# 2007 NATIONAL HOME IMPROVEMENT ESTIMATOR

## Edited by Ben Moselle

---

### Includes inside the back cover:

- An estimating CD with all the costs in this book, plus,
- An estimating program that makes it easy to use these costs,
- An interactive video guide to the National Estimator program,
- A program that converts your estimates into invoices,
- A program that exports your estimates to QuickBooks Pro.

**Monthly price updates on the Web are free** and automatic all during 2007. You'll be prompted when it's time to collect the next update. A connection to the Web is required.

Download all of Craftsman's most popular costbooks for one low price with the Craftsman Site License. http://CraftsmanSiteLicense.com

---

**Craftsman Book Company**
6058 Corte del Cedro / P.O. Box 6500 / Carlsbad, CA 92018

# Acknowledgments

Portions of chapters 8, 9 and 15 first appeared in the book *Renovating & Restyling Older Homes* by Lawrence Dworin. Mr. Dworin served as a resource in the development of this manuscript and contributed many valuable insights from his years of experience in the industry. An order form for *Renovating & Restyling Older Homes,* and his other construction reference *Profits in Buying & Renovating Homes* appears on the final pages of this manual.

Cover design by: *Bill Grote*
Cover photos by: *Ron South Photography*
Graphics by: *Nichole Campbell, D. Naomi Johnson and Devona Quindoy*
Layout by: *Nichole Campbell*
Software production: *Timothy Campbell*

© 2006 Craftsman Book Company
ISBN 978-1-57218-178-6
Published November 2006 for the year 2007.

# Contents

**1 Pricing Home Improvement Jobs** ......5
This book works two ways ....................8
Craft codes and crews ....................15
Area modification factors ....................17

**2 Demolition** ....................27
Before you begin ....................27
*Cost estimates*
Equipment rental ....................28
Sitework demolition ....................29
Building demolition ....................30
Wood framing demolition ....................33

**3 Foundations and Slabs** ....................37
Pre-construction checklist ....................38
*Cost estimates*
Equipment rental ....................39
Concrete flatwork ....................40
Concrete finishing ....................49

**4 Rough Carpentry** ....................53
Floor beams ....................53
Wood floor repairs ....................56
Wall framing ....................57
Roof repairs ....................58
*Cost estimates*
Framing repairs ....................62
Framing room additions ....................64
Subflooring ....................66
Wall and ceiling changes ....................70
Roof framing changes ....................75

**5 Insulation and Moisture Control** ......79
Ceiling insulation ....................79
Wall and floor insulation ....................80
Attic and roof ventilation ....................83
*Cost estimates*
Fiberglass insulation ....................84
Panel insulation ....................85
Vapor barrier ....................87
Vents and louvers ....................88
Roof ventilators ....................91

**6 Siding and Trim** ....................93
Vinyl siding installation ....................96
Wood siding installation ....................97
Estimating procedure for siding ....................104
Masonry fireplaces ....................104
*Cost estimates*
Siding and soffit repairs ....................105
Vinyl siding ....................107
Wood board siding ....................108
Texture 1-11 plywood siding ....................111
Fiber cement siding ....................112
Aluminum siding and soffit ....................113
Stucco and masonry siding ....................114

**7 Roofing and Flashing** ....................117
Flashing leaks ....................118
Roof repairs ....................121
Roofing material ....................122
*Cost estimates*
Roof repairs ....................129
Composition shingles ....................130
Built-up roofing ....................131
Roof patching ....................134
Wood shingle roofing ....................137
Slate and tile roofing ....................139
Roof flashing ....................141
Gutters and downspouts ....................146

**8 Windows** ....................147
Window repair ....................149
Sash repair ....................151
Counterweighted windows ....................152
Adding or moving a window ....................153
*Cost estimates*
Window repairs ....................156
Vinyl windows ....................157
Aluminum windows ....................161
Wood windows ....................164
Storm windows ....................166
Window accessories ....................166
Window glazing ....................170
Skylights ....................173

## 9 Doors . . . . . . . . . . . . . . . . . . . . . . . . . . .175
Door repair and replacement . . . . . . . . . . . . . . .175
Adding or moving a door . . . . . . . . . . . . . . . . . .182
Hanging a slab door . . . . . . . . . . . . . . . . . . . . . .184
Door terms . . . . . . . . . . . . . . . . . . . . . . . . . . . . .188
### Cost estimates
Exterior doors . . . . . . . . . . . . . . . . . . . . . . . .191
Prehung doors . . . . . . . . . . . . . . . . . . . . . . . .197
Screen and storm doors . . . . . . . . . . . . . . . .202
Interior doors . . . . . . . . . . . . . . . . . . . . . . . .206
Patio doors . . . . . . . . . . . . . . . . . . . . . . . . . .214
Closet doors . . . . . . . . . . . . . . . . . . . . . . . . .218
Garage doors . . . . . . . . . . . . . . . . . . . . . . . .223

## 10 Walls and Ceilings . . . . . . . . . . . . . . .227
Drywall installation . . . . . . . . . . . . . . . . . . . . . .228
Wood paneling . . . . . . . . . . . . . . . . . . . . . . . . . .235
Removing partitions . . . . . . . . . . . . . . . . . . . . . .239
### Cost estimates
Drywall . . . . . . . . . . . . . . . . . . . . . . . . . . . . .242
Textured finishes for drywall . . . . . . . . . . . . .244
Drywall bead and tools . . . . . . . . . . . . . . . . .247
Wall paneling . . . . . . . . . . . . . . . . . . . . . . . .251
Suspended ceilings . . . . . . . . . . . . . . . . . . . .255
Moldings . . . . . . . . . . . . . . . . . . . . . . . . . . . .259
Partition walls . . . . . . . . . . . . . . . . . . . . . . . .270

## 11 Floors and Tile . . . . . . . . . . . . . . . . . .271
Wood flooring installation . . . . . . . . . . . . . . . . .272
Ceramic tile installation . . . . . . . . . . . . . . . . . .276
### Cost estimates
Wood strip and plank flooring . . . . . . . . . . .278
Laminate plank flooring . . . . . . . . . . . . . . . .283
Vinyl flooring . . . . . . . . . . . . . . . . . . . . . . . . .286
Ceramic tile . . . . . . . . . . . . . . . . . . . . . . . . . .293
Natural stone and slate flooring . . . . . . . . . .298
Ceramic field wall tile . . . . . . . . . . . . . . . . . .301
Carpet . . . . . . . . . . . . . . . . . . . . . . . . . . . . . .308

## 12 Kitchens . . . . . . . . . . . . . . . . . . . . . . .309
Kitchen cabinet standards . . . . . . . . . . . . . . . . .312
### Cost estimates
Kitchen demolition . . . . . . . . . . . . . . . . . . . .315
Cabinets and countertops . . . . . . . . . . . . . . .317
Cabinet knobs, pulls and hinges . . . . . . . . . .324
Sinks and faucets . . . . . . . . . . . . . . . . . . . . .328
Appliances . . . . . . . . . . . . . . . . . . . . . . . . . .339

## 13 Bathrooms . . . . . . . . . . . . . . . . . . . . .345
Bathroom additions . . . . . . . . . . . . . . . . . . . . . .345

### Cost estimates
Tubs, enclosures and shower stalls . . . . . . . . .352
Tub and shower faucets . . . . . . . . . . . . . . . .360
Toilets . . . . . . . . . . . . . . . . . . . . . . . . . . . . . .362
Lavatory sinks . . . . . . . . . . . . . . . . . . . . . . . .367
Vanity cabinets and countertops . . . . . . . . . .374
Bath exhaust fans . . . . . . . . . . . . . . . . . . . . .378
Medicine cabinets . . . . . . . . . . . . . . . . . . . . .382
Bathroom accessories . . . . . . . . . . . . . . . . . .385

## 14 Plumbing and HVAC . . . . . . . . . . . . . .391
Heating types and considerations . . . . . . . . . . .391
Framing for plumbing and HVAC . . . . . . . . . . . .394
### Cost estimates
Furnace replacement . . . . . . . . . . . . . . . . . . .396
Chimney flue components . . . . . . . . . . . . . . .399
Water heaters . . . . . . . . . . . . . . . . . . . . . . . .405
DWV pipe replacement . . . . . . . . . . . . . . . . .407

## 15 Electrical . . . . . . . . . . . . . . . . . . . . . . .409
Wiring residences . . . . . . . . . . . . . . . . . . . . . . .411
Adding outlets and switches . . . . . . . . . . . . . . .414
### Cost estimates
Non-metallic cable . . . . . . . . . . . . . . . . . . . .417
PVC conduit and fittings . . . . . . . . . . . . . . . .420
EMT conduit and fittings . . . . . . . . . . . . . . . .422
Flex conduit and fittings . . . . . . . . . . . . . . . .426
Rigid and IMC conduit and fittings . . . . . . . .429
Switches and receptacles . . . . . . . . . . . . . . .434
Distribution panels . . . . . . . . . . . . . . . . . . . .436
Lighting . . . . . . . . . . . . . . . . . . . . . . . . . . . . .440

## 16 Porches and Decks . . . . . . . . . . . . . . .445
Adding a porch . . . . . . . . . . . . . . . . . . . . . . . . .446
Elevated decks . . . . . . . . . . . . . . . . . . . . . . . . .448
### Cost estimates
Deck lumber . . . . . . . . . . . . . . . . . . . . . . . . .452
Alternative deck materials . . . . . . . . . . . . . . .454
Porch and deck stairs . . . . . . . . . . . . . . . . . .457

## 17 Painting and Finishing . . . . . . . . . . . . .461
Paint application techniques . . . . . . . . . . . . . . .464
### Cost estimates
Labor estimates for painting . . . . . . . . . . . . .466
Stains and finishes . . . . . . . . . . . . . . . . . . . .487
Wood preservatives . . . . . . . . . . . . . . . . . . .490
Interior paints . . . . . . . . . . . . . . . . . . . . . . . .492
Exterior paints . . . . . . . . . . . . . . . . . . . . . . .494

## Index . . . . . . . . . . . . . . . . . . . . . . . . . . . . .497

# Pricing Home Improvement Jobs

Estimating home improvement costs requires specialized skills. You can't price home improvement work the same way you'd price new construction. The proportion of labor expense is greater. There's far more risk because there are far more variables and unknowns. An example will make this clear.

Hanging doors in new construction is a 1-2-3-affair. You know ahead of time exactly what's required. You built the wall, installed the frame and know each door will fit right the first time. There won't be any surprises. An experienced estimator can forecast the cost of hanging a door in a new home with a high degree of certainty.

Now let's look at the same task on a home improvement project. First, remember that you may be hanging only one door. There's no chance to improve the production rate on second or later doors. Installing this door is probably different from the last door you installed on a similar project and will be different from your next door installation.

Work starts with removing the old door. That's probably not too hard — once you break through six coats of accumulated paint on the hinges. Let's assume you work carefully and don't damage the casing or trim. The next step is to install a blind Dutchman where the original hinges were attached. If the frame's badly chewed up from years of neglect and abuse, you'll have to remove and replace the entire frame. Don't forget that the building has settled over the years. The frame has probably twisted out of plumb. More time will be needed to shim and level the frame.

Finally the door is installed and swings smoothly. But the job still isn't finished. You have to paint the door to match and will certainly have to haul away the old door and the debris.

Every home improvement cost estimate comes with dozens of chances to make an expensive mistake. Surprises are inevitable and nearly all surprises will add to the cost, not reduce the cost. Estimating new construction is a snap by comparison.

---

### All home improvement has similar problems:

- ❖ No chance for economies of scale (mass production),
- ❖ Difficulty in removing just enough of the old,
- ❖ Work doesn't follow the normal (from the ground, up) sequence of construction,
- ❖ Difficulty adding new materials to deteriorated or nonstandard existing materials,
- ❖ The need to match designs, colors and textures,
- ❖ Covering up for another contractor's mistakes,
- ❖ Struggling to get access to the place where the work is to be done,
- ❖ Protecting adjacent surfaces and pathway to the work area,
- ❖ Initial uncertainty about how much work is needed.

If you're an experienced home improvement contractor, you understand this already. Whether experienced or not, you know full well that accurate pricing is crucial to survival in the home improvement business. Unfortunately, most home improvement specialists have far less experience in pricing their work than in getting the work done. No wonder so many home improvement companies sink into obscurity.

But it doesn't have to be that way.

The chapters that follow explain how to price each type of home improvement work. Emphasis will be on avoiding risk — pricing pitfalls that can turn any home improvement contract into a financial nightmare.

## No Price Fits All Jobs

There's no single way to arrive at the correct price for home improvement work. Neither is there a single correct price for most home improvement projects. But there are both good and bad ways to estimate prices and there are good and bad prices for any proposed job. Your task as a home improvement estimator is to produce consistently good estimates on most jobs. If you're already doing that, congratulations. You don't need this book. Return it where you bought it and get a refund. But if you're not, information in the following chapters will make or save you many times what you paid for the book.

It's accepted wisdom among home improvement specialists that sales people need authority to quote prices when closing a sale. When trying to wrap up a deal, there's no substitute for having current cost information at your fingertips. That's especially true on larger jobs where you've prepared detailed plans and a written estimate. The owner will have questions and suggestions that change the job specs. If you want to close the sale then and there, you better know how much to add or subtract for each change the owner wants.

Most home improvement companies authorize salespeople to quote prices from an approved list in a price book. That simplifies the salesperson's job, eliminates most major errors, and saves the owner of the company from approving every item in every estimate. In our opinion, quoting from a price book is the only way to build sales volume in the home improvement business. That means every home improvement contractor needs a good price book.

Of course, the best price book for your company would be based on your actual cost experience — work done by your crews on your jobs with materials from your dealers and installed by your subcontractors. Since every contractor uses different crews, subcontractors and suppliers, every price book should be different. And, of course, prices in the company price book should be revised regularly to reflect current labor and material costs.

Having admitted that there's no substitute for developing your own price book, we'll suggest that you not bother. Most home improvement contractors don't have the time or patience to maintain current installed prices for thousands of repair and remodeling items. Even if you did, spending hundreds of hours a year keeping a price book current would be a waste of time. Prices for home improvement work are usually negotiated on site. There's little value in keeping a book of exact costs if you have to cut a special deal to close each sale.

Instead, we suggest that you let this manual serve as your company price book. Using prices in this book (and on the disk in the back of this manual) will eliminate most of the common estimating mistakes. If your labor costs are higher or if your crews aren't as skilled as most tradesmen, you may have to increase the selling prices listed here to make a reasonable profit. And, of course, sometimes you're going to have a job with costs that exceed what any reasonable estimate could have predicted. This manual isn't a substitute for the exercise of good judgment. That's always your job.

# Cost Plus Markup Equals Selling Price

The figures in this manual show both costs to the contractor and a recommended selling price. The difference between your cost and your selling price is your markup. Markup varies widely, far more than either material cost or labor cost. You're the final authority on markup. You decide what markup fits best — based on market conditions, your client, and the profit you feel is reasonable. Once you decide on the "right" markup, it's easy to plug that percentage into the bid, assuming you use National Estimator.

Every construction contractor needs to distinguish between "markup" and "profit." Markup is what you add to estimated labor and material costs (usually called "hard costs") to find the selling price. Profit is what you have left when all bills have been paid. The two are very different. Profit is just the frosting on the cake.

Markup is also different from "margin." Figure 1-1 shows percentages of markup for various levels of hard costs. For example, if your markup on a $20,000 job is 50 percent, the selling price will be $30,000. The margin on that job is 33 percent, $10,000 divided by the selling price of $30,000. (Note the bottom row in Figure 1-1.) Margin is what's left after hard costs are recovered, and includes profit.

| Markup vs. Margin | | | | |
|---|---|---|---|---|
| **Hard Cost** | **Price at 50% Markup** | **Price at 70% Markup** | **Price at 100% Markup** | **Price at 150% Markup** |
| $50 | $75 | $85 | $100 | $125 |
| $100 | $150 | $170 | $200 | $250 |
| $200 | $300 | $340 | $400 | $500 |
| $300 | $450 | $510 | $600 | $750 |
| $400 | $600 | $680 | $800 | $1,000 |
| $500 | $750 | $850 | $1,000 | $1,250 |
| $1,000 | $1,500 | $1,700 | $2,000 | $2,500 |
| $2,000 | $3,000 | $3,400 | $4,000 | $5,000 |
| $3,000 | $4,500 | $5,100 | $6,000 | $7,500 |
| $4,000 | $6,000 | $6,800 | $8,000 | $10,000 |
| $5,000 | $7,500 | $8,500 | $10,000 | $12,500 |
| $6,000 | $9,000 | $10,200 | $12,000 | $15,000 |
| $7,000 | $10,500 | $11,900 | $14,000 | $17,500 |
| $8,000 | $12,000 | $13,600 | $16,000 | $20,000 |
| $9,000 | $13,500 | $15,300 | $18,000 | $22,500 |
| $10,000 | $15,000 | $17,000 | $20,000 | $25,000 |
| $11,000 | $16,500 | $18,700 | $22,000 | $27,500 |
| $12,000 | $18,000 | $20,600 | $24,000 | $30,000 |
| $13,000 | $19,500 | $22,100 | $26,000 | $32,500 |
| $14,000 | $21,000 | $23,800 | $28,000 | $35,000 |
| $15,000 | $22,500 | $25,500 | $30,000 | $37,500 |
| $16,000 | $24,000 | $27,200 | $32,000 | $40,000 |
| $17,000 | $25,500 | $28,900 | $34,000 | $42,500 |
| $18,000 | $27,000 | $30,600 | $36,000 | $45,000 |
| $19,000 | $28,500 | $32,300 | $38,000 | $47,500 |
| $20,000 | $30,000 | $34,000 | $40,000 | $50,000 |
| $30,000 | $45,000 | $51,000 | $60,000 | $75,000 |
| $40,000 | $60,000 | $68,000 | $80,000 | $100,000 |
| $50,000 | $75,000 | $85,000 | $100,000 | $125,000 |
| **Hard Cost** | **33% Margin** | **41% Margin** | **50% Margin** | **60% Margin** |

**Figure 1-1**

*Markup (based on cost) vs. Margin (based on selling price)*

Many successful home improvement contractors find they can stay in business if hard costs are 59 percent of selling price. Hard costs include material expense, subcontract costs and labor (including taxes and insurance). The other 41 percent of selling price ("margin") compensates the sales staff, covers overhead, supervision and contingency, and should yield a modest profit. From Figure 1-1, you can see that markup on hard costs has to be 70 percent to yield a margin of 41 percent.

To put this formula to work in your home improvement company, add 70 percent to your hard costs to find the selling price. Of course, some jobs need more markup and others can get by on less. A smaller job with more risk done for a demanding client may require greater markup. A larger job done mostly by subcontractors can usually carry a smaller markup.

Home improvement contractors have all the overhead of any business: office rent, telephone, owner's salary, office salaries, legal and accounting expense, insurance, auto and truck expense, and more. But unlike other contractors, home improvement specialists routinely deal with high risk from both the unknown and the unknowable — at least until work actually begins. Since most of what you didn't or couldn't anticipate will inflate your costs, you're assuming significant risk. That's why markups for home improvement work have to be higher than for new construction.

Naturally, competition dictates markup. If you're not getting enough work at 70 percent markup, maybe 70 percent is too much for your area. But remember that you shouldn't have to bid remodeling work on a level playing field. New construction usually goes to the lowest responsible bidder. A creative salesman who follows the recommendations in this book has an advantage over lowball bidders who rely on price alone to sell their services.

If you have trouble using National Estimator, we'll be glad to help, and we don't charge you for it.

Free telephone assistance is available from 8 a.m. until 5 p.m. Pacific time Monday through Friday (except holidays). Call 760-438-7828.

# This Book Works Two Ways

The disk in the back of this book has all the information that appears in the printed book, but with one advantage. The National Estimator program makes it easy to copy and paste these costs into an estimate (or bid) and then add whatever markup you select. **Monthly price updates on the Web are free** and automatic all during 2007. You'll be prompted when it's time to collect the next update. A connection to the Web is required. To get started with the National Estimator disk, insert it into any computer that uses Microsoft Windows. Then follow instructions on the screen.

Before installation is complete, you'll be invited to watch ShowMe, an interactive video guide to the National Estimator program. ShowMe is designed to speed your mastery of National Estimator. You can also start ShowMe at any time while using National Estimator. Click the question mark button ⊙ at the right end of the toolbar. Then click **ShowMe Tutorial**. The disk has to be in your computer when ShowMe is running.

# Estimates That Work Two Ways

This manual is designed for use by both the owner of a home improvement business and by company sales staff. Prices shown in this manual include both hard costs (labor and material) and a recommended selling price (usually based on 70 percent markup). Sales staff will use the selling price.

Figure 1-2 shows the last page of an estimate for the Stillwel room addition, including 70 percent markup. The company estimator created this estimate from a job survey prepared by a sales representative. Figure 1-2 is for company use and was created in the National Estimator program.

National Estimator [STILLWEL.EST]

File  Edit  View  Utilities  Window  Help

| Qty | Craft@Hours | Unit | Material | Labor | Equipment | Total |
|---|---|---|---|---|---|---|
| Roll & brush 1 coat of water base undercoat on interior flush doors | | | | | | |
| Medium .3 MH/Door, 11.5 Doors/Gal | | | | | | |
| 3.00 | 2P@.9000 | Ea | 5.07 | 21.36 | 0.00 | 26.43 |
| Roll & brush 1st finish coat of water base enamel on interior flush doors | | | | | | |
| Medium .25 MH/Door, 12.5 Doors/Gal | | | | | | |
| 3.00 | 2P@.7500 | Ea | 5.39 | 17.80 | 0.00 | 23.18 |
| Roll & brush 1 coat water base undercoat on vanity cabinet | | | | | | |
| Medium 93 SF/MH, 250 SF/Gal | | | | | | |
| 0.90 | 2P@.9720 | CSF | 6.99 | 23.06 | 0.00 | 30.05 |
| Roll & brush 1st finish coat of water base enamel on vanity cabinet | | | | | | |
| Medium 103 SF/MH, 288 SF/Gal | | | | | | |
| 0.90 | 2P@.8740 | CSF | 7.02 | 20.70 | 0.00 | 27.72 |
| Roll 1st coat of water base stain on rough sawn or resawn wood siding | | | | | | |
| Medium 225 SF/MH, 213 SF/Gal | | | | | | |
| 17.00 | 2P@7.548 | CSF | 157.08 | 178.50 | 0.00 | 335.58 |
| Roll 2nd coat of water base stain on rough sawn or resawn wood siding | | | | | | |
| Medium 275 SF/MH, 273 SF/Gal | | | | | | |
| 17.00 | 2P@6.188 | CSF | 122.63 | 146.73 | 0.00 | 269.36 |
| **Subtotal: Painting | | | | | | |
| | 40.7 | | 1,014.54 | 973.46 | 0.00 | 1,988.00 |

| | | | | |
|---|---|---|---|---|
| Total Manhours, Material, Labor, and Equipment: | | | | |
| 456.0 | 18,641.82 | 11,675.42 | 540.29 | 30,857.53 |
| Total Only (Subcontract) Costs: | | | | 4,028.33 |
| | | Subtotal: | | 34,885.85 |
| | | 70.00% Markup: | | 24,420.10 |
| | | Estimate Total: | | 59,305.95 |

**Figure 1-2**

*Estimate with a 70 percent markup shown*

Figure 1-3 shows the last page of a proposal for the same job. This is what the customer sees. It was created by the Job Cost Wizard program. Both National Estimator and Job Cost Wizard are on the disk in the back of this book. Notice that the total is the same in both documents, $59,305.95. Figure 1-2, the estimate, shows 70 percent markup on hard costs. Figure 1-3, the proposal, doesn't show the markup at all. Instead, markup has been distributed proportionately throughout each cost item in the job. Job Cost Wizard does this distribution at the click of a button.

| Craftsman Construction | | | **Proposal** | |
|---|---|---|---|---|

6058 Corte del Cedro
Box 6500
Fairfield, GA 30456
856-3806

| Date | Estimate # |
|---|---|
| 8/3/08 | 494 |
| **Customer** | **Job** |
| Bill Stillwel | Room Addition |

| Description | Qty | Rate | Amount |
|---|---|---|---|
| Roll & brush 1st finish coat of water base enamel on vanity cabinet. Medium 103 SF/MH, 288 SF/Gal | | | |
|    Material, per CSF | 0.90 | 13.26 | 11.94 |
|    Labor, per CSF | 0.90 | 39.09 | 35.18 |
| Roll 1st coat of water base stain on rough sawn or resawn wood siding. Medium 225 SF/MH, 213 SF/Gal | | | |
|    Material, per CSF | 17 | 15.71 | 267.04 |
|    Labor, per CSF | 17 | 17.85 | 303.45 |
| Roll 2nd coat of water base stain on rough sawn or resawn wood siding. Medium 275 SF/MH, 273 SF/Gal | | | |
|    Material, per CSF | 17 | 12.26 | 208.47 |
|    Labor, per CSF | 17 | 14.67 | 249.44 |
|    *Painting subtotal | | | 3,379.60 |
| | | | |
| *Project Subtotal | | | 59,305.96 |
| *Project Total | | | 59,305.96 |

**Figure 1-3**

*Proposal for the same job not showing the markup*

Once an estimate is finished in the National Estimator program and saved to computer disk, press Ctrl on your keyboard and tap the letter J to convert the estimate into a proposal in Job Cost Wizard. You can't make changes in the Job Cost Wizard screen. But it's easy to toggle back to National Estimator program (press Alt and tap the Tab↹ key), make a change, and then press Ctrl-J once again.

Job Cost Wizard offers dozens of choices on what you show and don't show in written proposals. Your bids can be long (full description for everything in the estimate) or short (only a summary of each category). You can show or hide labor and material cost detail. You can show or hide markup and profit.

Once work begins, you'll want to monitor job expenses to be sure actual costs remain consistent with estimated costs. If you use QuickBooks Pro to pay bills and figure payroll, let QuickBooks do the comparisons for you. Job Cost Wizard exports the proposal to QuickBooks, where you can prepare progress invoices and track expenses against estimates.

The National Estimator program lets you change anything in the costbook or even add your own estimated costs. Make the *National Home Improvement Estimator* your collection point for all estimating and pricing information. Anything you add to the costbook shows up in red and can be migrated to later editions of this manual when available. The *National Home Improvement Estimator* starts out as your most useful estimating reference. What you add to the costbook will make it even more valuable.

# Job Survey (Scope of Work)

Of course, neither a good price book nor a computer estimating program will solve all of your estimating problems. Computers seldom make mistakes in addition or multiplication. But nothing prevents you from making expensive estimating mistakes on your own. By far the most common estimating mistake will be omitting something essential to the job.

Get in the habit of completing an exhaustive job survey before beginning any estimate for home improvement work. Some of your estimates on a job may be too high. Other estimates on a job may be too low. With any luck, your over-estimates will roughly balance with your under-estimates, leaving the job total about where it should be. But the estimated price for anything omitted from a job survey is always zero. That's a 100 percent miss. It's hard to balance a complete miss with anything. A few of those can create a financial disaster.

# Material Costs in This Book

Material costs for each item are listed in the column headed Material. These are neither retail nor wholesale prices. They are estimates of what most contractors who buy in moderate volume will pay suppliers as of mid-2007. Discounts may be available for purchases in larger volumes.

**Add Delivery Expense** to the material cost for other than local delivery of reasonably large quantities. Cost of delivery varies with the distance from source of supply, method of transportation, and quantity to be delivered. But most material dealers absorb the delivery cost on local delivery (5 to 15 miles) of larger quantities to good customers. Add the expense of job site delivery when it's a significant part of the material cost.

**Add Sales Tax** when sales tax will be charged to the contractor buying the materials. In some states, contractors have to collect sales tax based on the contract price. No matter what your state (or county) requires, National Estimator can handle the task. Click **Edit** on the National Estimator menu bar. Then click **Sales Tax**.

**Waste and Coverage Loss** is included in the installed material cost. The cost of many materials per unit after installation is greater than the purchase price for the same unit because of waste, shrinkage or coverage loss during installation. For example, about 120 square feet of nominal 1" x 4" square edge boards will be needed to cover 100 square feet of floor or wall. There's no coverage loss with plywood sheathing, but waste due to cutting and fitting will average about 6 percent.

Costs in the Material column of this book assume normal waste and coverage loss. Small and irregular jobs may require a greater waste allowance. Materials priced without installation (with no labor cost) don't include an allowance for waste and coverage except as noted.

**Labor Costs** for installing the material or doing the work described are listed in the column headed Labor. The labor cost per unit is the labor cost per hour multiplied by the manhours per unit shown after the @ sign in the Craft@Hrs column. Labor cost includes the basic wage, the employer's contribution to welfare, pension, vacation and apprentice funds, and all tax and insurance charges based on wages.

**Supervision Expense** to the general contractor isn't included in the labor cost. The cost of supervision and non-productive labor varies widely from job to job. Calculate the cost of supervision and non-productive labor and add this to the estimate.

**Payroll Taxes and Insurance** are included in the labor cost. See the following section for more on the "contractor's burden."

**Manhours per Unit and the Craft** performing the work are listed in the Craft@Hrs column. To find the units of work done per worker in an 8-hour day, divide 8 by the manhours per unit. To find the units done by a crew in an 8-hour day, multiply the units per worker per 8-hour day by the number of crew members.

**Manhours** include all productive labor normally associated with installing the materials described.

This will usually include tasks such as:

❖ Unloading and storing construction materials, tools and equipment on site.

❖ Moving tools and equipment from a storage area or truck on site at the beginning of the day.

❖ Returning tools and equipment to a storage area or truck on site at the end of the day.

❖ Normal time lost for work breaks.

❖ Planning and discussing the work to be performed.

❖ Normal handling, measuring, cutting and fitting.

❖ Keeping a record of the time spent and work done.

❖ Regular cleanup of construction debris.

❖ Infrequent correction or repairs required because of faulty installation.

**Adjust the Labor Cost** to the job you're figuring when your actual hourly labor cost is known or can be estimated. The labor costs listed in Figure 1-5 will apply within a few percent on many jobs. But labor costs may be much higher or much lower on the job you're estimating.

If the hourly wage rates listed in Figure 1-5 aren't accurate, divide your known or estimated cost per hour by the listed cost per hour. The result is your adjustment for any figure in the Labor column for that craft.

**Adjust for Unusual Labor Productivity**. Costs in the Labor column are for normal conditions: experienced craftsmen working on reasonably well-planned and managed home improvement projects with fair to good productivity. Labor estimates assume that materials are standard grade, appropriate tools are on hand, work done by other crafts is adequate, layout and installation are relatively uncomplicated, and working conditions don't slow progress.

Working conditions at the job site have a major effect on labor cost. Estimating experience and careful analysis can help you predict the effect of most changes in working conditions. Obviously, no single adjustment will apply on all jobs. But the adjustments that follow should help you produce more accurate labor estimates. More than one condition may apply on a job.

---

### *Jobsite conditions affect labor costs*

❖ Add 10% to 15% when working temperatures are below 40 degrees or above 95 degrees.

❖ Add 30% to 35% when temperatures are below 20 degrees. Materials and tools are hard to handle. Bulky clothing restricts freedom of movement.

❖ Add 15% to 25% for work on a ladder or a scaffold, in a 36" crawl space, in a congested area, or remote from the material storage point.

❖ Add 50% for work in an 18" to 36" crawl space. Allow extra time for cleaning out the area before work begins.

❖ Add 200% if portions of the crawl space are less than 18". Allow extra time for passing tools and materials. Few contractors bid on work like this.

❖ Deduct 10% when the work is in an open area with excellent access and good light.

❖ Add 1% for each 10 feet that materials must be lifted above ground level.

❖ Add 5% to 50% for tradesmen with below-average skills. Deduct 5% to 25% for highly-motivated, highly-skilled tradesmen.

❖ Deduct 10% to 20% when an identical task is repeated many times for several days at the same site.

❖ Add 20% to 50% on small jobs where fitting and matching of materials is required, adjacent surfaces have to be protected and the job site is occupied during construction.

❖ Add 25% to 50% for work done following a major flood, fire, earthquake, hurricane or tornado while skilled tradesmen are not readily available. Material costs may also be higher after a major disaster.

❖ Add 10% to 35% for demanding specs, rigid inspections, unreliable suppliers, a difficult owner or an inexperienced architect.

---

**Use an Area Modification Factor** from pages 18 through 26 if your material, hourly labor or equipment costs are unknown and can't be estimated. Here's how: Use the labor and material costs in this manual without modification. Then add or deduct the percentages shown for labor, material and equipment.

**Equipment Costs** for small tools and expendable supplies (such as saws and tape) are usually considered overhead expense and are covered by your markup. Equipment costs for larger equipment (such as a compressor or backhoe) should be based on the rental rate for the period needed. Equipment rental costs are included in sections where heavy equipment may be needed.

**Labor and Material Costs Change**. These costs were compiled in the fall of 2006 and projected to mid-2007 by adding a small percentage. This projection will be accurate for some materials and inaccurate for others. No one can predict material price changes accurately.

**How Accurate Are These Figures?** They're as accurate as possible considering that the estimators who wrote this book don't know your subcontractors or material suppliers, haven't seen the plans or specifications, don't know what building code applies or where the job is, had to project material costs at least six months into the future, and had no record of how much work the crew that will be assigned to the job can handle.

You wouldn't bid a job under those conditions. And we don't claim that all construction is done at these prices.

**Estimating Is an Art**, not a science. On many jobs the range between high and low bid will be 20 percent or more. There's room for legitimate disagreement on what the correct costs are, even when complete plans and specifications are available, the date and site are established, and labor and material costs are identical for all bidders.

No cost fits all jobs. Good estimates are custom-made for a particular project and a single contractor through judgment, analysis and experience.

This book isn't a substitute for judgment, analysis and sound estimating practice. It's an *aid* in developing an *informed opinion* of cost. If you're using this book as your sole cost authority for contract bids, you're reading more into these pages than the editors intend.

**Use These Figures** to compile preliminary estimates, when a "snap" bid is needed to close the deal, and when no actual costs are available. This book will reduce the chance of error or omission on bid estimates, speed "ball park" estimates, and be a good guide when there's no time to get a quote.

**Where Do We Get These Figures?** From the same sources all professional estimators use: material suppliers, material price services, analysis of plans, specifications, estimates and completed project costs, and both published and unpublished cost studies compiled by both private and government agencies. In addition, we conduct nationwide mail and phone surveys and have the use of several major national estimating databases.

**We'll Answer Your Questions** about any part of this book and explain how to apply these costs. Free telephone assistance is available from 8 a.m. until 5 p.m. Pacific time Monday through Friday except holidays. Phone 760-438-7828. We don't accept collect calls and can't estimate the job for you. But if you need clarification on something in this manual, we can help.

# Labor Costs and Crews

Throughout this manual you'll see a column headed *Craft@Hrs*. Letters and numbers in this column show our estimates of:

❖ Who will do the work (the craft code)

❖ An @ symbol which means at

❖ How long the work will take (manhours).

For example, in Chapter 4, page 70, you'll find estimates for installing BD plywood wall sheathing by the square foot. The Craft@Hrs column opposite ¹/₂" plywood wall sheathing shows:

B1@.016

That means we estimate the installation rate for crew B1 at 0.016 manhours per square foot. That's the same as 16 manhours per 1,000 square feet.

Figure 1-4 is a table that defines each of the craft codes used in this book. Notice that crew B1 is composed of two craftsmen: one laborer and one carpenter. To install 1,000 square feet of ¹/₂" BD wall sheathing at 0.016 manhours per square foot, that crew would need 16 manhours (one 8-hour day for a crew of two).

Notice also in the table that the cost per manhour for crew B1 is listed as $29.37. That's the average for a laborer (listed at $26.64 per hour) and a carpenter (listed at $32.09 per hour): $26.64 plus $32.09 is $58.73. Divide by 2 to get $29.37, the average cost per manhour for crew B1.

In the table, the cost per manhour is the sum of hourly costs of all crew members divided by the number of crew members. That's the average cost per manhour.

Costs in the Labor column in this book are the product of the installation time (in manhours) multiplied by the cost per manhour. For example, in Chapter 4, the labor cost listed for ¹/₂" BD wall sheathing is $0.47 per square foot. That's the installation time (0.016 manhours per square foot) multiplied by $29.37, the average cost per manhour for crew B1.

Figure 1-5 shows hourly labor costs components — base wage, typical fringe benefits, payroll taxes, insurance and the total hourly cost.

The labor costs shown in Column 6 were used to compute the manhour costs for crews and the figures in the Labor column.

## Craft Codes

| Craft Code | Cost Per Manhour | Crew Composition |
|---|---|---|
| B1 | $29.37 | 1 laborer, 1 carpenter |
| B2 | $30.27 | 1 laborer, 2 carpenters |
| B3 | $28.46 | 2 laborers, 1 carpenter |
| B4 | $31.17 | 1 laborer, 1 operating engineer, 1 reinforcing iron worker |
| B5 | $31.21 | 1 laborer, 1 carpenter, 1 cement mason, 1 operating engineer, 1 reinforcing iron worker |
| B6 | $28.54 | 1 laborer, 1 cement mason |
| B7 | $26.58 | 1 laborer, 1 truck driver |
| B8 | $31.31 | 1 laborer, 1 operating engineer |
| B9 | $28.74 | 1 bricklayer, 1 bricklayer's helper |
| BB | $32.55 | 1 bricklayer |
| BC | $32.09 | 1 carpenter |
| BE | $34.39 | 1 electrician |
| BF | $29.54 | 1 floor layer |
| BG | $29.78 | 1 glazier |
| BH | $24.93 | 1 bricklayer's helper |
| BL | $26.64 | 1 laborer |
| BR | $31.17 | 1 lather |
| BS | $27.87 | 1 marblesetter |
| CF | $30.44 | 1 cement mason |
| CT | $29.54 | 1 mosaic & terrazzo worker |
| D1 | $30.89 | 1 drywall installer, 1 drywall taper |
| DI | $30.91 | 1 drywall installer |
| DT | $30.86 | 1 drywall taper |
| HC | $25.26 | 1 plasterer helper |
| OE | $35.98 | 1 operating engineer |
| P1 | $31.50 | 1 laborer, 1 plumber |
| PM | $36.35 | 1 plumber |
| PP | $29.51 | 1 painter, 1 laborer |
| PR | $32.53 | 1 plasterer |
| PT | $32.38 | 1 painter |
| R1 | $31.21 | 1 roofer, 1 laborer |
| RI | $30.90 | 1 reinforcing iron worker |
| RR | $35.77 | 1 roofer |
| SW | $35.66 | 1 sheet metal worker |
| T1 | $28.72 | 1 tile layer, 1 laborer |
| TL | $30.79 | 1 tile layer |
| TR | $26.52 | 1 truck driver |

**Figure 1-4**

*Craft codes used in this manual*

| | 1 | 2 | 3 | 4 | 5 | 6 |
|---|---|---|---|---|---|---|
| **Hourly Labor Cost** | | | | | | |
| **Craft** | **Base wage per hour** | **Taxable fringe benefits (5.15% of base wage)** | **Insurance and employer taxes (%)** | **Insurance and employer taxes ($)** | **Non-Taxable fringe bene-fits (4.55% of base wage)** | **Total hourly cost used in this book** |
| Bricklayer | $23.84 | $1.23 | 25.54% | $6.40 | $1.08 | $32.55 |
| Bricklayer's Helper | 18.26 | 0.94 | 25.54 | 4.90 | 0.83 | 24.93 |
| Building Laborer | 18.46 | 0.95 | 32.93 | 6.39 | 0.84 | 26.64 |
| Carpenter | 22.42 | 1.15 | 31.83 | 7.50 | 1.02 | 32.09 |
| Cement Mason | 22.67 | 1.17 | 23.35 | 5.57 | 1.03 | 30.44 |
| Drywall installer | 22.95 | 1.18 | 23.77 | 5.74 | 1.04 | 30.91 |
| Drywall Taper | 22.91 | 1.18 | 23.77 | 5.73 | 1.04 | 30.86 |
| Electrician | 26.30 | 1.35 | 20.04 | 5.54 | 1.20 | 34.39 |
| Floor Layer | 21.88 | 1.13 | 24.02 | 5.53 | 1.00 | 29.54 |
| Glazier | 21.73 | 1.12 | 25.98 | 5.94 | 0.99 | 29.78 |
| Lather | 23.57 | 1.21 | 21.48 | 5.32 | 1.07 | 31.17 |
| Marble Setter | 21.06 | 1.08 | 21.56 | 4.77 | 0.96 | 27.87 |
| Millwright | 22.80 | 1.17 | 21.44 | 5.14 | 1.04 | 30.15 |
| Mosiac & Terrazzo Worker | 22.31 | 1.15 | 21.56 | 5.06 | 1.02 | 29.54 |
| Operating Engineer | 26.37 | 1.36 | 25.42 | 7.05 | 1.20 | 35.98 |
| Painter | 23.79 | 1.23 | 25.08 | 6.28 | 1.08 | 32.38 |
| Plasterer | 23.24 | 1.20 | 28.78 | 7.03 | 1.06 | 32.53 |
| Plasterer Helper | 18.05 | 0.93 | 28.78 | 5.46 | 0.82 | 25.26 |
| Plumber | 26.84 | 1.38 | 24.47 | 6.91 | 1.22 | 36.35 |
| Reinforcing Ironworker | 22.07 | 1.14 | 28.81 | 6.69 | 1.00 | 30.90 |
| Roofer | 22.88 | 1.18 | 44.34 | 10.67 | 1.04 | 35.77 |
| Sheet Metal Worker | 25.98 | 1.34 | 26.21 | 7.16 | 1.18 | 35.66 |
| Sprinkler Fitter | 26.37 | 1.36 | 25.28 | 7.01 | 1.20 | 35.94 |
| Tile Layer | 23.26 | 1.20 | 21.56 | 5.27 | 1.06 | 30.79 |
| Truck Driver | 19.29 | 0.99 | 26.42 | 5.36 | 0.88 | 26.52 |

**Figure 1-5**

*Components of hourly labor costs*

It's important that you understand what's included in the figures in each of the six columns in the table. Here's an explanation:

**Column 1,** the base wage per hour, is the craftsman's hourly wage. These figures are representative of what many contractors will be paying craftsmen working on home improvement jobs in 2007.

**Column 2,** taxable fringe benefits, includes vacation pay, sick leave and other taxable benefits. These fringe benefits average 5.15 percent of the base wage for many construction contractors. This benefit is in addition to the base wage.

**Column 3,** insurance and employer taxes in percent, shows the insurance and tax rate for construction trades. The cost of insurance in this column includes workers' compensation and contractor's casualty and liability coverage. Insurance rates vary widely from state to state and depend on a contractor's loss experience. Note that taxes and insurance increase the hourly labor cost by 30 to 35 percent for most trades. There is no legal way to avoid these costs.

**Column 4,** insurance and employer taxes in dollars, shows the hourly cost of taxes and insurance for each construction trade. Insurance and taxes are paid on the costs in both columns 1 and 2.

**Column 5,** non-taxable fringe benefits, includes employer paid non-taxable benefits such as medical coverage and tax-deferred pension and profit sharing plans. These fringe benefits average 4.55 percent of the base wage for many construction contractors. The employer pays no taxes or insurance on these benefits.

**Column 6,** the total hourly cost in dollars, is the sum of columns 1, 2, 4 and 5.

These hourly labor costs will apply within a few percent on many jobs. But wage rates may be much higher or lower in the area where you do business. We recommend using your actual labor cost rather than national averages. That's easy with the National Estimator program. When copying and pasting any cost to your estimate, adjust the assumed hourly labor cost to your actual cost. You need do this only once for each trade. And you can make this adjustment at any time. Any change you make is applied to that trade throughout the estimate.

## Abbreviations

| | | | | | |
|---|---|---|---|---|---|
| **4WD** | four-wheel drive | **EPA** | Environmental Protection Agency | **NEMA** | National Electrical Manufacturer's Assoc. |
| **ABS** | acrylonitrile butadiene styrene | **F** | Fahrenheit | **OC** | spacing from center to center |
| **AC** | alternating current | **GFCI** | ground fault circuit interrupter | **OD** | outside diameter |
| **ACQ** | Alkaline Copper Quat | **GRS** | galvanized rigid steel | **OSB** | oriented strand board |
| **ADA** | Americans with Disabilities Act | **H** | height | **OSHA** | Occupational Safety & Health Admin. |
| **ANSI** | American Nat. Standards Institute | **HOM** | higher order mode | **oz.** | ounce(s) |
| **APP** | atactic polypropylene | **HP** | horsepower | **PSI** | pounds per square inch |
| **ASTM** | Amer. Society for Testing Materials | **HVAC** | heating, ventilating & air cond. | **PVC** | polyvinyl chloride |
| **AWG** | American Wire Gauge | **ID** | inside diameter | **R** | thermal resistance |
| **BF** | board foot | **IMC** | intermediate metal conduit | **S4S** | surfaced 4 sides |
| **Btr** | better | **KD** | kiln dried | **SBS** | styrene butadiene styrene |
| **Btu(s)** | British thermal unit(s) | **KO** | knockout | **SF** | square foot |
| **C** | Celsius/Centigrade | **kW** | kilowatt(s) | **SFCA** | (per) square foot of contact area |
| **CF** | cubic foot | **L** | length | **Sq** | 100 square feet |
| **CFM** | cubic feet per minute | **Lb(s)** | pound(s) | **Std** | standard |
| **CLF** | 100 linear feet | **LF** | linear foot | **SY** | square yard |
| **CPM** | cycles per minute | **LS** | lump sum | **T** | thick |
| **CSF** | 100 square feet | **MBF** | 1,000 board feet | **T&G** | tongue & groove edge |
| **CY** | cubic yard | **MDF** | medium density fiberboard | **UL** | Underwriter's Laboratory |
| **d** | penny | **mm** | millimeter(s) | **USDA** | United States Dept. of Agriculture |
| **D** | depth | **Mo** | month | **UV** | ultraviolet |
| **DWV** | drain, waste & vent | **MPH** | miles per hour | **W** | width |
| **Ea** | each | **MSF** | 1,000 square feet | **X** | by or times |
| **EMT** | electric metallic tube | **NEC** | National Electrical Code | | |

**Figure 1-6**

*Abbreviations used in this manual*

# Area Modification Factors

Construction costs are higher in some areas than in other areas. Add or deduct the percentages shown on the following pages to adapt the costs in this book to your job site. Adjust your cost estimate by the appropriate percentages in this table to find the estimated cost for the site selected. Where 0% is shown, it means no modification is required.

Modification factors are listed by state and province. Areas within each state are listed by the first three digits of the postal zip code. For convenience, one representative city is identified in each three-digit zip or range of zips. Percentages are based on the average of all data points in the table. Factors listed for each state and province are the average of all data in that state or province. Figures for the zips are the average of all information in that area. And, figures in the Total column are the weighted average of factors for Labor, Material and Equipment.

The National Estimator program will apply an area modification factor for any five-digit zip you select. Click Utilities. Click Options. Then select the Area Modification Factors tab.

These percentages are composites of many costs and will not necessarily be accurate when estimating the cost of any particular part of a building. But when used to modify costs for an entire structure, they should improve the accuracy of your estimates.

## Area Modification Factors

| Location | Zip | Mat. | Lab. | Equip. | Total Wtd. Avg. |
|---|---|---|---|---|---|
| **Alabama Average** | | -1 | -15 | 0 | -7% |
| Auburn | 368 | -1 | -22 | 0 | -10% |
| Bellamy | 369 | -2 | -18 | -1 | -9% |
| Birmingham | 350-352 | -3 | 5 | -1 | 1% |
| Dothan | 363 | -1 | -20 | 0 | -10% |
| Evergreen | 364 | -2 | -30 | -1 | -15% |
| Gadsden | 359 | -4 | -20 | -1 | -11% |
| Huntsville | 358 | 1 | -10 | 0 | -4% |
| Jasper | 355 | -2 | -25 | -1 | -12% |
| Mobile | 365-366 | 4 | -1 | 1 | 1% |
| Montgomery | 360-361 | -1 | -11 | 0 | -5% |
| Scottsboro | 357 | 0 | -15 | 0 | -7% |
| Selma | 367 | -1 | -23 | 0 | -11% |
| Sheffield | 356 | -1 | -9 | 0 | -4% |
| Tuscaloosa | 354 | 1 | -16 | 0 | -7% |
| | | | | | |
| **Alaska Average** | | 13 | 49 | 4 | 30% |
| Anchorage | 995 | 16 | 67 | 5 | 40% |
| Fairbanks | 997 | 16 | 60 | 5 | 36% |
| Juneau | 998 | 17 | 45 | 6 | 29% |
| Ketchikan | 999 | 3 | 22 | 1 | 12% |
| King Salmon | 996 | 16 | 51 | 5 | 32% |

| Location | Zip | Mat. | Lab. | Equip. | Total Wtd. Avg. |
|---|---|---|---|---|---|
| **Arizona Average** | | 1 | -14 | 0 | -6% |
| Chambers | 865 | 2 | -39 | 1 | -17% |
| Douglas | 855 | 0 | -21 | 0 | -10% |
| Flagstaff | 860 | 2 | -24 | 1 | -10% |
| Kingman | 864 | 1 | -3 | 0 | -1% |
| Mesa | 852-853 | 1 | 6 | 0 | 3% |
| Phoenix | 850 | 1 | 8 | 0 | 4% |
| Prescott | 863 | 3 | -17 | 1 | -7% |
| Show Low | 859 | 2 | -22 | 1 | -9% |
| Tucson | 856-857 | 0 | -13 | 0 | -6% |
| | | | | | |
| **Arkansas Average** | | -2 | -20 | -1 | -10% |
| Batesville | 725 | 0 | -35 | 0 | -16% |
| Camden | 717 | -4 | -7 | -1 | -5% |
| Fayetteville | 727 | 0 | -11 | 0 | -5% |
| Fort Smith | 729 | -2 | -13 | -1 | -7% |
| Harrison | 726 | -1 | -39 | 0 | -18% |
| Hope | 718 | -4 | -23 | -1 | -12% |
| Hot Springs | 719 | -2 | -28 | -1 | -14% |
| Jonesboro | 724 | -1 | -25 | 0 | -12% |
| Little Rock | 720-722 | -1 | -7 | 0 | -4% |
| Pine Bluff | 716 | -4 | -10 | -1 | -7% |

# Area Modification Factors

| Location | Zip | Mat. | Lab. | Equip. | Total Wtd. Avg. |
|---|---|---|---|---|---|
| Russellville | 728 | 0 | -22 | 0 | -10% |
| West Memphis | 723 | -3 | -18 | -1 | -10% |
| | | | | | |
| **California Average** | | **2** | **20** | **1** | **10%** |
| Alhambra | 917-918 | 3 | 22 | 1 | 11% |
| Bakersfield | 932-933 | 0 | 1 | 0 | 1% |
| El Centro | 922 | 1 | 5 | 0 | 3% |
| Eureka | 955 | 1 | -3 | 0 | -1% |
| Fresno | 936-938 | 0 | 0 | 0 | 0% |
| Herlong | 961 | 2 | 7 | 1 | 4% |
| Inglewood | 902-905 | 3 | 22 | 1 | 12% |
| Irvine | 926-927 | 3 | 32 | 1 | 16% |
| Lompoc | 934 | 2 | 9 | 1 | 5% |
| Long Beach | 907-908 | 3 | 24 | 1 | 13% |
| Los Angeles | 900-901 | 3 | 20 | 1 | 11% |
| Marysville | 959 | 1 | 2 | 0 | 1% |
| Modesto | 953 | 1 | 1 | 0 | 1% |
| Mojave | 935 | 0 | 17 | 0 | 8% |
| Novato | 949 | 4 | 32 | 1 | 16% |
| Oakland | 945-947 | 3 | 42 | 1 | 21% |
| Orange | 928 | 3 | 29 | 1 | 15% |
| Oxnard | 930 | 3 | 10 | 1 | 6% |
| Pasadena | 910-912 | 4 | 23 | 1 | 12% |
| Rancho Cordova | 956-957 | 2 | 19 | 1 | 10% |
| Redding | 960 | 1 | 0 | 0 | 1% |
| Richmond | 948 | 2 | 40 | 1 | 20% |
| Riverside | 925 | 1 | 10 | 0 | 5% |
| Sacramento | 958 | 1 | 19 | 0 | 10% |
| Salinas | 939 | 3 | 12 | 1 | 7% |
| San Bernardino | 923-924 | 0 | 12 | 0 | 6% |
| San Diego | 919-921 | 3 | 20 | 1 | 11% |
| San Francisco | 941 | 3 | 57 | 1 | 28% |
| San Jose | 950-951 | 3 | 40 | 1 | 20% |
| San Mateo | 943-944 | 3 | 44 | 1 | 22% |
| Santa Barbara | 931 | 4 | 19 | 1 | 11% |
| Santa Rosa | 954 | 3 | 20 | 1 | 11% |
| Stockton | 952 | 2 | 13 | 1 | 7% |
| Sunnyvale | 940 | 3 | 44 | 1 | 22% |
| Van Nuys | 913-916 | 3 | 20 | 1 | 11% |
| Whittier | 906 | 3 | 22 | 1 | 11% |
| | | | | | |
| **Colorado Average** | | **2** | **0** | **1** | **1%** |
| Aurora | 800-801 | 3 | 17 | 1 | 9% |
| Boulder | 803-804 | 3 | 9 | 1 | 6% |
| Colorado Springs | 808-809 | 2 | 3 | 1 | 2% |
| Denver | 802 | 2 | 17 | 1 | 9% |
| Durango | 813 | 1 | -12 | 0 | -5% |
| Fort Morgan | 807 | 1 | -5 | 0 | -2% |
| Glenwood Springs | 816 | 2 | 10 | 1 | 6% |
| Grand Junction | 814-815 | 1 | -4 | 0 | -1% |
| Greeley | 806 | 2 | 8 | 1 | 5% |
| Longmont | 805 | 3 | 4 | 1 | 3% |
| Pagosa Springs | 811 | 0 | -16 | 0 | -7% |

| Location | Zip | Mat. | Lab. | Equip. | Total Wtd. Avg. |
|---|---|---|---|---|---|
| Pueblo | 810 | -1 | -14 | 0 | -7% |
| Salida | 812 | 1 | -19 | 0 | -8% |
| | | | | | |
| **Connecticut Average** | | **1** | **28** | **0** | **13%** |
| Bridgeport | 66 | 1 | 29 | 0 | 14% |
| Bristol | 60 | 1 | 33 | 0 | 16% |
| Fairfield | 64 | 2 | 33 | 1 | 16% |
| Hartford | 61 | 0 | 32 | 0 | 15% |
| New Haven | 65 | 1 | 28 | 0 | 13% |
| Norwich | 63 | 0 | 21 | 0 | 10% |
| Stamford | 068-069 | 4 | 35 | 1 | 18% |
| Waterbury | 67 | 1 | 27 | 0 | 13% |
| West Hartford | 62 | 0 | 13 | 0 | 6% |
| | | | | | |
| **Delaware Average** | | **1** | **11** | **0** | **6%** |
| Dover | 199 | 1 | -6 | 0 | -2% |
| Newark | 197 | 1 | 20 | 0 | 10% |
| Wilmington | 198 | 0 | 19 | 0 | 9% |
| | | | | | |
| **District of Columbia Average** | | **2** | **23** | **1** | **12%** |
| Washington | 200-205 | 2 | 23 | 1 | 12% |
| | | | | | |
| **Florida Average** | | **-1** | **-5** | **0** | **-3%** |
| Altamonte Springs | 327 | -1 | 1 | 0 | 0% |
| Bradenton | 342 | 0 | -3 | 0 | -1% |
| Brooksville | 346 | -1 | -12 | 0 | -6% |
| Daytona Beach | 321 | -2 | -18 | -1 | -9% |
| Fort Lauderdale | 333 | 2 | 10 | 1 | 6% |
| Fort Myers | 339 | 0 | 1 | 0 | 0% |
| Fort Pierce | 349 | -2 | -12 | -1 | -7% |
| Gainesville | 326 | -1 | -18 | 0 | -9% |
| Jacksonville | 322 | -1 | 4 | 0 | 1% |
| Lakeland | 338 | -3 | -11 | -1 | -7% |
| Melbourne | 329 | -2 | -3 | -1 | -3% |
| Miami | 330-332 | 2 | 8 | 1 | 5% |
| Naples | 341 | 3 | 6 | 1 | 5% |
| Ocala | 344 | -3 | -20 | -1 | -10% |
| Orlando | 328 | 0 | 9 | 0 | 4% |
| Panama City | 324 | -2 | -21 | -1 | -11% |
| Pensacola | 325 | -1 | -17 | 0 | -8% |
| Saint Augustine | 320 | -1 | -5 | 0 | -3% |
| Saint Cloud | 347 | -1 | 0 | 0 | -1% |
| St Petersburg | 337 | -1 | -5 | 0 | -3% |
| Tallahassee | 323 | 0 | -13 | 0 | -6% |
| Tampa | 335-336 | -1 | 2 | 0 | 0% |
| West Palm Beach | 334 | 1 | 4 | 0 | 2% |
| | | | | | |
| **Georgia Average** | | **-1** | **-9** | **0** | **-5%** |
| Albany | 317 | -2 | -23 | -1 | -12% |
| Athens | 306 | 0 | -12 | 0 | -5% |
| Atlanta | 303 | 3 | 38 | 1 | 19% |
| Augusta | 308-309 | -2 | -13 | -1 | -7% |
| Buford | 305 | 0 | -1 | 0 | 0% |

# Area Modification Factors

| Location | Zip | Mat. | Lab. | Equip. | Total Wtd. Avg. |
|---|---|---|---|---|---|
| Calhoun | 307 | -1 | -5 | 0 | -3% |
| Columbus | 318-319 | -1 | -21 | 0 | -11% |
| Dublin/Fort Valley | 310 | -3 | -19 | -1 | -10% |
| Hinesville | 313 | -2 | -10 | -1 | -5% |
| Kings Bay | 315 | -2 | -22 | -1 | -11% |
| Macon | 312 | -2 | -4 | -1 | -3% |
| Marietta | 300-302 | 1 | 15 | 0 | 7% |
| Savannah | 314 | -1 | -1 | 0 | -1% |
| Statesboro | 304 | -2 | -28 | -1 | -14% |
| Valdosta | 316 | -2 | -25 | -1 | -12% |
| **Hawaii Average** | | **17** | **48** | **6** | **31%** |
| Aliamanu | 968 | 17 | 53 | 6 | 33% |
| Ewa | 967 | 17 | 46 | 6 | 30% |
| Halawa Heights | 967 | 17 | 46 | 6 | 30% |
| Hilo | 967 | 17 | 46 | 6 | 30% |
| Honolulu | 968 | 17 | 53 | 6 | 33% |
| Kailua | 968 | 17 | 53 | 6 | 33% |
| Lualualei | 967 | 17 | 46 | 6 | 30% |
| Mililani Town | 967 | 17 | 46 | 6 | 30% |
| Pearl City | 967 | 17 | 46 | 6 | 30% |
| Wahiawa | 967 | 17 | 46 | 6 | 30% |
| Waianae | 967 | 17 | 46 | 6 | 30% |
| Wailuku (Maui) | 967 | 17 | 46 | 6 | 30% |
| **Idaho Average** | | **0** | **-16** | **0** | **-7%** |
| Boise | 837 | 1 | 1 | 0 | 1% |
| Coeur d'Alene | 838 | 0 | -17 | 0 | -7% |
| Idaho Falls | 834 | -1 | -24 | 0 | -11% |
| Lewiston | 835 | 0 | -20 | 0 | -9% |
| Meridian | 836 | 0 | -15 | 0 | -7% |
| Pocatello | 832 | -1 | -22 | 0 | -11% |
| Sun Valley | 833 | 0 | -14 | 0 | -7% |
| **Illinois Average** | | **-1** | **14** | **0** | **6%** |
| Arlington Heights | 600 | 2 | 47 | 1 | 22% |
| Aurora | 605 | 2 | 45 | 1 | 22% |
| Belleville | 622 | -3 | -2 | -1 | -2% |
| Bloomington | 617 | 0 | 5 | 0 | 3% |
| Carbondale | 629 | -3 | -9 | -1 | -6% |
| Carol Stream | 601 | 2 | 45 | 1 | 22% |
| Centralia | 628 | -3 | -14 | -1 | -8% |
| Champaign | 618 | -1 | 6 | 0 | 2% |
| Chicago | 606-608 | 2 | 45 | 1 | 22% |
| Decatur | 623 | -2 | -12 | -1 | -6% |
| Galesburg | 614 | -2 | -13 | -1 | -7% |
| Granite City | 620 | -3 | 0 | -1 | -2% |
| Green River | 612 | -2 | 3 | -1 | 0% |
| Joliet | 604 | 0 | 42 | 0 | 19% |
| Kankakee | 609 | -2 | 8 | -1 | 3% |
| Lawrenceville | 624 | -4 | -13 | -1 | -8% |
| Oak Park | 603 | 3 | 49 | 1 | 24% |
| Peoria | 615-616 | -1 | 16 | 0 | 7% |

| Location | Zip | Mat. | Lab. | Equip. | Total Wtd. Avg. |
|---|---|---|---|---|---|
| Peru | 613 | 0 | 6 | 0 | 3% |
| Quincy | 602 | 3 | 46 | 1 | 23% |
| Rockford | 610-611 | -2 | 21 | -1 | 9% |
| Springfield | 625-627 | -2 | 5 | -1 | 1% |
| Urbana | 619 | -3 | -4 | -1 | -3% |
| **Indiana Average** | | **-2** | **-2** | **-1** | **-2%** |
| Aurora | 470 | -1 | -16 | 0 | -8% |
| Bloomington | 474 | 1 | -6 | 0 | -2% |
| Columbus | 472 | 0 | -5 | 0 | -2% |
| Elkhart | 465 | -2 | 3 | -1 | 0% |
| Evansville | 476-477 | -2 | 7 | -1 | 2% |
| Fort Wayne | 467-468 | -2 | 4 | -1 | 0% |
| Gary | 463-464 | -4 | 22 | -1 | 8% |
| Indianapolis | 460-462 | -1 | 19 | 0 | 8% |
| Jasper | 475 | -2 | -15 | -1 | -8% |
| Jeffersonville | 471 | -1 | -8 | 0 | -4% |
| Kokomo | 469 | -3 | -7 | -1 | -5% |
| Lafayette | 479 | -1 | -9 | 0 | -5% |
| Muncie | 473 | -4 | -10 | -1 | -7% |
| South Bend | 466 | -4 | 6 | -1 | 1% |
| Terre Haute | 478 | -4 | -8 | -1 | -6% |
| **Iowa Average** | | **-1** | **-7** | **0** | **-4%** |
| Burlington | 526 | 0 | -8 | 0 | -4% |
| Carroll | 514 | -2 | -24 | -1 | -12% |
| Cedar Falls | 506 | -1 | -8 | 0 | -4% |
| Cedar Rapids | 522-524 | 0 | 6 | 0 | 3% |
| Cherokee | 510 | -2 | -5 | -1 | -4% |
| Council Bluffs | 515 | -2 | 11 | -1 | 4% |
| Creston | 508 | -3 | -15 | -1 | -8% |
| Davenport | 527-528 | -1 | 3 | 0 | 1% |
| Decorah | 521 | -2 | 6 | -1 | 1% |
| Des Moines | 500-503 | -2 | 10 | -1 | 4% |
| Dubuque | 520 | -1 | -1 | 0 | -1% |
| Fort Dodge | 505 | -2 | -8 | -1 | -5% |
| Mason City | 504 | 0 | -7 | 0 | -3% |
| Ottumwa | 525 | 0 | -15 | 0 | -7% |
| Sheldon | 512 | 0 | -22 | 0 | -10% |
| Shenandoah | 516 | -2 | -31 | -1 | -16% |
| Sioux City | 511 | -3 | 5 | -1 | 1% |
| Spencer | 513 | -1 | -22 | 0 | -11% |
| Waterloo | 507 | -4 | -11 | -1 | -7% |
| **Kansas Average** | | **-2** | **-19** | **-1** | **-10%** |
| Colby | 677 | 0 | -30 | 0 | -14% |
| Concordia | 669 | 0 | -31 | 0 | -15% |
| Dodge City | 678 | -2 | -13 | -1 | -7% |
| Emporia | 668 | -3 | -21 | -1 | -11% |
| Fort Scott | 667 | -3 | -15 | -1 | -8% |
| Hays | 676 | -2 | -29 | -1 | -14% |
| Hutchinson | 675 | -3 | -14 | -1 | -8% |
| Independence | 673 | -3 | -26 | -1 | -13% |

# Area Modification Factors

| Location | Zip | Mat. | Lab. | Equip. | Total Wtd. Avg. |
|---|---|---|---|---|---|
| Kansas City | 660-662 | 0 | 19 | 0 | 9% |
| Liberal | 679 | -2 | -56 | -1 | -27% |
| Salina | 674 | -2 | -15 | -1 | -8% |
| Topeka | 664-666 | -3 | -6 | -1 | -4% |
| Wichita | 670-672 | -2 | -8 | -1 | -5% |
| **Kentucky Average** | | **-1** | **-9** | **0** | **-5%** |
| Ashland | 411-412 | -3 | -11 | -1 | -7% |
| Bowling Green | 421 | 0 | -12 | 0 | -6% |
| Campton | 413-414 | -1 | 0 | 0 | -1% |
| Covington | 410 | -1 | 6 | 0 | 2% |
| Elizabethtown | 427 | -1 | -18 | 0 | -9% |
| Frankfort | 406 | 1 | -14 | 0 | -6% |
| Hazard | 417-418 | -2 | -2 | -1 | -2% |
| Hopkinsville | 422 | -2 | -16 | -1 | -9% |
| Lexington | 403-405 | 1 | -1 | 0 | 0% |
| London | 407-409 | -2 | -14 | -1 | -7% |
| Louisville | 400-402 | -1 | 3 | 0 | 1% |
| Owensboro | 423 | -2 | -17 | -1 | -9% |
| Paducah | 420 | -2 | -13 | -1 | -7% |
| Pikeville | 415-416 | -3 | -15 | -1 | -8% |
| Somerset | 425-426 | 0 | -26 | 0 | -12% |
| White Plains | 424 | -3 | 3 | -1 | 0% |
| **Louisiana Average** | | **1** | **-1** | **0** | **0%** |
| Alexandria | 713-714 | -3 | -14 | -1 | -8% |
| Baton Rouge | 707-708 | 5 | 4 | 2 | 4% |
| Houma | 703 | 3 | 17 | 1 | 10% |
| Lafayette | 705 | 0 | 6 | 0 | 3% |
| Lake Charles | 706 | -3 | -1 | -1 | -2% |
| Mandeville | 704 | 4 | 1 | 1 | 3% |
| Minden | 710 | -2 | -14 | -1 | -8% |
| Monroe | 712 | -2 | -19 | -1 | -10% |
| New Orleans | 700-701 | 4 | 17 | 1 | 10% |
| Shreveport | 711 | -2 | -11 | -1 | -6% |
| **Maine Average** | | **-1** | **-12** | **0** | **-6%** |
| Auburn | 42 | -1 | -12 | 0 | -6% |
| Augusta | 43 | -1 | -14 | 0 | -7% |
| Bangor | 44 | -2 | -11 | -1 | -6% |
| Bath | 45 | 0 | -15 | 0 | -7% |
| Brunswick | 039-040 | 1 | 0 | 0 | 0% |
| Camden | 48 | -1 | -17 | 0 | -8% |
| Cutler | 46 | -1 | -19 | 0 | -9% |
| Dexter | 49 | -2 | -11 | -1 | -6% |
| Northern Area | 47 | -2 | -26 | -1 | -13% |
| Portland | 41 | 2 | 7 | 1 | 4% |
| **Maryland Average** | | **1** | **7** | **0** | **3%** |
| Annapolis | 214 | 3 | 21 | 1 | 11% |
| Baltimore | 210-212 | -1 | 18 | 0 | 8% |
| Bethesda | 208-209 | 3 | 33 | 1 | 17% |
| Church Hill | 216 | 2 | -5 | 1 | -1% |
| Cumberland | 215 | -4 | -20 | -1 | -11% |
| Elkton | 219 | 2 | -8 | 1 | -3% |
| Frederick | 217 | 1 | 12 | 0 | 6% |
| Laurel | 206-207 | 2 | 21 | 1 | 11% |
| Salisbury | 218 | 1 | -13 | 0 | -6% |
| **Massachusetts Average** | | **2** | **30** | **1** | **15%** |
| Ayer | 015-016 | 1 | 21 | 0 | 10% |
| Bedford | 17 | 2 | 42 | 1 | 21% |
| Boston | 021-022 | 3 | 76 | 1 | 37% |
| Brockton | 023-024 | 3 | 49 | 1 | 24% |
| Cape Cod | 26 | 2 | 15 | 1 | 8% |
| Chicopee | 10 | 1 | 20 | 0 | 10% |
| Dedham | 19 | 2 | 38 | 1 | 19% |
| Fitchburg | 14 | 2 | 31 | 1 | 15% |
| Hingham | 20 | 3 | 51 | 1 | 25% |
| Lawrence | 18 | 2 | 39 | 1 | 19% |
| Nantucket | 25 | 2 | 26 | 1 | 13% |
| New Bedford | 27 | 2 | 16 | 1 | 8% |
| Northfield | 13 | 1 | 3 | 0 | 2% |
| Pittsfield | 12 | 0 | 3 | 0 | 2% |
| Springfield | 11 | -1 | 26 | 0 | 12% |
| **Michigan Average** | | **-2** | **8** | **-1** | **3%** |
| Battle Creek | 490-491 | -3 | 5 | -1 | 1% |
| Detroit | 481-482 | 0 | 29 | 0 | 13% |
| Flint | 484-485 | -3 | 3 | -1 | 0% |
| Grand Rapids | 493-495 | -1 | 7 | 0 | 3% |
| Grayling | 497 | 1 | -11 | 0 | -5% |
| Jackson | 492 | -3 | 3 | -1 | 0% |
| Lansing | 488-489 | 0 | 9 | 0 | 4% |
| Marquette | 498-499 | -1 | -2 | 0 | -2% |
| Pontiac | 483 | -2 | 37 | -1 | 16% |
| Royal Oak | 480 | -2 | 30 | -1 | 13% |
| Saginaw | 486-487 | -2 | -7 | -1 | -4% |
| Traverse City | 496 | -1 | -2 | 0 | -1% |
| **Minnesota Average** | | **0** | **4** | **0** | **2%** |
| Bemidji | 566 | -1 | -2 | 0 | -1% |
| Brainerd | 564 | 0 | -1 | 0 | -1% |
| Duluth | 556-558 | -2 | 7 | -1 | 2% |
| Fergus Falls | 565 | -1 | -12 | 0 | -6% |
| Magnolia | 561 | 0 | -19 | 0 | -9% |
| Mankato | 560 | 1 | 7 | 0 | 4% |
| Minneapolis | 553-555 | 1 | 32 | 0 | 15% |
| Rochester | 559 | 0 | 0 | 0 | 0% |
| St Cloud | 563 | -1 | 7 | 0 | 3% |
| St Paul | 550-551 | 1 | 30 | 0 | 14% |
| Thief River Falls | 567 | 0 | -4 | 0 | -2% |
| Willmar | 562 | -1 | 3 | 0 | 1% |
| **Mississippi Average** | | **-2** | **-21** | **-1** | **-11%** |
| Clarksdale | 386 | -3 | -15 | -1 | -8% |

# Area Modification Factors

| Location | Zip | Mat. | Lab. | Equip. | Total Wtd. Avg. |
|---|---|---|---|---|---|
| Columbus | 397 | -1 | -23 | 0 | -11% |
| Greenville | 387 | -4 | -33 | -1 | -17% |
| Greenwood | 389 | -3 | -25 | -1 | -13% |
| Gulfport | 395 | 3 | -7 | 1 | -2% |
| Jackson | 390-392 | -3 | -12 | -1 | -7% |
| Laurel | 394 | -3 | -20 | -1 | -11% |
| McComb | 396 | -2 | -33 | -1 | -16% |
| Meridian | 393 | -2 | -22 | -1 | -11% |
| Tupelo | 388 | -2 | -23 | -1 | -12% |
| | | | | | |
| **Missouri Average** | | **-1** | **-8** | **0** | **-4%** |
| Cape Girardeau | 637 | -2 | -13 | -1 | -7% |
| Caruthersville | 638 | -2 | -22 | -1 | -11% |
| Chillicothe | 646 | -2 | -22 | -1 | -11% |
| Columbia | 652 | 1 | -5 | 0 | -2% |
| East Lynne | 647 | -1 | -10 | 0 | -5% |
| Farmington | 636 | -3 | -22 | -1 | -12% |
| Hannibal | 634 | 1 | 6 | 0 | 3% |
| Independence | 640 | -1 | 17 | 0 | 7% |
| Jefferson City | 650-651 | 1 | -8 | 0 | -3% |
| Joplin | 648 | -2 | -23 | -1 | -12% |
| Kansas City | 641 | -1 | 21 | 0 | 9% |
| Kirksville | 635 | 0 | -36 | 0 | -17% |
| Knob Noster | 653 | 0 | -15 | 0 | -7% |
| Lebanon | 654-655 | -1 | -28 | 0 | -14% |
| Poplar Bluff | 639 | -1 | -21 | 0 | -10% |
| Saint Charles | 633 | 1 | 10 | 0 | 5% |
| Saint Joseph | 644-645 | -3 | 2 | -1 | -1% |
| Springfield | 656-658 | -2 | -18 | -1 | -9% |
| St Louis | 630-631 | -2 | 28 | -1 | 12% |
| | | | | | |
| **Montana Average** | | **0** | **-11** | **0** | **-5%** |
| Billings | 590-591 | 0 | -2 | 0 | -1% |
| Butte | 597 | 1 | -12 | 0 | -5% |
| Fairview | 592 | 0 | -16 | 0 | -8% |
| Great Falls | 594 | 0 | -2 | 0 | -1% |
| Havre | 595 | -1 | -25 | 0 | -12% |
| Helena | 596 | 0 | -9 | 0 | -4% |
| Kalispell | 599 | 2 | -14 | 1 | -6% |
| Miles City | 593 | 0 | -7 | 0 | -3% |
| Missoula | 598 | 1 | -15 | 0 | -6% |
| | | | | | |
| **Nebraska Average** | | **0** | **-19** | **0** | **-9%** |
| Alliance | 693 | -1 | -26 | 0 | -12% |
| Columbus | 686 | 0 | -14 | 0 | -7% |
| Grand Island | 688 | 0 | -20 | 0 | -9% |
| Hastings | 689 | 0 | -23 | 0 | -11% |
| Lincoln | 683-685 | 0 | -6 | 0 | -3% |
| McCook | 690 | 0 | -31 | 0 | -14% |
| Norfolk | 687 | -2 | -25 | -1 | -13% |
| North Platte | 691 | 0 | -26 | 0 | -12% |
| Omaha | 680-681 | -1 | 5 | 0 | 2% |
| Valentine | 692 | -1 | -24 | 0 | -12% |

| Location | Zip | Mat. | Lab. | Equip. | Total Wtd. Avg. |
|---|---|---|---|---|---|
| **Nevada Average** | | **2** | **8** | **1** | **5%** |
| Carson City | 897 | 2 | -3 | 1 | 0% |
| Elko | 898 | 0 | 30 | 0 | 14% |
| Ely | 893 | 2 | -19 | 1 | -8% |
| Fallon | 894 | 2 | 11 | 1 | 6% |
| Las Vegas | 889-891 | 2 | 20 | 1 | 10% |
| Reno | 895 | 2 | 10 | 1 | 6% |
| | | | | | |
| **New Hampshire Avg** | | **1** | **3** | **0** | **2%** |
| Charlestown | 36 | 1 | -4 | 0 | -2% |
| Concord | 34 | 1 | 7 | 0 | 4% |
| Dover | 38 | 0 | 10 | 0 | 5% |
| Lebanon | 37 | 2 | -3 | 1 | -1% |
| Littleton | 35 | -1 | -12 | 0 | -6% |
| Manchester | 032-033 | 0 | 10 | 0 | 5% |
| New Boston | 030-031 | 2 | 16 | 1 | 8% |
| | | | | | |
| **New Jersey Average** | | **1** | **35** | **0** | **17%** |
| Atlantic City | 080-084 | -2 | 23 | -1 | 10% |
| Brick | 87 | 2 | 21 | 1 | 11% |
| Dover | 78 | 1 | 38 | 0 | 18% |
| Edison | 088-089 | 0 | 41 | 0 | 19% |
| Hackensack | 76 | 3 | 31 | 1 | 16% |
| Monmouth | 77 | 3 | 43 | 1 | 21% |
| Newark | 071-073 | 1 | 36 | 0 | 17% |
| Passaic | 70 | 2 | 37 | 1 | 18% |
| Paterson | 074-075 | 2 | 31 | 1 | 15% |
| Princeton | 85 | -2 | 38 | -1 | 16% |
| Summit | 79 | 3 | 49 | 1 | 24% |
| Trenton | 86 | -3 | 34 | -1 | 14% |
| | | | | | |
| **New Mexico Average** | | **0** | **-22** | **0** | **-10%** |
| Alamogordo | 883 | 0 | -24 | 0 | -11% |
| Albuquerque | 870-871 | 2 | -5 | 1 | -1% |
| Clovis | 881 | -2 | -27 | -1 | -13% |
| Farmington | 874 | 2 | -14 | 1 | -6% |
| Fort Sumner | 882 | -3 | -14 | -1 | -8% |
| Gallup | 873 | 1 | -14 | 0 | -6% |
| Holman | 877 | 2 | -25 | 1 | -10% |
| Las Cruces | 880 | -1 | -28 | 0 | -13% |
| Santa Fe | 875 | 3 | -19 | 1 | -7% |
| Socorro | 878 | 1 | -31 | 0 | -14% |
| Truth or Consequences | 879 | -1 | -32 | 0 | -15% |
| Tucumcari | 884 | -1 | -30 | 0 | -14% |
| | | | | | |
| **New York Average** | | **0** | **22** | **0** | **10%** |
| Albany | 120-123 | -1 | 19 | 0 | 8% |
| Amityville | 117 | 2 | 30 | 1 | 15% |
| Batavia | 140 | -3 | 9 | -1 | 2% |
| Binghamton | 137-139 | -3 | 5 | -1 | 0% |
| Bronx | 104 | 2 | 34 | 1 | 17% |
| Brooklyn | 112 | 3 | 21 | 1 | 11% |
| Buffalo | 142 | -4 | 10 | -1 | 3% |

# Area Modification Factors

| Location | Zip | Mat. | Lab. | Equip. | Total Wtd. Avg. |
|---|---|---|---|---|---|
| Elmira | 149 | -4 | 1 | -1 | -2% |
| Flushing | 113 | 3 | 51 | 1 | 25% |
| Garden City | 115 | 3 | 40 | 1 | 20% |
| Hicksville | 118 | 3 | 36 | 1 | 18% |
| Ithaca | 148 | -4 | 0 | -1 | -2% |
| Jamaica | 114 | 3 | 48 | 1 | 24% |
| Jamestown | 147 | -4 | -8 | -1 | -6% |
| Kingston | 124 | 0 | -4 | 0 | -2% |
| Long Island | 111 | 3 | 78 | 1 | 37% |
| Montauk | 119 | 1 | 30 | 0 | 14% |
| New York (Manhattan) | 100-102 | 3 | 77 | 1 | 37% |
| New York City | 100-102 | 3 | 77 | 1 | 37% |
| Newcomb | 128 | -1 | 2 | 0 | 0% |
| Niagara Falls | 143 | -4 | -4 | -1 | -4% |
| Plattsburgh | 129 | 1 | -6 | 0 | -2% |
| Poughkeepsie | 125-126 | 1 | 8 | 0 | 4% |
| Queens | 110 | 4 | 48 | 1 | 24% |
| Rochester | 144-146 | -3 | 12 | -1 | 4% |
| Rockaway | 116 | 3 | 33 | 1 | 17% |
| Rome | 133-134 | -3 | -4 | -1 | -4% |
| Staten Island | 103 | 3 | 39 | 1 | 19% |
| Stewart | 127 | -1 | 4 | 0 | 1% |
| Syracuse | 130-132 | -3 | 9 | -1 | 3% |
| Tonawanda | 141 | -4 | 6 | -1 | 1% |
| Utica | 135 | -4 | -4 | -1 | -4% |
| Watertown | 136 | -3 | -6 | -1 | -4% |
| West Point | 109 | 1 | 26 | 0 | 13% |
| White Plains | 105-108 | 3 | 40 | 1 | 20% |
| **North Carolina Avg** | | **0** | **-8** | **0** | **-4%** |
| Asheville | 287-289 | 0 | -15 | 0 | -6% |
| Charlotte | 280-282 | 0 | 11 | 0 | 5% |
| Durham | 277 | 2 | 1 | 1 | 2% |
| Elizabeth City | 279 | 1 | -12 | 0 | -5% |
| Fayetteville | 283 | -1 | -19 | 0 | -9% |
| Goldsboro | 275 | 1 | -1 | 0 | 0% |
| Greensboro | 274 | 0 | -1 | 0 | 0% |
| Hickory | 286 | -1 | -17 | 0 | -8% |
| Kinston | 285 | -1 | -22 | 0 | -11% |
| Raleigh | 276 | 3 | 6 | 1 | 4% |
| Rocky Mount | 278 | -1 | -14 | 0 | -7% |
| Wilmington | 284 | 1 | -16 | 0 | -7% |
| Winston-Salem | 270-273 | 0 | -5 | 0 | -2% |
| **North Dakota Average** | | **-1** | **-7** | **0** | **-4%** |
| Bismarck | 585 | 0 | -7 | 0 | -3% |
| Dickinson | 586 | -1 | -9 | 0 | -4% |
| Fargo | 580-581 | 0 | 0 | 0 | 0% |
| Grand Forks | 582 | 0 | 0 | 0 | 0% |
| Jamestown | 584 | -1 | -13 | 0 | -6% |
| Minot | 587 | -1 | -2 | 0 | -1% |
| Nekoma | 583 | -1 | -18 | 0 | -9% |
| Williston | 588 | -1 | -9 | 0 | -5% |

| Location | Zip | Mat. | Lab. | Equip. | Total Wtd. Avg. |
|---|---|---|---|---|---|
| **Ohio Average** | | **-2** | **1** | **-1** | **0%** |
| Akron | 442-443 | -1 | 5 | 0 | 2% |
| Canton | 446-447 | -2 | -2 | -1 | -2% |
| Chillicothe | 456 | -2 | -8 | -1 | -5% |
| Cincinnati | 450-452 | -1 | 11 | 0 | 5% |
| Cleveland | 440-441 | -2 | 16 | -1 | 6% |
| Columbus | 432 | 0 | 18 | 0 | 8% |
| Dayton | 453-455 | -3 | 4 | -1 | 0% |
| Lima | 458 | -3 | -7 | -1 | -5% |
| Marietta | 457 | -3 | -7 | -1 | -5% |
| Marion | 433 | -3 | -11 | -1 | -7% |
| Newark | 430-431 | -1 | 14 | 0 | 6% |
| Sandusky | 448-449 | -2 | 1 | -1 | 0% |
| Steubenville | 439 | -3 | -9 | -1 | -6% |
| Toledo | 434-436 | -1 | 13 | 0 | 5% |
| Warren | 444 | -4 | -6 | -1 | -5% |
| Youngstown | 445 | -5 | 1 | -2 | -2% |
| Zanesville | 437-438 | -2 | -6 | -1 | -4% |
| **Oklahoma Average** | | **-3** | **-19** | **-1** | **-10%** |
| Adams | 739 | -2 | -28 | -1 | -14% |
| Ardmore | 734 | -3 | -20 | -1 | -11% |
| Clinton | 736 | -3 | -23 | -1 | -12% |
| Durant | 747 | -4 | -18 | -1 | -11% |
| Enid | 737 | -4 | -13 | -1 | -8% |
| Lawton | 735 | -3 | -26 | -1 | -13% |
| McAlester | 745 | -4 | -33 | -1 | -17% |
| Muskogee | 744 | -2 | -24 | -1 | -12% |
| Norman | 730 | -2 | -13 | -1 | -7% |
| Oklahoma City | 731 | -2 | -9 | -1 | -5% |
| Ponca City | 746 | -3 | -15 | -1 | -9% |
| Poteau | 749 | -2 | -20 | -1 | -10% |
| Pryor | 743 | -2 | -30 | -1 | -15% |
| Shawnee | 748 | -4 | -30 | -1 | -16% |
| Tulsa | 740-741 | -1 | -3 | 0 | -2% |
| Woodward | 738 | -4 | -3 | -1 | -3% |
| **Oregon Average** | | **1** | **-9** | **0** | **-4%** |
| Adrian | 979 | 0 | -35 | 0 | -16% |
| Bend | 977 | 1 | -6 | 0 | -2% |
| Eugene | 974 | 2 | -1 | 1 | 1% |
| Grants Pass | 975 | 2 | -13 | 1 | -5% |
| Klamath Falls | 976 | 3 | -26 | 1 | -11% |
| Pendleton | 978 | 0 | -12 | 0 | -6% |
| Portland | 970-972 | 2 | 20 | 1 | 10% |
| Salem | 973 | 2 | -1 | 1 | 0% |
| **Pennsylvania Average** | | **-3** | **1** | **-1** | **-1%** |
| Allentown | 181 | -2 | 19 | -1 | 8% |
| Altoona | 166 | -3 | -16 | -1 | -9% |
| Beaver Springs | 178 | -3 | -2 | -1 | -2% |
| Bethlehem | 180 | -1 | 20 | 0 | 9% |
| Bradford | 167 | -4 | -15 | -1 | -9% |

# Area Modification Factors

| Location | Zip | Mat. | Lab. | Equip. | Total Wtd. Avg. |
|---|---|---|---|---|---|
| Butler | 160 | -4 | -1 | -1 | -3% |
| Chambersburg | 172 | -1 | -9 | 0 | -5% |
| Clearfield | 168 | 1 | -4 | 0 | -1% |
| DuBois | 158 | -2 | -25 | -1 | -12% |
| East Stroudsburg | 183 | 0 | 0 | 0 | 0% |
| Erie | 164-165 | -3 | -7 | -1 | -5% |
| Genesee | 169 | -4 | -23 | -1 | -13% |
| Greensburg | 156 | -4 | -3 | -1 | -3% |
| Harrisburg | 170-171 | -2 | 11 | -1 | 4% |
| Hazleton | 182 | -3 | -5 | -1 | -4% |
| Johnstown | 159 | -4 | -18 | -1 | -11% |
| Kittanning | 162 | -4 | -13 | -1 | -8% |
| Lancaster | 175-176 | -3 | 22 | -1 | 9% |
| Meadville | 163 | -4 | -22 | -1 | -12% |
| Montrose | 188 | -3 | -16 | -1 | -9% |
| New Castle | 161 | -4 | -2 | -1 | -3% |
| Philadelphia | 190-191 | -3 | 36 | -1 | 15% |
| Pittsburgh | 152 | -4 | 16 | -1 | 5% |
| Pottsville | 179 | -4 | -11 | -1 | -7% |
| Punxsutawney | 157 | -4 | -18 | -1 | -11% |
| Reading | 195-196 | -4 | 12 | -1 | 4% |
| Scranton | 184-185 | -2 | -1 | -1 | -1% |
| Somerset | 155 | -4 | -15 | -1 | -9% |
| Southeastern | 193 | 0 | 24 | 0 | 11% |
| Uniontown | 154 | -4 | -11 | -1 | -8% |
| Valley Forge | 194 | -3 | 39 | -1 | 17% |
| Warminster | 189 | -1 | 34 | 0 | 15% |
| Warrendale | 150-151 | -4 | 14 | -1 | 4% |
| Washington | 153 | -4 | 12 | -1 | 4% |
| Wilkes Barre | 186-187 | -3 | -2 | -1 | -2% |
| Williamsport | 177 | -3 | -4 | -1 | -3% |
| York | 173-174 | -3 | 12 | -1 | 4% |
| **Rhode Island Average** | | **1** | **16** | **0** | **8%** |
| Bristol | 28 | 1 | 14 | 0 | 7% |
| Coventry | 28 | 1 | 14 | 0 | 7% |
| Cranston | 29 | 1 | 19 | 0 | 9% |
| Davisville | 28 | 1 | 14 | 0 | 7% |
| Narragansett | 28 | 1 | 14 | 0 | 7% |
| Newport | 28 | 1 | 14 | 0 | 7% |
| Providence | 29 | 1 | 19 | 0 | 9% |
| Warwick | 28 | 1 | 14 | 0 | 7% |
| **South Carolina Avg** | | **-1** | **-6** | **0** | **-3%** |
| Aiken | 298 | 0 | 10 | 0 | 4% |
| Beaufort | 299 | -1 | -2 | 0 | -2% |
| Charleston | 294 | -2 | -2 | -1 | -2% |
| Columbia | 290-292 | -1 | -10 | 0 | -5% |
| Greenville | 296 | 0 | -7 | 0 | -4% |
| Myrtle Beach | 295 | 0 | -18 | 0 | -8% |
| Rock Hill | 297 | -1 | -12 | 0 | -6% |
| Spartanburg | 293 | -2 | -9 | -1 | -5% |
| **South Dakota Average** | | **-1** | **-20** | **0** | **-10%** |
| Aberdeen | 574 | -1 | -14 | 0 | -7% |
| Mitchell | 573 | 0 | -22 | 0 | -10% |
| Mobridge | 576 | -1 | -41 | 0 | -20% |
| Pierre | 575 | -1 | -25 | 0 | -12% |
| Rapid City | 577 | -1 | -17 | 0 | -8% |
| Sioux Falls | 570-571 | 0 | -5 | 0 | -2% |
| Watertown | 572 | -1 | -19 | 0 | -9% |
| **Tennessee Average** | | **-1** | **-6** | **0** | **-3%** |
| Chattanooga | 374 | -1 | 1 | 0 | 0% |
| Clarksville | 370 | 1 | 5 | 0 | 3% |
| Cleveland | 373 | -1 | -4 | 0 | -3% |
| Columbia | 384 | -1 | -16 | 0 | -8% |
| Cookeville | 385 | 0 | -19 | 0 | -8% |
| Jackson | 383 | -1 | -12 | 0 | -6% |
| Kingsport | 376 | 0 | -14 | 0 | -7% |
| Knoxville | 377-379 | -1 | -6 | 0 | -3% |
| McKenzie | 382 | -2 | -17 | -1 | -9% |
| Memphis | 380-381 | -2 | 3 | -1 | 1% |
| Nashville | 371-372 | 1 | 14 | 0 | 7% |
| **Texas Average** | | **-2** | **-8** | **-1** | **-5%** |
| Abilene | 795-796 | -4 | -20 | -1 | -11% |
| Amarillo | 790-791 | -2 | -12 | -1 | -7% |
| Arlington | 760 | -1 | 7 | 0 | 3% |
| Austin | 786-787 | 0 | 10 | 0 | 5% |
| Bay City | 774 | -2 | 30 | -1 | 13% |
| Beaumont | 776-777 | -3 | 1 | -1 | -2% |
| Brownwood | 768 | -3 | -22 | -1 | -12% |
| Bryan | 778 | -1 | -18 | 0 | -9% |
| Childress | 792 | -3 | -42 | -1 | -21% |
| Corpus Christi | 783-784 | -3 | -8 | -1 | -5% |
| Dallas | 751-753 | -1 | 14 | 0 | 6% |
| Del Rio | 788 | -3 | -37 | -1 | -19% |
| El Paso | 798-799 | -3 | -25 | -1 | -13% |
| Fort Worth | 761-762 | -2 | 8 | -1 | 3% |
| Galveston | 775 | -3 | 14 | -1 | 5% |
| Giddings | 789 | 0 | -13 | 0 | -6% |
| Greenville | 754 | -3 | 7 | -1 | 2% |
| Houston | 770-772 | -1 | 17 | 0 | 7% |
| Huntsville | 773 | -2 | 15 | -1 | 6% |
| Longview | 756 | -3 | -8 | -1 | -5% |
| Lubbock | 793-794 | -3 | -22 | -1 | -12% |
| Lufkin | 759 | -3 | -23 | -1 | -12% |
| McAllen | 785 | -4 | -32 | -1 | -17% |
| Midland | 797 | -4 | -4 | -1 | -4% |
| Palestine | 758 | -3 | -20 | -1 | -10% |
| Plano | 750 | -1 | 14 | 0 | 6% |
| San Angelo | 769 | -4 | -16 | -1 | -9% |
| San Antonio | 780-782 | -3 | -3 | -1 | -3% |
| Texarkana | 755 | -3 | -16 | -1 | -9% |
| Tyler | 757 | -1 | -15 | 0 | -8% |

## Area Modification Factors

| Location | Zip | Mat. | Lab. | Equip. | Total Wtd. Avg. |
|---|---|---|---|---|---|
| Victoria | 779 | -3 | -2 | -1 | -2% |
| Waco | 765-767 | -3 | -12 | -1 | -7% |
| Wichita Falls | 763 | -3 | -19 | -1 | -10% |
| Woodson | 764 | -3 | -15 | -1 | -9% |
| **Utah Average** | | **2** | **-15** | **1** | **-6%** |
| Clearfield | 840 | 2 | -8 | 1 | -3% |
| Green River | 845 | 1 | -9 | 0 | -4% |
| Ogden | 843-844 | 1 | -26 | 0 | -12% |
| Provo | 846-847 | 2 | -24 | 1 | -10% |
| Salt Lake City | 841 | 2 | -6 | 1 | -1% |
| **Vermont Average** | | **0** | **-10** | **0** | **-4%** |
| Albany | 58 | 1 | -15 | 0 | -6% |
| Battleboro | 53 | 0 | -7 | 0 | -3% |
| Beecher Falls | 59 | 1 | -18 | 0 | -8% |
| Bennington | 52 | -1 | -10 | 0 | -5% |
| Burlington | 54 | 2 | 5 | 1 | 3% |
| Montpelier | 56 | 2 | -10 | 1 | -4% |
| Rutland | 57 | -1 | -15 | 0 | -7% |
| Springfield | 51 | -1 | -9 | 0 | -5% |
| White River Junction | 50 | 1 | -12 | 0 | -5% |
| **Virginia Average** | | **-1** | **-11** | **0** | **-5%** |
| Abingdon | 242 | -2 | -25 | -1 | -13% |
| Alexandria | 220-223 | 3 | 31 | 1 | 16% |
| Charlottesville | 229 | 1 | -15 | 0 | -6% |
| Chesapeake | 233 | -1 | -8 | 0 | -4% |
| Culpeper | 227 | 2 | -1 | 1 | 0% |
| Farmville | 239 | -2 | -30 | -1 | -15% |
| Fredericksburg | 224-225 | 1 | -5 | 0 | -2% |
| Galax | 243 | -2 | -28 | -1 | -14% |
| Harrisonburg | 228 | 1 | -17 | 0 | -7% |
| Lynchburg | 245 | -2 | -22 | -1 | -11% |
| Norfolk | 235-237 | 0 | -2 | 0 | -1% |
| Petersburg | 238 | -2 | -8 | -1 | -5% |
| Radford | 241 | -1 | -24 | 0 | -12% |
| Richmond | 232 | -1 | 13 | 0 | 5% |
| Roanoke | 240 | -1 | -19 | 0 | -10% |
| Staunton | 244 | 0 | -20 | 0 | -9% |
| Tazewell | 246 | -3 | -22 | -1 | -12% |
| Virginia Beach | 234 | 0 | -6 | 0 | -3% |
| Williamsburg | 230-231 | 0 | -2 | 0 | -1% |
| Winchester | 226 | 0 | -7 | 0 | -3% |
| **Washington Average** | | **1** | **0** | **0** | **1%** |
| Clarkston | 994 | -1 | -1 | 0 | -1% |
| Everett | 982 | 2 | 6 | 1 | 4% |
| Olympia | 985 | 2 | 1 | 1 | 1% |
| Pasco | 993 | 0 | -5 | 0 | -3% |
| Seattle | 980-981 | 3 | 26 | 1 | 14% |
| Spokane | 990-992 | 0 | -7 | 0 | -3% |
| Tacoma | 983-984 | 1 | 10 | 0 | 5% |
| Vancouver | 986 | 2 | 4 | 1 | 3% |

| Location | Zip | Mat. | Lab. | Equip. | Total Wtd. Avg. |
|---|---|---|---|---|---|
| Wenatchee | 988 | 1 | -17 | 0 | -7% |
| Yakima | 989 | 1 | -14 | 0 | -6% |
| **West Virginia Average** | | **-2** | **-16** | **-1** | **-9%** |
| Beckley | 258-259 | 0 | -14 | 0 | -6% |
| Bluefield | 247-248 | -2 | -19 | -1 | -10% |
| Charleston | 250-253 | 0 | 1 | 0 | 0% |
| Clarksburg | 263-264 | -3 | -16 | -1 | -9% |
| Fairmont | 266 | 0 | -22 | 0 | -10% |
| Huntington | 255-257 | -1 | -12 | 0 | -6% |
| Lewisburg | 249 | -2 | -22 | -1 | -11% |
| Martinsburg | 254 | -1 | -14 | 0 | -7% |
| Morgantown | 265 | -3 | -23 | -1 | -12% |
| New Martinsville | 262 | -2 | -26 | -1 | -13% |
| Parkersburg | 261 | -3 | -5 | -1 | -4% |
| Romney | 267 | -4 | -22 | -1 | -12% |
| Sugar Grove | 268 | -3 | -29 | -1 | -15% |
| Wheeling | 260 | -3 | -4 | -1 | -3% |
| **Wisconsin Average** | | **0** | **6** | **0** | **3%** |
| Amery | 540 | 1 | 8 | 0 | 4% |
| Beloit | 535 | 1 | 17 | 0 | 8% |
| Clam Lake | 545 | -1 | -5 | 0 | -3% |
| Eau Claire | 547 | 0 | 2 | 0 | 1% |
| Green Bay | 541-543 | 0 | 7 | 0 | 3% |
| La Crosse | 546 | -1 | -3 | 0 | -2% |
| Ladysmith | 548 | -2 | -9 | -1 | -5% |
| Madison | 537 | 2 | 23 | 1 | 12% |
| Milwaukee | 530-534 | 0 | 21 | 0 | 10% |
| Oshkosh | 549 | -1 | 10 | 0 | 4% |
| Portage | 539 | 0 | 9 | 0 | 4% |
| Prairie du Chien | 538 | -1 | -3 | 0 | -2% |
| Wausau | 544 | -1 | 5 | 0 | 2% |
| **Wyoming Average** | | **0** | **-10** | **0** | **-5%** |
| Casper | 826 | -2 | -3 | -1 | -2% |
| Cheyenne/Laramie | 820 | 1 | -14 | 0 | -6% |
| Gillette | 827 | -1 | 4 | 0 | 1% |
| Powell | 824 | 0 | -17 | 0 | -8% |
| Rawlins | 823 | -1 | -5 | 0 | -3% |
| Riverton | 825 | -1 | -16 | 0 | -8% |
| Rock Springs | 829-831 | 0 | 2 | 0 | 1% |
| Sheridan | 828 | 0 | -18 | 0 | -8% |
| Wheatland | 822 | 0 | -23 | 0 | -11% |
| **UNITED STATES TERRITORIES** | | | | | |
| Guam | | 53 | -21 | -5 | 18% |
| Puerto Rico | | 2 | -47 | -5 | -21% |
| **VIRGIN ISLANDS (U.S.)** | | | | | |
| St. Croix | | 18 | -15 | -4 | 2% |
| St. John | | 52 | -15 | -4 | 20% |
| St. Thomas | | 23 | -15 | -4 | 5% |

# Canadian Area Modification Factors

To find the cost in Canada in Canadian dollars, increase the costs in this book by the appropriate percentage listed below. These figures assume an exchange rate of $1.10 Canadian to $1.00 U.S.

| Location | Mat. | Lab. | Equip. | Total Wtd. Avg. |
|---|---|---|---|---|
| **Alberta Average** | **40.7** | **38.1** | **37.2** | **39.8** |
| Calgary | 39.4 | 33.7 | 34.1 | 37 |
| Edmonton | 40.9 | 39.6 | 36.4 | 40.6 |
| Ft. McMurray | 41.6 | 41.1 | 40.9 | 41.8 |
| | | | | |
| **British Columbia Avg** | **39.4** | **39.2** | **36.9** | **39.6** |
| Fraser Valley | 39.4 | 36 | 36.4 | 38.1 |
| Okanagan | 34.9 | 32.6 | 36.4 | 34.2 |
| Vancouver | 43.9 | 49.1 | 37.9 | 46.6 |
| | | | | |
| **Manitoba Average** | **42.2** | **43.4** | **36.8** | **43** |
| North Manitoba | 46.1 | 55.8 | 43.1 | 50.9 |
| South Manitoba | 42.2 | 43.1 | 34.1 | 42.8 |
| Selkirk | 39.4 | 34.2 | 35.6 | 37.3 |
| Winnipeg | 40.9 | 40.6 | 34.1 | 41 |
| | | | | |
| **New Brunswick Avg** | **36.4** | **36.4** | **39.4** | **36.9** |
| Moncton | 36.4 | 36.4 | 39.4 | 36.9 |

| Location | Mat. | Lab. | Equip. | Total Wtd. Avg. |
|---|---|---|---|---|
| **Nova Scotia Average** | **41.9** | **44.5** | **40.2** | **43.5** |
| Amherst | 41.6 | 43.3 | 40.2 | 42.8 |
| Nova Scotia | 40.2 | 37.9 | 40.2 | 39.6 |
| Sydney | 43.9 | 52.3 | 40.2 | 48.1 |
| | | | | |
| **Newfoundland/Labrador Average** | **37.9** | **35.6** | **41.6** | **37.4** |
| | | | | |
| **Ontario Average** | **41.4** | **42.2** | **34.9** | **42** |
| London | 42.4 | 46.5 | 34.9 | 44.5 |
| Thunder Bay | 37.9 | 33.7 | 34.1 | 36.3 |
| Toronto | 43.9 | 46.5 | 35.6 | 45.3 |
| | | | | |
| **Quebec Average** | **41.3** | **47.9** | **40.2** | **44.7** |
| Montreal | 42.4 | 53.2 | 40.2 | 47.8 |
| Quebec | 40.2 | 42.5 | 40.2 | 41.7 |
| | | | | |
| **Saskatchewan Avg** | **41.2** | **46.9** | **36.2** | **44.1** |
| La Ronge | 42.4 | 51.3 | 40.2 | 46.9 |
| Prince Albert | 40.9 | 45.8 | 34.1 | 43.4 |
| Saskatoon | 40.2 | 43.5 | 34.1 | 41.9 |

# *Demolition*

# 2

Nearly all home improvement projects require some breaking out and removing of existing materials and disposal of debris. Most will require protection of adjacent surfaces, taking safety precautions (such as setting up barricades), closing off doorways or windows, and normal cleanup. The demolition figures in this chapter include these tasks and will apply on most home improvement jobs that don't involve unusual conditions or complications. Costs will be higher when access to the work is limited, when your crew doesn't have complete control of the construction site, or when debris must be moved longer distances or around obstacles. All figures assume that debris is piled on site or in a bin. Add the cost of hauling to the nearest disposal site and tippage charges, if required. No salvage value is assumed except as noted.

Many of the following chapters include estimated costs to remove the old and replace with new materials of a similar description. For example, you'll find costs for removing shingles and replacing shingles in the roofing chapter. That's appropriate because they're part of the same task and usually performed by the same contractor. The estimates in this chapter are for demolition only and don't consider the cost of replacing what has been removed.

Of all construction tasks, demolition is probably the most difficult to estimate with any certainty. You're never sure what's in a wall to be demolished until the job begins. That's why it's best to exclude from your demolition bid what isn't in your estimate. Use the checklist to the right to help identify potential problems before they develop.

## Demolition Checklist

❒ Can you anticipate any problem getting a demolition permit for this job?

❒ Is there a noise ordinance which will limit hours of operation or selection of equipment?

❒ Can the debris box be located close to the work?

❒ Can a trencher get to the job site?

❒ Is the site accessible to wheeled equipment?

❒ Will the debris have to be hand-carried to the roll-off box or gondola?

❒ Is there a direct route from the demolition site to the debris box?

❒ Will you need to re-route any plumbing, electrical, phone or gas lines?

❒ Will you need to remove and replace lawn sprinklers?

❒ Will you need to cut back or remove any bushes or trees?

❒ Will you have to remove and re-install a fence or gate?

❒ Does a septic tank or drain field extend into the site of construction?

❒ Can (must) materials generated from your project be recycled or salvaged?

|  | Day | Week | Month |
|---|---|---|---|

**Equipment Rental for Demolition.**

Air compressors, wheel-mounted

| | Day | Week | Month |
|---|---|---|---|
| 16 CFM, shop type, electric | 45.00 | 136.00 | 408.00 |
| 30 CFM, shop type, electric | 63.00 | 187.00 | 562.00 |
| 80 CFM, shop type, electric | 73.00 | 221.00 | 663.00 |
| 100 CFM, gasoline unit | 91.00 | 272.00 | 816.00 |
| 125 CFM, gasoline unit | 113.00 | 340.00 | 1,020.00 |

Paving breakers (no bits included) hand-held, pneumatic

| | | | |
|---|---|---|---|
| To 40 lb | 30.00 | 85.00 | 230.00 |
| 41 - 55 lb | 25.00 | 70.00 | 210.00 |
| 56 - 70 lb | 40.00 | 110.00 | 310.00 |
| 71 - 90 lb | 40.00 | 110.00 | 315.00 |

Paving breakers jackhammer bits

| | | | |
|---|---|---|---|
| Moil points, 15" to 18" | 6.00 | 14.00 | 30.00 |
| Chisels, 3" | 7.00 | 17.00 | 44.00 |
| Clay spades, 5-1/2" | 10.00 | 23.00 | 62.00 |
| Asphalt cutters, 5" | 8.00 | 24.00 | 51.00 |

| | | | |
|---|---|---|---|
| Pneumatic chippers, medium weight, 10 lb | 33.00 | 100.00 | 275.00 |

Air hose rental, 50 LF section

| | | | |
|---|---|---|---|
| 5/8" air hose | 8.00 | 20.00 | 60.00 |
| 3/4" air hose | 8.00 | 20.00 | 55.00 |
| 1" air hose | 10.00 | 30.00 | 65.00 |
| 1-1/2" air hose | 18.00 | 55.00 | 135.00 |

Dump truck rental rate plus mileage

| | | | |
|---|---|---|---|
| 3 CY | 182.00 | 497.00 | 1,489.00 |
| 5 CY | 198.00 | 761.00 | 2,430.00 |

Hammer rental

| | | | |
|---|---|---|---|
| Electric brute breaker | 60.00 | 225.00 | 675.00 |
| Gas breaker | 63.00 | 252.00 | 756.00 |
| Demolition hammer, electric | 50.00 | 190.00 | 575.00 |
| Roto hammer, 7/8", electric | 52.00 | 190.00 | 550.00 |
| Roto hammer, 1-1/2", electric | 45.00 | 175.00 | 490.00 |

| | | | |
|---|---|---|---|
| Stump grinder, 9 HP | 110.00 | 400.00 | 1,200.00 |
| Brush chipper, trailer mounted 40 HP | 90.00 | 360.00 | 1,050.00 |
| Chain saw, 18", gasoline | 52.00 | 195.00 | 580.00 |
| Chop saw, 14", electric | 37.00 | 150.00 | 440.00 |

Backhoe/loader, wheel mounted, diesel or gasoline

| | | | |
|---|---|---|---|
| 1/2 CY bucket capacity, 55 HP | 185.00 | 600.00 | 1,750.00 |
| 1 CY bucket capacity, 65 HP | 235.00 | 690.00 | 1,900.00 |
| 1-1/4 CY bucket capacity, 75 HP | 300.00 | 720.00 | 2,100.00 |
| 1-1/2 CY bucket capacity, 100 HP | 255.00 | 800.00 | 2,300.00 |

Wheel loader, front-end load and dump, diesel

| | | | |
|---|---|---|---|
| 3/4 CY bucket, 4WD, articulated | 235.00 | 700.00 | 2,020.00 |
| 1 CY bucket, 4WD, articulated | 275.00 | 800.00 | 2,300.00 |
| 2 CY bucket, 4WD, articulated | 340.00 | 1,010.00 | 3,000.00 |

| | Craft@Hrs | Unit | Material | Labor | Total | Sell |
|---|---|---|---|---|---|---|

## Sitework Demolition

**Debris disposal and removal.** Tippage charges for solid waste disposal at the dump vary from $30 to $120 per ton. For planning purposes, estimate waste disposal at $75 per ton, plus the hauling cost. Call the dump or trash disposal company for actual charges. Typical costs are shown below.

| | Craft@Hrs | Unit | Material | Labor | Total | Sell |
|---|---|---|---|---|---|---|
| Dumpster, 3 CY trash bin, emptied weekly | — | Mo | — | — | 330.00 | — |
| Dumpster, 40 CY solid waste bin (lumber, drywall, roofing) | | | | | | |
|   Hauling cost, per load | — | Ea | — | — | 230.00 | — |
|   Add to per load charge, per ton | — | Ton | — | — | 55.00 | — |
| Low-boy, 14 CY solid waste container (asphalt, dirt, masonry, concrete) | | | | | | |
|   Hauling cost, per load (7 CY maximum load) | — | Ea | — | — | 230.00 | — |
|   Add to per load charge, per ton | — | Ton | — | — | 55.00 | — |

**Recycler fees.** Recycling construction waste materials can substantially reduce disposal costs. Recycling charges vary from $95 to $120 per load, depending on the type of material and the size of the load. Add the cost of hauling to a recycling facility.

| | Craft@Hrs | Unit | Material | Labor | Total | Sell |
|---|---|---|---|---|---|---|
| Greenwaste | — | Ton | — | — | 30.00 | — |
| Asphalt, per load (7 CY) | — | Ea | — | — | 100.00 | — |
| Concrete, masonry or rock, per load (7 CY) | — | Ea | — | — | 100.00 | — |
| Dirt, per load (7 CY) | — | Ea | — | — | 100.00 | — |
| Mixed loads, per load (7 CY) | — | Ea | — | — | 100.00 | — |

**Bush and tree removal.** Includes cutting into manageable pieces with power hand tools and dumping debris in a trash bin on site. Add the cost of power equipment rental, if needed.

Shrubs and bushes, including stump removal.

| | Craft@Hrs | Unit | Material | Labor | Total | Sell |
|---|---|---|---|---|---|---|
| 4' high, per each | BL@.750 | Ea | — | 20.00 | 20.00 | 34.00 |

Tree removal. Costs will vary widely depending on the condition, size, location and accessibility of the tree. Includes cutting into manageable pieces with power hand tools and dumping debris in a trash bin or chipper on site. Does not include stump removal or grinding. Add the cost of power equipment rental, bucket truck, and other specialized equipment if needed. Large shade tree removal may require a crew of 5 or more. Use $2000 as a minimum job charge for trees with a 36" diameter trunk and larger.

| | Craft@Hrs | Unit | Material | Labor | Total | Sell |
|---|---|---|---|---|---|---|
| 8" to 12" diameter trunk | B8@2.50 | Ea | — | 78.30 | 78.30 | 133.00 |
| 13" to 18" diameter trunk | B8@3.50 | Ea | — | 110.00 | 110.00 | 187.00 |
| 19" to 24" diameter trunk | B8@5.50 | Ea | — | 172.00 | 172.00 | 292.00 |
| 25" to 36" diameter trunk | B8@7.00 | Ea | — | 219.00 | 219.00 | 372.00 |

Stump grinding, using a 9 HP wheel-mounted stump grinder

| | Craft@Hrs | Unit | Material | Labor | Total | Sell |
|---|---|---|---|---|---|---|
| 6" to 10" diameter stump | B8@.800 | Ea | — | 25.00 | 25.00 | 42.50 |
| 11" to 14" diameter stump | B8@1.04 | Ea | — | 32.60 | 32.60 | 55.40 |
| 15" to 18" diameter stump | B8@1.30 | Ea | — | 40.70 | 40.70 | 69.20 |
| 19" to 24" diameter stump | B8@1.50 | Ea | — | 47.00 | 47.00 | 79.90 |

**Fencing demolition.** Remove chain link fence and cemented posts for disposal. These figures assume fencing is removed by cutting ties at posts and rails and rolling the fabric. These rolls can be heavy. A larger crew will be needed on larger jobs.

| | Craft@Hrs | Unit | Material | Labor | Total | Sell |
|---|---|---|---|---|---|---|
| To 6' high | BL@.100 | LF | — | 2.66 | 2.66 | 4.52 |
| Remove chain link fence and cemented posts for salvage, | | | | | | |
| To 6' high | BL@.120 | LF | — | 3.20 | 3.20 | 5.44 |
| Remove board fence and cemented posts for disposal, | | | | | | |
| To 6' high | BL@.100 | LF | — | 2.66 | 2.66 | 4.52 |

| | Craft@Hrs | Unit | Material | Labor | Total | Sell |
|---|---|---|---|---|---|---|

## Building Demolition

**Single-story room demolition.** Includes breaking out a concrete or masonry foundation using pneumatic tools, breaking materials into manageable pieces with hand tools and handling debris to a trash bin on site. Includes the cost of erecting a temporary vapor barrier to seal the wall opening in the existing structure. Add the cost of wall patching on the remaining structure. Add the cost of equipment rental.

Demolish a wood-frame addition built on a concrete slab, including grade beams. Use these figures to estimate the cost of demolishing an attached garage.

| | Craft@Hrs | Unit | Material | Labor | Total | Sell |
|---|---|---|---|---|---|---|
| Per square foot of floor area demolished | BL@.100 | SF | — | 2.66 | 2.66 | 4.52 |

Demolish a wood-frame addition built on a conventional foundation, including wood deck, stairs to grade, and the concrete or concrete block foundation.

| | Craft@Hrs | Unit | Material | Labor | Total | Sell |
|---|---|---|---|---|---|---|
| Per square foot of floor area demolished | BL@.120 | SF | — | 3.20 | 3.20 | 5.44 |

**Detached garage demolition.** Includes breaking materials into manageable pieces with hand tools and handling debris to a trash bin on site. No slab or foundation demolition included. No salvage of materials assumed.

| | Craft@Hrs | Unit | Material | Labor | Total | Sell |
|---|---|---|---|---|---|---|
| Frame garage with wood or aluminum siding | | | | | | |
| One-car garage, to 10' wide by 22' deep | BL@16.5 | Ea | — | 440.00 | 440.00 | 748.00 |
| Two-car garage, to 20' wide by 28' deep | BL@23.0 | Ea | — | 613.00 | 613.00 | 1,040.00 |
| Frame garage with stucco siding | | | | | | |
| One-car garage to 12' wide by 22' deep | BL@23.5 | Ea | — | 626.00 | 626.00 | 1,060.00 |
| One-car to 16' wide by 22' deep | BL@28.5 | Ea | — | 759.00 | 759.00 | 1,290.00 |
| Two-car garage to 22' wide by 28' deep | BL@42.5 | Ea | — | 1,130.00 | 1,130.00 | 1,920.00 |
| Brick or block garage | | | | | | |
| One-car garage to 12' wide by 22' deep | BL@23.0 | Ea | — | 613.00 | 613.00 | 1,040.00 |
| One-car garage to 16' wide by 22' deep | BL@28.0 | Ea | — | 746.00 | 746.00 | 1,270.00 |
| Two-car garage to 20' wide by 22' deep | BL@33.0 | Ea | — | 879.00 | 879.00 | 1,490.00 |

**Gutting a building.** Interior finish stripped back to the structural walls. Building structure to remain. No allowance for salvage value. These costs include loading and hauling up to 6 miles but no dump fees. Costs are per square foot of floor area based on 8' ceiling height. Add the cost of equipment rental. Figures in parentheses show the approximate "loose" volume of materials after demolition.

| | Craft@Hrs | Unit | Material | Labor | Total | Sell |
|---|---|---|---|---|---|---|
| Residential building (125 SF per CY) | BL@.100 | SF | — | 2.66 | 2.66 | 4.52 |

**Building Component Demolition.** Itemized costs for demolition of building components when building is being remodeled, repaired or rehabilitated and not completely demolished. Costs include protecting adjacent areas and normal clean-up. Costs are to break out the items listed and pile debris on site or in a bin.

**Brick wall demolition.** Cost per square foot removed measured on one face. Removed using a pneumatic breaker. Add the cost of equipment rental. Figures in parentheses show the volume before and after demolition and weight of the materials after demolition.

| | Craft@Hrs | Unit | Material | Labor | Total | Sell |
|---|---|---|---|---|---|---|
| 4" thick walls | | | | | | |
| (60 SF per CY and 36 lbs. per SF) | BL@.061 | SF | — | 1.63 | 1.63 | 2.77 |
| 8" thick walls | | | | | | |
| (30 SF per CY and 72 lbs. per SF) | BL@.110 | SF | — | 2.93 | 2.93 | 4.98 |

| | Craft@Hrs | Unit | Material | Labor | Total | Sell |
|---|---|---|---|---|---|---|

**Brick sidewalk demolition.** Cost per square foot removed. Add the cost of Bobcat rental. Larger jobs will require a crew of 2 or more. Figures in parentheses show the volume before and after demolition and weight of materials after demolition.

2-1/2" thick brick on sand base, no mortar bed, removed using hand tools

| | Craft@Hrs | Unit | Material | Labor | Total | Sell |
|---|---|---|---|---|---|---|
| (100 SF per CY and 28 lbs. per SF) | BL@.010 | SF | — | .27 | .27 | .46 |

Brick pavers up to 4-1/2" thick with mortar bed, removed using a pneumatic breaker

| | | | | | | |
|---|---|---|---|---|---|---|
| (55 SF per CY and 50 lbs. per SF) | BL@.050 | SF | — | 1.33 | 1.33 | 2.26 |

**Concrete masonry wall demolition.** Cost per square foot of wall removed measured on one side. Removed using a pneumatic breaker. Add the cost of equipment rental. Figures in parentheses show the volume before and after demolition and weight of materials after demolition.

4" thick walls

| | | | | | | |
|---|---|---|---|---|---|---|
| (60 SF per CY and 19 lbs. per SF) | BL@.066 | SF | — | 1.76 | 1.76 | 2.99 |

6" thick walls

| | | | | | | |
|---|---|---|---|---|---|---|
| (40 SF per CY and 28 lbs. per SF) | BL@.075 | SF | — | 2.00 | 2.00 | 3.40 |

8" thick walls

| | | | | | | |
|---|---|---|---|---|---|---|
| (30 SF per CY and 34 lbs. per SF) | BL@.098 | SF | — | 2.61 | 2.61 | 4.44 |

12" thick walls

| | | | | | | |
|---|---|---|---|---|---|---|
| (20 SF per CY and 46 lbs. per SF) | BL@.140 | SF | — | 3.73 | 3.73 | 6.34 |
| Reinforced or grouted walls add | — | % | — | — | 50.0 | — |

**Concrete foundation demolition (footings).** Steel reinforced, removed using a pneumatic breaker. Add the cost of equipment rental. Figures in parentheses show the volume before and after demolition.

| | Craft@Hrs | Unit | Material | Labor | Total | Sell |
|---|---|---|---|---|---|---|
| Cost per CY (.75 CY per CY) | BL@3.96 | CY | — | 105.00 | 105.00 | 179.00 |

Cost per LF with width and depth as shown

| | | | | | | |
|---|---|---|---|---|---|---|
| 6" W x 12" D  (35 LF per CY) | BL@.075 | LF | — | 2.00 | 2.00 | 3.40 |
| 8" W x 12" D  (30 LF per CY) | BL@.098 | LF | — | 2.61 | 2.61 | 4.44 |
| 8" W x 16" D  (20 LF per CY) | BL@.133 | LF | — | 3.54 | 3.54 | 6.02 |
| 8" W x 18" D  (18 LF per CY) | BL@.147 | LF | — | 3.92 | 3.92 | 6.66 |
| 10" W x 12" D  (21 LF per CY) | BL@.121 | LF | — | 3.22 | 3.22 | 5.47 |
| 10" W x 16" D  (16 LF per CY) | BL@.165 | LF | — | 4.40 | 4.40 | 7.48 |
| 10" W x 18" D  (14 LF per CY) | BL@.185 | LF | — | 4.93 | 4.93 | 8.38 |
| 12" W x 12" D  (20 LF per CY) | BL@.147 | LF | — | 3.92 | 3.92 | 6.66 |
| 12" W x 16" D  (13 LF per CY) | BL@.196 | LF | — | 5.22 | 5.22 | 8.87 |
| 12" W x 20" D  (11 LF per CY) | BL@.245 | LF | — | 6.53 | 6.53 | 11.10 |
| 12" W x 24" D  ( 9 LF per CY) | BL@.294 | LF | — | 7.83 | 7.83 | 13.30 |

**Concrete sidewalk demolition.** To 4" thick, cost per SF removed. Figures in parentheses show the volume before and after demolition.

Non-reinforced, removed by hand with sledge

| | | | | | | |
|---|---|---|---|---|---|---|
| (60 SF per CY) | BL@.050 | SF | — | 1.33 | 1.33 | 2.26 |

Reinforced with wire mesh. Mesh cut into manageable pieces, then removed using pneumatic breaker. Add the cost of equipment rental.

| | | | | | | |
|---|---|---|---|---|---|---|
| (55 SF per CY) | BL@.060 | SF | — | 1.60 | 1.60 | 2.72 |

| | Craft@Hrs | Unit | Material | Labor | Total | Sell |
|---|---|---|---|---|---|---|

**Porch and deck demolition.** Includes breaking materials into manageable pieces with hand tools and handling debris to a trash bin on site.

Demolish an enclosed wood-frame porch. Includes breaking out the concrete or masonry pier foundation. Per square foot of floor.

| | Craft@Hrs | Unit | Material | Labor | Total | Sell |
|---|---|---|---|---|---|---|
| One story | BL@.250 | SF | — | 6.66 | 6.66 | 11.30 |
| Two story | BL@.225 | SF | — | 5.99 | 5.99 | 10.20 |
| Three story | BL@.196 | SF | — | 5.22 | 5.22 | 8.87 |

Demolish a screened porch built on a concrete slab. Includes demolition of one concrete or wood frame step to grade. Per square foot of floor area.

| | Craft@Hrs | Unit | Material | Labor | Total | Sell |
|---|---|---|---|---|---|---|
| Porch and one step (no slab demolition) | BL@.100 | SF | — | 2.66 | 2.66 | 4.52 |

Demolish a wood deck with railing or kneewall. Includes demolition of the wood deck and up to 7 steps to grade.

| | Craft@Hrs | Unit | Material | Labor | Total | Sell |
|---|---|---|---|---|---|---|
| Deck to 150 SF | BL@18.5 | Ea | — | 493.00 | 493.00 | 838.00 |
| Add per SF for deck over 150 SF | BL@.100 | SF | — | 2.66 | 2.66 | 4.52 |

**Concrete slab demolition.** Break up and remove concrete slab using a compressor and pneumatic hammer. Includes loosening concrete with a pick and handling debris to a trash bin on site. Costs for demolition of reinforced concrete slab include burning off rebars with an acetylene torch. Add the rental cost of a compressor, pneumatic hammer and a torch, as required. Figures in parentheses show the volume before and after demolition.

| | Craft@Hrs | Unit | Material | Labor | Total | Sell |
|---|---|---|---|---|---|---|
| 4" non-reinforced concrete (60 SF per CY) | BL@.057 | SF | — | 1.52 | 1.52 | 2.58 |
| 4" reinforced concrete (60 SF per CY) | BL@.071 | SF | — | 1.89 | 1.89 | 3.21 |
| 6" non-reinforced concrete (45 SF per CY) | BL@.083 | SF | — | 2.21 | 2.21 | 3.76 |
| 6" reinforced concrete (45 SF per CY) | BL@.103 | SF | — | 2.74 | 2.74 | 4.66 |
| 8" non-reinforced concrete (30 SF per CY) | BL@.109 | SF | — | 2.90 | 2.90 | 4.93 |
| 8" reinforced concrete (30 SF per CY) | BL@.122 | SF | — | 3.25 | 3.25 | 5.53 |

**Asphalt pavement demolition.** Asphaltic concrete (bituminous) 3" thick, removed using a pneumatic spade. Add the cost of equipment rental and dump fees. Figures in parentheses show the volume before and after demolition and weight of debris.

| | Craft@Hrs | Unit | Material | Labor | Total | Sell |
|---|---|---|---|---|---|---|
| Break up and shovel on to a truck (4 SY per CY and 660 lbs. per SY) | BL@.140 | SY | — | 3.73 | 3.73 | 6.34 |

**Concrete wall demolition.** Steel reinforced, removed using a compressor and pneumatic breaker. Add the cost of equipment rental. Figures in parentheses show the volume before and after demolition.

| | Craft@Hrs | Unit | Material | Labor | Total | Sell |
|---|---|---|---|---|---|---|
| Cost per CY (1.33 CY per CY) | BL@4.76 | CY | — | 127.00 | 127.00 | 216.00 |

Cost per SF with thickness as shown

| | Craft@Hrs | Unit | Material | Labor | Total | Sell |
|---|---|---|---|---|---|---|
| 3" wall thickness (80 SF per CY) | BL@.045 | SF | — | 1.20 | 1.20 | 2.04 |
| 4" wall thickness (55 SF per CY) | BL@.058 | SF | — | 1.55 | 1.55 | 2.64 |
| 5" wall thickness (45 SF per CY) | BL@.075 | SF | — | 2.00 | 2.00 | 3.40 |
| 6" wall thickness (40 SF per CY) | BL@.090 | SF | — | 2.40 | 2.40 | 4.08 |
| 8" wall thickness (30 SF per CY) | BL@.120 | SF | — | 3.20 | 3.20 | 5.44 |
| 10" wall thickness (25 SF per CY) | BL@.147 | SF | — | 3.92 | 3.92 | 6.66 |
| 12" wall thickness (20 SF per CY) | BL@.176 | SF | — | 4.69 | 4.69 | 7.97 |

| | Craft@Hrs | Unit | Material | Labor | Total | Sell |
|---|---|---|---|---|---|---|

**Wood Framing Demolition.** Typical costs for demolition of wood frame structural components using hand tools. Normal center-to-center spacing (12" thru 24" OC) is assumed. Includes breaking materials into manageable pieces with hand tools and handling debris to a trash bin on site. Costs include removal of Romex cable as necessary, protecting adjacent areas, and normal clean-up. Costs do not include temporary shoring to support structural elements above or adjacent to the work area. (Add for demolition of flooring, roofing, sheathing, etc.) Figures in parentheses give the approximate "loose" volume and weight of the materials after being demolished.

**Ceiling joist demolition.** Per in-place SF of ceiling area removed.

| | Craft@Hrs | Unit | Material | Labor | Total | Sell |
|---|---|---|---|---|---|---|
| 2" x 4" (720 SF per CY and 1.18 lbs. per SF) | BL@.009 | SF | — | .24 | .24 | .41 |
| 2" x 6" (410 SF per CY and 1.76 lbs. per SF) | BL@.013 | SF | — | .35 | .35 | .60 |
| 2" x 8" (290 SF per CY and 2.34 lbs. per SF) | BL@.018 | SF | — | .48 | .48 | .82 |
| 2" x 10" (220 SF per CY and 2.94 lbs. per SF) | BL@.023 | SF | — | .61 | .61 | 1.04 |
| 2" x 12" (190 SF per CY and 3.52 lbs. per SF) | BL@.027 | SF | — | .72 | .72 | 1.22 |

**Floor joist demolition.** Per in-place SF of floor area removed.

| | Craft@Hrs | Unit | Material | Labor | Total | Sell |
|---|---|---|---|---|---|---|
| 2" x 6" (290 SF per CY and 2.04 lbs. per SF) | BL@.016 | SF | — | .43 | .43 | .73 |
| 2" x 8" (190 SF per CY and 2.72 lbs. per SF) | BL@.021 | SF | — | .56 | .56 | .95 |
| 2" x 10" (160 SF per CY and 3.42 lbs. per SF) | BL@.026 | SF | — | .69 | .69 | 1.17 |
| 2" x 12" (120 SF per CY and 4.10 lbs. per SF) | BL@.031 | SF | — | .83 | .83 | 1.41 |

**Rafter demolition.** Per in-place SF of actual roof area removed.

| | Craft@Hrs | Unit | Material | Labor | Total | Sell |
|---|---|---|---|---|---|---|
| 2" x 4" (610 SF per CY and 1.42 lbs. per SF) | BL@.011 | SF | — | .29 | .29 | .49 |
| 2" x 6" (360 SF per CY and 2.04 lbs. per SF) | BL@.016 | SF | — | .43 | .43 | .73 |
| 2" x 8" (270 SF per CY and 2.68 lbs. per SF) | BL@.020 | SF | — | .53 | .53 | .90 |
| 2" x 10" (210 SF per CY and 3.36 lbs. per SF) | BL@.026 | SF | — | .69 | .69 | 1.17 |
| 2" x 12" (180 SF per CY and 3.94 lbs. per SF) | BL@.030 | SF | — | .80 | .80 | 1.36 |

**Stud wall demolition.** Interior or exterior, includes allowance for plates and blocking, per in-place SF of wall area removed, measured on one face.

| | Craft@Hrs | Unit | Material | Labor | Total | Sell |
|---|---|---|---|---|---|---|
| 2" x 3" (430 SF per CY and 1.92 lbs. per SF) | BL@.013 | SF | — | .35 | .35 | .60 |
| 2" x 4" (310 SF per CY and 2.58 lbs. per SF) | BL@.017 | SF | — | .45 | .45 | .77 |
| 2" x 6" (190 SF per CY and 3.74 lbs. per SF) | BL@.025 | SF | — | .67 | .67 | 1.14 |

**Stud wall demolition, salvage condition.** Remove wall cover (nailed hardboard or fiberboard) from partition wall and salvage the stud framing. Per linear foot of stud wall.

| | Craft@Hrs | Unit | Material | Labor | Total | Sell |
|---|---|---|---|---|---|---|
| 2" x 4" x 8', 16" on center | BL@.156 | LF | — | 4.16 | 4.16 | 7.07 |

**Roof cover demolition.** Includes breaking materials into manageable pieces with hand tools and handling debris to a trash bin on site. Cost per Sq or "square". (1 square = 100 square feet). Figures in parentheses show the approximate "loose" volume and weight of materials after being demolished.

| | Craft@Hrs | Unit | Material | Labor | Total | Sell |
|---|---|---|---|---|---|---|
| Built-up, 5 ply | | | | | | |
| (2.5 Sq per CY and 250 lbs per Sq) | BL@1.50 | Sq | — | 40.00 | 40.00 | 68.00 |
| Shingles, asphalt, single layer | | | | | | |
| (2.5 Sq per CY and 240 lbs per Sq) | BL@1.33 | Sq | — | 35.40 | 35.40 | 60.20 |
| Shingles, asphalt, double layer | | | | | | |
| (1.2 Sq per CY and 480 lbs per Sq) | BL@2.00 | Sq | — | 53.30 | 53.30 | 90.60 |
| Shingles, slate. Weight ranges from 600 lbs. to 1200 lbs. per Sq | | | | | | |
| (1 Sq per CY and 900 lbs per Sq). | BL@1.79 | Sq | — | 47.70 | 47.70 | 81.10 |

| | Craft@Hrs | Unit | Material | Labor | Total | Sell |
|---|---|---|---|---|---|---|
| Shingles, wood | | | | | | |
|   (1.7 Sq per CY and 400 lbs per Sq) | BL@2.02 | Sq | — | 53.80 | 53.80 | 91.50 |
| Clay or concrete tile | | | | | | |
|   (.70 Sq per CY) | BL@1.65 | Sq | — | 44.00 | 44.00 | 74.80 |
| Remove gravel stop or flashing | BL@.070 | LF | — | 1.86 | 1.86 | 3.16 |

**Interior and exterior finish demolition.** Includes breaking materials into manageable pieces with hand tools and handling debris to a trash bin on site. Figures in parentheses show the approximate "loose" volume and weight of the materials after being demolished. Demolition for disposal. No salvage of materials included.

| | Craft@Hrs | Unit | Material | Labor | Total | Sell |
|---|---|---|---|---|---|---|
| Plywood sheathing, up to 1" thick, cost per in-place SF | | | | | | |
|   (200 SF per CY and 2 lbs. per SF) | BL@.017 | SF | — | .45 | .45 | .77 |
| Wood board sheathing | | | | | | |
|   (250 SF per CY and 2 lbs. per SF) | BL@.030 | SF | — | .80 | .80 | 1.36 |
| Metal siding, using hand tools | | | | | | |
|   (200 SF per CY and 2 lbs. per SF) | BL@.027 | SF | — | .72 | .72 | 1.22 |
| Stucco on walls or soffits, cost per in-place SF, removed using hand tools | | | | | | |
|   (150 SF per CY and 8 lbs. per SF) | BL@.036 | SF | — | .96 | .96 | 1.63 |

Wallboard 1/2" thick, cost per in-place SF, removed using hand tools. Add 50% to labor costs for 1" wallboard.

| | Craft@Hrs | Unit | Material | Labor | Total | Sell |
|---|---|---|---|---|---|---|
| Gypsum, walls or ceilings | | | | | | |
|   (250 SF per CY and 2.3 lbs. per SF) | BL@.010 | SF | — | .27 | .27 | .46 |
|     Including strip furring | BL@.015 | SF | — | .40 | .40 | .68 |
| Plywood or insulation board | | | | | | |
|   (200 SF per CY and 2 lbs. per SF) | BL@.018 | SF | — | .48 | .48 | .82 |
| Insulation, fiberglass batts or rolls, cost per square foot removed | | | | | | |
|   (500 SF per CY and .3 lb per SF) | BL@.003 | SF | — | .08 | .08 | .14 |
| Plaster on walls, cost per square foot removed by hand | | | | | | |
|   (150 SF per CY and 8 lbs. per SF) | BL@.015 | SF | — | .40 | .40 | .68 |

**Ceiling demolition.** Knock down with hand tools at heights to 9' and handle debris to a trash bin on site. Building structure to remain. No allowance for salvage value. Costs shown are per square foot of ceiling. Figures in parentheses show the approximate "loose" volume after demolition.

| | Craft@Hrs | Unit | Material | Labor | Total | Sell |
|---|---|---|---|---|---|---|
| Plaster ceiling (typically 175 to 200 SF per CY) | | | | | | |
|   Including lath and furring | BL@.025 | SF | — | .67 | .67 | 1.14 |
|   Including suspended grid | BL@.020 | SF | — | .53 | .53 | .90 |
| Acoustic tile ceiling (typically 200 to 250 SF per CY) | | | | | | |
|   Including suspended grid | BL@.010 | SF | — | .27 | .27 | .46 |
|   Including grid in salvage condition | BL@.019 | SF | — | .51 | .51 | .87 |
|   Tile glued or stapled to ceiling | BL@.015 | SF | — | .40 | .40 | .68 |
|   Tile on strip furring, including furring | BL@.025 | SF | — | .67 | .67 | 1.14 |
| Drywall ceiling (typically 250 to 300 SF per CY) | | | | | | |
|   Nailed or attached with screws | | | | | | |
|     to joists | BL@.010 | SF | — | .27 | .27 | .46 |

| | Craft@Hrs | Unit | Material | Labor | Total | Sell |
|---|---|---|---|---|---|---|

**Floor cover demolition.** Includes breaking materials into manageable pieces with hand tools and handling debris to a trash bin on site. Cost per square yard of floor removed. (1 square yard = 9 square feet). Figures in parentheses show the approximate "loose" volume and weight of the materials after demolition.

| | Craft@Hrs | Unit | Material | Labor | Total | Sell |
|---|---|---|---|---|---|---|
| Ceramic tile | | | | | | |
| (25 SY per CY and 34 lbs. per SY) | BL@.263 | SY | — | 7.01 | 7.01 | 11.90 |
| Hardwood, nailed | | | | | | |
| (25 SY per CY and 18 lbs. per SY) | BL@.290 | SY | — | 7.73 | 7.73 | 13.10 |
| Hardwood, glued | | | | | | |
| (25 SY per CY and 18 lbs. per SY) | BL@.503 | SY | — | 13.40 | 13.40 | 22.80 |
| Linoleum, sheet | | | | | | |
| (30 SY per CY and 3 lbs. per SY) | BL@.056 | SY | — | 1.49 | 1.49 | 2.53 |
| Resilient tile | | | | | | |
| (30 SY per CY and 3 lbs. per SY) | BL@.300 | SY | — | 7.99 | 7.99 | 13.60 |
| Terrazzo | | | | | | |
| (25 SY per CY and 34 lbs. per SY) | BL@.286 | SY | — | 7.62 | 7.62 | 13.00 |
| Carpet on tack strip | | | | | | |
| (40 SY per CY and 5 lbs. per SY) | BL@.028 | SY | — | .75 | .75 | 1.28 |
| Glue down carpet with rubber backing | | | | | | |
| (35 SY per CY and 1.7 lbs. per SY) | BL@.020 | SY | — | .53 | .53 | .90 |
| Add for each LF of room perimeter | BL@.008 | LF | — | .21 | .21 | .36 |
| Carpet pad | | | | | | |
| (35 SY per CY and 1.7 lbs. per SY) | BL@.014 | SY | — | .37 | .37 | .63 |

**Removal of interior items.** Includes breaking materials into manageable pieces with hand tools and handling debris to a trash bin on site. Items removed in salvageable condition do not include an allowance for salvage value. Figures in parentheses give the approximate "loose" volume of the materials after demolition. Removed in non-salvageable condition (demolished) except as noted.

| | Craft@Hrs | Unit | Material | Labor | Total | Sell |
|---|---|---|---|---|---|---|
| Windows, typical cost to remove window, frame and hardware, cost per SF of window | | | | | | |
| Metal (36 SF per CY) | BL@.058 | SF | — | 1.55 | 1.55 | 2.64 |
| Wood (36 SF per CY) | BL@.063 | SF | — | 1.68 | 1.68 | 2.86 |
| Wood stairs, cost per in-place SF or LF | | | | | | |
| Risers (25 SF per CY) | BL@.116 | SF | — | 3.09 | 3.09 | 5.25 |
| Landings (50 SF per CY) | BL@.021 | SF | — | .56 | .56 | .95 |
| Handrails (100 LF per CY) | BL@.044 | LF | — | 1.17 | 1.17 | 1.99 |
| Posts (200 LF per CY) | BL@.075 | LF | — | 2.00 | 2.00 | 3.40 |
| Hollow metal door and frame in a masonry wall. (2 doors per CY) | | | | | | |
| Single door to 4' x 7' | BL@1.00 | Ea | — | 26.60 | 26.60 | 45.20 |
| Two doors, per opening to 8' x 7' | BL@1.50 | Ea | — | 40.00 | 40.00 | 68.00 |
| Wood door and frame in a wood frame wall (2 doors per CY) | | | | | | |
| Single door to 4' x 7' | BL@.500 | Ea | — | 13.30 | 13.30 | 22.60 |
| Two doors, per opening to 8' x 7' | BL@.750 | Ea | — | 20.00 | 20.00 | 34.00 |
| Metal door and frame in masonry wall, salvage condition | | | | | | |
| Hollow metal door to 4' x 7' | BL@2.00 | Ea | — | 53.30 | 53.30 | 90.60 |
| Wood door to 4' x 7' | BL@1.00 | Ea | — | 26.60 | 26.60 | 45.20 |

# Foundations and Slabs

# 3

Many older homes have foundations that have cracked or settled. Check the foundation wall for deterioration that could allow water to enter the basement. Check both foundation walls and piers for settling. Windows or door frames out of square or loosened interior wall finish suggest that the foundation has settled. The next chapter explains how to correct minor settling by jacking and re-leveling beams and floor joists. Individual piers can be replaced. But if the pier has stopped settling, jack the supported girder or joist and add a block to the top of the pier.

Most concrete foundation walls develop minor hairline cracks that have no effect on the structure. However, open cracks may indicate a failure of the foundation that's getting progressively worse. To find out if a crack is active or dormant, scratch a line at the end of the crack and wedge a nail tightly into the crack. If the crack grows beyond the scratch mark or if the nail can be removed easily several months later, the crack is probably active.

If a crack is dormant, it can be repaired by routing and sealing. Enlarge the crack with a concrete saw or by chipping with hand tools. The crack should be routed $1^1/4"$ or more in width and about the same depth. Rinse the joint clean and let it dry. Then apply a joint sealer such as an epoxy-cement compound in accord with the manufacturer's instructions.

Active cracks require an elastic sealant. Again, follow the manufacturer's instructions. Good-quality concrete sealant will remain pliable for many years. The minimum routing depth and width for these sealants is $3/4"$. The elastic material will deform but maintain a tight seal as the crack moves. You could also apply a strip sealant over the crack. But these protrude from the surface and make a poor choice if the wall is visible from the building exterior.

Repair loose mortar by brushing thoroughly to remove dust and loose particles. Before applying new mortar, dampen the clean surface so that it won't absorb water during repair. You can buy premixed mortar with the consistency of putty. Apply mortar over the cavity as if you were filling a void with painter's caulk. For a good bond, force mortar into the crack. Then smooth the surface with a trowel. Cover with a vapor barrier for a few days to keep the mortar from drying too fast.

Damp or leaky basement walls are usually caused by clogged drain tile, clogged or broken downspouts, cracks in walls, or by water that puddles against the foundation. Look for downspouts that empty against the foundation wall or surface drainage channeling by the foundation. For a dry basement, keep water away from the foundation by proper grading.

## Foundation and Slab Pre-Construction Checklist

❐ Is the site accessible to wheeled equipment?

❐ Is the site accessible to a concrete truck?

❐ Will you need to re-route plumbing, electrical, gas or communication lines?

❐ Will you need to remove and replace lawn sprinklers?

❐ Will you need to cut back or remove bushes or trees?

❐ Will you have to remove and re-install a fence or gate?

❐ Does a septic tank or drain field extend into the site of construction?

❐ Can excess soil be spread on the premises?

❐ Will dirt need to be imported or exported for site balance?

❐ Will heavy equipment damage the roots of trees?

❐ Is a permit needed to move heavy equipment across city sidewalks?

❐ Do city sidewalks have to be planked?

❐ Will swales or culverts pose problems for equipment?

❐ Will a septic system, catch basins or wells have to be relocated?

❐ Will barricades and warning lights be needed?

A high water table is a more serious problem. A basement will never be completely dry if the water table extends above the basement floor. Heavy foundation waterproofing or footing drains may help, but they're unlikely to do more than minimize the problem.

|  | **Day** | **Week** | **Month** |
|---|---|---|---|
| **Equipment Rental for Foundations and Slabs.** | | | |
| Buggies, push type, 7 CF | 44.00 | 150.00 | 400.00 |
| Buggies, walking type, 12 CF | 65.00 | 190.00 | 535.00 |
| Buggies, riding type, 14 CF | 90.00 | 275.00 | 710.00 |
| Buggies, riding type, 18 CF | 105.00 | 310.00 | 835.00 |
| Vibrator, electric, 3 HP, flexible shaft | 39.00 | 125.00 | 340.00 |
| Vibrator, gasoline, 6 HP, flexible shaft | 60.00 | 175.00 | 525.00 |
| Troweling machine, 36", 4 paddle | 55.00 | 180.00 | 500.00 |
| Cement mixer, 6 CF | 60.00 | 185.00 | 535.00 |
| Towable mixer (10+ CF) | 105.00 | 295.00 | 795.00 |
| Power trowel 36", gas, with standard handle | 55.00 | 180.00 | 500.00 |
| Bull float | 13.00 | 40.00 | 110.00 |
| Concrete saws, gas powered, excluding blade cost | | | |
|     10 HP, push type | 60.00 | 195.00 | 550.00 |
|     20 HP, self propelled | 100.00 | 325.00 | 910.00 |
|     37 HP, self propelled | 140.00 | 475.00 | 1,330.00 |
| Concrete conveyor, belt type, portable, gas powered, | | | |
|     all sizes | 170.00 | 625.00 | 1,675.00 |
| Vibrating screed, 16', single beam | 75.00 | 225.00 | 685.00 |
| Column clamps, to 48", per set | — | — | 5.90 |
| Skid steer loaders | | | |
|     Bobcat 643, 1,000-lb. capacity | 155.00 | 520.00 | 1,500.00 |
|     Bobcat 753, 1,350-lb. capacity | 175.00 | 540.00 | 1,580.00 |
|     Bobcat 763, 1,750-lb. capacity | 190.00 | 565.00 | 1,650.00 |
|     Bobcat 863, 1,900-lb. capacity | 215.00 | 650.00 | 1,890.00 |
| Skid steer attachments | | | |
|     Auger | 80.00 | 250.00 | 690.00 |
|     Hydraulic breaker | 160.00 | 505.00 | 1,570.00 |
|     Backhoe | 110.00 | 345.00 | 1,060.00 |
|     Sweeper | 100.00 | 315.00 | 930.00 |
|     Grapple bucket | 60.00 | 170.00 | 510.00 |
| Trenchers, inclined chain boom type, pneumatic tired | | | |
|     15 HP, 12" wide, 48" max. depth, walking | 170.00 | 560.00 | 1,620.00 |
|     20 HP, 12" wide, 60" max. depth, riding | 220.00 | 740.00 | 2,200.00 |
| Wheel loaders, front-end load and dump, diesel | | | |
|     3/4 CY bucket, 4WD, articulated | 235.00 | 700.00 | 2,020.00 |
|     1 CY bucket, 4WD, articulated | 275.00 | 800.00 | 2,300.00 |
|     2 CY bucket, 4WD, articulated | 340.00 | 1,010.00 | 3,000.00 |
| Vibro plate, 300 lb, 24" plate width, gas | 80.00 | 250.00 | 700.00 |
| Vibro plate, 600 lb, 32" plate width, gas | 135.00 | 415.00 | 1,160.00 |
| Rammer, 60 CPM, 200 lb, gas powered | 70.00 | 215.00 | 600.00 |

|  | **Craft@Hrs** | **Unit** | **Material** | **Labor** | **Total** | **Sell** |
|---|---|---|---|---|---|---|
| **Concrete pumping.** Using a trailer-mounted gasoline pump. | | | | | | |
| 3/8" aggregate mix (pea gravel), using hose to 200' | | | | | | |
|     Cost per cubic yard (pumping only) | — | CY | — | — | 12.00 | — |
|     Add for hose over 200', per LF | — | LF | — | — | 1.25 | — |
| 3/4" aggregate mix, using hose to 200' | | | | | | |
|     Cost per cubic yard (pumping only) | — | CY | — | — | 13.20 | — |
|     Add for hose over 200', per LF | — | LF | — | — | 1.25 | — |

| | Craft@Hrs | Unit | Material | Labor | Total | Sell |
|---|---|---|---|---|---|---|

## Concrete Flatwork

Most concrete work on home improvement jobs will be slabs such as sidewalks, driveways and floors. All concrete flatwork requires fine grading, edge forms, concrete pouring, and finishing.

**4" thick concrete slab.** Typical cost for a 100 square foot slab, based on costs in this chapter. Use the figures that follow to adjust for other conditions: grading, base, forms, concrete thickness, concrete mix characteristics and finishing

| | Craft@Hrs | Unit | Material | Labor | Total | Sell |
|---|---|---|---|---|---|---|
| Fine grading and shaping, 100 SF | BL@.800 | LS | — | 21.30 | 21.30 | 36.20 |
| Aggregate 4" base (1.23 CY per CSF) | BL@.600 | LS | 30.00 | 16.00 | 46.00 | 78.20 |
| Lay out and set edge forms, 50 LF, 1 use | B2@2.50 | LS | 30.00 | 75.70 | 105.70 | 180.00 |
| .006" polyethylene vapor barrier | B5@.118 | LS | 7.00 | 3.68 | 10.68 | 18.20 |
| Place W2.9 x W2.9 x 6" x 6" mesh, 100 SF | RI@.400 | LS | 32.00 | 12.40 | 44.40 | 75.50 |
| Place ready-mix concrete, from chute 1.25 CY | B5@.540 | LS | 131.00 | 16.90 | 147.90 | 251.00 |
| Steel trowel finish | B6@1.00 | LS | — | 28.50 | 28.50 | 48.50 |
| Acrylic concrete cure and seal | B6@.300 | LS | 5.25 | 8.56 | 13.81 | 23.50 |
| Strip edge forms, 50 LF, no re-use | BL@.600 | LS | — | 16.00 | 16.00 | 27.20 |
| Total job cost for 4" thick 100 SF slab | — | LS | 235.25 | — | 235.25 | — |
| Cost per SF for 4" thick 100 SF job | — | SF | 2.35 | — | 2.35 | — |

**Excavation and Backfill by Hand.** Using hand tools.

General excavation, using a pick and shovel (loosening and one throw)

| | Craft@Hrs | Unit | Material | Labor | Total | Sell |
|---|---|---|---|---|---|---|
| Light soil | BL@1.10 | CY | — | 29.30 | 29.30 | 49.80 |
| Average soil | BL@1.70 | CY | — | 45.30 | 45.30 | 77.00 |
| Heavy soil or loose rock | BL@2.25 | CY | — | 59.90 | 59.90 | 102.00 |

Backfilling (one shovel throw from stockpile)

| | Craft@Hrs | Unit | Material | Labor | Total | Sell |
|---|---|---|---|---|---|---|
| Sand | BL@.367 | CY | — | 9.78 | 9.78 | 16.60 |
| Average soil | BL@.467 | CY | — | 12.40 | 12.40 | 21.10 |
| Rock or clay | BL@.625 | CY | — | 16.70 | 16.70 | 28.40 |
| Add for compaction, average soil or sand | BL@.400 | CY | — | 10.70 | 10.70 | 18.20 |
| Fine grading | BL@.008 | SF | — | .21 | .21 | .36 |

Footings, average soil, using a pick and shovel

| | Craft@Hrs | Unit | Material | Labor | Total | Sell |
|---|---|---|---|---|---|---|
| 6" deep x 12" wide (1.85 CY per CLF) | BL@.034 | LF | — | .91 | .91 | 1.55 |
| 8" deep x 12" wide (2.47 CY per CLF) | BL@.050 | LF | — | 1.33 | 1.33 | 2.26 |
| 8" deep x 16" wide (3.29 CY per CLF) | BL@.055 | LF | — | 1.47 | 1.47 | 2.50 |
| 8" deep x 18" wide (3.70 CY per CLF) | BL@.060 | LF | — | 1.60 | 1.60 | 2.72 |
| 10" deep x 12" wide (3.09 CY per CLF) | BL@.050 | LF | — | 1.33 | 1.33 | 2.26 |
| 10" deep x 16" wide (4.12 CY per CLF) | BL@.067 | LF | — | 1.78 | 1.78 | 3.03 |
| 10" deep x 18" wide (4.63 CY per CLF) | BL@.075 | LF | — | 2.00 | 2.00 | 3.40 |
| 12" deep x 12" wide (3.70 CY per CLF) | BL@.060 | LF | — | 1.60 | 1.60 | 2.72 |
| 12" deep x 16" wide (4.94 CY per CLF) | BL@.081 | LF | — | 2.16 | 2.16 | 3.67 |
| 12" deep x 20" wide (6.17 CY per CLF) | BL@.100 | LF | — | 2.66 | 2.66 | 4.52 |
| 12" deep x 24" wide (7.41 CY per CLF) | BL@.125 | LF | — | 3.33 | 3.33 | 5.66 |
| 16" deep x 16" wide (6.59 CY per CLF) | BL@.110 | LF | — | 2.93 | 2.93 | 4.98 |

Loading trucks (shoveling, one throw)

| | Craft@Hrs | Unit | Material | Labor | Total | Sell |
|---|---|---|---|---|---|---|
| Average soil | BL@1.45 | CY | — | 38.60 | 38.60 | 65.60 |
| Rock or clay | BL@2.68 | CY | — | 71.40 | 71.40 | 121.00 |

Pits to 5'. Pits over 5' deep require special consideration: shoring, liners and a way to lift the soil.

| | Craft@Hrs | Unit | Material | Labor | Total | Sell |
|---|---|---|---|---|---|---|
| Light soil | BL@1.34 | CY | — | 35.70 | 35.70 | 60.70 |
| Average soil | BL@2.00 | CY | — | 53.30 | 53.30 | 90.60 |
| Heavy soil | BL@2.75 | CY | — | 73.30 | 73.30 | 125.00 |

Shaping trench bottom for pipe

| | Craft@Hrs | Unit | Material | Labor | Total | Sell |
|---|---|---|---|---|---|---|
| To 10" pipe | BL@.024 | LF | — | .64 | .64 | 1.09 |
| 12" to 20" pipe | BL@.076 | LF | — | 2.02 | 2.02 | 3.43 |

| | Craft@Hrs | Unit | Material | Labor | Total | Sell |
|---|---|---|---|---|---|---|
| Shaping embankment slopes | | | | | | |
| Up to 1 in 4 slope | BL@.060 | SY | — | 1.60 | 1.60 | 2.72 |
| Over 1 in 4 slope | BL@.075 | SY | — | 2.00 | 2.00 | 3.40 |
| Add for top crown or toe | — | % | — | 50.0 | — | — |
| Add for swales | — | % | — | 90.0 | — | — |
| | | | | | | |
| Spreading material piled on site | | | | | | |
| Average soil | BL@.367 | CY | — | 9.78 | 9.78 | 16.60 |
| Stone or clay | BL@.468 | CY | — | 12.50 | 12.50 | 21.30 |
| Strip and pile top soil | BL@.024 | SF | — | .64 | .64 | 1.09 |
| Tamping, hand tamp only | BL@.612 | CY | — | 16.30 | 16.30 | 27.70 |
| Trenches to 5', soil piled beside trench | | | | | | |
| Light soil | BL@1.13 | CY | — | 30.10 | 30.10 | 51.20 |
| Average soil | BL@1.84 | CY | — | 49.00 | 49.00 | 83.30 |
| Heavy soil or loose rock | BL@2.86 | CY | — | 76.20 | 76.20 | 130.00 |
| Add for depth over 5' to 9' | BL@.250 | CY | — | 6.66 | 6.66 | 11.30 |

**Trenching and backfill with power equipment.** Using a 55 HP wheel loader with integral backhoe to excavate utility line trenches and continuous footings where soil is piled adjacent to the trench. Linear feet (LF) and cubic yards (CY) per hour shown are based on a 2-man crew. Reduce productivity by 10 to 25 percent when soil is loaded in trucks. Shoring, dewatering and unusual conditions will increase these costs. Add the cost of equipment rental.

| | Craft@Hrs | Unit | Material | Labor | Total | Sell |
|---|---|---|---|---|---|---|
| 12" wide bucket, for 12" wide trench. Depths 3' to 5' | | | | | | |
| Light soil (60 LF per hour) | B8@.033 | LF | — | 1.03 | 1.03 | 1.75 |
| Medium soil (55 LF per hour) | B8@.036 | LF | — | 1.13 | 1.13 | 1.92 |
| Heavy or wet soil (35 LF per hour) | B8@.057 | LF | — | 1.78 | 1.78 | 3.03 |
| 18" wide bucket, for 18" wide trench. Depths 3' to 5' | | | | | | |
| Light soil (55 LF per hour) | B8@.036 | LF | — | 1.13 | 1.13 | 1.92 |
| Medium soil (50 LF per hour) | B8@.040 | LF | — | 1.25 | 1.25 | 2.13 |
| Heavy or wet soil (30 LF per hour) | B8@.067 | LF | — | 2.10 | 2.10 | 3.57 |
| 24" wide bucket, for 24" wide trench. Depths 3' to 5' | | | | | | |
| Light soil (50 LF per hour) | B8@.040 | LF | — | 1.25 | 1.25 | 2.13 |
| Medium soil (45 LF per hour) | B8@.044 | LF | — | 1.38 | 1.38 | 2.35 |
| Heavy or wet soil (25 LF per hour) | B8@.080 | LF | — | 2.50 | 2.50 | 4.25 |

**Soil compaction.** Add the cost of equipment rental.

Compaction of soil in trenches in 8" layers.

| | Craft@Hrs | Unit | Material | Labor | Total | Sell |
|---|---|---|---|---|---|---|
| Pneumatic tampers | | | | | | |
| (40 CY per hour) | BL@.050 | CY | — | 1.33 | 1.33 | 2.26 |
| Vibrating rammers, gasoline powered "Jumping Jack" | | | | | | |
| (20 CY per hour) | BL@.100 | CY | — | 2.66 | 2.66 | 4.52 |

**Calculating slab area.** Patio slabs and driveways are often irregular in shape. For example, a driveway may be wider at the garage end and narrower at the street. The easy way to figure the area is to measure the width at both ends and then divide by two. Then multiply the result by the length of the driveway. For example, assume:

The length of the driveway is 30 feet
The driveway width at the apron end is 22 feet
The driveway width at the street tapers to 8 feet
Add width at the two ends: 20' + 8' = 28'
Then divide by two: 28 feet divided by 2 is 14 feet
Then multiply by the average width by the length: 14 feet times 30 feet is 420 square feet.

A cubic yard of concrete will cover the following area, assuming no waste:
At 2" thick, 1 cubic yard covers 162 square feet
At 3" thick, 1 cubic yard covers 108 square feet
At 4" thick, 1 cubic yard covers 81 square feet
At 5" thick, 1 cubic yard covers 64.8 square feet
At 6" thick, 1 cubic yard covers 54 square feet
At 7" thick, 1 cubic yard covers 46.2 square feet
At 8" thick, 1 cubic yard covers 40.5 square feet.

If the job requires 420 square feet of 4" concrete, divide 420 by 81 to determine that 5.18 cubic yards are needed. It's good estimating practice to allow 5 percent for waste and over-excavation (removing more soil than needed to maintain slab or footing depth). Add 5 percent to find the order quantity of 5.44 cubic yards.

**Garage slabs.** Slabs are poured monolithic and should be pitched from rear to the front opening so snow melt drains to the street. Garage slab floor drains are poor practice in colder climates. Melted snow freezes in the drain and tends to crack the slab. Keep the slab surface 4" above grade to provide good drainage. When grade beams are required by code, excavate the additional depth around the perimeter of the slab. No additional forming is needed. The beam is poured as an integral part of the garage floor. Wire mesh or rod reinforcement is good practice, even if not required by code. If the driveway includes a change in grade, be sure the incline, including any swale, won't be too steep for most cars. Be sure the grade at the street or alley matches the existing grade.

Consider the drainage pattern before quoting concrete work. Water draining around the perimeter of a slab can cause damage that's expensive to repair.

**Concrete forms.** Most concrete form lumber can be used more than once, either several times for concrete forming or once for forming and then on some other part of the job when forming is done. The figures that follow include options for multiple use of form lumber. These options assume that forms can be removed, partially disassembled and cleaned. Then 75 percent of the form material will be reused. Costs include assembly, oiling, setting, stripping and cleaning. Adjust material costs in this section to reflect your actual cost of lumber. Here's how: Divide your actual lumber cost per MBF by the assumed lumber cost. Then multiply the cost in the material column by this adjustment factor.

| | Craft@Hrs | Unit | Material | Labor | Total | Sell |
|---|---|---|---|---|---|---|

**Driveway and walkway edge forms.** Material costs include stakes, nails, form oil and 5 percent waste. Per LF of edge form. When forms are required on both sides of the concrete, include the length of each side plus the end widths.

| | Craft@Hrs | Unit | Material | Labor | Total | Sell |
|---|---|---|---|---|---|---|
| Using 2" x 4" lumber, .7 BF per LF | — | MBF | 622.00 | — | 622.00 | — |
| Using nails, ties and form oil, per SF | — | LF | .22 | — | .22 | — |
| 1 use | B2@.050 | LF | .60 | 1.51 | 2.11 | 3.59 |
| 3 use | B2@.050 | LF | .36 | 1.51 | 1.87 | 3.18 |
| 5 use | B2@.050 | LF | .30 | 1.51 | 1.81 | 3.08 |
| **2" x 6" edge form** | | | | | | |
| Using 2" x 6" lumber, 1.05 BF per LF | — | MBF | 641.00 | — | 641.00 | — |
| 1 use | B2@.050 | LF | .83 | 1.51 | 2.34 | 3.98 |
| 3 use | B2@.050 | LF | .46 | 1.51 | 1.97 | 3.35 |
| 5 use | B2@.050 | LF | .38 | 1.51 | 1.89 | 3.21 |
| **2" x 8" edge form** | | | | | | |
| Using 2" x 8" lumber, 1.4 BF per LF | — | MBF | 651.00 | — | 651.00 | — |
| 1 use | B2@.055 | LF | 1.12 | 1.66 | 2.78 | 4.73 |
| 3 use | B2@.055 | LF | .59 | 1.66 | 2.25 | 3.83 |
| 5 use | B2@.055 | LF | .48 | 1.66 | 2.14 | 3.64 |
| **2" x 10" edge form** | | | | | | |
| Using 2" x 10" lumber, 1.75 BF per LF | — | MBF | 683.00 | — | 683.00 | — |
| 1 use | B2@.055 | LF | 1.36 | 1.66 | 3.02 | 5.13 |
| 3 use | B2@.055 | LF | .70 | 1.66 | 2.36 | 4.01 |
| 5 use | B2@.055 | LF | .55 | 1.66 | 2.21 | 3.76 |
| **2" x 12" edge form** | | | | | | |
| Using 2" x 12" lumber, 2.1 BF per LF | — | MBF | 730.00 | — | 730.00 | — |
| 1 use | B2@.055 | LF | 1.69 | 1.66 | 3.35 | 5.70 |
| 3 use | B2@.055 | LF | .85 | 1.66 | 2.51 | 4.27 |
| 5 use | B2@.055 | LF | .67 | 1.66 | 2.33 | 3.96 |

**Plywood forming.** For foundation walls, building walls and retaining walls, using 3/4" plyform with 10 percent waste, and 2" bracing with 20 percent waste. All material costs include nails, ties, clamps, form oil and stripping. Costs shown are per square foot of contact area (SFCA). When forms are required on both sides of the concrete, include the contact surface area for each side.

| | Craft@Hrs | Unit | Material | Labor | Total | Sell |
|---|---|---|---|---|---|---|
| Using 3/4" plyform, per MSF | — | MSF | 1,410.00 | — | 1,410.00 | — |
| Using 2" bracing, per MBF | — | MBF | 631.00 | — | 631.00 | — |
| Using nails, ties, clamps & form oil, per SF | — | SFCA | .25 | — | .25 | — |
| **Walls up to 4' high (1.10 SF of plywood and .42 BF of bracing per SF of form area)** | | | | | | |
| 1 use | B2@.051 | SFCA | 2.04 | 1.54 | 3.58 | 6.09 |
| 3 use | B2@.051 | SFCA | 1.04 | 1.54 | 2.58 | 4.39 |
| 5 use | B2@.051 | SFCA | .82 | 1.54 | 2.36 | 4.01 |
| **Walls 4' to 8' high (1.10 SF of plywood and .60 BF of bracing per SF of form area)** | | | | | | |
| 1 use | B2@.060 | SFCA | 2.15 | 1.82 | 3.97 | 6.75 |
| 3 use | B2@.060 | SFCA | 1.09 | 1.82 | 2.91 | 4.95 |
| 5 use | B2@.060 | SFCA | .86 | 1.82 | 2.68 | 4.56 |
| **Walls 8' to 12' high (1.10 SF of plywood and .90 BF of bracing per SF of form area)** | | | | | | |
| 1 use | B2@.095 | SFCA | 2.34 | 2.88 | 5.22 | 8.87 |
| 3 use | B2@.095 | SFCA | 1.17 | 2.88 | 4.05 | 6.89 |
| 5 use | B2@.095 | SFCA | .92 | 2.88 | 3.80 | 6.46 |

| | Craft@Hrs | Unit | Material | Labor | Total | Sell |
|---|---|---|---|---|---|---|

**Form stripping.** Labor to remove forms and bracing, clean, and stack on job site.

Board forms at wall footings, grade beams and column footings. Per SF of contact area

| | Craft@Hrs | Unit | Material | Labor | Total | Sell |
|---|---|---|---|---|---|---|
| 1" thick lumber | BL@.010 | SF | — | .27 | .27 | .46 |
| 2" thick lumber | BL@.012 | SF | — | .32 | .32 | .54 |
| Removing keyways, 2" x 4" and 2" x 6" | BL@.006 | LF | — | .16 | .16 | .27 |

Slab edge forms. Per LF of edge form

| | Craft@Hrs | Unit | Material | Labor | Total | Sell |
|---|---|---|---|---|---|---|
| 2" x 4" to 2" x 6" | BL@.012 | LF | — | .32 | .32 | .54 |
| 2" x 8" to 2" x 12" | BL@.013 | LF | — | .35 | .35 | .60 |

Walls, plywood forms. Per SF of contact area stripped

| | Craft@Hrs | Unit | Material | Labor | Total | Sell |
|---|---|---|---|---|---|---|
| To 4' high | BL@.017 | SF | — | .45 | .45 | .77 |
| Over 4' to 8' high | BL@.018 | SF | — | .48 | .48 | .82 |
| Over 8' to 12' high | BL@.020 | SF | — | .53 | .53 | .90 |
| Over 12' to 16' high | BL@.027 | SF | — | .72 | .72 | 1.22 |
| Over 16' high | BL@.040 | SF | — | 1.07 | 1.07 | 1.82 |

**Board forming and stripping.** For wall footings, grade beams, column footings, site curbs and steps. Includes 5 percent waste. Per SF of contact area. When forms are required on both sides of the concrete, include the contact surface for each side.

2" thick forms and bracing, using 2.85 BF of form lumber per SF. Includes nails, ties, and form oil

| | Craft@Hrs | Unit | Material | Labor | Total | Sell |
|---|---|---|---|---|---|---|
| Using 2" lumber, per MBF | — | MBF | 631.00 | — | 631.00 | — |
| Using nails, ties and form oil, per SF | — | SFCA | .22 | — | .22 | — |
| 1 use | B2@.115 | SFCA | 2.09 | 3.48 | 5.57 | 9.47 |
| 3 use | B2@.115 | SFCA | 1.01 | 3.48 | 4.49 | 7.63 |
| 5 use | B2@.115 | SFCA | .79 | 3.48 | 4.27 | 7.26 |

Add for keyway beveled on two edges, one use. No stripping included

| | Craft@Hrs | Unit | Material | Labor | Total | Sell |
|---|---|---|---|---|---|---|
| 2" x 4" | B2@.027 | LF | .43 | .82 | 1.25 | 2.13 |
| 2" x 6" | B2@.027 | LF | .67 | .82 | 1.49 | 2.53 |

**Concrete Reinforcing Steel.** Steel reinforcing bars (rebar), ASTM A615 Grade 60. Material costs are for deformed steel reinforcing rebars, including 10 percent lap allowance, cutting and bending. Add for epoxy or galvanized coating of rebars, chairs and splicing, if required. Costs per pound (Lb) and per linear foot (LF) including tie wire and tying.

Reinforcing steel placed and tied in footings and grade beams

| | Craft@Hrs | Unit | Material | Labor | Total | Sell |
|---|---|---|---|---|---|---|
| 1/4" diameter, #2 rebar | RI@.015 | Lb | .78 | .46 | 1.24 | 2.11 |
| 1/4" diameter, #2 rebar (.17 lb per LF) | RI@.003 | LF | .14 | .09 | .23 | .39 |
| 3/8" diameter, #3 rebar | RI@.011 | Lb | .49 | .34 | .83 | 1.41 |
| 3/8" diameter, #3 rebar (.38 lb per LF) | RI@.004 | LF | .21 | .12 | .33 | .56 |
| 1/2" diameter, #4 rebar | RI@.010 | Lb | .47 | .31 | .78 | 1.33 |
| 1/2" diameter, #4 rebar (.67 lb per LF) | RI@.007 | LF | .35 | .22 | .57 | .97 |
| 5/8" diameter, #5 rebar | RI@.009 | Lb | .43 | .28 | .71 | 1.21 |
| 5/8" diameter, #5 rebar (1.04 lb per LF) | RI@.009 | LF | .49 | .28 | .77 | 1.31 |
| 3/4" diameter, #6 rebar | RI@.008 | Lb | .41 | .25 | .66 | 1.12 |
| 3/4" diameter, #6 rebar (1.50 lb per LF) | RI@.012 | LF | .67 | .37 | 1.04 | 1.77 |
| 7/8" diameter, #7 rebar | RI@.008 | Lb | .50 | .25 | .75 | 1.28 |
| 7/8" diameter, #7 rebar (2.04 lb per LF) | RI@.016 | LF | 1.12 | .49 | 1.61 | 2.74 |
| 1" diameter, #8 rebar | RI@.008 | Lb | .43 | .25 | .68 | 1.16 |
| 1" diameter, #8 rebar (2.67 lb per LF) | RI@.021 | LF | 1.26 | .65 | 1.91 | 3.25 |

Reinforcing steel placed and tied in structural slabs

| | Craft@Hrs | Unit | Material | Labor | Total | Sell |
|---|---|---|---|---|---|---|
| 1/4" diameter, #2 rebar | RI@.014 | Lb | .78 | .43 | 1.21 | 2.06 |
| 1/4" diameter, #2 rebar (.17 lb per LF) | RI@.002 | LF | .14 | .06 | .20 | .34 |
| 3/8" diameter, #3 rebar | RI@.010 | Lb | .49 | .31 | .80 | 1.36 |
| 3/8" diameter, #3 rebar (.38 lb per LF) | RI@.004 | LF | .21 | .12 | .33 | .56 |

|  | Craft@Hrs | Unit | Material | Labor | Total | Sell |
|---|---|---|---|---|---|---|
| 1/2" diameter, #4 rebar | RI@.009 | Lb | .47 | .28 | .75 | 1.28 |
| 1/2" diameter, #4 rebar (.67 lb per LF) | RI@.006 | LF | .35 | .19 | .54 | .92 |
| 5/8" diameter, #5 rebar | RI@.008 | Lb | .43 | .25 | .68 | 1.16 |
| 5/8" diameter, #5 rebar (1.04 lb per LF) | RI@.008 | LF | .49 | .25 | .74 | 1.26 |
| 3/4" diameter, #6 rebar | RI@.007 | Lb | .41 | .22 | .63 | 1.07 |
| 3/4" diameter, #6 rebar (1.50 lb per LF) | RI@.011 | LF | .67 | .34 | 1.01 | 1.72 |
| 7/8" diameter, #7 rebar | RI@.007 | Lb | .50 | .22 | .72 | 1.22 |
| 7/8" diameter, #7 rebar (2.04 lb per LF) | RI@.014 | LF | 1.12 | .43 | 1.55 | 2.64 |
| 1" diameter, #8 rebar | RI@.007 | Lb | .43 | .22 | .65 | 1.11 |
| 1" diameter, #8 rebar (2.67 lb per LF) | RI@.019 | LF | 1.26 | .59 | 1.85 | 3.15 |
| Reinforcing steel placed and tied in walls |  |  |  |  |  |  |
| 1/4" diameter, #2 rebar | RI@.017 | Lb | .78 | .53 | 1.31 | 2.23 |
| 1/4" diameter, #2 rebar (.17 lb per LF) | RI@.003 | LF | .14 | .09 | .23 | .39 |
| 3/8" diameter, #3 rebar | RI@.012 | Lb | .49 | .37 | .86 | 1.46 |
| 3/8" diameter, #3 rebar (.38 lb per LF) | RI@.005 | LF | .21 | .15 | .36 | .61 |
| 1/2" diameter, #4 rebar | RI@.011 | Lb | .47 | .34 | .81 | 1.38 |
| 1/2" diameter, #4 rebar (.67 lb per LF) | RI@.007 | LF | .35 | .22 | .57 | .97 |
| 5/8" diameter, #5 rebar | RI@.010 | Lb | .43 | .31 | .74 | 1.26 |
| 5/8" diameter, #5 rebar (1.04 lb per LF) | RI@.010 | LF | .49 | .31 | .80 | 1.36 |
| 3/4" diameter, #6 rebar | RI@.009 | Lb | .41 | .28 | .69 | 1.17 |
| 3/4" diameter, #6 rebar (1.50 lb per LF) | RI@.014 | LF | .67 | .43 | 1.10 | 1.87 |
| 7/8" diameter, #7 rebar | RI@.009 | Lb | .50 | .28 | .78 | 1.33 |
| 7/8" diameter, #7 rebar (2.04 lb per LF) | RI@.018 | LF | 1.12 | .56 | 1.68 | 2.86 |
| 1" diameter, #8 rebar | RI@.009 | Lb | .43 | .28 | .71 | 1.21 |
| 1" diameter, #8 rebar (2.67 lb per LF) | RI@.024 | LF | 1.26 | .74 | 2.00 | 3.40 |

Add for reinforcing steel coating. Coating applied prior to shipment to job site. Labor for placing coated rebar is the same as that shown for plain rebar above. Cost per pound of reinforcing steel.

|  | Craft@Hrs | Unit | Material | Labor | Total | Sell |
|---|---|---|---|---|---|---|
| Epoxy coating |  |  |  |  |  |  |
| Any size rebar, add | — | Lb | .28 | — | .28 | — |
| Galvanized coating |  |  |  |  |  |  |
| Any #2 rebar, #3 rebar or #4 rebar, add | — | Lb | .35 | — | .35 | — |
| Any #5 rebar or #6 rebar, add | — | Lb | .33 | — | .33 | — |
| Any #7 rebar or #8 rebar, add | — | Lb | .31 | — | .31 | — |

**Welded wire mesh.** Electric weld, including 15 percent waste and overlap.

|  | Craft@Hrs | Unit | Material | Labor | Total | Sell |
|---|---|---|---|---|---|---|
| 2" x 2" W.9 x W.9 (#12 x #12), slabs | RI@.004 | SF | .79 | .12 | .91 | 1.55 |
| 2" x 2" W.9 x W.9 (#12 x #12), beams and columns | RI@.020 | SF | .76 | .62 | 1.38 | 2.35 |
| 4" x 4" W1.4 x W1.4 (#10 x #10), slabs | RI@.003 | SF | .36 | .09 | .45 | .77 |
| 4" x 4" W2.0 x W2.0 (#8 x #8), slabs | RI@.004 | SF | .43 | .12 | .55 | .94 |
| 4" x 4" W2.9 x W2.9 (#6 x #6), slabs | RI@.005 | SF | .47 | .15 | .62 | 1.05 |
| 4" x 4" W4.0 x W4.0 (#4 x #4), slabs | RI@.006 | SF | .62 | .19 | .81 | 1.38 |
| 6" x 6" W1.4 x W1.4 (#10 x #10), slabs | RI@.003 | SF | .17 | .09 | .26 | .44 |
| 6" x 6" W2.0 x W2.0 (#8 x #8), slabs | RI@.004 | SF | .28 | .12 | .40 | .68 |
| 6" x 6" W2.9 x W2.9 (#6 x #6), slabs | RI@.004 | SF | .32 | .12 | .44 | .75 |
| 6" x 6" W4.0 x W4.0 (#4 x #4), slabs | RI@.005 | SF | .45 | .15 | .60 | 1.02 |
| Add for lengthwise cut, LF of cut | RI@.002 | LF | — | .06 | .06 | .10 |

| | Craft@Hrs | Unit | Material | Labor | Total | Sell |
|---|---|---|---|---|---|---|

**Concrete expansion joints.** Fiber or asphaltic-felt sided. Per linear foot.

1/4" wall thickness

| | Craft@Hrs | Unit | Material | Labor | Total | Sell |
|---|---|---|---|---|---|---|
| 3" deep | B1@.030 | LF | .17 | .88 | 1.05 | 1.79 |
| 3-1/2" deep | B1@.030 | LF | .21 | .88 | 1.09 | 1.85 |
| 4" deep | B1@.030 | LF | .41 | .88 | 1.29 | 2.19 |
| 6" deep | B1@.030 | LF | .57 | .88 | 1.45 | 2.47 |

3/8" wall thickness

| | Craft@Hrs | Unit | Material | Labor | Total | Sell |
|---|---|---|---|---|---|---|
| 3" deep | B1@.030 | LF | .22 | .88 | 1.10 | 1.87 |
| 3-1/2" deep | B1@.030 | LF | .28 | .88 | 1.16 | 1.97 |
| 4" deep | B1@.030 | LF | .55 | .88 | 1.43 | 2.43 |
| 6" deep | B1@.030 | LF | .77 | .88 | 1.65 | 2.81 |

1/2" wall thickness

| | Craft@Hrs | Unit | Material | Labor | Total | Sell |
|---|---|---|---|---|---|---|
| 3" deep | B1@.030 | LF | .28 | .88 | 1.16 | 1.97 |
| 3-1/2" deep | B1@.030 | LF | .35 | .88 | 1.23 | 2.09 |
| 4" deep | B1@.030 | LF | .69 | .88 | 1.57 | 2.67 |
| 6" deep | B1@.030 | LF | .96 | .88 | 1.84 | 3.13 |

**Concrete.** Ready-mix delivered by truck. Typical prices for most cities. Includes delivery up to 20 miles for 10 CY or more, 3" to 4" slump. Material cost only, no placing or pumping included. See forming and finishing costs on the following pages.

Footing and foundation concrete, 1-1/2" aggregate

| | Craft@Hrs | Unit | Material | Labor | Total | Sell |
|---|---|---|---|---|---|---|
| 2,000 PSI, 4.8 sack mix | — | CY | 99.20 | — | 99.20 | — |
| 2,500 PSI, 5.2 sack mix | — | CY | 102.00 | — | 102.00 | — |
| 3,000 PSI, 5.7 sack mix | — | CY | 104.00 | — | 104.00 | — |
| 3,500 PSI, 6.3 sack mix | — | CY | 107.00 | — | 107.00 | — |
| 4,000 PSI, 6.9 sack mix | — | CY | 110.00 | — | 110.00 | — |

Slab, sidewalk, and driveway concrete, 1" aggregate

| | Craft@Hrs | Unit | Material | Labor | Total | Sell |
|---|---|---|---|---|---|---|
| 2,000 PSI, 5.0 sack mix | — | CY | 101.00 | — | 101.00 | — |
| 2,500 PSI, 5.5 sack mix | — | CY | 104.00 | — | 104.00 | — |
| 3,000 PSI, 6.0 sack mix | — | CY | 105.00 | — | 105.00 | — |
| 3,500 PSI, 6.6 sack mix | — | CY | 107.00 | — | 107.00 | — |
| 4,000 PSI, 7.1 sack mix | — | CY | 108.00 | — | 108.00 | — |

Pea gravel pump mix / grout mix, 3/8" aggregate

| | Craft@Hrs | Unit | Material | Labor | Total | Sell |
|---|---|---|---|---|---|---|
| 2,000 PSI, 6.0 sack mix | — | CY | 109.00 | — | 109.00 | — |
| 2,500 PSI, 6.5 sack mix | — | CY | 111.00 | — | 111.00 | — |
| 3,000 PSI, 7.2 sack mix | — | CY | 114.00 | — | 114.00 | — |
| 3,500 PSI, 7.9 sack mix | — | CY | 118.00 | — | 118.00 | — |
| 5,000 PSI, 8.5 sack mix | — | CY | 125.00 | — | 125.00 | — |

Extra delivery costs for ready-mix concrete

| | Craft@Hrs | Unit | Material | Labor | Total | Sell |
|---|---|---|---|---|---|---|
| Add for delivery over 20 miles | — | Mile | 4.00 | — | 4.00 | — |
| Add for standby charge in excess of 5 minutes per CY delivered per minute of extra time | — | Ea | 1.75 | — | 1.75 | — |
| Add for less than 10 CY per load delivered | — | CY | 20.00 | — | 20.00 | — |

Extra costs for non-standard aggregates

| | Craft@Hrs | Unit | Material | Labor | Total | Sell |
|---|---|---|---|---|---|---|
| Add for lightweight aggregate, typical | — | CY | 42.50 | — | 42.50 | — |
| Add for lightweight aggregate, pump mix | — | CY | 44.00 | — | 44.00 | — |
| Add for granite aggregate, typical | — | CY | 6.50 | — | 6.50 | — |

| | Craft@Hrs | Unit | Material | Labor | Total | Sell |
|---|---|---|---|---|---|---|
| Extra costs for non-standard mix additives | | | | | | |
| High early strength 5 sack mix | — | CY | 12.00 | — | 12.00 | — |
| High early strength 6 sack mix | — | CY | 16.00 | — | 16.00 | — |
| Add for white cement (architectural) | — | CY | 55.00 | — | 55.00 | — |
| Add for 1% calcium chloride | — | CY | 1.50 | — | 1.50 | — |
| Add for chemical compensated shrinkage | — | CY | 20.00 | — | 20.00 | — |

Add for colored concrete, typical prices. The ready-mix supplier charges for cleanup of the ready-mix truck used for delivery of colored concrete. The usual practice is to provide one truck per day for delivery of all colored concrete for a particular job. Add the cost below to the cost per cubic yard for the design mix required. Also add the cost for truck cleanup.

| | Craft@Hrs | Unit | Material | Labor | Total | Sell |
|---|---|---|---|---|---|---|
| Adobe | — | CY | 19.00 | — | 19.00 | — |
| Black | — | CY | 30.00 | — | 30.00 | — |
| Blended red | — | CY | 17.00 | — | 17.00 | — |
| Brown | — | CY | 22.00 | — | 22.00 | — |
| Green | — | CY | 34.00 | — | 34.00 | — |
| Yellow | — | CY | 23.00 | — | 23.00 | — |
| Colored concrete, truck cleanup, per day | — | LS | 60.00 | — | 60.00 | — |

**Concrete for walls.** Cast-in-place concrete walls for buildings or retaining walls. Material costs for concrete placed direct from chute are based on a 2,000 PSI, 5.0 sack mix, with 1" aggregate, including 5 percent waste.

Pump mix cost includes an additional $10.00 per CY for the pump. Labor costs are for placing only. Add the cost of excavation, formwork, steel reinforcing, finishes and curing. Square foot costs are based on SF of wall measured on one face only. Costs don't include engineering, design or foundations.

| | Craft@Hrs | Unit | Material | Labor | Total | Sell |
|---|---|---|---|---|---|---|
| Using concrete (before 5% waste allowance) at | — | CY | 99.20 | — | 99.20 | — |
| 4" thick walls (1.23 CY per CSF) | | | | | | |
| To 4' high, direct from chute | B1@.013 | SF | 1.28 | .38 | 1.66 | 2.82 |
| 4' to 8' high, pumped | B3@.015 | SF | 1.43 | .43 | 1.86 | 3.16 |
| 8' to 12' high, pumped | B3@.017 | SF | 1.43 | .48 | 1.91 | 3.25 |
| 12' to 16' high, pumped | B3@.018 | SF | 1.43 | .51 | 1.94 | 3.30 |
| 16' high, pumped | B3@.020 | SF | 1.43 | .57 | 2.00 | 3.40 |
| 6" thick walls (1.85 CY per CSF) | | | | | | |
| To 4' high, direct from chute | B1@.020 | SF | 1.93 | .59 | 2.52 | 4.28 |
| 4' to 8' high, pumped | B3@.022 | SF | 2.15 | .63 | 2.78 | 4.73 |
| 8' to 12' high, pumped | B3@.025 | SF | 2.15 | .71 | 2.86 | 4.86 |
| 12' to 16' high, pumped | B3@.027 | SF | 2.15 | .77 | 2.92 | 4.96 |
| 16' high, pumped | B3@.030 | SF | 2.15 | .85 | 3.00 | 5.10 |
| 8" thick walls (2.47 CY per CSF) | | | | | | |
| To 4' high, direct from chute | B1@.026 | SF | 2.57 | .76 | 3.33 | 5.66 |
| 4' to 8' high, pumped | B3@.030 | SF | 2.87 | .85 | 3.72 | 6.32 |
| 8' to 12' high, pumped | B3@.033 | SF | 2.87 | .94 | 3.81 | 6.48 |
| 12' to 16' high, pumped | B3@.036 | SF | 2.87 | 1.02 | 3.89 | 6.61 |
| 16' high, pumped | B3@.040 | SF | 2.87 | 1.14 | 4.01 | 6.82 |
| 10" thick walls (3.09 CY per CSF) | | | | | | |
| To 4' high, direct from chute | B1@.032 | SF | 3.22 | .94 | 4.16 | 7.07 |
| 4' to 8' high, pumped | B3@.037 | SF | 3.59 | 1.05 | 4.64 | 7.89 |
| 8' to 12' high, pumped | B3@.041 | SF | 3.59 | 1.17 | 4.76 | 8.09 |
| 12' to 16' high, pumped | B3@.046 | SF | 3.59 | 1.31 | 4.90 | 8.33 |
| 16' high, pumped | B3@.050 | SF | 3.59 | 1.42 | 5.01 | 8.52 |
| 12" thick walls (3.70 CY per CSF) | | | | | | |
| To 4' high, placed direct from chute | B1@.040 | SF | 3.85 | 1.17 | 5.02 | 8.53 |
| 4' to 8' high, pumped | B3@.045 | SF | 4.30 | 1.28 | 5.58 | 9.49 |
| 8' to 12' high, pumped | B3@.050 | SF | 4.30 | 1.42 | 5.72 | 9.72 |
| 12' to 16' high, pumped | B3@.055 | SF | 4.30 | 1.57 | 5.87 | 9.98 |
| 16' high, pumped | B3@.060 | SF | 4.30 | 1.71 | 6.01 | 10.20 |

| | Craft@Hrs | Unit | Material | Labor | Total | Sell |
|---|---|---|---|---|---|---|

**Concrete Footings, Grade Beams and Stem Walls.** Use the figures below for preliminary estimates. Concrete costs are based on a 2,000 PSI, 5.0 sack mix with 1" aggregate placed directly from the chute of a ready-mix truck. Figures in parentheses show the cubic yards of concrete per linear foot of foundation including 5 percent waste. Costs shown include concrete, 60 pounds of reinforcing per CY of concrete, and typical excavation using a 3/4 CY backhoe with excess backfill spread on site.

**Concrete footings and grade beams.** Cast directly against the earth, including excavation, steel reinforcing and concrete but no forming or finishing. These costs assume the top of the foundation will be at finished grade. For scheduling purposes estimate that a crew of 3 can lay out, excavate, place and tie the reinforcing steel and place 13 CY of concrete in an 8-hour day. Add the cost of equipment rental. Use $900.00 as a minimum job charge.

| | Craft@Hrs | Unit | Material | Labor | Total | Sell |
|---|---|---|---|---|---|---|
| Using 2,000 PSI concrete, per CY | — | CY | 101.00 | — | 101.00 | — |
| Using reinforcing bars, per pound | — | Lb | .47 | — | .47 | — |
| Typical cost per CY | B4@1.80 | CY | 129.00 | 56.10 | 185.10 | 315.00 |
| 12" W x  6" D (0.019CY per LF) | B4@.034 | LF | 2.45 | 1.06 | 3.51 | 5.97 |
| 12" W x  8" D (0.025CY per LF) | B4@.045 | LF | 3.23 | 1.40 | 4.63 | 7.87 |
| 12" W x 10" D (0.032 CY per LF) | B4@.058 | LF | 4.13 | 1.81 | 5.94 | 10.10 |
| 12" W x 12" D (0.039 CY per LF) | B4@.070 | LF | 5.03 | 2.18 | 7.21 | 12.30 |
| 12" W x 18" D (0.058 CY per LF) | B4@.104 | LF | 7.48 | 3.24 | 10.72 | 18.20 |
| 16" W x  8" D (0.035 CY per LF) | B4@.063 | LF | 4.52 | 1.96 | 6.48 | 11.00 |
| 16" W x 10" D (0.043 CY per LF) | B4@.077 | LF | 5.55 | 2.40 | 7.95 | 13.50 |
| 16" W x 12" D (0.052 CY per LF) | B4@.094 | LF | 6.71 | 2.93 | 9.64 | 16.40 |
| 18" W x  8" D (0.037 CY per LF) | B4@.067 | LF | 4.77 | 2.09 | 6.86 | 11.70 |
| 18" W x 10" D (0.049 CY per LF) | B4@.088 | LF | 6.32 | 2.74 | 9.06 | 15.40 |
| 18" W x 24" D (0.117 CY per LF) | B4@.216 | LF | 15.10 | 6.73 | 21.83 | 37.10 |
| 20" W x 12" D (0.065 CY per LF) | B4@.117 | LF | 8.39 | 3.65 | 12.04 | 20.50 |
| 24" W x 12" D (0.078 CY per LF) | B4@.140 | LF | 10.10 | 4.36 | 14.46 | 24.60 |
| 24" W x 24" D (0.156 CY per LF) | B4@.281 | LF | 20.10 | 8.76 | 28.86 | 49.10 |
| 24" W x 30" D (0.195 CY per LF) | B4@.351 | LF | 25.20 | 10.90 | 36.10 | 61.40 |
| 24" W x 36" D (0.234 CY per LF) | B4@.421 | LF | 30.20 | 13.10 | 43.30 | 73.60 |
| 30" W x 36" D (0.292 CY per LF) | B4@.526 | LF | 37.70 | 16.40 | 54.10 | 92.00 |
| 30" W x 42" D (0.341 CY per LF) | B4@.614 | LF | 44.00 | 19.10 | 63.10 | 107.00 |
| 36" W x 48" D (0.467 CY per LF) | B4@.841 | LF | 60.20 | 26.20 | 86.40 | 147.00 |

To estimate the cost of footing or grade beam sizes not shown, multiply the width in inches by the depth in inches and divide the result by 3700. This is the CY of concrete per LF of footing including 5 percent waste. Multiply by the "Typical cost per CY" in the prior table to find the cost per LF.

**Continuous concrete footing with foundation stem wall.** These figures assume the foundation stem wall projects 24" above the finished grade and extends into the soil 18" to the top of the footing. Costs shown include typical excavation using a 3/4 CY backhoe with excess backfill spread on site, forming both sides of the foundation wall and the footing, based on three uses of the forms and 2 #4 rebar. Use $1,200.00 as a minimum cost for this type work. Add the cost of equipment rental.

| | Craft@Hrs | Unit | Material | Labor | Total | Sell |
|---|---|---|---|---|---|---|
| Typical cost per CY | B5@7.16 | CY | 145.00 | 223.00 | 368.00 | 626.00 |

**Concrete footing and stem wall for single-story structure.**
Typical single-story structure, footing 12" W x 8" D,

| | Craft@Hrs | Unit | Material | Labor | Total | Sell |
|---|---|---|---|---|---|---|
| wall 6" T x 42" D (.10 CY per LF) | B5@.716 | LF | 14.50 | 22.30 | 36.80 | 62.60 |

**Concrete footing and stem wall for two-story structure.**
Typical two-story structure, footing 18" W x 10" D,

| | Craft@Hrs | Unit | Material | Labor | Total | Sell |
|---|---|---|---|---|---|---|
| wall 8" T x 42" D (.14 CY per LF) | B5@1.00 | LF | 20.30 | 31.20 | 51.50 | 87.60 |

**Concrete footing and stem wall for three-story structure.**
Typical three-story structure, footing 24" W x 12" D,

| | Craft@Hrs | Unit | Material | Labor | Total | Sell |
|---|---|---|---|---|---|---|
| wall 10" T x 42" D (.19 CY per LF) | B5@1.36 | LF | 27.60 | 42.40 | 70.00 | 119.00 |

| | Craft@Hrs | Unit | Material | Labor | Total | Sell |
|---|---|---|---|---|---|---|

**Concrete slabs, walks and driveways.** Typical costs for reinforced concrete slabs-on-grade including fine grading, slab base, forms, vapor barrier, wire mesh, 3000 PSI concrete, finishing and curing. For thickened edge slabs, add the area of the thickened edge. Use 500 square feet as the minimum job size.

| | Craft@Hrs | Unit | Material | Labor | Total | Sell |
|---|---|---|---|---|---|---|
| 2" thick | B5@.067 | SF | 1.98 | 2.09 | 4.07 | 6.92 |
| 3" thick | B5@.068 | SF | 2.31 | 2.12 | 4.43 | 7.53 |
| 4" thick | B5@.069 | SF | 2.63 | 2.15 | 4.78 | 8.13 |
| 5" thick | B5@.070 | SF | 3.29 | 2.18 | 5.47 | 9.30 |
| 6" thick | B5@.071 | SF | 3.95 | 2.22 | 6.17 | 10.50 |

**Slab Base.** Aggregate base for slabs. No waste included. Labor costs are for spreading aggregate from piles only. Add for fine grading, using hand tools.

Crushed stone base 1.4 tons equal one cubic yard

| | Craft@Hrs | Unit | Material | Labor | Total | Sell |
|---|---|---|---|---|---|---|
| Using crushed stone, per CY | — | CY | 24.10 | — | 24.10 | — |
| 1" base ( .309 CY per CSF) | BL@.001 | SF | .07 | .03 | .10 | .17 |
| 2" base ( .617 CY per CSF) | BL@.003 | SF | .15 | .08 | .23 | .39 |
| 3" base ( .926 CY per CSF) | BL@.004 | SF | .22 | .11 | .33 | .56 |
| 4" base (1.23 CY per CSF) | BL@.006 | SF | .30 | .16 | .46 | .78 |
| 5" base (1.54 CY per CSF) | BL@.007 | SF | .37 | .19 | .56 | .95 |
| 6" base (1.85 CY per CSF) | BL@.008 | SF | .45 | .21 | .66 | 1.12 |

Sand fill base 1.40 tons equal one cubic yard

| | Craft@Hrs | Unit | Material | Labor | Total | Sell |
|---|---|---|---|---|---|---|
| Using sand, per CY | — | CY | 20.50 | — | 20.50 | — |
| 1" fill ( .309 CY per CSF) | BL@.001 | SF | .06 | .03 | .09 | .15 |
| 2" fill ( .617 CY per CSF) | BL@.002 | SF | .13 | .05 | .18 | .31 |
| 3" fill ( .926 CY per CSF) | BL@.003 | SF | .19 | .08 | .27 | .46 |
| 4" fill (1.23 CY per CSF) | BL@.004 | SF | .25 | .11 | .36 | .61 |
| 5" fill (1.54 CY per CSF) | BL@.006 | SF | .32 | .16 | .48 | .82 |
| 6" fill (1.85 CY per CSF) | BL@.007 | SF | .38 | .19 | .57 | .97 |
| Add for fine grading for slab on grade | BL@.008 | SF | — | .21 | .21 | .36 |

**Concrete finishing.**

Slab finishes

| | Craft@Hrs | Unit | Material | Labor | Total | Sell |
|---|---|---|---|---|---|---|
| Broom finish | B6@.012 | SF | — | .34 | .34 | .58 |
| Float finish | B6@.010 | SF | — | .29 | .29 | .49 |

Trowel finishing

| | Craft@Hrs | Unit | Material | Labor | Total | Sell |
|---|---|---|---|---|---|---|
| Steel, machine work | B6@.015 | SF | — | .43 | .43 | .73 |
| Steel, hand work | B6@.018 | SF | — | .51 | .51 | .87 |

Finish treads and risers

| | Craft@Hrs | Unit | Material | Labor | Total | Sell |
|---|---|---|---|---|---|---|
| No abrasives, no plastering, per LF of tread | B6@.042 | LF | — | 1.20 | 1.20 | 2.04 |
| With abrasives, plastered, per LF of tread | B6@.063 | LF | .57 | 1.80 | 2.37 | 4.03 |
| Slab finishesScoring concrete surface, hand work | B6@.005 | LF | — | .14 | .14 | .24 |
| Sweep, scrub and wash down | B6@.006 | SF | .02 | .17 | .19 | .32 |

Liquid curing and sealing compound, spray-on, acrylic concrete cure and seal,

| | Craft@Hrs | Unit | Material | Labor | Total | Sell |
|---|---|---|---|---|---|---|
| 400 SF per gallon | B6@.003 | SF | .05 | .09 | .14 | .24 |

Exposed aggregate (washed, including finishing),

| | Craft@Hrs | Unit | Material | Labor | Total | Sell |
|---|---|---|---|---|---|---|
| no disposal of slurry | B6@.017 | SF | .29 | .49 | .78 | 1.33 |

Non-metallic color and concrete hardener, troweled on, 2 applications, red, gray or black,

| | Craft@Hrs | Unit | Material | Labor | Total | Sell |
|---|---|---|---|---|---|---|
| 60 pounds per 100 SF | B6@.021 | SF | .41 | .60 | 1.01 | 1.72 |
| 100 pounds per 100 SF | B6@.025 | SF | .67 | .71 | 1.38 | 2.35 |

| | Craft@Hrs | Unit | Material | Labor | Total | Sell |
|---|---|---|---|---|---|---|
| **Concrete wall finish.** | | | | | | |
| Cut back ties and patch | B6@.011 | SF | .12 | .31 | .43 | .73 |
| Remove fins | B6@.008 | LF | .05 | .23 | .28 | .48 |
| Grind smooth | B6@.021 | SF | .09 | .60 | .69 | 1.17 |
| Sack, burlap grout rub | B6@.013 | SF | .06 | .37 | .43 | .73 |
| Wire brush, green | B6@.015 | SF | .05 | .43 | .48 | .82 |
| Wash with acid and rinse | B6@.004 | SF | .28 | .11 | .39 | .66 |
| Break fins, patch voids, carborundum rub | B6@.035 | SF | .07 | 1.00 | 1.07 | 1.82 |
| Break fins, patch voids, burlap grout rub | B6@.026 | SF | .07 | .74 | .81 | 1.38 |
| Specialty finishes | | | | | | |
| Monolithic natural aggregate topping | | | | | | |
|   3/16" | B6@.020 | SF | .25 | .57 | .82 | 1.39 |
|   1/2" | B6@.022 | SF | .29 | .63 | .92 | 1.56 |
| Integral colors | | | | | | |
| Primary colors, based on 25 lb sack | — | Lb | 4.79 | — | 4.79 | — |
| Blends, based on 25 lb sack | — | Lb | 2.42 | — | 2.42 | — |
| Colored release agent | — | Lb | 1.89 | — | 1.89 | — |
| Dry shake colored hardener | — | Lb | .42 | — | .42 | — |
| Stamped finish (embossed concrete) of joints, if required | | | | | | |
| Diamond, square, octagonal patterns | B6@.047 | SF | — | 1.34 | 1.34 | 2.28 |
| Spanish paver pattern | B6@.053 | SF | — | 1.51 | 1.51 | 2.57 |
| Add for grouting | B6@.023 | SF | .30 | .66 | .96 | 1.63 |

**Concrete slab sawing.** Typical rates quoted by concrete sawing subcontractors. Using a gasoline powered saw. Costs per linear foot for cured concrete assuming a level surface with good access and saw cut lines laid out and pre-marked by others. Costs include local travel time. Minimum cost will be $250. Electric powered slab sawing will be approximately 40 percent higher for the same depth.

| Depth | Unit | Under 200' | 200' to 1000' | Over 1000' |
|---|---|---|---|---|
| 1" deep | LF | 0.75 | 0.66 | 0.55 |
| 1-1/2" deep | LF | 1.10 | 1.02 | 0.85 |
| 2" deep | LF | 1.47 | 1.29 | 1.10 |
| 2-1/2" deep | LF | 1.85 | 1.66 | 1.38 |
| 3" deep | LF | 2.22 | 1.93 | 1.66 |
| 3-1/2" deep | LF | 2.58 | 2.31 | 1.93 |
| 4" deep | LF | 2.93 | 2.58 | 2.22 |
| 5" deep | LF | 3.68 | 3.23 | 2.76 |
| 6" deep | LF | 4.43 | 3.88 | 3.31 |
| 7" deep | LF | 5.35 | 4.59 | 3.97 |
| 8" deep | LF | 6.07 | 5.34 | 4.59 |
| 9" deep | LF | 6.91 | 6.07 | 5.26 |
| 10" deep | LF | 7.83 | 6.91 | 5.99 |
| 11" deep | LF | 8.74 | 7.73 | 6.72 |
| 12" deep | LF | 9.68 | 8.57 | 7.46 |

| | Craft@Hrs | Unit | Material | Labor | Total | Sell |
|---|---|---|---|---|---|---|

**Asphalt paving.** Small jobs and repairs. Using hand tools, except as noted. Add the cost of equipment rental (pneumatic hammer, compressor, front end loader, roller), hauling and dump fees, as required.

Break up and remove asphalt pavement up to 3" thick using a pneumatic hammer, loosen and load bituminous debris on truck with shovel

| | Craft@Hrs | Unit | Material | Labor | Total | Sell |
|---|---|---|---|---|---|---|
| Per square yard of pavement | B8@.147 | SY | — | 4.60 | 4.60 | 7.82 |

Trim area, break up pavement with a pneumatic hammer, load debris on truck with a front end loader, remove existing base to depth of 3"

| | Craft@Hrs | Unit | Material | Labor | Total | Sell |
|---|---|---|---|---|---|---|
| Per square yard of pavement | B8@.470 | SY | — | 14.70 | 14.70 | 25.00 |
| Add for job setup | B8@.053 | SY | — | 1.66 | 1.66 | 2.82 |

Spread hot mix 3" thick with shovel, rake smooth and hand tamp

| | Craft@Hrs | Unit | Material | Labor | Total | Sell |
|---|---|---|---|---|---|---|
| Per square yard of pavement | BL@.266 | SY | 12.20 | 7.09 | 19.29 | 32.80 |
| Add for job setup | BL@.026 | SY | — | .69 | .69 | 1.17 |

Shovel, rake and hand tamp 3" thick base material, sweep and tack coat, shovel, rake and hand tamp 3" bituminous pavement

| | Craft@Hrs | Unit | Material | Labor | Total | Sell |
|---|---|---|---|---|---|---|
| Per square yard of pavement | BL@.316 | SY | 15.20 | 8.42 | 23.62 | 40.20 |
| Add for extra 1" of thickness | BL@.024 | SY | 4.08 | .64 | 4.72 | 8.02 |
| Add for job setup | BL@.039 | SY | — | 1.04 | 1.04 | 1.77 |

Shovel, rake and machine roll 3" base material, sweep and tack coat area, shovel, rake and machine roll 3" bituminous mix

| | Craft@Hrs | Unit | Material | Labor | Total | Sell |
|---|---|---|---|---|---|---|
| Per square yard of pavement | B8@.407 | SY | 15.20 | 12.70 | 27.90 | 47.40 |
| Add for each 1" of thickness | B8@.022 | SY | 4.08 | .69 | 4.77 | 8.11 |
| Add for job setup | B8@.104 | SY | — | 3.26 | 3.26 | 5.54 |

# *Rough Carpentry*

# 4

Before bidding a home improvement job, spend a few minutes checking the integrity of the framing. While wood framing doesn't just wear out (some wood frame structures in Japan have been in daily use for over a thousand years), there are destructive organisms that attack wood. Framing that has serious decay, mold, or termite infestation may be deteriorated past the point of economical repair, and makes the home more a candidate for demolition than renovation. Financing for home repair may not be available from conventional lenders when a home has serious structural defects.

Fortunately, the organisms that attack wood can be controlled with wood preservative and by keeping wood dry. With a little care and attention, nearly any wood-frame structure will reach functional obsolescence long before the wood deteriorates.

## Floor Beams

Figure 4-1 shows the principal components of a framed floor. If the house has a basement, at least one wood or steel floor beam will support joists under the first floor.

Wood posts that support floor beams should rest on pedestals. Posts embedded in a concrete slab or foundation tend to absorb moisture and eventually decay. See Figure 4-2. Examine the base of wood support posts for decay, even if not in contact with concrete. Steel posts are normally supported on metal plates. Any wood in contact with concrete should be pressure-treated with a preservative such as ACQ (Alkaline Copper Quat), a product made from recycled copper waste. Use lumber treated with 0.25 pounds of ACQ per cubic foot for applications not in direct contact with the ground. For lumber in direct contact with the ground, use 0.40 ACQ treated lumber. Use 0.60 ACQ treated lumber for wood foundation systems.

Decayed beams and sills can be replaced, but it's slow, difficult work, especially when done in a crawl space. However difficult the repairs, they may be the only alternative to demolishing the home. Using temporary needle beams, most homes can be lifted enough to remove weight from the floor beam to make repairs. Insert a needle beam (6" x 6" or greater) through the foundation wall. Then lift the joists off the decayed beam with bottle jacks. All lifting in basements and crawl spaces should be done very slowly to minimize stress damage to the structure. A good rule is to lift no more than $1/4$" per day.

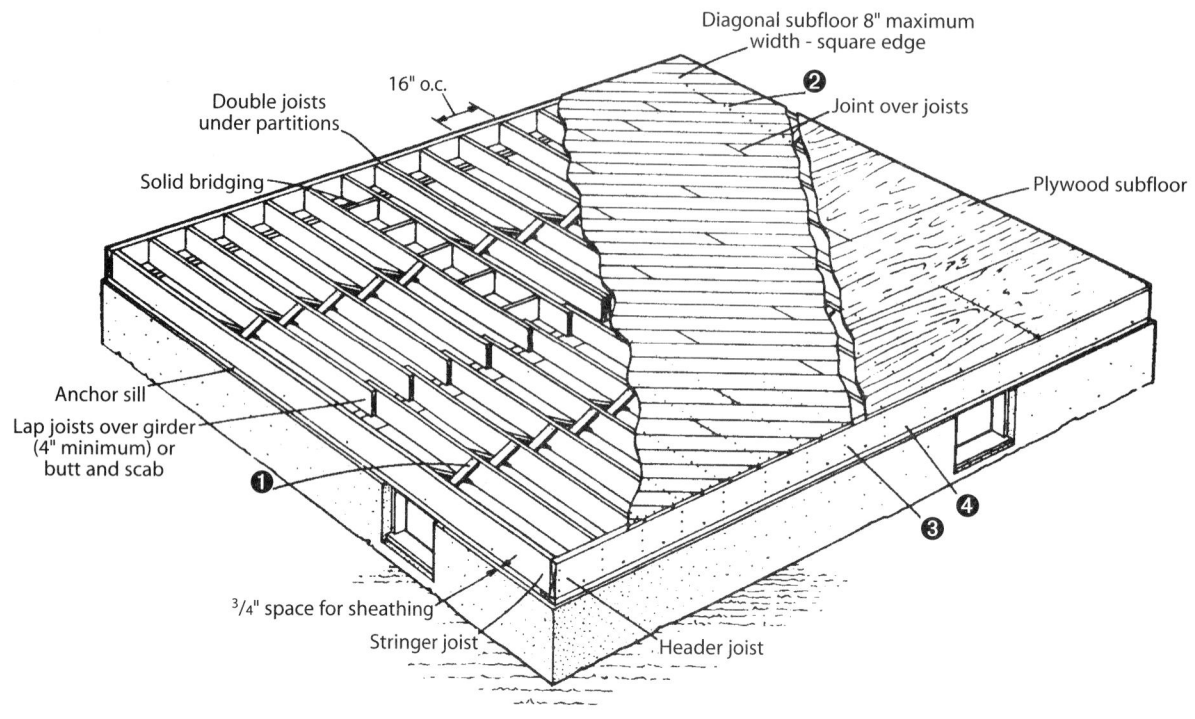

**Figure 4-1**

*Floor framing: (1) nailing bridging to joists; (2) nailing board subfloor to joists; (3) nailing header to joists; (4) toenailing header to sill*

When the weight is off the beam, cut out the decayed member and replace it. Then lower the needle beam slowly, again not more than $1/4$" per day. This work is easier if you set additional support piers or posts and replace the beam in sections shorter than the original member.

## Sagging Floors

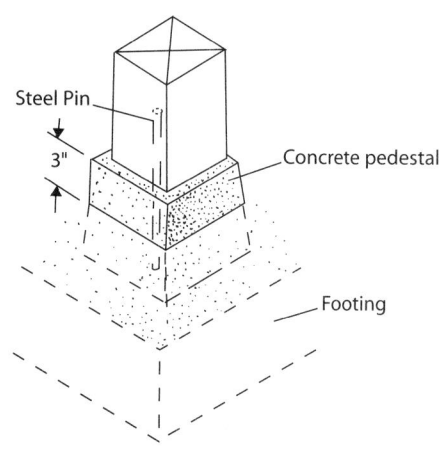

**Figure 4-2**

*Basement post on pedestal above the floor*

It's seldom practical to replace decayed floor joists, especially longer joists set with adhesive to the floor sheathing. It's much easier and just as effective to spike a new joist into position beside the old joist. First, remove the joist blocking to make room for the new joist. Then lift the section of decayed joists just enough to level the floor. A heavy 6" x 6" crossarm on top of a hydraulic jack will support a 4' to 6' width of a single-story house. Where you need additional support, add more jacks and crossarms. Raise the jacks slowly, over several days, and only just enough to allow you to insert a new pressure-treated joist beside the old joist. Excessive jacking will pull the building frame out of square. Insert the new joist, supporting it with joist hangers, a ledger, or by spiking it firmly to the old joist. Lower the jack. Then replace the joist bridges.

If only the joist end is decayed, brush wood preservative on the affected area. Then jack the existing joist up to the correct level and nail a short length of new material to the side of the existing joist. See Figure 4-3.

Some sag is normal in permanently-loaded wood beams. It's not a structural problem unless the walls above are obviously distorted. Deflection that doesn't exceed $^3/_8$" in 10' is generally acceptable for structural purposes.

Sagging floor joists aren't serious unless the foundation system has settled unevenly, causing excessive deflection in the floor. Look for deflection caused by poor support of heavy partition walls or by joist sections cut away by a plumbing, HVAC or electrical crew. If the floor appears to be reasonably level, joist sag is probably acceptable. Otherwise, jack the floor back to a level position and double the joists or add a floor beam supported at each end by concrete piers.

You can reinforce a sagging floor beam with jack posts. See Figures 4-4 and 4-5. Jack posts used to stiffen a springy floor can set directly on concrete. Where a jack post has to carry a heavy load, install a steel plate to distribute the load over a larger area of floor slab. Jack posts aren't made for heavy jacking. If heavy jacking is required, use cribbing and an automotive bottle jack to lift the load. Then slip the jack post into place.

If the floor bounces under foot traffic, check to be sure joist bridges are installed between joists every eight feet. If you see adequate bridging, consider adding extra joists alongside every second joist in the affected area. Be sure to place new joists with the crown up. It's more work, but you could also add a supplemental beam at mid-span under the joists. Support the beam on both ends by installing concrete piers or posts. Sagging or bouncy floors may also indicate one or more broken floor joists. Inspect the area around the furnace — furnace heat may have caused excessive drying and cracking of floor joists.

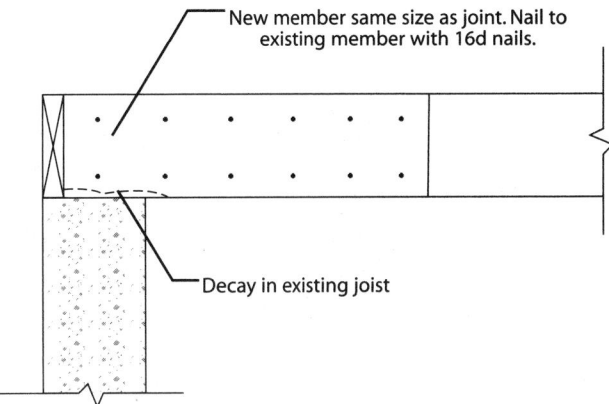

**Figure 4-3**

*Repair of joist with decay in end contacting the foundation*

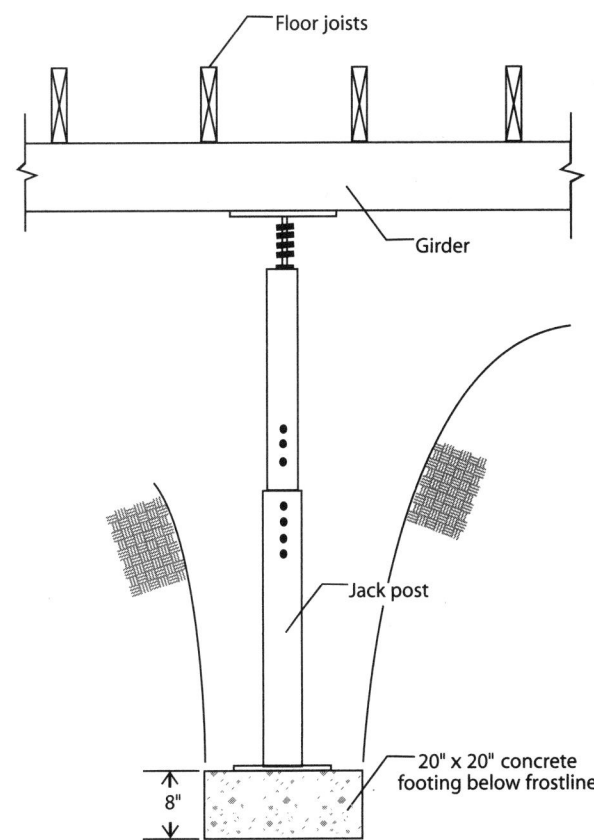

**Figure 4-4**

*Jack post supporting a sagging girder in a crawl space house*

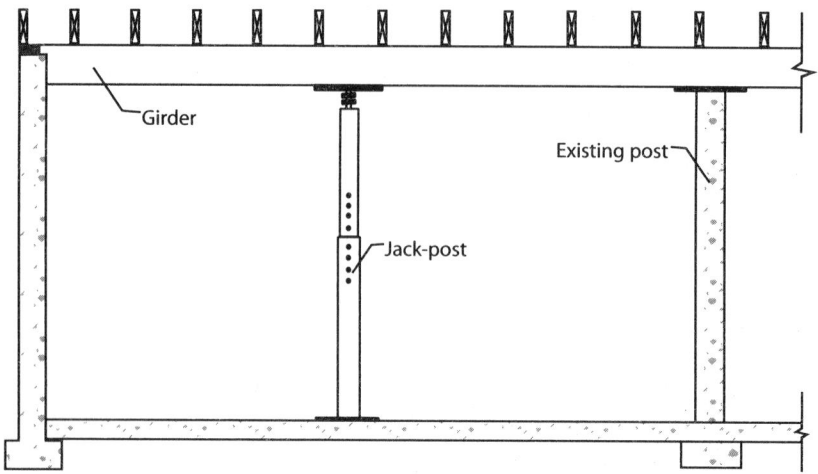

**Figure 4-5**

*Jack post used to level a sagging girder in a basement*

# Squeaky Wood Floors

Floor squeaks are usually caused by two edges rubbing together, such as the tongue of one flooring strip rubbing against the groove in the adjacent strip. Applying a small amount of mineral oil to the joint may solve the problem.

A sagging floor joist can pull away from the subfloor, leaving a small gap between the two. If you suspect this is the cause of squeaks, open a joint and squeeze construction mastic into the cavity. An alternate remedy is to drive small wedges into the space between the joists and subfloor. See Figure 4-6. Drive wedges only far enough for a snug fit. This repair only works for a small area. Too many wedges will make the finish floor look bumpy.

Floor joists that are too small for the span will squeak as the floor deflects under load. The best solution is to add a girder to shorten the joist span.

Strip flooring installed parallel to the joists usually deflects more under load than flooring installed perpendicular. If this is the cause of squeaks, solid blocking nailed between joists and fitted snugly against the subfloor (Figure 4-7) will limit deflection and help silence the floor.

Another common cause of squeaking is insufficient nails. If you decide to face nail the floor where it squeaks, countersink the nails and then fill each nail hole with wood putty. A better choice is to drill a small pilot hole through the tongue edge of a floor strip. Then nail through the pilot hole into a joist,

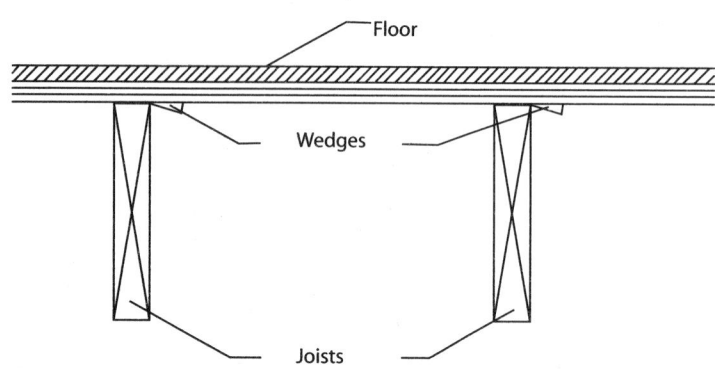

**Figure 4-6**

*Wedges driven between joists and subfloor to stop squeaks*

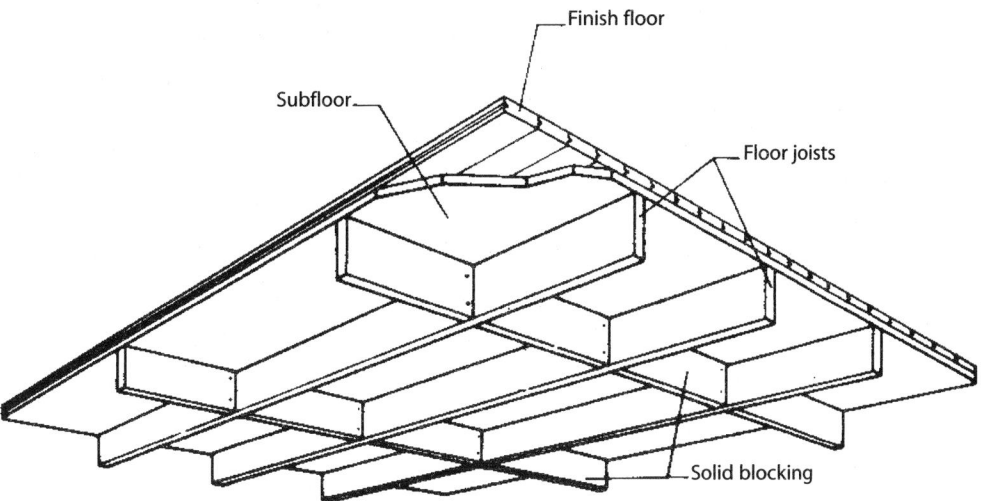

**Figure 4-7**

*Solid blocking between floor joists where finish floor is laid parallel to joists*

if possible. Even better, work from under the floor using screws driven through the subfloor into the finish floor. This method will also bring warped flooring flush with the subfloor.

# Wall Framing

Stud walls made of 2" x 4" lumber set 16" on center (Figure 4-8) and stiffened by fireblocks seldom need to be repaired. The studs are stiff enough to hold the roof up for many years. But individual studs may have to be replaced if they've been weakened, either by cuts made for plumbing or electrical lines, or by insect damage.

Attic conversions or adding a second story to a home generally require that the studs be doubled — another stud added between the original studs set 16" on center. Just installing ceramic tile in a second story bathroom or adding a bathtub on a second floor can increase the load carried by studs below by several thousand pounds. When installing new ceramic tile, you should also add solid blocking to the floor joists directly under the bathroom. A ceramic tile floor can't tolerate any movement. Any flexing of the floor will cause tile popping or cracking. Consider doubling studs in the first floor support wall when adding fixtures to the room above.

Occasionally you'll see a remodeled home where a partition wall, supported only by floor joists, is carrying a load from the second story. This is common when an attic has been converted to a storage room or an extra bedroom. The additional weight will tend to deform the main floor downward. Think of partition walls as curtains that simply divide up space. They aren't intended to support roof loads or any weight from the floor above. Doubling the support studs won't help if those

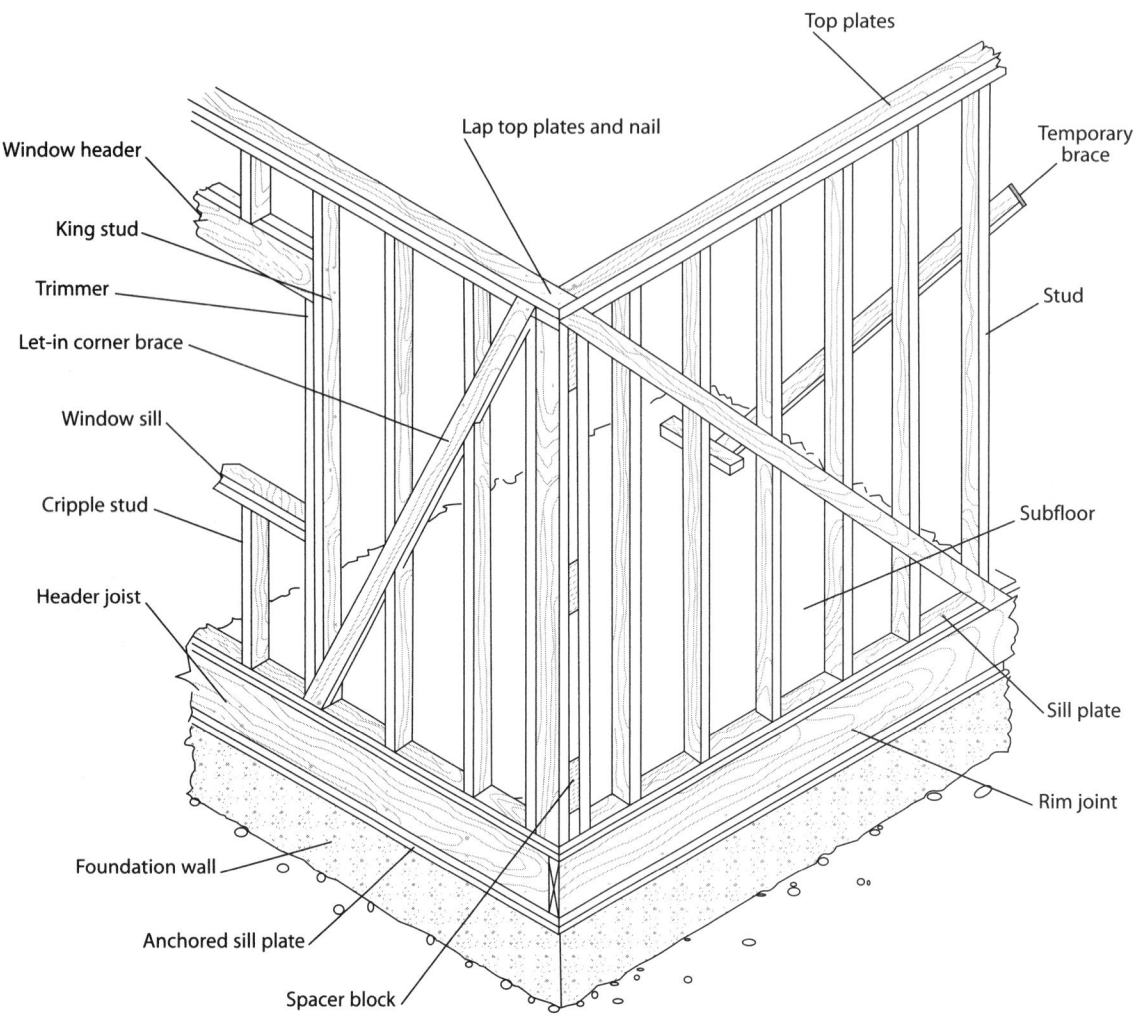

Window header

King stud

Trimmer

Let-in corner brace

Window sill

Cripple stud

Header joist

Foundation wall

Anchored sill plate

Spacer block

Top plates

Lap top plates and nail

Temporary brace

Stud

Subfloor

Sill plate

Rim joint

**Figure 4-8**

*Wall framing used with platform construction*

studs rest on a plate supported only by floor joists. You have to convert the partition wall to a load-bearing wall by adding a support beam and posts in the crawl space or basement below.

## Defects in the Roof

Roof leaks are usually caused by a failure of the roof cover, and can lead to rot damage to both the sheathing and the rafters. It would be poor practice to replace roof cover without making repairs to a sagging or decayed roof deck. Unlike problems with floor joists and floor beams, nearly all roof framing problems can be repaired economically. See Figure 4-9. Examine the roof for sagging at the ridge,

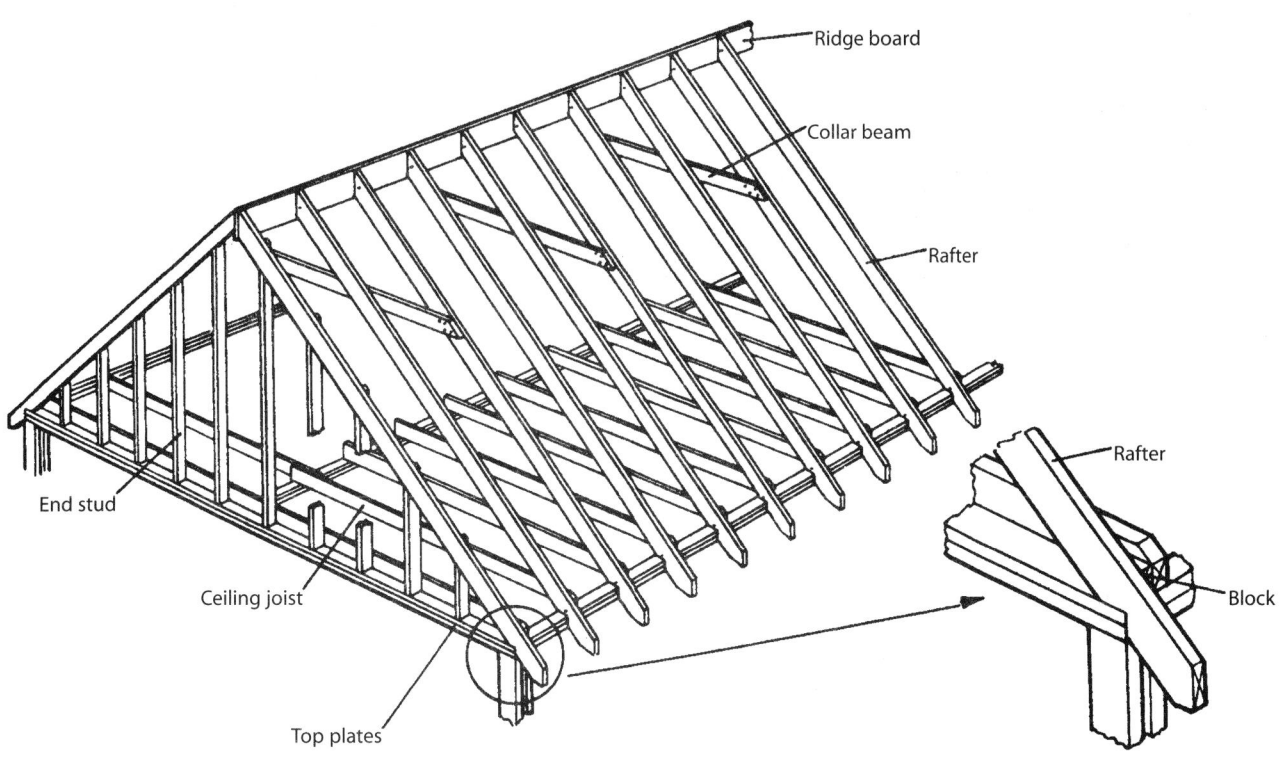

**Figure 4-9**

*Ceiling and roof framing: Overall gable roof framing*

rafters and in the sheathing. A ridge board can sag due to improper support, inadequate ties at the plate level, or even from sagging rafters. Rafters sag if the lumber wasn't well seasoned or if the lumber didn't have enough strength for the span. Sheathing sag is the result of rafter spacing that's too wide or sheathing material that's too thin. Occasionally, plywood sheathing will de-laminate and lose load-carrying capacity.

You can level a sagging ridge board or roof truss by jacking up points between supports and then installing braces to hold the member in place. When jacking any roof component, be sure to set the jack where the load will be transferred directly down through the structure to the foundation. Otherwise, pressure from the jack will deform the ceiling joists downward, damaging the ceiling below. If there's no bearing wall available, lay a beam in position where it can transfer loads to remote bearing points. When a ridge or truss chord is jacked into level position, cut a 2 x 4 just long enough to fit between the ceiling joist and ridge or chord. Then nail the prop at both ends. See Figure 4-10. For a short ridge board, one prop may be enough. Add more props as needed. If a rafter is sagging, jack the rafter into alignment. Then nail a new rafter into position at the side of the sagging rafter. That won't remove the sag from the old rafter, but it will transfer the load to a new, straight rafter.

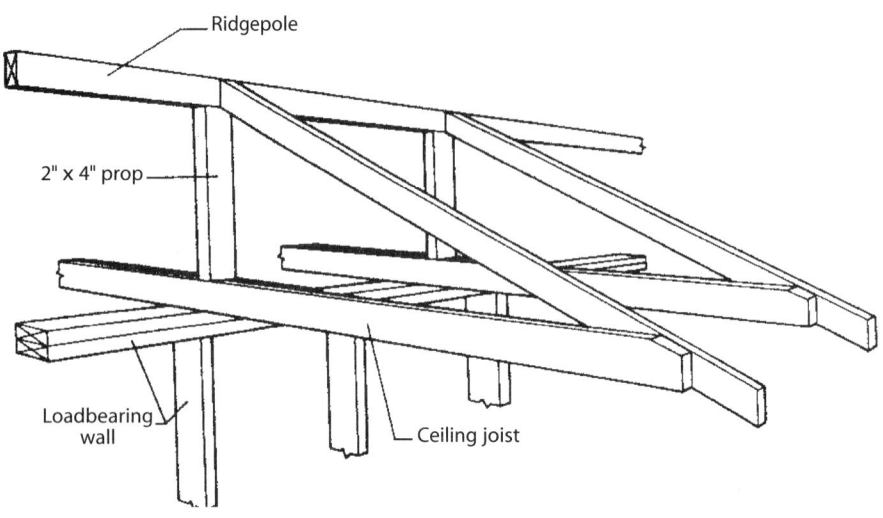

**Figure 4-10**

*Prop to hold sagging ridgepole in level position*

# Sheathing

A wavy roof surface is evidence that the sheathing has sagged. The only practical repair is to remove and replace the sheathing. However, you may be able to nail new sheathing right over the old. This saves both the labor of removing the old sheathing and a big cleanup job. Of course, you must remove old sheathing that shows signs of decay or mold.

New sheathing that's nailed over existing sheathing must be secured with longer nails than normally used on roof sheathing because the nails still need to penetrate the framing $1^1/4"$ to $1^1/2"$. You should nail the sheathing edges at 6" spacing and nail to the intermediate framing members at 12" spacing. Install sheathing with the long edge at right angles to the rafters. For a built-up roof, join panel edges with plywood clips where edges don't meet over a rafter. You can use $^3/8"$ thick sheathing for 16" rafter spacing, which is the minimum thickness required. However, $^1/2"$ sheathing plywood is better, and is required for heavy roofing materials such as clay or concrete tile. You'll find that Oriented Strand Board (OSB) sheathing costs less and works just as well as plywood sheathing.

Strip sheathing helps promote air circulation, which is an important consideration in any climate. Slate and tile roofs are commonly installed on furring strips rather than plywood sheathing. However, strip sheathing is not appropriate for asphalt shingle applications.

When replacing the roof cover, roof sheathing or roof supports, check the roof flashing for rust and physical damage. Inadequate or damaged flashing could result in leaks. Replace any flashing that shows signs of age. Then renew the caulking around chimneys and parapet walls.

# Adding a Roof Overhang

Many older homes have little or no roof overhang. Without an adequate overhang, rain and melting snow trickle down the wall finish, creating moisture problems in siding and trim, and at window and door openings, as well as increasing the moisture content at the foundation.

Extending the roof overhang will both improve the appearance of the house and reduce the maintenance required on siding and exterior trim. If you're adding new sheathing, consider extending the sheathing an inch or two beyond the edge of the existing roof. This is both an inexpensive and comparatively easy improvement.

| | Craft@Hrs | Unit | Material | Labor | Total | Sell |
|---|---|---|---|---|---|---|

## Framing Repairs

**Pricing for floor framing repairs.** Place temporary 6" x 6" x 10' needle beam under joists for floor jacking, using 4" x 4" pre-cut cribbing for end supports. Includes removing the needle beam when lifting is complete. All jacking materials to be salvaged. No concrete work included.

| | Craft@Hrs | Unit | Material | Labor | Total | Sell |
|---|---|---|---|---|---|---|
| Beam set in a basement or first floor | B1@1.72 | Ea | — | 50.50 | 50.50 | 85.90 |
| Beam set in a crawl space | B1@3.44 | Ea | — | 101.00 | 101.00 | 172.00 |

Remove a floor beam. Includes jacking and cutting out the old beam. Add the cost of repairs to sheathing, siding and interior finish. Assumes all jacking materials are salvaged. Add the cost of the beam from the section that follows. Add needle beam placement costs from the lines above. Lifting and lowering should be done slowly. A good rule is to lift no more than 1/4" per day.

| | Craft@Hrs | Unit | Material | Labor | Total | Sell |
|---|---|---|---|---|---|---|
| Per 8' section, basement or first floor | B1@6.87 | Ea | — | 202.00 | 202.00 | 343.00 |
| Per 16' section, basement or first floor | B1@9.34 | Ea | — | 274.00 | 274.00 | 466.00 |
| Per 8' section, work in a crawl space | B1@10.3 | Ea | — | 303.00 | 303.00 | 515.00 |
| Per 16' section, work in a crawl space | B1@14.0 | Ea | — | 411.00 | 411.00 | 699.00 |

Needle beam, purchase

| | Craft@Hrs | Unit | Material | Labor | Total | Sell |
|---|---|---|---|---|---|---|
| 6" x 6" x 10', green No. 1 Douglas fir | — | Ea | 40.00 | — | 40.00 | — |

Cribbing, 4" x 4" x 8' No. 2 fir

| | Craft@Hrs | Unit | Material | Labor | Total | Sell |
|---|---|---|---|---|---|---|
| Per crib, purchase | — | Ea | 9.63 | — | 9.63 | — |

Bottle (hydraulic) jack. Two-piece handle with internal overload valve, 9" to 18" lift

| | Craft@Hrs | Unit | Material | Labor | Total | Sell |
|---|---|---|---|---|---|---|
| 6 ton capacity, purchase | — | Ea | 21.60 | — | 21.60 | — |
| 12 ton capacity, purchase | — | Ea | 38.90 | — | 38.90 | — |
| 20 ton capacity, purchase | — | Ea | 54.00 | — | 54.00 | — |

**Beams and girders.** Std & Btr. First floor work. See beam removal costs in the previous section. Crawl space work assumes a minimum 24"-high reasonably accessible area. Work done in crawl spaces less than 24" high may cost up to 100% more. Figures in parentheses show board feet per linear foot of beam or girder, including 7% waste.

| | Craft@Hrs | Unit | Material | Labor | Total | Sell |
|---|---|---|---|---|---|---|
| 4" x 6", per MBF | B1@16.0 | MBF | 965.00 | 470.00 | 1,435.00 | 2,440.00 |
| 4" x 8", 10", 12", per MBF | B1@16.0 | MBF | 881.00 | 470.00 | 1,351.00 | 2,300.00 |
| 6" x 6", per MBF | B1@16.0 | MBF | 1,330.00 | 470.00 | 1,800.00 | 3,060.00 |
| 6" x 8", 10", 12", 8" x 8", per MBF | B1@16.0 | MBF | 1,330.00 | 470.00 | 1,800.00 | 3,060.00 |
| 4" x 6" (2.15 BF per LF) | B1@.034 | LF | 2.07 | 1.00 | 3.07 | 5.22 |
| 4" x 8" (2.85 BF per LF) | B1@.045 | LF | 2.51 | 1.32 | 3.83 | 6.51 |
| 4" x 10" (3.58 BF per LF) | B1@.057 | LF | 3.15 | 1.67 | 4.82 | 8.19 |
| 4" x 12" (4.28 BF per LF) | B1@.067 | LF | 3.77 | 1.97 | 5.74 | 9.76 |
| 6" x 6" (3.21 BF per LF) | B1@.051 | LF | 4.28 | 1.50 | 5.78 | 9.83 |
| 6" x 8" (4.28 BF per LF) | B1@.067 | LF | 5.71 | 1.97 | 7.68 | 13.10 |
| 6" x 10" (5.35 BF per LF) | B1@.083 | LF | 7.13 | 2.44 | 9.57 | 16.30 |
| 6" x 12" (6.42 BF per LF) | B1@.098 | LF | 8.56 | 2.88 | 11.44 | 19.40 |
| 8" x 8" (5.71 BF per LF) | B1@.088 | LF | 7.61 | 2.58 | 10.19 | 17.30 |
| Add if replacing a beam, basement or first floor | B1@.047 | LF | — | 1.38 | 1.38 | 2.35 |
| Add if replacing a beam in a crawl space | B1@.095 | LF | — | 2.79 | 2.79 | 4.74 |

**Shim up a floor beam.** Jack a floor beam into position with a crossarm mounted on a hydraulic jack. Insert a pressure-treated shim above the pier or column and remove the jack. Per jacking point. Work done on sloping or unstable ground will cost more. Assumes all jacking materials are salvaged and includes cost of removing the jack. Lifting and lowering should not exceed 1/4" per day.

| | Craft@Hrs | Unit | Material | Labor | Total | Sell |
|---|---|---|---|---|---|---|
| Jacking done in a basement | B1@.700 | Ea | 1.50 | 20.60 | 22.10 | 37.60 |
| Jacking done in a crawl space | B1@1.25 | Ea | 1.50 | 36.70 | 38.20 | 64.90 |

| | Craft@Hrs | Unit | Material | Labor | Total | Sell |
|---|---|---|---|---|---|---|

**Adjustable steel jack posts.** Temporary or permanent support for beams during home improvement work. Two height-adjustment pins seated in separate pockets. Top and bottom plates fit steel or wood beams. 8" turning bar. 5" screw height adjustment. 14,300 to 15,500 pound capacity. Set on an existing slab or pier.

| | Craft@Hrs | Unit | Material | Labor | Total | Sell |
|---|---|---|---|---|---|---|
| 6' 6" to 6' 10", 15,500 lb capacity | B1@.250 | Ea | 40.60 | 7.34 | 47.94 | 81.50 |
| 6' 9" to 7', 15,500 lb capacity | B1@.250 | Ea | 45.90 | 7.34 | 53.24 | 90.50 |
| 7' to 7' 4", 15,500 lb capacity | B1@.250 | Ea | 39.90 | 7.34 | 47.24 | 80.30 |
| 7' 3" to 7' 7", 15,500 lb capacity | B1@.250 | Ea | 46.30 | 7.34 | 53.64 | 91.20 |
| 7' 6" to 7' 10", 14,300 lb capacity | B1@.250 | Ea | 47.20 | 7.34 | 54.54 | 92.70 |
| 7' 9" to 8' 1", 14,300 lb capacity | B1@.250 | Ea | 49.30 | 7.34 | 56.64 | 96.30 |
| 8' to 8' 4", 14,300 lb capacity | B1@.250 | Ea | 50.90 | 7.34 | 58.24 | 99.00 |

**Remove basement post and replace concrete pier.** Add the cost of placing a temporary jack post from the previous section and setting the post itself from the following section.

| | Craft@Hrs | Unit | Material | Labor | Total | Sell |
|---|---|---|---|---|---|---|
| Remove an existing 4" x 4" x 7' wood basement post | | | | | | |
| Per post | B1@.250 | Ea | — | 7.34 | 7.34 | 12.50 |
| Break out an existing isolated concrete pier | | | | | | |
| Using hand tools, per cubic foot of concrete | B1@.435 | CF | — | 12.80 | 12.80 | 21.80 |
| Hand excavate for a concrete pad or pier, average soil | | | | | | |
| Per cubic foot of soil | B1@.100 | CF | — | 2.94 | 2.94 | 5.00 |
| Set and strip square concrete form for pier | | | | | | |
| Per square foot of contact area, 1 use | B1@.090 | SF | 1.60 | 2.64 | 4.24 | 7.21 |
| Mix, pour and finish concrete for an isolated pier | | | | | | |
| Per cubic foot of concrete | B1@.400 | CF | 6.46 | 11.70 | 18.16 | 30.90 |
| Place post anchor in concrete pier | | | | | | |
| Per anchor | B1@.132 | Ea | 4.79 | 3.88 | 8.67 | 14.70 |

**Wood support posts.** S4S, green. Set in basement or first floor. Material costs include 10% waste. For posts incorporated in wall framing, see stud wall costs.

| | Craft@Hrs | Unit | Material | Labor | Total | Sell |
|---|---|---|---|---|---|---|
| 4" x 4", 4" x 6" posts, per BF | — | BF | .95 | — | .95 | — |
| 4" x 8" to 4" x 12" posts, per BF | — | BF | .88 | — | .88 | — |
| 6" x 6" to 8" x 12" posts, per BF | — | BF | 1.33 | — | 1.33 | — |
| 4" x 4" per LF | B1@.110 | LF | 1.39 | 3.23 | 4.62 | 7.85 |
| 4" x 6" per LF | B1@.120 | LF | 2.09 | 3.52 | 5.61 | 9.54 |
| 4" x 8" per LF | B1@.140 | LF | 2.58 | 4.11 | 6.69 | 11.40 |
| 4" x 10" per LF | B1@.143 | LF | 3.23 | 4.20 | 7.43 | 12.60 |
| 4" x 12" per LF | B1@.145 | LF | 3.88 | 4.26 | 8.14 | 13.80 |
| 6" x 6" per LF | B1@.145 | LF | 4.40 | 4.26 | 8.66 | 14.70 |
| 6" x 8" per LF | B1@.145 | LF | 4.85 | 4.26 | 9.11 | 15.50 |
| 6" x 10" per LF | B1@.147 | LF | 7.33 | 4.32 | 11.65 | 19.80 |
| 6" x 12" per LF | B1@.150 | LF | 8.80 | 4.41 | 13.21 | 22.50 |
| 8" x 8" per LF | B1@.150 | LF | 7.82 | 4.41 | 12.23 | 20.80 |
| 8" x 10" per LF | B1@.155 | LF | 9.78 | 4.55 | 14.33 | 24.40 |

**Remove a sill plate.** Includes jacking and cutting out the old plate. Add the cost of repairs to sheathing, siding and interior finish. All jacking materials salvaged. Add the cost of placing a temporary needle beam from the previous section. Add the cost of setting the sill plate from the following section.

| | Craft@Hrs | Unit | Material | Labor | Total | Sell |
|---|---|---|---|---|---|---|
| Per 8' section, work in a basement | B1@6.87 | Ea | — | 202.00 | 202.00 | 343.00 |
| Per 16' section, work in a basement | B1@9.34 | Ea | — | 274.00 | 274.00 | 466.00 |
| Per 8' section, work in a crawl space | B1@10.3 | Ea | — | 303.00 | 303.00 | 515.00 |
| Per 16' section, work in a crawl space | B1@14.0 | Ea | — | 411.00 | 411.00 | 699.00 |

| | Craft@Hrs | Unit | Material | Labor | Total | Sell |
|---|---|---|---|---|---|---|

**Sill plate.** Mud sill at foundation, #2 pressure treated lumber, drilled and installed with foundation bolts at 48" OC, no bolts, nuts or washers included. Figures in parentheses show board feet per LF of foundation, including 5% waste.

| | Craft@Hrs | Unit | Material | Labor | Total | Sell |
|---|---|---|---|---|---|---|
| Sill plates, per MBF | B1@22.9 | MBF | 971.00 | 673.00 | 1,644.00 | 2,790.00 |
| 2" x 3" (.53 BF per LF) | B1@.020 | LF | .44 | .59 | 1.03 | 1.75 |
| 2" x 4" (.70 BF per LF) | B1@.023 | LF | .69 | .68 | 1.37 | 2.33 |
| 2" x 6" (1.05 BF per LF) | B1@.024 | LF | .99 | .70 | 1.69 | 2.87 |
| 2" x 8" (1.40 BF per LF) | B1@.031 | LF | 1.37 | .91 | 2.28 | 3.88 |
| Add for installation in a basement | B1@.095 | LF | — | 2.79 | 2.79 | 4.74 |
| Add for installation in a crawl space | B1@.158 | LF | — | 4.64 | 4.64 | 7.89 |

## Framing Room Additions

**Floor joists — room addition.** Per SF of area covered. Figures in parentheses indicate board feet per square foot of floor including box or band joist, typical double joists (under partition walls), and 6% waste. No beams, blocking or bridging included. Deduct for openings over 25 SF. Costs shown are based on a job with 1,000 SF of area covered. For scheduling purposes, estimate that a crew of two can complete 750 SF of area per 8-hour day for 12" center-to-center framing; 925 SF for 16" OC; 1,100 SF for 20" OC; or 1,250 SF for 24" OC. See the following section for costs of repairing floor joists in an existing building. For more complete coverage, see National Framing and Finish Carpentry Estimator at http://CraftsmanSiteLicense.com.

| | Craft@Hrs | Unit | Material | Labor | Total | Sell |
|---|---|---|---|---|---|---|
| 2" x 6" Std & Btr | | | | | | |
| 2" x 6" floor joists, per MBF | — | MBF | 641.00 | — | 641.00 | — |
| 12" centers (1.28 BF per SF) | B1@.021 | SF | .82 | .62 | 1.44 | 2.45 |
| 16" centers (1.02 BF per SF) | B1@.017 | SF | .65 | .50 | 1.15 | 1.96 |
| 20" centers (.88 BF per SF) | B1@.014 | SF | .56 | .41 | .97 | 1.65 |
| 24" centers (.73 BF per SF) | B1@.012 | SF | .47 | .35 | .82 | 1.39 |
| 2" x 8" Std & Btr | | | | | | |
| 2" x 8" floor joists, per MBF | — | MBF | 652.00 | — | 652.00 | — |
| 12" centers (1.71 BF per SF) | B1@.023 | SF | 1.11 | .68 | 1.79 | 3.04 |
| 16" centers (1.36 BF per SF) | B1@.018 | SF | .89 | .53 | 1.42 | 2.41 |
| 20" centers (1.17 BF per SF) | B1@.015 | SF | .76 | .44 | 1.20 | 2.04 |
| 24" centers (1.03 BF per SF) | B1@.013 | SF | .67 | .38 | 1.05 | 1.79 |
| 2" x 10" Std & Btr | | | | | | |
| 2" x 10" floor joists, per MBF | — | MBF | 683.00 | — | 683.00 | — |
| 12" centers (2.14 BF per SF) | B1@.025 | SF | 1.46 | .73 | 2.19 | 3.72 |
| 16" centers (1.71 BF per SF) | B1@.020 | SF | 1.17 | .59 | 1.76 | 2.99 |
| 20" centers (1.48 BF per SF) | B1@.016 | SF | 1.01 | .47 | 1.48 | 2.52 |
| 24" centers (1.30 BF per SF) | B1@.014 | SF | .89 | .41 | 1.30 | 2.21 |
| 2" x 12" Std & Btr | | | | | | |
| 2" x 12" floor joists, per MBF | — | MBF | 730.00 | — | 730.00 | — |
| 12" centers (2.56 BF per SF) | B1@.026 | SF | 1.87 | .76 | 2.63 | 4.47 |
| 16" centers (2.05 BF per SF) | B1@.021 | SF | 1.50 | .62 | 2.12 | 3.60 |
| 20" centers (1.77 BF per SF) | B1@.017 | SF | 1.29 | .50 | 1.79 | 3.04 |
| 24" centers (1.56 BF per SF) | B1@.015 | SF | 1.14 | .44 | 1.58 | 2.69 |

**Add floor joists under an existing building.** Per linear foot of new joist set in joist hangers or "doubled" by spiking to an existing joist.

Jack two or three adjacent joists into position with a crossarm mounted on a hydraulic jack. Per jacking point. Work done on sloping or unstable ground will cost more. Includes the cost of removing the jack.

| | Craft@Hrs | Unit | Material | Labor | Total | Sell |
|---|---|---|---|---|---|---|
| Jacking done in a basement | B1@.700 | Ea | — | 20.60 | 20.60 | 35.00 |
| Jacking done in a crawl space | B1@1.25 | Ea | — | 36.70 | 36.70 | 62.40 |
| Set 2" x 6" floor joist | | | | | | |
| Work done in a basement | B1@.042 | LF | .44 | 1.23 | 1.67 | 2.84 |
| Work done in a crawl space | B1@.070 | LF | .44 | 2.06 | 2.50 | 4.25 |

|  | Craft@Hrs | Unit | Material | Labor | Total | Sell |
|---|---|---|---|---|---|---|
| Set 2" x 8" floor joist |  |  |  |  |  |  |
|    Work done in a basement | B1@.046 | LF | .60 | 1.35 | 1.95 | 3.32 |
|    Work done in a crawl space | B1@.077 | LF | .60 | 2.26 | 2.86 | 4.86 |
| Set 2" x 10" floor joist |  |  |  |  |  |  |
|    Work done in a basement | B1@.050 | LF | .84 | 1.47 | 2.31 | 3.93 |
|    Work done in a crawl space | B1@.083 | LF | .84 | 2.44 | 3.28 | 5.58 |
| Set 2" x 12" floor joist |  |  |  |  |  |  |
|    Work done in a basement | B1@.052 | LF | 1.01 | 1.53 | 2.54 | 4.32 |
|    Work done in a crawl space | B1@.087 | LF | 1.01 | 2.56 | 3.57 | 6.07 |

**Floor joist wood, TJI truss type.** Suitable for residential use, 50 PSF floor load design. Costs shown are per square foot (SF) of floor area, based on joists at 16" OC, for a job with 1,000 SF of floor area. Figure 1.22 SF of floor area for each LF of joist. Add the cost of beams, supports and blocking. For scheduling purposes, estimate that a crew of two can install 900 to 950 SF of joists in an 8-hour day.

|  | Craft@Hrs | Unit | Material | Labor | Total | Sell |
|---|---|---|---|---|---|---|
|    9-1/2" TJI/15 | B1@.017 | SF | 1.93 | .50 | 2.43 | 4.13 |
|    11-7/8" TJI/15 | B1@.017 | SF | 2.12 | .50 | 2.62 | 4.45 |
|    14" TJI/35 | B1@.018 | SF | 3.06 | .53 | 3.59 | 6.10 |
|    16" TJI/35 | B1@.018 | SF | 3.34 | .53 | 3.87 | 6.58 |

**Bridging or blocking.** Installed between 2" x 6" thru 2" x 12" joists. Costs shown are per each set of cross bridges or per each block for solid bridging and include normal waste. Spacing between the bridging or blocking, sometimes called a "bay", depends on job requirements. Labor costs assume bridging is cut to size on site.

|  | Craft@Hrs | Unit | Material | Labor | Total | Sell |
|---|---|---|---|---|---|---|
| 1" x 4" cross |  |  |  |  |  |  |
|    Joist bridging, per MBF | — | MBF | 893.00 | — | 893.00 | — |
|    Joists on 12" centers | B1@.034 | Ea | .45 | 1.00 | 1.45 | 2.47 |
|    Joists on 16" centers | B1@.034 | Ea | .61 | 1.00 | 1.61 | 2.74 |
|    Joists on 20" centers | B1@.034 | Ea | .77 | 1.00 | 1.77 | 3.01 |
|    Joists on 24" centers | B1@.034 | Ea | .94 | 1.00 | 1.94 | 3.30 |
|    Add for work in a basement or first floor | B1@.068 | Ea | — | 2.00 | 2.00 | 3.40 |
|    Add for work in a crawl space | B1@.115 | Ea | — | 3.38 | 3.38 | 5.75 |
| 2" x 6" solid, Std & Btr |  |  |  |  |  |  |
|    2" x 6" blocking, per MBF | — | MBF | 641.00 | — | 641.00 | — |
|    Joists on 12" centers | B1@.042 | Ea | .71 | 1.23 | 1.94 | 3.30 |
|    Joists on 16" centers | B1@.042 | Ea | .94 | 1.23 | 2.17 | 3.69 |
|    Joists on 20" centers | B1@.042 | Ea | 1.18 | 1.23 | 2.41 | 4.10 |
|    Joists on 24" centers | B1@.042 | Ea | 1.41 | 1.23 | 2.64 | 4.49 |
|    Add for work in a basement or first floor | B1@.084 | Ea | — | 2.47 | 2.47 | 4.20 |
|    Add for work in a crawl space | B1@.140 | Ea | — | 4.11 | 4.11 | 6.99 |
| 2" x 8" solid, Std & Btr |  |  |  |  |  |  |
|    2" x 8" blocking, per MBF | — | MBF | 652.00 | — | 652.00 | — |
|    Joists on 12" centers | B1@.042 | Ea | .96 | 1.23 | 2.19 | 3.72 |
|    Joists on 16" centers | B1@.042 | Ea | 1.27 | 1.23 | 2.50 | 4.25 |
|    Joists on 20" centers | B1@.042 | Ea | 1.59 | 1.23 | 2.82 | 4.79 |
|    Joists on 24" centers | B1@.042 | Ea | 1.91 | 1.23 | 3.14 | 5.34 |
|    Add for work in a basement or first floor | B1@.084 | Ea | — | 2.47 | 2.47 | 4.20 |
|    Add for work in a crawl space | B1@.140 | Ea | — | 4.11 | 4.11 | 6.99 |
| 2" x 10" solid, Std & Btr |  |  |  |  |  |  |
|    2" x 10" blocking, per MBF | — | MBF | 683.00 | — | 683.00 | — |
|    Joists on 12" centers | B1@.057 | Ea | 1.25 | 1.67 | 2.92 | 4.96 |
|    Joists on 16" centers | B1@.057 | Ea | 1.67 | 1.67 | 3.34 | 5.68 |
|    Joists on 20" centers | B1@.057 | Ea | 2.09 | 1.67 | 3.76 | 6.39 |
|    Joists on 24" centers | B1@.057 | Ea | 2.50 | 1.67 | 4.17 | 7.09 |
|    Add for work in a basement or first floor | B1@.114 | Ea | — | 3.35 | 3.35 | 5.70 |
|    Add for work in a crawl space | B1@.190 | Ea | — | 5.58 | 5.58 | 9.49 |

| | Craft@Hrs | Unit | Material | Labor | Total | Sell |
|---|---|---|---|---|---|---|
| 2" x 12" solid, Std & Btr | | | | | | |
| 2" x 12" blocking, per MBF | — | MBF | 730.00 | — | 730.00 | — |
| Joists on 12" centers | B1@.057 | Ea | 1.61 | 1.67 | 3.28 | 5.58 |
| Joists on 16" centers | B1@.057 | Ea | 2.14 | 1.67 | 3.81 | 6.48 |
| Joists on 20" centers | B1@.057 | Ea | 2.67 | 1.67 | 4.34 | 7.38 |
| Joists on 24" centers | B1@.057 | Ea | 3.21 | 1.67 | 4.88 | 8.30 |
| Add for work in a basement or first floor | B1@.114 | Ea | — | 3.35 | 3.35 | 5.70 |
| Add for work in a crawl space | B1@.190 | Ea | — | 5.58 | 5.58 | 9.49 |
| Steel, no nail type, cross | | | | | | |
| Wood joists on 12" centers | B1@.020 | Ea | 2.46 | .59 | 3.05 | 5.19 |
| Wood joists on 16" centers | B1@.020 | Ea | 2.56 | .59 | 3.15 | 5.36 |
| Wood joists on 20" centers | B1@.020 | Ea | 2.80 | .59 | 3.39 | 5.76 |
| Wood joists on 24" centers | B1@.020 | Ea | 3.20 | .59 | 3.79 | 6.44 |

**Subflooring.**

Plywood sheathing, CD standard exterior grade. Material costs shown include fasteners and 5% for waste.

| | Craft@Hrs | Unit | Material | Labor | Total | Sell |
|---|---|---|---|---|---|---|
| 3/8" | B1@.011 | SF | .38 | .32 | .70 | 1.19 |
| 1/2" | B1@.012 | SF | .44 | .35 | .79 | 1.34 |
| 5/8" | B1@.012 | SF | .55 | .35 | .90 | 1.53 |
| 3/4" | B1@.013 | SF | .74 | .38 | 1.12 | 1.90 |
| Remove plywood floor sheathing | B1@.020 | SF | — | .59 | .59 | 1.00 |

OSB sheathing, Material costs shown include fasteners and 5% for waste.

| | Craft@Hrs | Unit | Material | Labor | Total | Sell |
|---|---|---|---|---|---|---|
| 3/8" | B1@.011 | SF | .30 | .32 | .62 | 1.05 |
| 1/2" | B1@.012 | SF | .36 | .35 | .71 | 1.21 |
| 5/8" | B1@.012 | SF | .52 | .35 | .87 | 1.48 |
| 3/4" | B1@.013 | SF | .73 | .38 | 1.11 | 1.89 |
| 7/8" | B1@.014 | SF | .96 | .41 | 1.37 | 2.33 |
| 1" | B1@.014 | SF | 1.11 | .41 | 1.52 | 2.58 |
| Remove OSB floor sheathing | B1@.020 | SF | — | .59 | .59 | 1.00 |

Board sheathing, laid diagonal, 1" x 6" #3 & Btr, 1.18 board feet per square foot. Costs shown include nails, 12% shrinkage, and 5% waste.

| | Craft@Hrs | Unit | Material | Labor | Total | Sell |
|---|---|---|---|---|---|---|
| Board sheathing, per MBF | — | MBF | 1,410.00 | — | 1,410.00 | — |
| 1" x 6", joists 16" on center | B1@.027 | SF | 1.67 | .79 | 2.46 | 4.18 |
| 1" x 6", joists 24" on center | B1@.030 | SF | 1.67 | .88 | 2.55 | 4.34 |
| 1" x 8", joists 16" on center | B1@.016 | SF | 1.67 | .47 | 2.14 | 3.64 |
| 1" x 8", joists 24" on center | B1@.017 | SF | 1.67 | .50 | 2.17 | 3.69 |
| Remove board floor sheathing | B1@.030 | SF | — | .88 | .88 | 1.50 |

**Wall studs — room addition.** Per square foot of wall area. Do not subtract for openings less than 16' wide. Figures in parentheses indicate typical board feet per SF of wall area, measured on one side, and include normal waste. Add for each corner and partition from below. Costs include studding and nails. Add for plates from the sections above and below. Add for door and window opening framing, backing, let-in bracing, fire blocking, and sheathing for shear walls from the sections that follow. Labor includes layout, plumb and align.

| | Craft@Hrs | Unit | Material | Labor | Total | Sell |
|---|---|---|---|---|---|---|
| 2" x 3", Std & Btr | | | | | | |
| 2" x 3" studs, 16" centers, per MBF | — | MBF | 697.00 | — | 697.00 | — |
| 12" centers (.55 BF per SF) | B1@.024 | SF | .38 | .70 | 1.08 | 1.84 |
| 16" centers (.41 BF per SF) | B1@.018 | SF | .29 | .53 | .82 | 1.39 |
| 20" centers (.33 BF per SF) | B1@.015 | SF | .23 | .44 | .67 | 1.14 |
| 24" centers (.28 BF per SF) | B1@.012 | SF | .20 | .35 | .55 | .94 |
| Add for each corner or partition | B1@.083 | Ea | 9.05 | 2.44 | 11.49 | 19.50 |
| 2" x 4", Std & Btr | | | | | | |
| 2" x 4" studs, 16" centers, per MBF | — | MBF | 622.00 | — | 622.00 | — |
| 12" centers (.73 BF per SF) | B1@.031 | SF | .45 | .91 | 1.36 | 2.31 |
| 16" centers (.54 BF per SF) | B1@.023 | SF | .34 | .68 | 1.02 | 1.73 |

| | Craft@Hrs | Unit | Material | Labor | Total | Sell |
|---|---|---|---|---|---|---|
| 20" centers (.45 BF per SF) | B1@.020 | SF | .28 | .59 | .87 | 1.48 |
| 24" centers (.37 BF per SF) | B1@.016 | SF | .23 | .47 | .70 | 1.19 |
| Add for each corner or partition | B1@.083 | Ea | 10.60 | 2.44 | 13.04 | 22.20 |
| 2" x 6", Std & Btr | | | | | | |
| 2" x 6" studs, 24" centers, per MBF | — | MBF | 641.00 | — | 641.00 | — |
| 12" centers (1.10 BF per SF) | B1@.044 | SF | .71 | 1.29 | 2.00 | 3.40 |
| 16" centers (.83 BF per SF) | B1@.033 | SF | .53 | .97 | 1.50 | 2.55 |
| 20" centers (.67 BF per SF) | B1@.027 | SF | .43 | .79 | 1.22 | 2.07 |
| 24" centers (.55 BF per SF) | B1@.022 | SF | .35 | .65 | 1.00 | 1.70 |
| Add for each corner or partition | B1@.083 | Ea | 15.30 | 2.44 | 17.74 | 30.20 |
| 4" x 4", Std & Btr | | | | | | |
| Installed in wall framing, with 10% waste | B1@.064 | LF | 1.39 | 1.88 | 3.27 | 5.56 |

**Stud wall repair.** Costs per stud added or repaired. Costs will be higher when plumbing or electrical lines run through the stud wall affected.

| | Craft@Hrs | Unit | Material | Labor | Total | Sell |
|---|---|---|---|---|---|---|
| Straighten stud by cutting saw kerf. Drive shim or screw into the kerf | | | | | | |
| Reinforce with a plywood scab | BC@.300 | Ea | — | 9.63 | 9.63 | 16.40 |
| Remove and replace a badly warped stud | BC@.300 | Ea | — | 9.63 | 9.63 | 16.40 |

**Add stud to an existing stud wall.** Remove blocking, insert an additional stud in an existing wall, replace blocking.

| | Craft@Hrs | Unit | Material | Labor | Total | Sell |
|---|---|---|---|---|---|---|
| 2" x 3" x 8' stud | B1@.300 | Ea | 2.71 | 8.81 | 11.52 | 19.60 |
| 2" x 4" x 8' stud | B1@.346 | Ea | 3.31 | 10.20 | 13.51 | 23.00 |
| 2" x 6" x 8' stud | B1@.392 | Ea | 4.96 | 11.50 | 16.46 | 28.00 |

**Plates.** (Wall plates), Std & Btr, untreated. For pressure-treated plates see also Sill Plates in this chapter. Figures in parentheses indicate board feet per LF. Costs shown include 10% for waste and nails.

| | Craft@Hrs | Unit | Material | Labor | Total | Sell |
|---|---|---|---|---|---|---|
| 2" x 3" Wall plates, per MBF | — | MBF | 697.00 | — | 697.00 | — |
| 2" x 3" (.55 BF per LF) | B1@.010 | LF | .38 | .29 | .67 | 1.14 |
| 2" x 4" Wall plates, per MBF | — | MBF | 622.00 | — | 622.00 | — |
| 2" x 4" (.73 BF per LF) | B1@.012 | LF | .45 | .35 | .80 | 1.36 |
| 2" x 6" Wall plates, per MBF | — | MBF | 641.00 | — | 641.00 | — |
| 2" x 6" (1.10 BF per LF) | B1@.018 | LF | .71 | .53 | 1.24 | 2.11 |

**Wall Bracing.** See also sheathing for plywood bracing and shear panels.

| | Craft@Hrs | Unit | Material | Labor | Total | Sell |
|---|---|---|---|---|---|---|
| Let-in wall bracing, using Std & Btr lumber | | | | | | |
| 1" x 4" | B1@.021 | LF | .53 | .62 | 1.15 | 1.96 |
| 1" x 6" | B1@.027 | LF | .77 | .79 | 1.56 | 2.65 |
| 2" x 4" | B1@.035 | LF | .41 | 1.03 | 1.44 | 2.45 |
| Steel strap bracing, 1-1/4" wide | | | | | | |
| 9'6" or 11'6" lengths | B1@.010 | LF | .48 | .29 | .77 | 1.31 |
| Steel "V" bracing, 3/4" x 3/4" | | | | | | |
| 9'6" or 11'6" lengths | B1@.010 | LF | .66 | .29 | .95 | 1.62 |
| Temporary wood frame wall bracing, assumes salvage at 50% and 3 uses | | | | | | |
| 1" x 4" Std & Btr | B1@.006 | LF | .13 | .18 | .31 | .53 |
| 1" x 6" Std & Btr | B1@.010 | LF | .19 | .29 | .48 | .82 |
| 2" x 4" utility | B1@.012 | LF | .10 | .35 | .45 | .77 |
| 2" x 6" utility | B1@.018 | LF | .16 | .53 | .69 | 1.17 |

| | Craft@Hrs | Unit | Material | Labor | Total | Sell |
|---|---|---|---|---|---|---|

**Fireblocks.** Installed in wood frame walls, per LF of wall to be blocked. Figures in parentheses indicate board feet of fire blocking per linear foot of wall, including 10% cutting waste. See also Bridging and Backing and Nailers.

2" x 3" blocking, Std & Btr

| | Craft@Hrs | Unit | Material | Labor | Total | Sell |
|---|---|---|---|---|---|---|
| 2" x 3" fireblocks, 16" centers, per MBF | — | MBF | 697.00 | — | 697.00 | — |
| 12" OC members (.48 BF per LF) | B1@.026 | LF | .33 | .76 | 1.09 | 1.85 |
| 16" OC members (.50 BF per LF) | B1@.025 | LF | .35 | .73 | 1.08 | 1.84 |
| 20" OC members (.51 BF per LF) | B1@.022 | LF | .36 | .65 | 1.01 | 1.72 |
| 24" OC members (.52 BF per LF) | B1@.020 | LF | .36 | .59 | .95 | 1.62 |
| Add for removing and replacing, per block | B1@.026 | LF | — | .76 | .76 | 1.29 |

2" x 4" blocking, Std & Btr

| | Craft@Hrs | Unit | Material | Labor | Total | Sell |
|---|---|---|---|---|---|---|
| 2" x 4" fireblocks, 16" centers, per MBF | — | MBF | 622.00 | — | 622.00 | — |
| 12" OC members (.64 BF per LF) | B1@.030 | LF | .40 | .88 | 1.28 | 2.18 |
| 16" OC members (.67 BF per LF) | B1@.025 | LF | .42 | .73 | 1.15 | 1.96 |
| 20" OC members (.68 BF per LF) | B1@.022 | LF | .42 | .65 | 1.07 | 1.82 |
| 24" OC members (.69 BF per LF) | B1@.020 | LF | .43 | .59 | 1.02 | 1.73 |
| Add to remove and replace, per block | B1@.030 | LF | — | .88 | .88 | 1.50 |

2" x 6" blocking, Std & Btr

| | Craft@Hrs | Unit | Material | Labor | Total | Sell |
|---|---|---|---|---|---|---|
| 2" x 6" fireblocks, 24" centers, per MBF | — | MBF | 641.00 | — | 641.00 | — |
| 12" OC members (.96 BF per LF) | B1@.031 | LF | .62 | .91 | 1.53 | 2.60 |
| 16" OC members (1.00 BF per LF) | B1@.026 | LF | .64 | .76 | 1.40 | 2.38 |
| 20" OC members (1.02 BF per LF) | B1@.022 | LF | .65 | .65 | 1.30 | 2.21 |
| 24" OC members (1.03 BF per LF) | B1@.020 | LF | .66 | .59 | 1.25 | 2.13 |
| Add to remove and replace, per block | B1@.034 | LF | — | 1.00 | 1.00 | 1.70 |

**Ceiling beams — room addition.** Std & Btr. Installed over wall openings and around floor, ceiling and roof openings or where a flush beam is called out in the plans, including 10% waste. Don't use these cost estimates for door or window headers in framed walls. See the door or window opening framing assemblies costs that follow.

| | Craft@Hrs | Unit | Material | Labor | Total | Sell |
|---|---|---|---|---|---|---|
| 2" x 6" beams, per MBF | B1@24.5 | MBF | 641.00 | 720.00 | 1,361.00 | 2,310.00 |
| 2" x 8" beams, per MBF | B1@25.2 | MBF | 652.00 | 740.00 | 1,392.00 | 2,370.00 |
| 2" x 10" beams, per MBF | B1@25.1 | MBF | 683.00 | 737.00 | 1,420.00 | 2,410.00 |
| 2" x 12" beams, per MBF | B1@25.9 | MBF | 730.00 | 761.00 | 1,491.00 | 2,530.00 |
| 4" x 6" beams, per MBF | B1@25.9 | MBF | 965.00 | 761.00 | 1,726.00 | 2,930.00 |
| 4" x 8", 10", 12", 14" beams, per MBF | B1@25.5 | MBF | 881.00 | 749.00 | 1,630.00 | 2,770.00 |
| 6" x 6", 8", 10", 12", 14" beams, per MBF | B1@19.1 | MBF | 1,330.00 | 561.00 | 1,891.00 | 3,210.00 |
| 2" x 6" (1.10 BF per LF) | B1@.028 | LF | .71 | .82 | 1.53 | 2.60 |
| 2" x 8" (1.47 BF per LF) | B1@.037 | LF | .96 | 1.09 | 2.05 | 3.49 |
| 2" x 10" (1.83 BF per LF) | B1@.046 | LF | 1.25 | 1.35 | 2.60 | 4.42 |
| 2" x 12" (2.20 BF per LF) | B1@.057 | LF | 1.61 | 1.67 | 3.28 | 5.58 |
| 4" x 6" (2.20 BF per LF) | B1@.057 | LF | 2.12 | 1.67 | 3.79 | 6.44 |
| 4" x 8" (2.93 BF per LF) | B1@.073 | LF | 2.58 | 2.14 | 4.72 | 8.02 |
| 4" x 10" (3.67 BF per LF) | B1@.094 | LF | 3.23 | 2.76 | 5.99 | 10.20 |
| 4" x 12" (4.40 BF per LF) | B1@.112 | LF | 3.88 | 3.29 | 7.17 | 12.20 |
| 4" x 14" (5.13 BF per LF) | B1@.115 | LF | 4.52 | 3.38 | 7.90 | 13.40 |
| 6" x 6" (3.30 BF per LF) | B1@.060 | LF | 4.40 | 1.76 | 6.16 | 10.50 |
| 6" x 8" (4.40 BF per LF) | B1@.080 | LF | 5.87 | 2.35 | 8.22 | 14.00 |
| 6" x 10" (5.50 BF per LF) | B1@.105 | LF | 7.33 | 3.08 | 10.41 | 17.70 |
| 6" x 12" (6.60 BF per LF) | B1@.115 | LF | 8.80 | 3.38 | 12.18 | 20.70 |
| 6" x 14" (7.70 BF per LF) | B1@.120 | LF | 10.30 | 3.52 | 13.82 | 23.50 |

| | Craft@Hrs | Unit | Material | Labor | Total | Sell |
|---|---|---|---|---|---|---|

**Window opening framing.** In 2" x 4" stud wall, based on 8' wall height. Costs shown are per window opening framed into a new stud wall and include a header of appropriate size, sub-sill plate (double sub-sill if opening is 8' wide or wider), double vertical studs each side of openings less than 8' wide (triple vertical studs each side of openings 8' wide or wider), top and bottom cripples, blocking, nails, and normal waste. Figures in parentheses indicate header size.

| | Craft@Hrs | Unit | Material | Labor | Total | Sell |
|---|---|---|---|---|---|---|
| To 2' wide (4" x 4" header) | B1@1.00 | Ea | 17.50 | 29.40 | 46.90 | 79.70 |
| Over 2' to 3' wide (4" x 4" header) | B1@1.17 | Ea | 21.90 | 34.40 | 56.30 | 95.70 |
| Over 3' to 4' wide (4" x 6" header) | B1@1.45 | Ea | 27.50 | 42.60 | 70.10 | 119.00 |
| Over 4' to 5' wide (4" x 6" header) | B1@1.73 | Ea | 30.90 | 50.80 | 81.70 | 139.00 |
| Over 5' to 6' wide (4" x 8" header) | B1@2.01 | Ea | 36.50 | 59.00 | 95.50 | 162.00 |
| Over 6' to 7' wide (4" x 8" header) | B1@2.29 | Ea | 46.30 | 67.30 | 113.60 | 193.00 |
| Over 7' to 8' wide (4" x 10" header) | B1@2.57 | Ea | 55.90 | 75.50 | 131.40 | 223.00 |
| Over 8' to 10' wide (4" x 12" header) | B1@2.57 | Ea | 70.20 | 75.50 | 145.70 | 248.00 |
| Over 10' to 12' wide (4" x 14" header) | B1@2.85 | Ea | 91.60 | 83.70 | 175.30 | 298.00 |
| Add per foot of height for walls over 8' high | — | Ea | 2.49 | — | 2.49 | — |

In 2" x 6" stud wall, based on 8' wall height. Costs shown are per window opening framed into a new stud wall and include a header of appropriate size, sub-sill plate (double sub-sill if opening is 8' wide or wider), double vertical studs each side of openings less than 8' wide (triple vertical studs each side of openings 8' wide or wider), top and bottom cripples, blocking, nails, and normal waste. Figures in parentheses indicate header size.

| | Craft@Hrs | Unit | Material | Labor | Total | Sell |
|---|---|---|---|---|---|---|
| To 2' wide (4" x 4" header) | B1@1.17 | Ea | 22.60 | 34.40 | 57.00 | 96.90 |
| Over 2' to 3' wide (6" x 4" header) | B1@1.45 | Ea | 29.20 | 42.60 | 71.80 | 122.00 |
| Over 3' to 4' wide (6" x 6" header) | B1@1.73 | Ea | 41.60 | 50.80 | 92.40 | 157.00 |
| Over 4' to 5' wide (6" x 6" header) | B1@2.01 | Ea | 49.40 | 59.00 | 108.40 | 184.00 |
| Over 5' to 6' wide (6" x 8" header) | B1@2.29 | Ea | 60.50 | 67.30 | 127.80 | 217.00 |
| Over 6' to 8' wide (6" x 8" header) | B1@2.85 | Ea | 94.40 | 83.70 | 178.10 | 303.00 |
| Over 8' to 10' wide (6" x 12" header) | B1@2.85 | Ea | 129.00 | 83.70 | 212.70 | 362.00 |
| Over 8' to 12' wide (6" x 14" header) | B1@3.13 | Ea | 164.00 | 91.90 | 255.90 | 435.00 |
| Add per foot of height for walls over 8' high | — | Ea | 3.85 | — | 3.85 | — |

**Door opening framing.** In 2" x 4" stud wall, based on 8' wall height. Costs shown are per each door opening framed into a new stud wall and include a header of appropriate size, double vertical studs each side of the opening less than 8' wide (triple vertical studs each side of openings 8' wide or wider), cripples, blocking, nails, and normal waste. Width shown is size of finished opening. Figures in parentheses indicate header size.

| | Craft@Hrs | Unit | Material | Labor | Total | Sell |
|---|---|---|---|---|---|---|
| To 3' wide (4" x 4" header) | B1@.830 | Ea | 19.30 | 24.40 | 43.70 | 74.30 |
| Over 3' to 4' wide (4" x 6" header) | B1@1.11 | Ea | 24.10 | 32.60 | 56.70 | 96.40 |
| Over 4' to 5' wide (4" x 6" header) | B1@1.39 | Ea | 26.60 | 40.80 | 67.40 | 115.00 |
| Over 5' to 6' wide (4" x 8" header) | B1@1.66 | Ea | 31.50 | 48.80 | 80.30 | 137.00 |
| Over 6' to 8' wide (4" x 10" header) | B1@1.94 | Ea | 49.30 | 57.00 | 106.30 | 181.00 |
| Over 8' to 10' wide (4" x 12" header) | B1@1.94 | Ea | 62.00 | 57.00 | 119.00 | 202.00 |
| Over 10' to 12' wide (4" x 14" header) | B1@2.22 | Ea | 76.70 | 65.20 | 141.90 | 241.00 |
| Add per foot of height for walls over 8' high | — | Ea | 2.49 | — | 2.49 | — |

In 2" x 6" stud wall, based on 8' wall height. Costs shown are per each door opening framed into a new stud wall and include a header of appropriate size, double vertical studs each side of the opening less than 8' wide (triple vertical studs each side of openings 8' wide or wider), cripples, blocking, nails, and normal waste. Width shown is size of finished opening. Figures in parentheses indicate header size.

| | Craft@Hrs | Unit | Material | Labor | Total | Sell |
|---|---|---|---|---|---|---|
| To 3' wide (6" x 4" header) | B1@1.11 | Ea | 32.40 | 32.60 | 65.00 | 111.00 |
| Over 3' to 4' wide (4" x 6" header) | B1@1.39 | Ea | 38.10 | 40.80 | 78.90 | 134.00 |
| Over 4' to 5' wide (4" x 6" header) | B1@1.66 | Ea | 42.80 | 48.80 | 91.60 | 156.00 |
| Over 5' to 6' wide (4" x 8" header) | B1@1.94 | Ea | 55.30 | 57.00 | 112.30 | 191.00 |
| Over 6' to 8' wide (4" x 10" header) | B1@2.22 | Ea | 87.50 | 65.20 | 152.70 | 260.00 |
| Over 8' to 10' wide (4" x 12" header) | B1@2.22 | Ea | 116.00 | 65.20 | 181.20 | 308.00 |
| Over 10' to 12' wide (4" x 14" header) | B1@2.50 | Ea | 149.00 | 73.40 | 222.40 | 378.00 |
| Add per foot of height for walls over 8' high | — | Ea | 3.85 | — | 3.85 | — |

| | Craft@Hrs | Unit | Material | Labor | Total | Sell |
|---|---|---|---|---|---|---|

**OSB sheathing.** Oriented strand board. Compressed wood strands bonded by phenolic resin. By nominal thickness.

| | Craft@Hrs | Unit | Material | Labor | Total | Sell |
|---|---|---|---|---|---|---|
| 3/8" x 4' x 8' | B1@.014 | SF | .28 | .41 | .69 | 1.17 |
| 7/16" x 4' x 8' | B1@.015 | SF | .28 | .44 | .72 | 1.22 |
| 15/32" x 4' x 8' | B1@.016 | SF | .32 | .47 | .79 | 1.34 |
| 1/2" x 4' x 8' | B1@.016 | SF | .34 | .47 | .81 | 1.38 |
| 5/8" x 4' x 8', square edge | B1@.018 | SF | .48 | .53 | 1.01 | 1.72 |
| 5/8" x 4' x 8', tongue and groove | B1@.018 | SF | .52 | .53 | 1.05 | 1.79 |
| 3/4" x 4' x 8', square edge | B1@.018 | SF | .63 | .53 | 1.16 | 1.97 |
| 3/4" x 4' x 8', tongue and groove | B1@.018 | SF | .61 | .53 | 1.14 | 1.94 |
| Remove OSB sheathing | B1@.017 | SF | — | .50 | .50 | .85 |

**Wall sheathing.** Per square foot of wall surface, including 5% waste. See also OSB sheathing.

BD plywood wall sheathing, plugged and touch sanded, interior grade

| | Craft@Hrs | Unit | Material | Labor | Total | Sell |
|---|---|---|---|---|---|---|
| 1/4" | B1@.013 | SF | .43 | .38 | .81 | 1.38 |
| 3/8" | B1@.015 | SF | .52 | .44 | .96 | 1.63 |
| 1/2" | B1@.016 | SF | .61 | .47 | 1.08 | 1.84 |
| 5/8" | B1@.018 | SF | .84 | .53 | 1.37 | 2.33 |
| 3/4" | B1@.020 | SF | .92 | .59 | 1.51 | 2.57 |
| Remove plywood wall sheathing | B1@.017 | SF | — | .50 | .50 | .85 |

Board wall sheathing 1" x 6" or 1" x 8" utility T&G

| | Craft@Hrs | Unit | Material | Labor | Total | Sell |
|---|---|---|---|---|---|---|
| 1" x 6" or 8" utility T&G, per MBF | — | MBF | 953.00 | — | 953.00 | — |
| (1.13 BF per SF) | B1@.020 | SF | 1.08 | .59 | 1.67 | 2.84 |
| Add for diagonal patterns | B1@.006 | SF | — | .18 | .18 | .31 |
| Remove board sheathing | B1@.030 | SF | — | .88 | .88 | 1.50 |

**Patch wall sheathing.** Studs 16" on center, remove and replace panels in several places on wall surface.

| | Craft@Hrs | Unit | Material | Labor | Total | Sell |
|---|---|---|---|---|---|---|
| 1" x 8" x 12' long, per board | B1@.248 | Ea | 10.70 | 7.28 | 17.98 | 30.60 |
| 1" x 8", per square foot | B1@.031 | SF | 1.34 | .91 | 2.25 | 3.83 |
| 1/2", 4-ply plywood, per 4' x 8' sheet | B1@.960 | Ea | 16.20 | 28.20 | 44.40 | 75.50 |
| 1/2", 4-ply plywood, per square foot | B1@.030 | SF | .51 | .88 | 1.39 | 2.36 |

## Changes to Walls and Ceilings

**Remove and replace interior partition wall.** Use 16 linear feet as the minimum job size. No interior floor finish included.

2" x 4" stud walls with 1/2" gypsum drywall both sides, ready for painting

| | Craft@Hrs | Unit | Material | Labor | Total | Sell |
|---|---|---|---|---|---|---|
| Cost per square foot of wall | B1@.105 | SF | 1.17 | 3.08 | 4.25 | 7.23 |
| Cost per running foot, 8' high wall | B1@.842 | LF | 9.33 | 24.70 | 34.03 | 57.90 |

2" x 4" stud walls with 5/8" gypsum fire-rated drywall both sides, ready for painting

| | Craft@Hrs | Unit | Material | Labor | Total | Sell |
|---|---|---|---|---|---|---|
| Cost per square foot of wall | B1@.108 | SF | 1.33 | 3.17 | 4.50 | 7.65 |
| Cost per running foot, 8' high wall | B1@.864 | LF | 10.60 | 25.40 | 36.00 | 61.20 |

2" x 6" stud walls with 1/2" gypsum drywall both sides, ready for painting

| | Craft@Hrs | Unit | Material | Labor | Total | Sell |
|---|---|---|---|---|---|---|
| Cost per square foot of wall | B1@.112 | SF | 1.46 | 3.29 | 4.75 | 8.08 |
| Cost per running foot, 8' high wall | B1@.896 | LF | 11.70 | 26.30 | 38.00 | 64.60 |

2" x 6" stud walls with 5/8" gypsum fire-rated drywall both sides, ready for painting

| | Craft@Hrs | Unit | Material | Labor | Total | Sell |
|---|---|---|---|---|---|---|
| Cost per square foot of wall | B1@.116 | SF | 1.62 | 3.41 | 5.03 | 8.55 |
| Cost per running foot, 8' high wall | B1@.928 | LF | 12.90 | 27.30 | 40.20 | 68.30 |

| | Craft@Hrs | Unit | Material | Labor | Total | Sell |
|---|---|---|---|---|---|---|

**Remove and replace exterior wall.** Add temporary joist support costs from below. Use 16 linear feet as the minimum job size. No interior floor finish included.

2" x 4" stud walls with drywall interior, wood siding exterior, 1/2" gypsum drywall inside face ready for painting, over 3-1/2" R-11 insulation with 5/8" thick rough sawn T-1-11 exterior grade plywood siding on the outside face

| | Craft@Hrs | Unit | Material | Labor | Total | Sell |
|---|---|---|---|---|---|---|
| Using 5/8" rough sawn T-1-11 siding at | — | MSF | 809.00 | — | 809.00 | — |
| Cost per square foot of wall | B1@.108 | SF | 1.95 | 3.17 | 5.12 | |
| Cost per running foot, 8' high wall | B1@.864 | LF | 15.60 | 25.40 | 41.00 | 69.70 |

2" x 6" stud walls with drywall interior, wood siding exterior, same construction as above, except with 6-1/4" R-19 insulation.

| | Craft@Hrs | Unit | Material | Labor | Total | Sell |
|---|---|---|---|---|---|---|
| Cost per square foot of wall | B1@.117 | SF | 2.38 | 3.44 | 5.82 | 9.89 |
| Cost per running foot, 8' high wall | B1@.936 | LF | 19.00 | 27.50 | 46.50 | 79.10 |

2" x 4" stud walls with drywall interior, 1/2" gypsum drywall inside face ready for painting, over 3-1/2" R-11 insulation with .042" vinyl lap siding on the outside face.

| | Craft@Hrs | Unit | Material | Labor | Total | Sell |
|---|---|---|---|---|---|---|
| Using .042" vinyl lap siding at | — | MSF | 809.00 | — | 809.00 | — |
| Cost per square foot of wall | B1@.112 | SF | 2.35 | 3.29 | 5.64 | 9.59 |
| Cost per running foot, 8' high wall | B1@.896 | LF | 18.80 | 26.30 | 45.10 | 76.70 |

2" x 4" stud walls with drywall interior, 1/2" gypsum drywall on inside face ready for painting, over 3-1/2" R-11 insulation with 1" x 6" southern yellow pine drop siding, D grade, 1.19 BF per SF at 5-1/4" exposure on the outside face.

| | Craft@Hrs | Unit | Material | Labor | Total | Sell |
|---|---|---|---|---|---|---|
| Using D grade yellow pine drop siding at | — | MSF | 2,310.00 | — | 2,310.00 | — |
| Cost per square foot of wall | B1@.114 | SF | 3.77 | 3.35 | 7.12 | 12.10 |
| Cost per running foot, 8' high wall | B1@.912 | LF | 30.20 | 26.80 | 57.00 | 96.90 |

2" x 6" stud walls with drywall interior, 1" x 6" drop siding exterior, same construction as above, except with 6-1/4" R-19 insulation

| | Craft@Hrs | Unit | Material | Labor | Total | Sell |
|---|---|---|---|---|---|---|
| Cost per square foot of wall | B1@.123 | SF | 4.20 | 3.61 | 7.81 | 13.30 |
| Cost per running foot, 8' high wall | B1@.984 | LF | 33.60 | 28.90 | 62.50 | 106.00 |

2" x 4" stud walls with drywall interior, stucco exterior, 1/2" gypsum drywall on inside face ready for painting, over 3-1/2" R-11 insulation and a three-coat exterior plaster (stucco) finish with integral color on the outside face

| | Craft@Hrs | Unit | Material | Labor | Total | Sell |
|---|---|---|---|---|---|---|
| Cost per square foot of wall | B1@.090 | SF | 3.74 | 2.64 | 6.38 | 10.80 |
| Cost per running foot, 8' high wall | B1@.720 | LF | 30.00 | 21.10 | 51.10 | 86.90 |

2" x 6" stud walls with drywall interior, stucco exterior, same construction as above, except with 6-1/4" R-19 insulation

| | Craft@Hrs | Unit | Material | Labor | Total | Sell |
|---|---|---|---|---|---|---|
| Cost per square foot of wall | B1@.099 | SF | 4.17 | 2.91 | 7.08 | 12.00 |
| Cost per running foot, 8' high wall | B1@.792 | LF | 33.40 | 23.30 | 56.70 | 96.40 |

Add for other types of gypsum board

| | Craft@Hrs | Unit | Material | Labor | Total | Sell |
|---|---|---|---|---|---|---|
| 1/2" or 5/8" moisture-resistant greenboard | | | | | | |
| Cost per SF, greenboard per side, add | — | SF | .10 | — | .10 | — |
| 1/2" or 5/8" moisture-resistant greenboard | | | | | | |
| Cost per running foot per side 8' high | — | LF | .99 | — | .99 | — |
| 5/8" thick fire-rated type X gypsum drywall | | | | | | |
| Cost per SF, per side, add | — | SF | .09 | — | .09 | — |

5/8" thick fire-rated type X gypsum drywall

| | Craft@Hrs | Unit | Material | Labor | Total | Sell |
|---|---|---|---|---|---|---|
| Cost per running foot per side 8' high | — | LF | .69 | — | .69 | — |

Support ceiling joists during replacement of a stud wall, using one 4" x 4" post, 2" x 8" pads and wedges, 4" x 4" strong back. All materials salvaged. Allow one post per each 5' of wall.

| | Craft@Hrs | Unit | Material | Labor | Total | Sell |
|---|---|---|---|---|---|---|
| Per support post set and removed | B1@.567 | Ea | — | 16.70 | 16.70 | 28.40 |

| | Craft@Hrs | Unit | Material | Labor | Total | Sell |
|---|---|---|---|---|---|---|

**Breakthrough wall opening.** Includes breaking an opening in an existing interior or exterior wall, cutting out the existing wall frame and cover, installing a header where studs or brick are removed, and trimming and casing the opening. Add the cost of floor and wall finish to match the existing surfaces on both sides of the new opening. In multi-story construction or in wide openings, a steel lintel may be required to support the framing above the opening. Add the cost of engineering, the lintel, and one hour of carpenter time for each foot of lintel length.

Door or cased opening to 3' wide x 7' high. Includes the door frame. Add the cost of the door and jamb, as required.

| | Craft@Hrs | Unit | Material | Labor | Total | Sell |
|---|---|---|---|---|---|---|
| Frame wall | B1@5.25 | Ea | — | 154.00 | 154.00 | 262.00 |
| Add for each foot of width over 3' | B1@.700 | LF | — | 20.60 | 20.60 | 35.00 |
| Brick veneer over frame wall | B1@6.25 | Ea | — | 184.00 | 184.00 | 313.00 |
| Add for each foot of width over 3' | B1@.750 | LF | — | 22.00 | 22.00 | 37.40 |
| Block or brick wall | B1@6.75 | Ea | — | 198.00 | 198.00 | 337.00 |
| Add for each foot of width over 3' | B1@.850 | LF | — | 25.00 | 25.00 | 42.50 |

Wall breakthrough for window or wall air conditioner, to 3' wide x 4' high. Includes the window frame and cripple stud fill below the new opening. Add the cost of the window, as required.

| | Craft@Hrs | Unit | Material | Labor | Total | Sell |
|---|---|---|---|---|---|---|
| Frame wall | B1@6.50 | Ea | — | 191.00 | 191.00 | 325.00 |
| Brick veneer over frame wall | B1@10.5 | Ea | — | 308.00 | 308.00 | 524.00 |
| Block or brick | B1@11.8 | Ea | — | 347.00 | 347.00 | 590.00 |

**Furring.** Per SF of surface area to be covered, including typical wedging. Figures in parentheses show coverage, including 7% waste, typical job.

Over masonry, utility grade

| | Craft@Hrs | Unit | Material | Labor | Total | Sell |
|---|---|---|---|---|---|---|
| 1" furring, per MBF | — | MBF | 893.00 | — | 893.00 | — |
| 12" OC, 1" x 2" (.24 BF per SF) | B1@.025 | SF | .21 | .73 | .94 | 1.60 |
| 16" OC, 1" x 2" (.20 BF per SF) | B1@.020 | SF | .18 | .59 | .77 | 1.31 |
| 20" OC, 1" x 2" (.17 BF per SF) | B1@.018 | SF | .15 | .53 | .68 | 1.16 |
| 24" OC, 1" x 2" (.15 BF per SF) | B1@.016 | SF | .13 | .47 | .60 | 1.02 |
| 12" OC, 1" x 3" (.36 BF per SF) | B1@.025 | SF | .32 | .73 | 1.05 | 1.79 |
| 16" OC, 1" x 3" (.29 BF per SF) | B1@.020 | SF | .26 | .59 | .85 | 1.45 |
| 20" OC, 1" x 3" (.25 BF per SF) | B1@.018 | SF | .22 | .53 | .75 | 1.28 |
| 24" OC, 1" x 3" (.22 BF per SF) | B1@.016 | SF | .20 | .47 | .67 | 1.14 |

Over wood frame, utility grade

| | Craft@Hrs | Unit | Material | Labor | Total | Sell |
|---|---|---|---|---|---|---|
| 1" furring, per MBF | — | MBF | 893.00 | — | 893.00 | — |
| 12" OC, 1" x 2" (.24 BF per SF) | B1@.016 | SF | .21 | .47 | .68 | 1.16 |
| 16" OC, 1" x 2" (.20 BF per SF) | B1@.013 | SF | .18 | .38 | .56 | .95 |
| 20" OC, 1" x 2" (.17 BF per SF) | B1@.011 | SF | .15 | .32 | .47 | .80 |
| 24" OC, 1" x 2" (.15 BF per SF) | B1@.010 | SF | .13 | .29 | .42 | .71 |
| 12" OC, 1" x 3" (.36 BF per SF) | B1@.016 | SF | .32 | .47 | .79 | 1.34 |
| 16" OC, 1" x 3" (.29 BF per SF) | B1@.013 | SF | .26 | .38 | .64 | 1.09 |
| 20" OC, 1" x 3" (.25 BF per SF) | B1@.011 | SF | .22 | .32 | .54 | .92 |
| 24" OC, 1" x 3" (.22 BF per SF) | B1@.010 | SF | .20 | .29 | .49 | .83 |

Over wood subfloor

| | Craft@Hrs | Unit | Material | Labor | Total | Sell |
|---|---|---|---|---|---|---|
| 1" furring, per MBF | — | MBF | 893.00 | — | 893.00 | — |
| 12" OC, 1" x 2" utility (.24 BF per SF) | B1@.033 | SF | .21 | .97 | 1.18 | 2.01 |
| 16" OC, 1" x 2" utility (.20 BF per SF) | B1@.028 | SF | .18 | .82 | 1.00 | 1.70 |
| 20" OC, 1" x 2" utility (.17 BF per SF) | B1@.024 | SF | .15 | .70 | .85 | 1.45 |
| 24" OC, 1" x 2" utility (.15 BF per SF) | B1@.021 | SF | .13 | .62 | .75 | 1.28 |
| 2" furring, per MBF | — | MBF | 622.00 | — | 622.00 | — |
| 12" OC, 2" x 2" Std & Btr (.48 BF per SF) | B1@.033 | SF | .30 | .97 | 1.27 | 2.16 |
| 16" OC, 2" x 2" Std & Btr (.39 BF per SF) | B1@.028 | SF | .24 | .82 | 1.06 | 1.80 |
| 20" OC, 2" x 2" Std & Btr (.34 BF per SF) | B1@.024 | SF | .21 | .70 | .91 | 1.55 |
| 24" OC, 2" x 2" Std & Btr (.30 BF per SF) | B1@.021 | SF | .19 | .62 | .81 | 1.38 |

| | Craft@Hrs | Unit | Material | Labor | Total | Sell |
|---|---|---|---|---|---|---|

Remove 1" furring. All nails pulled or driven flush. Per linear foot of 1" furring

| | Craft@Hrs | Unit | Material | Labor | Total | Sell |
|---|---|---|---|---|---|---|
| Furring installed on masonry | B1@.030 | LF | — | .88 | .88 | 1.50 |
| Furring installed on wood frame | B1@.020 | LF | — | .59 | .59 | 1.00 |
| Furring installed on subfloor | B1@.035 | LF | — | 1.03 | 1.03 | 1.75 |

**Sleepers.** Std & Btr pressure treated lumber, laid on concrete, including 5% waste.

| | Craft@Hrs | Unit | Material | Labor | Total | Sell |
|---|---|---|---|---|---|---|
| 2" x 6" sleepers, per MBF | — | MBF | 943.00 | — | 943.00 | — |
| 2" x 6" sleepers, per LF | B1@.017 | LF | .99 | .50 | 1.49 | 2.53 |
| Add for taper cuts on sleepers, per cut | B1@.050 | Ea | — | 1.47 | 1.47 | 2.50 |

**Backing and nailers.** Std & Btr, for wall finishes, "floating" backing for drywall ceilings, trim, z-bar, appliances and fixtures, etc. Figures in parentheses show board feet per LF, including 10% waste.

| | Craft@Hrs | Unit | Material | Labor | Total | Sell |
|---|---|---|---|---|---|---|
| 1" x 4" (.37 BF per LF) | B1@.011 | LF | .33 | .32 | .65 | 1.11 |
| 1" x 6" (.55 BF per LF) | B1@.017 | LF | .52 | .50 | 1.02 | 1.73 |
| 1" x 8" (.73 BF per LF) | B1@.022 | LF | .70 | .65 | 1.35 | 2.30 |
| 2" x 4" (.73 BF per LF) | B1@.023 | LF | .45 | .68 | 1.13 | 1.92 |
| 2" x 6" (1.10 BF per LF) | B1@.034 | LF | .71 | 1.00 | 1.71 | 2.91 |
| 2" x 8" (1.47 BF per LF) | B1@.045 | LF | .96 | 1.32 | 2.28 | 3.88 |
| 2" x 10" (1.83 BF per LF) | B1@.057 | LF | 1.25 | 1.67 | 2.92 | 4.96 |

**Ledger strips.** Std & Btr, nailed to faces of studs, beams, girders, joists, etc. See also Ribbons in this section for let-in type. Figures in parentheses indicate board feet per LF, including 10% waste.

| | Craft@Hrs | Unit | Material | Labor | Total | Sell |
|---|---|---|---|---|---|---|
| 1" x 2", 3" or 4" ledger, per MBF | — | MBF | 893.00 | — | 893.00 | — |
| 1" x 2" (.18 BF per LF) | B1@.010 | LF | .16 | .29 | .45 | .77 |
| 1" x 3" (.28 BF per LF) | B1@.010 | LF | .25 | .29 | .54 | .92 |
| 1" x 4" (.37 BF per LF) | B1@.010 | LF | .33 | .29 | .62 | 1.05 |
| 2" x 2", 3" or 4" ledger, per MBF | — | MBF | 622.00 | — | 622.00 | — |
| 2" x 2" (.37 BF per LF) | B1@.010 | LF | .23 | .29 | .52 | .88 |
| 2" x 3" (.55 BF per LF) | B1@.010 | LF | .34 | .29 | .63 | 1.07 |
| 2" x 4" (.73 BF per LF) | B1@.010 | LF | .45 | .29 | .74 | 1.26 |

**Ribbons** (Ribbands). Let-in to wall framing. Figures in parentheses indicate board feet per LF, including 10% waste.

| | Craft@Hrs | Unit | Material | Labor | Total | Sell |
|---|---|---|---|---|---|---|
| 1" x 3", 4" or 6" ribbon, per MBF | — | MBF | 893.00 | — | 893.00 | — |
| 1" x 3", Std & Btr (.28 BF per LF) | B1@.020 | LF | .25 | .59 | .84 | 1.43 |
| 1" x 4", Std & Btr (.37 BF per LF) | B1@.020 | LF | .33 | .59 | .92 | 1.56 |
| 1" x 6", Std & Btr (.55 BF per LF) | B1@.030 | LF | .49 | .88 | 1.37 | 2.33 |
| 2" x 3", 4" ribbon, per MBF | — | MBF | 622.00 | — | 622.00 | — |
| 2" x 3", Std & Btr (.55 BF per LF) | B1@.041 | LF | .34 | 1.20 | 1.54 | 2.62 |
| 2" x 4", Std & Btr (.73 BF per LF) | B1@.041 | LF | .45 | 1.20 | 1.65 | 2.81 |
| 2" x 6" ribbon, per MBF | — | MBF | 641.00 | — | 641.00 | — |
| 2" x 6", Std & Btr (1.10 BF per LF) | B1@.045 | LF | .71 | 1.32 | 2.03 | 3.45 |
| 2" x 8" ribbon, per MBF | — | MBF | 652.00 | — | 652.00 | — |
| 2" x 8", Std & Btr (1.47 BF per LF) | B1@.045 | LF | .96 | 1.32 | 2.28 | 3.88 |

**Ceiling joists — room addition.** Per SF of area covered. Figures in parentheses indicate board feet per square foot of ceiling, including end joists, header joists, and 5% waste. No beams, bridging, blocking, or ledger strips included. Deduct for openings over 25 SF. For scheduling purposes, estimate that a crew of two can complete 650 SF of area per 8-hour day for 12" center-to-center framing; 800 SF for 16" OC; 950 SF for 20" OC; or 1,100 SF for 24" OC.

2" x 4", Std & Btr grade

| | Craft@Hrs | Unit | Material | Labor | Total | Sell |
|---|---|---|---|---|---|---|
| 2" x 4" ceiling joists, 16" centers, per MBF | — | MBF | 622.00 | — | 622.00 | — |
| 12" centers (.78 BF per SF) | B1@.026 | SF | .48 | .76 | 1.24 | 2.11 |
| 16" centers (.59 BF per SF) | B1@.020 | SF | .37 | .59 | .96 | 1.63 |
| 20" centers (.48 BF per SF) | B1@.016 | SF | .30 | .47 | .77 | 1.31 |
| 24" centers (.42 BF per SF) | B1@.014 | SF | .26 | .41 | .67 | 1.14 |

| | Craft@Hrs | Unit | Material | Labor | Total | Sell |
|---|---|---|---|---|---|---|
| 2" x 6", Std & Btr grade | | | | | | |
|   2" x 6" ceiling joists, 16" centers, per MBF | — | MBF | 641.00 | — | 641.00 | — |
|   12" centers (1.15 BF per SF) | B1@.026 | SF | .74 | .76 | 1.50 | 2.55 |
|   16" centers (.88 BF per SF) | B1@.020 | SF | .56 | .59 | 1.15 | 1.96 |
|   20" centers (.72 BF per SF) | B1@.016 | SF | .46 | .47 | .93 | 1.58 |
|   24" centers (.63 BF per SF) | B1@.014 | SF | .40 | .41 | .81 | 1.38 |
| 2" x 8", Std & Btr grade | | | | | | |
|   2" x 8" ceiling joists, 16" centers, per MBF | — | MBF | 651.00 | — | 651.00 | — |
|   12" centers (1.53 BF per SF) | B1@.028 | SF | 1.00 | .82 | 1.82 | 3.09 |
|   16" centers (1.17 BF per SF) | B1@.022 | SF | .76 | .65 | 1.41 | 2.40 |
|   20" centers (.96 BF per SF) | B1@.018 | SF | .63 | .53 | 1.16 | 1.97 |
|   24" centers (.84 BF per SF) | B1@.016 | SF | .55 | .47 | 1.02 | 1.73 |
| 2" x 10", Std & Btr grade | | | | | | |
|   2" x 10" ceiling joists, 16" centers, per MBF | — | MBF | 683.00 | — | 683.00 | — |
|   12" centers (1.94 BF per SF) | B1@.030 | SF | 1.32 | .88 | 2.20 | 3.74 |
|   16" centers (1.47 BF per SF) | B1@.023 | SF | 1.00 | .68 | 1.68 | 2.86 |
|   20" centers (1.21 BF per SF) | B1@.019 | SF | .83 | .56 | 1.39 | 2.36 |
|   24" centers (1.04 BF per SF) | B1@.016 | SF | .71 | .47 | 1.18 | 2.01 |
| 2" x 12", Std & Btr grade | | | | | | |
|   2" x 12" ceiling joists, 16" centers, per MBF | — | MBF | 730.00 | — | 730.00 | — |
|   12" centers (2.30 BF per SF) | B1@.033 | SF | 1.68 | .97 | 2.65 | 4.51 |
|   16" centers (1.76 BF per SF) | B1@.025 | SF | 1.28 | .73 | 2.01 | 3.42 |
|   20" centers (1.44 BF per SF) | B1@.020 | SF | 1.05 | .59 | 1.64 | 2.79 |
|   24" centers (1.26 BF per SF) | B1@.018 | SF | .92 | .53 | 1.45 | 2.47 |

**Adding ceiling joists.** Per linear foot of new joist set in joist hangers or "doubled" by spiking to an existing joist. Work done from the floor below. Add the cost of removing ceiling cover to expose the joists.

Jack two or three adjacent joists into position with a crossarm mounted on a hydraulic jack. Includes the cost of removing the jack.

| | Craft@Hrs | Unit | Material | Labor | Total | Sell |
|---|---|---|---|---|---|---|
|   Per jacking point | B1@.700 | Ea | — | 20.60 | 20.60 | 35.00 |
| Add 2" x 6" ceiling joist | | | | | | |
|   Per linear foot of joist | B1@.042 | LF | .62 | 1.23 | 1.85 | 3.15 |
| Add 2" x 8" ceiling joist | | | | | | |
|   Per linear foot of joist | B1@.046 | LF | .64 | 1.35 | 1.99 | 3.38 |
| Add 2" x 10" ceiling joist | | | | | | |
|   Per linear foot of joist | B1@.050 | LF | .66 | 1.47 | 2.13 | 3.62 |
| Add 2" x 12" ceiling joist | | | | | | |
|   Per linear foot of joist | B1@.052 | LF | .74 | 1.53 | 2.27 | 3.86 |

| | Craft@Hrs | Unit | Material | Labor | Total | Sell |
|---|---|---|---|---|---|---|

## Changes to Roof Framing

**Rafters.** Flat, shed, or gable roofs, up to 5 in 12 slope (5/24 pitch), maximum 25' span. Figures in parentheses indicate board feet per square foot of actual roof surface area (not roof plan area), including rafters, ridge boards, collar beams and normal waste, but no blocking, bracing, purlins, curbs, or gable walls.

| | Craft@Hrs | Unit | Material | Labor | Total | Sell |
|---|---|---|---|---|---|---|
| 2" x 4", Std & Btr | | | | | | |
| 2" x 4" rafters, 16" centers, per MBF | — | MBF | 622.00 | — | 622.00 | — |
| 12" centers (.89 BF per SF) | B1@.021 | SF | .55 | .62 | 1.17 | 1.99 |
| 16" centers (.71 BF per SF) | B1@.017 | SF | .44 | .50 | .94 | 1.60 |
| 24" centers (.53 BF per SF) | B1@.013 | SF | .33 | .38 | .71 | 1.21 |
| Remove and replace 2" x 4" rafter, per LF | B1@.050 | LF | .62 | 1.47 | 2.09 | 3.55 |
| 2" x 6", Std & Btr | | | | | | |
| 2" x 6" rafters, 16" centers, per MBF | — | MBF | 641.00 | — | 641.00 | — |
| 12" centers (1.29 BF per SF) | B1@.029 | SF | .83 | .85 | 1.68 | 2.86 |
| 16" centers (1.02 BF per SF) | B1@.023 | SF | .65 | .68 | 1.33 | 2.26 |
| 24" centers (.75 BF per SF) | B1@.017 | SF | .48 | .50 | .98 | 1.67 |
| Remove and replace 2" x 6" rafter, per LF | B1@.068 | LF | .62 | 2.00 | 2.62 | 4.45 |
| 2" x 8", Std & Btr | | | | | | |
| 2" x 8" rafters, 16" centers, per MBF | — | MBF | 651.00 | — | 651.00 | — |
| 12" centers (1.71 BF per SF) | B1@.036 | SF | 1.11 | 1.06 | 2.17 | 3.69 |
| 16" centers (1.34 BF per SF) | B1@.028 | SF | .87 | .82 | 1.69 | 2.87 |
| 24" centers (1.12 BF per SF) | B1@.024 | SF | .73 | .70 | 1.43 | 2.43 |
| Remove and replace 2" x 8" rafter, per LF | B1@.087 | LF | .64 | 2.56 | 3.20 | 5.44 |
| 2" x 10", Std & Btr | | | | | | |
| 2" x 10" rafters, 16" centers, per MBF | — | MBF | 683.00 | — | 683.00 | — |
| 12" centers (2.12 BF per SF) | B1@.039 | SF | 1.45 | 1.15 | 2.60 | 4.42 |
| 16" centers (1.97 BF per SF) | B1@.036 | SF | 1.35 | 1.06 | 2.41 | 4.10 |
| 24" centers (1.21 BF per SF) | B1@.022 | SF | .83 | .65 | 1.48 | 2.52 |
| Remove and replace 2" x 10" rafter, per LF | B1@.098 | LF | .66 | 2.88 | 3.54 | 6.02 |
| 2" x 12", Std & Btr | | | | | | |
| 2" x 12" rafters, 16" centers, per MBF | — | MBF | 730.00 | — | 730.00 | — |
| 12" centers (2.52 BF per SF) | B1@.045 | SF | 1.84 | 1.32 | 3.16 | 5.37 |
| 16" centers (1.97 BF per SF) | B1@.035 | SF | 1.44 | 1.03 | 2.47 | 4.20 |
| 24" centers (1.43 BF per SF) | B1@.026 | SF | 1.04 | .76 | 1.80 | 3.06 |
| Remove and replace 2" x 12" rafter, per LF | B1@.113 | LF | .74 | 3.32 | 4.06 | 6.90 |
| Add for hip roof | B1@.007 | SF | — | .21 | .21 | .36 |
| Add for slope over 5 in 12 | B1@.015 | SF | — | .44 | .44 | .75 |

**Roof frame jacking.** Repair sag in ridge board by jacking and installing shoring supported by bearing walls.

| | Craft@Hrs | Unit | Material | Labor | Total | Sell |
|---|---|---|---|---|---|---|
| Per support installed | B1@7.27 | Ea | 4.50 | 214.00 | 218.50 | 371.00 |

**Trimmers and curbs.** At stairwells, skylights, dormers, etc. Figures in parentheses show board feet per LF, Std & Btr grade, including 10% waste.

| | Craft@Hrs | Unit | Material | Labor | Total | Sell |
|---|---|---|---|---|---|---|
| 2" x 4" (.73 BF per LF) | B1@.018 | LF | .45 | .53 | .98 | 1.67 |
| 2" x 6" (1.10 BF per LF) | B1@.028 | LF | .71 | .82 | 1.53 | 2.60 |
| 2" x 8" (1.47 BF per LF) | B1@.038 | LF | .96 | 1.12 | 2.08 | 3.54 |
| 2" x 10" (1.83 BF per LF) | B1@.047 | LF | 1.25 | 1.38 | 2.63 | 4.47 |
| 2" x 12" (2.20 BF per LF) | B1@.057 | LF | 1.61 | 1.67 | 3.28 | 5.58 |

**Collar beams & collar ties.** Std & Btr grade, including 10% waste.

| | Craft@Hrs | Unit | Material | Labor | Total | Sell |
|---|---|---|---|---|---|---|
| Collar beams, 2" x 6" | B1@.013 | LF | .71 | .38 | 1.09 | 1.85 |
| Collar ties, 1" x 6" | B1@.006 | LF | .52 | .18 | .70 | 1.19 |

**Purlins** (Purling). Std & Btr, installed below roof rafters. Figures in parentheses indicate board feet per LF including 5% waste.

| | | | | | | |
|---|---|---|---|---|---|---|
| 2" x 4" (.70 BF per LF) | B1@.012 | LF | .44 | .35 | .79 | 1.34 |
| 2" x 6" (1.05 BF per LF) | B1@.017 | LF | .67 | .50 | 1.17 | 1.99 |
| 2" x 8" (1.40 BF per LF) | B1@.023 | LF | .91 | .68 | 1.59 | 2.70 |
| 2" x 10" (1.75 BF per LF) | B1@.027 | LF | 1.20 | .79 | 1.99 | 3.38 |
| 2" x 12" (2.10 BF per LF) | B1@.030 | LF | 1.53 | .88 | 2.41 | 4.10 |
| 4" x 6" (2.10 BF per LF) | B1@.034 | LF | 2.03 | 1.00 | 3.03 | 5.15 |
| 4" x 8" (2.80 BF per LF) | B1@.045 | LF | 2.48 | 1.32 | 3.80 | 6.46 |

**Dormer studs.** Std & Btr, per square foot of wall area, including 10% waste.

| | | | | | | |
|---|---|---|---|---|---|---|
| 2" x 4", (.84 BF per SF) | B1@.033 | SF | .52 | .97 | 1.49 | 2.53 |

**Roof trusses.** 24" OC, any slope from 3 in 12 to 12 in 12, total height not to exceed 12' high from bottom chord to highest point on truss. Prices for trusses over 12' high will be up to 100% higher. Square foot (SF) costs, where shown, are per square foot of roof area to be covered.

Scissor trusses

| | | | | | | |
|---|---|---|---|---|---|---|
| 2" x 4" top and bottom chords | | | | | | |
| Up to 38' span | B1@.022 | SF | 2.48 | .65 | 3.13 | 5.32 |
| 40' to 50' span | B1@.028 | SF | 3.12 | .82 | 3.94 | 6.70 |

Fink truss "W" (conventional roof truss)

| | | | | | | |
|---|---|---|---|---|---|---|
| 2" x 4" top and bottom chords | | | | | | |
| Up to 38' span | B1@.017 | SF | 2.16 | .50 | 2.66 | 4.52 |
| 40' to 50' span | B1@.022 | SF | 2.57 | .65 | 3.22 | 5.47 |
| 2" x 6" top and bottom chords | | | | | | |
| Up to 38' span | B1@.020 | SF | 2.62 | .59 | 3.21 | 5.46 |
| 40' to 50' span | B1@.026 | SF | 3.17 | .76 | 3.93 | 6.68 |

Truss with gable fill at 16" OC

| | | | | | | |
|---|---|---|---|---|---|---|
| 28' span, 5 in 12 slope | B1@.958 | Ea | 177.00 | 28.10 | 205.10 | 349.00 |
| 32' span, 5 in 12 slope | B1@1.26 | Ea | 219.00 | 37.00 | 256.00 | 435.00 |
| 40' span, 5 in 12 slope | B1@1.73 | Ea | 309.00 | 50.80 | 359.80 | 612.00 |

**Roof truss repairs.**

Repair splits in 2" x 6" diagonal or 2" x 6" vertical member by installing five prefabricated clamps

| | | | | | | |
|---|---|---|---|---|---|---|
| per truss | B1@.492 | Ea | — | 14.50 | 14.50 | 24.70 |

Repair splits at various locations by installing splice plates

| | | | | | | |
|---|---|---|---|---|---|---|
| per splice plate | B1@.104 | Ea | — | 3.05 | 3.05 | 5.19 |

Repair splits in bottom chord by installing two prefabricated clamps near joints, drill holes to arrest split

| | | | | | | |
|---|---|---|---|---|---|---|
| per truss | B1@.280 | Ea | — | 8.22 | 8.22 | 14.00 |

Repair split ends in 2" x 6" bottom chord by installing splice plate and four 1/2" stitch bolts

| | | | | | | |
|---|---|---|---|---|---|---|
| per truss | B1@.617 | Ea | — | 18.10 | 18.10 | 30.80 |

Repair vertical compression 2" x 4" member by scabbing pieces to each side and bolting

| | | | | | | |
|---|---|---|---|---|---|---|
| per scab repair | B1@1.00 | Ea | — | 29.40 | 29.40 | 50.00 |

**Fascia board.** Installed on roof eaves working at 8' height from ladders.

Pry off nailed fascia and drop debris to the ground

| | | | | | | |
|---|---|---|---|---|---|---|
| per linear foot of fascia | B1@.007 | LF | — | .21 | .21 | .36 |

Install 1" x 6" fascia board by nailing at roof eaves

| | | | | | | |
|---|---|---|---|---|---|---|
| per linear foot of fascia | B1@.044 | LF | .35 | 1.29 | 1.64 | 2.79 |

Remove and replace fascia board at roof eaves

| | | | | | | |
|---|---|---|---|---|---|---|
| per linear foot of fascia | B1@.051 | LF | .35 | 1.50 | 1.85 | 3.15 |
| Add for heights to 20', per linear foot | B1@.015 | LF | .35 | .44 | .79 | 1.34 |

| | Craft@Hrs | Unit | Material | Labor | Total | Sell |
|---|---|---|---|---|---|---|
| **Roof sheathing.** | | | | | | |
| CDX plywood sheathing, rough, power nailed, including normal waste. | | | | | | |
|   1/2" | B1@.013 | SF | .41 | .38 | .79 | 1.34 |
|   1/2", 4-ply | B1@.013 | SF | .46 | .38 | .84 | 1.43 |
|   1/2", 5-ply | B1@.013 | SF | .51 | .38 | .89 | 1.51 |
|   5/8", 4-ply | B1@.013 | SF | .57 | .38 | .95 | 1.62 |
|   5/8", 5-ply | B1@.013 | SF | .67 | .38 | 1.05 | 1.79 |
|   3/4" | B1@.013 | SF | .78 | .38 | 1.16 | 1.97 |
|   Add for hip roof | B1@.007 | SF | — | .21 | .21 | .36 |
|   Add for steep pitch or cut-up roof | B1@.015 | SF | — | .44 | .44 | .75 |
|   Remove existing sheathing | B1@.017 | SF | — | .50 | .50 | .85 |
| OSB sheathing, rough, power nailed, including normal waste. | | | | | | |
|   1/2" | B1@.013 | SF | .37 | .38 | .75 | 1.28 |
|   5/8" | B1@.013 | SF | .54 | .38 | .92 | 1.56 |
|   3/4" | B1@.013 | SF | .77 | .38 | 1.15 | 1.96 |
|   7/8" | B1@.014 | SF | 1.01 | .41 | 1.42 | 2.41 |
|   1" | B1@.015 | SF | 1.17 | .44 | 1.61 | 2.74 |
|   Add for hip roof | B1@.007 | SF | — | .21 | .21 | .36 |
|   Add for steep pitch or cut-up roof | B1@.015 | SF | — | .44 | .44 | .75 |
|   Remove existing sheathing | B1@.017 | SF | — | .50 | .50 | .85 |
| Board sheathing, 1" x 6" or 1" x 8" utility T&G laid diagonal | | | | | | |
|   1" utility T&G lumber, per MBF | — | MBF | 1,040.00 | — | 1,040.00 | — |
|   (1.13 BF per SF) | B1@.026 | SF | 1.30 | .76 | 2.06 | 3.50 |
|   Add for hip roof | B1@.007 | SF | — | .21 | .21 | .36 |
|   Add for steep pitch or cut-up roof | B1@.015 | SF | — | .44 | .44 | .75 |
|   Remove board sheathing | B1@.030 | SF | — | .88 | .88 | 1.50 |
|   Remove deteriorated boards | B1@.034 | SF | — | 1.00 | 1.00 | 1.70 |
|   Remove boards in salvage condition | B1@.036 | SF | — | 1.06 | 1.06 | 1.80 |
| Remove and patch roof sheathing, rafters 24" on center, 3 in 12 pitch. | | | | | | |
|   1" x 8" x 12' long, per board | B1@.400 | Ea | 10.70 | 11.70 | 22.40 | 38.10 |
|   1" x 8", per square foot | B1@.050 | SF | 1.34 | 1.47 | 2.81 | 4.78 |
|   1/2", 4-ply plywood, per 4' x 8' sheet | B1@.960 | Ea | 16.20 | 28.20 | 44.40 | 75.50 |
|   1/2", 4-ply plywood, per square foot | B1@.030 | SF | .51 | .88 | 1.39 | 2.36 |

# *Insulation and Moisture Control*

5

**H**omes built before the early 1970's were designed in an era when gasoline was less than 30 cents a gallon, heating oil was cheap and natural gas supplies were plentiful. Few people worried about energy efficiency because there was energy to spare. Builders and homeowners didn't spend money on insulating houses because it was cheaper just to turn up the heat. But times have changed. Modern insulation standards are considerably higher.

## Ceiling Insulation

Most homes have an accessible attic with exposed ceiling framing. That makes it easy to check the depth of insulation and apply more when needed. Before about 1970, the most common attic insulation, if there was any, was reflective foil (in warmer climates) or a few inches of rock wool (in colder climates).

Optimum insulation still varies with location. But homes in milder climates can benefit from 6" of fiberglass above the ceiling. In colder climates, you need 12". If the home is heated by electricity or a means more expensive than gas, propane or heating oil, extra insulation usually makes good economic sense.

Batt and roll blanket insulation is made to fit between joists and studs installed either 16" or 24" on center. When you see coverage figures for insulation, those numbers include framing area. For example a roll of insulation described as "covers 80 SF" will fill a wall area measuring 8' by 10'.

If a vapor barrier is already in place, or if you're installing a separate vapor barrier, use unfaced rolls or batts. Otherwise, install Kraft-faced insulation. However, there's one exception when adding more insulation in an attic space. Even if the existing ceiling insulation doesn't have a vapor barrier, install unfaced insulation.

Loose fill insulation is just as effective as rolls or batts and it's easy to install in an attic. Simply pour insulation between the joists and screed it off to the right thickness (Figure 5-1).

> The effectiveness of insulation is measured in R-value, which is the resistance of the material to heat transfer. The higher the rating, the more insulating value. For example, 12" of fiberglass insulation usually carries an R-38 rating. Six inches of the same fiberglass will be rated at R-19.

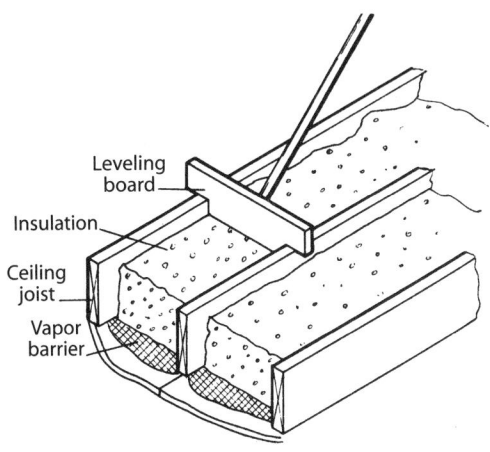

**Figure 5-1**

*Installation of loose-fill ceiling insulation*

# Wall Insulation

Any time you remove interior wall finish, check the insulation. Walls in mild climates need R-13 insulation. That's $3^1/2$" of fiberglass, which fits perfectly in the cavity of a 2 x 4 stud wall. R-19 wall insulation is appropriate in severely cold climates. But that requires a 6" wall cavity and 2 x 6 studs. If you need more than R-19 from an existing 2 x 4 stud wall, 1" polyurethane sheathing can add R-6.

Adding insulation is easy when the wall cavity is open. When the cavity is closed, you can blow cellulose insulation through a 1"-diameter hole cut in siding and sheathing. Open a hole at the top of each stud space. Lower a plumb bob into the cavity to find where the fire block is installed. Then drill another hole in the same cavity below the fire block. Blow cellulose into the stud space both above and below the block. When the cavity is filled, patch entry holes with either plastic or wood plugs.

# Floor Insulation

Houses with a furnace in the basement don't need insulation between the floor joists. But homes should have floor insulation equivalent to the wall insulation if the floors are located above an unheated crawl space or any other unheated space. Fit insulation batts between the floor joists. Hold the batts in place with wood or wire strips cut slightly longer than the joist space so they spring into place (Figure 5-2). You could also staple wire netting under the floor joists to hold either batt or blanket insulation tight against the underside of the floor sheathing. A vapor barrier has to be on the top side of insulation placed between floor joists. That presents a problem when installing insulation above a crawl space. Insulation manufacturers now offer roll insulation with inverse tabs so insulation can be stapled between floor joists with the poly face-up against the floor sheathing.

# Board Insulation

You can insulate brick, block and concrete walls by applying insulation board to the interior surface. The thicker the board, the higher the R-value. Polystyrene insulated sheathing board has an R-value of 3.5 per inch of thickness. Polyurethane board carries a 6.25 rating per inch of thickness. Many types of insulated sheathing can be installed over the existing siding. No tear-off is necessary, though additional edge trim will be needed around windows and doors. If the owner wants 2" of foam insulation on a basement wall, consider attaching 2" x 2" furring strips to the wall at 16" on center. Then install 2" insulation board between the strips and hang drywall or other wall cover on the furring. Most insulation boards, such as polystyrene, must be covered with drywall. They're fragile, and won't hold paint or wallpaper. Also, most insulation boards aren't fire-resistant. Fire codes require that interior wall surfaces be covered with fire-resistant material, such as drywall.

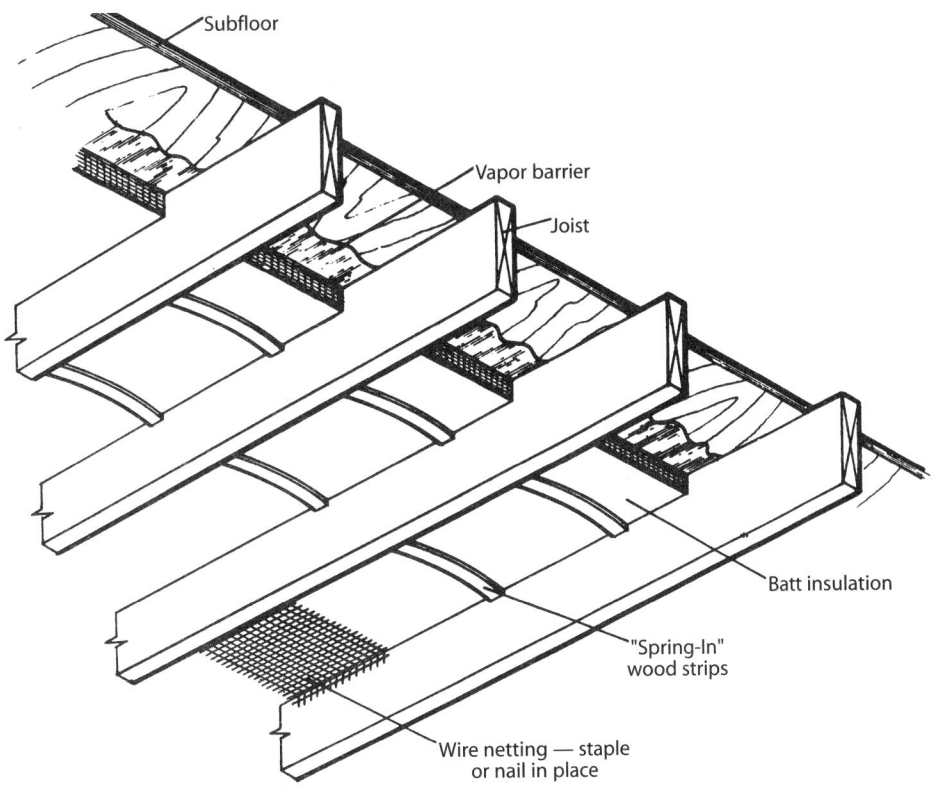

**Figure 5-2**

*Installing insulating batts in floor*

# Vapor Barrier

Good insulation cuts heating costs and adds to comfort by making the temperature in the house more uniform. But simply adding insulation to an old home can create more problems than it solves. In poorly-insulated older homes, water vapor escapes to the exterior without interference. But in a tighter home with fewer air leaks and better insulation, water vapor tends to collect in walls. There, it cools and condenses into liquid. The result can be saturated insulation and siding. Running a humidifier, cooking, bathing, and simply breathing aggravate the problem. Liquid moisture in a wall reduces the effectiveness of insulation, promotes decay and supports the growth of mold. To control moisture problems, be sure there is a vapor barrier on the (winter) warm side of walls and ceilings.

Most homes built before the mid-1930's don't have a vapor barrier in either the walls or the ceilings. If attic insulation has been upgraded since construction, you'll probably find batts or blankets with a Kraft face that resist the passage of moisture. The Kraft face should face down, against the ceiling finish. But if the ceiling insulation is loose fill, there should be a separate vapor barrier of coated paper, aluminum foil, or plastic film below the insulation.

Older homes develop thousands of tiny air leaks around doors and windows, at the sill, around electrical outlets and where interior and exterior walls have cracked.

Vapor movement through any barrier is measured in perms. The lower the perm rating, the more effective the barrier. Materials with good perm ratings include polyethylene at least 2 mils thick, asphalt-impregnated and surface-coated Kraft papers, and asphalt-laminated paper.

To be truly effective, the vapor barrier should be continuous over the inside of studs and joists. That's easy in new construction. In home improvement work, applying continuous vapor barrier is practical only when interior wall cover is being replaced.

If removing the ceiling or wall finish is part of the job, staple vapor barrier directly to the studs or joists. Lay vapor barrier on a subfloor directly under the new floor finish. Be sure to lap joints at least 2" and avoid unnecessary punctures. Nails or screws driven snug against the vapor barrier don't do any harm.

## Blanket Insulation With Vapor Barrier

Most roll (or "blanket") insulation has a vapor barrier on one side that extends beyond the edges of the roll to form a stapling tab. Lap these tabs over the 2" thickness of studs and joists. (Figure 5-3). Tabs on adjacent blankets should overlap. Your drywall crew probably prefers to see these tabs stapled to the inside face of studs or joists. That makes it easier to hang drywall with a smooth, even surface. But stapling tabs inside the stud or joist cavity will also reduce the effectiveness of the vapor barrier. Moisture will seep through openings between the tabs and framing members. It's better to make the drywall finishers work a little harder and protect the studs and joists from decay, mold and termites.

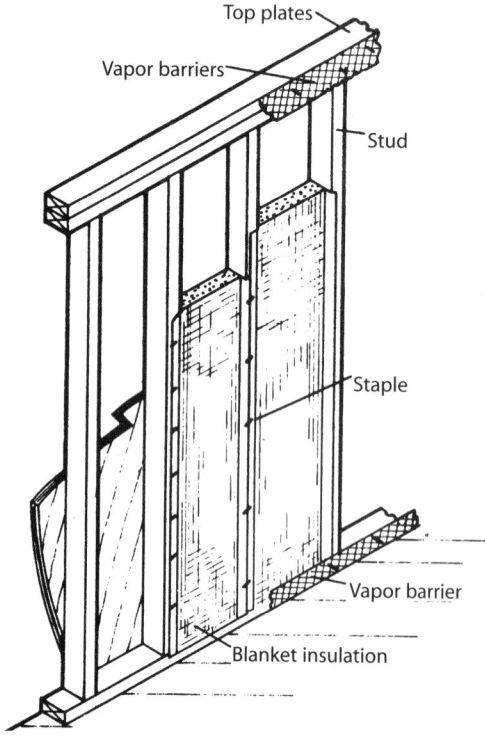

**Figure 5-3**

*Installation of blanket insulation with vapor barriers on one side*

## Vapor-Resistant Coating

If walls and ceilings don't have a vapor barrier, the next best thing is a vapor-resistant interior coating such as aluminum primer. Two coats of aluminum primer covered with acrylic latex offer moderate vapor resistance, though not as much as a true membrane.

## Crawl Space Ventilation

Damp soil in a crawl space will transfer moisture to the floor framing and even to occupied rooms above. Good ventilation helps keep the crawl space dry. Promote good cross-ventilation by installing a vent at each corner of the foundation. Taken together, foundation vent "free circulation area" should be $1/_{150}$th of the crawl space surface (10 square feet of vent for a 1,500 square foot crawl space). Free vent area excludes space occupied by the vent frame, screen wire and louvers. Most manufactured vents are identified by the overall dimensions and the net free area. If there's a partial basement, the crawl space can be vented to the basement instead of the exterior.

If soil in the crawl space is covered with a vapor barrier, foundation vent area can be as little as $1/1500$ of the ground area (1 square foot for a 1,500 square foot crawl space). It's usually much cheaper to cover the soil in a crawl space with membrane than to add more foundation vents. Ground cover should lap at least 2" at all joints (Figure 5-4). Lay brick or stones on top of the membrane to hold it down and prevent curling.

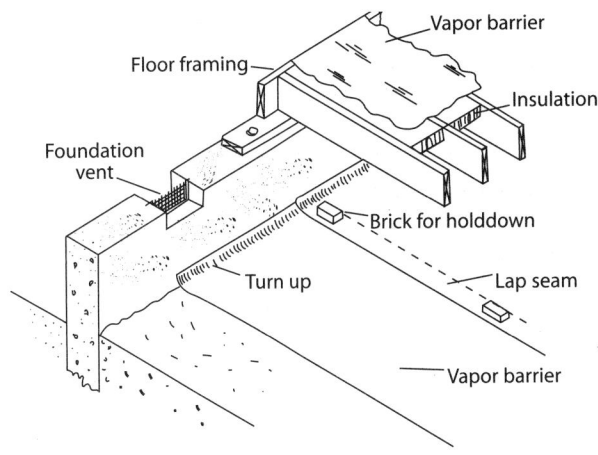

**Figure 5-4**

*Vapor barrier for crawl space (ground cover)*

## Attic and Roof Ventilation

Insulation laid between ceiling joists helps hold heat inside the home. But above that insulation, air should circulate freely. Excess moisture accumulation will damage insulation and other attic materials. Good air circulation is especially important in colder climates and when the roof cover is less permeable. With built-up roofing or asphalt shingles, the only way to get moisture out of an attic is with vents.

Inadequately-vented attics can reach temperatures above 140 degrees F on a sunny day. Some of the heat will work its way into the house, overburdening the air conditioning system. Properly-vented attics will stay cooler and can save up to 30 percent on cooling costs. In addition, a cooler attic will keep asphalt shingles from degrading prematurely in the heat.

Air circulation works best with soffit vents around the edge of the roof and outlet vents located high, near the ridge. Warm air rises in the attic, exits at the peak, and draws fresh air through soffit vents at the perimeter. Under those conditions, inlet vents should be $1/900$ of attic floor area. Outlet vents should have the same area. A home with only gable vents needs more free vent area. Air circulation through gable vents depends on wind blowing the right direction. Provide one square foot of free vent area for each 300' of attic area.

Hip roofs don't have gables and flat roofs don't have peaks. That reduces your venting options. The estimating section that follows includes power ventilators that can be installed on nearly any roof.

Flat roofs with no attic require some type of ventilation above the ceiling insulation. If this space is divided by joists, each joist space should be ventilated. A continuous soffit vent strip is the best choice. Drill through all headers that impede passage of air to the opposite eave.

A post and beam roof with a roof-plank ceiling has no attic and needs no ventilation. But the roof surface requires insulation board coated with a walkable sprayed-on insulator such as urethane.

| | Craft@Hrs | Unit | Material | Labor | Total | Sell |
|---|---|---|---|---|---|---|

## Fiberglass Insulation

**Kraft-faced fiberglass roll insulation.** Stapled between studs or joists. Coverage includes framing area.

| | Craft@Hrs | Unit | Material | Labor | Total | Sell |
|---|---|---|---|---|---|---|
| R-11, 15" x 70'6" x 3-1/2" thick | BC@.010 | SF | .30 | .32 | .62 | 1.05 |
| R-11, 23" x 70'6" x 3-1/2" thick | BC@.009 | SF | .30 | .29 | .59 | 1.00 |
| R-13, 15" x 32' x 3-1/2" thick | BC@.010 | SF | .29 | .32 | .61 | 1.04 |
| R-19, 15" x 39' x 6-1/4" thick | BC@.010 | SF | .48 | .32 | .80 | 1.36 |
| R-19, 23" x 39' x 6-1/4" thick | BC@.009 | SF | .41 | .29 | .70 | 1.19 |

**Unfaced fiberglass roll insulation.** Laid between ceiling joists. Coverage includes framing area.

| | Craft@Hrs | Unit | Material | Labor | Total | Sell |
|---|---|---|---|---|---|---|
| R-6.7, 15" x 48" x 2" thick handy roll | BC@.005 | SF | .86 | .16 | 1.02 | 1.73 |
| R-19, 15" x 39' x 6-1/4" thick | BC@.005 | SF | .42 | .16 | .58 | .99 |
| R-19, 23" x 39' x 6-1/4" thick | BC@.005 | SF | .42 | .16 | .58 | .99 |
| R-25, 23" x 17' x 8" thick | BC@.005 | SF | .59 | .16 | .75 | 1.28 |
| R-25, 15" x 18' x 8" thick | BC@.005 | SF | .63 | .16 | .79 | 1.34 |
| R-25, 15" x 26' x 8" thick | BC@.005 | SF | .66 | .16 | .82 | 1.39 |
| R-25, 23" x 26' x 8" thick | BC@.005 | SF | .64 | .16 | .80 | 1.36 |
| Add for rolls in floor joists above a crawl space | BC@.008 | SF | .04 | .26 | .30 | .51 |

**Kraft-faced fiberglass batt insulation.** Stapled between studs and joists. 16" widths and 24" widths are laid over ceiling joists.

| | Craft@Hrs | Unit | Material | Labor | Total | Sell |
|---|---|---|---|---|---|---|
| R-11, 15" x 93" x 3" thick | BC@.012 | SF | .31 | .39 | .70 | 1.19 |
| R-13, 23" x 89" x 3-1/2" thick | BC@.013 | SF | .40 | .42 | .82 | 1.39 |
| R-13, 15" x 93" x 3-1/2" thick | BC@.012 | SF | .40 | .39 | .79 | 1.34 |
| R-13, 23" x 80" x 3-1/2" thick | BC@.013 | SF | .41 | .42 | .83 | 1.41 |
| R-15, 15" x 93" x 4" thick | BC@.012 | SF | .77 | .39 | 1.16 | 1.97 |
| R-19, 15" x 93" x 6-1/4" thick | BC@.009 | SF | .48 | .29 | .77 | 1.31 |
| R-19, 23" x 93" x 6-1/4" thick | BC@.013 | SF | .48 | .42 | .90 | 1.53 |
| R-30, 16" x 48" x 10" thick | BC@.005 | SF | .79 | .16 | .95 | 1.62 |
| R-30, 24" x 48" x 10" thick | BC@.005 | SF | .79 | .16 | .95 | 1.62 |
| R-38, 15" x 48" x 10-1/4" high performance | BC@.009 | SF | 1.46 | .29 | 1.75 | 2.98 |
| R-38, 16" x 12-1/2" thick | BC@.007 | SF | .78 | .22 | 1.00 | 1.70 |
| R-38, 24" x 12-1/2" thick | BC@.005 | SF | .93 | .16 | 1.09 | 1.85 |

**Unfaced fiberglass batt insulation.** Fit between studs and joists. 16" and 24" widths are laid over ceiling joists. Coverage includes framing area.

| | Craft@Hrs | Unit | Material | Labor | Total | Sell |
|---|---|---|---|---|---|---|
| R-11, 15" x 94" x 3-1/2" thick | BC@.007 | SF | .29 | .22 | .51 | .87 |
| R-11, 16" x 96" x 3-1/2" thick | BC@.006 | SF | .29 | .19 | .48 | .82 |
| R-13, 15" x 93" x 3-1/2" thick | BC@.007 | SF | .39 | .22 | .61 | 1.04 |
| R-19, 15" x 93" x 6-1/4" thick | BC@.007 | SF | .48 | .22 | .70 | 1.19 |
| R-19, 23" x 93" x 6-1/4" thick | BC@.006 | SF | .48 | .19 | .67 | 1.14 |
| R-21, 15" x 93" x 5-1/2" thick | BC@.007 | SF | .62 | .22 | .84 | 1.43 |
| R-21, 23" x 93" x 5-1/2" thick | BC@.006 | SF | .63 | .19 | .82 | 1.39 |
| R-30, 16" x 48" x 9-1/2" thick | BC@.005 | SF | .78 | .16 | .94 | 1.60 |
| R-30, 16" x 48" x 10-1/4" thick | BC@.005 | SF | .83 | .16 | .99 | 1.68 |
| R-30, 24" x 48" x 9-1/2" thick | BC@.005 | SF | .78 | .16 | .94 | 1.60 |
| R-30, 24" x 48" x 10-1/4" thick | BC@.005 | SF | .85 | .16 | 1.01 | 1.72 |
| R-38, 16" x 48" x 13" thick | BC@.007 | SF | .98 | .22 | 1.20 | 2.04 |
| R-38, 24" x 48" x 13" thick | BC@.007 | SF | .98 | .22 | 1.20 | 2.04 |
| Add for batts in floor joists above a crawl space | BC@.008 | SF | .04 | .26 | .30 | .51 |

| | Craft@Hrs | Unit | Material | Labor | Total | Sell |
|---|---|---|---|---|---|---|

**Encapsulated roll fiberglass insulation.** Stapled between studs and joists. Coverage includes framing area. Fibers are coated to reduce irritation.

| | Craft@Hrs | Unit | Material | Labor | Total | Sell |
|---|---|---|---|---|---|---|
| R-13, 15" x 37'6" x 3-1/2" thick | BC@.010 | SF | .32 | .32 | .64 | 1.09 |
| R-13, 15" x 32' x 3-1/2" thick | BC@.010 | SF | .32 | .32 | .64 | 1.09 |
| R-25, 15" x 22' x 8-1/4" thick | BC@.010 | SF | .78 | .32 | 1.10 | 1.87 |
| R-25, 23" x 22' x 8-1/4" thick | BC@.009 | SF | .79 | .29 | 1.08 | 1.84 |

**Encapsulated batt fiberglass insulation.** Fit between studs and joists. Coverage includes framing area. Fibers are coated to reduce irritation. 16" widths and 24" widths are laid over ceiling joists.

| | Craft@Hrs | Unit | Material | Labor | Total | Sell |
|---|---|---|---|---|---|---|
| R-13, 15" x 93" x 3-1/2" thick | BC@.007 | SF | .46 | .22 | .68 | 1.16 |
| R-19, 15" x 93" x 6-1/2" thick | BC@.007 | SF | .56 | .22 | .78 | 1.33 |
| R-19, 23" x 8' x 6-1/2" thick | BC@.005 | SF | .56 | .16 | .72 | 1.22 |
| R-30, 16" x 48" x 10-1/4" thick | BC@.005 | SF | .82 | .16 | .98 | 1.67 |
| R-30, 24" x 48" x 10-1/4" thick | BC@.005 | SF | .84 | .16 | 1.00 | 1.70 |

**Blow-in cellulose insulation.** One bag covers 40 square feet at 6" deep and provides an R-19 rating.

| | Craft@Hrs | Unit | Material | Labor | Total | Sell |
|---|---|---|---|---|---|---|
| 40 square foot bag of stabilized cellulose | BC@.400 | Ea | 7.04 | 12.80 | 19.84 | 33.70 |
| Blower and hose, daily rental | — | Ea | 27.00 | — | 27.00 | — |
| Hose connector | — | Ea | 6.50 | — | 6.50 | — |
| 2" to 1" hose reducer and nozzle | — | Ea | 36.60 | — | 36.60 | — |
| 1" wood plugs, bag of 25 | BC@.500 | Ea | 8.07 | 16.00 | 24.07 | 40.90 |
| 1" plastic plugs, bag of 25 | BC@.500 | Ea | 7.30 | 16.00 | 23.30 | 39.60 |

## Panel Insulation

**Insulation board, 4' x 8' panels.** Owens Corning Foamular XAE rigid extruded polystyrene foam board. Film-faced, stapled in place, including taped joints and 5% waste. 4' x 8' or 9' panels, tongue and groove or square edge.

| | Craft@Hrs | Unit | Material | Labor | Total | Sell |
|---|---|---|---|---|---|---|
| 1/2" thick (R-2.5), $9.66 per panel | BC@.010 | SF | .30 | .32 | .62 | 1.05 |
| 3/4" thick (R-3.8), $11.97 per panel | BC@.011 | SF | .37 | .35 | .72 | 1.22 |
| 1" thick (R-5.0), $17.06 per panel | BC@.011 | SF | .53 | .35 | .88 | 1.50 |
| 1-1/2" thick (R-7.5), $17.67 per panel | BC@.015 | SF | .55 | .48 | 1.03 | 1.75 |
| 2" thick (R-10), $24.20 per panel | BC@.015 | SF | .76 | .48 | 1.24 | 2.11 |

**Extruded polystyrene insulated sheathing.** Dow Blue Board, water resistant. Blue Board has a density of 1.6 pounds per cubic foot. Gray Board has a density of 1.3 pounds per cubic foot. Include 5% waste.

| | Craft@Hrs | Unit | Material | Labor | Total | Sell |
|---|---|---|---|---|---|---|
| 3/4" x 2' x 8', tongue & groove, R-3.8 | BC@.011 | SF | .37 | .35 | .72 | 1.22 |
| 1" x 2' x 8', R-5.0 | BC@.011 | SF | .54 | .35 | .89 | 1.51 |
| 1" x 2' x 8', tongue & groove, R-5.0 | BC@.011 | SF | .38 | .35 | .73 | 1.24 |
| 1" x 4' x 8', R-5.0 | BC@.011 | SF | .53 | .35 | .88 | 1.50 |
| 1" x 4' x 8', tongue & groove, R-5.0 | BC@.011 | SF | .44 | .35 | .79 | 1.34 |
| 1-1/2" x 2' x 8', R-7.5 | BC@.015 | SF | .79 | .48 | 1.27 | 2.16 |
| 1-1/2" x 2' x 8', tongue & groove, R-7.5 | BC@.015 | SF | .88 | .48 | 1.36 | 2.31 |
| 1-1/2" 4' x 8', square edge, R-7.5 | BC@.015 | SF | .76 | .48 | 1.24 | 2.11 |
| 2" x 2' x 8', R-10 | BC@.015 | SF | 1.06 | .48 | 1.54 | 2.62 |
| 2" x 4' x 8', R-10 | BC@.015 | SF | 1.08 | .48 | 1.56 | 2.65 |
| 1" x 2' x 8' gray (1.3 Lbs. per CF) | BC@.015 | SF | .50 | .48 | .98 | 1.67 |
| 1-1/2" x 2' x 8' gray board | BC@.015 | SF | .70 | .48 | 1.18 | 2.01 |
| 2" x 2' x 8' gray board | BC@.015 | SF | .94 | .48 | 1.42 | 2.41 |

**Rigid foam insulated sheathing.** Fiberglass polymer faced two sides, Celotex Tuff-R, 4' x 8' panels, including 5% waste.

| | Craft@Hrs | Unit | Material | Labor | Total | Sell |
|---|---|---|---|---|---|---|
| 1/2" (R-3.2) | BC@.010 | SF | .27 | .32 | .59 | 1.00 |
| 3/4" (R-5.6) | BC@.011 | SF | .36 | .35 | .71 | 1.21 |
| 1" (R-7.2) | BC@.011 | SF | .50 | .35 | .85 | 1.45 |
| 2" (R-14.4) | BC@.015 | SF | .92 | .48 | 1.40 | 2.38 |

| | Craft@Hrs | Unit | Material | Labor | Total | Sell |
|---|---|---|---|---|---|---|
| **Fanfold extruded polystyrene insulation.** 4' x 50' panels, including 5% waste. | | | | | | |
| 1/4" (R-1) | BC@.011 | SF | 2.13 | .35 | 2.48 | 4.22 |
| 3/8" (R-1) | BC@.011 | SF | 2.41 | .35 | 2.76 | 4.69 |
| Add for nails, per 50-pound carton (large square heads) | | | | | | |
| 2-1/2", for 1-1/2" boards | — | LS | 116.00 | — | 116.00 | — |
| 3", for 2" boards | — | LS | 116.00 | — | 116.00 | — |
| **Foil-faced urethane sheathing.** Including 5% waste. 4' x 8' panels. Including 5% waste. | | | | | | |
| 1" (R-7.2) | BC@.011 | SF | .58 | .35 | .93 | 1.58 |
| **Polystyrene foam underlay.** Polyethylene covered, R-Gard. Including 5% waste. | | | | | | |
| 3/8" 4' x 24' fanfold | BC@.011 | SF | .25 | .35 | .60 | 1.02 |
| 1/2" x 4' x 8' | BC@.011 | SF | .29 | .35 | .64 | 1.09 |
| 1" x 4' x 8' | BC@.011 | SF | .41 | .35 | .76 | 1.29 |
| 1-1/2" x 2' x 8' | BC@.015 | SF | .78 | .48 | 1.26 | 2.14 |
| 2" x 2' x 8' | BC@.015 | SF | .82 | .48 | 1.30 | 2.21 |
| **Perlite roof insulation.** 24" x 48" board. | | | | | | |
| 3/4" thick (R-2.08) | RR@.007 | SF | .34 | .25 | .59 | 1.00 |
| 1" thick (R-2.78) | RR@.007 | SF | .45 | .25 | .70 | 1.19 |
| 1-1/2" thick (R-4.17) | RR@.007 | SF | .62 | .25 | .87 | 1.48 |
| 2" thick (R-5.26) | RR@.007 | SF | .76 | .25 | 1.01 | 1.72 |
| **Cant strips for insulated roof board.** One piece wood, tapered. | | | | | | |
| 2" strip | RR@.018 | LF | .16 | .64 | .80 | 1.36 |
| 3" strip | RR@.020 | LF | .15 | .72 | .87 | 1.48 |
| 4" strip | RR@.018 | LF | .14 | .64 | .78 | 1.33 |
| **Sill plate gasket.** Fills gaps between the foundation and the plate. Also used for sealing around windows and doors. | | | | | | |
| 1/4" x 50', 3-1/2" wide | BC@.003 | LF | .09 | .10 | .19 | .32 |
| 1/4" x 50', 5-1/2" wide | BC@.003 | LF | .12 | .10 | .22 | .37 |
| **Perimeter foundation insulation.** Expanded polystyrene with polyethylene skinned surface. Low moisture retention. Minimum compressive strength of 1440 pounds per SF. | | | | | | |
| 3/4" x 4' x 8' | BC@.011 | SF | .27 | .35 | .62 | 1.05 |
| 1-1/2" x 2' x 4' | BC@.015 | SF | .59 | .48 | 1.07 | 1.82 |
| 2" x 2' x 4' | BC@.015 | SF | .78 | .48 | 1.26 | 2.14 |
| 2" x 4' x 4' | BC@.015 | SF | .64 | .48 | 1.12 | 1.90 |
| **FBX 1240 industrial board insulation.** Rockwool, semi-rigid, no waste included. 24" on-center framing members. | | | | | | |
| 2" (R-8) | BC@.004 | SF | .47 | .13 | .60 | 1.02 |
| 3" (R-12) | BC@.004 | SF | .69 | .13 | .82 | 1.39 |
| 3-1/2" (R-14) | BC@.004 | SF | .80 | .13 | .93 | 1.58 |
| **Sound control insulation.** Fiberglass acoustically designed to absorb sound vibration. For use between interior walls, floors and ceilings. Quietzone™ batts. Coverage allows for framing. | | | | | | |
| 3-1/2", 16" on-center framing | BC@.006 | SF | .31 | .19 | .50 | .85 |
| 3-1/2", 24" on-center framing | BC@.006 | SF | .32 | .19 | .51 | .87 |
| **Sound insulation board.** 4' x 8' panels installed on walls. | | | | | | |
| 1/2" Homasote | BC@.013 | SF | .68 | .42 | 1.10 | 1.87 |
| 1/2" sound deadening board | BC@.013 | SF | .28 | .42 | .70 | 1.19 |
| Add for installation on ceilings | BC@.004 | SF | — | .13 | .13 | .22 |

| | Craft@Hrs | Unit | Material | Labor | Total | Sell |
|---|---|---|---|---|---|---|

**Sound Attenuation Fire Batt Insulation** (SAFB). Rockwool, semi-rigid, no waste included, pressed between framing members.

16" on-center framing members

| | Craft@Hrs | Unit | Material | Labor | Total | Sell |
|---|---|---|---|---|---|---|
| 2" (R-8) | BC@.004 | SF | .39 | .13 | .52 | .88 |
| 3" (R-12) | BC@.004 | SF | .57 | .13 | .70 | 1.19 |
| 4" (R-16) | BC@.005 | SF | .76 | .16 | .92 | 1.56 |

24" on-center framing members

| | Craft@Hrs | Unit | Material | Labor | Total | Sell |
|---|---|---|---|---|---|---|
| 2" (R-8) | BC@.003 | SF | .39 | .10 | .49 | .83 |
| 3" (R-12) | BC@.003 | SF | .58 | .10 | .68 | 1.16 |
| 4" (R-16) | BC@.004 | SF | .76 | .13 | .89 | 1.51 |

## Loose Insulation

**Vermiculite insulation.** Poured over ceilings.

| | Craft@Hrs | Unit | Material | Labor | Total | Sell |
|---|---|---|---|---|---|---|
| Vermiculite, 25 lb sack (3 CF) | — | Ea | 32.00 | — | 32.00 | — |
| At 3" depth (60 sacks per 1,000 SF) | BL@.007 | SF | 2.00 | .19 | 2.19 | 3.72 |
| At 4" depth (72 sacks per 1,000 SF) | BL@.007 | SF | 2.67 | .19 | 2.86 | 4.86 |

**Masonry fill insulation.** Poured in concrete block cores.

| | Craft@Hrs | Unit | Material | Labor | Total | Sell |
|---|---|---|---|---|---|---|
| Using 3 CF bags of perlite, per bag | — | Ea | 29.00 | — | 29.00 | — |
| 4" wall, 8.1 SF per CF | B9@.006 | SF | .90 | .17 | 1.07 | 1.82 |
| 6" wall, 5.4 SF per CF | B9@.006 | SF | 1.34 | .17 | 1.51 | 2.57 |
| 8" wall, 3.6 SF per CF | B9@.006 | SF | 2.01 | .17 | 2.18 | 3.71 |

**Natural fiber building insulation.** EPA registered anti-microbial agent that offers protection from mold, mildew, fungi and pests. Made from all natural cotton fibers. Friction fit between framing 16" or 24" on center. Based on Ultra Touch. http://www.bondedlogic.com/

| | Craft@Hrs | Unit | Material | Labor | Total | Sell |
|---|---|---|---|---|---|---|
| 16" x 94", R-13 | BC@.005 | SF | .58 | .16 | .74 | 1.26 |
| 24" x 94", R-13 | BC@.006 | SF | .58 | .19 | .77 | 1.31 |
| 16" x 94", R-19 | BC@.005 | SF | .86 | .16 | 1.02 | 1.73 |
| 24" x 94", R-19 | BC@.006 | SF | .86 | .19 | 1.05 | 1.79 |

## Vapor Barrier

**Polyethylene film.** Clear or black. Includes 5% for waste and 10% for laps.

4 mil (.004" thick)

| | Craft@Hrs | Unit | Material | Labor | Total | Sell |
|---|---|---|---|---|---|---|
| 50 LF rolls, 3' to 20' wide | BL@.003 | SF | .06 | .08 | .14 | .24 |

6 mil (.006" thick)

| | Craft@Hrs | Unit | Material | Labor | Total | Sell |
|---|---|---|---|---|---|---|
| 50 LF rolls, 3' to 40' wide | BL@.003 | SF | .07 | .08 | .15 | .26 |

**Vapor barrier paper.** Including 12% for overlap and waste.

| | Craft@Hrs | Unit | Material | Labor | Total | Sell |
|---|---|---|---|---|---|---|
| 15 lb. asphalt felt, 432 SF roll, 36" x 144' | BC@.003 | SF | .05 | .10 | .15 | .26 |
| 30 lb. asphalt felt, 216 SF roll, 36" x 72' | BC@.003 | SF | .07 | .10 | .17 | .29 |
| SuperJumbo Tex black building paper, asphalt saturated, (162 SF roll | BC@.003 | SF | .09 | .10 | .19 | .32 |
| Red rosin sized sheathing paper (duplex sheathing) 36" wide, (500 SF roll) | BC@.003 | SF | .02 | .10 | .12 | .20 |
| Double Kraft Aquabar, class A, 36" wide, 30-50-30, (500 SF roll) | BC@.003 | SF | .06 | .10 | .16 | .27 |

| | Craft@Hrs | Unit | Material | Labor | Total | Sell |
|---|---|---|---|---|---|---|
| Double Kraft Aquabar, class B, 36" wide, 30-30-30,<br>  (500 SF roll) | BC@.003 | SF | .05 | .10 | .15 | .26 |
| Vaporstop 298, fiberglass reinforcing and asphaltic adhesive between<br>  2 layers of Kraft,<br>  32" x 405' roll (1,080 SF roll) | BC@.003 | SF | .08 | .10 | .18 | .31 |
| Pyro-Kure 600, 2 layers of heavy Kraft with fire-retardant adhesive<br>  edge reinforced with fiberglass,<br>  32" x 405' roll (1,080 SF roll) | BC@.003 | SF | .16 | .10 | .26 | .44 |
| Foil Barrier 718, fiberglass-reinforced aluminum foil and<br>  adhesive between 2 layers of Kraft,<br>  52" x 231' roll (1,000 SF roll) | BC@.004 | SF | .27 | .13 | .40 | .68 |
| Seekure, fiberglass reinforcing strands and non-staining adhesive<br>  between 2 layers of Kraft,<br>  48" x 300' roll, (1,200 SF roll) | BC@.003 | SF | .07 | .10 | .17 | .29 |
| Moistop, fiberglass-reinforced Kraft between 2 layers of polyethylene,<br>  8' x 250' roll (2,000 SF roll) | BC@.003 | SF | .08 | .10 | .18 | .31 |

**Ice and water shield.** Self-adhesive rubberized asphalt and poly. Vycor.

| | Craft@Hrs | Unit | Material | Labor | Total | Sell |
|---|---|---|---|---|---|---|
| 225 SF roll at $101.00 | BC@.006 | SF | .48 | .19 | .67 | 1.14 |

**Roof flashing paper.** Seals around skylights, dormers, vents, valleys and eaves. Rubberized, fiberglass reinforced, self-adhesive. GAF StormGuard.

| | Craft@Hrs | Unit | Material | Labor | Total | Sell |
|---|---|---|---|---|---|---|
| Roll covers 200 SF at $68.40 | RR@.006 | SF | .37 | .21 | .58 | .99 |

**Weatherproof roof underlay.** For use under shingles. Granular surface.

| | Craft@Hrs | Unit | Material | Labor | Total | Sell |
|---|---|---|---|---|---|---|
| Roll covers 400 SF at $59.00 | RR@.006 | SF | .16 | .21 | .37 | .63 |

**Tyvek™ House Wrap.** Air infiltration barrier. High-density polyethylene fiber sheet.

| | Craft@Hrs | Unit | Material | Labor | Total | Sell |
|---|---|---|---|---|---|---|
| 3' x 165' rolls or 9' x 100' rolls | BC@.003 | SF | .13 | .10 | .23 | .39 |
| House Wrap tape, 2" x 165' | BC@.005 | LF | .07 | .16 | .23 | .39 |

## Vents and Louvers

**Rafter Bay Vent Channel.** Keeps attic insulation away from soffit vents to promote good air circulation. Extruded polystyrene. 48" x 1-1/4".

| | Craft@Hrs | Unit | Material | Labor | Total | Sell |
|---|---|---|---|---|---|---|
| Rafters 16" on center,<br>  14" wide, 15" net-free vent area | BC@.220 | Ea | 1.92 | 7.06 | 8.98 | 15.30 |
| Rafters 24" on center,<br>  22" wide, 26" net-free vent area | BC@.220 | Ea | 2.45 | 7.06 | 9.51 | 16.20 |

**Replacement automatic foundation vent.** Fits 8" x 16" concrete block opening. Aluminum mesh on front and molded plastic screen on back. Opens at 72 degrees F and closes at and 38 degrees F. 1" frame each side.

| | Craft@Hrs | Unit | Material | Labor | Total | Sell |
|---|---|---|---|---|---|---|
| Automatic open and close | SW@.255 | Ea | 14.30 | 9.09 | 23.39 | 39.80 |

**Foundation vent with manual damper.** Sheet metal.

| | Craft@Hrs | Unit | Material | Labor | Total | Sell |
|---|---|---|---|---|---|---|
| 4" x 16" | SW@.255 | Ea | 5.70 | 9.09 | 14.79 | 25.10 |
| 6" x 16" | SW@.255 | Ea | 6.24 | 9.09 | 15.33 | 26.10 |
| 8" x 16" | SW@.255 | Ea | 6.55 | 9.09 | 15.64 | 26.60 |

**Foundation access door.** Replacement cost only.

| | Craft@Hrs | Unit | Material | Labor | Total | Sell |
|---|---|---|---|---|---|---|
| 24" x 24" | BC@.363 | Ea | 25.90 | 11.60 | 37.50 | 63.80 |
| 32" x 24" | BC@.363 | Ea | 32.40 | 11.60 | 44.00 | 74.80 |

| | Craft@Hrs | Unit | Material | Labor | Total | Sell |
|---|---|---|---|---|---|---|
| **Foundation screen vent.** | | | | | | |
| 14" x 6" | BC@.255 | Ea | 2.21 | 8.18 | 10.39 | 17.70 |
| 16" x 4" | BC@.255 | Ea | 2.24 | 8.18 | 10.42 | 17.70 |
| 16" x 6" | BC@.255 | Ea | 2.69 | 8.18 | 10.87 | 18.50 |
| 16" x 8" | BC@.255 | Ea | 2.91 | 8.18 | 11.09 | 18.90 |
| **Under eave vent.** Decomesh vent strips. 26 square inches free air per linear foot. 3/32" diameter perforations. 0.016-gauge aluminum. | | | | | | |
| 4-1/2" x 8', white | BC@.040 | LF | .93 | 1.28 | 2.21 | 3.76 |
| 4-1/2" x 8', brown | BC@.040 | LF | 1.00 | 1.28 | 2.28 | 3.88 |
| 6" x 8', white | BC@.040 | LF | 1.50 | 1.28 | 2.78 | 4.73 |
| **Continuous louvered soffit vent.** Reversible for flush or recessed mounting. | | | | | | |
| 2-5/8" x 8', mill finish aluminum | BC@.360 | Ea | 4.76 | 11.60 | 16.36 | 27.80 |
| 2-5/8" x 8', white finish aluminum | BC@.360 | Ea | 4.70 | 11.60 | 16.30 | 27.70 |
| **Wood soffit vent.** | | | | | | |
| 8" x 16" | BC@.220 | Ea | 1.40 | 7.06 | 8.46 | 14.40 |
| **Louvered soffit vent.** Finned louvers with galvanized steel frame and wire mesh. Replacement cost only. For new installations, add the cost of cutting a hole in an existing soffit. | | | | | | |
| 14" x 4" | BC@.220 | Ea | 2.54 | 7.06 | 9.60 | 16.30 |
| 14" x 5" | BC@.220 | Ea | 2.99 | 7.06 | 10.05 | 17.10 |
| 14" x 6" | BC@.220 | Ea | 3.14 | 7.06 | 10.20 | 17.30 |
| 14" x 8" | BC@.220 | Ea | 2.97 | 7.06 | 10.03 | 17.10 |
| 22" x 3" | BC@.220 | Ea | 3.02 | 7.06 | 10.08 | 17.10 |
| **Plastic gable louver vent.** With double baffle and screen. Replacement cost only. For new installations, add the cost of cutting a hole in an existing gable. | | | | | | |
| Half round, 16" x 24", white | BC@.500 | Ea | 40.50 | 16.00 | 56.50 | 96.10 |
| Half round, 22" x 34", paintable | BC@.500 | Ea | 56.00 | 16.00 | 72.00 | 122.00 |
| Half round, 22" x 34", white | BC@.500 | Ea | 43.60 | 16.00 | 59.60 | 101.00 |
| Octagon, 18", #01 white | BC@.500 | Ea | 32.40 | 16.00 | 48.40 | 82.30 |
| Octagon, 22", paintable | BC@.500 | Ea | 42.00 | 16.00 | 58.00 | 98.60 |
| Octagon, 22", white | BC@.500 | Ea | 41.40 | 16.00 | 57.40 | 97.60 |
| Rectangular, 12" x 18", ivory | BC@.500 | Ea | 29.70 | 16.00 | 45.70 | 77.70 |
| Rectangular, 12" x 18", paintable | BC@.500 | Ea | 29.70 | 16.00 | 45.70 | 77.70 |
| Rectangular, 18" x 24", paintable | BC@.500 | Ea | 32.40 | 16.00 | 48.40 | 82.30 |
| Rectangular, 12" x 18", white | BC@.500 | Ea | 22.40 | 16.00 | 38.40 | 65.30 |
| Rectangular, 18" x 24", white | BC@.500 | Ea | 32.40 | 16.00 | 48.40 | 82.30 |
| Replace-A-Vent, 12" x 18", white | BC@.500 | Ea | 32.40 | 16.00 | 48.40 | 82.30 |
| Round, 22", paintable | BC@.500 | Ea | 38.80 | 16.00 | 54.80 | 93.20 |
| Round, 22", white | BC@.500 | Ea | 41.90 | 16.00 | 57.90 | 98.40 |
| Square, 12" x 12", white | BC@.500 | Ea | 20.30 | 16.00 | 36.30 | 61.70 |

| | Craft@Hrs | Unit | Material | Labor | Total | Sell |
|---|---|---|---|---|---|---|

**Metal gable louver vent.** Replacement cost only. For new installations, add the cost of cutting a hole in an existing gable.

| | Craft@Hrs | Unit | Material | Labor | Total | Sell |
|---|---|---|---|---|---|---|
| 12" x 12", brown | BC@.500 | Ea | 14.70 | 16.00 | 30.70 | 52.20 |
| 12" x 12", galvanized | BC@.500 | Ea | 12.20 | 16.00 | 28.20 | 47.90 |
| 12" x 12", white | BC@.500 | Ea | 14.60 | 16.00 | 30.60 | 52.00 |
| 12" x 18", brown | BC@.500 | Ea | 18.70 | 16.00 | 34.70 | 59.00 |
| 12" x 18", galvanized | BC@.500 | Ea | 16.30 | 16.00 | 32.30 | 54.90 |
| 12" x 18", white | BC@.500 | Ea | 17.80 | 16.00 | 33.80 | 57.50 |
| 14" x 12", white | BC@.500 | Ea | 12.90 | 16.00 | 28.90 | 49.10 |
| 14" x 18", brown | BC@.500 | Ea | 18.90 | 16.00 | 34.90 | 59.30 |
| 14" x 18", galvanized | BC@.500 | Ea | 17.80 | 16.00 | 33.80 | 57.50 |
| 14" x 18", white | BC@.500 | Ea | 18.90 | 16.00 | 34.90 | 59.30 |
| 14" x 24", brown | BC@.500 | Ea | 23.30 | 16.00 | 39.30 | 66.80 |
| 14" x 24", galvanized | BC@.500 | Ea | 20.20 | 16.00 | 36.20 | 61.50 |
| 14" x 24", white | BC@.500 | Ea | 22.10 | 16.00 | 38.10 | 64.80 |

**Redwood gable vent.** Replacement cost only. For new installations, add the cost of cutting a hole in an existing gable.

| | Craft@Hrs | Unit | Material | Labor | Total | Sell |
|---|---|---|---|---|---|---|
| 14" x 18" rectangle | BC@.500 | Ea | 25.90 | 16.00 | 41.90 | 71.20 |
| 14" x 24" rectangle | BC@.500 | Ea | 30.20 | 16.00 | 46.20 | 78.50 |
| 18" x 24" rectangle | BC@.500 | Ea | 47.50 | 16.00 | 63.50 | 108.00 |
| 24" octagon | BC@.500 | Ea | 51.80 | 16.00 | 67.80 | 115.00 |
| 24" full round | BC@.500 | Ea | 69.10 | 16.00 | 85.10 | 145.00 |

**Cedar louver vent.** Replacement cost only.

| | Craft@Hrs | Unit | Material | Labor | Total | Sell |
|---|---|---|---|---|---|---|
| Arch top, 16" x 24" | BC@.500 | Ea | 48.60 | 16.00 | 64.60 | 110.00 |
| Octagon, 18" | BC@.500 | Ea | 48.60 | 16.00 | 64.60 | 110.00 |
| Round, 18" | BC@.500 | Ea | 48.60 | 16.00 | 64.60 | 110.00 |
| Square, 16" x 24" | BC@.500 | Ea | 36.70 | 16.00 | 52.70 | 89.60 |

**Dormer louver vent.** With galvanized wire mesh to keep birds out. Replacement cost only.

| | Craft@Hrs | Unit | Material | Labor | Total | Sell |
|---|---|---|---|---|---|---|
| 19" x 3", low profile | BC@.500 | Ea | 22.40 | 16.00 | 38.40 | 65.30 |

**Galvanized under eave vent.**

| | Craft@Hrs | Unit | Material | Labor | Total | Sell |
|---|---|---|---|---|---|---|
| 14" x 5" | BC@.400 | Ea | 2.56 | 12.80 | 15.36 | 26.10 |
| 14" x 6" | BC@.400 | Ea | 3.79 | 12.80 | 16.59 | 28.20 |
| 22" x 3" | BC@.400 | Ea | 3.56 | 12.80 | 16.36 | 27.80 |
| 22" x 6" | BC@.400 | Ea | 3.70 | 12.80 | 16.50 | 28.10 |
| 22" x 7" | BC@.400 | Ea | 4.18 | 12.80 | 16.98 | 28.90 |

**Aluminum ridge vent.**

| | Craft@Hrs | Unit | Material | Labor | Total | Sell |
|---|---|---|---|---|---|---|
| 10' long, black | BC@.850 | Ea | 13.60 | 27.30 | 40.90 | 69.50 |
| 10' long, brown | BC@.850 | Ea | 13.90 | 27.30 | 41.20 | 70.00 |
| 10' long, white | BC@.850 | Ea | 13.70 | 27.30 | 41.00 | 69.70 |
| Joint strap | BC@.010 | Ea | 1.22 | .32 | 1.54 | 2.62 |
| End connector plug | BC@.010 | Ea | 1.94 | .32 | 2.26 | 3.84 |
| 4' long, hinged for steep roof | BC@.355 | Ea | 10.80 | 11.40 | 22.20 | 37.70 |

**Ridge vent coil.**

| | Craft@Hrs | Unit | Material | Labor | Total | Sell |
|---|---|---|---|---|---|---|
| 10.5" wide, 20' coil | BC@1.00 | Ea | 50.80 | 32.10 | 82.90 | 141.00 |
| 10.5" wide, 50' coil | BC@2.50 | Ea | 104.00 | 80.20 | 184.20 | 313.00 |

| | Craft@Hrs | Unit | Material | Labor | Total | Sell |
|---|---|---|---|---|---|---|
| **Slant back aluminum roof louver.** 60-square inch net-free venting area. | | | | | | |
| 18" x 20', black | BC@1.00 | Ea | 10.90 | 32.10 | 43.00 | 73.10 |
| 18" x 20', brown | BC@1.00 | Ea | 10.90 | 32.10 | 43.00 | 73.10 |
| 18" x 20', mill | BC@1.00 | Ea | 9.62 | 32.10 | 41.72 | 70.90 |
| 18" x 20', shingle match weathered wood | BC@1.00 | Ea | 11.60 | 32.10 | 43.70 | 74.30 |
| **Aluminum wall louver vent.** | | | | | | |
| 12" x 12" | BC@.450 | Ea | 8.02 | 14.40 | 22.42 | 38.10 |
| 12" x 18" | BC@.450 | Ea | 10.70 | 14.40 | 25.10 | 42.70 |
| 14" x 24" | BC@.450 | Ea | 13.50 | 14.40 | 27.90 | 47.40 |
| 18" x 24" | BC@.450 | Ea | 15.70 | 14.40 | 30.10 | 51.20 |
| **Galvanized wall louver vent.** | | | | | | |
| 12" x 12" | BC@.450 | Ea | 7.09 | 14.40 | 21.49 | 36.50 |
| 12" x 18" | BC@.450 | Ea | 8.95 | 14.40 | 23.35 | 39.70 |
| 14" x 24" | BC@.450 | Ea | 11.90 | 14.40 | 26.30 | 44.70 |
| **Plastic wall louver vent.** | | | | | | |
| 8" x 8" | BC@.450 | Ea | 4.59 | 14.40 | 18.99 | 32.30 |
| 12" x 12" | BC@.450 | Ea | 5.39 | 14.40 | 19.79 | 33.60 |
| 12" x 18" | BC@.450 | Ea | 7.50 | 14.40 | 21.90 | 37.20 |
| 18" x 24" | BC@.450 | Ea | 10.80 | 14.40 | 25.20 | 42.80 |
| **Aluminum midget louver.** For venting eaves and soffits. | | | | | | |
| 2", vent area 1-3/16" | BC@.440 | Ea | 1.74 | 14.10 | 15.84 | 26.90 |
| 3", vent area 1-3/4" | BC@.440 | Ea | 2.41 | 14.10 | 16.51 | 28.10 |
| 4", vent area 3-1/2" | BC@.440 | Ea | 2.98 | 14.10 | 17.08 | 29.00 |
| **Combination attic vent.** | | | | | | |
| 12" x 12", brown | BC@.450 | Ea | 8.63 | 14.40 | 23.03 | 39.20 |
| 12" x 12", galvanized | BC@.450 | Ea | 7.00 | 14.40 | 21.40 | 36.40 |
| 12" x 12", white | BC@.450 | Ea | 8.63 | 14.40 | 23.03 | 39.20 |
| 12" x 18", brown | BC@.450 | Ea | 9.68 | 14.40 | 24.08 | 40.90 |
| 12" x 18", white | BC@.450 | Ea | 9.69 | 14.40 | 24.09 | 41.00 |
| 14" x 24", brown | BC@.450 | Ea | 10.80 | 14.40 | 25.20 | 42.80 |
| 14" x 24", galvanized | BC@.450 | Ea | 9.69 | 14.40 | 24.09 | 41.00 |
| 14" x 24", white | BC@.450 | Ea | 10.70 | 14.40 | 25.10 | 42.70 |

## Ventilators

| | Craft@Hrs | Unit | Material | Labor | Total | Sell |
|---|---|---|---|---|---|---|
| **Roof turbine vent with base.** | | | | | | |
| 12", galvanized | SW@1.00 | Ea | 30.40 | 35.70 | 66.10 | 112.00 |
| 12", weathered wood | SW@1.00 | Ea | 33.80 | 35.70 | 69.50 | 118.00 |
| 12", white aluminum | SW@1.00 | Ea | 32.30 | 35.70 | 68.00 | 116.00 |
| Add for steep pitch base | SW@.440 | Ea | 10.20 | 15.70 | 25.90 | 44.00 |
| Add for weather cap | — | Ea | 9.20 | — | 9.20 | — |
| **Belt drive attic exhaust fan.** 1/3 HP, with aluminum shutter and plenum boards. | | | | | | |
| 30" diameter | SW@3.00 | Ea | 237.00 | 107.00 | 344.00 | 585.00 |
| 36" diameter | SW@3.00 | Ea | 230.00 | 107.00 | 337.00 | 573.00 |
| 12-hour timer switch | BE@.950 | Ea | 23.60 | 32.70 | 56.30 | 95.70 |

|  | Craft@Hrs | Unit | Material | Labor | Total | Sell |
|---|---|---|---|---|---|---|
| **Power attic gable vent.** | | | | | | |
| 1,280 CFM | SW@1.65 | Ea | 37.20 | 58.80 | 96.00 | 163.00 |
| 1,540 CFM | SW@1.65 | Ea | 58.20 | 58.80 | 117.00 | 199.00 |
| 1,600 CFM | SW@1.65 | Ea | 82.20 | 58.80 | 141.00 | 240.00 |
| Shutter for gable vent | SW@.450 | Ea | 31.20 | 16.00 | 47.20 | 80.20 |
| Humidistat control | BE@.550 | Ea | 25.00 | 18.90 | 43.90 | 74.60 |
| Thermostat control | BE@.550 | Ea | 18.30 | 18.90 | 37.20 | 63.20 |
| | | | | | | |
| **Roof-mount power ventilator.** | | | | | | |
| 1130 CFM | SW@1.65 | Ea | 42.70 | 58.80 | 101.50 | 173.00 |
| 1250 CFM | SW@1.65 | Ea | 71.90 | 58.80 | 130.70 | 222.00 |
| 1600 CFM | SW@1.65 | Ea | 95.20 | 58.80 | 154.00 | 262.00 |

# *Siding and Trim*

**6**

$M$ost older homes have wood siding, often referred to as *clapboard*. The term is dated now. But the functional equivalent is still around as *bevel*, *bungalow* or *lap* siding. All consist of tapered boards installed horizontally. The top edge is thinner than the bottom edge so moisture falling down the wall is diverted away from the joint between boards, keeping the wall interior dry. At least that was the theory. Figure 6-1 shows bevel (or clapboard) siding installed over lumber sheathing.

Unlike vinyl and aluminum siding or stucco, board siding needs to be painted about every ten years. But heavily deteriorated board siding, usually as a result of prolonged exposure to moisture, may not be worth painting. Look for signs of water coming from either the inside or outside of the building. Exterior moisture damage is probably caused by lack of a roof overhang or by siding in contact with damp concrete or soil. Damage from moisture originating inside the home is probably caused by an inadequate vapor barrier.

> Check siding for decay where two boards are butted together end to end, at corners, and around window and door openings. Look for gaps between horizontal siding boards by sighting along the wall.

## Repair or Replace?

Fill gaps and cracks in the siding with stainable latex wood filler, and re-nail warped boards. Simple repairs like these can go along way towards extending the life of board siding. When board siding is seriously damaged or decayed, it's fairly easy to replace a few boards. Slip a pry bar into the bevel and carefully pry up to loosen the nails. When the nails are loose, slip the damaged board down and off the wall. Then cut a replacement strip, leaving a $^1/_{16}$" gap at each end. If your local lumberyard doesn't stock exactly the pattern you need, a few minutes work with a tabletop router will yield boards indistinguishable from what's being replaced. Repaint, and your repair will be invisible.

Aluminum siding and mineral siding don't decay like wood siding. That's an advantage. However, both aluminum and mineral siding can be damaged fairly easily by impact, such as from a foul ball. Vinyl siding has nearly replaced aluminum siding in many areas. Finding an aluminum replacement section may be a problem — and matching mineral siding may be even more difficult. If only a small section of aluminum siding is damaged, cut it out and patch with sheet aluminum set in construction adhesive. Then repaint. GAF offers a fiber-cement lookalike product sold under the name *Purity* that you may be able to use to match mineral siding.

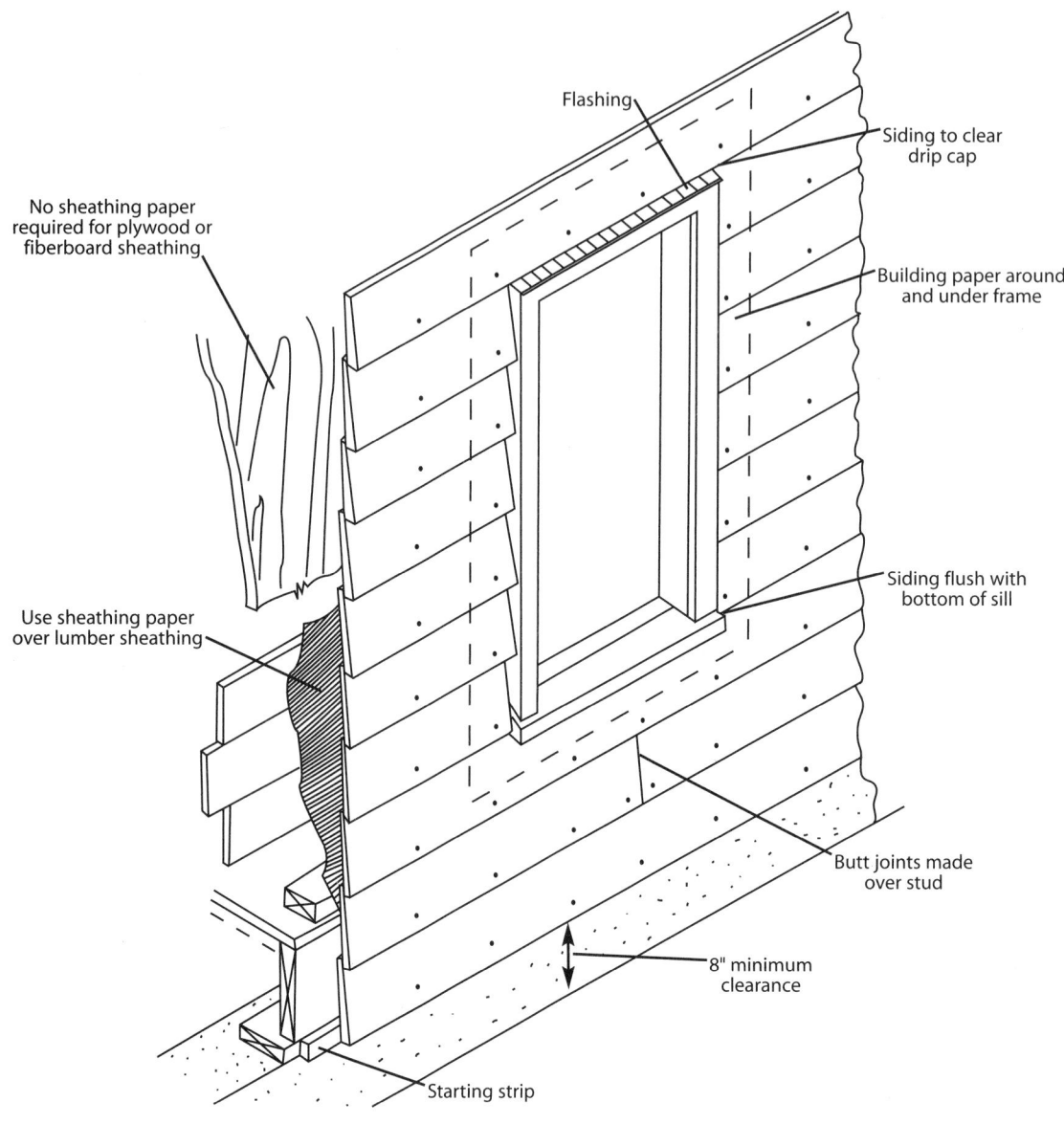

**Figure 6-1**

*Application of bevel siding to coincide with window sill and drip cap*

In a pinch, scavenge a few pieces from an inconspicuous location on the same building to make your repair. Then repair the inconspicuous area with a non-matching material. Both aluminum and mineral siding can be painted.

Incidentally, mineral siding was formerly sold as *asbestos-cement* siding. Some asbestos materials are known to pose a health risk and require special handling by qualified experts. But the asbestos in most cement-asbestos siding is non-friable. That means asbestos fibers aren't normally released into the air. If you're in doubt, have the siding checked for friable content. To find a testing laboratory, search for *asbestos testing* in your Web browser. If the friable content is about 1 percent, removing the siding presents little or no danger. If friable asbestos content is high, removal is a task for the experts. Even better, leave asbestos siding undisturbed or cover it with another type of siding. Asbestos siding left alone presents no danger, though the owner may have to make disclosure to any prospective buyer.

# New Siding Over Old?

If the building has only one layer of siding, and if that siding is sound (no decay problem), consider applying new siding over the old. This offers several advantages. First, you avoid the cost of demolition and disposal. Second, the extra layer of wall cover adds at least some insulation value. And third, installing siding over siding isn't much more effort than installing siding over sheathing.

In some cases, removing the old siding is the best choice. If wood siding has decayed to the point where it won't hold a nail, you could try using longer nails to attach it right into the sheathing, but it's probably a candidate for tear-off. Mineral siding fractures into pieces when nailed. It's easier to remove mineral siding than to try to nail through it. Also, if the home already has two layers of siding, it's probably best to do a tear-off.

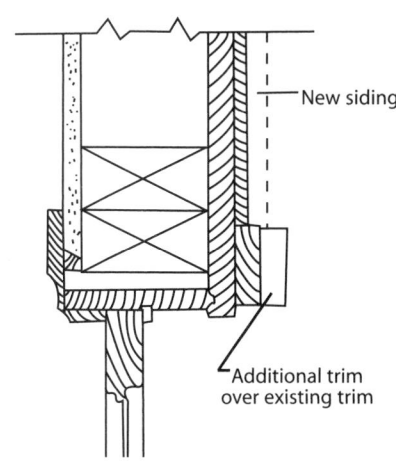

**Figure 6-2**

*Top view of window casing extended by adding trim over existing trim*

Trim around windows and doors needs special attention when applying new siding over old. Extra trim is needed to compensate for the added wall thickness. Window sills aren't a problem. Generally, sills extend well beyond the existing siding in older homes. Look back at Figure 6-1. But if the window casing is nearly flush with existing siding, some type of extension molding will be needed. Build up the casing with an extra thickness of trim, as shown in Figure 6-2. You'll probably need a wider drip cap at the top of the window. If the drip cap is in good condition and can be removed in one piece, extend the cap by fitting a block where the cap was. Then re-nail the cap over the block. Compare Figure 6-3 A (before) and Figure 6-3 B (after). Building out trim on windows and doors is usually easier when vertical siding is applied over horizontal siding.

Rather than build up the depth of window casing, you could add a wider trim piece at the edge of existing trim. See Figure 6-4. A wider drip cap will also be required. Extend exterior door trim in the same way.

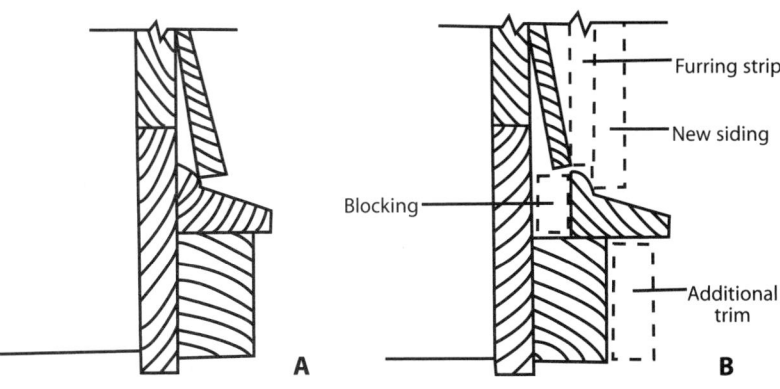

**Figure 6-3**

*Change in drip cap with new siding: **A**, existing drip cap and trim; **B**, drip cap blocked out to extend beyond new siding and added trim*

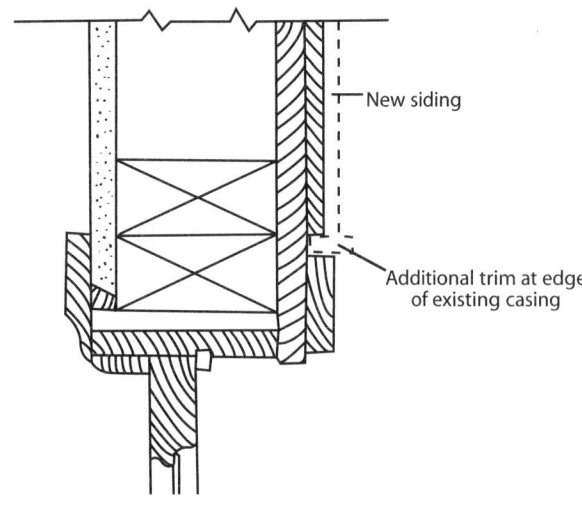

New siding

Additional trim at edge
of existing casing

**Figure 6-4**

*Top view of window casing extended for new siding by adding trim at the edge of existing casing*

# Vinyl Siding

Vinyl siding is by far the most popular siding in use today. It's relatively inexpensive, and comes in a wide variety of styles and colors. Vinyl siding is maintenance-free and resistant to most damage. It's also easy to install. With a little care, your first installation will look great. You can install vinyl siding directly over horizontal board siding if the wood surface is firm and flat. Or, even better, install fanfold polystyrene insulating board over the wood siding first. That provides extra insulation and assures a smooth, even surface over which to install the vinyl siding. Remove and replace any deteriorated boards. Re-nail loose boards. If the existing siding is irregular, install furring strips so the new siding is supported by a flush surface. Then re-caulk around doors and windows and renew the window flashing, if needed.

Follow the manufacturer's installation instructions. But note that all vinyl siding expands and contracts with heat and cold, as much as $1/2$" over the length of a 12'6" panel. If nails or screws are driven tight against the nailing slot, the panel will buckle noticeably in warm weather. Use aluminum, galvanized steel or other corrosion-resistant nails or screws and leave $1/32$" (the thickness of a dime) clearance between the head and the vinyl. Drive fasteners through the center of the nailing slot in the hem so horizontal siding can slide $1/2$" left and right after installation. Drive a fastener every 16" on horizontal panels, every 12" on vertical panels and every 8" to 10" on accessories such as J-channel and corner posts.

Adjacent vinyl panels lock together to form a snug joint. Where panel ends meet, overlap the ends by about 1". Avoid end overlap near doors or windows where regular use is likely to put stress on the joint.

All vinyl siding manufacturers offer trim pieces designed for use with their siding materials. These include starter strip, drip cap, J-channel, molding, corner posts and soffits. Anything projecting from the wall, such as a doorbell, a faucet or a porch light, should be surrounded by a vinyl *mounting block*. The block forms a collar that surrounds the siding penetration. Setting a mounting block is much easier and neater than cutting a hole in a full-length vinyl siding panel. Mounting blocks also offer better protection from the elements.

The trickiest part of any vinyl siding job is dealing with existing windows and doors. Most new windows are designed with vinyl siding in mind. They're contoured so that vinyl siding will trim out neatly around them. Older existing windows, however, are a serious problem. If you just trim around the window frame, you'll leave exposed wood, which will need to be painted. Most siding installers will cover the existing window frames with aluminum, which needs to be custom-bent on-site using an aluminum brake. This is very time consuming. Also,

Vinyl siding can be installed over stucco. But first install 1" x 3" furring secured to the studs. Then install the siding, driving either ring-shank nails or screws into the furring. Nails or screws driven into the stucco alone will not secure the siding.

the new siding will stand out further from the house than the existing siding, causing the windows to be slightly recessed. This alters the look of the house, which the homeowner may not like. In that case, you'll need to build out the window frames to restore the original depth.

If the homeowner is considering having new windows installed in his house, this would be an excellent time to do it. New windows will eliminate all these problems. You might want to offer the homeowner a package deal on new windows to go along with the new vinyl siding. It will save them money, and save you a lot of time and aggravation.

# Horizontal Wood Siding

Horizontal wood siding patterns include bevel, Dolly Varden, drop, and channel siding. Common widths are 4", 6", 8", 10" and even 12". Smooth-finish wood siding can be stained or painted. Rough-sawn wood siding is usually stained.

Horizontal siding has to be applied over a smooth surface. If you decide to leave the existing siding in place, either cover it with OSB (oriented strand board panels) or nail furring strips over each stud. Nail siding at each stud with corrosion-resistant nails. Use 6d nails for siding less than $1/2$" thick and 8d nails for thicker siding. Try to avoid nailing through the brittle top edge of siding boards. Bevel siding joints should overlap no less than 1". Be sure the butt joints of horizontal siding boards fall over a stud. If possible, adjust the lap so butt edges coincide with the bottom of the sill and the top of the drip cap at window frames. Refer back to Figure 6-1. Finish interior corners by butting the siding against a corner board at least $1^{1}/8$" square. See Figure 6-5. Exterior corners can be mitered, butted against corner boards at least $1^{1}/8$" square and $1^{1}/2$" wide (Figure 6-6), or covered with metal corners.

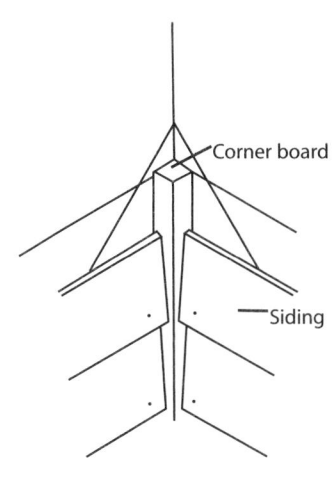

**Figure 6-5**

*Corner board for application of horizontal siding at interior corner*

Pre-finished hardboard plank siding is sold in widths from 6" to 16" and in lengths to 16'. Matching trim, corners and connectors are available. Nail hardboard plank siding at each vertical joint, the same as bevel siding.

# Vertical Wood Siding

Vertical board siding applied over horizontal board siding makes a good combination. Probably the most popular vertical pattern is matched (tongue and groove) boards. Nail through the existing siding and into the sheathing. If you tear off the existing siding, you may need nailing strips. Sheathing that's $1/2$" or $3/8$" thick doesn't have enough nail-holding power. When the existing sheathing is thinner than $5/8$", apply 1" x 4" nailing strips horizontally across the wall, spaced 16" to 24" apart vertically. Then nail

> Wood siding can also be installed diagonally. Add 15 percent to the labor estimate and allow 5 percent more waste of materials.

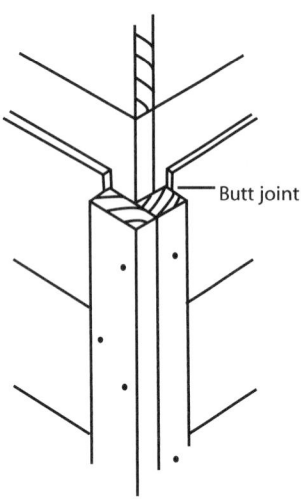

**Figure 6-6**

*Corner boards for application of horizontal siding at exterior corner*

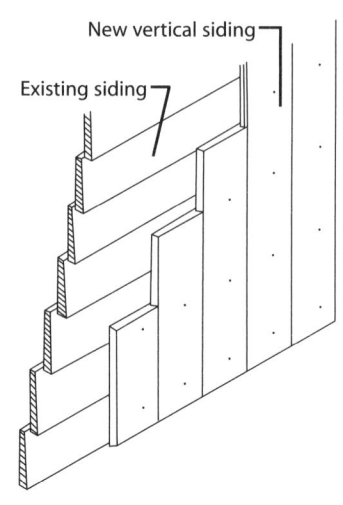

**Figure 6-7**

*Application of vertical siding*

the vertical siding to the strips. Blind-nail through the siding tongue at each strip with galvanized 7d finishing nails. When boards are nominal 6" or wider, also face-nail the middle of each board with an 8d galvanized nail. See Figure 6-7.

Board and batten (*board 'n bat*) is another popular choice for vertical siding. Again, apply horizontal nailing strips if sheathing is less than $^5/_8$" thick. Nail each board at the top and bottom with a galvanized 8d nail. See Figure 6-8. Wider boards need two nails at the top and bottom, spaced 1" each side of the center line. Close spacing prevents splitting as the boards shrink. Cover the $^1/_2$" gap between boards with the batten, secured with 12d finishing nails. Be careful to miss the underboard when nailing batts. Use only corrosion-resistant nails. Figure 6-8 shows two variations of board and batt siding.

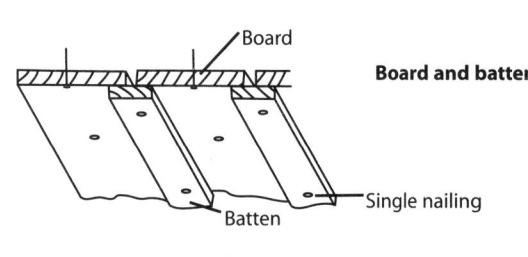

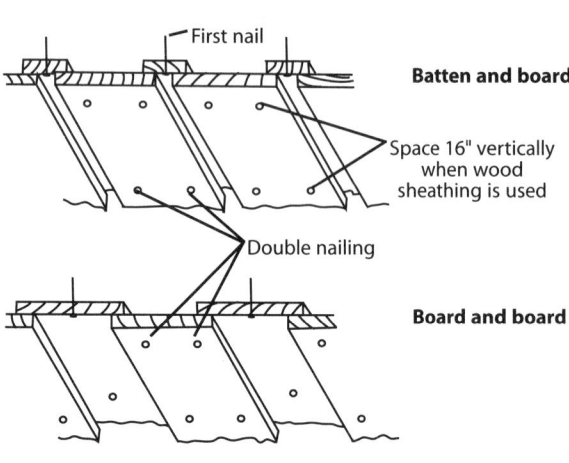

**Figure 6-8**

*Application of vertical wood siding*

# Panel Siding

If existing siding is uneven, panel siding (plywood, hardboard or particleboard) is the best choice. Panel siding tends to smooth out any unevenness in the existing surface. And panel siding is probably the easiest to install. It can be applied over nearly any surface.

If you use plywood, it's best to select a plywood siding product that's specifically made for use as siding. Hardboard used for siding must be tempered. Most panel siding products are sold in both 4' x 8' and 4' x 9' sheets. Rough-textured plywood (such as Texture 1-11) soaks up water-repellent preservative stain better than smooth-finish plywood. Smooth plywood can be stained, but the finish isn't as durable as saw-texture plywood. Fiber-overlaid plywood (usually called MDO, for medium-density overlay) is particularly good if you plan to paint the surface. The resin-treated fiber overlay presents a very smooth surface that both bonds securely to paint and minimizes expansion and contraction due to moisture changes.

> Panel siding with vertical grooves 4" or 8" on center adds visual appeal. Or, get a similar effect by nailing 1" x 2" vertical battens over each panel joint and stud.

# Plywood Siding

In new construction, plywood siding can be installed directly over the studs without any sheathing. If there is no sheathing, the plywood should be at least $3/8$" thick for 16" stud spacing and $1/2$" thick for 24" stud spacing. Grooved plywood is normally $5/8$" thick with $1/4$" deep grooves.

Nail plywood siding every 7" to 8" around the perimeter of the panel and at each intermediate stud. Use galvanized or other rust-resistant nails. When installing panel siding over existing siding, use nails long enough to penetrate $3/4$" into the surface below.

Some plywood siding has shiplap joints. Treat the joint edges with a water-repellent preservative and then nail the siding on both sides of the joint. See Figure 6-9A. With shiplap joints, no caulking is needed. When hanging square-edge panels, you'll need to caulk the joints with sealant, as shown in Figure 6-9B. Again, nail the panels on both sides of the joint. If you install battens over panel joints and at intermediate studs, nail the battens with 8d galvanized nails spaced 12" apart. Use longer nails if you have to penetrate through existing siding to the sheathing below.

> You'll pay a premium for pre-finished plywood siding, but the finish lasts longer and requires little or no maintenance. When installing pre-finished panel siding, use nails and connectors recommended by the manufacturer.

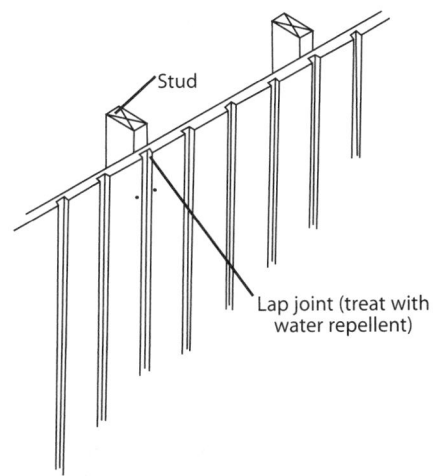

**Figure 6-9A**

*Joint of plywood panel siding: shiplap joint*

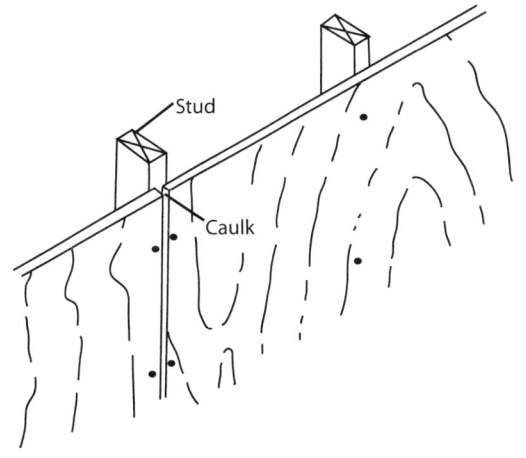

**Figure 6-9B**

*Joint of plywood panel siding: square-edge joint*

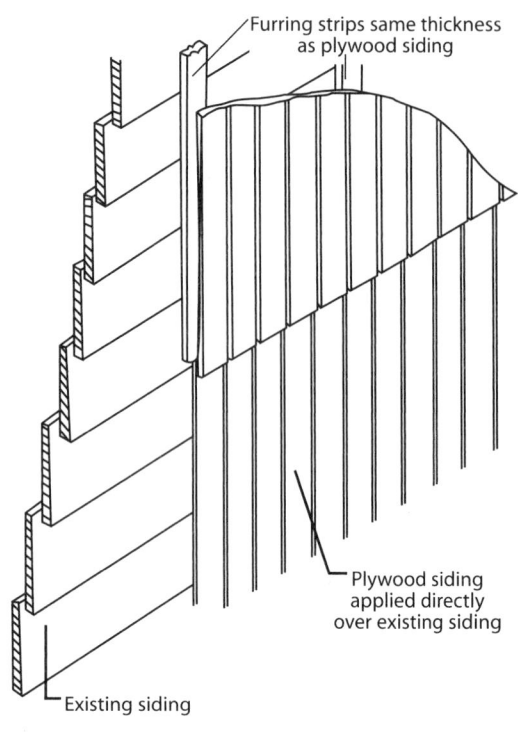

Furring strips same thickness as plywood siding

Plywood siding applied directly over existing siding

Existing siding

**Figure 6-10**

*Application of plywood siding at gable end*

Gable ends present a problem because panels aren't long enough to extend the full wall height. Horizontal butt joints in panels are an invitation to water damage, even when covered with trim. Be sure the joints are as snug as possible. Alternatively, you can lap siding panels over the top of panels below — although this may detract from the overall appearance. See Figure 6-10. Install furring strips on the gable that are the same thickness as the new siding below. Nail a furring strip over the existing siding or sheathing at each stud. Then install siding on the strips the way you'd apply siding directly to studs.

## Hardboard Siding

Hardboard panel siding is available in lengths up to 16'. It's usually $^1/_4$" thick, but may be thicker when grooved. Hardboard is usually factory primed. Apply a finish coat after installation. Hang hardboard siding the same way you install plywood siding.

Finish corners with corner boards, the same as for horizontal board siding. Use $1^1/_8$" x $1^1/_8$" corner boards at inside corners and $1^1/_8$" x $1^1/_2$" or $2^1/_2$" boards at outside corners. Apply caulking where siding butts against corner boards, window or door casings, and trim boards at gable ends.

## Shingle and Shake Siding

It's not always easy to spot shingle siding that's reached the end of its useful life, especially if it's been painted. Sometimes rotted shingles appear fine visually, but will crumble if you touch them. Look closely for broken, warped, and upturned shingles.

It's common to find houses in which some areas of shingles are decayed, but the rest are fine. This is because exposure to the weather varies from one part of the house to the other. Areas that are more exposed, or are kept damp by overhanging trees, will tend to decay first. Removing damaged shingles can be tricky. Prying them up tends to damage the shingle directly above the bad one. You may find yourself replacing a vertical strip of shingles that extends substantially past the bad shingles.

Repairs made to individual shingles, or small sections of shingles, will be obvious for several years. Old shingles are bleached nearly gray by sun and weather. New shingles have the rich tone of freshly-cut cedar. It would be nicer to re-side the entire house so that all the shingles match, but budgets don't always allow for this.

### Estimating Furring

| Shingle Exposure | Linear Feet per 100 SF of Siding |
|---|---|
| 4" exposure | 300 feet |
| $4^1/_2$" exposure | 270 feet |
| 5" exposure | 240 feet |
| $5^1/_2$" exposure | 220 feet |
| 6" exposure | 200 feet |
| $6^1/_2$" exposure | 180 feet |
| 7" exposure | 171 feet |
| 8" exposure | 150 feet |

**Figure 6-11**

*Linear feet of furring per 100 square feet of shingle siding*

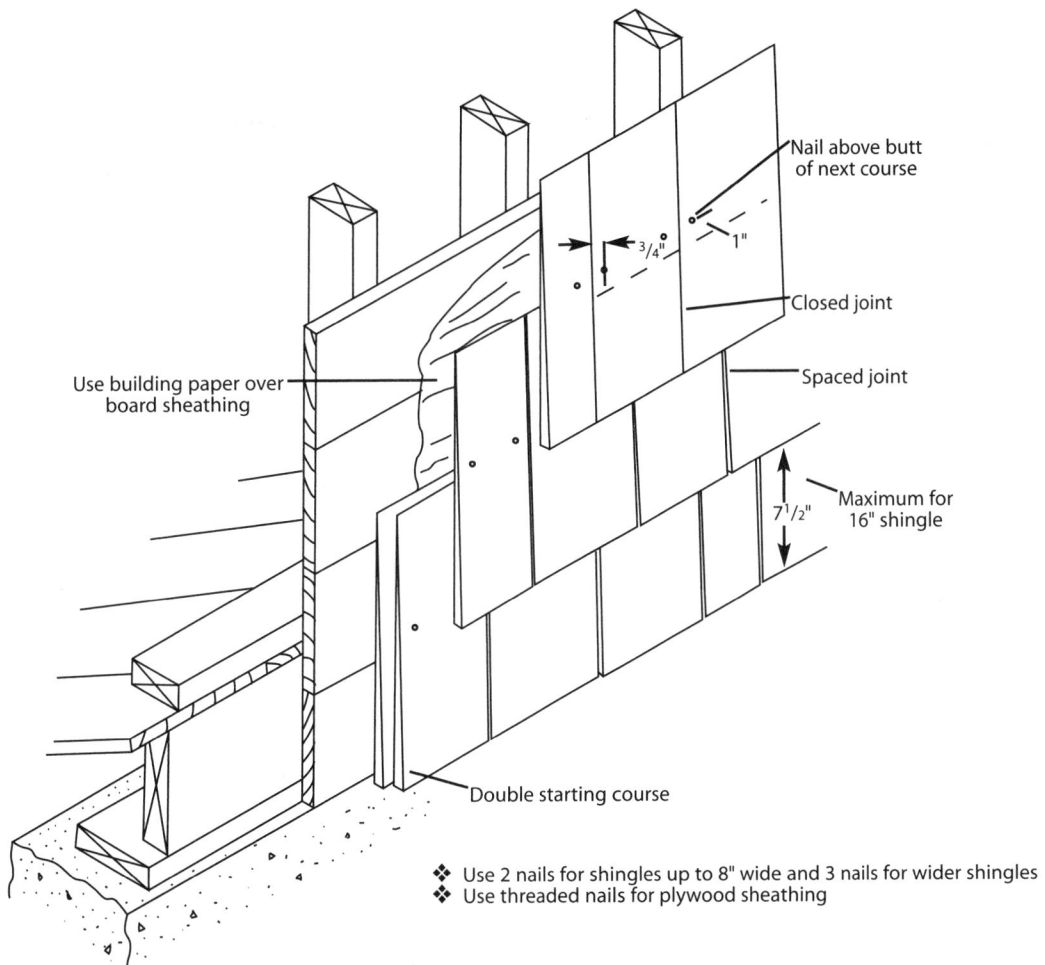

Nail above butt
of next course

3/4"

1"

Closed joint

Use building paper over
board sheathing

Spaced joint

7¹/₂"

Maximum for
16" shingle

Double starting course

❖ Use 2 nails for shingles up to 8" wide and 3 nails for wider shingles
❖ Use threaded nails for plywood sheathing

**Figure 6-12**

*Single-course application of shingle siding*

Siding shingles can be applied over either wood or plywood sheathing or over existing board siding. When installed over sheathing, apply 15-pound asphalt felt between the sheathing and the shingles. If the surface is irregular, apply horizontal furring strips of 1" x 3" lumber or 1" x 4" lumber. Spacing has to match the exposed shingle length. Use Figure 6-11 to estimate the furring required for various exposures per 100 square feet of shingle siding. Leave a gap of ¹/₈" to ¹/₄" between shingles to allow for expansion during wet weather.

Single-course shingles are applied like bevel siding. Each top course laps over the course below with all joints staggered. See Figure 6-12. Use second-grade shingles (rather than first or third grade) because only one-half or less of the butt portion is exposed.

Compare the double-course shingles in Figure 6-13. The first grade (No. 1) shingles cover the undercourse (third grade) shingles. The butt ends of top course shingles should project ¹/₄" to ¹/₂" beyond the butt ends of undercourse shingles.

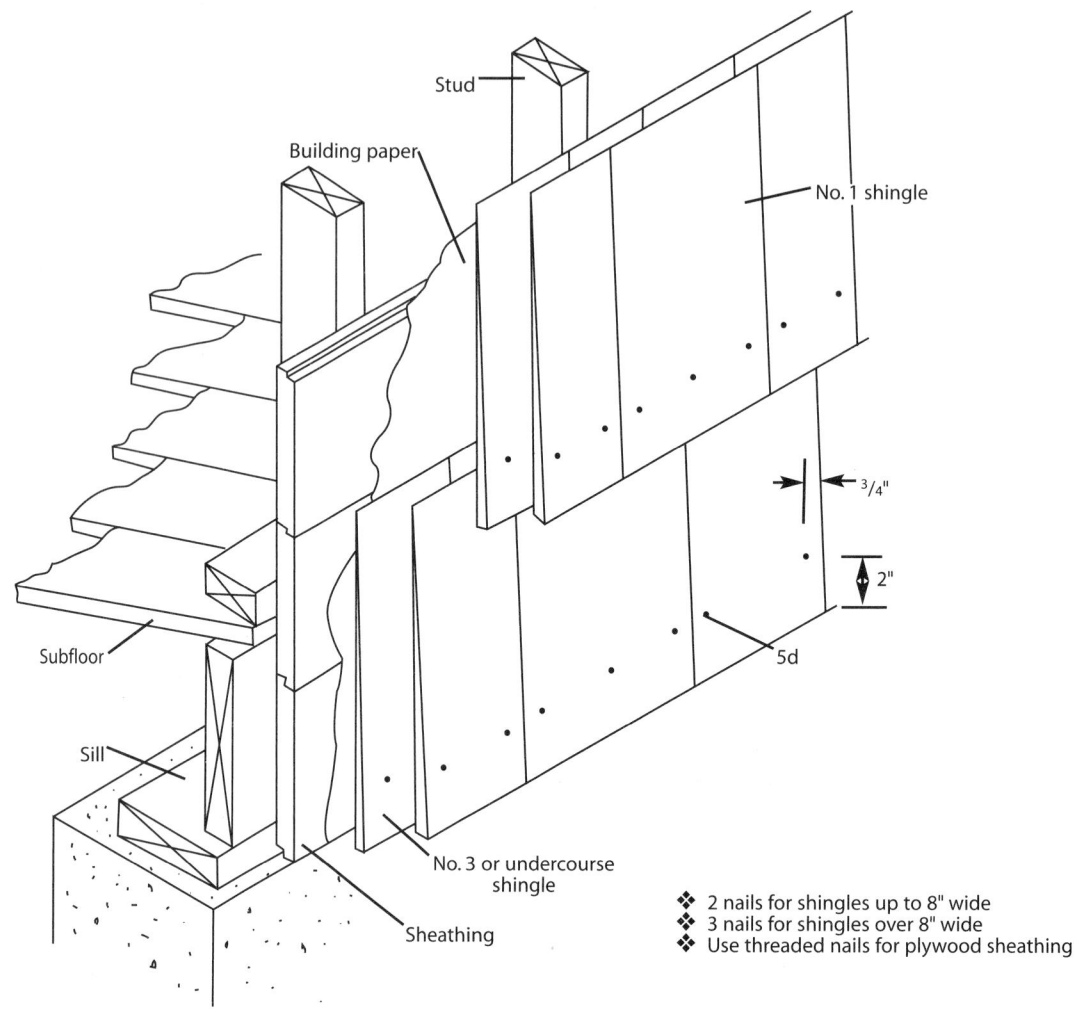

Stud

Building paper

No. 1 shingle

3/4"

2"

5d

Subfloor

Sill

No. 3 or undercourse
shingle

Sheathing

❖ 2 nails for shingles up to 8" wide
❖ 3 nails for shingles over 8" wide
❖ Use threaded nails for plywood sheathing

**Figure 6-13**

*Double-course application of shingle siding*

With double coursing, shingle exposure of the top course can be greater. Figure 6-14 shows recommended exposure distances for shingles.

Whether single-course or double-course, all joints should be staggered. Vertical joints of the upper shingle should be at least $1^1/2$" from any joint in the lower shingle.

Use rust-resistant, anti-stain nails when installing shingles. Two nails are enough on shingles up to 8" wide. Use three nails on shingles over 8" wide. For single-coursing, 3d or 4d shingle nails are best. Nails with small flat heads are best for double-coursing because the nail heads are exposed. Use 5d nails for the top course and 3d or 4d nails on the lower course. When plywood sheathing is less than $3/4$" thick, use threaded nails for increased holding power. Keep nails at least $3/4$" from the shingle edge. When single-coursing, keep nails 1" above the horizontal butt line of the next higher course. See Figure 6-12. When double-coursing, nails should be 2" above the bottom butt.

| Maximum Exposure for Shingles | | | | |
|---|---|---|---|---|
| Material | Length | Single Coursing | Double Coursing No. 1 grade | Double Coursing No. 2 grade |
| **Shingles** | 16" | 7$\frac{1}{2}$" | 12" | 10" |
| | 18" | 8$\frac{1}{2}$" | 14" | 11" |
| | 24" | 11$\frac{1}{2}$" | 16" | 14" |
| **Shakes (handsplit and resawn)** | 18" | 8$\frac{1}{2}$" | 14" | — |
| | 24" | 11$\frac{1}{2}$" | 20" | — |
| | 32" | 15" | — | — |

**Figure 6-14**

*Exposure distances for wood shingles and shakes on sidewalls*

# Masonry Veneer

Brick or stone veneer siding won't wear out. But mortar can crumble and settling will cause cracks. In either case, apply new mortar to keep moisture out of the wall and to improve the overall appearance. Scratch out all loose mortar. Brush the joint to remove dust and loose particles. Then dampen the surface and apply mortar to fill the joint. Tamp mortar into the joint for a tight bond. Use pre-mixed joint compound and a jointing tool to keep the joint depths uniform. When repointing red brick, any mortar left on the face of individual brick will cause a permanent stain.

Washing or brushing the brick after repointing tends to erode the joint you've just patched. Consider masking the face of brick when making joint repairs, especially when the brick is going to be left its natural color rather than painted. Let the mortar dry. Then pull off the masking tape. Finally, consider coating the entire surface with transparent waterproofing seal.

Stained or discolored masonry can be cleaned with a water blaster or by brushing with diluted (22 percent) muriatic acid. Common household bleach will remove green mold from masonry.

# Stucco Repair

Hairline cracks in stucco are normal as the stucco dries and ages. Simply recoat the surface with masonry paint. Wider cracks, especially cracks that run diagonally, are usually the result of the structure settling. To repair any crack too wide for paint, scribe out $\frac{1}{4}$" of stucco along the crack. A screwdriver makes a good scribing tool. Then fill the scribed line with stucco patch or caulk and recoat with masonry paint.

Occasionally, cracks will be the result of defects in the stucco itself. For example, a grid of horizontal and vertical cracks may indicate that the lath is working loose from the studs. A spider web of cracks indicates that the scratch or brown coat is too thin or dried too fast. Under those circumstances, there may be no alternative to removing and replacing the affected section.

When stucco is exposed to water for long periods, the color coat flakes off and turns to powder. Remove the source of the moisture. Then wire-brush the surface and apply a new color coat.

# Estimating Procedure for Siding

When measuring quantities for siding jobs, it's good practice to ignore wall openings (such as windows and doors). Window and door openings result in only a modest saving of materials, but do increase the installation cost. Each wall corner will also increase the labor costs. The figures in this chapter assume a typical job consisting of six corners and 2,500 square feet of siding. Allow one additional installation hour for each corner over six. The accessories needed for most siding jobs are a substantial part of the costs: J-channel, starting strip, L-channel, undersill trim, corners, nails, and colored caulk to match the siding. These figures also assume that you'll be working with vinyl siding-ready windows, or that you'll simply be trimming around existing doors and windows. If it will be necessary to cover windows with aluminum trim bent on-site, add figures from the section Aluminum siding trim in this chapter.

On soffits, assume that 30 percent of the soffit will be vented. Trim on fascia and soffit jobs can include utility trim, starter, cove molding, divider for miter cuts, frieze molding, reversible frieze, overhanging drip edge, brick trim, matching nails and matching caulk.

The figures that follow assume that the work is done on the exterior of a one- or two-story dwelling. Work at higher levels requires scaffolding. Add the cost of scaffold or lifting equipment and about $1/2$ hour of extra time per 100 square feet of siding installed.

# Masonry Fireplaces and Chimneys

Cracks in a masonry chimney will usually be the result of someone attaching an antenna or holiday decorations at the roof level. Cracks can also be the result of settlement of the fireplace foundation. A well-built chimney has a terra-cotta flue lining which keeps the chimney air-tight, even when the masonry exterior develops cracks. If the chimney does not have a fireproof lining, cracks can be a fire hazard. Chip out and replace any fractured brick. Seal all deep cracks and regrout (tuck point) the masonry joints. The combustion chamber (firebox) in a fireplace will be lined with firebrick. With age and frequent use, firebrick can spall and flake, leaving the common brick backing exposed to excessive heat. Chip out and reset any deteriorated firebrick. House framing should neither support nor be supported by the chimney. The fireplace and chimney should be supported on a separate foundation. If the fireplace has developed stress fractures due to settling of either the foundation or the house, demolition may be the only practical alternative.

Prefabricated metal fireplaces are a good choice for home improvement because they can be set on a wood-frame floor without a separate foundation and without any masonry.

| | Craft@Hrs | Unit | Material | Labor | Total | Sell |
|---|---|---|---|---|---|---|

## Siding and Soffit Repairs

**Repair of board siding.** Using hand tools and working from a ladder.

| | Craft@Hrs | Unit | Material | Labor | Total | Sell |
|---|---|---|---|---|---|---|
| Cut out 3" x 3" or smaller damaged area with a chisel, fill with stainable latex wood filler, sand smooth, per area cut and filled | BC@.133 | Ea | — | 4.27 | 4.27 | 7.26 |
| Stainable latex wood filler, 16 oz. tube | — | Ea | 9.70 | — | 9.70 | — |
| Re-nail split or warped siding board, per board repaired | BC@.067 | Ea | 1.90 | 2.15 | 4.05 | 6.89 |
| Pry out 8' length of siding, cut replacement siding board to fit, nail new siding board in place, per board replaced | BC@.416 | Ea | 11.90 | 13.30 | 25.20 | 42.80 |
| Remove and replace 8' vertical siding board or batt, per board or batt | BC@.350 | Ea | 11.90 | 11.20 | 23.10 | 39.30 |
| Replace 1" x 2" x 8' siding corner board | BC@.400 | Ea | 1.97 | 12.80 | 14.77 | 25.10 |
| Touch-up painting to match | PT@.333 | Ea | 4.16 | 10.80 | 14.96 | 25.40 |

**Repair of vinyl or aluminum siding.** Using hand tools and working from a ladder.

| | Craft@Hrs | Unit | Material | Labor | Total | Sell |
|---|---|---|---|---|---|---|
| Cut out damaged panel with a siding removal tool, loosen the panel above, remove damaged panel nails, insert replacement panel, per panel | BC@1.15 | Ea | 6.32 | 36.90 | 43.22 | 73.50 |
| Siding removal tool | — | Ea | 5.33 | — | 5.33 | — |
| Scab patch vinyl or aluminum siding, trim away damaged material, cut patch from scrap, set patch in construction adhesive, per patch | BC@.300 | Ea | .50 | 9.63 | 10.13 | 17.20 |
| Remove and replace aluminum corner cap, pry off or trim off damaged cap, set new cap in panel adhesive, per corner cap | BC@.167 | Ea | .69 | 5.36 | 6.05 | 10.30 |
| Touch-up painting to match | PT@.333 | Ea | 4.16 | 10.80 | 14.96 | 25.40 |

**Remove and replace a siding shake or shingle.** Using hand tools and working from a ladder. Per section of 2-3 shingles removed and replaced.

| | Craft@Hrs | Unit | Material | Labor | Total | Sell |
|---|---|---|---|---|---|---|
| Single course. Chip out damaged shingle, hacksaw old nails, trim a new shingle and place 1/2" below correct alignment, nail at top of course, tap shingle into alignment | BC@.416 | Ea | 1.89 | 13.30 | 15.19 | 25.80 |
| Double course. Chip out damaged shingle in two courses, pull old nails, cut undercourse and top course shingles to fit, place and nail both courses, per set of two shingles | BC@.500 | Ea | 3.78 | 16.00 | 19.78 | 33.60 |
| Shingle removal tool | — | Ea | 27.30 | — | 27.30 | — |
| Touch-up stain to match | PT@.333 | Ea | 4.16 | 10.80 | 14.96 | 25.40 |

**Stucco repair.** Using hand tools and working from a ladder.

| | Craft@Hrs | Unit | Material | Labor | Total | Sell |
|---|---|---|---|---|---|---|
| Fill crack by scribing out the crack to 1/4" depth, fill crack with elastomeric caulk, per linear foot filled | CF@.033 | Ea | — | 1.00 | 1.00 | 1.70 |
| Elastomeric patch and caulk, 10 oz. tube | — | Ea | 2.80 | — | 2.80 | — |
| Patch hole in exterior stucco. Chip and wire brush loose stucco, moisten surface, apply stucco patch with a trowel, per 4" x 4" patch | CF@.250 | Ea | .26 | 7.61 | 7.87 | 13.40 |
| Premixed stucco patch, quart | — | Ea | 5.38 | — | 5.38 | — |
| Touch-up painting to match | PT@.333 | Ea | 4.16 | 10.80 | 14.96 | 25.40 |

| | Craft@Hrs | Unit | Material | Labor | Total | Sell |
|---|---|---|---|---|---|---|

**Siding tear off.** Remove the old siding, break debris into convenient size and pile on site. Using hand tools. Includes the cost of driving or pulling out nails to prepare the surface for installation of new siding. Add the cost of caulking and flashing, if needed. No salvage value assumed.

| | Craft@Hrs | Unit | Material | Labor | Total | Sell |
|---|---|---|---|---|---|---|
| Aluminum siding, horizontal | BL@.027 | SF | — | .72 | .72 | 1.22 |
| Aluminum siding, vertical | BL@.031 | SF | — | .83 | .83 | 1.41 |
| Asphalt siding, panel or roll | BL@.021 | SF | — | .56 | .56 | .95 |
| Mineral siding | BL@.027 | SF | — | .72 | .72 | 1.22 |
| Plywood siding | BL@.017 | SF | — | .45 | .45 | .77 |
| Plywood siding with battens | BL@.019 | SF | — | .51 | .51 | .87 |
| Stucco on metal lath | BL@.035 | SF | — | .93 | .93 | 1.58 |
| Stucco on wood lath | BL@.040 | SF | — | 1.07 | 1.07 | 1.82 |
| Vinyl siding, horizontal | BL@.021 | SF | — | .56 | .56 | .95 |
| Vinyl siding, vertical | BL@.024 | SF | — | .64 | .64 | 1.09 |
| Wood siding, horizontal | BL@.030 | SF | — | .80 | .80 | 1.36 |
| Wood siding, vertical | BL@.033 | SF | — | .88 | .88 | 1.50 |
| Wood shingles or shakes, single course | BL@.021 | SF | — | .56 | .56 | .95 |
| Wood shingles or shakes, double course | BL@.026 | SF | — | .69 | .69 | 1.17 |
| Add for work above 9' high | BL@.008 | SF | — | .21 | .21 | .36 |
| Remove 1" wood furring, per linear foot | BL@.020 | LF | — | .53 | .53 | .90 |

**Soffit tear off.** Remove board, panel, vinyl or aluminum soffit and trim and pile debris on site. Using hand tools and working from a ladder at 8' to 10' height. Includes the cost of driving or pulling out nails to prepare the surface for installation of a new soffit. No salvage value assumed. Per linear foot of soffit.

| | Craft@Hrs | Unit | Material | Labor | Total | Sell |
|---|---|---|---|---|---|---|
| To 12" wide soffit | BL@.016 | SF | — | .43 | .43 | .73 |
| Over 12" wide soffit | BL@.020 | SF | — | .53 | .53 | .90 |
| Add for heights over 10' to 20' | BL@.008 | SF | — | .21 | .21 | .36 |

**Remove and replace for siding work.** Remove in salvage condition using hand tools. Then replace when siding work is complete.

| | Craft@Hrs | Unit | Material | Labor | Total | Sell |
|---|---|---|---|---|---|---|
| Awning or hood over door or window | BC@1.75 | Ea | — | 56.20 | 56.20 | 95.50 |
| Patio canopy, per square foot | BC@.088 | SF | — | 2.82 | 2.82 | 4.79 |
| Fixed shutters, per pair | BC@1.00 | Pair | — | 32.10 | 32.10 | 54.60 |
| Operable shutters, per pair | BC@1.50 | Pair | — | 48.10 | 48.10 | 81.80 |
| Storm windows, each | BC@.625 | Ea | — | 20.10 | 20.10 | 34.20 |
| Storm doors, each | BC@2.50 | Ea | — | 80.20 | 80.20 | 136.00 |
| Gutters or downspouts, per linear foot | BC@.080 | LF | — | 2.57 | 2.57 | 4.37 |

|  | Craft@Hrs | Unit | Material | Labor | Total | Sell |
|---|---|---|---|---|---|---|

## Vinyl Siding

**Vinyl lap siding.** Solid vinyl double traditional lap profile. When calculating quantities, do not deduct for openings (such as doors and windows) less than 6' wide. Labor and material needed for accessories will be a large part of the total cost. Accessories include J-channel, starting strip, L-channel, undersill trim, corners, nails, and colored caulking to match the siding. Each corner adds about an hour of work to the job. These figures assume six walls are being covered. Add the cost of soffit, fascia and extra trim needed over doors and windows. These figures assume 5% waste of materials. Per 100 square feet (Square).

|  | Craft@Hrs | Unit | Material | Labor | Total | Sell |
|---|---|---|---|---|---|---|
| White, .042" x 8" exposure x 12'6" | B1@3.34 | Sq | 76.60 | 98.10 | 174.70 | 297.00 |
| Colors, .042" x 8" exposure x 12'6" | B1@3.34 | Sq | 76.60 | 98.10 | 174.70 | 297.00 |
| White, .042" x 9" exposure x 12'1" | B1@3.34 | Sq | 76.60 | 98.10 | 174.70 | 297.00 |
| Colors, .042" x 9" exposure x 12'1" | B1@3.34 | Sq | 76.60 | 98.10 | 174.70 | 297.00 |
| White, .042" x 10" exposure x 12' | B1@3.34 | Sq | 76.60 | 98.10 | 174.70 | 297.00 |
| Colors, .042" x 10" exposure x 12' | B1@3.34 | Sq | 76.60 | 98.10 | 174.70 | 297.00 |
| White, .040" x 8" exposure x 12'6" | B1@3.34 | Sq | 51.70 | 98.10 | 149.80 | 255.00 |
| Colors, .040" x 8" exposure x 12'6" | B1@3.34 | Sq | 65.30 | 98.10 | 163.40 | 278.00 |
| White, .040" x 9" exposure x 12'1" | B1@3.34 | Sq | 54.60 | 98.10 | 152.70 | 260.00 |
| Colors, .040" x 9" exposure x 12'1" | B1@3.34 | Sq | 54.60 | 98.10 | 152.70 | 260.00 |
| White, .040" x 10" exposure x 12' | B1@3.34 | Sq | 52.90 | 98.10 | 151.00 | 257.00 |
| Colors, .040" x 10" exposure x 12' | B1@3.34 | Sq | 52.90 | 98.10 | 151.00 | 257.00 |
| Starter strip, 5" wide | B1@.025 | LF | .34 | .73 | 1.07 | 1.82 |
| J-channel trim at wall openings, 3/4" | B1@.033 | LF | .36 | .97 | 1.33 | 2.26 |
| Outside corner post, 3/4" x 3/4" | B1@.033 | LF | 1.56 | .97 | 2.53 | 4.30 |
| Inside corner post, 3/4" x 3/4" | B1@.033 | LF | 1.03 | .97 | 2.00 | 3.40 |
| Under sill trim, 1-1/2" | B1@.038 | LF | .39 | 1.12 | 1.51 | 2.57 |
| Casing and utility trim | B1@.038 | LF | .39 | 1.12 | 1.51 | 2.57 |
| Vinyl siding 1-1/4" nails, per pound | — | Lb | 8.41 | — | 8.41 | — |
| Add for R-3 insulated vinyl siding | B1@.350 | Sq | 27.25 | 10.30 | 37.55 | 63.80 |

**Siding fixture mounting block.** For mounting accessories such as electrical outlets and hose bibs on wood, aluminum, vinyl, stucco, shake, or brick siding. Snap-on finishing ring. No caulking or J-channel required. 1-1/4" diameter fixture cutout.

|  | Craft@Hrs | Unit | Material | Labor | Total | Sell |
|---|---|---|---|---|---|---|
| Recessed, 6-1/4" x 4-3/4" | B1@.350 | Ea | 5.60 | 10.30 | 15.90 | 27.00 |
| Surface mount, 6-3/4" x 6-3/4" | B1@.350 | Ea | 4.94 | 10.30 | 15.24 | 25.90 |
| Jumbo, 12" x 8" | B1@.500 | Ea | 8.59 | 14.70 | 23.29 | 39.60 |

**Thermoplastic resin siding.** Nailite™ 9/10" thick panels installed over sheathing. Add for trim board, corners, starter strips, J-channels and mortar fill. Based on five nails per piece. Includes 10% waste.

|  | Craft@Hrs | Unit | Material | Labor | Total | Sell |
|---|---|---|---|---|---|---|
| Hand-split shake, 41-3/8" x 18-3/4" | B1@.043 | SF | 2.18 | 1.26 | 3.44 | 5.85 |
| Hand-laid brick, 44-1/4" x 18-5/8" | B1@.043 | SF | 2.32 | 1.26 | 3.58 | 6.09 |
| Hand-cut stone, 44-1/4" x 18-5/8" | B1@.043 | SF | 2.32 | 1.26 | 3.58 | 6.09 |
| Rough sawn cedar, 59-1/4" x 15" | B1@.043 | SF | 2.42 | 1.26 | 3.68 | 6.26 |
| Perfection plus cedar, 36-1/8" x 15" | B1@.043 | SF | 2.42 | 1.26 | 3.68 | 6.26 |
| Ledge trim |  |  |  |  |  |  |
| Brick ledge trim | B1@.033 | LF | 5.04 | .97 | 6.01 | 10.20 |
| Stone ledge trim | B1@.033 | LF | 5.04 | .97 | 6.01 | 10.20 |
| Corners |  |  |  |  |  |  |
| Hand-split shake, 4" x 18" | B1@.030 | Ea | 9.52 | .88 | 10.40 | 17.70 |
| Rough sawn cedar, 3" x 26" | B1@.030 | Ea | 10.10 | .88 | 10.98 | 18.70 |
| Hand-laid brick, 4" x 18" | B1@.030 | Ea | 10.10 | .88 | 10.98 | 18.70 |
| Hand-cut stone, 4" x 18" | B1@.030 | Ea | 10.10 | .88 | 10.98 | 18.70 |
| Perfection plus cedar, 3" x 13" | B1@.030 | Ea | 8.40 | .88 | 9.28 | 15.80 |
| Starter strips |  |  |  |  |  |  |
| 12' universal | B1@.025 | LF | .50 | .73 | 1.23 | 2.09 |

| | Craft@Hrs | Unit | Material | Labor | Total | Sell |
|---|---|---|---|---|---|---|
| J-channels | | | | | | |
| Hand-split shake, 3/4" opening | B1@.025 | LF | .77 | .73 | 1.50 | 2.55 |
| Hand-split shake, 1-1/4" flexible opening | B1@.027 | LF | 1.52 | .79 | 2.31 | 3.93 |
| Rough sawn cedar, 3/4" opening | B1@.025 | LF | .57 | .73 | 1.30 | 2.21 |
| Rough sawn cedar, 1-1/4" flexible opening | B1@.027 | LF | 1.52 | .79 | 2.31 | 3.93 |
| Hand-laid brick, 3/4" opening | B1@.025 | LF | .57 | .73 | 1.30 | 2.21 |
| Hand-cut stone, 3/4" opening | B1@.025 | LF | .57 | .73 | 1.30 | 2.21 |
| Perfection plus cedar, 3/4" opening | B1@.025 | LF | .57 | .73 | 1.30 | 2.21 |
| Thermoplastic siding accessories | | | | | | |
| Mortar fill, 12 tubes, 200 LF per carton | — | Ea | 49.90 | — | 49.90 | — |
| Touch up paint, 6 oz. aerosol can | — | Ea | 4.20 | — | 4.20 | — |

**Vinyl fascia.** Installed on rafter tails with small tools, working from a ladder. Per linear foot, including 5% waste.

| | Craft@Hrs | Unit | Material | Labor | Total | Sell |
|---|---|---|---|---|---|---|
| 8" wide x 12' long panels | B1@.025 | LF | .90 | .73 | 1.63 | 2.77 |

**Vinyl soffit.** Panel conceals the under side of roof overhang. Per linear foot including 5% waste.

| | Craft@Hrs | Unit | Material | Labor | Total | Sell |
|---|---|---|---|---|---|---|
| Soffit, 12" x 12' | B1@.040 | LF | .90 | 1.17 | 2.07 | 3.52 |
| J-channel support, 1/2" x 12' | B1@.033 | LF | .36 | .97 | 1.33 | 2.26 |
| F-channel support, 1/2" x 12' | B1@.033 | LF | .54 | .97 | 1.51 | 2.57 |
| H-bar double channel support, 1/2" x 12' | B1@.025 | LF | .38 | .73 | 1.11 | 1.89 |

## Wood Board Siding

**Pine bevel siding.** 1/2" thick by 6" wide. Each linear foot of 6"-wide bevel siding covers .38 square feet of wall. Do not deduct for wall openings less than 10 square feet. Add 5% for typical waste and end cutting.

| | Craft@Hrs | Unit | Material | Labor | Total | Sell |
|---|---|---|---|---|---|---|
| 8' lengths | B1@.038 | SF | 1.03 | 1.12 | 2.15 | 3.66 |
| 12' lengths | B1@.036 | SF | 1.02 | 1.06 | 2.08 | 3.54 |
| 16' lengths | B1@.033 | SF | 1.03 | .97 | 2.00 | 3.40 |

**Western red cedar bevel siding.** A grade and better, 1/2" thick by 6" wide. Each linear foot of 6"-wide bevel siding covers .38 square feet of wall. Do not deduct for wall openings less than 10 square feet. Add 5% for typical waste and end cutting.

| | Craft@Hrs | Unit | Material | Labor | Total | Sell |
|---|---|---|---|---|---|---|
| 8' lengths | B1@.038 | SF | 2.18 | 1.12 | 3.30 | 5.61 |
| 12' lengths | B1@.036 | SF | 2.33 | 1.06 | 3.39 | 5.76 |
| 16' lengths | B1@.033 | SF | 2.35 | .97 | 3.32 | 5.64 |

**Clear cedar 1/2" bevel siding.** Vertical grain, 1/2" thick by 6" wide, kiln dried. Each linear foot of 6"-wide bevel siding covers .38 square feet of wall. Do not deduct for wall openings less than 10 square feet. Add 5% for typical waste and end cutting.

| | Craft@Hrs | Unit | Material | Labor | Total | Sell |
|---|---|---|---|---|---|---|
| 8' lengths | B1@.038 | SF | 2.42 | 1.12 | 3.54 | 6.02 |
| 12' lengths | B1@.036 | SF | 2.42 | 1.06 | 3.48 | 5.92 |
| 16' lengths | B1@.033 | SF | 2.42 | .97 | 3.39 | 5.76 |

**Cedar 3/4" bevel siding.** 3/4" thick by 8" wide. Each linear foot of 8"-wide bevel siding covers .54 square feet of wall. Do not deduct for wall openings less than 10 square feet. Add 5% for typical waste and end cutting.

| | Craft@Hrs | Unit | Material | Labor | Total | Sell |
|---|---|---|---|---|---|---|
| 8' lengths | B1@.036 | SF | 2.13 | 1.06 | 3.19 | 5.42 |
| 12' lengths | B1@.034 | SF | 2.13 | 1.00 | 3.13 | 5.32 |
| 16' lengths | B1@.031 | SF | 2.13 | .91 | 3.04 | 5.17 |

**Primed pine V rustic siding.** 1" thick by 6" wide. Each linear foot of 6"-wide channel siding covers .38 square feet of wall. Do not deduct for wall openings less than 10 square feet. Add 5% for typical waste and end cutting.

| | Craft@Hrs | Unit | Material | Labor | Total | Sell |
|---|---|---|---|---|---|---|
| 12' lengths | B1@.034 | SF | 3.90 | 1.00 | 4.90 | 8.33 |

| | Craft@Hrs | Unit | Material | Labor | Total | Sell |
|---|---|---|---|---|---|---|

**Tongue and groove cedar siding.** 1" thick by 6" wide. Kiln dried, select. Each linear foot of 6"-wide tongue and groove siding covers .43 square feet of wall. Do not deduct for wall openings less than 10 square feet. Add 5% for typical waste and end cutting.

| | Craft@Hrs | Unit | Material | Labor | Total | Sell |
|---|---|---|---|---|---|---|
| 8' lengths | B1@.036 | SF | 2.88 | 1.06 | 3.94 | 6.70 |
| 10' lengths | B1@.035 | SF | 2.95 | 1.03 | 3.98 | 6.77 |
| 12' lengths | B1@.034 | SF | 2.85 | 1.00 | 3.85 | 6.55 |

**Redwood 5/8" shiplap siding.** B grade. 5/8" thick by 5-3/8" wide. Each linear foot of 5-3/8"-wide shiplap siding covers .38 square feet of wall. Do not deduct for wall openings less than 10 square feet. Add 5% for typical waste and end cutting.

| | Craft@Hrs | Unit | Material | Labor | Total | Sell |
|---|---|---|---|---|---|---|
| 8' lengths | B1@.038 | SF | 4.46 | 1.12 | 5.58 | 9.49 |
| 10' lengths | B1@.037 | SF | 4.46 | 1.09 | 5.55 | 9.44 |
| 12' lengths | B1@.036 | SF | 4.44 | 1.06 | 5.50 | 9.35 |

**Southern yellow pine 6" pattern 105 siding.** 1" thick by 6" wide. Select grade except as noted. Each linear foot of 6"-wide siding covers .38 square feet of wall. Do not deduct for wall openings less than 10 square feet. Add 5% for typical waste and end cutting.

| | Craft@Hrs | Unit | Material | Labor | Total | Sell |
|---|---|---|---|---|---|---|
| 8' lengths | B1@.038 | SF | 3.18 | 1.12 | 4.30 | 7.31 |
| 10' lengths | B1@.037 | SF | 3.18 | 1.09 | 4.27 | 7.26 |
| 12' lengths, No. 2 | B1@.036 | SF | 1.82 | 1.06 | 2.88 | 4.90 |
| 12' lengths | B1@.036 | SF | 3.02 | 1.06 | 4.08 | 6.94 |

**Southern yellow pine 8" pattern 105 siding.** 1" thick by 8" wide. Select grade except as noted. Each linear foot of 8"-wide siding covers .54 square feet of wall. Do not deduct for wall openings less than 10 square feet. Add 5% for typical waste and end cutting.

| | Craft@Hrs | Unit | Material | Labor | Total | Sell |
|---|---|---|---|---|---|---|
| 8' lengths | B1@.036 | SF | 2.94 | 1.06 | 4.00 | 6.80 |
| 10' lengths | B1@.034 | SF | 2.85 | 1.00 | 3.85 | 6.55 |
| 12' lengths, No. 2 | B1@.031 | SF | 1.58 | .91 | 2.49 | 4.23 |
| 12' lengths | B1@.031 | SF | 2.83 | .91 | 3.74 | 6.36 |

**Southern yellow pine 6" pattern 116 siding.** 1" thick by 6" wide. Select grade except as noted. Each linear foot of 6"-wide siding covers .38 square feet of wall. Do not deduct for wall openings less than 10 square feet. Add 5% for typical waste and end cutting.

| | Craft@Hrs | Unit | Material | Labor | Total | Sell |
|---|---|---|---|---|---|---|
| 8' lengths | B1@.038 | SF | 3.18 | 1.12 | 4.30 | 7.31 |
| 10' lengths | B1@.037 | SF | 3.18 | 1.09 | 4.27 | 7.26 |
| 12' lengths | B1@.036 | SF | 3.18 | 1.06 | 4.24 | 7.21 |

**Southern yellow pine 6" pattern 117 siding.** 1" thick by 6" wide. Select grade except as noted. Each linear foot of 6"-wide siding covers .38 square feet of wall. Do not deduct for wall openings less than 10 square feet. Add 5% for typical waste and end cutting.

| | Craft@Hrs | Unit | Material | Labor | Total | Sell |
|---|---|---|---|---|---|---|
| 8' lengths | B1@.038 | SF | 3.18 | 1.12 | 4.30 | 7.31 |
| 10' lengths | B1@.037 | SF | 3.18 | 1.09 | 4.27 | 7.26 |
| 12' lengths | B1@.036 | SF | 3.18 | 1.06 | 4.24 | 7.21 |

| | Craft@Hrs | Unit | Material | Labor | Total | Sell |
|---|---|---|---|---|---|---|

**Southern yellow pine 8" pattern 131 siding.** 1" thick by 8" wide. Select grade except as noted. Each linear foot of 8"-wide siding covers .54 square feet of wall. Do not deduct for wall openings less than 10 square feet. Add 5% for typical waste and end cutting.

| | Craft@Hrs | Unit | Material | Labor | Total | Sell |
|---|---|---|---|---|---|---|
| 8' lengths | B1@.036 | SF | 2.89 | 1.06 | 3.95 | 6.72 |
| 10' lengths | B1@.034 | SF | 2.85 | 1.00 | 3.85 | 6.55 |
| 12' lengths | B1@.031 | SF | 2.83 | .91 | 3.74 | 6.36 |

**Southern yellow pine 8" pattern 122 siding.** 1" thick by 8" wide. V-joint. Select grade except as noted. Each linear foot of 8"-wide siding covers .54 square feet of wall. Do not deduct for wall openings less than 10 square feet. Add 5% for typical waste and end cutting.

| | Craft@Hrs | Unit | Material | Labor | Total | Sell |
|---|---|---|---|---|---|---|
| 8' lengths | B1@.036 | SF | 2.89 | 1.06 | 3.95 | 6.72 |
| 10' lengths | B1@.034 | SF | 2.85 | 1.00 | 3.85 | 6.55 |
| 12' lengths | B1@.031 | SF | 2.83 | .91 | 3.74 | 6.36 |

**Southern yellow pine 6" pattern 122 siding.** 1" thick by 6" wide. Select grade except as noted. Each linear foot of 6"-wide siding covers .38 square feet of wall. Do not deduct for wall openings less than 10 square feet. Add 5% for typical waste and end cutting.

| | Craft@Hrs | Unit | Material | Labor | Total | Sell |
|---|---|---|---|---|---|---|
| 8' lengths | B1@.038 | SF | 2.23 | 1.12 | 3.35 | 5.70 |
| 10' lengths | B1@.037 | SF | 2.23 | 1.09 | 3.32 | 5.64 |
| 12' lengths | B1@.036 | SF | 2.23 | 1.06 | 3.29 | 5.59 |

**Southern yellow pine 8" shiplap siding.** 1" thick by 8" wide. No. 2 grade. Each linear foot of 8"-wide siding covers .54 square feet of wall. Do not deduct for wall openings less than 10 square feet. Add 5% for typical waste and end cutting.

| | Craft@Hrs | Unit | Material | Labor | Total | Sell |
|---|---|---|---|---|---|---|
| 8' lengths | B1@.036 | SF | 1.21 | 1.06 | 2.27 | 3.86 |
| 10' lengths | B1@.034 | SF | 1.19 | 1.00 | 2.19 | 3.72 |
| 12' lengths | B1@.031 | SF | 1.22 | .91 | 2.13 | 3.62 |

**Rustic 8" pine siding.** 1" thick by 8" wide. V-grooved. Each linear foot of 8"-wide siding covers .54 square feet of wall. Do not deduct for wall openings less than 10 square feet. Add 5% for typical waste and end cutting.

| | Craft@Hrs | Unit | Material | Labor | Total | Sell |
|---|---|---|---|---|---|---|
| 16' lengths | B1@.031 | SF | 3.50 | .91 | 4.41 | 7.50 |
| 20' lengths | B1@.031 | SF | 3.50 | .91 | 4.41 | 7.50 |

**Log cabin siding.** 1-1/2" thick. Each linear foot of 6"-wide siding covers .41 square feet of wall. Each linear foot of 8"-wide siding covers .55 square feet of wall. Do not deduct for wall openings less than 10 square feet. Add 5% for typical waste and end cutting.

| | Craft@Hrs | Unit | Material | Labor | Total | Sell |
|---|---|---|---|---|---|---|
| 6" x 12', T&G latia | B1@.036 | SF | 2.41 | 1.06 | 3.47 | 5.90 |
| 8" x 12' | B1@.031 | SF | 2.76 | .91 | 3.67 | 6.24 |
| 8" x 16' | B1@.030 | SF | 2.82 | .88 | 3.70 | 6.29 |

**Log lap spruce siding.** 1-1/2" thick. Each linear foot of 6"-wide siding covers .41 square feet of wall. Each linear foot of 8"-wide siding covers .55 square feet of wall. Each linear foot of 10"-wide siding covers .72 square feet of wall. Do not deduct for wall openings less than 10 square feet. Add 5% for typical waste and end cutting.

| | Craft@Hrs | Unit | Material | Labor | Total | Sell |
|---|---|---|---|---|---|---|
| 8" width | B1@.031 | SF | 3.24 | .91 | 4.15 | 7.06 |
| 10" width | B1@.030 | SF | 3.24 | .88 | 4.12 | 7.00 |

|  | Craft@Hrs | Unit | Material | Labor | Total | Sell |
|---|---|---|---|---|---|---|

## Texture 1-11 plywood siding

**Texture 1-11 plywood siding.** Per square foot including 5% waste. Exterior grade 4' x 8' panels.

| | Craft@Hrs | Unit | Material | Labor | Total | Sell |
|---|---|---|---|---|---|---|
| 3/8" southern yellow pine, grooved 4" OC | B1@.022 | SF | .57 | .65 | 1.22 | 2.07 |
| 3/8" plain fir | B1@.022 | SF | .89 | .65 | 1.54 | 2.62 |
| 3/8" ACQ treated | B1@.022 | SF | .84 | .65 | 1.49 | 2.53 |
| 5/8" borate treated | B1@.022 | SF | 1.65 | .65 | 2.30 | 3.91 |
| 5/8" fir, grooved 8" OC | B1@.022 | SF | 1.20 | .65 | 1.85 | 3.15 |
| 5/8" x 4' x 8' southern yellow pine, grooved 8" OC | B1@.022 | SF | .83 | .65 | 1.48 | 2.52 |
| 5/8" ACQ treated southern, grooved 8" OC | B1@.022 | SF | 1.29 | .65 | 1.94 | 3.30 |
| 5/8" borate treated, grooved 4" OC | B1@.022 | SF | 1.67 | .65 | 2.32 | 3.94 |
| Add for 4' x 9' panels | — | SF | .16 | — | .16 | — |

**Plywood panel siding.** Per square foot including 5% waste. Exterior grade 4' x 8' panels.

| | Craft@Hrs | Unit | Material | Labor | Total | Sell |
|---|---|---|---|---|---|---|
| 3/8" plain southern yellow pine | B1@.022 | SF | .67 | .65 | 1.32 | 2.24 |
| 5/8" southern yellow pine, reverse board and batt 12" OC | B1@.022 | SF | .86 | .65 | 1.51 | 2.57 |
| 5/8" ACQ treated southern yellow pine, grooved 12 OC | B1@.022 | SF | 1.38 | .65 | 2.03 | 3.45 |

**OSB panel siding.** Oriented strand board ("Waferboard"). Compressed wood strands bonded by phenolic resin. Per square foot including 5% waste. Exterior grade 4' x 8' panels.

| | Craft@Hrs | Unit | Material | Labor | Total | Sell |
|---|---|---|---|---|---|---|
| 3/8" Smart Panel II | B1@.022 | SF | .66 | .65 | 1.31 | 2.23 |
| 7/16" textured and grooved 8" OC | B1@.022 | SF | .90 | .65 | 1.55 | 2.64 |
| 5/8" borate treated, grooved 8" OC | B1@.022 | SF | 1.37 | .65 | 2.02 | 3.43 |
| Add for 4' x 9' panels | — | SF | .14 | — | .14 | — |

**OSB plank lap siding.** Oriented strand board ("Waferboard"). Compressed wood strands bonded by phenolic resin. Per square foot including 8% waste. Textured.

| | Craft@Hrs | Unit | Material | Labor | Total | Sell |
|---|---|---|---|---|---|---|
| 7/16" x 6" x 16' | B1@.043 | SF | .98 | 1.26 | 2.24 | 3.81 |
| 7/16" x 8" x 16' | B1@.040 | SF | .94 | 1.17 | 2.11 | 3.59 |

## Hardboard Panel

**Hardboard panel siding.** Cost per square foot including 5% waste. 4' x 8' panels.

| | Craft@Hrs | Unit | Material | Labor | Total | Sell |
|---|---|---|---|---|---|---|
| 7/16" plain panel | B1@.022 | SF | .70 | .65 | 1.35 | 2.30 |
| 7/16" grooved 8" OC | B1@.022 | SF | .85 | .65 | 1.50 | 2.55 |
| 7/16" Sturdi-Panel, grooved 8" OC | B1@.022 | SF | .62 | .65 | 1.27 | 2.16 |
| 7/16" textured board and batt 12" OC | B1@.022 | SF | .60 | .65 | 1.25 | 2.13 |
| 7/16" cedar, grooved 8" OC | B1@.022 | SF | .63 | .65 | 1.28 | 2.18 |
| 7/16" cedar, plain, textured | B1@.022 | SF | .71 | .65 | 1.36 | 2.31 |
| 15/32" Duratemp, grooved 8" OC 1/8" hardboard face, plywood back | BL@.022 | SF | .89 | .59 | 1.48 | 2.52 |
| Add for 4' x 9' panels | — | SF | .14 | — | .14 | — |

| | Craft@Hrs | Unit | Material | Labor | Total | Sell |
|---|---|---|---|---|---|---|

**Hardboard lap siding.** Primed plank siding. With 8% waste. 6" x 16' covers 5.6 square feet, 8" x 16' covers 7.4 square feet, 16" x 16' covers 14.8 square feet.

| | Craft@Hrs | Unit | Material | Labor | Total | Sell |
|---|---|---|---|---|---|---|
| 7/16" x 6" x 16', smooth lap | B1@.033 | SF | .87 | .97 | 1.84 | 3.13 |
| 7/16" x 6" x 16', textured cedar | B1@.033 | SF | .91 | .97 | 1.88 | 3.20 |
| 7/16" x 8" x 16', textured cedar | B1@.031 | SF | 1.02 | .91 | 1.93 | 3.28 |
| 7/16" x 8" x 16', smooth | B1@.031 | SF | .86 | .91 | 1.77 | 3.01 |
| 1/2" x 8" x 16', self-aligning | B1@.031 | SF | 1.45 | .91 | 2.36 | 4.01 |
| 1/2" x 8" x 16', old mill sure lock | B1@.031 | SF | 1.58 | .91 | 2.49 | 4.23 |
| 1/2" x 16" x 16', cottage beveled 5" OC | B1@.030 | SF | 1.68 | .88 | 2.56 | 4.35 |
| 7/16" x 8", textured joint cover | — | Ea | .70 | — | .70 | — |
| 4/4" x 4" x 16' reversible trim | — | LF | .50 | — | .50 | — |
| 4/4" x 6" x 16' reversible trim | — | LF | .68 | — | .68 | — |
| 4/5" x 4" x 16' reversible trim | — | LF | .74 | — | .74 | — |
| 4/5" x 6" x 16' reversible trim | — | LF | 1.01 | — | 1.01 | — |

## Fiber Cement Siding

**Fiber cement lap siding, Hardiplank®.** 6-1/4" width requires 20 planks per 100 square feet of wall. 7-1/4" width requires 17 planks per 100 square feet of wall. 8-1/4" width requires 15 planks per 100 square feet of wall. 9-1/4" width requires 13 planks per 100 square feet of wall. 12" width requires 10 planks per 100 square feet of wall.

| | Craft@Hrs | Unit | Material | Labor | Total | Sell |
|---|---|---|---|---|---|---|
| 6-1/4" x 12', cedarmill | B1@.048 | SF | .98 | 1.41 | 2.39 | 4.06 |
| 7-1/4" x 12', cedarmill | B1@.048 | SF | .94 | 1.41 | 2.35 | 4.00 |
| 8-1/4" x 12', colonial roughsawn | B1@.048 | SF | .96 | 1.41 | 2.37 | 4.03 |
| 8-1/4" x 12', cedarmill | B1@.048 | SF | .93 | 1.41 | 2.34 | 3.98 |
| 8-1/4" x 12', beaded cedarmill | B1@.048 | SF | 1.04 | 1.41 | 2.45 | 4.17 |
| 8-1/4" x 12', beaded | B1@.048 | SF | .94 | 1.41 | 2.35 | 4.00 |
| 8-1/4" x 12', smooth | B1@.048 | SF | 1.03 | 1.41 | 2.44 | 4.15 |
| 9-1/4" x 12', cedarmill | B1@.046 | SF | 1.12 | 1.35 | 2.47 | 4.20 |
| 12" x 12', smooth | B1@.042 | SF | .94 | 1.23 | 2.17 | 3.69 |
| 12" x 12', cedarmill | B1@.042 | SF | .96 | 1.23 | 2.19 | 3.72 |

**Fiber cement panel siding, Hardipanel®.** 5/16" thick. Cost per square foot including 5% waste.

| | Craft@Hrs | Unit | Material | Labor | Total | Sell |
|---|---|---|---|---|---|---|
| 4' x 8', Sierra, grooved 8" OC | B1@.042 | SF | .82 | 1.23 | 2.05 | 3.49 |
| 4' x 8', Stucco | B1@.042 | SF | .85 | 1.23 | 2.08 | 3.54 |
| 4' x 8', cedarmill, grooved 8" OC | B1@.042 | SF | .89 | 1.23 | 2.12 | 3.60 |
| 4' x 9', Sierra, grooved 8" OC | B1@.042 | SF | .73 | 1.23 | 1.96 | 3.33 |

**Trim for fiber cement siding, Hardipanel®.** Per linear foot including 10% waste.

| | Craft@Hrs | Unit | Material | Labor | Total | Sell |
|---|---|---|---|---|---|---|
| 5/4" x 4" x 10', smooth | B1@.035 | LF | .97 | 1.03 | 2.00 | 3.40 |
| 5/4" x 8" x 10', smooth | B1@.035 | LF | 2.05 | 1.03 | 3.08 | 5.24 |
| 5/4" x 12" x 12', smooth XLD | B1@.035 | LF | 2.60 | 1.03 | 3.63 | 6.17 |
| 5/4" x 6" x 12', smooth XLD | B1@.035 | LF | 1.30 | 1.03 | 2.33 | 3.96 |

**Fiber cement soffit, Hardisoffit®.** 1/4" thick. Per square foot including 10% waste. Vented.

| | Craft@Hrs | Unit | Material | Labor | Total | Sell |
|---|---|---|---|---|---|---|
| 2' x 8', smooth | B1@.040 | SF | .90 | 1.17 | 2.07 | 3.52 |
| 4' x 8', cedarmill | B1@.040 | SF | .80 | 1.17 | 1.97 | 3.35 |
| 4' x 8', smooth | B1@.040 | SF | .95 | 1.17 | 2.12 | 3.60 |
| 12" x 12', smooth | B1@.044 | SF | 1.48 | 1.29 | 2.77 | 4.71 |
| 12" x 12', cedarmill | B1@.044 | SF | 1.28 | 1.29 | 2.57 | 4.37 |
| 16" x 12', cedarmill | B1@.044 | SF | 1.31 | 1.29 | 2.60 | 4.42 |

| | Craft@Hrs | Unit | Material | Labor | Total | Sell |
|---|---|---|---|---|---|---|

**Fiber-cement siding shingle.** 18 pieces cover 33.3 square feet at 11" exposure. Three cartons cover 100 square feet. Price per square foot.

| | Craft@Hrs | Unit | Material | Labor | Total | Sell |
|---|---|---|---|---|---|---|
| Purity shingle | B1@.030 | SF | 1.97 | .88 | 2.85 | 4.85 |

## Shingle Siding

**Cedar sidewall shingle panels.** Shakertown. Clear all-heart, KD. Applied over building paper to solid or spaced nailable sheathing. For use on walls or mansard roofs with minimum 20 in 12 pitch. Use rust resistant, 7d nails for penetration of studs at least 1/2".

| | Craft@Hrs | Unit | Material | Labor | Total | Sell |
|---|---|---|---|---|---|---|
| 1 course shingle panels, 8' long, 7" exposure | B1@.033 | SF | 3.92 | .97 | 4.89 | 8.31 |
| 2 course shingle panels, 8' long, 7" exposure | B1@.031 | SF | 4.05 | .91 | 4.96 | 8.43 |
| 3 course shingle panels, 8' long, 7" exposure | B1@.028 | SF | 4.59 | .82 | 5.41 | 9.20 |
| Designer shingles, 7-1/2" exposure | B1@.040 | SF | 6.29 | 1.17 | 7.46 | 12.70 |
| Striated sidewall shake, 18" with natural groove | B1@.030 | SF | 2.71 | .88 | 3.59 | 6.10 |
| 1 or 2 course corner unit, 7" exposure | B1@.030 | LF | 5.79 | .88 | 6.67 | 11.30 |
| 3 course corner unit, 7" exposure | B1@.030 | LF | 7.91 | .88 | 8.79 | 14.90 |

**Shingle siding.** Cedar rebutted sidewall shingles, Western red cedar.

| | Craft@Hrs | Unit | Material | Labor | Total | Sell |
|---|---|---|---|---|---|---|
| No. 1 western red cedar rebutted and rejointed shingles. Carton covers 25 square feet at 7" exposure, $112 per carton | R1@.035 | SF | 4.50 | 1.09 | 5.59 | 9.50 |

Red cedar perfects 5x green, 16" long, 7-1/2" exposure, 5 shingles are 2" thick (5/2), 3 bundles cover 100 square feet (one square) including waste. Costs include fasteners.

| | Craft@Hrs | Unit | Material | Labor | Total | Sell |
|---|---|---|---|---|---|---|
| #1 | B1@.039 | SF | 2.08 | 1.15 | 3.23 | 5.49 |
| #2 red label | B1@.039 | SF | 1.16 | 1.15 | 2.31 | 3.93 |
| #3 | B1@.039 | SF | .84 | 1.15 | 1.99 | 3.38 |

Panelized shingles, Shakertown, KD, regraded shingle on plywood backing, no waste.

| | Craft@Hrs | Unit | Material | Labor | Total | Sell |
|---|---|---|---|---|---|---|
| 8' long x 7" wide single course | | | | | | |
|   Colonial 1 | B1@.028 | SF | 4.14 | .82 | 4.96 | 8.43 |
|   Cascade | B1@.028 | SF | 4.01 | .82 | 4.83 | 8.21 |
| 8' long x 14" wide, single course | | | | | | |
|   Colonial 1 | B1@.028 | SF | 4.14 | .82 | 4.96 | 8.43 |
| 8' long x 14" wide, double course | | | | | | |
|   Colonial 2 | B1@.028 | SF | 4.53 | .82 | 5.35 | 9.10 |
|   Cascade Classic | B1@.028 | SF | 4.01 | .82 | 4.83 | 8.21 |

Decorative shingles, 5" wide x 18" long, Shakertown, fancy cuts, 9 hand-shaped patterns.

| | Craft@Hrs | Unit | Material | Labor | Total | Sell |
|---|---|---|---|---|---|---|
|   7-1/2" exposure | B1@.030 | SF | 4.84 | .88 | 5.72 | 9.72 |
|   Deduct for 10" exposure | — | SF | - .52 | — | - .52 | — |

## Aluminum Siding and Soffit

Aluminum corrugated 4-V x 2-1/2", plain or embossed finish. Includes 15% waste.

| | Craft@Hrs | Unit | Material | Labor | Total | Sell |
|---|---|---|---|---|---|---|
|   17 gauge, 26" x 6' to 24' | B1@.034 | SF | 2.56 | 1.00 | 3.56 | 6.05 |
|   19 gauge, 26" x 6' to 24' | B1@.034 | SF | 2.65 | 1.00 | 3.65 | 6.21 |
|   Rubber filler strip, 3/4" x 7/8" x 6' | — | LF | .39 | — | .39 | — |
| Flashing for corrugated aluminum siding, embossed. | | | | | | |
|   End wall, 10" x 52" | B1@.048 | Ea | 4.58 | 1.41 | 5.99 | 10.20 |
|   Side wall, 7-1/2" x 10' | B1@.015 | LF | 1.97 | .44 | 2.41 | 4.10 |

| | Craft@Hrs | Unit | Material | Labor | Total | Sell |
|---|---|---|---|---|---|---|
| Aluminum smooth 24 gauge, horizontal patterns, non-insulated. | | | | | | |
| 8" or double 4" widths, acrylic finish | B1@.028 | SF | 1.82 | .82 | 2.64 | 4.49 |
| 12" widths, bonded vinyl finish | B1@.028 | SF | 1.85 | .82 | 2.67 | 4.54 |
| Add for foam backing | — | SF | .38 | — | .38 | — |
| Starter strip | B1@.030 | LF | .44 | .88 | 1.32 | 2.24 |
| Inside and outside corners | B1@.033 | LF | 1.31 | .97 | 2.28 | 3.88 |
| Casing and trim | B1@.033 | LF | .44 | .97 | 1.41 | 2.40 |
| Drip cap | B1@.044 | LF | .45 | 1.29 | 1.74 | 2.96 |
| Aluminum siding trim, fabrication and installation labor only. | | | | | | |
| Cover frieze board, part of a siding job | B1@.055 | LF | — | 1.62 | 1.62 | 2.75 |
| Cover frieze board only | B1@.061 | LF | — | 1.79 | 1.79 | 3.04 |
| Cover gable trim | B1@.075 | LF | — | 2.20 | 2.20 | 3.74 |
| Cover fascia board | B1@.061 | LF | — | 1.79 | 1.79 | 3.04 |
| Cover porch ceiling, part of siding job | B1@.058 | SF | — | 1.70 | 1.70 | 2.89 |
| Cover porch ceiling only | B1@.082 | SF | — | 2.41 | 2.41 | 4.10 |
| Cover porch post and framework | B1@.073 | SF | — | 2.14 | 2.14 | 3.64 |
| Cover window trim, masonry wall | B1@.610 | Ea | — | 17.90 | 17.90 | 30.40 |
| Cover window trim, frame wall | B1@.720 | Ea | — | 21.10 | 21.10 | 35.90 |
| Cover door trim | B1@1.00 | Ea | — | 29.40 | 29.40 | 50.00 |
| Cover trim on one car garage, service door | B1@3.25 | Ea | — | 95.50 | 95.50 | 162.00 |
| Cover trim on two car garage, service door | B1@5.00 | Ea | — | 147.00 | 147.00 | 250.00 |

**Aluminum fascia.** Installed on rafter tails with small tools, working from a ladder. Per linear foot, including 5% waste.

| | Craft@Hrs | Unit | Material | Labor | Total | Sell |
|---|---|---|---|---|---|---|
| 4" x 12' | B1@.025 | LF | .66 | .73 | 1.39 | 2.36 |
| 6" x 12' | B1@.025 | LF | .89 | .73 | 1.62 | 2.75 |
| 8" x 12' | B1@.025 | LF | .94 | .73 | 1.67 | 2.84 |
| 12" x 12' | B1@.025 | LF | .99 | .73 | 1.72 | 2.92 |
| Frieze molding, 1-1/2" x 12' | B1@.030 | LF | .52 | .88 | 1.40 | 2.38 |
| Inside fascia, 12" corner | B1@.333 | Ea | 1.20 | 9.78 | 10.98 | 18.70 |
| Outside fascia, 12" corner | B1@.333 | Ea | 1.20 | 9.78 | 10.98 | 18.70 |

**Aluminum soffit.** Installed on rafter tails with small tools, working from a ladder. Per linear foot, including 5% waste.

| | Craft@Hrs | Unit | Material | Labor | Total | Sell |
|---|---|---|---|---|---|---|
| 12" x 12' non-vented | B1@.031 | LF | .97 | .91 | 1.88 | 3.20 |
| 12" x 12' perforated | B1@.031 | LF | .98 | .91 | 1.89 | 3.21 |
| 12' miter divider | B1@.042 | LF | .75 | 1.23 | 1.98 | 3.37 |
| J-channel support | B1@.033 | LF | .37 | .97 | 1.34 | 2.28 |
| Aluminum nails, 1-1/4', 1/4 pound, 250 nails | — | Ea | 3.98 | — | 3.98 | — |

## Stucco and Masonry Siding

**Portland cement stucco.** Including paperback lath, 3/8" scratch coat, 3/8" brown coat and 1/8" color coat. Per square yard covered.

| | Craft@Hrs | Unit | Material | Labor | Total | Sell |
|---|---|---|---|---|---|---|
| Stucco on walls, | | | | | | |
| Natural gray, sand float finish | PR@.567 | SY | 5.76 | 18.40 | 24.16 | 41.10 |
| Natural gray, trowel finish | PR@.640 | SY | 5.76 | 20.80 | 26.56 | 45.20 |
| White cement, sand float finish | PR@.675 | SY | 6.61 | 22.00 | 28.61 | 48.60 |
| White cement, trowel finish | PR@.735 | SY | 6.81 | 23.90 | 30.71 | 52.20 |
| Stucco on soffits, | | | | | | |
| Natural gray, sand float finish | PR@.678 | SY | 5.49 | 22.10 | 27.59 | 46.90 |
| Natural gray, trowel finish | PR@.829 | SY | 5.49 | 27.00 | 32.49 | 55.20 |
| White cement, sand float finish | PR@.878 | SY | 6.44 | 28.60 | 35.04 | 59.60 |
| White cement, trowel finish | PR@1.17 | SY | 6.62 | 38.10 | 44.72 | 76.00 |

| | Craft@Hrs | Unit | Material | Labor | Total | Sell |
|---|---|---|---|---|---|---|

**Cleaning and pointing masonry.** These costs assume the masonry surface is in fair to good condition, with no unusual damage. Add the cost of protecting adjacent surfaces such as trim or the base of the wall, and the cost of scaffolding, if required. Labor required to presoak or saturate the area cleaned is included in the labor cost. Work more than 12' above floor level will increase costs.

Brushing (hand cleaning) masonry, includes the cost of detergent or chemical solution.

| | Craft@Hrs | Unit | Material | Labor | Total | Sell |
|---|---|---|---|---|---|---|
| Light cleanup (100 SF per manhour) | B9@.010 | SF | .02 | .29 | .31 | .53 |
| Medium scrub (75 SF per manhour) | B9@.013 | SF | .03 | .37 | .40 | .68 |
| Heavy (50 SF per manhour) | B9@.020 | SF | .04 | .57 | .61 | 1.04 |

Water blasting masonry, using rented 400 to 700 PSI power washer with 3 to 8 gallon per minute flow rate, includes blaster rental at $50 per day.

| | Craft@Hrs | Unit | Material | Labor | Total | Sell |
|---|---|---|---|---|---|---|
| Smooth face (250 SF per manhour) | B9@.004 | SF | .05 | .11 | .16 | .27 |
| Rough face (200 SF per manhour) | B9@.005 | SF | .07 | .14 | .21 | .36 |

Sandblasting masonry, using 150 PSI compressor (with accessories and sand) at $150.00 per day.

| | Craft@Hrs | Unit | Material | Labor | Total | Sell |
|---|---|---|---|---|---|---|
| Smooth face (50 SF per manhour) | B9@.020 | SF | .25 | .57 | .82 | 1.39 |
| Rough face (40 SF per manhour) | B9@.025 | SF | .28 | .72 | 1.00 | 1.70 |

Steam cleaning masonry, using rented steam cleaning rig (with accessories) at $52.50 per day.

| | Craft@Hrs | Unit | Material | Labor | Total | Sell |
|---|---|---|---|---|---|---|
| Smooth face (75 SF per manhour) | B9@.013 | SF | .04 | .37 | .41 | .70 |
| Rough face (55 SF per manhour) | B9@.018 | SF | .06 | .52 | .58 | .99 |
| Add for masking adjacent surfaces | — | % | — | 5.0 | — | — |
| Add for difficult stain removal | — | % | 50.0 | 50.0 | — | — |
| Add for working from scaffold | — | % | — | 20.0 | — | — |

Repointing brick, cut out joint, mask (blend-in), and regrout (tuck pointing).

| | Craft@Hrs | Unit | Material | Labor | Total | Sell |
|---|---|---|---|---|---|---|
| 30 SF per manhour | B9@.033 | SF | .05 | .95 | 1.00 | 1.70 |

**Masonry cleaning.** Costs to clean masonry surfaces using commercial cleaning agents, add for pressure washing and scaffolding equipment.

| | Craft@Hrs | Unit | Material | Labor | Total | Sell |
|---|---|---|---|---|---|---|
| Typical cleaning of surfaces, Granite, sandstone, terra cotta, brick | BL@.015 | SF | .28 | .40 | .68 | 1.16 |
| Cleaning surfaces of heavily carbonated Limestone or cast stone | BL@.045 | SF | .38 | 1.20 | 1.58 | 2.69 |
| Typical wash with acid and rinse | BL@.004 | SF | .44 | .11 | .55 | .94 |

## Masonry fireplaces

**Masonry fireplace demolition.** Includes breaking a concrete block or brick fireplace and chimney into manageable pieces using a pneumatic chipper and piling on site. No fireplace foundation included. Add the cost of debris disposal.

| | Craft@Hrs | Unit | Material | Labor | Total | Sell |
|---|---|---|---|---|---|---|
| Demolish interior fireplace and chimney, to 12' high | BL@8.00 | Ea | — | 213.00 | 213.00 | 362.00 |
| Demolish fireplace and exterior chimney, to 12' high | BL@10.7 | Ea | — | 285.00 | 285.00 | 485.00 |
| Demolish chimney over 12' high, per linear foot of height over 12' | BL@.700 | Ea | — | 18.60 | 18.60 | 31.60 |

**Masonry fireplace repairs.** Chip out using a pneumatic chipper.

| | Craft@Hrs | Unit | Material | Labor | Total | Sell |
|---|---|---|---|---|---|---|
| Remove firebrick in a firebox wall | B9@.388 | SF | — | 11.20 | 11.20 | 19.00 |
| Remove firebrick in a firebox floor | B9@.204 | SF | — | 5.86 | 5.86 | 9.96 |
| Replace firebrick firebox wall or floor | B9@.186 | SF | 5.00 | 5.35 | 10.35 | 17.60 |
| Remove and replace nonadjacent fireplace or chimney brick, per brick removed and replaced | B9@.360 | Ea | 1.00 | 10.30 | 11.30 | 19.20 |
| Remove and replace brick in 4" fireplace or chimney wall, per square foot of wall | B9@.675 | SF | 4.00 | 19.40 | 23.40 | 39.80 |

| | Craft@Hrs | Unit | Material | Labor | Total | Sell |
|---|---|---|---|---|---|---|

**Masonry fireplaces.** Fire box lined with refractory firebrick, backed with common brick and enclosed in a common brick wall. Includes fireplace throat and smoke chamber. Overall height of the fireplace and throat is 5'. Chimney (flue) is made from common brick and a flue liner. Including lintels, damper and 12" thick reinforced concrete foundation.

| | Craft@Hrs | Unit | Material | Labor | Total | Sell |
|---|---|---|---|---|---|---|
| 30" wide x 29" high x 16" deep fire box | B9@39.7 | LS | 451.00 | 1,140.00 | 1,591.00 | 2,700.00 |
| 36" wide x 29" high x 16" deep fire box | B9@43.5 | LS | 476.00 | 1,250.00 | 1,726.00 | 2,930.00 |
| 42" wide x 32" high x 16" deep fire box | B9@48.3 | LS | 528.00 | 1,390.00 | 1,918.00 | 3,260.00 |
| 48" wide x 32" high x 18" deep fire box | B9@52.8 | LS | 578.00 | 1,520.00 | 2,098.00 | 3,570.00 |
| 12" x 12" chimney to 12' high | B9@9.50 | LS | 270.00 | 273.00 | 543.00 | 923.00 |
| Add for higher chimney, per LF | B9@1.30 | LF | 36.80 | 37.40 | 74.20 | 126.00 |
| Add for 8" x 10" ash cleanout door | B9@.500 | SF | 16.00 | 14.40 | 30.40 | 51.70 |
| Add for combustion air inlet kit | B9@.750 | SF | 25.00 | 21.60 | 46.60 | 79.20 |
| Add for common brick veneer interior face | B9@.451 | SF | 3.85 | 13.00 | 16.85 | 28.60 |
| Add for rubble or cobble stone veneer face | B9@.480 | SF | 10.50 | 13.80 | 24.30 | 41.30 |
| Add for limestone veneer face or hearth | B9@.451 | SF | 20.00 | 13.00 | 33.00 | 56.10 |
| Add for 1-1/4" marble face or hearth | B9@.451 | SF | 30.00 | 13.00 | 43.00 | 73.10 |
| Add for modern design 11" x 77" pine mantel | B1@3.81 | Ea | 600.00 | 112.00 | 712.00 | 1,210.00 |
| Add for gas valve, log lighter and key | PM@.750 | Ea | 68.00 | 27.30 | 95.30 | 162.00 |

**Prefabricated fireplaces.** Zero clearance factory-built fireplaces, including metal fireplace body, refractory interior and fuel grate. UL listed.

| | Craft@Hrs | Unit | Material | Labor | Total | Sell |
|---|---|---|---|---|---|---|
| Radiant heating, open front only, with fuel grate | | | | | | |
|    38" wide | B9@5.00 | Ea | 549.00 | 144.00 | 693.00 | 1,178.00 |
|    43" wide | B9@5.00 | Ea | 667.00 | 144.00 | 811.00 | 1,380.00 |
| Radiant heating, open front and one end | | | | | | |
|    38" wide | B9@5.00 | Ea | 831.00 | 144.00 | 975.00 | 1,660.00 |
| Radiant heating, three sides open, with brass and glass doors and refractory floor | | | | | | |
|    36" wide (long side) | B9@6.00 | Ea | 1,330.00 | 172.00 | 1,502.00 | 2,550.00 |
| Radiant heating, air circulating, open front only, recessed screen panels | | | | | | |
|    38" screen size | B9@5.50 | Ea | 748.00 | 158.00 | 906.00 | 1,540.00 |
|    43" screen size | B9@5.50 | Ea | 865.00 | 158.00 | 1,023.00 | 1,740.00 |
| Radiant heating simulated masonry fireplace | | | | | | |
|    45" wide | B9@5.00 | Ea | 1,250.00 | 144.00 | 1,394.00 | 2,370.00 |
| Convection-type heat circulating prefabricated fireplace, open front | | | | | | |
|    38" screen size | B9@5.50 | Ea | 624.00 | 158.00 | 782.00 | 1,330.00 |
|    43" screen size | B9@5.50 | Ea | 767.00 | 158.00 | 925.00 | 1,570.00 |
| Forced air heat circulating prefabricated fireplace, open front | | | | | | |
|    35" screen size | B9@5.50 | Ea | 549.00 | 158.00 | 707.00 | 1,200.00 |
|    45" screen size | B9@5.50 | Ea | 831.00 | 158.00 | 989.00 | 1,680.00 |
| Radiant, see-through with brass and glass doors, designer model | | | | | | |
|    38" screen size | B9@6.50 | Ea | 1,330.00 | 187.00 | 1,517.00 | 2,580.00 |
| Header plate, | | | | | | |
|    Connects venting system to gas fireplace | B9@.167 | Ea | 76.30 | 4.80 | 81.10 | 138.00 |
| Fireplace flue (chimney), double wall, typical cost for straight vertical installation | | | | | | |
|    8" inside diameter | B9@.167 | LF | 23.00 | 4.80 | 27.80 | 47.30 |
|    10" inside diameter | B9@.167 | LF | 32.80 | 4.80 | 37.60 | 63.90 |
| Add for flue spacers, for passing through ceiling | | | | | | |
|    1" or 2" clearance | B9@.167 | Ea | 13.30 | 4.80 | 18.10 | 30.80 |
| Add for typical flue offset | B9@.250 | Ea | 84.00 | 7.19 | 91.19 | 155.00 |
| Add for flashing and storm collar, one required per roof penetration | | | | | | |
|    Flat to 6/12 pitch roofs | B9@3.25 | Ea | 53.40 | 93.40 | 146.80 | 250.00 |
|    6/12 to 12/12 pitch roofs | B9@3.25 | Ea | 77.60 | 93.40 | 171.00 | 291.00 |
| Add for spark arrestor top | B9@.500 | Ea | 109.00 | 14.40 | 123.40 | 210.00 |

# Roofing and Flashing

**7**

**D**amage from a leaky roof is usually obvious. However, the precise source of the leak won't always be clear. The hole in the roof is seldom directly above the drip stains on the ceiling. Water gets through the roof surface and runs along the top of sheathing until it finds a panel edge. Then it may run along a rafter or the underside of the panel for a foot or so before dropping down to the attic where it puddles on the vapor barrier. The puddle expands downhill, finds a seam, and then begins saturating the drywall or plaster below. The distance between the hole in the roof and the wet spot on the ceiling can be several feet or more.

So where do you start patching? If the attic crawl space is readily accessible, start your search in the attic. Look for recent water stains. Once you have that information, you can go up on the roof and look for:

❖ *Missing shingles.* Leaks are nearly certain when roof components get blown away.

❖ *Physical damage.* Foot traffic can break seals and dislodge shingles. Was anyone working on the roof or in the attic before the last rain? Check for a tree branch that could be hitting the roof's surface during a high wind.

❖ *Deteriorated caulking.* Caulked joints harden with age and then fracture along the joint with movement.

❖ *Debris.* Anything resting on a roof can collect water and speed deterioration. Debris tends to accumulate on the high side of the roof next to penetrations such as a chimney or skylight.

❖ *Shiners.* Shiners are nails not covered by the shingle course above. If left exposed for many years, the nails can rust out, leaving a hole in the roof.

❖ *Broken mortar.* Any gap or opening in a chimney or parapet wall offers an opening to the attic.

❖ *Obstructed gutters.* Be sure rain runoff gear is working properly. Water accumulating in a gutter can back up over the top of the fascia, run along the soffit, and then down the wall interior.

❖ *Open roll roofing seams.* Even if you see bitumen or adhesive sticking out under a lap, the seam may be wicking moisture to the interior. Try running a putty knife or pocketknife blade under the lap. If the blade slides into the lap more than an inch, assume the seam needs to be resealed.

❖ *Roof blisters.* Blistering (air pockets trapped between the layers of roofing material) is a failure of the bond. Punctured blisters are an obvious source of leaks.

❖ *Splits in the membrane.* Walk the area with your feet close together, taking many small steps. Twist your feet as you walk. If a split is developing, you'll see roof membrane separate between your feet.

❖ *Obstructed roof drains.* Drains are made to carry water one way. When backed up, the roof will leak.

❖ *Deteriorated expansion joints.* Seams on expansion joints may have to be replaced more often than the roof itself.

❖ *A leak when it isn't raining.* Condensate from roof-mounted coolers should be piped to a legal drain, not discharged on the roof.

❖ *Hundred year flood.* It's probably OK to ignore leaks first noticed after the heaviest rain of the century or when rain is driven by unusually heavy wind. Neither you nor the present owner may be around the next time that happens.

# Flashing Leaks

If the roof surface appears sound, look for problems with the flashing. The term flashing is used to describe two types of roofing material. One is metal flashing, which is used at roof joints, such as in the valley of a hip roof where one surface meets another, or the metal counterflashing that extends down a parapet wall and over the roof surface below. The other type is non-metallic flashing. It's usually made from asphalt-coated materials and adds an extra layer of protection under the roof cover, especially along the eave line. In cold climates, roll-roof flashing helps prevent damage from ice dams, as shown in Figure 7-1. Ice dams form when melting snow runs down the roof and re-freezes at the cornice. The ice forms a barrier, causing water to back up under the shingles. A layer of 45-pound smooth-surface roll roofing (flashing) won't prevent the dam from forming, but it will minimize the chance of water entering the attic and the wall below. Be sure this flashing extends 36" inside the warm wall. If two strips are needed, seal the joint with mastic. Also seal the end joints with mastic. Many building departments require self-adhesive type roll roofing on eaves and gables.

Metal counterflashing is lapped over the top of bituminous base flashing to keep water above the roof surface. The most common metal counterflashing problems are:

❖ Corrosion and deterioration of the metal.

❖ Open joints between pieces of metal flashing.

❖ Separation of counterflashing from vertical surfaces.

❖ Air gaps between the counterflashing and base flashing. Counter-flashing that ends too high above the base flashing.

Common problems with non-metallic base flashing include:

❖ Flashing that's too thin, such as 15-pound roofing felt used where 45-pound is needed.

❖ Flashing that's not wide enough, such as an 18" strip of felt used where 36" felt is needed.

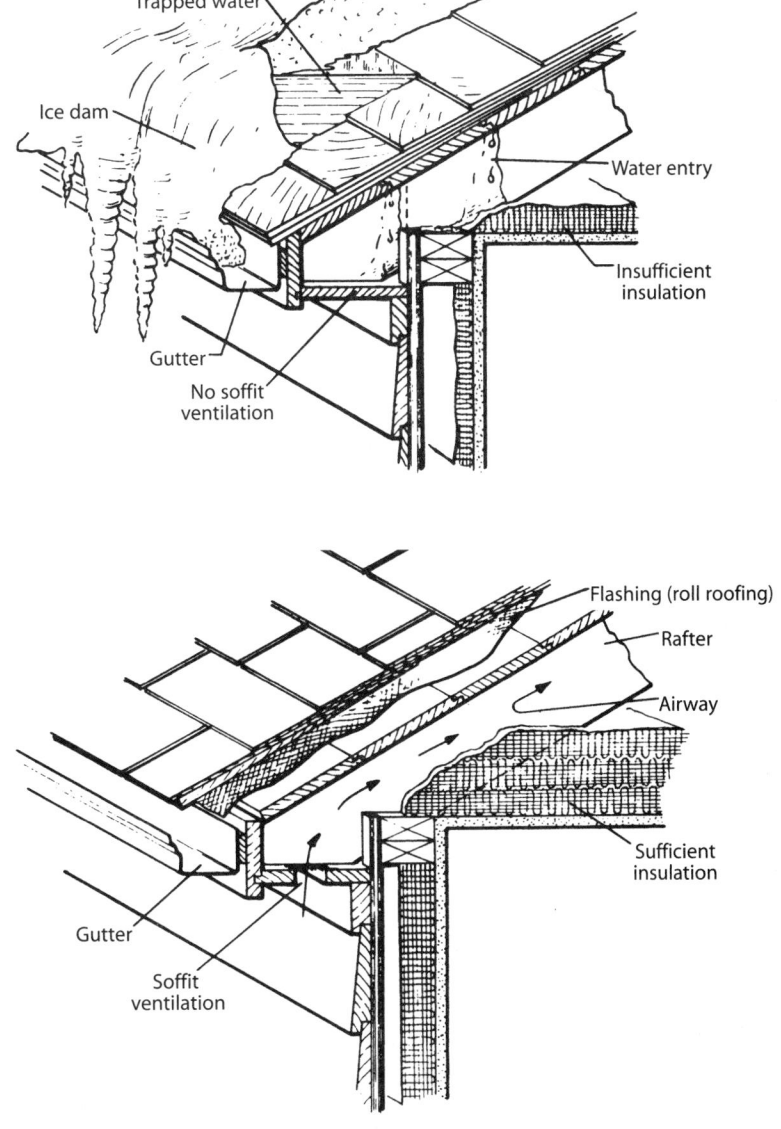

**Figure 7-1**

*Flashing and ventilation to prevent damage from ice dams.*

❖ Lack of a 45-degree cant strip where a roof joins the wall, leaving the flashing unsupported at the base of the wall.

❖ Poorly sealed seams and end laps.

❖ Poorly fastened base flashing at walls or curbs.

❖ Decay of the sheathing below.

Both metallic and non-metallic flashing can be used on the roof surface (*base*) with most types of roof cover. Built-up roofing is an exception. As the surface heats and cools during the day, metal flashing expands and contracts much more than roll roofing. The result will be ripples, tearing and eventually separation of the built-up roofing at the edges of the metal flashing. Use roll-roofing materials as flashing under a built-up roof.

> Attaching an antenna, flagpole, sign or bracing directly to the roof membrane is asking for trouble. Leaks are almost certain. The best repair is to remove the attachment. However, if it must be mounted on top of the roof, mount it using a raised curb-type support. Then install flashing to keep the roof watertight.

All penetrations in the roof surface need flashing. Plumbing stacks and small vents should be flashed with a roof jack or flange placed directly on the last ply of roofing material and stripped in with felt and mastic or felt and bitumen. Prefabricated vinyl or plastic seals for plumbing stacks are effective, and are available in a wide variety of styles and sizes. Larger penetrations should be made through a roof-mounted curb built from 2 x 6 or larger framing lumber. Bituminous base flashing should run up the curb, and metal counterflashing should run down the curb from above.

Leaks around a roof penetration are usually caused by:

❖ Open or broken seams in a metal curb due to expansion and contraction.

❖ Standing water behind a penetration.

❖ Sagging or separating base flashing.

❖ Missing or deteriorated counterflashing.

❖ Splitting or separation of felt stripping over the edge of metal flanges.

❖ Improper priming and stripping of metal surfaces.

❖ Fasteners working loose around the flashing.

❖ Movement between stack vents or pipes and flashing.

## Can't Find the Leak? If All Else Fails . . .

If you're truly stumped about the source of a leak, or have been called back several times to make unsuccessful repairs of the same leak, try this; take a garden hose up on the roof and direct a steady stream of water at every source likely to leak. When the leak is active again, trace the path of the leak from the building interior to the roof surface. Be sure to brief the owner before you test for leaks — and get the owner's help in reporting an active leak.

# Roof Repairs

Damaged or missing shingles will need to be replaced. In other cases, once you've found the source of a leak, use trowel-grade wet patch plastic roofing cement to make repairs. First, wire brush or scrape away everything that's loose, including gravel, granules, membrane, dirt, dust and debris. Then trowel on plastic roofing cement so the patch extends at least 6" beyond the edge of the repair. Press the roofing cement firmly into the membrane. Then test the patch by laying the trowel flat on the cement surface. Plastic cement shouldn't lift off when you pull the trowel away. If it does, work the cement deeper into the membrane. Next, cut fiberglass-reinforcing fabric to fit the patch. Press the fabric into the first layer of plastic roof cement. Fiberglass reinforces the plastic cement and extends the useful life of the repair. Over the embedded fabric, trowel on another layer of plastic roof cement. Be sure to cover all the exposed fabric. Finally, dust the patch with mineral granules that match the existing surface. Granules make the patch more durable by adding sun and weather protection.

On a smooth surface roof, such as built-up asphalt, E.P.D.M. or PVC thermoplastic, make emergency repairs with roof tape membrane. Start by cleaning the damaged surface with alcohol or a household cleaning solution, such as window cleaner. Then wipe the surface with a clean cloth saturated with splice cleaner or white gas. (Use rubber gloves to protect your skin.) Apply roof tape to the damaged roofing material. Be sure to press the tape firmly over the entire surface. If you can't locate a source for roof tape, follow the procedure described here, but apply a smooth coat of butyl or polyurethane caulk.

Most flashing can be patched with plastic roofing cement, but on a new roof the useful life of the repair will usually be less than for the roof surface. Make more permanent flashing repairs by stripping off old flashing down to the roof surface. Then build up new flashing in layers of felt and adhesive. You can make flashing around scuppers and drains watertight with fresh caulk.

After high winds, on homes with a flat, gravel-covered roof, it's common to find roof gravel scattered, leaving some areas bare and unprotected. Before redistributing the gravel, inspect the membrane for open seams, punctures, and tears from flying debris. If you find a leak that's the result of a failed expansion joint or deteriorated caulk, remove and replace the defective material.

# Life Expectancy of Roofing Materials

The life expectancy of roofing materials will vary considerably with the type and how well it's maintained. Common 3-tab asphalt shingles last about 20 years. Under favorable conditions, a wood shingle roof will last 30 years. A 3-ply built-up roof should last 15 years, and you can expect 20 years from a 4-ply or 5-ply built-up roof. Some architectural grade laminated shingles are rated at 45 years. And a well-maintained slate or tile roof can last 150 years.

Premature roof failure is usually caused by fungus, high winds, roof deck deterioration or failure of the flashing. In some climates, portions of a roof can develop fungus that discolors and eventually deteriorates the shingles. After a high wind (over 50 miles per hour), it's common to see shingles blown off. Replacing a shingle takes only minutes, but that's seldom the extent of damage. High winds create a vacuum above portions of the roof, lifting and flexing the surface like the wing of an airplane. Such movement of the roof deck will cause seals to break and fasteners to work loose. It's this unseen damage you have to watch out for — it may not be apparent until the next rain.

When a roof is approaching the end of its useful life, you'll see obvious signs of weathering. Asphalt shingles lose granules, curl on the edges and get brittle. Wood shingles break, warp, curl and blow off in high winds. A good wood shingle roof forms a perfect mosaic. A worn wood shingle roof looks ragged. Excessively worn roofing loses the ability to shed water, making it a candidate for replacement — even if it hasn't begun to leak.

# Choosing a New Roof Material

The roof is a major design element of most homes. If the roof cover has to be replaced, be sure to select material that fits the character of the building. You don't want to put Spanish tile on a Tudor house.

A major consideration in choosing a new roof covering is the weight of the material. You can't replace lighter roofing materials (such as wood shingles) with heavier roofing materials (such as concrete tile or slate) unless the owner is prepared to reinforce the roof to support the additional load. Select materials that are similar in weight to what you're replacing.

Wood shingles and composition shingles are still the most popular roofing materials for pitched roofs, though in an increasing number of areas in the U.S., wood shingles untreated with fire-retardant may not be installed. Both of these can be applied directly over some types of roofing without doing a tear-off. But there are limits. First, ask your customer if the house has been previously reroofed. When in doubt, you'll need to get up on a ladder and peel back a corner of the existing shingles to see what's under them. Be sure you understand your local building department's requirements regarding layering and tear-offs.

Here are some other tips to help you decide whether or not you can avoid a tear-off. You can't install fiberglass strip shingles over hexagon, Dutch lap or T-lock shingles. However, you can install laminated fiberglass strip shingles over asphalt shingles without cutouts and over wood shingles, assuming the old surface is smooth or can be made smooth. T-lock asphalt shingles can't be installed over hexagonal, Dutch lap or wood shingles, or on a roof with 2 in 12 or 3 in 12 pitch. But T-locks can be installed over 3-tab shingles and shingles without cutouts. Slate, tile, wood shakes and cement fiber roofing must always be removed before reroofing.

Before signing a contract for reroofing over an old roof, consider:

❖ *Building codes.* Some building codes prohibit installing a new roof over old. Others allow either two or three layers of roofing. A few allow a one-layer tear-off — remove and replace just the top layer of a two- or three-layer roof. If in doubt, place a call to your local building department.

❖ *Bad decking.* Walk the deck. If you see sags between the rafters or trusses, or if the deck feels weak under foot, the sheathing may be deteriorating or damaged. You may find that only a small section of the deck needs to be replaced. This is often caused by a long-term leak. If possible, go into the attic space and inspect the deck from underneath. This may show you the extent of the problem. Do a tear-off and plan on repairing the deck. The deck under a wood shingle roof may use skip sheathing: 1" x 2" to 6" boards laid with a 1" to 3" gap between boards. Skip sheathing is fine for wood shakes and shingles, but asphalt shingles need a solid board or plywood deck.

❖ *Ice dams.* A full tear-off is best if there's a history of damage from ice dams.

❖ *Incompatible shingles.* Installing heavyweight architectural shingles over lightweight 3-tabs is nearly always successful. A lightweight shingle installed over a heavyweight shingle may leave obvious bumps and ridges.

❖ *Poor condition.* If the existing roof is in really bad shape, such as with curled tabs and crooked rows, a tear-off is in order.

❖ *Shorter life span.* Most roofing professionals agree that the life expectancy of a new roof is reduced by 10 to 20 percent when it's installed as an overlay on an old roof.

When installing new shingles over old wood or asphalt shingles, manufacturers may recommend removing a 6" wide strip of the old shingles along eaves and gables. Then nail down a nominal l" x 6" board where the shingles were cut away. If the old shingles were asphalt 3-tabs, use thinner boards. Remove the old covering from ridges and hips and place a strip of lumber over each valley to separate old metal flashing from new.

# Wood Shingles

If you decide on wood shingles as your new roofing material, use Number 1 grade shingles. Number 1 shingles are all heartwood, all edge grain and tapered. Western red cedar shingles and redwood shingles offer the best decay resistance and shrink very little as they age. Wood shingle widths vary. Narrower shingles are classified as either Number 2 or Number 3. Figure 7-2 shows common shingle sizes and recommended exposures.

| Maximum Exposure for Wood Shingles | | | |
|---|---|---|---|
| Length | Thickness (green) | Slope less [1] than 4 in 12 | Slope 4 in 12 or over |
| 16" | 5 butts in 2" | 3³/₄" | 5" |
| 18" | 5 butts in 2¹/₄" | 4¹/₄" | 5¹/₂" |
| 24" | 4 butts in 2" | 5³/₄" | 7¹/₂" |

[1] Minimum slope for main roofs — 4 in 12.
Minimum slopes for porch roofs — 3 in 12.

**Figure 7-2**

*Recommended exposure for wood shingles*

Follow the standards illustrated in Figure 7-3 when you apply wood shingles:

**1.** Extend the shingles 1¹/₂" to 2" beyond the eave line and about ³/₄" beyond the rake (gable) edge.

**2.** Nail each shingle with two rust-resistant nails spaced about ³/₄" from the edge and 1¹/₂" above the butt line of the next course. Use 3d nails for 16" and 18" shingles and 4d nails for 24" shingles. If you're applying new shingles over old wood shingles, you'll need longer nails to penetrate through the old roofing and into the sheathing. Use ring-shank (threaded) nails on plywood roof sheathing that's less than ¹/₂" thick.

**3.** Allow a ¹/₈" to ¹/₄" expansion space between shingles. Stagger the joints in succeeding courses so the joint in one course isn't in line with either of the two courses above.

**4**. Shingle away from valleys, using the widest shingles in the valleys. The valley should be 4" wide at the top and increase in width at the rate of ¹/₈" per foot from the top to the bottom. Valley flashing with a standing seam (W valley) works best. Don't nail shingles through the metal. Valley flashing should be a minimum of 24" wide for roof slopes under 4 in 12; 18" wide for roof slopes of 4 in 12 to 7 in 12; and 12" wide for roof slopes of 7 in 12 and over.

**5.** Place metal edging along the gable end to direct water away from the end walls.

You can use the same instructions for applying wood shakes, except you'll need longer nails because shakes are thicker. Shakes are also longer and the exposure distances can be greater: 8" for 18" shakes, 10" for 24" shakes, and 13" for 32" shakes. For a rustic appearance, lay the butts unevenly.

Install an 18"-wide strip of 30-pound asphalt felt between each course to protect against wind-driven snow. The lower edge of the shake felt should be above the butt end by twice the exposure distance. If the exposure distance is less than one-third the total length, most shingle roofers don't bother with shake felt.

# Asphalt-Fiberglass (Composition) Shingles

Asphalt shingles are made from a fiberglass mat saturated with asphalt. Fiberglass adds strength and durability, and asphalt adds moisture protection. But asphalt doesn't bind very well with fiberglass, so mineral fibers are added to the fiberglass before coating. The result is a "composition" shingle that is both durable and effective.

The most common type of asphalt shingle is the square-butt strip shingle. It measures 12" by 36", has three tabs, and is usually laid with 5" exposed to the weather. Top-quality laminated asphalt shingles weigh as much as 450 pounds per 100 square feet (a square) and can be expected to last up to 45 years. Heavyweight

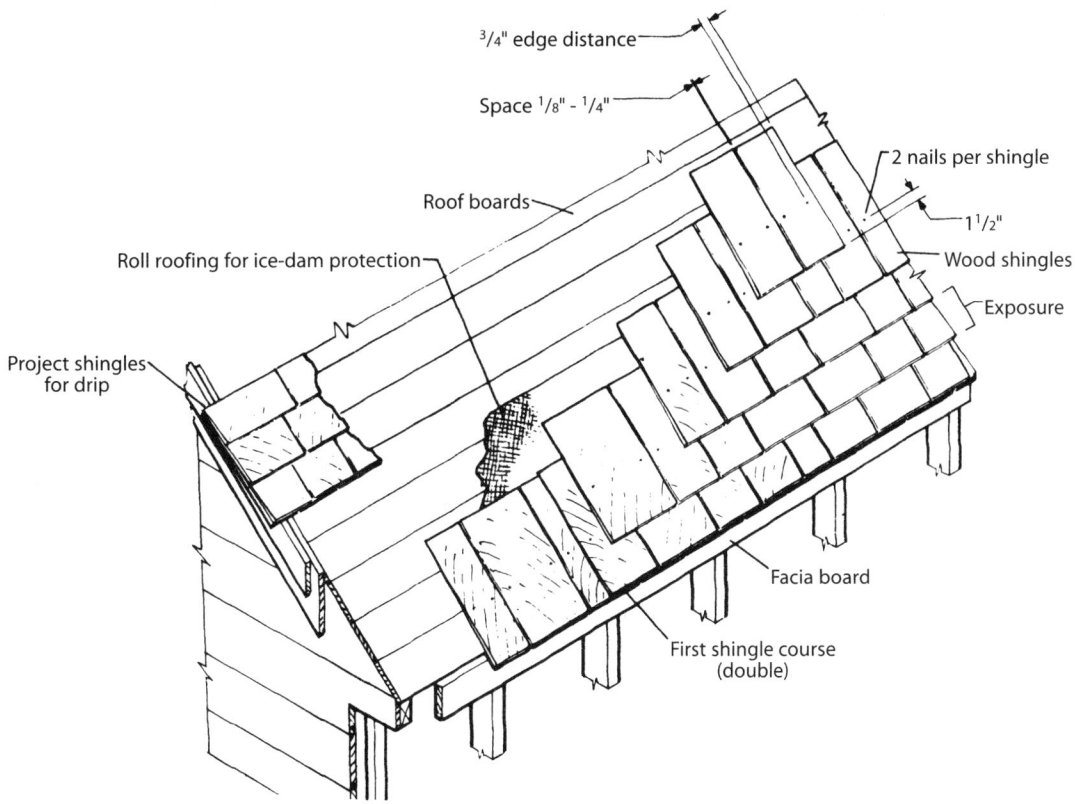

**Figure 7-3**

*Application of wood-shingle roofing over boards*

laminated asphalt shingles are sold in four bundles to the square. Regular weight asphalt shingles are sold in three bundles to the square. Store the bundles flat so the strips don't curl when they're installed. Avoid storing them in the sun, which will activate the self-sealing tabs, causing the top shingles in each bundle to stick together.

Install 15-pound saturated felt before laying asphalt shingles on a roof with a 4 in 12 to 7 in 12 pitch. Start by laying the felt at the roof eave and lap each successive course by several inches. Stagger the felt courses to separate joints. Apply a double layer of 15-pound felt on a roof with less than a 4 in 12 pitch. Doubling the felt will take more time because it's harder to keep wrinkles out.

Begin application with metal edging along the eave line. Cut the exposure tabs off the first course of asphalt shingles and lay them along the roof edge. See Figure 7-4 A. Then apply another course on top of the first, extending the shingle exposure 1/2" over the metal edging at the rake to prevent water from backing up under the shingles (Figure 7-4 B). Apply the next course with the recommended exposure. Snap horizontal chalk lines on the felt to help keep shingles in straight lines. Follow the manufacturer's recommendation on nailing. When a nail penetrates a crack or knothole, remove the nail, seal the hole, and drive another nail into sound wood. Any nail that doesn't hit sound wood will gradually work out and cause a hump in the shingle above. Use seal-tab or T-lock shingles in high wind areas.

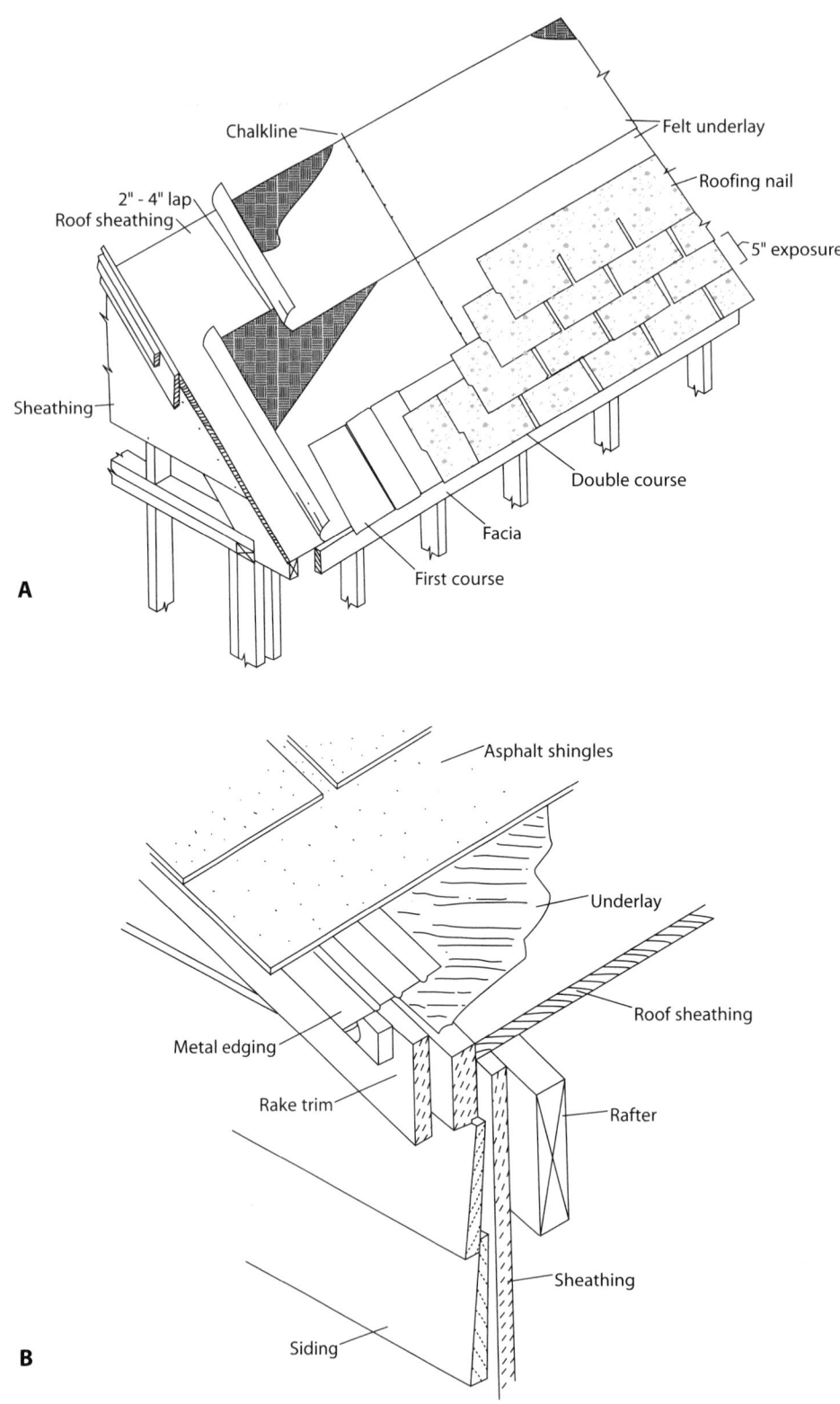

**Figure 7-4**

*Application of asphalt-shingle roofing over plywood: **A**, with strip shingles ; **B**, metal edging at gable end.*

# Built-up Roofing

A built-up roof that isn't worth repairing must be stripped off before the new roofing is installed. Once the deck is clean, pound flush or pull all exposed nails and make repairs to the deck and flashing as needed.

The base sheet for a built-up roof can be either resin-coated or asphalt-coated organic felt or fiberglass. On a wood deck, nail the base sheet to the deck. Use a Type I or Type II venting base sheet on a concrete deck.

The felt layers are applied on top of the base sheet. Type I (15-pound) and Type II (30-pound) asphalt-saturated organic felts are the most common. Asphalt-impregnated fiberglass felts are Types III, IV and V. Type III fiberglass felt is lower quality than Type IV. Coal tar-saturated felts are popular in some areas. Polyester felts are stronger than fiberglass felts, and fiberglass felts are stronger than organic felts.

Lay each 100 square feet of felt in 25 to 30 pounds of hot asphalt, 20 to 25 pounds of coal tar or 2 to 4 gallons of cold-applied roofing adhesive (lap cement). If the last layer is a mineral surface cap sheet, use the same quantity of asphalt, coal tar or adhesive. For a topping of gravel or slag, flood each 100 square feet of surface with 60 pounds of hot asphalt or 70 pounds of coal tar or 3 gallons of emulsion. Then install 400 to 500 pounds of gravel or 300 to 400 pounds of slag per 100 square feet of roof surface. If gravel or slag is set in emulsion, apply an aluminum reflective coating to protect against UV damage.

# Modified Bitumen Roofing (APP and SBS)

In the 1970s, chemists in Europe discovered that asphalt took on rubber-like properties when combined with certain additives. These additives include APP (atactic polypropylene) and SBS (styrene butadiene styrene). Adding about 30 percent APP to asphalt creates a product that can be stretched up to 50 percent without breaking. SBS modified roofing was developed by French and German chemists. They found that asphalt took on the character of rubber when 10 to 15 percent SBS was added. SBS modified asphalt will stretch up to six times its original length without breaking. Unlike APP, SBS modified asphalt returns to its original size when released. This elasticity makes asphalt roofing more durable. The roof surface can shift, or expand and contract, without breaking.

APP roofing uses a polyester mat that matches the pliability of APP modified asphalt better than a woven glass mat. You apply APP membranes using a torch. The back of the sheet has extra asphalt that bonds to the surface below when heated. That's convenient on smaller jobs with limited working area, because you need very little equipment.

You can use many mat materials with SBS, including fiberglass and polyester. Fiberglass SBS mats weigh from 1.0 to 2.5 pounds per 100 square feet. Polyester SBS mats weigh from 3.5 to 5.0 pounds per 100 square feet. You can apply SBS membranes with hot asphalt, a torch, or cold adhesive.

Modified bitumen membrane roofing comes in 100-square-foot rolls. Use one, two, or three plies to create a finished roof. Or, combine modified bitumen membrane with fiberglass felt to form a hybrid roof system.

The minimum roof slope should be not less than $1/4$" of rise per 12" of horizontal run. The surface can be finished with mineral granules or an aggregate, such as gravel or slag.

# How to Measure Roof Area

Flat roofs are easy to measure. Just divide the area into rectangles, calculate the area of each, and add up the total. You can probably do that from the ground. Sloping roofs are a little more complex. The steeper the slope, the more the roof area differs from the horizontal (plane) area. To calculate the roof surface accurately, you have to multiply the rafter lengths by the width of each pitch. However, if the roof pitch is no more than 5 in 12, a simple rule of thumb will yield nearly the same result in a fraction of the time: Multiply the length and width of the building, including eaves and overhang. Divide by 100 to find the number of roofing "squares." Then add 10 percent for a gable roof, 15 percent for a hip roof and 20 percent for a roof with dormers.

| | Craft@Hrs | Unit | Material | Labor | Total | Sell |
|---|---|---|---|---|---|---|

## Roof Repairs

**Tear-off roofing.** Includes removing the existing surface and throwing debris into a bin below a one-story or two-story building. Also includes pulling or driving all protruding nails but no repairs to the roof deck, flashing or rafters. Add the cost of hauling away debris and dump or recycling fees. Roofing tear-off will take longer when the pitch is steeper than 6 in 12.

| | Craft@Hrs | Unit | Material | Labor | Total | Sell |
|---|---|---|---|---|---|---|
| Asphalt or fiberglass shingles, single layer | BL@1.00 | Sq | — | 26.60 | 26.60 | 45.20 |
| Asphalt or fiberglass shingles, double layer | BL@1.45 | Sq | — | 38.60 | 38.60 | 65.60 |
| Wood shingles or shakes | BL@1.50 | Sq | — | 40.00 | 40.00 | 68.00 |
| Slate shingles | BL@1.79 | Sq | — | 47.70 | 47.70 | 81.10 |
| Clay or concrete tile | BL@1.65 | Sq | — | 44.00 | 44.00 | 74.80 |
| Remove gravel stop or flashing, per LF | BL@.070 | LF | — | 1.86 | 1.86 | 3.16 |
| Remove chimney flashing, per chimney | BL@1.50 | Ea | — | 40.00 | 40.00 | 68.00 |
| Roof insulation board | BL@.650 | Sq | — | 17.30 | 17.30 | 29.40 |
| Gypsum plank | BL@3.16 | Sq | — | 84.20 | 84.20 | 143.00 |
| Metal decking | BL@3.52 | Sq | — | 93.80 | 93.80 | 159.00 |
| Remove vent flashing, per vent | BL@.400 | Sq | — | 10.70 | 10.70 | 18.20 |
| Remove and reset skylight to 10 SF | BL@2.20 | Ea | — | 58.60 | 58.60 | 99.60 |

**Tear-off built-up roofing.** Includes removing the existing surface and throwing debris into a bin below a one-story or two-story building. Add the cost of hauling away debris and dump or recycling fees.

| | Craft@Hrs | Unit | Material | Labor | Total | Sell |
|---|---|---|---|---|---|---|
| Roofing on wood deck | BL@1.08 | Sq | — | 28.80 | 28.80 | 49.00 |
| Roofing and insulation on wood deck | BL@1.73 | Sq | — | 46.10 | 46.10 | 78.40 |
| Roofing on concrete deck | BL@1.31 | Sq | — | 34.90 | 34.90 | 59.30 |
| Roofing and insulation on concrete deck | BL@1.96 | Sq | — | 52.20 | 52.20 | 88.70 |

**Repair roof deck — remove and replace.** Includes removing the existing surface, throwing debris into a bin below a one-story or two-story building, pulling or driving all protruding nails but no repairs to flashing or rafters. Add the cost of hauling away debris and dump or recycling fees. Per square foot removed and replaced.

| | Craft@Hrs | Unit | Material | Labor | Total | Sell |
|---|---|---|---|---|---|---|
| 2" x 6" T&G boards | B1@.032 | SF | .53 | .94 | 1.47 | 2.50 |
| 2" x 10" T&G boards | B1@.029 | SF | .57 | .85 | 1.42 | 2.41 |
| 1" x 6" square edge boards | B1@.029 | SF | .24 | .85 | 1.09 | 1.85 |
| 1" x 8" square edge boards | B1@.027 | SF | .24 | .79 | 1.03 | 1.75 |
| 1" x 12" square edge boards | B1@.026 | SF | .26 | .76 | 1.02 | 1.73 |
| Plywood 4' x 8' sheets or partial sheets | B1@.025 | SF | .44 | .73 | 1.17 | 1.99 |

**Repair roof flashing — remove and replace.**

| | Craft@Hrs | Unit | Material | Labor | Total | Sell |
|---|---|---|---|---|---|---|
| Roof drain or vent | R1@.918 | Ea | 13.00 | 28.70 | 41.70 | 70.90 |
| Flashing felt and counterflashing | R1@.160 | LF | 1.30 | 4.99 | 6.29 | 10.70 |
| Pitch pocket, set of two | R1@.400 | Set | 16.20 | 12.50 | 28.70 | 48.80 |
| 1-ply membrane at headwall | R1@.053 | LF | .24 | 1.65 | 1.89 | 3.21 |
| 2-ply membrane at headwall | R1@.076 | LF | .48 | 2.37 | 2.85 | 4.85 |
| 1-ply felt and flashing at headwall | R1@.129 | LF | .60 | 4.03 | 4.63 | 7.87 |
| 2-ply felt and flashing at headwall | R1@.152 | LF | .82 | 4.74 | 5.56 | 9.45 |

**Repair roof waterproofing.** Install membrane on smooth and clean vertical concrete or masonry surface, such as a parapet wall. Per square foot.

| | Craft@Hrs | Unit | Material | Labor | Total | Sell |
|---|---|---|---|---|---|---|
| 1-ply | R1@.025 | SF | .46 | .78 | 1.24 | 2.11 |
| 2-ply | R1@.041 | SF | .86 | 1.28 | 2.14 | 3.64 |
| 3-ply | R1@.056 | SF | 1.42 | 1.75 | 3.17 | 5.39 |
| Vertical joint, 2-ply | R1@.046 | LF | .28 | 1.44 | 1.72 | 2.92 |

**Non-fibered roof and foundation coat.** Moisture barrier on exterior concrete and masonry surfaces or heavy-duty roof primer.

| | Craft@Hrs | Unit | Material | Labor | Total | Sell |
|---|---|---|---|---|---|---|
| Gallon, covers 80 square feet | R1@.006 | SF | .27 | .19 | .46 | .78 |

| | Craft@Hrs | Unit | Material | Labor | Total | Sell |
|---|---|---|---|---|---|---|

**Remove and replace roof waterproofing.** Chip away damaged membrane on a vertical surface, such as a parapet wall. Clean with wire brush. Replace with new waterproof membrane. Per square foot.

| | Craft@Hrs | Unit | Material | Labor | Total | Sell |
|---|---|---|---|---|---|---|
| 1-ply | R1@.042 | SF | .46 | 1.31 | 1.77 | 3.01 |
| 2-ply | R1@.058 | SF | .86 | 1.81 | 2.67 | 4.54 |
| 3-ply | R1@.074 | SF | 1.42 | 2.31 | 3.73 | 6.34 |

**Waterproof roof vent.** Per vent. Based on vents averaging 24" diameter.

| | Craft@Hrs | Unit | Material | Labor | Total | Sell |
|---|---|---|---|---|---|---|
| 2-ply, using hot asphalt | R1@.797 | Ea | 1.65 | 24.90 | 26.55 | 45.10 |
| 4-ply, using hot asphalt | R1@1.59 | Ea | 2.23 | 49.60 | 51.83 | 88.10 |
| 2-ply, using plastic roofing cement | R1@1.28 | Ea | 2.79 | 39.90 | 42.69 | 72.60 |
| 4-ply, using plastic roofing cement | R1@2.54 | Ea | 3.92 | 79.30 | 83.22 | 141.00 |

**Repair roof vent waterproofing.** Chip away damaged membrane around a roof vent. Clean with a wire brush. Replace with new waterproof membrane. Per vent. Based on vents averaging 24" diameter.

| | Craft@Hrs | Unit | Material | Labor | Total | Sell |
|---|---|---|---|---|---|---|
| 2-ply, using hot asphalt | R1@1.56 | Ea | 1.65 | 48.70 | 50.35 | 85.60 |
| 4-ply, using hot asphalt | R1@2.35 | Ea | 2.23 | 73.30 | 75.53 | 128.00 |
| 1-ply, using plastic roofing cement | R1@.764 | Ea | 1.65 | 23.80 | 25.45 | 43.30 |
| 2-ply, using plastic roofing cement | R1@2.04 | Ea | 2.79 | 63.70 | 66.49 | 113.00 |
| 4-ply, using plastic roofing cement | R1@3.31 | Ea | 3.92 | 103.00 | 106.92 | 182.00 |

**Waterproof masonry wall.** Waterproof masonry such as a parapet wall. Membrane set in asphaltic waterproofing. Per square foot of concrete or concrete block wall.

| | Craft@Hrs | Unit | Material | Labor | Total | Sell |
|---|---|---|---|---|---|---|
| 1-ply, vertical wall | R1@.034 | SF | .46 | 1.06 | 1.52 | 2.58 |
| 2-ply, vertical wall | R1@.041 | SF | .86 | 1.28 | 2.14 | 3.64 |
| 3-ply, vertical wall | R1@.056 | SF | 1.42 | 1.75 | 3.17 | 5.39 |

**Remove and replace asphalt shingle.**

| | Craft@Hrs | Unit | Material | Labor | Total | Sell |
|---|---|---|---|---|---|---|
| 5 square foot patch (3 shingles) | R1@.171 | Ea | 1.81 | 5.34 | 7.15 | 12.20 |
| 12 square foot patch (10 shingles) | R1@.367 | Ea | 4.35 | 11.50 | 15.85 | 26.90 |
| 31 square foot patch (25 shingles) | R1@.804 | Ea | 11.20 | 25.10 | 36.30 | 61.70 |
| Re-seal shingle with roofing cement, per shingle | R1@.033 | Ea | .17 | 1.03 | 1.20 | 2.04 |

## Composition Shingles

**Architectural grade laminated shingles, Owens Corning.** Three bundles per square (100 square feet). Class A fire rating. 70 MPH wind resistance.

| | Craft@Hrs | Unit | Material | Labor | Total | Sell |
|---|---|---|---|---|---|---|
| Oakridge, 30 year | R1@1.83 | Sq | 51.40 | 57.10 | 108.50 | 184.00 |
| Oakridge, 30 year, algae-resistant | R1@1.83 | Sq | 54.40 | 57.10 | 111.50 | 190.00 |
| Oakridge, 50 year, algae-resistant | R1@1.83 | Sq | 61.60 | 57.10 | 118.70 | 202.00 |

**Hip and ridge shingles with sealant, Owens Corning.** Per linear foot of hip or ridge.

| | Craft@Hrs | Unit | Material | Labor | Total | Sell |
|---|---|---|---|---|---|---|
| High ridge | R1@.028 | LF | 1.67 | .87 | 2.54 | 4.32 |
| High ridge, algae-resistant | R1@.028 | LF | 1.70 | .87 | 2.57 | 4.37 |
| High style | R1@.028 | LF | 1.38 | .87 | 2.25 | 3.83 |

**3-tab asphalt fiberglass shingles, Owens Corning.** Three bundles per square (100 square feet). Class A fire rating. 60 MPH wind resistance.

| | Craft@Hrs | Unit | Material | Labor | Total | Sell |
|---|---|---|---|---|---|---|
| Supreme, 25 year | R1@1.83 | Sq | 43.90 | 57.10 | 101.00 | 172.00 |
| Supreme, 25 year, algae-resistant | R1@1.83 | Sq | 43.00 | 57.10 | 100.10 | 170.00 |
| Classic, 20 year | R1@1.83 | Sq | 37.60 | 57.10 | 94.70 | 161.00 |
| Classic, 20 year, algae-resistant | R1@1.83 | Sq | 37.30 | 57.10 | 94.40 | 160.00 |

**Hip and ridge shingles with sealant, Owens Corning.** Per linear foot of hip or ridge. 41 linear feet per carton (22 pieces).

| | Craft@Hrs | Unit | Material | Labor | Total | Sell |
|---|---|---|---|---|---|---|
| Asphalt-fiberglass | R1@.028 | LF | .97 | .87 | 1.84 | 3.13 |

| | Craft@Hrs | Unit | Material | Labor | Total | Sell |
|---|---|---|---|---|---|---|

**Glaslock® interlocking shingles, Owens Corning.** Three bundles per square (100 square feet). Class A fire rating. 60 MPH wind resistance.

| | Craft@Hrs | Unit | Material | Labor | Total | Sell |
|---|---|---|---|---|---|---|
| 25 year rating | R1@1.83 | Sq | 48.70 | 57.10 | 105.80 | 180.00 |

**Architectural grade laminated shingles, Timberline® Ultra®, GAF.** Class A fire rating. 110 MPH wind resistance. Four bundles per square (100 square feet).

| | Craft@Hrs | Unit | Material | Labor | Total | Sell |
|---|---|---|---|---|---|---|
| Ultra, lifetime warranty | R1@1.83 | Sq | 80.00 | 57.10 | 137.10 | 233.00 |
| Select, 40 year | R1@1.83 | Sq | 79.90 | 57.10 | 137.00 | 233.00 |
| Select, 40 year, algae-resistant | R1@1.83 | Sq | 82.40 | 57.10 | 139.50 | 237.00 |
| Select, 30 year | R1@1.83 | Sq | 72.90 | 57.10 | 130.00 | 221.00 |
| Select, 30 year, blend, fungus-resist. | R1@1.83 | Sq | 78.20 | 57.10 | 135.30 | 230.00 |

**Hip and ridge shingles, GAF.** Per linear foot of hip or ridge.

| | Craft@Hrs | Unit | Material | Labor | Total | Sell |
|---|---|---|---|---|---|---|
| Timbertex® | R1@.028 | LF | 1.78 | .87 | 2.65 | 4.51 |
| Pacific Ridge | R1@.028 | LF | 1.43 | .87 | 2.30 | 3.91 |
| Timber Ridge | R1@.028 | LF | 1.32 | .87 | 2.19 | 3.72 |

**3-tab fiberglass asphalt shingles, GAF.** Class A fire rating. 60 MPH wind warranty. Three bundles cover 100 square feet at 5" exposure. Per square (100 square feet).

| | Craft@Hrs | Unit | Material | Labor | Total | Sell |
|---|---|---|---|---|---|---|
| Royal Sovereign®, 25 year | R1@1.83 | Sq | 44.60 | 57.10 | 101.70 | 173.00 |
| Royal Sovereign®, 25 year, algae-resistant | R1@1.83 | Sq | 46.40 | 57.10 | 103.50 | 176.00 |
| Sentinel®, 20 year | R1@1.83 | Sq | 35.60 | 57.10 | 92.70 | 158.00 |
| Sentinel®, 20 year, fungus- and algae-resistant | R1@1.83 | Sq | 43.10 | 57.10 | 100.20 | 170.00 |

**Shingle starter strip.** Peel 'N Stick SBS membrane. Rubberized to seal around nails and resist wind lift. Per roll.

| | Craft@Hrs | Unit | Material | Labor | Total | Sell |
|---|---|---|---|---|---|---|
| 7" x 33' | R1@.460 | Ea | 10.65 | 14.40 | 25.05 | 42.60 |

**Asphalt roofing starter strip.**

| | Craft@Hrs | Unit | Material | Labor | Total | Sell |
|---|---|---|---|---|---|---|
| 9" x 36' | R1@.500 | Ea | 7.61 | 15.60 | 23.21 | 39.50 |

## Built-up Roofing

**Patch built-up roofing — concrete deck.** Includes removing a section of the existing built-up roof on a concrete deck and replacing with built-up roofing. Per square (100 square feet) patched.

| | Craft@Hrs | Unit | Material | Labor | Total | Sell |
|---|---|---|---|---|---|---|
| 3-ply roof and gravel | R1@2.94 | Sq | 50.60 | 91.80 | 142.40 | 242.00 |
| 4-ply roof and gravel | R1@3.11 | Sq | 58.40 | 97.10 | 155.50 | 264.00 |
| 5-ply roof and gravel | R1@3.28 | Sq | 70.80 | 102.00 | 172.80 | 294.00 |

**Patch built-up roofing and insulation — concrete deck.** Includes removing a section of the existing built-up roofing and insulation on a concrete deck and replacing with built-up roofing and 1" polystyrene insulation board. Per square (100 square feet) patched.

| | Craft@Hrs | Unit | Material | Labor | Total | Sell |
|---|---|---|---|---|---|---|
| 3-ply roof, gravel and insulation | R1@3.18 | Sq | 94.10 | 99.20 | 193.30 | 329.00 |
| 4-ply roof, gravel and insulation | R1@3.35 | Sq | 102.00 | 105.00 | 207.00 | 352.00 |
| 5-ply roof, gravel and insulation | R1@3.52 | Sq | 114.00 | 110.00 | 224.00 | 381.00 |

| | Craft@Hrs | Unit | Material | Labor | Total | Sell |
|---|---|---|---|---|---|---|

**Patch built-up roofing — wood deck.** Includes removing a section of the existing built-up roof on a wood deck and replacing with built-up roofing. Per square (100 square feet) patched.

| | Craft@Hrs | Unit | Material | Labor | Total | Sell |
|---|---|---|---|---|---|---|
| 2-ply roof and cap sheet | R1@2.66 | Sq | 44.30 | 83.00 | 127.30 | 216.00 |
| 3-ply roof and cap sheet | R1@2.84 | Sq | 46.20 | 88.60 | 134.80 | 229.00 |
| 4-ply roof and cap sheet | R1@3.01 | Sq | 52.30 | 93.90 | 146.20 | 249.00 |

**Patch built-up roofing and insulation — wood deck.** Includes removing a section of the existing built-up roofing and insulation on a wood deck and replacing with built-up roofing and 1" roof insulation board. Per square (100 square feet) patched.

| | Craft@Hrs | Unit | Material | Labor | Total | Sell |
|---|---|---|---|---|---|---|
| 2-ply roof,cap sheet and insulation | R1@2.91 | Sq | 90.80 | 90.80 | 181.60 | 309.00 |
| 3-ply roof,cap sheet and insulation | R1@3.08 | Sq | 92.70 | 96.10 | 188.80 | 321.00 |
| 4-ply roof,cap sheet and insulation | R1@3.25 | Sq | 98.90 | 101.00 | 199.90 | 340.00 |

**Built-up roofing.** Installed over existing suitable substrate. Add the cost of equipment rental (the kettle) and consumable supplies (mops, brooms, gloves). Typical cost for equipment and consumable supplies is $15 per square applied.

Type 1: 2-ply felt and 1-ply 90-lb. cap sheet, including 2 plies of hot mop asphalt.

| | Craft@Hrs | Unit | Material | Labor | Total | Sell |
|---|---|---|---|---|---|---|
| Base sheet, nailed down, per 100 SF | R1@.250 | Sq | 4.19 | 7.80 | 11.99 | 20.40 |
| Hot mop asphalt, 30 lbs. per 100 SF | R1@.350 | Sq | 10.40 | 10.90 | 21.30 | 36.20 |
| Asphalt felt, 15 lbs. per 100 SF | R1@.100 | Sq | 4.19 | 3.12 | 7.31 | 12.40 |
| Hot mop asphalt, 30 lbs. per 100 SF | R1@.350 | Sq | 10.40 | 10.90 | 21.30 | 36.20 |
| Mineral surface, 90 lbs. per 100 SF | R1@.200 | Sq | 23.00 | 6.24 | 29.24 | 49.70 |
| Type 1, total per 100 SF | R1@1.25 | Sq | 52.18 | 39.00 | 91.18 | 155.00 |

Type 2: 3-ply asphalt, 1-ply 30 pound felt, 2 plies 15 pound felt, 3 coats hot mop asphalt and gravel.

| | Craft@Hrs | Unit | Material | Labor | Total | Sell |
|---|---|---|---|---|---|---|
| Asphalt felt, 30 lbs., nailed down | R1@.300 | Sq | 8.37 | 9.36 | 17.73 | 30.10 |
| Hot mop asphalt, 30 lbs. per 100 SF | R1@.350 | Sq | 10.40 | 10.90 | 21.30 | 36.20 |
| Asphalt felt, 15 lbs. per 100 SF | R1@.100 | Sq | 4.19 | 3.12 | 7.31 | 12.40 |
| Hot mop asphalt, 30 lbs. per 100 SF | R1@.350 | Sq | 10.40 | 10.90 | 21.30 | 36.20 |
| Asphalt felt, 15 lbs. per 100 SF | R1@.100 | Sq | 4.19 | 3.12 | 7.31 | 12.40 |
| Hot mop asphalt, 60 lbs. per 100 SF | R1@.350 | Sq | 10.40 | 10.90 | 21.30 | 36.20 |
| Gravel, 400 lbs. per 100 SF | R1@.600 | Sq | 26.50 | 18.70 | 45.20 | 76.80 |
| Type 2, total per 100 SF | R1@2.15 | Sq | 74.45 | 67.10 | 141.55 | 241.00 |

Type 3: 4-ply asphalt, 4 plies 15 pound felt including 4 coats hot mop and gravel.

| | Craft@Hrs | Unit | Material | Labor | Total | Sell |
|---|---|---|---|---|---|---|
| Base sheet, nailed down, per 100 SF | R1@.250 | Sq | 4.19 | 7.80 | 11.99 | 20.40 |
| Hot mop asphalt, 30 lbs. per 100 SF | R1@.350 | Sq | 10.40 | 10.90 | 21.30 | 36.20 |
| Asphalt felt, 15 lbs. per 100 SF | R1@.100 | Sq | 4.19 | 3.12 | 7.31 | 12.40 |
| Hot mop asphalt, 30 lbs. per 100 SF | R1@.350 | Sq | 10.40 | 10.90 | 21.30 | 36.20 |
| Asphalt felt, 15 lbs. per 100 SF | R1@.100 | Sq | 4.19 | 3.12 | 7.31 | 12.40 |
| Hot mop asphalt, 30 lbs. per 100 SF | R1@.350 | Sq | 10.40 | 10.90 | 21.30 | 36.20 |
| Asphalt felt, 15 lbs. per 100 SF | R1@.100 | Sq | 4.19 | 3.12 | 7.31 | 12.40 |
| Hot mop asphalt, 60 lbs. per 100 SF | R1@.350 | Sq | 10.40 | 10.90 | 21.30 | 36.20 |
| Gravel, 400 lbs. per 100 SF | R1@.600 | Sq | 26.50 | 18.70 | 45.20 | 76.80 |
| Type 3, total per 100 square feet | R1@2.55 | Sq | 84.86 | 79.60 | 164.46 | 280.00 |

**Roofing asphalt.** 100 pound keg.

| | Craft@Hrs | Unit | Material | Labor | Total | Sell |
|---|---|---|---|---|---|---|
| Type III, to 3 in 12 roof, 195-205 degree | — | Keg | 29.50 | — | 29.50 | — |
| Type III, at 30 pounds per 100 SF | R1@.350 | Sq | 8.85 | 10.90 | 19.75 | 33.60 |
| Type IV, to 6 in 12 roof, 210-225 degree | — | Keg | 28.00 | — | 28.00 | — |
| Type IV, at 30 pounds per 100 SF | R1@.350 | Sq | 8.40 | 10.90 | 19.30 | 32.80 |

| | Craft@Hrs | Unit | Material | Labor | Total | Sell |
|---|---|---|---|---|---|---|

**Mineral surface roll roofing.** Roll covers one square (100 square feet).

| | Craft@Hrs | Unit | Material | Labor | Total | Sell |
|---|---|---|---|---|---|---|
| 90 pound, fiberglass mat reinforced | R1@.200 | Sq | 22.40 | 6.24 | 28.64 | 48.70 |
| GAF Mineral Guard | R1@.200 | Sq | 23.00 | 6.24 | 29.24 | 49.70 |
| SBS-modified bitumen, Flintlastic | R1@.200 | Sq | 45.90 | 6.24 | 52.14 | 88.60 |
| Add for mopping in asphalt | R1@.350 | Sq | 4.50 | 10.90 | 15.40 | 26.20 |
| Add for adhesive application | R1@.350 | Sq | 8.50 | 10.90 | 19.40 | 33.00 |

**Roofing membrane.** Asphalt saturated. Per square. 15-pound roll covers 400 square feet. 30-pound roll covers 200 square feet. Nailed down.

| | Craft@Hrs | Unit | Material | Labor | Total | Sell |
|---|---|---|---|---|---|---|
| 15-pound, Type I | R1@.140 | Sq | 3.93 | 4.37 | 8.30 | 14.10 |
| 15-pound, ASTM D226 | R1@.140 | Sq | 5.03 | 4.37 | 9.40 | 16.00 |
| 30-pound, Type II | R1@.150 | Sq | 7.84 | 4.68 | 12.52 | 21.30 |
| 30-pound, ASTM D226 | R1@.150 | Sq | 10.14 | 4.68 | 14.82 | 25.20 |
| Add for mopping in asphalt | R1@.350 | Sq | 4.50 | 10.90 | 15.40 | 26.20 |
| Add for adhesive application | R1@.350 | Sq | 8.50 | 10.90 | 19.40 | 33.00 |

**Fiberglass Type IV roll roofing.** 500 square foot roll. ASTM D-2178. Fiberglass mat coated with asphalt and surfaced with a liquid parting agent that prevents rolls from sticking.

| | Craft@Hrs | Unit | Material | Labor | Total | Sell |
|---|---|---|---|---|---|---|
| Per 100 square feet | R1@.140 | Sq | 4.15 | 4.37 | 8.52 | 14.50 |
| Add for mopping in asphalt | R1@.350 | Sq | 4.50 | 10.90 | 15.40 | 26.20 |
| Add for adhesive application | R1@.350 | Sq | 8.50 | 10.90 | 19.40 | 33.00 |

**Fiberglass base sheet.** ASTM D-4601. 300-square-foot roll. Nailed down.

| | Craft@Hrs | Unit | Material | Labor | Total | Sell |
|---|---|---|---|---|---|---|
| Per 100 square feet | R1@.300 | Sq | 5.48 | 9.36 | 14.84 | 25.20 |

**Modified bitumen adhesive.** Rubberized. Use to adhere SBS modified bitumen membranes and rolled roofing. Blind nailing cement. Brush grade.

| | Craft@Hrs | Unit | Material | Labor | Total | Sell |
|---|---|---|---|---|---|---|
| 2 gallons per 100 square feet | R1@.350 | Sq | 13.80 | 10.90 | 24.70 | 42.00 |

**Cold process and lap cement.** Bonds plies of built-up roofing and gravel.

| | Craft@Hrs | Unit | Material | Labor | Total | Sell |
|---|---|---|---|---|---|---|
| 2 gallons per 100 square feet | R1@.350 | Sq | 8.62 | 10.90 | 19.52 | 33.20 |

**Modified bitumen cold process adhesive.** For placing shingles and cementing asphalt-coated roofing sheets together. Use for bonding felts, smooth and granule-surfaced roll roofing. Quick-setting, waterproof adhesive with strong bonding action.

| | Craft@Hrs | Unit | Material | Labor | Total | Sell |
|---|---|---|---|---|---|---|
| 2 gallons per 100 square feet | R1@.350 | Sq | 12.30 | 10.90 | 23.20 | 39.40 |

**Non-fibered asphalt emulsion.** For metal and smooth-surface built-up roofs. Type III, Class I.

| | Craft@Hrs | Unit | Material | Labor | Total | Sell |
|---|---|---|---|---|---|---|
| 3 gallons cover 100 square feet | R1@.350 | Sq | 14.60 | 10.90 | 25.50 | 43.40 |

**Cold-Ap® built-up roofing cement, Henry.** Bonds layers of conventional built-up roofing or SBS base and cap sheets. Replaces hot asphalt in built-up roofing. For embedding polyester fabric, gravel and granules. Not recommended for saturated felt. ASTM D3019 Type III.

| | Craft@Hrs | Unit | Material | Labor | Total | Sell |
|---|---|---|---|---|---|---|
| 2 gallons per 100 square feet | R1@.350 | Sq | 14.50 | 10.90 | 25.40 | 43.20 |

**PBA™ Permanent Bond Adhesive, Henry.** Cold-process waterproofing roofing adhesive. ASTM D3019 Type III.

| | Craft@Hrs | Unit | Material | Labor | Total | Sell |
|---|---|---|---|---|---|---|
| 2 gallons per 100 square feet | R1@.350 | Sq | 14.70 | 10.90 | 25.60 | 43.50 |

|  | Craft@Hrs | Unit | Material | Labor | Total | Sell |
|---|---|---|---|---|---|---|

**GAFGLAS torch and mop system built-up roofing, GAF.** Low-slope roofing. Cap sheet made of asphalt-coated glass fiber mat. Cap sheet surfaced with mineral granules. Use as a surfacing ply in the application of hot-applied built-up roofs, and as a top ply in base flashing construction. Glass base resists moisture, and granule surface provides an ultraviolet protective surface. One 33.5' x 39.6" roll covers 1 square (100 square feet). One roll of felt covers 5 squares (500 square feet). Cost per 100 square feet.

| | Craft@Hrs | Unit | Material | Labor | Total | Sell |
|---|---|---|---|---|---|---|
| Cap sheet, white, | | | | | | |
| 33.5' x 39.6" roll | RR@.550 | Sq | 26.00 | 19.70 | 45.70 | 77.70 |
| Torch application, granular, black, | | | | | | |
| 33.5' x 39.6" roll | RR@.350 | Sq | 57.80 | 12.50 | 70.30 | 120.00 |
| Torch application, granular, white, | | | | | | |
| 33.5' x 39.6" roll | RR@.350 | Sq | 57.50 | 12.50 | 70.00 | 119.00 |
| Torch application, smooth, white, | | | | | | |
| 33.5' x 39.6" roll | RR@.350 | Sq | 56.30 | 12.50 | 68.80 | 117.00 |
| Mop SBS polyester mat, granular, black, | | | | | | |
| 100 square foot roll | RR@.350 | Sq | 56.30 | 12.50 | 68.80 | 117.00 |
| Mop SBS polyester mat, granular, white, | | | | | | |
| 100 square foot roll | RR@.350 | Sq | 57.80 | 12.50 | 70.30 | 120.00 |
| #75 glass mat base sheet, | | | | | | |
| 300 square foot roll | RR@.500 | Sq | 28.60 | 17.90 | 46.50 | 79.10 |

**3-ply SBS field adhesive, GAF.** Low-slope roofing. GAF Torch and Mop System.

| | | Unit | Material | Labor | Total | Sell |
|---|---|---|---|---|---|---|
| 5 gallons | — | Ea | 29.30 | — | 29.30 | — |

**GAFGLAS Type IV glass felt, GAF.** Glass fiber mat coated with asphalt. Use as a ply felt or base sheet in built-up roof or as a flashing membrane. Class A, B, or C roofs. 39.4" x 161.8' roll covers 500 square feet.

| | Craft@Hrs | Unit | Material | Labor | Total | Sell |
|---|---|---|---|---|---|---|
| Per 100 square feet | R1@.350 | Ea | 5.79 | 10.90 | 16.69 | 28.40 |

**Ruberoid Mop SBS granular white roll roofing.** Non-woven polyester mat coated with SBS polymer-modified asphalt. For new roofing, reroofing, flashing, and built-up roofing. Roll size, 1 square (100 square feet).

| | Craft@Hrs | Unit | Material | Labor | Total | Sell |
|---|---|---|---|---|---|---|
| Granular, white, 0.160" thick | R1@.350 | Sq | 61.50 | 10.90 | 72.40 | 123.00 |

**Building paper.** Moisture barrier installed on a vertical or horizontal surface over sheathing. Per 100 square feet (square).

| | | Unit | Material | Labor | Total | Sell |
|---|---|---|---|---|---|---|
| 1-ply, black, 40" x 97' roll | — | Sq | 3.02 | — | 3.02 | — |
| 2-ply, 162 SF roll | — | Sq | 6.70 | — | 6.70 | — |
| 2-ply Jumbo Tex, 162 SF roll | — | Sq | 8.69 | — | 8.69 | — |
| Rosin paper, 500 SF roll | — | Sq | 2.48 | — | 2.48 | — |

## Roof Patching

**Self-adhesive roof flashing.** Self-adhesive ice and water shield. For use as flashing around valleys, vents, chimneys and skylights. Adhesive allows for one-time repositioning but aggressive bond over time. Per 100 square feet (square).

| | Craft@Hrs | Unit | Material | Labor | Total | Sell |
|---|---|---|---|---|---|---|
| GAF StormGuard | R1@.500 | Sq | 37.20 | 15.60 | 52.80 | 89.80 |
| GAF leak barrier | R1@.500 | Sq | 31.90 | 15.60 | 47.50 | 80.80 |
| Owens Corning Weatherlock® | R1@.500 | Sq | 47.60 | 15.60 | 63.20 | 107.00 |
| Tarco ice and water armor | R1@.500 | Sq | 22.70 | 15.60 | 38.30 | 65.10 |
| Grace ice and water shield | R1@.500 | Sq | 54.70 | 15.60 | 70.30 | 120.00 |

**Remove roof flashing.** Chip off damaged waterproofing on a wood or concrete surface and clean with a wire brush.

| | Craft@Hrs | Unit | Material | Labor | Total | Sell |
|---|---|---|---|---|---|---|
| Per linear foot of joint | R1@.030 | LF | — | .94 | .94 | 1.60 |

| | Craft@Hrs | Unit | Material | Labor | Total | Sell |
|---|---|---|---|---|---|---|

**Peel & Patch™ waterproofing.** Aluminum foil, polymer film and rubberized asphalt with self-sticking back. For metal roof repair, gutter patch, valley patch, flashing and waterproofing.

| | | | | | | |
|---|---|---|---|---|---|---|
| 4" x 10' | — | Ea | 7.77 | — | 7.77 | — |
| 6" x 10' | — | Ea | 10.93 | — | 10.93 | — |

**Leak Stopper™ rubberized roof patch, Gardner.** Stops roof leaks on wet or dry surfaces. Fiber-reinforced. Trowel grade. One gallon covers 12 square feet at 1/8" thick.

| | | | | | | |
|---|---|---|---|---|---|---|
| 10.1 ounces | — | Ea | 3.30 | — | 3.30 | — |
| 1 gallon | — | Ea | 8.65 | — | 8.65 | — |

**SEBS (styrene-ethylene-butylene-styrene) asphalt.** Seals openings around chimneys, vents, pipes, air conditioning equipment, skylights, drains, flashings, and sheet metal. One gallon covers 12 square feet at 1/8" thick.

| | | | | | | |
|---|---|---|---|---|---|---|
| All-weather, per gallon | — | Ea | 9.47 | — | 9.47 | — |
| Reflective, per gallon | — | Ea | 13.20 | — | 13.20 | — |

**Quick Roof, Cofair Products.** Self-adhesive, reinforced, aluminum surface for flashing repairs on flat and metal roofs.

| | | | | | | |
|---|---|---|---|---|---|---|
| 3" x 16' roll | — | Ea | 9.05 | — | 9.05 | — |
| 6" x 16' roll | — | Ea | 13.90 | — | 13.90 | — |

**Flashing cement.** For repairing chimneys, shingle cracks, skylights, pipe penetrations and gutters.

| | | | | | | |
|---|---|---|---|---|---|---|
| 1 gallon | — | Ea | 9.65 | — | 9.65 | — |
| 3 gallons | — | Ea | 24.60 | — | 24.60 | — |
| 5 gallons | — | Ea | 37.60 | — | 37.60 | — |
| Rubberized, 3 gallons | — | Ea | 31.00 | — | 31.00 | — |

**Patch roof or flashing.** Apply roofing cement and reinforce with embedded fiberglass mat. Per linear foot of patch.

| | | | | | | |
|---|---|---|---|---|---|---|
| One-ply cement and fiberglass | R1@.098 | LF | 1.25 | 3.06 | 4.31 | 7.33 |
| Two-ply cement and fiberglass | R1@.121 | LF | 2.00 | 3.78 | 5.78 | 9.83 |

**Rubber wet patch roof cement.** For sealing leaks on wet or dry surfaces. One gallon covers 12 square feet at 1/8" thick.

| | | | | | | |
|---|---|---|---|---|---|---|
| 11 ounces | — | Ea | 3.46 | — | 3.46 | — |
| 1 gallon | — | Ea | 12.10 | — | 12.10 | — |
| 3-1/2 gallons | — | Ea | 37.10 | — | 37.10 | — |

**Wet Patch® roof cement, Henry.** Seals around chimneys, vents, skylights, and flashings.

| | | | | | | |
|---|---|---|---|---|---|---|
| 11 ounce cartridge | — | Ea | 2.84 | — | 2.84 | — |
| 1 gallon | — | Ea | 8.94 | — | 8.94 | — |
| 3-1/2 gallons | — | Ea | 26.90 | — | 26.90 | — |
| 5 gallons | — | Ea | 33.40 | — | 33.40 | — |

**Wet-Stick asphalt plastic roof cement, DeWitts.** Adheres to wood, felt, plastics, glass, concrete, and all metals. One gallon covers 15 to 20 square feet when applied 1/8" thick.

| | | | | | | |
|---|---|---|---|---|---|---|
| 10-1/2 ounces | — | Ea | 2.13 | — | 2.13 | — |
| 1 quart | — | Ea | 3.21 | — | 3.21 | — |
| 1 gallon | — | Ea | 6.48 | — | 6.48 | — |
| 5 gallons | — | Ea | 24.80 | — | 24.80 | — |

**Kool Patch™ white acrylic patching cement, Kool Seal.** Stops leaks on damp or dry surfaces. Blends in on existing white roofs.

| | | | | | | |
|---|---|---|---|---|---|---|
| 11 ounce cartridge | — | Ea | 3.20 | — | 3.20 | — |
| 1 gallon | — | Ea | 15.70 | — | 15.70 | — |

| | Craft@Hrs | Unit | Material | Labor | Total | Sell |
|---|---|---|---|---|---|---|

**Fiberglass reinforcing membrane.** Fabric membrane for repairing cracks, holes, seams, and joints with reinforcing roof cements and coatings.

| | | | | | | |
|---|---|---|---|---|---|---|
| Black, 6" wide, 50' long | — | Ea | 7.00 | — | 7.00 | — |
| Black, 6" wide, 150' long | — | Ea | 13.00 | — | 13.00 | — |

**Polyester stress-block fabric, Gardner.** Asphalt-saturated. Repairs cracks, holes, seams, and joints with reinforcing roof cements and coatings.

| | | | | | | |
|---|---|---|---|---|---|---|
| White, 4" wide, 25' long | — | Ea | 7.33 | — | 7.33 | — |

**Yellow resin coated glass fabric, Henry.** Woven, open-mesh fiberglass. 20 x 10 thread count. ASTM D1668.

| | | | | | | |
|---|---|---|---|---|---|---|
| 4" wide, 150' long | — | Ea | 10.33 | — | 10.33 | — |
| 6" wide, 150' long | — | Ea | 6.30 | — | 6.30 | — |

**Asphalt saturated cotton roof patch.** Cotton fabric membrane for patching roof seams and holes.

| | | | | | | |
|---|---|---|---|---|---|---|
| 4" wide, 150' long | — | Ea | 9.25 | — | 9.25 | — |
| 6" wide, 150' long | — | Ea | 13.60 | — | 13.60 | — |

**Elastotape, Henry.** Non-woven polyester mat for use as mastic reinforcement. Lightweight, heat set, 100% spunbonded.

| | | | | | | |
|---|---|---|---|---|---|---|
| 4" wide, 150' long | — | Ea | 10.42 | — | 10.42 | — |

**Roofing granules.**

| | | | | | | |
|---|---|---|---|---|---|---|
| Colors, 60 pound pail | — | Ea | 23.10 | — | 23.10 | — |
| White, 100 pound bag | — | Ea | 29.10 | — | 29.10 | — |

**Siliconizer crack filler.** Seals cracks and stops leaks in concrete, metal, asphalt, brick and polyurethane surfaces.

| | | | | | | |
|---|---|---|---|---|---|---|
| 10.1 ounce tube | — | Ea | 2.16 | — | 2.16 | — |
| 1 quart | — | Ea | 7.80 | — | 7.80 | — |
| 1 gallon | — | Ea | 14.30 | — | 14.30 | — |

**Elastocaulk®, Henry.** White elastomeric acrylic roof patch. Apply by brush or trowel.

| | | | | | | |
|---|---|---|---|---|---|---|
| 11 ounce tube | — | Ea | 2.98 | — | 2.98 | — |
| 1 gallon | — | Ea | 21.50 | — | 21.50 | — |

**Elastomastic, Henry.** SEBS-modified sealant for roof areas subject to movement. Seals metal to metal joints. Also used to fill pitch pockets.

| | | | | | | |
|---|---|---|---|---|---|---|
| 11 ounce tube | — | Ea | 4.57 | — | 4.57 | — |
| 1 gallon | — | Ea | 21.50 | — | 21.50 | — |

**Roof caulk, Karnak.** Seals shingle tabs, flashing and joints at chimneys, skylights, pipe penetrations and gutters.

| | | | | | | |
|---|---|---|---|---|---|---|
| Flashing cement, 10 ounces | — | Ea | 1.89 | — | 1.89 | — |
| Wet or dry roof caulk, 10 ounces | — | Ea | 2.05 | — | 2.05 | — |
| Rubberized wet/dry caulk, 10 ounces | — | Ea | 3.18 | — | 3.18 | — |

**PondPatch™, Henry.** Fills roof ponds. Trowel on over or under roof membrane. For use on built-up, SBS or APP surfaces.

| | | | | | | |
|---|---|---|---|---|---|---|
| 6 Gallonpail covers 12 SF at 3/4" thick | R1@.085 | SF | 5.04 | 2.65 | 7.69 | 13.10 |

**Roof repair tools.**

| | | | | | | |
|---|---|---|---|---|---|---|
| Roof mop, cotton | — | Ea | 13.10 | — | 13.10 | — |
| Roof cement trowel, 9" | — | Ea | 2.95 | — | 2.95 | — |
| Roof shingle remover | — | Ea | 27.30 | — | 27.30 | — |
| Poly roofing broom with scraper | — | Ea | 31.30 | — | 31.30 | — |

| | Craft@Hrs | Unit | Material | Labor | Total | Sell |
|---|---|---|---|---|---|---|

**Fibered roof coating.** Moisture barrier for use over metal, composition and built-up roofs. Seals worn roof surfaces.

| | Craft@Hrs | Unit | Material | Labor | Total | Sell |
|---|---|---|---|---|---|---|
| On built-up roof, 50 SF per gallon | R1@.600 | Sq | 5.17 | 18.70 | 23.87 | 40.60 |
| On metal roof, 80 SF per gallon | R1@.600 | Sq | 5.18 | 18.70 | 23.88 | 40.60 |

**SolarFlex® white acrylic roof coating, Henry.** Non-fibered elastomeric. Two gallons per 100 square feet using two coats.

| | Craft@Hrs | Unit | Material | Labor | Total | Sell |
|---|---|---|---|---|---|---|
| White, two coats | R1@1.00 | Sq | 24.40 | 31.20 | 55.60 | 94.50 |
| Tan, two coats | R1@1.00 | Sq | 30.21 | 31.20 | 61.41 | 104.00 |

**Fibered aluminum roof coating.** For metal, smooth or mineral-surfaced built-up, composition, SBS and APP roofs.

| | Craft@Hrs | Unit | Material | Labor | Total | Sell |
|---|---|---|---|---|---|---|
| 2 gallons cover 100 square feet | R1@.600 | Sq | 25.90 | 18.70 | 44.60 | 75.80 |

**Snow Roof.** White liquid roof coating for built-up, modified bitumen, composition, spray urethane foam, metal, asphalt, fiberglass, and polystyrene foam roofing.

| | Craft@Hrs | Unit | Material | Labor | Total | Sell |
|---|---|---|---|---|---|---|
| 2 gallons cover 100 square feet | R1@.600 | Sq | 34.30 | 18.70 | 53.00 | 90.10 |

**Ultra White elastomeric roof coating.** White acrylic roof coating for built-up, modified bitumen, urethane foam, concrete, and previously coated roofs.

| | Craft@Hrs | Unit | Material | Labor | Total | Sell |
|---|---|---|---|---|---|---|
| 60 square feet per gallon | R1@.600 | Ea | 17.30 | 18.70 | 36.00 | 61.20 |

**Wet-Stick fibered roof coat.** For coating old or new metal, built-up, composition or gravel roofs.

| | Craft@Hrs | Unit | Material | Labor | Total | Sell |
|---|---|---|---|---|---|---|
| 2 gallons cover 100 square feet | R1@.600 | Ea | 6.48 | 18.70 | 25.18 | 42.80 |

**Cant strip.** 3" x 3" x 3-5/8" x 4' long.

| | Craft@Hrs | Unit | Material | Labor | Total | Sell |
|---|---|---|---|---|---|---|
| Per 4' piece | R1@.080 | Ea | .56 | 2.50 | 3.06 | 5.20 |
| 30-piece bundle, 120 linear feet | R1@2.40 | Ea | 18.10 | 74.90 | 93.00 | 158.00 |

## Wood Shingle Roofing

**Patch wood shingle roof.** Remove wood shingles by chipping out with a chisel. Trim new shingles to fit. Nail down and tap shingle into alignment.

| | Craft@Hrs | Unit | Material | Labor | Total | Sell |
|---|---|---|---|---|---|---|
| 3 shingle patch (1 square foot) | R1@.256 | Ea | 1.42 | 7.99 | 9.41 | 16.00 |
| 12 shingle patch (4 square feet) | R1@.551 | Ea | 5.70 | 17.20 | 22.90 | 38.90 |
| 30 shingle patch (10 square feet) | R1@1.20 | Ea | 14.20 | 37.50 | 51.70 | 87.90 |

**Wood shakes and shingles.** Red cedar. No pressure treating. Labor includes flashing and assumes a gable or hip roof to 5 in 12. Add 6% to the labor costs for a cut-up roof. Add 17% to the labor cost for a steeper roof or more complex hip roof. Add 25% to the labor cost when shakes or shingles are laid with staggered butts or in an irregular pattern such as thatch, serrated weave or ocean.

Taper split No. 1 medium cedar shakes, 4 bundles cover 100 square feet (1 square) at 10" exposure,

| | Craft@Hrs | Unit | Material | Labor | Total | Sell |
|---|---|---|---|---|---|---|
| No. 1 medium shakes, $35.60 per bundle | R1@3.52 | Sq | 142.00 | 110.00 | 252.00 | 428.00 |
| Add for pressure treated shakes | — | % | 30.0 | — | 30.0 | — |
| Deduct in the Pacific Northwest | — | % | -20.0 | — | -20.0 | — |
| Shake felt, 30 lb. (180 SF roll at $15.40) | R1@.500 | Sq | 15.40 | 15.60 | 31.00 | 52.70 |

Hip and ridge shakes, medium, 20 per bundle, covers 16.7 LF at 10" and 20 LF at 12" exposure,

| | Craft@Hrs | Unit | Material | Labor | Total | Sell |
|---|---|---|---|---|---|---|
| Per bundle | R1@1.00 | Ea | 36.80 | 31.20 | 68.00 | 116.00 |

Sawn shakes, sawn 1 side, Class "C" fire retardant,

| | Craft@Hrs | Unit | Material | Labor | Total | Sell |
|---|---|---|---|---|---|---|
| 1/2" to 3/4" x 24" (4 bundle/sq at 10" exp.) | R1@3.52 | Sq | 194.00 | 110.00 | 304.00 | 517.00 |
| 3/4" to 5/4" x 24" (5 bundle/sq at 10" exp.) | R1@4.16 | Sq | 269.00 | 130.00 | 399.00 | 678.00 |

|  | Craft@Hrs | Unit | Material | Labor | Total | Sell |
|---|---|---|---|---|---|---|

Cedar roofing shingles, No. 1 grade.

Four bundles of Perfection grade 18" shingles cover 100 square feet (1 square) at 5-1/2" exposure. Five Perfections are 2-1/4" thick. Four bundles of 5x Perfect grade shingles cover 100 square feet (1 square) at 5" exposure. Five Perfects are 2" thick.

| | Craft@Hrs | Unit | Material | Labor | Total | Sell |
|---|---|---|---|---|---|---|
| Perfections, $53.90 per bundle (Florida) | R1@3.52 | Sq | 231.00 | 110.00 | 341.00 | 580.00 |
| Perfects, $47.40 per bundle | R1@3.52 | Sq | 202.00 | 110.00 | 312.00 | 530.00 |
| No. 2 Perfections, $36 per bundle | R1@3.52 | Sq | 163.00 | 110.00 | 273.00 | 464.00 |
| Add for pitch over 6 in 12 | — | % | — | 40.0 | — | — |

No. 2 cedar ridge shingles, medium, bundle covers 16 linear feet,

| | | | | | | |
|---|---|---|---|---|---|---|
| Per bundle | R1@1.00 | Ea | 34.00 | 31.20 | 65.20 | 111.00 |

Fire-treated cedar shingles, 18", 5-1/2" exposure, 5 shingles are 2-1/4" thick (5/2-1/4),

| | | | | | | |
|---|---|---|---|---|---|---|
| No. 1 (houses) | R1@3.52 | Sq | 250.00 | 110.00 | 360.00 | 612.00 |
| No. 2, red label (houses or garages) | R1@2.00 | Sq | 208.00 | 62.40 | 270.40 | 460.00 |
| No. 3 (garages) | R1@2.00 | Sq | 179.00 | 62.40 | 241.40 | 410.00 |

Eastern white cedar shingles, smooth butt edge, four bundles cover 100 square feet (1 square) at 5" exposure,

| | | | | | | |
|---|---|---|---|---|---|---|
| White extras, $26.25 per bundle | R1@2.00 | Sq | 125.00 | 62.40 | 187.40 | 319.00 |
| White clears, $42.00 per bundle | R1@2.00 | Sq | 99.60 | 62.40 | 162.00 | 275.00 |
| White No. 2 clears, $13.50 per bundle | R1@2.00 | Sq | 67.60 | 62.40 | 130.00 | 221.00 |

Cedar shim shingles, tapered Western red cedar,

| | | | | | | |
|---|---|---|---|---|---|---|
| Builder's 12-pack, per pack | — | Ea | 3.75 | — | 3.75 | — |

## Sheet Metal Roofing

### Roofing sheets.

Metal sheet roofing, utility gauge.

26" wide, 5-V crimp (includes 15% coverage loss),

| | | | | | | |
|---|---|---|---|---|---|---|
| 6' to 12' lengths | R1@.027 | SF | .64 | .84 | 1.48 | 2.52 |

26" wide, corrugated (includes 15% coverage loss),

| | | | | | | |
|---|---|---|---|---|---|---|
| 6' to 12' lengths | R1@.027 | SF | .58 | .84 | 1.42 | 2.41 |

27-1/2" wide, corrugated (includes 20% coverage loss),

| | | | | | | |
|---|---|---|---|---|---|---|
| 6' to 12' lengths | R1@.027 | SF | .57 | .84 | 1.41 | 2.40 |

Ridge roll, plain,

| | | | | | | |
|---|---|---|---|---|---|---|
| 10" wide | R1@.030 | LF | 1.95 | .94 | 2.89 | 4.91 |

Ridge cap, formed, plain, 12",

| | | | | | | |
|---|---|---|---|---|---|---|
| 12" x 10' roll | R1@.061 | LF | .95 | 1.90 | 2.85 | 4.85 |

Sidewall flashing, plain,

| | | | | | | |
|---|---|---|---|---|---|---|
| 3" x 4" x 10' | R1@.035 | LF | .70 | 1.09 | 1.79 | 3.04 |

5-V crimp closure strips,

| | | | | | | |
|---|---|---|---|---|---|---|
| 24" long | R1@.035 | Ea | 1.05 | 1.09 | 2.14 | 3.64 |

Endwall flashing, 2-1/2" corrugated,

| | | | | | | |
|---|---|---|---|---|---|---|
| 10" x 28" | R1@.035 | LF | .99 | 1.09 | 2.08 | 3.54 |

Wood filler strip, corrugated,

| | | | | | | |
|---|---|---|---|---|---|---|
| 7/8" x 7/8" x 6' (at $1.65 each) | R1@.035 | LF | .28 | 1.09 | 1.37 | 2.33 |

| | Craft@Hrs | Unit | Material | Labor | Total | Sell |
|---|---|---|---|---|---|---|

## Slate and Tile Roofing

**Roofing slate.** Local delivery included. Costs will be higher where slate is not mined. Add freight cost at 800 to 1,000 pounds per 100 square feet. Includes 20" long random width slate, 3/16" thick, 7-1/2" exposure. Meets Standard SS-S-451.

| | Craft@Hrs | Unit | Material | Labor | Total | Sell |
|---|---|---|---|---|---|---|
| Semi-weathering green and gray | R1@11.3 | Sq | 404.00 | 353.00 | 757.00 | 1,290.00 |
| Vermont black and gray black | R1@11.3 | Sq | 480.00 | 353.00 | 833.00 | 1,420.00 |
| China black or gray | R1@11.3 | Sq | 545.00 | 353.00 | 898.00 | 1,530.00 |
| Unfading and variegated purple | R1@11.3 | Sq | 545.00 | 353.00 | 898.00 | 1,530.00 |
| Unfading mottled green and purple | R1@11.3 | Sq | 545.00 | 353.00 | 898.00 | 1,530.00 |
| Unfading green | R1@11.3 | Sq | 446.00 | 353.00 | 799.00 | 1,360.00 |
| Red slate | R1@13.6 | Sq | 1,400.00 | 424.00 | 1,824.00 | 3,100.00 |
| Add for other specified widths and lengths | — | % | 20.0 | — | 20.0 | — |

**Fiber-cement slate roofing.** Eternit™, Stonit™, or Thrutone™, various colors. Non-asbestos formulation. Laid over 1/2" sheathing. Includes 30-pound felt and one copper storm anchor per slate. Add for sheathing, flashing, hip, ridge and valley units.

| | Craft@Hrs | Unit | Material | Labor | Total | Sell |
|---|---|---|---|---|---|---|
| 23-5/8" x 11-7/8", English (420 lb, 30 year) | | | | | | |
|   2" head lap (113 pieces per Sq) | R1@5.50 | Sq | 346.00 | 172.00 | 518.00 | 881.00 |
|   3" head lap (119 pieces per Sq) | R1@5.75 | Sq | 364.00 | 179.00 | 543.00 | 923.00 |
|   4" head lap (124 pieces per Sq) | R1@6.00 | Sq | 378.00 | 187.00 | 565.00 | 961.00 |
|   Add for hip and ridge units, in mastic | R1@0.10 | LF | 7.08 | 3.12 | 10.20 | 17.30 |
| 15-3/4" x 10-5/8", Continental (420 lb, 30 year) | | | | | | |
|   2" head lap (197 pieces per Sq) | R1@6.50 | Sq | 358.00 | 203.00 | 561.00 | 954.00 |
|   3" head lap (214 pieces per Sq) | R1@6.75 | Sq | 390.00 | 211.00 | 601.00 | 1,020.00 |
|   4" head lap (231 pieces per Sq) | R1@7.00 | Sq | 422.00 | 218.00 | 640.00 | 1,090.00 |
|   Add for hip and ridge units, in mastic | R1@0.10 | LF | 6.50 | 3.12 | 9.62 | 16.40 |
| Copper storm anchors, box of 1,000 ($36.50) | — | Ea | .04 | — | .04 | — |
| Stainless steel slate hooks | — | Ea | .25 | — | .25 | — |
| Add for extra felt under 5 in 12 pitch | R1@0.20 | Sq | 15.30 | 6.24 | 21.54 | 36.60 |

**Roofing tile, clay.** Material costs include felt and flashing. No freight or waste included. Normal waste is 5%. Roofing tiles are very heavy and may have to be transported long distances to the job site. Many roofing supply dealers don't stock the tile and quote prices F.O.B. at the factory. Cost of delivery to the job site can be a very substantial item.

Spanish tile, "S" shaped, 88 pieces per square, 800 pounds per square at 11" centers and 15" exposure.

| | Craft@Hrs | Unit | Material | Labor | Total | Sell |
|---|---|---|---|---|---|---|
|   Red clay tile | R1@3.46 | Sq | 112.00 | 108.00 | 220.00 | 374.00 |
|   Add for coloring | — | Sq | 12.30 | — | 12.30 | — |
|   Red hip and ridge units | R1@.047 | LF | .93 | 1.47 | 2.40 | 4.08 |
|   Color hip and ridge units | R1@.047 | LF | 1.33 | 1.47 | 2.80 | 4.76 |
|   Red rake units | R1@.047 | LF | 1.77 | 1.47 | 3.24 | 5.51 |
|   Color rake units | R1@.047 | LF | 1.94 | 1.47 | 3.41 | 5.80 |

Red clay mission tile, 2-piece, 86 pans and 86 tops per square, 7-1/2" x 18" x 8-1/2" tiles at 11" centers and 15" exposure.

| | Craft@Hrs | Unit | Material | Labor | Total | Sell |
|---|---|---|---|---|---|---|
|   Red clay tile | R1@5.84 | Sq | 207.00 | 182.00 | 389.00 | 661.00 |
|   Add for coloring (costs vary widely) | — | Sq | 46.60 | — | 46.60 | — |
|   Red hip and ridge units | R1@.047 | LF | .98 | 1.47 | 2.45 | 4.17 |
|   Color hip and ridge units | R1@.047 | LF | 1.33 | 1.47 | 2.80 | 4.76 |
|   Red rake units | R1@.047 | LF | 1.77 | 1.47 | 3.24 | 5.51 |
|   Color rake units | R1@.047 | LF | 1.94 | 1.47 | 3.41 | 5.80 |

| | Craft@Hrs | Unit | Material | Labor | Total | Sell |
|---|---|---|---|---|---|---|
| **Concrete roof tile.** Material includes felt, nails, and flashing. Approximately 90 pieces per square. Monier Lifetile. | | | | | | |
| Shake, slurry color | R1@3.25 | Sq | 117.00 | 101.00 | 218.00 | 371.00 |
| Slate, thru color | R1@3.25 | Sq | 141.00 | 101.00 | 242.00 | 411.00 |
| Espana, slurry coated | R1@3.25 | Sq | 141.00 | 101.00 | 242.00 | 411.00 |
| Monier 2000 | R1@3.25 | Sq | 152.00 | 101.00 | 253.00 | 430.00 |
| Vignette | R1@3.25 | Sq | 132.00 | 101.00 | 233.00 | 396.00 |
| Collage | R1@3.25 | Sq | 132.00 | 101.00 | 233.00 | 396.00 |
| Tapestry | R1@3.25 | Sq | 145.00 | 101.00 | 246.00 | 418.00 |
| Split shake | R1@3.25 | Sq | 159.00 | 101.00 | 260.00 | 442.00 |
| Trim tile | | | | | | |
| Mansard "V" ridge or rake, slurry coated | — | Ea | 1.87 | — | 1.87 | — |
| Mansard, ridge or rake, thru color | — | Ea | 1.95 | — | 1.95 | — |
| Hipstarters, slurry coated | — | Ea | 15.90 | — | 15.90 | — |
| Hipstarters, thru color | — | Ea | 20.30 | — | 20.30 | — |
| Accessories | | | | | | |
| Eave closure or birdstop | R1@.030 | LF | 1.16 | .94 | 2.10 | 3.57 |
| Hurricane or wind clips | R1@.030 | Ea | .39 | .94 | 1.33 | 2.26 |
| Underlayment or felt | R1@.050 | Sq | 15.30 | 1.56 | 16.86 | 28.70 |
| Metal flashing and nails | R1@.306 | Sq | 8.94 | 9.55 | 18.49 | 31.40 |
| Pre-formed plastic flashing | | | | | | |
| Hip (13" long) | R1@.020 | Ea | 30.60 | .62 | 31.22 | 53.10 |
| Ridge (39" long) | R1@.030 | Ea | 30.60 | .94 | 31.54 | 53.60 |
| Anti-ponding foam | R1@.030 | LF | .25 | .94 | 1.19 | 2.02 |
| Batten extenders | R1@.030 | Ea | 1.02 | .94 | 1.96 | 3.33 |
| Roof loading | | | | | | |
| Add to load tile and accessories on roof | R1@.822 | Sq | — | 25.70 | 25.70 | 43.70 |

## Elastomeric Decking and Roofing

**Roof walking deck.** Includes cleaning of the existing wood, masonry or coated deck, filling fractures, masking adjacent surfaces, primer at .3 gallons per 100 SF, Deckgard elastomeric polyurethane texture coat at 1 gallon per 100 square feet, polyurethane top coat at 1 gallon per 100 square feet. Per 100 square feet (square). Includes the cost of spray equipment.

| | Craft@Hrs | Unit | Material | Labor | Total | Sell |
|---|---|---|---|---|---|---|
| Based on 800 SF job | R1@5.00 | Sq | 58.30 | 156.00 | 214.30 | 364.00 |

**Urethane spray foam roof.** Applied over an existing low-slope urethane roof. Includes cleaning roof with a broom and air blast, caulking fractures with polyurethane, masking adjacent surfaces, emulsion primer at 2 gallons per 100 SF, 3.0-pound density urethane foam, 20 mil elastomeric acrylic top coat covered with 40 pounds of mineral granules per square. Per 100 square feet (square). Includes the cost of spray equipment.

| | Craft@Hrs | Unit | Material | Labor | Total | Sell |
|---|---|---|---|---|---|---|
| Based on 2500 SF job | R1@2.80 | Sq | 97.60 | 87.40 | 185.00 | 315.00 |

## Fiberglass Roofing

**Fiberglass panels.** Nailed or screwed onto wood frame.

Corrugated, 8', 10', 12' panels, 2-1/2" corrugations, standard colors. Costs include 10% for waste,

| | Craft@Hrs | Unit | Material | Labor | Total | Sell |
|---|---|---|---|---|---|---|
| Polycarbonate, 26"-wide panels | BC@.012 | SF | 1.07 | .39 | 1.46 | 2.48 |
| Fiberglass, 26"-wide panels | BC@.012 | SF | .65 | .39 | 1.04 | 1.77 |
| PVC, 26"-wide panels | BC@.012 | SF | .63 | .39 | 1.02 | 1.73 |
| Flat panels, clear, green or white, | | | | | | |
| .06" flat sheets, 4' x 8', 10', 12' | BC@.012 | SF | .95 | .39 | 1.34 | 2.28 |
| .03" rolls, 24", 36", 48" x 50' | BC@.012 | SF | 1.07 | .39 | 1.46 | 2.48 |
| .037" rolls, 24", 36", 48" x 50' | BC@.012 | SF | 1.23 | .39 | 1.62 | 2.75 |
| Nails, ring shank with rubber washer, 1-3/4" (labor included with panels), | | | | | | |
| 100 per box (covers 130 SF) | — | SF | .05 | — | .05 | — |

|  | Craft@Hrs | Unit | Material | Labor | Total | Sell |
|---|---|---|---|---|---|---|
| Self-tapping screws, 1-1/4", rust-proof (labor included with panels), 100 per box (covers 130 SF) | — | SF | .20 | — | .20 | — |
| Horizontal closure strips, corrugated (labor included with panels), | | | | | | |
| Redwood, 2-1/2" x 1-1/2" | — | LF | .30 | — | .30 | — |
| Poly-foam, 1" x 1" | — | LF | .52 | — | .52 | — |
| Rubber, 1" x 1" | — | LF | .43 | — | .43 | — |
| Vertical crown molding, | | | | | | |
| Redwood, 1-1/2" x 1" or poly-foam, 1" x 1" | — | LF | .30 | — | .30 | — |
| Rubber, 1" x 1" | — | LF | .81 | — | .81 | — |

## Roof Flashing

**Aluminum roll flashing.** Installs under shingles, along roof valley. Provides a waterproof valley at intersecting roof sections. 0.010" economy gauge. Mill finish.

| | Craft@Hrs | Unit | Material | Labor | Total | Sell |
|---|---|---|---|---|---|---|
| 6" wide, 10' long | RR@.200 | Ea | 4.42 | 7.15 | 11.57 | 19.70 |
| 6" wide, 50' long | RR@1.00 | Ea | 13.25 | 35.80 | 49.05 | 83.40 |
| 8" wide, 10' long | RR@.200 | Ea | 5.54 | 7.15 | 12.69 | 21.60 |
| 8" wide, 50' long | RR@1.00 | Ea | 17.40 | 35.80 | 53.20 | 90.40 |
| 10" wide, 10' long | RR@.200 | Ea | 6.90 | 7.15 | 14.05 | 23.90 |
| 10" wide, 50' long | RR@1.00 | Ea | 21.30 | 35.80 | 57.10 | 97.10 |
| 14" wide, 10' long | RR@.200 | Ea | 8.77 | 7.15 | 15.92 | 27.10 |
| 14" wide, 50' long | RR@1.00 | Ea | 28.70 | 35.80 | 64.50 | 110.00 |
| 20" wide, 10' long | RR@.200 | Ea | 11.30 | 7.15 | 18.45 | 31.40 |
| 20" wide, 50' long | RR@1.00 | Ea | 39.00 | 35.80 | 74.80 | 127.00 |

**Aluminum roll valley flashing.** Use for valley, hip and ridge flashing.

| | Craft@Hrs | Unit | Material | Labor | Total | Sell |
|---|---|---|---|---|---|---|
| 6" wide, 25' long | RR@.500 | Ea | 11.80 | 17.90 | 29.70 | 50.50 |
| 8" wide, 25' long | RR@.500 | Ea | 15.20 | 17.90 | 33.10 | 56.30 |
| 10" wide, 10' long | RR@.200 | Ea | 9.82 | 7.15 | 16.97 | 28.80 |
| 10" wide, 25' long | RR@.500 | Ea | 17.20 | 17.90 | 35.10 | 59.70 |
| 14" wide, 10' long | RR@.200 | Ea | 11.40 | 7.15 | 18.55 | 31.50 |
| 14" wide, 25' long | RR@.500 | Ea | 20.10 | 17.90 | 38.00 | 64.60 |
| 20" wide, 25' long | RR@.500 | Ea | 30.40 | 17.90 | 48.30 | 82.10 |

**Aluminum window drip cap.** Commercial gauge 0.0175". Metal edge cap over door and window frames.

| | Craft@Hrs | Unit | Material | Labor | Total | Sell |
|---|---|---|---|---|---|---|
| 1-1/4" x 10' long, white | BC@.400 | Ea | 3.72 | 12.80 | 16.52 | 28.10 |
| 1-5/8" x 10' long, white | BC@.350 | Ea | 4.64 | 11.20 | 15.84 | 26.90 |

**Aluminum drip edge.**

| | Craft@Hrs | Unit | Material | Labor | Total | Sell |
|---|---|---|---|---|---|---|
| 2-3/4" x 1-3/4" x 10', brown | RR@.200 | Ea | 4.32 | 7.15 | 11.47 | 19.50 |
| 2-3/4" x 1-3/4" x 10', white | RR@.200 | Ea | 4.32 | 7.15 | 11.47 | 19.50 |

**Copper roll flashing.**

| | Craft@Hrs | Unit | Material | Labor | Total | Sell |
|---|---|---|---|---|---|---|
| 6" x 10' roll | RR@.500 | Ea | 35.60 | 17.90 | 53.50 | 91.00 |
| 6" x 25' roll | RR@1.25 | Ea | 85.30 | 44.70 | 130.00 | 221.00 |
| 8" x 10' roll | RR@.500 | Ea | 45.30 | 17.90 | 63.20 | 107.00 |
| 8" x 150' roll | RR@3.00 | Ea | 485.00 | 107.00 | 592.00 | 1,010.00 |
| 10" x 10' roll | RR@.500 | Ea | 58.30 | 17.90 | 76.20 | 130.00 |
| 12" x 100' roll | RR@2.50 | Ea | 491.00 | 89.40 | 580.40 | 987.00 |
| 24" x 50' roll | RR@2.25 | Ea | 501.00 | 80.50 | 581.50 | 989.00 |
| 3' x 4' flat sheet | — | Ea | 58.30 | — | 58.30 | — |

| | Craft@Hrs | Unit | Material | Labor | Total | Sell |
|---|---|---|---|---|---|---|
| **Copper step flashing.** | | | | | | |
| 5" x 7" | RR@.040 | Ea | 2.32 | 1.43 | 3.75 | 6.38 |
| 4" x 6" x 14" | RR@.040 | Ea | 8.62 | 1.43 | 10.05 | 17.10 |
| **Copper formed flashing.** | | | | | | |
| 1-1/2" x 2", roof edge | RR@.350 | Ea | 20.50 | 12.50 | 33.00 | 56.10 |
| 2" x 2", roof edge | RR@.350 | Ea | 23.70 | 12.50 | 36.20 | 61.50 |
| 4" x 6", roof-to-wall | RR@.250 | Ea | 59.40 | 8.94 | 68.34 | 116.00 |
| 5" x 7", shingle | — | Ea | 1.59 | — | 1.59 | — |
| 8" x 12", shingle | — | Ea | 4.84 | — | 4.84 | — |
| **Galvanized roll flashing.** Ridge and valley flashing. 26 gauge. | | | | | | |
| 6" wide, 10' long | RR@.200 | Ea | 4.89 | 7.15 | 12.04 | 20.50 |
| 6" wide, 50' long | RR@1.00 | Ea | 17.30 | 35.80 | 53.10 | 90.30 |
| 8" wide, 50' long | RR@1.00 | Ea | 17.30 | 35.80 | 53.10 | 90.30 |
| 10" wide, 50' long | RR@1.00 | Ea | 26.40 | 35.80 | 62.20 | 106.00 |
| 14" wide, 10' long | RR@.200 | Ea | 7.50 | 7.15 | 14.65 | 24.90 |
| 14" wide, 50' long | RR@1.25 | Ea | 35.00 | 44.70 | 79.70 | 135.00 |
| 20" wide, 10' long | RR@.250 | Ea | 12.10 | 8.94 | 21.04 | 35.80 |
| 20" wide, 50' long | RR@1.40 | Ea | 48.10 | 50.10 | 98.20 | 167.00 |
| **Galvanized angle flashing.** 10' lengths. 26 gauge. | | | | | | |
| 3" x 5", 90-degree | RR@.200 | Ea | 6.20 | 7.15 | 13.35 | 22.70 |
| 4" x 4" | RR@.200 | Ea | 7.54 | 7.15 | 14.69 | 25.00 |
| 4" x 4", hemmed | RR@.200 | Ea | 6.56 | 7.15 | 13.71 | 23.30 |
| 4" x 5" | RR@.200 | Ea | 7.52 | 7.15 | 14.67 | 24.90 |
| 4" x 5" x 1/2", turnback flashing | RR@.200 | Ea | 7.00 | 7.15 | 14.15 | 24.10 |
| 6" x 6" | RR@.200 | Ea | 8.35 | 7.15 | 15.50 | 26.40 |
| 7-1/2", apron | RR@.200 | Ea | 4.58 | 7.15 | 11.73 | 19.90 |
| **Galvanized drip edge.** Used along roof edge to direct water away from the fascia or into rain gutter. | | | | | | |
| 1-1/4" x 1-1/4" x 10', white | RR@.200 | Ea | 4.20 | 7.15 | 11.35 | 19.30 |
| 2" x 2" x 10' | RR@.200 | Ea | 2.73 | 7.15 | 9.88 | 16.80 |
| 2-1/8" x 2" x 10', white | RR@.200 | Ea | 4.08 | 7.15 | 11.23 | 19.10 |
| 2-1/2" x 1" x 10', white | RR@.200 | Ea | 3.23 | 7.15 | 10.38 | 17.60 |
| 3" x 1" x 10', heavy duty | RR@.200 | Ea | 3.65 | 7.15 | 10.80 | 18.40 |
| 10' long, vented | RR@.200 | Ea | 16.20 | 7.15 | 23.35 | 39.70 |
| **Gravel stop.** Raised 26-gauge metal bead nailed around the roof perimeter to keep gravel on the roof. | | | | | | |
| 2" x 4" x 1/2" x 10', bonderized | RR@.300 | Ea | 6.29 | 10.70 | 16.99 | 28.90 |
| 2" x 4" x 1/2" x 10', galvanized | RR@.330 | Ea | 4.17 | 11.80 | 15.97 | 27.10 |
| 3-1/2" x 3-1/2" x 10', galvanized | RR@.300 | Ea | 5.07 | 10.70 | 15.77 | 26.80 |
| 3" x 4" x 1/4" x 10', galvanized | RR@.330 | Ea | 5.73 | 11.80 | 17.53 | 29.80 |
| 4" x 4" x 1/2" x 10', bonderized | RR@.330 | Ea | 6.75 | 11.80 | 18.55 | 31.50 |

| | Craft@Hrs | Unit | Material | Labor | Total | Sell |
|---|---|---|---|---|---|---|
| **L-flashing.** Galvanized 28 gauge, except as noted. | | | | | | |
| 1" x 1-1/2" x 10', galvanized | RR@.200 | Ea | 2.45 | 7.15 | 9.60 | 16.30 |
| 1" x 2" x 10', galvanized | RR@.200 | Ea | 2.97 | 7.15 | 10.12 | 17.20 |
| 1-1/2" x 1-1/2", galvanized | RR@.200 | Ea | 2.97 | 7.15 | 10.12 | 17.20 |
| 2" x 2" x 10', dark brown | RR@.240 | Ea | 4.84 | 8.58 | 13.42 | 22.80 |
| 2" x 2" x 10', galvanized | RR@.240 | Ea | 3.62 | 8.58 | 12.20 | 20.70 |
| 2" x 3" x 10', dark brown | RR@.240 | Ea | 5.93 | 8.58 | 14.51 | 24.70 |
| 2" x 3" x 10', galvanized | RR@.240 | Ea | 4.86 | 8.58 | 13.44 | 22.80 |
| 3" x 3" x 10', bonderized | RR@.250 | Ea | 7.89 | 8.94 | 16.83 | 28.60 |
| 3" x 3" x 10', galvanized | RR@.250 | Ea | 6.30 | 8.94 | 15.24 | 25.90 |
| 4" x 4" x 10', galvanized | RR@.250 | Ea | 7.67 | 8.94 | 16.61 | 28.20 |
| 4" x 6" x 10', white | RR@.270 | Ea | 10.78 | 9.66 | 20.44 | 34.70 |
| **Pipe flashing.** Hard base. | | | | | | |
| 3" to 1" | RR@.046 | Ea | 4.77 | 1.65 | 6.42 | 10.90 |
| 3" to 4" | RR@.046 | Ea | 5.89 | 1.65 | 7.54 | 12.80 |
| **Lead flashing roll.** | | | | | | |
| 8" x 30' | RR@.750 | Ea | 71.60 | 26.80 | 98.40 | 167.00 |
| 10" x 24' | RR@.672 | Ea | 72.60 | 24.00 | 96.60 | 164.00 |
| 12" x 20' | RR@.600 | Ea | 74.50 | 21.50 | 96.00 | 163.00 |
| **Galvanized roll valley flashing.** Laid in the valley of a hip roof. | | | | | | |
| 7" x 10', galvanized | RR@.200 | Ea | 6.62 | 7.15 | 13.77 | 23.40 |
| 7" x 25', galvanized | RR@.500 | Ea | 12.50 | 17.90 | 30.40 | 51.70 |
| 7" x 50', galvanized | RR@1.00 | Ea | 29.00 | 35.80 | 64.80 | 110.00 |
| 8" x 10', brown | RR@.220 | Ea | 10.75 | 7.87 | 18.62 | 31.70 |
| 8" x 10', white | RR@.220 | Ea | 10.80 | 7.87 | 18.67 | 31.70 |
| 10" x 10', galvanized | RR@.250 | Ea | 8.91 | 8.94 | 17.85 | 30.30 |
| 10" x 25', galvanized | RR@.625 | Ea | 20.98 | 22.40 | 43.38 | 73.70 |
| 10" x 50', galvanized | RR@1.25 | Ea | 23.70 | 44.70 | 68.40 | 116.00 |
| 12" x 12', galvanized | RR@2.50 | Ea | 16.50 | 89.40 | 105.90 | 180.00 |
| 12" x 25', galvanized | RR@.500 | Ea | 27.00 | 17.90 | 44.90 | 76.30 |
| 14" x 10', galvanized | RR@.270 | Ea | 11.40 | 9.66 | 21.06 | 35.80 |
| 14" x 25', galvanized | RR@.675 | Ea | 23.90 | 24.10 | 48.00 | 81.60 |
| 20" x 10', galvanized | RR@.300 | Ea | 16.30 | 10.70 | 27.00 | 45.90 |
| 20" x 12', galvanized | RR@.500 | Ea | 19.30 | 17.90 | 37.20 | 63.20 |
| 20" x 25', green | RR@.625 | Ea | 74.50 | 22.40 | 96.90 | 165.00 |
| 20" x 25', galvanized | RR@.600 | Ea | 34.40 | 21.50 | 55.90 | 95.00 |
| 20" x 25', red | RR@.625 | Ea | 48.50 | 22.40 | 70.90 | 121.00 |
| **Galvanized roof apron flashing.** Slips under shingles to extend drip beyond fascia board. | | | | | | |
| 10' long, brown | RR@.400 | Ea | 3.31 | 14.30 | 17.61 | 29.90 |
| 10' long, white | RR@.400 | Ea | 3.40 | 14.30 | 17.70 | 30.10 |
| **Galvanized roof edge flashing.** Installs under shingles along roof edge to retard decay. | | | | | | |
| 1-1/2" x 1-1/2" x 10' | RR@.250 | Ea | 2.72 | 8.94 | 11.66 | 19.80 |
| 1-1/2" x 1-1/2" x 10', dark brown | RR@.250 | Ea | 3.55 | 8.94 | 12.49 | 21.20 |
| 2" x 2" x 10', white | RR@.250 | Ea | 4.07 | 8.94 | 13.01 | 22.10 |
| 2" x 3" x 10', bonderized | RR@.250 | Ea | 5.61 | 8.94 | 14.55 | 24.70 |
| 2" x 3" x 10' | RR@.250 | Ea | 3.95 | 8.94 | 12.89 | 21.90 |
| 2" x 8" x 10' | RR@.350 | Ea | 8.36 | 12.50 | 20.86 | 35.50 |
| 6" x 6" x 10' | RR@.350 | Ea | 11.04 | 12.50 | 23.54 | 40.00 |

|  | Craft@Hrs | Unit | Material | Labor | Total | Sell |
|---|---|---|---|---|---|---|
| **Galvanized roof-to-wall flashing.** | | | | | | |
| 2" x 3" x 10' | RR@.250 | Ea | 4.63 | 8.94 | 13.57 | 23.10 |
| 2" x 3" x 10' | RR@.250 | Ea | 4.35 | 8.94 | 13.29 | 22.60 |
| 2" x 4" x 10' | RR@.250 | Ea | 5.60 | 8.94 | 14.54 | 24.70 |
| 2" x 6" x 10', bonderized | RR@.250 | Ea | 7.40 | 8.94 | 16.34 | 27.80 |
| 2" x 6" x 10' | RR@.250 | Ea | 7.80 | 8.94 | 16.74 | 28.50 |
| 3" x 5" x 10', dark brown | RR@.250 | Ea | 9.74 | 8.94 | 18.68 | 31.80 |
| 3" x 5" x 10' | RR@.250 | Ea | 8.13 | 8.94 | 17.07 | 29.00 |
| 4" x 5" x 10' | RR@.250 | Ea | 10.20 | 8.94 | 19.14 | 32.50 |
| 4" x 6" x 10' | RR@.250 | Ea | 8.75 | 8.94 | 17.69 | 30.10 |

**Galvanized step flashing.** Sheet metal formed into two leafs meeting at a right angle. Placed so each step overlaps the step below. Used to waterproof the meeting point of two surfaces, such as joint between chimney and roof surface.

|  | Craft@Hrs | Unit | Material | Labor | Total | Sell |
|---|---|---|---|---|---|---|
| 3" x 4" x 7" | RR@.150 | Ea | .48 | 5.37 | 5.85 | 9.95 |
| 4" x 4" x 8" | RR@.150 | Ea | .69 | 5.37 | 6.06 | 10.30 |
| 4" x 4" x 8" | RR@.150 | Ea | .39 | 5.37 | 5.76 | 9.79 |
| 4" x 4" x 12" | RR@.150 | Ea | .92 | 5.37 | 6.29 | 10.70 |
| 4" x 6" x 14" | RR@.150 | Ea | 1.18 | 5.37 | 6.55 | 11.10 |

**Tin repair shingles.** Galvanized.

|  | Craft@Hrs | Unit | Material | Labor | Total | Sell |
|---|---|---|---|---|---|---|
| 5" x 7" | — | Ea | .46 | — | .46 | — |
| 8" x 12" | — | Ea | .84 | — | .84 | — |
| 8" x 12", dark brown | — | Ea | .99 | — | .99 | — |

**Flashing shingles.**

|  | Craft@Hrs | Unit | Material | Labor | Total | Sell |
|---|---|---|---|---|---|---|
| 5" x 8", dark brown | — | Ea | .76 | — | .76 | — |
| 5" x 8", galvanized | — | Ea | .52 | — | .52 | — |
| 6" x 8", gray | — | Ea | .37 | — | .37 | — |

**W-valley flashing.** 10' lengths. 28 gauge. Sheet metal formed into a flattened "W" shape and laid into a roof valley. Channels water flow better than a smooth valley. Roof finish is applied over the valley from both sides.

|  | Craft@Hrs | Unit | Material | Labor | Total | Sell |
|---|---|---|---|---|---|---|
| 18", brown | RR@.250 | Ea | 22.10 | 8.94 | 31.04 | 52.80 |
| 18", copper | RR@.250 | Ea | 81.00 | 8.94 | 89.94 | 153.00 |
| 18", galvanized | RR@.250 | Ea | 13.40 | 8.94 | 22.34 | 38.00 |
| 18", white | RR@.250 | Ea | 28.10 | 8.94 | 37.04 | 63.00 |
| 24", dark brown | RR@.250 | Ea | 21.60 | 8.94 | 30.54 | 51.90 |
| 24", galvanized | RR@.250 | Ea | 15.70 | 8.94 | 24.64 | 41.90 |

**Counterflashing.**

|  | Craft@Hrs | Unit | Material | Labor | Total | Sell |
|---|---|---|---|---|---|---|
| 6" x 10', galvanized | RR@.500 | Ea | 7.00 | 17.90 | 24.90 | 42.30 |
| 6" x 10', grey | RR@.500 | Ea | 7.00 | 17.90 | 24.90 | 42.30 |

**Galvanized dormer vent.** Wire mesh screen keeps out birds and large insects. 28-gauge galvanized steel.

|  | Craft@Hrs | Unit | Material | Labor | Total | Sell |
|---|---|---|---|---|---|---|
| 18" x 18", low profile | RR@.871 | Ea | 27.00 | 31.20 | 58.20 | 98.90 |
| 19" x 3", low profile | RR@.871 | Ea | 22.40 | 31.20 | 53.60 | 91.10 |

**Ridge vent.** For attic and rafter-bay ventilation.

|  | Craft@Hrs | Unit | Material | Labor | Total | Sell |
|---|---|---|---|---|---|---|
| 1-1/4" x 11" x 4' | R1@.255 | Ea | 9.85 | 7.96 | 17.81 | 30.30 |

**VentSure™ Ridge Vent, Owens Corning.**

|  | Craft@Hrs | Unit | Material | Labor | Total | Sell |
|---|---|---|---|---|---|---|
| Rigid roll | R1@.028 | LF | 2.06 | .87 | 2.93 | 4.98 |

| | Craft@Hrs | Unit | Material | Labor | Total | Sell |
|---|---|---|---|---|---|---|
| **Cobra® Rigid Vent II ridge vent, GAF.** Polyester composite. | | | | | | |
| 4' long, .75" thick | R1@.255 | Ea | 10.75 | 7.96 | 18.71 | 31.80 |
| **Shingle-over ridge vent.** Narrow opening keeps out insects and debris. | | | | | | |
| 4' vent | R1@.255 | Ea | 9.73 | 7.96 | 17.69 | 30.10 |
| Vent end plug | R1@.025 | Ea | 4.00 | .78 | 4.78 | 8.13 |
| **Galvanized roof jack.** | | | | | | |
| 4" pipe, 4 in 12 pitch | RR@.311 | Ea | 12.90 | 11.10 | 24.00 | 40.80 |
| 6" pipe, 4 in 12 pitch | RR@.311 | Ea | 18.30 | 11.10 | 29.40 | 50.00 |
| 8" pipe, 4 in 12 pitch | RR@.311 | Ea | 31.60 | 11.10 | 42.70 | 72.60 |
| **Aluminum base roof jack.** | | | | | | |
| 1" to 3" pipe | RR@.311 | Ea | 7.00 | 11.10 | 18.10 | 30.80 |
| 3" to 4" pipe | RR@.311 | Ea | 8.16 | 11.10 | 19.26 | 32.70 |
| **3-N-1 rubber gasket roof jack.** Roof pitch to 12 in 12. | | | | | | |
| 1" to 3" pipe | RR@.311 | Ea | 6.33 | 11.10 | 17.43 | 29.60 |
| 3" to 4" pipe | RR@.311 | Ea | 7.45 | 11.10 | 18.55 | 31.50 |
| **All rubber roof jack.** | | | | | | |
| 1-1/4" to 4" pipe | R1@.311 | Ea | 19.30 | 9.71 | 29.01 | 49.30 |
| **Roof jack lead flashing.** | | | | | | |
| 1-1/2" pipe | RR@.271 | Ea | 21.60 | 9.69 | 31.29 | 53.20 |
| 2" pipe | RR@.271 | Ea | 24.80 | 9.69 | 34.49 | 58.60 |
| 4" pipe | RR@.271 | Ea | 27.00 | 9.69 | 36.69 | 62.40 |
| **Tile roof pipe flashing.** For plumbing, electrical, and heating lines exiting a Spanish tile roof with pitch from 5 in 12 to 12 in 12. Aluminum base. | | | | | | |
| 1-1/2" pipe | RR@.311 | Ea | 5.85 | 11.10 | 16.95 | 28.80 |
| 2" pipe | RR@.311 | Ea | 6.04 | 11.10 | 17.14 | 29.10 |
| 3" pipe | RR@.311 | Ea | 7.20 | 11.10 | 18.30 | 31.10 |
| 4" pipe | RR@.311 | Ea | 7.54 | 11.10 | 18.64 | 31.70 |
| **Square roof vent.** Duct outlet for kitchen or bathroom exhaust fan. 50-square-inch net free vent area. | | | | | | |
| Aluminum | RR@.440 | Ea | 6.58 | 15.70 | 22.28 | 37.90 |
| Plastic | RR@.440 | Ea | 8.10 | 15.70 | 23.80 | 40.50 |
| **Square slant back aluminum roof vent.** 60-square-inch free vent area. 18" x 20" flashing. | | | | | | |
| Aluminum | RR@.440 | Ea | 10.90 | 15.70 | 26.60 | 45.20 |
| Plastic, low profile | RR@.440 | Ea | 9.05 | 15.70 | 24.75 | 42.10 |
| **T-top roof exhaust vent.** For bath, dryer, kitchen or other exhaust ducts terminating through the roof. | | | | | | |
| 4", bonderized | RR@.440 | Ea | 10.00 | 15.70 | 25.70 | 43.70 |
| 4", galvanized | RR@.440 | Ea | 8.00 | 15.70 | 23.70 | 40.30 |
| 7", bonderized | RR@.440 | Ea | 10.70 | 15.70 | 26.40 | 44.90 |
| 7", galvanized | RR@.440 | Ea | 10.13 | 15.70 | 25.83 | 43.90 |
| 4", T-top sub-base flashing | RR@.100 | Ea | 6.45 | 3.58 | 10.03 | 17.10 |
| 7", T-top sub-base flashing | RR@.100 | Ea | 7.54 | 3.58 | 11.12 | 18.90 |

| | Craft@Hrs | Unit | Material | Labor | Total | Sell |
|---|---|---|---|---|---|---|
| **Chimney cap.** | | | | | | |
| 9" x 9" | RR@.440 | Ea | 25.80 | 15.70 | 41.50 | 70.60 |
| 9" x 13" | RR@.440 | Ea | 28.00 | 15.70 | 43.70 | 74.30 |
| 8" x 8" to 13" x 13", adjustable | RR@.440 | Ea | 27.50 | 15.70 | 43.20 | 73.40 |
| 13" x 13" | RR@.440 | Ea | 39.10 | 15.70 | 54.80 | 93.20 |
| 13" x 18" | RR@.440 | Ea | 32.30 | 15.70 | 48.00 | 81.60 |
| 13" x 13", stainless steel | RR@.440 | Ea | 43.20 | 15.70 | 58.90 | 100.00 |
| | | | | | | |
| **Roof turbine vent with base.** Externally braced. Stainless steel ball bearings. | | | | | | |
| 12", galvanized base | RR@.694 | Ea | 30.40 | 24.80 | 55.20 | 93.80 |
| 12", wood base | RR@.694 | Ea | 33.80 | 24.80 | 58.60 | 99.60 |
| 12", combination aluminum base | RR@.694 | Ea | 39.50 | 24.80 | 64.30 | 109.00 |

## Rain Handling Gear

### Gutters and downspouts.

| | Craft@Hrs | Unit | Material | Labor | Total | Sell |
|---|---|---|---|---|---|---|
| Galvanized steel, 4" box type, | | | | | | |
|    Gutter, 5", 10' lengths. .015" | B1@.070 | LF | .64 | 2.06 | 2.70 | 4.59 |
|    Inside or outside corner 566950 | B1@.091 | Ea | 4.62 | 2.67 | 7.29 | 12.40 |
|    Fascia support brackets | — | Ea | 1.31 | — | 1.31 | — |
|    Spike and ferrule support, 7" | — | Ea | .54 | — | .54 | — |
|    Add per 2" x 3" galvanized downspout, with drop outlet, 3 ells and straps, | | | | | | |
|      10' downspout | B1@.540 | Ea | 18.90 | 15.90 | 34.80 | 59.20 |
| Aluminum rain gutter, 5" box type, primed, | | | | | | |
|    Gutter, 5", 10' lengths | B1@.070 | LF | .60 | 2.06 | 2.66 | 4.52 |
|    Inside or outside corner | B1@.091 | Ea | 4.85 | 2.67 | 7.52 | 12.80 |
|    Fascia support brackets | — | Ea | 1.31 | — | 1.31 | — |
|    Spike and ferrule support, 7" | — | Ea | .54 | — | .54 | — |
|    Section joint kit | — | Ea | 3.90 | — | 3.90 | — |
|    Add per 2" x 3" aluminum downspout, with drop outlet, strainer, 3 ells and straps, | | | | | | |
|      10' downspout | B1@.540 | Ea | 24.50 | 15.90 | 40.40 | 68.70 |
|      20' downspout, with connector | B1@.810 | Ea | 17.60 | 23.80 | 41.40 | 70.40 |
| Vinyl rain gutter, white PVC, (Plastmo Vinyl), half round gutter, (10' lengths), | | | | | | |
|    Gutter, 4" | B1@.054 | LF | 1.19 | 1.59 | 2.78 | 4.73 |
|      Add for fittings (connectors, hangers, etc) | — | LF | 1.25 | — | 1.25 | — |
|      Add for inside or outside corner | B1@.091 | Ea | 6.95 | 2.67 | 9.62 | 16.40 |
|      Add for bonding kit (covers 150 LF) | — | Ea | 5.64 | — | 5.64 | — |
|    Add for 5" gutter | — | % | 15.0 | — | 15.0 | — |
|    Downspouts, 3" round | B1@.022 | LF | .92 | .65 | 1.57 | 2.67 |
|      Add for 3 ells and 2 clamps, per downspout | — | Ea | 31.20 | — | 31.20 | — |

# Windows

# 8

Windows can be an important (and expensive) part of any home improvement project. Loose-fitting, single-glazed windows are a major source of heat loss and gain. Wood windows in older homes often look similar to the eyesore in Figure 8-1. If your first instinct is to replace the window, keep reading. That may not be the best choice. This chapter explains why.

About one-quarter of the heating and cooling load in a well-insulated home is the result of heat transfer through windows. Old window frames and sash tend to be very porous, admitting streams of outside air during both heating and cooling seasons. Modern windows have superior weatherstripping and are much tighter. In fact, modern windows are so tight that a home can develop condensation problems and a musty odor after leaky, old windows are replaced. Modern windows have insulated frames and dual glazing with about $1/2"$ of dead air space between the panes. In better-quality insulated windows, this dead space is filled with argon gas, which transfers even less heat than dead air. Low-emissivity (Low-E) glass has two invisible layers of microscopically-thin silver sandwiched between anti-reflective layers of metal oxide. Low-E glass admits sunlight but blocks radiant

**Figure 8-1**

*Signs of excessive water damage are evident in the paint peeling off this window sill and sash and broken caulking around the window*

heat from escaping during the winter. During the cooling season, Low-E glass reflects back to the exterior as much as 25 percent of unwanted solar heat. Low-E coatings also block more than 80 percent of the ultraviolet light that tends to fade fabrics. Glass edges in better-quality insulated windows are sealed with "warm edge" spacers that transfer even less heat around the glass perimeter.

Modern windows have clear advantages over windows installed in homes before about 1970. But these advantages come at a price, especially if windows have to be custom-made to fit the opening. Before recommending replacement, take a few minutes to examine the existing windows.

Check the tightness of fit. A window sash that rattles in the frame probably leaks cold air in the winter, hot air in the summer and water during heavy rain. Check the operation of the window. Old double-hung and single-hung wood windows seldom operate smoothly. You may find many that are painted shut. Years of re-painting has sealed and immobilized the sash in the frame. That may reduce leaking but it also eliminates venting as an option. Old casement and awning windows tend to warp at the top and bottom. Dark stains on the sash and sill are usually caused by condensation running down the glass. If you see condensation stains, check for softening of the molding and sill.

Brown or black discoloration near joints is a sign of decay. Untreated wood windows are particularly vulnerable to insect attack, either by termites, powder-post beetles or carpenter ants. Termites usually leave sand-like pellets outside the wood they're destroying. Powder-post beetles make the wood look as though it was hit by birdshot. Carpenter ants deposit neat piles of coarse sawdust where they are nesting. Fumigation by a professional exterminator is usually required to eliminate termites. Use spray insecticide to eliminate carpenter ants and powder post beetles.

Although aluminum windows don't decay, they are still candidates for replacement. Aluminum is highly conductive to heat and cold. Aluminum windows can get so cold in winter that frost will form on the inside of them, wasting heat while creating unpleasant drafts and moisture problems. Modern windows offer better insulating value against heat, cold, UV radiation and noise. They protect better against forced entry, and resist both wind and rain better. And if that's not enough, modern windows are easier to keep clean. Many tilt inward so both sides can be cleaned from the interior. Perhaps you're old enough to remember your mother or grandmother sitting on the window sills of the upstairs bedrooms, facing in, with her legs dangling inside the room, cleaning the outside of the windows. It wasn't a good idea then, and it's not a good idea now.

# Windows and the Building Code

Most states and counties have energy codes that set energy-efficiency standards for both new construction and alteration work. To get a building permit for a significant home improvement project, you may have to submit energy calculations that demonstrate energy code compliance. State, city and county standards vary. But the U.S. Department of Energy offers free computer software to help you meet requirements of most model energy codes. The D.O.E. Web site is http://www.energycodes.gov. Naturally, the size and orientation of windows affect energy use. Too many large windows in the wrong places increase both the heating and cooling load beyond what the energy code allows.

Building codes require that sleeping rooms below the fourth floor have at least one operable window or door approved for an emergency exit (egress). If your building department requires upgrading the home to modern egress standards, you'll have to install at least one egress window (or door) in each bedroom. An egress window measures at least 20" wide by at least 24" high and has a clear opening of at least 5.7 square feet. The sill can't be more than 44" above the floor.

Many building codes also require that some windows have tempered glass. These include windows within 18" of the floor, windows on a stairway landing and windows used as a sidelight beside an entry door. Tempered glass fractures into pebble-like shards less likely to cause injury.

# Storm Windows

In colder climates, storm windows (storms) reduce the heating load and prevent damage from condensation on window interiors. Storms also protect the windows from extreme weather. The damaged window in Figure 8-2 wasn't protected by a storm window. The sash joints are weak and need to be reinforced; the glazing putty has to be replaced; the swollen, rough wood has to be sanded smooth so the window will open and close without sticking; and the window needs paint. Most of this work could have been avoided if a storm window had been installed during winter months.

When evaluating windows, note whether the house has removable storms and screens. Most homes built in cold climates between 1920 and 1980 have (or had) storm windows. Newer homes usually have insulated (double pane) glazing that offers weather protection similar to storm windows.

Galvanized self-storing storm-and-screen windows were introduced in the 1940s. Aluminum storms and screens became popular in the 1950s. Older aluminum storms will be oxidized and pitted and rubber glazing strips will be deteriorated. If you elect to replace windows, storms may not be needed. If you plan to repair windows, storms will be required in colder climates. It's usually better to replace (rather than repair) deteriorated storms, especially for standard size windows. Most storm windows have a 2" skirt that can be trimmed off to make minor adjustments in width and height. Storms can be custom fabricated for odd-size windows at moderate cost. For a little extra money, consider high-performance storms with Low-E glass.

**Figure 8-2**

*Weather-damaged window*

# Repair of Existing Wood Windows

Repair is easy when the only problem with a wood window is minor decay. A coat of water-repellent preservative on the exterior will stop further deterioration. This is a three-step process:

1. Use a heat gun or paint stripper to remove the existing paint.

2. Brush on preservative.

3. Repaint the window.

Paint can't be used over some types of wood preservative, so be sure to select a paintable preservative.

Next, consider aesthetics. This is especially important on older or historic homes. If you install modern windows, the house will have a new look. Is that new look going to be in harmony with the exterior trim? With the siding? With the roofing? With other windows in the house? If just a few windows need replacing, can you replace just those few without creating an eyesore? In such cases, repairing a window may be a much better choice than replacement, both in the money saved and in the preservation of a consistent style for the home.

Reproducing the leaded glass windows in Figure 8-3 would cost thousands today. Instead, repair fine-quality windows like these. Replacing the windows in Figure 8-4 with modern vinyl windows would be a mistake. The vinyl window would stick out like a sore thumb.

Weigh the repair work needed against the cost of a replacement window. Nearly all old windows need new glazing putty. Cracked panes have to be replaced. Corners may need reinforcement. And you'll probably have to scrape off layers of paint so the window opens and closes properly. These are relatively minor repairs. More extensive decay may require that the window be rebuilt rather than repaired.

**Figure 8-3**

*Fancy Victorian windows should be saved*

**Figure 8-4**

*These windows are historically accurate*

# Replacing Glazing Putty

Glazing putty lasts 40 to 50 years when protected by a storm window. Without protection from severe weather, glazing putty cracks in 5 to 10 years. Sections of putty fall out, leaving glass loose in the window frame. If a window pane is cracked and needs to be replaced, you'll have to remove the old putty to get the broken glass out.

When re-glazing, select putty that's smooth, creamy, and easy to work with. Avoid glazing putty that's hard to spread. You can apply glazing putty with a putty knife, but it's better to use a glazing tool that both applies new putty and removes old. Most of the old putty will come out easily. If some refuses to budge, use the glazing tool to cut under the old putty, removing a little sliver of wood, putty and all.

Allow about 15 minutes to remove and replace putty for one pane of glass, whether large or small. So figure four hours to renew putty for a pair of windows like the ones in Figure 8-5. If a pane is broken, take the sash out of the frame. Lay the window down flat on a newspaper. Then break the old glass out into the newspaper. When you're done, just fold up the newspaper, broken glass and all, and throw it away. This keeps the job site free of broken glass. Measure carefully for replacement glass, including the $1/4"$ of glass area under the glazing putty.

# Sash Repair

Joints tend to work loose and decay as windows age. Use wood glue and an angle brace to reinforce a weak corner. Figure 8-6 shows this type of repair on the weather-damaged window in Figure 8-2. When the window is painted, the braces won't be noticeable.

If a single- or double-hung wood window doesn't operate smoothly, try applying paste or paraffin wax to parts of the stop, jamb, and parting strip that come in contact during operation. If that doesn't solve the problem, abrasion marks on the sash will usually reveal where the sash is binding. Move nailed stops (Figure 8-7) away from the sash slightly. If stops are fastened with screws, remove the stop and sand or plane it lightly on the face in contact with the sash. But be careful not to plane away too much of the stop. Opening up too much space creates a path for air infiltration. If the sash is binding against the jamb, remove the sash and then plane the vertical edges slightly.

**Figure 8-5**

*Multi-paned windows*

If the sash rattles in the frame, glue a thin strip of wood veneer to one side of the sash. Keep veneer strips narrow enough so they're hidden by the stops.

The most common cause of window jamming is paint build-up. This is particularly true of the top sash, which is opened less than the bottom sash. Remove paint build-up with a heat gun. Start by removing the interior stop. Then lift out both top and bottom sash. Use a heat gun and a scraper to remove paint from the casing. Remove paint from the sash as well. Build-up may be jamming the sash against the stop. Be careful to keep the heat gun away from glass. Hot air from the gun can crack cold glass.

Fit the windows back into the frame and check operation. If the sash is still tight, a little sanding should solve the problem.

Once the window is working properly, it's OK to repaint the casing and sash. A single coat of paint won't do any harm. This entire job usually takes about two hours per window, including touch-up painting.

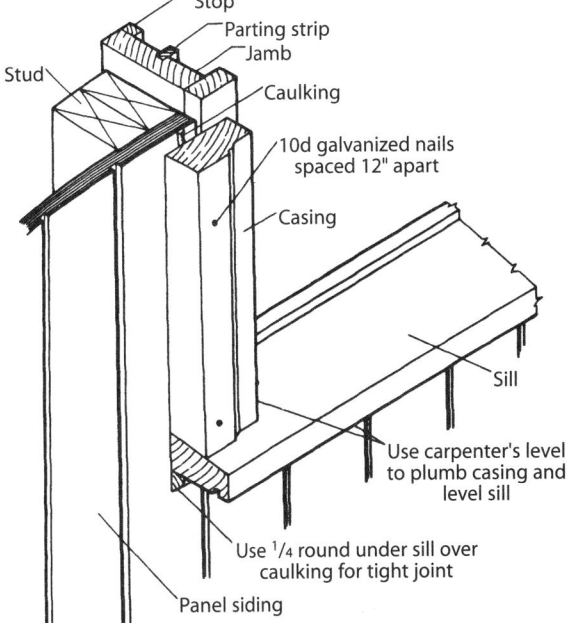

**Figure 8-7**

*Installation of double-hung window frame*

**Figure 8-6**

*Steel angle brace used to reinforce sash*

# Counterweighted Windows

Quality wood windows built before about 1930 had sash balances inside the wall, usually two weights, one on each side of the window. See Figure 8-8. The weights and the window were perfectly balanced so the window would stay put when opened to any height. Most likely, at least one pulley and rope failed many years ago. Now, the only way to keep the window open is with a stick supporting the sash. Even worse, the balance mechanism may still be working on just one side. That immobilizes the sash at nearly any height.

If sash balances have failed on an otherwise sound window, consider installing a new spring balance or replacing the broken sash cord. Installing a spring balance only requires removing the parting strip and stops. Follow the spring balance manufacturer's installation instructions. Then replace the stops.

Replacing broken balance ropes is simple, but not easy. The weights are difficult to get at. Working from inside the house, remove the inside stop. Back out screws or carefully pry out the small nails anchoring the stop. Once loose, lift the sash and swivel it out of the window frame. With luck, you'll only need to remove the stop on one side of the window.

If you see a little access cover under the inside stop on each side of the window, remove the cover to expose the channel where sash weights are supposed to operate. If you don't see an access cover, pry off the outer casing boards on each side of the window. Be careful not to crack the casing. Casing is usually attached with 4" trim-head nails. If these have rusted, they can be very hard to get loose. Don't put too much pressure on one spot. You'll either break the trim board or knock a hole in the wall. Work all around the board until you get it loose.

When the trim board is off, you won't see any weights. But have faith, the weights are in there. The window was installed before the furring strips and plaster went on. The outer casing boards were put on top of that. The weights are under the furring strips.

You'll need to cut through the plaster and furring strips at the bottom of the window frame. Don't worry, the hole won't show once you put the trim board back on. Cut carefully to avoid breaking any big chunks of plaster loose.

You may not see the weight at first. It's recessed a little below the bottom edge of the window frame. That's to give it someplace to go

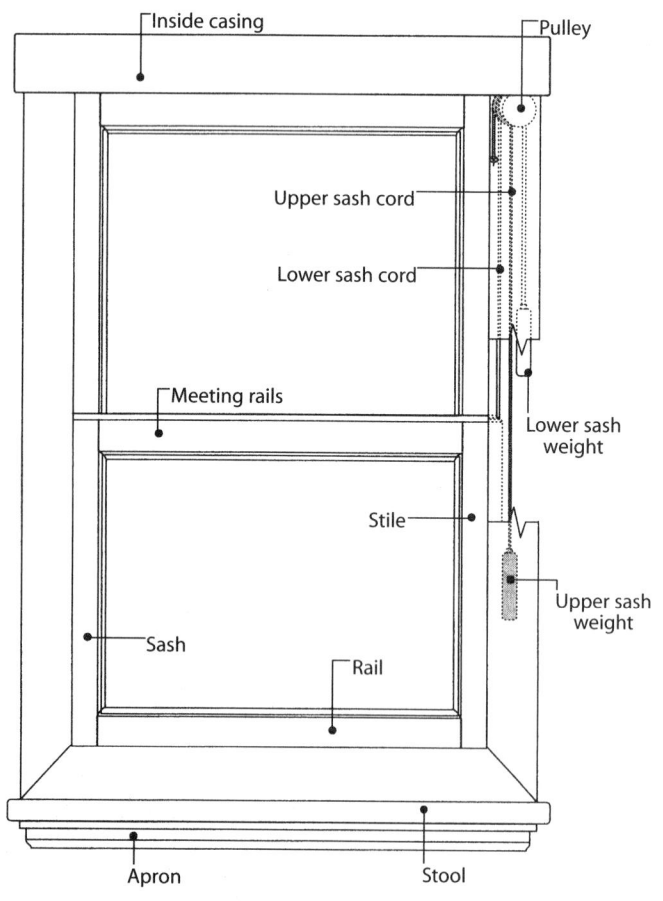

**Figure 8-8**

*Weight assembly in counterweighted windows*

when the window is all the way up. If you stick your fingers into the hole, you'll feel the top of the sash weight. It's usually a rough cast-iron cylinder with a loop on top for attaching the sash cord.

Now you need to replace the sash cord. You can use cotton clothesline, because it looks just like the cord that was there originally. Feed an end over the pulley until it appears in the hole. Tie the end to the loop on top of the weight.

If you look at the outside edge of the sash frame on each side, you'll see keyhole-shaped cutouts where the cord is supposed to go. Cut new cord to the proper length and tie a knot in the end. The knot goes in the round-hole part of the keyhole-shaped cutout. The long thin part of the cutout acts as a guide for the rest of the rope. Once you fit the sash back in the frame, the cord will be held in place by pressure.

Before you put the window back together, check and see that it operates properly. If the repair has been done correctly, the window will slide smoothly and stay put when released in any position. If the window doesn't slide easily, do a little cleaning or sanding.

Replacing a set of four sash cords requires about two hours per window, assuming you don't break any trim boards or knock chunks out of the plaster. But those two hours will save the several hundred dollars a new custom-made insulated vinyl window would cost. Since counterweighted windows were made from high-quality materials like oak, a vinyl window might not look right in this location anyway. And a new window probably won't cut heating or cooling cost by more than a dollar or so a month. When in doubt, repair the existing window and buy new storms.

## Adding or Moving a Window

Adding or moving a window will usually require both structural changes (studs and headers) and changes to the interior finish (wallboard, painting and trim). If the home improvement budget is modest, money should be spent elsewhere. But a strategically-placed window can add both livability and resale value to an older home. Figures 8-9 and 8-10 are before and after shots of a window re-sizing job. To take good advantage of a beautiful view, both openings were enlarged considerably. This job required about 12 hours per window. Chapter 3 includes estimates for cutting new wall openings for windows and doors.

Here are some rules to observe when adding, moving or re-sizing a window:

❖ Glass area should be equivalent to about 10 percent of floor area.

❖ Locate windows to promote good cross-ventilation.

❖ Group windows to eliminate undesirable contrasts in brightness.

❖ Window venting area should be about half of the window area.

❖ Provide screens for all sash that open for ventilation.

**Figure 8-9**
*Original windows*

**Figure 8-10**
*New larger windows take advantage of view*

❖ Sill height can be lower in rooms used mostly for seating (living room).

❖ Sill heights can be higher in rooms used mostly for standing (kitchen).

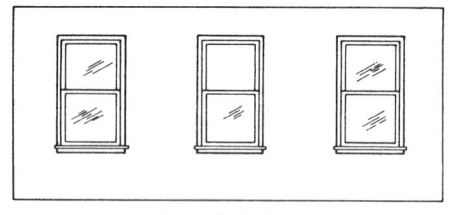

Spaced windows

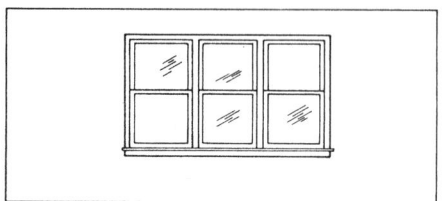

Grouped windows

**Figure 8-11**
*Window spacing*

Avoid small windows spaced evenly across a wall. Scattered windows cut up the wall space, making it unusable for larger pieces of furniture, such as a book case. It's better to cluster windows into one or two areas and leave other wall space undisturbed. See Figure 8-11. If you have a choice, put windows in south-facing walls, especially in colder climates. Winter sun is lower on the horizon, projecting sunlight deeper into the room through south-facing windows. Summer sun is higher in the sky and will be shaded even by a small overhang on the south side. In very warm climates, favor windows in north walls to avoid heat gain. Avoid west-facing windows to limit solar heating during the hottest part of the day. Even a large overhang is ineffective against a setting sun.

If you decide to close up a window opening, install vertical framing members spaced to match stud spacing below the window. Keeping the stud spacing consistent simplifies the nailing of both interior and exterior wall cover. Toenail new framing to the old window header and to the sill using three 8d or 10d nails at each joint. Install sheathing of the same thickness as the existing sheathing. Then add insulation and apply a vapor barrier on the inside face of the framing.

Make sure the new vapor barrier laps over the vapor barrier in the remainder of the wall. Finally, apply interior and exterior wall covering to match the existing wall cover.

# Replacement Windows

Figure 8-12 shows a double-casement window too decayed for repair. The replacement window (Figure 8-13) preserves the appearance of the old window but adds modern comfort and convenience.

Windows made for new construction have a nailing flange that laps over the exterior of the rough opening. For replacement purposes, that requires removal of the siding or stucco around the window perimeter and either a lot of patching or some very wide casing. Replacement windows fit into the existing jamb and have only a slightly wider casing to hide what's left of the old window. Be sure to specify the type of window you're replacing, either brick mold frame, wood frame or metal frame.

No special tools or equipment are needed to install a replacement window. But the window has to fit the opening exactly. If you're lucky, a stock size replacement window will fit. The tables that follow in this chapter list stock sizes for both replacement windows and new construction windows. If a stock size doesn't fit the opening, consider adding blocks and cripples so the opening fits a standard size replacement window. Of course, you'll need wider casing to cover the extra framing width at the head and jamb. If re-framing the window opening isn't practical, order a replacement window custom-made to size. For a price quote on the Web, go to http://www.thewindowsite.com.

**Figure 8-12**

*This rotted window had to be replaced*

**Figure 8-13**

*New thermopane window retains the look of the original*

|  | Craft@Hrs | Unit | Material | Labor | Total | Sell |
|---|---|---|---|---|---|---|

## Window Repairs

**Close up window opening, wood frame wall.** Remove and dispose of window and frame, install studs, sheathing, insulation, hang drywall interior (not taped or finished), install exterior wall finish and job clean-up. Near (not perfect) match of exterior wall finish. Per opening.

| | | | | | | |
|---|---|---|---|---|---|---|
| Stucco, wood, or vinyl siding exterior | B1@6.85 | Ea | 120.00 | 201.00 | 321.00 | 546.00 |
| Brick veneer toothed into existing brick | B9@9.50 | Ea | 230.00 | 273.00 | 503.00 | 855.00 |

**Close up window opening, masonry wall.** Remove and dispose of window and frame, install brick or block backup, tooth face brick into the existing brick, install insulation, hang drywall interior (not taped or finished) and job clean-up. Near (not perfect) match of materials. Per opening.

| | | | | | | |
|---|---|---|---|---|---|---|
| Brick or brick and block | B9@12.0 | Ea | 262.00 | 345.00 | 607.00 | 1,030.00 |

**Change window opening size, wood frame wall.** Including cripples, shims and blocks as needed. Add the cost of patching the interior and exterior wall surfaces.

| | | | | | | |
|---|---|---|---|---|---|---|
| Opening to 20 square feet, per opening | B1@1.53 | Ea | 16.40 | 44.90 | 61.30 | 104.00 |
| Opening over 20 square feet, per SF | B1@.077 | SF | .82 | 2.26 | 3.08 | 5.24 |

**Change window opening size, masonry wall.** Includes masonry patching and toothing of brick as needed.

| | | | | | | |
|---|---|---|---|---|---|---|
| Opening to 20 square feet, per opening | B6@1.95 | Ea | 16.40 | 55.70 | 72.10 | 123.00 |
| Opening over 20 square feet, per SF | B6@.098 | SF | 2.08 | 2.80 | 4.88 | 8.30 |
| Add for new cut stone sill, if needed | B6@.080 | LF | 4.10 | 2.28 | 6.38 | 10.80 |
| Add for new steel lintel, if needed | B6@.250 | Ea | 7.00 | 7.14 | 14.14 | 24.00 |

**Cut window opening and install new window.** Cut window opening in an exterior frame wall, frame a new window opening, install new window and patch interior and exterior surfaces. Based on wall with 1" x 6" drop siding, diagonal 1" x 12" sheathing, 2" x 4" studs 18" on center, and drywall interior. Assumes a double 2" x 6" header for a 3' opening, double 2" x 8" header for a 5' opening, double 2" x 10" header for a 6' opening or a double 2" x 12" header for an 8' opening.

| | | | | | | |
|---|---|---|---|---|---|---|
| Window to 10 square feet | BC@9.50 | Ea | 18.00 | 305.00 | 323.00 | 549.00 |
| Add per square foot for window over 10 SF | B1@.950 | SF | 1.80 | 27.90 | 29.70 | 50.50 |
| Add per linear foot of 4" wide vinyl sill | BC@.050 | LF | 2.75 | 1.60 | 4.35 | 7.40 |

**Remove and replace wood window.** Pry off trim. Remove stops and sash. Loosen siding to expose jamb. Cut or pull nails securing window to wall framing. Lift out window. Level and install same-size jamb. Re-install trim. Add the cost of patching and finishing interior and exterior walls.

| | | | | | | |
|---|---|---|---|---|---|---|
| Window to 10 square feet | BC@3.12 | Ea | — | 100.00 | 100.00 | 170.00 |
| Add per square foot for window over 10 SF | B1@.310 | SF | — | 9.10 | 9.10 | 15.50 |

**Remove, repair and replace window sash.** Wood double-hung window. Remove stops and sash. Plane and sand sash. Apply paraffin. Re-install sash and stops.

| | | | | | | |
|---|---|---|---|---|---|---|
| Per window sash | BC@.400 | Ea | — | 12.80 | 12.80 | 21.80 |

**Break paint seal to free window sash.** Using a hammer, wood block and putty knife blade. Add the cost of touch-up painting, if required.

| | | | | | | |
|---|---|---|---|---|---|---|
| Per sash | BC@.200 | Ea | — | 6.42 | 6.42 | 10.90 |

**Remove and replace sash cord.** Wood double-hung window. Remove window stop, sash and weight pocket covers. Remove weights and cords. Replace and attach new cords. Replace window sash and stop.

| | | | | | | |
|---|---|---|---|---|---|---|
| Per sash (2 cords) | BC@1.00 | Ea | 1.10 | 32.10 | 33.20 | 56.40 |

**Remove exterior window trim.** Working from a ladder.

| | | | | | | |
|---|---|---|---|---|---|---|
| 12' trim length | BC@.220 | Ea | — | 7.06 | 7.06 | 12.00 |

| | Craft@Hrs | Unit | Material | Labor | Total | Sell |
|---|---|---|---|---|---|---|

**Caulk window exterior.** Caulk up to 12 linear feet of window perimeter with 5 ounces of paintable silicone sealant.

| | Craft@Hrs | Unit | Material | Labor | Total | Sell |
|---|---|---|---|---|---|---|
| Per window to 12 linear feet | BC@.104 | Ea | 2.85 | 3.34 | 6.19 | 10.50 |
| Add for each additional LF | BC@.009 | LF | .24 | .29 | .53 | .90 |

**Remove and replace window caulk.** Working from a ladder. Chip out old caulk around window perimeter and replace with 12 linear feet of paintable silicone sealant.

| | Craft@Hrs | Unit | Material | Labor | Total | Sell |
|---|---|---|---|---|---|---|
| Per window to 12 linear feet | BC@.178 | Ea | 2.85 | 5.71 | 8.56 | 14.60 |
| Add for each additional LF | BC@.015 | LF | .24 | .48 | .72 | 1.22 |

## Vinyl Windows

**Custom-made vinyl replacement windows.** Typical costs. Prices vary with window specifications and distance from the point of manufacture. Allow two weeks to a month for delivery. Insulating glass. Base cost per unit plus additional cost per square foot of opening size. Add the cost of Low-E glazing, grilles and screens below.

| | Craft@Hrs | Unit | Material | Labor | Total | Sell |
|---|---|---|---|---|---|---|
| Fixed (picture) vinyl window, base cost | BC@1.00 | Ea | 303.00 | 32.10 | 335.10 | 570.00 |
| Add per square foot of opening | B1@.050 | SF | 6.07 | 1.47 | 7.54 | 12.80 |
| Double-hung vinyl window, base cost | BC@1.00 | Ea | 326.00 | 32.10 | 358.10 | 609.00 |
| Add per square foot of opening | B1@.050 | SF | 5.51 | 1.47 | 6.98 | 11.90 |
| Two-lite sliding vinyl window, base cost | BC@1.00 | Ea | 348.00 | 32.10 | 380.10 | 646.00 |
| Add per square foot of opening | B1@.050 | SF | 5.51 | 1.47 | 6.98 | 11.90 |
| Three-lite sliding vinyl window, base cost | BC@1.00 | Ea | 375.00 | 32.10 | 407.10 | 692.00 |
| Add per square foot of opening | B1@.050 | SF | 5.51 | 1.47 | 6.98 | 11.90 |
| Add for Low-E glazing, per SF of opening | — | SF | 1.84 | — | 1.84 | — |
| Add for colonial grille, per SF of opening | — | SF | 1.79 | — | 1.79 | — |
| Add for screen, per screen | — | Ea | 16.60 | — | 16.60 | — |

**Vinyl double-hung stock replacement windows, 7/8" insulating Low-E glass.** Fusion-welded white vinyl frame. By nominal (opening) size, width x height. Actual size is 1/2" less in both width and height. Tilt sash. With screen. Obscure glass and colonial grilles are available on some stock sizes. Labor includes removing the old stops and sash and setting new J-trim as needed.

| | Craft@Hrs | Unit | Material | Labor | Total | Sell |
|---|---|---|---|---|---|---|
| 24" x 36" | BC@1.00 | Ea | 150.00 | 32.10 | 182.10 | 310.00 |
| 24" x 38" | BC@1.00 | Ea | 139.00 | 32.10 | 171.10 | 291.00 |
| 24" x 46" | BC@1.00 | Ea | 156.00 | 32.10 | 188.10 | 320.00 |
| 24" x 54" | BC@1.00 | Ea | 159.00 | 32.10 | 191.10 | 325.00 |
| 28" x 37" | BC@1.00 | Ea | 156.00 | 32.10 | 188.10 | 320.00 |
| 28" x 38" | BC@1.00 | Ea | 152.00 | 32.10 | 184.10 | 313.00 |
| 28" x 46" | BC@1.00 | Ea | 157.00 | 32.10 | 189.10 | 321.00 |
| 28" x 53" | BC@1.00 | Ea | 185.00 | 32.10 | 217.10 | 369.00 |
| 28" x 54" | B1@1.50 | Ea | 184.00 | 44.10 | 228.10 | 388.00 |
| 28" x 58" | B1@1.50 | Ea | 184.00 | 44.10 | 228.10 | 388.00 |
| 28" x 62" | B1@1.50 | Ea | 185.00 | 44.10 | 229.10 | 389.00 |
| 30" x 38" | B1@1.50 | Ea | 151.00 | 44.10 | 195.10 | 332.00 |
| 30" x 54" | B1@1.50 | Ea | 179.00 | 44.10 | 223.10 | 379.00 |
| 30" x 58" | B1@1.50 | Ea | 184.00 | 44.10 | 228.10 | 388.00 |
| 30" x 62" | B1@1.50 | Ea | 200.00 | 44.10 | 244.10 | 415.00 |
| 30" x 66" | B1@1.50 | Ea | 200.00 | 44.10 | 244.10 | 415.00 |
| 31" x 53" | B1@1.50 | Ea | 187.00 | 44.10 | 231.10 | 393.00 |
| 31" x 54" | B1@1.50 | Ea | 187.00 | 44.10 | 231.10 | 393.00 |
| 31" x 57" | B1@1.50 | Ea | 190.00 | 44.10 | 234.10 | 398.00 |
| 31" x 58" | B1@1.50 | Ea | 190.00 | 44.10 | 234.10 | 398.00 |
| 31" x 61" | B1@1.50 | Ea | 193.00 | 44.10 | 237.10 | 403.00 |
| 32" x 38" | B1@1.50 | Ea | 158.00 | 44.10 | 202.10 | 344.00 |

| | Craft@Hrs | Unit | Material | Labor | Total | Sell |
|---|---|---|---|---|---|---|
| 32" x 46" | B1@1.50 | Ea | 160.00 | 44.10 | 204.10 | 347.00 |
| 32" x 50" | B1@1.50 | Ea | 179.00 | 44.10 | 223.10 | 379.00 |
| 32" x 54" | B1@1.50 | Ea | 190.00 | 44.10 | 234.10 | 398.00 |
| 32" x 58" | B1@1.50 | Ea | 191.00 | 44.10 | 235.10 | 400.00 |
| 32" x 61" | B1@1.50 | Ea | 193.00 | 44.10 | 237.10 | 403.00 |
| 32" x 62" | B1@1.50 | Ea | 202.00 | 44.10 | 246.10 | 418.00 |
| 32" x 66" | B1@1.50 | Ea | 200.00 | 44.10 | 244.10 | 415.00 |
| 32" x 70" | B1@1.50 | Ea | 208.00 | 44.10 | 252.10 | 429.00 |
| 34" x 38" | B1@1.50 | Ea | 141.00 | 44.10 | 185.10 | 315.00 |
| 34" x 46" | B1@1.50 | Ea | 177.00 | 44.10 | 221.10 | 376.00 |
| 34" x 54" | B1@1.50 | Ea | 190.00 | 44.10 | 234.10 | 398.00 |
| 34" x 62" | B1@1.50 | Ea | 195.00 | 44.10 | 239.10 | 406.00 |
| 36" x 38" | B1@1.50 | Ea | 157.00 | 44.10 | 201.10 | 342.00 |
| 36" x 46" | B1@1.50 | Ea | 187.00 | 44.10 | 231.10 | 393.00 |
| 36" x 50" | B1@1.50 | Ea | 190.00 | 44.10 | 234.10 | 398.00 |
| 36" x 54" | B1@1.50 | Ea | 190.00 | 44.10 | 234.10 | 398.00 |
| Add for colonial grille between the lites | — | % | 15.0 | — | — | — |

**Vinyl single-hung stock replacement windows, insulating glass.** Tilt-in sash. Front flange frame for installation in masonry block openings. Lower sash opens. Upper sash is fixed. Includes half screen. By opening size, width x height. Grille pattern shows lites in upper and lower sash, such as 6/6. Labor includes removing the old stops and sash and setting new J-trim as needed.

| | Craft@Hrs | Unit | Material | Labor | Total | Sell |
|---|---|---|---|---|---|---|
| 35-7/8" x 24-7/8" | B1@1.00 | Ea | 105.00 | 29.40 | 134.40 | 228.00 |
| 35-7/8" x 37-1/4" | B1@1.00 | Ea | 116.00 | 29.40 | 145.40 | 247.00 |
| 35-7/8" x 49-1/2" | B1@1.00 | Ea | 129.00 | 29.40 | 158.40 | 269.00 |
| 35-7/8" x 61-7/8" | B1@1.50 | Ea | 139.00 | 44.10 | 183.10 | 311.00 |
| 52" x 37-1/4" | B1@1.50 | Ea | 137.00 | 44.10 | 181.10 | 308.00 |
| 52" x 49-1/2" | B1@1.50 | Ea | 152.00 | 44.10 | 196.10 | 333.00 |
| 35-7/8" x 37-1/4", 6/6 grille | B1@1.50 | Ea | 132.00 | 44.10 | 176.10 | 299.00 |
| 35-7/8" x 49-1/2", 6/6 grille | B1@1.50 | Ea | 145.00 | 44.10 | 189.10 | 321.00 |
| 35-7/8" x 61-7/8", 6/6 grille | B1@1.50 | Ea | 157.00 | 44.10 | 201.10 | 342.00 |
| 52" x 37-1/4", 8/8 grille | B1@1.50 | Ea | 158.00 | 44.10 | 202.10 | 344.00 |
| 52" x 49-1/2", 8/8 grille | B1@1.50 | Ea | 160.00 | 44.10 | 204.10 | 347.00 |

**Vinyl casement bay windows, 7/8" insulating glass.** Fixed center lite and two casement flankers. White vinyl. Factory assembled. By nominal (opening) size, width x height. Actual size is 1/2" less in both width and height. Includes two screens. Low-E glass is available on some stock sizes.

| | Craft@Hrs | Unit | Material | Labor | Total | Sell |
|---|---|---|---|---|---|---|
| 69" x 50" | B1@4.00 | Ea | 972.00 | 117.00 | 1,089.00 | 1,850.00 |
| 74" x 50" | B1@4.00 | Ea | 980.00 | 117.00 | 1,097.00 | 1,860.00 |
| 93" x 50" | B1@4.50 | Ea | 1,115.00 | 132.00 | 1,247.00 | 2,120.00 |
| 98" x 50" | B1@4.50 | Ea | 1,174.00 | 132.00 | 1,306.00 | 2,220.00 |

**Vinyl garden windows, 7/8" insulating glass.** Single front awning vent. White vinyl construction with wood frame. Tempered glass shelf. 5/4" vinyl laminated head and seat board. With hardware and screen. By nominal (opening) size, width x height. Replacement window. Labor includes removing the old stops and sash and setting new J-trim as needed.

| | Craft@Hrs | Unit | Material | Labor | Total | Sell |
|---|---|---|---|---|---|---|
| 36" x 36" | B1@2.75 | Ea | 528.00 | 80.80 | 608.80 | 1,030.00 |
| 36" x 36", Low-E glass | B1@2.75 | Ea | 566.00 | 80.80 | 646.80 | 1,100.00 |
| 36" x 48" | B1@2.75 | Ea | 580.00 | 80.80 | 660.80 | 1,120.00 |
| 36" x 48", Low-E glass | B1@2.75 | Ea | 699.00 | 80.80 | 779.80 | 1,330.00 |

| | Craft@Hrs | Unit | Material | Labor | Total | Sell |
|---|---|---|---|---|---|---|

**Half-circle fixed vinyl windows.** Installed in an opening that's already framed to the correct size. Dual glazed, white frame. Low-E insulating glass. Stock units for new construction. By width x height.

| | Craft@Hrs | Unit | Material | Labor | Total | Sell |
|---|---|---|---|---|---|---|
| 20" x 10" | B1@2.75 | Ea | 450.00 | 80.80 | 530.80 | 902.00 |
| 30" x 16" | B1@2.75 | Ea | 450.00 | 80.80 | 530.80 | 902.00 |
| 40" x 20" | B1@2.75 | Ea | 600.00 | 80.80 | 680.80 | 1,160.00 |
| 50" x 26" | B1@2.75 | Ea | 710.00 | 80.80 | 790.80 | 1,340.00 |
| 60" x 30" | B1@2.75 | Ea | 806.00 | 80.80 | 886.80 | 1,510.00 |
| 80" x 40" | B1@2.75 | Ea | 1,170.00 | 80.80 | 1,250.80 | 2,130.00 |

**Vinyl jalousie windows.** Installed in an opening that's already framed to the correct size. Positive locking handle for security. By opening size, width x height.

| | Craft@Hrs | Unit | Material | Labor | Total | Sell |
|---|---|---|---|---|---|---|
| 24" x 35-1/2", clear | BC@1.00 | Ea | 249.00 | 32.10 | 281.10 | 478.00 |
| 24" x 35-1/2", obscure glass | BC@1.00 | Ea | 286.00 | 32.10 | 318.10 | 541.00 |
| 30" x 35-1/2", clear | BC@1.00 | Ea | 265.00 | 32.10 | 297.10 | 505.00 |
| 30" x 35-1/2", obscure glass | BC@1.00 | Ea | 312.00 | 32.10 | 344.10 | 585.00 |
| 36" x 35-1/2", clear | BC@1.00 | Ea | 276.00 | 32.10 | 308.10 | 524.00 |
| 36" x 35-1/2", obscure glass | BC@1.00 | Ea | 332.00 | 32.10 | 364.10 | 619.00 |

**Hopper vinyl windows.** White. Installed in an opening that's already framed to the correct size. Does not include finish or trim work. Low-E insulating glass. Argon filled. Frame depth 3-5/16". With screen. Stock units for new construction. By opening size, width x height.

| | Craft@Hrs | Unit | Material | Labor | Total | Sell |
|---|---|---|---|---|---|---|
| 32" x 15" | B1@.500 | Ea | 80.00 | 14.70 | 94.70 | 161.00 |
| 32" x 17" | B1@.500 | Ea | 83.70 | 14.70 | 98.40 | 167.00 |
| 32" x 19" | B1@.500 | Ea | 87.20 | 14.70 | 101.90 | 173.00 |
| 32" x 23" | B1@.500 | Ea | 91.00 | 14.70 | 105.70 | 180.00 |

## Vinyl-Clad Wood Windows

**Vinyl-clad double-hung low E insulated glass wood windows.** Prefinished pine interior. White vinyl exterior. Tilt sash. For new construction. By nominal (opening) size, width x height. Actual width is 1-5/8" more. Actual height is 3-1/4" more. Includes factory-applied extension jamb for 4-9/16" wall. Add the cost of screens and colonial grilles.

| | Craft@Hrs | Unit | Material | Labor | Total | Sell |
|---|---|---|---|---|---|---|
| 2'0" x 3'0" | B1@.500 | Ea | 156.00 | 14.70 | 170.70 | 290.00 |
| 2'0" x 3'6" | B1@.500 | Ea | 174.00 | 14.70 | 188.70 | 321.00 |
| 2'4" x 3'0" | B1@.500 | Ea | 158.00 | 14.70 | 172.70 | 294.00 |
| 2'4" x 3'6" | B1@.500 | Ea | 184.00 | 14.70 | 198.70 | 338.00 |
| 2'4" x 4'0" | B1@.750 | Ea | 200.00 | 22.00 | 222.00 | 377.00 |
| 2'4" x 4'6" | B1@.750 | Ea | 213.00 | 22.00 | 235.00 | 400.00 |
| 2'4" x 4'9" | B1@.750 | Ea | 224.00 | 22.00 | 246.00 | 418.00 |
| 2'8" x 3'0" | B1@.750 | Ea | 182.00 | 22.00 | 204.00 | 347.00 |
| 2'8" x 3'6" | B1@.750 | Ea | 194.00 | 22.00 | 216.00 | 367.00 |
| 2'8" x 4'0" | B1@.750 | Ea | 208.00 | 22.00 | 230.00 | 391.00 |
| 2'8" x 4'6" | B1@.750 | Ea | 225.00 | 22.00 | 247.00 | 420.00 |
| 2'8" x 4'9" | B1@.750 | Ea | 242.00 | 22.00 | 264.00 | 449.00 |
| 3'0" x 3'0" | B1@.750 | Ea | 262.00 | 22.00 | 284.00 | 483.00 |
| 3'0" x 3'6" | B1@.750 | Ea | 204.00 | 22.00 | 226.00 | 384.00 |
| 3'0" x 4'0" | B1@.750 | Ea | 219.00 | 22.00 | 241.00 | 410.00 |
| 3'0" x 4'6" | B1@.750 | Ea | 234.00 | 22.00 | 256.00 | 435.00 |
| 3'0" x 4'9" | B1@.750 | Ea | 242.00 | 22.00 | 264.00 | 449.00 |
| 3'0" x 5'6" | B1@.750 | Ea | 264.00 | 22.00 | 286.00 | 486.00 |
| 3'0" x 6'0" | B1@.750 | Ea | 279.00 | 22.00 | 301.00 | 512.00 |
| Deduct for non low-E glazing, per SF of opening | — | SF | -1.70 | — | -1.70 | — |

| | Craft@Hrs | Unit | Material | Labor | Total | Sell |
|---|---|---|---|---|---|---|

**Screens for vinyl-clad wood windows.** White. By window nominal size.

| | Craft@Hrs | Unit | Material | Labor | Total | Sell |
|---|---|---|---|---|---|---|
| 2'0" x 3'0" | B1@.050 | Ea | 16.70 | 1.47 | 18.17 | 30.90 |
| 2'0" x 3'6" | B1@.050 | Ea | 18.90 | 1.47 | 20.37 | 34.60 |
| 2'4" x 3'0" | B1@.050 | Ea | 18.00 | 1.47 | 19.47 | 33.10 |
| 2'4" x 3'6" | B1@.050 | Ea | 20.20 | 1.47 | 21.67 | 36.80 |
| 2'4" x 4'0" | B1@.050 | Ea | 21.30 | 1.47 | 22.77 | 38.70 |
| 2'4" x 4'6" | B1@.050 | Ea | 23.20 | 1.47 | 24.67 | 41.90 |
| 2'4" x 4'9" | B1@.050 | Ea | 23.60 | 1.47 | 25.07 | 42.60 |
| 2'8" x 3'0" | B1@.050 | Ea | 19.30 | 1.47 | 20.77 | 35.30 |
| 2'8" x 3'6" | B1@.050 | Ea | 21.30 | 1.47 | 22.77 | 38.70 |
| 2'8" x 4'0" | B1@.050 | Ea | 22.60 | 1.47 | 24.07 | 40.90 |
| 2'8" x 4'6" | B1@.050 | Ea | 24.50 | 1.47 | 25.97 | 44.10 |
| 2'8" x 4'9" | B1@.050 | Ea | 25.40 | 1.47 | 26.87 | 45.70 |
| 3'0" x 3'0" | B1@.050 | Ea | 20.40 | 1.47 | 21.87 | 37.20 |
| 3'0" x 3'6" | B1@.050 | Ea | 22.30 | 1.47 | 23.77 | 40.40 |
| 3'0" x 4'0" | B1@.050 | Ea | 23.60 | 1.47 | 25.07 | 42.60 |
| 3'0" x 4'6" | B1@.050 | Ea | 24.50 | 1.47 | 25.97 | 44.10 |
| 3'0" x 4'9" | B1@.050 | Ea | 26.40 | 1.47 | 27.87 | 47.40 |
| 3'0" x 5'6" | B1@.050 | Ea | 24.40 | 1.47 | 25.87 | 44.00 |

**Vinyl-clad wood awning windows, insulating glass.** Prefinished pine interior. White vinyl exterior. For new construction. Add the cost of screens. By nominal (opening) size, width x height.

| | Craft@Hrs | Unit | Material | Labor | Total | Sell |
|---|---|---|---|---|---|---|
| 2'1" x 2'0" | B1@.500 | Ea | 183.00 | 14.70 | 197.70 | 336.00 |
| 3'1" x 2'0" | B1@.500 | Ea | 215.00 | 14.70 | 229.70 | 390.00 |
| 4'1" x 2'0" | B1@.500 | Ea | 242.00 | 14.70 | 256.70 | 436.00 |

**Vinyl-clad wood casement windows, insulating glass.** Prefinished pine interior. White or tan vinyl exterior. Tilt sash. For new construction. By nominal (opening) size, width x height. Add the cost of screens and hardware. Hinge operation as viewed from exterior.

| | Craft@Hrs | Unit | Material | Labor | Total | Sell |
|---|---|---|---|---|---|---|
| 2'1" x 3'0", right hinge | B1@.500 | Ea | 204.00 | 14.70 | 218.70 | 372.00 |
| 2'1" x 3'0", left hinge | B1@.500 | Ea | 204.00 | 14.70 | 218.70 | 372.00 |
| 2'1" x 3'6", right hinge | B1@.500 | Ea | 228.00 | 14.70 | 242.70 | 413.00 |
| 2'1" x 3'6", left hinge | B1@.500 | Ea | 228.00 | 14.70 | 242.70 | 413.00 |
| 2'1" x 4'0", right hinge | B1@.750 | Ea | 237.00 | 22.00 | 259.00 | 440.00 |
| 2'1" x 4'0", left hinge | B1@.750 | Ea | 237.00 | 22.00 | 259.00 | 440.00 |
| 2'1" x 5'0", right hinge | B1@.750 | Ea | 269.00 | 22.00 | 291.00 | 495.00 |
| 2'1" x 5'0", left hinge | B1@.750 | Ea | 269.00 | 22.00 | 291.00 | 495.00 |
| 4'1" x 3'0", left and right hinge | B1@.750 | Ea | 301.00 | 22.00 | 323.00 | 549.00 |
| 4'1" x 3'5", left and right hinge | B1@.750 | Ea | 420.00 | 22.00 | 442.00 | 751.00 |
| 4'1" x 4'0", left and right hinge | B1@.750 | Ea | 463.00 | 22.00 | 485.00 | 825.00 |
| Add for casement hardware | — | Ea | 7.98 | — | 7.98 | — |

## Aluminum Windows

**Aluminum single-hung vertical sliding white windows, insulating glass.** White frame. Tilt-in removable sash. Dual sash weatherstripping. Includes screen. By nominal (opening) size, width x height. Nailing flange for new construction.

| | Craft@Hrs | Unit | Material | Labor | Total | Sell |
|---|---|---|---|---|---|---|
| 2' x 3' | B1@.500 | Ea | 48.60 | 14.70 | 63.30 | 108.00 |
| 2' x 5' | B1@1.00 | Ea | 71.30 | 29.40 | 100.70 | 171.00 |
| 2'8" x 5' | B1@1.00 | Ea | 77.80 | 29.40 | 107.20 | 182.00 |
| 3' x 3' | B1@1.00 | Ea | 62.60 | 29.40 | 92.00 | 156.00 |
| 3' x 4' | B1@1.00 | Ea | 73.40 | 29.40 | 102.80 | 175.00 |
| 3' x 5' | B1@1.25 | Ea | 84.20 | 36.70 | 120.90 | 206.00 |
| 3' x 6' | B1@1.25 | Ea | 94.00 | 36.70 | 130.70 | 222.00 |
| 4' x 3' | B1@1.00 | Ea | 81.70 | 29.40 | 111.10 | 189.00 |

| | Craft@Hrs | Unit | Material | Labor | Total | Sell |
|---|---|---|---|---|---|---|
| 4' x 4' | B1@1.50 | Ea | 99.80 | 44.10 | 143.90 | 245.00 |
| 5' x 3' | B1@1.50 | Ea | 96.30 | 44.10 | 140.40 | 239.00 |
| 5' x 4' | B3@2.00 | Ea | 114.00 | 56.90 | 170.90 | 291.00 |
| 6' x 4' | B3@2.00 | Ea | 135.00 | 56.90 | 191.90 | 326.00 |
| 8' x 4' | B3@2.00 | Ea | 174.00 | 56.90 | 230.90 | 393.00 |
| Add for obscure glass in 2' windows | — | Ea | 3.15 | — | 3.15 | — |

**Aluminum single-hung vertical sliding bronze windows, insulating glass.** Bronze frame. Better quality removable sash. Dual sash weatherstripping. Includes screen. By nominal (opening) size, width x height. Includes a window grid between the panes. Nailing flange for new construction.

| | Craft@Hrs | Unit | Material | Labor | Total | Sell |
|---|---|---|---|---|---|---|
| 2' x 3' | B1@.500 | Ea | 75.60 | 14.70 | 90.30 | 154.00 |
| 2' 8" x 3' | B1@.500 | Ea | 84.00 | 14.70 | 98.70 | 168.00 |
| 2' 8" x 4' 4" | B1@1.00 | Ea | 107.00 | 29.40 | 136.40 | 232.00 |
| 2' 8" x 5' | B1@1.00 | Ea | 117.00 | 29.40 | 146.40 | 249.00 |
| 3' x 3' | B1@1.00 | Ea | 91.00 | 29.40 | 120.40 | 205.00 |
| 3' x 4' | B1@1.00 | Ea | 116.00 | 29.40 | 145.40 | 247.00 |
| 3' x 4' 4" | B1@1.00 | Ea | 119.00 | 29.40 | 148.40 | 252.00 |
| 3' x 5' | B1@1.25 | Ea | 120.00 | 36.70 | 156.70 | 266.00 |
| 3' x 6' | B1@1.25 | Ea | 136.00 | 36.70 | 172.70 | 294.00 |

**Aluminum single-hung vertical sliding windows, single glazed.** Mill finish. Includes screen. By nominal (opening) size, width x height. Clear glass except as noted. Nailing flange for new construction.

| | Craft@Hrs | Unit | Material | Labor | Total | Sell |
|---|---|---|---|---|---|---|
| 2' x 1' | B1@.500 | Ea | 21.50 | 14.70 | 36.20 | 61.50 |
| 2' x 1', obscure glass | B1@.500 | Ea | 24.80 | 14.70 | 39.50 | 67.20 |
| 2' x 2' | B1@.500 | Ea | 31.30 | 14.70 | 46.00 | 78.20 |
| 2' x 2', obscure glass | B1@.500 | Ea | 33.50 | 14.70 | 48.20 | 81.90 |
| 2' x 3', obscure glass | B1@.500 | Ea | 47.50 | 14.70 | 62.20 | 106.00 |
| 3' x 2' | B1@.500 | Ea | 35.60 | 14.70 | 50.30 | 85.50 |
| 3' x 3' | B1@1.00 | Ea | 46.40 | 29.40 | 75.80 | 129.00 |
| 4' x 3' | B1@1.00 | Ea | 52.90 | 29.40 | 82.30 | 140.00 |
| 5' x 3' | B1@1.50 | Ea | 57.20 | 44.10 | 101.30 | 172.00 |
| 6' x 3' | B3@1.50 | Ea | 62.40 | 42.70 | 105.10 | 179.00 |

**Aluminum horizontal sliding windows, single glazed.** Right sash is operable when viewed from the exterior. With screen. 7/8" nailing fin setback for 3-coat stucco or wood siding. By nominal (opening) size, width x height. For new construction. Obscure glass is usually available in smaller (1' and 2') windows at no additional cost.

| | Craft@Hrs | Unit | Material | Labor | Total | Sell |
|---|---|---|---|---|---|---|
| 2' x 2', mill finish | B1@.500 | Ea | 64.00 | 14.70 | 78.70 | 134.00 |
| 2' x 2', bronze or white finish | B1@.500 | Ea | 66.00 | 14.70 | 80.70 | 137.00 |
| 3' x 1', mill finish | B1@.750 | Ea | 60.00 | 22.00 | 82.00 | 139.00 |
| 3' x 2', mill finish | B1@.750 | Ea | 79.00 | 22.00 | 101.00 | 172.00 |
| 3' x 3', mill finish | B1@1.00 | Ea | 92.00 | 29.40 | 121.40 | 206.00 |
| 3' x 3', bronze or white finish | B1@1.00 | Ea | 96.00 | 29.40 | 125.40 | 213.00 |
| 3' x 4', mill finish | B1@1.00 | Ea | 107.00 | 29.40 | 136.40 | 232.00 |
| 3' x 4', bronze or white finish | B1@1.00 | Ea | 111.00 | 29.40 | 140.40 | 239.00 |
| 4' x 3', mill finish | B1@1.00 | Ea | 100.00 | 29.40 | 129.40 | 220.00 |
| 4' x 3', bronze or white finish | B1@1.00 | Ea | 102.00 | 29.40 | 131.40 | 223.00 |
| 4' x 4', mill finish | B1@1.50 | Ea | 120.00 | 44.10 | 164.10 | 279.00 |
| 4' x 4', bronze or white finish | B1@1.50 | Ea | 130.00 | 44.10 | 174.10 | 296.00 |
| 5' x 3', mill finish | B3@1.50 | Ea | 126.00 | 42.70 | 168.70 | 287.00 |
| 5' x 4', mill finish | B3@1.50 | Ea | 139.00 | 42.70 | 181.70 | 309.00 |
| 5' x 4', bronze or white finish | B3@1.50 | Ea | 143.00 | 42.70 | 185.70 | 316.00 |
| 6' x 3', mill finish | B3@1.50 | Ea | 148.00 | 42.70 | 190.70 | 324.00 |
| 6' x 4', mill finish | B3@1.50 | Ea | 160.00 | 42.70 | 202.70 | 345.00 |
| 6' x 4', bronze or white finish | B3@1.50 | Ea | 170.00 | 42.70 | 212.70 | 362.00 |

| | Craft@Hrs | Unit | Material | Labor | Total | Sell |
|---|---|---|---|---|---|---|

**Aluminum horizontal sliding windows, insulated glass.** Right sash is operable when viewed from the exterior. With screen. Nailing fin for new construction. By nominal (opening) size, width x height. Obscure glass is usually available in smaller (1' and 2') windows at no additional cost. White or bronze finish.

| | Craft@Hrs | Unit | Material | Labor | Total | Sell |
|---|---|---|---|---|---|---|
| 2' x 2' | B1@.500 | Ea | 58.00 | 14.70 | 72.70 | 124.00 |
| 3' x 1' | B1@.750 | Ea | 59.30 | 22.00 | 81.30 | 138.00 |
| 3' x 2' | B1@.750 | Ea | 72.00 | 22.00 | 94.00 | 160.00 |
| 3' x 3' | B1@1.00 | Ea | 86.00 | 29.40 | 115.40 | 196.00 |
| 3' x 4' | B1@1.00 | Ea | 100.00 | 29.40 | 129.40 | 220.00 |
| 4' x 3' | B1@1.00 | Ea | 110.00 | 29.40 | 139.40 | 237.00 |
| 4' x 4' | B1@1.50 | Ea | 112.00 | 44.10 | 156.10 | 265.00 |
| Add for colonial grid between the lites | — | % | 15.0 | — | 15.0 | — |
| Add for Low-E glass | — | % | 15.0 | — | 15.0 | — |

**Aluminum awning windows, insulated glass.** With hardware and screen. Natural mill finish. For new construction. Nominal (opening) sizes, width x height.

| | Craft@Hrs | Unit | Material | Labor | Total | Sell |
|---|---|---|---|---|---|---|
| 19" x 26", two lites high | BC@.500 | Ea | 73.60 | 16.00 | 89.60 | 152.00 |
| 19" x 38-1/2", three lites high | BC@.750 | Ea | 88.80 | 24.10 | 112.90 | 192.00 |
| 19" x 50-1/2", four lites high | BC@.750 | Ea | 106.00 | 24.10 | 130.10 | 221.00 |
| 19" x 63", five lites high | BC@1.00 | Ea | 125.00 | 32.10 | 157.10 | 267.00 |
| 26" x 26", two lites high | BC@.750 | Ea | 91.40 | 24.10 | 115.50 | 196.00 |
| 26" x 38-1/2", three lites high | BC@.750 | Ea | 107.00 | 24.10 | 131.10 | 223.00 |
| 26" x 50-1/2", four lites high | BC@1.00 | Ea | 138.00 | 32.10 | 170.10 | 289.00 |
| 26" x 63", five lites high | BC@1.00 | Ea | 162.00 | 32.10 | 194.10 | 330.00 |
| 37" x 26", two lites high | BC@.750 | Ea | 106.00 | 24.10 | 130.10 | 221.00 |
| 37" x 38-1/2", three lites high | BC@1.00 | Ea | 123.00 | 32.10 | 155.10 | 264.00 |
| 37" x 50-1/2", four lites high | BC@1.00 | Ea | 141.00 | 32.10 | 173.10 | 294.00 |
| 37" x 63", five lites high | BC@1.00 | Ea | 190.00 | 32.10 | 222.10 | 378.00 |
| 53" x 26", two lites high | BC@1.00 | Ea | 126.00 | 32.10 | 158.10 | 269.00 |
| 53" x 38-1/2", three lites high | BC@1.00 | Ea | 143.00 | 32.10 | 175.10 | 298.00 |
| 53" x 50-1/2", four lites high | B3@1.50 | Ea | 186.00 | 42.70 | 228.70 | 389.00 |
| 53" x 63", five lites high | B3@1.50 | Ea | 218.00 | 42.70 | 260.70 | 443.00 |
| Add for bronze finish | — | % | 15.0 | — | — | — |
| Add for enamel finish | — | % | 10.0 | — | — | — |
| Add for special order replacement window | — | % | 100.0 | — | — | — |

**Fixed aluminum windows (picture windows), insulating glass.** Bronze or white enamel finish. For new construction. Wall opening (nominal) size, width x height.

| | Craft@Hrs | Unit | Material | Labor | Total | Sell |
|---|---|---|---|---|---|---|
| 24" x 20" | BC@.500 | Ea | 99.80 | 16.00 | 115.80 | 197.00 |
| 24" x 30" | BC@.500 | Ea | 117.00 | 16.00 | 133.00 | 226.00 |
| 24" x 40" | BC@.500 | Ea | 138.00 | 16.00 | 154.00 | 262.00 |
| 30" x 20" | BC@.750 | Ea | 122.00 | 24.10 | 146.10 | 248.00 |
| 30" x 30" | BC@.750 | Ea | 143.00 | 24.10 | 167.10 | 284.00 |
| 30" x 40" | BC@.750 | Ea | 169.00 | 24.10 | 193.10 | 328.00 |
| 30" x 50" | BC@1.00 | Ea | 206.00 | 32.10 | 238.10 | 405.00 |
| 30" x 60" | BC@1.00 | Ea | 223.00 | 32.10 | 255.10 | 434.00 |
| 36" x 20" | BC@.750 | Ea | 147.00 | 24.10 | 171.10 | 291.00 |
| 36" x 30" | BC@.750 | Ea | 173.00 | 24.10 | 197.10 | 335.00 |
| 36" x 40" | BC@1.00 | Ea | 215.00 | 32.10 | 247.10 | 420.00 |

| | Craft@Hrs | Unit | Material | Labor | Total | Sell |
|---|---|---|---|---|---|---|
| 36" x 50" | B1@1.50 | Ea | 187.00 | 44.10 | 231.10 | 393.00 |
| 36" x 60" | B1@1.50 | Ea | 307.00 | 44.10 | 351.10 | 597.00 |
| 40" x 20" | BC@.750 | Ea | 173.00 | 24.10 | 197.10 | 335.00 |
| 40" x 30" | BC@.750 | Ea | 216.00 | 24.10 | 240.10 | 408.00 |
| 40" x 40" | BC@1.00 | Ea | 253.00 | 32.10 | 285.10 | 485.00 |
| 48" x 20" | BC@.750 | Ea | 207.00 | 24.10 | 231.10 | 393.00 |
| 48" x 30" | BC@1.00 | Ea | 258.00 | 32.10 | 290.10 | 493.00 |
| 48" x 40" | BC@1.00 | Ea | 284.00 | 32.10 | 316.10 | 537.00 |
| Add for storm sash, per square foot of glass | — | SF | 7.27 | — | 7.27 | — |

**Louvered aluminum (jalousie) windows.** Single glazed, 4" clear glass louvers, including hardware. Jalousie windows are usually made to order and can be fabricated to any size.

| | Craft@Hrs | Unit | Material | Labor | Total | Sell |
|---|---|---|---|---|---|---|
| 18" x 24" | BC@1.00 | Ea | 55.50 | 32.10 | 87.60 | 149.00 |
| 18" x 36" | BC@1.00 | Ea | 80.20 | 32.10 | 112.30 | 191.00 |
| 18" x 48" | BC@1.00 | Ea | 98.80 | 32.10 | 130.90 | 223.00 |
| 18" x 60" | BC@1.00 | Ea | 129.00 | 32.10 | 161.10 | 274.00 |
| 24" x 24" | BC@1.00 | Ea | 62.50 | 32.10 | 94.60 | 161.00 |
| 24" x 36" | BC@1.00 | Ea | 90.90 | 32.10 | 123.00 | 209.00 |
| 24" x 48" | BC@1.00 | Ea | 114.00 | 32.10 | 146.10 | 248.00 |
| 24" x 60" | BC@1.50 | Ea | 150.00 | 48.10 | 198.10 | 337.00 |
| 30" x 24" | BC@1.00 | Ea | 74.80 | 32.10 | 106.90 | 182.00 |
| 30" x 36" | BC@1.00 | Ea | 107.00 | 32.10 | 139.10 | 236.00 |
| 30" x 48" | BC@1.50 | Ea | 135.00 | 48.10 | 183.10 | 311.00 |
| 30" x 60" | BC@1.50 | Ea | 173.00 | 48.10 | 221.10 | 376.00 |
| 36" x 24" | BC@1.00 | Ea | 81.00 | 32.10 | 113.10 | 192.00 |
| 36" x 36" | BC@1.50 | Ea | 117.00 | 48.10 | 165.10 | 281.00 |
| 36" x 48" | BC@1.50 | Ea | 150.00 | 48.10 | 198.10 | 337.00 |
| 36" x 60" | BC@1.50 | Ea | 194.00 | 48.10 | 242.10 | 412.00 |
| 40" x 24" | BC@1.00 | Ea | 94.10 | 32.10 | 126.20 | 215.00 |
| 40" x 36" | BC@1.50 | Ea | 133.00 | 48.10 | 181.10 | 308.00 |
| 40" x 48" | BC@1.50 | Ea | 169.00 | 48.10 | 217.10 | 369.00 |
| 40" x 60" | BC@1.50 | Ea | 216.00 | 48.10 | 264.10 | 449.00 |

## Wood Windows

**Wood double-hung windows.** Treated western pine. Natural wood interior. 4-9/16" jamb width. 5/8" insulating glass. Weatherstripped. Stock units for new construction. Tilt-in, removable sash. With hardware. Nominal sizes, width x height.

| | Craft@Hrs | Unit | Material | Labor | Total | Sell |
|---|---|---|---|---|---|---|
| 2'0" x 3'2" | BC@.750 | Ea | 107.00 | 24.10 | 131.10 | 223.00 |
| 2'4" x 3'2" | BC@.750 | Ea | 124.00 | 24.10 | 148.10 | 252.00 |
| 2'4" x 4'6" | BC@.750 | Ea | 145.00 | 24.10 | 169.10 | 287.00 |
| 2'8" x 3'2" | BC@.750 | Ea | 133.00 | 24.10 | 157.10 | 267.00 |
| 2'8" x 3'10" | BC@.750 | Ea | 139.00 | 24.10 | 163.10 | 277.00 |
| 2'8" x 4'6" | BC@1.25 | Ea | 152.00 | 40.10 | 192.10 | 327.00 |
| 2'8" x 5'2" | BC@1.25 | Ea | 161.00 | 40.10 | 201.10 | 342.00 |
| 3'0" x 3'2" | BC@.750 | Ea | 140.00 | 24.10 | 164.10 | 279.00 |
| 3'0" x 4'6" | BC@1.25 | Ea | 161.00 | 40.10 | 201.10 | 342.00 |
| 3'0" x 5'2" | BC@1.25 | Ea | 172.00 | 40.10 | 212.10 | 361.00 |
| 2'8" x 4'6", twin | BC@1.25 | Ea | 295.00 | 40.10 | 335.10 | 570.00 |

**Fixed octagon wood windows.** Stock units for new construction. Installed in a pre-cut and framed opening.

| | Craft@Hrs | Unit | Material | Labor | Total | Sell |
|---|---|---|---|---|---|---|
| 24", insulated glass | BC@1.50 | Ea | 80.50 | 48.10 | 128.60 | 219.00 |
| 24", single glazed | BC@1.50 | Ea | 57.20 | 48.10 | 105.30 | 179.00 |
| 24", venting insulated glass | BC@1.50 | Ea | 123.00 | 48.10 | 171.10 | 291.00 |

| | Craft@Hrs | Unit | Material | Labor | Total | Sell |
|---|---|---|---|---|---|---|

**Angle bay windows.** Fixed center lite and two double-hung flankers. 30 degree angle. Pine with aluminum cladding, insulated glass, screens, white or bronze finish, 4-9/16" wall thickness. Stock units for new construction. Nominal (opening) sizes, width x height. Add the cost of interior trim and installing or framing the bay window roof. See the figures that follow.

| | Craft@Hrs | Unit | Material | Labor | Total | Sell |
|---|---|---|---|---|---|---|
| 6'4" x 4'8" | B1@4.00 | Ea | 1,380.00 | 117.00 | 1,497.00 | 2,540.00 |
| 7'2" x 4'8" | B1@4.00 | Ea | 1,390.00 | 117.00 | 1,507.00 | 2,560.00 |
| 7'10" x 4'8" | B1@4.50 | Ea | 1,510.00 | 132.00 | 1,642.00 | 2,790.00 |
| 8'6" x 4'8" | B1@4.50 | Ea | 1,730.00 | 132.00 | 1,862.00 | 3,170.00 |
| 8'10" x 4'8" | B1@4.50 | Ea | 1,760.00 | 132.00 | 1,892.00 | 3,220.00 |
| 9'2" x 4'8" | B1@5.00 | Ea | 1,700.00 | 147.00 | 1,847.00 | 3,140.00 |
| 9'10" x 4'8" | B1@5.00 | Ea | 1,790.00 | 147.00 | 1,937.00 | 3,290.00 |
| Add for 45 degree angle windows | — | Ea | 71.00 | — | 71.00 | — |
| Add for roof framing kit | — | Ea | 126.00 | — | 126.00 | — |

**Bay window roof cover.** Architectural copper roof system for standard angle bay windows. Includes hardware. Stock units for new construction.

| | Craft@Hrs | Unit | Material | Labor | Total | Sell |
|---|---|---|---|---|---|---|
| Up to 48" width | SW@.500 | Ea | 635.00 | 17.80 | 652.80 | 1,110.00 |
| 49" to 60" width | SW@.500 | Ea | 722.00 | 17.80 | 739.80 | 1,260.00 |
| 61" to 72" width | SW@.600 | Ea | 809.00 | 21.40 | 830.40 | 1,410.00 |
| 73" to 84" width | SW@.600 | Ea | 897.00 | 21.40 | 918.40 | 1,560.00 |
| 85" to 96" width | SW@.600 | Ea | 985.00 | 21.40 | 1,006.40 | 1,710.00 |
| 97" to 108" width | SW@.700 | Ea | 1,070.00 | 25.00 | 1,095.00 | 1,860.00 |
| 109" to 120" width | SW@.700 | Ea | 1,160.00 | 25.00 | 1,185.00 | 2,010.00 |
| 121" to 132" width | SW@.800 | Ea | 1,330.00 | 28.50 | 1,358.50 | 2,310.00 |
| 133" to 144" width | SW@.800 | Ea | 1,420.00 | 28.50 | 1,448.50 | 2,460.00 |
| 145" to 156" width | SW@1.00 | Ea | 1,510.00 | 35.70 | 1,545.70 | 2,630.00 |
| 157" to 168" width | SW@1.00 | Ea | 1,590.00 | 35.70 | 1,625.70 | 2,760.00 |
| Add for aged copper, brass or aluminum finish | — | % | 20.0 | — | — | — |
| Add for 24" to 26" projection | — | % | 10.0 | — | — | — |
| Add for 26" to 28" projection | — | % | 15.0 | — | — | — |
| Add for 28" to 30" projection | — | % | 20.0 | — | — | — |
| Add for brick soffit kit | — | % | 45.0 | — | — | — |
| Add for 90-degree bays | — | Ea | 105.00 | — | 105.00 | — |

**Casement bow windows.** Insulated glass, assembled, with screens, weatherstripped, unfinished pine, 16" projection, 22" x 56" lites, 4-7/16" jambs. Add the cost of interior trim and installing or framing the bow window roof. Stock units for new construction. Generally a roof kit is required for projections of 12" or more. See the figures that follow.

| | Craft@Hrs | Unit | Material | Labor | Total | Sell |
|---|---|---|---|---|---|---|
| 8'1" wide x 4'8" (4 lites, 2 venting) | B1@4.75 | Ea | 1,470.00 | 140.00 | 1,610.00 | 2,740.00 |
| 8'1" wide x 5'4" (4 lites, 2 venting) | B1@4.75 | Ea | 1,590.00 | 140.00 | 1,730.00 | 2,940.00 |
| 8'1" wide x 6'0" (4 lites, 2 venting) | B1@4.75 | Ea | 1,700.00 | 140.00 | 1,840.00 | 3,130.00 |
| 10' wide x 4'8" (5 lites, 2 venting) | B1@4.75 | Ea | 1,790.00 | 140.00 | 1,930.00 | 3,280.00 |
| 10' wide x 5'4" (5 lites, 2 venting) | B1@4.75 | Ea | 1,930.00 | 140.00 | 2,070.00 | 3,520.00 |
| 10' wide x 6'0" (5 lites, 2 venting) | B1@4.75 | Ea | 1,070.00 | 140.00 | 1,210.00 | 2,060.00 |
| Add for roof framing kit, 5 lites | — | Ea | 116.00 | — | 116.00 | — |

**Bow window wood roof kit.** Framed wood roof kit for four panel bow window. Built with a 10 in 12 pitch, roof overhangs bay or bow 1" to 3". Includes hardware.

| | Craft@Hrs | Unit | Material | Labor | Total | Sell |
|---|---|---|---|---|---|---|
| Up to 72" width | B1@.500 | Ea | 307.00 | 14.70 | 321.70 | 547.00 |
| 73" to 96" width | B1@.500 | Ea | 360.00 | 14.70 | 374.70 | 637.00 |
| 97" to 120" width | B1@.500 | Ea | 416.00 | 14.70 | 430.70 | 732.00 |
| 121" to 144" width | B1@.600 | Ea | 470.00 | 17.60 | 487.60 | 829.00 |
| 145" to 168" width | B1@.600 | Ea | 648.00 | 17.60 | 665.60 | 1,130.00 |
| Add for 5 panel bow | — | Ea | 47.80 | — | 47.80 | — |

| | Craft@Hrs | Unit | Material | Labor | Total | Sell |
|---|---|---|---|---|---|---|

**Bow or bay window roof, framed on-site.** 1/2" plywood sheathing, fiberglass batt insulation, fascia board, cove molding, roof flashing, 15-pound felt, 240-pound seal-tab asphalt shingles, and cleanup.

| | Craft@Hrs | Unit | Material | Labor | Total | Sell |
|---|---|---|---|---|---|---|
| Up to 120" width | B1@3.50 | Ea | 69.50 | 103.00 | 172.50 | 293.00 |
| Add per linear foot of width over 10' | B1@.350 | LF | 6.95 | 10.30 | 17.25 | 29.30 |

## Storm Windows

**Basement storm windows.** White aluminum frame with removable plexiglass window. Clips on over aluminum or wood basement window. Screen included. Nominal window sizes, width x height.

| | Craft@Hrs | Unit | Material | Labor | Total | Sell |
|---|---|---|---|---|---|---|
| 32" x 14" | BC@.420 | Ea | 20.50 | 13.50 | 34.00 | 57.80 |
| 32" x 18" | BC@.420 | Ea | 20.50 | 13.50 | 34.00 | 57.80 |
| 32" x 22" | BC@.420 | Ea | 20.50 | 13.50 | 34.00 | 57.80 |

**Aluminum storm windows.** Removable glass inserts. Removable screen. Double-track. Includes weatherstripping. Adaptable to aluminum single-hung windows.

| | Craft@Hrs | Unit | Material | Labor | Total | Sell |
|---|---|---|---|---|---|---|
| 24" x 39", mill finish | BC@.420 | Ea | 31.30 | 13.50 | 44.80 | 76.20 |
| 24" x 39", white finish | BC@.420 | Ea | 35.60 | 13.50 | 49.10 | 83.50 |
| 24" x 55", mill finish | BC@.420 | Ea | 37.80 | 13.50 | 51.30 | 87.20 |
| 24" x 55", white finish | BC@.420 | Ea | 45.40 | 13.50 | 58.90 | 100.00 |
| 28" x 39", mill finish | BC@.420 | Ea | 32.40 | 13.50 | 45.90 | 78.00 |
| 28" x 39", white finish | BC@.420 | Ea | 37.80 | 13.50 | 51.30 | 87.20 |
| 28" x 47", white finish | BC@.420 | Ea | 42.10 | 13.50 | 55.60 | 94.50 |
| 28" x 55", mill finish | BC@.420 | Ea | 37.90 | 13.50 | 51.40 | 87.40 |
| 28" x 55", white finish | BC@.420 | Ea | 45.40 | 13.50 | 58.90 | 100.00 |
| 32" x 39", mill finish | BC@.420 | Ea | 33.50 | 13.50 | 47.00 | 79.90 |
| 32" x 39", white finish | BC@.420 | Ea | 42.10 | 13.50 | 55.60 | 94.50 |
| 32" x 47", white finish | BC@.420 | Ea | 47.50 | 13.50 | 61.00 | 104.00 |
| 32" x 55", mill finish | BC@.420 | Ea | 37.90 | 13.50 | 51.40 | 87.40 |
| 32" x 55", white finish | BC@.420 | Ea | 47.50 | 13.50 | 61.00 | 104.00 |
| 32" x 63", mill finish | BC@.420 | Ea | 47.90 | 13.50 | 61.40 | 104.00 |
| 32" x 63", white finish | BC@.420 | Ea | 47.50 | 13.50 | 61.00 | 104.00 |
| 32" x 75", white finish | BC@.420 | Ea | 64.00 | 13.50 | 77.50 | 132.00 |
| 36" x 39", mill finish | BC@.420 | Ea | 33.60 | 13.50 | 47.10 | 80.10 |
| 36" x 39", white finish | BC@.420 | Ea | 42.10 | 13.50 | 55.60 | 94.50 |
| 36" x 55", mill finish | BC@.420 | Ea | 40.10 | 13.50 | 53.60 | 91.10 |
| 36" x 55", white finish | BC@.420 | Ea | 47.50 | 13.50 | 61.00 | 104.00 |
| 36" x 63", mill finish | BC@.420 | Ea | 48.20 | 13.50 | 61.70 | 105.00 |
| 36" x 63", white finish | BC@.420 | Ea | 63.70 | 13.50 | 77.20 | 131.00 |

## Window Accessories

**Window sills, EasySills.** Solid vinyl. Set with latex or solvent base adhesive. Hard scratch-resistant surface. Ultraviolet protection to prevent yellowing. Gloss finish.

| | Craft@Hrs | Unit | Material | Labor | Total | Sell |
|---|---|---|---|---|---|---|
| 3-1/2" x 3' | BC@.129 | Ea | 12.30 | 4.14 | 16.44 | 27.90 |
| 3-1/2" x 4' | BC@.172 | Ea | 15.30 | 5.52 | 20.82 | 35.40 |
| 3-1/2" x 5' | BC@.215 | Ea | 18.10 | 6.90 | 25.00 | 42.50 |
| 3-1/2" x 6' | BC@.258 | Ea | 21.10 | 8.28 | 29.38 | 49.90 |
| 4-1/4" x 3' | BC@.129 | Ea | 13.10 | 4.14 | 17.24 | 29.30 |
| 4-1/4" x 4' | BC@.172 | Ea | 16.40 | 5.52 | 21.92 | 37.30 |
| 4-1/4" x 5' | BC@.215 | Ea | 21.00 | 6.90 | 27.90 | 47.40 |
| 4-1/4" x 6' | BC@.258 | Ea | 24.20 | 8.28 | 32.48 | 55.20 |
| 5-1/2" x 3' | BC@.129 | Ea | 16.30 | 4.14 | 20.44 | 34.70 |
| 5-1/2" x 4' | BC@.172 | Ea | 20.70 | 5.52 | 26.22 | 44.60 |
| 5-1/2" x 5' | BC@.215 | Ea | 24.70 | 6.90 | 31.60 | 53.70 |
| 5-1/2" x 6' | BC@.258 | Ea | 29.20 | 8.28 | 37.48 | 63.70 |
| 6-1/4" x 3' | BC@.129 | Ea | 17.10 | 4.14 | 21.24 | 36.10 |

| | Craft@Hrs | Unit | Material | Labor | Total | Sell |
|---|---|---|---|---|---|---|
| 6-1/4" x 4' | BC@.172 | Ea | 23.20 | 5.52 | 28.72 | 48.80 |
| 6-1/4" x 5' | BC@.215 | Ea | 27.70 | 6.90 | 34.60 | 58.80 |
| 6-1/4" x 6' | BC@.258 | Ea | 34.20 | 8.28 | 42.48 | 72.20 |

**Aluminum window mullions.** White.

| | | | | | | |
|---|---|---|---|---|---|---|
| 38-1/4" | — | Ea | 41.00 | — | 41.00 | — |
| 50-1/2" | — | Ea | 50.76 | — | 50.76 | — |
| 62-7/8" | — | Ea | 57.20 | — | 57.20 | — |

**Replacement casement window operator handle.** 9" arm. Casement opening direction when viewed from the exterior.

| | | | | | | |
|---|---|---|---|---|---|---|
| Left, aluminum window | BC@.250 | Ea | 8.60 | 8.02 | 16.62 | 28.30 |
| Right, aluminum window | BC@.250 | Ea | 8.60 | 8.02 | 16.62 | 28.30 |
| Left, wood window | BC@.250 | Ea | 16.80 | 8.02 | 24.82 | 42.20 |
| Right, wood window | BC@.250 | Ea | 16.80 | 8.02 | 24.82 | 42.20 |

**Casement window lock handle.** Universal casement window lock for aluminum or steel windows. Accepts a 1/8" security pin. Screws included.

| | | | | | | |
|---|---|---|---|---|---|---|
| Aluminum | BC@.250 | Ea | 3.85 | 8.02 | 11.87 | 20.20 |
| Bronze finish, 5/16" bore size | BC@.250 | Ea | 3.74 | 8.02 | 11.76 | 20.00 |
| White | BC@.250 | Ea | 3.90 | 8.02 | 11.92 | 20.30 |

**Awning window operator handle.**

| | | | | | | |
|---|---|---|---|---|---|---|
| Left | BC@.250 | Ea | 8.60 | 8.02 | 16.62 | 28.30 |
| Right | BC@.250 | Ea | 8.60 | 8.02 | 16.62 | 28.30 |

**Jalousie window operator handle.**

| | | | | | | |
|---|---|---|---|---|---|---|
| Crank handle, 5/16" | BC@.250 | Ea | 2.14 | 8.02 | 10.16 | 17.30 |
| Crank handle, 3/8" | BC@.250 | Ea | 2.28 | 8.02 | 10.30 | 17.50 |
| T-handle, 5/16" | BC@.250 | Ea | 2.14 | 8.02 | 10.16 | 17.30 |
| T-handle, 3/8" | BC@.250 | Ea | 2.26 | 8.02 | 10.28 | 17.50 |

**Wood window sash spring.** Replaces broken sash cords and eliminates window rattle. Includes two sash springs and two nail fasteners.

| | | | | | | |
|---|---|---|---|---|---|---|
| Per pair | BC@.750 | Ea | 4.30 | 24.10 | 28.40 | 48.30 |

**Sash balance.**

| | | | | | | |
|---|---|---|---|---|---|---|
| 4-1/2-pound | BC@.360 | Ea | 20.48 | 11.60 | 32.08 | 54.50 |
| 6-pound | BC@.360 | Ea | 21.20 | 11.60 | 32.80 | 55.80 |
| 8-pound | BC@.360 | Ea | 22.10 | 11.60 | 33.70 | 57.30 |
| 10-pound | BC@.360 | Ea | 24.30 | 11.60 | 35.90 | 61.00 |
| 12-pound | BC@.360 | Ea | 25.30 | 11.60 | 36.90 | 62.70 |

**Sash lift pull.**

| | | | | | | |
|---|---|---|---|---|---|---|
| Oil rubbed bronze | BC@.060 | Ea | 9.93 | 1.93 | 11.86 | 20.20 |
| Satin nickel | BC@.060 | Ea | 9.93 | 1.93 | 11.86 | 20.20 |

**Double-hung sash lock.**

| | | | | | | |
|---|---|---|---|---|---|---|
| Bronze | BC@.350 | Ea | 10.77 | 11.20 | 21.97 | 37.30 |
| Oil rubbed bronze | BC@.350 | Ea | 10.03 | 11.20 | 21.23 | 36.10 |
| White | BC@.350 | Ea | 10.77 | 11.20 | 21.97 | 37.30 |
| Satin nickel | BC@.350 | Ea | 9.50 | 11.20 | 20.70 | 35.20 |

**Double-hung window bolt.**

| | | | | | | |
|---|---|---|---|---|---|---|
| Brass | BC@.350 | Ea | 2.56 | 11.20 | 13.76 | 23.40 |

| | Craft@Hrs | Unit | Material | Labor | Total | Sell |
|---|---|---|---|---|---|---|

**Arch top louvered exterior shutters.** Copolymer construction with molded-through color. Available in colors and paintable. Price per pair.

| | Craft@Hrs | Unit | Material | Labor | Total | Sell |
|---|---|---|---|---|---|---|
| 15" x 36" | BC@.200 | Ea | 23.20 | 6.42 | 29.62 | 50.40 |
| 15" x 39" | BC@.200 | Ea | 24.50 | 6.42 | 30.92 | 52.60 |
| 15" x 43" | BC@.200 | Ea | 25.70 | 6.42 | 32.12 | 54.60 |
| 15" x 48" | BC@.200 | Ea | 26.80 | 6.42 | 33.22 | 56.50 |
| 15" x 52" | BC@.320 | Ea | 30.15 | 10.30 | 40.45 | 68.80 |
| 15" x 55" | BC@.320 | Ea | 31.70 | 10.30 | 42.00 | 71.40 |
| 15" x 60" | BC@.320 | Ea | 32.90 | 10.30 | 43.20 | 73.40 |
| 15" x 64" | BC@.320 | Ea | 36.80 | 10.30 | 47.10 | 80.10 |
| 15" x 67" | BC@.320 | Ea | 35.80 | 10.30 | 46.10 | 78.40 |
| 15" x 72" | BC@.320 | Ea | 40.25 | 10.30 | 50.55 | 85.90 |
| 15" x 75" | BC@.320 | Ea | 44.00 | 10.30 | 54.30 | 92.30 |
| Decorative shutter hardware | — | Ea | 8.26 | — | 8.26 | — |
| Lock fasteners | BC@.125 | Ea | 5.20 | 4.01 | 9.21 | 15.70 |

**Classic window guards.** Welded steel 1/2" square tubes with spear points and ornamentation. Add screws and brackets below. Width x height.

| | Craft@Hrs | Unit | Material | Labor | Total | Sell |
|---|---|---|---|---|---|---|
| 24" x 24" | BC@.500 | Ea | 16.05 | 16.00 | 32.05 | 54.50 |
| 24" x 36" | BC@.500 | Ea | 18.10 | 16.00 | 34.10 | 58.00 |
| 24" x 54" | BC@.500 | Ea | 23.30 | 16.00 | 39.30 | 66.80 |
| 30" x 54" | BC@.500 | Ea | 23.00 | 16.00 | 39.00 | 66.30 |
| 36" x 24" | BC@.500 | Ea | 18.70 | 16.00 | 34.70 | 59.00 |
| 36" x 36" | BC@.500 | Ea | 23.80 | 16.00 | 39.80 | 67.70 |
| 36" x 42" | BC@.750 | Ea | 24.90 | 24.10 | 49.00 | 83.30 |
| 36" x 48" | BC@.750 | Ea | 22.90 | 24.10 | 47.00 | 79.90 |
| 36" x 54" | BC@.750 | Ea | 30.30 | 24.10 | 54.40 | 92.50 |
| 42" x 42" | BC@.750 | Ea | 30.00 | 24.10 | 54.10 | 92.00 |
| 48" x 36" | BC@.750 | Ea | 24.07 | 24.10 | 48.17 | 81.90 |
| 48" x 48" | BC@.750 | Ea | 30.10 | 24.10 | 54.20 | 92.10 |
| 60" x 36" | BC@.750 | Ea | 27.20 | 24.10 | 51.30 | 87.20 |
| 60" x 48" | BC@.750 | Ea | 30.10 | 24.10 | 54.20 | 92.10 |
| 72" x 36" | BC@.750 | Ea | 38.70 | 24.10 | 62.80 | 107.00 |
| 1" one-way screws | — | Ea | 2.81 | — | 2.81 | — |
| 3" one-way screws | — | Ea | 3.36 | — | 3.36 | — |
| 4" one-way screws | — | Ea | 3.95 | — | 3.95 | — |
| Flush brackets | — | Ea | 7.55 | — | 7.55 | — |
| 3" projection brackets | — | Ea | 8.77 | — | 8.77 | — |
| 3" tee brackets | — | Ea | 9.48 | — | 9.48 | — |
| Connector pins | — | Ea | 5.46 | — | 5.46 | — |

**Aluminum window screen frame kits.** For rescreening windows. White or bronze finish. Matching color corners snap into place. Add the cost of screen wire and spline. Labor includes removing the old screen, measuring, cutting and assembling the frame, measuring and cutting screen wire, pressing screen and spline into the frame channel, re-hanging the screen.

| | Craft@Hrs | Unit | Material | Labor | Total | Sell |
|---|---|---|---|---|---|---|
| 5/16" frame, to 24" x 36" screen | BC@.750 | Ea | 9.96 | 24.10 | 34.06 | 57.90 |
| 5/16" frame, to 36" x 48" screen | BC@1.00 | Ea | 9.70 | 32.10 | 41.80 | 71.10 |
| 5/16" frame, to 48" x 60" screen | BC@1.25 | Ea | 11.00 | 40.10 | 51.10 | 86.90 |
| 7/16" frame, to 24" x 36" screen | BC@.750 | Ea | 9.56 | 24.10 | 33.66 | 57.20 |
| 7/16" frame, to 48" x 60" screen | BC@1.25 | Ea | 10.84 | 40.10 | 50.94 | 86.60 |

| | Craft@Hrs | Unit | Material | Labor | Total | Sell |
|---|---|---|---|---|---|---|

**Window screens, aluminum.** Aluminum mill finish frame. Fiberglass mesh. For double hung windows. Labor includes removing the old screen and re-hanging the new screen.

| | Craft@Hrs | Unit | Material | Labor | Total | Sell |
|---|---|---|---|---|---|---|
| 2'0" x 3'2" | BC@.350 | Ea | 12.00 | 11.20 | 23.20 | 39.40 |
| 2'4" x 3'2" | BC@.350 | Ea | 13.00 | 11.20 | 24.20 | 41.10 |
| 2'4" x 4'6" | BC@.350 | Ea | 15.00 | 11.20 | 26.20 | 44.50 |
| 2'8" x 3'1" | BC@.350 | Ea | 15.00 | 11.20 | 26.20 | 44.50 |
| 2'8" x 3'2" | BC@.350 | Ea | 14.00 | 11.20 | 25.20 | 42.80 |
| 2'8" x 4'6" | BC@.350 | Ea | 16.50 | 11.20 | 27.70 | 47.10 |
| 2'8" x 5'2" | BC@.350 | Ea | 17.25 | 11.20 | 28.45 | 48.40 |
| 2'8" x 6'2" | BC@.350 | Ea | 19.50 | 11.20 | 30.70 | 52.20 |
| 3'0" x 3'1" | BC@.350 | Ea | 16.50 | 11.20 | 27.70 | 47.10 |
| 3'0" x 3'2" | BC@.350 | Ea | 15.00 | 11.20 | 26.20 | 44.50 |
| 3'0" x 4'6" | BC@.350 | Ea | 17.00 | 11.20 | 28.20 | 47.90 |
| 3'0" x 5'2" | BC@.350 | Ea | 17.50 | 11.20 | 28.70 | 48.80 |
| 3'0" x 6'2" | BC@.350 | Ea | 19.50 | 11.20 | 30.70 | 52.20 |

**Aluminum screen rolls.** Replaces worn-out screening on windows, patio doors and screened-in porches. Bright or charcoal.

| | | Unit | Material | Labor | Total | Sell |
|---|---|---|---|---|---|---|
| 36" x 7' | — | Ea | 7.47 | — | 7.47 | — |
| 36" x 25' | — | Ea | 21.60 | — | 21.60 | — |
| 48" x 7' | — | Ea | 9.66 | — | 9.66 | — |
| 48" x 25' | — | Ea | 28.70 | — | 28.70 | — |

**Fiberglass screen rolls.** Charcoal or gray.

| | | Unit | Material | Labor | Total | Sell |
|---|---|---|---|---|---|---|
| 36" x 7' | — | Ea | 8.10 | — | 8.10 | — |
| 36" x 25' | — | Ea | 13.60 | — | 13.60 | — |
| 48" x 25' | — | Ea | 16.90 | — | 16.90 | — |
| 60" x 25' | — | Ea | 21.40 | — | 21.40 | — |
| 72" x 25' | — | Ea | 25.70 | — | 25.70 | — |
| 84" x 25' | — | Ea | 27.00 | — | 27.00 | — |
| 96" x 25' | — | Ea | 30.20 | — | 30.20 | — |

**Screen patch.** Self-stick.

| | | Unit | Material | Labor | Total | Sell |
|---|---|---|---|---|---|---|
| 3" x 3", aluminum | — | Ea | 2.27 | — | 2.27 | — |
| 3" x 3", charcoal | — | Ea | 2.27 | — | 2.27 | — |

**Screen replacement splines.** Black or gray.

| | | Unit | Material | Labor | Total | Sell |
|---|---|---|---|---|---|---|
| 0.125" x 25' | — | Ea | 3.74 | — | 3.74 | — |
| 0.125" x 100' | — | Ea | 7.21 | — | 7.21 | — |
| 0.140" x 25' | — | Ea | 3.77 | — | 3.77 | — |
| 0.140" x 100' | — | Ea | 5.37 | — | 5.37 | — |
| 0.160" x 25' | — | Ea | 3.66 | — | 3.66 | — |
| 0.160" x 100' | — | Ea | 7.21 | — | 7.21 | — |
| 0.175" x 25' | — | Ea | 3.73 | — | 3.73 | — |
| 0.175" x 100' | — | Ea | 7.13 | — | 7.13 | — |
| 0.190" x 25' | — | Ea | 3.77 | — | 3.77 | — |
| 0.190" x 100' | — | Ea | 8.10 | — | 8.10 | — |
| 0.210" x 25' | — | Ea | 3.77 | — | 3.77 | — |
| 0.210" x 100', flat | — | Ea | 8.10 | — | 8.10 | — |
| 0.220" x 25' | — | Ea | 3.77 | — | 3.77 | — |
| 0.220" x 100' | — | Ea | 9.17 | — | 9.17 | — |
| 0.250" x 25' | — | Ea | 3.77 | — | 3.77 | — |

**Roller spline tool.**

| | | Unit | Material | Labor | Total | Sell |
|---|---|---|---|---|---|---|
| Plastic handle | — | Ea | 2.81 | — | 2.81 | — |

| | Craft@Hrs | Unit | Material | Labor | Total | Sell |
|---|---|---|---|---|---|---|

**Tilt sash window grille sets.** Removable window grille system fits double-hung wood windows.

| | Craft@Hrs | Unit | Material | Labor | Total | Sell |
|---|---|---|---|---|---|---|
| 2'0" x 3'2", 4 x 4 grille | BC@.250 | Ea | 11.60 | 8.02 | 19.62 | 33.40 |
| 2'4" x 3'2", 6 x 6 grille | BC@.250 | Ea | 17.60 | 8.02 | 25.62 | 43.60 |
| 2'4" x 4'6", 6 x 6 grille | BC@.250 | Ea | 17.60 | 8.02 | 25.62 | 43.60 |
| 2'8" x 3'2", 6 x 6 grille | BC@.250 | Ea | 17.60 | 8.02 | 25.62 | 43.60 |
| 2'8" x 3'10", 6 x 6 grille | BC@.250 | Ea | 17.60 | 8.02 | 25.62 | 43.60 |
| 2'8" x 4'6", 6 x 6 grille | BC@.250 | Ea | 17.60 | 8.02 | 25.62 | 43.60 |
| 2'8" x 5'2", 6 x 6 grille | BC@.250 | Ea | 17.60 | 8.02 | 25.62 | 43.60 |
| 2'8" x 6'2", 9 x 9 grille | BC@.250 | Ea | 20.20 | 8.02 | 28.22 | 48.00 |
| 3'0" x 3'2", 6 x 6 grille | BC@.250 | Ea | 17.60 | 8.02 | 25.62 | 43.60 |
| 3'0" x 3'10", 6 x 6 grille | BC@.250 | Ea | 17.60 | 8.02 | 25.62 | 43.60 |
| 3'0" x 4'6", 6 x 6 grille | BC@.250 | Ea | 17.60 | 8.02 | 25.62 | 43.60 |
| 3'0" x 5'2", 6 x 6 grille | BC@.250 | Ea | 17.60 | 8.02 | 25.62 | 43.60 |

## Window Glazing

**Single-strength replacement glass.**

| | Craft@Hrs | Unit | Material | Labor | Total | Sell |
|---|---|---|---|---|---|---|
| 8" x 10", clear | BG@.246 | Ea | 1.81 | 7.33 | 9.14 | 15.50 |
| 9" x 12", clear | BG@.333 | Ea | 1.53 | 9.92 | 11.45 | 19.50 |
| 10" x 12", clear | BG@.370 | Ea | 1.64 | 11.00 | 12.64 | 21.50 |
| 12" x 14", clear | BG@.280 | Ea | 1.89 | 8.34 | 10.23 | 17.40 |
| 12" x 16", clear | BG@.322 | Ea | 3.07 | 9.59 | 12.66 | 21.50 |
| 12" x 18", clear | BG@.363 | Ea | 2.47 | 10.80 | 13.27 | 22.60 |
| 12" x 24", clear | BG@.296 | Ea | 5.27 | 8.81 | 14.08 | 23.90 |
| 12" x 30", clear | BG@.370 | Ea | 3.47 | 11.00 | 14.47 | 24.60 |
| 12" x 36", clear | BG@.444 | Ea | 5.51 | 13.20 | 18.71 | 31.80 |
| 12" x 36", obscure | BG@.444 | Ea | 6.77 | 13.20 | 19.97 | 33.90 |
| 14" x 20", clear | BG@.369 | Ea | 2.92 | 11.00 | 13.92 | 23.70 |
| 16" x 20", clear | BG@.329 | Ea | 4.47 | 9.80 | 14.27 | 24.30 |
| 16" x 28", clear | BG@.460 | Ea | 4.20 | 13.70 | 17.90 | 30.40 |
| 16" x 32", clear | BG@.525 | Ea | 4.85 | 15.60 | 20.45 | 34.80 |
| 16" x 36", clear | BG@.592 | Ea | 5.72 | 17.60 | 23.32 | 39.60 |
| 18" x 24", clear | BG@.444 | Ea | 3.59 | 13.20 | 16.79 | 28.50 |
| 18" x 36", clear | BG@.666 | Ea | 6.99 | 19.80 | 26.79 | 45.50 |
| 18" x 36", obscure | BG@.666 | Ea | 16.90 | 19.80 | 36.70 | 62.40 |
| 18" x 52", clear | BG@.962 | Ea | 10.30 | 28.60 | 38.90 | 66.10 |
| 20" x 24", clear | BG@.492 | Ea | 5.26 | 14.70 | 19.96 | 33.90 |
| 24" x 30", clear | BG@.555 | Ea | 9.34 | 16.50 | 25.84 | 43.90 |
| 24" x 36", clear | BG@.666 | Ea | 8.32 | 19.80 | 28.12 | 47.80 |
| 24" x 52", clear | BG@.961 | Ea | 13.50 | 28.60 | 42.10 | 71.60 |
| 28" x 36", clear | BG@.777 | Ea | 9.36 | 23.10 | 32.46 | 55.20 |
| 30" x 32", clear | BG@.739 | Ea | 8.96 | 22.00 | 30.96 | 52.60 |
| 30" x 36", clear | BG@.833 | Ea | 14.30 | 24.80 | 39.10 | 66.50 |
| 32" x 40", clear | BG@.985 | Ea | 12.00 | 29.30 | 41.30 | 70.20 |
| 36" x 48", clear | BG@1.30 | Ea | 19.00 | 38.70 | 57.70 | 98.10 |

**Jalousie glass.** Replacement horizontal panes or slats for a jalousie window.

| | Craft@Hrs | Unit | Material | Labor | Total | Sell |
|---|---|---|---|---|---|---|
| 4" x 36-1/4", clear | BG@.200 | Ea | 6.04 | 5.96 | 12.00 | 20.40 |
| 4-1/2" x 36-1/4", clear | BG@.200 | Ea | 6.04 | 5.96 | 12.00 | 20.40 |
| 4" x 24-7/32", obscure | BG@.200 | Ea | 4.98 | 5.96 | 10.94 | 18.60 |
| 4" x 36-7/32", obscure | BG@.200 | Ea | 6.11 | 5.96 | 12.07 | 20.50 |
| 4-1/2" x 36-7/32", obscure | BG@.200 | Ea | 7.37 | 5.96 | 13.33 | 22.70 |

| | Craft@Hrs | Unit | Material | Labor | Total | Sell |
|---|---|---|---|---|---|---|

**Remove and replace glazing putty.** Using the existing glass pane. Heat old putty with a heat gun. Scrape out old putty. Refill the channel with putty.

| | Craft@Hrs | Unit | Material | Labor | Total | Sell |
|---|---|---|---|---|---|---|
| Pane to 4 square feet | BG@.250 | Ea | — | 7.45 | 7.45 | 12.70 |
| Add per square foot over 4 SF | BG@.063 | SF | — | 1.88 | 1.88 | 3.20 |

**Measure and cut glass to size.** Single or double strength window glass.

| | Craft@Hrs | Unit | Material | Labor | Total | Sell |
|---|---|---|---|---|---|---|
| Per pane, to 4 square feet | BG@.167 | Ea | 6.12 | 4.97 | 11.09 | 18.90 |
| Add for panes over 4 SF, per SF | BG@.042 | SF | 1.52 | 1.25 | 2.77 | 4.71 |

**Remove and replace window glass set in putty.** Remove sash. Hack out glass fragments. Heat old putty with a heat gun. Scrape out old putty. Pull glazing points. Set glass in putty bead. Reset glazing points. Fill channel with putty

| | Craft@Hrs | Unit | Material | Labor | Total | Sell |
|---|---|---|---|---|---|---|
| Remove sash from frame and replace | BG@.400 | Ea | — | 11.90 | 11.90 | 20.20 |
| Hack out glass fragments and putty | BG@.250 | Ea | — | 7.45 | 7.45 | 12.70 |
| Set glass and putty, pane to 4 square feet | BG@.250 | Ea | 1.00 | 7.45 | 8.45 | 14.40 |
| Set glass and putty, per SF over 4 SF | BG@.063 | SF | .25 | 1.88 | 2.13 | 3.62 |

**Remove window glass set in vinyl stops.** Remove sash. Hack out glass fragments. No glass included. Per pane, width plus height. For example, the width plus height of a window pane measuring 24" x 36" is 60".

| | Craft@Hrs | Unit | Material | Labor | Total | Sell |
|---|---|---|---|---|---|---|
| To 40" width plus height | BG@.580 | Ea | — | 17.30 | 17.30 | 29.40 |
| 41" to 50" width plus height | BG@.640 | Ea | — | 19.10 | 19.10 | 32.50 |
| 51" to 60" width plus height | BG@.710 | Ea | — | 21.10 | 21.10 | 35.90 |
| 61" to 70" width plus height | BG@.900 | Ea | — | 26.80 | 26.80 | 45.60 |
| 81" to 90" width plus height | BG@1.00 | Ea | — | 29.80 | 29.80 | 50.70 |
| Over 91" width plus height | BG@1.10 | Ea | — | 32.80 | 32.80 | 55.80 |

**Set window glass in vinyl stops.** No glass included. Add the cost of new vinyl bead, if required. Per pane, width plus height. For example, the width plus height of a window pane measuring 24" x 36" is 60".

| | Craft@Hrs | Unit | Material | Labor | Total | Sell |
|---|---|---|---|---|---|---|
| To 40" width plus height | BG@.580 | Ea | — | 17.30 | 17.30 | 29.40 |
| 41" to 50" width plus height | BG@.640 | Ea | — | 19.10 | 19.10 | 32.50 |
| 51" to 60" width plus height | BG@.710 | Ea | — | 21.10 | 21.10 | 35.90 |
| 61" to 70" width plus height | BG@.900 | Ea | — | 26.80 | 26.80 | 45.60 |
| 81" to 90" width plus height | BG@1.00 | Ea | — | 29.80 | 29.80 | 50.70 |
| Over 91" width plus height | BG@1.10 | Ea | — | 32.80 | 32.80 | 55.80 |

**Vinyl glazing bead.**

| | Craft@Hrs | Unit | Material | Labor | Total | Sell |
|---|---|---|---|---|---|---|
| 7/32" x 6', brown | — | Ea | 1.93 | — | 1.93 | — |
| 7/32" x 6', gray | — | Ea | 1.89 | — | 1.89 | — |
| 7/32" x 6', white | — | Ea | 1.89 | — | 1.89 | — |

**Glazing compound, 33®, DAP®.** For face-glazing wood or metal window frames. Weather resistant. Smooth and ready mixed.

| | Craft@Hrs | Unit | Material | Labor | Total | Sell |
|---|---|---|---|---|---|---|
| 1/2-pint | — | Ea | 3.20 | — | 3.20 | — |
| Pint | — | Ea | 4.29 | — | 4.29 | — |
| Quart | — | Ea | 5.30 | — | 5.30 | — |

**Glazing compound, "1012", DAP®.** For face-glazing metal sash. Requires no painting. Gray.

| | Craft@Hrs | Unit | Material | Labor | Total | Sell |
|---|---|---|---|---|---|---|
| Quart | — | Ea | 6.98 | — | 6.98 | — |
| Gallon | — | Ea | 18.30 | — | 18.30 | — |

**Latex glazing compound, DAP®.** Resists cracking, sagging, and chalking.

| | Craft@Hrs | Unit | Material | Labor | Total | Sell |
|---|---|---|---|---|---|---|
| 10.3-ounce tube | — | Ea | 4.60 | — | 4.60 | — |

**Glazing points.** Secures glass to sash before glazing putty is applied. Push "T" style.

| | Craft@Hrs | Unit | Material | Labor | Total | Sell |
|---|---|---|---|---|---|---|
| Bundle | — | Ea | 2.13 | — | 2.13 | — |

| | Craft@Hrs | Unit | Material | Labor | Total | Sell |
|---|---|---|---|---|---|---|
| **Wheel glass cutter.** | | | | | | |
| Steel wheel | — | Ea | 3.93 | — | 3.93 | — |

**Glazing tools.** 2-in-1 glazing tool has a slotted V-blade on one end to apply a smooth strip of putty. The other end has a heavy duty 1-1/4" chisel edge to remove old putty and drive push style glazing points.

| | Craft@Hrs | Unit | Material | Labor | Total | Sell |
|---|---|---|---|---|---|---|
| 2-in-1 glazing tool | — | Ea | 5.57 | — | 5.57 | — |
| Pushmate for glazing points | — | Ea | 9.15 | — | 9.15 | — |
| Acrylic sheet cutter | — | Ea | 3.12 | — | 3.12 | — |
| 6" nipping/running pliers | — | Ea | 10.88 | — | 10.88 | — |

## Glass Substitutes

**Acrylic sheet.**

| | Craft@Hrs | Unit | Material | Labor | Total | Sell |
|---|---|---|---|---|---|---|
| .093", 11" x 14" | — | Ea | 2.38 | — | 2.38 | — |
| .093", 18" x 24" | — | Ea | 6.46 | — | 6.46 | — |
| .093", 20" x 32" | — | Ea | 9.21 | — | 9.21 | — |
| .093", 24" x 48" | — | Ea | 16.30 | — | 16.30 | — |
| .093", 30" x 36" | — | Ea | 14.40 | — | 14.40 | — |
| .093", 30" x 60" | — | Ea | 25.90 | — | 25.90 | — |
| .093", 36" x 48" | — | Ea | 21.60 | — | 21.60 | — |
| .118", 36" x 72" | — | Ea | 42.30 | — | 42.30 | — |
| .230", 48" x 96" | — | Ea | 107.00 | — | 107.00 | — |
| .236", 18" x 24" | — | Ea | 14.10 | — | 14.10 | — |
| .236", 24" x 48" | — | Ea | 39.05 | — | 39.05 | — |
| .236", 30" x 36" | — | Ea | 34.70 | — | 34.70 | — |

**Fabback acrylic mirror sheet.** Ideal for ceilings, doors, children's rooms.

| | Craft@Hrs | Unit | Material | Labor | Total | Sell |
|---|---|---|---|---|---|---|
| .118", 24" x 48" | — | Ea | 32.00 | — | 32.00 | — |

**Polycarbonate sheet Plexiglas®, Lexan® XL-10.** Impact-strength and long service life. UV-resistant surface. Suitable for vertical-glazing and overhead applications.

| | Craft@Hrs | Unit | Material | Labor | Total | Sell |
|---|---|---|---|---|---|---|
| .093" thick, 18" x 24" | — | Ea | 14.00 | — | 14.00 | — |
| .093" thick, 30" x 36" | — | Ea | 30.24 | — | 30.24 | — |
| .093" thick, 32" x 44" | — | Ea | 32.40 | — | 32.40 | — |
| .093" thick, 36" x 48" | — | Ea | 66.90 | — | 66.90 | — |
| .093" thick, 36" x 72" | — | Ea | 82.00 | — | 82.00 | — |

**White silicon plastic adhesive.**

| | Craft@Hrs | Unit | Material | Labor | Total | Sell |
|---|---|---|---|---|---|---|
| White adhesive | — | Ea | 7.62 | — | 7.62 | — |

## Window Components

**Sliding window track lock with steel locking plate.** For horizontal or vertical metal sliding windows.

| | Craft@Hrs | Unit | Material | Labor | Total | Sell |
|---|---|---|---|---|---|---|
| Aluminum key lock | BC@.250 | Ea | 3.76 | 8.02 | 11.78 | 20.00 |
| Aluminum lever lock | BC@.250 | Ea | 1.62 | 8.02 | 9.64 | 16.40 |
| Aluminum sash lock | BC@.250 | Ea | 2.30 | 8.02 | 10.32 | 17.50 |
| White lever lock | BC@.250 | Ea | 2.12 | 8.02 | 10.14 | 17.20 |

**Sliding window thumbscrew bar lock.** Anti-lift-out bar. Securely locks window in closed or ventilating position.

| | Craft@Hrs | Unit | Material | Labor | Total | Sell |
|---|---|---|---|---|---|---|
| Reversible, aluminum | BC@.250 | Ea | 3.74 | 8.02 | 11.76 | 20.00 |
| Reversible, white | BC@.250 | Ea | 4.61 | 8.02 | 12.63 | 21.50 |

**Wood window flip lock.** Mounts on vertical track. Stops double-hung window in vent position. Wedge flips out of the way to allow opening. Pack of two.

| | Craft@Hrs | Unit | Material | Labor | Total | Sell |
|---|---|---|---|---|---|---|
| Brass | BC@.250 | Ea | 2.34 | 8.02 | 10.36 | 17.60 |

| | Craft@Hrs | Unit | Material | Labor | Total | Sell |
|---|---|---|---|---|---|---|

**Window weatherstrip spring brass seal sets.** Peel 'N Stick®, per linear foot per side.

| | Craft@Hrs | Unit | Material | Labor | Total | Sell |
|---|---|---|---|---|---|---|
| Double-hung window, head and sill | BC@.008 | LF | 1.48 | .26 | 1.74 | 2.96 |
| Add to remove and replace window sash | BC@.250 | Ea | — | 8.02 | 8.02 | 13.60 |
| Casement window, head and sill | BC@.008 | LF | .69 | .26 | .95 | 1.62 |
| Gray or black pile gasket | BC@.008 | LF | .55 | .26 | .81 | 1.38 |
| Gray or black pile gasket with adhesive | BC@.006 | LF | .76 | .19 | .95 | 1.62 |

## Specialty Windows and Skylights

**Window well, ScapeWEL.** Meets code requirements for basement egress window. Snaps into place on-site. For new construction and remodeling, attaches directly to foundation. Two tier. 41" projection from foundation. 48" high.

| | Craft@Hrs | Unit | Material | Labor | Total | Sell |
|---|---|---|---|---|---|---|
| 42" wide window, 3' window buck | B1@1.50 | Ea | 599.00 | 44.10 | 643.10 | 1,090.00 |
|   Polycarbonate well cover | — | Ea | 285.00 | — | 285.00 | — |
| 54" wide window, 4' window buck | B1@1.50 | Ea | 635.00 | 44.10 | 679.10 | 1,150.00 |
|   Polycarbonate well cover | — | Ea | 310.00 | — | 310.00 | — |
| 66" wide window, 5' window buck | B1@1.50 | Ea | 655.00 | 44.10 | 699.10 | 1,190.00 |
|   Polycarbonate well cover | — | Ea | 325.00 | — | 325.00 | — |

**Tubular skylights with install kit.** Acrylic dome on the roof reflects sunlight down a light shaft providing natural light for bathrooms and hallways. Use a 10" diameter tube for rooms to 150 square feet and a 14" diameter tube for rooms to 300 square feet. Use the composition flashing kit for asphalt shingle, slate, or wood shake roofs. Aluminum flashing kit is suitable for other surfaces with pitch from level to 12 in 12. Adjustable tubes make it easy to maneuver around obstacles. Kit contains material to complete a 48" installation from roof surface to interior ceiling surface. Fits between 16" or 24" on center rafters.

| | Craft@Hrs | Unit | Material | Labor | Total | Sell |
|---|---|---|---|---|---|---|
| 10" tube, aluminum flash kit | B1@3.00 | Ea | 196.00 | 88.10 | 284.10 | 483.00 |
| 10" tube, composition flash kit | B1@3.00 | Ea | 161.00 | 88.10 | 249.10 | 423.00 |
| 10" tube, tile or flat roof | B1@3.00 | Ea | 193.00 | 88.10 | 281.10 | 478.00 |
| 14" tube, aluminum flash kit | B1@3.00 | Ea | 295.00 | 88.10 | 383.10 | 651.00 |
| 14" tube, composition flash kit | B1@3.00 | Ea | 256.00 | 88.10 | 344.10 | 585.00 |
| 14" tube, tile or flat roof | B1@3.00 | Ea | 289.00 | 88.10 | 377.10 | 641.00 |

**Remove an existing poly dome skylight.** No salvage of materials. Includes disposal.

| | Craft@Hrs | Unit | Material | Labor | Total | Sell |
|---|---|---|---|---|---|---|
| To 10 square feet | BL@2.65 | Ea | — | 70.60 | 70.60 | 120.00 |
| Add per square foot over 10 square feet | BL@.210 | SF | — | 5.59 | 5.59 | 9.50 |

**Cut roof opening for a skylight.** Cut shingles and sheathing, install L-flashing. Self-flashing flush-mount skylights can be installed directly on a roof with a pitch of 3 in 12 or greater. Curb mounted skylights require a roof curb. Add the cost of roof framing (doubled rafters and two headers) if required. Add the cost of framing a light well through the attic, if required. Add the cost of cutting and finishing a hole in the ceiling (including doubled joists and two headers) if required.

| | Craft@Hrs | Unit | Material | Labor | Total | Sell |
|---|---|---|---|---|---|---|
| Measure and cut roof opening for skylight | BC@3.50 | Ea | 15.00 | 112.00 | 127.00 | 216.00 |
| Frame roof curb | BC@2.00 | LF | 15.00 | 64.20 | 79.20 | 135.00 |
| Curb flashing for skylight | BC@1.25 | Ea | 62.60 | 40.10 | 102.70 | 175.00 |

**Fixed flush-mount skylights, Velux.** Low-E insulating laminated and tempered glass. Wood interior frame. Exterior aluminum cladding covers gaskets and sealants. Pre-mounted brackets allows adjustment for high or low profile roofing materials.

| | Craft@Hrs | Unit | Material | Labor | Total | Sell |
|---|---|---|---|---|---|---|
| 15-5/16" x 46-3/8" | B1@3.42 | Ea | 178.00 | 100.00 | 278.00 | 473.00 |
| 21-1/2" x 27-1/2" | B1@3.42 | Ea | 150.00 | 100.00 | 250.00 | 425.00 |
| 21-1/2" x 38-1/2" | B1@3.42 | Ea | 182.00 | 100.00 | 282.00 | 479.00 |
| 21-1/2" x 46-3/8" | B1@3.42 | Ea | 203.00 | 100.00 | 303.00 | 515.00 |
| 30-5/8" x 38-1/2", type 74 | B1@3.42 | Ea | 219.00 | 100.00 | 319.00 | 542.00 |
| 30-5/8" x 38-1/2", type 75 | B1@3.42 | Ea | 177.00 | 100.00 | 277.00 | 471.00 |
| 30-5/8" x 46-7/8" | B1@3.42 | Ea | 249.00 | 100.00 | 349.00 | 593.00 |
| 30-5/8" x 55" | B1@3.42 | Ea | 279.00 | 100.00 | 379.00 | 644.00 |

| | Craft@Hrs | Unit | Material | Labor | Total | Sell |
|---|---|---|---|---|---|---|

**Fixed curb-mount skylights, Velux.** Low-E insulating laminated and tempered glass. Adapts to curb size and roofing material. Anodized aluminum exterior frame with gasket. UV-resistant ABS inner frame.

| | Craft@Hrs | Unit | Material | Labor | Total | Sell |
|---|---|---|---|---|---|---|
| 22-1/2" x 22-1/2" | B1@3.00 | Ea | 135.00 | 88.10 | 223.10 | 379.00 |
| 22-1/2" x 46-1/2" | B1@3.42 | Ea | 176.00 | 100.00 | 276.00 | 469.00 |
| 22-1/2" x 46-1/2", impact glass | B1@3.42 | Ea | 278.00 | 100.00 | 378.00 | 643.00 |
| 46-1/2" x 46-1/2" | B1@3.42 | Ea | 296.00 | 100.00 | 396.00 | 673.00 |

**Venting flush-mount glass skylights, Velux.** Low-E tempered insulating glass. Wood interior frame. Aluminum cladding. Thru-screen allows skylight operation without removing screen. Stainless steel scissor operator allows easy opening for maximum ventilation. Removable, integrated insect screen.

| | Craft@Hrs | Unit | Material | Labor | Total | Sell |
|---|---|---|---|---|---|---|
| 21-1/2" x 27-1/2" | B1@3.00 | Ea | 292.00 | 88.10 | 380.10 | 646.00 |
| 21-1/2" x 38-1/2" | B1@3.42 | Ea | 316.00 | 100.00 | 416.00 | 707.00 |
| 21-1/2" x 46-3/8" | B1@3.42 | Ea | 343.00 | 100.00 | 443.00 | 753.00 |
| 30-1/2" x 39" | B1@3.52 | Ea | 346.00 | 103.00 | 449.00 | 763.00 |
| 30-1/2" x 55-1/2" | B1@3.52 | Ea | 417.00 | 103.00 | 520.00 | 884.00 |
| 44-3/4" x 46-3/8 | B1@3.52 | Ea | 465.00 | 103.00 | 568.00 | 966.00 |

**Venting flush-mount laminated glass skylights, Velux.** Low-E tempered insulating glass. Wood interior frame. Aluminum cladding. Thru-screen allows skylight operation without removing screen. Stainless steel scissor operator allows easy opening for maximum ventilation. Removable, integrated insect screen.

| | Craft@Hrs | Unit | Material | Labor | Total | Sell |
|---|---|---|---|---|---|---|
| 21-1/2" x 27-1/2" | B1@3.00 | Ea | 289.00 | 88.10 | 377.10 | 641.00 |
| 21-1/2" x 38-1/2" | B1@3.42 | Ea | 312.00 | 100.00 | 412.00 | 700.00 |
| 21-1/2" x 46-3/8" | B1@3.42 | Ea | 371.00 | 100.00 | 471.00 | 801.00 |
| 30-5/8" x 38-1/2" | B1@3.42 | Ea | 392.00 | 100.00 | 492.00 | 836.00 |
| 30-5/8" x 46-3/8", type 75 | B1@3.52 | Ea | 211.00 | 103.00 | 314.00 | 534.00 |
| 30-5/8" x 46-3/8", type 74 | B1@3.52 | Ea | 373.00 | 103.00 | 476.00 | 809.00 |
| 30-5/8" x 55" | B1@3.52 | Ea | 471.00 | 103.00 | 574.00 | 976.00 |
| 44-3/8" x 46-3/8" | B1@3.52 | Ea | 523.00 | 103.00 | 626.00 | 1,060.00 |
| Telescoping control rod, 6' to10' | — | Ea | 32.70 | — | 32.70 | — |

**Fixed skylights.** Folded corners. Watertight seal. 360 degree thermal break frame.

| | Craft@Hrs | Unit | Material | Labor | Total | Sell |
|---|---|---|---|---|---|---|
| 2' x 2', acrylic | B1@3.00 | Ea | 113.00 | 88.10 | 201.10 | 342.00 |
| 2' x 2', glass | B1@3.00 | Ea | 124.00 | 88.10 | 212.10 | 361.00 |
| 2' x 4', acrylic | B1@3.42 | Ea | 152.00 | 100.00 | 252.00 | 428.00 |
| 2' x 4', glass | B1@3.42 | Ea | 155.00 | 100.00 | 255.00 | 434.00 |
| 4' x 4', glass | B1@3.42 | Ea | 294.00 | 100.00 | 394.00 | 670.00 |

# *Doors* 9

The front entry door is the most important door in any home. It's usually visible from the street and always makes a first impression on visitors. Whether new or old, the front door should both look good and complement the style of the house.

If only cosmetic repairs are needed to an original door, that's the place to start. If the door is original, it's probably in perfect harmony with the exterior décor. Repairing the door saves the time and expense of finding exactly the right replacement door and hauling away the old door. Finding historically correct replacement doors for older homes can be difficult, and they're likely to be expensive. But if the door is already a replacement, then replacing it again probably won't do any harm.

How can you tell if the entry door is original? Replacement doors can be beautiful, but seldom harmonize with the home décor. For example, a light colored oak flush door (popular in the 1960s) installed on a home built in the 1920s is an obvious replacement. Hanging a 21st century door doesn't modernize a 19th century home.

## Repair or Replace

Balance the cost of repair against the cost of buying and hanging a new door of similar quality. Obviously, repair makes sense when the cost is substantially less than the cost of replacement. Figure 9-1 shows extra-wide French doors set with thick, beveled glass. These doors were very dirty and the bottoms scarred and damaged. Brushing them with detergent removed the grime, and the damage on the bottoms was taken care of by simply covering them with brass kickplates. Two hours of work and a $50 investment salvaged doors that might cost thousands to duplicate.

The most common problem with entry doors is a deteriorated finish. Peeling paint or alligatored varnish will turn an otherwise attractive door into an eyesore. A bad finish alone isn't an adequate reason to replace an entry door. If the door has a varnish finish that has darkened, wash it before you make a decision about stripping the finish. If the door has been waxed, the finish may be nothing more than dirty. Wax collects dirt and darkens as it ages. That makes the door ugly. But dirty wax also helps protect the finish below. You may find a perfect gem under that wax.

**Figure 9-1**
*Repaired French doors*

Scrub a small area with a powerful detergent such as Formula 409® or Fantastic®. If you see the color changing dramatically, dirt and wax buildup could be the source of the problem. In that case, it's worth cleaning the entire door. Refinishing may not be needed.

# Refinishing an Entry Door

If a door has to be refinished, a heat gun is the best tool for stripping the finish. Heat softens the old paint or varnish, making it easy to scrape off. Stripping like this is easier if the door is laid flat across two sawhorses. Once the paint is off, you can apply either paint or varnish. If it's a quality wood door, varnishing will bring out the beauty of the wood.

If you elect to varnish, begin by washing the door with acetone to remove residue that escaped the heat gun. Residue won't show through paint, but it will show through varnish. When using acetone, wear rubber or neoprene gloves to avoid skin irritation, and work in a well-ventilated area to reduce the risk from fumes.

Pour some acetone into a jar with a screw-on lid. Dip a piece of extra fine steel wool in the acetone (synthetic steel wool may be an even better choice). Then gently scrub the door surface. Work only on a few square inches at a time. Don't scrub hard; let the acetone do the work. You'll see residue dissolve and the original wood surface appear. After a few minutes of scrubbing, the steel wool will get clogged with old paint or varnish. Rinse the steel wool in acetone. Then continue scrubbing. Be sure to keep the jar of acetone capped between dips. Acetone evaporates very quickly.

Before applying varnish, examine the wood tone. Do you like the color? If it's too light, you might want to stain it. If it's too dark, consider bleaching. Light sanding may also lighten the tone a little. Expect to spend about six hours on a door that requires paint stripping, acetone wash and a coat of varnish. But when you're done, the door will probably be as good as new.

If the door is damaged, painting it may be a better choice. Paint will hide repairs that varnish and stain won't. Varnish will work over minor repairs, such as nail holes filled with wood putty, but larger repairs, like a filled-in hole from an old deadbolt, will show through varnish. Wood putty doesn't absorb stain the same way wood does.

If staining the door is essential, buy colored wood putty that matches the stain selected. The patch will still be visible, but not as apparent.

## *Antiquing*

An antique finish is a painted-on wood grain. When done correctly, the results can be surprisingly believable. You have to look very closely to notice that the finish isn't actually wood grain. Antiquing kits come with complete instructions and all the materials needed, including a special tool for making the fake wood grain.

Antiquing is time-consuming. Any antiqued surface requires three coats: a base coat, the antique wood-grain coat, and a finish coat of varnish. Each coat has to dry overnight before the next coat is applied. Figure about three hours to antique an entry door, not including any other repairs.

Antique finishes are fragile and not recommended for high traffic surfaces. But if the door is too damaged for varnish and painting isn't an option, antiquing will replicate the look and feel of wood grain.

## Repairing Holes

Holes are common in older doors. They range from small nail holes, such as used to hang holiday decorations, to abandoned bore holes as big as your fist. Holes of up to about 1/2" can be filled with wood putty. From a foot or two away, the repair won't be noticeable. Repairing larger holes requires a different strategy.

Don't consider a door ruined just because an amateur butchered it trying to install a security lock — or a thief smashed it trying to defeat the same lock. Large holes are fairly easy to fix if you know how. Even a ragged hole the size of a softball can be fixed.

Rather than replace the door, buy a gallon of auto body filler, such as Bondo®. It's strong, weatherproof, workable, paintable, and cheap. A gallon goes a long way. Auto body filler is a gooey, clay-like material. The hardener comes in a separate tube. Mix body filler and hardener on a flat surface such as a piece of scrap wood or heavy cardboard. Do the mixing with a putty knife. Be careful not to get Bondo on your skin, as it burns. And keep anything valuable that's nearby well protected, as it's almost impossible to remove once it's hardened. The more hardener you add to the mix, the faster it hardens. Add too much hardener and it sets in seconds.

Body filler is semi-liquid and will ooze and run when applied to a vertical surface. So lay the door down flat across two sawhorses. If the hole you're repairing goes completely through the door, form the underside of the patch with a piece of cardboard. The form material will bond to the door, so if you use wood for a form, you'll end up having to chisel it off. Cardboard will easily peel off the door. Then you can just sand away the residue.

Fill the hole completely with the first layer of body filler. If you have hills and valleys, let the filler harden, then fill in the valleys. In auto body work, three layers are considered the minimum. When repairing doors, two coats will usually do the job.

Use a Surform plane to smooth out hardened body filler. A Surform plane has hundreds of little cutting teeth. Each tooth trims away a little material. You can work at any angle and not have to worry about cutting too deep. The Surform cuts away hills and blobs in seconds, leaving a flat, although rough, surface. When you're done planing, the surface will look like someone clawed the material with fingernails. Use a power sander with coarse sandpaper to smooth out the marks left by the Surform. The result will be a smooth, strong, solid surface.

**Figure 9-2**

*Door has been patched with Bondo®*

You'll have to paint this type of repair. It won't take a stain. However, once the door is painted, the repair will be completely invisible. Don't try to bore or drill through the body filler to install a lockset. It's likely to break loose. Bore for a new lockset either above or below the repair, never in exactly the same place.

If you have to install the lock at the same height, turn the door around (outside to inside) so the hinge edge becomes the lock edge and vice versa. Since the hinge edge has never been bored for a lock, you'll have clean wood to work with. The problem, of course, is that the hinge edge has been mortised for hinges. Fill the old hinge mortise cuts with body filler and make new mortise cuts for hinges on the other edge. This adds about a half hour to the task.

Body filler comes in handy when replacing a mortise lock with a modern cylinder lockset. You can buy a lock conversion kit for this. The kit includes an escutcheon plate that fastens over and covers the old mortise hole. The new cylinder lockset mounts in the plate. Unfortunately, these plates aren't very strong and don't offer much protection against forced entry. Although it's more work to plug the mortise void with body filler and bore for a new circular lockset, the finished product is stronger and far more attractive.

Figure 9-2 shows a door repaired with Bondo® after it was kicked in. The old mortise lock burst out, leaving a huge, ragged hole. Figure 9-3 shows the same door, flipped inside to out, painted, and hung with new hinges and lockset. The complete job, including repair, painting, and hanging, took approximately four hours and cost about $20, not including the lock.

## Rebuilding an Entry Door

Most wood doors are assembled from panels. Figure 9-4 shows an exploded view of door panels. The top and bottom rails are each one piece, and each side is one panel. Each divider between panels is a single piece. The more panels a door has, the more pieces there are. All these pieces are held together with glue. Weather and moisture cause the glue to deteriorate, usually along the lower half of the door.

If an entry door is coming apart but is otherwise worthy of repair, you'll have to disassemble the door and rebuild it. Estimate the time required at about six hours for most doors. The door will be off the hinges for a day or two, so you'll need to board up the opening.

Rebuilding a door with bad joints is a lot like repairing an old piece of furniture that has come unglued. First, take it completely apart. Pry the joints apart carefully, working from the edges to minimize tool marks on the more visible door parts. Most joints will come apart easily. All joints have small, interlocking pieces that are glued together. Try not to break

**Figure 9-3**

*Repaired door, painted and installed*

these joints. If a joint can't be pried apart easily, look for small nails driven into the joint at right angles. Remove these nails very carefully to avoid damaging the joint.

Outside rails are the hardest to get apart. That's where the strength of the door is — or rather, was. Once you get the outside rails apart, the center panels and rails will usually fall right out.

Usually, most of the glue in failing joints has turned to powder. In order for the new glue to form a tight bond in the joint, you'll need to thoroughly clean out all the old, powdery glue. Wipe the visible powder away; then remove any remaining glue residue with a sanding disc. You may need to do a little hand sanding where the sanding disc can't reach.

### Reassembling the Door

When you've removed the old glue, reglue and reassemble the door. Lay the door flat and make sure that all the corners are straight and true. Once the glue has set, you won't be able to straighten anything. Use a good carpenter's glue and bar clamps to hold the joints firmly in place. Be careful to wipe up any glue drips — once dry, these are hard to remove. Leave the door undisturbed overnight.

If any joints were damaged in the disassembly process, add strength by driving screws through the joint. Use long, thin 8" to 10" wood screws. Working from the edge of the door, pre-drill a hole through the weak joint. The screw should be long enough to go through one rail and into the next. Dip the screw in glue and drive it snug. Since you're driving from the edge of the door, the screw heads will only show when the door is open. Adding screws will substantially strengthen the door.

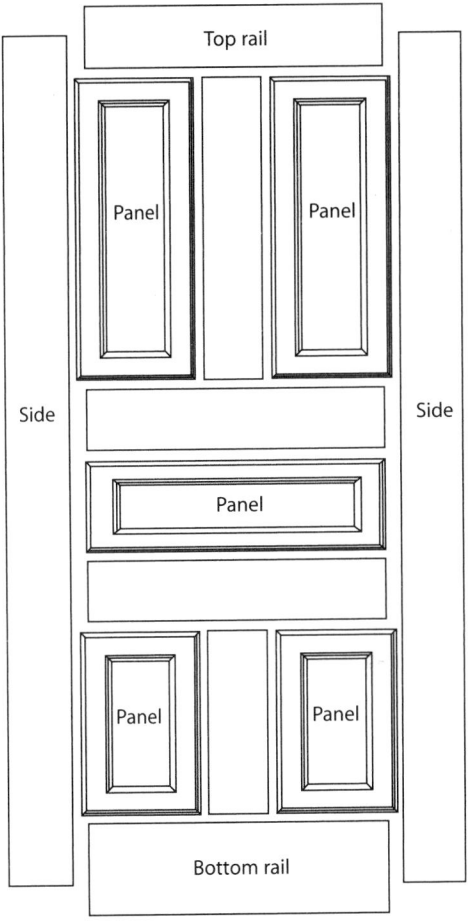

**Figure 9-4**

*Diagram of a panel door*

# Framing and Replacing Exterior Doors

Entry doors usually measure 6'8" high by 3' wide by $1^3/4$" thick. Side and rear doors should be at least 2'6" wide. Rough opening height should be the height of the door plus $2^1/4$" above the finish floor. The width of the rough opening should be the width of the door plus $2^1/2$". For a 3' door opening, make the header out of two lengths of 2" x 6" set with two 16d nails driven into the studs on each side. If the stud space on each side of the door isn't accessible, toenail the header to studs. Cripple studs (door bucks) support the header on each side of the opening.

If you're framing a new door opening, install a prehung door assembly complete with jamb, hinges and threshold. Prehung doors cost a little more than slab doors, but reduce the labor cost.

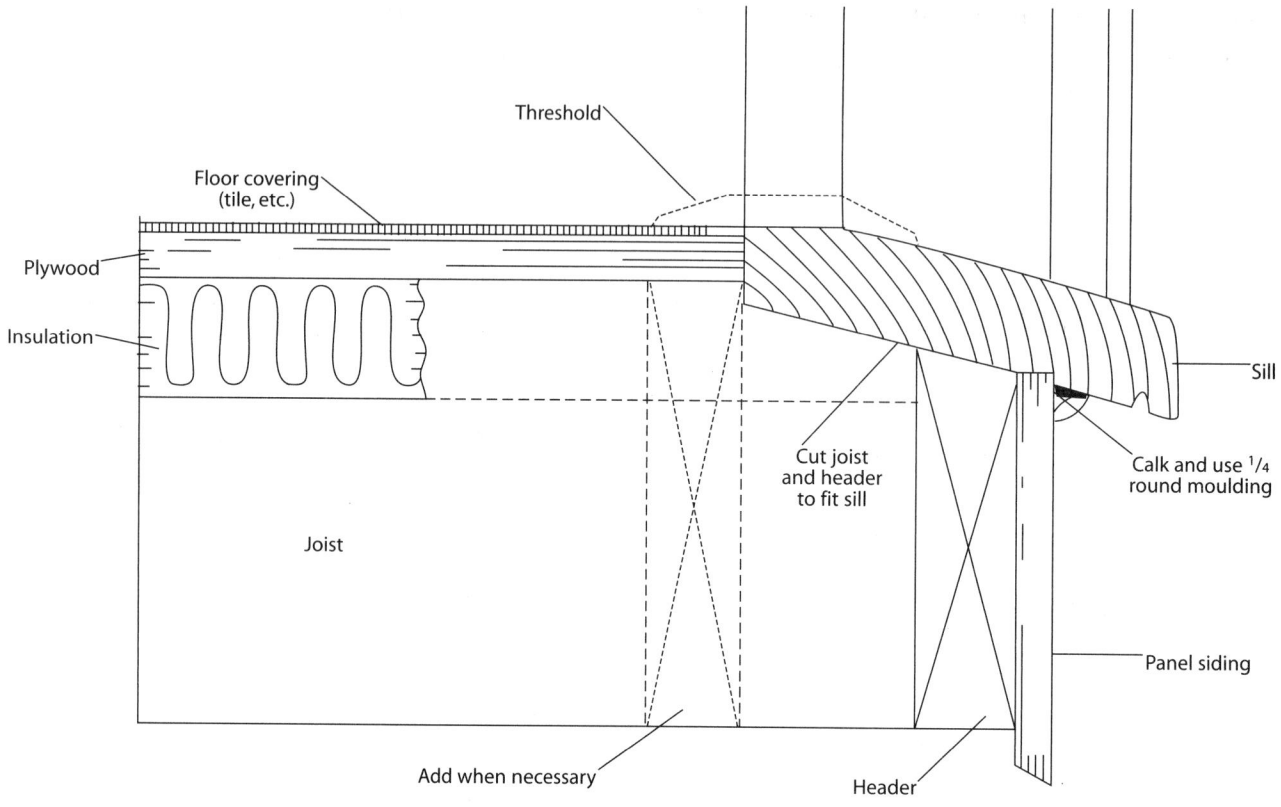

**Figure 9-5**

*Door installation at sill*

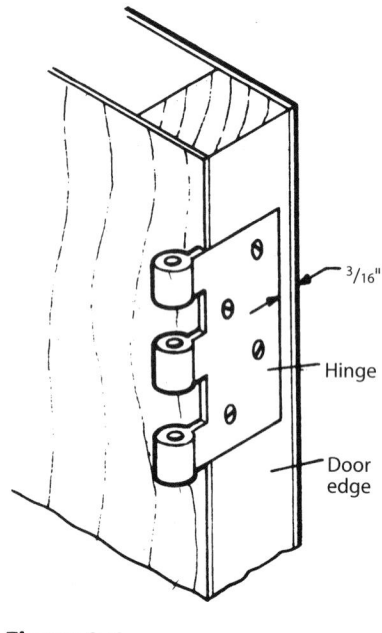

**Figure 9-6**

*Installation of door hinges*

In a new installation, you may have to trim floor joists and the joist header to receive the sill. The top of the sill should be the same height as the finish floor so the threshold can be installed over the joint. See Figure 9-5. Shim the sill when necessary so it's fully supported by floor framing. When joists run parallel to the sill, install headers and a short support member at the edge of the sill. Install the threshold over the joint by nailing with finishing nails.

When the opening is ready for installation, apply a ribbon of caulking sealer down each side and above the door opening. This caulk should seal the gap between the wall and the door casing. Set the door frame in the opening and secure it by nailing through the side and head casing. Nail the hinge side first.

Exterior doors should have a solid core. Hollow-core doors offer little security and warp too much during cold weather to use on the exterior. If a door has to be trimmed to fit the frame, cut on the hinge edge. Mortise hinges into the edge of the door with about $3/16"$ or $1/4"$ back spacing. See Figure 9-6. Use $3^1/2"$ by $3^1/2"$ loose-pin hinges on exterior doors that swing in. Use non-removable pins on out-swinging doors. Use three hinges (rather than two) on solid core doors to minimize warping.

Figure 9-7 shows standard door clearances. Measure the opening width and plane the edge for proper side clearances. Slightly bevel each door edge that fits against the stop. If necessary, square and trim the top of the door for proper fit. In cold climates, exterior doors should have weather seals.

## Interior Doors

Most interior doors get heavy use. But they don't have many moving parts and tend to be durable. Sticking or binding with the jamb and failure to latch are the most common problems. Sticking or binding will rarely be the result of heavy use or abuse. The most common cause is movement in the building frame. Settling at the foundation, shrinkage of framing lumber or heavy floor loads can warp the frame out of square and cause a door to bind. In most cases, minor adjustments will solve the problem.

To remedy a binding door, first find the point of contact. You'll usually see scuff marks where the door and jamb are rubbing together. Here are some simple repairs.

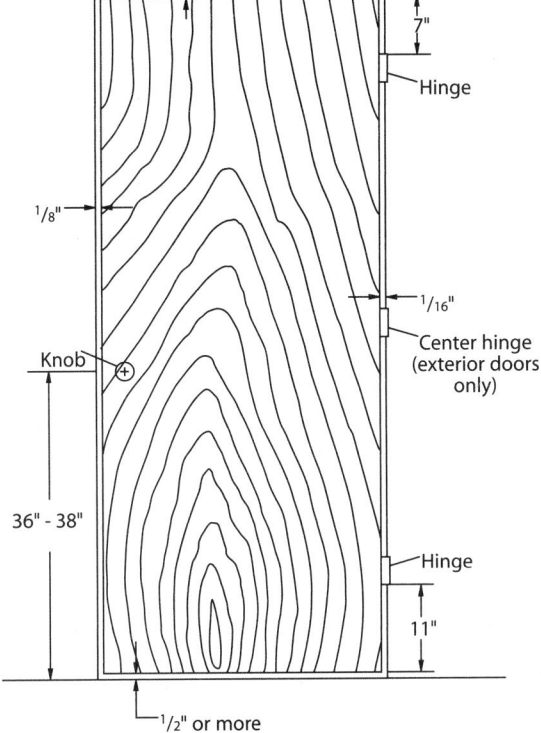

**Figure 9-7**

*Door clearances*

- ❖ *Binding on the top or lock edge.* Use a hand plane to remove ¹⁄₈" of door where you see scuff marks. Some paint touch-up will be needed where the door edge was sanded or planed off.

- ❖ *Binding on the hinge side.* The hinges may be routed too deeply, causing the door to "bounce" open if the latch doesn't catch. Loosen the hinge leaf enough to add a shim under the leaf. Then retighten the leaf screws.

- ❖ *Scraping at the bottom edge.* The frame may be out of square. Or, a thicker floor cover has been installed. Plane or saw off enough door edge to provide floor clearance.

- ❖ *The latch doesn't close — interior door.* Remove the strike plate and adjust the position slightly or shim it out slightly. To replace the latch, you may need longer screws. Before replacing the old screws, try inserting a filler (such as a wood matchstick) into the old screw hole. Then drive the screw in the same hole. If the latch is secure, you're done.

Doors will also bind when they absorb excess moisture during humid weather. This is a symptom of the door not being completely sealed at the edges. Sealing the edges with paint or varnish should solve the problem.

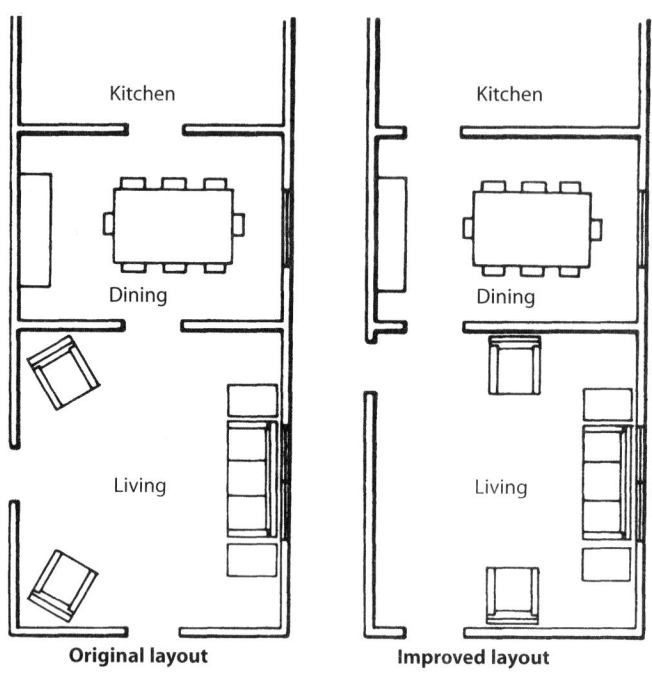

**Figure 9-8**

*Relocation of doors to direct traffic to one side of the rooms*

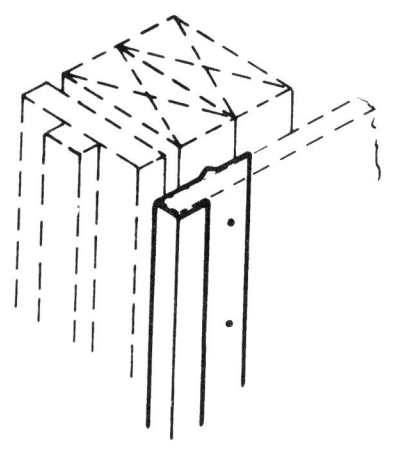

**Figure 9-9**

*Metal casing used with drywall*

# Making Changes

Adding or moving a door can make a home more convenient. Traffic through a room should be at the edges, not through the middle. Many older homes have doors centered in a wall, which cuts the wall space in two and makes furniture arrangement difficult. Moving a door opening from the middle of a wall space to the corner of the room will usually add livability to a home. See Figure 9-8. In some cases, simply closing up a doorway is the best choice. If you relocate a door to a corner of the room, be sure the new door swings against the nearest wall (rather than into the open room) and away from the light switch (so the switch is just inside the doorway). Also keep in mind that entire walls (affecting each adjoining room) may need to be re-drywalled when relocating a door. If it's a loadbearing wall, you'll need to allow room for installation of the header. Be sure you and your customer are aware of the additional costs and complexities involved before deciding to relocate a door.

Rough opening dimensions for interior doors are the same as rough opening dimensions for exterior doors. The opening width is the door width plus $2^1/2$". The opening height is the door height plus $2^1/4$" above the finished floor. The head jamb and two side jambs are the same width as the overall wall thickness where wood casing is used. Where metal casing is used with drywall (Figure 9-9), the jamb width is the same as the stud depth.

Installation goes faster if you hang the door in the jamb before setting the frame in the wall. Let the door act as a jig around which you assemble the jambs. To save time, buy jambs in precut sets. The jamb should be about $1/8$" wider than the thickness of the finished wall. That's probably $4^9/16$" for a 2" x 4" stud wall, or $6^9/16$" for a 2" x 6" stud wall. The jamb height is $3/8$" less than the rough opening height. Fit the door to the frame, using the clearances shown earlier in Figure 9-7. Bevel the edges slightly where the door meets the stops. Cutting and assembling the jamb will take about a half hour.

Before installing the door, temporarily set the narrow wood strips used as stops. Stops are usually $7/16$" thick and from $1^1/2$" to $2^1/4$" wide. See Figure 9-10. Cut a mitered joint where the stops meet at the head and side jambs. A 45-degree bevel cut 1" to $1^1/2$" above the finish floor (as in Figure 9-10) will eliminate a dirt pocket and make cleaning easier. This is called a sanitary stop.

Rout or mortise the hinges into the edge of the door with a $3/16$" to $1/4$" back spacing. Adjust the hinge, if necessary, so screws penetrate into solid wood. For interior doors, use two 3" by 3" loose-pin hinges. If a router isn't available, mark the hinge outline and depth of cut. Then remove just enough wood with a chisel so that the surface of the hinge is flush with the remainder of the door edge.

Place the door in the opening. Block the door in place with proper clearances all around. Then mark the location of the door hinges on the jamb. Remove the door from the opening and rout or chisel the jamb to the thickness of the hinge half. Install the hinge halves on the jamb.

Setting hinges on the door and jamb will require about an hour if you're using hand tools. If a router and door jig are available, reduce that time by about half.

Place the door in the opening and insert the hinge pins. Plumb and fasten the hinge side of the frame first. Drive shingle wedge sets between the side jamb and the rough door buck to plumb the jamb. Place a wedge set opposite the latch, each hinge and at intermediate locations. Nail the jamb with pairs of 8d nails at each wedge. Then fasten the opposite jamb the same way. When the door jamb is secure, cut off the shingle wedges flush with the wall. Allow about a half hour to set the door and jamb in the frame and nail the assembly in place.

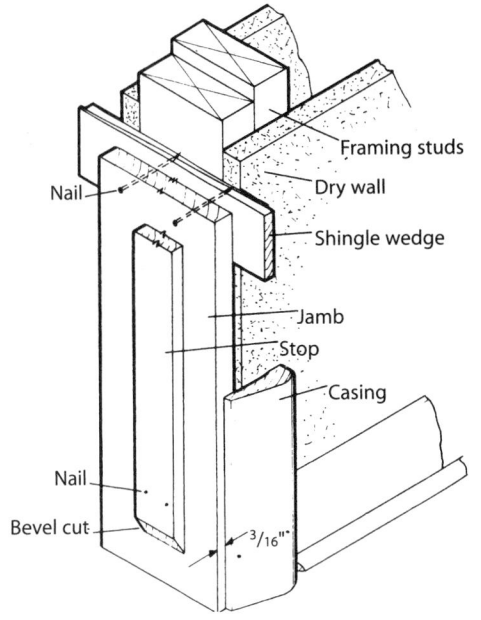

**Figure 9-10**

*Installation of sanitary stop*

## Door Locksets

Many doors come pre-bored for the lockset. There are four types of locksets:

1. Decorative entry keyed locksets;

2. Common keyed locks;

3. Privacy locksets with an inside lock button and a safety slot that allows opening from the outside; and

4. Passage latch sets with no lock.

There are also dummy knobs that merely attach to the face of the door.

After the lockset has been installed, mark the location of the latch on the jamb. Do this with the door nearly closed. Then mark an outline for the strike plate at this position. Rout the jamb so the strike plate will be flush with the face of the jamb. See Figure 9-11. Installing the lockset and setting the strike will take about 15 minutes, assuming the door is already bored for a lockset. If you have to drill the door for a cylinder lock, allow 45 minutes. Routing for a mortise lockset will take over an hour.

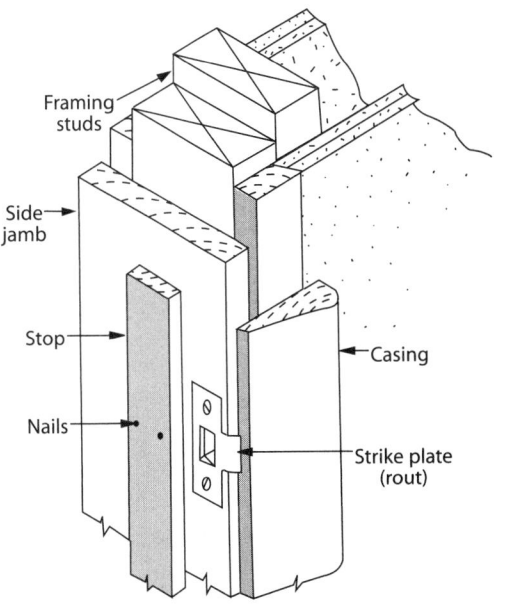

**Figure 9-11**

*Installation of strike plate*

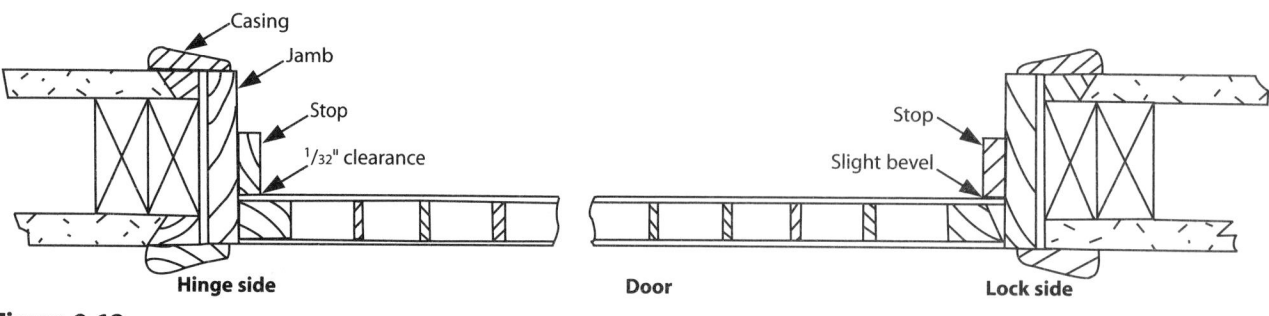

**Figure 9-12**

*Location of stops*

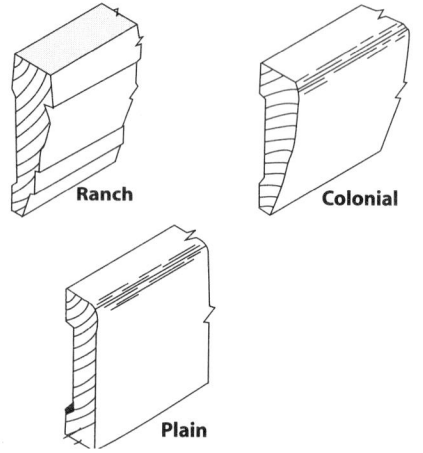

**Figure 9-13**

*Styles of door casings*

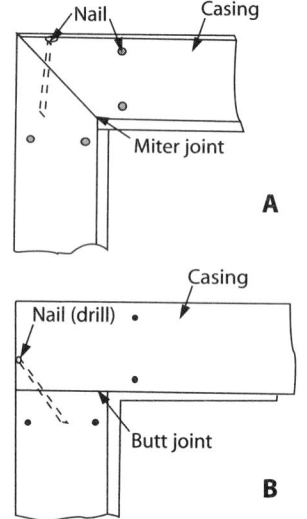

**Figure 9-14**

*Casing joints*

The stops, which were temporarily tacked in place, can now be permanently installed. Nail the stop on the lock side first, setting it against the door face when the door is fully latched. Nail the stops with $1^1/2$" long finishing nails or brads spaced in pairs about 16" apart. The stop at the hinge side of the door should have a clearance of $^1/32$" (Figure 9-12). Allow 15 minutes to set the stops in place.

## Setting the Trim

Casing is the trim set around a door or window opening. Most casing is from $^1/2$" to $^3/4$" thick and from $2^1/4$" to $3^1/2$" wide. Figure 9-13 shows three popular casing profiles.

Set the casing back about $^3/16$" from the face of the jamb. (Refer back to Figure 9-10.) Nail casing with 6d or 7d finishing nails, depending on the thickness of the casing. Space nails in pairs about 16" apart. Casings with molded profiles need a mitered joint where the head and side casings join (Figure 9-14 A). Rectangular casings can be butt-joined (Figure 9-14 B). With metal casing, fit the casing over the drywall, position the sheet properly, then nail through the drywall and casing into the stud behind. See Figure 9-15. Use the same type of nails and spacing as for drywall alone. Setting casing on both sides of the door will take about an hour.

## Hanging a Slab Door

In round numbers, allow about two hours for routine assembly and hanging of a slab door in a rough wall opening. Here's a breakdown:

❖ Cut and fit the jamb, 30 minutes

❖ Set hinges on the door and jamb, 15 minutes

❖ Install jambs and hang the door, 15 minutes

❖ Install the lockset and strike, 15 minutes

❖ Set door stop on the jamb, 15 minutes

❖ Set casing on two sides, 30 minutes

❖ Add 15 minutes for a French door

❖ Add 30 minutes for setting a threshold

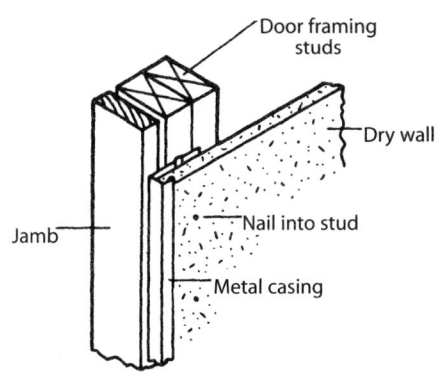

An experienced finish carpenter using a pneumatic nailer, router and door jig may be able to cut this time in half.

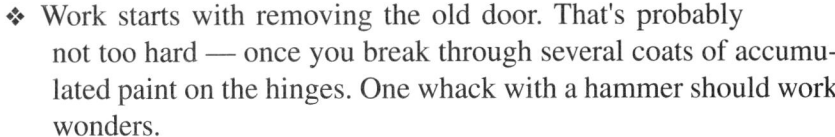

**Figure 9-15**
*Metal casing*

But as mentioned at the start of Chapter 1, your typical door replacement job may be anything but routine.

❖ Work starts with removing the old door. That's probably not too hard — once you break through several coats of accumulated paint on the hinges. One whack with a hammer should work wonders.

❖ Check the opening for plumb and square. Allow an extra hour or more if you have to align the framing. If the wall is way out of plumb, you may have a serious structural problem that requires reframing the entire wall.

❖ If hinge butts are loose, stripped out or poorly aligned, the next step may be installing a blind Dutchman on the jamb where the original hinges were attached.

❖ If the jamb is badly chewed up from years of neglect and abuse, you'll have to remove and replace the entire frame.

❖ In any case, you'll have to refinish the door and repair cosmetic defects in the interior and exterior wall finish.

❖ Finally, you have haul away the old door and debris.

The pages that follow show a labor estimate of 1.45 hours for hanging a new exterior slab door. That estimate includes setting three hinges on the door and jamb, fitting and hanging the door in an existing cased opening, checking the operation and making minor adjustments as required to the stop, jamb or strike. Add the cost of removing the old door, reframing the opening, repairs to the existing jamb, resetting the latch and strike, and replacing interior casing or exterior brick mold, if required.

The labor estimate for hanging an interior slab door is 1.15 hours and includes setting two hinges on the door and jamb and hanging the door in an existing cased opening. Add the cost of removing the old door, reframing the opening, repairs to the existing jamb, resetting the latch and strike, and replacing casing on two sides, if required.

Prehung doors are a better choice for most jobs and offer obvious advantages.

# Prehung Doors

Prehung doors (door, hinges, jamb and casing) cost a little more than slab doors (the door alone) but save money on labor because they require less cutting and fitting. Prehung doors come with the hinges attached, casing set, jamb assembled and door bored for a lockset. Exterior prehung doors usually include an adjustable threshold. The main jamb is attached to the door and is installed from one side of the opening. The other half of the jamb installs from the other side of the opening. When assembled, the two halves fit snugly together. Split jamb prehung doors have a tongue-and-groove joint behind the jamb stop that adjusts to fit the actual wall thickness. The casing will extend about 2" beyond the door edge at both the head and side jambs.

Prehung doors are easy to install. Allow about 45 minutes for an interior prehung door and 60 minutes for an exterior prehung door, assuming the opening is plumb, square and flush. But the door you order has to be exactly the door you need. It's impractical to make changes at the job site. All of the following are set at the mill:

> Identify the "hinge" of a prehung door by opening it away from you. Right-hand (or right hinge) doors swing out of the way to the right when fully open. Left hand doors swing to the left. Inward swing is most common for exterior doors.

❖ Door width and height

❖ Left or right hinge

❖ Inward or outward swing

❖ Jamb depth that matches the finished wall thickness

❖ Bore diameter and setback.

# Closet Doors

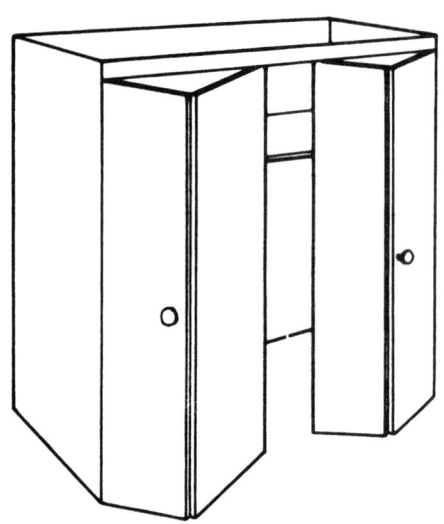

**Figure 9-16**

*Double-hinged door set for full-width opening closet*

Old homes tend to have larger rooms, higher ceilings and completely inadequate closet space, at least by 21st century standards. Even if there's enough closet space, it may not be well-arranged for good utility. Adding or rearranging closet space is a popular home improvement project.

One easy closet improvement is to replace a conventional single closet door with doors that open the full width of the closet. You'll need to remove the wall finish and enough studs to provide the opening needed for accordion, bypass or bi-fold doors (Figure 9-16). If the closet wall is a loadbearing wall, install a header across the new opening: two 2 x 8s for a 5' opening; two 2 x 10s for a 6' opening; or two 2 x 12s for an 8' opening. No header is needed for non-loadbearing walls. Frame the opening the same as for any new door opening. The rough opening should be 2$^1/_2$" wider than the door or set of doors.

You can also make closets more useful by adding or altering shelves and clothes rods. The usual closet has one rod with a single shelf above. If hanging space is limited, install a second clothes rod about halfway between the existing rod and the floor. This type of space is good for children's clothing, shirts and folded pants. Support a new shelf or rod with 1" x 4" cleats nailed to the closet end walls. Nail these cleats with three 6d nails at each end of the closet and at the intermediate stud. Shelf ends can rest directly on these cleats. Attach clothes rods to the same cleats that support the shelf. A long closet rod may sag in the middle when supported with nothing but end cleats. In that case, add steel support brackets fastened to studs at the back of the closet.

# New Closet Space

A plywood wardrobe cabinet is an economical and practical alternative to building a conventional closet with studs, drywall and casing. Figure 9-17 shows a simple plywood wardrobe built against a wall. For extra convenience, install additional shelves, drawers and doors.

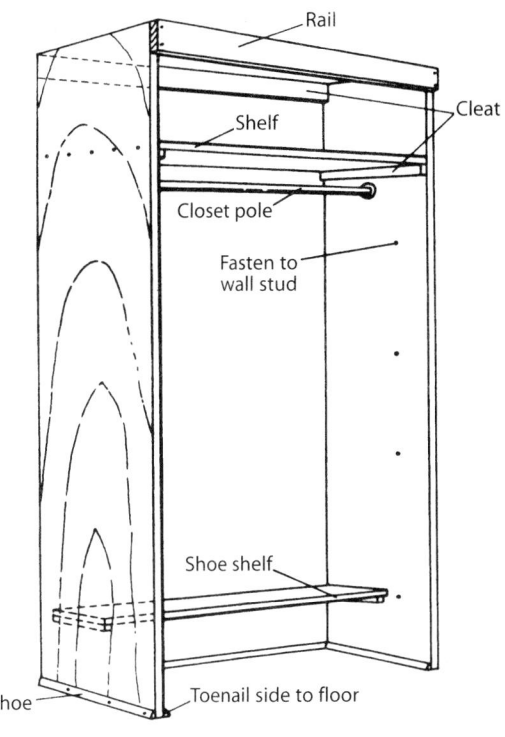

**Figure 9-17**

*Wardrobe closet*

Build wardrobe closets from $^5/_8$" or $^3/_4$" plywood or particleboard supported on cleats. Use a 1" x 4" top rail and back cleat. Fasten the cleat to the back wall. Fasten the sidewalls to a wall stud. Toenail base shoe molding to the floor to hold the bottoms of the sidewalls in place. Add shelves and closet poles as needed. Then enclose the space with bi-fold doors.

Plan for a coat closet near both the front and rear entrances. There should be a cleaning closet in the work area, and a linen closet in the bedroom area. Each bedroom needs a closet. In most real estate listings, you can't refer to a room as a "bedroom" unless it has a closet. If bedrooms are large, it's a simple matter to build a closet across one end of the room. Build partition walls with 2 x 3 or 2 x 4 framing lumber covered with gypsum board. Provide a cased opening for the closet doors.

In a small house, look for wasted space at the end of a hallway or at a wall offset. If the front door opens directly into the living room, consider building a coat closet beside or in front of the door to form an entry (Figure 9-18). In a story-and-a-half house, build closets in attic space where headroom is too limited for occupancy.

Closets used for hanging clothes should be about 24" deep. Shallower closets can be useful with appropriate hanging fixtures. Deeper closets need rollout hanging rods. To make the best use of closet space, plan for a full-front opening.

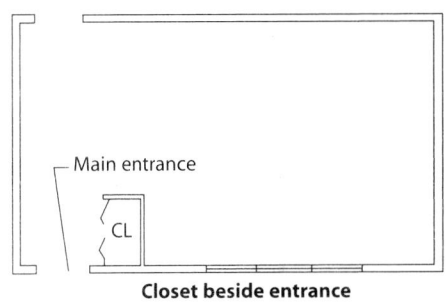

**Closet beside entrance**

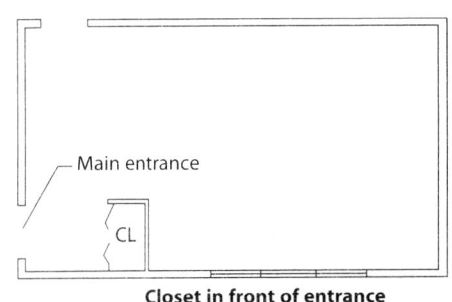

**Closet in front of entrance**

**Figure 9-18**

*Entry formed by coat closet*

**Hollow core door**

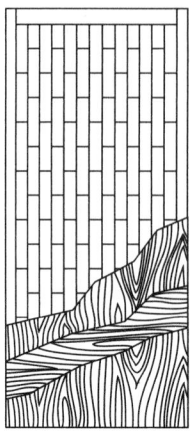

**Solid core door**

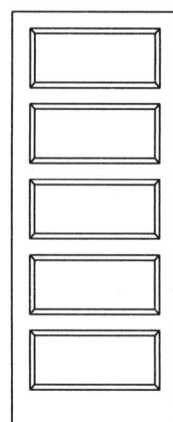

**French door, 5 lites**

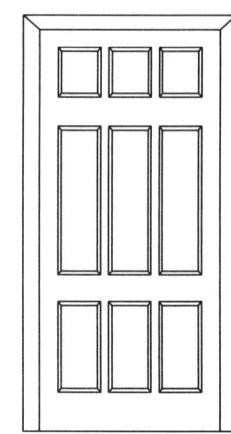

**Colonial 9-panel door**

# Door Terms

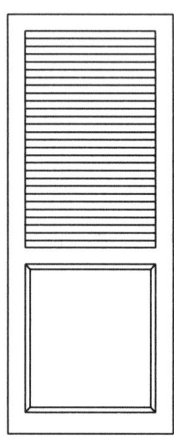

**Half louver door**

**Flush door**

❖ Interior doors are usually $1^3/_8$" thick and have a "hollow" core. A wood frame is filled with a matrix of corrugated paper and covered with a thin veneer.

❖ Exterior doors are usually $1^3/_4$" thick and solid core — again a thin veneer laid over a frame, but with some other material filling the door cavity. This filler could be foam or composition board or even lumber. Other popular exterior door choices include solid wood, steel and fiberglass.

❖ French doors are mostly glass, with one, to as many as fifteen panes set in wood sash.

❖ Insulated French doors have dual layers of glass, usually with a dividing grille installed between layers.

❖ Colonial doors have raised decorative wood panels that resemble doors popular in colonial America. Six or nine panels are most common.

❖ Flush doors are smooth and flat, with no decorative treatment.

❖ Louver doors include wood louvers that allow air to circulate but obscure vision, an advantage for enclosing a closet, pantry or water heater.

❖ Grilles (grids) are laid over the glass in a single-pane French door to create the illusion of multiple glass panes.

❖ Paint grade doors are usually primed at the factory and are designed to be painted after installation.

❖ Unfinished doors can be stained or given a clear coating, such as urethane, to enhance the beauty of the wood.

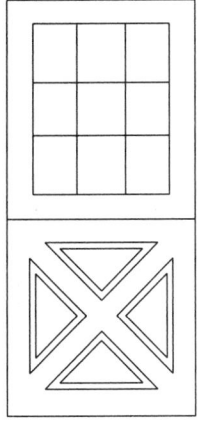

**Crossbuck door**

**Bar doors**

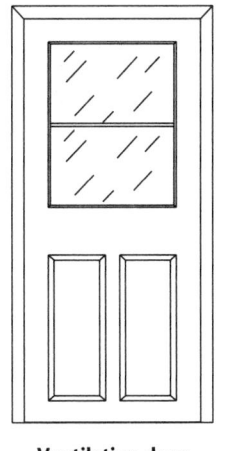

**Ventilating door**

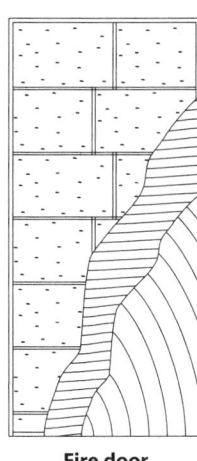

**Fire door**

❖ Entry doors are decorative exterior doors, often with glass panels set in a border of brass caming.

❖ Slab doors are the door alone, as distinguished from prehung doors which are sold with jamb, hinges, trim and threshold attached.

❖ Prehung doors are sold as a package with the door, jamb, hinges, and trim already assembled.

❖ Brick mold (also called stucco mold) is the casing around each side and head on the exterior side of an exterior door. Prehung doors are sold either with or without brick mold.

❖ Crossbuck doors have a raised X design in the bottom half of the door.

❖ Bar (café) doors are familiar to everyone who has seen a Western movie.

❖ Ventilating doors include a window that can be opened.

❖ Fire doors carry a fire rating and are designed to meet building code requirements for specialized occupancies, such as hotel rooms or dormitories. Code also requires that fire doors be used between a house and an attached garage, and that an automatic door closer is installed.

❖ Bi-fold doors on closets fold in half when open, revealing nearly the full interior.

❖ Bypass closet doors slide left or right, never revealing more than one-half of the interior.

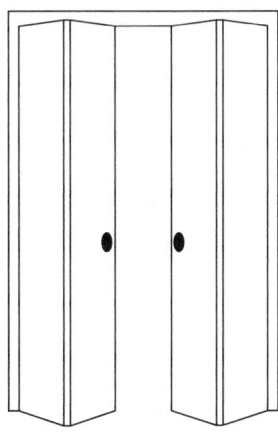

**Bi-fold doors**

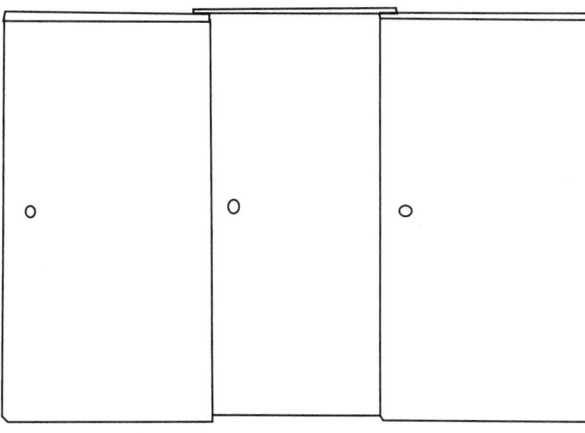

**Bypass doors**

**Security door**

**³/₄ oval lite door**

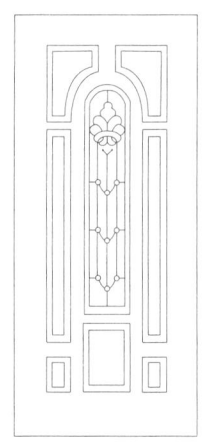

**Arch lite panel door**

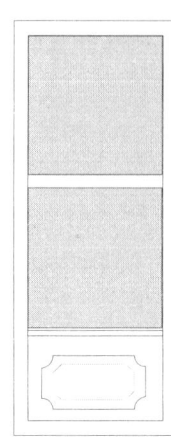

**Self-storing storm door**

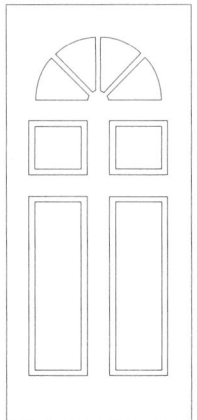

**Fan lite panel door**

❖ Hardboard doors are made of composition wood veneer, the least expensive material available.

❖ Casing is the trim around each side and head of an interior door. Casing covers the joint between jamb and the wall.

❖ Finger joint trim is made from two or more lengths of wood joined together with a finger-like joint that yields nearly as much strength as trim made of a single, solid piece.

❖ Hardwood doors are finished with a thin layer of true hardwood veneer.

❖ Lauan doors are finished with a veneer of Philippine hardwood from the mahogany family.

❖ Stiles and rails form the framework of a flush door. Stiles are vertical and every door has two; a lock stile and the hinge stile. Rails run horizontally at the top and bottom of the door.

❖ Molded face doors have decorative molding applied to at least one side.

| | Craft@Hrs | Unit | Material | Labor | Total | Sell |
|---|---|---|---|---|---|---|

## Exterior Doors

**Cut wall and frame exterior door opening, install slab door and trim.** Includes temporary wall bracing and closure, header, trimmer studs, exterior flush slab door ($52), frame, threshold, three hinges, stop, brick mold, casing, cylinder lockset ($74.50), and weatherstrip. Add the cost of debris removal, painting, and floor and wall finish if needed.

| | Craft@Hrs | Unit | Material | Labor | Total | Sell |
|---|---|---|---|---|---|---|
| 36" x 80" slab exterior door | BC@10.7 | Ea | 354.00 | 343.00 | 697.00 | 1,180.00 |

**Remove and replace single exterior slab door and trim.** Includes removing the existing door, trim and jamb, aligning the rough frame, installing exterior flush slab door ($52), frame, threshold, three hinges, stop, brick mold, casing, cylinder lockset ($74.50), and weatherstrip. Add the cost of debris removal and painting.

| | Craft@Hrs | Unit | Material | Labor | Total | Sell |
|---|---|---|---|---|---|---|
| Single door, stud wall | BC@6.40 | Ea | 325.00 | 205.00 | 530.00 | 901.00 |
| Single door, masonry wall | BC@7.35 | Ea | 325.00 | 236.00 | 561.00 | 954.00 |

**Remove and replace double exterior slab doors and trim.** Includes removing the existing doors, trim and jamb, aligning the rough frame, installing two exterior 6-panel slab doors ($104), frame, threshold, six hinges, stop, brick mold, casing, cylinder lockset ($75), two surface slide bolts, and weatherstrip.

| | Craft@Hrs | Unit | Material | Labor | Total | Sell |
|---|---|---|---|---|---|---|
| Double door, stud wall, per pair | BC@8.75 | Ea | 461.00 | 281.00 | 742.00 | 1,260.00 |
| Double door, masonry wall, per pair | BC@9.75 | Ea | 461.00 | 313.00 | 774.00 | 1,320.00 |

**Remove, clean, refinish and reinstall exterior slab door.** Includes removing door finish on all six sides with heat gun and solvent, filling small voids, sanding and refinishing with varnish or stain and sealer, and rehanging the door.

| | Craft@Hrs | Unit | Material | Labor | Total | Sell |
|---|---|---|---|---|---|---|
| Single door | PT@6.00 | Ea | 10.90 | 194.00 | 204.90 | 348.00 |
| Double door | PT@11.0 | Ea | 21.80 | 356.00 | 377.80 | 642.00 |
| Refinish jamb and trim, per opening | PT@2.00 | Ea | 10.90 | 64.80 | 75.70 | 129.00 |
| Add for antique finish, per door | PT@1.50 | Ea | 16.40 | 48.60 | 65.00 | 111.00 |

**Door repairs.**

| | Craft@Hrs | Unit | Material | Labor | Total | Sell |
|---|---|---|---|---|---|---|
| Remove set of three hinge pins | BC@.067 | Ea | — | 2.15 | 2.15 | 3.66 |
| Install set of three hinge pins | BC@.091 | Ea | — | 2.92 | 2.92 | 4.96 |
| Patch 6" hole with body filler | BC@4.00 | Ea | 21.00 | 128.00 | 149.00 | 253.00 |
| Align strike plate with latch bolt | BC@.500 | Ea | — | 16.00 | 16.00 | 27.20 |
| Plane 1/8" off door edge | BC@.250 | Ea | — | 8.02 | 8.02 | 13.60 |
| Loosen and reset one hinge butt | BC@.333 | Ea | — | 10.70 | 10.70 | 18.20 |
| Touch-up painting | PT@.250 | Ea | 1.10 | 8.10 | 9.20 | 15.60 |

**Install exterior slab door in an existing framed and cased opening.** Includes minor repair to the jamb (blind Dutchman), three hinges, flush slab door ($52), stop, cylinder lockset ($74.50), and weatherstrip.

| | Craft@Hrs | Unit | Material | Labor | Total | Sell |
|---|---|---|---|---|---|---|
| 36" x 80" exterior door | BC@3.10 | Ea | 191.00 | 99.50 | 290.50 | 494.00 |

**Remove and replace exterior door casing, jamb and stop.** Includes frame, stop, and casing on one side only. Per door opening.

| | Craft@Hrs | Unit | Material | Labor | Total | Sell |
|---|---|---|---|---|---|---|
| Door opening to 36" x 80" | BC@1.50 | Ea | 108.00 | 48.10 | 156.10 | 265.00 |

**Remove and replace exterior slab door, on existing hinges.** Includes exterior 6-panel slab door ($52) and cylinder lockset ($74.00).

| | Craft@Hrs | Unit | Material | Labor | Total | Sell |
|---|---|---|---|---|---|---|
| 36" x 80" exterior door | BC@1.90 | Ea | 126.00 | 61.00 | 187.00 | 318.00 |

| | Craft@Hrs | Unit | Material | Labor | Total | Sell |
|---|---|---|---|---|---|---|

## Wood Exterior Slab Doors

**Flush hardboard exterior doors.** Solid core. 1-3/4" thick. Primed and ready to paint. Labor includes setting three hinges on the door and jamb and hanging the door in an existing cased opening.

| | Craft@Hrs | Unit | Material | Labor | Total | Sell |
|---|---|---|---|---|---|---|
| 30" x 80" | BC@1.45 | Ea | 47.00 | 46.50 | 93.50 | 159.00 |
| 32" x 80" | BC@1.45 | Ea | 49.00 | 46.50 | 95.50 | 162.00 |
| 34" x 80" | BC@1.45 | Ea | 54.70 | 46.50 | 101.20 | 172.00 |
| 36" x 80" | BC@1.45 | Ea | 52.90 | 46.50 | 99.40 | 169.00 |

**Flush hardwood exterior doors.** Solid core. 1-3/4" thick. Labor includes setting three hinges on the door and jamb and hanging the door in an existing cased opening.

| | Craft@Hrs | Unit | Material | Labor | Total | Sell |
|---|---|---|---|---|---|---|
| 28" x 80" | BC@1.45 | Ea | 50.90 | 46.50 | 97.40 | 166.00 |
| 30" x 80" | BC@1.45 | Ea | 54.60 | 46.50 | 101.10 | 172.00 |
| 32" x 80" | BC@1.45 | Ea | 55.60 | 46.50 | 102.10 | 174.00 |
| 36" x 80" | BC@1.45 | Ea | 58.80 | 46.50 | 105.30 | 179.00 |
| 36" x 80" | BC@1.45 | Ea | 63.70 | 46.50 | 110.20 | 187.00 |
| 30" x 80", diamond light | BC@1.45 | Ea | 63.70 | 46.50 | 110.20 | 187.00 |
| 30" x 80", square light | BC@1.45 | Ea | 63.70 | 46.50 | 110.20 | 187.00 |
| 32" x 80", square light | BC@1.45 | Ea | 66.00 | 46.50 | 112.50 | 191.00 |
| 36" x 80", diamond light | BC@1.45 | Ea | 69.10 | 46.50 | 115.60 | 197.00 |
| 36" x 80", square light | BC@1.45 | Ea | 69.10 | 46.50 | 115.60 | 197.00 |

**Flush birch exterior doors.** Solid core. 1-3/4" thick. Stainable. Labor includes setting three hinges on the door and jamb and hanging the door in an existing cased opening.

| | Craft@Hrs | Unit | Material | Labor | Total | Sell |
|---|---|---|---|---|---|---|
| 32" x 80" | BC@1.45 | Ea | 52.20 | 46.50 | 98.70 | 168.00 |
| 34" x 80" | BC@1.45 | Ea | 60.80 | 46.50 | 107.30 | 182.00 |
| 36" x 80" | BC@1.45 | Ea | 54.90 | 46.50 | 101.40 | 172.00 |

**Flush lauan exterior doors.** Solid core. 1-3/4" thick. Stainable. Labor includes setting three hinges on the door and jamb and hanging the door in an existing cased opening.

| | Craft@Hrs | Unit | Material | Labor | Total | Sell |
|---|---|---|---|---|---|---|
| 30" x 80" | BC@1.45 | Ea | 48.60 | 46.50 | 95.10 | 162.00 |
| 30" x 84" | BC@1.45 | Ea | 50.70 | 46.50 | 97.20 | 165.00 |
| 32" x 80" | BC@1.45 | Ea | 49.50 | 46.50 | 96.00 | 163.00 |
| 36" x 80" | BC@1.45 | Ea | 52.20 | 46.50 | 98.70 | 168.00 |

**Flush fire-rated doors.** 1-3/4" thick. 20-minute fire rating. Labor includes setting three hinges on the door and jamb and hanging the door in an existing cased opening.

| | Craft@Hrs | Unit | Material | Labor | Total | Sell |
|---|---|---|---|---|---|---|
| 32" x 80", hardboard veneer | BC@1.45 | Ea | 55.70 | 46.50 | 102.20 | 174.00 |
| 36" x 80", hardboard veneer | BC@1.45 | Ea | 57.40 | 46.50 | 103.90 | 177.00 |
| 32" x 80", birch veneer | BC@1.45 | Ea | 60.80 | 46.50 | 107.30 | 182.00 |
| 36" x 80", birch veneer | BC@1.45 | Ea | 63.50 | 46.50 | 110.00 | 187.00 |

**Red oak exterior doors.** Wood stave core. 1-3/4" thick. Labor includes setting three hinges on the door and jamb and hanging the door in an existing cased opening.

| | Craft@Hrs | Unit | Material | Labor | Total | Sell |
|---|---|---|---|---|---|---|
| 30" x 80" | BC@1.45 | Ea | 58.70 | 46.50 | 105.20 | 179.00 |
| 32" x 80" | BC@1.45 | Ea | 61.00 | 46.50 | 107.50 | 183.00 |
| 34" x 80" | BC@1.45 | Ea | 64.80 | 46.50 | 111.30 | 189.00 |
| 36" x 80" | BC@1.45 | Ea | 64.60 | 46.50 | 111.10 | 189.00 |

**6-panel fir exterior doors.** 1-3/4" thick. Beveled, hip-raised panels. Doweled stile and rail joints. 4-1/2" wide stiles. Unfinished clear fir. Ready to paint or stain. Labor includes setting three hinges on the door and jamb and hanging the door in an existing cased opening.

| | Craft@Hrs | Unit | Material | Labor | Total | Sell |
|---|---|---|---|---|---|---|
| 30" x 80" | BC@1.45 | Ea | 163.00 | 46.50 | 209.50 | 356.00 |
| 32" x 80" | BC@1.45 | Ea | 158.00 | 46.50 | 204.50 | 348.00 |
| 36" x 80" | BC@1.45 | Ea | 158.00 | 46.50 | 204.50 | 348.00 |

| | Craft@Hrs | Unit | Material | Labor | Total | Sell |
|---|---|---|---|---|---|---|

**6-panel Douglas fir entry doors.** 1-3/4" thick. Solid core. Labor includes setting three hinges on the door and jamb and hanging the door in an existing cased opening.

| | Craft@Hrs | Unit | Material | Labor | Total | Sell |
|---|---|---|---|---|---|---|
| 30" x 80" | BC@1.45 | Ea | 127.00 | 46.50 | 173.50 | 295.00 |
| 32" x 80" | BC@1.45 | Ea | 124.00 | 46.50 | 170.50 | 290.00 |
| 36" x 80" | BC@1.45 | Ea | 129.00 | 46.50 | 175.50 | 298.00 |

**Mahogany entry doors.** 1-3/4" thick. Beveled tempered glass lite. Unfinished solid wood. Ready to paint or stain. Labor includes setting three hinges on the door and jamb and hanging the door in an existing cased opening.

| | Craft@Hrs | Unit | Material | Labor | Total | Sell |
|---|---|---|---|---|---|---|
| 30" x 80", arch lite | BC@1.45 | Ea | 438.00 | 46.50 | 484.50 | 824.00 |
| 30" x 80", moon lite | BC@1.45 | Ea | 438.00 | 46.50 | 484.50 | 824.00 |
| 30" x 80", oval lite | BC@1.45 | Ea | 445.00 | 46.50 | 491.50 | 836.00 |

**4-panel 4-lite fir exterior doors.** Vertical grain Douglas fir. 1-3/4" thick. 3/4" thick panels. Unfinished. Veneered and doweled construction. 1/8" tempered safety glass. Labor includes setting three hinges on the door and jamb and hanging the door in an existing cased opening.

| | Craft@Hrs | Unit | Material | Labor | Total | Sell |
|---|---|---|---|---|---|---|
| 36" x 80" | BC@1.70 | Ea | 198.00 | 54.60 | 252.60 | 429.00 |

**2-panel 9-lite entry doors.** 1-3/4" thick. Single-pane glass. Solid core with doweled joints. Unfinished. Suitable for stain or paint. Labor includes setting three hinges on the door and jamb and hanging the door in an existing cased opening.

| | Craft@Hrs | Unit | Material | Labor | Total | Sell |
|---|---|---|---|---|---|---|
| 30" x 80" | BC@1.70 | Ea | 163.00 | 54.60 | 217.60 | 370.00 |
| 32" x 80" | BC@1.70 | Ea | 164.00 | 54.60 | 218.60 | 372.00 |
| 36" x 80" | BC@1.70 | Ea | 164.00 | 54.60 | 218.60 | 372.00 |

**1-lite clear fir entry doors.** 1-3/4" thick. Clear fir surface ready to stain or paint. Insulated glass. Labor includes setting three hinges on the door and jamb and hanging the door in an existing cased opening.

| | Craft@Hrs | Unit | Material | Labor | Total | Sell |
|---|---|---|---|---|---|---|
| 32" x 80" | BC@1.70 | Ea | 165.00 | 54.60 | 219.60 | 373.00 |
| 36" x 80" | BC@1.70 | Ea | 166.00 | 54.60 | 220.60 | 375.00 |

**Fir 10-lite exterior doors.** 1-3/4" thick. Unfinished. Labor includes setting three hinges on the door and jamb and hanging the door in an existing cased opening.

| | Craft@Hrs | Unit | Material | Labor | Total | Sell |
|---|---|---|---|---|---|---|
| 30" x 80" x 1-3/4" | BC@1.70 | Ea | 144.00 | 54.60 | 198.60 | 338.00 |
| 32" x 80" x 1-3/4" | BC@1.70 | Ea | 152.00 | 54.60 | 206.60 | 351.00 |
| 36" x 80" x 1-3/4" | BC@1.70 | Ea | 153.00 | 54.60 | 207.60 | 353.00 |

**Fir 15-lite entry doors.** Single-pane glass. Ready for stain or paint. 1-3/4" thick. Doweled stile and rail joints. Labor includes setting three hinges on the door and jamb and hanging the door in an existing cased opening.

| | Craft@Hrs | Unit | Material | Labor | Total | Sell |
|---|---|---|---|---|---|---|
| 30" x 80" x 1-3/4" | BC@1.70 | Ea | 159.00 | 54.60 | 213.60 | 363.00 |
| 32" x 80" x 1-3/4" | BC@1.70 | Ea | 159.00 | 54.60 | 213.60 | 363.00 |
| 36" x 80" x 1-3/4" | BC@1.70 | Ea | 159.00 | 54.60 | 213.60 | 363.00 |

**Fir fan lite entry doors.** Vertical grain Douglas fir. 1-3/4" thick. 3/4" thick panels. Unfinished. Veneer and doweled construction. 1/8" tempered safety glass. Labor includes setting three hinges on the door and jamb and hanging the door in an existing cased opening.

| | Craft@Hrs | Unit | Material | Labor | Total | Sell |
|---|---|---|---|---|---|---|
| 32" x 80" | BC@1.45 | Ea | 195.00 | 46.50 | 241.50 | 411.00 |
| 30" x 80" | BC@1.45 | Ea | 189.00 | 46.50 | 235.50 | 400.00 |

**Deco fan lite entry doors.** 1-3/4" thick. 3/4" thick panels. Unfinished solid wood. Ready to paint or stain. Beveled insulated glass. Labor includes setting three hinges on the door and jamb and hanging the door in an existing cased opening.

| | Craft@Hrs | Unit | Material | Labor | Total | Sell |
|---|---|---|---|---|---|---|
| 36" x 80" | BC@1.45 | Ea | 226.00 | 46.50 | 272.50 | 463.00 |

| | Craft@Hrs | Unit | Material | Labor | Total | Sell |
|---|---|---|---|---|---|---|

**Flush oak entry doors.** 1-3/4" thick. Prefinished. Labor includes setting three hinges on the door and jamb and hanging the door in an existing cased opening.

| | Craft@Hrs | Unit | Material | Labor | Total | Sell |
|---|---|---|---|---|---|---|
| 36" x 80" | BC@1.45 | Ea | 864.00 | 46.50 | 910.50 | 1,550.00 |

**Reversible hollow metal fire doors.**

| | Craft@Hrs | Unit | Material | Labor | Total | Sell |
|---|---|---|---|---|---|---|
| 34" x 80" | BC@.721 | Ea | 136.00 | 23.10 | 159.10 | 270.00 |
| 36" x 80" | BC@.721 | Ea | 157.00 | 23.10 | 180.10 | 306.00 |

**Knock down steel door frames.** Bored for (3) 4-1/2" hinges, 4-7/8" strike plate and 2-3/4" backset. 90-minute fire label manufactured to N.Y.C. M.E.A. procedures. Hollow core. For drywall and masonry applications. Galvanized steel.

| | Craft@Hrs | Unit | Material | Labor | Total | Sell |
|---|---|---|---|---|---|---|
| 30" x 80" | BC@.944 | Ea | 67.00 | 30.30 | 97.30 | 165.00 |
| 32" x 80" | BC@.944 | Ea | 68.90 | 30.30 | 99.20 | 169.00 |
| 34" x 80" | BC@.944 | Ea | 69.10 | 30.30 | 99.40 | 169.00 |
| 36" x 80" | BC@.944 | Ea | 70.60 | 30.30 | 100.90 | 172.00 |
| 36" x 84" | BC@.944 | Ea | 72.00 | 30.30 | 102.30 | 174.00 |

**90-minute fire-rated door frame parts.**

| | Craft@Hrs | Unit | Material | Labor | Total | Sell |
|---|---|---|---|---|---|---|
| 4-7/8" x 80" legs, pack of 2 | BC@.778 | Ea | 66.30 | 25.00 | 91.30 | 155.00 |
| 4-7/8" x 36" head | BC@.166 | Ea | 17.30 | 5.33 | 22.63 | 38.50 |

**Steel cellar doors.** Durable heavy gauge steel.

| | Craft@Hrs | Unit | Material | Labor | Total | Sell |
|---|---|---|---|---|---|---|
| 57" x 45" x 24-1/2" | BC@1.00 | Ea | 285.00 | 32.10 | 317.10 | 539.00 |
| 63" x 49" x 22" | BC@1.00 | Ea | 291.00 | 32.10 | 323.10 | 549.00 |
| 71" x 53" x 19-1/2" | BC@1.00 | Ea | 301.00 | 32.10 | 333.10 | 566.00 |

## Door Trim

**Install door trim.**

| | Craft@Hrs | Unit | Material | Labor | Total | Sell |
|---|---|---|---|---|---|---|
| Door casing, two sides, hand nailed | BC@.500 | Ea | — | 16.00 | 16.00 | 27.20 |
| Door casing, two sides, power nailed | BC@.333 | Ea | — | 10.70 | 10.70 | 18.20 |
| Install mail slot in door | BC@.529 | Ea | — | 17.00 | 17.00 | 28.90 |

**T-astragal molding.** Covers the vertical joint where a pair of doors meet.

| | Craft@Hrs | Unit | Material | Labor | Total | Sell |
|---|---|---|---|---|---|---|
| 1-1/4" x 2" x 7', pine | BC@.529 | Ea | 22.10 | 17.00 | 39.10 | 66.50 |
| Remove and replace 80" long astragal | BC@.680 | Ea | — | 21.80 | 21.80 | 37.10 |

**Primed exterior door jamb.** Primed finger joint pine. 1-1/4" thick. Set for a 3' door consists of two 6'8" legs and one 3' head.

| | Craft@Hrs | Unit | Material | Labor | Total | Sell |
|---|---|---|---|---|---|---|
| 4-5/8" wide, per 6'8" leg | BC@.204 | Ea | 26.50 | 6.55 | 33.05 | 56.20 |
| 5-1/4" wide, per 6'8" leg | BC@.204 | Ea | 29.70 | 6.55 | 36.25 | 61.60 |
| 6-9/16" wide, per 6'8" leg | BC@.204 | Ea | 30.90 | 6.55 | 37.45 | 63.70 |
| 4-5/8" wide, per set | BC@.500 | Ea | 71.60 | 16.00 | 87.60 | 149.00 |
| 5-1/4" wide, per set | BC@.500 | Ea | 29.70 | 16.00 | 45.70 | 77.70 |
| 6-9/16" wide, per set | BC@.500 | Ea | 30.90 | 16.00 | 46.90 | 79.70 |

**Primed kerfed exterior door jamb.** 1-1/4" thick. Set for a 3' door consists of two 6'8" legs and one 3' head.

| | Craft@Hrs | Unit | Material | Labor | Total | Sell |
|---|---|---|---|---|---|---|
| 4-5/8" x 7' leg | BC@.204 | Ea | 24.60 | 6.55 | 31.15 | 53.00 |
| 4-5/8" x 3' head | BC@.092 | Ea | 13.40 | 2.95 | 16.35 | 27.80 |
| 6-5/8" x 7' leg | BC@.204 | Ea | 26.50 | 6.55 | 33.05 | 56.20 |
| 6-5/8" x 3' head | BC@.092 | Ea | 14.70 | 2.95 | 17.65 | 30.00 |

| | Craft@Hrs | Unit | Material | Labor | Total | Sell |
|---|---|---|---|---|---|---|
| **Aluminum thermal door sill.** With vinyl insert. 36" x 5-5/8", mill finish. | | | | | | |
| Install 30" to 36" aluminum threshold | BC@.518 | Ea | 14.40 | 16.60 | 31.00 | 52.70 |
| Install 56" to 72" aluminum threshold | BC@.623 | Ea | 28.90 | 20.00 | 48.90 | 83.10 |
| Remove and replace 30" to 36" threshold | BC@.850 | Ea | 14.40 | 27.30 | 41.70 | 70.90 |
| Remove and replace 56" to 72" threshold | BC@1.20 | Ea | 28.90 | 38.50 | 67.40 | 115.00 |
| **Oak high boy threshold.** Fits standard door sizes. | | | | | | |
| 5/8" x 3-1/2" x 36" | BC@.518 | Ea | 14.00 | 16.60 | 30.60 | 52.00 |
| **Stop molding.** Nails to face of door jamb to prevent door from swinging through. Solid pine. | | | | | | |
| 3/8" x 1-1/4" | BC@.015 | LF | .68 | .48 | 1.16 | 1.97 |
| 7/16" x 3/4" | BC@.015 | LF | .50 | .48 | .98 | 1.67 |
| 7/16" x 1-3/8" | BC@.015 | LF | .77 | .48 | 1.25 | 2.13 |
| 7/16" x 1-5/8" | BC@.015 | LF | .78 | .48 | 1.26 | 2.14 |
| 7/16" x 2-1/8" | BC@.015 | LF | 1.23 | .48 | 1.71 | 2.91 |
| **Solid pine ranch casing.** | | | | | | |
| 9/16" x 2-1/4" | BC@.015 | LF | .95 | .48 | 1.43 | 2.43 |
| 9/16" x 2-1/4" x 7' | BC@.105 | Ea | 6.65 | 3.37 | 10.02 | 17.00 |
| 11/16" x 3-1/2" | BC@.015 | LF | 1.66 | .48 | 2.14 | 3.64 |
| 11/16" x 2-1/2" x 7' | BC@.105 | Ea | 8.62 | 3.37 | 11.99 | 20.40 |
| **Primed finger joint ranch door casing.** | | | | | | |
| 9/16" x 3-1/4" | BC@.015 | LF | 1.23 | .48 | 1.71 | 2.91 |
| 11/16" x 2-1/4" | BC@.015 | LF | .89 | .48 | 1.37 | 2.33 |
| 11/16" x 2-1/4" x 7' | BC@.105 | Ea | 6.20 | 3.37 | 9.57 | 16.30 |
| 11/16" x 2-1/2" | BC@.015 | LF | .96 | .48 | 1.44 | 2.45 |
| 11/16" x 2-1/2" x 7' | BC@.105 | Ea | 6.73 | 3.37 | 10.10 | 17.20 |
| **Colonial door casing.** Solid pine. | | | | | | |
| 1/2" x 2-1/4" | BC@.015 | LF | .80 | .48 | 1.28 | 2.18 |
| 9/16" x 2-1/4" x 7' | BC@.105 | Ea | 6.30 | 3.37 | 9.67 | 16.40 |
| 9/16" x 2-1/4" | BC@.015 | LF | .95 | .48 | 1.43 | 2.43 |
| 11/16" x 2-1/2" | BC@.015 | LF | 1.25 | .48 | 1.73 | 2.94 |
| 11/16" x 2-1/2" x 7' | BC@.105 | Ea | 7.15 | 3.37 | 10.52 | 17.90 |
| 11/16" x 3-1/2" | BC@.015 | LF | 1.85 | .48 | 2.33 | 3.96 |
| 2-1/2", pre-mitered, 3' opening | BC@.250 | Ea | 25.90 | 8.02 | 33.92 | 57.70 |
| **Fluted pine casing and rosette set.** Includes 1 header piece, 2 side pieces, 2 rosettes and 2 base blocks. | | | | | | |
| Set | BC@.250 | Ea | 31.60 | 8.02 | 39.62 | 67.40 |
| **Brick mold set.** Trim for the exterior side of an exterior door. Two legs and one head. Primed finger joint pine. 36" x 80" door. 1-1/4" x 2". | | | | | | |
| Finger joint pine | BC@.250 | Ea | 36.10 | 8.02 | 44.12 | 75.00 |
| Pressure treated pine | BC@.250 | Ea | 22.70 | 8.02 | 30.72 | 52.20 |
| Remove and replace brick mold | BC@.350 | Ea | — | 11.20 | 11.20 | 19.00 |
| **Brick mold.** | | | | | | |
| 1-3/16" x 2", hemlock | BC@.015 | LF | 2.00 | .48 | 2.48 | 4.22 |
| 1-1/4" x 2", fir | BC@.015 | LF | 1.37 | .48 | 1.85 | 3.15 |
| 1-1/4" x 2", solid pine | BC@.015 | LF | 2.08 | .48 | 2.56 | 4.35 |

| | Craft@Hrs | Unit | Material | Labor | Total | Sell |
|---|---|---|---|---|---|---|
| **Stucco molding.** Redwood. | | | | | | |
| 13/16" x 1-1/2" | BC@.015 | LF | .90 | .48 | 1.38 | 2.35 |
| 1" x 1-3/8" | BC@.015 | LF | 1.04 | .48 | 1.52 | 2.58 |
| | | | | | | |
| **Remove door hinge butt.** Remove four screws per 3-1/2" hinge butt half. Per hinge. | | | | | | |
| Door or jamb half hinge butt only | BC@.089 | Ea | — | 2.86 | 2.86 | 4.86 |
| Door and jamb half hinge butts | BC@.135 | Ea | — | 4.33 | 4.33 | 7.36 |
| Remove and reset door hinge in existing mortise | BC@.275 | Ea | — | 8.82 | 8.82 | 15.00 |

**Door hinges.** Solid brass. Mortise type. Non-rising removable tip. Includes screws. Installation labor is included with cost of the door.

| | | Unit | Material | | Total | |
|---|---|---|---|---|---|---|
| 3" x 3", square corners | — | Ea | 10.30 | — | 10.30 | — |
| 3-1/2" x 3-1/2", round corners | — | Ea | 10.80 | — | 10.80 | — |
| 3-1/2" x 3-1/2", square corners | — | Ea | 10.90 | — | 10.90 | — |
| 4" x 4", round corners | — | Ea | 12.70 | — | 12.70 | — |
| 4" x 4", square corners | — | Ea | 12.10 | — | 12.10 | — |

## Door Weatherstrip

**Remove and replace vinyl door weatherstrip.**

| | Craft@Hrs | Unit | Material | Labor | Total | Sell |
|---|---|---|---|---|---|---|
| 36" x 80" exterior door | BC@.589 | Ea | 12.70 | 18.90 | 31.60 | 53.70 |

**Aluminum and vinyl adjustable door weatherstrip.** Seals out drafts and moisture. Fits top and sides of 36" x 80" door. Slotted holes for easy alignment.

| | Craft@Hrs | Unit | Material | Labor | Total | Sell |
|---|---|---|---|---|---|---|
| 3/4" x 17', brown or white | BC@.381 | Ea | 12.70 | 12.20 | 24.90 | 42.30 |
| 3/4" x 17', grey | BC@.381 | Ea | 10.30 | 12.20 | 22.50 | 38.30 |

**Gray foam weatherstrip.** Self-adhesive. Resilient. Open-cell. Compresses flat for a tight seal. Sticks to any dry clean surface. Width x thickness x length.

| | Craft@Hrs | Unit | Material | Labor | Total | Sell |
|---|---|---|---|---|---|---|
| 1/4" x 1/2" x 17' | BC@.250 | Ea | 3.11 | 8.02 | 11.13 | 18.90 |
| 3/8" x 1/2" x 17' | BC@.250 | Ea | 3.41 | 8.02 | 11.43 | 19.40 |

**Aluminum and vinyl door bottom.** Heavy gauge aluminum with vinyl sweep. Seals out drafts, rain and dust. Screws included. 36" x 1-5/8".

| | Craft@Hrs | Unit | Material | Labor | Total | Sell |
|---|---|---|---|---|---|---|
| Brown | BC@.250 | Ea | 5.85 | 8.02 | 13.87 | 23.60 |
| Brushed chrome | BC@.250 | Ea | 10.25 | 8.02 | 18.27 | 31.10 |

**Drip cap door bottom seal.** Drip cap diverts rain away. Seals out drafts, dust and insects. 36" long.

| | Craft@Hrs | Unit | Material | Labor | Total | Sell |
|---|---|---|---|---|---|---|
| 1-1/2", brite gold | BC@.250 | Ea | 11.40 | 8.02 | 19.42 | 33.00 |

**Magnetic weatherstrip.** Provides airtight, refrigerator-like seal. Hinge compression tear strip adjusts after installation.

| | Craft@Hrs | Unit | Material | Labor | Total | Sell |
|---|---|---|---|---|---|---|
| 36" long | BC@.250 | Ea | 14.60 | 8.02 | 22.62 | 38.50 |

**Flex-O-Matic door sweep.** As the door opens, the spring-action door sweep rises to clear the carpet. When the door is closed, the door sweep automatically lowers to firmly press against floor or carpet. Keeps out dirt and dust.

| | Craft@Hrs | Unit | Material | Labor | Total | Sell |
|---|---|---|---|---|---|---|
| Brite gold | BC@.250 | Ea | 5.37 | 8.02 | 13.39 | 22.80 |

| | Craft@Hrs | Unit | Material | Labor | Total | Sell |
|---|---|---|---|---|---|---|

## Prehung Wood Exterior Doors

**Cut wall and frame exterior door opening, install prehung door.** Includes cutting stud wall, temporary wall bracing and closure, header, trimmer studs, exterior 6-panel prehung steel door ($139) with frame, threshold, hinges, stop, brick mold, casing, weatherstrip and cylinder lockset ($74.50). Add the cost of debris removal, painting, floor and wall finish if needed.

| | Craft@Hrs | Unit | Material | Labor | Total | Sell |
|---|---|---|---|---|---|---|
| 36" x 80" slab exterior door | BC@8.20 | Ea | 243.00 | 263.00 | 506.00 | 860.00 |

**Remove door and frame and replace with an exterior prehung steel door.** Includes removing the existing door, trim and jamb, aligning the rough frame, installing one exterior 6-panel prehung steel door ($139) with frame, threshold, hinges, stop, brick mold, casing, weatherstrip and cylinder lockset ($74.50). Add the cost of debris removal and painting.

| | Craft@Hrs | Unit | Material | Labor | Total | Sell |
|---|---|---|---|---|---|---|
| Single door, stud wall | BC@3.90 | Ea | 214.00 | 125.00 | 339.00 | 576.00 |
| Single door, masonry wall | BC@4.85 | Ea | 214.00 | 156.00 | 370.00 | 629.00 |

**Remove double doors and frame and replace with double exterior prehung steel doors.** Includes removing an existing double door, trim and jamb, aligning the rough frame, installing double exterior 6-panel prehung steel doors ($291), with frame, threshold, hinges, stop, brick mold, casing, and weatherstrip. Add cylinder lockset ($74.50) and surface slide bolts ($15).

| | Craft@Hrs | Unit | Material | Labor | Total | Sell |
|---|---|---|---|---|---|---|
| Double door, stud wall, per pair | BC@6.25 | Ea | 381.00 | 201.00 | 582.00 | 989.00 |
| Double door, masonry wall, per pair | BC@7.25 | Ea | 381.00 | 233.00 | 614.00 | 1,040.00 |

**Flush hardwood prehung exterior doors.** 1-3/4" door. Solid particleboard core. Lauan veneer. Weatherstripped. Includes oak sill and threshold with vinyl insert. Finger joint 4-9/16" jamb and finger joint brick mold. 2-1/8" bore for lockset with 2-3/8" backset. Labor includes hanging the door in an existing framed opening.

| | Craft@Hrs | Unit | Material | Labor | Total | Sell |
|---|---|---|---|---|---|---|
| 32" x 80" | BC@1.00 | Ea | 104.00 | 32.10 | 136.10 | 231.00 |
| 36" x 80" | BC@1.00 | Ea | 107.00 | 32.10 | 139.10 | 236.00 |

**Flush birch prehung exterior doors.** 1-3/4" door. Solid particleboard core. Weatherstripped. Includes oak sill and threshold. Finger joint 4-9/16" jamb and finger joint brick mold. 2-1/8" bore for lockset with 2-3/8" backset. Labor includes hanging the door in an existing framed opening.

| | Craft@Hrs | Unit | Material | Labor | Total | Sell |
|---|---|---|---|---|---|---|
| 36" x 80" | BC@1.00 | Ea | 99.00 | 32.10 | 131.10 | 223.00 |

**Flush hardboard prehung exterior doors.** Primed. Solid core. Weatherstripped. Includes sill and threshold. Finger joint 4-9/16" jamb and finger joint brick mold. 2-1/8" bore for lockset with 2-3/8" backset. Labor includes hanging the door in an existing framed opening.

| | Craft@Hrs | Unit | Material | Labor | Total | Sell |
|---|---|---|---|---|---|---|
| 32" x 80" | BC@1.00 | Ea | 101.00 | 32.10 | 133.10 | 226.00 |
| 36" x 80" | BC@1.00 | Ea | 104.00 | 32.10 | 136.10 | 231.00 |

**Fan lite prehung hemlock exterior doors.** Weatherstripped. Includes sill and threshold. Finger joint 4-9/16" jamb and finger joint brick mold. 2-1/8" bore for lockset with 2-3/8" backset. Labor includes hanging the door in an existing framed opening.

| | Craft@Hrs | Unit | Material | Labor | Total | Sell |
|---|---|---|---|---|---|---|
| 36" x 80" | BC@1.00 | Ea | 301.00 | 32.10 | 333.10 | 566.00 |

**9-lite prehung fir exterior doors.** Inward swing. No brick mold. Solid core. Weatherstripped. Includes sill and threshold. Finger joint 4-9/16" jamb and finger joint brick mold. 2-1/8" bore for lockset with 2-3/8" backset. Labor includes hanging the door in an existing framed opening.

| | Craft@Hrs | Unit | Material | Labor | Total | Sell |
|---|---|---|---|---|---|---|
| 36" x 80" | BC@1.00 | Ea | 239.00 | 32.10 | 271.10 | 461.00 |

**Prehung hardboard fire-rated doors.** 20-minute fire rating. 80" high. Labor includes hanging the door in an existing framed opening.

| | Craft@Hrs | Unit | Material | Labor | Total | Sell |
|---|---|---|---|---|---|---|
| 30" x 80" | BC@1.00 | Ea | 138.00 | 32.10 | 170.10 | 289.00 |
| 32" x 80" | BC@1.00 | Ea | 144.00 | 32.10 | 176.10 | 299.00 |
| 36" x 80" | BC@1.00 | Ea | 144.00 | 32.10 | 176.10 | 299.00 |

| | Craft@Hrs | Unit | Material | Labor | Total | Sell |
|---|---|---|---|---|---|---|

## Prehung Steel Exterior Doors

**Utility flush prehung steel exterior doors.** No brick mold. 26-gauge galvanized steel. Foam core. Compression weatherstrip. Triple fin sweep. Non-thermal threshold. Single bore.

| | Craft@Hrs | Unit | Material | Labor | Total | Sell |
|---|---|---|---|---|---|---|
| 32" x 80" | BC@1.00 | Ea | 113.00 | 32.10 | 145.10 | 247.00 |
| 36" x 80" | BC@1.00 | Ea | 113.00 | 32.10 | 145.10 | 247.00 |

**Premium flush prehung steel exterior doors.** No brick mold. 24-gauge galvanized steel. Foam core. Compression weatherstrip. Triple fin sweep. Thermal threshold.

| | Craft@Hrs | Unit | Material | Labor | Total | Sell |
|---|---|---|---|---|---|---|
| 30" x 80" | BC@1.00 | Ea | 147.00 | 32.10 | 179.10 | 304.00 |
| 32" x 80" | BC@1.00 | Ea | 147.00 | 32.10 | 179.10 | 304.00 |
| 36" x 80" | BC@1.00 | Ea | 147.00 | 32.10 | 179.10 | 304.00 |

**6-panel prehung steel exterior doors.** With brick mold. Inward swing. Non-thermal threshold. Bored for lockset with 2-3/4" backset. Includes compression weatherstrip and thermal-break threshold.

| | Craft@Hrs | Unit | Material | Labor | Total | Sell |
|---|---|---|---|---|---|---|
| 32" x 80" | BC@1.00 | Ea | 113.00 | 32.10 | 145.10 | 247.00 |
| 36" x 80" | BC@1.00 | Ea | 113.00 | 32.10 | 145.10 | 247.00 |

**6-panel premium prehung steel doors.** 4-9/16" jamb. Adjustable thermal-break sill. Inward swing. 24-gauge galvanized steel. Bored for 2-3/8" lockset. 12" lock block. Impact-resistant laminated glass. Polyurethane core. Triple sweep and compression weatherstripping. Factory primed.

| | Craft@Hrs | Unit | Material | Labor | Total | Sell |
|---|---|---|---|---|---|---|
| 32" x 80", no brick mold | BC@1.00 | Ea | 139.00 | 32.10 | 171.10 | 291.00 |
| 32" x 80", with brick mold | BC@1.00 | Ea | 149.00 | 32.10 | 181.10 | 308.00 |
| 36" x 80", no brick mold | BC@1.00 | Ea | 139.00 | 32.10 | 171.10 | 291.00 |
| 36" x 80", with brick mold | BC@1.00 | Ea | 149.00 | 32.10 | 181.10 | 308.00 |
| 72" x 80", no brick mold | BC@1.50 | Ea | 291.00 | 48.10 | 339.10 | 576.00 |

**6-panel prehung steel exterior doors.** 24-gauge galvanized steel. Foam core. Caming matches hinge and sill finishes. Jamb guard security plate. No brick mold. Fixed sill. 4-9/16" primed jamb. Triple sweep compression weatherstripping. Factory primed. Thermal-break threshold. 20-minute fire rating. 10-year limited warranty.

| | Craft@Hrs | Unit | Material | Labor | Total | Sell |
|---|---|---|---|---|---|---|
| 30" x 80" | BC@1.00 | Ea | 139.00 | 32.10 | 171.10 | 291.00 |
| 32" x 80" | BC@1.00 | Ea | 139.00 | 32.10 | 171.10 | 291.00 |
| 36" x 80" | BC@1.00 | Ea | 139.00 | 32.10 | 171.10 | 291.00 |

**6-panel prehung steel entry door with sidelites.** Two 10"-wide sidelites. Sidelites have 5 divided lites.

| | Craft@Hrs | Unit | Material | Labor | Total | Sell |
|---|---|---|---|---|---|---|
| 36" x 80" | BC@2.00 | Ea | 453.00 | 64.20 | 517.20 | 879.00 |

**Fan lite premium prehung steel exterior doors.** With brick mold. 24-gauge galvanized steel. Foam core. Fixed sill. Compression weatherstripping. Factory primed. Thermal-break threshold.

| | Craft@Hrs | Unit | Material | Labor | Total | Sell |
|---|---|---|---|---|---|---|
| 32" x 80", 4-9/16" jamb | BC@1.00 | Ea | 172.00 | 32.10 | 204.10 | 347.00 |
| 32" x 80", 6-9/16" jamb | BC@1.00 | Ea | 174.00 | 32.10 | 206.10 | 350.00 |
| 36" x 80", 4-9/16" jamb | BC@1.00 | Ea | 172.00 | 32.10 | 204.10 | 347.00 |
| 36" x 80", 6-9/16" jamb | BC@1.00 | Ea | 108.00 | 32.10 | 140.10 | 238.00 |

**Half lite prehung steel exterior doors.** 4-9/16" jamb. 22" x 36" light. Adjustable sill. With brick mold.

| | Craft@Hrs | Unit | Material | Labor | Total | Sell |
|---|---|---|---|---|---|---|
| 32" x 80" | BC@1.00 | Ea | 279.00 | 32.10 | 311.10 | 529.00 |
| 36" x 80" | BC@1.00 | Ea | 279.00 | 32.10 | 311.10 | 529.00 |

**Half lite prehung steel exterior doors with mini-blind.** 22" x 36" light. Adjustable sill. 24-gauge galvanized steel. Non-yellowing, white, warp-resistant, paintable, high-performance light frames. Brick mold applied.

| | Craft@Hrs | Unit | Material | Labor | Total | Sell |
|---|---|---|---|---|---|---|
| 32" x 80", 4-9/16" jamb | BC@1.00 | Ea | 261.00 | 32.10 | 293.10 | 498.00 |
| 36" x 80", 4-9/16" jamb | BC@1.00 | Ea | 261.00 | 32.10 | 293.10 | 498.00 |

| | Craft@Hrs | Unit | Material | Labor | Total | Sell |
|---|---|---|---|---|---|---|

**Full lite prehung steel exterior doors.** 22" x 64" light. 4-9/16" jamb. Adjustable sill. No brick mold. 24-gauge galvanized steel.

| | | | | | | |
|---|---|---|---|---|---|---|
| 32" x 80" | BC@1.00 | Ea | 235.00 | 32.10 | 267.10 | 454.00 |
| 36" x 80" | BC@1.00 | Ea | 235.00 | 32.10 | 267.10 | 454.00 |

**2-lite, 4-panel prehung steel doors.** 4-9/16" jamb. Adjustable sill. Ready-to-install door and jamb system. 24-gauge galvanized steel. Non-yellowing, white, warp-resistant, paintable, high-performance light frames. Brick mold applied.

| | | | | | | |
|---|---|---|---|---|---|---|
| 30" x 80" | BC@1.00 | Ea | 194.00 | 32.10 | 226.10 | 384.00 |
| 32" x 80" | BC@1.00 | Ea | 194.00 | 32.10 | 226.10 | 384.00 |
| 36" x 80" | BC@1.00 | Ea | 194.00 | 32.10 | 226.10 | 384.00 |

**9-lite prehung steel exterior doors.** Insulating glass with internal 9-lite grille. Adjustable sill. Ready-to-install door and jamb system. 24-gauge galvanized steel. Non-yellowing, white, warp-resistant, paintable, high-performance light frames. Brick mold applied.

| | | | | | | |
|---|---|---|---|---|---|---|
| 32" x 80", 4-9/16" jamb | BC@1.00 | Ea | 204.00 | 32.10 | 236.10 | 401.00 |
| 32" x 80", 6-9/16" jamb | BC@1.00 | Ea | 183.00 | 32.10 | 215.10 | 366.00 |

**11-lite prehung steel exterior doors.** 24-gauge, high-profile, deep embossed steel skins. Impact-resistant glass. Double-bored for deadbolt and doorknob. Brass adjustable threshold. 4-9/16" primed frame. No brick mold.

| | | | | | | |
|---|---|---|---|---|---|---|
| 32" x 80" left | BC@1.00 | Ea | 237.00 | 32.10 | 269.10 | 457.00 |
| 36" x 80" right | BC@1.00 | Ea | 237.00 | 32.10 | 269.10 | 457.00 |

**15-lite prehung steel exterior doors.** Multi-paned insulated glass. 24-gauge hot-dipped galvanized steel. Polyurethane foam core. Compression weatherstrip. Thermal-break construction. No brick mold. Adjustable seal. Aluminum sill with composite adjustable threshold.

| | | | | | | |
|---|---|---|---|---|---|---|
| 32" x 80" | BC@1.00 | Ea | 215.00 | 32.10 | 247.10 | 420.00 |
| 36" x 80" | BC@1.00 | Ea | 238.00 | 32.10 | 270.10 | 459.00 |

**Flush ventlite prehung steel exterior doors.** No brick mold. 4-9/16" primed jamb. 24-gauge galvanized steel. Foam core. Magnetic weatherstrip. Triple seal sweep. Thermal-break threshold. Bored for lockset and deadbolt. 20-minute fire rating. 10-year limited warranty.

| | | | | | | |
|---|---|---|---|---|---|---|
| 30" x 80" | BC@1.00 | Ea | 215.00 | 32.10 | 247.10 | 420.00 |
| 32" x 80" | BC@1.00 | Ea | 215.00 | 32.10 | 247.10 | 420.00 |
| 36" x 80" | BC@1.00 | Ea | 215.00 | 32.10 | 247.10 | 420.00 |

## Decorative Steel Entry Doors

**Decorative 3/4 oval lite prehung steel entry doors.** 24-gauge galvanized steel. Non-yellowing, white, warp-resistant, paintable, high-performance light frames. Decorative impact-resistant glass. Rot-resistant frame. Polyurethane insulated core. With brick mold. Adjustable sill.

| | | | | | | |
|---|---|---|---|---|---|---|
| 36" x 80", brass caming, 4-9/16" jamb | BC@1.00 | Ea | 312.00 | 32.10 | 344.10 | 585.00 |
| 36" x 80", zinc caming, 4-9/16" jamb | BC@1.00 | Ea | 301.00 | 32.10 | 333.10 | 566.00 |
| 36" x 80", zinc caming, 6-9/16" jamb | BC@1.00 | Ea | 310.00 | 32.10 | 342.10 | 582.00 |

**Camber top prehung steel entry doors.** 4-9/16" jamb.

| | | | | | | |
|---|---|---|---|---|---|---|
| 36" x 80", with brick mold | BC@1.00 | Ea | 269.00 | 32.10 | 301.10 | 512.00 |
| 36" x 80", no brick mold | BC@1.00 | Ea | 269.00 | 32.10 | 301.10 | 512.00 |

| | Craft@Hrs | Unit | Material | Labor | Total | Sell |
|---|---|---|---|---|---|---|

**Decorative central arch lite prehung steel entry door.** Decorative laminated glass with brass caming to match hinge and sill finish. 24-gauge galvanized steel. Polyurethane core. Adjustable thermal-break sill. Triple sweep compression weatherstripping. Non-yellowing and warp-resistant light frame. Factory primed.
36" x 80", with brick mold,

| | Craft@Hrs | Unit | Material | Labor | Total | Sell |
|---|---|---|---|---|---|---|
| 4-9/16" jamb | BC@1.00 | Ea | 315.00 | 32.10 | 347.10 | 590.00 |

36" x 80", with no brick mold,

| | Craft@Hrs | Unit | Material | Labor | Total | Sell |
|---|---|---|---|---|---|---|
| 4-9/16" jamb | BC@1.00 | Ea | 312.00 | 32.10 | 344.10 | 585.00 |

**Decorative fan lite prehung steel entry doors.** Polyurethane core. Adjustable thermal-break sill with composite saddle. 24-gauge galvanized steel. Laminated glass. Brass caming matches hinge and sill. Triple sweep and compression weatherstrip. Non-yellowing and warp-resistant light frame. Factory primed. Includes brick mold.

| | Craft@Hrs | Unit | Material | Labor | Total | Sell |
|---|---|---|---|---|---|---|
| 32" x 80" | BC@1.00 | Ea | 225.00 | 32.10 | 257.10 | 437.00 |
| 36" x 80" | BC@1.00 | Ea | 217.00 | 32.10 | 249.10 | 423.00 |

**Decorative half lite prehung steel entry doors.** 22" x 36" laminated safety glass. Brass caming matches sill and hinges. 24-gauge galvanized insulated steel.

| | Craft@Hrs | Unit | Material | Labor | Total | Sell |
|---|---|---|---|---|---|---|
| 36" x 80", no brick mold | BC@1.00 | Ea | 290.00 | 32.10 | 322.10 | 548.00 |
| 36" x 80", with brick mold | BC@1.00 | Ea | 290.00 | 32.10 | 322.10 | 548.00 |

## Fiberglass Prehung Entry Doors

**6-panel fiberglass prehung exterior doors.** Polyurethane core. PVC stiles and rails. Double bore. 4-9/16" prefinished jambs. Adjustable thermal-break brass threshold. Weatherstripped. With brick mold.

| | Craft@Hrs | Unit | Material | Labor | Total | Sell |
|---|---|---|---|---|---|---|
| 32" x 80" | BC@1.00 | Ea | 158.00 | 32.10 | 190.10 | 323.00 |
| 36" x 80" | BC@1.00 | Ea | 158.00 | 32.10 | 190.10 | 323.00 |

**Fan lite prehung light oak fiberglass entry doors.** With brick mold. Factory prefinished. Triple pane insulated glass with brass caming. Adjustable brass thermal-break threshold and fully weatherstripped. 4-9/16" jamb.

| | Craft@Hrs | Unit | Material | Labor | Total | Sell |
|---|---|---|---|---|---|---|
| 36" x 80" | BC@1.00 | Ea | 377.00 | 32.10 | 409.10 | 695.00 |

**3/4 oval lite prehung fiberglass entry doors.** With brick mold. Ready to finish.

| | Craft@Hrs | Unit | Material | Labor | Total | Sell |
|---|---|---|---|---|---|---|
| 36" x 80", 4-9/16" jamb | BC@1.00 | Ea | 323.00 | 32.10 | 355.10 | 604.00 |
| 36" x 80", 6-9/16" jamb | BC@1.00 | Ea | 345.00 | 32.10 | 377.10 | 641.00 |

**3/4 oval lite prehung light oak fiberglass doors.** With brick mold. Prefinished. Polyurethane core. Triple pane insulated glass. Glue chip and clear 30" x 18" beveled glass surrounded by brass caming. Double bore. Extended wood lock block. 4-9/16" primed jamb. Thermal-break brass threshold with adjustable oak cap.

| | Craft@Hrs | Unit | Material | Labor | Total | Sell |
|---|---|---|---|---|---|---|
| 36" x 80", 4-9/16" jamb | BC@1.00 | Ea | 517.32 | 32.10 | 549.42 | 934.00 |
| 36" x 80", 6-9/16" jamb | BC@1.00 | Ea | 538.92 | 32.10 | 571.02 | 971.00 |

**Full height oval lite prehung medium oak fiberglass entry doors.**

| | Craft@Hrs | Unit | Material | Labor | Total | Sell |
|---|---|---|---|---|---|---|
| 36" x 80", 4-9/16" jamb | BC@1.00 | Ea | 593.00 | 32.10 | 625.10 | 1,060.00 |
| 36" x 80", 6-9/16" jamb | BC@1.00 | Ea | 615.00 | 32.10 | 647.10 | 1,100.00 |

|  | Craft@Hrs | Unit | Material | Labor | Total | Sell |
|---|---|---|---|---|---|---|

**Center arch lite medium oak prehung fiberglass doors.** Factory prefinished. Triple pane insulated glass with brass caming. Adjustable brass thermal-break threshold and fully weatherstripped. 4-9/16" jamb except as noted. PVC stiles and rails.

| | | | | | | |
|---|---|---|---|---|---|---|
| 36" x 80", no brick mold | BC@1.00 | Ea | 455.00 | 32.10 | 487.10 | 828.00 |
| 36" x 80", with brick mold | BC@1.00 | Ea | 463.00 | 32.10 | 495.10 | 842.00 |
| 36" x 80", with brick mold, 6-9/16" jamb | BC@1.00 | Ea | 485.00 | 32.10 | 517.10 | 879.00 |

**Fan lite prehung fiberglass entry doors.** With brick mold. Ready to finish. Insulated glass. Adjustable thermal-break brass threshold. 4-9/16" jamb except as noted.

| | | | | | | |
|---|---|---|---|---|---|---|
| 32" x 80" | BC@1.00 | Ea | 269.00 | 32.10 | 301.10 | 512.00 |
| 36" x 80" | BC@1.00 | Ea | 269.00 | 32.10 | 301.10 | 512.00 |
| 36" x 80", 6-9/16" jamb | BC@1.00 | Ea | 291.00 | 32.10 | 323.10 | 549.00 |

**9-lite fiberglass prehung exterior doors.** Inward swing. With brick mold. Smooth face.

| | | | | | | |
|---|---|---|---|---|---|---|
| 32" x 80" | BC@1.00 | Ea | 224.00 | 32.10 | 256.10 | 435.00 |
| 36" x 80" | BC@1.00 | Ea | 224.00 | 32.10 | 256.10 | 435.00 |

**15-lite smooth prehung fiberglass entry doors.** No brick mold. Factory prefinished. Left or right hinge. Triple pane, insulated glass with brass caming. Adjustable brass thermal-break threshold and fully weatherstripped. 4-9/16" jamb.

| | | | | | | |
|---|---|---|---|---|---|---|
| 32" x 80" | BC@1.00 | Ea | 247.00 | 32.10 | 279.10 | 474.00 |
| 32" x 80" | BC@1.00 | Ea | 247.00 | 32.10 | 279.10 | 474.00 |

**Full lite prehung smooth fiberglass doors.** With brick mold. Inward swing.

| | | | | | | |
|---|---|---|---|---|---|---|
| 36" x 80", 4-9/16" jamb | BC@1.00 | Ea | 539.00 | 32.10 | 571.10 | 971.00 |
| 36" x 80", 6-9/16" jamb | BC@1.00 | Ea | 561.00 | 32.10 | 593.10 | 1,010.00 |

**Smooth prehung fiberglass ventlite exterior doors.** Ready to finish. No brick mold.

| | | | | | | |
|---|---|---|---|---|---|---|
| 32" x 80" | BC@1.00 | Ea | 228.00 | 32.10 | 260.10 | 442.00 |
| 36" x 80" | BC@1.00 | Ea | 228.00 | 32.10 | 260.10 | 442.00 |

## Entry Locksets

**Door lockset repairs.**

| | | | | | | |
|---|---|---|---|---|---|---|
| Adjust strike plate to align with latch bolt | BC@.200 | Ea | — | 6.42 | 6.42 | 10.90 |
| Remove cylinder or deadlock | BC@.137 | Ea | — | 4.40 | 4.40 | 7.48 |
| Reinstall same cylinder or deadlock | BC@.183 | Ea | — | 5.87 | 5.87 | 9.98 |
| Install cylinder lock and strike in pre-bored door | BC@.250 | Ea | — | 8.02 | 8.02 | 13.60 |
| Bore door, install cylinder lock, strike | BC@.750 | Ea | — | 24.10 | 24.10 | 41.00 |
| Remove and replace cylinder or deadlock in existing bore | BC@.300 | Ea | — | 9.63 | 9.63 | 16.40 |
| Remove and replace combination cylinder lock and deadbolt in existing bore | BC@1.52 | Ea | — | 48.80 | 48.80 | 83.00 |
| Remove mortise lock | BC@.146 | Ea | — | 4.69 | 4.69 | 7.97 |
| Install mortise lock | BC@1.31 | Ea | — | 42.00 | 42.00 | 71.40 |
| Remove and replace mortise lock | BC@1.45 | Ea | — | 46.50 | 46.50 | 79.10 |

**Keyed entry locksets.** Front or back door. Keyed exterior, turn button interior. Labor cost assumes the door is already bored for a lockset.

| | | | | | | |
|---|---|---|---|---|---|---|
| Antique brass | BC@.250 | Ea | 18.50 | 8.02 | 26.52 | 45.10 |
| Polished brass | BC@.250 | Ea | 12.80 | 8.02 | 20.82 | 35.40 |
| Satin chrome | BC@.250 | Ea | 14.00 | 8.02 | 22.02 | 37.40 |

| | Craft@Hrs | Unit | Material | Labor | Total | Sell |
|---|---|---|---|---|---|---|

**Keyed entry locksets.** Meets ANSI Grade 2 standards for residential security. One-piece knob. Triple option faceplate fits square, radius-round, and drive-in door preps. 10-year finish warranty, lifetime mechanical warranty. Labor cost assumes the door is already bored for a lockset.

| | Craft@Hrs | Unit | Material | Labor | Total | Sell |
|---|---|---|---|---|---|---|
| Antique brass | BC@.250 | Ea | 23.90 | 8.02 | 31.92 | 54.30 |
| Antique pewter | BC@.250 | Ea | 31.70 | 8.02 | 39.72 | 67.50 |
| Bright brass | BC@.250 | Ea | 23.80 | 8.02 | 31.82 | 54.10 |
| Satin chrome | BC@.250 | Ea | 21.40 | 8.02 | 29.42 | 50.00 |

**Lever locksets.** Dual torque springs eliminate wobble. Labor cost assumes the door is already bored for a lockset.

| | | | | | | |
|---|---|---|---|---|---|---|
| Bright brass | BC@.250 | Ea | 45.00 | 8.02 | 53.02 | 90.10 |
| Satin nickel | BC@.250 | Ea | 54.30 | 8.02 | 62.32 | 106.00 |

**Egg knob entry locksets.** Labor cost assumes the door is already bored for a lockset.

| | | | | | | |
|---|---|---|---|---|---|---|
| Satin nickel finish | BC@.250 | Ea | 44.40 | 8.02 | 52.42 | 89.10 |

**Egg knob entry locksets — solid forged brass.** 5-pin cylinder. Full lip strike. Adjustable backset. Labor cost assumes the door is already bored for a lockset.

| | | | | | | |
|---|---|---|---|---|---|---|
| Polished brass | BC@.250 | Ea | 75.60 | 8.02 | 83.62 | 142.00 |

**Colonial knob entry locksets.** Labor cost assumes the door is already bored for a lockset and frame is already bored for a strike plate.

| | | | | | | |
|---|---|---|---|---|---|---|
| Polished brass | BC@.250 | Ea | 74.60 | 8.02 | 82.62 | 140.00 |

**Double cylinder entry door handle set.** Forged brass handle with thumb latch. Separate grade 1 security double cylinder deadbolt. Labor cost assumes the door is already bored for a lockset, but frame is not.

| | | | | | | |
|---|---|---|---|---|---|---|
| Satin nickel finish | BC@.820 | Ea | 91.00 | 26.30 | 117.30 | 199.00 |
| Bronze finish | BC@.820 | Ea | 95.00 | 26.30 | 121.30 | 206.00 |

**Deadbolt and lockset combo.** Includes keyed alike entry knob and grade 2 security deadbolt with 4 keys. Labor cost assumes the door is already bored for a lockset.

| | | | | | | |
|---|---|---|---|---|---|---|
| Brass, single cylinder | BC@.750 | Ea | 44.80 | 24.10 | 68.90 | 117.00 |
| Brass, double cylinder | BC@.750 | Ea | 53.20 | 24.10 | 77.30 | 131.00 |

**Deadbolts.** All metal and brass. ANSI Grade 2 rating. 1" throw deadbolt. Includes 2 keys. Polished brass.

| | | | | | | |
|---|---|---|---|---|---|---|
| Single cylinder | BC@.500 | Ea | 20.60 | 16.00 | 36.60 | 62.20 |
| Double cylinder | BC@.500 | Ea | 28.80 | 16.00 | 44.80 | 76.20 |

## Screen Doors

**Install wood screen door.** Complete with three hinge butts, pneumatic door closer, latch lock and guard. Labor cost includes adjustments to existing screen door frame.

| | | | | | | |
|---|---|---|---|---|---|---|
| Per screen door | BC@2.17 | Ea | — | 69.60 | 69.60 | 118.00 |

**Install aluminum or vinyl screen door.** Complete with pneumatic door closer, chain and spring restrainer and latch lock.

| | | | | | | |
|---|---|---|---|---|---|---|
| Per screen door | BC@1.35 | Ea | — | 43.30 | 43.30 | 73.60 |

**Remove and replace wood screen door.** Using existing hinge butts. With pneumatic door closer, latch lock and guard. Labor cost assumes new door is an exact replacement for the old door.

| | | | | | | |
|---|---|---|---|---|---|---|
| Per screen door | BC@1.00 | Ea | — | 32.10 | 32.10 | 54.60 |

**Remove and replace aluminum or vinyl screen door.** Complete with pneumatic door closer, chain and spring restrainer and latch lock.

| | | | | | | |
|---|---|---|---|---|---|---|
| Per screen door | BC@1.97 | Ea | — | 63.20 | 63.20 | 107.00 |

|  | Craft@Hrs | Unit | Material | Labor | Total | Sell |
|---|---|---|---|---|---|---|
| **Remove wood and replace with aluminum or vinyl screen door.** With pneumatic door closer, chain spring retainer and latch lock. | | | | | | |
| Per screen door | BC@1.70 | Ea | — | 54.60 | 54.60 | 92.80 |
| **Screen door hardware.** | | | | | | |
| Remove and replace pneumatic closer | BC@.296 | Ea | — | 9.50 | 9.50 | 16.20 |
| Remove and replace latch lock | BC@.185 | Ea | — | 5.94 | 5.94 | 10.10 |
| Remove and replace chain spring retainer | BC@.203 | Ea | — | 6.51 | 6.51 | 11.10 |
| Remove and replace door bottom wiper | BC@.187 | Ea | — | 6.00 | 6.00 | 10.20 |
| **Wood screen doors.** 3-3/4"-wide stile and rail with 7-1/4" bottom rail. Removable screen. Mortise and tenon solid construction. Wide center push bar. Ready to finish. | | | | | | |
| 32" x 80" | BC@2.17 | Ea | 66.90 | 69.60 | 136.50 | 232.00 |
| 36" x 80" | BC@2.17 | Ea | 69.00 | 69.60 | 138.60 | 236.00 |
| **T-style wood screen doors.** 3-3/4"-wide stile and rail with 7-1/4" bottom rail. Removable screen. Mortise and tenon solid construction. All joints sealed with water-resistant adhesive. Wide center push bar. | | | | | | |
| 30" x 80" | BC@2.17 | Ea | 21.30 | 69.60 | 90.90 | 155.00 |
| 32" x 80" | BC@2.17 | Ea | 21.60 | 69.60 | 91.20 | 155.00 |
| 36" x 80" | BC@2.17 | Ea | 21.60 | 69.60 | 91.20 | 155.00 |
| **Metal screen doors.** Includes hardware and fiberglass screening. 1-piece steel frame. Heavy-duty pneumatic closer. 5" kickplate. | | | | | | |
| 32" x 80" | BC@1.35 | Ea | 37.00 | 43.30 | 80.30 | 137.00 |
| 36" x 80" | BC@1.35 | Ea | 37.00 | 43.30 | 80.30 | 137.00 |
| **Five-bar solid vinyl screen doors.** Accepts standard screen door hardware. | | | | | | |
| 32" x 80" | BC@1.35 | Ea | 80.00 | 43.30 | 123.30 | 210.00 |
| 36" x 80" | BC@1.35 | Ea | 80.00 | 43.30 | 123.30 | 210.00 |
| **Solid pine screen doors.** Right or left side installation. 1-1/4" thick. Unfinished. Hardware included. | | | | | | |
| 32" x 80" | BC@1.35 | Ea | 160.00 | 43.30 | 203.30 | 346.00 |
| 36" x 80" | BC@1.35 | Ea | 162.00 | 43.30 | 205.30 | 349.00 |
| **Steel security screen doors.** Class I rated to withstand 300 pounds. All welded steel construction for strength and durability. Double lockbox with extension plate. Baked-on powder coating. | | | | | | |
| 36" x 80" | BC@1.35 | Ea | 69.00 | 43.30 | 112.30 | 191.00 |
| **Replacement aluminum screen doors.** Includes all hardware. 2-piece protective grille. 3" x 1-1/4" heavy-duty rails. | | | | | | |
| 36" wide | BC@1.35 | Ea | 89.00 | 43.30 | 132.30 | 225.00 |
| **Replacement roll-formed screen doors.** 7/8" 1-piece unitized roll-formed frame, mechanically secured corners, epoxy-painted finish over electro-galvanizing, 5" heavy-duty kickplate. | | | | | | |
| 32" x 80", gray | BC@1.35 | Ea | 53.00 | 43.30 | 96.30 | 164.00 |
| 36" x 80", white | BC@1.35 | Ea | 87.50 | 43.30 | 130.80 | 222.00 |
| 6" x 80", bronze | BC@1.35 | Ea | 83.00 | 43.30 | 126.30 | 215.00 |
| **Replacement steel screen doors.** Powder-coated finish. With hardware and closer. | | | | | | |
| 36" x 80" | BC@1.35 | Ea | 69.00 | 43.30 | 112.30 | 191.00 |

| | Craft@Hrs | Unit | Material | Labor | Total | Sell |
|---|---|---|---|---|---|---|
| **Storm Doors** | | | | | | |
| **Remove existing storm door.** | | | | | | |
| Wood or aluminum | BC@.353 | Ea | — | 11.30 | 11.30 | 19.20 |
| Remove and replace storm door | BC@1.80 | Ea | — | 57.80 | 57.80 | 98.30 |

**Self-storing aluminum storm and screen doors.** Pneumatic closer and sweep. Tempered safety glass. Maintenance-free finish. Push-button hardware. 1" x 2-1/8" frame size. Bronze or white. Includes screen.

| | Craft@Hrs | Unit | Material | Labor | Total | Sell |
|---|---|---|---|---|---|---|
| 30" x 80" x 1-1/4" | BC@1.45 | Ea | 96.00 | 46.50 | 142.50 | 242.00 |
| 32" x 80" x 1-1/4" | BC@1.45 | Ea | 95.00 | 46.50 | 141.50 | 241.00 |
| 36" x 80" x 1-1/4" | BC@1.45 | Ea | 95.00 | 46.50 | 141.50 | 241.00 |

**Self-storing vinyl-covered wood storm doors.** White vinyl-covered wood core. Push-button handle set. Self-storing window and screen. Single black closer.

| | Craft@Hrs | Unit | Material | Labor | Total | Sell |
|---|---|---|---|---|---|---|
| 30" x 80" | BC@1.45 | Ea | 102.00 | 46.50 | 148.50 | 252.00 |
| 32" x 80" | BC@1.45 | Ea | 102.00 | 46.50 | 148.50 | 252.00 |
| 34" x 80" | BC@1.45 | Ea | 102.00 | 46.50 | 148.50 | 252.00 |
| 36" x 80" | BC@1.45 | Ea | 102.00 | 46.50 | 148.50 | 252.00 |

**Store-in-Door™ storm doors.** 1-1/2"-thick polypropylene frame. Reversible hinge. Slide window or screen, completely concealed when not in use. Brass-finished keylock hardware system with separate color-matched deadbolt. Triple-fin door sweep. 2 color-matched closers. Full-length piano hinge. Flexible weather seal. White.

| | Craft@Hrs | Unit | Material | Labor | Total | Sell |
|---|---|---|---|---|---|---|
| 30" x 80", crossbuck style | BC@1.45 | Ea | 258.00 | 46.50 | 304.50 | 518.00 |
| 32" x 80", crossbuck style | BC@1.45 | Ea | 258.00 | 46.50 | 304.50 | 518.00 |
| 36" x 80", crossbuck style | BC@1.45 | Ea | 258.00 | 46.50 | 304.50 | 518.00 |
| 30" x 80", traditional style | BC@1.45 | Ea | 258.00 | 46.50 | 304.50 | 518.00 |
| 36" x 80", traditional style | BC@1.45 | Ea | 258.00 | 46.50 | 304.50 | 518.00 |

**Aluminum and wood storm doors.** Aluminum over wood core. Self-storing window and screen. Black lever handle set. Separate deadbolt. Single black closer. White finish. Traditional panel style.

| | Craft@Hrs | Unit | Material | Labor | Total | Sell |
|---|---|---|---|---|---|---|
| 30" x 80" | BC@1.45 | Ea | 144.00 | 46.50 | 190.50 | 324.00 |
| 32" x 80" | BC@1.45 | Ea | 144.00 | 46.50 | 190.50 | 324.00 |
| 36" x 80" | BC@1.45 | Ea | 144.00 | 46.50 | 190.50 | 324.00 |

**Triple-track storm doors.** Low-maintenance aluminum over solid wood core. Ventilates from top, bottom or both. Solid brass exterior handle. Color-matched interior handle. Separate deadbolt. 80" high. White, almond or bronze.

| | Craft@Hrs | Unit | Material | Labor | Total | Sell |
|---|---|---|---|---|---|---|
| 30" x 80" | BC@1.45 | Ea | 199.00 | 46.50 | 245.50 | 417.00 |
| 32" x 80" | BC@1.45 | Ea | 225.00 | 46.50 | 271.50 | 462.00 |
| 34" x 80" | BC@1.45 | Ea | 199.00 | 46.50 | 245.50 | 417.00 |
| 36" x 80" | BC@1.45 | Ea | 199.00 | 46.50 | 245.50 | 417.00 |

**All-season storm doors.** 1-1/4" thick frame. 1-piece construction with window molding built in. Self-storing tempered safety glass and screen for easy ventilation. Includes solid-brass lockset with deadbolt security. Brass-finished sweep for tight threshold seal. Color-matched screw covers, and two closers. Reversible hinge opens to 180 degrees. White, almond or sandstone.

| | Craft@Hrs | Unit | Material | Labor | Total | Sell |
|---|---|---|---|---|---|---|
| 32" x 80" | BC@1.45 | Ea | 208.00 | 46.50 | 254.50 | 433.00 |
| 36" x 80" | BC@1.45 | Ea | 208.00 | 46.50 | 254.50 | 433.00 |

**All-season triple track storm doors.** Ventilates from top or bottom, or both. 1" composite frame. Brass handle set with deadbolt security. Wood-grain finish. Color-matched screws. 1-piece construction with window molding built in. Self-storing window and screen. White, almond or sandstone.

| | Craft@Hrs | Unit | Material | Labor | Total | Sell |
|---|---|---|---|---|---|---|
| 32" x 80" | BC@1.45 | Ea | 215.00 | 46.50 | 261.50 | 445.00 |
| 36" x 80" | BC@1.45 | Ea | 215.00 | 46.50 | 261.50 | 445.00 |

| | Craft@Hrs | Unit | Material | Labor | Total | Sell |
|---|---|---|---|---|---|---|

**Colonial triple-track storm doors.** 1-piece composite frame. 12-lite grille. Vents from top, bottom, or both. Solid-brass handle set with deadbolt security. Brass-finished sweep ensures tight seal across the entire threshold. Wood-grain finish. White, almond or sandstone.

| | Craft@Hrs | Unit | Material | Labor | Total | Sell |
|---|---|---|---|---|---|---|
| 32" x 80" | BC@1.45 | Ea | 276.00 | 46.50 | 322.50 | 548.00 |
| 36" x 80" | BC@1.45 | Ea | 266.00 | 46.50 | 312.50 | 531.00 |

**Full-view brass all-season storm doors.** Water-diverting rain cap. Brass-finished, color-matched sweep. One closer. Window and insect screen snap in and out. Reversible. Opens 180 degrees when needed. Separate deadbolt for added security. Tempered glass. Heavy-gauge one-piece 1" aluminum construction with reinforced corners and foam insulation.

| | Craft@Hrs | Unit | Material | Labor | Total | Sell |
|---|---|---|---|---|---|---|
| 36" x 80", green | BC@1.45 | Ea | 322.00 | 46.50 | 368.50 | 626.00 |
| 36" x 80", white | BC@1.45 | Ea | 257.00 | 46.50 | 303.50 | 516.00 |

**Full-view aluminum storm doors.** 1-1/2"-thick heavy-gauge aluminum frame with foam insulation. Extra perimeter weatherstripping and water-diverting rain cap. Solid brass handle set and double-throw deadbolt. Brass-finished sweep. Full-length, piano-style hinge. Reversible hinge for left- or right-side entry door handle. Two heavy-duty, color-matched closers. Solid brass perimeter locks. Insect screen. White, brown, green or almond.

| | Craft@Hrs | Unit | Material | Labor | Total | Sell |
|---|---|---|---|---|---|---|
| 30" x 80" | BC@1.45 | Ea | 268.00 | 46.50 | 314.50 | 535.00 |
| 32" x 80" | BC@1.45 | Ea | 269.00 | 46.50 | 315.50 | 536.00 |
| 36" x 80" | BC@1.45 | Ea | 269.00 | 46.50 | 315.50 | 536.00 |

**Full-view woodcore storm doors.** 1-1/2-inch thick, heavy-gauge aluminum frame over foam insulation. Perimeter weatherstripping and water-diverting rain cap. Full-length, piano-style hinge. Reversible hinge for left- or right-side entry door handle. Two heavy-duty closers.

| | Craft@Hrs | Unit | Material | Labor | Total | Sell |
|---|---|---|---|---|---|---|
| 30" x 80", White | BC@1.45 | Ea | 268.00 | 46.50 | 314.50 | 535.00 |
| 32" x 80", White | BC@1.45 | Ea | 269.00 | 46.50 | 315.50 | 536.00 |
| 36" x 80", White | BC@1.45 | Ea | 272.00 | 46.50 | 318.50 | 541.00 |

## Security Doors

**Security storm doors.** 16-gauge steel. Brass deadbolt lock. Tempered safety glass. Interchangeable screen panel. Heavy-gauge all-welded steel frame with mitered top. Double weatherstripped jamb and sweep. Tamper-resistant hinges with concealed interior screws. Heavy-duty pneumatic closer and safety wind chain. Prehung with reversible hinge.

| | Craft@Hrs | Unit | Material | Labor | Total | Sell |
|---|---|---|---|---|---|---|
| 30" x 80" | BC@1.45 | Ea | 323.00 | 46.50 | 369.50 | 628.00 |
| 32" x 80" | BC@1.45 | Ea | 323.00 | 46.50 | 369.50 | 628.00 |
| 36" x 80" | BC@1.45 | Ea | 323.00 | 46.50 | 369.50 | 628.00 |

**Class II security doors.** Rated to withstand 400 pounds. Double lockbox with extension plate. Decorative geometric ornaments. All welded steel construction. Expanded, galvanized metal screen. 3/4" square frame design.

| | Craft@Hrs | Unit | Material | Labor | Total | Sell |
|---|---|---|---|---|---|---|
| 32" x 80" | BC@1.45 | Ea | 114.00 | 46.50 | 160.50 | 273.00 |
| 36" x 80" | BC@1.45 | Ea | 108.00 | 46.50 | 154.50 | 263.00 |

**Heavy-duty security door with screen.** Right- or left-hand exterior mounting. All welded steel construction. Baked-on powder coating. Double lockbox with tamper-resistant extension plate. Uses standard 2-3/8" backset lock (not included). Galvanized perforated metal security screen. 1" x 1" door frame. 1" x 1-1/2" jambs, prehung on hinge side.

| | Craft@Hrs | Unit | Material | Labor | Total | Sell |
|---|---|---|---|---|---|---|
| 32" x 80" | BC@1.45 | Ea | 138.00 | 46.50 | 184.50 | 314.00 |
| 36" x 80" | BC@1.45 | Ea | 140.00 | 46.50 | 186.50 | 317.00 |

**Paradise style security door.** Five semi-concealed hinges. Heavy-duty deadbolt. All steel frame.

| | Craft@Hrs | Unit | Material | Labor | Total | Sell |
|---|---|---|---|---|---|---|
| 30" x 80" | BC@1.45 | Ea | 406.00 | 46.50 | 452.50 | 769.00 |
| 32" x 80" | BC@1.45 | Ea | 406.00 | 46.50 | 452.50 | 769.00 |
| 36" x 80" | BC@1.45 | Ea | 406.00 | 46.50 | 452.50 | 769.00 |

**Folding security gates.** Riveted 3/4" top and bottom channel. Gate pivots and folds flat to one side. Provides air circulation and positive visibility.

| | Craft@Hrs | Unit | Material | Labor | Total | Sell |
|---|---|---|---|---|---|---|
| 48" max width, 79" high | BC@1.25 | Ea | 118.00 | 40.10 | 158.10 | 269.00 |

| | Craft@Hrs | Unit | Material | Labor | Total | Sell |
|---|---|---|---|---|---|---|

## Interior Slab Doors

**Cut wall and frame interior door opening, install slab door and trim.** Includes header, trimmer studs, interior flush lauan slab door ($25.20), jamb, two hinges, stop, casing two sides, and passage lockset ($28). Add the cost of debris removal, painting, and floor and wall finish, as needed.

| | Craft@Hrs | Unit | Material | Labor | Total | Sell |
|---|---|---|---|---|---|---|
| 32" to 36" slab interior door | BC@6.30 | Ea | 153.00 | 202.00 | 355.00 | 604.00 |

**Install interior slab door in an existing framed and cased opening.** Includes minor repair to the jamb (blind Dutchman), two hinges, interior flush lauan slab door ($25.20), stop, and passage lockset ($28).

| | | | | | | |
|---|---|---|---|---|---|---|
| 28" to 36" x 80" door | BC@2.93 | Ea | 79.00 | 94.00 | 173.00 | 294.00 |

**Remove and replace interior door casing, jamb and stop.** Includes jamb, stop, and casing one side only. Per door opening.

| | | | | | | |
|---|---|---|---|---|---|---|
| Door opening to 36" x 80" | BC@1.20 | Ea | 65.40 | 38.50 | 103.90 | 177.00 |

**Remove and replace interior slab door, on existing hinges.** Includes interior flush lauan slab door ($25.20), and passage lockset ($28).

| | | | | | | |
|---|---|---|---|---|---|---|
| 36" x 80" exterior door | BC@2.00 | Ea | 53.20 | 64.20 | 117.40 | 200.00 |

**Install interior slab door in an existing cased opening.** Includes interior flush lauan slab door ($25.20), two hinges and stop. Add the cost of additional work to the jamb (if needed), installing the lockset and strike, setting or adjusting the stop, and setting the door casing.

| | | | | | | |
|---|---|---|---|---|---|---|
| 36" x 80" interior door | BC@1.15 | Ea | 50.80 | 36.90 | 87.70 | 149.00 |

**Flush hardboard interior doors.** Hollow core. Bored for lockset. Ready to paint or stain. 1-3/8" thick.

| | | | | | | |
|---|---|---|---|---|---|---|
| 24" x 80" | BC@1.15 | Ea | 22.00 | 36.90 | 58.90 | 100.00 |
| 28" x 80" | BC@1.15 | Ea | 23.50 | 36.90 | 60.40 | 103.00 |
| 30" x 80" | BC@1.15 | Ea | 24.50 | 36.90 | 61.40 | 104.00 |
| 32" x 80" | BC@1.15 | Ea | 25.60 | 36.90 | 62.50 | 106.00 |
| 36" x 80" | BC@1.15 | Ea | 27.00 | 36.90 | 63.90 | 109.00 |

**Flush birch interior doors.** Hollow core. 1-3/8" thick. Wood veneer. Stainable and paintable. All wood stile and rail.

| | | | | | | |
|---|---|---|---|---|---|---|
| 18" x 80" | BC@1.15 | Ea | 26.50 | 36.90 | 63.40 | 108.00 |
| 24" x 80" | BC@1.15 | Ea | 29.00 | 36.90 | 65.90 | 112.00 |
| 28" x 80" | BC@1.15 | Ea | 31.70 | 36.90 | 68.60 | 117.00 |
| 30" x 80" | BC@1.15 | Ea | 33.00 | 36.90 | 69.90 | 119.00 |
| 32" x 80" | BC@1.15 | Ea | 34.90 | 36.90 | 71.80 | 122.00 |
| 36" x 80" | BC@1.15 | Ea | 36.60 | 36.90 | 73.50 | 125.00 |

**Flush lauan interior doors.** Hollow core. 1-3/8" thick. Wood veneer. Stainable and paintable. All wood stile and rail.

| | | | | | | |
|---|---|---|---|---|---|---|
| 18" x 80" | BC@1.15 | Ea | 19.50 | 36.90 | 56.40 | 95.90 |
| 24" x 80" | BC@1.15 | Ea | 21.00 | 36.90 | 57.90 | 98.40 |
| 28" x 80" | BC@1.15 | Ea | 22.50 | 36.90 | 59.40 | 101.00 |
| 30" x 80" | BC@1.15 | Ea | 23.50 | 36.90 | 60.40 | 103.00 |
| 32" x 80" | BC@1.15 | Ea | 23.30 | 36.90 | 60.20 | 102.00 |
| 36" x 80" | BC@1.15 | Ea | 26.70 | 36.90 | 63.60 | 108.00 |

**Flush oak interior doors.** Hollow core. 1-3/8" thick. Wood veneer. Stainable and paintable. All wood stile and rail.

| | | | | | | |
|---|---|---|---|---|---|---|
| 18" x 80" | BC@1.15 | Ea | 28.40 | 36.90 | 65.30 | 111.00 |
| 24" x 80" | BC@1.15 | Ea | 30.30 | 36.90 | 67.20 | 114.00 |
| 28" x 80" | BC@1.15 | Ea | 33.40 | 36.90 | 70.30 | 120.00 |
| 30" x 80" | BC@1.15 | Ea | 34.60 | 36.90 | 71.50 | 122.00 |
| 32" x 80" | BC@1.15 | Ea | 36.60 | 36.90 | 73.50 | 125.00 |
| 36" x 80" | BC@1.15 | Ea | 39.50 | 36.90 | 76.40 | 130.00 |

| | Craft@Hrs | Unit | Material | Labor | Total | Sell |
|---|---|---|---|---|---|---|

**Flush hardboard solid core interior doors.** Primed. 1-3/8" thick.

| | Craft@Hrs | Unit | Material | Labor | Total | Sell |
|---|---|---|---|---|---|---|
| 30" x 80" | BC@1.15 | Ea | 43.90 | 36.90 | 80.80 | 137.00 |
| 32" x 80" | BC@1.15 | Ea | 45.40 | 36.90 | 82.30 | 140.00 |
| 36" x 80" | BC@1.15 | Ea | 49.00 | 36.90 | 85.90 | 146.00 |

**Flush lauan solid core interior doors.** 1-3/8" thick. Stainable.

| | Craft@Hrs | Unit | Material | Labor | Total | Sell |
|---|---|---|---|---|---|---|
| 28" x 80" | BC@1.15 | Ea | 46.40 | 36.90 | 83.30 | 142.00 |
| 32" x 80" | BC@1.15 | Ea | 50.40 | 36.90 | 87.30 | 148.00 |
| 36" x 80" | BC@1.15 | Ea | 50.30 | 36.90 | 87.20 | 148.00 |
| 30" x 84" | BC@1.15 | Ea | 52.90 | 36.90 | 89.80 | 153.00 |
| 32" x 84" | BC@1.15 | Ea | 55.00 | 36.90 | 91.90 | 156.00 |
| 34" x 84" | BC@1.15 | Ea | 59.40 | 36.90 | 96.30 | 164.00 |
| 36" x 84" | BC@1.15 | Ea | 59.40 | 36.90 | 96.30 | 164.00 |

**Flush birch solid core interior doors.** 1-3/8" thick. Solid core.

| | Craft@Hrs | Unit | Material | Labor | Total | Sell |
|---|---|---|---|---|---|---|
| 24" x 80" | BC@1.15 | Ea | 76.40 | 36.90 | 113.30 | 193.00 |
| 32" x 80" | BC@1.15 | Ea | 59.00 | 36.90 | 95.90 | 163.00 |
| 36" x 80" | BC@1.15 | Ea | 62.40 | 36.90 | 99.30 | 169.00 |

**Flush hardboard interior doors with lite.** 1-3/8" thick.

| | Craft@Hrs | Unit | Material | Labor | Total | Sell |
|---|---|---|---|---|---|---|
| 30" x 80" x 1-3/8" | BC@1.15 | Ea | 99.20 | 36.90 | 136.10 | 231.00 |
| 32" x 80" x 1-3/8" | BC@1.15 | Ea | 105.00 | 36.90 | 141.90 | 241.00 |

**Hardboard 6-panel colonist interior doors.** Hollow core. Raised panels. 1-3/8" thick. Masonite high-density fiberboard with CraftMaster door facings. Embossed simulated woodgrain panels. Primed.

| | Craft@Hrs | Unit | Material | Labor | Total | Sell |
|---|---|---|---|---|---|---|
| 18" x 80" | BC@1.15 | Ea | 29.80 | 36.90 | 66.70 | 113.00 |
| 24" x 80" | BC@1.15 | Ea | 32.40 | 36.90 | 69.30 | 118.00 |
| 28" x 80" | BC@1.15 | Ea | 34.40 | 36.90 | 71.30 | 121.00 |
| 30" x 80" | BC@1.15 | Ea | 35.40 | 36.90 | 72.30 | 123.00 |
| 32" x 80" | BC@1.15 | Ea | 36.70 | 36.90 | 73.60 | 125.00 |
| 36" x 80" | BC@1.15 | Ea | 38.50 | 36.90 | 75.40 | 128.00 |

**Hardboard 6-panel molded face interior doors.** 1-3/8" thick. Molded and primed hardboard. Embossed simulated woodgrain. Hollow core. Ready to paint. Bored for lockset.

| | Craft@Hrs | Unit | Material | Labor | Total | Sell |
|---|---|---|---|---|---|---|
| 24" x 80" | BC@1.15 | Ea | 34.40 | 36.90 | 71.30 | 121.00 |
| 28" x 80" | BC@1.15 | Ea | 36.00 | 36.90 | 72.90 | 124.00 |
| 30" x 80" | BC@1.15 | Ea | 36.50 | 36.90 | 73.40 | 125.00 |
| 32" x 80" | BC@1.15 | Ea | 38.20 | 36.90 | 75.10 | 128.00 |
| 36" x 80" | BC@1.15 | Ea | 40.20 | 36.90 | 77.10 | 131.00 |

**Lauan 6-panel molded face interior doors.** Bored for latchset. 1-3/8" thick.

| | Craft@Hrs | Unit | Material | Labor | Total | Sell |
|---|---|---|---|---|---|---|
| 24" x 80" | BC@1.15 | Ea | 37.00 | 36.90 | 73.90 | 126.00 |
| 28" x 80" | BC@1.15 | Ea | 42.00 | 36.90 | 78.90 | 134.00 |
| 30" x 80" | BC@1.15 | Ea | 42.00 | 36.90 | 78.90 | 134.00 |
| 32" x 80" | BC@1.15 | Ea | 42.20 | 36.90 | 79.10 | 134.00 |
| 36" x 80" | BC@1.15 | Ea | 44.50 | 36.90 | 81.40 | 138.00 |

| | Craft@Hrs | Unit | Material | Labor | Total | Sell |
|---|---|---|---|---|---|---|

**Pine 6-panel molded face interior doors.** Radiata pine veneer. Primed. Hollow core. 1-3/8" thick. 3/4" double hip raised panels. Unfinished. Ready to paint, stain, or varnish.

| | Craft@Hrs | Unit | Material | Labor | Total | Sell |
|---|---|---|---|---|---|---|
| 24" x 80" | BC@1.15 | Ea | 45.00 | 36.90 | 81.90 | 139.00 |
| 30" x 80" | BC@1.15 | Ea | 48.50 | 36.90 | 85.40 | 145.00 |
| 32" x 80" | BC@1.15 | Ea | 49.20 | 36.90 | 86.10 | 146.00 |
| 36" x 80" | BC@1.15 | Ea | 50.80 | 36.90 | 87.70 | 149.00 |

**Pine 6-panel stile and rail interior doors.** Ready to paint or stain. 1-3/8" thick.

| | Craft@Hrs | Unit | Material | Labor | Total | Sell |
|---|---|---|---|---|---|---|
| 24" x 80" | BC@1.15 | Ea | 80.70 | 36.90 | 117.60 | 200.00 |
| 28" x 80" | BC@1.15 | Ea | 84.60 | 36.90 | 121.50 | 207.00 |
| 30" x 80" | BC@1.15 | Ea | 87.30 | 36.90 | 124.20 | 211.00 |
| 32" x 80" | BC@1.15 | Ea | 90.30 | 36.90 | 127.20 | 216.00 |
| 36" x 80" | BC@1.15 | Ea | 96.80 | 36.90 | 133.70 | 227.00 |

**Fir 1-panel stile and rail interior doors.** 1-3/8" thick.

| | Craft@Hrs | Unit | Material | Labor | Total | Sell |
|---|---|---|---|---|---|---|
| 28" x 80" | BC@1.15 | Ea | 169.00 | 36.90 | 205.90 | 350.00 |
| 30" x 80" | BC@1.15 | Ea | 170.00 | 36.90 | 206.90 | 352.00 |
| 32" x 80" | BC@1.15 | Ea | 171.00 | 36.90 | 207.90 | 353.00 |

**Fir 3-panel stile and rail interior doors.** 1-3/8" thick.

| | Craft@Hrs | Unit | Material | Labor | Total | Sell |
|---|---|---|---|---|---|---|
| 28" x 80" | BC@1.15 | Ea | 211.00 | 36.90 | 247.90 | 421.00 |
| 30" x 80" | BC@1.15 | Ea | 206.00 | 36.90 | 242.90 | 413.00 |
| 32" x 80" | BC@1.15 | Ea | 207.00 | 36.90 | 243.90 | 415.00 |

**Fir 6-panel stile and rail interior doors.** Select grade. 1-3/8" thick.

| | Craft@Hrs | Unit | Material | Labor | Total | Sell |
|---|---|---|---|---|---|---|
| 24" x 80" | BC@1.15 | Ea | 130.00 | 36.90 | 166.90 | 284.00 |
| 28" x 80" | BC@1.15 | Ea | 133.00 | 36.90 | 169.90 | 289.00 |
| 30" x 80" | BC@1.15 | Ea | 137.00 | 36.90 | 173.90 | 296.00 |
| 32" x 80" | BC@1.15 | Ea | 139.00 | 36.90 | 175.90 | 299.00 |
| 36" x 80" | BC@1.15 | Ea | 142.00 | 36.90 | 178.90 | 304.00 |

**Oak 6-panel stile and rail interior doors.** Red oak. Double bevel hip raised panels. 1-3/8" thick.

| | Craft@Hrs | Unit | Material | Labor | Total | Sell |
|---|---|---|---|---|---|---|
| 24" x 80" | BC@1.15 | Ea | 129.00 | 36.90 | 165.90 | 282.00 |
| 28" x 80" | BC@1.15 | Ea | 134.00 | 36.90 | 170.90 | 291.00 |
| 30" x 80" | BC@1.15 | Ea | 139.00 | 36.90 | 175.90 | 299.00 |
| 32" x 80" | BC@1.15 | Ea | 143.00 | 36.90 | 179.90 | 306.00 |
| 36" x 80" | BC@1.15 | Ea | 148.00 | 36.90 | 184.90 | 314.00 |

**French stile and rail interior doors.** Clear pine. Ready to paint or stain. 1-3/8" thick. Pre-masked tempered glass.

| | Craft@Hrs | Unit | Material | Labor | Total | Sell |
|---|---|---|---|---|---|---|
| 24" x 80", 10 lite | BC@1.40 | Ea | 112.00 | 44.90 | 156.90 | 267.00 |
| 28" x 80", 10 lite | BC@1.40 | Ea | 117.00 | 44.90 | 161.90 | 275.00 |
| 30" x 80", 15 lite | BC@1.40 | Ea | 123.00 | 44.90 | 167.90 | 285.00 |
| 32" x 80", 15 lite | BC@1.40 | Ea | 127.00 | 44.90 | 171.90 | 292.00 |
| 36" x 80", 15 lite | BC@1.40 | Ea | 133.00 | 44.90 | 177.90 | 302.00 |

**Half louver stile and rail interior doors.** 1-3/8" thick.

| | Craft@Hrs | Unit | Material | Labor | Total | Sell |
|---|---|---|---|---|---|---|
| 24" x 80" | BC@1.15 | Ea | 147.00 | 36.90 | 183.90 | 313.00 |
| 28" x 80" | BC@1.15 | Ea | 152.00 | 36.90 | 188.90 | 321.00 |
| 30" x 80" | BC@1.15 | Ea | 158.00 | 36.90 | 194.90 | 331.00 |
| 32" x 80" | BC@1.15 | Ea | 163.00 | 36.90 | 199.90 | 340.00 |
| 36" x 80" | BC@1.15 | Ea | 166.00 | 36.90 | 202.90 | 345.00 |

| | Craft@Hrs | Unit | Material | Labor | Total | Sell |
|---|---|---|---|---|---|---|

**Full louver cafe (bar) doors.** Per pair of doors. With hardware.

| | Craft@Hrs | Unit | Material | Labor | Total | Sell |
|---|---|---|---|---|---|---|
| 24" wide, 42" high | BC@1.15 | Ea | 80.70 | 36.90 | 117.60 | 200.00 |
| 30" wide, 42" high | BC@1.15 | Ea | 88.50 | 36.90 | 125.40 | 213.00 |
| 32" wide, 42" high | BC@1.15 | Ea | 97.00 | 36.90 | 133.90 | 228.00 |
| 36" wide, 42" high | BC@1.15 | Ea | 99.40 | 36.90 | 136.30 | 232.00 |

**Stile and rail cafe doors.** Per pair of doors. With hardware.

| | Craft@Hrs | Unit | Material | Labor | Total | Sell |
|---|---|---|---|---|---|---|
| 30" x 42" | BC@1.15 | Ea | 50.50 | 36.90 | 87.40 | 149.00 |
| 32" x 42" | BC@1.15 | Ea | 52.00 | 36.90 | 88.90 | 151.00 |
| 36" x 42" | BC@1.15 | Ea | 53.10 | 36.90 | 90.00 | 153.00 |

## Interior Door Jambs

**Finger joint pine interior jamb set.** Unfinished. Two 6'8" legs and one 3' head.

| | Craft@Hrs | Unit | Material | Labor | Total | Sell |
|---|---|---|---|---|---|---|
| 4-9/16" x 11/16" | BC@.500 | Ea | 30.20 | 16.00 | 46.20 | 78.50 |
| 5-1/4" x 11/16" | BC@.500 | Ea | 63.30 | 16.00 | 79.30 | 135.00 |

**Solid clear pine interior jamb.** Varnish or stain. Jamb set includes two 6'8" legs and one 3' head.

| | Craft@Hrs | Unit | Material | Labor | Total | Sell |
|---|---|---|---|---|---|---|
| 11/16" x 4-9/16" x 7' leg | BC@.210 | Ea | 15.60 | 6.74 | 22.34 | 38.00 |
| 11/16" x 4-9/16" x 3' head | BC@.090 | Ea | 10.10 | 2.89 | 12.99 | 22.10 |
| 11/16" x 4-9/16" jamb set | BC@.500 | Ea | 39.70 | 16.00 | 55.70 | 94.70 |

**Door jamb set with hinges.**

| | Craft@Hrs | Unit | Material | Labor | Total | Sell |
|---|---|---|---|---|---|---|
| 80" x 4-9/16", per set | BC@.500 | Ea | 48.60 | 16.00 | 64.60 | 110.00 |
| 80" x 6-9/16", per set | BC@.500 | Ea | 52.90 | 16.00 | 68.90 | 117.00 |

## Prehung Interior Doors

**Cut wall and frame interior door opening, install prehung door and trim.** Includes header, trimmer studs, interior flush prehung door ($59.40) with jamb, two hinges, stop, casing two sides, and passage lockset ($28). Add the cost of debris removal, painting, floor and wall finish, as needed.

| | Craft@Hrs | Unit | Material | Labor | Total | Sell |
|---|---|---|---|---|---|---|
| 32" to 36" prehung interior door | BC@5.08 | Ea | 96.80 | 163.00 | 259.80 | 442.00 |

**Install interior prehung door in an existing wall opening.** Includes labor setting the door, casing, jamb and stops. Add the cost of setting the lockset.

| | Craft@Hrs | Unit | Material | Labor | Total | Sell |
|---|---|---|---|---|---|---|
| 32" to 36" x 80" interior door | BC@.750 | Ea | — | 24.10 | 24.10 | 41.00 |

**Flush prehung hardboard interior doors.** 1-3/8" thick. Hollow core. Primed. Ready to paint.

| | Craft@Hrs | Unit | Material | Labor | Total | Sell |
|---|---|---|---|---|---|---|
| 24" x 80" | BC@.750 | Ea | 51.00 | 24.10 | 75.10 | 128.00 |
| 28" x 80" | BC@.750 | Ea | 57.70 | 24.10 | 81.80 | 139.00 |
| 30" x 80" | BC@.750 | Ea | 59.80 | 24.10 | 83.90 | 143.00 |
| 32" x 80" | BC@.750 | Ea | 62.00 | 24.10 | 86.10 | 146.00 |
| 36" x 80" | BC@.750 | Ea | 64.20 | 24.10 | 88.30 | 150.00 |

|  | Craft@Hrs | Unit | Material | Labor | Total | Sell |
|---|---|---|---|---|---|---|
| **Flush prehung lauan interior doors.** 1-3/8" thick. 4-9/16" jamb. Bored for lockset. | | | | | | |
| 18" x 78" | BC@.750 | Ea | 54.00 | 24.10 | 78.10 | 133.00 |
| 24" x 78" | BC@.750 | Ea | 54.00 | 24.10 | 78.10 | 133.00 |
| 28" x 78" | BC@.750 | Ea | 55.10 | 24.10 | 79.20 | 135.00 |
| 30" x 78" | BC@.750 | Ea | 56.20 | 24.10 | 80.30 | 137.00 |
| 32" x 78" | BC@.750 | Ea | 58.30 | 24.10 | 82.40 | 140.00 |
| 36" x 78" | BC@.750 | Ea | 60.50 | 24.10 | 84.60 | 144.00 |
| 18" x 80" | BC@.750 | Ea | 52.90 | 24.10 | 77.00 | 131.00 |
| 24" x 80" | BC@.750 | Ea | 52.90 | 24.10 | 77.00 | 131.00 |
| 28" x 80" | BC@.750 | Ea | 54.00 | 24.10 | 78.10 | 133.00 |
| 30" x 80" | BC@.750 | Ea | 57.30 | 24.10 | 81.40 | 138.00 |
| 32" x 80" | BC@.750 | Ea | 59.40 | 24.10 | 83.50 | 142.00 |
| 36" x 80" | BC@.750 | Ea | 61.60 | 24.10 | 85.70 | 146.00 |
| **Flush prehung lauan interior doors.** 1-3/8" thick. Hollow core. Bored for 2-1/8" lockset with 2-3/8" backset. Casing and lockset sold separately. Wood stile construction. Can be stained or finished naturally. Two brass hinges. Paint grade pine split jambs and 2-1/4" casings. 4" to 4-5/8" adjustable jamb. | | | | | | |
| 24" x 80" | BC@.750 | Ea | 68.80 | 24.10 | 92.90 | 158.00 |
| 28" x 80" | BC@.750 | Ea | 70.20 | 24.10 | 94.30 | 160.00 |
| 30" x 80" | BC@.750 | Ea | 71.70 | 24.10 | 95.80 | 163.00 |
| 32" x 80" | BC@.750 | Ea | 73.60 | 24.10 | 97.70 | 166.00 |
| 36" x 80" | BC@.750 | Ea | 76.00 | 24.10 | 100.10 | 170.00 |
| **Flush prehung birch interior doors.** Stain grade. 1-3/8" thick. Wood stiles. Bored for lockset. Multi-finger jointed door jamb 11/16" x 4-9/16". 3 brass finish hinges 3-1/2" x 3-1/2". | | | | | | |
| 24" x 80" | BC@.750 | Ea | 67.60 | 24.10 | 91.70 | 156.00 |
| 28" x 80" | BC@.750 | Ea | 71.30 | 24.10 | 95.40 | 162.00 |
| 30" x 80" | BC@.750 | Ea | 73.60 | 24.10 | 97.70 | 166.00 |
| 32" x 80" | BC@.750 | Ea | 75.80 | 24.10 | 99.90 | 170.00 |
| 36" x 80" | BC@.750 | Ea | 78.00 | 24.10 | 102.10 | 174.00 |
| **Flush prehung oak interior doors.** Stain grade. 1-3/8" thick. Wood stiles. Bored for lockset. Multi-finger jointed door jamb 11/16" x 4-9/16". 3 brass finish hinges 3-1/2" x 3-1/2". | | | | | | |
| 24" x 80" | BC@.750 | Ea | 60.60 | 24.10 | 84.70 | 144.00 |
| 28" x 80" | BC@.750 | Ea | 63.70 | 24.10 | 87.80 | 149.00 |
| 30" x 80" | BC@.750 | Ea | 64.70 | 24.10 | 88.80 | 151.00 |
| 32" x 80" | BC@.750 | Ea | 68.20 | 24.10 | 92.30 | 157.00 |
| 36" x 80" | BC@.750 | Ea | 69.90 | 24.10 | 94.00 | 160.00 |
| **Flush prehung red oak interior doors.** Hollow core. 1-3/8" thick. Stain grade. | | | | | | |
| 24" x 80" | BC@.750 | Ea | 69.70 | 24.10 | 93.80 | 159.00 |
| 28" x 80" | BC@.750 | Ea | 72.40 | 24.10 | 96.50 | 164.00 |
| 30" x 80" | BC@.750 | Ea | 74.50 | 24.10 | 98.60 | 168.00 |
| 32" x 80" | BC@.750 | Ea | 77.20 | 24.10 | 101.30 | 172.00 |
| 36" x 80" | BC@.750 | Ea | 79.90 | 24.10 | 104.00 | 177.00 |
| **Interior prehung lauan heater closet doors.** | | | | | | |
| 24" x 60" | BC@.750 | Ea | 72.30 | 24.10 | 96.40 | 164.00 |

| | Craft@Hrs | Unit | Material | Labor | Total | Sell |
|---|---|---|---|---|---|---|

**6-panel colonial prehung interior doors.** 11/16" x 4-9/16" jamb. Bored for lockset. Casing and hardware sold separately. Hollow core. 1-3/8" thick.

| | Craft@Hrs | Unit | Material | Labor | Total | Sell |
|---|---|---|---|---|---|---|
| 18" x 80", 3-panel | BC@.750 | Ea | 63.80 | 24.10 | 87.90 | 149.00 |
| 24" x 80" | BC@.750 | Ea | 64.50 | 24.10 | 88.60 | 151.00 |
| 28" x 80" | BC@.750 | Ea | 66.20 | 24.10 | 90.30 | 154.00 |
| 30" x 80" | BC@.750 | Ea | 68.30 | 24.10 | 92.40 | 157.00 |
| 32" x 80" | BC@.750 | Ea | 70.50 | 24.10 | 94.60 | 161.00 |
| 36" x 80" | BC@.750 | Ea | 72.10 | 24.10 | 96.20 | 164.00 |

**6-panel colonist prehung interior doors.** 6-9/16" jamb. Hollow core.

| | Craft@Hrs | Unit | Material | Labor | Total | Sell |
|---|---|---|---|---|---|---|
| 24" x 80" | BC@.750 | Ea | 103.00 | 24.10 | 127.10 | 216.00 |
| 30" x 80" | BC@.750 | Ea | 107.00 | 24.10 | 131.10 | 223.00 |
| 32" x 80" | BC@.750 | Ea | 110.00 | 24.10 | 134.10 | 228.00 |
| 36" x 80" | BC@.750 | Ea | 112.00 | 24.10 | 136.10 | 231.00 |

**6-panel colonist prehung interior doors.** Ready to paint or stain. Split flush jambs. Colonial casing attached. Bored for lockset.

| | Craft@Hrs | Unit | Material | Labor | Total | Sell |
|---|---|---|---|---|---|---|
| 18" x 78" | BC@.750 | Ea | 85.30 | 24.10 | 109.40 | 186.00 |
| 24" x 78" | BC@.750 | Ea | 86.40 | 24.10 | 110.50 | 188.00 |
| 28" x 78" | BC@.750 | Ea | 87.50 | 24.10 | 111.60 | 190.00 |
| 30" x 78" | BC@.750 | Ea | 83.90 | 24.10 | 108.00 | 184.00 |
| 32" x 78" | BC@.750 | Ea | 91.80 | 24.10 | 115.90 | 197.00 |
| 36" x 78" | BC@.750 | Ea | 92.90 | 24.10 | 117.00 | 199.00 |
| 18" x 80" | BC@.750 | Ea | 85.30 | 24.10 | 109.40 | 186.00 |
| 24" x 80" | BC@.750 | Ea | 86.40 | 24.10 | 110.50 | 188.00 |
| 28" x 80" | BC@.750 | Ea | 87.50 | 24.10 | 111.60 | 190.00 |
| 30" x 80" | BC@.750 | Ea | 88.50 | 24.10 | 112.60 | 191.00 |
| 32" x 80" | BC@.750 | Ea | 91.80 | 24.10 | 115.90 | 197.00 |
| 36" x 80" | BC@.750 | Ea | 82.90 | 24.10 | 107.00 | 182.00 |

**6-panel hardboard prehung interior doors.** Wood stiles. Bored for lockset. Primed finger joint 11/16" x 4-9/16" jamb. 3 brass finish hinges 3-1/2" x 3-1/2". Hollow core.

| | Craft@Hrs | Unit | Material | Labor | Total | Sell |
|---|---|---|---|---|---|---|
| 24" x 80" | BC@.750 | Ea | 52.40 | 24.10 | 76.50 | 130.00 |
| 28" x 80" | BC@.750 | Ea | 55.10 | 24.10 | 79.20 | 135.00 |
| 30" x 80" | BC@.750 | Ea | 56.90 | 24.10 | 81.00 | 138.00 |
| 32" x 80" | BC@.750 | Ea | 58.70 | 24.10 | 82.80 | 141.00 |
| 36" x 80" | BC@.750 | Ea | 60.50 | 24.10 | 84.60 | 144.00 |

**6-panel hardboard prehung interior doors.** 4-9/16" adjustable split jambs. Hollow core. Paint grade casing applied. Pre-drilled for lockset. Ready to paint or gel stain.

| | Craft@Hrs | Unit | Material | Labor | Total | Sell |
|---|---|---|---|---|---|---|
| 24" x 80" | BC@.750 | Ea | 73.00 | 24.10 | 97.10 | 165.00 |
| 28" x 80" | BC@.750 | Ea | 75.00 | 24.10 | 99.10 | 168.00 |
| 30" x 80" | BC@.750 | Ea | 78.00 | 24.10 | 102.10 | 174.00 |
| 32" x 80" | BC@.750 | Ea | 79.00 | 24.10 | 103.10 | 175.00 |
| 36" x 80" | BC@.750 | Ea | 82.00 | 24.10 | 106.10 | 180.00 |

**6-panel hardboard prehung double interior doors.** 1-3/8" thick. Flat jamb.

| | Craft@Hrs | Unit | Material | Labor | Total | Sell |
|---|---|---|---|---|---|---|
| 48" x 80" | BC@1.25 | Ea | 149.00 | 40.10 | 189.10 | 321.00 |
| 60" x 80" | BC@1.25 | Ea | 157.00 | 40.10 | 197.10 | 335.00 |

**6-panel prehung fire-rated interior doors.** 4-9/16" jamb. 20-minute fire rating.

| | Craft@Hrs | Unit | Material | Labor | Total | Sell |
|---|---|---|---|---|---|---|
| 32" x 80" | BC@.750 | Ea | 204.00 | 24.10 | 228.10 | 388.00 |
| 36" x 80" | BC@.750 | Ea | 208.00 | 24.10 | 232.10 | 395.00 |

| | Craft@Hrs | Unit | Material | Labor | Total | Sell |
|---|---|---|---|---|---|---|

**6-panel oak prehung interior doors.** Hollow core. 1-3/8" thick.

| | Craft@Hrs | Unit | Material | Labor | Total | Sell |
|---|---|---|---|---|---|---|
| 24" x 80" | BC@.750 | Ea | 134.00 | 24.10 | 158.10 | 269.00 |
| 28" x 80" | BC@.750 | Ea | 138.00 | 24.10 | 162.10 | 276.00 |
| 30" x 80" | BC@.750 | Ea | 142.00 | 24.10 | 166.10 | 282.00 |
| 32" x 80" | BC@.750 | Ea | 149.00 | 24.10 | 173.10 | 294.00 |
| 36" x 80" | BC@.750 | Ea | 152.00 | 24.10 | 176.10 | 299.00 |

**6-panel prehung interior double doors.** Bored for lockset. Ready to paint or stain. 1-3/8" thick. 4-9/16" jambs.

| | Craft@Hrs | Unit | Material | Labor | Total | Sell |
|---|---|---|---|---|---|---|
| 4'0" x 6'8" | BC@1.50 | Ea | 147.00 | 48.10 | 195.10 | 332.00 |
| 6'0" x 6'8" | BC@1.50 | Ea | 166.00 | 48.10 | 214.10 | 364.00 |

**6-panel hemlock prehung interior doors.** Bored for lockset. Ready to paint or stain. 1-3/8" thick. 4-9/16" jambs.

| | Craft@Hrs | Unit | Material | Labor | Total | Sell |
|---|---|---|---|---|---|---|
| 24" x 80" | BC@.750 | Ea | 166.00 | 24.10 | 190.10 | 323.00 |
| 28" x 80" | BC@.750 | Ea | 171.00 | 24.10 | 195.10 | 332.00 |
| 30" x 80" | BC@.750 | Ea | 174.00 | 24.10 | 198.10 | 337.00 |
| 32" x 80" | BC@.750 | Ea | 175.00 | 24.10 | 199.10 | 338.00 |
| 36" x 80" | BC@.750 | Ea | 178.00 | 24.10 | 202.10 | 344.00 |

**2-panel knotty alder prehung interior doors.** Bored for lockset. 1-3/8" thick.

| | Craft@Hrs | Unit | Material | Labor | Total | Sell |
|---|---|---|---|---|---|---|
| 24" x 80" | BC@.750 | Ea | 291.00 | 24.10 | 315.10 | 536.00 |
| 28" x 80" | BC@.750 | Ea | 296.00 | 24.10 | 320.10 | 544.00 |
| 30" x 80" | BC@.750 | Ea | 301.00 | 24.10 | 325.10 | 553.00 |
| 32" x 80" | BC@.750 | Ea | 308.00 | 24.10 | 332.10 | 565.00 |
| 36" x 80" | BC@.750 | Ea | 312.00 | 24.10 | 336.10 | 571.00 |

**Pine full louver prehung interior doors.** Ready to install. Provides ventilation while maintaining privacy. 11/16" x 4-9/16" clear face pine door jamb. 1-3/8" thick. Includes (3) 3-1/2" x 3-1/2" hinges with brass finish. Bored for lockset.

| | Craft@Hrs | Unit | Material | Labor | Total | Sell |
|---|---|---|---|---|---|---|
| 24" x 80" | BC@.750 | Ea | 111.00 | 24.10 | 135.10 | 230.00 |
| 28" x 80" | BC@.750 | Ea | 129.00 | 24.10 | 153.10 | 260.00 |
| 30" x 80" | BC@.750 | Ea | 111.00 | 24.10 | 135.10 | 230.00 |
| 32" x 80" | BC@.750 | Ea | 115.00 | 24.10 | 139.10 | 236.00 |
| 36" x 80" | BC@.750 | Ea | 116.00 | 24.10 | 140.10 | 238.00 |

**Pine louver over louver prehung interior doors.** Bored for lockset. 1-3/8" thick.

| | Craft@Hrs | Unit | Material | Labor | Total | Sell |
|---|---|---|---|---|---|---|
| 32" x 80" | BC@.750 | Ea | 189.00 | 24.10 | 213.10 | 362.00 |

**Prehung pantry doors.** Finger joint flat jamb. Bored for lockset. 1-3/8" thick.

| | Craft@Hrs | Unit | Material | Labor | Total | Sell |
|---|---|---|---|---|---|---|
| 28" x 80" | BC@.750 | Ea | 215.00 | 24.10 | 239.10 | 406.00 |
| 30" x 80" | BC@.750 | Ea | 215.00 | 24.10 | 239.10 | 406.00 |

**10-lite wood French prehung double interior doors.** Pre-masked tempered glass for easy finishing. True divided light design. Ponderosa pine. 1-3/8" thick.

| | Craft@Hrs | Unit | Material | Labor | Total | Sell |
|---|---|---|---|---|---|---|
| 48" x 80", unfinished | BC@1.25 | Ea | 314.00 | 40.10 | 354.10 | 602.00 |
| 60" x 80", unfinished | BC@1.25 | Ea | 335.00 | 40.10 | 375.10 | 638.00 |
| 48" x 80", primed | BC@1.25 | Ea | 268.00 | 40.10 | 308.10 | 524.00 |
| 60" x 80", primed | BC@1.25 | Ea | 285.00 | 40.10 | 325.10 | 553.00 |

**15-lite wood French prehung double interior doors.** Pre-masked tempered glass for easy finishing. True divided light design. Ponderosa pine. 1-3/8" thick.

| | Craft@Hrs | Unit | Material | Labor | Total | Sell |
|---|---|---|---|---|---|---|
| 48" x 80" | BC@1.25 | Ea | 215.00 | 40.10 | 255.10 | 434.00 |
| 60" x 80" | BC@1.25 | Ea | 215.00 | 40.10 | 255.10 | 434.00 |

|  | Craft@Hrs | Unit | Material | Labor | Total | Sell |
|---|---|---|---|---|---|---|
| **Oak French prehung double interior doors.** Bored for lockset. 1-3/8" thick. | | | | | | |
| 48" x 80" | BC@1.25 | Ea | 461.00 | 40.10 | 501.10 | 852.00 |
| 60" x 80" | BC@1.25 | Ea | 502.00 | 40.10 | 542.10 | 922.00 |
| 72" x 80" | BC@1.25 | Ea | 353.00 | 40.10 | 393.10 | 668.00 |

## Interior Latchsets

| **Knob passage latch.** | | | | | | |
|---|---|---|---|---|---|---|
| Dummy, brass | BC@.250 | Ea | 12.20 | 8.02 | 20.22 | 34.40 |
| Dummy, satin nickel | BC@.250 | Ea | 13.00 | 8.02 | 21.02 | 35.70 |
| Hall or closet, brass | BC@.250 | Ea | 24.00 | 8.02 | 32.02 | 54.40 |
| Hall or closet, satin nickel | BC@.250 | Ea | 28.00 | 8.02 | 36.02 | 61.20 |

| **Egg knob privacy latchset.** | | | | | | |
|---|---|---|---|---|---|---|
| Bright brass | BC@.250 | Ea | 26.00 | 8.02 | 34.02 | 57.80 |
| Satin nickel | BC@.250 | Ea | 30.20 | 8.02 | 38.22 | 65.00 |

**Privacy lever latchset.** Both levers locked or unlocked by turn button. Outside lever can be unlocked by emergency key (included).

|  | Craft@Hrs | Unit | Material | Labor | Total | Sell |
|---|---|---|---|---|---|---|
| Antique brass | BC@.250 | Ea | 21.50 | 8.02 | 29.52 | 50.20 |
| Polished brass | BC@.250 | Ea | 20.50 | 8.02 | 28.52 | 48.50 |
| Satin nickel | BC@.250 | Ea | 24.00 | 8.02 | 32.02 | 54.40 |

## Door Hardware

| **Door knockers.** Polished brass. | | | | | | |
|---|---|---|---|---|---|---|
| 3-1/2" x 6", knocker and eye viewer | BC@.250 | Ea | 16.20 | 8.02 | 24.22 | 41.20 |
| 6-1/2" x 3-1/2", colonial, ornate | BC@.250 | Ea | 42.10 | 8.02 | 50.12 | 85.20 |

| **Door pull plates.** | | | | | | |
|---|---|---|---|---|---|---|
| 4" x 16", satin chrome | BC@.333 | Ea | 27.70 | 10.70 | 38.40 | 65.30 |

| **Door push plates.** | | | | | | |
|---|---|---|---|---|---|---|
| 3" x 12", bright brass | BC@.333 | Ea | 9.70 | 10.70 | 20.40 | 34.70 |
| 3-1/2" x 15" anodized aluminum | BC@.333 | Ea | 5.40 | 10.70 | 16.10 | 27.40 |
| 3-1/2" x 15", solid brass | BC@.333 | Ea | 10.80 | 10.70 | 21.50 | 36.60 |

| **Door kick plates.** | | | | | | |
|---|---|---|---|---|---|---|
| 6" x 30", anodized aluminum | BC@.333 | Ea | 14.10 | 10.70 | 24.80 | 42.20 |
| 6" x 30", lifetime brass finish | BC@.333 | Ea | 28.80 | 10.70 | 39.50 | 67.20 |
| 6" x 30", mirror brass finish | BC@.333 | Ea | 44.30 | 10.70 | 55.00 | 93.50 |
| 6" x 34", polished brass finish | BC@.333 | Ea | 47.80 | 10.70 | 58.50 | 99.50 |
| 8" x 34", .050" stainless | BC@.333 | Ea | 20.50 | 10.70 | 31.20 | 53.00 |
| 8" x 34", anodized aluminum | BC@.333 | Ea | 21.60 | 10.70 | 32.30 | 54.90 |
| 8" x 34", lifetime brass finish | BC@.333 | Ea | 37.20 | 10.70 | 47.90 | 81.40 |
| 8" x 34", satin nickel finish | BC@.333 | Ea | 36.20 | 10.70 | 46.90 | 79.70 |
| 8" x 34", magnetic, brass finish | BC@.333 | Ea | 38.90 | 10.70 | 49.60 | 84.30 |
| 8" x 34", polished brass finish | BC@.333 | Ea | 52.00 | 10.70 | 62.70 | 107.00 |

| **Door stops.** | | | | | | |
|---|---|---|---|---|---|---|
| 2" spring wall stop | BC@.084 | Ea | 1.90 | 2.70 | 4.60 | 7.82 |
| 2" wall stop, bright brass | BC@.084 | Ea | 3.75 | 2.70 | 6.45 | 11.00 |
| 2" wall stop, satin stainless | BC@.084 | Ea | 4.30 | 2.70 | 7.00 | 11.90 |
| 4" kick down stop, aluminum | BC@.120 | Ea | 4.25 | 3.85 | 8.10 | 13.80 |
| 4" kick down stop, bright brass | BC@.120 | Ea | 4.30 | 3.85 | 8.15 | 13.90 |
| 4" kick down door holder, brass | BC@.166 | Ea | 17.60 | 5.33 | 22.93 | 39.00 |
| Floor mounted door bumper | BC@.166 | Ea | 7.55 | 5.33 | 12.88 | 21.90 |

| | Craft@Hrs | Unit | Material | Labor | Total | Sell |
|---|---|---|---|---|---|---|
| **Door peep sights.**, Brass with optical lens. | | | | | | |
| 160 Degree, 1/2" diameter | BC@.300 | Ea | 4.40 | 9.63 | 14.03 | 23.90 |
| 190 Degree, 1/2" diameter | BC@.300 | Ea | 7.50 | 9.63 | 17.13 | 29.10 |
| 160 Degree, 1" diameter | BC@.300 | Ea | 8.90 | 9.63 | 18.53 | 31.50 |
| **House numbers.** Brass or black coated. | | | | | | |
| 4" high, brass | BC@.103 | Ea | 2.70 | 3.31 | 6.01 | 10.20 |
| 4" high, stainless steel | BC@.103 | Ea | 6.50 | 3.31 | 9.81 | 16.70 |
| 5" high, bronze | BC@.103 | Ea | 7.00 | 3.31 | 10.31 | 17.50 |

## Patio Doors

**Install a sliding or hinged patio door.** Setting a door in an existing framed opening.

| | Craft@Hrs | Unit | Material | Labor | Total | Sell |
|---|---|---|---|---|---|---|
| To 6' wide x 8' high | B1@3.58 | Ea | — | 105.00 | 105.00 | 179.00 |
| Over 6' wide x 8' high to 12' x 8' | B1@4.60 | Ea | — | 135.00 | 135.00 | 230.00 |

**Remove and replace a sliding or hinged patio door.** Includes removing the existing door and frame down to the rough opening, trim the siding as needed, caulk the opening, shim and nail the door in place, set exterior trim. Add the cost of painting, debris removal, wall and floor finish, as required.

| | Craft@Hrs | Unit | Material | Labor | Total | Sell |
|---|---|---|---|---|---|---|
| To 6' wide x 8' high | B1@5.50 | Ea | — | 162.00 | 162.00 | 275.00 |
| Over 6' wide x 8' high to 12' x 8' | B1@6.95 | Ea | — | 204.00 | 204.00 | 347.00 |

**10-lite swinging patio double doors.** White finish. Dual-glazed with clear tempered glass. Fully weatherstripped. Pre-drilled lock-bore.

| | Craft@Hrs | Unit | Material | Labor | Total | Sell |
|---|---|---|---|---|---|---|
| 6'0" wide x 6'8" high | B1@3.58 | Ea | 592.00 | 105.00 | 697.00 | 1,180.00 |

**Vinyl-clad hinged patio double doors, Frenchwood®, Andersen.** White exterior. Perma-Shield® low-maintenance finish. Clear pine interior. 1-lite. 3-point locking system. Adjustable hinges with ball-bearing pivots. Low-E2 tempered insulating glass. Mortise and tenon joints.

| | Craft@Hrs | Unit | Material | Labor | Total | Sell |
|---|---|---|---|---|---|---|
| 6'0" x 6'8" | B1@3.58 | Ea | 1295.00 | 105.00 | 1,400.00 | 2,380.00 |

**Center hinge 15-lite aluminum patio double doors.** Sized for replacement. Unfinished clear pine interior. White extruded aluminum-clad exterior. Weatherstripped frame. Self-draining sill. Bored for lockset. 3/4" Low-E insulated safety glass.

| | Craft@Hrs | Unit | Material | Labor | Total | Sell |
|---|---|---|---|---|---|---|
| 6'0" x 6'8" | B1@3.58 | Ea | 848.00 | 105.00 | 953.00 | 1,620.00 |

**Aluminum inward swing 1-lite patio double doors.** Low-E insulated safety glass. Clear wood interior ready to paint or stain.

| | Craft@Hrs | Unit | Material | Labor | Total | Sell |
|---|---|---|---|---|---|---|
| 6'0" x 6'8" | B1@3.58 | Ea | 756.00 | 105.00 | 861.00 | 1,460.00 |

**Aluminum inward swing 10-lite patio double doors.** Unfinished interior. Low-E2 argon-filled insulated glass. Thermal-break sill. Bored for lockset. With head and foot bolts.

| | Craft@Hrs | Unit | Material | Labor | Total | Sell |
|---|---|---|---|---|---|---|
| 6'0" x 6'8" | B1@3.58 | Ea | 827.00 | 105.00 | 932.00 | 1,580.00 |

**Vinyl-clad gliding patio doors, Andersen.** White. Tempered 1" Low-E insulating glass. Anodized aluminum track.

| | Craft@Hrs | Unit | Material | Labor | Total | Sell |
|---|---|---|---|---|---|---|
| 6'0" x 6'8" | B1@3.58 | Ea | 648.00 | 105.00 | 753.00 | 1,280.00 |

**Vinyl-clad gliding patio doors.** Pine core. Dual-pane Low-E tempered insulating glass. Stainless-steel tracks. Ball-bearing rollers. Combination weatherstrip and interlock.

| | Craft@Hrs | Unit | Material | Labor | Total | Sell |
|---|---|---|---|---|---|---|
| 6'0" x 6'8" | B1@3.58 | Ea | 1,030.00 | 105.00 | 1,135.00 | 1,930.00 |

**10-lite sliding patio doors.** Solid pine core. Weatherstripped on 4 sides of operating panels. Butcher block 4-1/2"-wide stiles and 9-1/2" bottom rail. 4-9/16" jamb width. White. Actual dimensions: 71-1/4" wide x 79-1/2" high. Pre-drilled lock-bore.

| | Craft@Hrs | Unit | Material | Labor | Total | Sell |
|---|---|---|---|---|---|---|
| 6'0" wide x 6'8" high | B1@3.58 | Ea | 598.00 | 105.00 | 703.00 | 1,200.00 |

| | Craft@Hrs | Unit | Material | Labor | Total | Sell |
|---|---|---|---|---|---|---|

**Aluminum sliding patio doors.** Unfinished interior. White aluminum-clad exterior. With lockset. Insulating Low-E2 glass.

| | | | | | | |
|---|---|---|---|---|---|---|
| 6'0" x 6'8", 1 lite | B1@3.58 | Ea | 744.00 | 105.00 | 849.00 | 1,440.00 |
| 6'0" x 6'8", 15 lites | B1@3.58 | Ea | 744.00 | 105.00 | 849.00 | 1,440.00 |

**Aluminum sliding patio doors.** Insulating glass. Reversible, self-aligning ball bearing rollers. Zinc finish lock. With screen. By nominal width x height. Actual height and width are 1/2" less.

**Aluminum sliding patio doors.** With screen. Single glazed. Reversible, self-aligning ball bearing rollers. Zinc finish lock. By nominal width x height. Actual height and width are 1/2" less.

| | | | | | | |
|---|---|---|---|---|---|---|
| 5' x 6'8", mill finish | B1@3.58 | Ea | 182.00 | 105.00 | 287.00 | 488.00 |
| 6' x 6'8" mill finish | B1@3.58 | Ea | 327.00 | 105.00 | 432.00 | 734.00 |
| 6' x 6'8" white finish | B1@3.58 | Ea | 350.00 | 105.00 | 455.00 | 774.00 |
| 6' x 6'8" bronze finish | B1@3.58 | Ea | 360.00 | 105.00 | 465.00 | 791.00 |
| 6' x 6'8" white finish, solar gray glass | B1@3.58 | Ea | 417.00 | 105.00 | 522.00 | 887.00 |
| 8' x 8'0", white finish | B1@4.60 | Ea | 348.00 | 135.00 | 483.00 | 821.00 |

**Vinyl sliding patio doors.** Single glazed. By opening size. Actual width and height are 1/2" less.

| | | | | | | |
|---|---|---|---|---|---|---|
| 5'0" x 6'8", with grille | B1@3.58 | Ea | 409.00 | 105.00 | 514.00 | 874.00 |
| 6'0" x 6'8" | B1@3.58 | Ea | 420.00 | 105.00 | 525.00 | 893.00 |
| 6'0" x 6'8", with grille | B1@3.58 | Ea | 453.00 | 105.00 | 558.00 | 949.00 |
| 8'0" x 6'8" | B1@3.58 | Ea | 528.00 | 105.00 | 633.00 | 1,080.00 |
| 8'0" x 6'8", with grille | B1@3.58 | Ea | 561.00 | 105.00 | 666.00 | 1,130.00 |

**1-lite swinging prehung steel patio double doors.** Polyurethane core. High-performance weatherstrip. Adjustable thermal-break sill. Non-yellowing frame and grille. Inward swing.

| | | | | | | |
|---|---|---|---|---|---|---|
| 6'0" x 6'8" | B1@3.58 | Ea | 323.00 | 105.00 | 428.00 | 728.00 |

**10-lite swinging prehung steel patio double doors.** 24-gauge galvanized steel. Double bored for deadbolt. Compression weatherstripping. Aluminum T-astragal. Single glazed.

| | | | | | | |
|---|---|---|---|---|---|---|
| 6'0" x 6'8" | B1@3.58 | Ea | 282.00 | 105.00 | 387.00 | 658.00 |

**10-lite swinging steel patio double doors.** Inward swing. Insulated Low-E glass.

| | | | | | | |
|---|---|---|---|---|---|---|
| 5'0" x 6'8" | B1@3.58 | Ea | 455.00 | 105.00 | 560.00 | 952.00 |
| 6'0" x 6'8" | B1@3.58 | Ea | 448.00 | 105.00 | 553.00 | 940.00 |

**10-lite venting steel patio double doors.** Inward swing. Insulated glass.

| | | | | | | |
|---|---|---|---|---|---|---|
| 6'0" x 6'8" | B1@3.58 | Ea | 571.00 | 105.00 | 676.00 | 1,150.00 |
| 8'0" x 6'8" | B1@3.58 | Ea | 712.00 | 105.00 | 817.00 | 1,390.00 |

**15-lite swinging steel patio double doors.** Inward swing.

| | | | | | | |
|---|---|---|---|---|---|---|
| 5'0" x 6'8", insulated glass | B1@3.58 | Ea | 323.00 | 105.00 | 428.00 | 728.00 |
| 6'0" x 6'8", Low-E insulated glass | B1@3.58 | Ea | 432.00 | 105.00 | 537.00 | 913.00 |
| 6'0" x 6'8", single glazed | B1@3.58 | Ea | 323.00 | 105.00 | 428.00 | 728.00 |

**15-lite prehung swinging fiberglass patio double doors.** Outward swing. Insulated glass. Adjustable mill finish thermal-break threshold.

| | | | | | | |
|---|---|---|---|---|---|---|
| 6'0" x 6'8" | B1@3.58 | Ea | 636.00 | 105.00 | 741.00 | 1,260.00 |

| | Craft@Hrs | Unit | Material | Labor | Total | Sell |
|---|---|---|---|---|---|---|

## Patio Door Replacement Parts

**Replacement sliding patio door screen and track.**

| | Craft@Hrs | Unit | Material | Labor | Total | Sell |
|---|---|---|---|---|---|---|
| 2'6" x 6'8" screen | BC@.335 | Ea | 111.00 | 10.80 | 121.80 | 207.00 |
| 3'0" x 6'8" screen | BC@.335 | Ea | 120.00 | 10.80 | 130.80 | 222.00 |

**Replacement sliding patio door screens.** Adjustable 78-1/2" to 80" height. Self-latching. Fiberglass screening. Steel frame. Powder-coated finish.

| | Craft@Hrs | Unit | Material | Labor | Total | Sell |
|---|---|---|---|---|---|---|
| 30" wide, bronze | BC@.335 | Ea | 32.00 | 10.80 | 42.80 | 72.80 |
| 36" wide, bronze | BC@.335 | Ea | 33.30 | 10.80 | 44.10 | 75.00 |
| 48" wide, gray | BC@.335 | Ea | 40.70 | 10.80 | 51.50 | 87.60 |

**Patio door hardware set, Frenchwood® and Perma-Shield®, Andersen.** Fits Andersen patio doors.

| | Craft@Hrs | Unit | Material | Labor | Total | Sell |
|---|---|---|---|---|---|---|
| Bright brass, gliding doors | — | Ea | 190.00 | — | 190.00 | — |
| Bright brass, hinged doors | — | Ea | 201.00 | — | 201.00 | — |
| Metro, stone, hinged doors | — | Ea | 57.80 | — | 57.80 | — |
| Metro, white, gliding doors | — | Ea | 57.40 | — | 57.40 | — |

**Sliding aluminum patio door hardware kit.**

| | Craft@Hrs | Unit | Material | Labor | Total | Sell |
|---|---|---|---|---|---|---|
| 6'0" x 6'8", bronze | BC@.450 | Ea | 43.90 | 14.40 | 58.30 | 99.10 |
| 6'0" x 6'8", white | BC@.450 | Ea | 43.90 | 14.40 | 58.30 | 99.10 |

**Sliding patio door replacement handle set.**

| | Craft@Hrs | Unit | Material | Labor | Total | Sell |
|---|---|---|---|---|---|---|
| White | BC@.250 | Ea | 27.50 | 8.02 | 35.52 | 60.40 |

**Sliding patio door replacement locks.**

| | Craft@Hrs | Unit | Material | Labor | Total | Sell |
|---|---|---|---|---|---|---|
| Auxiliary footlock with screws | BC@.250 | Ea | 28.60 | 8.02 | 36.62 | 62.30 |
| Reachout dead lock, white | BC@.250 | Ea | 30.80 | 8.02 | 38.82 | 66.00 |

**Sliding patio door replacement rollers.**

| | Craft@Hrs | Unit | Material | Labor | Total | Sell |
|---|---|---|---|---|---|---|
| Lower screen roller, per pair | BC@.450 | Ea | 11.70 | 14.40 | 26.10 | 44.40 |
| Steel tandem rollers | BC@.450 | Ea | 11.30 | 14.40 | 25.70 | 43.70 |

**Patio door window grilles.** For 6'0" x 6' 8" door.

| | Craft@Hrs | Unit | Material | Labor | Total | Sell |
|---|---|---|---|---|---|---|
| 15 lite | — | Ea | 64.80 | — | 64.80 | — |

**Patio door jamb extension kit.**

| | Craft@Hrs | Unit | Material | Labor | Total | Sell |
|---|---|---|---|---|---|---|
| 6-9/16", white | — | Ea | 64.80 | — | 64.80 | — |

**Swinging patio door threshold.**

| | Craft@Hrs | Unit | Material | Labor | Total | Sell |
|---|---|---|---|---|---|---|
| 6'0", oak | BC@.500 | Ea | 30.75 | 16.00 | 46.75 | 79.50 |

**Cat patio pet door.**

| | Craft@Hrs | Unit | Material | Labor | Total | Sell |
|---|---|---|---|---|---|---|
| 6" x 6", mill finish | BC@.800 | Ea | 138.00 | 25.70 | 163.70 | 278.00 |

**Pet door.** Mill finish aluminum.

| | Craft@Hrs | Unit | Material | Labor | Total | Sell |
|---|---|---|---|---|---|---|
| Small | BC@.800 | Ea | 25.00 | 25.70 | 50.70 | 86.20 |
| Medium | BC@.800 | Ea | 35.60 | 25.70 | 61.30 | 104.00 |
| Large | BC@.800 | Ea | 44.90 | 25.70 | 70.60 | 120.00 |
| Extra Large | BC@.800 | Ea | 54.00 | 25.70 | 79.70 | 135.00 |
| Super Large | BC@.800 | Ea | 73.20 | 25.70 | 98.90 | 168.00 |

| | Craft@Hrs | Unit | Material | Labor | Total | Sell |
|---|---|---|---|---|---|---|

## Door Closers

**Remove and replace door closers.**
Interior or exterior door — BC@.540 Ea — 17.30 17.30 29.40

**Pneumatic door closers.** For wood or metal out swing doors to 90 degrees. Adjustable closing speed. Internal spring keeps rod from bending. 2-hole end plug for adjusting latching power. Bronze, black or white.
Closer — BC@.462 Ea 9.20 14.80 24.00 40.80

**Light duty door closers.** For free-swinging doors up to 85 pounds and 30" wide. Parallel arm mount. Adjustable closing speed. 125-degree maximum opening angle. Brown or ivory.
Non hold-open — BC@.462 Ea 31.00 14.80 45.80 77.90
Hold-open — BC@.462 Ea 30.40 14.80 45.20 76.80

**Medium duty door closers.** Adjustable closing and latching speeds. Designed for free-swinging residential and light commercial doors up to 140 pounds. Brown or ivory.
Closer — BC@.462 Ea 37.90 14.80 52.70 89.60

**Heavy duty door closers.**
No cover, painted steel finish — BC@.462 Ea 58.70 14.80 73.50 125.00
No cover, aluminum finish — BC@.462 Ea 58.70 14.80 73.50 125.00
With cover, aluminum finish — BC@.462 Ea 81.70 14.80 96.50 164.00

**Commercial grade hydraulic door closers.** For free-swinging doors up to 176 pounds. Includes brackets for three possible mounting applications. Adjustable back check. Adjustable closing and latching speeds. Bronze or silver.
Closer — BC@.462 Ea 53.20 14.80 68.00 116.00

**Adjustable hinge pin door closers.** Includes screws.
Brass — BC@.250 Ea 13.89 8.02 21.91 37.20

## Accordion Folding Doors

**Accordion folding doors.** Solid PVC slats with flexible hinges. Vertical embossed surface. Includes hardware. Fully assembled. 32" x 80". Hung in a cased opening.
Light weight, white — BC@.750 Ea 18.60 24.10 42.70 72.60
Standard weight, gray — BC@.750 Ea 36.70 24.10 60.80 103.00
Standard weight, white — BC@.750 Ea 36.70 24.10 60.80 103.00

**Premium accordion folding doors.** Vertical embossed surface. Preassembled. Prefinished. Adjustable width and height. Flexible vinyl hinges. Hung in a cased opening.
8" x 80" section — BC@.750 Ea 18.90 24.10 43.00 73.10
32" x 80" — BC@.750 Ea 57.30 24.10 81.40 138.00

**Folding door lock.**
Door lock — BC@.250 Ea 4.20 8.02 12.22 20.80

| | Craft@Hrs | Unit | Material | Labor | Total | Sell |
|---|---|---|---|---|---|---|

## Bypass Closet Doors

**White bypass closet doors.** Prefinished white vinyl-covered hardboard panels instead of mirrors. White steel frame. Jump-proof bottom track. 1-1/2" dual race ball bearing wheels. Injection-molded top guides. Dimensions are width x height. Per pair of doors.

| | Craft@Hrs | Unit | Material | Labor | Total | Sell |
|---|---|---|---|---|---|---|
| 48" x 80" | BC@.750 | Ea | 57.30 | 24.10 | 81.40 | 138.00 |
| 60" x 80" | BC@1.00 | Ea | 67.00 | 32.10 | 99.10 | 168.00 |
| 72" x 80" | BC@1.00 | Ea | 78.90 | 32.10 | 111.00 | 189.00 |
| 48" x 96" | BC@.750 | Ea | 65.90 | 24.10 | 90.00 | 153.00 |
| 60" x 96" | BC@1.00 | Ea | 81.00 | 32.10 | 113.10 | 192.00 |
| 72" x 96" | BC@1.00 | Ea | 96.10 | 32.10 | 128.20 | 218.00 |

**Economy mirrored bypass closet doors.** Gold finished steel. Plate mirror with double strength glass. Bottom rail with jump-proof track. Dimensions are width x height. Per pair of doors.

| | Craft@Hrs | Unit | Material | Labor | Total | Sell |
|---|---|---|---|---|---|---|
| 48" x 80" | BC@.750 | Ea | 76.70 | 24.10 | 100.80 | 171.00 |
| 60" x 80" | BC@1.00 | Ea | 96.10 | 32.10 | 128.20 | 218.00 |
| 72" x 80" | BC@1.00 | Ea | 108.00 | 32.10 | 140.10 | 238.00 |
| 96" x 80" | BC@1.25 | Ea | 178.00 | 40.10 | 218.10 | 371.00 |

**Good quality mirrored bypass closet doors.** White frame. Safety-backed mirror. Dimensions are width x height. Per pair of doors.

| | Craft@Hrs | Unit | Material | Labor | Total | Sell |
|---|---|---|---|---|---|---|
| 48" x 80", white | BC@.750 | Ea | 93.00 | 24.10 | 117.10 | 199.00 |
| 60" x 80", white | BC@1.00 | Ea | 114.00 | 32.10 | 146.10 | 248.00 |
| 72" x 80", white | BC@1.00 | Ea | 133.00 | 32.10 | 165.10 | 281.00 |
| 96" x 80", white | BC@1.25 | Ea | 207.00 | 40.10 | 247.10 | 420.00 |
| 48" x 80", brushed nickel | BC@.750 | Ea | 118.00 | 24.10 | 142.10 | 242.00 |
| 60" x 80", brushed nickel | BC@1.00 | Ea | 139.00 | 32.10 | 171.10 | 291.00 |
| 72" x 80", brushed nickel | BC@1.00 | Ea | 161.00 | 32.10 | 193.10 | 328.00 |
| 48" x 96", white | BC@1.00 | Ea | 144.00 | 32.10 | 176.10 | 299.00 |
| 60" x 96", white | BC@1.00 | Ea | 156.00 | 32.10 | 188.10 | 320.00 |
| 72" x 96", white | BC@1.25 | Ea | 190.00 | 40.10 | 230.10 | 391.00 |
| 96" x 96", white | BC@1.25 | Ea | 263.00 | 40.10 | 303.10 | 515.00 |

**Better quality mirrored bypass closet doors.** White steel frame. Select 3mm plate mirror. Jump-proof bottom track. 1-1/2" dual race ball bearing wheels. Injection-molded top guides. Dimensions are width x height. Per pair of doors.

| | Craft@Hrs | Unit | Material | Labor | Total | Sell |
|---|---|---|---|---|---|---|
| 48" x 80" | BC@.750 | Ea | 94.00 | 24.10 | 118.10 | 201.00 |
| 60" x 80" | BC@1.00 | Ea | 118.00 | 32.10 | 150.10 | 255.00 |
| 72" x 80" | BC@1.00 | Ea | 138.00 | 32.10 | 170.10 | 289.00 |
| 96" x 80" | BC@1.25 | Ea | 195.00 | 40.10 | 235.10 | 400.00 |
| 48" x 96" | BC@1.00 | Ea | 133.00 | 32.10 | 165.10 | 281.00 |
| 60" x 96" | BC@1.00 | Ea | 154.00 | 32.10 | 186.10 | 316.00 |
| 72" x 96" | BC@1.25 | Ea | 176.00 | 40.10 | 216.10 | 367.00 |
| 96" x 96" | BC@1.25 | Ea | 254.00 | 40.10 | 294.10 | 500.00 |

**Beveled mirror bypass closet doors.** 1/2" bevel on both sides of glass. Safety-backed mirror. Includes hardware. Dimensions are width x height. Per pair of doors.

| | Craft@Hrs | Unit | Material | Labor | Total | Sell |
|---|---|---|---|---|---|---|
| 48" x 80" | BC@.750 | Ea | 123.00 | 24.10 | 147.10 | 250.00 |
| 60" x 80" | BC@1.00 | Ea | 145.00 | 32.10 | 177.10 | 301.00 |
| 72" x 80" | BC@1.00 | Ea | 162.00 | 32.10 | 194.10 | 330.00 |
| 96" x 80" | BC@1.25 | Ea | 212.00 | 40.10 | 252.10 | 429.00 |

| | Craft@Hrs | Unit | Material | Labor | Total | Sell |
|---|---|---|---|---|---|---|

**Premium mirrored bypass closet doors.** Bright gold aluminum frame with integrated handle. Select quality 3mm plate mirror and heavy-duty 1-1/2" wide commercial-grade tubular side frame molding. Ultraglide® 2-1/4" deep felt-lined top channel, jump-proof bottom track and precision 1-1/2" dual race ball bearing wheels. Dimensions are width x height. Per pair of doors.

| | Craft@Hrs | Unit | Material | Labor | Total | Sell |
|---|---|---|---|---|---|---|
| 48" x 81" | BC@.750 | Ea | 187.00 | 24.10 | 211.10 | 359.00 |
| 60" x 81" | BC@1.00 | Ea | 208.00 | 32.10 | 240.10 | 408.00 |
| 72" x 81" | BC@1.00 | Ea | 240.00 | 32.10 | 272.10 | 463.00 |

**Bypass door accessories.**

| | Craft@Hrs | Unit | Material | Labor | Total | Sell |
|---|---|---|---|---|---|---|
| Bumper | BC@.175 | Ea | 1.95 | 5.62 | 7.57 | 12.90 |
| Carpet riser | BC@.175 | Ea | 1.62 | 5.62 | 7.24 | 12.30 |
| Guide | BC@.175 | Ea | 1.65 | 5.62 | 7.27 | 12.40 |

## Bi-Fold Closet Doors

**Lauan bi-fold flush closet doors.** 1-3/8" thick. 2 prehinged panels. Ready to paint, stain, or varnish.

| | Craft@Hrs | Unit | Material | Labor | Total | Sell |
|---|---|---|---|---|---|---|
| 24" x 80" | BC@.700 | Ea | 28.00 | 22.50 | 50.50 | 85.90 |
| 30" x 80" | BC@.700 | Ea | 31.30 | 22.50 | 53.80 | 91.50 |
| 32" x 80" | BC@.700 | Ea | 32.40 | 22.50 | 54.90 | 93.30 |
| 36" x 80" | BC@.700 | Ea | 35.30 | 22.50 | 57.80 | 98.30 |

**Primed hardboard bi-fold flush closet doors.** 1-3/8" thick. 2 prehinged panels.

| | Craft@Hrs | Unit | Material | Labor | Total | Sell |
|---|---|---|---|---|---|---|
| 24" x 80" | BC@.700 | Ea | 28.30 | 22.50 | 50.80 | 86.40 |
| 30" x 80" | BC@.700 | Ea | 31.00 | 22.50 | 53.50 | 91.00 |
| 32" x 80" | BC@.700 | Ea | 34.50 | 22.50 | 57.00 | 96.90 |
| 36" x 80" | BC@.700 | Ea | 36.50 | 22.50 | 59.00 | 100.00 |

**Clear mahogany bi-fold flush doors.** 1-1/8" thick. 2 prehinged panels. Unfinished.

| | Craft@Hrs | Unit | Material | Labor | Total | Sell |
|---|---|---|---|---|---|---|
| 24" x 80" | BC@.700 | Ea | 45.90 | 22.50 | 68.40 | 116.00 |
| 30" x 80" | BC@.700 | Ea | 54.10 | 22.50 | 76.60 | 130.00 |
| 32" x 80" | BC@.700 | Ea | 57.80 | 22.50 | 80.30 | 137.00 |
| 36" x 80" | BC@.700 | Ea | 59.30 | 22.50 | 81.80 | 139.00 |

**2-door colonist molded face bi-fold closet doors.** 1-3/8" thick. Primed. Two panels.

| | Craft@Hrs | Unit | Material | Labor | Total | Sell |
|---|---|---|---|---|---|---|
| 24" x 78" | BC@.700 | Ea | 32.50 | 22.50 | 55.00 | 93.50 |
| 30" x 78" | BC@.700 | Ea | 37.00 | 22.50 | 59.50 | 101.00 |
| 32" x 78" | BC@.700 | Ea | 39.00 | 22.50 | 61.50 | 105.00 |
| 36" x 78" | BC@.700 | Ea | 41.00 | 22.50 | 63.50 | 108.00 |

**4-panel birch bi-fold flush closet doors.** Wood stile and rail construction. 1-3/8" thick. Open left or right. Includes track and hardware.

| | Craft@Hrs | Unit | Material | Labor | Total | Sell |
|---|---|---|---|---|---|---|
| 24" x 80" | BC@.700 | Ea | 36.80 | 22.50 | 59.30 | 101.00 |
| 30" x 80" | BC@.700 | Ea | 40.50 | 22.50 | 63.00 | 107.00 |
| 36" x 80" | BC@.700 | Ea | 45.50 | 22.50 | 68.00 | 116.00 |
| 48" x 80" | BC@.800 | Ea | 63.80 | 25.70 | 89.50 | 152.00 |

**4-panel lauan flush bi-fold closet doors.** Wood stile and rail construction. 1-3/8" thick. Open left or right. Includes track and hardware.

| | Craft@Hrs | Unit | Material | Labor | Total | Sell |
|---|---|---|---|---|---|---|
| 48" x 80" | BC@.800 | Ea | 48.60 | 25.70 | 74.30 | 126.00 |

**4-panel red oak flush bi-fold closet doors.** Wood stile and rail construction. 1-3/8" thick. Open left or right. Includes track and hardware.

| | Craft@Hrs | Unit | Material | Labor | Total | Sell |
|---|---|---|---|---|---|---|
| 24" x 80" | BC@.700 | Ea | 36.40 | 22.50 | 58.90 | 100.00 |
| 30" x 80" | BC@.700 | Ea | 40.80 | 22.50 | 63.30 | 108.00 |
| 36" x 80" | BC@.700 | Ea | 45.10 | 22.50 | 67.60 | 115.00 |
| 48" x 80" | BC@.800 | Ea | 65.60 | 25.70 | 91.30 | 155.00 |

| | Craft@Hrs | Unit | Material | Labor | Total | Sell |
|---|---|---|---|---|---|---|

**4-panel molded face bi-fold closet doors.** 1-3/8" thick. Primed. Two doors.

| | Craft@Hrs | Unit | Material | Labor | Total | Sell |
|---|---|---|---|---|---|---|
| 24" x 80" | BC@.700 | Ea | 32.40 | 22.50 | 54.90 | 93.30 |
| 30" x 80" | BC@.700 | Ea | 36.60 | 22.50 | 59.10 | 100.00 |
| 36" x 80" | BC@.700 | Ea | 41.30 | 22.50 | 63.80 | 108.00 |

**6-panel molded face bi-fold closet doors.** 1-3/8" thick. Unfinished. Molded and primed hardboard. Two doors.

| | Craft@Hrs | Unit | Material | Labor | Total | Sell |
|---|---|---|---|---|---|---|
| 24" x 80" | BC@.700 | Ea | 34.10 | 22.50 | 56.60 | 96.20 |
| 30" x 80" | BC@.700 | Ea | 38.80 | 22.50 | 61.30 | 104.00 |
| 32" x 80" | BC@.700 | Ea | 40.40 | 22.50 | 62.90 | 107.00 |
| 36" x 80" | BC@.700 | Ea | 43.50 | 22.50 | 66.00 | 112.00 |
| 48" x 80" | BC@.800 | Ea | 68.90 | 25.70 | 94.60 | 161.00 |
| 60" x 80" | BC@.900 | Ea | 73.10 | 28.90 | 102.00 | 173.00 |
| 72" x 80" | BC@1.00 | Ea | 81.80 | 32.10 | 113.90 | 194.00 |

**Colonial 6-panel pine bi-fold doors.** 1-3/8" thick. Includes track and hardware. Unfinished.

| | Craft@Hrs | Unit | Material | Labor | Total | Sell |
|---|---|---|---|---|---|---|
| 24" x 78" | BC@.700 | Ea | 51.00 | 22.50 | 73.50 | 125.00 |
| 30" x 78" | BC@.700 | Ea | 60.50 | 22.50 | 83.00 | 141.00 |
| 36" x 78" | BC@.700 | Ea | 71.00 | 22.50 | 93.50 | 159.00 |
| 24" x 80" | BC@.700 | Ea | 52.00 | 22.50 | 74.50 | 127.00 |
| 30" x 80" | BC@.700 | Ea | 61.00 | 22.50 | 83.50 | 142.00 |
| 32" x 80" | BC@.700 | Ea | 68.00 | 22.50 | 90.50 | 154.00 |
| 36" x 80" | BC@.700 | Ea | 72.00 | 22.50 | 94.50 | 161.00 |

**Clear pine 2-panel colonial bi-fold doors.** Clear pine. Recessed rails. Unfinished. Open left or right. Includes all hardware and track. Two hinged panels.

| | Craft@Hrs | Unit | Material | Labor | Total | Sell |
|---|---|---|---|---|---|---|
| 24" x 80" | BC@.700 | Ea | 91.80 | 22.50 | 114.30 | 194.00 |
| 30" x 80" | BC@.700 | Ea | 97.20 | 22.50 | 119.70 | 203.00 |
| 32" x 80" | BC@.700 | Ea | 89.00 | 22.50 | 111.50 | 190.00 |
| 36" x 80" | BC@.700 | Ea | 96.80 | 22.50 | 119.30 | 203.00 |

**6-panel oak bi-fold doors.** Includes track and hardware. 1-1/8" thick.

| | Craft@Hrs | Unit | Material | Labor | Total | Sell |
|---|---|---|---|---|---|---|
| 24" x 80" | BC@.700 | Ea | 168.00 | 22.50 | 190.50 | 324.00 |
| 30" x 80" | BC@.700 | Ea | 179.00 | 22.50 | 201.50 | 343.00 |
| 36" x 80" | BC@.700 | Ea | 196.00 | 22.50 | 218.50 | 371.00 |

**Metal louver over louver bi-fold doors.** Includes track and hardware. 1-3/8" thick. Two prehinged panels.

| | Craft@Hrs | Unit | Material | Labor | Total | Sell |
|---|---|---|---|---|---|---|
| 24" x 80" | BC@.700 | Ea | 51.00 | 22.50 | 73.50 | 125.00 |
| 30" x 80" | BC@.700 | Ea | 61.50 | 22.50 | 84.00 | 143.00 |
| 36" x 80" | BC@.700 | Ea | 64.00 | 22.50 | 86.50 | 147.00 |

**Oak louver over louver bi-fold doors.** Includes track and hardware. 1-3/8" thick. Two hinged panels.

| | Craft@Hrs | Unit | Material | Labor | Total | Sell |
|---|---|---|---|---|---|---|
| 24" x 96" | BC@.700 | Ea | 65.00 | 22.50 | 87.50 | 149.00 |
| 30" x 96" | BC@.700 | Ea | 81.00 | 22.50 | 103.50 | 176.00 |
| 36" x 96" | BC@.700 | Ea | 84.00 | 22.50 | 106.50 | 181.00 |

**Pine louver over louver bi-fold doors.** 1-1/8" thick ponderosa pine. 3-11/16" wide rails. 1-1/4" wide stiles. Two prehinged panels. Unfinished.

| | Craft@Hrs | Unit | Material | Labor | Total | Sell |
|---|---|---|---|---|---|---|
| 24" x 80" | BC@.700 | Ea | 46.50 | 22.50 | 69.00 | 117.00 |
| 30" x 80" | BC@.700 | Ea | 52.80 | 22.50 | 75.30 | 128.00 |
| 32" x 80" | BC@.700 | Ea | 58.00 | 22.50 | 80.50 | 137.00 |
| 36" x 80" | BC@.700 | Ea | 63.70 | 22.50 | 86.20 | 147.00 |

| | Craft@Hrs | Unit | Material | Labor | Total | Sell |
|---|---|---|---|---|---|---|

**Louver over panel white bi-fold doors.** With track and hardware. Two hinged panels.

| | Craft@Hrs | Unit | Material | Labor | Total | Sell |
|---|---|---|---|---|---|---|
| 24" x 80" | BC@.700 | Ea | 48.60 | 22.50 | 71.10 | 121.00 |
| 30" x 80" | BC@.700 | Ea | 51.90 | 22.50 | 74.40 | 126.00 |
| 36" x 80" | BC@.700 | Ea | 57.30 | 22.50 | 79.80 | 136.00 |

**Pine louver over panel bi-fold doors.** Recessed rails. Open left or right. 1/2" swing space. Two prehinged panels. Includes track and hardware. Unfinished stain grade.

| | Craft@Hrs | Unit | Material | Labor | Total | Sell |
|---|---|---|---|---|---|---|
| 24" x 80" | BC@.700 | Ea | 50.70 | 22.50 | 73.20 | 124.00 |
| 30" x 80" | BC@.700 | Ea | 67.00 | 22.50 | 89.50 | 152.00 |
| 36" x 80" | BC@.700 | Ea | 76.20 | 22.50 | 98.70 | 168.00 |

**Bi-fold door hardware.**

| | Craft@Hrs | Unit | Material | Labor | Total | Sell |
|---|---|---|---|---|---|---|
| Bi-fold hinge | BC@.175 | Ea | 2.87 | 5.62 | 8.49 | 14.40 |
| Carpet riser | BC@.175 | Ea | 1.74 | 5.62 | 7.36 | 12.50 |
| Hardware set | BC@.175 | Ea | 7.00 | 5.62 | 12.62 | 21.50 |
| 48" track | — | Ea | 20.50 | — | 20.50 | — |
| 60" track | — | Ea | 21.40 | — | 21.40 | — |
| 72" track | — | Ea | 23.50 | — | 23.50 | — |

## Mirrored Bi-Fold Doors

**Framed bi-fold mirror doors.** White frame. Top-hung with single wheel top hanger. Snap-in bottom guide and fascia. Low rise bottom track. Non-binding door operation. Reversible top and bottom track. 3mm safety-backed mirror. Two hinged panels.

| | Craft@Hrs | Unit | Material | Labor | Total | Sell |
|---|---|---|---|---|---|---|
| 24" x 80" | BC@.700 | Ea | 85.30 | 22.50 | 107.80 | 183.00 |
| 30" x 80" | BC@.700 | Ea | 96.10 | 22.50 | 118.60 | 202.00 |
| 36" x 80" | BC@.700 | Ea | 107.00 | 22.50 | 129.50 | 220.00 |

**Bevel edge bi-fold mirror doors.** 1/2" vertical bevel with bearing hinges. White finished steel frame with safety-backed mirror. Includes hardware. Full access to closet opening. Reversible high- and low-profile top and bottom tracks. Two hinged panels.

| | Craft@Hrs | Unit | Material | Labor | Total | Sell |
|---|---|---|---|---|---|---|
| 24" x 80" | BC@.700 | Ea | 95.80 | 22.50 | 118.30 | 201.00 |
| 30" x 80" | BC@.700 | Ea | 105.00 | 22.50 | 127.50 | 217.00 |
| 36" x 80" | BC@.700 | Ea | 124.00 | 22.50 | 146.50 | 249.00 |

**Frameless bi-fold mirror doors.** Frameless style maximizes the mirror surface. Beveled edge glass. All preassembled hinges. Safety-reinforced mirror backing. Two hinged panels. Includes hardware.

| | Craft@Hrs | Unit | Material | Labor | Total | Sell |
|---|---|---|---|---|---|---|
| 24" x 80" | BC@.700 | Ea | 95.80 | 22.50 | 118.30 | 201.00 |
| 30" x 80" | BC@.700 | Ea | 106.60 | 22.50 | 129.10 | 219.00 |
| 36" x 80" | BC@.700 | Ea | 117.40 | 22.50 | 139.90 | 238.00 |

**Chrome bi-fold mirror doors.** Two hinged panels. Includes hardware.

| | Craft@Hrs | Unit | Material | Labor | Total | Sell |
|---|---|---|---|---|---|---|
| 24" x 80" | BC@.700 | Ea | 150.10 | 22.50 | 172.60 | 293.00 |
| 30" x 80" | BC@.700 | Ea | 160.90 | 22.50 | 183.40 | 312.00 |
| 36" x 80" | BC@.700 | Ea | 171.70 | 22.50 | 194.20 | 330.00 |

| | Craft@Hrs | Unit | Material | Labor | Total | Sell |
|---|---|---|---|---|---|---|

## Pocket Door Hardware

**Pocket door frame and track.** Fully assembled frame. Designed to hold 150-pound standard 80" door. Door sold separately.

| | Craft@Hrs | Unit | Material | Labor | Total | Sell |
|---|---|---|---|---|---|---|
| 24" x 80" | BC@1.00 | Ea | 60.60 | 32.10 | 92.70 | 158.00 |
| 28" x 80" | BC@1.00 | Ea | 60.70 | 32.10 | 92.80 | 158.00 |
| 30" x 80" | BC@1.00 | Ea | 60.60 | 32.10 | 92.70 | 158.00 |
| 32" x 80" | BC@1.00 | Ea | 60.70 | 32.10 | 92.80 | 158.00 |
| 36" x 80" | BC@1.00 | Ea | 60.50 | 32.10 | 92.60 | 157.00 |

**Universal pocket door jamb kit.**

| | Craft@Hrs | Unit | Material | Labor | Total | Sell |
|---|---|---|---|---|---|---|
| Clear pine | — | Ea | 35.10 | — | 35.10 | — |

**Pocket door edge pulls.**

| | Craft@Hrs | Unit | Material | Labor | Total | Sell |
|---|---|---|---|---|---|---|
| 2-1/8" round brass | BC@.300 | Ea | 3.38 | 9.63 | 13.01 | 22.10 |
| 2-1/8" round brass dummy | BC@.300 | Ea | 4.88 | 9.63 | 14.51 | 24.70 |
| 1-3/8" x 3" oval brass | BC@.300 | Ea | 3.34 | 9.63 | 12.97 | 22.00 |
| 1-3/8" x 3" rectangular brass | BC@.300 | Ea | 3.34 | 9.63 | 12.97 | 22.00 |
| 3/4" round brass | BC@.300 | Ea | 2.11 | 9.63 | 11.74 | 20.00 |
| Brass plate | BC@.300 | Ea | 4.48 | 9.63 | 14.11 | 24.00 |

**Ball bearing pocket door hangers.**

| | Craft@Hrs | Unit | Material | Labor | Total | Sell |
|---|---|---|---|---|---|---|
| Hanger | BC@.175 | Ea | 9.70 | 5.62 | 15.32 | 26.00 |

**Converging pocket door kit.**

| | Craft@Hrs | Unit | Material | Labor | Total | Sell |
|---|---|---|---|---|---|---|
| Door kit | — | Ea | 3.76 | — | 3.76 | — |

**Pocket door hangers.**

| | Craft@Hrs | Unit | Material | Labor | Total | Sell |
|---|---|---|---|---|---|---|
| 3/8" offset, one wheel | BC@.175 | Ea | 2.55 | 5.62 | 8.17 | 13.90 |
| 1/16" offset, one wheel | BC@.175 | Ea | 2.55 | 5.62 | 8.17 | 13.90 |
| 7/8" offset, two wheel | BC@.175 | Ea | 3.97 | 5.62 | 9.59 | 16.30 |

**2-door bypass pocket door track hardware.**

| | Craft@Hrs | Unit | Material | Labor | Total | Sell |
|---|---|---|---|---|---|---|
| 48" width | BC@.350 | Ea | 9.82 | 11.20 | 21.02 | 35.70 |
| 60" width | BC@.350 | Ea | 11.80 | 11.20 | 23.00 | 39.10 |
| 72" width | BC@.350 | Ea | 13.80 | 11.20 | 25.00 | 42.50 |

## Mobile Home Doors and Windows

**Exterior mobile home combo doors.** Inswinging door with outswinging storm and screen door.

| | Craft@Hrs | Unit | Material | Labor | Total | Sell |
|---|---|---|---|---|---|---|
| 32" x 72" | BC@1.00 | Ea | 230.00 | 32.10 | 262.10 | 446.00 |
| 32" x 76" | BC@1.00 | Ea | 223.00 | 32.10 | 255.10 | 434.00 |
| 34" x 76" | BC@1.00 | Ea | 236.00 | 32.10 | 268.10 | 456.00 |

**Exterior mobile home doors.** Left hinge. Outswinging.

| | Craft@Hrs | Unit | Material | Labor | Total | Sell |
|---|---|---|---|---|---|---|
| 32" x 72" | BC@1.00 | Ea | 110.00 | 32.10 | 142.10 | 242.00 |
| 32" x 76" | BC@1.00 | Ea | 110.00 | 32.10 | 142.10 | 242.00 |
| 34" x 76" | BC@1.00 | Ea | 111.00 | 32.10 | 143.10 | 243.00 |

|  | Craft@Hrs | Unit | Material | Labor | Total | Sell |
|---|---|---|---|---|---|---|

## Garage Doors

**Install overhead rollup garage door sections only.** Assemble and install rollup door. Excludes trolley and springs.

|  | Craft@Hrs | Unit | Material | Labor | Total | Sell |
|---|---|---|---|---|---|---|
| 4 sections high, 8' to 9' wide | B1@2.73 | Set | — | 80.20 | 80.20 | 136.00 |
| 4 sections high, 12' to 16' wide | B1@4.27 | Set | — | 125.00 | 125.00 | 213.00 |
| 6 or 7 sections high, 12' to 16' wide | B1@5.19 | Set | — | 152.00 | 152.00 | 258.00 |

**Install overhead rollup garage door.** Assemble and install rollup door. Includes trolley and torsion spring counter-balance installed on wood frame building.

|  | Craft@Hrs | Unit | Material | Labor | Total | Sell |
|---|---|---|---|---|---|---|
| 4 sections high, 8' to 9' wide | B1@6.46 | Ea | — | 190.00 | 190.00 | 323.00 |
| 4 sections high, 12' to 16' wide | B1@8.33 | Ea | — | 245.00 | 245.00 | 417.00 |
| 6 or 7 sections high, 12' to 16' wide | B1@9.60 | Ea | — | 282.00 | 282.00 | 479.00 |

**Remove and replace overhead rollup garage door sections.** 4-section-high door. No painting included.

|  | Craft@Hrs | Unit | Material | Labor | Total | Sell |
|---|---|---|---|---|---|---|
| One bottom section, 8' to 9' wide | B1@2.20 | Ea | — | 64.60 | 64.60 | 110.00 |
| One bottom section, 12' to 16' wide | B1@3.00 | Ea | — | 88.10 | 88.10 | 150.00 |

**Remove and replace overhead rollup garage door sections.** 6- or 7-section-high door. No painting included.

|  | Craft@Hrs | Unit | Material | Labor | Total | Sell |
|---|---|---|---|---|---|---|
| One bottom section, 12' to 16' wide | B1@3.84 | Ea | — | 113.00 | 113.00 | 192.00 |

**Install overhead rollup garage door torsion spring counter-balance.** Per set of two torsion springs.

|  | Craft@Hrs | Unit | Material | Labor | Total | Sell |
|---|---|---|---|---|---|---|
| 4 section door 8' to 9' wide | B1@1.02 | Ea | — | 30.00 | 30.00 | 51.00 |
| 4 section door 12' to 16' wide | B1@1.33 | Ea | — | 39.10 | 39.10 | 66.50 |
| 6 or 7 section door 12' to 16' wide | B1@2.01 | Ea | — | 59.00 | 59.00 | 100.00 |

**Remove and replace torsion springs on overhead garage door.** Remove set of two old torsion springs and replace with set of two new torsion springs.

|  | Craft@Hrs | Unit | Material | Labor | Total | Sell |
|---|---|---|---|---|---|---|
| 4 section door 8' to 9' wide | B1@1.10 | Ea | — | 32.30 | 32.30 | 54.90 |
| 4 section door 12' to 16' wide | B1@1.40 | Ea | — | 41.10 | 41.10 | 69.90 |
| 6 or 7 section door 12' to 16' wide | B1@2.10 | Ea | — | 61.70 | 61.70 | 105.00 |

**Install trollies for overhead rollup garage door.** Installation on a frame building. Includes trollies but no torsion springs.

|  | Craft@Hrs | Unit | Material | Labor | Total | Sell |
|---|---|---|---|---|---|---|
| 4 section door, 8' to 16' wide | B1@2.70 | Ea | — | 79.30 | 79.30 | 135.00 |
| 6 to 7 section door, 12' to 16' wide | B1@3.32 | Ea | — | 97.50 | 97.50 | 166.00 |

**Install garage door bottom seal.**

|  | Craft@Hrs | Unit | Material | Labor | Total | Sell |
|---|---|---|---|---|---|---|
| 8' to 9' wide | B1@.143 | Ea | — | 4.20 | 4.20 | 7.14 |
| 12' wide | B1@.204 | Ea | — | 5.99 | 5.99 | 10.20 |
| 16' wide | B1@.259 | Ea | — | 7.61 | 7.61 | 12.90 |

**Steel rollup garage door.** Torsion springs. 2-layer insulated construction. 7/8" insulation. Steel frame with overlapped section joints. With hardware and lock.

|  | Craft@Hrs | Unit | Material | Labor | Total | Sell |
|---|---|---|---|---|---|---|
| 8' x 7' | B1@6.46 | Ea | 247.00 | 190.00 | 437.00 | 743.00 |
| 9' x 7' | B1@6.46 | Ea | 291.00 | 190.00 | 481.00 | 818.00 |
| 16' x 7' | B1@8.33 | Ea | 499.00 | 245.00 | 744.00 | 1,260.00 |

**Raised panel steel garage door.** 2-layer construction. 7/8" thick polystyrene insulation.

|  | Craft@Hrs | Unit | Material | Labor | Total | Sell |
|---|---|---|---|---|---|---|
| 8' x 7' | B1@6.46 | Ea | 304.00 | 190.00 | 494.00 | 840.00 |
| 9' x 7' | B1@6.46 | Ea | 328.00 | 190.00 | 518.00 | 881.00 |
| 16' x 7' | B1@8.33 | Ea | 554.00 | 245.00 | 799.00 | 1,360.00 |

| | Craft@Hrs | Unit | Material | Labor | Total | Sell |
|---|---|---|---|---|---|---|

**Bonded steel garage door.** 3-layer steel exterior. 2"-thick insulation. With torsion springs, hardware and lock.

| | Craft@Hrs | Unit | Material | Labor | Total | Sell |
|---|---|---|---|---|---|---|
| 8' x 7' | B1@6.46 | Ea | 397.00 | 190.00 | 587.00 | 998.00 |
| 8' x 7', molding face | B1@6.46 | Ea | 301.00 | 190.00 | 491.00 | 835.00 |
| 9' x 7' | B1@6.46 | Ea | 419.00 | 190.00 | 609.00 | 1,040.00 |
| 9' x 7', molding face | B1@6.46 | Ea | 341.00 | 190.00 | 531.00 | 903.00 |
| 16' x 7' | B1@6.46 | Ea | 702.00 | 190.00 | 892.00 | 1,520.00 |
| 16' x 7', molding face | B1@8.33 | Ea | 743.00 | 245.00 | 988.00 | 1,680.00 |
| 8' x 6'6", with glass window | B1@6.46 | Ea | 617.00 | 190.00 | 807.00 | 1,370.00 |

**2-layer steel garage door.** 7/8"-thick insulation. With torsion springs, hardware and lock.

| | Craft@Hrs | Unit | Material | Labor | Total | Sell |
|---|---|---|---|---|---|---|
| 16' x 7', raised panel face | B1@8.33 | Ea | 500.00 | 245.00 | 745.00 | 1,270.00 |

**Non-insulated steel garage door.** With rollup hardware and lock.

| | Craft@Hrs | Unit | Material | Labor | Total | Sell |
|---|---|---|---|---|---|---|
| 8' x 7' | B1@6.46 | Ea | 239.00 | 190.00 | 429.00 | 729.00 |
| 9' x 7' | B1@6.46 | Ea | 240.00 | 190.00 | 430.00 | 731.00 |
| 16' x 7' | B1@8.33 | Ea | 437.00 | 245.00 | 682.00 | 1,160.00 |

**Wood garage doors.** Unfinished, jamb hinge, with hardware.

| | Craft@Hrs | Unit | Material | Labor | Total | Sell |
|---|---|---|---|---|---|---|
| Plain panel un-insulated hardboard door | | | | | | |
|   9' wide x 7' high | B1@4.42 | Ea | 300.00 | 130.00 | 430.00 | 731.00 |
|   16' wide x 7' high | B1@5.90 | Ea | 521.00 | 173.00 | 694.00 | 1,180.00 |
| Styrofoam core hardboard door | | | | | | |
|   9' wide x 7' high | B1@4.42 | Ea | 324.00 | 130.00 | 454.00 | 772.00 |
|   16' wide x 7' high | B1@5.90 | Ea | 570.00 | 173.00 | 743.00 | 1,260.00 |
| Redwood door, raised panels | | | | | | |
|   9' wide x 7' high | B1@4.42 | Ea | 598.00 | 130.00 | 728.00 | 1,240.00 |
|   16' wide x 7' high | B1@5.90 | Ea | 1,320.00 | 173.00 | 1,493.00 | 2,540.00 |
| Hardboard door with two small lites | | | | | | |
|   9' wide x 7' high | B1@4.42 | Ea | 330.00 | 130.00 | 460.00 | 782.00 |
|   16' wide x 7' high | B1@5.90 | Ea | 298.00 | 173.00 | 471.00 | 801.00 |
| Hardboard door with sunburst lites | | | | | | |
|   9' wide x 7' high | B1@4.42 | Ea | 480.00 | 130.00 | 610.00 | 1,040.00 |
|   16' wide x 7' high | B1@5.90 | Ea | 928.00 | 173.00 | 1,101.00 | 1,870.00 |
| Hardboard door with cathedral-style lites | | | | | | |
|   9' wide x 7' high | B1@4.42 | Ea | 424.00 | 130.00 | 554.00 | 942.00 |
|   16' wide x 7' high | B1@5.90 | Ea | 893.00 | 173.00 | 1,066.00 | 1,810.00 |

**Garage door jamb hinge**. With carriage bolts and nuts.

| | Craft@Hrs | Unit | Material | Labor | Total | Sell |
|---|---|---|---|---|---|---|
| No. 1, 14 gauge | BC@.333 | Ea | 3.35 | 10.70 | 14.05 | 23.90 |
| No. 2, 14 gauge | BC@.333 | Ea | 3.40 | 10.70 | 14.10 | 24.00 |
| No. 3, 14 gauge | BC@.333 | Ea | 3.50 | 10.70 | 14.20 | 24.10 |

**Steel garage door weather seal.** Cellular vinyl.

| | Craft@Hrs | Unit | Material | Labor | Total | Sell |
|---|---|---|---|---|---|---|
| 7/16" x 2" x 7' | BC@.400 | Ea | 5.79 | 12.80 | 18.59 | 31.60 |
| 7/16" x 2" x 9' | BC@.500 | Ea | 7.49 | 16.00 | 23.49 | 39.90 |
| 7/16" x 2" x 16' | BC@.750 | Ea | 14.33 | 24.10 | 38.43 | 65.30 |

| | Craft@Hrs | Unit | Material | Labor | Total | Sell |
|---|---|---|---|---|---|---|
| **Garage door hardware.** | | | | | | |
| Adjustable kicker | BC@.250 | Ea | 10.50 | 8.02 | 18.52 | 31.50 |
| Door cantilever bolts | BC@.250 | Ea | 3.85 | 8.02 | 11.87 | 20.20 |
| Garage door handle | BC@.500 | Ea | 4.20 | 16.00 | 20.20 | 34.30 |
| Garage door wall pushbutton | BC@.250 | Ea | 8.65 | 8.02 | 16.67 | 28.30 |
| Garage door rubber bumpers | BC@.250 | Ea | 3.85 | 8.02 | 11.87 | 20.20 |
| 16' Door truss rod kit | BC@.500 | Ea | 27.90 | 16.00 | 43.90 | 74.60 |
| 9' Door hardware kit | BC@1.00 | Ea | 91.60 | 32.10 | 123.70 | 210.00 |
| 16' Door hardware kit | BC@1.00 | Ea | 72.00 | 32.10 | 104.10 | 177.00 |
| Containment C hook | — | Ea | 2.80 | -- | 2.80 | — |
| Spring connector C Hook | — | Ea | 5.50 | -- | 5.50 | — |
| Double door low headroom kit | BC@.500 | Ea | 63.35 | 16.00 | 79.35 | 135.00 |
| Track hanger kit | — | Ea | 17.90 | — | 17.40 | — |
| **Garage door replacement hardware.** | | | | | | |
| 3" replacement pulley | BC@.500 | Ea | 3.00 | 16.00 | 19.00 | 32.30 |
| 4" replacement pulley | BC@.500 | Ea | 5.91 | 16.00 | 21.91 | 37.20 |
| 7' lift cable set | BC@.500 | Ea | 3.95 | 16.00 | 19.95 | 33.90 |
| 8' lift cable set | BC@.500 | Ea | 4.69 | 16.00 | 20.69 | 35.20 |
| Safety lift cable set | BC@.500 | Ea | 4.54 | 16.00 | 20.54 | 34.90 |
| Window insert | BC@.250 | Ea | 45.40 | 8.02 | 53.42 | 90.80 |
| Track, bolt and flange nuts | — | Ea | 2.02 | — | 2.02 | — |
| Replacement nylon rollers | BC@.500 | Ea | 2.45 | 16.00 | 18.45 | 31.40 |
| G-top bracket | — | Ea | 1.98 | — | 1.98 | — |
| Lift handles | BC@.250 | Ea | 2.02 | 8.02 | 10.04 | 17.10 |

**Garage door openers.** Radio controlled, electric, single or double door, includes typical electric hookup to existing adjacent 110 volt outlet. 1/2 HP opener, wireless receiver and two multiple frequency transmitters.

| | Craft@Hrs | Unit | Material | Labor | Total | Sell |
|---|---|---|---|---|---|---|
| Chain drive glide opener, | BC@3.90 | Ea | 175.00 | 125.00 | 300.00 | 510.00 |
| Screw drive glide opener | BC@3.90 | Ea | 200.00 | 125.00 | 325.00 | 553.00 |
| Add for additional transmitter | — | Ea | 32.60 | — | 32.60 | — |
| Add for operator reinforcement kit | — | Ea | 16.40 | — | 16.40 | — |
| Add for low headroom kit | — | Ea | 63.40 | — | 63.40 | — |
| Add for garage door monitor | — | Ea | 27.00 | — | 27.00 | — |
| Add for monitor add-on sensor | — | Ea | 15.10 | — | 15.10 | — |
| Add for 8' chain drive extension kit | — | Ea | 32.40 | — | 32.40 | — |
| Add for 8' belt drive extension kit | — | Ea | 20.50 | — | 20.50 | — |
| Add for remote light control | — | Ea | 21.00 | — | 21.00 | — |
| Add for wireless keyless entry control | — | Ea | 37.80 | — | 37.80 | — |
| Add for motion detecting wall control | — | Ea | 21.60 | — | 21.60 | — |
| Add for opener surge protector | — | Ea | 16.20 | — | 16.20 | — |
| Add for replacement safety beam kit | — | Ea | 32.40 | — | 32.40 | — |
| **Garage door single extension spring.** | | | | | | |
| 70 pounds | BC@.500 | Ea | 10.70 | 16.00 | 26.70 | 45.40 |
| 110 pounds | BC@.500 | Ea | 11.30 | 16.00 | 27.30 | 46.40 |
| 120 pounds | BC@.500 | Ea | 12.00 | 16.00 | 28.00 | 47.60 |
| 130 pounds | BC@.500 | Ea | 12.70 | 16.00 | 28.70 | 48.80 |
| 140 pounds | BC@.500 | Ea | 13.60 | 16.00 | 29.60 | 50.30 |
| 150 pounds | BC@.500 | Ea | 14.50 | 16.00 | 30.50 | 51.90 |
| 160 pounds | BC@.500 | Ea | 15.60 | 16.00 | 31.60 | 53.70 |
| U-bolt spring anchor | BC@.100 | Ea | 10.80 | 3.21 | 14.01 | 23.80 |
| 8-link spring anchor chain | BC@.100 | Ea | 4.32 | 3.21 | 7.53 | 12.80 |

| | Craft@Hrs | Unit | Material | Labor | Total | Sell |
|---|---|---|---|---|---|---|

## Emergency Exit Hardware

**Install rim panic bar on metal door.** Includes measure and mark, drill holes, cut bar to length, assemble bar, set strike plate.

| | Craft@Hrs | Unit | Material | Labor | Total | Sell |
|---|---|---|---|---|---|---|
| Per bar, no exit bolts | BC@.685 | Ea | 117.75 | 22.00 | 139.75 | 238.00 |

**Touch bar exit hardware.** For use with 1-3/4"-thick hollow metal and wood doors. Extruded anodized aluminum with stainless steel spring. 3/4" throw stainless steel latch bolts. Strike for 5/8" stop; shim for 1/2" stop; 2-3/4" backset, 4-1/4" minimum stile. Includes standard mounting with sheet metal and machine screws. Recommended mounting height: 40-5/16" from ceiling to finished floor. 36" maximum width. Rim-type cylinders with Schlage "C" keyway. Includes measure and mark, drill holes, assemble, set strike plate.

| | Craft@Hrs | Unit | Material | Labor | Total | Sell |
|---|---|---|---|---|---|---|
| Install, single door | BC@1.45 | Ea | 111.00 | 46.50 | 157.50 | 268.00 |
| Install, per set of double doors | BC@2.97 | Ea | 220.00 | 95.30 | 315.30 | 536.00 |
| Remove, single door | BC@.369 | Ea | — | 11.80 | 11.80 | 20.10 |
| Remove, double doors and mortise lock | BC@1.38 | Ea | — | 44.30 | 44.30 | 75.30 |

# Walls and Ceilings 10

**B**efore about 1950, most homes were built with interior walls and ceilings finished in lath and plaster. During the 1950s, gypsum wallboard became generally available and most owners and industry professionals switched over. While plaster and gypsum wallboard are equally durable (when protected from moisture and impact) and both make attractive wall and ceiling coverings, gypsum wallboard is preferred because it's so much less expensive to install. Lath and plaster is very labor-intensive, and because it requires days or weeks to dry out, it tends to delay construction. Today, it's unlikely you'll be called on to do a lath and plaster job. But you may have to make repairs on one.

Nearly all plaster develops hairline cracks. These cracks and small holes must be patched with patching plaster or joint compound or spackle before repainting. Plan on covering or replacing the surface if plaster is uneven, bulging, loose, or there are many large holes. If there are only a few large holes, you can patch them. If there are more than a few, it's easier to replace the entire wall. Loose or bulging plaster is a sign of water damage. When you find damage caused by water, your first step will be eliminating the source of moisture. Otherwise, the repair will be worthless.

Gypsum wallboard is usually called drywall, though you'll hear the same material referred to as wallboard, gypboard, Sheetrock or simply rock. The term drywall is a little misleading. First, drywall isn't just for walls. It adapts equally well to installation on ceilings. Second, drywall is dry only by comparison with plaster, which goes on wet. The cement (sometimes called mud) used to cover drywall joints is mixed with water and needs at least a few hours to dry completely.

Drywall does have its problems. Unlike plaster walls and ceilings, it can develop nail pops — nail heads that work loose and rise above the surface. Also, the paper tape used to cover panel joints can de-laminate and curl if not bedded correctly. Loose tape is a sign of a poor drywall job.

## Choices for Renewing Wall and Ceiling Cover

**1.** Patch and smooth what's there. This is the least expensive choice. But it's not practical when walls or ceilings are seriously damaged.

**2.** Install furring strips as a base for new wall cover, or apply new wall and ceiling cover directly over the old. This option works when the existing surface is flat and firm. I don't generally recommend this, as the wall will be

| Minimum Material Thickness for On Center Spacing of Studs and Joists | | |
|---|---|---|
| **Material** | 16" OC | 24" OC |
| **Drywall** | $3/8$" | $1/2$" |
| **Plywood Paneling** | $1/4$" | $3/8$" |
| **Hardboard Paneling** | $1/4$" | — |
| **Tongue & Groove Plank** | $3/8$" | $1/2$" |

**Note 1.** New wall cover set with nails or adhesive directly on existing wall cover can be any thickness because it's supported continuously on the existing surface.
**Note 2.** $1/4$" plywood or hardboard set on framing members 16" on center may be slightly wavy unless applied over at least $3/8$" drywall.

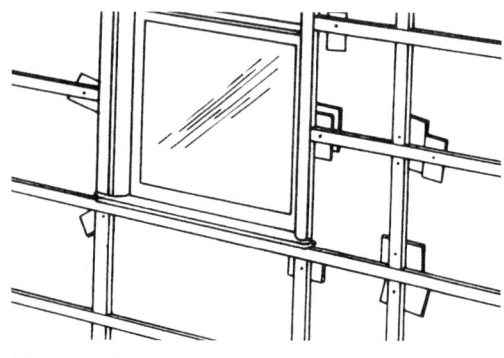

**Figure 10-1**

*Shingle shims behind furring to produce a smooth vertical surface*

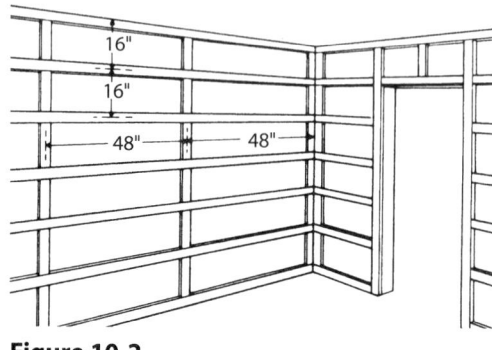

**Figure 10-2**

*Application of horizontal furring to interior wall*

significantly thicker, requiring moldings and casings to be redone, and new trim around wall penetrations. Plus, setting furring strips requires a great deal of work, and can result in an inferior job. Generally it's easier to tear off the existing wall cover and replace it, but you may come across a situation where this may be the best choice — perhaps when it's just one wall in a room and there are no openings. It's a lot less messy.

If new wall cover is going to be nailed over old or set on furring strips, the first step will be finding the studs. They're usually installed 16" on center and at the edge of doors and windows. Where you see a nail head in drywall or baseboard, assume there's a stud under the nail. If stud locations aren't obvious, put a stud finder to work. Stud finders locate dense material in the wall cavity — usually a stud. Mark the center of each stud at the top and bottom of the wall. Then snap a chalk line between the two marks.

Check walls for flatness by holding the straight side of a 2 × 4 against the surface. Furring will correct a bulging or indented wall. Mark wall locations that are uneven. Also check for true vertical alignment by holding a large carpenter's level against a straight 2 × 4 held against the wall. Use shingles to shim out furring and create a flat, vertical surface. See Figure 10-1.

Figure 10-2 shows 1" × 2" furring strips set 16" on center. Nail the furring at each stud. Remove existing base trim, window casings, and door casings. Apply furring around all the openings. Be sure there will be a vertical furring strip under each vertical joint in the new wall cover.

**3.** Your third choice for renewing wall and ceiling cover is to strip off what's covering the studs and joists and start over. The result is a wall or ceiling that's as good as new. In fact, it is new. This is by far the best option for seriously damaged walls, although stripping off the existing wall covering is messy, and it creates a disposal problem.

Before applying the new wall cover, be sure studs are aligned. Run a string across the front of the studs, holding it taut. The studs should all touch the string equally. Those that don't touch the string, or push the string out, are out of line with the rest and need to be re-aligned or replaced.

# Hanging Drywall

You can hang drywall either with the long edges horizontal (Figure 10-3A) or vertical (Figure 10-3B). Hanging the drywall horizontally reduces the number of vertical joints at the middle

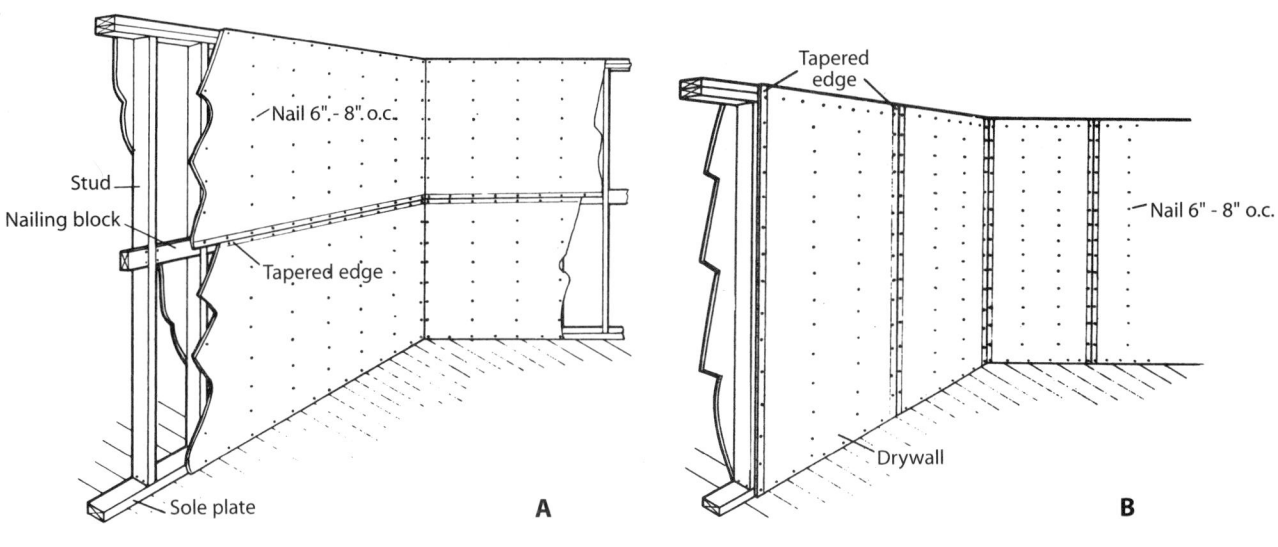

**Figure 10-3**

*Installing drywall on walls: **A**, horizontal application; **B** vertical application*

of walls. If you install drywall horizontally, be sure all joints are supported at the ends. If you install drywall vertically, you won't need nailing blocks — the stud acts as the nailing block. All drywall joints must be supported, or they'll crack. Whether hung vertically or horizontally, proper fastening is essential. If you have a drywall screw gun, drive screws every 12" on ceiling joists and every 16" on wall studs. Adjust the gun so it sets the screw just slightly into the board without breaking the paper. Some drywall specialists set the board initially with a few nails driven at the edges and then secure the board in place with screws.

If you nail the board in place, drive a nail every 8" along wall studs and every 6" along ceiling joists. Use 4d ring shank nails on $^3/_8$" board and 5d ring shank nails on $^1/_2$" board. Keep nails at least $^3/_8$" away from the edges of the board. Use a drywall hammer with a slightly convex head that leaves a dimple at each nail location. Be careful not to break the surface of the paper. See Figure 10-4. The dimple will be covered later with drywall joint compound (mud).

Also be aware of where you're driving fasteners. There may be pipe runs through the studs, and copper pipe is thin, soft and easily punctured with a drywall nail or a screw. Copper pipes and wire are supposed to be protected by metal plates to prevent punctures. But don't count on it. Take care when working around pocket door frames. Don't drive any nails or screw points into the pocket door cavity. When in doubt, use shorter nails or screws.

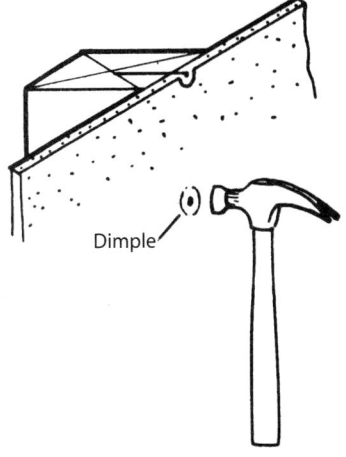

**Figure 10-4**

*Drive nails in "dimple" fashion*

Drywall is heavy. If you're hanging full 4' x 8' sheets on ceilings, you'll need a crew of at least two, and three would be better. One person working alone can drywall ceilings if it's an emergency, but it's not recommended. Start with the ceiling panels. Center the panel edges on the ceiling joist. Plan the layout so that any cut edges will wind up in the corners, and not in the middle of the ceiling. All joints must center on a joist. If there's no ceiling joist where you need one, you'll have to add one. Don't cut the drywall sheets to make the joint center on a joist — the seam will show.

When the ceiling is done, begin hanging drywall on the walls. The standard ceiling height is $97^1/8''$ between the floor and the bottom of the ceiling joists. Subtracting $^1/2''$ for the thickness of ceiling panels leaves a $96^5/8''$ wall height. Two panel widths total 96". Hang wall panels $^5/8''$ above the finished floor for a snug fit at the ceiling. The $^5/8''$ at the floor will be covered with baseboard. Use a drywall foot lift to hold the panel $^5/8''$ above the floor while driving nails or screws.

Regular core $^1/2''$ drywall panels are recommended for single-layer wall application in new residential construction. $^3/8''$ panels are recommended for ceilings in residential repair and remodeling in single or double-layers. Type X drywall is designed to meet requirements for fire safety. Greenboard is water-resistant for use behind tile. Brownboard is designed for exterior sheathing or soffits.

Joint compound comes in both powder and pre-mixed forms. Home improvement specialists generally use pre-mixed compound. Dry mix has to be used right away after adding water. Pre-mixed cement will last for weeks if kept in a sealed container. Regardless of which you use, the mix should have a soft, putty-like consistency that spreads easily with a trowel or wide putty knife. Mud that runs off the knife is too thin.

Most drywall has a tapered, or beveled edge. Joint compound and tape fill this recess, leaving a smooth, flat surface. If a sheet has been cut to fit, the edge won't be tapered. You can tape a square edge the same way you tape beveled edges, but the joint compound will rise slightly above the finished surface. The extra depth won't be as obvious if you feather out the joint compound at least 4" beyond the joint but one located in the middle of the ceiling is going to be noticeable. In the corners, a taped cut edge won't show. Taping and feathering cut joints will slow your job down to a crawl. Avoid this situation whenever possible.

Taping and finishing joints takes three or four days. Figure 10-5 illustrates the taping sequence. Paper joint tape is the most economical, about $2.50 per thousand square feet of board hung. Fiberglass tape costs more, about $6.00 per thousand square feet of board. But self-adhesive fiberglass joint mesh needs no embedding coat and is more durable than paper tape. It flexes rather than curling or tearing if the joint moves.

### Taping and Finishing Drywall

**1.** Start with the ceiling and work down the walls. If you've selected paper tape rather than fiberglass, spread joint compound over panel edges with a 5"-wide taping knife. Don't skimp on the mud. If you're using self-adhesive fiberglass tape, press tape over the joints and skip to step three below.

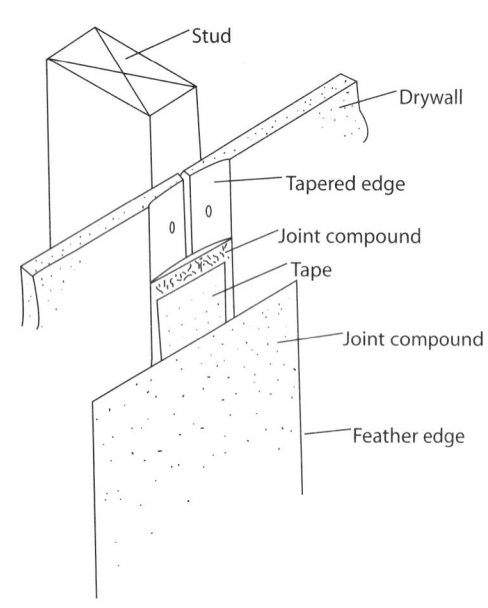

**Figure 10-5**

*Detail of joint treatment*

**2.** Press the paper tape into the mud with a drywall knife, not your hand. Then smooth the surface with the knife. Press hard enough to force joint compound through the small perforations in the tape, if the tape has perforations. But don't press too hard. Some mud should remain under the tape. When you're done taping and embedding the tape, let everything dry overnight before beginning the finish coats.

**3.** Cover the tape with cement, feathering the outer edges at least 2" on each side of the paper tape. Feather an additional 2" when covering a cut joint, as there's no taper. Then let the cement dry overnight.

**4.** When dry, sand lightly. A pole sander speeds this work. Then apply a thin second finish coat, feathering the edges a little past the edge of the prior coat. Use drywall topping compound designed for finish coats to create a smooth joint that's easy to sand. You can buy "all-purpose" compound, which can be used both for bedding and topping. But it's better to use bedding compound for bedding, and topping compound for finishing. Use a wider drywall knife for this finish coat, up to 12" wide. To save sanding time, keep this finish coat as smooth as possible. For top quality work, apply a third coat of mud after the second coat has dried and been sanded.

**5.** When the last coat of cement is dry, sand smooth.

**6.** Fill all nail and screw dimples with at least one coat of joint compound. Sand the surface after each coat is dry.

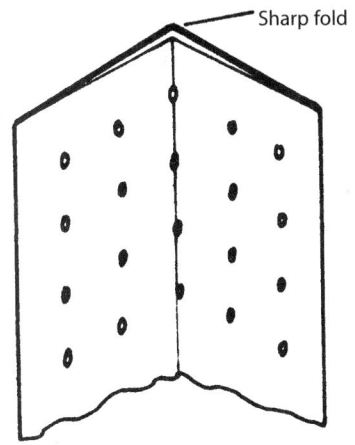

**Figure 10-6**
*Corner tape*

Use folded perforated tape on interior corners. See Figure 10-6. Fold tape down the center to form a right angle. Tape designed for this purpose already has a crease down the middle. Then apply cement on each side of the corner and press tape in place with the putty knife. Use a drywall corner knife to embed the tape on both intersecting walls at the same time. Finish the corner with a coat of joint compound. Smooth the cement on both surfaces of a corner at the same time with a corner knife. Let the corner dry overnight and then sand the surface smooth.

For exterior corners, use metal or plastic drywall corner bead. This makes a more durable corner, able to withstand impacts that are likely at external corners. Also apply paper drywall tape over the edge of the metal bead. Nail or screw outside corners to the board every 8" and finish with joint compound. When you're finished applying tape, bead and mud at both internal and external corners, you should have a 4" strip of mud on each side of every corner. Don't worry if this strip isn't smooth. Sanding and more finishing will follow.

Drywall mud shrinks as it dries. So apply a little more than actually needed to make a smooth finish, especially over fastener heads. Bear in mind that drywall sanding is very messy! It creates huge, choking clouds of dust. Wear a dust mask to avoid inhaling this stuff. In an unoccupied house, the mess isn't as much of a problem. However, if you're working in an occupied home, you'll need to control the dust. Homeowners don't appreciate having everything in their home covered with a thick layer of white powder. You'll either need to seal off your work area with plastic sheeting, or use a dustless sanding system, such as a wet sanding sponge.

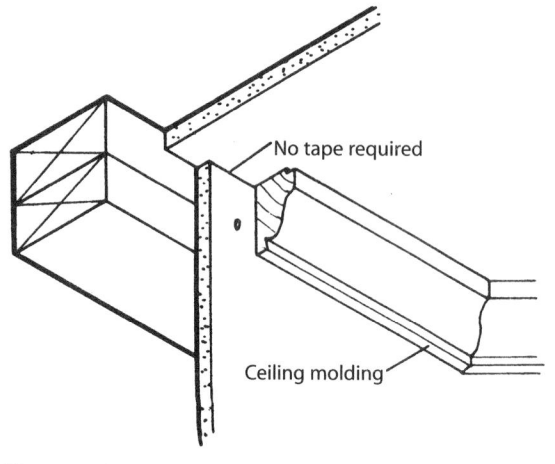

**Figure 10-7**
*Ceiling molding*

If you plan to install crown molding, no taping is needed where walls meet the ceiling. See Figure 10-7. Set the trim with 8d finishing nails spaced 12" to 16" apart. Be sure to nail into the top wall plate.

## Drywall Ceilings

You can apply new drywall directly over old plaster or on furring strips nailed over an uneven plaster ceiling. Applying furring strips on a ceiling won't create the problems that furring would on a wall. But furring out the ceiling will usually be more work than tearing down the ceiling cover and starting over. Use 2" x 2" or 2" x 3" furring strips nailed perpendicular to the joists. Space these furring strips 16" on center for $^3/_8$" drywall or 24" on center for $^1/_2$" drywall. Nail the furring strips with two 8d nails at each joist. Stagger board end joints. Be sure board edges end on a joist or furring strip. Don't jam the boards tightly together. It's best if there's only light contact at each edge.

Hanging ceiling panels is easier with a drywall lift that allows precise positioning while leaving two hands free for driving nails or screws. If you don't have a drywall lift, cut two braces like the one shown in Figure 10-8. Make them slightly longer than the ceiling height. Nail or screw the drywall to all supporting members, spacing the fasteners 7" to 8" apart. If you use nails, select 5d ($1^1/_4$") ring-shank nails for $^1/_2$" drywall and 4d (1") ring shank nails for $^3/_8$" drywall. Again, don't break the surface of the paper when driving fasteners. Finish ceiling joints the same way you finish wall joints.

## Textured Finishes for Drywall

Wall finish is usually smooth to make cleaning easier. But the ceiling finish may be textured, usually with some form of joint compound. Texture hides ridges and bumps in the ceiling and improves acoustics by eliminating echo off the ceiling. But textured ceilings are also an admission that there's something to hide. Many owners don't like textured ceilings and know that texture is used to hide defects. The first thing they'll say when they see a textured ceiling is, "What's wrong with your ceiling?"

*Orange peel* texture consists of thinned joint compound applied with a long-nap paint roller. In an emergency, you can make ceiling texture by thinning out joint compound with water until it reaches a consistency similar to that of paint. But it's better to buy mix that's specifically made for texturing. It's much easier than trying to make your own. To ensure proper consistency, try applying some mixture to a scrap piece of drywall held upright. Adjust the consistency as necessary by adding

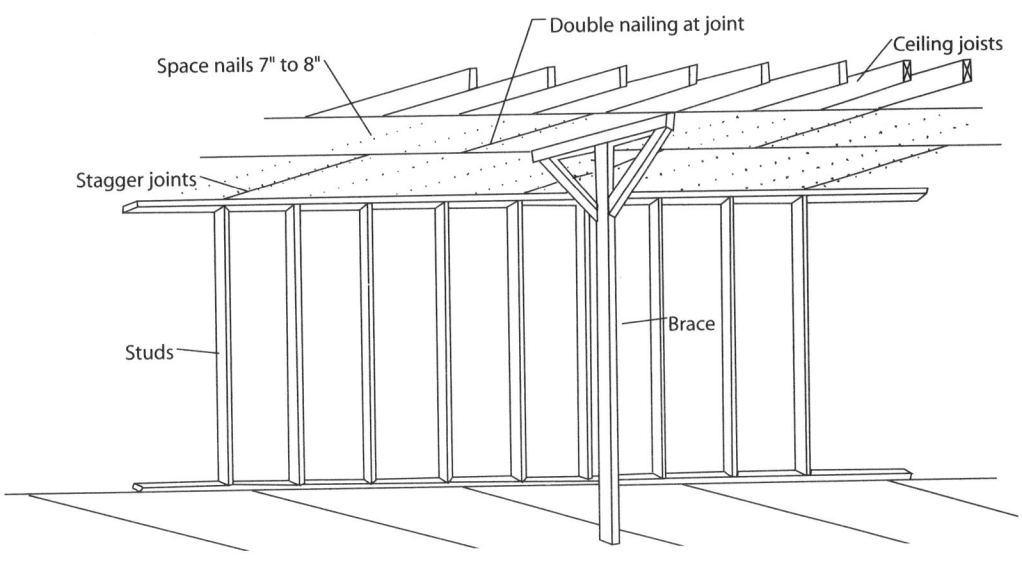

**Figure 10-8**

*Installing drywall on ceiling*

more water or joint compound to the bucket of mixture. When you've got the mixture just right, roll it onto the ceiling or wall. Keep the rolling pattern uniform so the texture appears to have a grain. When the mixture dries, avoid the temptation to sand the surface. The texture is very fragile. Sanding can knock off too much of the desirable surface.

*Spatter finish* is done with a compressor-operated spatter gun that shoots globules of thinned drywall mud on the ceiling at random. Scrape overspray off the walls. Other mixes are available to create different effects. You can get nearly the same spatter effect by dipping a stiff-bristle brush in thinned drywall mud and slinging mud on the ceiling with a snap of the wrist. This takes practice. Don't count on getting this right on the first attempt. Control the size of the spatters by making the mix thinner or thicker. Obviously, this is messy work. But it's an effective technique when you have to match only a few square feet of spatter-finished ceiling. For minor patching, you can also buy spatter finish in an aerosol can.

*Knockdown finish* uses the same technique — spatter blown or snapped on the ceiling. But the mix should be stiffer so spatters are between the size of a pea and a grape. Let the globules dry for a few minutes. Then knock the tops off with a masonry trowel. Work the trowel in all directions to avoid creating an obvious grain in the texture.

*Skip trowel or imperial texture* is like a knockdown finish, only more so. Apply mud to the ceiling or wall in a random pattern. Then smooth out what's there, leaving irregular patterns of texture in some areas and no texture in others. When done, it should look like Spanish stucco.

*Veneer plaster* is used in one or two $1/8''$ coats over a veneer plaster base such as blueboard. Blueboard is similar to drywall, with a paper surface designed to bond well with the veneer plaster. Apply enough plaster to trowel a smooth, even finish over the entire surface. It's a lot of work, but veneer plaster hides imperfections and joints better than regular plaster, and provides a hard coat that protects the paper surface below.

*Cottage cheese or popcorn texture* is applied with a compressor, hopper and applicator gun. Although popcorn ceiling texture isn't currently in fashion, and you're not likely to be installing it, you may still be called on to do a repair job. Aerosol sprays are available to match the existing cottage cheese texture if you're only patching a small area.

No matter what finish you apply, the job isn't done until the surface has been primed and painted.

# Repairing Drywall

Fill nail holes and small cracks in board by applying a smooth coat of drywall compound. Let it dry. Then sand the surface smooth. Repairing larger holes in drywall isn't as easy. There's nothing but wall cavity behind a full penetration of the board. Drywall compound will fall into holes wider than about 1/2". Cover larger cracks and small holes with self-adhesive fiberglass tape. Then press stiff drywall compound into the mesh. When the first coat is dry, apply a finish coat. With a patch like this, feather the drywall compound 12" on each side of the crack to avoid leaving an obvious ridge. Again, don't count on getting this right the first time.

Holes larger than a golf ball need some type of backing to hold the drywall mud until it sets. You can buy a drywall repair kit with clips that support drywall cut to cover nearly any size hole. If these drywall clips create lumps or otherwise don't work for you, make a patch kit with cardboard, string and a short length of dowel. Cut a piece of stiff cardboard slightly larger than the hole. Loop a short length of string through the center of the cardboard patch. Then fold the cardboard in half and insert it into the cavity. Pull the string tight, flattening the cardboard against the cavity side of the board and closing off the hole. Tie the loose end of the string around a short dowel laid across the hole. Then apply a coat of drywall compound over the hole and against the cardboard backing. Leave the patch slightly concave. When dry, cut the string and remove the dowel.

Many experienced drywall experts use neither clips nor cardboard. Instead, they cut a piece of scrap wood that will fit through the hole and extend about 2" to either side. They screw this in place with drywall screws on either side of the hole. This puts a firm foundation behind a portion of the hole. Then they cut a piece of drywall to fit the hole and screw it to the scrap wood. Once in place, they lay lengths of self-adhesive fiberglass drywall tape over the patch so it laps several inches onto firm wallboard. Then they apply a finish coat of joint compound and feather out several inches beyond the patch. When dry, they sand the patch smooth. Once primed and painted, there should be no evidence of the repair.

A hole more than 12" across is probably too large for a cardboard-backed patch. Instead, mark and cut out a rectangular section of wallboard all the way to the middle of the studs at both sides. Cut two nailing blocks to fit horizontally between the studs. Insert the blocks into the cutout and toenail them at the top and bottom of the rectangular cutout. Then cut drywall to fit in the cutout. Tape and finish the perimeter of the cutout as with any drywall joint.

# Wood Paneling

Plywood paneling is sold in many grains and species. Hardboard imprinted with a wood grain pattern is generally less expensive. Better hardboard paneling has a realistic wood grain pattern. Both plywood and hardboard paneling are sold with a hard, plastic finish that's easily wiped clean. Hardboard is also available with vinyl coatings in many patterns and colors, including some that have the appearance of ceramic tile.

Wood paneling should be delivered to the site a few days before application. Panels need time to adapt to room temperature and humidity before application. Stack panels in the room separated by full length furring strips so air can circulate to panel faces and backs. Figure 10-9.

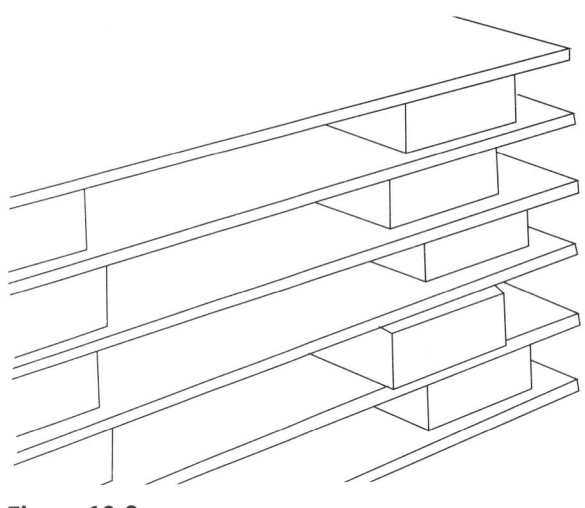

**Figure 10-9**

*Stacking panels for conditioning to room environment prior to use*

Always start a panel application with a truly vertical edge. If a corner is straight and vertical, butt the first panel into that corner. Cut subsequent panels so they lap on studs. If you don't have a vertical corner, tack a panel perfectly vertical and 2" from the starting corner. Use an art compass to scribe the outline of the corner on the panel edge. See Figure 10-10. Cut the panel along this line and move it into the corner. Butt the next panel against the first, being careful to keep the long edges truly vertical. Use the same art compass to scribe a line for panel top edges.

Fasten the panels with nails or adhesive. Adhesive saves filling nail holes on the panel surface. Use adhesive that provides "work time" before forming a tight bond. That makes it easier to adjust panels for a good fit. If panels are nailed, use small finishing nails (brads). Use $1^{1}/_{2}$" long brads for $^{1}/_{4}$" or $^{3}/_{8}$" thick materials. Drive a brad each 8" to 10" along edges and at intermediate supports. Most panels are grooved to simulate hardboard panels. Drive brads in these grooves. Set brads slightly below the surface

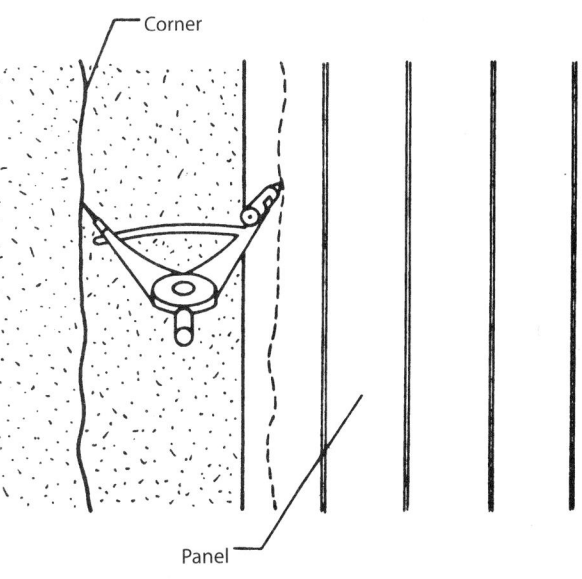

**Figure 10-10**

*Scribing of cut at panel edge to provide exact fit in a corner or at ceiling*

with a nail set. Many vendors of prefinished paneling also sell matching nails that require no putty to fill nail holes. Other vendors sell wood-filler putty to match their panels.

### Hardwood Paneling

Most hardwood paneling is 8" wide or less. Hardwood paneling needs several days to adapt to room temperature and moisture conditions before being applied. Most paneling is applied with the long edges running vertically. But rustic patterns may be applied horizontally or diagonally to achieve a special effect.

Nail vertical paneling to horizontal furring strips or to nailing blocks set between studs. Use $1^1/_2$" to 2" finishing or casing nails. Blind nail through the tongue on narrow strips. For 8" boards, face nail near each edge.

# Ceiling Tile

Tile attached to the ceiling is usually 12" x 12". Suspended ceiling panels are usually 2' x 2' or 2' x 4'. Ceiling tile can be set with adhesive if the surface is smooth, level and firm. Dab a small spot of adhesive at the center and at each corner of the tile. Edge-matched tile can be stapled if the backing is wood.

You can set tile on furring strips to cover unsightly defects. But it's usually faster, cheaper and results in a better job if you tear off the existing cover and start over. If you want to try setting tile over the existing ceiling, use 1" x 3" or 1" x 4" furring strips where ceiling joists are 16" or 24" on center. Fasten the furring with two 7d or 8d nails at each joist. Where trusses or ceiling joists are spaced up to 48" apart, fasten 2" x 2" or 2" x 3" furring strips with two 10d nails at each joist. The furring should be a low-density wood, such as a soft pine, if tile is to be stapled to the furring.

Lay furring strips from the center of the room to the edges. Find the center by snapping chalk lines from opposite corners. The ceiling center is where the diagonal lines cross. Place the first furring strip at the room center and at a right angle to the joists. Run parallel furring strips each 12" to both edges of the room. See Figure 10-11 A. Edge courses on opposite walls should be equal in width. Plan spacing perpendicular to joists the same way. End courses should also be equal in width. Install tile the same way, working from the center to the edges. Set edge tile last so you get a close fit. Ceiling tile usually has a tongue on two adjacent edges and grooves on the other edges. Keep the tongue edges on the open side so they can be stapled to furring strips. Attach edge tile on the groove side with finishing nails or adhesive. Use one staple at each furring strip on the leading edge and two staples along the side, as in Figure 10-11 B. Drive a small finishing nail or use adhesive to set edge tile against the wall.

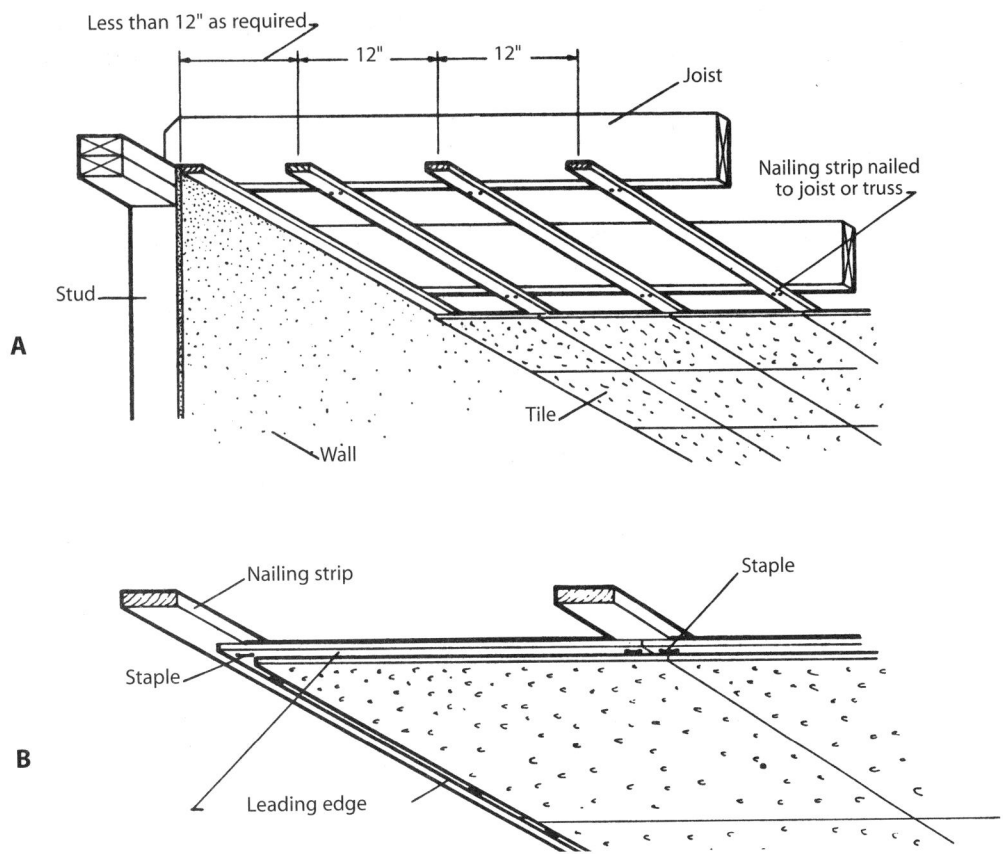

**Figure 10-11**

*Ceiling tile installation: **A**, nailing strip location; **B**, stapling*

Be careful not to soil the tile surface. Grease will leave permanent stains on ceiling tile. Professional tile installers rub corn meal between their palms to keep their hands oil-free.

# Suspended Ceilings

These are fine for basements, or other informal areas. Using them in "formal" rooms, like a living room or dining room, creates a "low-income" effect that may not be exactly what your client intends. Suspended ceilings cover imperfections, lower the ceiling to a more practical height, and add a plenum for running new electrical, plumbing, and HVAC lines.

The ceiling grid is suspended from wires or straps attached to joists. Panels drop into the completed grid. Ceiling height can be any level. Hanger wires may be only 2" or 3" long if the primary purpose is to cover fractured plaster. In earthquake zones, seismic bracing may be required by the building code. Your building department will have more information on this.

# Interior Trim

Many older homes have trim styles no longer available at building material dealers. Matching trim exactly may require expensive custom fabrication. Try to remove trim in salvage condition so it can be re-installed. If trim is damaged or if you have to move doors or windows, it may be easier to replace all the trim in the room rather than try to match existing trim.

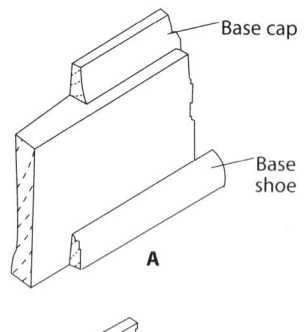

Keep in mind that trim work requires a very high level of carpentry skill. Trim needs to be essentially perfect, sloppy joints and visible nail heads won't do. Don't ask a rough carpenter to do trim work — the results will be a disappointment If trim is going to be painted, select a trim made of extruded polymer, ponderosa pine or northern white pine, or primed MDF. Highly decorative cast trim is another good choice if trim will be painted. Most natural finish trim in modern homes is pine or oak. These woods can be very attractive if they're nicely finished.

## Casing

Casing is the interior edge trim for door and window openings. Modern casing patterns vary in width from $2^1/4"$ to $3^1/2"$ and in thickness from $1/2"$ to $3/4"$. Install casing about $3/16"$ back from the face of the door or window jamb. Nail with 6d or 7d casing or finishing nails, depending on thickness of the casing. Space nails in pairs about 16" apart, nailing to both jambs and framing. Casing with molded forms requires mitered joints, while rectangular casing can be butt-joined.

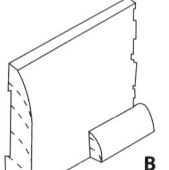

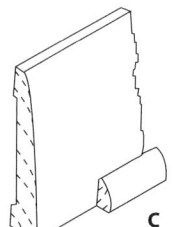

**Figure 10-12**

*Baseboard: **A**, two-piece; **B**, narrow; **C**, medium width*

## Baseboard

Finish the joint between the wall and floor with baseboard. Figure 10-12 shows several sizes and forms of baseboard. Two-piece base consists of a baseboard topped with a small base cap. The cap covers any gap caused by irregularities in the wall finish. Base shoe is nailed into the subfloor and covers irregularities in the finished floor. Drywall walls seldom need a base cap. Carpeted floors hide variations in the floor and make base shoe unnecessary.

Install square-edged baseboard with butt joints at inside corners and mitered joints at outside corners. Nail at each stud with two 8d finishing nails. Molded base, base cap, and base shoe require a coped joint at inside corners and a mitered joint at outside corners. See Figure 10-13.

## Other Molding

Ceiling molding may be strictly decorative or may be used to hide the joint where the wall and ceiling meet. Use crown molding to cover the gap where wood paneling meets the ceiling. Attach crown molding with finishing nails driven into upper wall plates. Wide crown molding should be nailed both to the wall plate and the ceiling joists.

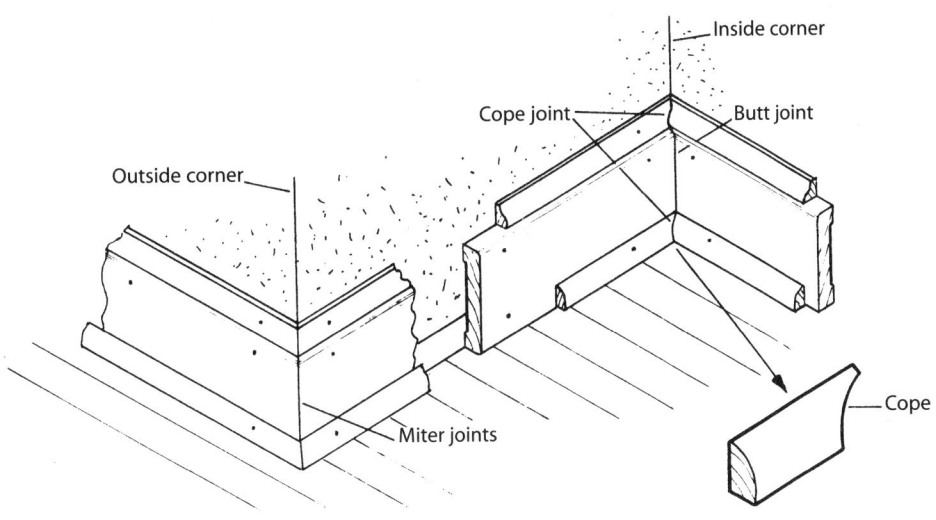

**Figure 10-13**

*Installation of base molding*

# Removing Partitions

Modern taste favors more open space in homes. For example, new homes often have the family room and kitchen combined as one open area. Many older homes have a formal dining room or kitchen dining area enclosed by four walls and with a door that can be closed. Removing a nonbearing wall (partition) can add livability to an older home.

Nonbearing partitions support neither the roof nor a floor above. Breaking out a partition is only a cosmetic change. If wall cover is plaster or drywall, there's no salvage value in the partition. But save the trim, if possible. You may need it later.

Many partition walls include plumbing, electric or HVAC lines. Plan how those will be handled before you begin demolition.

When the partition is gone, there will be a strip of exposed floor, ceiling and wall where bottom, top and side edges of the partition had been. Finish the ceiling and walls with strips of drywall, tape, joint compound and paint. Filling the strip in the floor isn't as easy. Usually the best you can do is level the surface and cover the area with carpet or vinyl. With oak strip floors, it's possible to patch holes by weaving in new oak strips. However, this is a tricky job, and the entire floor may have to be refinished to get a perfect match of colors.

Removing a loadbearing partition requires the same patching of walls, ceiling, and floor. But you also have to add support for ceiling joists. If there's attic space above the partition, install a support beam above the ceiling joists. If it's a loadbearing wall on the first floor of a two-story house, it's holding up the second floor. If you remove it, the upstairs rooms can collapse onto the first floor. You'll need a large beam and posts to carry the weight that the wall was carrying. In this case, the beam will have to be below the ceiling joists, since the ceiling joists of the first

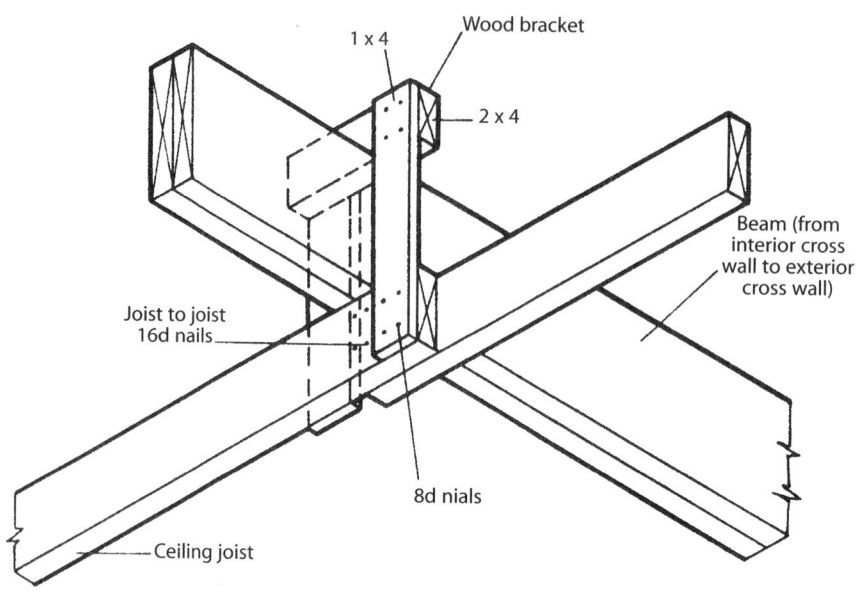

**Figure 10-14**

*Framing for flush ceiling with wood brackets*

floor are also the floor joists of the second floor. Be sure both ends of the beam are well supported on a bearing wall or post that is supported by the foundation. Support joists with metal framing anchors or wood brackets, as illustrated in Figure 10-14. To eliminate the need for temporary support, install the new beam before demolition begins.

If there's no attic space for a concealed beam, substitute an exposed beam at least 6'8" above the floor. Support ceiling joists temporarily with jacks and blocking while the partition is demolished and until the new beam is in place.

Figure 10-15 illustrates another option if an exposed beam is objectionable and there's no attic space for a concealed beam. Existing joists are supported by a new beam inserted where the top of the bearing wall had been. Place temporary joist supports on both sides of the bearing wall. Then remove the bearing wall and cut the joists as needed. Insert the new beam and install a hanger for each joist. Posts will also be needed to support this new beam.

The size of the beam required will vary with the span, load and lumber grade. Beam sizing like this is work for a civil engineer. In some communities, you'll need the approval of a licensed engineer before a permit is issued. Your building department or lumber yard probably has span tables for beams and load tables for posts that cover the most common residential situations. You probably won't need an engineer unless you're spanning a huge opening. For example, licensed contractors are often allowed to do simple engineering like this for houses up

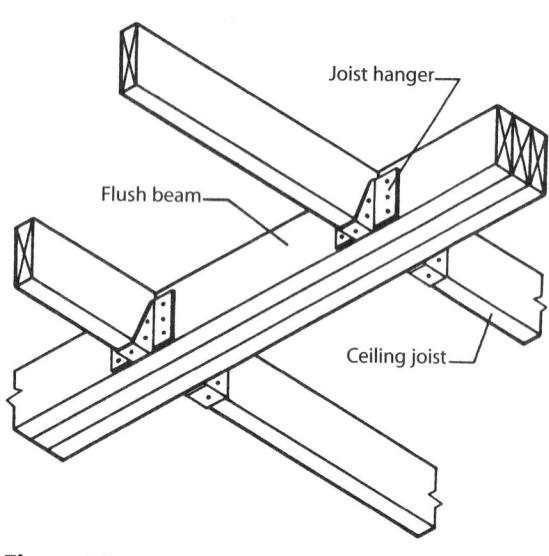

**Figure 10-15**

*Flush beam with joist hangers*

to 3,000 square feet. Houses this size and smaller are fairly straightforward. Larger houses and commercial buildings are more likely to present more complex engineering problems. When in doubt, make the framing much stronger than necessary. Building departments never have a problem with this.

# Adding Partitions

Partition walls support nothing but their own weight and can be framed from 2" x 3" lumber, though 2" x 4" studs and plates are more common. The first step is to install the top plate. If ceiling joists are perpendicular to the partition, nail the top plate to each joist using 16d nails. If ceiling joists are parallel to the top plate and the partition is not directly under a joist, install solid blocking between joists. Blocks should be no more than 24" on center. See Figure 10-16. Nail the top plate to the blocks.

To be sure the new partition will be vertical, hold a plumb line along the side of the top plate at several points. Mark where the plumb bob touches the floor. Nail the sole plate to the floor joist at that position. If there's no joist where needed, nail solid blocking between joists. Blocks should be no more than 24" on center. Cut studs to fit snugly between the plates every 16" on center. Stud lengths may vary, so measure for each stud. Toenail the studs to the plates using 8d nails.

If you have enough space, assemble the wall on the floor and tilt it into place. Nail the top plate to the studs first. Tilt the assembled wall into place. Then toenail the studs to the bottom plate as described above.

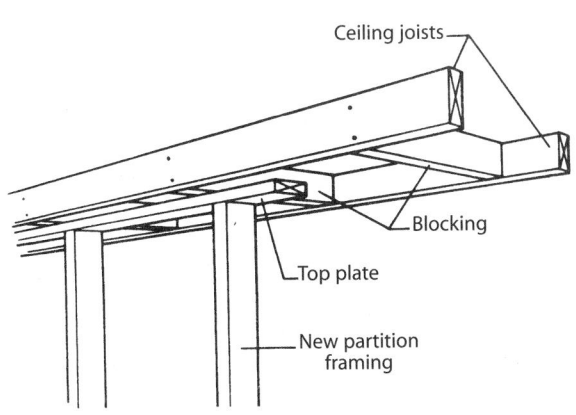

**Figure 10-16**
*Blocking between joists to which the top plate of a new partition is nailed*

| | Craft@Hrs | Unit | Material | Labor | Total | Sell |
|---|---|---|---|---|---|---|

## Drywall

**Ceiling and wall demolition.** Knock down with hand tools at heights to 9' and handle debris to a trash bin on site. Building structure to remain. Includes the cost of breaking out old ceiling or wall cover, pulling or driving the old fasteners and cleaning up the debris. Add the cost of hauling debris off the site and dump fees. These figures assume demolition with a crowbar. Knock a hole with a crowbar, hook the crowbar in the hole, pull, and get out of the way! Ceilings come down fast. Add extra time if you're planning to save the floor, and put down floor protection. Plaster will fall in big chunks that can gouge holes in the floor. Figures in parentheses show the volume and weight of materials after demolition.

| | Craft@Hrs | Unit | Material | Labor | Total | Sell |
|---|---|---|---|---|---|---|
| Plaster on ceiling (175 to 200 SF per CY and 8 pounds per SF) | | | | | | |
| Lath and plaster only | BL@.010 | SF | — | .27 | .27 | .46 |
| Lath, plaster and furring | BL@.015 | SF | — | .40 | .40 | .68 |
| Suspended lath and plaster | BL@.010 | SF | — | .27 | .27 | .46 |
| Plaster on walls (150 SF per CY and 8 pounds per SF) | | | | | | |
| Lath and plaster only | BL@.011 | SF | — | .29 | .29 | .49 |
| Lath, plaster and furring | BL@.015 | SF | — | .40 | .40 | .68 |

**Remove drywall.** Includes the cost of breaking out old board, pulling or driving the old fasteners and cleaning up the debris. Add the cost of hauling debris off the site and dump fees.

| | Craft@Hrs | Unit | Material | Labor | Total | Sell |
|---|---|---|---|---|---|---|
| Remove full panels on ceilings | BL@.009 | SF | — | .24 | .24 | .41 |
| Remove full panels on walls | BL@.008 | SF | — | .21 | .21 | .36 |
| Remove full panels and furring strips | BL@.028 | SF | — | .75 | .75 | 1.28 |
| Remove and install drywall, no joint treatment | | | | | | |
| per square foot | BC@.022 | SF | — | .71 | .71 | 1.21 |
| Add for job setup, per room | BC@.200 | Ea | — | 6.42 | 6.42 | 10.90 |

**Repair drywall.** Includes joint tape, three coats of joint compound and finish sanding.

| | Craft@Hrs | Unit | Material | Labor | Total | Sell |
|---|---|---|---|---|---|---|
| Cut out section, to 4' x 4' | D1@.288 | Ea | — | 8.90 | 8.90 | 15.10 |
| Remove and replace section, to 4' x 4' | D1@.840 | Ea | 5.80 | 25.90 | 31.70 | 53.90 |
| Taping and finishing only, ceilings | DT@.010 | SF | .01 | .31 | .32 | .54 |
| Taping and finishing only, walls | DT@.008 | SF | .01 | .25 | .26 | .44 |
| Tape and finish cracks, ceilings | DT@.060 | LF | .04 | 1.85 | 1.89 | 3.21 |
| Tape and finish cracks, walls | DT@.050 | LF | .04 | 1.54 | 1.58 | 2.69 |

**Patch hole in drywall.** Cut back drywall, insert backing in wall cavity, apply three coats of joint compound, sand smooth.

| | Craft@Hrs | Unit | Material | Labor | Total | Sell |
|---|---|---|---|---|---|---|
| Per patch | BC@1.00 | Ea | 1.76 | 32.10 | 33.86 | 57.60 |

**Regular core drywall.** Tapered edges. Cut ends. Labor includes cutting board around electrical boxes and obstacles, installing on wall studs or ceiling joists 8' to 12' above floor level, joint tape, three coats of joint compound and finish sanding. Material includes 1/2 gallon of premixed joint compound per 100 square feet, 38 linear feet of 2" perforated joint tape per 100 square feet, 1/2 pound of drywall screws per 100 square feet and 10% waste.

| | Craft@Hrs | Unit | Material | Labor | Total | Sell |
|---|---|---|---|---|---|---|
| 3/8" x 4' x 8', ceilings | D1@.023 | SF | .38 | .71 | 1.09 | 1.85 |
| 3/8" x 4' x 8', walls | D1@.017 | SF | .38 | .53 | .91 | 1.55 |
| 1/2" x 4' x 8', ceilings | D1@.024 | SF | .38 | .74 | 1.12 | 1.90 |
| 1/2" x 4' x 8', walls | D1@.018 | SF | .38 | .56 | .94 | 1.60 |
| 1/2" x 4' x 9', ceilings | D1@.024 | SF | .38 | .74 | 1.12 | 1.90 |
| 1/2" x 4' x 9', walls | D1@.018 | SF | .38 | .56 | .94 | 1.60 |
| 1/2" x 4' x 10', ceilings | D1@.024 | SF | .37 | .74 | 1.11 | 1.89 |
| 1/2" x 4' x 10', walls | D1@.018 | SF | .37 | .56 | .93 | 1.58 |
| 1/2" x 4' x 12', ceilings | D1@.024 | SF | .38 | .74 | 1.12 | 1.90 |
| 1/2" x 4' x 12', walls | D1@.018 | SF | .38 | .56 | .94 | 1.60 |
| 1/2" x 54" x 12', ceilings | D1@.024 | SF | .40 | .74 | 1.14 | 1.94 |
| 1/2" x 54" x 12', walls | D1@.018 | SF | .40 | .56 | .96 | 1.63 |

| | Craft@Hrs | Unit | Material | Labor | Total | Sell |
|---|---|---|---|---|---|---|
| 1/2" x 4' x 12', ceilings, sag resistant | D1@.024 | SF | .37 | .74 | 1.11 | 1.89 |
| 1/2" x 4' x 12', soffit board | D1@.035 | SF | .46 | 1.08 | 1.54 | 2.62 |
| 1/2" panels on ceilings, no tape or finish | D1@.012 | SF | .34 | .37 | .71 | 1.21 |
| 1/2" panels on walls, no tape or finish | D1@.008 | SF | .34 | .25 | .59 | 1.00 |
| Arches, soffits, recesses, columns, angles | D1@.035 | SF | — | 1.08 | 1.08 | 1.84 |

**Firecode type X drywall.** Fire-resistant gypsum core. Labor Includes cutting board around electrical boxes and obstacles, installing on wall studs or ceiling joists 8' to 12' above floor level, joint tape, three coats of joint compound and finish sanding. Material includes 1/2 gallon of premixed joint compound per 100 square feet, 38 linear feet of 2" perforated joint tape per 100 square feet, 1/2 pound of drywall screws per 100 square feet and 10% waste.

| | Craft@Hrs | Unit | Material | Labor | Total | Sell |
|---|---|---|---|---|---|---|
| 5/8" x 4' x 6', ceilings | D1@.025 | SF | .51 | .77 | 1.28 | 2.18 |
| 5/8" x 4' x 6', walls | D1@.019 | SF | .51 | .59 | 1.10 | 1.87 |
| 5/8" x 4' x 8', ceilings | D1@.025 | SF | .44 | .77 | 1.21 | 2.06 |
| 5/8" x 4' x 8', walls | D1@.019 | SF | .44 | .59 | 1.03 | 1.75 |
| 5/8" x 4' x 9', ceilings | D1@.025 | SF | .43 | .77 | 1.20 | 2.04 |
| 5/8" x 4' x 9', walls | D1@.019 | SF | .43 | .59 | 1.02 | 1.73 |
| 5/8" x 4' x 10', ceilings | D1@.025 | SF | .46 | .77 | 1.23 | 2.09 |
| 5/8" x 4' x 10', walls | D1@.019 | SF | .46 | .59 | 1.05 | 1.79 |
| 5/8" x 4' x 12', ceilings | D1@.025 | SF | .44 | .77 | 1.21 | 2.06 |
| 5/8" x 4' x 12', walls | D1@.019 | SF | .44 | .59 | 1.03 | 1.75 |
| 5/8" panels on ceilings, no tape or finish | D1@.013 | SF | .35 | .40 | .75 | 1.28 |
| 5/8" panels on walls, no tape or finish | D1@.009 | SF | .35 | .28 | .63 | 1.07 |

**Water-resistant drywall.** Moisture-resistant gypsum core and paper as a base for tile or plastic-faced wall panels in bathrooms and kitchens. Not designed for use in high-moisture areas such as tub and shower surrounds. Includes cutting board around obstacles and 10% waste but no taping or finishing.

| | Craft@Hrs | Unit | Material | Labor | Total | Sell |
|---|---|---|---|---|---|---|
| 1/2" x 4' x 8' | D1@.008 | SF | .45 | .25 | .70 | 1.19 |
| 1/2" x 4' x 12' | D1@.008 | SF | .50 | .25 | .75 | 1.28 |
| 5/8" x 4' x 8' | D1@.009 | SF | .56 | .28 | .84 | 1.43 |
| 5/8" x 4' x 12' | D1@.009 | SF | .57 | .28 | .85 | 1.45 |

**Mold and mildew-resistant drywall.** DensArmor has a glass mat backing. DensArmor Plus has glass matt on both front and back. Includes cutting board around obstacles and 10% waste but no taping or finishing.

| | Craft@Hrs | Unit | Material | Labor | Total | Sell |
|---|---|---|---|---|---|---|
| 1/2" x 4' x 8' DensArmor | D1@.008 | SF | .42 | .25 | .67 | 1.14 |
| 1/2" x 4' x 8' DensArmor Plus | D1@.008 | SF | .45 | .25 | .70 | 1.19 |
| 1/2" x 4' x 12' DensArmor Plus | D1@.009 | SF | .52 | .28 | .80 | 1.36 |
| 5/8" x 4' x 8' DensArmor Plus | D1@.009 | SF | .58 | .28 | .86 | 1.46 |
| 5/8" x 4' x 12' DensArmor Plus | D1@.009 | SF | .58 | .28 | .86 | 1.46 |

**Moisture- and fire-resistant drywall.** For use under masonry veneer; aluminum, steel and vinyl siding; wood and mineral shingles; and stucco. Includes cutting board around obstacles and 10% waste, but no taping or finishing. Used to meet fire code in some types of multi-unit residential applications.

| | Craft@Hrs | Unit | Material | Labor | Total | Sell |
|---|---|---|---|---|---|---|
| 1/2" x 2' x 8' | D1@.008 | SF | .34 | .25 | .59 | 1.00 |
| 5/8" x 2' x 8' | D1@.009 | SF | .52 | .28 | .80 | 1.36 |

**Sound-deadening drywall.** For use as a base layer under drywall finish. Interior applications only. Must be kept dry. Includes cutting board around electrical boxes and obstacles, hanging on wall studs or ceiling joists 8' to 12' above floor level and 10% waste.

| | Craft@Hrs | Unit | Material | Labor | Total | Sell |
|---|---|---|---|---|---|---|
| 1/2" x 4' x 8', ceilings | D1@.012 | Ea | .35 | .37 | .72 | 1.22 |
| 1/2" x 4' x 8', walls | D1@.008 | Ea | .35 | .25 | .60 | 1.02 |

| | Craft@Hrs | Unit | Material | Labor | Total | Sell |
|---|---|---|---|---|---|---|

**Flexible drywall.** For covering curved surfaces. Includes cutting board around electrical boxes and obstacles, hanging board on wall studs or ceiling joists 8' to 12' above floor level, joint tape, three coats of joint compound and finish sanding. Material includes 1/2 gallon of premixed joint compound per 100 square feet, 38 linear feet of 2" perforated joint tape per 100 square feet, 1/2 pound of drywall screws per 100 square feet and 10% waste.

| | Craft@Hrs | Unit | Material | Labor | Total | Sell |
|---|---|---|---|---|---|---|
| 1/4" x 4' x 8', ceilings | D1@.035 | SF | .41 | 1.08 | 1.49 | 2.53 |
| 1/4" x 4' x 8', walls | D1@.035 | SF | .41 | 1.08 | 1.49 | 2.53 |

**Drywall repair sheet.** Includes taping and finishing.

| | Craft@Hrs | Unit | Material | Labor | Total | Sell |
|---|---|---|---|---|---|---|
| 1/2" x 16" x 16", greenboard | D1@1.00 | Ea | 3.03 | 30.90 | 33.93 | 57.70 |
| 1/2" x 16" x 16", regular | D1@1.00 | Ea | 1.97 | 30.90 | 32.87 | 55.90 |
| 5/8" x 16" x 16", regular | D1@1.00 | Ea | 2.24 | 30.90 | 33.14 | 56.30 |

**Drywall texture.** Finish applied to gypsum wallboard.

| | Craft@Hrs | Unit | Material | Labor | Total | Sell |
|---|---|---|---|---|---|---|
| Orange peel, rolled on, one coat | PT@.006 | SF | .03 | .19 | .22 | .37 |
| Spatter finish, one coat | PT@.007 | SF | .04 | .23 | .27 | .46 |
| Knockdown finish, one coat | PT@.008 | SF | .04 | .26 | .30 | .51 |
| Skip trowel finish, one coat | PT@.009 | SF | .07 | .29 | .36 | .61 |
| Smooth finish veneer plaster, one coat | PR@.018 | SF | .12 | .59 | .71 | 1.21 |

**Wall and ceiling spray texture, unaggregated.** Creates spatter, knock down and orange peel texture. Powder mixes with water. Can be spray applied over drywall, concrete or plaster.

| | Craft@Hrs | Unit | Material | Labor | Total | Sell |
|---|---|---|---|---|---|---|
| 40 pound bag, walls and ceilings | — | Ea | 11.05 | — | 11.05 | — |
| 50 pound bag, walls and ceilings | — | Ea | 12.00 | — | 12.00 | — |

**Wall and ceiling spray texture, aggregated.** Polystyrene aggregate mixes with water for spray application. Produces a white, simulated acoustical ceiling finish. Coarser mix conceals minor surface defects better. For gypsum, plaster and concrete ceilings. One pound of dry mix covers 8 square feet.

| | Craft@Hrs | Unit | Material | Labor | Total | Sell |
|---|---|---|---|---|---|---|
| Regular, 40 pound bag | PT@1.60 | Ea | 10.60 | 51.80 | 62.40 | 106.00 |
| Medium, 32 pound bag | PT@1.28 | Ea | 10.20 | 41.40 | 51.60 | 87.70 |
| Medium, 40 pound bag | PT@1.60 | Ea | 12.30 | 51.80 | 64.10 | 109.00 |
| Coarse, 40 pound bag | PT@1.60 | Ea | 11.70 | 51.80 | 63.50 | 108.00 |

**Spray texture touch-up kit.** Hand-operated sprayer for small repairs and touch-up work. Pump body adjusts to spray orange peel, spatter, or knockdown textures. Reusable and refillable. Kit covers 15 square feet. Refill covers 10 square feet for orange peel and medium spatter or 6 square feet for heavy spatter/knockdown finish.

| | Craft@Hrs | Unit | Material | Labor | Total | Sell |
|---|---|---|---|---|---|---|
| Spray texture touch-up kit | PT@.250 | Ea | 17.50 | 8.10 | 25.60 | 43.50 |
| Premixed refill kit | PT@.167 | Ea | 6.52 | 5.41 | 11.93 | 20.30 |

**Popcorn ceiling texture.** Finish applied to gypsum ceiling board. Applied with a compressed air hopper gun. Add the cost of a compressor.

| | Craft@Hrs | Unit | Material | Labor | Total | Sell |
|---|---|---|---|---|---|---|
| Blown on polystyrene texture | PT@.005 | SF | .09 | .16 | .25 | .43 |
| Pneumatic hopper gun, purchase | — | Ea | 75.00 | — | 75.00 | — |
| Shop-type compressor rental, per day | — | Ea | 40.00 | — | 40.00 | — |

**Popcorn ceiling spray mix.** Polystyrene. Mix with water and spray. 13-pound bag.

| | Craft@Hrs | Unit | Material | Labor | Total | Sell |
|---|---|---|---|---|---|---|
| Bag, titanium white | PT@1.00 | Ea | 9.68 | 32.40 | 42.08 | 71.50 |

**Popcorn ceiling texture touch-up kit.** Hand-operated sprayer for small repair and touch-up work. Pump body sprays acoustic popcorn texture. Reusable and refillable. Includes texture with polystyrene chip material for matching. Covers two square feet.

| | Craft@Hrs | Unit | Material | Labor | Total | Sell |
|---|---|---|---|---|---|---|
| Touch-up kit | PT@.500 | Ea | 17.50 | 16.20 | 33.70 | 57.30 |
| Pack of 2 dry refills | PT@1.00 | Ea | 4.73 | 32.40 | 37.13 | 63.10 |

| | Craft@Hrs | Unit | Material | Labor | Total | Sell |
|---|---|---|---|---|---|---|

**Popcorn ceiling patch.** For repair of popcorn or acoustical rough-textured ceilings. Blends with most existing textures to cover holes, scuffs and scars. Apply with a paint roller or putty knife. One gallon covers 200 square feet.

| | | | | | | |
|---|---|---|---|---|---|---|
| 13 ounce spray | PT@.250 | Ea | 7.60 | 8.10 | 15.70 | 26.70 |
| 5 gallons | PT@3.50 | Ea | 20.70 | 113.00 | 133.70 | 227.00 |

**Acousti-Tex™ ceiling texture, Homax®.** Mix one 20-ounce bag with one gallon of paint. Apply with a brush or roller. Creates a simulated cottage cheese ceiling texture. Covers cracks, flaws and scrapes.

| | | | | | | |
|---|---|---|---|---|---|---|
| 20 ounce bag | — | Ea | 3.77 | — | 3.77 | — |

**Ceiling spray texture.** Premixed. Bottle attaches to spray texture hopper gun and covers 200 square feet.

| | | | | | | |
|---|---|---|---|---|---|---|
| 2.2 liter bottle | PT@1.00 | Ea | 12.80 | 32.40 | 45.20 | 76.80 |

**Textured ceiling finish aerosol.** 10-ounce aerosol can. Repair for small patches and cracks. Water cleanup.

| | | | | | | |
|---|---|---|---|---|---|---|
| Popcorn, white | PT@.167 | Ea | 9.70 | 5.41 | 15.11 | 25.70 |
| Knockdown or heavy spatter | PT@.167 | Ea | 9.68 | 5.41 | 15.09 | 25.70 |
| Orange peel | PT@.167 | Ea | 9.68 | 5.41 | 15.09 | 25.70 |

**Imperial veneer plaster base, USG.** Rigid plaster base resists sag and sound transmission. Face paper resist plaster slide. Includes 10% waste.

| | | | | | | |
|---|---|---|---|---|---|---|
| 1/2" x 4' x 8', square edge | BC@.012 | SF | .40 | .39 | .79 | 1.34 |
| 1/2" x 4' x 12', square edge | BC@.012 | SF | .33 | .39 | .72 | 1.22 |
| 5/8" x 4' x 8', fire-resistant type X | BC@.013 | SF | .40 | .42 | .82 | 1.39 |
| 5/8" x 4' x 12', fire-resistant type X | BC@.013 | SF | .33 | .42 | .75 | 1.28 |

**Cameo Veneer Plaster Base Board, Georgia-Pacific.** High-suction face paper. Use as a base for 1- or 2-coat Cameo Veneer Plaster. Includes 10% waste.

| | | | | | | |
|---|---|---|---|---|---|---|
| 3/8" x 4' x 8', tapered edge | BC@.012 | SF | .36 | .39 | .75 | 1.28 |
| 1/2" x 4' x 12' | BC@.012 | SF | .35 | .39 | .74 | 1.26 |

**DensShield® Tile Backer, Georgia-Pacific.** Glass mat facing acrylic coating that blocks moisture. For walls, floors, countertops, showers, laboratories. Includes 10% waste.

| | | | | | | |
|---|---|---|---|---|---|---|
| 1/4" x 4' x 4' | TL@.042 | SF | .71 | 1.29 | 2.00 | 3.40 |
| 1/2" x 32" x 5' | TL@.042 | SF | .84 | 1.29 | 2.13 | 3.62 |

**Durock® Cement Board, USG.** Tile base for tub and shower areas, underlayment for tile on floors and countertops. Use the smooth side for ceramic tile mastic and the rough side for thin-set mortar and Portland cement. Hang with Durock® screws or galvanized roofing nails. Includes taping and sealing seams with Durock tape and 10% waste.

| | | | | | | |
|---|---|---|---|---|---|---|
| 5/16" x 4' x 4' | TL@.037 | SF | .76 | 1.14 | 1.90 | 3.23 |
| 1/2" x 3' x 5' | TL@.042 | SF | .77 | 1.29 | 2.06 | 3.50 |
| 1/2" x 4' x 8', interior | TL@.042 | SF | .83 | 1.29 | 2.12 | 3.60 |
| 5/8" x 3' x 5' | TL@.047 | SF | 1.05 | 1.45 | 2.50 | 4.25 |
| 5/8" x 4' x 8', interior | TL@.047 | SF | 1.03 | 1.45 | 2.48 | 4.22 |

## Drywall Joint Compound

**Lightweight ready-mix joint compound.** For embedding drywall joint tape and finishing drywall joints. One-half gallon finishes three 4' x 8' drywall panels. Installation cost is included in the cost of hanging drywall.

| | | | | | | |
|---|---|---|---|---|---|---|
| 3.5 gallon carton | — | Ea | 7.87 | — | 7.87 | — |
| 4.5 gallon carton, tinted | — | Ea | 11.40 | — | 11.40 | — |
| 4.5 gallon pail | — | Ea | 13.06 | — | 13.06 | — |

| | Craft@Hrs | Unit | Material | Labor | Total | Sell |
|---|---|---|---|---|---|---|

**All-purpose ready-mix joint compound.** For embedding joint tape, finishing drywall joints, repairing small cracks and holes in drywall and plaster surfaces, and simple hand-applied texturing. One-half gallon finishes three 4' x 8' drywall panels. Installation cost is included in the cost of hanging drywall.

| | Craft@Hrs | Unit | Material | Labor | Total | Sell |
|---|---|---|---|---|---|---|
| 3 pound, 1 quart tub | — | Ea | 3.05 | — | 3.05 | — |
| 12 pound pail | — | Ea | 5.76 | — | 5.76 | — |
| 48 pound carton | — | Ea | 7.34 | — | 7.34 | — |
| 50 pound carton | — | Ea | 8.51 | — | 8.51 | — |
| 61.7 pound carton | — | Ea | 10.47 | — | 10.47 | — |
| 61.7 pound pail | — | Ea | 12.70 | — | 12.70 | — |

**Lightweight setting-type joint compound powder.** 20, 45, and 90 minute setting ranges allow one-day drywall joint finishing with next-day texturing. After a coat has set, apply another coat. No need to wait for each coat to dry completely. Weighs 25% less than all-purpose compound. Sands as easily as ready-mix joint compound. Ideal for patching drywall and plaster surfaces. 18-pound bag.

| | Craft@Hrs | Unit | Material | Labor | Total | Sell |
|---|---|---|---|---|---|---|
| 20-minute set time | — | Ea | 8.73 | — | 8.73 | — |
| 45-minute set time | — | Ea | 8.80 | — | 8.80 | — |
| 90-minute set time | — | Ea | 8.75 | — | 8.75 | — |

**Ready-mix joint topping compound.** For filling, leveling and finishing coats over drywall joints. Sands easier and faster than all-purpose compound, and weighs 35% less. Not suitable for embedding joint tape, skim coating, or texturing.

| | Craft@Hrs | Unit | Material | Labor | Total | Sell |
|---|---|---|---|---|---|---|
| 3.5 gallon carton | — | Ea | 7.87 | — | 7.87 | — |

**Drywall ready-mix primer.** Primes new drywall panel walls and ceilings before texturing or painting. Brush, roller, or spray application. Dries white in 30 minutes. Topcoat in 1 hour. 300 to 500 square feet per gallon.

| | Craft@Hrs | Unit | Material | Labor | Total | Sell |
|---|---|---|---|---|---|---|
| 5 gallon pail | — | Ea | 32.70 | — | 32.70 | — |

**Tuf-Tex hopper gun respray texture.** For use with hopper gun or trailer-mounted spray rig. Approximately 500 square feet per bag.

| | Craft@Hrs | Unit | Material | Labor | Total | Sell |
|---|---|---|---|---|---|---|
| 40 pound bag | — | Ea | 7.70 | — | 7.70 | — |

**Plaster of Paris.** For patching interior drywall or plaster walls and ceilings. Sets hard in 30 minutes.

| | Craft@Hrs | Unit | Material | Labor | Total | Sell |
|---|---|---|---|---|---|---|
| 25 pounds | — | Ea | 10.07 | — | 10.07 | — |

**One coat veneer plaster.** High resistance to cracking, nail-popping, impact and abrasion failure. Mill-mixed plaster components help assure uniform installation performance and finished job quality. Can be applied directly to concrete block.

| | Craft@Hrs | Unit | Material | Labor | Total | Sell |
|---|---|---|---|---|---|---|
| 50 pounds | — | Ea | 11.70 | — | 11.70 | — |

## Drywall Joint Tape

**Drywall joint tape.** Center-creased paper tape. Six 4' x 8' drywall panels require about 75 linear feet of tape. Tape installation cost is included in the cost of hanging drywall. 2-1/16" wide. Perforated.

| | Craft@Hrs | Unit | Material | Labor | Total | Sell |
|---|---|---|---|---|---|---|
| 250' roll | — | Ea | 2.21 | — | 2.21 | — |
| 500' roll | — | Ea | 3.39 | — | 3.39 | — |

**Fiberglass drywall tape.** For drywall joints and repairs, veneer plastering, stucco and tile backer board. Self-adhesive 100% fiberglass mesh. Apply directly to drywall joint to eliminate embedding coat. 1-7/8" wide. Six 4' x 8' drywall panels require about 75 linear feet of tape. Cost of joint tape application is included in the Drywall section above.

| | Craft@Hrs | Unit | Material | Labor | Total | Sell |
|---|---|---|---|---|---|---|
| 50' long, white | — | Ea | 2.16 | — | 2.16 | — |
| 150' long, white | — | Ea | 4.18 | — | 4.18 | — |
| 150' long, yellow | — | Ea | 3.81 | — | 3.81 | — |
| 300' long, white | — | Ea | 6.12 | — | 6.12 | — |
| 300' long, yellow | — | Ea | 5.57 | — | 5.57 | — |
| 500' long, white | — | Ea | 9.44 | — | 9.44 | — |
| 500' long, yellow | — | Ea | 8.33 | — | 8.33 | — |

| | Craft@Hrs | Unit | Material | Labor | Total | Sell |
|---|---|---|---|---|---|---|

## Drywall Bead

**Galvanized bullnose corner bead.** All-metal galvanized steel reinforcement for protecting external drywall corners. Rounded 3/4" radius outside corner on 1/2" or 5/8" drywall. Attached with nails or crimping tool. Finish with USG joint compounds.

| | Craft@Hrs | Unit | Material | Labor | Total | Sell |
|---|---|---|---|---|---|---|
| 3/4" x 8' long | DI@.128 | Ea | 2.76 | 3.96 | 6.72 | 11.40 |
| 3/4" x 10' long | DI@.128 | Ea | 3.13 | 3.96 | 7.09 | 12.10 |

**Plastic bullnose corner bead.** For use where smooth or rounded corners are specified. Will not dent or rust. Paint and drywall mud adhere well. Finish compatible with drywall mud and all paint finishes.

| | Craft@Hrs | Unit | Material | Labor | Total | Sell |
|---|---|---|---|---|---|---|
| 3/4" x 8' | DI@.128 | Ea | 2.60 | 3.96 | 6.56 | 11.20 |

**135-degree open-angle bullnose drywall arch.** Plastic. For hard-to-do arches. No shiner. Wide flange.

| | Craft@Hrs | Unit | Material | Labor | Total | Sell |
|---|---|---|---|---|---|---|
| 3/4" x 8' | DI@.128 | Ea | 4.82 | 3.96 | 8.78 | 14.90 |

**Bullnose open plastic corner.** Used where three exterior corners intersect. Treated for maximum joint adhesion and rust protection. Wide flange.

| | Craft@Hrs | Unit | Material | Labor | Total | Sell |
|---|---|---|---|---|---|---|
| 3-way | DI@.160 | Ea | 2.72 | 4.95 | 7.67 | 13.00 |

**Paper-faced metal bullnose outside corner, tape-on.** Smooth, rounded radius outside corner for 1/2" or 5/8" drywall. Paper tape cover is laminated to a rust-resistant metal profile for good adhesion of joint compound. Held in place with joint compound instead of nails to bond the bead to the drywall surface. Bead is covered with joint compound.

| | Craft@Hrs | Unit | Material | Labor | Total | Sell |
|---|---|---|---|---|---|---|
| 3/4" x 8' long | DI@.128 | Ea | 2.54 | 3.96 | 6.50 | 11.10 |
| 9/16" x 13/16" x 10' | DI@.160 | Ea | 2.16 | 4.95 | 7.11 | 12.10 |
| 11/16" x 15/16" x 8' | DI@.216 | Ea | 2.40 | 6.68 | 9.08 | 15.40 |
| 1-1/16" x 1-1/16" x 8' | DI@.216 | Ea | 2.09 | 6.68 | 8.77 | 14.90 |
| 1-1/16" x 1-1/16" x 10' | DI@.270 | Ea | 2.59 | 8.35 | 10.94 | 18.60 |

**Galvanized bullnose corner bead.** Protects corners from damage with a more rounded edge for curves. Textured flanges allow for better bonding.

| | Craft@Hrs | Unit | Material | Labor | Total | Sell |
|---|---|---|---|---|---|---|
| 1-1/4" x 8' | DI@.128 | Ea | 1.59 | 3.96 | 5.55 | 9.44 |
| 1-1/4" x 9' | DI@.144 | Ea | 1.75 | 4.45 | 6.20 | 10.50 |
| 1-1/4" x 10' | DI@.160 | Ea | 1.68 | 4.95 | 6.63 | 11.30 |

**Beadex® paper-faced metal L-shaped tape-on trim.** Edge reinforcement for drywall panels where panels abut other materials such as suspended ceilings, beams, plaster, masonry, concrete, door or window jambs. Paper tape covering laminated to an L-shaped rust-resistant metal profile provides excellent adhesion of joint compound, texture and paint. Applied using joint compound instead of nails to bond trim to the drywall surface. Trim is finished with joint compound. For use with 1/2" drywall.

| | Craft@Hrs | Unit | Material | Labor | Total | Sell |
|---|---|---|---|---|---|---|
| 1/2" x 8' | DI@.160 | Ea | 1.91 | 4.95 | 6.86 | 11.70 |
| 1/2" x 10' | DI@.160 | Ea | 2.07 | 4.95 | 7.02 | 11.90 |

**Paper-faced metal inside corner bead.** Tape-on trim used to protect drywall panel assembly edges, corners and sides. Forms a true, inner (90-degree) corner. No fasteners required, adheres to compound. Eliminates edge cracking, nail pops and chipping.

| | Craft@Hrs | Unit | Material | Labor | Total | Sell |
|---|---|---|---|---|---|---|
| 8' long | DI@.216 | Ea | 2.65 | 6.68 | 9.33 | 15.90 |

**Drywall flex corner bead.** For use where curve or arch is formed. Pre-slotted flange allows for perfect arches. Tapered flanges are perforated and striated for adhesion. No need to tape the flange. Finish is compatible with drywall mud and all paint finishes.

| | Craft@Hrs | Unit | Material | Labor | Total | Sell |
|---|---|---|---|---|---|---|
| 1-1/4" x 10' | DI@.270 | Ea | 2.55 | 8.35 | 10.90 | 18.50 |

| | Craft@Hrs | Unit | Material | Labor | Total | Sell |
|---|---|---|---|---|---|---|

**Beadex® tape-on flexible metal corner.** Provides straight, strong inside or outside corners of any angle. Paper tape is 2-1/16" wide laminated to two 7/16" wide galvanized steel strips with a 1/16" gap between the strips. The strength of steel with the superior bond of joint compound on paper. Strong, chip-resistant, smooth finish. Applied with joint compound instead of nails. Guaranteed against edge cracking. Bead is finished with joint compound.

| | Craft@Hrs | Unit | Material | Labor | Total | Sell |
|---|---|---|---|---|---|---|
| 25' roll | DI@.675 | Ea | 4.42 | 20.90 | 25.32 | 43.00 |

**Drywall metal J bead.** Protects drywall edges around doors and windows. Textured flanges allow for better bonding.

| | Craft@Hrs | Unit | Material | Labor | Total | Sell |
|---|---|---|---|---|---|---|
| 1/2" x 8' | DI@.216 | Ea | 2.88 | 6.68 | 9.56 | 16.30 |
| 1/2" x 10' | DI@.270 | Ea | 3.08 | 8.35 | 11.43 | 19.40 |

**Drywall plastic J bead.** For capping vertical and horizontal edges of raw drywall. Helps keep moisture out of board. Finish is compatible with drywall mud and paint finishes.

| | Craft@Hrs | Unit | Material | Labor | Total | Sell |
|---|---|---|---|---|---|---|
| 1/2" x 10' | DI@.270 | Ea | 1.34 | 8.35 | 9.69 | 16.50 |
| 5/8" x 10' | DI@.270 | Ea | 1.58 | 8.35 | 9.93 | 16.90 |

**Drywall metal J trim.** Galvanized steel. Provides maximum protection and neat finished edges where panels join window and door jambs, and at internal angles. Install with nails or screws.

| | Craft@Hrs | Unit | Material | Labor | Total | Sell |
|---|---|---|---|---|---|---|
| 1/2" x 8' | DI@.216 | Ea | 1.43 | 6.68 | 8.11 | 13.80 |
| 1/2" x 10' | DI@.270 | Ea | 1.80 | 8.35 | 10.15 | 17.30 |
| 5/8" x 8' | DI@.216 | Ea | 2.12 | 6.68 | 8.80 | 15.00 |
| 5/8" x 10' | DI@.270 | Ea | 2.44 | 8.35 | 10.79 | 18.30 |

**Vinyl drywall outside corner bead.** Rust- and impact-resistant. Tapered legs are perforated and striated for better adhesion. No need to tape flange. Prevents electrolysis. Finish compatible with drywall mud and all paint finishes.

| | Craft@Hrs | Unit | Material | Labor | Total | Sell |
|---|---|---|---|---|---|---|
| 1-1/4" x 8' outside corner | DI@.216 | Ea | 1.43 | 6.68 | 8.11 | 13.80 |
| 1-1/4" x 10' outside corner | DI@.270 | Ea | 1.77 | 8.35 | 10.12 | 17.20 |
| 1/2" x 10' zip corner | DI@.270 | Ea | 1.99 | 8.35 | 10.34 | 17.60 |
| 10' bullnose corner | DI@.270 | Ea | 2.36 | 8.35 | 10.71 | 18.20 |
| 10' arch bullnose | DI@.270 | Ea | 3.21 | 8.35 | 11.56 | 19.70 |
| 1/2" x 8' J trim | DI@.216 | Ea | 1.07 | 6.68 | 7.75 | 13.20 |
| 5/8" x 8' J trim | DI@.216 | Ea | 1.11 | 6.68 | 7.79 | 13.20 |

**Drywall L trim.** Metal covered with paper. Trims drywall where it abuts with other building components such as suspended ceilings, beams, plaster, masonry, or concrete walls, and untrimmed door or window jambs.

| | Craft@Hrs | Unit | Material | Labor | Total | Sell |
|---|---|---|---|---|---|---|
| 1/2" x 8' | DI@.216 | Ea | 2.15 | 6.68 | 8.83 | 15.00 |
| 1/2" x 10' | DI@.270 | Ea | 3.00 | 8.35 | 11.35 | 19.30 |

**Utility access door.** For plumbing and electrical access. 28-gauge galvanized steel. Piano hinge. Pressure fit door. Drywall frame with nailing flange.

| | Craft@Hrs | Unit | Material | Labor | Total | Sell |
|---|---|---|---|---|---|---|
| 14" x 14" | DI@.500 | Ea | 11.80 | 15.50 | 27.30 | 46.40 |

**Drywall repair kit.** Includes 1-pound bag of setting compound, 5' of drywall, fiberglass drywall tape, 120-grit sandpaper, plastic spreader, four drywall repair clips, and eight drywall screws.

| | Craft@Hrs | Unit | Material | Labor | Total | Sell |
|---|---|---|---|---|---|---|
| Kit | — | Ea | 9.90 | — | 9.90 | — |

**Drywall repair patch.** No-rust aluminum plate provides rigid backing. Fiberglass mesh eliminates the need to embed tape.

| | Craft@Hrs | Unit | Material | Labor | Total | Sell |
|---|---|---|---|---|---|---|
| 4" x 4" | — | Ea | 3.50 | — | 3.50 | — |
| 8" x 8" | — | Ea | 4.50 | — | 4.50 | — |

**Crack repair mesh, Quik-Tape™.** Self-adhesive 100% fiberglass mesh tape. Provides tensile strength and dimensional stability.

| | Craft@Hrs | Unit | Material | Labor | Total | Sell |
|---|---|---|---|---|---|---|
| 6" wide, 25' long | — | Ea | 3.27 | — | 3.27 | — |

|  | Craft@Hrs | Unit | Material | Labor | Total | Sell |
|---|---|---|---|---|---|---|

**Drywall repair clips.** For repair of holes over 2". Suitable for both 1/2" and 5/8" thick drywall. Clips hold and reinforce the new piece of drywall sized to fit the hole being repaired.

| Kit with 8 clips | — | Ea | 2.25 | — | 2.25 | — |

**Tile backer tape.** Pressure sensitive, anti-corrosion coating.

| 2" x 50' | — | Ea | 3.20 | — | 3.20 | — |

**Flexible metal corner tape.** For inside and outside corners.

| 2" x 25' | DI@.175 | Ea | 4.25 | 5.41 | 9.66 | 16.40 |
| 2" x 100' | DI@.700 | Ea | 10.80 | 21.60 | 32.40 | 55.10 |

**Corner bead spray adhesive, 3M Company.** Quick grab so corner bead can be joined without delay. Longer open time allows assembly flexibility. Covers up to 1200 linear feet per can. Tinted adhesive helps provide better wall coverage in hard-to-see areas.

| Adhesive | — | Ea | 6.12 | — | 6.12 | — |

## Drywall Fasteners

**Drywall screws, Twinfast.** Sharp, non-walking point. Dual lead. Each 100 square feet of board requires approximately 170 screws (1/2 pound). The cost of fasteners needed to hang drywall is included in the previous section, Drywall.

| #6 x 1", 5 pounds, 2,000 screws | — | Ea | 19.35 | — | 19.35 | — |
| #6 x 1-1/4", 5 pounds, 1,700 screws | — | Ea | 21.50 | — | 21.50 | — |
| #6 x 1-5/8", 5 pounds, 1,550 screws | — | Ea | 19.30 | — | 19.30 | — |

**Drywall nails.** Approximately 300 nails per pound. Each 100 square feet of board requires approximately 200 nails (2/3 of a pound). The cost of fasteners needed to hang drywall is included in the previous section, Drywall.

| 1-3/8", 30 pounds | — | Ea | 42.20 | — | 42.20 | — |
| 1-1/2", 1 pound | — | Ea | 2.66 | — | 2.66 | — |
| 1-1/2", coated, 1 pound | — | Ea | 2.32 | — | 2.32 | — |
| 1-1/2", galvanized, 1 pound | — | Ea | 3.48 | — | 3.48 | — |

## Drywall Tools

**Electronic stud sensor.** Detects wood at 3/4" depth.

| Wall-Tech | — | Ea | 5.10 | — | 5.10 | — |

**Drywall screwdriver, 6-amp, DeWalt.** Helical-cut steel, heat-treated gears. Quiet clutch. Depth-sensitive nosepiece. Tool weight: 2.9 pounds. UL Listed and OSHA approved. 60 inch-pounds of torque.

| 0 To 4000 RPM | — | Ea | 86.30 | — | 86.30 | — |

**Battery-powered drywall screw gun, DuraSpin, Senco.** Drives 600 1" to 2" drywall screws an hour. Adjustable nosepiece. Includes: 2 batteries, 2 chargers, spare drive bits, and case.

| 14.4 volts | — | Ea | 161.00 | — | 161.00 | — |

**Drywall to wood collated screws, Senco.** Collated pack of 1,000.

| #6 x 1-1/4" | — | Ea | 15.20 | — | 15.20 | — |
| #6 x 1-5/8" | — | Ea | 17.30 | — | 17.30 | — |
| #7 x 2" | — | Ea | 20.20 | — | 20.20 | — |

**Drywall to steel collated screws, Senco.** Collated pack of 1,000.

| #6 x 1-1/4", light steel | — | Ea | 15.10 | — | 15.10 | — |
| #6 x 1-1/4", heavy steel | — | Ea | 23.50 | — | 23.50 | — |

| | Craft@Hrs | Unit | Material | Labor | Total | Sell |
|---|---|---|---|---|---|---|
| **Phillips-head drywall screwdriver bit, Hilti.** | | | | | | |
| #1 insert bit, pack of 10 | — | Ea | 8.70 | — | 8.70 | — |
| #2 insert bit, pack of 10 | — | Ea | 8.66 | — | 8.66 | — |

**Drywall hammer, Estwing.** All solid steel construction. Head and handle forged in one piece. Fully polished. Shock-reduction nylon vinyl grip. Crowned and scored face.

| | | | | | | |
|---|---|---|---|---|---|---|
| 13-1/2" long, 14 ounce | — | Ea | 17.20 | — | 17.20 | — |

**Drywall T-square.** Head riveted to blade. Silver anodized aluminum. 1/8" and 1/16" graduations.

| | | | | | | |
|---|---|---|---|---|---|---|
| 48" blade | — | Ea | 11.30 | — | 11.30 | — |

**Stainless steel curved drywall trowel.** Soft handle.

| | | | | | | |
|---|---|---|---|---|---|---|
| 11" trowel | — | Ea | 18.30 | — | 18.30 | — |
| 14" trowel | — | Ea | 18.80 | — | 18.80 | — |

**Drywall corner trowel.** Blade set at 80-degree angle that flexes to 90 degrees during use.

| | | | | | | |
|---|---|---|---|---|---|---|
| Hardwood handle | — | Ea | 9.00 | — | 9.00 | — |

**Single-texture drywall brush.** Horsehair plastic, round for swirled or sponged effects on ceilings and walls. Standard threaded hole in center can be used with any standard broom handle.

| | | | | | | |
|---|---|---|---|---|---|---|
| 2-1/2" trim, 4-3/4" block | — | Ea | 14.30 | — | 14.30 | — |
| Stippling brush | — | Ea | 8.58 | — | 8.58 | — |

**Crow's foot texturing brush.** Tampico hemp bristles set in a wooden base. Can be hand held or center-hole pole mounted.

| | | | | | | |
|---|---|---|---|---|---|---|
| 13" x 9" | — | Ea | 8.50 | — | 8.50 | — |

**Drywall stilts.** High-strength aluminum alloy with adjustable heights. Spring wishbone locks securely to prevent legs from sliding. Rubber soles prevent slipping. To 225 pounds.

| | | | | | | |
|---|---|---|---|---|---|---|
| 18" to 28" lift | — | Ea | 236.00 | — | 236.00 | — |

**Drywall panel lift.** Lifts and holds drywall in place on the ceiling.

| | | | | | | |
|---|---|---|---|---|---|---|
| Daily rental | — | Ea | 25.00 | — | 25.00 | — |
| Panel lift deposit | — | Ea | 300.00 | — | 300.00 | — |

**Drywall sander kit.** For drywall sanding without dust. Includes 6' hose with swivel ends, 2 hose adapters, 1 medium grit sandscreen, 1-piece high-impact polystyrene plastic sanding head, and instructions.

| | | | | | | |
|---|---|---|---|---|---|---|
| Drywall sander kit | — | Ea | 16.20 | — | 16.20 | — |

**Drywall sanding fiberglass respirator.** Strengthened outer shell helps prevent collapse due to moisture buildup. Soft inner shell for comfort and durability. Adjustable metal nosepiece with foam seal fits variety of face sizes. Heavy-duty pre-stretched straps.

| | | | | | | |
|---|---|---|---|---|---|---|
| Pack of 2 | — | Ea | 5.38 | — | 5.38 | — |

**Drywall sanding sponge.** For sanding drywall and plaster during spot repair. Resists clogging and buildup. Fine and medium sanding grits in one sponge.

| | | | | | | |
|---|---|---|---|---|---|---|
| Small | — | Ea | 2.10 | — | 2.10 | — |
| Large | — | Ea | 3.09 | — | 3.09 | — |

**Plastic drywall hand sander.** Lightweight, durable plastic base with multiple sandpaper usage. Use with die-cut or 1/2-sheet standard sandpaper.

| | | | | | | |
|---|---|---|---|---|---|---|
| 3-1/4" x 9-1/4" | — | Ea | 6.44 | — | 6.44 | — |

| | Craft@Hrs | Unit | Material | Labor | Total | Sell |
|---|---|---|---|---|---|---|

**Pole and hand drywall sander.**
Sander — Ea 6.85 — 6.85 —

**Sand & Kleen™ dustless drywall sanding system, Magna.** 20' of hose to sand 7 drywall panels between water changes. 5-gallon aquair filter, 1-1/4", 2-1/4", and 2-1/2" vacuum adapters.
No. MT800 dustless sander — Ea 43.50 — 43.50 —

## 4 x 8 Paneling

**Embossed tileboard, ABTCO.** Printed hardboard. Including 10% waste.

| | Craft@Hrs | Unit | Material | Labor | Total | Sell |
|---|---|---|---|---|---|---|
| Aquatile, 4' x 8' panels, 1/8" thick | BC@.020 | SF | 1.25 | .64 | 1.89 | 3.21 |
| Glaztile III, 4' x 8' panels, 1/8" thick | BC@.020 | SF | 1.10 | .64 | 1.74 | 2.96 |

**Embossed tileboard.** Water-resistant surface. The look of real ceramic tile. 1/8" x 4' x 8' panels. Including 10% waste.

| | Craft@Hrs | Unit | Material | Labor | Total | Sell |
|---|---|---|---|---|---|---|
| Alicante | BC@.034 | SF | .83 | 1.09 | 1.92 | 3.26 |
| Bisque plain | BC@.034 | SF | .56 | 1.09 | 1.65 | 2.81 |
| Blue floral | BC@.034 | SF | .80 | 1.09 | 1.89 | 3.21 |
| Florableu | BC@.034 | SF | .69 | 1.09 | 1.78 | 3.03 |
| Golden lace | BC@.034 | SF | .67 | 1.09 | 1.76 | 2.99 |
| Morning glories | BC@.034 | SF | .70 | 1.09 | 1.79 | 3.04 |
| Scored white | BC@.034 | SF | .65 | 1.09 | 1.74 | 2.96 |
| Silver quartz | BC@.034 | SF | .82 | 1.09 | 1.91 | 3.25 |
| Stone white | BC@.034 | SF | .72 | 1.09 | 1.81 | 3.08 |
| Thrifty white | BC@.034 | SF | .38 | 1.09 | 1.47 | 2.50 |
| Tuscan marble | BC@.034 | SF | .63 | 1.09 | 1.72 | 2.92 |
| White | BC@.034 | SF | .45 | 1.09 | 1.54 | 2.62 |

**Primed fiberboard paneling.** Medium-density fiberboard 4' x 8' panels. Per square foot including 10% waste. 3/16" thick. White.

| | Craft@Hrs | Unit | Material | Labor | Total | Sell |
|---|---|---|---|---|---|---|
| Beaded | BC@.020 | Ea | .58 | .64 | 1.22 | 2.07 |
| Beaded, grooved 3" on center | BC@.020 | Ea | .58 | .64 | 1.22 | 2.07 |

**Paneling, Georgia-Pacific.** Durable wood fiber backing. Simulated woodgrain face. Can be used in basements. 4' x 8' panels. Per square foot, including 10% waste.

| | Craft@Hrs | Unit | Material | Labor | Total | Sell |
|---|---|---|---|---|---|---|
| Blanc polare lauan, 5/32" | BC@.020 | SF | .56 | .64 | 1.20 | 2.04 |
| Moonlight, 1/8" | BC@.020 | SF | .42 | .64 | 1.06 | 1.80 |

**Mount Vernon paneling, Georgia-Pacific.** Simulated wood grain face on medium-density fiberboard backing. Can be used in basements. 4' x 8' panels. Per square foot, including 10% waste.

| | Craft@Hrs | Unit | Material | Labor | Total | Sell |
|---|---|---|---|---|---|---|
| Kimberly oak, 1/8" | BC@.020 | SF | .34 | .64 | .98 | 1.67 |

**Jubilee® beaded paneling, Georgia-Pacific.** 3/16" thick. Ready to paint. Simulated woodgrain and decorative finishes. Acryglas® topcoat. 4' x 8' panels. Per square foot including 10% waste.

| | Craft@Hrs | Unit | Material | Labor | Total | Sell |
|---|---|---|---|---|---|---|
| Beaded white ice, 3/16" | BC@.020 | SF | .73 | .64 | 1.37 | 2.33 |
| Jubilee®, 3/16" | BC@.020 | SF | .71 | .64 | 1.35 | 2.30 |

**Hardboard paneling.** Printed wood grain. 4' x 8' panels. Per square foot, including 10% waste.

| | Craft@Hrs | Unit | Material | Labor | Total | Sell |
|---|---|---|---|---|---|---|
| Alaskan oak, 1/8" | BC@.020 | SF | .48 | .64 | 1.12 | 1.90 |
| Beaded finished oak, 5/32" | BC@.020 | SF | 1.30 | .64 | 1.94 | 3.30 |
| Charterhouse oak, 1/8" | BC@.020 | SF | .70 | .64 | 1.34 | 2.28 |
| Knotty cedar, 5/32" | BC@.020 | SF | .58 | .64 | 1.22 | 2.07 |
| Norwegian pine, 1/8" | BC@.020 | SF | .63 | .64 | 1.27 | 2.16 |
| Spartan oak, 1/8" | BC@.020 | SF | .31 | .64 | .95 | 1.62 |
| Valencia birch, 1/8" | BC@.020 | SF | .60 | .64 | 1.24 | 2.11 |

| | Craft@Hrs | Unit | Material | Labor | Total | Sell |
|---|---|---|---|---|---|---|
| Water resistant, 1/4" | BC@.020 | SF | .73 | .64 | 1.37 | 2.33 |
| Westminster red brick, 1/4" | BC@.020 | SF | .85 | .64 | 1.49 | 2.53 |
| Vintage maple, 1/8" | BC@.020 | SF | .53 | .64 | 1.17 | 1.99 |
| Traditional oak, 1/8" | BC@.020 | SF | .63 | .64 | 1.27 | 2.16 |
| Honey oak, 1/8" | BC@.020 | SF | .63 | .64 | 1.27 | 2.16 |
| Cherry oak, 1/8" | BC@.020 | SF | .63 | .64 | 1.27 | 2.16 |
| Arabella pine, 5/32" | BC@.020 | SF | .48 | .64 | 1.12 | 1.90 |
| Heather oak, 5/32" | BC@.020 | SF | .63 | .64 | 1.27 | 2.16 |
| Juliet cherry, beaded, 5/32" | BC@.020 | SF | .63 | .64 | 1.27 | 2.16 |
| Kristen leather, 5/32" | BC@.020 | SF | .74 | .64 | 1.38 | 2.35 |

**Designer paneling, Fashion Series.** 3.2 millimeter backing. Texture of wallpaper. Not recommended for floors, ceilings, and high-moisture areas. 4' x 8' panels. Per square foot, including 10% waste.

| | Craft@Hrs | Unit | Material | Labor | Total | Sell |
|---|---|---|---|---|---|---|
| Sculptured stripe, 3.2mm thick | BC@.034 | SF | .65 | 1.09 | 1.74 | 2.96 |

## Hardwood Paneling

**Ann Arbor birch prefinished paneling, Georgia-Pacific.** For use in basements and below ground level. Rotary cut face. Plywood backing. Permagard® topcoat finish. 4' x 8' panels. Per square foot, including 10% waste.

| | Craft@Hrs | Unit | Material | Labor | Total | Sell |
|---|---|---|---|---|---|---|
| 5.0mm thick | BC@.034 | SF | 1.10 | 1.09 | 2.19 | 3.72 |

**Bridgeport Bryant birch paneling, Georgia-Pacific.** Offers the beauty of real wood with a durable hardwood plywood backing. Can be used in basements and below ground level. 4' x 8' panels. Per square foot, including 10% waste.

| | Craft@Hrs | Unit | Material | Labor | Total | Sell |
|---|---|---|---|---|---|---|
| 5.0mm thick | BC@.034 | SF | 1.06 | 1.09 | 2.15 | 3.66 |

**Brookstone birch paneling, Georgia-Pacific.** May be used in basements and below ground level. Rotary cut face and prefinished hardwood plywood substrate. Permagard® topcoat finish and unfinished. 4' x 8' panels. Per square foot, including 10% waste.

| | Craft@Hrs | Unit | Material | Labor | Total | Sell |
|---|---|---|---|---|---|---|
| Unfinished, 5.0mm thick | BC@.034 | SF | 1.21 | 1.09 | 2.30 | 3.91 |

**Remove tongue-and-groove plank paneling.** Remove 1/4" x 8" x 4' to 8' pieces in salvage condition. Per square foot of wall covered.

| | Craft@Hrs | Unit | Material | Labor | Total | Sell |
|---|---|---|---|---|---|---|
| Working from a ladder | BC@.040 | SF | — | 1.28 | 1.28 | 2.18 |

**Beaded Cape Cod plank.** Tongue-and-groove. Medium-density fiberboard backing.

| | Craft@Hrs | Unit | Material | Labor | Total | Sell |
|---|---|---|---|---|---|---|
| 1/4" x 4" x 8' | BC@.040 | SF | .62 | 1.28 | 1.90 | 3.23 |
| 9/16" x 8', trim kit | BC@.250 | Ea | 13.60 | 8.02 | 21.62 | 36.80 |

**Edge V plank.** 4" wide x 8' long. Including 10% waste.

| | Craft@Hrs | Unit | Material | Labor | Total | Sell |
|---|---|---|---|---|---|---|
| Beaded Adirondack oak, 5/16" | BC@.040 | SF | 1.28 | 1.28 | 2.56 | 4.35 |
| Beaded knotty pine, 3/4" | BC@.040 | SF | .63 | 1.28 | 1.91 | 3.25 |
| Knotty cedar, 5/16" | BC@.040 | SF | .70 | 1.28 | 1.98 | 3.37 |

**Edge V plank trim pack.** Tongue-and-groove.

| | Craft@Hrs | Unit | Material | Labor | Total | Sell |
|---|---|---|---|---|---|---|
| Adirondack oak, 5/16" x 6' | BC@.250 | Ea | 31.00 | 8.02 | 39.02 | 66.30 |

**Pine planking.** 4" x 8' x 5/16". Tongue-and-groove. Including 10% waste.

| | Craft@Hrs | Unit | Material | Labor | Total | Sell |
|---|---|---|---|---|---|---|
| Beaded knotty pine | BC@.040 | SF | .54 | 1.28 | 1.82 | 3.09 |
| Knotty pine 8' trim pack | BC@.250 | Ea | 16.03 | 8.02 | 24.05 | 40.90 |
| V-groove pine | BC@.040 | SF | .49 | 1.28 | 1.77 | 3.01 |

|  | Craft@Hrs | Unit | Material | Labor | Total | Sell |
|---|---|---|---|---|---|---|

## Wainscot Paneling

**Hardboard wainscot paneling, Masonite Corp.** Paneling covers only the lower 32" of walls. Hardboard with the classic look of fine raised wood. Cost per panel.

|  | Craft@Hrs | Unit | Material | Labor | Total | Sell |
|---|---|---|---|---|---|---|
| 8" x 32" | BC@.035 | Ea | 4.30 | 1.12 | 5.42 | 9.21 |
| 12" x 32", Pack of 2 | BC@.105 | Ea | 17.90 | 3.37 | 21.27 | 36.20 |
| 16" x 32", Pack of 2 | BC@.140 | Ea | 18.40 | 4.49 | 22.89 | 38.90 |
| 48" x 32", Pack of 2 | BC@.420 | Ea | 39.30 | 13.50 | 52.80 | 89.80 |
| Dual outlet cover | BC@.250 | Ea | 4.35 | 8.02 | 12.37 | 21.00 |
| Universal outlet cover | BC@.250 | Ea | 4.31 | 8.02 | 12.33 | 21.00 |

**Birch wainscot paneling, Bedford Village, Georgia-Pacific.** Precut to 32" high. Unfinished. Stain to match an existing wainscot room, or customize. Beaded grooves every 1-1/2". 48" wide x 32" high. 5mm thick. Including 10% waste.

|  | Craft@Hrs | Unit | Material | Labor | Total | Sell |
|---|---|---|---|---|---|---|
| Ann Arbor | BC@.040 | SF | .93 | 1.28 | 2.21 | 3.76 |
| Unfinished | BC@.040 | SF | 1.22 | 1.28 | 2.50 | 4.25 |

**Pine wainscot paneling.** 4" x 32". Tongue-and-groove. Including 10% waste.

|  | Craft@Hrs | Unit | Material | Labor | Total | Sell |
|---|---|---|---|---|---|---|
| 1/4", paintable white | BC@.040 | Ea | .90 | 1.28 | 2.18 | 3.71 |
| 5/16", beaded knotty pine | BC@.040 | Ea | 1.08 | 1.28 | 2.36 | 4.01 |
| 3/4" x 8', knotty pine trim pack | BC@.250 | Ea | 16.60 | 8.02 | 24.62 | 41.90 |

## Fiberglass Reinforced Plastic Panels

**FRP wall and ceiling panels, Structoglas®, Sequentia.** For residential construction and remodeling. Won't rust, rot, corrode, stain, dent, peel, or splinter. One side textured. Surface won't support mold, mildew, or other bacterial growth. Impermeable to cooking fumes and grease. Cleans with household detergents. Meets all major model code requirements. Class C flame spread. 4' x 8', .090 thick panels. Including 10% waste.

|  | Craft@Hrs | Unit | Material | Labor | Total | Sell |
|---|---|---|---|---|---|---|
| Almond | BC@.032 | SF | 1.17 | 1.03 | 2.20 | 3.74 |
| White | BC@.032 | SF | 1.10 | 1.03 | 2.13 | 3.62 |

**FRP panel molding.** Fiberglass reinforced plastic. White. 10' long.

|  | Craft@Hrs | Unit | Material | Labor | Total | Sell |
|---|---|---|---|---|---|---|
| Division bar | BC@.160 | Ea | 3.61 | 5.13 | 8.74 | 14.90 |
| Inside corner | BC@.160 | Ea | 3.61 | 5.13 | 8.74 | 14.90 |
| Outside corner | BC@.160 | Ea | 3.61 | 5.13 | 8.74 | 14.90 |
| End cap, 8' long | BC@.160 | Ea | 2.53 | 5.13 | 7.66 | 13.00 |

**Nylon panel rivets.** Non-staining. Maintains moisture seal and provides mechanical fastening for FRP panels when used with silicone sealants. 1/16" minimum grip, 1/2" maximum grip. Pack of 100.

|  | Craft@Hrs | Unit | Material | Labor | Total | Sell |
|---|---|---|---|---|---|---|
| Almond | — | Ea | 13.60 | — | 13.60 | — |
| White | — | Ea | 14.00 | — | 14.00 | — |

## Ceiling Tile

**Acoustical ceiling tile.** 12" x 12" tile. Working from a ladder at heights 8' to 12' above floor level.

|  | Craft@Hrs | Unit | Material | Labor | Total | Sell |
|---|---|---|---|---|---|---|
| Remove nailed or glued tile ceiling, per square foot of ceiling | BC@.009 | SF | — | .29 | .29 | .49 |
| Remove and install glued or stapled tile at various locations, per tile | BC@.100 | Ea | — | 3.21 | 3.21 | 5.46 |
| Remove and install border tile by slipping tile under molding, per tile replaced | BC@.110 | Ea | — | 3.53 | 3.53 | 6.00 |
| Staple tile directly on furring, no molding included, per square foot of tile | BC@.020 | SF | — | .64 | .64 | 1.09 |
| Add for job setup, per room | BC@.150 | Ea | — | 4.81 | 4.81 | 8.18 |

| | Craft@Hrs | Unit | Material | Labor | Total | Sell |
|---|---|---|---|---|---|---|
| Staple tile to 1" x 2" furring strips, including strips, | | | | | | |
| per square foot of tile | BC@.042 | SF | — | 1.35 | 1.35 | 2.30 |
| Add for job setup, per room | BC@.150 | Ea | — | 4.81 | 4.81 | 8.18 |
| Set tile directly on ceiling with adhesive, including | | | | | | |
| cove molding, per square foot of tile | BC@.020 | SF | — | .64 | .64 | 1.09 |
| Add for job setup, per room | BC@.150 | Ea | — | 4.81 | 4.81 | 8.18 |
| Secure loose tile by nailing or driving screws, | | | | | | |
| per tile | BC@.035 | Ea | — | 1.12 | 1.12 | 1.90 |

**Ceiling tile, Armstrong.** Covers cracked or stained drywall ceilings. Fire-retardant. Stapled or set with adhesive.

| | Craft@Hrs | Unit | Material | Labor | Total | Sell |
|---|---|---|---|---|---|---|
| 12" x 12", Glenwood | BC@.020 | Ea | .77 | .64 | 1.41 | 2.40 |
| 12" x 12", Grenoble | BC@.020 | Ea | .62 | .64 | 1.26 | 2.14 |
| 12" x 12", White, washable | BC@.020 | Ea | .55 | .64 | 1.19 | 2.02 |

**Tin design ceiling tile, Armstrong.** Hidden seams, durable, vinyl-coated, fire-retardant.

| | Craft@Hrs | Unit | Material | Labor | Total | Sell |
|---|---|---|---|---|---|---|
| 12" x 12" | BC@.020 | Ea | 1.66 | .64 | 2.30 | 3.91 |

**Wood fiber ceiling tile, Advantage Series, USG.** Class C panels for non-acoustical and non fire-rated applications. Tongue-and-groove edging hides staples for a smooth, clean look. 12" x 12" x 1/2" with staple flange. Stapled or set with adhesive.

| | Craft@Hrs | Unit | Material | Labor | Total | Sell |
|---|---|---|---|---|---|---|
| Lace | BC@.020 | Ea | .82 | .64 | 1.46 | 2.48 |
| Tivoli | BC@.020 | Ea | .93 | .64 | 1.57 | 2.67 |
| White wood fiber | BC@.020 | Ea | .94 | .64 | 1.58 | 2.69 |

**Acoustical ceiling tile adhesive, 237, Henry.** For use on 12" x 12" or 12" x 24" tile. Bonds to concrete, concrete block, drywall, plaster and brick. 30-minute working time. One gallon covers 60 square feet.

| | Craft@Hrs | Unit | Material | Labor | Total | Sell |
|---|---|---|---|---|---|---|
| Per square foot | — | Ea | .25 | — | .25 | — |

## Suspended Ceilings

**Remove suspended tile ceiling.** (200 to 250 SF of ceiling yields one CY of debris.)

| | Craft@Hrs | Unit | Material | Labor | Total | Sell |
|---|---|---|---|---|---|---|
| Including suspended grid | BL@.010 | SF | — | .27 | .27 | .46 |
| Including grid in salvage condition | BL@.019 | SF | — | .51 | .51 | .87 |

**Install suspended ceiling.** Working from a ladder at heights 8' to 12' above floor level. No light fixtures included. Suspended from wires attached to wood joists through holes broken in drywall, plaster or fiberboard ceiling.

| | Craft@Hrs | Unit | Material | Labor | Total | Sell |
|---|---|---|---|---|---|---|
| 24" x 24" ceiling grid with panels | BC@.042 | SF | — | 1.35 | 1.35 | 2.30 |
| 24" x 48" ceiling grid with panels | BC@.028 | SF | — | .90 | .90 | 1.53 |
| 48" x 48" ceiling grid with panels | BC@.025 | SF | — | .80 | .80 | 1.36 |
| Add for job setup, per room | BC@.183 | Ea | — | 5.87 | 5.87 | 9.98 |

**Remove and replace 24" x 48" ceiling panels on an existing grid.**

| | Craft@Hrs | Unit | Material | Labor | Total | Sell |
|---|---|---|---|---|---|---|
| No cutting of panels, per panel | BC@.052 | Ea | — | 1.67 | 1.67 | 2.84 |
| 10% of panels cut, per panel | BC@.059 | Ea | — | 1.89 | 1.89 | 3.21 |
| 20% of panels cut, per panel | BC@.064 | Ea | — | 2.05 | 2.05 | 3.49 |
| 30% of panels cut, per panel | BC@.070 | Ea | — | 2.25 | 2.25 | 3.83 |
| 40% of panels cut, per panel | BC@.076 | Ea | — | 2.44 | 2.44 | 4.15 |
| 50% of panels cut, per panel | BC@.082 | Ea | — | 2.63 | 2.63 | 4.47 |

| | Craft@Hrs | Unit | Material | Labor | Total | Sell |
|---|---|---|---|---|---|---|

**Prelude® suspended ceiling system cross tee, Armstrong.** Intermediate-duty performance. Meets all building codes for non-residential construction. Hot-dipped galvanized, rust-resistant steel. Double-web construction. Requires only 2-1/2" clearance.

| | Craft@Hrs | Unit | Material | Labor | Total | Sell |
|---|---|---|---|---|---|---|
| 2', stab-in, white | BC@.033 | Ea | .94 | 1.06 | 2.00 | 3.40 |
| 4', stab-in, white | BC@.033 | Ea | 1.83 | 1.06 | 2.89 | 4.91 |

**Prelude® fire guard suspended ceiling system cross tee, Armstrong.** Fire rated intermediate-duty performance. Meets all building codes for non-residential construction. Hot-dipped galvanized, rust-resistant steel. Double-web construction. Requires only 2-1/2" clearance.

| | Craft@Hrs | Unit | Material | Labor | Total | Sell |
|---|---|---|---|---|---|---|
| 2', stab-in | BC@.033 | Ea | .85 | 1.06 | 1.91 | 3.25 |
| 4', stab-in, white | BC@.033 | Ea | 2.25 | 1.06 | 3.31 | 5.63 |

**Prelude® suspended ceiling system main beam, Armstrong.** Intermediate-duty performance. Hot-dipped galvanized, rust-resistant steel. Double-web construction. PeakForm™ end. 12.6 pound per foot carrying capacity.

| | Craft@Hrs | Unit | Material | Labor | Total | Sell |
|---|---|---|---|---|---|---|
| 12', white | BC@.167 | Ea | 5.37 | 5.36 | 10.73 | 18.20 |
| 12', fire guard, white | BC@.167 | Ea | 6.13 | 5.36 | 11.49 | 19.50 |

**Prelude® suspended ceiling system wall molding, Armstrong.** Fire rated performance. Hemmed edges. Hot-dipped galvanized, rust-resistant steel, 12' x 7/8" x 7/8", 0.018" thick. Meets all building codes for nonresidential construction.

| | Craft@Hrs | Unit | Material | Labor | Total | Sell |
|---|---|---|---|---|---|---|
| 12', white | BC@.167 | Ea | 3.54 | 5.36 | 8.90 | 15.10 |

**CeilingMAX™ zero-clearance ceiling tile grid system.** Surface mounting for covering old plaster, drywall or paste-up ceilings. No wires to hang or leveling required. Top hanger fastens directly to joists or existing ceiling. Use any 2' x 2' or 4' x 4' acoustic ceiling tile. Eliminates demolition and re-installation. Runner snaps into top hanger to lock entire grid system and ceiling panels. Cross tee is installed perpendicular to top hanger to support ceiling panels. Wall bracket is installed around perimeter of room. Grid snaps out to access plenum above the grid. White vinyl.

| | Craft@Hrs | Unit | Material | Labor | Total | Sell |
|---|---|---|---|---|---|---|
| 2' cross tee | BC@.033 | Ea | 1.35 | 1.06 | 2.41 | 4.10 |
| 8' runner insert | BC@.333 | Ea | 5.30 | 10.70 | 16.00 | 27.20 |
| 8' top hanger | BC@.333 | Ea | 6.40 | 10.70 | 17.10 | 29.10 |
| 8' wall bracket | BC@.111 | Ea | 3.18 | 3.56 | 6.74 | 11.50 |

**Drill tip suspension lag screws.**

| | Craft@Hrs | Unit | Material | Labor | Total | Sell |
|---|---|---|---|---|---|---|
| Package of 100 | — | Ea | 32.00 | — | 32.00 | — |

**White rivets.**

| | Craft@Hrs | Unit | Material | Labor | Total | Sell |
|---|---|---|---|---|---|---|
| Package of 100 | — | Ea | 3.97 | — | 3.97 | — |

**Ceiling wire hanging kit.** Contains nine 12-gauge support wires and nine J-hook nails. Suspends 256 square feet with main tees supported on 4' centers.

| | Craft@Hrs | Unit | Material | Labor | Total | Sell |
|---|---|---|---|---|---|---|
| Ceiling wire kit | BC@1.67 | Ea | 2.80 | 53.60 | 56.40 | 95.90 |

**Suspended ceiling installation kit, Armstrong.** Kit includes everything needed to install 64 square feet of suspended ceiling grid. White.

| | Craft@Hrs | Unit | Material | Labor | Total | Sell |
|---|---|---|---|---|---|---|
| 2' x 2' grid | BC@1.25 | Ea | 44.50 | 40.10 | 84.60 | 144.00 |
| 2' x 4' grid | BC@1.25 | Ea | 34.10 | 40.10 | 74.20 | 126.00 |

**Random textured suspended ceiling panel, Armstrong.** Square lay-in panel. Recommended for use with Prelude 15/16" exposed-tee grid. Class A fire-resistance. Reduces sound by 55%. 5-year limited warranty.

| | Craft@Hrs | Unit | Material | Labor | Total | Sell |
|---|---|---|---|---|---|---|
| 2' x 2' x 5/8" thick | BC@.027 | Ea | 1.78 | .87 | 2.65 | 4.51 |
| 2' x 4' x 5/8" thick | BC@.056 | Ea | 4.23 | 1.80 | 6.03 | 10.30 |

| | Craft@Hrs | Unit | Material | Labor | Total | Sell |
|---|---|---|---|---|---|---|

**Contractor suspended ceiling panel, Armstrong.** Fire-retardant. Acoustical NRC .55. Reduces sound by 55%.

| | Craft@Hrs | Unit | Material | Labor | Total | Sell |
|---|---|---|---|---|---|---|
| 2' x 2', random textured | BC@.027 | Ea | 3.00 | .87 | 3.87 | 6.58 |

**Plain white suspended ceiling panel, Armstrong.** Fire-retardant. 10-year limited warranty.

| | Craft@Hrs | Unit | Material | Labor | Total | Sell |
|---|---|---|---|---|---|---|
| 2' x 4', square edge | BC@.056 | Ea | 5.06 | 1.80 | 6.86 | 11.70 |

**Classic fine-textured suspended ceiling panel, Armstrong.** 3-dimensional look. Reduces sound by 60%. High light reflectance. BioBlock paint inhibits mold and mildew. HumiGuard Plus for extra sag resistance. Fire-retardant.

| | Craft@Hrs | Unit | Material | Labor | Total | Sell |
|---|---|---|---|---|---|---|
| 2' x 2' | BC@.027 | Ea | 5.76 | .87 | 6.63 | 11.30 |

**Acoustical suspended ceiling panel, Armstrong.** Square edge. Light reflectance minimum LR 0.80. Reduces sound by 55%.

| | Craft@Hrs | Unit | Material | Labor | Total | Sell |
|---|---|---|---|---|---|---|
| 2' x 4', random textured | BC@.056 | Ea | 2.90 | 1.80 | 4.70 | 7.99 |

**Textured suspended ceiling panel, Armstrong.** Good for basements to hide plumbing and electrical fixtures while allowing access. Acoustical (NRC .55, CAC min. 35). Fire retardant. Square edge. 5-year limited warranty.

| | Craft@Hrs | Unit | Material | Labor | Total | Sell |
|---|---|---|---|---|---|---|
| 2' x 4' | BC@.056 | Ea | 2.87 | 1.80 | 4.67 | 7.94 |

**Sahara suspended ceiling panel, Armstrong.** Regular edge panel. HumiGuard™ Plus and Bioguard™ paint. Class A fire-resistant. 3-dimensional look. Reduces sound by 50%.

| | Craft@Hrs | Unit | Material | Labor | Total | Sell |
|---|---|---|---|---|---|---|
| 2' x 2' | BC@.027 | Ea | 5.00 | .87 | 5.87 | 9.98 |

**HomeStyle™ suspended ceiling panel, Armstrong.** Grid-hiding designs. Fire-retardant. 10-year limited warranty.

| | Craft@Hrs | Unit | Material | Labor | Total | Sell |
|---|---|---|---|---|---|---|
| 2' x 2', Brighton | BC@.027 | Ea | 2.60 | .87 | 3.47 | 5.90 |
| 2' x 2', Royal oak | BC@.027 | Ea | 4.65 | .87 | 5.52 | 9.38 |

**Fiberglass suspended ceiling panel, Armstrong.** Square-edge panel with HumiGuard Plus for extra sag-resistance for high humidity.

| | Craft@Hrs | Unit | Material | Labor | Total | Sell |
|---|---|---|---|---|---|---|
| 2' x 4' x 5/8" thick,<br>    Random fissured | BC@.056 | Ea | 3.88 | 1.80 | 5.68 | 9.66 |
| 2' x 4' x 5/8" thick, Esprit | BC@.056 | Ea | 5.13 | 1.80 | 6.93 | 11.80 |

**Fiberglass suspended ceiling panel, Armstrong.** Acoustical. UL-approved fire-retardant. Light reflectance minimum LR 0.80. Class A surface-burning characteristics for Flame Spread 25 or under. Meets Class A requirements of ASTM E 1264.

| | Craft@Hrs | Unit | Material | Labor | Total | Sell |
|---|---|---|---|---|---|---|
| 2' x 4', reveal-edge design | BC@.056 | Ea | 5.13 | 1.80 | 6.93 | 11.80 |

**Fissured fire guard suspended ceiling panel, Armstrong.** Square lay-in panel. Recommended for use with Prelude 15/16" exposed-tee grid. Class A fire-resistance. UL approved for fire rated ceiling assemblies. Reduces sound by 55%.

| | Craft@Hrs | Unit | Material | Labor | Total | Sell |
|---|---|---|---|---|---|---|
| 2' x 4' x 5/8" | BC@.056 | Ea | 5.40 | 1.80 | 7.20 | 12.20 |

**Prestige suspended ceiling panel, Armstrong.** Architectural detail. Grid-blending pattern. Fire-retardant.

| | Craft@Hrs | Unit | Material | Labor | Total | Sell |
|---|---|---|---|---|---|---|
| 2' x 2' | BC@.027 | Ea | 3.79 | .87 | 4.66 | 7.92 |

**Grenoble suspended ceiling panel, Armstrong.** HomeStyle™ ceiling panel. Fire-retardant.

| | Craft@Hrs | Unit | Material | Labor | Total | Sell |
|---|---|---|---|---|---|---|
| 2' x 4' | BC@.056 | Ea | 4.34 | 1.80 | 6.14 | 10.40 |

|  | Craft@Hrs | Unit | Material | Labor | Total | Sell |
|---|---|---|---|---|---|---|

**RADAR™ Illusion suspended ceiling panel, Professional Series, USG Interiors.** Non-directional pattern for fast installation. Made with a perforated, water-felted surface for good sound absorption. Face-scored to create the illusion of a smaller-scaled ceiling system without compromising accessibility and speed of installation. Surface cleans with a soft brush or vacuum.

| | | | | | | |
|---|---|---|---|---|---|---|
| 2' x 4' x 3/4", slant edge | BC@.056 | Ea | 5.70 | 1.80 | 7.50 | 12.80 |
| 2' x 2' x 5/8", square edge | BC@.027 | Ea | 2.16 | .87 | 3.03 | 5.15 |
| 2' x 4' x 5/8", square edge | BC@.056 | Ea | 3.63 | 1.80 | 5.43 | 9.23 |
| 2' x 2' x 5/8", shadowline tapered edge | BC@.027 | Ea | 2.94 | .87 | 3.81 | 6.48 |

**Tundra fire guard suspended ceiling panel, Armstrong** Fine textured ceiling panels with high light reflectance and scratch resistance. Washable. Sag-resistant. Resists growth of mold and mildew.

| | | | | | | |
|---|---|---|---|---|---|---|
| 2' x 2' x 1/2", square edge | BC@.056 | Ea | 4.70 | 1.80 | 6.50 | 11.10 |

**Fifth Avenue suspended ceiling panel, Professional Series, USG Interiors.** Medium-textured acoustical panels with directional pattern. Made with a water-felted manufacturing process for good sound absorption. Surface cleans easily with a soft brush or vacuum. For home and light commercial applications. Square edge.

| | | | | | | |
|---|---|---|---|---|---|---|
| 2' x 2' x 5/8" | BC@.027 | Ea | 1.92 | .87 | 2.79 | 4.74 |
| 2' x 4' x 5/8" | BC@.056 | Ea | 3.62 | 1.80 | 5.42 | 9.21 |

**Firecode™ Fifth Avenue suspended ceiling panel, Professional Series, USG Interiors.** Medium-textured acoustical panels with directional fissures. Made with a water-felted manufacturing process for good sound absorption. Cleans easily with a soft brush or vacuum. Washable. Class A fire-resistant.

| | | | | | | |
|---|---|---|---|---|---|---|
| 2' x 4' x 5/8", square edge | BC@.056 | Ea | 4.85 | 1.80 | 6.65 | 11.30 |

**Sheetrock® gypsum lay-in ceiling panel, Professional Series, USG Interiors.** Vinyl-faced panels for interior or exterior applications. Fire-rated construction up to 1-1/2 hours. USDA accepted for food processing areas.

| | | | | | | |
|---|---|---|---|---|---|---|
| 2' x 4' x 1/2", square edge | BC@.056 | Ea | 6.26 | 1.80 | 8.06 | 13.70 |

**Stonehurst mineral fiber suspended ceiling panel, Advantage Series, USG Interiors.** Square edge design for quick installation.

| | | | | | | |
|---|---|---|---|---|---|---|
| 2' x 4' x 9/16" | BC@.056 | Ea | 3.43 | 1.80 | 5.23 | 8.89 |

**Luna™ suspended ceiling panel, Elite Series, USG Interiors.** Fine-textured acoustical panels with Climaplus performance to prevent visible sag. High sound absorption and sound blocking. High light reflectance values. Cleans easily with a soft brush or vacuum. High humidity resistance.

| | | | | | | |
|---|---|---|---|---|---|---|
| 2' x 2' x 3/4", slant edge | BC@.027 | Ea | 6.04 | .87 | 6.91 | 11.70 |

**Alpine suspended ceiling panel, USG Interiors.** Medium-textured acoustical panels made with a water-felted manufacturing process for good sound absorption. Perforated. Excellent light reflectance. Cleans easily with a soft brush or vacuum.

| | | | | | | |
|---|---|---|---|---|---|---|
| 2' x 2' x 5/8", slant edge | BC@.027 | Ea | 2.71 | .87 | 3.58 | 6.09 |

**Astro textured suspended ceiling panel, Professional Series, USG Interiors.** Ceiling panels with textured surface and good sound absorption.

| | | | | | | |
|---|---|---|---|---|---|---|
| 2' x 2' x 5/8", square edge | BC@.027 | Ea | 4.32 | .87 | 5.19 | 8.82 |

**Saville Row suspended ceiling panel, Professional Series, USG Interiors.** Medium, natural random texture with a sculptured, upscale appearance at an economy price. Made with a water-felted manufacturing process for good sound absorption. Perforated for good sound absorption. Excellent light reflectance. Cleans easily with a soft brush or vacuum.

| | | | | | | |
|---|---|---|---|---|---|---|
| 2' x 2' x 3/4", slant edge | BC@.027 | Ea | 5.57 | .87 | 6.44 | 10.90 |

| | Craft@Hrs | Unit | Material | Labor | Total | Sell |
|---|---|---|---|---|---|---|
| **Acrylic suspended lighting panel.** | | | | | | |
| 2' x 4', cracked ice, clear | BC@.056 | Ea | 8.58 | 1.80 | 10.38 | 17.60 |
| 2' x 4', cracked ice, white | BC@.056 | Ea | 8.52 | 1.80 | 10.32 | 17.50 |
| 2' x 4', prismatic, clear | BC@.056 | Ea | 8.58 | 1.80 | 10.38 | 17.60 |
| 2' x 4', prismatic, white | BC@.056 | Ea | 8.49 | 1.80 | 10.29 | 17.50 |
| | | | | | | |
| **Styrene suspended lighting panel.** | | | | | | |
| 2' x 2', cracked ice, clear | BC@.027 | Ea | 3.00 | .87 | 3.87 | 6.58 |
| 2' x 2', cracked ice, white | BC@.027 | Ea | 3.00 | .87 | 3.87 | 6.58 |
| 2' x 2', prismatic, clear | BC@.027 | Ea | 3.00 | .87 | 3.87 | 6.58 |
| 2' x 2', prismatic, white | BC@.027 | Ea | 3.00 | .87 | 3.87 | 6.58 |
| 2' x 4', cracked ice, clear | BC@.056 | Ea | 4.61 | 1.80 | 6.41 | 10.90 |
| 2' x 4', cracked ice, white | BC@.056 | Ea | 4.61 | 1.80 | 6.41 | 10.90 |
| 2' x 4', eggcrate, metallic | BC@.056 | Ea | 25.30 | 1.80 | 27.10 | 46.10 |
| 2' x 4', eggcrate, white | BC@.056 | Ea | 10.30 | 1.80 | 12.10 | 20.60 |
| 2' x 4', flat mist, white | BC@.056 | Ea | 8.00 | 1.80 | 9.80 | 16.70 |
| 2' x 4', prismatic, clear | BC@.056 | Ea | 4.61 | 1.80 | 6.41 | 10.90 |
| 2' x 4', prismatic, white | BC@.056 | Ea | 4.61 | 1.80 | 6.41 | 10.90 |
| 2' x 4', Victorian pattern | BC@.056 | Ea | 14.00 | 1.80 | 15.80 | 26.90 |

**Fluorescent troffer light fixture.** For suspending in a ceiling grid. Labor includes fixture hookup only. Add the cost of rough-in electrical and lamps. 32 watt, T-8.

| | Craft@Hrs | Unit | Material | Labor | Total | Sell |
|---|---|---|---|---|---|---|
| 2' x 2', 2 light | BE@.800 | Ea | 45.30 | 27.50 | 72.80 | 124.00 |
| 2' x 4', 3 light | BE@.800 | Ea | 42.70 | 27.50 | 70.20 | 119.00 |
| 2' x 4', 4 light | BE@.800 | Ea | 53.80 | 27.50 | 81.30 | 138.00 |

**Parabolic troffer fluorescent fixture.** For suspending in a ceiling grid. Labor includes fixture hookup only. Add the cost of rough-in electrical and lamps. 32 watt, T-8. Contoured louvers reduce glare. Contains electronic ballast for maximum efficiency.

| | Craft@Hrs | Unit | Material | Labor | Total | Sell |
|---|---|---|---|---|---|---|
| 2' x 4', 3 light | BE@.800 | Ea | 74.50 | 27.50 | 102.00 | 173.00 |

**U-lamp troffer fluorescent fixture.** For suspending in a ceiling grid. Labor includes fixture hookup only. Add the cost of rough-in electrical and lamps. Uses two U6 lamps. Energy saving ballast. White unibody steel housing, reinforced flat white steel door, and mitered corners. Hinged and mitered doorframe. Comes with cam latches for easy opening. 40 or 34 watt bulbs.

| | Craft@Hrs | Unit | Material | Labor | Total | Sell |
|---|---|---|---|---|---|---|
| 2' x 2', 2 light | BE@.800 | Ea | 37.80 | 27.50 | 65.30 | 111.00 |

| | Craft@Hrs | Unit | Material | Labor | Total | Sell |
|---|---|---|---|---|---|---|

## Ornamental Wood Molding

**Hardwood Victorian base molding.** Labor for moldings is best estimated by the piece rather than by the foot. The hard part is cutting lengths precisely to size. The more cuts needed, the more time required. For example, one 12' run of molding requires two cuts — one on each end. Twelve 1' lengths of molding would require 24 cuts, even though it's the same 12 linear feet. The estimates that follow are based on linear feet, but assume that full 8' lengths can be used to good advantage.

| | Craft@Hrs | Unit | Material | Labor | Total | Sell |
|---|---|---|---|---|---|---|
| 5/8" x 4", per foot | BC@.032 | LF | 3.00 | 1.03 | 4.03 | 6.85 |

**Beaded baseboard molding.** Solid wood. Embossed. Stain or paint to match paneling, cabinets, hardwood floors, or walls. Baseboard connectors provide architectural detail. Protects corners. Traditional period-style detail. White hardwood oak. Per foot.

| | | | | | | |
|---|---|---|---|---|---|---|
| 3/4" x 6-1/2", Victorian | BC@.032 | LF | 3.55 | 1.03 | 4.58 | 7.79 |

**Embossed molding, basswood.**

| | | | | | | |
|---|---|---|---|---|---|---|
| No. 630, 3/4" x 1/2" x 8' | BC@.290 | Ea | 7.75 | 9.31 | 17.06 | 29.00 |

**Base corner molding.** Per corner.

| | | | | | | |
|---|---|---|---|---|---|---|
| No. 623, 7/8" x 7/8" x 6-3/4", inside | BC@.050 | Ea | 2.45 | 1.60 | 4.05 | 6.89 |
| No. 623, 7/8" x 7/8" x 6-3/4", outside | BC@.050 | Ea | 4.07 | 1.60 | 5.67 | 9.64 |

**Rosette block molding.** Per block.

| | | | | | | |
|---|---|---|---|---|---|---|
| 2-3/4" x 2-3/4" x 1" | BC@.050 | Ea | 1.90 | 1.60 | 3.50 | 5.95 |
| 3-3/4" x 3-3/4" x 11/16" | BC@.050 | Ea | 1.89 | 1.60 | 3.49 | 5.93 |

**Square rosette molding, solid pine.**

| | | | | | | |
|---|---|---|---|---|---|---|
| 7/8" x 2-1/2" x 2-1/2" | BC@.050 | Ea | 1.98 | 1.60 | 3.58 | 6.09 |
| 7/8" x 3-1/2" x 3-1/2" | BC@.050 | Ea | 3.07 | 1.60 | 4.67 | 7.94 |

**Plinth block.**

| | | | | | | |
|---|---|---|---|---|---|---|
| 3-3/4" x 5-1/2", solid pine | BC@.050 | Ea | 5.09 | 1.60 | 6.69 | 11.40 |
| 3-3/4" x 5-3/4", solid pine | BC@.050 | Ea | 4.23 | 1.60 | 5.83 | 9.91 |

**Ornamental crown molding.** Per foot.

| | | | | | | |
|---|---|---|---|---|---|---|
| 3-3/4", dentil, pine | BC@.044 | LF | 4.26 | 1.41 | 5.67 | 9.64 |
| 3-3/4", rope, pine | BC@.044 | LF | 4.91 | 1.41 | 6.32 | 10.70 |
| 4-7/16", embossed | BC@.044 | LF | 3.94 | 1.41 | 5.35 | 9.10 |
| 4-7/8", pine | BC@.044 | LF | 2.97 | 1.41 | 4.38 | 7.45 |
| 5-1/2", dentil, pine | BC@.044 | LF | 6.43 | 1.41 | 7.84 | 13.30 |

**Embossed decorative hardwood molding.**

| | | | | | | |
|---|---|---|---|---|---|---|
| Base, 5/8" x 4" x 8' | BC@.128 | Ea | 19.50 | 4.11 | 23.61 | 40.10 |
| Egg and Dart, 1/2" x 3/4" x 8' | BC@.128 | Ea | 8.57 | 4.11 | 12.68 | 21.60 |
| Egg and Dart, 5/8" x 1" x 8' | BC@.128 | Ea | 8.91 | 4.11 | 13.02 | 22.10 |
| Flower, 5/8" x 1" x 8' | BC@.128 | Ea | 10.77 | 4.11 | 14.88 | 25.30 |
| Flower, 3/4" x 1-15/16" x 8' | BC@.128 | Ea | 8.60 | 4.11 | 12.71 | 21.60 |
| Greek Key, 3/8" x 3/4" x 8' | BC@.128 | Ea | 6.42 | 4.11 | 10.53 | 17.90 |
| Leaf, 3/8" x 3/4" x 8' | BC@.128 | Ea | 6.42 | 4.11 | 10.53 | 17.90 |
| Molding, 3/8" x 3/4" x 8' | BC@.128 | Ea | 6.37 | 4.11 | 10.48 | 17.80 |
| Repeating Flower, 3/8" x 1-15/16" x 8' | BC@.128 | Ea | 8.56 | 4.11 | 12.67 | 21.50 |
| Rope, 1/4" x 1" x 8' | BC@.128 | Ea | 6.92 | 4.11 | 11.03 | 18.80 |
| Rope Flute, 1/2" x 2-1/4" x 7' | BC@.112 | Ea | 10.50 | 3.59 | 14.09 | 24.00 |
| Rope Flute, 1/2" x 3" x 7' | BC@.112 | Ea | 15.80 | 3.59 | 19.39 | 33.00 |
| Sunburst, 3/4" x 1-15/16" x 8' | BC@.128 | Ea | 8.56 | 4.11 | 12.67 | 21.50 |
| Victorian, 5/8" x 3" x 8' | BC@.128 | Ea | 19.30 | 4.11 | 23.41 | 39.80 |
| Vine, 3/8" x 3/4" x 8' | BC@.128 | Ea | 6.30 | 4.11 | 10.41 | 17.70 |
| Vine, 3/8" x 1-3/4" x 8' | BC@.128 | Ea | 10.22 | 4.11 | 14.33 | 24.40 |

|  | Craft@Hrs | Unit | Material | Labor | Total | Sell |
|---|---|---|---|---|---|---|

## Primed Finger Joint Molding

**Primed finger joint base molding.** Per foot.

|  | Craft@Hrs | Unit | Material | Labor | Total | Sell |
|---|---|---|---|---|---|---|
| No. S, 7/16" x 3" | BC@.032 | LF | .74 | 1.03 | 1.77 | 3.01 |
| No. B, 7/16" x 3" | BC@.032 | LF | .75 | 1.03 | 1.78 | 3.03 |
| No. 663, 9/16" x 3-1/4" | BC@.032 | LF | .92 | 1.03 | 1.95 | 3.32 |
| No. 753, 9/16" x 3-1/4" | BC@.032 | LF | .94 | 1.03 | 1.97 | 3.35 |
| No. 750, 9/16" x 4-1/4" | BC@.032 | LF | 1.36 | 1.03 | 2.39 | 4.06 |
| No. 620, 9/16" x 4-1/4" | BC@.032 | LF | 1.33 | 1.03 | 2.36 | 4.01 |
| No. LB-11, 9/16" x 4-1/2" | BC@.032 | LF | 1.35 | 1.03 | 2.38 | 4.05 |
| No. 618, 9/16" x 5-1/4" | BC@.032 | LF | 1.61 | 1.03 | 2.64 | 4.49 |
| No. 5180, 9/16" x 5-1/4" | BC@.032 | LF | 1.60 | 1.03 | 2.63 | 4.47 |
| No. 5163, 9/16" x 5-1/4" | BC@.032 | LF | 1.65 | 1.03 | 2.68 | 4.56 |
| No. A314, 5/8" x 3-1/4" | BC@.032 | LF | .89 | 1.03 | 1.92 | 3.26 |
| No. B322, 5/8" x 3-1/4" | BC@.032 | LF | .87 | 1.03 | 1.90 | 3.23 |

**Primed finger joint base cap molding.** Per foot.

|  | Craft@Hrs | Unit | Material | Labor | Total | Sell |
|---|---|---|---|---|---|---|
| No. 1207, 11/16" x 7/8" | BC@.018 | LF | .71 | .58 | 1.29 | 2.19 |
| No. 167, 11/16" x 1-1/8" | BC@.018 | LF | .68 | .58 | 1.26 | 2.14 |
| No. 163, 11/16" x 1-3/8" | BC@.018 | LF | .71 | .58 | 1.29 | 2.19 |

**Primed finger joint base shoe molding.** Per foot.

|  | Craft@Hrs | Unit | Material | Labor | Total | Sell |
|---|---|---|---|---|---|---|
| No. 129, 7/16" x 11/16" | BC@.020 | LF | .38 | .64 | 1.02 | 1.73 |
| No. WM126, 1/2" x 3/4" | BC@.020 | LF | .48 | .64 | 1.12 | 1.90 |

**Primed finger joint bead molding.** Per foot.

|  | Craft@Hrs | Unit | Material | Labor | Total | Sell |
|---|---|---|---|---|---|---|
| No. WM75, 9/16" x 1-5/8" | BC@.032 | LF | .68 | 1.03 | 1.71 | 2.91 |
| No. WM74, 9/16" x 1-3/4" | BC@.032 | LF | .82 | 1.03 | 1.85 | 3.15 |

**Primed finger joint brick molding.** Pine.

|  | Craft@Hrs | Unit | Material | Labor | Total | Sell |
|---|---|---|---|---|---|---|
| No. 180WMMPA, 1-1/4" x 2", set | BC@.250 | LF | 36.10 | 8.02 | 44.12 | 75.00 |
| No. WM180, 1-1/4" x 2" x 7' | BC@.105 | LF | 11.90 | 3.37 | 15.27 | 26.00 |

**Primed finger joint ranch casing.**

|  | Craft@Hrs | Unit | Material | Labor | Total | Sell |
|---|---|---|---|---|---|---|
| No. C101, 1/2" x 2-1/8" x 7' | BC@.105 | Ea | 4.60 | 3.37 | 7.97 | 13.50 |
| No. 713, 9/16" x 3-1/4" | BC@.015 | LF | 1.23 | .48 | 1.71 | 2.91 |
| No. 327, 11/16" x 2-1/4" | BC@.015 | LF | .89 | .48 | 1.37 | 2.33 |
| No. 327, 11/16" x 2-1/4" x 7' | BC@.105 | Ea | 6.20 | 3.37 | 9.57 | 16.30 |
| No. 315, 11/16" x 2-1/2" x 7' | BC@.105 | Ea | 6.73 | 3.37 | 10.10 | 17.20 |

**Primed finger joint colonial casing.** Pine. Per foot.

|  | Craft@Hrs | Unit | Material | Labor | Total | Sell |
|---|---|---|---|---|---|---|
| No. WM445, 5/8" x 3-1/4" | BC@.015 | LF | 1.04 | .48 | 1.52 | 2.58 |
| No. WM444, 5/8" x 3-1/2" | BC@.015 | LF | 1.14 | .48 | 1.62 | 2.75 |
| No. 442, 11/16" x 2-1/4" | BC@.015 | LF | .98 | .48 | 1.46 | 2.48 |

**Primed finger joint chair rail molding.** Per foot.

|  | Craft@Hrs | Unit | Material | Labor | Total | Sell |
|---|---|---|---|---|---|---|
| No. 297H, 5/8" x 3" | BC@.030 | LF | 1.13 | .96 | 2.09 | 3.55 |
| No. CR-7, 11/16" x 2-5/8" | BC@.030 | LF | 1.64 | .96 | 2.60 | 4.42 |
| No. AIL300, 1-1/16" x 2-3/4" | BC@.030 | LF | 1.65 | .96 | 2.61 | 4.44 |

| | Craft@Hrs | Unit | Material | Labor | Total | Sell |
|---|---|---|---|---|---|---|
| **Primed finger joint crown molding.** Per foot. | | | | | | |
| No. WM53, 9/16" x 2-5/8" | BC@.044 | LF | 1.30 | 1.41 | 2.71 | 4.61 |
| No. AMC-R-47, 9/16" x 4-5/8" | BC@.044 | LF | 2.47 | 1.41 | 3.88 | 6.60 |
| No. L47-PFJ, 9/16" x 4-5/8" | BC@.044 | LF | 1.84 | 1.41 | 3.25 | 5.53 |
| **Primed finger joint reversible flute and reed molding.** Per foot. | | | | | | |
| No. 286-PFJ, 11/16" x 3-1/4" | BC@.023 | LF | 1.46 | .74 | 2.20 | 3.74 |
| No. 286-POM, 11/16" x 3-1/4" | BC@.023 | LF | 2.00 | .74 | 2.74 | 4.66 |
| **Primed finger joint quarter round molding.** Per foot. | | | | | | |
| No. WM106, 11/16" x 11/16" | BC@.020 | LF | .64 | .64 | 1.28 | 2.18 |
| No. WM105, 3/4" x 3/4" | BC@.020 | LF | .79 | .64 | 1.43 | 2.43 |
| **Primed finger joint stool molding.** Per foot. | | | | | | |
| No. LWM1021, 11/16" x 5-1/4" | BC@.037 | LF | 2.60 | 1.19 | 3.79 | 6.44 |
| **Round edge stop molding.** Per foot. | | | | | | |
| No. P433FJ, 3/8" x 1-1/4" | BC@.015 | LF | .52 | .48 | 1.00 | 1.70 |
| No. P349PR, 7/16" x 1-1/4" | BC@.015 | LF | .66 | .48 | 1.14 | 1.94 |
| No. P435PR, 1/2" x 1-5/8" | BC@.015 | LF | .69 | .48 | 1.17 | 1.99 |

## Medium Density Fiberboard (MDF) Molding

| | Craft@Hrs | Unit | Material | Labor | Total | Sell |
|---|---|---|---|---|---|---|
| **MDF colonial base molding.** Medium density fiberboard. Per foot. | | | | | | |
| 7/16" x 3" | BC@.032 | LF | .80 | 1.03 | 1.83 | 3.11 |
| No. WM-663, 9/16" x 3-1/4" | BC@.032 | LF | .98 | 1.03 | 2.01 | 3.42 |
| No. WM662, 9/16" x 3-1/2" | BC@.032 | LF | .94 | 1.03 | 1.97 | 3.35 |
| 9/16" x 5-1/8" | BC@.032 | LF | 1.65 | 1.03 | 2.68 | 4.56 |
| No. 620-MDF, 4-1/4" | BC@.032 | LF | 1.20 | 1.03 | 2.23 | 3.79 |
| No. L163E, 5-1/8" | BC@.032 | LF | 1.45 | 1.03 | 2.48 | 4.22 |
| **MDF primed base corner.** Per corner piece. | | | | | | |
| No. MDF409A-711, 7/16" x 2-1/2" | BC@.050 | Ea | 1.91 | 1.60 | 3.51 | 5.97 |
| No. MDF413A, 7/16" x 2-1/2", 3 step | BC@.050 | Ea | 1.91 | 1.60 | 3.51 | 5.97 |
| No. MDF410A-711, 7/16" x 3-1/2" | BC@.050 | Ea | 2.43 | 1.60 | 4.03 | 6.85 |
| No. MDF24A, 5/8" x 6", imperial | BC@.050 | Ea | 4.20 | 1.60 | 5.80 | 9.86 |
| No. MDF28A, 5/8" x 6-1/2", vintage | BC@.050 | Ea | 4.20 | 1.60 | 5.80 | 9.86 |
| **MDF casing.** Medium density fiberboard. | | | | | | |
| No. AMH52, 1" x 4" | BC@.015 | LF | 1.90 | .48 | 2.38 | 4.05 |
| No. AMH56, 1" x 3-1/2" | BC@.015 | Ea | 1.47 | .48 | 1.95 | 3.32 |
| No. C322, 5/8" x 2-1/2", per foot | BC@.015 | LF | .78 | .48 | 1.26 | 2.14 |
| No. 1684, 12mm x 3-3/8", per foot | BC@.015 | LF | .84 | .48 | 1.32 | 2.24 |
| **MDF chair rail.** Medium density fiberboard. Per foot. | | | | | | |
| 9/16" x 3", primed | BC@.030 | LF | 1.30 | .96 | 2.26 | 3.84 |
| 11/16" x 2-1/2" | BC@.030 | LF | 1.03 | .96 | 1.99 | 3.38 |
| 5/8" x 2-5/8" | BC@.030 | LF | 1.06 | .96 | 2.02 | 3.43 |

| | Craft@Hrs | Unit | Material | Labor | Total | Sell |
|---|---|---|---|---|---|---|

**MDF crown molding.** Medium density fiberboard. Per foot.

| | Craft@Hrs | Unit | Material | Labor | Total | Sell |
|---|---|---|---|---|---|---|
| No. 45, 9/16" x 5-1/4" | BC@.044 | LF | 1.76 | 1.41 | 3.17 | 5.39 |
| No. L47, 9/16" x 4-5/8" x 16' | BC@.044 | LF | 1.52 | 1.41 | 2.93 | 4.98 |

**MDF ultralite crown molding.** Per foot.

| | Craft@Hrs | Unit | Material | Labor | Total | Sell |
|---|---|---|---|---|---|---|
| No. 40, 11/16" x 4-1/4" | BC@.044 | LF | 2.21 | 1.41 | 3.62 | 6.15 |
| No. 49, 15mm x 3-5/8" | BC@.044 | LF | 1.05 | 1.41 | 2.46 | 4.18 |
| No. 47, 15mm x 4-5/8" | BC@.044 | LF | 1.30 | 1.41 | 2.71 | 4.61 |
| No. 43, 18mm x 7-1/4" | BC@.044 | LF | 3.25 | 1.41 | 4.66 | 7.92 |

**MDF primed plinth block.**

| | Craft@Hrs | Unit | Material | Labor | Total | Sell |
|---|---|---|---|---|---|---|
| 1" x 3-3/4" x 8" | BC@.050 | Ea | 2.75 | 1.60 | 4.35 | 7.40 |
| 1" x 4-1/8" x 9", Victorian | BC@.050 | Ea | 3.69 | 1.60 | 5.29 | 8.99 |
| 1-1/4" x 2-1/2" x 5-3/4" | BC@.050 | Ea | 2.51 | 1.60 | 4.11 | 6.99 |
| 1-1/4" x 3-1/2" x 6-1/2" | BC@.050 | Ea | 3.18 | 1.60 | 4.78 | 8.13 |

## Polyurethane Cast Molding

**Cross head polyurethane cast molding.**

| | Craft@Hrs | Unit | Material | Labor | Total | Sell |
|---|---|---|---|---|---|---|
| 6" x 60", dentil | BC@.250 | Ea | 48.80 | 8.02 | 56.82 | 96.60 |
| 6" x 60", Tuscan | BC@.250 | Ea | 48.80 | 8.02 | 56.82 | 96.60 |

**Frieze polyurethane cast molding, American Molding.** Horizontal flat band, usually installed just below cornice molding.

| | Craft@Hrs | Unit | Material | Labor | Total | Sell |
|---|---|---|---|---|---|---|
| 4-7/8" x 96", rococo | BC@.250 | Ea | 21.70 | 8.02 | 29.72 | 50.50 |
| 7-1/2" x 92-1/4", grand palmetto | BC@.250 | Ea | 43.40 | 8.02 | 51.42 | 87.40 |
| 7-1/2" x 92-1/4", grand Tuscan | BC@.250 | Ea | 43.40 | 8.02 | 51.42 | 87.40 |

**Polyurethane cast molding.**

| | Craft@Hrs | Unit | Material | Labor | Total | Sell |
|---|---|---|---|---|---|---|
| 4-1/8" x 8', dentil | BC@.200 | Ea | 21.70 | 6.42 | 28.12 | 47.80 |
| 4-1/8" x 8', egg and dart | BC@.200 | Ea | 21.70 | 6.42 | 28.12 | 47.80 |
| 4-1/8" x 8', Grecian | BC@.200 | Ea | 21.70 | 6.42 | 28.12 | 47.80 |
| 5-7/8" x 8', dentil | BC@.200 | Ea | 28.20 | 6.42 | 34.62 | 58.90 |
| 5-7/8" x 8', egg and dart | BC@.200 | Ea | 28.20 | 6.42 | 34.62 | 58.90 |
| 5-7/8" x 8', rope crown | BC@.200 | Ea | 28.20 | 6.42 | 34.62 | 58.90 |

**Polyurethane corner block.**

| | Craft@Hrs | Unit | Material | Labor | Total | Sell |
|---|---|---|---|---|---|---|
| 4-1/8", inside | BC@.050 | Ea | 7.58 | 1.60 | 9.18 | 15.60 |
| 4-1/8", outside | BC@.050 | Ea | 7.56 | 1.60 | 9.16 | 15.60 |
| 5-7/8", inside | BC@.050 | Ea | 8.64 | 1.60 | 10.24 | 17.40 |
| 5-7/8", outside | BC@.050 | Ea | 8.64 | 1.60 | 10.24 | 17.40 |

**Rope polyurethane cast molding.** Rope molding is a bead or Torus molding carved to imitate rope. Per foot.

| | Craft@Hrs | Unit | Material | Labor | Total | Sell |
|---|---|---|---|---|---|---|
| 3-1/2" x 8', casing | BC@.400 | Ea | 20.60 | 12.80 | 33.40 | 56.80 |
| 3-1/2" x 8', chair rail | BC@.300 | Ea | 20.62 | 9.63 | 30.25 | 51.40 |
| 5-1/2" x 8', base | BC@.250 | Ea | 28.20 | 8.02 | 36.22 | 61.60 |

**Chair rail polyurethane cast molding.** Per foot.

| | Craft@Hrs | Unit | Material | Labor | Total | Sell |
|---|---|---|---|---|---|---|
| 3-1/2" x 8', dentil | BC@.300 | Ea | 21.90 | 9.63 | 31.53 | 53.60 |
| 3-1/2" x 8', Greek key | BC@.300 | Ea | 21.80 | 9.63 | 31.43 | 53.40 |

|  | Craft@Hrs | Unit | Material | Labor | Total | Sell |
|---|---|---|---|---|---|---|
| **Poly molding.** | | | | | | |
| Bead, brite white, 8' | BC@.128 | Ea | 2.47 | 4.11 | 6.58 | 11.20 |
| Bead, cherry/mahogany, 8' | BC@.128 | Ea | 2.47 | 4.11 | 6.58 | 11.20 |
| Bead, cinnamon chestnut, 8' | BC@.128 | Ea | 2.47 | 4.11 | 6.58 | 11.20 |
| Bed mold, brite white, 8' | BC@.128 | Ea | 4.80 | 4.11 | 8.91 | 15.10 |
| Bed mold, cinnamon chestnut, 8' | BC@.128 | Ea | 4.82 | 4.11 | 8.93 | 15.20 |
| Bed mold, cherry/mahogany, 8' | BC@.128 | Ea | 4.81 | 4.11 | 8.92 | 15.20 |
| Casing and base, brite white, 8' | BC@.184 | Ea | 4.80 | 5.90 | 10.70 | 18.20 |
| Casing and base, cherry/mahogany, 8' | BC@.184 | Ea | 4.81 | 5.90 | 10.71 | 18.20 |
| Casing and base, cinnamon chestnut, 8' | BC@.184 | Ea | 4.83 | 5.90 | 10.73 | 18.20 |
| Outside corner, brite white, 8' | BC@.128 | Ea | 2.47 | 4.11 | 6.58 | 11.20 |
| Outside corner, cherry/mahogany, 8' | BC@.128 | Ea | 2.47 | 4.11 | 6.58 | 11.20 |
| Outside corner, cinnamon chestnut, 8' | BC@.128 | Ea | 2.47 | 4.11 | 6.58 | 11.20 |
| Plywood cap, brite white, 8' | BC@.128 | Ea | 3.01 | 4.11 | 7.12 | 12.10 |
| Shoe, brite white, 8' | BC@.128 | Ea | 2.46 | 4.11 | 6.57 | 11.20 |
| Shoe, cherry/mahogany, 8' | BC@.128 | Ea | 2.47 | 4.11 | 6.58 | 11.20 |
| Shoe, cinnamon chestnut, 8' | BC@.128 | Ea | 2.47 | 4.11 | 6.58 | 11.20 |
| Stop, brite white, 8' | BC@.200 | Ea | 2.47 | 6.42 | 8.89 | 15.10 |
| Stop, cinnamon chestnut, 8' | BC@.200 | Ea | 2.47 | 6.42 | 8.89 | 15.10 |

**White extruded polymer molding.** Durable in high traffic areas. Easy to clean. Paintable and stainable. Unaffected by moisture, will not mold, mildew, swell, or rot. Termite proof. No sanding needed. Per linear foot.

|  | Craft@Hrs | Unit | Material | Labor | Total | Sell |
|---|---|---|---|---|---|---|
| No. 49, 1/2" x 3-5/8", crown | BC@.044 | LF | 1.43 | 1.41 | 2.84 | 4.83 |
| No. 623, 9/16" x 3-1/4", base | BC@.032 | LF | 1.05 | 1.03 | 2.08 | 3.54 |
| No. 356, 11/16" x 2-1/4", casing | BC@.050 | LF | .82 | 1.60 | 2.42 | 4.11 |
| No. 105, 3/4" x 3/4", quarter round | BC@.032 | LF | .62 | 1.03 | 1.65 | 2.81 |
| 1-3/8", closet pole | BC@.032 | LF | 1.37 | 1.03 | 2.40 | 4.08 |
| No. 390, 2-5/8", chair rail | BC@.032 | LF | 1.60 | 1.03 | 2.63 | 4.47 |

## Prefinished Molding

**Embossed polystyrene base molding, Heirloom™, Royal Mouldings.** Deep embossed ornate design. Won't crack, warp, split, or bow. Paint process finish. Stainable.

|  | Craft@Hrs | Unit | Material | Labor | Total | Sell |
|---|---|---|---|---|---|---|
| 5" x 9/16" x 8', Bone white | BC@.250 | Ea | 21.50 | 8.02 | 29.52 | 50.20 |
| 5" x 9/16" x 8', Clearwood | BC@.250 | Ea | 19.60 | 8.02 | 27.62 | 47.00 |
| 8', Highlands oak | BC@.250 | Ea | 31.30 | 8.02 | 39.32 | 66.80 |
| 8', Imperial oak | BC@.250 | Ea | 31.70 | 8.02 | 39.72 | 67.50 |

**Colonial base molding, Royal Mouldings.** Pre-sealed and pre-sanded. Non-repetitive woodgrain finish for consistent appearance. Stain or paint with any oil or latex paint applied by brush, roller or sprayer. Stain or paint absorbs uniformly without discoloration. Clean by wiping with paint thinner, naphtha or ammonia.

|  | Craft@Hrs | Unit | Material | Labor | Total | Sell |
|---|---|---|---|---|---|---|
| 7/16" x 3-1/4" x 8', Bone white | BC@.250 | Ea | 8.12 | 8.02 | 16.14 | 27.40 |
| 7/16" x 3-1/4" x 8', Clearwood | BC@.250 | Ea | 7.32 | 8.02 | 15.34 | 26.10 |

**Colonial base paperwrap molding, Royal Collection™, Royal Mouldings.**

|  | Craft@Hrs | Unit | Material | Labor | Total | Sell |
|---|---|---|---|---|---|---|
| 7/16" x 3-1/4" x 8', Highlands oak | BC@.250 | Ea | 10.14 | 8.02 | 18.16 | 30.90 |
| 7/16" x 3-1/4" x 8', Imperial oak | BC@.250 | Ea | 10.06 | 8.02 | 18.08 | 30.70 |
| 7/16" x 3-1/4" x 8', Natural maple | BC@.250 | Ea | 10.00 | 8.02 | 18.02 | 30.60 |

| | Craft@Hrs | Unit | Material | Labor | Total | Sell |
|---|---|---|---|---|---|---|

**Colonial base polystyrene molding, Royal Mouldings.** Non-repetitive woodgrain finish. Pre-sealed and pre-sanded. Stain or paint with any oil or latex paint. Stain and paint absorb uniformly.

| | Craft@Hrs | Unit | Material | Labor | Total | Sell |
|---|---|---|---|---|---|---|
| 7/16" x 3-1/4" x 8', | | | | | | |
|     Bone white | BC@.250 | Ea | 11.50 | 8.02 | 19.52 | 33.20 |
| 7/16" x 3-1/4" x 8', | | | | | | |
|     Dominion oak | BC@.250 | Ea | 9.15 | 8.02 | 17.17 | 29.20 |

**Combination base molding, Royal Mouldings.** 11/16" x 5-1/4".

| | Craft@Hrs | Unit | Material | Labor | Total | Sell |
|---|---|---|---|---|---|---|
| 8', Bone white | BC@.250 | Ea | 17.60 | 8.02 | 25.62 | 43.60 |
| 8', Clearwood | BC@.250 | Ea | 18.10 | 8.02 | 26.12 | 44.40 |

**Ranch base paperwrap molding, Royal Collection™, Royal Mouldings.**

| | Craft@Hrs | Unit | Material | Labor | Total | Sell |
|---|---|---|---|---|---|---|
| 7/16" x 3-1/4" x 8', | | | | | | |
|     Clearwood | BC@.250 | Ea | 7.33 | 8.02 | 15.35 | 26.10 |
| 1/2" x 2-7/16" x 8', | | | | | | |
|     Dominion oak | BC@.250 | Ea | 8.27 | 8.02 | 16.29 | 27.70 |
| 1/2" x 2-7/16" x 8', | | | | | | |
|     Highlands oak | BC@.250 | Ea | 8.60 | 8.02 | 16.62 | 28.30 |
| 1/2" x 2-7/16" x 8', | | | | | | |
|     Imperial oak | BC@.250 | Ea | 8.56 | 8.02 | 16.58 | 28.20 |

**Base shoe molding, Royal Mouldings.** Pre-sealed and pre-sanded. Stainable or paintable with any oil or latex paint applied by brush, roller or sprayer. Non-repetitive woodgrain finish for consistent appearance. Stain or paint absorbs uniformly without discoloration. Clean by wiping with paint thinner, naphtha or ammonia.

| | Craft@Hrs | Unit | Material | Labor | Total | Sell |
|---|---|---|---|---|---|---|
| 3/8" x 11/16" x 8', | | | | | | |
|     Clearwood | BC@.128 | Ea | 3.11 | 4.11 | 7.22 | 12.30 |

**Colonial cap paperwrap molding, Royal Collection™, Royal Mouldings.** Non-repetitive woodgrain finish. Pre-sealed and pre-sanded. Stainable or paintable with any oil or latex paint. Stain and paint absorb uniformly.

| | Craft@Hrs | Unit | Material | Labor | Total | Sell |
|---|---|---|---|---|---|---|
| 9/16" x 1-1/8" x 8', | | | | | | |
|     Highlands oak | BC@.250 | Ea | 6.56 | 8.02 | 14.58 | 24.80 |
| 9/16" x 1-1/8" x 8', | | | | | | |
|     Imperial oak | BC@.250 | Ea | 6.56 | 8.02 | 14.58 | 24.80 |
| 9/16" x 1-1/8" x 8', | | | | | | |
|     Natural maple | BC@.250 | Ea | 6.61 | 8.02 | 14.63 | 24.90 |

**Colonial casing paperwrap molding, Royal Collection™, Royal Mouldings.**

| | Craft@Hrs | Unit | Material | Labor | Total | Sell |
|---|---|---|---|---|---|---|
| 9/16" x 2-1/4" x 7', | | | | | | |
|     Highlands oak | BC@.105 | Ea | 8.31 | 3.37 | 11.68 | 19.90 |
| 9/16" x 2-1/4" x 7', | | | | | | |
|     Natural maple | BC@.105 | Ea | 8.08 | 3.37 | 11.45 | 19.50 |

**Colonial casing polystyrene molding, Royal Mouldings.** Non-repetitive woodgrain finish. Pre-sealed and pre-sanded. Stainable or paintable with any oil or latex paint. Stain and paint absorb uniformly.

| | Craft@Hrs | Unit | Material | Labor | Total | Sell |
|---|---|---|---|---|---|---|
| 7/16" x 1-9/16" x 7' | BC@.105 | Ea | 4.80 | 3.37 | 8.17 | 13.90 |
| 9/16" x 2-1/4" x 7', | | | | | | |
|     Bone white | BC@.105 | Ea | 6.93 | 3.37 | 10.30 | 17.50 |
| 9/16" x 2-1/4" x 7', | | | | | | |
|     Dominion oak | BC@.105 | Ea | 7.11 | 3.37 | 10.48 | 17.80 |

**Colonial casing prefinished molding, Royal Mouldings.**

| | Craft@Hrs | Unit | Material | Labor | Total | Sell |
|---|---|---|---|---|---|---|
| No. CWPS,7', "Creations" | BC@.105 | Ea | 8.54 | 3.37 | 11.91 | 20.20 |
| 9/16" x 2-1/4" x 7', "Reflections" | BC@.105 | Ea | 6.13 | 3.37 | 9.50 | 16.20 |
| Highlands oak, 7', "Heirloom" | BC@.105 | Ea | 25.50 | 3.37 | 28.87 | 49.10 |
| Imperial oak, 7', "Heirloom" | BC@.105 | Ea | 25.50 | 3.37 | 28.87 | 49.10 |

|  | Craft@Hrs | Unit | Material | Labor | Total | Sell |
|---|---|---|---|---|---|---|
| **Economy casing molding, Royal Mouldings.** 9/16" x 2-1/4" x 7', | | | | | | |
| Highlands oak | BC@.105 | Ea | 4.56 | 3.37 | 7.93 | 13.50 |
| Imperial oak | BC@.105 | Ea | 4.56 | 3.37 | 7.93 | 13.50 |

**Fluted casing embossed polystyrene molding, Heirloom™, Royal Mouldings.** Stainable or paintable finish. Deep embossed ornate design. Won't crack, warp, split, or bow. Paint process finish.

| | | | | | | |
|---|---|---|---|---|---|---|
| 11/16" x 4" x 8', Bone white | BC@.105 | Ea | 19.50 | 3.37 | 22.87 | 38.90 |
| 11/16" x 4" x 8', Clearwood | BC@.105 | Ea | 18.90 | 3.37 | 22.27 | 37.90 |

**Fluted casing prefinished molding, Royal Mouldings.**

| | | | | | | |
|---|---|---|---|---|---|---|
| 2-1/8", Bone white, 8' | BC@.105 | Ea | 10.70 | 3.37 | 14.07 | 23.90 |
| 3-1/8", Bone white, 8' | BC@.105 | Ea | 12.30 | 3.37 | 15.67 | 26.60 |

**Ranch casing molding, Clearwood PS™, Royal Mouldings.** Pre-sealed and pre-sanded. Stainable or paintable with any oil or latex paint applied by brush, roller, or sprayer. Non-repetitive woodgrain finish for consistent appearance. Stain or paint absorbs uniformly without discoloration. Clean by wiping with paint thinner, naphtha, or ammonia.

| | | | | | | |
|---|---|---|---|---|---|---|
| 9/16" x 2-1/4" x 7', Bone white | BC@.105 | Ea | 5.71 | 3.37 | 9.08 | 15.40 |

**Ranch casing molding, Royal Mouldings.** Pre-sealed and pre-sanded. Stainable or paintable with any oil or latex paint applied by brush, roller or sprayer. Non-repetitive woodgrain finish for consistent appearance. Stain or paint absorbs uniformly without discoloration. Clean by wiping with paint thinner, naphtha or ammonia.

| | | | | | | |
|---|---|---|---|---|---|---|
| 9/16" x 2-1/4" x 7', Clearwood | BC@.105 | Ea | 4.62 | 3.37 | 7.99 | 13.60 |

**Ranch casing paperwrap molding, Royal Collection™, Royal Mouldings.**

| | | | | | | |
|---|---|---|---|---|---|---|
| 9/16" x 2-1/4" x 7', Highlands oak | BC@.105 | Ea | 6.77 | 3.37 | 10.14 | 17.20 |
| 9/16" x 2-1/4" x 7', Imperial oak | BC@.105 | Ea | 6.76 | 3.37 | 10.13 | 17.20 |

**Ranch casing polystyrene molding, Royal Mouldings.** Non-repetitive woodgrain finish. Pre-sealed and pre-sanded. Stainable or paintable with any oil or latex paint. Stain and paint absorb uniformly.

| | | | | | | |
|---|---|---|---|---|---|---|
| 13/32" x 1-9/16" x 7' | BC@.105 | Ea | 4.85 | 3.37 | 8.22 | 14.00 |
| 9/16" x 2-1/4" x 7', Dominion oak | BC@.105 | Ea | 7.10 | 3.37 | 10.47 | 17.80 |
| 5/8" x 1-5/8" x 7', Clearwood | BC@.105 | Ea | 7.44 | 3.37 | 10.81 | 18.40 |

| | Craft@Hrs | Unit | Material | Labor | Total | Sell |
|---|---|---|---|---|---|---|

**Reversible polystyrene casing, Decorator Lykewood® RB3, Royal Mouldings.** Won't crack, warp, split, or bow. Paint process finish.

1" x 3-1/2" x 7',
| | | | | | | |
|---|---|---|---|---|---|---|
| Bone white | BC@.105 | Ea | 11.90 | 3.37 | 15.27 | 26.00 |

1" x 3-1/2" x 7',
| | | | | | | |
|---|---|---|---|---|---|---|
| Clearwood | BC@.105 | Ea | 11.80 | 3.37 | 15.17 | 25.80 |

**Chair rail embossed polystyrene molding, Heirloom™, Royal Mouldings.** Stainable or paintable finish. Deep embossed ornate design. Won't crack, warp, split, or bow. Paint process finish.

4-1/2" x 1-1/4" x 8',
| | | | | | | |
|---|---|---|---|---|---|---|
| Bone white | BC@.300 | Ea | 18.50 | 9.63 | 28.13 | 47.80 |

4-1/2" x 1-1/4" x 8',
| | | | | | | |
|---|---|---|---|---|---|---|
| Clearwood | BC@.300 | Ea | 18.50 | 9.63 | 28.13 | 47.80 |

**Chair rail embossed prefinished molding, Heirloom™, Royal Mouldings.**

| | | | | | | |
|---|---|---|---|---|---|---|
| Highlands oak, 8' | BC@.300 | Ea | 34.90 | 9.63 | 44.53 | 75.70 |
| Imperial oak, 8' | BC@.300 | Ea | 34.90 | 9.63 | 44.53 | 75.70 |

**Chair rail molding, Clearwood PS™, Royal Mouldings.** Pre-sealed and pre-sanded. Stainable or paintable with any oil or latex paint applied by brush, roller, or sprayer. Non-repetitive woodgrain finish for consistent appearance. Stain or paint absorbs uniformly without discoloration. Clean by wiping with paint thinner, naphtha, or ammonia.

11/16" x 2-5/8" x 8',
| | | | | | | |
|---|---|---|---|---|---|---|
| Bone white | BC@.300 | Ea | 9.15 | 9.63 | 18.78 | 31.90 |

**Chair rail paperwrap molding, Royal Collection™, Royal Mouldings.**

11/16" x 2-5/8" x 8',
| | | | | | | |
|---|---|---|---|---|---|---|
| Highlands oak | BC@.300 | Ea | 10.23 | 9.63 | 19.86 | 33.80 |

11/16" x 2-5/8" x 8',
| | | | | | | |
|---|---|---|---|---|---|---|
| Imperial oak | BC@.300 | Ea | 10.24 | 9.63 | 19.87 | 33.80 |

11/16" x 2-5/8" x 8',
| | | | | | | |
|---|---|---|---|---|---|---|
| Natural maple | BC@.300 | Ea | 10.21 | 9.63 | 19.84 | 33.70 |

**Chair rail polystyrene molding, Royal Mouldings.** Non-repetitive woodgrain finish. Pre-sealed and pre-sanded. Stainable or paintable with any oil or latex paint. Stain and paint absorb uniformly.

11/16" x 2-5/8" x 8',
| | | | | | | |
|---|---|---|---|---|---|---|
| Bone white | BC@.300 | Ea | 9.18 | 9.63 | 18.81 | 32.00 |

11/16" x 2-5/8" x 8',
| | | | | | | |
|---|---|---|---|---|---|---|
| Clearwood | BC@.300 | Ea | 7.35 | 9.63 | 16.98 | 28.90 |

**Chair rail prefinished molding, Reflections™, Royal Mouldings.**

| | | | | | | |
|---|---|---|---|---|---|---|
| 11/16" x 2-5/8" x 8', White marble | BC@.300 | Ea | 14.00 | 9.63 | 23.63 | 40.20 |

**Inside corner and cove molding, Royal Mouldings.** Pre-sealed and pre-sanded. Stainable or paintable with any oil or latex paint applied by brush, roller or sprayer. Non-repetitive woodgrain finish for consistent appearance. Stain or paint absorbs uniformly without discoloration. Clean by wiping with paint thinner, naphtha or ammonia.

11/16" x 11/16" x 8',
| | | | | | | |
|---|---|---|---|---|---|---|
| Clearwood | BC@.300 | Ea | 3.20 | 9.63 | 12.83 | 21.80 |

**Inside corner and cove paperwrap molding, Royal Collection™, Royal Mouldings.**

5/16" x 15/16" x 8',
| | | | | | | |
|---|---|---|---|---|---|---|
| Highlands oak | BC@.300 | Ea | 4.64 | 9.63 | 14.27 | 24.30 |

5/16" x 15/16" x 8',
| | | | | | | |
|---|---|---|---|---|---|---|
| Imperial oak | BC@.300 | Ea | 4.65 | 9.63 | 14.28 | 24.30 |

5/16" x 15/16" x 8',
| | | | | | | |
|---|---|---|---|---|---|---|
| Natural maple | BC@.300 | Ea | 4.48 | 9.63 | 14.11 | 24.00 |

| | Craft@Hrs | Unit | Material | Labor | Total | Sell |
|---|---|---|---|---|---|---|

**Inside corner and cove prefinished molding, Reflections™, Royal Mouldings.**

| 11/16" x 11/16" x 8', White marble | BC@.300 | Ea | 3.55 | 9.63 | 13.18 | 22.40 |

**Inside corner molding, Clearwood PS™, Royal Mouldings.** Pre-sealed and pre-sanded. Stainable or paintable with any oil or latex paint applied by brush, roller, or sprayer. Non-repetitive woodgrain finish for consistent appearance. Stain or paint absorbs uniformly without discoloration. Clean by wiping with paint thinner, naphtha, or ammonia.

| 5/16" x 1" x 8', Bone white | BC@.300 | Ea | 3.45 | 9.63 | 13.08 | 22.20 |

**Inside corner polystyrene molding, Royal Mouldings.** Non-repetitive woodgrain finish. Pre-sealed and pre-sanded. Stainable or paintable with any oil or latex paint. Stain and paint absorb uniformly.

| 5/16" x 1" x 8', Dominion oak | BC@.300 | Ea | 3.54 | 9.63 | 13.17 | 22.40 |

**Inside corner prefinished molding, Royal Mouldings.**

| Natural pine | BC@.300 | Ea | 3.84 | 9.63 | 13.47 | 22.90 |
| Perfect corner, bone white | BC@.300 | Ea | 5.71 | 9.63 | 15.34 | 26.10 |

**Outside corner molding, Royal Mouldings.** Pre-sealed and pre-sanded. Stainable or paintable with any oil or latex paint applied by brush, roller or sprayer. Non-repetitive woodgrain finish for consistent appearance. Stain or paint absorbs uniformly without discoloration. Clean by wiping with paint thinner, naphtha or ammonia.

| 11/16" x 11/16" x 8', Clearwood | BC@.300 | Ea | 3.87 | 9.63 | 13.50 | 23.00 |
| 7/8" x 7/8" x 8', Bone white | BC@.300 | Ea | 3.50 | 9.63 | 13.13 | 22.30 |
| 1" x 1" x 8', Bone white | BC@.300 | Ea | 5.40 | 9.63 | 15.03 | 25.60 |
| 1-1/8" x 1-1/8" x 8', Clearwood | BC@.300 | Ea | 7.09 | 9.63 | 16.72 | 28.40 |

**Outside corner polystyrene molding, Royal Mouldings.** Non-repetitive woodgrain finish. Pre-sealed and pre-sanded. Stainable or paintable with any oil or latex paint. Stain and paint absorb uniformly.

| 7/8" x 7/8" x 8', Dominion oak | BC@.300 | Ea | 4.41 | 9.63 | 14.04 | 23.90 |

**Outside corner prefinished molding, Reflections™, Royal Mouldings.**

| 7/8" x 7/8" x 8', white marble | BC@.300 | Ea | 4.09 | 9.63 | 13.72 | 23.30 |
| 8', Perfect corner, bone white | BC@.300 | Ea | 5.71 | 9.63 | 15.34 | 26.10 |

**Colonial outside corner paperwrap molding, Royal Collection™, Royal Mouldings.**

| 1" x 1" x 8', Highlands | BC@.300 | Ea | 5.53 | 9.63 | 15.16 | 25.80 |
| 1" x 1" x 8', Imperial oak | BC@.300 | Ea | 5.53 | 9.63 | 15.16 | 25.80 |
| 1" x 1" x 8', Natural maple | BC@.300 | Ea | 5.49 | 9.63 | 15.12 | 25.70 |

**Colonial outside corner polystyrene molding, Royal Mouldings.** Non-repetitive woodgrain finish. Pre-sealed and pre-sanded. Stainable or paintable with any oil or latex paint. Stain and paint absorb uniformly.

| 1" x 1" x 8', Dominion oak | BC@.300 | Ea | 5.55 | 9.63 | 15.18 | 25.80 |

| | Craft@Hrs | Unit | Material | Labor | Total | Sell |
|---|---|---|---|---|---|---|

**Oak counter edge, Royal Mouldings.** Molding specially shaped for edge of tile countertops. Provides durable protection from everyday impacts that can chip ceramics. Wood adds a warm, natural accent to tile.

| | Craft@Hrs | Unit | Material | Labor | Total | Sell |
|---|---|---|---|---|---|---|
| 13/16" x 1-3/4" x 5', Oak trim | BC@.300 | Ea | 17.02 | 9.63 | 26.65 | 45.30 |
| 13/16" x 1-3/4" x 8', Oak trim | BC@.480 | Ea | 27.00 | 15.40 | 42.40 | 72.10 |

**Cove molding, Royal Mouldings.** Pre-sealed and pre-sanded. Stainable or paintable with any oil or latex paint applied by brush, roller or sprayer. Non-repetitive woodgrain finish for consistent appearance. Stain or paint absorbs uniformly without discoloration. Clean by wiping with paint thinner, naphtha or ammonia.

| | Craft@Hrs | Unit | Material | Labor | Total | Sell |
|---|---|---|---|---|---|---|
| 9/16" x 1-5/8" x 8', Clearwood | BC@.300 | Ea | 5.38 | 9.63 | 15.01 | 25.50 |

**Spring cove molding, Clearwood PS™, Royal Mouldings.** Pre-sealed and pre-sanded. Stainable or paintable with any oil or latex paint applied by brush, roller, or sprayer. Non-repetitive woodgrain finish for consistent appearance. Stain or paint absorbs uniformly without discoloration. Clean by wiping with paint thinner, naphtha, or ammonia.

| | Craft@Hrs | Unit | Material | Labor | Total | Sell |
|---|---|---|---|---|---|---|
| 9/16" x 1-5/8" x 8', Bone white | BC@.300 | Ea | 6.00 | 9.63 | 15.63 | 26.60 |

**Spring cove polystyrene molding, Royal Mouldings.** Non-repetitive woodgrain finish. Pre-sealed and pre-sanded. Stainable or paintable with any oil or latex paint. Stain and paint absorb uniformly.

| | Craft@Hrs | Unit | Material | Labor | Total | Sell |
|---|---|---|---|---|---|---|
| 9/16" x 1-5/8" x 8', Dominion oak | BC@.300 | Ea | 5.90 | 9.63 | 15.53 | 26.40 |

**Colonial crown paperwrap molding, Royal Collection™, Royal Mouldings.**

| | Craft@Hrs | Unit | Material | Labor | Total | Sell |
|---|---|---|---|---|---|---|
| 9/16" x 3-5/8" x 8', Highlands oak | BC@.352 | Ea | 14.50 | 11.30 | 25.80 | 43.90 |
| 9/16" x 3-5/8" x 8', Imperial oak | BC@.352 | Ea | 12.60 | 11.30 | 23.90 | 40.60 |
| 11/16" x 3-5/8" x 8', Natural maple | BC@.352 | Ea | 14.80 | 11.30 | 26.10 | 44.40 |

**Crown embossed polystyrene molding, Heirloom™, Royal Mouldings.** Stainable or paintable finish. Deep embossed ornate design. Won't crack, warp, split, or bow. Paint process finish.

| | Craft@Hrs | Unit | Material | Labor | Total | Sell |
|---|---|---|---|---|---|---|
| 5" x 11/16" x 8', Bone white | BC@.352 | Ea | 21.70 | 11.30 | 33.00 | 56.10 |
| 5" x 11/16" x 8', Clearwood | BC@.352 | Ea | 21.60 | 11.30 | 32.90 | 55.90 |

**Crown polystyrene molding, Royal Mouldings.** Non-repetitive woodgrain finish. Pre-sealed and pre-sanded. Stainable or paintable with any oil or latex paint. Stain and paint absorb uniformly.

| | Craft@Hrs | Unit | Material | Labor | Total | Sell |
|---|---|---|---|---|---|---|
| 11/16" x 3-5/8" x 8', Bone white | BC@.350 | Ea | 17.10 | 11.20 | 28.30 | 48.10 |

**Crown prefinished molding, Heirloom™, Royal Mouldings.**

| | Craft@Hrs | Unit | Material | Labor | Total | Sell |
|---|---|---|---|---|---|---|
| 8', Highlands oak | BC@.352 | Ea | 32.70 | 11.30 | 44.00 | 74.80 |
| 8', Oak | BC@.352 | Ea | 32.70 | 11.30 | 44.00 | 74.80 |

**Crown prefinished molding, Royal Mouldings.**

| | Craft@Hrs | Unit | Material | Labor | Total | Sell |
|---|---|---|---|---|---|---|
| 11/16" x 3-5/8" x 8', Clearwood | BC@.352 | Ea | 9.17 | 11.30 | 20.47 | 34.80 |
| 8', Bone white, smooth | BC@.352 | Ea | 18.07 | 11.30 | 29.37 | 49.90 |

| | Craft@Hrs | Unit | Material | Labor | Total | Sell |
|---|---|---|---|---|---|---|

**Lattice molding, Royal Mouldings.** Pre-sealed and pre-sanded. Stainable or paintable with any oil or latex paint applied by brush, roller or sprayer. Non-repetitive woodgrain finish for consistent appearance. Stain or paint absorbs uniformly without discoloration. Clean by wiping with paint thinner, naphtha or ammonia.

| | Craft@Hrs | Unit | Material | Labor | Total | Sell |
|---|---|---|---|---|---|---|
| 5/32" x 1-1/8" x 8',<br>Bone white | BC@.250 | Ea | 3.94 | 8.02 | 11.96 | 20.30 |
| 1/4" x 1-1/8" x 8',<br>Clearwood | BC@.250 | Ea | 3.39 | 8.02 | 11.41 | 19.40 |
| 1/4" x 1-3/4" x 8',<br>Clearwood | BC@.250 | Ea | 3.95 | 8.02 | 11.97 | 20.30 |

**Lattice polystyrene molding, Royal Mouldings.** Non-repetitive woodgrain finish. Pre-sealed and pre-sanded. Stainable or paintable with any oil or latex paint. Stain and paint absorb uniformly.

| | Craft@Hrs | Unit | Material | Labor | Total | Sell |
|---|---|---|---|---|---|---|
| 5/32" x 1-1/8" x 8',<br>Dominion oak | BC@.250 | Ea | 4.08 | 8.02 | 12.10 | 20.60 |

**Parting stop molding, Royal Mouldings.** Pre-sealed and pre-sanded. Stainable or paintable with any oil or latex paint applied by brush, roller or sprayer. Non-repetitive woodgrain finish for consistent appearance. Stain or paint absorbs uniformly without discoloration. Clean by wiping with paint thinner, naphtha or ammonia.

| | Craft@Hrs | Unit | Material | Labor | Total | Sell |
|---|---|---|---|---|---|---|
| 1/2" x 3/4" x 8',<br>Clearwood | BC@.200 | Ea | 3.75 | 6.42 | 10.17 | 17.30 |

**Colonial stop molding, Royal Mouldings.** Pre-sealed and pre-sanded. Stainable or paintable with any oil or latex paint applied by brush, roller or sprayer. Non-repetitive woodgrain finish for consistent appearance. Stain or paint absorbs uniformly without discoloration. Clean by wiping with paint thinner, naphtha or ammonia.

| | Craft@Hrs | Unit | Material | Labor | Total | Sell |
|---|---|---|---|---|---|---|
| 3/8" x 1-1/4" x 7',<br>Clearwood | BC@.175 | Ea | 3.56 | 5.62 | 9.18 | 15.60 |

**Colonial stop paperwrap molding, Royal Collection™, Royal Mouldings.**

| | Craft@Hrs | Unit | Material | Labor | Total | Sell |
|---|---|---|---|---|---|---|
| 3/8" x 1-1/4" x 7',<br>Highlands oak | BC@.175 | Ea | 5.07 | 5.62 | 10.69 | 18.20 |
| 3/8" x 1-1/4" x 7',<br>Imperial oak | BC@.175 | Ea | 5.09 | 5.62 | 10.71 | 18.20 |
| 3/8" x 1-1/4" x 7',<br>Natural maple | BC@.175 | Ea | 5.70 | 5.62 | 11.32 | 19.20 |

**Colonial stop polystyrene molding, Royal Mouldings.** Non-repetitive woodgrain finish. Pre-sealed and pre-sanded. Stainable or paintable with any oil or latex paint. Stain and paint absorb uniformly.

| | Craft@Hrs | Unit | Material | Labor | Total | Sell |
|---|---|---|---|---|---|---|
| 3/8" x 1-1/4" x 7',<br>Bone white | BC@.175 | Ea | 4.87 | 5.62 | 10.49 | 17.80 |
| 3/8" x 1-1/4" x 7',<br>Dominion oak | BC@.175 | Ea | 4.80 | 5.62 | 10.42 | 17.70 |
| 3/8" x 1-1/4" x 7',<br>Washed oak | BC@.175 | Ea | 4.85 | 5.62 | 10.47 | 17.80 |

**Ranch stop molding, Royal Mouldings.** Pre-sealed and pre-sanded. Stainable or paintable with any oil or latex paint applied by brush, roller or sprayer. Non-repetitive woodgrain finish for consistent appearance. Stain or paint absorbs uniformly without discoloration. Clean by wiping with paint thinner, naphtha or ammonia.

| | Craft@Hrs | Unit | Material | Labor | Total | Sell |
|---|---|---|---|---|---|---|
| 3/8" x 1-1/4" x 7',<br>Clearwood | BC@.175 | Ea | 3.06 | 5.62 | 8.68 | 14.80 |

**Ranch stop polystyrene molding, Royal Mouldings.** Non-repetitive woodgrain finish. Pre-sealed and pre-sanded. Stainable or paintable with any oil or latex paint. Stain and paint absorb uniformly.

| | Craft@Hrs | Unit | Material | Labor | Total | Sell |
|---|---|---|---|---|---|---|
| 3/8" x 1-1/4" x 7',<br>Bone white | BC@.175 | Ea | 4.86 | 5.62 | 10.48 | 17.80 |

| | Craft@Hrs | Unit | Material | Labor | Total | Sell |
|---|---|---|---|---|---|---|
| **Vinyl tileboard molding, prefinished rigid divider, Royal Mouldings.** 1/8" to 4mm insert. | | | | | | |
| 8', white | BC@.200 | Ea | 2.38 | 6.42 | 8.80 | 15.00 |
| 8', white | BC@.200 | Ea | 2.31 | 6.42 | 8.73 | 14.80 |
| 8', white | BC@.200 | Ea | 2.35 | 6.42 | 8.77 | 14.90 |
| 8', white | BC@.200 | Ea | 2.35 | 6.42 | 8.77 | 14.90 |

## Interior Partition Walls

**Remove non-bearing drywall stud walls.** Includes allowance for plates, blocking and wall cover. Per SF of wall area demolished, measured one side. Add the cost of patching at floor, wall and ceiling joints, hauling away debris, and dump fees. No salvage of material assumed.

| | Craft@Hrs | Unit | Material | Labor | Total | Sell |
|---|---|---|---|---|---|---|
| 2" x 3" framing, drywall both sides | BL@.022 | SF | — | .59 | .59 | 1.00 |
| 2" x 4" framing, drywall both sides | BL@.026 | SF | — | .69 | .69 | 1.17 |
| 2" x 6" framing, drywall both sides | BL@.034 | SF | — | .91 | .91 | 1.55 |

**Remove plastered stud walls.** Includes allowance for plates, blocking and wall cover. Per SF of wall area demolished, measured one side. Add the cost of patching at floor, wall and ceiling joints, hauling away debris, and dump fees. No salvage of material assumed.

| | Craft@Hrs | Unit | Material | Labor | Total | Sell |
|---|---|---|---|---|---|---|
| 2" x 3" framing, lath and plaster both sides | BL@.038 | SF | — | 1.01 | 1.01 | 1.72 |
| 2" x 4" framing, lath and plaster both sides | BL@.042 | SF | — | 1.12 | 1.12 | 1.90 |

**Add 2" x 4" interior stud partition with 1/2" drywall both sides.** Add the cost of patching, finishing and trim.

| | Craft@Hrs | Unit | Material | Labor | Total | Sell |
|---|---|---|---|---|---|---|
| Cost per square foot of wall | B1@.064 | SF | 1.21 | 1.88 | 3.09 | 5.25 |
| Cost per running foot, 8' high walls | B1@.512 | LF | 9.65 | 15.00 | 24.65 | 41.90 |

**Add 2" x 4" interior stud partition with 5/8" drywall both sides.** Includes drywall taping. Add the cost of patching, finishing and trim.

| | Craft@Hrs | Unit | Material | Labor | Total | Sell |
|---|---|---|---|---|---|---|
| Cost per square foot of wall | B1@.068 | SF | 1.35 | 2.00 | 3.35 | 5.70 |
| Cost per running foot, 8' high walls | B1@.544 | LF | 10.80 | 16.00 | 26.80 | 45.60 |

**Add 2" x 6" interior stud partition with 1/2" drywall both sides.** Add the cost of patching, finishing and trim.

| | Craft@Hrs | Unit | Material | Labor | Total | Sell |
|---|---|---|---|---|---|---|
| Cost per square foot of wall | B1@.072 | SF | 1.50 | 2.11 | 3.61 | 6.14 |
| Cost per running foot, 8' high walls | B1@.576 | LF | 12.00 | 16.90 | 28.90 | 49.10 |

**Add 2" x 6" interior stud partition with 5/8" drywall both sides.** Add the cost of patching, finishing and trim.

| | Craft@Hrs | Unit | Material | Labor | Total | Sell |
|---|---|---|---|---|---|---|
| Cost per square foot of wall | B1@.076 | SF | 1.64 | 2.23 | 3.87 | 6.58 |
| Cost per running foot, 8' high walls | B1@.608 | LF | 13.10 | 17.90 | 31.00 | 52.70 |

# Floors and Tile

**11**

Some types of floor cover are more durable than others. Vinyl lasts longer than carpet. Ceramic tile lasts longer than wood block or strip flooring. But no floor material has a life expectancy equal to that of the house itself. That makes flooring a popular focus in home improvement work.

All floor cover requires a base that's structurally sound, clean, level (to $1/4$" per 10' span) and dry (moisture content of the subfloor should not exceed 13 percent). Concrete makes a good base, assuming the surface is smooth and incorporates a good vapor barrier. Untempered hardboard, plywood, and particleboard also make a good base for flooring. Use either $1/4$"- or $3/8$"-thick sheets. Underlayment needs a $1/32$" gap at the edges and the ends to allow for expansion. Underlayment-grade plywood has a sanded, C-plugged or better face. If moisture isn't a problem, use interior type plywood. Otherwise use either exterior or interior grade plywood with exterior glue. Trowel on a smooth coat of cement-based underlay to prepare nearly any floor surface for resilient flooring.

## Repair, Recover or Remove?

Adhesive used to secure resilient flooring tends to deteriorate when moisture comes up through the subfloor. If resilient tile comes loose, try resetting the tile in new adhesive that's designed for use below grade. If the resilient tile is cracked, broken or has chipped edges, it's usually better to install new flooring. Matching new tile with old tile isn't practical. Resilient tile changes color with age. But it may not be necessary to remove the old surface when installing new. If the old surface is scarred, stained, abraded or has been embossed by the weight of furniture, apply a liquid leveler, or trowel on a cement-based underlayment to smooth the surface. Then install the new floor cover. If unevenness in the under-layment is showing through, remove the old surface and do some leveling before installing the new floor cover. Remove resilient tile if the new floor cover is also to be resilient tile.

If a wood floor is smooth and free of large cracks, refinishing may put the floor back in like-new condition. Most wood flooring can be sanded and refinished several times. Softwood flooring with no subfloor is an exception. Even one sanding might weaken the floor too much. Plywood block flooring can some-times be sanded and refinished. Thin wood flooring and wood flooring with wide cracks usually has to be replaced — any patch would be obvious.

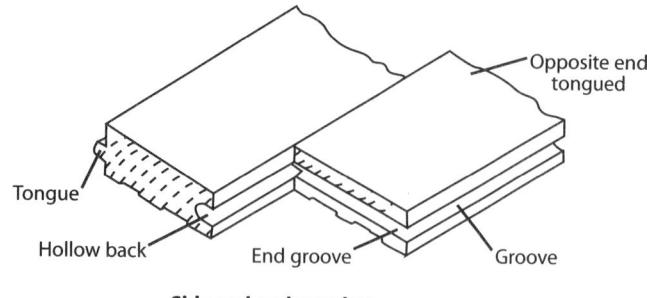

Opposite end
tongued

Tongue

Hollow back

End groove

Groove

**Side and end matches**

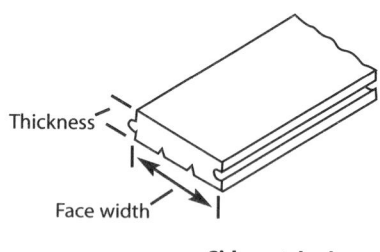

Thickness

Face width

**Side matched**

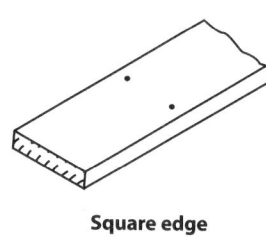

**Square edge**

**Figure 11-1**

*Strip flooring*

Wood flooring, sheet vinyl with resilient backing, and carpeting can be installed directly over an existing hardwood floor, assuming any voids have been filled and the surface isn't loose. Shrinkage cracks are more common where boards are wide. Be sure to check for boards that are buckling, cupping or cracking due to moisture. If there's a moisture problem, solve that before you lay new flooring.

You can install laminated wood flooring over ceramic, wood or resilient flooring, so long as the surface is firm and dry.

# New Wood Flooring

Hardwood flooring is available in tongue-and-groove strips and blocks (parquet). Some thinner patterns of strip flooring are square-edged. See Figure 11-1. The most common hardwood strip flooring is $^{25}/_{32}$" thick by $2^1/_4$" wide and has a hollow back. Strips are random lengths and vary from 2' to 16' long. The face is slightly wider than the bottom so joints will be tight on the surface.

Softwood flooring is also available in strips and blocks. Most softwood strip flooring has tongue-and-groove edges, although some types are end matched. Softwood flooring costs less than most hardwood species, but it's also less wear-resistant and shows surface abrasions more readily. Use softwood flooring in light traffic areas such as closets. No matter which type of flooring you select, give the material a few days to reach the moisture content of the room where it will be installed.

Strip flooring is normally laid at right angles to the floor joists. When new strip flooring is installed over old, lay the new strips at right angles to the old, no matter what direction the floor joists run. Use 8d flooring nails for $^{25}/_{32}$" thick flooring, 6d flooring nails for $^1/_2$" flooring, and 4d casing nails for $^3/_8$" flooring. Some manufacturers recommend ring-shank or screw-shank nails. To help prevent splitting the tongue, use flooring brads with blunted points.

Begin installation of tongue-and-groove flooring by placing the first strip $^1/_2$" to $^5/_8$" away from the wall. That allows for expansion and prevents buckling when the moisture content increases. Nail straight down through the face of the first strip, as in Figure 11-2. The nail should be close enough to the wall to be covered by the base or shoe molding. Try to nail into a joist if the new flooring is laid at right

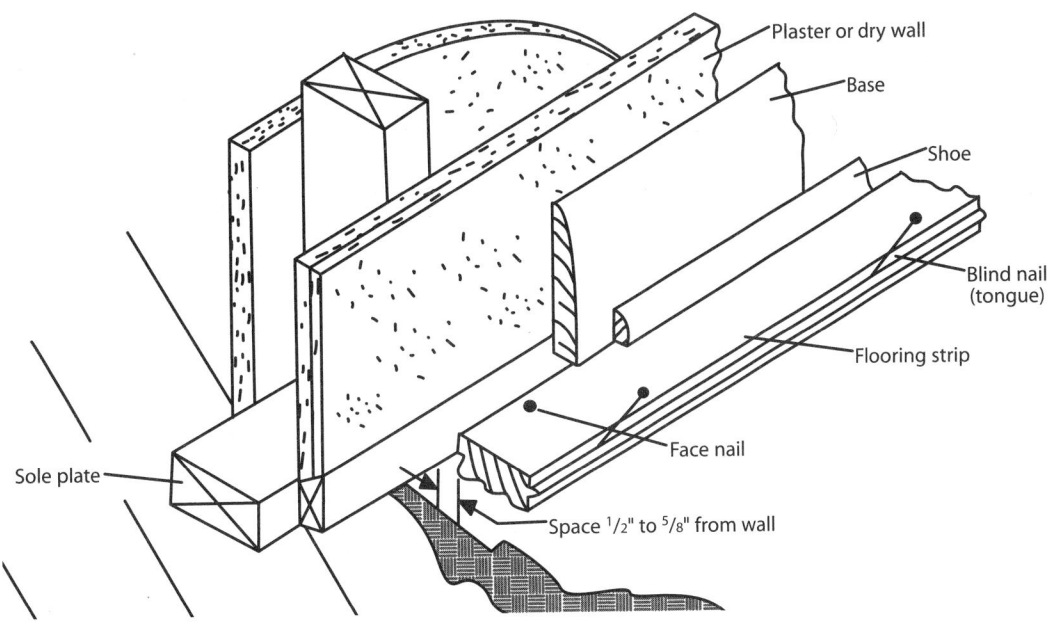

**Figure 11-2**

*Installation of first strip of flooring*

angles to the joists. Drive a second nail through the tongue of this first strip. All other strips are nailed through the tongue only. Drive these nails at an angle of 45 to 50 degrees. But leave the head just above the surface to avoid damaging the strip with your hammer. Use a large nail set to drive nails the last quarter inch. You can lay the nail set flat against the flooring when setting these nails, See Figure 11-3.

Stagger the end joints of strip flooring so butts are separated in adjacent courses. Install each new strip tightly against the previously installed strip. Use shorter strips and crooked strips at the end of courses or in closets. Leave a $1/2$" to $5/8$" space between the last course of flooring and the wall, just as with the first course. Face-nail the last course where the base or shoe will cover the nail head.

Square-edged strip flooring must be installed over a substantial subfloor and should be face-nailed. Other than that, the installation procedure is the same as for matched (tongue-and-groove) flooring. Wood strip flooring is always nailed.

Parquet tile is made from narrow wood slats formed into a square. Parquet block flooring can be applied with adhesive over a concrete floor protected from moisture with a vapor barrier. Spread adhesive on the slab or underlayment with a notched trowel. Then lay parquet in the adhesive. If you elect to nail parquet flooring to wood underlayment, nail through the tongue, the same as with wood strip flooring. Minimize problems associated with shrinkage and swelling by changing the grain direction of alternate blocks.

You can install particleboard tile over underlayment the same way you install parquet tile — except particleboard tile shouldn't be installed directly over concrete.

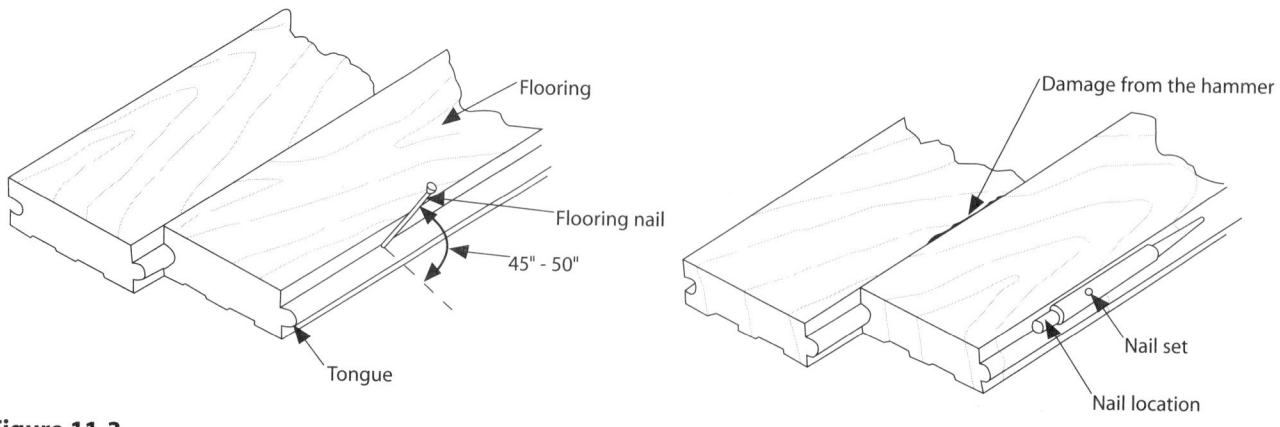

**Figure 11-3**

*Nailing strip flooring*

Follow the manufacturer's installation instructions. Particleboard tile is usually 9" x 9" and 3/8" thick, with tongue-and-groove edges. The back is often marked with small saw kerfs to stabilize the tile and provide a better grip for adhesive.

## Laminate Wood Flooring

Laminate flooring strips are made from layers of wood and finished with a hard synthetic surface. Pergo® is one popular name. Most laminate flooring is loose lay; neither nails nor adhesive are used. Instead, the flooring floats on a cushioning material designed to reduce noise from foot traffic. Laminate flooring can be installed over nearly any firm, flat flooring material. Install strips parallel to the longest wall in the room. Keep the strips about 1/4" away from the side wall and end wall so the floor can expand with changes in temperature and moisture. Cut laminate flooring with the finish side down, using a carbide-tip blade.

Lay the tongue side of the first strip against the wall. Continue laying boards along that wall, fitting ends snug against the previous board. Use spacers to maintain a 1/4" gap between the flooring and the wall. Avoid short lengths of flooring at the end of a course. If the last board in any course is less than 8", trim that amount off the first board in the course and move the entire course down by that distance. Second and later courses lock into the previous course. Stagger end joints in adjacent courses. Finish the job with base molding that covers the 1/4" gap at side and end walls.

## Resilient Flooring

Sheet vinyl with resilient backing smoothes out minor surface imperfections. Some sheet vinyl is designated *loose lay* and doesn't require adhesive. But use double-faced tape at joints and around edges to keep the covering in place. Manufacturers recommend spreading adhesive under all parts of the sheet for most products. Others can be bonded at the perimeter and seams only. Minimize the number of joints needed by using wider sheets — some sheet vinyl comes in widths up to 15'.

Both resilient sheet flooring and resilient tile require a smooth surface for proper adhesive bonding. You can repair an irregular surface with an embossing leveler or a masonry leveling compound. When the surface is dry, spread adhesive with a notched trowel, following the adhesive manufacturers instructions. Lay the tile so joints don't coincide with the joints in the underlayment.

Seamless flooring, consisting of resin chips combined with a urethane binder, can be applied over any stable base, including old floor tile. Apply this liquid in several coats, allowing each coat to dry. A complete application may take several days, depending on the brand. You can repair a seamless floor by applying another coat. Damaged spots are easy to patch by adding more chips and binder.

# Cork Tile

Cork is a natural sound absorber and insulator. It is quiet underfoot, and can last for decades when properly maintained. Cork will expand and contract based on humidity, although to a lesser degree than wood. Cork tiles should be given time to acclimate to the environment before installation. Remove tiles from their packaging and store them in the room where they will be installed for at least 48 hours prior to installation.

Most manufacturers recommend using a water-based contact cement adhesive for cork installation. Cork is porous, allowing the water in the adhesive to evaporate and create a strong bond. It's a good idea to test for proper adhesion before proceeding with the installation. Excessive moisture can damage cork flooring. For kitchen, bathroom or other high-risk applications, follow the manufacturer's guidelines for sealing cork floors with urethane or floor wax.

# Granite and Marble Tile

Common granite and marble surface finishes include *polished*, *honed*, and *flamed*. A polished surface is highly reflective, and is best suited for low-traffic areas. A honed surface has a duller, more slip-resistant finish that's less likely to show scratches. Flamed tiles have a deeply textured surface that's useful for applications requiring additional slip-resistance.

Marble is softer and more porous than granite, so it's more susceptible to scratches, but it can be repolished when necessary. Marble is also susceptible to damage from alcohols, oils and acids commonly found in the home. A penetrative sealer is generally recommended when installing marble in high-risk areas such as kitchens and bathrooms.

Colors and grains will vary from tile to tile and batch to batch. To assure the installation will have a uniform look, be sure you have enough material to finish a complete area. Consider using tiles randomly from different boxes when allocating material for the job. This will more-evenly distribute irregularities, and may actually create a more homogeneous look.

# Flooring Over Concrete

A concrete floor that stays dry in all seasons probably has a good vapor barrier under the slab. If the surface is also smooth and level, nearly any type of resilient flooring or carpet can be installed directly over the slab. If a basement slab is both uneven and moist to the touch, one remedy is to lay a vapor barrier over the existing slab, then cover the entire surface with a 2"- to 3"-thick concrete topping. Another approach is to lay a good-quality vapor barrier directly on the slab, then anchor furring strips or sleepers to the slab with concrete nails or shot fasteners. You can then install hardwood strip flooring directly over the sleepers. See Figure 11-4. For tile or sheet vinyl, nail underlayment or plywood to the sleepers before you install the finish floor.

# Ceramic Tile

Ceramic tile can be set in either mortar (*thin-set or thick set*) or applied with adhesive. Adhesive is more convenient because no mixing is required, though cleanup takes a little longer. Tile is set on backerboard, cement board reinforced with polymer-coated glass mesh. Common names are Durock®, WonderBoard®, RhinoBoard® and Hardibacker®. For floors and counters, set the backerboard in adhesive on $3/4$" exterior grade plywood. For walls, affix backerboard to the studs with cement board screws every 8". On ceilings, drive cement board screws every 6". Regular drywall screws don't have enough holding strength for use on backerboard. Cover panel joints with fiberglass mesh and joint cement. One side of backerboard is rough for use with tile in thin-set mortar. The other side is smooth for use with tile adhesive.

## Ceramic Tile Definitions

❖ Field tiles make up most of the job, the "field".

❖ Border tiles are trim pieces set around the edge of the field.

❖ Listello tiles have a decorative design different from field tile and are generally used on the edge of the field, like the frame of a picture.

❖ Rope tiles, as you might expect, have a rope design, usually in raised relief, and are used on the border.

Ceramic tile sizes range from 1" square mosaic to 12" x 12" and even larger. Mosaic tile are usually sold in 12" x 12" squares held together with a mesh backing. The most popular tile size for walls and counters is $4^1/4$" x $4^1/4$".

Avoid using tile with a bright glaze finish on floors. A highly-reflective finish tends to be slippery and offers less resistance to wear. Vitrified porcelain tiles are hard to cut accurately with a tile cutter and may require a circular ceramic wet saw. You also have to apply adhesive to both the tile and the floor when you're installing porcelain tile.

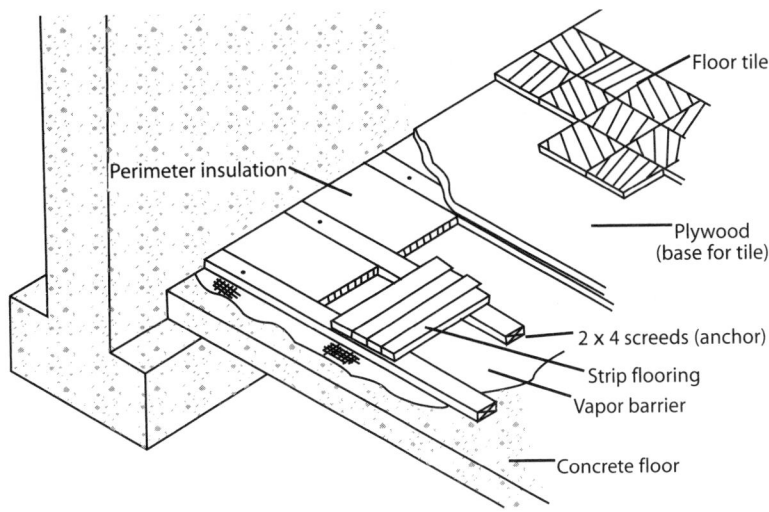

**Figure 11-4**

*Installation of wood floors in a basement*

## Tips on Ceramic Tile

Most ceramic tile carries a PEI (Porcelain Enamel Institute) wear rating:

❖ Class 1, no foot traffic. Interior residential and commercial walls only.

❖ Class 2, light traffic. Interior residential and commercial walls and residential bathroom floors.

❖ Class 3, light to moderate traffic. Residential floors, countertops, and walls.

❖ Class 4, moderate to heavy traffic. Residential, medium commercial and light institutional floors and walls.

❖ Class 5, heavy to extra-heavy traffic. Residential, commercial and institutional floors and walls.

## Indoor vs. Outdoor Tile

Tile that absorbs water will crack when exposed to freezing and thawing. Tile with an absorption rating of 3 percent or less is usually considered acceptable for outdoor use. That includes vitrified and porcelain ceramic tiles. Outdoor tile is very dense and doesn't break easily. Use thin-set mortar with a latex admix.

## Matching Styles and Batches

Tile colors and glazes can vary from batch to batch. To make matching easier, many tile manufacturers emboss batch numbers into the back of each tile.

| | Craft@Hrs | Unit | Material | Labor | Total | Sell |
|---|---|---|---|---|---|---|

**Remove flooring.** Per 100 square feet of floor. Using a long handle spudding spade. Includes removing baseboard as required. Debris piled on site. No salvage of materials or furniture moving included.

| | Craft@Hrs | Unit | Material | Labor | Total | Sell |
|---|---|---|---|---|---|---|
| Remove nailed hardwood floor and prepare surface for new flooring | BF@3.22 | CSF | — | 95.10 | 95.10 | 162.00 |
| Remove glued hardwood floor and prepare surface for new flooring | BF@5.58 | CSF | — | 165.00 | 165.00 | 281.00 |
| Remove sheet vinyl set in adhesive | BF@.620 | CSF | — | 18.30 | 18.30 | 31.10 |
| Remove sheet vinyl and adhesive residue for new flooring | BF@1.37 | CSF | — | 40.50 | 40.50 | 68.90 |
| Remove loose lay sheet flooring | BF@.138 | CSF | — | 4.08 | 4.08 | 6.94 |
| Remove resilient tile, no floor prep | BF@.777 | CSF | — | 23.00 | 23.00 | 39.10 |
| Remove resilient tile and adhesive residue for new flooring | BF@1.53 | CSF | — | 45.20 | 45.20 | 76.80 |
| Remove vinyl cove base | BF@2.50 | CSF | — | 73.90 | 73.90 | 126.00 |
| Remove rubber-back glue-down carpet, no scraping of adhesive included | BF@.217 | CSF | — | 6.41 | 6.41 | 10.90 |
| Remove rubber-back glue-down carpet, and prepare for new flooring | BF@.967 | CSF | — | 28.60 | 28.60 | 48.60 |
| Remove ceramic tile, air hammer | BL@2.90 | CSF | — | 77.30 | 77.30 | 131.00 |
| Remove ceramic tile and tile backer for new flooring | BL@5.80 | CSF | — | 155.00 | 155.00 | 264.00 |

## Wood Strip and Plank Flooring

**Bruce oak solid hardwood strip flooring.** These estimates assume work is done by skilled tradespeople, but not flooring specialists. Highly skilled flooring specialists may be able to install strip and plank flooring in 50% less time. Prefinished solid oak strip. Eased edge, square ends. 2-1/4" wide x 3/4" thick. Includes 10% cutting waste.

| | Craft@Hrs | Unit | Material | Labor | Total | Sell |
|---|---|---|---|---|---|---|
| Gunstock (western) | BF@.050 | SF | 4.71 | 1.48 | 6.19 | 10.50 |
| Gunstock (eastern) | BF@.050 | SF | 4.42 | 1.48 | 5.90 | 10.00 |
| Natural (western) | BF@.050 | SF | 4.71 | 1.48 | 6.19 | 10.50 |
| Natural (eastern) | BF@.050 | SF | 4.41 | 1.48 | 5.89 | 10.00 |
| Butterscotch | BF@.050 | SF | 4.35 | 1.48 | 5.83 | 9.91 |
| Marsh | BF@.050 | SF | 3.26 | 1.48 | 4.74 | 8.06 |

**Bruce hardwood molding.** Prefinished.

| | Craft@Hrs | Unit | Material | Labor | Total | Sell |
|---|---|---|---|---|---|---|
| 3'11" reducer | BF@.160 | Ea | 21.50 | 4.73 | 26.23 | 44.60 |
| 3'11" stair nose | BF@.160 | Ea | 21.50 | 4.73 | 26.23 | 44.60 |
| 3'11" T-molding | BF@.160 | Ea | 21.50 | 4.73 | 26.23 | 44.60 |
| 7'10" quarter round | BF@.320 | Ea | 16.10 | 9.45 | 25.55 | 43.40 |

**Bruce touch-up kit.** Includes two touch-up markers, three filler sticks and 1/2 ounce of Dura Luster urethane.

| | Craft@Hrs | Unit | Material | Labor | Total | Sell |
|---|---|---|---|---|---|---|
| Kit | — | Ea | 21.30 | — | 21.30 | — |

**Northwoods hardwood strip flooring, Harris-Tarkett.** These estimates assume work is done by skilled tradespeople, but not flooring specialists. Highly skilled flooring specialists may be able to install strip and plank flooring in 50% less time. Alumide® enhanced urethane finish resists stains and cleans easily. Precise tongue-and-groove for precise fit. Can be sanded and refinished up to three times. Nail-down installation only, on or above grade. 3/4" thick x 2-1/4" wide. Random lengths. Includes 10% cutting waste.

| | Craft@Hrs | Unit | Material | Labor | Total | Sell |
|---|---|---|---|---|---|---|
| Oak Toffee | BF@.050 | SF | 4.51 | 1.48 | 5.99 | 10.20 |
| Red Oak natural | BF@.050 | SF | 4.51 | 1.48 | 5.99 | 10.20 |

|  | Craft@Hrs | Unit | Material | Labor | Total | Sell |
|---|---|---|---|---|---|---|
| **Northwoods hardwood molding.** Prefinished toffee. | | | | | | |
| 7'10" quarter round | BF@.320 | Ea | 16.20 | 9.45 | 25.65 | 43.60 |
| 3'11" reducer strip, carpet to hard surface | BF@.160 | Ea | 19.90 | 4.73 | 24.63 | 41.90 |
| 3'11" stair nose | BF@.160 | Ea | 21.50 | 4.73 | 26.23 | 44.60 |
| 3'11" T-molding | BF@.160 | Ea | 21.50 | 4.73 | 26.23 | 44.60 |

**Rosewood hardwood strip flooring.** These estimates assume work is done by skilled tradespeople, but not flooring specialists. Highly skilled flooring specialists may be able to install strip and plank flooring in 50% less time. Prefinished one-piece strips. Nail or staple to wood subfloor. 3/4" thick. Random lengths except as noted. Includes 10% cutting waste.

| | Craft@Hrs | Unit | Material | Labor | Total | Sell |
|---|---|---|---|---|---|---|
| 2-1/4" wide | BF@.050 | SF | 4.12 | 1.48 | 5.60 | 9.52 |
| 2-1/4" wide, 7' long | BF@.048 | SF | 6.30 | 1.42 | 7.72 | 13.10 |
| 3-1/2" wide, 7' long | BF@.047 | SF | 6.00 | 1.39 | 7.39 | 12.60 |
| 4-1/2" wide, select | BF@.047 | SF | 9.40 | 1.39 | 10.79 | 18.30 |
| 5-1/2" wide, premium | BF@.045 | SF | 9.54 | 1.33 | 10.87 | 18.50 |
| 5-1/2" wide, select | BF@.045 | SF | 9.36 | 1.33 | 10.69 | 18.20 |
| 5-1/2" wide, select, 6' long | BF@.045 | SF | 8.94 | 1.33 | 10.27 | 17.50 |

**Rosewood hardwood molding.** 72" long. Prefinished.

| | Craft@Hrs | Unit | Material | Labor | Total | Sell |
|---|---|---|---|---|---|---|
| Grooved reducer | BF@.360 | Ea | 22.80 | 10.60 | 33.40 | 56.80 |
| Quarter round | BF@.360 | Ea | 17.80 | 10.60 | 28.40 | 48.30 |
| Stair nosing | BF@.360 | Ea | 50.25 | 10.60 | 60.85 | 103.00 |
| T-molding | BF@.360 | Ea | 42.20 | 10.60 | 52.80 | 89.80 |

**Maple hardwood flooring, Harris-Tarkett.** These estimates assume work is done by skilled tradespeople, but not flooring specialists. Highly skilled flooring specialists may be able to install strip and plank flooring in 50% less time. Light colored with less distinct grain. Extremely hard. 3/4" thick x 2-1/4" wide. Random lengths. Includes 10% cutting waste. Prefinished.

| | Craft@Hrs | Unit | Material | Labor | Total | Sell |
|---|---|---|---|---|---|---|
| Ginger | BF@.050 | Ea | 4.85 | 1.48 | 6.33 | 10.80 |
| Natural | BF@.050 | Ea | 5.40 | 1.48 | 6.88 | 11.70 |

**Maple hardwood molding.** Prefinished.

| | Craft@Hrs | Unit | Material | Labor | Total | Sell |
|---|---|---|---|---|---|---|
| 3'11" floor to carpet reducer strip | BF@.160 | Ea | 20.00 | 4.73 | 24.73 | 42.00 |
| 3'11" reducer strip | BF@.160 | Ea | 20.00 | 4.73 | 24.73 | 42.00 |
| 3'11" stair nose | BF@.160 | Ea | 20.00 | 4.73 | 24.73 | 42.00 |
| 3'11" T-molding | BF@.160 | Ea | 20.00 | 4.73 | 24.73 | 42.00 |

**Longstrip hardwood flooring.** These estimates assume work is done by skilled tradespeople, but not flooring specialists. Highly skilled flooring specialists may be able to install strip and plank flooring in 50% less time. Random lengths. Cost per square foot including 10% cutting waste. Prefinished.

| | Craft@Hrs | Unit | Material | Labor | Total | Sell |
|---|---|---|---|---|---|---|
| 3/4" x 3-1/4", Medium brown teak | BF@.047 | SF | 3.71 | 1.39 | 5.10 | 8.67 |
| 3/8" x 5", Rustic antique oak | BF@.045 | SF | 6.16 | 1.33 | 7.49 | 12.70 |

**Pre-finished natural flooring strips, Armstrong.** These estimates assume work is done by skilled tradespeople, but not flooring specialists. Highly skilled flooring specialists may be able to install strip and plank flooring in 50% less time. 3/4" thick x 2-1/4" wide. Includes 10% cutting waste.

| | Craft@Hrs | Unit | Material | Labor | Total | Sell |
|---|---|---|---|---|---|---|
| Natural strips | BF@.050 | SF | 3.87 | 1.48 | 5.35 | 9.10 |

| | Craft@Hrs | Unit | Material | Labor | Total | Sell |
|---|---|---|---|---|---|---|

**Vanguard TapTight hardwood plank flooring, Harris-Tarkett.** These estimates assume work is done by skilled tradespeople, but not flooring specialists. Highly skilled flooring specialists may be able to install strip and plank flooring in 50% less time. Factory pre-glued tongue-and-groove. Aluminide® enhanced urethane finish. Thick top layer can be sanded and refinished up to 3 times. Cross-directional construction for added strength and stability. Floating, glue-down or staple-down installation on grade or above grade. 9/16" x 7.5" x 4' planks. Case covers 13.25 square feet with 10% end-cutting waste.

| | Craft@Hrs | Unit | Material | Labor | Total | Sell |
|---|---|---|---|---|---|---|
| Heritage Maple | BF@.045 | SF | 4.80 | 1.33 | 6.13 | 10.40 |
| Oak Wheat | BF@.045 | SF | 4.87 | 1.33 | 6.20 | 10.50 |
| Red Oak natural | BF@.045 | SF | 4.87 | 1.33 | 6.20 | 10.50 |

**TapTight hardwood moldings.** Prefinished.

| | Craft@Hrs | Unit | Material | Labor | Total | Sell |
|---|---|---|---|---|---|---|
| 3'11" carpet reducer | BF@.160 | Ea | 19.40 | 4.73 | 24.13 | 41.00 |
| 3'11" stair nose | BF@.160 | Ea | 15.90 | 4.73 | 20.63 | 35.10 |
| 3'11" T-molding | BF@.160 | Ea | 19.40 | 4.73 | 24.13 | 41.00 |
| 7'10" quarter round | BF@.160 | Ea | 19.40 | 4.73 | 24.13 | 41.00 |

**Vanguard TapTight floating floor installation kit, Harris-Tarkett.** Includes pull tool, spacers and tapping block.

| | Craft@Hrs | Unit | Material | Labor | Total | Sell |
|---|---|---|---|---|---|---|
| Kit | — | Ea | 18.30 | — | 18.30 | — |

**Bamboo plank flooring.** These estimates assume work is done by skilled tradespeople, but not flooring specialists. Highly skilled flooring specialists may be able to install strip and plank flooring in 50% less time. 5/8" thick x 3-5/8" wide. Includes 10% waste. Prefinished.

| | Craft@Hrs | Unit | Material | Labor | Total | Sell |
|---|---|---|---|---|---|---|
| Natural | BF@.050 | SF | 4.49 | 1.48 | 5.97 | 10.10 |
| Spice | BF@.050 | SF | 4.49 | 1.48 | 5.97 | 10.10 |

**Bamboo flooring molding.** Prefinished.

| | Craft@Hrs | Unit | Material | Labor | Total | Sell |
|---|---|---|---|---|---|---|
| Quarter round, 78" | BF@.360 | Ea | 10.42 | 10.60 | 21.02 | 35.70 |
| Reducer, 47" | BF@.240 | Ea | 24.40 | 7.09 | 31.49 | 53.50 |
| Stair nose, 47" | BF@.240 | Ea | 32.40 | 7.09 | 39.49 | 67.10 |
| T-mold, 47" | BF@.240 | Ea | 19.10 | 7.09 | 26.19 | 44.50 |

**Wood plank flooring.** These estimates assume work is done by skilled tradespeople, but not flooring specialists. Highly skilled flooring specialists may be able to install strip and plank flooring in 50% less time. Prefinished. 3/8" thick x 3" wide. Includes 10% cutting waste.

| | Craft@Hrs | Unit | Material | Labor | Total | Sell |
|---|---|---|---|---|---|---|
| Butterscotch | BF@.050 | SF | 3.93 | 1.48 | 5.41 | 9.20 |
| Gunstock | BF@.050 | SF | 3.93 | 1.48 | 5.41 | 9.20 |

**Rustic oak plank flooring.** These estimates assume work is done by skilled tradespeople, but not flooring specialists. Highly skilled flooring specialists may be able to install strip and plank flooring in 50% less time. Prefinished. 3/4" thick x 2-1/4" wide. Includes 10% cutting waste.

| | Craft@Hrs | Unit | Material | Labor | Total | Sell |
|---|---|---|---|---|---|---|
| Toffee | BF@.050 | SF | 3.55 | 1.48 | 5.03 | 8.55 |
| Umber | BF@.050 | SF | 3.55 | 1.48 | 5.03 | 8.55 |

**Heart pine plank flooring.** These estimates assume work is done by skilled tradespeople, but not flooring specialists. Highly skilled flooring specialists may be able to install strip and plank flooring in 50% less time. Century-old heart pine timber milled to 3/4" thick. Tongue-and-groove. Random lengths. Includes 10% cutting waste. Prefinished.

| | Craft@Hrs | Unit | Material | Labor | Total | Sell |
|---|---|---|---|---|---|---|
| 3/4" x 2-1/4" x 7' | BF@.050 | SF | 5.30 | 1.48 | 6.78 | 11.50 |
| 3/4" x 5-1/2", premium | BF@.045 | SF | 6.90 | 1.33 | 8.23 | 14.00 |
| 3/4" x 5-1/2", select | BF@.045 | SF | 7.65 | 1.33 | 8.98 | 15.30 |

| | Craft@Hrs | Unit | Material | Labor | Total | Sell |
|---|---|---|---|---|---|---|
| **Heart pine molding.** 72" long. Prefinished. | | | | | | |
| Baby threshold | BF@.360 | Ea | 21.30 | 10.60 | 31.90 | 54.20 |
| Grooved reducer | BF@.360 | Ea | 22.10 | 10.60 | 32.70 | 55.60 |
| Quarter round | BF@.360 | Ea | 11.10 | 10.60 | 21.70 | 36.90 |
| Stair nose | BF@.360 | Ea | 44.60 | 10.60 | 55.20 | 93.80 |
| T-molding | BF@.360 | Ea | 34.20 | 10.60 | 44.80 | 76.20 |
| Threshold | BF@.360 | Ea | 34.20 | 10.60 | 44.80 | 76.20 |

**Unfinished red oak strip flooring, Harris-Tarkett.** These estimates assume work is done by skilled tradespeople, but not flooring specialists. Highly skilled flooring specialists may be able to install strip and plank flooring in 50% less time. Random lengths. Nested bundle covers 17.6 square feet including 10% end-cutting waste. Add the cost of finishing.

| | Craft@Hrs | Unit | Material | Labor | Total | Sell |
|---|---|---|---|---|---|---|
| 3/4" x 2-1/4", select | BF@.050 | SF | 3.05 | 1.48 | 4.53 | 7.70 |
| 3/4" x 2-1/4", #1 common | BF@.050 | SF | 2.63 | 1.48 | 4.11 | 6.99 |
| 3/4" x 2-1/4", #2 common | BF@.050 | SF | 2.35 | 1.48 | 3.83 | 6.51 |
| 3/4" x 3-1/4", select | BF@.047 | SF | 3.76 | 1.39 | 5.15 | 8.76 |
| 3/4" x 3-1/4", #1 common | BF@.047 | SF | 2.79 | 1.39 | 4.18 | 7.11 |
| 3/4" x 3-1/4", #2 common | BF@.047 | SF | 2.09 | 1.39 | 3.48 | 5.92 |

**Unfinished white oak strip flooring, Harris-Tarkett.** These estimates assume work is done by skilled tradespeople, but not flooring specialists. Highly skilled flooring specialists may be able to install strip and plank flooring in 50% less time. Random lengths. Nested bundle covers 17.6 square feet including 10% end-cutting waste.

| | Craft@Hrs | Unit | Material | Labor | Total | Sell |
|---|---|---|---|---|---|---|
| 3/4" x 2-1/4", select | BF@.050 | SF | 2.73 | 1.48 | 4.21 | 7.16 |
| 3/4" x 2-1/4", #1 common | BF@.050 | SF | 2.30 | 1.48 | 3.78 | 6.43 |

**Unfinished oak molding.**

| | Craft@Hrs | Unit | Material | Labor | Total | Sell |
|---|---|---|---|---|---|---|
| 3/4" x 2" x 39", reducer | BF@.160 | Ea | 10.69 | 4.73 | 15.42 | 26.20 |
| 3/4" x 3/4" x 78", quarter round | BF@.360 | Ea | 12.50 | 10.60 | 23.10 | 39.30 |
| 3/4" x 78", stair nose | BF@.360 | Ea | 27.70 | 10.60 | 38.30 | 65.10 |

**Unfinished maple strip flooring.** These estimates assume work is done by skilled tradespeople, but not flooring specialists. Highly skilled flooring specialists may be able to install strip and plank flooring in 50% less time. Random lengths. Including 10% end-cutting waste.

| | Craft@Hrs | Unit | Material | Labor | Total | Sell |
|---|---|---|---|---|---|---|
| 3/4" x 2-1/4", 1st grade | BF@.050 | SF | 3.73 | 1.48 | 5.21 | 8.86 |

**Acrylic wood filler.** Fills nicks and dents in hardwood flooring.

| | | Unit | Material | Labor | Total | Sell |
|---|---|---|---|---|---|---|
| 3-1/4 ounce tube | — | Ea | 5.36 | — | 5.36 | — |

**Floor installation tools.** Heavy duty.

| | | Unit | Material | Labor | Total | Sell |
|---|---|---|---|---|---|---|
| Pull bar | — | Ea | 8.40 | — | 8.40 | — |
| Tapping block | — | Ea | 10.28 | — | 10.28 | — |

**Flooring installation clamp.** Use to install most wood and laminate flooring. Adjustable cinching system. Recommended use: every 3' to 5'.

| | | Unit | Material | Labor | Total | Sell |
|---|---|---|---|---|---|---|
| 18' maximum span | — | Ea | 33.50 | — | 33.50 | — |

**Sand hardwood floor.** Sanding floors requires skill. An inexperienced operator can ruin a hardwood floor with a power sander. Using a vibrating or drum sander. Per 100 square feet of floor. No furniture moving included. Add the cost of equipment.

| | Craft@Hrs | Unit | Material | Labor | Total | Sell |
|---|---|---|---|---|---|---|
| Sand only, 3 passes | BF@1.08 | CSF | — | 31.90 | 31.90 | 54.20 |
| Sand and scrape, 4 passes | BF@1.31 | CSF | — | 38.70 | 38.70 | 65.80 |
| Small room or closet, 3 passes | BF@2.80 | CSF | — | 82.70 | 82.70 | 141.00 |
| Small room or closet, 4 passes | BF@3.40 | CSF | — | 100.00 | 100.00 | 170.00 |

| | Craft@Hrs | Unit | Material | Labor | Total | Sell |
|---|---|---|---|---|---|---|

## Wood Parquet Flooring

**Hevea parquet flooring, OakCrest Products.** Solid hardwood. Locking tongue-and-groove on all five sides with Tru-Square™ edges. 7.5mm thick. Wire back. UV-cured polyurethane finish, no-wax and stain-resistant. Can be sanded and refinished several times over the life of the floor. Pack of 10 covers 9 square feet with 10% waste. Add the cost of adhesive.

| | Craft@Hrs | Unit | Material | Labor | Total | Sell |
|---|---|---|---|---|---|---|
| 12" x 12" | BC@.055 | SF | 1.08 | 1.76 | 2.84 | 4.83 |

**Bruce oak parquet flooring.** 12" x 12" x 5/16". Pack of 10 covers 9 square feet with 10% waste. Add the cost of adhesive.

| | Craft@Hrs | Unit | Material | Labor | Total | Sell |
|---|---|---|---|---|---|---|
| Chestnut | BF@.055 | SF | 1.95 | 1.62 | 3.57 | 6.07 |
| Desert | BF@.055 | SF | 1.95 | 1.62 | 3.57 | 6.07 |

**Parquet floor tile.** Urethane finish. 12" x 12" x 5/16". Case of 25 covers 22.5 square feet with 10% waste. Add the cost of adhesive.

| | Craft@Hrs | Unit | Material | Labor | Total | Sell |
|---|---|---|---|---|---|---|
| Mellow | BF@.055 | SF | 1.17 | 1.62 | 2.79 | 4.74 |
| Sorrento Chestnut | BF@.055 | SF | 1.17 | 1.62 | 2.79 | 4.74 |
| Sorrento Quartz | BF@.055 | SF | 1.17 | 1.62 | 2.79 | 4.74 |

**Bruce oak self-stick parquet flooring.** Dura-Luster® urethane finish. 12" x 12" x 7/16". Pack of 10 covers 9 square feet with 10% waste.

| | Craft@Hrs | Unit | Material | Labor | Total | Sell |
|---|---|---|---|---|---|---|
| Chestnut | BF@.050 | SF | 2.17 | 1.48 | 3.65 | 6.21 |
| Desert | BF@.050 | SF | 2.17 | 1.48 | 3.65 | 6.21 |

**Hardwood moldings for Bruce parquet.**

| | Craft@Hrs | Unit | Material | Labor | Total | Sell |
|---|---|---|---|---|---|---|
| 3'11" reducer | BF@.160 | Ea | 21.50 | 4.73 | 26.23 | 44.60 |
| 3'11" stair nose | BF@.160 | Ea | 21.50 | 4.73 | 26.23 | 44.60 |
| 3'11" T-molding | BF@.160 | Ea | 21.50 | 4.73 | 26.23 | 44.60 |
| 7'10" quarter round | BF@.360 | Ea | 16.10 | 10.60 | 26.70 | 45.40 |

## Wood Flooring Adhesives

**Adhesive remover, Jasco®.** Nonflammable and water rinseable. Softens old adhesive in 5 to 15 minutes. One gallon removes adhesive from 100 to 200 square feet. Semi-paste clings to vertical surfaces. Use with neoprene or rubber gloves and adequate ventilation.

| | Craft@Hrs | Unit | Material | Labor | Total | Sell |
|---|---|---|---|---|---|---|
| Quart | — | Ea | 8.10 | — | 8.10 | — |
| Gallon | — | Ea | 22.04 | — | 22.04 | — |

**Adhesive primer, Jasco®.** For use on plywood, concrete, and glasscrete. Primes floors before applying adhesive. Prevents dry out when adhesive is drawn below the surface. Ensures adhesion of self-stick floor tile. Water cleanup. Covers 300 to 500 square feet per gallon.

| | Craft@Hrs | Unit | Material | Labor | Total | Sell |
|---|---|---|---|---|---|---|
| Quart | — | Ea | 5.15 | — | 5.15 | — |
| Gallon | — | Ea | 11.90 | — | 11.90 | — |

**Wood flooring adhesive, Roberts 1404.** Synthetic latex-based adhesive for use on engineered laminated wood plank and wood parquet flooring. May be used over interior grade plywood subflooring and concrete on and above grade. Not recommended for foam backed, solid wood flooring or high-pressure laminate flooring (e.g., Wilsonart, Pergo, Formica, etc.). Excellent early grab and develops good water resistance. Covers 40 to 50 square feet per gallon when applied with a 1/8" x 1/8" x 1/8" square-notch trowel for parquet or a 1/4" x 1/4" x 1/4" V-notch trowel for laminated plank.

| | Craft@Hrs | Unit | Material | Labor | Total | Sell |
|---|---|---|---|---|---|---|
| Gallon | — | Ea | 19.80 | — | 19.80 | — |
| 4 gallons | — | Ea | 59.90 | — | 59.90 | — |
| Pro trowel | — | Ea | 8.05 | — | 8.05 | — |

| | Craft@Hrs | Unit | Material | Labor | Total | Sell |
|---|---|---|---|---|---|---|

## Laminate Plank Flooring

**Pergo Presto laminate wood flooring.** Tongue-and-groove joints lock together for a tight fit. Guaranteed secure joint. Case covers 19.69 square feet before end-cutting waste.

| | Craft@Hrs | Unit | Material | Labor | Total | Sell |
|---|---|---|---|---|---|---|
| Beech, blocked | BF@.051 | SF | 3.21 | 1.51 | 4.72 | 8.02 |
| Cherry, blocked | BF@.051 | SF | 3.21 | 1.51 | 4.72 | 8.02 |
| Red oak, blocked | BF@.051 | SF | 3.21 | 1.51 | 4.72 | 8.02 |
| Salem oak | BF@.051 | SF | 2.75 | 1.51 | 4.26 | 7.24 |
| Colby walnut | BF@.051 | SF | 2.75 | 1.51 | 4.26 | 7.24 |

**Pergo floor molding.**

| | Craft@Hrs | Unit | Material | Labor | Total | Sell |
|---|---|---|---|---|---|---|
| Carpet transition, 94" | BF@.320 | Ea | 28.10 | 9.45 | 37.55 | 63.80 |
| T-molding, 94" | BF@.320 | Ea | 31.60 | 9.45 | 41.05 | 69.80 |
| Wall base, 94" | BF@.320 | Ea | 16.80 | 9.45 | 26.25 | 44.60 |

**Pergo flooring installation spacers.** For maintaining proper spacing along walls and fixed objects. Minimizes movement of floor during installation. Package covers about 48 square feet.

| | Craft@Hrs | Unit | Material | Labor | Total | Sell |
|---|---|---|---|---|---|---|
| Pack of 48 | — | Ea | 7.33 | — | 7.33 | — |

**SilentStep™ laminate flooring underlayment, Pergo®.** Sound reduction for laminate flooring. Flexible. Cuts without leaving debris or dust. Roll covers 100 square feet.

| | Craft@Hrs | Unit | Material | Labor | Total | Sell |
|---|---|---|---|---|---|---|
| Flooring underlayment | BF@.010 | SF | .74 | .30 | 1.04 | 1.77 |

**Soundbloc™ underlayment foam, Pergo®.** Evens out minor irregularities in subfloor. Provides sound and heat insulation. Makes floor comfortable underfoot. Roll covers 100 square feet.

| | Craft@Hrs | Unit | Material | Labor | Total | Sell |
|---|---|---|---|---|---|---|
| Foam underlayment | BF@.006 | SF | .20 | .18 | .38 | .65 |

**SoftSeal™ underlayment, Pergo®.** Vapor barrier and sound reducing underlayment in one. For concrete subfloors only. 8" band of film along edges for overlap. Alkaline resistant. Roll covers 100 square feet.

| | Craft@Hrs | Unit | Material | Labor | Total | Sell |
|---|---|---|---|---|---|---|
| Foam and film pack | — | SF | .32 | — | .32 | — |

**Pergo laminate glue.** Penetrates the core material to create a super-strong joint. Pergo warranty is valid only if Pergo glue is used for installation.

| | Craft@Hrs | Unit | Material | Labor | Total | Sell |
|---|---|---|---|---|---|---|
| 16.9-ounce tube | — | Ea | 6.83 | — | 6.83 | — |

**Laminate flooring sealant, Pergo®.** Acrylic-based sealing compound for moisture resistance. For use in bathrooms, kitchens, laundry rooms and other wet areas. Formulated to maintain high level of flexibility and elasticity.

| | Craft@Hrs | Unit | Material | Labor | Total | Sell |
|---|---|---|---|---|---|---|
| 11-ounce tube | — | Ea | 8.59 | — | 8.59 | — |

**Finishing putty, Pergo®.** Water-based putty formulated to fill nail and screw holes in Pergo moldings and wall base. Repairs small chips and dents in the laminate layer.

| | Craft@Hrs | Unit | Material | Labor | Total | Sell |
|---|---|---|---|---|---|---|
| Tube | — | Ea | 8.83 | — | 8.83 | — |

**Pergo® universal installation kit.** Contains underseal tapping block, 48 installation spacers, pull bar and glue scraper.

| | Craft@Hrs | Unit | Material | Labor | Total | Sell |
|---|---|---|---|---|---|---|
| Kit | — | Ea | 18.14 | — | 18.14 | — |

| | Craft@Hrs | Unit | Material | Labor | Total | Sell |
|---|---|---|---|---|---|---|

**Hampton TrafficMaster laminate flooring.** Engineered "longstrip" wood floors with premium oak veneers. Cross-ply, multiple layer construction for long-term dimensional stability and structural integrity. Each plank is 7.5" wide, 72" long, 7/16" thick.

| | Craft@Hrs | Unit | Material | Labor | Total | Sell |
|---|---|---|---|---|---|---|
| Baltic beech, 22 SF per case | BF@.051 | SF | 1.58 | 1.51 | 3.09 | 5.25 |
| Southern oak, 22 SF per case | BF@.051 | SF | 1.58 | 1.51 | 3.09 | 5.25 |
| Light red oak, 19.5 SF per case | BF@.051 | SF | 1.95 | 1.51 | 3.46 | 5.88 |
| Medium red oak, 19.5 SF per case | BF@.051 | SF | 1.95 | 1.51 | 3.46 | 5.88 |
| Dark red oak, 19.5 SF per case | BF@.051 | SF | 1.95 | 1.51 | 3.46 | 5.88 |

**Hampton floor molding.**

| | Craft@Hrs | Unit | Material | Labor | Total | Sell |
|---|---|---|---|---|---|---|
| 3'11" carpet reducer | BF@.160 | Ea | 11.80 | 4.73 | 16.53 | 28.10 |
| 3'11" reducer | BF@.160 | Ea | 11.80 | 4.73 | 16.53 | 28.10 |
| 3'11" stair nose | BF@.160 | Ea | 9.80 | 4.73 | 14.53 | 24.70 |
| 3'11" T-molding | BF@.160 | Ea | 11.80 | 4.73 | 16.53 | 28.10 |
| 7'10" quarter round | BF@.320 | Ea | 7.00 | 9.45 | 16.45 | 28.00 |
| 7'10" wall base | BF@.320 | Ea | 9.52 | 9.45 | 18.97 | 32.20 |

**Legends strip laminate flooring** Case covers 20 square feet before cutting waste.

| | Craft@Hrs | Unit | Material | Labor | Total | Sell |
|---|---|---|---|---|---|---|
| Kempas | BF@.051 | SF | 2.15 | 1.51 | 3.66 | 6.22 |
| Brazilian cherry | BF@.051 | SF | 2.15 | 1.51 | 3.66 | 6.22 |
| Wheat | BF@.051 | SF | 2.15 | 1.51 | 3.66 | 6.22 |
| Cognac | BF@.051 | SF | 2.15 | 1.51 | 3.66 | 6.22 |
| Coffee | BF@.051 | SF | 2.15 | 1.51 | 3.66 | 6.22 |

**Legends molding.** Molding used to transition from laminate flooring to carpet or at the base of a wall.

| | Craft@Hrs | Unit | Material | Labor | Total | Sell |
|---|---|---|---|---|---|---|
| 3'11" reducer | BF@.160 | Ea | 11.80 | 4.73 | 16.53 | 28.10 |
| 3'11" stair nose | BF@.160 | Ea | 18.30 | 4.73 | 23.03 | 39.20 |
| 3'11" T-molding | BF@.160 | Ea | 11.80 | 4.73 | 16.53 | 28.10 |
| 8' quarter round | BF@.320 | Ea | 7.06 | 9.45 | 16.51 | 28.10 |
| 8' wall base | BF@.320 | Ea | 16.10 | 9.45 | 25.55 | 43.40 |

**Wood and laminate flooring underlayment, Shaw HardSurfaces.** Roll covers 50 square feet.

| | Craft@Hrs | Unit | Material | Labor | Total | Sell |
|---|---|---|---|---|---|---|
| 3-in-1 underlayment | BF@.010 | SF | .73 | .30 | 1.03 | 1.75 |

**Foam underlayment, Shaw HardSurfaces.** Evens out minor irregularities in subfloor. Provides sound and heat insulation. Makes floor comfortable underfoot. Roll covers 100 square feet.

| | Craft@Hrs | Unit | Material | Labor | Total | Sell |
|---|---|---|---|---|---|---|
| 2-in-1 underlayment | BF@.006 | SF | .27 | .18 | .45 | .77 |

**Moisturbloc™ polyethylene film.** For installation over concrete subfloors. Lasts for the life of the floor. 6 mil polyethylene film with increased alkaline resistance and 0.2% HALS additive. 120 square foot roll.

| | Craft@Hrs | Unit | Material | Labor | Total | Sell |
|---|---|---|---|---|---|---|
| Poly film | BF@.006 | SF | .12 | .18 | .30 | .51 |

**Laminate flooring installation system.** Adjustable cinching system pulls laminate floor together and provides a perfect, tight seam.

| | Craft@Hrs | Unit | Material | Labor | Total | Sell |
|---|---|---|---|---|---|---|
| Cinch system | — | Ea | 20.50 | — | 20.50 | — |

## Floor Surface Prep

**Prepare surface for new flooring.** Per 100 square feet of floor.

| | Craft@Hrs | Unit | Material | Labor | Total | Sell |
|---|---|---|---|---|---|---|
| Scrape concrete floor to remove adhesive residue | BF@.750 | CSF | — | 22.20 | 22.20 | 37.70 |
| Apply skim-coat to level surface | BF@.563 | CSF | 71.50 | 16.60 | 88.10 | 150.00 |

**Sentinel 810 floor prep and adhesive remover.** Removes latex, acrylic and pressure sensitive adhesives from concrete, wood, sheet vinyl, tile, ceramic and terrazzo substrates. Highly concentrated cleaning solution.

| | Craft@Hrs | Unit | Material | Labor | Total | Sell |
|---|---|---|---|---|---|---|
| Quart | — | Ea | 5.37 | — | 5.37 | — |

| | Craft@Hrs | Unit | Material | Labor | Total | Sell |
|---|---|---|---|---|---|---|

**Sentinel 747 flooring adhesive remover.** For cleaning cutback, emulsion and outdoor flooring mastics. Coverage is 75 to 150 square feet per gallon.

| | | | | | | |
|---|---|---|---|---|---|---|
| Quart | — | Ea | 6.10 | — | 6.10 | — |
| Gallon | — | Ea | 19.30 | — | 19.30 | — |

**Henry 550 embossing leveler.** Levels embossed areas in existing resilient flooring prior to installing new residential flooring. One-part latex polymer resin. No mixing required. Can be used directly over urethane wear surfaces with no special preparation. Eliminates the need to remove the old floor. Prevents pattern show-through. Coverage: approximately 54 square feet per quart over heavy embossed floors or 108 square feet per quart over lightly embossed floors. Working time 15 to 20 minutes.

| | | | | | | |
|---|---|---|---|---|---|---|
| Quart | — | Ea | 26.90 | — | 26.90 | — |

**Liquid underlayment, Henry.** For residential use as an embossing leveler. One-part latex polymer resin. Can be used directly over CleanSweep surfaces with no special preparation. Eliminates the need to remove a damaged floor. Prevents pattern show-through. Coverage is approximately 216 square feet per gallon over heavily embossed floors or 432 square feet per gallon over lightly embossed floors. Working time is 15 to 20 minutes.

| | | | | | | |
|---|---|---|---|---|---|---|
| Quart | — | Ea | 27.00 | — | 27.00 | — |

**Latex primer and additive, S-185, Armstrong.** For priming wood and concrete before installing resilient floors. Improves adhesion. Fast-drying, ready for adhesive in 60 minutes. Gallon covers 400 square feet.

| | | | | | | |
|---|---|---|---|---|---|---|
| Quart | — | Ea | 5.35 | — | 5.35 | — |
| Gallon | — | Ea | 10.75 | — | 10.75 | — |

**Henry 336 floor primer.** Improves adhesion on dry, dusty and porous substrates. Bonds to concrete, poured-in-place gypsum subfloors, wood and wood underlayments, radiant-heated subfloors where the surface temperature does not exceed 85 degrees F (29 degrees C), wall surfaces such as wood, plaster, drywall, and masonry.

| | | | | | | |
|---|---|---|---|---|---|---|
| Quart | — | Ea | 5.35 | — | 5.35 | — |
| Gallon | — | Ea | 10.70 | — | 10.70 | — |

**Fast-setting patch and underlayment, S-184, Armstrong.** Cement-based smooth finish under resilient flooring. Ready for adhesive in 60 minutes. Can be used to skim coat over old cutback adhesive.

| | | | | | | |
|---|---|---|---|---|---|---|
| 3 pounds | — | Ea | 6.45 | — | 6.45 | — |
| 10 pounds | — | Ea | 14.00 | — | 14.00 | — |
| 25 pounds | — | Ea | 16.80 | — | 16.80 | — |

**Cement-based underlayment, Armstrong.** Fast-setting cement-based patch and underlayment for use over wood and concrete in resilient flooring application. Ready for most flooring in 1-2 hours. 10-pound bag covers 10 square feet at 1/8" thickness.

| | | | | | | |
|---|---|---|---|---|---|---|
| 10 pounds | — | Ea | 10.70 | — | 10.70 | — |
| 40 pounds | — | Ea | 20.33 | — | 20.33 | — |

**Henry 331 patch.** Ready-to-use gypsum based patch – just add water. Fills voids and levels floors from particle thin to 1" thick in a single application. Non-shrinking. Fast-setting.

| | | | | | | |
|---|---|---|---|---|---|---|
| 25 pounds | — | Ea | 8.40 | — | 8.40 | — |

| | Craft@Hrs | Unit | Material | Labor | Total | Sell |
|---|---|---|---|---|---|---|

**Henry 547 universal underlayment and floor patch.** Use as a patch, underlayment or embossing leveler. For patching, leveling and skim coating; covering existing "cutback" adhesive residue; and leveling the embossing of existing resilient flooring. Sets for most flooring in 1 to 2 hours. Excellent fill characteristics allows for installations from as little as a featheredge to depths of 1" thick. Non-sanded formulation allows product to be troweled to a smooth finish. Compressive strength exceeds 5,000 PSI. Non-shrinking formulation. Bonds to concrete, wood substrates, steel, stainless steel, brass, lead, ceramic, terrazzo, marble, existing "cutback" adhesive residue and existing resilient flooring (as an embossing leveler). Metal substrates must be clean and abraded. 10-year limited warranty. One pound covers one square foot at 1/8" thick.

| | | | | | | |
|---|---|---|---|---|---|---|
| 3 pounds | — | Ea | 6.86 | — | 6.86 | — |
| 10 pounds | — | Ea | 14.00 | — | 14.00 | — |
| 25 pounds | — | Ea | 23.60 | — | 23.60 | — |

**Henry 546 feather edge additive.** Improves the bond and enhances flexural strength when used as a mix instead of water with Henry 547. Extends the pot life for added working time. Not to be used when installing over asphalt "cutback" adhesive residue.

| | | | | | | |
|---|---|---|---|---|---|---|
| Gallon | — | Ea | 16.20 | — | 16.20 | — |

**Skimcoat white cover patch and underlayment.** Covers and levels defects in concrete, wood, metal, vinyl, ceramic tile and other subfloors. Smooth-spreading, quick-setting and fast-drying. High compressive strength and adhesion-resistance. Non-staining. Non-alkaline, adhesive-friendly. Non-shrinking, dust-free in thick or thin applications. Use on virtually any flooring surface from particle thin to over 1" thick. One pound covers one square foot at 1/8" thick.

| | | | | | | |
|---|---|---|---|---|---|---|
| 10 pounds | — | Ea | 6.30 | — | 6.30 | — |
| 25 pounds | — | Ea | 10.55 | — | 10.55 | — |
| 35 pounds | — | Ea | 17.20 | — | 17.20 | — |
| 50 pounds | — | Ea | 23.40 | — | 23.40 | — |

## Sheet Vinyl Flooring

**Medley™ sheet vinyl flooring, Armstrong.** CleanSweep® surface. ToughGuard® durability — guaranteed to not rip, tear or gouge. 15-year warranty. 6' wide roll. 0.08" thick. Urethane no-wax wear layer. Includes 10% waste.

| | | | | | | |
|---|---|---|---|---|---|---|
| Chalk White | BF@.300 | SY | 16.90 | 8.86 | 25.76 | 43.80 |
| Edinburgh White | BF@.300 | SY | 17.10 | 8.86 | 25.96 | 44.10 |
| Sage Stone | BF@.300 | SY | 18.20 | 8.86 | 27.06 | 46.00 |

**Caspian™ sheet vinyl flooring, Armstrong.** CleanSweep® surface. ToughGuard® durability — guaranteed to not rip, tear or gouge. 15-year limited warranty. 12' wide rolls. 0.07" thick. Urethane no-wax wear layer. 3.8 Performance Appearance Rating. Includes 10% waste.

| | | | | | | |
|---|---|---|---|---|---|---|
| Black Dot | BF@.300 | SY | 14.50 | 8.86 | 23.36 | 39.70 |
| Corner Dot, Hunter | BF@.300 | SY | 14.80 | 8.86 | 23.66 | 40.20 |
| Corner Dot, Navy | BF@.300 | SY | 14.80 | 8.86 | 23.66 | 40.20 |
| Corner Dot, Taupe | BF@.300 | SY | 15.70 | 8.86 | 24.56 | 41.80 |
| Decker Station, Honey Spice | BF@.300 | SY | 14.60 | 8.86 | 23.46 | 39.90 |
| Mt. Bethel Park, Moss | BF@.300 | SY | 14.60 | 8.86 | 23.46 | 39.90 |
| Shavano Park, Spice | BF@.300 | SY | 14.50 | 8.86 | 23.36 | 39.70 |
| Walkway, Green | BF@.300 | SY | 14.50 | 8.86 | 23.36 | 39.70 |
| Walkway, Taupe | BF@.300 | SY | 14.90 | 8.86 | 23.76 | 40.40 |
| Walkway, Warm Beige | BF@.300 | SY | 14.70 | 8.86 | 23.56 | 40.10 |

| | Craft@Hrs | Unit | Material | Labor | Total | Sell |
|---|---|---|---|---|---|---|

**Themes sheet vinyl flooring, Armstrong.** Easy to clean. ToughGuard® durability. Guaranteed to not rip, tear or gouge. 10-year limited warranty. Urethane no-wax wear layer. 12' wide rolls. 0.08" thick. 2.9 Performance Appearance Rating. Includes adhesive and 10% waste.

| | Craft@Hrs | Unit | Material | Labor | Total | Sell |
|---|---|---|---|---|---|---|
| Most colors and designs | BF@.300 | SY | 12.00 | 8.86 | 20.86 | 35.50 |
| Sandstone | BF@.300 | SY | 12.10 | 8.86 | 20.96 | 35.60 |
| Marble Wisp, White Crystal | BF@.300 | SY | 11.15 | 8.86 | 20.01 | 34.00 |
| Prescott, Natural White | BF@.300 | SY | 11.15 | 8.86 | 20.01 | 34.00 |

**Sundial™ sheet vinyl flooring, Armstrong.** Urethane wear layer. Modified loose lay. No adhesive needed. 10-year warranty. 12' wide rolls. 0.098" thick. Includes 10% waste.

| | Craft@Hrs | Unit | Material | Labor | Total | Sell |
|---|---|---|---|---|---|---|
| Most colors and designs | BF@.300 | SY | 12.00 | 8.86 | 20.86 | 35.50 |
| Chestnut Corner, Rust | BF@.300 | SY | 13.20 | 8.86 | 22.06 | 37.50 |
| Myrtlewood, Light Oak | BF@.300 | SY | 13.30 | 8.86 | 22.16 | 37.70 |
| Quinault, Light Limestone | BF@.300 | SY | 13.70 | 8.86 | 22.56 | 38.40 |

**Sentinel sheet vinyl flooring, Armstrong.** 5-year limited warranty. 12' wide rolls. Includes adhesive and 10% waste.

| | Craft@Hrs | Unit | Material | Labor | Total | Sell |
|---|---|---|---|---|---|---|
| Most colors and designs | BF@.300 | SY | 9.44 | 8.86 | 18.30 | 31.10 |
| Darker colors, intricate designs | BF@.300 | SY | 11.90 | 8.86 | 20.76 | 35.30 |

**Metro™ sheet vinyl flooring, Armstrong.** ToughGuard® durability. Guaranteed to not rip, tear or gouge. 5-year limited warranty. 12' wide rolls. 0.045" thick. Vinyl no-wax wear layer. 2.3 Performance Appearance Rating. Includes adhesive and 10% waste.

| | Craft@Hrs | Unit | Material | Labor | Total | Sell |
|---|---|---|---|---|---|---|
| Adobe Tan | BF@.300 | SY | 7.10 | 8.86 | 15.96 | 27.10 |
| Bluelake | BF@.300 | SY | 7.12 | 8.86 | 15.98 | 27.20 |
| Cocoa | BF@.300 | SY | 7.05 | 8.86 | 15.91 | 27.00 |
| Media Heights, Sandstone | BF@.300 | SY | 7.60 | 8.86 | 16.46 | 28.00 |
| Navy | BF@.300 | SY | 9.35 | 8.86 | 18.21 | 31.00 |
| Sand | BF@.300 | SY | 7.07 | 8.86 | 15.93 | 27.10 |

**Cambray™ sheet vinyl flooring, Armstrong.** Vinyl ToughGuard® durability. FHA spec. Guaranteed to not rip, tear or gouge. 5-year limited warranty. 12' wide rolls. 1.7 Performance Appearance Rating. Includes 10% waste.

| | Craft@Hrs | Unit | Material | Labor | Total | Sell |
|---|---|---|---|---|---|---|
| Aegean Mist, White | BF@.300 | SY | 7.12 | 8.86 | 15.98 | 27.20 |
| Country Flair, Rose | BF@.300 | SY | 7.11 | 8.86 | 15.97 | 27.10 |
| Hampton House, White | BF@.300 | SY | 7.13 | 8.86 | 15.99 | 27.20 |
| Middlesex, Toast | BF@.300 | SY | 7.10 | 8.86 | 15.96 | 27.10 |

**Royelle® sheet vinyl flooring, Armstrong.** No-wax surface. Easy to install. 12' wide rolls. One-year limited warranty. 0.05" thick. Vinyl no-wax wear layer. 1.3 Performance Appearance Rating. Includes 10% waste.

| | Craft@Hrs | Unit | Material | Labor | Total | Sell |
|---|---|---|---|---|---|---|
| Augusta, Dark Oak | BF@.300 | SY | 4.80 | 8.86 | 13.66 | 23.20 |
| Kaley's Korner, Seashell | BF@.300 | SY | 4.80 | 8.86 | 13.66 | 23.20 |
| Sheffley, Black and White | BF@.300 | SY | 4.70 | 8.86 | 13.56 | 23.10 |

|  | Craft@Hrs | Unit | Material | Labor | Total | Sell |
|---|---|---|---|---|---|---|

## Vinyl Floor Tile

**Metro Series™ vinyl floor tile, Armstrong.** 12" x 12". Resists stains, scratches, scuffs and indentations. 5-year warranty. 0.045" thick. Vinyl no-wax wear layer. 2.3 Performance Appearance Rating. Per square foot, including adhesive and 10% waste.

| | | | | | | |
|---|---|---|---|---|---|---|
| Avlana, Golden Mosaic | BF@.023 | SF | .77 | .68 | 1.45 | 2.47 |
| Salina, Beige | BF@.023 | SF | .67 | .68 | 1.35 | 2.30 |
| Salina, Gray | BF@.023 | SF | .72 | .68 | 1.40 | 2.38 |
| Toledo, Slate Blue | BF@.023 | SF | .74 | .68 | 1.42 | 2.41 |
| Tetherow Provincial, Cherry | BF@.023 | SF | .72 | .68 | 1.40 | 2.38 |

**Elegant Images™ Series vinyl floor tile, Armstrong.** 12" x 12". ToughGuard® durability. Guaranteed to not rip, tear or gouge. Resists stains, scratches, scuffs and indentations. 25-year warranty. Per square foot, including adhesive and 10% waste.

| | | | | | | |
|---|---|---|---|---|---|---|
| Cornwall, Burnt Almond | BF@.023 | SF | 2.32 | .68 | 3.00 | 5.10 |
| Modern, Rose/Green Inset | BF@.023 | SF | 1.65 | .68 | 2.33 | 3.96 |

**Vernay™ Series vinyl floor tile, Armstrong.** 12" x 12". ToughGuard® durability. Guaranteed not rip, tear or gouge. Resists stains, scratches, scuffs and indentations. 25-year warranty. Self-adhesive. Per square foot, including 10% waste.

| | | | | | | |
|---|---|---|---|---|---|---|
| Allenbury, Shadow Blue | BF@.021 | SF | .58 | .62 | 1.20 | 2.04 |
| Classic Marble, Blue | BF@.021 | SF | .64 | .62 | 1.26 | 2.14 |
| Englewood, Oak | BF@.021 | SF | .62 | .62 | 1.24 | 2.11 |
| Floral Focus, Rose Petal | BF@.021 | SF | .64 | .62 | 1.26 | 2.14 |
| Montelena, Cinnamon | BF@.021 | SF | .63 | .62 | 1.25 | 2.13 |
| Montelena, White | BF@.021 | SF | .62 | .62 | 1.24 | 2.11 |
| Snapshot, Bianco White | BF@.021 | SF | .71 | .62 | 1.33 | 2.26 |

**Chelsea Collection™ vinyl floor tile, Armstrong.** 12" x 12". ToughGuard® durability. Guaranteed to not rip, tear or gouge. Resists stains, scratches, scuffs and indentations. 10-year warranty. Self-adhesive. Per square foot, including 10% waste.

| | | | | | | |
|---|---|---|---|---|---|---|
| Marble Beauty, Verde Agate | BF@.021 | SF | 2.12 | .62 | 2.74 | 4.66 |

**Themes Collection™ vinyl floor tile, Armstrong.** 12" x 12". ToughGuard® durability. Guaranteed not to rip, tear or gouge. Resists stains, scratches, scuffs and indentations. 10-year warranty. Per square foot, including adhesive and 10% waste.

| | | | | | | |
|---|---|---|---|---|---|---|
| Center Piece, Black | BF@.023 | SF | 1.19 | .68 | 1.87 | 3.18 |
| Charmayne, Sandstone | BF@.023 | SF | 1.19 | .68 | 1.87 | 3.18 |
| Classic Mosaic, Deep Blue/Natural | BF@.023 | SF | 1.19 | .68 | 1.87 | 3.18 |
| Classic Mosaic, Burgundy/Natural | BF@.023 | SF | 1.19 | .68 | 1.87 | 3.18 |
| Classic Mosaic, Hunter Green/Almond | BF@.023 | SF | 1.20 | .68 | 1.88 | 3.20 |
| Colebrook, Fieldstone | BF@.023 | SF | 1.19 | .68 | 1.87 | 3.18 |
| Midway, Beige | BF@.023 | SF | 1.19 | .68 | 1.87 | 3.18 |
| Midway, Bisque | BF@.023 | SF | 1.19 | .68 | 1.87 | 3.18 |
| Naples, Black | BF@.023 | SF | 1.19 | .68 | 1.87 | 3.18 |
| Naples, Green | BF@.023 | SF | 1.19 | .68 | 1.87 | 3.18 |
| Naples, Natural | BF@.023 | SF | 1.17 | .68 | 1.85 | 3.15 |
| Senegal, Light Beige | BF@.023 | SF | 1.16 | .68 | 1.84 | 3.13 |
| Stonehaven, Orange Bisque | BF@.023 | SF | 1.18 | .68 | 1.86 | 3.16 |
| Tripoli, Blue/Natural | BF@.023 | SF | 1.19 | .68 | 1.87 | 3.18 |

|  | Craft@Hrs | Unit | Material | Labor | Total | Sell |
|---|---|---|---|---|---|---|

**Safety Zone vinyl floor tile, Armstrong.** Slip-retardant vinyl composition tile for use where slips and falls are a concern. Meets or exceeds ADA slip-retardant performance ranges. Low profile for easy maintenance. Styled for all commercial interiors. Per square foot, including adhesive and 10% waste.

|  | Craft@Hrs | Unit | Material | Labor | Total | Sell |
|---|---|---|---|---|---|---|
| Earth Stone | BF@.023 | SF | 2.48 | .68 | 3.16 | 5.37 |
| Slate Black | BF@.023 | SF | 2.48 | .68 | 3.16 | 5.37 |
| Stone Beige | BF@.023 | SF | 2.48 | .68 | 3.16 | 5.37 |
| Weathered Alabaster | BF@.023 | SF | 2.48 | .68 | 3.16 | 5.37 |

**Stylistik® II vinyl floor tile, Armstrong.** 12" x 12". Resists stains, scratches, scuffs and indentations. 5-year warranty. 0.065" thick. Vinyl no-wax wear layer. Self-adhesive. Per square foot, including 10% waste.

|  | Craft@Hrs | Unit | Material | Labor | Total | Sell |
|---|---|---|---|---|---|---|
| Bayville, Bisque | BF@.021 | SF | 1.18 | .62 | 1.80 | 3.06 |
| Criswood, Russet Oak | BF@.021 | SF | 1.09 | .62 | 1.71 | 2.91 |
| Criswood, Vintage Oak | BF@.021 | SF | 1.06 | .62 | 1.68 | 2.86 |
| Firenza, Sapphire | BF@.021 | SF | 1.05 | .62 | 1.67 | 2.84 |
| Gladstone, Blue/Natural | BF@.021 | SF | 1.18 | .62 | 1.80 | 3.06 |
| Oakland, Dusty Brick | BF@.021 | SF | 1.06 | .62 | 1.68 | 2.86 |
| Pequea Park, Canyon Multi | BF@.021 | SF | 1.03 | .62 | 1.65 | 2.81 |
| Terramora, Black | BF@.021 | SF | 1.06 | .62 | 1.68 | 2.86 |
| Terramora, White | BF@.021 | SF | 1.03 | .62 | 1.65 | 2.81 |

**Chesapeake Collection™ vinyl floor tile, Armstrong.** 12" x 12". 25-year warranty. 0.08" thick. Urethane no-wax wear layer. Per square foot, including adhesive and 10% waste.

|  | Craft@Hrs | Unit | Material | Labor | Total | Sell |
|---|---|---|---|---|---|---|
| Leesport, Sienna/Gray | BF@.023 | SF | 1.77 | .68 | 2.45 | 4.17 |
| Millport, Wheat | BF@.023 | SF | 1.70 | .68 | 2.38 | 4.05 |
| Steinway, White Essence | BF@.023 | SF | 1.54 | .68 | 2.22 | 3.77 |

**Natural Images™ vinyl floor tile, Armstrong.** 12" x 12". ToughGuard® durability. 25-year warranty. Per square foot including adhesive and 10% waste.

|  | Craft@Hrs | Unit | Material | Labor | Total | Sell |
|---|---|---|---|---|---|---|
| Stonegate, Antique Ivory | BF@.023 | SF | 2.34 | .68 | 3.02 | 5.13 |
| Stonegate, Blue Slate | BF@.023 | SF | 2.34 | .68 | 3.02 | 5.13 |

**Harbour Collection™ vinyl floor tile, Armstrong.** 12" x 12". ToughGuard® durability. Guaranteed not to rip, tear or gouge. Resists stains, scratches, scuffs and indentations. 25-year warranty. Per square foot including adhesive and 10% waste.

|  | Craft@Hrs | Unit | Material | Labor | Total | Sell |
|---|---|---|---|---|---|---|
| Andora, Spring Green | BF@.023 | SF | 1.60 | .68 | 2.28 | 3.88 |
| Arrington, Dusty Green | BF@.023 | SF | 1.60 | .68 | 2.28 | 3.88 |
| Arrington, Trail Beige | BF@.023 | SF | 1.60 | .68 | 2.28 | 3.88 |
| Matheson Park, Emerald | BF@.023 | SF | 1.60 | .68 | 2.28 | 3.88 |
| Matheson Park, Sapphire | BF@.023 | SF | 1.60 | .68 | 2.28 | 3.88 |

**Place n Press floor tile, Armstrong.** 12" x 12". No-wax easy to clean finish. Household gauge. Self-adhesive. Per square foot, including 10% waste.

|  | Craft@Hrs | Unit | Material | Labor | Total | Sell |
|---|---|---|---|---|---|---|
| Lockeport, Slate Blue | BF@.021 | SF | .35 | .62 | .97 | 1.65 |
| Parkson, Light Oak | BF@.021 | SF | .35 | .62 | .97 | 1.65 |

**Excelon® Imperial Texture vinyl composition tile, Armstrong.** 1/8"-thick tile. 12" x 12". Color pattern goes through the tile. Commercial traffic rated. Per square foot, including 10% waste and adhesive.

|  | Craft@Hrs | Unit | Material | Labor | Total | Sell |
|---|---|---|---|---|---|---|
| Blue Cloud | BF@.023 | SF | .73 | .68 | 1.41 | 2.40 |
| Blue/Gray | BF@.023 | SF | .73 | .68 | 1.41 | 2.40 |
| Caribbean Blue | BF@.023 | SF | .69 | .68 | 1.37 | 2.33 |
| Cherry Red | BF@.023 | SF | .69 | .68 | 1.37 | 2.33 |
| Classic Black | BF@.023 | SF | .73 | .68 | 1.41 | 2.40 |
| Classic White | BF@.023 | SF | .73 | .68 | 1.41 | 2.40 |
| Cool White | BF@.023 | SF | .73 | .68 | 1.41 | 2.40 |
| Cottage Tan | BF@.023 | SF | .72 | .68 | 1.40 | 2.38 |

| | Craft@Hrs | Unit | Material | Labor | Total | Sell |
|---|---|---|---|---|---|---|
| Fortress White | BF@.023 | SF | .72 | .68 | 1.40 | 2.38 |
| Hazelnut | BF@.023 | SF | .71 | .68 | 1.39 | 2.36 |
| Marina Blue | BF@.023 | SF | .71 | .68 | 1.39 | 2.36 |
| Pearl White | BF@.023 | SF | .70 | .68 | 1.38 | 2.35 |
| Sandrift White | BF@.023 | SF | .73 | .68 | 1.41 | 2.40 |
| Sea Green | BF@.023 | SF | .73 | .68 | 1.41 | 2.40 |
| Shelter White | BF@.023 | SF | .73 | .68 | 1.41 | 2.40 |
| Sterling | BF@.023 | SF | .73 | .68 | 1.41 | 2.40 |
| Teal | BF@.023 | SF | .73 | .68 | 1.41 | 2.40 |

**Excelon® Civic Square vinyl composition tile, Armstrong.** Color pattern goes through the tile. 1/8"-thick tile. 12" x 12". Per square foot, including adhesive and 10% waste.

| | Craft@Hrs | Unit | Material | Labor | Total | Sell |
|---|---|---|---|---|---|---|
| Oyster White | BF@.023 | SF | .57 | .68 | 1.25 | 2.13 |
| Stone Tan | BF@.023 | SF | .57 | .68 | 1.25 | 2.13 |

**Rubber studded tile.** 12" x 12". Highly resilient to reduce fatigue. 10-year wear warranty. Slip-resistant. Sound absorbent. High-abrasion resistance. Recommended in kitchens and active use rooms.

| | Craft@Hrs | Unit | Material | Labor | Total | Sell |
|---|---|---|---|---|---|---|
| Black | BF@.025 | SF | 6.83 | .74 | 7.57 | 12.90 |
| Indigo | BF@.025 | SF | 6.83 | .74 | 7.57 | 12.90 |
| Taupe | BF@.025 | SF | 6.83 | .74 | 7.57 | 12.90 |

## Vinyl Flooring Adhesive

**Sheet flooring adhesive, Armstrong.** For residential felt-backed floors. Water-based rubber-resin. Can be used on all grade levels of concrete, existing resilient floors, ceramic, terrazzo, marble, and wood floors. Solvent-free and low odor. Non-staining. Long working time. Gallon covers 350 to 400 square feet.

| | Craft@Hrs | Unit | Material | Labor | Total | Sell |
|---|---|---|---|---|---|---|
| Quart | — | Ea | 5.41 | — | 5.41 | — |
| Gallon | — | Ea | 13.50 | — | 13.50 | — |

**Henry 430 clear thin-spread vinyl tile adhesive.** For installing vinyl composition tile (VCT). 24-hour working time. Dries clear so chalk lines show through. Moisture and alkali resistance allows installation above or below grade. Bonds to concrete, existing asphalt "cutback" adhesive residue, underlayments, wood substrates, terrazzo, clean and abraded steel, stainless steel, aluminum, lead, copper, brass and bronze, and existing resilient flooring. Gallon covers 350 to 400 square feet.

| | Craft@Hrs | Unit | Material | Labor | Total | Sell |
|---|---|---|---|---|---|---|
| Quart | — | Ea | 4.27 | — | 4.27 | — |
| Gallon | — | Ea | 12.50 | — | 12.50 | — |
| 4 gallons | — | Ea | 33.10 | — | 33.10 | — |

**Resilient tile adhesive, Armstrong.** For Excelon®, dry-back vinyl, and residential dry-back tile. Used on all grade levels of concrete, ceramic, terrazzo, marble and polymeric poured floors. Water-based. Up to 6-hour working time. Low odor and solvent-free. Gallon covers 350 to 400 square feet.

| | Craft@Hrs | Unit | Material | Labor | Total | Sell |
|---|---|---|---|---|---|---|
| Quart | — | Ea | 6.73 | — | 6.73 | — |
| Gallon | — | Ea | 18.85 | — | 18.85 | — |
| 4 gallons | — | Ea | 53.40 | — | 53.40 | — |

**Rubber tile adhesive.** Gallon covers 350 to 400 square feet.

| | Craft@Hrs | Unit | Material | Labor | Total | Sell |
|---|---|---|---|---|---|---|
| Gallon | — | Ea | 21.30 | — | 21.30 | — |

**Henry 440 cove base adhesive.** Wet set adhesive for rubber and vinyl cove base. Aggressive initial grab prevents slip during installation. Cleans up with water.

| | Craft@Hrs | Unit | Material | Labor | Total | Sell |
|---|---|---|---|---|---|---|
| 11 ounces | — | Ea | 2.35 | — | 2.35 | — |
| 1 quart | — | Ea | 4.68 | — | 4.68 | — |
| 30 ounces | — | Ea | 4.17 | — | 4.17 | — |
| 1 gallon | — | Ea | 13.50 | — | 13.50 | — |

| | Craft@Hrs | Unit | Material | Labor | Total | Sell |
|---|---|---|---|---|---|---|

**Acrylic tape, Cal-Flor.** Double-face. For Armstrong Sundial sheet vinyl flooring. Allows flooring to be repositioned during installation.

| | | | | | | |
|---|---|---|---|---|---|---|
| 2" x 50' roll | — | Ea | 10.35 | — | 10.35 | — |

**Low-gloss seam coating kit, S-564, Armstrong.** Protects seams from dirt and wear. Includes seam cleaner, coating, deglosser, and applicator. Covers approximately 100 linear feet per kit.

| | | | | | | |
|---|---|---|---|---|---|---|
| Seam coat kit | — | Ea | 11.80 | — | 11.80 | — |

**High-gloss seam coating kit, S-595, Armstrong.** For use on high-gloss residential sheet vinyl floors. Includes coating, accelerator, and cleaner. Coverage is 250 linear feet.

| | | | | | | |
|---|---|---|---|---|---|---|
| 3 ounces | — | Ea | 12.50 | — | 12.50 | — |

## Vinyl Flooring Tools

**Extendable roller.** Extension tube with firm grip handle. Extends from 17" to 27". For rolling vinyl floor covering, carpet, wall covering and cove base.

| | | | | | | |
|---|---|---|---|---|---|---|
| 7-1/2" wide | — | Ea | 37.80 | — | 37.80 | — |

**Margin notch trowel.** 2" x 6" flexible steel trowel for patch jobs and hard-to-reach areas.

| | | | | | | |
|---|---|---|---|---|---|---|
| 1/4" x 1/4" x 1/4" square notch | — | Ea | 3.88 | — | 3.88 | — |
| 5/16" x 1/4" x 1/16" V-notch | — | Ea | 3.88 | — | 3.88 | — |

**Notched trowels.** Large, foam comfort grip. Flexible spring steel blade.

| | | | | | | |
|---|---|---|---|---|---|---|
| 1/8" x 1/8" V-notch | — | Ea | 8.04 | — | 8.04 | — |
| 3/16" x 1/4" U-notch | — | Ea | 13.20 | — | 13.20 | — |
| 1/2" x 1/4" U-notch | — | Ea | 13.20 | — | 13.20 | — |
| 3/8" x 1/4" U-notch | — | Ea | 13.20 | — | 13.20 | — |
| 2-3/4" x 8" square notch | — | Ea | 2.60 | — | 2.60 | — |

**Underlay finish trowel.** Constructed of welded tool steel. Comfortable hardwood handle.

| | | | | | | |
|---|---|---|---|---|---|---|
| 2" x 5" margin trowel | — | Ea | 4.42 | — | 4.42 | — |
| 6" pointing trowel | — | Ea | 4.38 | — | 4.38 | — |

**Chalk line.** 100" self-chalking line with aluminum casing and plumb bob.

| | | | | | | |
|---|---|---|---|---|---|---|
| 100" chalk line | — | Ea | 8.04 | — | 8.04 | — |
| Chalk refill, 8 oz. plastic bottle | — | Ea | 1.53 | — | 1.53 | — |

**White rubber mallet.** Will not mark coping or tile. Rubber head securely mounted on hickory handle.

| | | | | | | |
|---|---|---|---|---|---|---|
| Mallet | — | Ea | 18.20 | — | 18.20 | — |

**Vinyl tile cutter.** Cuts vinyl tiles to 12" x 12". Hardened steel blade with die-cast aluminum base. Cushioned rubber handle. Ball bearings at stress points with "guillotine" action. Built-in measuring gauge. Works equally well with peel-and-stick tiles.

| | | | | | | |
|---|---|---|---|---|---|---|
| Tile cutter | — | Ea | 54.10 | — | 54.10 | — |

**Cove base adhesive nozzle.** Fits all cove base caulking guns. Spreads adhesive evenly for cove base.

| | | | | | | |
|---|---|---|---|---|---|---|
| 1-3/4" long, 3" wide, 4-1/2" high | — | Ea | 3.22 | — | 3.22 | — |

**Scribing felt, Armstrong.** For transferring the pattern of the floor to the flooring material.

| | | | | | | |
|---|---|---|---|---|---|---|
| Per square yard | — | Ea | 1.58 | — | 1.58 | — |
| 50 square yard roll | — | Ea | 54.00 | — | 54.00 | — |

|  | Craft@Hrs | Unit | Material | Labor | Total | Sell |
|---|---|---|---|---|---|---|

## Flooring Transition Strips

**Vinyl flooring transition reducer strip, Roppe.** Beveled edge for resilient flooring material. Butting edge is 1/8".

| | | | | | | |
|---|---|---|---|---|---|---|
| 3' long, black | BF@.250 | Ea | 1.61 | 7.39 | 9.00 | 15.30 |
| 3' long, brown | BF@.250 | Ea | 1.61 | 7.39 | 9.00 | 15.30 |
| 3' long, gray | BF@.250 | Ea | 1.61 | 7.39 | 9.00 | 15.30 |

**Carpet-to-tile transition strip, Roppe.** Joins 1/4" carpet to 1/8" tile.

| | | | | | | |
|---|---|---|---|---|---|---|
| 4' long, black | BF@.250 | Ea | 5.17 | 7.39 | 12.56 | 21.40 |
| 4' long, brown | BF@.250 | Ea | 5.17 | 7.39 | 12.56 | 21.40 |
| 4' long, gray | BF@.250 | Ea | 5.17 | 7.39 | 12.56 | 21.40 |

**Oak-to-ceramic tile transition strip.**

| | | | | | | |
|---|---|---|---|---|---|---|
| 3', high pile | BF@.250 | Ea | 13.30 | 7.39 | 20.69 | 35.20 |
| 3', laminate | BF@.250 | Ea | 13.70 | 7.39 | 21.09 | 35.90 |
| 3', vinyl | BF@.250 | Ea | 12.90 | 7.39 | 20.29 | 34.50 |
| 6', high pile | BF@.500 | Ea | 23.70 | 14.80 | 38.50 | 65.50 |
| 6', laminate | BF@.500 | Ea | 23.70 | 14.80 | 38.50 | 65.50 |
| 6', vinyl | BF@.500 | Ea | 23.70 | 14.80 | 38.50 | 65.50 |

## Cove Base for Vinyl Flooring

**Dryback vinyl wall base.** Resists scratches, scuffs and whitening. Flexible. 0.080 gauge. Per linear foot.

| | | | | | | |
|---|---|---|---|---|---|---|
| 4", blue gray | BF@.025 | LF | .56 | .74 | 1.30 | 2.21 |
| 4", burnt umber | BF@.025 | LF | .57 | .74 | 1.31 | 2.23 |
| 4", mauve | BF@.025 | LF | .58 | .74 | 1.32 | 2.24 |
| 4", mid gray | BF@.025 | LF | .58 | .74 | 1.32 | 2.24 |
| 4", walnut | BF@.025 | LF | .65 | .74 | 1.39 | 2.36 |
| 4", white | BF@.025 | LF | .50 | .74 | 1.24 | 2.11 |

**Thermoplastic dryback wall base, Roppe.** Type TP thermoplastic rubber. More flexible than vinyl due to its rubber content. Thickness hides many wall irregularities. Good stability. Rib back for positive adhesion. Per foot.

| | | | | | | |
|---|---|---|---|---|---|---|
| 2-1/2" x 120' x 1/8", almond | BF@.025 | LF | .52 | .74 | 1.26 | 2.14 |
| 2-1/2" x 120' x 1/8", black | BF@.025 | LF | .49 | .74 | 1.23 | 2.09 |
| 2-1/2" x 120' x 1/8", brown | BF@.025 | LF | .49 | .74 | 1.23 | 2.09 |
| 2-1/2" x 120' x 1/8", dark gray | BF@.025 | LF | .52 | .74 | 1.26 | 2.14 |
| 2-1/2" x 120' x 1/8", fawn | BF@.025 | LF | .52 | .74 | 1.26 | 2.14 |
| 2-1/2" x 120' x 1/8", hunter green | BF@.025 | LF | .63 | .74 | 1.37 | 2.33 |
| 2-1/2" x 120' x 1/8", white | BF@.025 | LF | .52 | .74 | 1.26 | 2.14 |
| 4" x 120' x 1/8", almond | BF@.025 | LF | .59 | .74 | 1.33 | 2.26 |
| 4" x 120' x 1/8", black | BF@.025 | LF | .59 | .74 | 1.33 | 2.26 |
| 4" x 120' x 1/8", brown | BF@.025 | LF | .59 | .74 | 1.33 | 2.26 |
| 4" x 120' x 1/8", fawn | BF@.025 | LF | .63 | .74 | 1.37 | 2.33 |
| 4" x 120' x 1/8", hunter green | BF@.025 | LF | .52 | .74 | 1.26 | 2.14 |
| 4" x 120' x 1/8", white | BF@.025 | LF | .59 | .74 | 1.33 | 2.26 |

**Rubber dryback wall base.** 100% synthetic rubber. Extremely flexible. Won't shrink, grab or separate. Resists scuffing, gouging and most chemicals. Low gloss stain finish. Flexible with shade control. 1/8" thick. Per foot.

| | | | | | | |
|---|---|---|---|---|---|---|
| 2-1/2", almond | BF@.025 | LF | .73 | .74 | 1.47 | 2.50 |
| 2-1/2", black | BF@.025 | LF | .71 | .74 | 1.45 | 2.47 |
| 2-1/2", brown | BF@.025 | LF | .72 | .74 | 1.46 | 2.48 |
| 2-1/2", dark gray | BF@.025 | LF | .72 | .74 | 1.46 | 2.48 |
| 2-1/2", fawn | BF@.025 | LF | .74 | .74 | 1.48 | 2.52 |
| 2-1/2", white | BF@.025 | LF | .72 | .74 | 1.46 | 2.48 |
| 4", almond | BF@.025 | LF | .85 | .74 | 1.59 | 2.70 |

| | Craft@Hrs | Unit | Material | Labor | Total | Sell |
|---|---|---|---|---|---|---|
| 4", black | BF@.025 | LF | .85 | .74 | 1.59 | 2.70 |
| 4", brown | BF@.025 | LF | .84 | .74 | 1.58 | 2.69 |
| 4", dark gray | BF@.025 | LF | .84 | .74 | 1.58 | 2.69 |
| 4", fawn | BF@.025 | LF | .90 | .74 | 1.64 | 2.79 |
| 4", white | BF@.025 | LF | .84 | .74 | 1.58 | 2.69 |

**Self-stick wall base, Roppe.** Full-cover adhesive for sure-stick installation. Top lip design for tight fit. Special coating resists scuffs and scratches. 1/8" thick. Per foot.

| | Craft@Hrs | Unit | Material | Labor | Total | Sell |
|---|---|---|---|---|---|---|
| 2-1/2", almond | BF@.016 | LF | .80 | .47 | 1.27 | 2.16 |
| 2-1/2", black | BF@.016 | LF | .81 | .47 | 1.28 | 2.18 |
| 2-1/2", brown | BF@.016 | LF | .81 | .47 | 1.28 | 2.18 |
| 2-1/2", dark gray | BF@.016 | LF | .81 | .47 | 1.28 | 2.18 |
| 2-1/2", snow | BF@.016 | LF | .81 | .47 | 1.28 | 2.18 |
| 4", almond | BF@.016 | LF | 1.02 | .47 | 1.49 | 2.53 |
| 4", black | BF@.016 | LF | 1.02 | .47 | 1.49 | 2.53 |
| 4", brown | BF@.016 | LF | 1.02 | .47 | 1.49 | 2.53 |
| 4", dark gray | BF@.016 | LF | 1.02 | .47 | 1.49 | 2.53 |
| 4", snow | BF@.016 | LF | 1.02 | .47 | 1.49 | 2.53 |

**Residential self-stick vinyl wall base, Armstrong.** Per foot.

| | Craft@Hrs | Unit | Material | Labor | Total | Sell |
|---|---|---|---|---|---|---|
| 4", almond high gloss | BF@.016 | LF | 1.08 | .47 | 1.55 | 2.64 |
| 4", architectural white | BF@.016 | LF | 1.09 | .47 | 1.56 | 2.65 |
| 4", black high gloss | BF@.016 | LF | 1.08 | .47 | 1.55 | 2.64 |
| 4", gray | BF@.016 | LF | 1.10 | .47 | 1.57 | 2.67 |
| 4", walnut high gloss | BF@.016 | LF | 1.07 | .47 | 1.54 | 2.62 |
| 4", white high gloss | BF@.016 | LF | 1.08 | .47 | 1.55 | 2.64 |

## Cork Wall and Floor Cover

**Natural cork tile, Boone International.** Case of 18 tiles covers 16.2 square feet with 10% waste.

| | Craft@Hrs | Unit | Material | Labor | Total | Sell |
|---|---|---|---|---|---|---|
| 12" x 12" x 3/8" | BF@.030 | SF | 1.50 | .89 | 2.39 | 4.06 |

**Cork roll.**

| | | Unit | Material | Labor | Total | Sell |
|---|---|---|---|---|---|---|
| 4' wide, 1/4" thick, per linear foot | — | Ea | 3.15 | — | 3.15 | — |
| 4' x 50' x 1/4" roll | — | Ea | 155.00 | — | 155.00 | — |

## Ceramic Tile Installation

**Ceramic tile adhesive.** Premixed Type 1, Acryl-4000. Gallon covers 70 square feet applied with #1 trowel (3/16" x 5/32" V-notch), 50 square feet applied with #2 trowel (3-/16" x 1/4" V-notch) or 40 square feet applied with #3 trowel (1/4" x 1/4" square-notch)

| | | Unit | Material | Labor | Total | Sell |
|---|---|---|---|---|---|---|
| 1/2 pint | — | Ea | 2.93 | — | 2.93 | — |
| 1 gallon | — | Ea | 11.40 | — | 11.40 | — |
| 3-1/2 gallons | — | Ea | 31.50 | — | 31.50 | — |
| Tile adhesive applied with #2 trowel | — | SF | .23 | — | .23 | — |

**Thin-set tile mortar**. 50 pound bag covers 100 square feet applied with #1 trowel (1/4" x 1/4" square-notch), 80 square feet applied with #2 trowel (1/4" x 3/8" square-notch), 45 square feet applied with #3 trowel (1/2" x 1/2" square-notch). Cost per 50 pound bag.

| | | Unit | Material | Labor | Total | Sell |
|---|---|---|---|---|---|---|
| FlexBond®, gray | — | Ea | 27.60 | — | 27.60 | — |
| Marble and granite mix | — | Ea | 22.60 | — | 22.60 | — |
| MasterBlend™, white | — | Ea | 10.60 | — | 10.60 | — |
| Standard, gray | — | Ea | 8.96 | — | 8.96 | — |
| MasterBlend™ applied with #3 trowel | — | SF | .22 | — | .22 | — |

| | Craft@Hrs | Unit | Material | Labor | Total | Sell |
|---|---|---|---|---|---|---|

**Tile grout.** Polymer-modified dry grout for joints 1/16" to 1/2". Coverage varies with tile and joint size. Cost per 25 pound bag.

| | Craft@Hrs | Unit | Material | Labor | Total | Sell |
|---|---|---|---|---|---|---|
| Sanded, dark colors | — | Ea | 12.00 | — | 12.00 | — |
| Sanded, light colors | — | Ea | 16.40 | — | 16.40 | — |
| Unsanded | — | Ea | 16.40 | — | 16.40 | — |

**Setting counter, wall and floor tile in adhesive.** Costs are for adhesive and grout. Add the cost of surface preparation (backerboard) and the cost of tile.

| | Craft@Hrs | Unit | Material | Labor | Total | Sell |
|---|---|---|---|---|---|---|
| Countertops | | | | | | |
| Ceramic mosaic field tile | TL@.203 | SF | .27 | 6.25 | 6.52 | 11.10 |
| 4-1/4" x 4-1/4" to 6" x 6" glazed field tile | TL@.180 | SF | .27 | 5.54 | 5.81 | 9.88 |
| Countertop trim pieces and edge tile | TL@.180 | LF | .05 | 5.54 | 5.59 | 9.50 |
| Floors | | | | | | |
| Ceramic mosaic field tile | TL@.121 | SF | .27 | 3.73 | 4.00 | 6.80 |
| 4-1/4" x 4-1/4" to 6" x 6" glazed field tile | TL@.110 | SF | .27 | 3.39 | 3.66 | 6.22 |
| 8" x 8" and larger field tile | TL@.100 | SF | .27 | 3.08 | 3.35 | 5.70 |
| Floor trim pieces and edge tile | TL@.200 | LF | .05 | 6.16 | 6.21 | 10.60 |
| Walls | | | | | | |
| Ceramic mosaic field tile | TL@.143 | SF | .27 | 4.40 | 4.67 | 7.94 |
| 4-1/4" x 4-1/4" to 6" x 6" glazed field tile | TL@.131 | SF | .27 | 4.03 | 4.30 | 7.31 |
| 8" x 8" and larger field tile | TL@.120 | SF | .27 | 3.69 | 3.96 | 6.73 |
| Wall trim pieces and edge tile | TL@.200 | LF | .05 | 6.16 | 6.21 | 10.60 |

**Setting counter, wall and floor tile in thin-set mortar.** Costs are for mortar and grout. Add the cost of surface preparation (backerboard) and tile.

| | Craft@Hrs | Unit | Material | Labor | Total | Sell |
|---|---|---|---|---|---|---|
| Countertops | | | | | | |
| Ceramic mosaic field tile | TL@.300 | SF | .32 | 9.24 | 9.56 | 16.30 |
| 4-1/4" x 4-1/4" to 6" x 6" glazed field tile | TL@.250 | SF | .32 | 7.70 | 8.02 | 13.60 |
| Countertop trim pieces and edge tile | TL@.400 | LF | .05 | 12.30 | 12.35 | 21.00 |
| Floors | | | | | | |
| Ceramic mosaic field tile | TL@.238 | SF | .32 | 7.33 | 7.65 | 13.00 |
| 4-1/4" x 4-1/4" to 6" x 6" field tile | TL@.210 | SF | .32 | 6.47 | 6.79 | 11.50 |
| 8" x 8" and larger field tile | TL@.200 | SF | .32 | 6.16 | 6.48 | 11.00 |
| Quarry or paver tile | TL@.167 | SF | .32 | 5.14 | 5.46 | 9.28 |
| Marble, granite or stone tile | TL@.354 | SF | .32 | 10.90 | 11.22 | 19.10 |
| Floor trim pieces and edge tile | TL@.210 | LF | .05 | 6.47 | 6.52 | 11.10 |
| Walls | | | | | | |
| Ceramic mosaic field tile | TL@.315 | SF | .32 | 9.70 | 10.02 | 17.00 |
| 4-1/4" x 4-1/4" to 6" x 6" glazed field tile | TL@.270 | SF | .32 | 8.31 | 8.63 | 14.70 |
| 8" x 8" and larger field tile | TL@.260 | SF | .32 | 8.01 | 8.33 | 14.20 |
| Wall trim pieces and edge tile | TL@.270 | LF | .05 | 8.31 | 8.36 | 14.20 |

| | Craft@Hrs | Unit | Material | Labor | Total | Sell |
|---|---|---|---|---|---|---|

**Tile backerboard.** (Durock™ or Wonderboard™). Water-resistant underlayment for ceramic tile on floors, walls, countertops and other interior wet areas. Material cost for 100 square feet (CSF) of board includes waterproofing membrane, the backerboard (and 10% waste), 50 pounds of job mixed latex-fortified mortar for the joints and surface skim coat, 75 linear feet of fiberglass joint tape, and 200 1-1/4" backerboard screws. For scheduling purposes, estimate that a crew of 2 can install, tape and apply the skim coat on the following quantity of backerboard in an 8-hour day: countertops 180 SF, floors 525 SF and walls 350 SF. Use $250.00 as a minimum charge for this type work

| | Craft@Hrs | Unit | Material | Labor | Total | Sell |
|---|---|---|---|---|---|---|
| Using 15 lb. felt waterproofing, per 400 SF, roll | — | Ea | 16.10 | — | 16.10 | — |
| Using 1/4" or 1/2" backerboard, per 100 SF | — | SF | 77.10 | — | 77.10 | — |
| Using 50 lbs. of latex-fortified mortar each 100 SF, per 50 lb. sack | — | Ea | 10.80 | — | 10.80 | — |
| Using 75 LF of backerboard tape for each 100 SF, per 50 LF roll | — | Ea | 3.20 | — | 3.20 | — |
| Using 200 backerboard screws for each 100 SF, per pack of 200 screws | — | Ea | 6.88 | — | 6.88 | — |
| Backerboard with waterproofing, mortar, tape and screws. | | | | | | |
| Countertops | T1@.090 | SF | 1.06 | 2.58 | 3.64 | 6.19 |
| Floors | T1@.030 | SF | 1.06 | .86 | 1.92 | 3.26 |
| Walls | T1@.045 | SF | 1.06 | 1.29 | 2.35 | 4.00 |

**Wonderboard® backerboard.** Underlayment for tile floors in kitchens and baths and for countertops. 1/4" thickness reduces subfloor modifications to adjacent floors, thresholds, carpets and cabinets. 1/2" thickness matches up with surrounding 1/2" (12 mm) drywall without the need to use shims or spacers. Easy to score, snap, cut and nail. For installation cost, see Tile backerboard. Includes 10% waste.

| | | | | | | |
|---|---|---|---|---|---|---|
| 1/4" x 3' x 5' | — | SF | .77 | — | .77 | — |
| 1/2" x 3' x 5' | — | SF | .77 | — | .77 | — |

**RhinoBoard™.** Lightweight fiber-cement backerboard for interior floors and countertops. Used under ceramic tile, natural stone, wood, vinyl and other resilient flooring. Easy to score, snap, cut and nail. For installation costs, see Tile backerboard. Includes 10% waste.

| | | | | | | |
|---|---|---|---|---|---|---|
| 1/4" x 3' x 5' | — | SF | .76 | — | .76 | — |

**Hardibacker® backerboard, James Hardie.** Water-resistant, cement-based underlayment for floors, walls and countertops. Bonds with mastic or latex modified thin-set tile adhesive. Non-abrasive to enamel or porcelain surfaces. Easy to score and snap. For installation costs, see Tile backerboard. Includes 10% waste.

| | | | | | | |
|---|---|---|---|---|---|---|
| 1/4" x 3' x 5' | — | SF | .74 | — | .74 | — |
| 1/2" x 3' x 5' | — | SF | .79 | — | .79 | — |
| 1/2" x 4' x 8' | — | SF | .77 | — | .77 | — |

**Aqua-Tough™ tile underlayment, USG Interiors.** Underlayment for ceramic tile, resilient flooring, laminated or hardwood flooring in both wet and dry areas. For installation costs, see Tile backerboard. Includes 10% waste.

| | | | | | | |
|---|---|---|---|---|---|---|
| 1/4" x 4' x 4' | — | SF | .47 | — | .47 | — |

## 8" x 8" Ceramic Floor Tile

**8" ceramic floor tile.** Glazed ceramic, residential indoor tile. Class 3, light to moderate traffic. Case of 25 tiles covers 10 square feet, including 10% waste. See Ceramic Tile Installation for tile setting costs, including tile backer, thin-set mortar or adhesive and grout.

| | | | | | | |
|---|---|---|---|---|---|---|
| Ivory | — | SF | 1.29 | — | 1.29 | — |
| White | — | SF | 1.29 | — | 1.29 | — |

| | Craft@Hrs | Unit | Material | Labor | Total | Sell |
|---|---|---|---|---|---|---|

**8" Ragione ceramic floor tile.** Satin finish for ease of maintenance. Subtle shading provides a warm, natural, earthy feel. Stained finish makes tile durable, practical and easy to maintain. Case of 25 tiles covers 10 square feet, including 10% waste. See Ceramic Tile Installation for tile setting costs, including tile backer, thin-set mortar or adhesive and grout.

| | | | | | | |
|---|---|---|---|---|---|---|
| Almond | — | SF | 1.33 | — | 1.33 | — |
| White | — | SF | 1.19 | — | 1.19 | — |

**8" ceramic floor tile.** Case of 36 tiles covers 15 square feet, including 10% waste. See Ceramic Tile Installation for tile setting costs, including tile backer, thin-set mortar or adhesive and grout.

| | | | | | | |
|---|---|---|---|---|---|---|
| Cape Gray | — | SF | 1.26 | — | 1.26 | — |

**8" Natura® ceramic floor tile.** A natural stone-look ceramic tile with warm earth tones. Suitable for all interior residential and light commercial applications. Case of 25 tiles covers 10 square feet, including 10% waste. See Ceramic Tile Installation for tile setting costs, including tile backer, thin-set mortar or adhesive and grout.

| | | | | | | |
|---|---|---|---|---|---|---|
| Stone | — | SF | 1.86 | — | 1.86 | — |

**8" x 8" ceramic floor tile.** Class 3, light to moderate traffic for bathrooms, kitchens, foyers, dining rooms, and recreation areas. Case of 30 tiles covers 12.92 square feet, including 10% waste. See Ceramic Tile Installation for tile setting costs, including tile backer, thin-set mortar or adhesive and grout.

| | | | | | | |
|---|---|---|---|---|---|---|
| Pietra, Alabaster | — | SF | .78 | — | .78 | — |
| Pietra, Gem | — | SF | .79 | — | .79 | — |

**8" glazed ceramic floor tile.** Extra durable matte glaze. Non-slip finish. Exceptional wear and scratch resistance. Case of 25 tiles covers 10 square feet, including 10% waste. See Ceramic Tile Installation for tile setting costs, including tile backer, thin-set mortar or adhesive and grout.

| | | | | | | |
|---|---|---|---|---|---|---|
| Blue | — | SF | 2.34 | — | 2.34 | — |
| White | — | SF | 2.34 | — | 2.34 | — |

## 12" x 12" Ceramic Floor Tile

**12" x 12" ceramic residential floor tile.** Class 3, light to moderate traffic. For bathrooms, kitchens, foyers, dining rooms and recreation areas. Case of 15 tiles covers 15 square feet, including 10% waste. See Ceramic Tile Installation for tile setting costs, including tile backer, thin-set mortar or adhesive and grout.

| | | | | | | |
|---|---|---|---|---|---|---|
| Carrara, Beige | — | SF | 2.12 | — | 2.12 | — |
| Carrara, Gray | — | SF | 1.93 | — | 1.93 | — |
| Pietra, Alabaster | — | SF | 2.17 | — | 2.17 | — |
| Pietra, Gem | — | SF | 2.04 | — | 2.04 | — |
| Saltillo, Monterey | — | SF | 1.90 | — | 1.90 | — |

**12" x 12" glazed porcelain residential and commercial floor tile, American Marazzi.** For residential, commercial and light institutional applications, including restaurant dining rooms, shopping malls, offices, lobbies, showrooms and corridors. Class 4, moderate to heavy traffic. Case of 15 tiles covers 15 square feet, including 10% waste. This product does not meet the minimum coefficient of friction to be considered a slip-resistant tile. See Ceramic Tile Installation for tile setting costs, including tile backer, thin-set mortar or adhesive and grout.

| | | | | | | |
|---|---|---|---|---|---|---|
| Rapolano, Noce | — | SF | 2.23 | — | 2.23 | — |
| Salento Sabbia, Beige | — | SF | 2.15 | — | 2.15 | — |
| Salento Sabbia, Gray | — | SF | 2.14 | — | 2.14 | — |
| Vermont, Caledonia Porcelain | — | SF | 2.15 | — | 2.15 | — |

**12" x 12" ceramic floor tile.** For all residential applications. Class 3, light to moderate traffic including bathrooms, kitchens, foyers, dining rooms and recreation areas. Case of 15 tiles covers 15 square feet, including 10% waste. This product does not meet the minimum coefficient of friction to be considered a slip-resistant tile. See Ceramic Tile Installation for tile setting costs, including tile backer, thin-set mortar or adhesive and grout.

| | | | | | | |
|---|---|---|---|---|---|---|
| Verona, Beige | — | SF | 1.92 | — | 1.92 | — |
| Verona, White | — | SF | 1.93 | — | 1.93 | — |

| | Craft@Hrs | Unit | Material | Labor | Total | Sell |
|---|---|---|---|---|---|---|

**12" x 12" ceramic floor tile, Dal-Tile.** Field tile. For interior floors, walls, and counter tops. Case of 11 tiles covers 10 square feet, including 10% waste. See Ceramic Tile Installation for tile setting costs, including tile backer, thin-set mortar or adhesive and grout.

| | Craft@Hrs | Unit | Material | Labor | Total | Sell |
|---|---|---|---|---|---|---|
| Dakota™, Rushmore Gray | — | SF | 1.96 | — | 1.96 | — |
| French Quarter®, Cobblestone | — | SF | 2.69 | — | 2.69 | — |
| French Quarter®, Mardi Gras | — | SF | 2.90 | — | 2.90 | — |
| Gold Rush®, California Sand | — | SF | 2.09 | — | 2.09 | — |
| Gold Rush®, Golden Nugget | — | SF | 2.10 | — | 2.10 | — |
| Gold Rush®, Goldust | — | SF | 2.26 | — | 2.26 | — |
| Edgefield™, Beige | — | SF | 1.70 | — | 1.70 | — |

**12" x 12" ceramic floor and wall tile, Eliane Ceramic Tile.** Marbleized look floor ceramic designed for bathrooms. Can be used as wall coverings. Rated PEI 4 and Mohs scale 3. Eleven tiles cover 10 square feet, including 10% waste. See Ceramic Tile Installation for tile setting costs, including tile backer, thin-set mortar or adhesive and grout.

| | Craft@Hrs | Unit | Material | Labor | Total | Sell |
|---|---|---|---|---|---|---|
| Autumn, White | — | SF | 1.77 | — | 1.77 | — |
| Illusione, Beige | — | SF | 2.00 | — | 2.00 | — |
| Illusione, Caramel | — | SF | 2.00 | — | 2.00 | — |
| Illusione, Ice | — | SF | 2.00 | — | 2.00 | — |

## Large Ceramic Floor Tile

**Home Series 16" x 16" ceramic floor tile, Megatrade.** Smooth stone finish. PEI rating of 4. Indoor use only. Case of 9 tiles covers 15.5 square feet, including 10% waste. See Ceramic Tile Installation for tile setting costs, including tile backer, thin-set mortar or adhesive and grout.

| | Craft@Hrs | Unit | Material | Labor | Total | Sell |
|---|---|---|---|---|---|---|
| Beige | — | SF | 1.71 | — | 1.71 | — |
| Ivory | — | SF | 1.78 | — | 1.78 | — |
| White | — | SF | 1.78 | — | 1.78 | — |

**Fossile Cotto 16" x 16" ceramic floor tile, Eliane Ceramic Tile.** Rated PEI 4 and Mohs scale 5. Indoor. Suitable for offices, living rooms, bedrooms, bathrooms, halls and kitchens. Rustic look. Can also be used as wall covering. Case of 9 tiles covers 16 square feet, including 10% waste. See Ceramic Tile Installation for tile setting costs, including tile backer, thin-set mortar or adhesive and grout.

| | Craft@Hrs | Unit | Material | Labor | Total | Sell |
|---|---|---|---|---|---|---|
| Rustic look | — | SF | 1.91 | — | 1.91 | — |

**Illusione 16" x 16" ceramic floor tile, Eliane Ceramic Tile.** Rated PEI 4 and Mohs scale 3. Marbleized look. Designed for bathrooms. Can also be used as wall coverings. Case of 9 tiles covers 14.4 square feet, including 10% waste. See Ceramic Tile Installation for tile setting costs, including tile backer, thin-set mortar or adhesive and grout.

| | Craft@Hrs | Unit | Material | Labor | Total | Sell |
|---|---|---|---|---|---|---|
| Beige | — | SF | 2.86 | — | 2.86 | — |
| Caramel | — | SF | 2.86 | — | 2.86 | — |
| Ice | — | SF | 2.87 | — | 2.87 | — |

**16" x 16" ceramic residential and medium commercial floor tile, American Marazzi.** PEI class 4, moderate to heavy traffic. For residential, medium commercial and light institutional applications, including restaurant dining rooms, shopping malls, offices, lobbies, showrooms and corridors. Case of 8 tiles covers 13.77 square feet, including 10% waste. Does not meet the minimum coefficient of friction to be considered a slip-resistant tile. See Ceramic Tile Installation for tile setting costs, including tile backer, thin-set mortar or adhesive and grout.

| | Craft@Hrs | Unit | Material | Labor | Total | Sell |
|---|---|---|---|---|---|---|
| Andes, Peru | — | SF | 2.03 | — | 2.03 | — |
| Andes, Bolivia | — | SF | 2.03 | — | 2.03 | — |

| | Craft@Hrs | Unit | Material | Labor | Total | Sell |
|---|---|---|---|---|---|---|

**16" x 16" glazed porcelain residential and medium commercial floor tile, American Marazzi.** PEI class 4, moderate to heavy traffic. For residential, medium commercial and light institutional applications, including restaurant dining rooms, shopping malls, offices, lobbies, showrooms and corridors. Case of 9 tiles covers 15.5 square feet. Does not meet the minimum coefficient of friction to be considered a slip-resistant tile. See Ceramic Tile Installation for tile setting costs, including tile backer, thin-set mortar or adhesive and grout.

| | | | | | | |
|---|---|---|---|---|---|---|
| Romance, Almond | — | SF | 2.15 | — | 2.15 | — |

## Natural Stone and Slate Flooring

**Botticino Americana marble floor tile.** Highly polished genuine marble. See Ceramic Tile Installation for tile setting costs, including tile backer, thin-set mortar or adhesive and grout.

| | | | | | | |
|---|---|---|---|---|---|---|
| 12" x 12" | — | Ea | 7.32 | — | 7.32 | — |

**Grigio Venato marble tile.** Gray with natural shade variations. For interior and exterior floors and walls. See Ceramic Tile Installation for tile setting costs, including tile backer, thin-set mortar or adhesive and grout.

| | | | | | | |
|---|---|---|---|---|---|---|
| 12" x 12" | — | Ea | 2.37 | — | 2.37 | — |

**Coral Venato marble tile.** Polished marble with soft hues of rose and gray to match all decors. For interior and exterior floors, wall fireplaces and bathroom countertops. Gloss finish. High durability. See Ceramic Tile Installation for tile setting costs, including tile backer, thin-set mortar or adhesive and grout.

| | | | | | | |
|---|---|---|---|---|---|---|
| 12" x 12" | — | Ea | 2.12 | — | 2.12 | — |

**Maron Venato marble tile.** For floors, fireplaces, and countertops. Residential and commercial applications. High sheen finish. See Ceramic Tile Installation for tile setting costs, including tile backer, thin-set mortar or adhesive and grout.

| | | | | | | |
|---|---|---|---|---|---|---|
| 12" x 12" | — | Ea | 3.66 | — | 3.66 | — |

**White Gioia marble floor tile.** First-quality Italian marble. Each tile has unique polished surface. Residential and commercial applications. Rich classic look. High sheen finish. See Ceramic Tile Installation for tile setting costs, including tile backer, thin-set mortar or adhesive and grout.

| | | | | | | |
|---|---|---|---|---|---|---|
| 12" x 12" | — | Ea | 5.40 | — | 5.40 | — |

**Nero Black marble floor tile.** Tile size is approximate. See Ceramic Tile Installation for tile setting costs, including tile backer, thin-set mortar or adhesive and grout.

| | | | | | | |
|---|---|---|---|---|---|---|
| 12" x 12" | — | Ea | 3.98 | — | 3.98 | — |

**Marble floor tile, Bestview Ltd.** 12" x 12" and standard thickness of 10mm. Suitable for both interior and exterior wall and floor applications. Highly polished surface and beveled on the edges. See Ceramic Tile Installation for tile setting costs, including tile backer, thin-set mortar or adhesive and grout.

| | | | | | | |
|---|---|---|---|---|---|---|
| Botticino Beige | — | Ea | 4.73 | — | 4.73 | — |
| Emperador Lite | — | Ea | 4.30 | — | 4.30 | — |
| Cream Jade | — | Ea | 2.04 | — | 2.04 | — |

**Travertine stone tile, Emser International.** Rustic look with natural color variations. For floors or fireplaces. See Ceramic Tile Installation for tile setting costs, including tile backer, thin-set mortar or adhesive and grout.

| | | | | | | |
|---|---|---|---|---|---|---|
| 12" x 12", Dorato, regular | — | Ea | 4.41 | — | 4.41 | — |
| 12" x 12", Coliseum, honed | — | Ea | 6.45 | — | 6.45 | — |
| 12" x 12", Medium beige | — | Ea | 3.23 | — | 3.23 | — |

| | Craft@Hrs | Unit | Material | Labor | Total | Sell |
|---|---|---|---|---|---|---|

**Travertine floor tile.** See Ceramic Tile Installation for tile setting costs, including tile backer, thin-set mortar or adhesive and grout.

| | | | | | | |
|---|---|---|---|---|---|---|
| 12" x 12", Noce | — | Ea | 6.45 | — | 6.45 | — |
| 12" x 12", Rojo | — | Ea | 5.63 | — | 5.63 | — |
| 16" x 16", Desert Sand | — | Ea | 8.03 | — | 8.03 | — |
| 18" x 18", Desert Sand | — | Ea | 9.94 | — | 9.94 | — |

**Granite floor tile.** For bathrooms, entryways, fireplaces or countertops. See Ceramic Tile Installation for tile setting costs, including tile backer, thin-set mortar or adhesive and grout.

| | | | | | | |
|---|---|---|---|---|---|---|
| 12" x 12", Mauve | — | Ea | 4.30 | — | 4.30 | — |
| 12" x 12", Absolute Black | — | Ea | 9.45 | — | 9.45 | — |
| 12" x 12", Mystique | — | Ea | 3.21 | — | 3.21 | — |
| 12" x 12", Pacific Gray | — | Ea | 7.54 | — | 7.54 | — |
| 12" x 12", Sonoma Montecruz | — | Ea | 2.97 | — | 2.97 | — |

**Natural granite tile.** Pattern and color varies. For interior, exterior walls, heavy-duty interior floors and polished, honed, or flamed surfaces. Slightly eased edges. 3/8" thick. See Ceramic Tile Installation for tile setting costs, including tile backer, thin-set mortar or adhesive and grout.

| | | | | | | |
|---|---|---|---|---|---|---|
| 12" x 12", Black | — | Ea | 10.79 | — | 10.79 | — |
| 12" x 12", Balmoral Red | — | Ea | 10.70 | — | 10.70 | — |

**Natural gauged slate floor tile.** Rustic earth tones. Indoor or outdoor use. See Ceramic Tile Installation for tile setting costs, including tile backer, thin-set mortar or adhesive and grout.

| | | | | | | |
|---|---|---|---|---|---|---|
| 12" x 12", Autumn | — | Ea | 2.03 | — | 2.03 | — |
| 12" x 12", Earth | — | Ea | 2.18 | — | 2.18 | — |
| 12" x 12", Golden White Quartz | — | Ea | 2.50 | — | 2.50 | — |
| 16" x 16", Autumn | — | Ea | 3.40 | — | 3.40 | — |
| 16" x 16", Earth | — | Ea | 3.79 | — | 3.79 | — |

**Natural stone floor tile.** See Ceramic Tile Installation for tile setting costs, including tile backer, thin-set mortar or adhesive and grout.

| | | | | | | |
|---|---|---|---|---|---|---|
| 12" x 12", Absolute Black | — | Ea | 5.72 | — | 5.72 | — |
| 12" x 12", Caffe Mist | — | Ea | 3.21 | — | 3.21 | — |
| 12" x 12", Malaga Beige | — | Ea | 5.15 | — | 5.15 | — |
| 12" x 12", Panetella Brown | — | Ea | 3.21 | — | 3.21 | — |

## Paver Tile

**Saltillo floor tile.** 12" x 12". Refined natural Saltillo clay tile. Slight rustic appearance with subtle color variations. Rounded edges. Indoor or outdoor use. See Ceramic Tile Installation for tile setting costs, including tile backer, thin-set mortar or adhesive and grout.

| | | | | | | |
|---|---|---|---|---|---|---|
| Handcrafted, regular | — | Ea | 1.27 | — | 1.27 | — |
| Handcrafted, super | — | Ea | 1.50 | — | 1.50 | — |
| Oil sealed | — | Ea | 2.17 | — | 2.17 | — |

**Quarrybasics® Mayflower Red quarry tile, U.S. Ceramic Tile.** Indoor or outdoor use. 6" x 6" x 1/2" thick. Relieved edges. Four tiles per square foot before cutting waste and breakage. Low absorption. See Ceramic Tile Installation for tile setting costs, including tile backer, thin-set mortar or adhesive and grout.

| | | | | | | |
|---|---|---|---|---|---|---|
| 6" x 6" x 1/2" flat tile | — | SF | .35 | — | .35 | — |
| 6" bullnose trim | — | Ea | 1.70 | — | 1.70 | — |
| 6" bullnose corner | — | Ea | 3.03 | — | 3.03 | — |
| 6" round top cove base | — | Ea | 1.56 | — | 1.56 | — |

|  | Craft@Hrs | Unit | Material | Labor | Total | Sell |
|---|---|---|---|---|---|---|

**Brick paver, Magnolia Brick and Tile.** 4" x 8" x 1/2". Magnolia Brick Pavers install like ceramic tile, but look and perform like real brick floors. Case of pavers covers approximately 10 square feet. See Ceramic Tile Installation for tile setting costs, including tile backer, thin-set mortar or adhesive and grout.

| | Craft@Hrs | Unit | Material | Labor | Total | Sell |
|---|---|---|---|---|---|---|
| Old Kentucky | — | SF | 1.78 | — | 1.78 | — |
| Old South | — | SF | 2.10 | — | 2.10 | — |
| Savannah | — | SF | 2.10 | — | 2.10 | — |
| St. Louis | — | SF | 2.11 | — | 2.11 | — |

## Tumbled Marble Tile Flooring

**4" x 4" tumbled stone tile, Jeffrey Court.** Case of 9 tiles covers 1 square foot. See Ceramic Tile Installation for tile setting costs, including tile backer, thin-set mortar or adhesive and grout.

| | | | | | | |
|---|---|---|---|---|---|---|
| Botticino | — | SF | 6.45 | — | 6.45 | — |
| Light Travertine | — | SF | 6.45 | — | 6.45 | — |
| Roso Perlina | — | SF | 6.46 | — | 6.46 | — |
| Travertino Noce | — | SF | 6.46 | — | 6.46 | — |

**6" x 6" tumbled slate tile, Jeffrey Court.** Case of 4 tiles covers 1 square foot. See Ceramic Tile Installation for tile setting costs, including either thin-set mortar or adhesive and grout.

| | | | | | | |
|---|---|---|---|---|---|---|
| Sequoia | — | SF | 5.30 | — | 5.30 | — |

**Tumbled marble listello tile, Westminster Ceramics.** For walls or floors. Stone marble from Italy. Per linear foot of tile before waste and breakage. See Ceramic Tile Installation for tile setting costs, including either thin-set mortar or adhesive and grout.

| | | | | | | |
|---|---|---|---|---|---|---|
| 3" x 12", Chiaro Rosso | — | Ea | 5.35 | — | 5.35 | — |
| 3" x 12", Chiaro Rosso Verde | — | Ea | 5.36 | — | 5.36 | — |
| 3" x 12", Noce Chiaro | — | Ea | 5.35 | — | 5.35 | — |
| 4" x 12", Bottichino, Scabas | — | Ea | 5.92 | — | 5.92 | — |
| 4" x 12", Chiaro Rosso | — | Ea | 5.94 | — | 5.94 | — |
| 4" x 12", Chiaro Rosso Verde | — | Ea | 5.97 | — | 5.97 | — |
| 4" x 12", Noce | — | Ea | 5.94 | — | 5.94 | — |

## Ceramic Tile Threshold, Saddle

**Carrera sill and saddle.** 5/8" thick.

| | | | | | | |
|---|---|---|---|---|---|---|
| 4" x 24", double Hollywood saddle | TL@.687 | Ea | 9.72 | 21.20 | 30.92 | 52.60 |
| 5" x 30", Hollywood saddle | TL@.687 | Ea | 19.70 | 21.20 | 40.90 | 69.50 |
| 5" x 36", double Hollywood saddle | TL@.687 | Ea | 29.40 | 21.20 | 50.60 | 86.00 |
| 5" x 37", sill | TL@.687 | Ea | 15.40 | 21.20 | 36.60 | 62.20 |
| 5" x 73", sill | TL@.687 | Ea | 28.80 | 21.20 | 50.00 | 85.00 |
| 6" x 37", sill | TL@.687 | Ea | 20.00 | 21.20 | 41.20 | 70.00 |

**Marble threshold.** A natural marble threshold provides a transition from one flooring surface to another or from one room to another.

| | | | | | | |
|---|---|---|---|---|---|---|
| 36" x 2" x 3/8", gray | TL@.687 | Ea | 4.43 | 21.20 | 25.63 | 43.60 |
| 36" x 2" x 3/8", white | TL@.687 | Ea | 7.44 | 21.20 | 28.64 | 48.70 |
| 36" x 2" x 5/8", white | TL@.687 | Ea | 8.52 | 21.20 | 29.72 | 50.50 |
| 36" x 4" x 3/8", white | TL@.687 | Ea | 8.52 | 21.20 | 29.72 | 50.50 |
| 36" x 4" x 5/8", white, gray veins | TL@.687 | Ea | 8.62 | 21.20 | 29.82 | 50.70 |

| | Craft@Hrs | Unit | Material | Labor | Total | Sell |
|---|---|---|---|---|---|---|
| **Ceramic threshold.** Off white. | | | | | | |
| 1/4" x 36" | TL@.515 | Ea | 14.00 | 15.90 | 29.90 | 50.80 |
| 1/4" x 72" | TL@.730 | Ea | 31.05 | 22.50 | 53.55 | 91.00 |
| 1/2" x 36" | TL@.515 | Ea | 15.40 | 15.90 | 31.30 | 53.20 |
| 1/2" x 72" | TL@.730 | Ea | 32.20 | 22.50 | 54.70 | 93.00 |
| 3/4" x 36" | TL@.515 | Ea | 24.80 | 15.90 | 40.70 | 69.20 |
| 3/4" x 72" | TL@.730 | Ea | 44.80 | 22.50 | 67.30 | 114.00 |
| **Ceramic window sill.** | | | | | | |
| 4" x 37" x 3/4" | TL@.687 | Ea | 13.60 | 21.20 | 34.80 | 59.20 |
| 4" x 54" x 3/4" | TL@.687 | Ea | 23.70 | 21.20 | 44.90 | 76.30 |
| 4" x 74" x 3/4" | TL@.730 | Ea | 32.30 | 22.50 | 54.80 | 93.20 |
| 5" x 37" x 3/4" | TL@.687 | Ea | 20.50 | 21.20 | 41.70 | 70.90 |
| 5" x 54" x 3/4" | TL@.687 | Ea | 29.10 | 21.20 | 50.30 | 85.50 |
| 5" x 74" x 3/4" | TL@.730 | Ea | 34.10 | 22.50 | 56.60 | 96.20 |
| 6" x 37" x 3/4" | TL@.687 | Ea | 23.70 | 21.20 | 44.90 | 76.30 |
| 6" x 54" x 3/4" | TL@.687 | Ea | 35.90 | 21.20 | 57.10 | 97.10 |
| 6" x 74" x 3/4" | TL@.730 | Ea | 43.10 | 22.50 | 65.60 | 112.00 |

## Ceramic Field Wall Tile

**4-1/4" x 4-1/4" ceramic wall tile, U.S. Ceramic Tile.** For full-wall interior residential, commercial, and institutional applications. Matte glaze suitable for residential countertops and light-duty floors. Designer's choice glaze. A case of 80 tiles covers 10 square feet, and a case of 120 tiles covers 15 square feet. Add for cutting waste and breakage. See Ceramic Tile Installation for tile setting costs, including tile backer, thin-set mortar or adhesive and grout.

| | Craft@Hrs | Unit | Material | Labor | Total | Sell |
|---|---|---|---|---|---|---|
| Black | — | SF | 3.11 | — | 3.11 | — |
| Burgundy | — | SF | 2.85 | — | 2.85 | — |
| Crystal Bone | — | SF | 2.42 | — | 2.42 | — |
| Crystal White | — | SF | 2.42 | — | 2.42 | — |
| Fawn Beige, bright glaze | — | SF | 1.65 | — | 1.65 | — |
| Flaxseed | — | SF | 1.65 | — | 1.65 | — |
| Gold Dust, bright glaze | — | SF | 1.65 | — | 1.65 | — |
| Granite | — | SF | 1.73 | — | 1.73 | — |
| Kelly | — | SF | 2.85 | — | 2.85 | — |
| Medium Blue, bright glaze | — | SF | 1.65 | — | 1.65 | — |
| Moonstone Bone, matte glaze | — | SF | 1.55 | — | 1.55 | — |
| Navy | — | SF | 3.45 | — | 3.45 | — |
| Pink | — | SF | 1.55 | — | 1.55 | — |
| Royal Cobalt | — | SF | 2.85 | — | 2.85 | — |
| Snow White, bright glaze | — | SF | 1.21 | — | 1.21 | — |
| Sterling, bright glaze | — | SF | 1.65 | — | 1.65 | — |
| White, matte glaze | — | SF | 1.64 | — | 1.64 | — |
| Windrift White, matte glaze | — | SF | 2.12 | — | 2.12 | — |
| Yellow | — | SF | 1.65 | — | 1.65 | — |

**4-1/4" x 4-1/4" ceramic single bullnose tile, U.S. Ceramic Tile.** Per trim piece. Add for cutting waste and breakage. See Ceramic Tile Installation for tile setting costs, including tile backer, thin-set mortar or adhesive and grout.

| | Craft@Hrs | Unit | Material | Labor | Total | Sell |
|---|---|---|---|---|---|---|
| Fawn Beige | — | Ea | .78 | — | .78 | — |
| Flaxseed | — | Ea | .76 | — | .76 | — |
| Gold Dust, bright glaze | — | Ea | .89 | — | .89 | — |
| Granite | — | Ea | .61 | — | .61 | — |
| Medium Blue, bright glaze | — | Ea | .78 | — | .78 | — |
| Moonstone, matte glaze | — | Ea | .78 | — | .78 | — |
| Pink | — | Ea | .78 | — | .78 | — |

| | Craft@Hrs | Unit | Material | Labor | Total | Sell |
|---|---|---|---|---|---|---|
| Sterling | — | Ea | .77 | — | .77 | — |
| White, matte glaze | — | Ea | .83 | — | .83 | — |
| Yellow | — | Ea | .75 | — | .75 | — |

**4-1/4" x 4-1/4" ceramic bullnose corner tile, U.S. Ceramic Tile.** Per trim piece. Add for cutting waste and breakage. See Ceramic Tile Installation for tile setting costs, including tile backer, thin-set mortar or adhesive and grout.

| | Craft@Hrs | Unit | Material | Labor | Total | Sell |
|---|---|---|---|---|---|---|
| Crystal White | — | Ea | 1.68 | — | 1.68 | — |
| Fawn Beige | — | Ea | 1.47 | — | 1.47 | — |
| Flaxseed | — | Ea | 1.47 | — | 1.47 | — |
| Gold Dust, bright glaze | — | Ea | 1.55 | — | 1.55 | — |
| Granite | — | Ea | 1.47 | — | 1.47 | — |
| Medium Blue, bright glaze | — | Ea | 1.47 | — | 1.47 | — |
| Moonstone, matte glaze | — | Ea | 1.43 | — | 1.43 | — |
| Pink | — | Ea | 1.47 | — | 1.47 | — |
| Sterling | — | Ea | 1.47 | — | 1.47 | — |
| White, matte glaze | — | Ea | 1.50 | — | 1.50 | — |
| Yellow | — | Ea | 1.45 | — | 1.45 | — |

**Ceramic sink rail, U.S. Ceramic Tile.** Per trim piece. Add for cutting waste and breakage. See Ceramic Tile Installation for tile setting costs, including tile backer, thin-set mortar or adhesive and grout.

| | Craft@Hrs | Unit | Material | Labor | Total | Sell |
|---|---|---|---|---|---|---|
| 1-5/8" x 2-1/4", corner, Moonstone | — | Ea | 3.41 | — | 3.41 | — |
| 2" x 2", corner, Bone, bright glaze | — | Ea | 3.42 | — | 3.42 | — |
| 2" x 6", Bone, bright glaze | — | Ea | 1.68 | — | 1.68 | — |
| 2" x 6", Moonstone | — | Ea | 1.92 | — | 1.92 | — |
| 2" x 6", White, matte glaze | — | Ea | 1.95 | — | 1.95 | — |

**2" x 2" ceramic surface cap corner, U.S. Ceramic Tile.** Per trim piece. Add for cutting waste and breakage. See Ceramic Tile Installation for tile setting costs, including tile backer, thin-set mortar or adhesive and grout.

| | Craft@Hrs | Unit | Material | Labor | Total | Sell |
|---|---|---|---|---|---|---|
| Black | — | Ea | .92 | — | .92 | — |
| Bone, bright glaze | — | Ea | .65 | — | .65 | — |
| Burgundy | — | Ea | .94 | — | .94 | — |
| Fawn Beige | — | Ea | .65 | — | .65 | — |
| Flaxseed | — | Ea | .67 | — | .67 | — |
| Gold Dust, bright glaze | — | Ea | .67 | — | .67 | — |
| Granite | — | Ea | .66 | — | .66 | — |
| Kelly | — | Ea | .91 | — | .91 | — |
| Medium Blue, bright glaze | — | Ea | .64 | — | .64 | — |
| Moonstone, matte glaze | — | Ea | .67 | — | .67 | — |
| Navy | — | Ea | .90 | — | .90 | — |
| Pink | — | Ea | .46 | — | .46 | — |
| Sterling, bright glaze | — | Ea | .67 | — | .67 | — |
| White, bright glaze | — | Ea | .51 | — | .51 | — |
| White, matte glaze | — | Ea | .70 | — | .70 | — |
| Yellow | — | Ea | .62 | — | .62 | — |

**2" x 6" ceramic surface cap, U.S. Ceramic Tile.** Per trim piece. Add for cutting waste and breakage. See Ceramic Tile Installation for tile setting costs, including tile backer, thin-set mortar or adhesive and grout.

| | Craft@Hrs | Unit | Material | Labor | Total | Sell |
|---|---|---|---|---|---|---|
| Black, Designer's Choice glaze | — | Ea | .92 | — | .92 | — |
| Bone, bright glaze | — | Ea | .67 | — | .67 | — |
| Burgundy, Designer's Choice glaze | — | Ea | .96 | — | .96 | — |
| Fawn Beige | — | Ea | .67 | — | .67 | — |
| Flaxseed | — | Ea | .67 | — | .67 | — |
| Gold Dust, bright glaze | — | Ea | .66 | — | .66 | — |
| Granite | — | Ea | .49 | — | .49 | — |

| | Craft@Hrs | Unit | Material | Labor | Total | Sell |
|---|---|---|---|---|---|---|
| Kelly, Designer's Choice glaze | — | Ea | .96 | — | .96 | — |
| Medium Blue, bright glaze | — | Ea | .67 | — | .67 | — |
| Moonstone, matte glaze | — | Ea | .67 | — | .67 | — |
| Navy, Designer's Choice glaze | — | Ea | .93 | — | .93 | — |
| Pink | — | Ea | .64 | — | .64 | — |
| White, matte glaze | — | Ea | .69 | — | .69 | — |
| Yellow | — | Ea | .67 | — | .67 | — |

**Gold Rush® glazed ceramic tile, Dal-Tile.** Glazed paver with unique textured surface. Durable and easy to maintain. Ideal for floors, walls and countertops. Provides good slip, wear and stain resistance. No sealers or waxes needed. One case covers 11 square feet.

| | Craft@Hrs | Unit | Material | Labor | Total | Sell |
|---|---|---|---|---|---|---|
| 6" x 6", Goldust | — | SF | 3.13 | — | 3.13 | — |
| 6" x 12", Goldust | — | SF | 3.60 | — | 3.60 | — |

**6" x 6" ceramic wall tile, U.S. Ceramic Tile.** For full-wall interior residential, commercial, and institutional applications. Bright glaze suitable for residential countertops and light-duty floors. Case of 50 tiles covers 12.5 square feet. Add for cutting waste and breakage. See Ceramic Tile Installation for tile setting costs, including tile backer, thin-set mortar or adhesive and grout.

| | Craft@Hrs | Unit | Material | Labor | Total | Sell |
|---|---|---|---|---|---|---|
| Biscuit | — | SF | 1.69 | — | 1.69 | — |
| White | — | SF | 1.69 | — | 1.69 | — |

**6" x 6" ceramic residential and commercial wall tile, American Marazzi.** For interior residential and commercial wall applications and for residential bathroom floor applications only. PEI class 2, light traffic. Case of 48 tiles covers 12 square feet.

| | Craft@Hrs | Unit | Material | Labor | Total | Sell |
|---|---|---|---|---|---|---|
| Explorer Atlantis, 6" x 6", per SF | — | SF | 2.50 | — | 2.50 | — |
| 2" x 2" angle, each | — | Ea | 2.33 | — | 2.33 | — |
| 2" x 6" single bullnose, each | — | Ea | 1.97 | — | 1.97 | — |
| Explorer Columbia, 6" x 6", per SF | — | SF | 2.55 | — | 2.55 | — |
| 2" x 2" angle, each | — | Ea | 2.26 | — | 2.26 | — |
| 2" x 6" single bullnose, each | — | Ea | 2.02 | — | 2.02 | — |
| Explorer Gemini, 6" x 6", per SF | — | SF | 1.99 | — | 1.99 | — |
| 2" x 2" angle, each | — | Ea | 2.26 | — | 2.26 | — |
| 2" x 6" single bullnose, each | — | Ea | 1.07 | — | 1.07 | — |
| Explorer Voyager, 6" x 6", per SF | — | SF | 2.55 | — | 2.55 | — |
| 2" x 2" angle, each | — | Ea | 2.26 | — | 2.26 | — |
| 2" x 6" single bullnose, each | — | Ea | 2.12 | — | 2.12 | — |

**6" x 8" ceramic wall tile, U.S. Ceramic Tile.** For full-wall interior residential, commercial, and institutional applications. Matte glaze suitable for residential countertops and light-duty floors.

| | Craft@Hrs | Unit | Material | Labor | Total | Sell |
|---|---|---|---|---|---|---|
| Beige | — | SF | 2.20 | — | 2.20 | — |
| Biscuit, bright glaze | — | SF | 1.72 | — | 1.72 | — |
| Blush | — | SF | 2.20 | — | 2.20 | — |
| Sierra Gray, bright glaze | — | SF | 2.17 | — | 2.17 | — |
| Sierra Sand, bright glaze | — | SF | 2.17 | — | 2.17 | — |
| White, bright glaze | — | SF | 1.52 | — | 1.52 | — |

**Tile edge protection, Schluter Systems.** Finishes and protects the edges of tiled surfaces. Provides a smooth transition between surfaces of the same height or different heights. Per 8'2-1/2" length.

| | Craft@Hrs | Unit | Material | Labor | Total | Sell |
|---|---|---|---|---|---|---|
| 1/4", aluminum/bright brass | — | Ea | 15.03 | — | 15.03 | — |
| 5/16", aluminum | — | Ea | 7.82 | — | 7.82 | — |
| 5/16", aluminum/brass, round edge | — | Ea | 17.20 | — | 17.20 | — |
| 5/16", aluminum/chrome, round edge | — | Ea | 16.70 | — | 16.70 | — |
| 5/16", aluminum/gold | — | Ea | 12.20 | — | 12.20 | — |
| 3/8", aluminum | — | Ea | 8.37 | — | 8.37 | — |
| 3/8", aluminum/brass, round edge | — | Ea | 19.40 | — | 19.40 | — |

| | Craft@Hrs | Unit | Material | Labor | Total | Sell |
|---|---|---|---|---|---|---|
| 3/8", aluminum/gold | — | Ea | 13.70 | — | 13.70 | — |
| 3/8", brass | — | Ea | 26.00 | — | 26.00 | — |
| 3/8", solid brass | — | Ea | 14.00 | — | 14.00 | — |
| 1/2", aluminum | — | Ea | 8.96 | — | 8.96 | — |

**Tile reducer strip, Schluter Systems.** Protective edge profile with a sloped transition. Use for higher tiled surfaces to lower carpet, vinyl or wood.

| | Craft@Hrs | Unit | Material | Labor | Total | Sell |
|---|---|---|---|---|---|---|
| 5/16" x 4', aluminum/bright gold | — | Ea | 6.37 | — | 6.37 | — |
| 5/16" x 8', aluminum/bright brass | — | Ea | 13.70 | — | 13.70 | — |
| 5/16" x 8', anodized aluminum | — | Ea | 14.00 | — | 14.00 | — |
| 3/8" x 4', aluminum/bright brass | — | Ea | 17.70 | — | 17.70 | — |
| 3/8" x 8', anodized aluminum | — | Ea | 13.70 | — | 13.70 | — |
| 3/8" x 4', aluminum/bright gold | — | Ea | 6.80 | — | 6.80 | — |
| 3/8" x 4', brass | — | Ea | 21.90 | — | 21.90 | — |

## Mosaic Tile

**Octagon mosaic dot ceramic tile, Dal-Tile.** Face-mounted 12" x 12" sheets. One sheet covers one square foot before cutting waste and breakage. See Ceramic Tile Installation for tile setting costs, including tile backer, thin-set mortar or grout.

| | Craft@Hrs | Unit | Material | Labor | Total | Sell |
|---|---|---|---|---|---|---|
| 2", burgundy matte with white | — | SF | 2.55 | — | 2.55 | — |
| 2", white with black | — | SF | 2.53 | — | 2.53 | — |
| 2", white with white | — | SF | 2.56 | — | 2.56 | — |

**Octagon mosaic dot ceramic tile.** Face-mounted 12" x 12" sheets. One sheet covers one square foot before cutting waste and breakage. See Ceramic Tile Installation for tile setting costs, including tile backer, thin-set mortar or grout.

| | Craft@Hrs | Unit | Material | Labor | Total | Sell |
|---|---|---|---|---|---|---|
| 2-1/4", white with black | — | SF | 2.09 | — | 2.09 | — |
| 2-1/4", white with green | — | SF | 1.80 | — | 1.80 | — |
| 2-1/4", white with white | — | SF | 2.15 | — | 2.15 | — |

**Accent mosaic ceramic tile, Dal-Tile.** Face-mounted 12" x 12" sheets. One sheet covers one square foot before cutting waste and breakage. See Ceramic Tile Installation for tile setting costs, including tile backer, thin-set mortar or grout.

| | Craft@Hrs | Unit | Material | Labor | Total | Sell |
|---|---|---|---|---|---|---|
| 2" x 2", mottled sandalwood | — | SF | 7.00 | — | 7.00 | — |

## Ceramic Tile Fixtures

**Ceramic tile corner shelf, Professional, Lenape.** Genuine porcelain. Mastic installation.

| | Craft@Hrs | Unit | Material | Labor | Total | Sell |
|---|---|---|---|---|---|---|
| 4-1/2" x 4-1/2" x 2-1/4", bone | TL@.500 | Ea | 11.50 | 15.40 | 26.90 | 45.70 |
| 4-1/2" x 4-1/2" x 2-1/4", white | TL@.500 | Ea | 11.53 | 15.40 | 26.93 | 45.80 |
| 7-1/4" x 7-1/4" x 3", bone | TL@.500 | Ea | 20.40 | 15.40 | 35.80 | 60.90 |
| 7-1/4" x 7-1/4" x 3", white | TL@.500 | Ea | 20.45 | 15.40 | 35.85 | 60.90 |

**Ceramic tile soap dish, Dal-Tile.** Mastic installation.

| | Craft@Hrs | Unit | Material | Labor | Total | Sell |
|---|---|---|---|---|---|---|
| 4" x 6", almond | TL@.250 | Ea | 10.41 | 7.70 | 18.11 | 30.80 |
| 4" x 6", white | TL@.250 | Ea | 10.42 | 7.70 | 18.12 | 30.80 |

**Ceramic tile toilet tissue paper holder, Professional, Lenape.** 4-1/4" x 6", Genuine porcelain. Mastic installation.

| | Craft@Hrs | Unit | Material | Labor | Total | Sell |
|---|---|---|---|---|---|---|
| Bone | BC@.300 | Ea | 7.67 | 9.63 | 17.30 | 29.40 |
| White | BC@.300 | Ea | 7.67 | 9.63 | 17.30 | 29.40 |

| | Craft@Hrs | Unit | Material | Labor | Total | Sell |
|---|---|---|---|---|---|---|

**Ceramic tile toothbrush and tumbler holder, Professional, Lenape.** 4-1/4" x 4-1/4". Genuine porcelain. Mastic installation.

| | Craft@Hrs | Unit | Material | Labor | Total | Sell |
|---|---|---|---|---|---|---|
| Bone | TL@.250 | Ea | 7.06 | 7.70 | 14.76 | 25.10 |
| White | TL@.250 | Ea | 7.06 | 7.70 | 14.76 | 25.10 |

**Ceramic tile 24" towel bar assembly, Professional, Lenape.** Includes bar and posts. High-fired porcelain. Coordinates with most ceramic tile. Mastic installation.

| | Craft@Hrs | Unit | Material | Labor | Total | Sell |
|---|---|---|---|---|---|---|
| Bone | BC@.400 | Ea | 10.55 | 12.80 | 23.35 | 39.70 |
| White | BC@.400 | Ea | 10.53 | 12.80 | 23.33 | 39.70 |

**5-piece ceramic tile porcelain bath fixture set, Lenape.** Includes soap dish, toothbrush holder, tub soap dish, toilet paper holder and 24" towel bar and posts. Genuine porcelain. Made in USA. Lifetime warranty. Mastic installation.

| | Craft@Hrs | Unit | Material | Labor | Total | Sell |
|---|---|---|---|---|---|---|
| Bone | BC@.500 | Ea | 23.20 | 16.00 | 39.20 | 66.60 |
| White | BC@.500 | Ea | 22.90 | 16.00 | 38.90 | 66.10 |

**SunTouch® ceramic tile floor warming system, Watts Radiant.** Electric floor warming for tile and stone floors. Dual-insulated heating cable woven into thin tile warming mats that install in thin-set mortar below tile and stone. All mats are 120 VAC and deliver 12 watts per square foot. All sizes are 30" wide by 4', 6', 8', and 10' long. Warms the floor and keeps it dry. Adds warmth to the floor, as well as the entire room. The SunTouch® system consists of an electric mat (or mats) and a control (either programmable or non-programmable). The system comes with an in-depth installation manual and installation video. The mat installs easily — just roll it out on the subfloor or backerboard and secure it with the double-sided tape provided, or just staple it down. Then apply the mortar.

| | Craft@Hrs | Unit | Material | Labor | Total | Sell |
|---|---|---|---|---|---|---|
| 4' x 30" mat | — | Ea | 136.00 | — | 136.00 | — |
| 6' x 30" mat | — | Ea | 168.00 | — | 168.00 | — |
| 8' x 30" mat | — | Ea | 188.00 | — | 188.00 | — |
| 10' x 30" mat | — | Ea | 204.00 | — | 204.00 | — |

**Floor warming system thermostat, Watts Radiant.** Multi-function thermostat. Senses the floor temperature via a remote floor sensor. Includes a large digital display showing floor temperature, comfort temperature, and set-back temperature. Built-in 15 amp relay can control up to 150 square feet of SunTouch (15 amp load). With ground fault circuit interrupter. Select both comfort setting and set-back setting at the touch of a button.

| | Craft@Hrs | Unit | Material | Labor | Total | Sell |
|---|---|---|---|---|---|---|
| Non-programmable | — | Ea | 107.00 | — | 107.00 | — |
| Programmable | — | Ea | 150.00 | — | 150.00 | — |

## Tile Adhesive and Grout

**Semco anti-fracture tile membrane.** Protects tile installations from minor substrate movement. Provides crack-resistance. Can set tile directly over membrane. Easy-to-use, one-step system. Covers 40 to 45 square feet per gallon when applied with a 3/16" x 1/4" V-notch trowel.

| | Craft@Hrs | Unit | Material | Labor | Total | Sell |
|---|---|---|---|---|---|---|
| Gallon | — | Ea | 39.20 | — | 39.20 | — |

**RedGard™ waterproofing and anti-fracture membrane, Custom® Building Products.** Elastomeric waterproofing and anti-fracture membrane for interior or exterior commercial and residential tile and stone installations. Can also be used as a slab-on-grade moisture barrier under resilient or wood flooring. Applied with roller, trowel or sprayer. Reduces crack transmission in ceramic tile and stone floors.

| | Craft@Hrs | Unit | Material | Labor | Total | Sell |
|---|---|---|---|---|---|---|
| Gallon | — | Ea | 40.50 | — | 40.50 | — |

**VersaBond® bonding mortar, Custom® Building Products.** All-purpose tile mortar for walls and floors. Trowels smoothly and cures fast, even in cold weather. Just add water and mix; no additives needed.

| | Craft@Hrs | Unit | Material | Labor | Total | Sell |
|---|---|---|---|---|---|---|
| Gray, 50 pounds | — | Ea | 14.20 | — | 14.20 | — |
| White, 50 pounds | — | Ea | 16.05 | — | 16.05 | — |

| | Craft@Hrs | Unit | Material | Labor | Total | Sell |
|---|---|---|---|---|---|---|

**Medium bed mortar, Custom® Building Products.** For installing large tile over 12" x 12" tile or natural stone. To 3/4" thick on horizontal applications. Meets ANSI A118.4.

| | | | | | | |
|---|---|---|---|---|---|---|
| Gray, 50 pounds | — | Ea | 11.70 | — | 11.70 | — |

**CustomBlend® thin-set mortar, Custom® Building Products.** Economical bonding adhesive for most basic job tile installations. Good for floor installations and Saltillo or clay pavers. Offers good open time, pot life and adjustment time to apply and adjust tiles. Interior or exterior use. Mix with acrylic mortar admix for best results when setting dense tile or setting over hard-to-bond-to surfaces such as laminates, existing ceramic tile and plywood.

| | | | | | | |
|---|---|---|---|---|---|---|
| Gray, 50 pounds | — | Ea | 5.50 | — | 5.50 | — |
| White, 50 pounds | — | Ea | 7.90 | — | 7.90 | — |

**100 percent solids epoxy mortar, Custom® Building Products.** For use where high chemical resistance is required as in commercial kitchens, dairies, and animal hospitals. Pure, 3-component epoxy mortar with high bond and compressive strengths. Provides exceptional resistance to chemicals and staining. Water washable. Also for installation of true green marble.

| | | | | | | |
|---|---|---|---|---|---|---|
| 5-gallon pail | — | Ea | 63.70 | — | 63.70 | — |

**FlexBond® premium flexible bonding mortar, Custom® Building Products.** For tiling over surfaces subject to minor movement and hard-to-bond-to surfaces such as laminates, ceramic tile and plywood. Allows tiling over cracks up to 1/16" (2 mm) without repairs. Just add water and mix. No additives needed.

| | | | | | | |
|---|---|---|---|---|---|---|
| Gray, 50 pounds | — | Ea | 27.50 | — | 27.50 | — |
| White, 50 pounds | — | Ea | 28.20 | — | 28.20 | — |

**CustomFloat® bedding mortar, Custom® Building Products.** A lightweight Portland cement-based, pre-blended mortar for use as a bedding or brown coat. Very low shrinkage, exceptional bond strength. For floors, walls and countertops. Excellent for overhead work such as showers, arches and coves.

| | | | | | | |
|---|---|---|---|---|---|---|
| 50 pounds | — | Ea | 14.30 | — | 14.30 | — |

**Porcelain mortar, Custom® Building Products.** For setting impervious porcelain, glass mosaics and large tile to concrete. Also used over radiant heating systems.

| | | | | | | |
|---|---|---|---|---|---|---|
| 50 pounds | — | Ea | 21.30 | — | 21.30 | — |

**Marble and granite mortar mix, Custom® Building Products.** For setting marble, granite, other natural stone, large modular tile, Saltillo pavers and other tiles requiring a medium-bed mortar. Can be applied up to 3/4" (19 mm) thick. Polymer-modified for high bond strength. Interior or exterior.

| | | | | | | |
|---|---|---|---|---|---|---|
| White, 50 pounds | — | Ea | 22.60 | — | 22.60 | — |

**Marble and granite stone setting adhesive, Custom® Building Products.** Formulated specifically for setting moisture-sensitive stones such as green and black marble and granite. Fast-setting, self-drying technology inhibits warping and staining that can occur with traditional thin-set mortars. Grout in two hours. No additives required. Interior use only.

| | | | | | | |
|---|---|---|---|---|---|---|
| 12.5 pounds | — | Ea | 14.20 | — | 14.20 | — |

**Glass block mortar mix, Custom® Building Products.** For installing all types of glass block. Suitable for new installations or to repair old mortar joints. Pre-mixed.

| | | | | | | |
|---|---|---|---|---|---|---|
| White, 50 pounds | — | Ea | 10.90 | — | 10.90 | — |

**PremiumPlus® thin-set mortar, Custom® Building Products.** Extra high bond strength for most basic job tile installations. Offers maximum open time, pot life and extra long working time.

| | | | | | | |
|---|---|---|---|---|---|---|
| Gray, 50 pounds | — | Ea | 12.10 | — | 12.10 | — |
| White, 50 pounds | — | Ea | 14.30 | — | 14.30 | — |

**Tile repair mortar, Custom® Building Products.** A polymer-modified mortar with maximum bond strength. Use to replace broken interior or exterior tile. Ready to grout in 24 hours.

| | | | | | | |
|---|---|---|---|---|---|---|
| White, 1 pound | — | Ea | 4.32 | — | 4.32 | — |

| | Craft@Hrs | Unit | Material | Labor | Total | Sell |
|---|---|---|---|---|---|---|

**Underwater tile repair mortar, Custom® Building Products.** Replace broken tile underwater without draining fountains, pools or spas. Mix with water for immediate use.

| | | | | | | |
|---|---|---|---|---|---|---|
| 1 pound | — | Ea | 8.66 | — | 8.66 | — |

**Acrylic mortar admix, Custom® Building Products.** Greatly increases bond strength, water-resistance and impact-resistance of non-polymer modified thin-sets, mortars and grouts. Use with PremiumPlus®, MasterBlend™ and CustomBlend® thin-sets, Portland-cement mortar mixes and non-polymer modified grouts. Recommended to use with non-polymer modified thin-set mortars when setting vitreous, dense tile or setting over hard-to-bond-to surfaces such as laminates, existing ceramic tile and plywood.

| | | | | | | |
|---|---|---|---|---|---|---|
| 3 quarts | — | Ea | 9.75 | — | 9.75 | — |
| 2-1/2 gallons | — | Ea | 20.80 | — | 20.80 | — |

**AFM® anti-fracture membrane kit, ProtectoWrap.** Peel-and-stick 40 mm membrane. For indoor use under thin-set tile, stone and marble for crack suppression. Includes primer, knife, and instruction sheet.

| | | | | | | |
|---|---|---|---|---|---|---|
| 12" x 25' roll, per square foot | — | SF | 1.24 | — | 1.24 | — |
| 5' wide roll, per square foot | — | SF | .95 | — | .95 | — |

**Ceramic tile adhesive caulk, Tile Perfect.** For tile, tubs, showers, and countertops. Mold and mildew resistant. Remains flexible and paintable.

| | | | | | | |
|---|---|---|---|---|---|---|
| 5.5 ounce tube | — | Ea | 3.24 | — | 3.24 | — |

**White dry tile grout (non-sanded), Custom® Building Products.** Portland cement-based grout for interior/exterior floor and wall tiles with narrow joints up to 1/8" (3 mm). For soft-glazed, high-glazed and marble tile subject to scratching by sanded grouts.

| | | | | | | |
|---|---|---|---|---|---|---|
| 1 pound | — | Ea | 1.26 | — | 1.26 | — |
| 5 pounds | — | Ea | 4.77 | — | 4.77 | — |
| 10 pounds | — | Ea | 7.78 | — | 7.78 | — |
| 25 pounds | — | Ea | 9.08 | — | 9.08 | — |

**Grout colorant, Custom® Building Products.** Renews or changes grout color, interior or exterior. Seals grout joints and evens color. Water-based and fade-resistant. Available in 10 popular Polyblend® grout colors.

| | | | | | | |
|---|---|---|---|---|---|---|
| 1/2 pint, most colors | — | Ea | 9.70 | — | 9.70 | |

**Saltillo grout mix, Custom® Building Products.** Pre-blended mix of Portland cement, silica sand and pigments. For any interior or exterior tile with wide grout joints, 1/2" to 1-1/4". Recommend mixing with acrylic mortar admix diluted 1:1 with water for higher strength.

| | | | | | | |
|---|---|---|---|---|---|---|
| Gray, tan or red, 50 pounds | — | Ea | 11.20 | — | 11.20 | — |

**Sanded tile grout, Weco.** Provides a smooth, colorfast joint for wall and floor installations.

| | | | | | | |
|---|---|---|---|---|---|---|
| Most colors, 25 pounds | — | Ea | 9.90 | — | 9.90 | — |

**Polyblend® ceramic tile caulk, Custom® Building Products.** Fills and repairs joints around sinks, tubs, showers and countertops. Siliconate acrylic formula.

| | | | | | | |
|---|---|---|---|---|---|---|
| Sanded, various colors, 5.5 ounces | — | Ea | 3.61 | — | 3.61 | — |

**High-gloss finish/sealer, Miracle Sealants Co.** Medium-to-high gloss interior surface sealer for Saltillo tile, terra cotta tile, unglazed ceramic tile, brick, slate and concrete. One gallon covers 500 to 1000 square feet.

| | | | | | | |
|---|---|---|---|---|---|---|
| Quart | — | Ea | 9.13 | — | 9.13 | — |
| Gallon | — | Ea | 27.60 | — | 27.60 | — |

**Sealer's Choice 15 Gold, Aqua Mix.** Water-based sealer leaves a no-sheen natural look. Maximum stain protection in food preparation areas, on porous tile and natural stone. One gallon covers 300 to 1500 square feet.

| | | | | | | |
|---|---|---|---|---|---|---|
| Pint | — | Ea | 21.60 | — | 21.60 | — |
| Quart | — | Ea | 32.70 | — | 32.70 | — |
| Gallon | — | Ea | 98.50 | — | 98.50 | — |

| | Craft@Hrs | Unit | Material | Labor | Total | Sell |
|---|---|---|---|---|---|---|

**Tile gloss and seal, Jasco®.** Two coats protect from grease, food and oil. For unglazed Mexican paver, terrazzo, flagstone, and Saltillo. Clear interior and exterior tile sealer. Non-yellowing, non-flammable, low odor. Dries in 8 to 24 hours. Water clean up. Wax can be applied over product. One gallon covers 125 to 300 square feet.

| | Craft@Hrs | Unit | Material | Labor | Total | Sell |
|---|---|---|---|---|---|---|
| Gallon | — | Ea | 21.80 | — | 21.80 | — |
| 2.5 gallons | — | Ea | 48.60 | — | 48.60 | — |

**Grout sealer, Aqua Mix.** Water-based sealer resists water, oil and acid-based contaminants. Repels food, dirt and grease. Does not change the appearance of grout. Inhibits mildew and bacteria. For interior and exterior use.

| | Craft@Hrs | Unit | Material | Labor | Total | Sell |
|---|---|---|---|---|---|---|
| Pint | — | Ea | 8.56 | — | 8.56 | — |

**Ditra detaching membrane, Schluter Systems.** Polyethylene membrane with a grid structure of square, cut-back cavities on a base of anchoring fleece. Forms an uncoupling, waterproofing, and vapor pressure equalization layer. Installed in binding mortar spread with a notched trowel. Apply tile over Ditra using thin-set mortar.

| | Craft@Hrs | Unit | Material | Labor | Total | Sell |
|---|---|---|---|---|---|---|
| Detaching membrane | — | SF | 1.27 | — | 1.27 | — |

## Carpet

Includes sweeping a prepared surface, setting carpet tack strip, rebond pad, unrolling 12' to 15' wide carpet, measuring, marking, cutting, trimming one edge, hot melt tape on seams and disposal of debris.

Minimum quality olefin, 25 to 35 ounce, with 1/2" (3 lb. density) pad

| | Craft@Hrs | Unit | Material | Labor | Total | Sell |
|---|---|---|---|---|---|---|
| | BF@.150 | SY | 15.50 | 4.43 | 19.93 | 33.90 |

Standard quality polyester or olefin, 35 to 50 ounce, with 1/2" (6 lb. density) pad

| | | | | | | |
|---|---|---|---|---|---|---|
| | BF@.160 | SY | 20.60 | 4.73 | 25.33 | 43.10 |

Better quality nylon, 50 (plus) ounce, with 1/2" (6 lb. density) pad

| | Craft@Hrs | Unit | Material | Labor | Total | Sell |
|---|---|---|---|---|---|---|
| | BF@.170 | SY | 30.90 | 5.02 | 35.92 | 61.10 |
| Wool carpet, 36 ounce medium traffic | BF@.300 | SY | 36.10 | 8.86 | 44.96 | 76.40 |
| Wool carpet, 42 ounce heavy traffic | BF@.300 | SY | 46.40 | 8.86 | 55.26 | 93.90 |
| Sisal carpet, no pad | BF@.130 | SY | 10.30 | 3.84 | 14.14 | 24.00 |
| Foam-backed olefin carpet, no pad | BF@.130 | SY | 7.88 | 3.84 | 11.72 | 19.90 |
| Add for typical furniture moving | BF@.060 | SY | — | 1.77 | 1.77 | 3.01 |
| Add for waterfall (box steps) stairways | BF@.190 | Step | — | 5.61 | 5.61 | 9.54 |
| Add for wrapped steps (open riser), sewn | BF@.250 | Step | — | 7.39 | 7.39 | 12.60 |
| Add for steps with sewn edge on one side | BF@.310 | Step | — | 9.16 | 9.16 | 15.60 |
| Add for circular stair steps, typical cost | BF@.440 | Step | — | 13.00 | 13.00 | 22.10 |
| Add for carpet edge binding | BF@.062 | LF | — | 1.83 | 1.83 | 3.11 |
| Rebond 1/2" thick 4-pound pad only | — | SY | 2.30 | — | 2.30 | — |
| Rebond 1/2" thick 6-pound pad only | — | SY | 3.71 | — | 3.71 | — |
| Carpet tile, 22" x 23", with adhesive | BF@.022 | SF | 2.77 | .65 | 3.42 | 5.81 |
| Peel and stick carpet tile, 12" x 12" | BF@.020 | SF | .98 | .59 | 1.57 | 2.67 |
| Indoor/outdoor carpet, 16 ounce olefin | BF@.025 | SF | .56 | .74 | 1.30 | 2.21 |

# Kitchens

**12**

Living area in a home can be divided into three categories: the private area (bedrooms and bathrooms), the relaxing area (living room, dining room, den) and the work area (kitchen and laundry).

This work area is more efficient when the kitchen is like the hub of the wheel. The dining area should be only a few steps away. The entrance off the garage should be nearby in another direction to make unloading groceries easier. The laundry room should be another spoke off the hub. Anyone preparing meals in the kitchen should be able to monitor the washer and dryer. Storage space (or a pantry) should be close at hand. But remember that a kitchen is a work area. Traffic from the rear entry or laundry shouldn't pass through the kitchen work triangle — range-to-refrigerator-to-sink.

In the 1960s and 1970s, designers favored small kitchens on the theory that smaller work triangles save steps and reduce fatigue. The proliferation of appliances at the end of the 20th century made small kitchens seem cramped. Modern kitchen counters need space for juicers, extractors, grinders, mixers, toasters, dispensers, coffee makers, microwave ovens, electric pots, grills, griddles and more.

Avoid setting a door that swings *into* a kitchen. Doors interfere with use of appliances, cabinets and countertops. If a door is essential, consider a pocket or folding door. Provide enough glass area to make the kitchen a light and cheerful place. Time spent in this work area is more enjoyable when there's a window over the kitchen sink with a view. To make food preparation easier for outdoor living, install a large sliding window over the sink or counter. Extend the window sill 14" beyond the window to form a level shelf for passage of plates or a food tray between interior and exterior.

Every kitchen is also a family meeting-place. Consider combining work with pleasure. Merge the kitchen (work area) with the family room (relaxing area) by removing a partition or by adding extra living space. Figure 12-1 shows a kitchen and family room combined with an island cook center that doubles as a breakfast bar for casual dining.

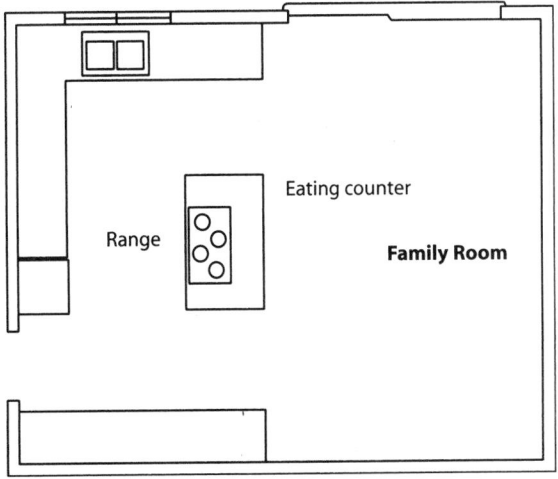

**Figure 12-1**

*Island counter dividing kitchen and family room*

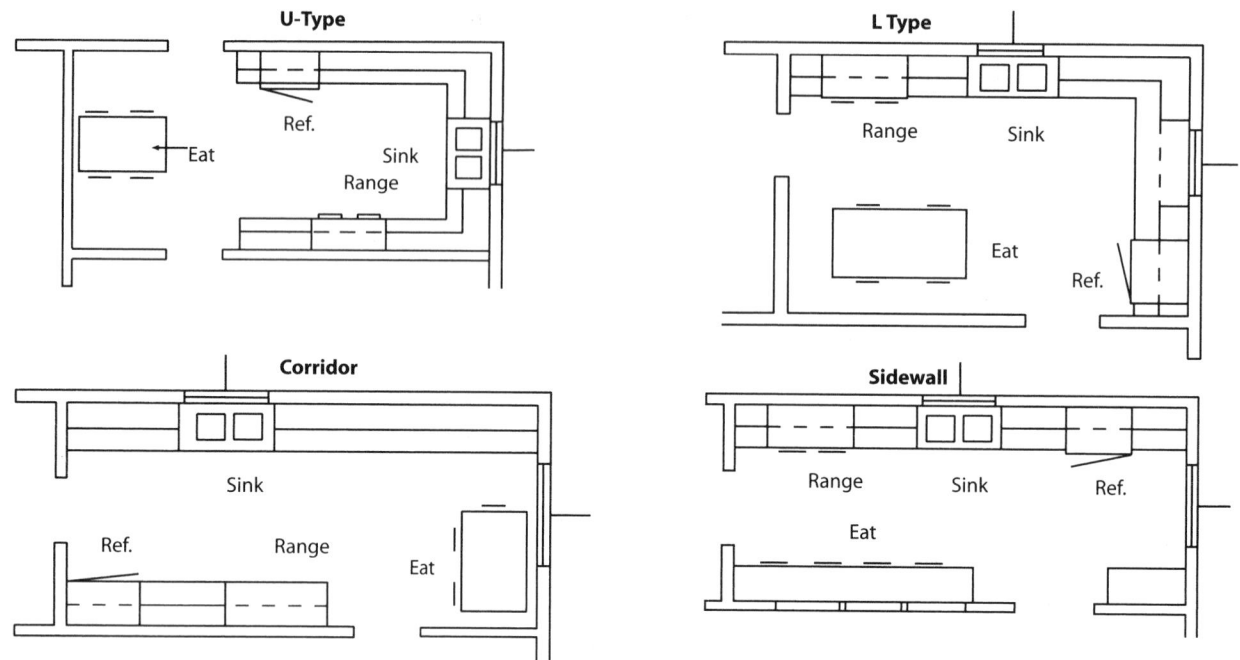

**Figure 12-2**

*Kitchen arrangements*

Food usually moves from the refrigerator to the sink to the range. That's the work triangle. Figure 12-2 shows four common kitchen floor plans: "U", "L", corridor and sidewall kitchens. The work triangle is smallest in "U" and corridor kitchens. Use a sidewall kitchen when space is very limited. The "L" arrangement is best for a square kitchen that includes an eating area with a table.

# Cabinets and Counters

Here's a yardstick to judge whether there's enough cabinet and counter space in a home. Figures are in linear feet of counter or cabinet face per thousand square feet of floor in the living area (everything under the roof except the garage, soffit, patio and porch). Allow a little more counter and cabinet space in a small home (under 1200 square feet). A very large home (over 4,000 square feet) may need less space than indicated below. Measure linear feet of counter at the front or back edge, whichever is shorter.

❖ *Counter.* 8 linear feet per 1000 square feet of living area

❖ *Base cabinet.* 7 linear feet per 1000 square feet of living area

❖ *Wall cabinet.* 6 linear feet per 1000 square feet of living area

**Notes:** Base cabinet length assumes one sink base cabinet. Wall cabinets can include vertical storage cabinets (such as for trays and cookie sheets) above a built-in oven. Count utility cabinets (6' high or more) as both base and wall cabinet. Credit a corner rotating shelf (Lazy Susan) base cabinet with 2 linear feet of base cabinet, regardless of the actual width.

❖ *Drawer base cabinet.* one per 10 linear feet of base cabinet

❖ *Counter width and height.* 25" wide ✕ 36" high

❖ *Dishwasher.* 24" wide ✕ 35" high

❖ *Trash compactor.* 15" wide ✕ 35" high

Modern kitchens include a work desk with connections for electric service, telephone, computer and TV. Minimum desk width is about 3' with a book shelf (for recipe books) above the desk. Minimum desk cutout (seating area) is 24". Install a 14"- to 20"-wide drawer base cabinet at the right of the cutout. Desk depth can be 25", the same as normal counter depth. Desk height should be 28". Provide task lighting above the desk.

# Refreshing the Look

Kitchen cabinets seldom wear out before they go out of style. If kitchen cabinet space is adequate and well arranged, refinishing cabinets and adding new hardware may be enough. Replace roller catches with magnetic hardware. The magnet attaches to the cabinet interior and the complementary metal plate fits on the door interior. New knobs or pulls (handles) complete the makeover.

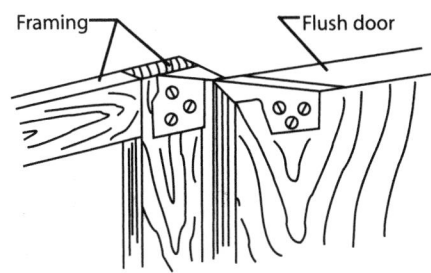

**Figure 12-3**
*Concealed hinge used with flush cabinet door*

For a more modern look, replace cabinet doors and drawer fronts on the existing cabinet frames. Consider flush doors with concealed (European) hinges, as in Figure 12-3. Cabinet doors fit edge to edge for a continuous panel effect. Finger slots at the edge of doors and drawers make cabinet knobs and pulls unnecessary.

Countertops are subject to considerable wear and often need to be replaced before the cabinets. Installing a simulated stone countertop will modernize any kitchen. Figure 12-4 shows standard counter and cabinet dimensions. Counter height is usually 36". Counter depth is usually 25".

# Cabinet Costs

Cabinets are like furniture. Prices vary widely. For example, a drawer with hardwood rails, plywood bottom and dovetail joints will cost considerably more than a particleboard drawer with stapled butt joints. Better-quality custom and semi-custom cabinets are made from 3/4" or 1/2" furniture-grade plywood covered with hardwood veneer. Drawers have full-extension roller hardware. Less-expensive cabinets are made from 3/8" or 1/2" particleboard, usually with a melamine coating.

"Ready to assemble" cabinets cost the least and are sold primarily in D-I-Y outlets. Options are limited and units require about an hour of assembly before they are ready to install.

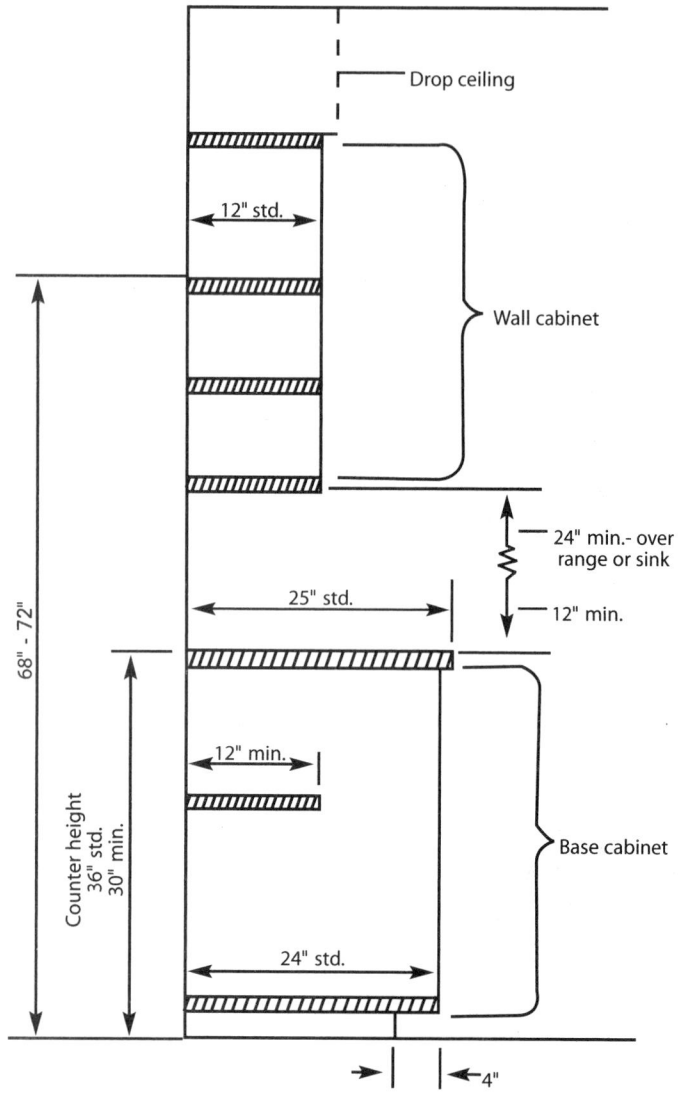

**Figure 12-4**

*Kitchen cabinet standards*

Cabinet costs listed in this chapter are based on good quality stock units such as those offered by American Woodmark, IXL (Triangle Pacific), Kabinart, Merillat, Mills Pride, Prestige, and Thomasville. Semi-custom cabinets will cost about 50 percent more. Manufacturers of semi-custom cabinets include Brandom, Decora, Diamond, Kemper, KraftMaid, Schrock, Shenandoah and Yorktowne. Costs for true custom cabinets will be about double the figures listed later in this chapter. Vendors of true custom cabinets include Crystal, J.H. Brubaker, DBS, Fieldstone, Neff, Omega, Plato, Poggenpohl, Rutt, Snaidero, SieMatic, Studio Becker, Wood-Mode, and any cabinet shop in your community.

The advantage of custom-made cabinets is flexibility in style, finish, size and design. For example, custom cabinets can be made in any width or height to fit any kitchen. Stock cabinets come in widths that increment 3 inches at a time, such as 15", 18", 21" and 24". Most installations will require a filler strip to extend the line of cabinets to exactly the right length. The disadvantage of custom cabinets is price — about double the cost of stock cabinets. Incidentally, nearly all cabinets are priced with the screws, hinges, rails and guides needed to finish the job. But door and drawer pulls and knobs are generally left to the discretion of the homeowner.

Cabinet prices in this chapter reflect what most home improvement contractors install — stock cabinets with flat panel faces, picture frame molding or a simple design. Doors and drawers with grooves, raised panels, bead or elaborate molding will cost more. Full overlay doors (installed with Euro hinges mortised into the interior cabinet wall) hide nearly all the cabinet frame and will cost more than traditional doors with hinges mounted on the exterior. Doors and drawers set inside the cabinet face frame (*inset*) will cost even more than full overlay doors and drawers.

About 60 percent of the cost of most cabinets is the wood itself. So you can expect to pay more for cabinets made with exotic wood veneers such as cherry, hickory, alder, redwood or teak on a plywood base. Oak, birch, maple and pine are the most common wood species used for cabinets. The least expensive cabinets have a melamine or plastic laminate surface on a particleboard core. Cabinets surfaced in stainless steel are among the most expensive. Some older homes have steel cabinets that were popular in the middle of the 20th century. Steel cabinets are still made, but their primary use is in hospitals and laboratories rather than in homes.

Wood cabinets are stained to bring out the warmth of the wood grain and then sealed for moisture protection. Install unfinished cabinets if your client wants to match existing woodwork in the home. But be prepared to do a lot of sanding and hand rubbing to produce a finish equal to the best of stock cabinets. Many cabinet vendors offer special custom glazes and layered finishes at extra cost — usually about 10 percent more than standard stained cabinets.

# Kitchen Appliances

---

### Kitchen Remodeling Checklist

❏ *Cabinets* — Soffits, countertops, size of back splash, accommodating a dishwasher?

❏ *Ceiling* — Repaired, acoustic tile, suspended ceiling, drywall

❏ *Electrical* — See the Kitchen Electrical Service Checklist for new outlets needed

❏ *Floor* — Underlayment, vinyl tile, sheet vinyl, ceramic or clay tile

❏ *Heating and cooling* — Ducts or radiators relocated

❏ *Doors* — Change in location or swing. New trim

❏ *Plumbing* — New venting, gas line relocated, change sink location, dishwasher

❏ *Structural changes* — Partitions removed or moved

❏ *Wall repairs* — Ceramic tile, paint, wallpaper, paneling

❏ *Windows* — Relocated, increase or decrease in size. Storm windows

❏ *Ventilation* — Kitchen exhaust hood, ducted or ductless

---

Most home improvement contractors avoid reselling kitchen appliances that only need to be plugged in. Every homeowner knows where to get a good deal on a refrigerator. Instead of quoting prices, include a cash allowance for appliances in the contract price. For example, your contract might include the following language:

> *This agreement includes an allowance of $ _____ for the following appliances:*
>
> _____
>
> *Owner agrees to bear the cost of appliances that exceed this allowance. Contractor agrees to furnish and install all gas, water, drain, vent and electrical lines required for operation of these appliances. Contractor is not responsible for installation, service or maintenance of these appliances.*

Built-in appliances, such as garbage disposers, cooktops and wall ovens are an exception to the rule on appliances. These are fixtures, a part of the home itself, and should be included in your bid as a courtesy to the owner. Be sure to specify in the contract the brand and model the customer chose so that's what you price.

You'll never see an old kitchen with adequate electrical service. Upgrading the electrical service is a prime reason for remodeling most kitchens. Consider the following checklist when planning extra runs from the electrical service panel. Note that some building codes require as many as three ground-fault receptacles in the kitchen. If you're adding base cabinets, figure which electrical outlets have to be moved.

### Kitchen Electrical Service Checklist

- ❏ Ceiling fixture
- ❏ Ceiling paddle fan
- ❏ Clock in soffit
- ❏ Dishwasher
- ❏ Disposer
- ❏ Electric range
- ❏ Hood over the stove
- ❏ Soffit lighting (fluorescent strip)
- ❏ Light fixture recessed over the sink
- ❏ Light fixture over the desk
- ❏ Microwave oven
- ❏ Oven
- ❏ Refrigerator
- ❏ Small appliance outlets
- ❏ Three ground-fault receptacles
- ❏ Trash compactor
- ❏ T.V.
- ❏ Wall or ceiling exhaust fan
- ❏ Wall switches

# Other Cost Considerations

Extensive kitchen remodeling will usually require HVAC work such as moving duct, registers, grilles, hydronic piping or radiators. Consider also the cost of patching walls or ceilings after HVAC materials have been moved or added. Figure a half-day of work ($150) as the minimum charge for drywall hanging, taping and finishing.

| | Craft@Hrs | Unit | Material | Labor | Total | Sell |
|---|---|---|---|---|---|---|

## Kitchen Demolition

**Remove cabinets.** Per linear foot of cabinet face or back (whichever is longer). The cost of removing base cabinets includes the cost of removing the countertop. Add the cost of moving plumbing, electrical and HVAC lines. These figures assume debris is piled on site. No salvage of materials or fixture moving is included.

| | Craft@Hrs | Unit | Material | Labor | Total | Sell |
|---|---|---|---|---|---|---|
| Wall cabinets, wood | BL@.250 | LF | — | 6.66 | 6.66 | 11.30 |
| Wall cabinets, metal | BL@.350 | LF | — | 9.32 | 9.32 | 15.80 |
| Base cabinets, wood | BL@.400 | LF | — | 10.70 | 10.70 | 18.20 |
| Base cabinets, metal | BL@.550 | LF | — | 14.70 | 14.70 | 25.00 |
| Remove cabinet door only | BC@.250 | Ea | — | 8.02 | 8.02 | 13.60 |
| Remove and replace cabinet door | BC@.580 | Ea | — | 18.60 | 18.60 | 31.60 |

**Remove kitchen fixtures.** These figures include turning off the water, disconnecting the drain, capping the lines, disconnecting faucets and fittings, removing the sink from the base cabinet and piling debris on site. No salvage value assumed. Reduce these costs by 25% if countertop and base cabinets are also being demolished. If a new floor surface has been laid over the original floor, there may be no way to remove undercounter appliances without damaging either the floor or the countertop.

| | Craft@Hrs | Unit | Material | Labor | Total | Sell |
|---|---|---|---|---|---|---|
| Self-rimming sink | BL@1.00 | Ea | — | 26.60 | 26.60 | 45.20 |
| Ledge-type or huddee ring sink | BL@1.65 | Ea | — | 44.00 | 44.00 | 74.80 |
| Self-rimming sink with disposer | BL@1.50 | Ea | — | 40.00 | 40.00 | 68.00 |
| Ledge-type or huddee ring sink with disposer | BL@2.00 | Ea | — | 53.30 | 53.30 | 90.60 |
| Garbage disposer | BL@.500 | Ea | — | 13.30 | 13.30 | 22.60 |
| Undercounter dishwasher | BL@1.00 | Ea | — | 26.60 | 26.60 | 45.20 |
| Undercounter trash compactor | BL@.800 | Ea | — | 21.30 | 21.30 | 36.20 |

**Remove ceiling cover.** Per square foot of ceiling. Add the cost of removing electrical fixtures. Debris piled on site. No salvage of materials included except as noted.

| | Craft@Hrs | Unit | Material | Labor | Total | Sell |
|---|---|---|---|---|---|---|
| Plaster ceiling | | | | | | |
|     Including lath and furring | BL@.025 | SF | — | .67 | .67 | 1.14 |
|     Including suspended grid | BL@.020 | SF | — | .53 | .53 | .90 |
| Acoustic tile ceiling | | | | | | |
|     Including suspended grid | BL@.010 | SF | — | .27 | .27 | .46 |
|     Including grid in salvage condition | BL@.019 | SF | — | .51 | .51 | .87 |
|     Tile glued or stapled to ceiling | BL@.015 | SF | — | .40 | .40 | .68 |
|     Tile on strip furring, including furring | BL@.025 | SF | — | .67 | .67 | 1.14 |
| Drywall ceiling | | | | | | |
|     Nailed or attached with screws to joists | BL@.010 | SF | — | .27 | .27 | .46 |
|     Dropped drywall ceiling on wood or metal grid, including grid | BL@.022 | SF | — | .59 | .59 | 1.00 |
|     Drywall on plaster and lath, including lath and plaster | BL@.041 | SF | — | 1.09 | 1.09 | 1.85 |

**Remove floor cover.** Per square yard of floor cover removed. Debris piled on site. No salvage of materials included.

| | Craft@Hrs | Unit | Material | Labor | Total | Sell |
|---|---|---|---|---|---|---|
| Ceramic tile, pneumatic breaker | BL@.263 | SY | — | 7.01 | 7.01 | 11.90 |
| Hardwood, nailed | BL@.290 | SY | — | 7.73 | 7.73 | 13.10 |
| Hardwood, glued | BL@.503 | SY | — | 13.40 | 13.40 | 22.80 |
| Linoleum or sheet vinyl | BL@.056 | SY | — | 1.49 | 1.49 | 2.53 |
| Resilient tile | BL@.300 | SY | — | 7.99 | 7.99 | 13.60 |
| Terrazzo | BL@.286 | SY | — | 7.62 | 7.62 | 13.00 |
| Carpet on tack strip | BL@.028 | SY | — | .75 | .75 | 1.28 |
| Bonded floor, scraped | BL@.135 | SY | — | 3.60 | 3.60 | 6.12 |

| | Craft@Hrs | Unit | Material | Labor | Total | Sell |
|---|---|---|---|---|---|---|

**Remove wall cover.** Per square foot of wall. Debris piled on site.

| | Craft@Hrs | Unit | Material | Labor | Total | Sell |
|---|---|---|---|---|---|---|
| Drywall nailed or attached with screws to joists | BL@.010 | SF | — | .27 | .27 | .46 |
| Drywall on plaster and lath, including lath and plaster | BL@.041 | SF | — | 1.09 | 1.09 | 1.85 |
| Plywood or insulation board | BL@.018 | SF | — | .48 | .48 | .82 |
| Plaster and lath | BL@.025 | SF | — | .67 | .67 | 1.14 |
| Beaded partitioning dado | BL@.025 | SF | — | .67 | .67 | 1.14 |

**Soffits.** These figures assume a 12" wide by 12" high soffit or drop ceiling (as in Figure 12-4) installed above wall cabinets. Per linear foot of soffit measured on the longest side. Demolition figures assume debris is piled on site. Figure a half-day of work ($150) as the minimum charge for drywall or plaster finish.

| | Craft@Hrs | Unit | Material | Labor | Total | Sell |
|---|---|---|---|---|---|---|
| Remove wood frame soffit with drywall or plaster finish | BL@.059 | LF | — | 1.57 | 1.57 | 2.67 |
| Install wood frame soffit with drywall finish | BC@.471 | LF | 2.62 | 15.10 | 17.72 | 30.10 |
| Install wood frame soffit with rock lath and plaster finish | BC@.600 | LF | 2.74 | 19.30 | 22.04 | 37.50 |
| Illuminated soffit with ceiling strips, battens and luminous ceiling panel | BC@.318 | LF | 3.00 | 10.20 | 13.20 | 22.40 |
| Undercounter refrigerator | BL@1.00 | Ea | — | 26.60 | 26.60 | 45.20 |
| Pendent, wall or recessed light fixture | BL@.590 | Ea | — | 15.70 | 15.70 | 26.70 |

## Kitchen Cabinets

For detailed coverage of custom-built cabinets, see *National Framing & Finish Carpentry Estimator*, http://CraftsmanSiteLicense.com.

**Unfinished oak cabinet door only.** Ready to stain. With hardware. Labor cost assumes exact fit replacement door with standard hinges.

| | Craft@Hrs | Unit | Material | Labor | Total | Sell |
|---|---|---|---|---|---|---|
| 10" x 22" | BC@.250 | Ea | 18.30 | 8.02 | 26.32 | 44.70 |
| 10" x 28" | BC@.250 | Ea | 20.00 | 8.02 | 28.02 | 47.60 |
| 13" x 13" | BC@.250 | Ea | 14.60 | 8.02 | 22.62 | 38.50 |
| 13" x 22" | BC@.250 | Ea | 20.00 | 8.02 | 28.02 | 47.60 |
| 13" x 28" | BC@.250 | Ea | 22.30 | 8.02 | 30.32 | 51.50 |
| 14-1/2" x 22" | BC@.250 | Ea | 21.20 | 8.02 | 29.22 | 49.70 |
| 14" x 28" | BC@.250 | Ea | 23.50 | 8.02 | 31.52 | 53.60 |
| 16" x 13" | BC@.250 | Ea | 17.60 | 8.02 | 25.62 | 43.60 |
| 16" x 22" | BC@.250 | Ea | 18.90 | 8.02 | 26.92 | 45.80 |
| 16" x 28" | BC@.250 | Ea | 21.20 | 8.02 | 29.22 | 49.70 |
| 22" x 22" | BC@.250 | Ea | 25.90 | 8.02 | 33.92 | 57.70 |
| 22" x 28" | BC@.250 | Ea | 29.30 | 8.02 | 37.32 | 63.40 |

**Unfinished oak cabinet drawer front only.** Ready to stain. With hardware.

| | Craft@Hrs | Unit | Material | Labor | Total | Sell |
|---|---|---|---|---|---|---|
| 10" x 5-3/4" | BC@.250 | Ea | 6.50 | 8.02 | 14.52 | 24.70 |
| 13" x 5-3/4" | BC@.250 | Ea | 7.00 | 8.02 | 15.02 | 25.50 |
| 16" x 5-3/4" | BC@.250 | Ea | 7.60 | 8.02 | 15.62 | 26.60 |
| 22" x 5-3/4" | BC@.250 | Ea | 10.00 | 8.02 | 18.02 | 30.60 |

**Unfinished oak cabinet door trim.** Ready to stain.

| | Craft@Hrs | Unit | Material | Labor | Total | Sell |
|---|---|---|---|---|---|---|
| 3" x 30" filler strip | BC@.500 | Ea | 8.50 | 16.00 | 24.50 | 41.70 |
| 3" x 36" valance | BC@.250 | Ea | 15.80 | 8.02 | 23.82 | 40.50 |
| 4-1/2" x 96" toe kick | BC@.750 | Ea | 7.59 | 24.10 | 31.69 | 53.90 |

**Unfinished oak cabinet end panel.** Ready to stain.

| | Craft@Hrs | Unit | Material | Labor | Total | Sell |
|---|---|---|---|---|---|---|
| 11" x 30", wall cabinet | BC@.200 | Ea | 5.45 | 6.42 | 11.87 | 20.20 |
| 24" x 34-1/2", base cabinet | BC@.200 | Ea | 10.50 | 6.42 | 16.92 | 28.80 |
| 24" x 84", utility cabinet | BC@.200 | Ea | 21.90 | 6.42 | 28.32 | 48.10 |

| | Craft@Hrs | Unit | Material | Labor | Total | Sell |
|---|---|---|---|---|---|---|

## Unfinished Birch Kitchen Cabinets

**Unfinished birch base cabinets.** 34-1/2" high, 24" deep. Birch face frame and door frame. Ready to stain. With hardware.

| | Craft@Hrs | Unit | Material | Labor | Total | Sell |
|---|---|---|---|---|---|---|
| 12" wide, 1 door | BC@.640 | Ea | 38.00 | 20.50 | 58.50 | 99.50 |
| 15" wide, 1 door | BC@.640 | Ea | 51.00 | 20.50 | 71.50 | 122.00 |
| 18" wide, 1 door | BC@.770 | Ea | 58.20 | 24.70 | 82.90 | 141.00 |
| 18" wide, 3 drawer | BC@.770 | Ea | 62.40 | 24.70 | 87.10 | 148.00 |
| 24" wide, 2 doors | BC@.770 | Ea | 70.00 | 24.70 | 94.70 | 161.00 |
| 24" wide, 3 drawer | BC@.910 | Ea | 97.50 | 29.20 | 126.70 | 215.00 |
| 30" wide, 2 doors | BC@.910 | Ea | 80.60 | 29.20 | 109.80 | 187.00 |
| 36" wide, 2 doors | BC@.910 | Ea | 90.20 | 29.20 | 119.40 | 203.00 |
| 36" wide, blind corner base, 1 door | BC@.910 | Ea | 122.00 | 29.20 | 151.20 | 257.00 |
| 36" wide, sink base, 1 door | BC@.770 | Ea | 86.60 | 24.70 | 111.30 | 189.00 |
| 48" wide, sink base, 2 doors | BC@.770 | Ea | 107.00 | 24.70 | 131.70 | 224.00 |
| 60" wide, sink base, 2 doors | BC@.910 | Ea | 124.00 | 29.20 | 153.20 | 260.00 |

**Unfinished birch wall cabinets.** 12" deep. Birch face frame and door frame. Ready to stain. With hardware.

| | Craft@Hrs | Unit | Material | Labor | Total | Sell |
|---|---|---|---|---|---|---|
| 12" wide x 30" high, 1 door | BC@.640 | Ea | 25.00 | 20.50 | 45.50 | 77.40 |
| 15" wide x 30" high, 1 door | BC@.640 | Ea | 33.80 | 20.50 | 54.30 | 92.30 |
| 18" wide x 30" high, 1 door | BC@.770 | Ea | 39.20 | 24.70 | 63.90 | 109.00 |
| 24" wide x 30" high, corner, 1 door | BC@.770 | Ea | 77.30 | 24.70 | 102.00 | 173.00 |
| 24" wide x 30" high, 2 doors | BC@.770 | Ea | 50.60 | 24.70 | 75.30 | 128.00 |
| 30" wide x 30" high, 2 doors | BC@.910 | Ea | 57.00 | 29.20 | 86.20 | 147.00 |
| 36" wide x 30" high, 2 doors | BC@.910 | Ea | 69.60 | 29.20 | 98.80 | 168.00 |
| 30" wide x 15" high, 2 doors | BC@.770 | Ea | 42.90 | 24.70 | 67.60 | 115.00 |
| 36" wide x 15" high, 2 doors | BC@.770 | Ea | 47.00 | 24.70 | 71.70 | 122.00 |
| 42" wide x 21" high, 2 doors | BC@.910 | Ea | 59.00 | 29.20 | 88.20 | 150.00 |
| 54" wide x 24" high, laundry, 2 doors | BC@.910 | Ea | 76.80 | 29.20 | 106.00 | 180.00 |

**Unfinished birch utility cabinet.** 2 doors and 2 drawers. 24" deep. Birch face frame and door frame. Ready to stain. With hardware.

| | Craft@Hrs | Unit | Material | Labor | Total | Sell |
|---|---|---|---|---|---|---|
| 18" wide x 84" high | BC@.910 | Ea | 162.00 | 29.20 | 191.20 | 325.00 |

**Unfinished birch cabinet end panels.**

| | Craft@Hrs | Unit | Material | Labor | Total | Sell |
|---|---|---|---|---|---|---|
| 12" wide x 30" high | BC@.200 | Ea | 4.60 | 6.42 | 11.02 | 18.70 |
| 24" wide x 35" high | BC@.200 | Ea | 6.80 | 6.42 | 13.22 | 22.50 |
| 24" wide x 84" high | BC@.200 | Ea | 16.40 | 6.42 | 22.82 | 38.80 |

**Unfinished birch filler strip.** Ready to stain.

| | Craft@Hrs | Unit | Material | Labor | Total | Sell |
|---|---|---|---|---|---|---|
| 3" x 32" | BC@.500 | Ea | 3.20 | 16.00 | 19.20 | 32.60 |

| | Craft@Hrs | Unit | Material | Labor | Total | Sell |
|---|---|---|---|---|---|---|

## Unfinished Red Oak Kitchen Cabinets

**Unfinished red oak base cabinets.** Red oak face frames and flat panel. Picture frame doors. 34-1/2" high, 24" deep. Ready to stain. With hardware.

| | Craft@Hrs | Unit | Material | Labor | Total | Sell |
|---|---|---|---|---|---|---|
| 12" wide, 1 door | BC@.640 | Ea | 57.30 | 20.50 | 77.80 | 132.00 |
| 15" wide, 1 door | BC@.640 | Ea | 64.80 | 20.50 | 85.30 | 145.00 |
| 18" wide, 1 door | BC@.770 | Ea | 74.00 | 24.70 | 98.70 | 168.00 |
| 18" wide 3-drawer base | BC@.770 | Ea | 118.00 | 24.70 | 142.70 | 243.00 |
| 24" wide, 2 doors | BC@.770 | Ea | 84.00 | 24.70 | 108.70 | 185.00 |
| 24" wide 3-drawer base | BC@.770 | Ea | 138.00 | 24.70 | 162.70 | 277.00 |
| 30" wide, 2 doors | BC@.910 | Ea | 114.80 | 29.20 | 144.00 | 245.00 |
| 36" Lazy Susan corner base | BC@.910 | Ea | 200.00 | 29.20 | 229.20 | 390.00 |
| 36" wide, 2 doors | BC@.910 | Ea | 129.00 | 29.20 | 158.20 | 269.00 |
| 36" wide blind base | BC@.910 | Ea | 121.00 | 29.20 | 150.20 | 255.00 |
| 36" wide corner base | BC@.910 | Ea | 151.00 | 29.20 | 180.20 | 306.00 |
| 36" wide sink base, 1 door | BC@.770 | Ea | 111.00 | 24.70 | 135.70 | 231.00 |
| 60" wide sink base, 2 doors | BC@.910 | Ea | 175.00 | 29.20 | 204.20 | 347.00 |

**Unfinished red oak wall cabinets.** Red oak face frames and flat panel. 12" deep. Ready to stain. With hardware.

| | Craft@Hrs | Unit | Material | Labor | Total | Sell |
|---|---|---|---|---|---|---|
| 12" wide x 30" high, 1 door | BC@.640 | Ea | 41.00 | 20.50 | 61.50 | 105.00 |
| 15" wide x 30" high, 1 door | BC@.640 | Ea | 45.90 | 20.50 | 66.40 | 113.00 |
| 18" wide x 30" high, 1 door | BC@.770 | Ea | 50.00 | 24.70 | 74.70 | 127.00 |
| 21" wide x 30" high, 1 door | BC@.770 | Ea | 64.00 | 24.70 | 88.70 | 151.00 |
| 24" wide x 30" high, 2 doors | BC@.770 | Ea | 73.00 | 24.70 | 97.70 | 166.00 |
| 24" wide x 30" high, diagonal corner | BC@.770 | Ea | 103.00 | 24.70 | 127.70 | 217.00 |
| 30" wide x 30" high, 2 doors | BC@.910 | Ea | 81.00 | 29.20 | 110.20 | 187.00 |
| 36" wide x 30" high, 2 doors | BC@.910 | Ea | 98.00 | 29.20 | 127.20 | 216.00 |
| 30" wide x 15" high, 2 doors | BC@.770 | Ea | 55.00 | 24.70 | 79.70 | 135.00 |
| 33" wide x 15" high, 2 doors | BC@.770 | Ea | 62.70 | 24.70 | 87.40 | 149.00 |
| 36" wide x 15" high, 2 doors | BC@.770 | Ea | 66.00 | 24.70 | 90.70 | 154.00 |
| 54" wide x 24" high, 2 doors, laundry | BC@.910 | Ea | 107.00 | 29.20 | 136.20 | 232.00 |

**Unfinished red oak utility cabinet.** Red oak face frames and flat panel. Ready to stain. With hardware.

| | Craft@Hrs | Unit | Material | Labor | Total | Sell |
|---|---|---|---|---|---|---|
| 18" wide x 84" high, 24" deep, 1 door | BC@.910 | Ea | 184.00 | 29.20 | 213.20 | 362.00 |
| 24" wide x 84" high, 18" deep, 1 door | BC@.910 | Ea | 203.00 | 29.20 | 232.20 | 395.00 |

**Unfinished red oak cabinet end panel.** Ready to stain.

| | Craft@Hrs | Unit | Material | Labor | Total | Sell |
|---|---|---|---|---|---|---|
| 11-3/8" wide x 30" high | BC@.200 | Ea | 9.40 | 6.42 | 15.82 | 26.90 |
| 23" wide x 34-1/2" high | BC@.200 | Ea | 14.00 | 6.42 | 20.42 | 34.70 |

**Unfinished red oak filler strip.** Ready to stain.

| | Craft@Hrs | Unit | Material | Labor | Total | Sell |
|---|---|---|---|---|---|---|
| 3" x 32" | BC@.500 | Ea | 7.10 | 16.00 | 23.10 | 39.30 |
| 6" x 32" | BC@.500 | Ea | 10.60 | 16.00 | 26.60 | 45.20 |

## Unfinished Natural Oak Kitchen Cabinets

**Unfinished natural oak base cabinets.** 34-1/2" high, 24" deep. Solid oak face frame and door frame. Ready to stain. With hardware.

| | Craft@Hrs | Unit | Material | Labor | Total | Sell |
|---|---|---|---|---|---|---|
| 12" wide, 1 door | BC@.640 | Ea | 61.30 | 20.50 | 81.80 | 139.00 |
| 15" wide, 1 door | BC@.640 | Ea | 67.50 | 20.50 | 88.00 | 150.00 |
| 15" wide, 3-drawer base | BC@.770 | Ea | 100.00 | 24.70 | 124.70 | 212.00 |
| 18" wide, 1 door | BC@.770 | Ea | 71.00 | 24.70 | 95.70 | 163.00 |
| 18" wide, 3-drawer base | BC@.770 | Ea | 111.00 | 24.70 | 135.70 | 231.00 |
| 21" wide, 1 door | BC@.770 | Ea | 91.60 | 24.70 | 116.30 | 198.00 |
| 24" wide, 2 doors | BC@.770 | Ea | 91.60 | 24.70 | 116.30 | 198.00 |
| 24" wide, 3-drawer base | BC@.910 | Ea | 117.00 | 29.20 | 146.20 | 249.00 |
| 30" wide, 2 doors | BC@.910 | Ea | 122.00 | 29.20 | 151.20 | 257.00 |

| | Craft@Hrs | Unit | Material | Labor | Total | Sell |
|---|---|---|---|---|---|---|
| 36" wide, 2 doors | BC@.910 | Ea | 156.00 | 29.20 | 185.20 | 315.00 |
| 36" wide, Lazy Susan base | BC@.910 | Ea | 217.00 | 29.20 | 246.20 | 419.00 |
| 36" wide sink base, 1 door | BC@.770 | Ea | 103.00 | 24.70 | 127.70 | 217.00 |
| 48" wide sink base, 2 doors, oak | BC@.910 | Ea | 162.00 | 29.20 | 191.20 | 325.00 |
| 60" wide sink base, 2 doors, oak | BC@.910 | Ea | 193.00 | 29.20 | 222.20 | 378.00 |
| Corner sink base | BC@.770 | Ea | 130.00 | 24.70 | 154.70 | 263.00 |

**Unfinished natural oak wall cabinets.** 12" deep. Solid oak face frame and door frame. Ready to stain. With hardware.

| | Craft@Hrs | Unit | Material | Labor | Total | Sell |
|---|---|---|---|---|---|---|
| 12" wide x 30" high, 1 door | BC@.640 | Ea | 34.60 | 20.50 | 55.10 | 93.70 |
| 15" wide x 30" high, 1 door | BC@.640 | Ea | 43.30 | 20.50 | 63.80 | 108.00 |
| 18" wide x 30" high, 1 door | BC@.770 | Ea | 48.50 | 24.70 | 73.20 | 124.00 |
| 18" wide x 36" high, 1 door | BC@.910 | Ea | 77.10 | 29.20 | 106.30 | 181.00 |
| 24" wide x 30" high diagonal corner | BC@.770 | Ea | 87.70 | 24.70 | 112.40 | 191.00 |
| 24" wide x 36" high, 2 doors | BC@.910 | Ea | 77.40 | 29.20 | 106.60 | 181.00 |
| 30" wide x 36" high, 2 doors | BC@.910 | Ea | 70.80 | 29.20 | 100.00 | 170.00 |
| 24" wide x 15" high, 2 doors | BC@.640 | Ea | 61.10 | 20.50 | 81.60 | 139.00 |
| 30" wide x 15" high, 2 doors | BC@.770 | Ea | 55.40 | 24.70 | 80.10 | 136.00 |
| 33" wide x 15" high, 2 doors | BC@.770 | Ea | 58.40 | 24.70 | 83.10 | 141.00 |
| 36" wide x 15" high, 2 doors | BC@.770 | Ea | 61.00 | 24.70 | 85.70 | 146.00 |
| 36" wide blind corner | BC@.910 | Ea | 96.40 | 29.20 | 125.60 | 214.00 |

**Unfinished natural oak utility cabinet.** 2 doors and 2 drawers. 24" deep. Solid oak face frame and door frame. Ready to stain. With hardware.

| | Craft@Hrs | Unit | Material | Labor | Total | Sell |
|---|---|---|---|---|---|---|
| 18" wide x 84" high | BC@.910 | Ea | 172.00 | 29.20 | 201.20 | 342.00 |

**Unfinished natural oak washer-dryer wall cabinet.** Oak face frame and door frame. Ready to stain. With hardware. 12" deep.

| | Craft@Hrs | Unit | Material | Labor | Total | Sell |
|---|---|---|---|---|---|---|
| 24" high x 54" wide, 2 doors | BC@.910 | Ea | 96.40 | 29.20 | 125.60 | 214.00 |

**Unfinished natural oak cabinet end panel.** Ready to stain.

| | Craft@Hrs | Unit | Material | Labor | Total | Sell |
|---|---|---|---|---|---|---|
| 12" wide x 15" high | BC@.200 | Ea | 3.90 | 6.42 | 10.32 | 17.50 |
| 12" wide x 24" high | BC@.200 | Ea | 6.10 | 6.42 | 12.52 | 21.30 |
| 12" wide x 30" high | BC@.200 | Ea | 5.00 | 6.42 | 11.42 | 19.40 |
| 23" wide x 84" high | BC@.200 | Ea | 29.20 | 6.42 | 35.62 | 60.60 |

**Unfinished natural oak trim.** Ready to stain.

| | Craft@Hrs | Unit | Material | Labor | Total | Sell |
|---|---|---|---|---|---|---|
| 3" x 32" filler strip | BC@.500 | Ea | 6.30 | 16.00 | 22.30 | 37.90 |
| 6" x 32" filler strip | BC@.500 | Ea | 13.00 | 16.00 | 29.00 | 49.30 |
| 36" x 4" valance | BC@.250 | Ea | 14.00 | 8.02 | 22.02 | 37.40 |
| 36" x 8" valance | BC@.250 | Ea | 19.30 | 8.02 | 27.32 | 46.40 |

| | Craft@Hrs | Unit | Material | Labor | Total | Sell |
|---|---|---|---|---|---|---|

## Finished Kitchen Cabinets

For detailed coverage of finished custom-built cabinets, see *National Framing & Finish Carpentry Estimator*, http://CraftsmanSiteLicense.com.

**Arctic white finished base cabinets.** Pre-assembled. 3/4"-thick medium-density fiberboard. Door face and drawer fronts vacuum-formed with white thermofoil. Door back is white melamine laminate. 34-1/2" high x 24" deep. With hardware.

| | Craft@Hrs | Unit | Material | Labor | Total | Sell |
|---|---|---|---|---|---|---|
| 12" wide, 1 door | BC@.640 | Ea | 108.00 | 20.50 | 128.50 | 218.00 |
| 15" wide, 1 door | BC@.640 | Ea | 112.00 | 20.50 | 132.50 | 225.00 |
| 15" wide, 3-drawer base | BC@.770 | Ea | 177.00 | 24.70 | 201.70 | 343.00 |
| 18" wide, 1 door | BC@.770 | Ea | 122.00 | 24.70 | 146.70 | 249.00 |
| 18" wide, 3-drawer base | BC@.770 | Ea | 205.00 | 24.70 | 229.70 | 390.00 |
| 24" wide, 2 doors | BC@.770 | Ea | 143.00 | 24.70 | 167.70 | 285.00 |
| 30" wide, 2 doors | BC@.910 | Ea | 189.00 | 29.20 | 218.20 | 371.00 |
| 36" wide, 2 doors | BC@.910 | Ea | 231.00 | 29.20 | 260.20 | 442.00 |
| 36" wide, easy reach corner | BC@.910 | Ea | 247.00 | 29.20 | 276.20 | 470.00 |
| 36" wide, sink base | BC@.770 | Ea | 168.00 | 24.70 | 192.70 | 328.00 |
| 45" wide, blind corner base | BC@.910 | Ea | 208.00 | 29.20 | 237.20 | 403.00 |

**Arctic white finished wall cabinets.** Pre-assembled. 3/4"-thick medium-density fiberboard. Door face vacuum formed with white thermofoil. Door back is white melamine laminate. 12" deep. With hardware.

| | Craft@Hrs | Unit | Material | Labor | Total | Sell |
|---|---|---|---|---|---|---|
| 12" wide x 30" high | BC@.640 | Ea | 83.00 | 20.50 | 103.50 | 176.00 |
| 15" wide x 30" high | BC@.640 | Ea | 90.60 | 20.50 | 111.10 | 189.00 |
| 18" wide x 30" high | BC@.770 | Ea | 95.00 | 24.70 | 119.70 | 203.00 |
| 24" wide x 30" high | BC@.770 | Ea | 109.00 | 24.70 | 133.70 | 227.00 |
| 24" wide x 30" high, corner angle cabinet | BC@.770 | Ea | 174.00 | 24.70 | 198.70 | 338.00 |
| 27" wide x 30" high, blind corner cabinet | BC@.910 | Ea | 141.00 | 29.20 | 170.20 | 289.00 |
| 30" wide x 15" high | BC@.770 | Ea | 91.70 | 24.70 | 116.40 | 198.00 |
| 30" wide x 18" high | BC@.770 | Ea | 109.00 | 24.70 | 133.70 | 227.00 |
| 30" wide x 30" high | BC@.910 | Ea | 134.00 | 29.20 | 163.20 | 277.00 |
| 36" wide x 15" high | BC@.910 | Ea | 104.00 | 29.20 | 133.20 | 226.00 |
| 36" wide x 30" high | BC@.910 | Ea | 151.00 | 29.20 | 180.20 | 306.00 |

**Arctic white finished utility cabinet.** Pre-assembled. 3/4"-thick medium-density fiberboard. Door face vacuum formed with white thermofoil. Door back is white melamine laminate. With hardware.

| | Craft@Hrs | Unit | Material | Labor | Total | Sell |
|---|---|---|---|---|---|---|
| 18" wide x 48" high x 24" deep | BC@.910 | Ea | 169.00 | 29.20 | 198.20 | 337.00 |

**Arctic white cabinet trim.**

| | Craft@Hrs | Unit | Material | Labor | Total | Sell |
|---|---|---|---|---|---|---|
| Crown mold, 8' long | BC@.750 | Ea | 44.20 | 24.10 | 68.30 | 116.00 |
| Filler strip, 3" x 30" | BC@.500 | Ea | 15.10 | 16.00 | 31.10 | 52.90 |
| Toe tick, 8' long | BC@.750 | Ea | 18.90 | 24.10 | 43.00 | 73.10 |

**Oak finished base cabinets.** Pre-assembled. Raised veneer center panel framed with 3/4"-thick solid oak. Arched top rail. Drawer fronts are 3/4"-thick solid oak. 34-1/2" high, 24" deep. With hardware.

| | Craft@Hrs | Unit | Material | Labor | Total | Sell |
|---|---|---|---|---|---|---|
| 12" wide, 1 door | BC@.640 | Ea | 120.00 | 20.50 | 140.50 | 239.00 |
| 15" wide, 1 door | BC@.640 | Ea | 127.00 | 20.50 | 147.50 | 251.00 |
| 15" wide, 3-drawer base | BC@.640 | Ea | 183.00 | 20.50 | 203.50 | 346.00 |
| 18" wide, 1 door | BC@.770 | Ea | 131.00 | 24.70 | 155.70 | 265.00 |
| 18" wide, 3-drawer base | BC@.770 | Ea | 217.00 | 24.70 | 241.70 | 411.00 |
| 24" wide, 2 doors | BC@.770 | Ea | 159.00 | 24.70 | 183.70 | 312.00 |
| 30" wide, 2 doors | BC@.910 | Ea | 200.00 | 29.20 | 229.20 | 390.00 |
| 36" wide, 2 doors | BC@.910 | Ea | 238.00 | 29.20 | 267.20 | 454.00 |
| 36" wide, corner base | BC@.770 | Ea | 170.00 | 24.70 | 194.70 | 331.00 |
| 36" wide, easy-reach base | BC@.910 | Ea | 255.00 | 29.20 | 284.20 | 483.00 |
| 45" wide, blind corner base | BC@.910 | Ea | 212.00 | 29.20 | 241.20 | 410.00 |

|  | Craft@Hrs | Unit | Material | Labor | Total | Sell |
|---|---|---|---|---|---|---|

**Oak finished wall cabinets.** Pre-assembled. Raised veneer center panel framed with 3/4"-thick solid oak. 12" deep. Arched top rail. With hardware.

| | Craft@Hrs | Unit | Material | Labor | Total | Sell |
|---|---|---|---|---|---|---|
| 12" wide x 30" high, 1 door | BC@.640 | Ea | 88.90 | 20.50 | 109.40 | 186.00 |
| 15" wide x 30" high, 1 door | BC@.640 | Ea | 97.60 | 20.50 | 118.10 | 201.00 |
| 18" wide x 30" high, 1 door | BC@.770 | Ea | 105.00 | 24.70 | 129.70 | 220.00 |
| 24" wide x 30" high, 2 doors | BC@.770 | Ea | 124.00 | 24.70 | 148.70 | 253.00 |
| 27" wide x 30" high, corner | BC@.770 | Ea | 140.00 | 24.70 | 164.70 | 280.00 |
| 30" wide x 15" high, 2 doors | BC@.770 | Ea | 96.80 | 24.70 | 121.50 | 207.00 |
| 30" wide x 18" high, 2 doors | BC@.770 | Ea | 113.00 | 24.70 | 137.70 | 234.00 |
| 30" wide x 30" high, 2 doors | BC@.910 | Ea | 143.00 | 29.20 | 172.20 | 293.00 |
| 36" wide x 15" high, 2 doors | BC@.910 | Ea | 115.00 | 29.20 | 144.20 | 245.00 |
| 36" wide x 30" high, 2 doors | BC@.910 | Ea | 163.00 | 29.20 | 192.20 | 327.00 |

**Oak finished utility top cabinet.** Pre-assembled. Raised veneer center panel framed with 3/4"-thick solid oak. 24" deep. Two doors. Arched top rail. With hardware.

| | Craft@Hrs | Unit | Material | Labor | Total | Sell |
|---|---|---|---|---|---|---|
| 18" wide x 48" high | BC@.910 | Ea | 192.00 | 29.20 | 221.20 | 376.00 |

**Finished oak cabinet trim.**

| | Craft@Hrs | Unit | Material | Labor | Total | Sell |
|---|---|---|---|---|---|---|
| Crown mold, 8' long | BC@.750 | Ea | 49.50 | 24.10 | 73.60 | 125.00 |
| Toe kick, 8' long | BC@.750 | Ea | 19.30 | 24.10 | 43.40 | 73.80 |
| Universal fill strip, 3" x 30" | BC@.500 | Ea | 15.10 | 16.00 | 31.10 | 52.90 |

**Maple finished base cabinets.** Pre-assembled. Veneer arched raised center panel framed with 2"-wide, 3/4"-thick solid maple. Base cabinet doors have a square raised center panel. Drawer fronts are 3/4"-thick solid maple. 34-1/2" high x 24" deep. With hardware.

| | Craft@Hrs | Unit | Material | Labor | Total | Sell |
|---|---|---|---|---|---|---|
| 12" wide, 1 door | BC@.640 | Ea | 131.00 | 20.50 | 151.50 | 258.00 |
| 15" wide, 1 door | BC@.640 | Ea | 138.00 | 20.50 | 158.50 | 269.00 |
| 15" wide, 3-drawer base | BC@.640 | Ea | 191.00 | 20.50 | 211.50 | 360.00 |
| 18" wide, 1 door | BC@.770 | Ea | 146.00 | 24.70 | 170.70 | 290.00 |
| 18" wide, 3-drawer base | BC@.770 | Ea | 219.00 | 24.70 | 243.70 | 414.00 |
| 24" wide, 2 doors | BC@.770 | Ea | 180.00 | 24.70 | 204.70 | 348.00 |
| 30" wide, 2 doors | BC@.910 | Ea | 214.00 | 29.20 | 243.20 | 413.00 |
| 36" sink base | BC@.770 | Ea | 203.00 | 24.70 | 227.70 | 387.00 |
| 36" wide, 2 doors | BC@.910 | Ea | 246.00 | 29.20 | 275.20 | 468.00 |
| 45" wide blind base | BC@.910 | Ea | 225.00 | 29.20 | 254.20 | 432.00 |

**Maple finished wall cabinets.** Pre-assembled. Veneer arched raised center panel framed with 2"-wide, 3/4"-thick solid maple. The top rail has a gentle arch. 12" deep. With hardware.

| | Craft@Hrs | Unit | Material | Labor | Total | Sell |
|---|---|---|---|---|---|---|
| 12" wide x 30" high, 1 door | BC@.640 | Ea | 95.00 | 20.50 | 115.50 | 196.00 |
| 15" wide x 30" high, 1 door | BC@.640 | Ea | 106.00 | 20.50 | 126.50 | 215.00 |
| 18" wide x 30" high, 1 door | BC@.770 | Ea | 112.00 | 24.70 | 136.70 | 232.00 |
| 24" wide x 30" high, 1 door | BC@.770 | Ea | 135.00 | 24.70 | 159.70 | 271.00 |
| 24" wide x 30" high, angle corner | BC@.770 | Ea | 188.00 | 24.70 | 212.70 | 362.00 |
| 27" wide x 30" high, blind corner | BC@.770 | Ea | 141.00 | 24.70 | 165.70 | 282.00 |
| 30" wide x 30" high, 2 doors | BC@.910 | Ea | 152.00 | 29.20 | 181.20 | 308.00 |
| 36" wide x 30" high, 2 doors | BC@.910 | Ea | 174.00 | 29.20 | 203.20 | 345.00 |
| 36" wide x 30" high, easy reach corner | BC@.910 | Ea | 269.00 | 29.20 | 298.20 | 507.00 |
| 30" wide x 15" high, 2 doors | BC@.770 | Ea | 105.00 | 24.70 | 129.70 | 220.00 |
| 36" wide x 15" high, 2 doors | BC@.770 | Ea | 127.00 | 24.70 | 151.70 | 258.00 |
| 30" wide x 18" high, 2 doors | BC@.770 | Ea | 122.00 | 24.70 | 146.70 | 249.00 |

**Maple finished utility top cabinet.** Pre-assembled. Veneer arched raised center panel framed with 2"-wide, 3/4"-thick solid maple. 24" deep. With hardware.

| | Craft@Hrs | Unit | Material | Labor | Total | Sell |
|---|---|---|---|---|---|---|
| 18" x 48" high | BC@.910 | Ea | 200.00 | 29.20 | 229.20 | 390.00 |

| | Craft@Hrs | Unit | Material | Labor | Total | Sell |
|---|---|---|---|---|---|---|
| **Maple finished cabinet trim.** | | | | | | |
| Crown molding, 8' long | BC@.750 | Ea | 49.50 | 24.10 | 73.60 | 125.00 |
| Toe kick, 8' long | BC@.750 | Ea | 19.40 | 24.10 | 43.50 | 74.00 |
| Universal filler, 3" x 30" | BC@.500 | Ea | 15.10 | 16.00 | 31.10 | 52.90 |

## Custom Countertops

**Laminated plastic countertops**. Custom-fabricated straight, U- or L-shaped tops (such as Formica, Textolite or Wilsonart) on a particleboard base. 22" to 25" wide. Cost per LF of longest edge (either back or front) in solid colors with 3-1/2" to 4" backsplash.

| | Craft@Hrs | Unit | Material | Labor | Total | Sell |
|---|---|---|---|---|---|---|
| Accent color front edge | B1@.181 | LF | 28.00 | 5.32 | 33.32 | 56.60 |
| Full wrap (180 degree) front edge | B1@.181 | LF | 22.00 | 5.32 | 27.32 | 46.40 |
| Inlaid front and back edge | B1@.181 | LF | 33.00 | 5.32 | 38.32 | 65.10 |
| Rolled drip edge (post formed) | B1@.181 | LF | 20.50 | 5.32 | 25.82 | 43.90 |
| Square edge, separate backsplash | B1@.181 | LF | 14.50 | 5.32 | 19.82 | 33.70 |
| Additional costs for laminated countertops | | | | | | |
| Add for contour end splash | — | Ea | 25.00 | — | 25.00 | — |
| Add for drilling 3 plumbing fixture holes | — | LS | 10.40 | — | 10.40 | — |
| Add for half round corner | — | Ea | 29.80 | — | 29.80 | — |
| Add for mitered corners | — | Ea | 20.70 | — | 20.70 | — |
| Add for quarter round corner | — | Ea | 20.80 | — | 20.80 | — |
| Add for seamless tops | — | Ea | 30.00 | — | 30.00 | — |
| Add for sink, range or chop block cutout | — | Ea | 8.10 | — | 8.10 | — |
| Add for square end splash | — | Ea | 17.50 | — | 17.50 | — |
| Add for textures, patterns | — | % | 20.0 | — | — | — |

**Solid surface countertops**. Many brands available, including Corian, Wilsonart and Avenite. Tops are cut from acrylic or polyester-acrylic sheets measuring 1/2" or 3/4" thick, 30" to 36" wide and 12' long. The estimates below include cutting, polishing, sink and range cutouts and a flat polish edge. Color groups are usually offered with distinct edge treatments. Group 1 colors are light shades with little texture and a matte finish. Group 2 colors include darker colors and stone textures. Group 3 colors are extra dark or bright and include high-gloss surfaces. Solid surface countertops are set directly on cabinets with silicone adhesive.

| | Craft@Hrs | Unit | Material | Labor | Total | Sell |
|---|---|---|---|---|---|---|
| Color group 1, Basic edge | B1@.128 | SF | 49.00 | 3.76 | 52.76 | 89.70 |
| Color group 1, Custom edge | B1@.128 | SF | 57.00 | 3.76 | 60.76 | 103.00 |
| Color group 1, Premium edge | B1@.128 | SF | 53.00 | 3.76 | 56.76 | 96.50 |
| Color group 2, Basic edge | B1@.128 | SF | 55.00 | 3.76 | 58.76 | 99.90 |
| Color group 2, Custom edge | B1@.128 | SF | 64.00 | 3.76 | 67.76 | 115.00 |
| Color group 2, Premium edge | B1@.128 | SF | 60.00 | 3.76 | 63.76 | 108.00 |
| Color group 3, Basic edge | B1@.128 | SF | 64.00 | 3.76 | 67.76 | 115.00 |
| Color group 3, Custom edge | B1@.128 | SF | 73.00 | 3.76 | 76.76 | 130.00 |
| Color group 3, Premium edge | B1@.128 | SF | 69.00 | 3.76 | 72.76 | 124.00 |
| Add for delivery (typical) | — | Ea | 150.00 | — | 150.00 | — |
| Add for integral sink, one basin | — | Ea | 350.00 | — | 350.00 | — |
| Add for integral sink, two basins | — | Ea | 450.00 | — | 450.00 | — |
| Add for radius or beveled edge | — | LF | 7.00 | — | 7.00 | — |

| | Craft@Hrs | Unit | Material | Labor | Total | Sell |
|---|---|---|---|---|---|---|

**Granite countertops.** Granite countertop work is generally done by a specialist. Custom made to order from a template. 22" to 25" wide. Per square foot of top and backsplash overall dimension, including measuring, cutting, and flat-edge polishing. Add the cost of edge detail as itemized below and jobsite delivery. 3/4" granite weighs about 30 pounds per square foot. 1-1/4" granite weighs about 45 pounds per square foot. 3/4" tops should be set on 3/4" plywood and set in silicon adhesive every two feet. 1-1/4" granite tops can be set directly on base cabinets with silicon adhesive.

| | Craft@Hrs | Unit | Material | Labor | Total | Sell |
|---|---|---|---|---|---|---|
| Most 3/4" (2cm) granite tops | B1@.128 | SF | 39.40 | 3.76 | 43.16 | 73.40 |
| Most 1-1/4" (3cm) granite tops | B1@.128 | SF | 51.50 | 3.76 | 55.26 | 93.90 |
| Add for 1/2" beveled edge | — | LF | 7.07 | — | 7.07 | — |
| Add for backsplash seaming and edging | — | LF | 13.00 | — | 13.00 | — |
| Add for cooktop overmount cutout | — | Ea | 131.00 | — | 131.00 | — |
| Add for delivery (typical) | — | Ea | 152.00 | — | 152.00 | — |
| Add for eased edge | — | LF | 5.05 | — | 5.05 | — |
| Add for electrical outlet cutout | — | Ea | 35.50 | — | 35.50 | — |
| Add for faucet holes | — | Ea | 26.50 | — | 26.50 | — |
| Add for full bullnose edge | — | LF | 20.20 | — | 20.20 | — |
| Add for half bullnose edge | — | LF | 10.10 | — | 10.10 | — |
| Add to remove existing non-ceramic top | — | LF | 6.50 | — | 6.50 | — |
| Add to remove existing ceramic tile top | — | LF | 9.00 | — | 9.00 | — |
| Add for rounded corners | — | Ea | 30.00 | — | 30.00 | — |
| Add for undermount sink, polished | — | Ea | 300.00 | — | 300.00 | — |

**Engineered stone countertops.** Custom made from a template. Quartz particles with acrylic or epoxy binder. Trade names include Crystalite, Silestone and Cambria. Per square foot of top and backsplash surface.

| | Craft@Hrs | Unit | Material | Labor | Total | Sell |
|---|---|---|---|---|---|---|
| Small light chips, 3/4" square edge | B1@.128 | SF | 49.50 | 3.76 | 53.26 | 90.50 |
| Large dark chips, 3/4" square edge | B1@.128 | SF | 55.60 | 3.76 | 59.36 | 101.00 |
| Add for 1-1/2" bullnose edge | — | LF | 48.50 | — | 48.50 | — |
| Add for 1-1/2" ogee edge | — | LF | 84.80 | — | 84.80 | — |
| Add for 3/4" bullnose edge | — | LF | 25.30 | — | 25.30 | — |
| Add for 3/4" ogee edge | — | LF | 33.30 | — | 33.30 | — |
| Add for delivery (typical) | — | Ea | 152.00 | — | 152.00 | — |
| Add for electrical outlet cutout | — | Ea | 30.30 | — | 30.30 | — |
| Add for radius corner or end | — | Ea | 187.00 | — | 187.00 | — |
| Add to remove existing laminated top | — | Ea | 280.00 | — | 280.00 | — |
| Add for sink cutout | — | Ea | 250.00 | — | 250.00 | — |

## Stock Kitchen Countertops

**Post-formed double-radius plastic laminate countertop.** Standard laminate. 3" backsplash. 25" width. Substrate is 3/4" industrial grade particleboard.

| | Craft@Hrs | Unit | Material | Labor | Total | Sell |
|---|---|---|---|---|---|---|
| 5' long | BC@1.00 | Ea | 49.70 | 32.10 | 81.80 | 139.00 |
| 6' long | BC@1.20 | Ea | 59.70 | 38.50 | 98.20 | 167.00 |
| 6' long, left-hand miter | BC@1.20 | Ea | 67.20 | 38.50 | 105.70 | 180.00 |
| 6' long, right-hand miter | BC@1.20 | Ea | 67.20 | 38.50 | 105.70 | 180.00 |
| 8' long | BC@1.60 | Ea | 79.60 | 51.30 | 130.90 | 223.00 |
| 10' long | BC@2.00 | Ea | 99.50 | 64.20 | 163.70 | 278.00 |
| 10' long, left-hand miter | BC@1.60 | Ea | 107.00 | 51.30 | 158.30 | 269.00 |
| 10' long, right-hand miter | BC@1.60 | Ea | 107.00 | 51.30 | 158.30 | 269.00 |
| End cap kit | BC@.250 | Ea | 10.80 | 8.02 | 18.82 | 32.00 |
| End splash | BC@.250 | Ea | 14.00 | 8.02 | 22.02 | 37.40 |

| | Craft@Hrs | Unit | Material | Labor | Total | Sell |
|---|---|---|---|---|---|---|

**Post-formed double-radius premium plastic laminate countertop.** 3" backsplash. 25" width. Substrate is 3/4" industrial grade particleboard. Exceeds ANSI A161.2 standard for countertops.

| | Craft@Hrs | Unit | Material | Labor | Total | Sell |
|---|---|---|---|---|---|---|
| 5' long | BC@1.00 | Ea | 54.20 | 32.10 | 86.30 | 147.00 |
| 6' long | BC@1.20 | Ea | 64.80 | 38.50 | 103.30 | 176.00 |
| 6' long, left-hand miter | BC@1.20 | Ea | 70.70 | 38.50 | 109.20 | 186.00 |
| 6' long, right-hand miter | BC@1.20 | Ea | 70.70 | 38.50 | 109.20 | 186.00 |
| 8' long | BC@1.60 | Ea | 88.50 | 51.30 | 139.80 | 238.00 |
| 10' long | BC@2.00 | Ea | 109.00 | 64.20 | 173.20 | 294.00 |
| 10' long, left-hand miter | BC@2.00 | Ea | 113.00 | 64.20 | 177.20 | 301.00 |
| 10' long, right-hand miter | BC@2.00 | Ea | 113.00 | 64.20 | 177.20 | 301.00 |
| End cap kit | BC@.250 | Ea | 11.80 | 8.02 | 19.82 | 33.70 |
| End splash | BC@.250 | Ea | 13.30 | 8.02 | 21.32 | 36.20 |

**Miter bolt kit.** Joins 2 mitered tops. Includes 4 miter bolts, packet of glue, and instructions.

| | Craft@Hrs | Unit | Material | Labor | Total | Sell |
|---|---|---|---|---|---|---|
| Kit | — | Ea | 5.92 | — | 5.92 | — |

## Cabinet Knobs, Pulls and Hinges

**Round ceramic knob.**

| | Craft@Hrs | Unit | Material | Labor | Total | Sell |
|---|---|---|---|---|---|---|
| 1-3/8", Almond | BC@.060 | Ea | 2.02 | 1.93 | 3.95 | 6.72 |
| 1-3/8", White with white flower | BC@.060 | Ea | 2.42 | 1.93 | 4.35 | 7.40 |

**Round ceramic knob.** Quality porcelain. Classic style. Fasteners included.

| | Craft@Hrs | Unit | Material | Labor | Total | Sell |
|---|---|---|---|---|---|---|
| 1", white | BC@.060 | Ea | 1.73 | 1.93 | 3.66 | 6.22 |
| 1-1/4", white | BC@.060 | Ea | 1.93 | 1.93 | 3.86 | 6.56 |
| 1-3/8", white | BC@.060 | Ea | 1.94 | 1.93 | 3.87 | 6.58 |
| 1-1/2", white | BC@.060 | Ea | 2.15 | 1.93 | 4.08 | 6.94 |
| 1-1/2", white with apple decal | BC@.060 | Ea | 2.35 | 1.93 | 4.28 | 7.28 |
| 1-1/2", white with pear decal | BC@.060 | Ea | 2.43 | 1.93 | 4.36 | 7.41 |
| 1-1/2", white with plum decal | BC@.060 | Ea | 2.38 | 1.93 | 4.31 | 7.33 |

**Classic cabinet knob.** 30mm diameter.

| | Craft@Hrs | Unit | Material | Labor | Total | Sell |
|---|---|---|---|---|---|---|
| Black nickel | BC@.060 | Ea | 3.26 | 1.93 | 5.19 | 8.82 |
| Brushed red antique | BC@.060 | Ea | 3.46 | 1.93 | 5.39 | 9.16 |
| Matte blue | BC@.060 | Ea | 3.31 | 1.93 | 5.24 | 8.91 |
| Matte red | BC@.060 | Ea | 3.31 | 1.93 | 5.24 | 8.91 |
| Pearl gold | BC@.060 | Ea | 3.11 | 1.93 | 5.04 | 8.57 |
| Satin chrome | BC@.060 | Ea | 3.12 | 1.93 | 5.05 | 8.59 |

**Brushed antique satin cabinet knob.** Striking brushed; lacquered antique finish. 30mm diameter. Fasteners included.

| | Craft@Hrs | Unit | Material | Labor | Total | Sell |
|---|---|---|---|---|---|---|
| Bronze | BC@.060 | Ea | 3.15 | 1.93 | 5.08 | 8.64 |
| Copper, red | BC@.060 | Ea | 3.27 | 1.93 | 5.20 | 8.84 |
| Nickel | BC@.060 | Ea | 3.25 | 1.93 | 5.18 | 8.81 |

**Decorative star cabinet knob.**

| | Craft@Hrs | Unit | Material | Labor | Total | Sell |
|---|---|---|---|---|---|---|
| Black | BC@.060 | Ea | 3.23 | 1.93 | 5.16 | 8.77 |
| Nickel | BC@.060 | Ea | 4.31 | 1.93 | 6.24 | 10.60 |
| Pewter | BC@.060 | Ea | 3.23 | 1.93 | 5.16 | 8.77 |

**Round cabinet knob.** Lacquered black chrome finish. Fasteners included.

| | Craft@Hrs | Unit | Material | Labor | Total | Sell |
|---|---|---|---|---|---|---|
| 1-1/4" diameter | BC@.060 | Ea | 3.26 | 1.93 | 5.19 | 8.82 |

|  | Craft@Hrs | Unit | Material | Labor | Total | Sell |
|---|---|---|---|---|---|---|
| **Round knob.** | | | | | | |
| 1-3/8", almond | BC@.060 | Ea | 1.13 | 1.93 | 3.06 | 5.20 |
| 1-3/8", black | BC@.060 | Ea | 1.06 | 1.93 | 2.99 | 5.08 |
| 1-3/8", green | BC@.060 | Ea | 1.05 | 1.93 | 2.98 | 5.07 |
| 1-3/8", speckled black | BC@.060 | Ea | 1.05 | 1.93 | 2.98 | 5.07 |
| **Satin nickel knob.** Lacquered, brushed satin finish. Traditional styling. Fasteners included. | | | | | | |
| 1-1/4", round | BC@.060 | Ea | 3.24 | 1.93 | 5.17 | 8.79 |
| 1-3/4" x 1-1/16", oval | BC@.060 | Ea | 1.83 | 1.93 | 3.76 | 6.39 |
| **Brass-plated cabinet knob.** Decorative knob. Lacquered; polished brass plate. Classic style. Fasteners included. | | | | | | |
| 1" diameter | BC@.060 | Ea | 2.70 | 1.93 | 4.63 | 7.87 |
| 1-1/4" diameter | BC@.060 | Ea | 3.25 | 1.93 | 5.18 | 8.81 |
| 1-1/4" diameter, ringed design | BC@.060 | Ea | 1.91 | 1.93 | 3.84 | 6.53 |
| **Brass-plated knob with insert.** Decorative knob. Lacquered; polished brass plate. Decorative beveled edge. Fasteners included. 1-1/4" diameter. | | | | | | |
| Frost maple insert | BC@.060 | Ea | 2.26 | 1.93 | 4.19 | 7.12 |
| High-gloss white ceramic insert | BC@.060 | Ea | 2.15 | 1.93 | 4.08 | 6.94 |
| Medium oak wood insert | BC@.060 | Ea | 2.12 | 1.93 | 4.05 | 6.89 |
| **Natural wood round knob.** Unfinished solid birch. Pre-sanded. Perfect for refinishing. Fasteners included. Limited lifetime warranty. | | | | | | |
| 1" diameter | BC@.060 | Ea | .85 | 1.93 | 2.78 | 4.73 |
| 1-1/4" diameter | BC@.060 | Ea | .96 | 1.93 | 2.89 | 4.91 |
| 1-1/2" diameter | BC@.060 | Ea | 1.06 | 1.93 | 2.99 | 5.08 |
| 1-3/4" diameter | BC@.060 | Ea | 1.32 | 1.93 | 3.25 | 5.53 |
| 2" diameter | BC@.060 | Ea | 1.39 | 1.93 | 3.32 | 5.64 |
| **Circle swirl knob.** Distressed finish. Fasteners included. Limited lifetime warranty. | | | | | | |
| 4/5" x 1-1/3", black finish | BC@.060 | Ea | 3.30 | 1.93 | 5.23 | 8.89 |
| 33mm, rustic pewter patina finish | BC@.060 | Ea | 3.32 | 1.93 | 5.25 | 8.93 |
| **Solid brass and chrome knob.** Polished brass with bright chrome for a contemporary look. High-quality lacquered solid brass. Classic style. Fasteners included. | | | | | | |
| 1-1/4", 2-tone | BC@.060 | Ea | 5.44 | 1.93 | 7.37 | 12.50 |
| **Solid brass knob.** High-quality lacquered solid brass. Fasteners included. | | | | | | |
| 1-3/16" x 3/4", oval | BC@.060 | Ea | 4.46 | 1.93 | 6.39 | 10.90 |
| 1-1/4", round | BC@.060 | Ea | 4.32 | 1.93 | 6.25 | 10.60 |
| 1-3/8", round | BC@.060 | Ea | 5.30 | 1.93 | 7.23 | 12.30 |
| 1-1/2", round, rope design | BC@.060 | Ea | 4.66 | 1.93 | 6.59 | 11.20 |
| **Small forked branch knob.** | | | | | | |
| Black | BC@.060 | Ea | 3.51 | 1.93 | 5.44 | 9.25 |
| **Antique brass target knob.** Round. Provincial ringed design with antiqued finish. Fasteners included. | | | | | | |
| 1-1/8" diameter | BC@.060 | Ea | .98 | 1.93 | 2.91 | 4.95 |
| 1-5/16" diameter | BC@.060 | Ea | 1.05 | 1.93 | 2.98 | 5.07 |

|  | Craft@Hrs | Unit | Material | Labor | Total | Sell |
|---|---|---|---|---|---|---|
| **Round backplate.** For knobs. | | | | | | |
| 1-1/4", chrome | BC@.020 | Ea | 1.61 | .64 | 2.25 | 3.83 |
| 1-1/4", polished brass | BC@.020 | Ea | 1.61 | .64 | 2.25 | 3.83 |
| | | | | | | |
| **Birdcage knob.** 50mm diameter. | | | | | | |
| Antique pewter | BC@.060 | Ea | 4.33 | 1.93 | 6.26 | 10.60 |
| Flat black | BC@.060 | Ea | 4.32 | 1.93 | 6.25 | 10.60 |
| | | | | | | |
| **Birdcage bail pull.** 96mm post spacing. | | | | | | |
| Antique pewter | BC@.120 | Ea | 7.14 | 3.85 | 10.99 | 18.70 |
| Flat black | BC@.120 | Ea | 7.55 | 3.85 | 11.40 | 19.40 |
| | | | | | | |
| **Brass pull with oak insert.** Medium oak insert. Traditional die-cast spoon-foot design. Matching knob sold separately. Fasteners included. 3" post spacing. | | | | | | |
| Oak and antique brass | BC@.120 | Ea | 3.43 | 3.85 | 7.28 | 12.40 |
| Oak and polished brass | BC@.120 | Ea | 2.99 | 3.85 | 6.84 | 11.60 |
| | | | | | | |
| **Birdcage wire pull.** 3" post spacing. | | | | | | |
| Flat black | BC@.120 | Ea | 5.40 | 3.85 | 9.25 | 15.70 |
| Pewter | BC@.120 | Ea | 5.49 | 3.85 | 9.34 | 15.90 |
| | | | | | | |
| **Bow pull.** Versatile design. Classic style for kitchen or bath. Fasteners included. 96 mm post spacing. | | | | | | |
| Lacquered black nickel | BC@.120 | Ea | 2.60 | 3.85 | 6.45 | 11.00 |
| Lacquered brass | BC@.120 | Ea | 2.71 | 3.85 | 6.56 | 11.20 |
| | | | | | | |
| **Fusilli pull.** Shallow S design. Fasteners included. Limited lifetime warranty. 96mm post spacing. | | | | | | |
| Black nickel | BC@.120 | Ea | 4.48 | 3.85 | 8.33 | 14.20 |
| Bright chrome | BC@.120 | Ea | 4.35 | 3.85 | 8.20 | 13.90 |
| Brushed bronze antique | BC@.120 | Ea | 4.31 | 3.85 | 8.16 | 13.90 |
| Brushed satin antique red | BC@.120 | Ea | 4.15 | 3.85 | 8.00 | 13.60 |
| Brushed satin chrome | BC@.120 | Ea | 4.33 | 3.85 | 8.18 | 13.90 |
| Brushed satin nickel | BC@.120 | Ea | 4.34 | 3.85 | 8.19 | 13.90 |
| Flat black | BC@.120 | Ea | 4.32 | 3.85 | 8.17 | 13.90 |
| Pearl gold | BC@.120 | Ea | 5.54 | 3.85 | 9.39 | 16.00 |
| Satin gold | BC@.120 | Ea | 4.31 | 3.85 | 8.16 | 13.90 |
| | | | | | | |
| **White decorative cabinet pull.** Durable epoxy coating. Classic style for kitchen or bath. Fasteners included. 3" post spacing. | | | | | | |
| Minaret | BC@.120 | Ea | 2.43 | 3.85 | 6.28 | 10.70 |
| Spoon-foot | BC@.120 | Ea | 2.37 | 3.85 | 6.22 | 10.60 |

| | Craft@Hrs | Unit | Material | Labor | Total | Sell |
|---|---|---|---|---|---|---|

**Chrome decorative cabinet pull.** Fasteners included. Limited lifetime warranty. 3" post spacing.
Tapered bow,
    bright lacquered, polished chrome — BC@.120 — Ea — 1.62 — 3.85 — 5.47 — 9.30
Die-cast, spoon-foot,
    bright polished chrome — BC@.120 — Ea — 2.32 — 3.85 — 6.17 — 10.50
Ornate beaded,
    lacquered, polished chrome — BC@.120 — Ea — 2.15 — 3.85 — 6.00 — 10.20

**Satin nickel decorative cabinet pull.** Lacquered, brushed satin nickel finish. Fasteners included. 3" post spacing.
Die-cast spoon-foot — BC@.120 — Ea — 2.31 — 3.85 — 6.16 — 10.50
Fan design foot — BC@.120 — Ea — 2.28 — 3.85 — 6.13 — 10.40

**Brass-plated pull with ceramic insert.** High-gloss non-porous ceramic insert. Traditional die-cast spoon-foot design. 3" post spacing.
Almond insert — BC@.120 — Ea — 3.29 — 3.85 — 7.14 — 12.10
White insert — BC@.120 — Ea — 3.23 — 3.85 — 7.08 — 12.00

**S swirl pull.** Distressed die-cast rust-resistant finish. Fasteners included. Limited lifetime warranty. 3" post spacing.
Pewter, small — BC@.120 — Ea — 4.04 — 3.85 — 7.89 — 13.40

**Solid brass wire pull.** 96mm post spacing.
2-tone — BC@.120 — Ea — 5.69 — 3.85 — 9.54 — 16.20

**Brass-plated spoon-foot pull.** Lacquered, polished brass finish. Traditional die-cast spoon-foot design. Fasteners included. 3" post spacing.
Smooth — BC@.120 — Ea — 2.38 — 3.85 — 6.23 — 10.60

**Wire cabinet pull.** Fasteners included. By post spacing.

| | Craft@Hrs | Unit | Material | Labor | Total | Sell |
|---|---|---|---|---|---|---|
| 3", black nickel | BC@.120 | Ea | 2.58 | 3.85 | 6.43 | 10.90 |
| 3", flat black | BC@.120 | Ea | 1.90 | 3.85 | 5.75 | 9.78 |
| 3", polished chrome | BC@.120 | Ea | 2.55 | 3.85 | 6.40 | 10.90 |
| 3", satin chrome | BC@.120 | Ea | 2.67 | 3.85 | 6.52 | 11.10 |
| 3", solid brass | BC@.120 | Ea | 2.46 | 3.85 | 6.31 | 10.70 |
| 3", white | BC@.120 | Ea | 1.95 | 3.85 | 5.80 | 9.86 |
| 3-1/2", satin chrome | BC@.120 | Ea | 2.67 | 3.85 | 6.52 | 11.10 |
| 3-1/2", white | BC@.120 | Ea | 1.94 | 3.85 | 5.79 | 9.84 |
| 4", white | BC@.120 | Ea | 2.07 | 3.85 | 5.92 | 10.10 |

**35mm Euro face frame overlay cabinet hinge.** For face frame cabinets with 5/8" overlay cabinet doors. Nickel-plated stamped steel cup and zinc die-cast mounting plate. Three-way adjustable. Door opening angle 100 degrees. Fasteners included. Labor assumes cabinet is already cut for this type of hinge.
Per hinge — BC@.250 — Ea — 6.98 — 8.02 — 15.00 — 25.50
Pack of 5 — BC@1.25 — Ea — 19.50 — 40.10 — 59.60 — 101.00

| | Craft@Hrs | Unit | Material | Labor | Total | Sell |
|---|---|---|---|---|---|---|

**Semi-concealed cabinet hinge.** For face frame cabinets using 3/8" inset doors. Fasteners included. Pack of two.

| | Craft@Hrs | Unit | Material | Labor | Total | Sell |
|---|---|---|---|---|---|---|
| Antique brass | BC@.250 | Ea | 2.37 | 8.02 | 10.39 | 17.70 |
| Chrome | BC@.250 | Ea | 2.26 | 8.02 | 10.28 | 17.50 |
| Polished brass | BC@.250 | Ea | 2.37 | 8.02 | 10.39 | 17.70 |

**Semi-concealed self-closing cabinet hinge.** For face frame cabinets using 3/8" inset doors. Wrought steel. Self-closing spring. Fasteners included.

| | Craft@Hrs | Unit | Material | Labor | Total | Sell |
|---|---|---|---|---|---|---|
| Antique brass | BC@.250 | Ea | 2.04 | 8.02 | 10.06 | 17.10 |
| Black nickel | BC@.250 | Ea | 2.93 | 8.02 | 10.95 | 18.60 |
| Chrome | BC@.250 | Ea | 3.02 | 8.02 | 11.04 | 18.80 |
| Polished brass | BC@.250 | Ea | 2.04 | 8.02 | 10.06 | 17.10 |
| White | BC@.250 | Ea | 3.03 | 8.02 | 11.05 | 18.80 |
| White, pack of 20 | BC@5.00 | Ea | 23.40 | 160.00 | 183.40 | 312.00 |

## Stainless Steel Sinks

**Stainless steel single bowl sink, Neptune Series, Elkay.** Heavy-gauge stainless steel. Four-hole faucet drilling. Faucet and accessories sold separately. Bright satin finish. 25" long x 22" wide.

| | Craft@Hrs | Unit | Material | Labor | Total | Sell |
|---|---|---|---|---|---|---|
| 5-1/2" deep | P1@.600 | Ea | 33.50 | 18.90 | 52.40 | 89.10 |
| 6" deep | P1@.600 | Ea | 40.00 | 18.90 | 58.90 | 100.00 |
| 7" deep | P1@.600 | Ea | 62.60 | 18.90 | 81.50 | 139.00 |
| 8" deep | P1@.600 | Ea | 83.70 | 18.90 | 102.60 | 174.00 |

**Stainless steel double bowl sink, Neptune Series, Elkay.** Heavy-gauge nickel bearing stainless steel. Full undercoating to deaden sound and prevent condensation. Exclusive hand-mounting fasteners. 3-1/2" drain openings. Brilliant finish with highlights. 33" x 22". Four-hole faucet drilling. Faucet and accessories sold separately.

| | Craft@Hrs | Unit | Material | Labor | Total | Sell |
|---|---|---|---|---|---|---|
| 5-1/2" deep, double bowl | P1@.600 | Ea | 37.40 | 18.90 | 56.30 | 95.70 |
| 6" deep, double bowl | P1@.600 | Ea | 47.70 | 18.90 | 66.60 | 113.00 |
| 8" deep, double bowl | P1@.600 | Ea | 94.60 | 18.90 | 113.50 | 193.00 |
| 7" deep, small and large bowls | P1@.600 | Ea | 99.90 | 18.90 | 118.80 | 202.00 |

**Stainless steel single bowl sink, Signature Series, Elkay.** Drop-in single bowl sink. Made from nickel bearing type 302 stainless steel. 3" radius vertical corners. 8-1/4"-deep bowl with mirror-finished interior deck. Machine-ground satin-finished bowls with satin highlighted outer rim edge and bowl radius. Fully undercoated for sound deadening. Self-rimming installation. Four holes. Drain openings are 3-1/2".

| | Craft@Hrs | Unit | Material | Labor | Total | Sell |
|---|---|---|---|---|---|---|
| 25" x 22", 20 gauge | P1@.600 | Ea | 102.00 | 18.90 | 120.90 | 206.00 |
| 25" x 22", 18 gauge | P1@.600 | Ea | 162.00 | 18.90 | 180.90 | 308.00 |
| 33" x 22", 18 gauge | P1@.600 | Ea | 183.00 | 18.90 | 201.90 | 343.00 |

**Stainless steel double bowl sink, Signature Series, Elkay.** Drop-in double bowl sink. Self-rimming installation. 3" radius vertical coved corners. Machine-ground satin-finished bowls and satin highlighted outer rim edge and bowl radius. Mirror-finished interior deck. Sound Guard undercoat prevents condensation and ensures quiet operation. Four holes. 3-1/2" drain openings. 33" long x 22" wide. 8-1/4" deep.

| | Craft@Hrs | Unit | Material | Labor | Total | Sell |
|---|---|---|---|---|---|---|
| 20 gauge | P1@.600 | Ea | 161.00 | 18.90 | 179.90 | 306.00 |
| 18 gauge | P1@.600 | Ea | 215.00 | 18.90 | 233.90 | 398.00 |
| Offset appearance | P1@.600 | Ea | 159.00 | 18.90 | 177.90 | 302.00 |
| Dual level, 7-1/2" deep | P1@.600 | Ea | 198.00 | 18.90 | 216.90 | 369.00 |

**Stainless steel kitchen sink, Kohler.** 20-gauge with sound deadening. Undercoating reduces disposal noise, minimizes condensation and maintains water temperature. Stain- and corrosion-resistant. Self-rimming, four-hole installation. Faucet sold separately.

| | Craft@Hrs | Unit | Material | Labor | Total | Sell |
|---|---|---|---|---|---|---|
| 25" x 22", single bowl | P1@.600 | Ea | 114.00 | 18.90 | 132.90 | 226.00 |
| 33" x 22", double bowl | P1@.600 | Ea | 171.00 | 18.90 | 189.90 | 323.00 |

|  | Craft@Hrs | Unit | Material | Labor | Total | Sell |
|---|---|---|---|---|---|---|

**Stainless steel double bowl sink, Staccato, Kohler.** 8" basin depth. Double equal basins. Four-hole punched. 18 gauge. Faucet sold separately.

| 33" x 22" | P1@.600 | Ea | 264.00 | 18.90 | 282.90 | 481.00 |

**Stainless steel kitchen sink, Banner.** Standard-gauge stainless steel. Rounded corners. Self-rimming design. Includes sink clips. Faucet, sprayer, and strainers sold separately. Five-year limited warranty.

| 25" x 22", 7" depth, single bowl | P1@.600 | Ea | 73.60 | 18.90 | 92.50 | 157.00 |
| 33" x 22", 6" depth, double buffed bowl | P1@.600 | Ea | 45.50 | 18.90 | 64.40 | 109.00 |
| 33" x 22", 7" depth, double bowl | P1@.600 | Ea | 67.10 | 18.90 | 86.00 | 146.00 |
| 33" x 22", 8" depth, double bowl | P1@.600 | Ea | 98.70 | 18.90 | 117.60 | 200.00 |

**Sink mounting clips and screws, Elkay.** U-channel. For use with Elkay Signature and Neptune sinks.

| Clips | — | Ea | 3.21 | — | 3.21 | — |

## Enameled Steel Sinks

**Charleston double bowl kitchen sink.** White enameled steel with four faucet holes. Self-rimming installation. 33" x 22" x 7-1/2".

| White | P1@.750 | Ea | 85.30 | 23.60 | 108.90 | 185.00 |

**Emory kitchen sink.** White enameled steel with three faucet holes. Rimless installation – requires use of metal mounting frame (sold separately). Double bowl. Faucet and sprayer not included. 32" x 21".

| White | P1@.750 | Ea | 44.80 | 23.60 | 68.40 | 116.00 |

## Cast Iron Sinks

**Dumont double bowl cast iron sink, Eljer.** White enameled cast iron. 33" x 22" x 8" deep.

| Self-rimming | P1@.750 | Ea | 161.00 | 23.60 | 184.60 | 314.00 |
| Tile-in installation | P1@.750 | Ea | 204.00 | 23.60 | 227.60 | 387.00 |

**Risotto double bowl cast iron sink, Eljer.** White enameled cast iron. Self-rimming. Offset drains. Elliptical bowl design with right-hand offset faucet shelf. Faucet, sprayer, and strainers sold separately.

| 28" x 22" x 9", white | P1@.750 | Ea | 212.00 | 23.60 | 235.60 | 401.00 |
| 33" x 22" x 9", bisque | P1@.750 | Ea | 198.00 | 23.60 | 221.60 | 377.00 |

**Mayfield cast iron single bowl sink, Kohler.** Self-rimming. 25" x 22" x 8". Four-hole drilling.

| Almond | P1@.750 | Ea | 217.00 | 23.60 | 240.60 | 409.00 |
| White | P1@.750 | Ea | 204.00 | 23.60 | 227.60 | 387.00 |

**Efficiency™ double bowl cast iron sink, Kohler.** Self-rimming installation. Offset faucet ledge to maximize basin area. 33" x 22" x 6-3/4".

| White | P1@.750 | Ea | 161.00 | 23.60 | 184.60 | 314.00 |

**Ashland cast iron double bowl sink, Kohler.** 33" x 22". Four-hole faucet drilling. Large and medium double basins. Self-rimming.

| White, 7" deep | P1@.750 | Ea | 172.00 | 23.60 | 195.60 | 333.00 |

| | Craft@Hrs | Unit | Material | Labor | Total | Sell |
|---|---|---|---|---|---|---|

**Hartland™ double bowl cast iron sink, Kohler.** Unique design lines. Four-hole. Self-rimming. Deep, glossy color. 33" long x 22" wide x 7-3/4" deep.

| | Craft@Hrs | Unit | Material | Labor | Total | Sell |
|---|---|---|---|---|---|---|
| Almond | P1@.750 | Ea | 388.00 | 23.60 | 411.60 | 700.00 |
| Bisque | P1@.750 | Ea | 257.00 | 23.60 | 280.60 | 477.00 |
| White | P1@.750 | Ea | 206.00 | 23.60 | 229.60 | 390.00 |

**Brookfield™ double bowl cast iron sink, Kohler.** Deep, glossy color. 33" long x 22" wide. Faucet, sprayer, and strainers sold separately. Self-rimming. Four holes.

| | Craft@Hrs | Unit | Material | Labor | Total | Sell |
|---|---|---|---|---|---|---|
| White, 6-3/4" deep | P1@.750 | Ea | 250.00 | 23.60 | 273.60 | 465.00 |
| White, 8" deep | P1@.750 | Ea | 247.00 | 23.60 | 270.60 | 460.00 |

**Clarity cast iron double bowl sink, Kohler.** 33" x 22". Four-hole faucet drilling. 9" deep. Interior corners are nearly square. Large and medium bowls.

| | Craft@Hrs | Unit | Material | Labor | Total | Sell |
|---|---|---|---|---|---|---|
| White, tile-in | P1@.750 | Ea | 289.00 | 23.60 | 312.60 | 531.00 |
| White, under counter | P1@.750 | Ea | 272.00 | 23.60 | 295.60 | 503.00 |

## Composite Sinks

**Silhouette polycast single bowl sink, American Standard.** Porcelain enameled Americast. Lighter than cast iron. Silicone sealant supplied. Four faucet holes. Faucet, sprayer, and strainer sold separately. Self-rimming, tile edge or under the counter installation. 25" x 22" x 9-1/2" deep.

| | Craft@Hrs | Unit | Material | Labor | Total | Sell |
|---|---|---|---|---|---|---|
| Bisque | P1@.750 | Ea | 172.00 | 23.60 | 195.60 | 333.00 |
| Black | P1@.750 | Ea | 215.00 | 23.60 | 238.60 | 406.00 |
| Bone | P1@.750 | Ea | 170.00 | 23.60 | 193.60 | 329.00 |
| White heat | P1@.750 | Ea | 152.00 | 23.60 | 175.60 | 299.00 |

**Silhouette polycast double bowl sink, American Standard.** Porcelain enameled Americast. Lighter than cast iron. Silicone sealant supplied. Four faucet holes. Faucet, sprayer, and strainer sold separately. Self-rimming, tile edge or under the counter installation. 33" x 22" x 9-1/2" deep. Insulating material holds water temperature and reduces noise of running water and garbage disposal. Lead-free porcelain enamel finish and scratch-resistant. Faucets, sprayers and strainers sold separately.

| | Craft@Hrs | Unit | Material | Labor | Total | Sell |
|---|---|---|---|---|---|---|
| Bisque | P1@.750 | Ea | 247.00 | 23.60 | 270.60 | 460.00 |
| Bone | P1@.750 | Ea | 279.00 | 23.60 | 302.60 | 514.00 |
| White heat | P1@.750 | Ea | 214.00 | 23.60 | 237.60 | 404.00 |

**Single bowl composite sink, CorStone.** Four-hole faucet drilling. Scratch- and stain-resistant. Acrylic, self-rimming design. Faucets, drains and accessories sold separately.

| | Craft@Hrs | Unit | Material | Labor | Total | Sell |
|---|---|---|---|---|---|---|
| 25" x 22", black | P1@.600 | Ea | 154.00 | 18.90 | 172.90 | 294.00 |

**Double bowl composite sink, CorStone.** Large bowl stretches from the front of the sink to the back rim. A large 8-1/2" deep offset bowl in front of the faucet deck is separated by a lowered dam that helps avoid chipping and banging of dishes. For 33" x 22" sink cut-out. Faucets, drains and accessories sold separately. Equal bowls.

| | Craft@Hrs | Unit | Material | Labor | Total | Sell |
|---|---|---|---|---|---|---|
| Kendall, white | P1@.750 | Ea | 107.00 | 23.60 | 130.60 | 222.00 |
| Kennesaw, biscuit | P1@.750 | Ea | 172.00 | 23.60 | 195.60 | 333.00 |

**Forsyth hi/lo double bowl composite sink, CorStone.** 33" x 22" x 8-1/2" deep. Four-hole faucet drilling. Scratch- and stain-resistant. Acrylic, self-rimming design. Faucets, drains and accessories sold separately.

| | Craft@Hrs | Unit | Material | Labor | Total | Sell |
|---|---|---|---|---|---|---|
| White | P1@.750 | Ea | 187.00 | 23.60 | 210.60 | 358.00 |

**Greenwich double bowl composite sink, CorStone.** 33" x 22" x 8-1/2" deep. Four-hole faucet drilling. Scratch- and stain-resistant. Acrylic, self-rimming design. Faucets, drains and accessories sold separately. Equal bowls.

| | Craft@Hrs | Unit | Material | Labor | Total | Sell |
|---|---|---|---|---|---|---|
| White | P1@.750 | Ea | 145.00 | 23.60 | 168.60 | 287.00 |

| | Craft@Hrs | Unit | Material | Labor | Total | Sell |
|---|---|---|---|---|---|---|

**Providence 60/40 bowl composite sink, CorStone.** 10" main bowl depth. 33" x 22" sink for use with disposals. Easy installation and easy cleaning.

| | | | | | | |
|---|---|---|---|---|---|---|
| Royston white | P1@.750 | Ea | 198.00 | 23.60 | 221.60 | 377.00 |
| Toccoa black | P1@.750 | Ea | 216.00 | 23.60 | 239.60 | 407.00 |

**Inverness single bowl composite sink, International Thermocast.** 25" x 22" x 9" deep. Cutout dimensions 23" x 20" with 1" radius corners. Rich high gloss finish that is easy to clean with non-abrasive cleaners. Beauty of porcelain with strength of cast iron, yet lighter weight. Insulated construction to absorb waste disposal noise. Large basins for more workspace. Resists stain, rust, oxidation, chipping, and scratches. Surface can be polished with ordinary liquid sink, countertop, or auto polish. Drop-in ready. Faucets, drains and accessories sold separately.

| | | | | | | |
|---|---|---|---|---|---|---|
| Inverness, white, 4" deck | P1@.750 | Ea | 106.00 | 23.60 | 129.60 | 220.00 |
| Inverness, bone, 4" deck | P1@.750 | Ea | 139.00 | 23.60 | 162.60 | 276.00 |
| Wellington, white, 5" deck | P1@.750 | Ea | 129.00 | 23.60 | 152.60 | 259.00 |

**Beaumont double bowl composite kitchen sink, International Thermocast.** Equal bowls. The beauty of porcelain with strength of cast iron yet lighter weight. Insulated construction to absorb waste disposal noise. Large basins for more workspace. Resists stain, rust, oxidation, chipping, and scratches. Surface can be polished with ordinary liquid sink, countertop, or auto polish. Drop-in ready. No mounting clips required. Certified by the National Association of Home Builders Research Center. 33" x 22" x 8-1/4" deep.

| | | | | | | |
|---|---|---|---|---|---|---|
| Bone | P1@.750 | Ea | 129.00 | 23.60 | 152.60 | 259.00 |
| White | P1@.750 | Ea | 102.00 | 23.60 | 125.60 | 214.00 |

**Breckenridge double bowl composite sink, International Thermocast.** 33" x 22". Large bowl 9" deep. Small bowl 7" deep. Cutout dimensions 31" x 20" with 1" radius corners. Drains positioned to maximize workspace. Insulated construction to absorb waste disposal noise. Resists stain, rust, oxidation, chipping, and scratches. Drop-in ready. No mounting clips required.

| | | | | | | |
|---|---|---|---|---|---|---|
| Black | P1@.750 | Ea | 216.00 | 23.60 | 239.60 | 407.00 |
| White | P1@.750 | Ea | 183.00 | 23.60 | 206.60 | 351.00 |

**Newport double bowl composite sink, International Thermocast.** 33" x 22" x 9" deep. Equal bowls. 4-1/4" deck. Four-hole faucet drilling. Self-rimming. Faucets, drains and accessories sold separately.

| | | | | | | |
|---|---|---|---|---|---|---|
| Biscuit | P1@.750 | Ea | 172.00 | 23.60 | 195.60 | 333.00 |
| Bone | P1@.750 | Ea | 183.00 | 23.60 | 206.60 | 351.00 |
| White | P1@.750 | Ea | 146.00 | 23.60 | 169.60 | 288.00 |

**Cambridge offset double bowl composite sink, International Thermocast.** 33" x 22". Large bowl 10-1/2" deep. Small bowl 8-1/2" deep. Self-rimming. Four-hole faucet drilling. Cutout dimensions 31" x 20" with 1" radius corners. Insulated construction to absorb waste disposal noise. Faucets, drains and accessories sold separately.

| | | | | | | |
|---|---|---|---|---|---|---|
| Black | P1@.750 | Ea | 215.00 | 23.60 | 238.60 | 406.00 |
| White | P1@.750 | Ea | 172.00 | 23.60 | 195.60 | 333.00 |

## Pull-Out Kitchen Faucets

**Kitchen faucet, single-handle, pull-out spout.** Deck mount. One- or three-hole installation. Loop handle. Labor includes fasten in place only, and assumes sink is already in place. Installation times will be reduced by half if installing faucet before sink is in place. Add the cost of plumbing rough-in as required. With soap dispenser.

| | | | | | | |
|---|---|---|---|---|---|---|
| Chrome finish | P1@1.10 | Ea | 49.80 | 34.70 | 84.50 | 144.00 |

**Kitchen faucet, single-handle, pull-out spout, Aqua-Touch.** Deck mount. One- or three-hole installation. Lever handle. Labor includes fasten in place only, and assumes sink is already in place. Installation times will be reduced by half if installing faucet before sink is in place. Add the cost of plumbing rough-in as required.

| | | | | | | |
|---|---|---|---|---|---|---|
| Chrome finish | P1@1.00 | Ea | 87.50 | 31.50 | 119.00 | 202.00 |

| | Craft@Hrs | Unit | Material | Labor | Total | Sell |
|---|---|---|---|---|---|---|

**Kitchen faucet, single-handle, pull-out spout, Bacharach, Glacier Bay.** Deck mount. One- or three-hole installation. Loop handle. Ceramic disc cartridge. 20-year warranty. Labor includes fasten in place only, and assumes sink is already in place. Installation times will be reduced by half if installing faucet before sink is in place. Add the cost of plumbing rough-in as required.

| | Craft@Hrs | Unit | Material | Labor | Total | Sell |
|---|---|---|---|---|---|---|
| Polished chrome finish | P1@1.00 | Ea | 91.00 | 31.50 | 122.50 | 208.00 |

**Kitchen faucet, single-handle, pull-out spout, Reliant+™, American Standard.** Deck mount. One- or three-hole installation. Lever handle. Ceramic disc cartridge. Spray or aerated stream. 4-1/2" spout. ADA compliant. Labor includes fasten in place only, and assumes sink is already in place. Installation times will be reduced by half if installing faucet before sink is in place. Add the cost of plumbing rough-in as required.

| | Craft@Hrs | Unit | Material | Labor | Total | Sell |
|---|---|---|---|---|---|---|
| Bone finish | P1@1.00 | Ea | 118.00 | 31.50 | 149.50 | 254.00 |
| Chrome with brass accents | P1@1.00 | Ea | 107.00 | 31.50 | 138.50 | 235.00 |
| Stainless steel | P1@1.00 | Ea | 151.00 | 31.50 | 182.50 | 310.00 |
| Velvet finish | P1@1.00 | Ea | 140.00 | 31.50 | 171.50 | 292.00 |
| White finish | P1@1.00 | Ea | 117.00 | 31.50 | 148.50 | 252.00 |

**Kitchen faucet, single-handle, pull-out spout, Signature, Delta.** Deck mount. One- or three-hole installation. Lever handle. Spray or aerated stream. Arched spout provides 9" of reach and 6-1/2" of height. Touch-Clean® wand makes it easy to wipe away residue and mineral deposits, ensuring proper spraying. Lifetime limited warranty. Labor includes fasten in place only, and assumes sink is already in place. Installation times will be reduced by half if installing faucet before sink is in place. Add the cost of plumbing rough-in as required.

| | Craft@Hrs | Unit | Material | Labor | Total | Sell |
|---|---|---|---|---|---|---|
| Chrome | P1@1.00 | Ea | 145.00 | 31.50 | 176.50 | 300.00 |
| Stainless steel | P1@1.00 | Ea | 214.00 | 31.50 | 245.50 | 417.00 |
| White | P1@1.00 | Ea | 195.00 | 31.50 | 226.50 | 385.00 |

**Kitchen faucet, single-handle, pull-out hi-arc spout, Glacier Extensa®, Moen.** Deck mount. One- or three-hole installation. Lever handle. Spray or aerated stream. Washerless cartridge. IntuiTouch™ wand. Lifetime limited warranty. Labor includes fasten in place only, and assumes sink is already in place. Installation times will be reduced by half if installing faucet before sink is in place. Add the cost of plumbing rough-in as required.

| | Craft@Hrs | Unit | Material | Labor | Total | Sell |
|---|---|---|---|---|---|---|
| Chrome finish | P1@1.00 | Ea | 192.00 | 31.50 | 223.50 | 380.00 |
| Matte black finish | P1@1.00 | Ea | 271.00 | 31.50 | 302.50 | 514.00 |
| Sand finish | P1@1.00 | Ea | 214.00 | 31.50 | 245.50 | 417.00 |
| Stainless steel finish | P1@1.00 | Ea | 219.00 | 31.50 | 250.50 | 426.00 |
| White finish | P1@1.00 | Ea | 237.00 | 31.50 | 268.50 | 456.00 |

**Kitchen faucet, single-handle, pull-out spout, Colonnade, Moen.** Deck mount. One- or three-hole installation. Lever handle. Integral vacuum breaker. Labor includes fasten in place only, and assumes sink is already in place. Installation times will be reduced by half if installing faucet before sink is in place. Add the cost of plumbing rough-in as required.

| | Craft@Hrs | Unit | Material | Labor | Total | Sell |
|---|---|---|---|---|---|---|
| Chrome finish | P1@1.00 | Ea | 186.00 | 31.50 | 217.50 | 370.00 |
| White finish | P1@1.00 | Ea | 203.00 | 31.50 | 234.50 | 399.00 |

**Kitchen faucet, single-handle, pull-out spout, Genesis, Price Pfister.** Deck mount. One- or three-hole installation. Loop handle. Spray or aerated stream. Ceramic disc cartridge. Pforever Pfaucet™ limited lifetime warranty. Labor includes fasten in place only, and assumes sink is already in place. Installation times will be reduced by half if installing faucet before sink is in place. Add the cost of plumbing rough-in as required.

| | Craft@Hrs | Unit | Material | Labor | Total | Sell |
|---|---|---|---|---|---|---|
| Stainless steel | P1@1.00 | Ea | 151.00 | 31.50 | 182.50 | 310.00 |
| White finish | P1@1.00 | Ea | 150.00 | 31.50 | 181.50 | 309.00 |

**Kitchen faucet, single-handle, pull-out spout, Forte, Kohler.** Labor includes fasten in place only and assumes sink is already in place. Installation times will be reduced by half if installing faucet before sink is in place. Add the cost of plumbing rough-in as required.

| | Craft@Hrs | Unit | Material | Labor | Total | Sell |
|---|---|---|---|---|---|---|
| Polished chrome | P1@1.00 | Ea | 193.00 | 31.50 | 224.50 | 382.00 |
| Stainless steel | P1@1.00 | Ea | 234.00 | 31.50 | 265.50 | 451.00 |

| | Craft@Hrs | Unit | Material | Labor | Total | Sell |
|---|---|---|---|---|---|---|

**Kitchen faucet, high arch single-handle, Forte, Kohler.** Remote valve and matching deck mounted spray head. Labor includes fasten in place only and assumes sink is already in place. Installation times will be reduced by half if installing faucet before sink is in place. Add the cost of plumbing rough-in as required.

| | Craft@Hrs | Unit | Material | Labor | Total | Sell |
|---|---|---|---|---|---|---|
| Polished chrome | P1@1.10 | Ea | 183.00 | 34.70 | 217.70 | 370.00 |
| Brushed nickel | P1@1.10 | Ea | 236.00 | 34.70 | 270.70 | 460.00 |
| Polished chrome with soap dispenser | P1@1.20 | Ea | 195.00 | 37.80 | 232.80 | 396.00 |
| Brushed nickel with soap dispenser | P1@1.20 | Ea | 247.00 | 37.80 | 284.80 | 484.00 |

## Two-Handle Kitchen Faucets

**Wall-mount kitchen faucet, two-handle, American Standard.** 8" center set. Porcelain lever handles. Lifetime limited warranty on function and finish. Labor includes fasten in place only. Add the cost of plumbing rough-in as required.

| | Craft@Hrs | Unit | Material | Labor | Total | Sell |
|---|---|---|---|---|---|---|
| Chrome finish | P1@1.00 | Ea | 138.00 | 31.50 | 169.50 | 288.00 |

**Kitchen faucet, two-handle, Cadet, American Standard.** Deck mount. 8": center set. Labor includes fasten in place only, and assumes sink is already in place. Installation times will be reduced by half if installing faucet before sink is in place. Add the cost of plumbing rough-in as required.

| | Craft@Hrs | Unit | Material | Labor | Total | Sell |
|---|---|---|---|---|---|---|
| Chrome finish | P1@1.00 | Ea | 46.50 | 31.50 | 78.00 | 133.00 |

**Kitchen faucet, two-handle, Savannah, Price Pfister.** Solid brass construction. Ceramic disc valving. High arc spout. Dual porcelain handles. Lifetime warranty. Labor assumes sink is already in place. Installation times will be reduced by half if installing faucet before sink is in place.

| | Craft@Hrs | Unit | Material | Labor | Total | Sell |
|---|---|---|---|---|---|---|
| No spray, chrome | P1@1.00 | Ea | 78.50 | 31.50 | 110.00 | 187.00 |
| No spray, gooseneck spout, chrome | P1@1.00 | Ea | 86.40 | 31.50 | 117.90 | 200.00 |
| With spray, chrome | P1@1.10 | Ea | 100.00 | 34.70 | 134.70 | 229.00 |
| With spray, stainless steel | P1@1.10 | Ea | 156.00 | 34.70 | 190.70 | 324.00 |

**Kitchen faucet, two-handle, Redford, Price Pfister.** Deck mount. 8" center set. Three-hole installation. Metal handles. Labor includes fasten in place only, and assumes sink is already in place. Installation times will be reduced by half if installing faucet before sink is in place. Add the cost of plumbing rough-in as required.

| | Craft@Hrs | Unit | Material | Labor | Total | Sell |
|---|---|---|---|---|---|---|
| Chrome finish | P1@1.00 | Ea | 48.40 | 31.50 | 79.90 | 136.00 |

**Kitchen faucet, two-handle, deck spray, Waterfall, Delta.** Deck mount. Chrome finish. 3-hole installation. Washerless design. Lifetime function and finish warranty. Labor includes fasten in place only, and assumes sink is already in place. Installation times will be reduced by half if installing faucet before sink is in place. Add the cost of plumbing rough-in as required.

| | Craft@Hrs | Unit | Material | Labor | Total | Sell |
|---|---|---|---|---|---|---|
| Chrome | P1@1.10 | Ea | 150.00 | 34.70 | 184.70 | 314.00 |
| White | P1@1.10 | Ea | 172.00 | 34.70 | 206.70 | 351.00 |

**Kitchen faucet, two-handle, deck spray, Fairfax, Kohler.** Deck mount. Labor includes fasten in place only, and assumes sink is already in place. Installation times will be reduced by half if installing faucet before sink is in place. Add the cost of plumbing rough-in as required.

| | Craft@Hrs | Unit | Material | Labor | Total | Sell |
|---|---|---|---|---|---|---|
| Polished chrome finish | P1@1.10 | Ea | 129.00 | 34.70 | 163.70 | 278.00 |

**Kitchen faucet, two-handle, Glacier Bay.** Deck mount. 8" center set. Four-hole installation. Acrylic handles. 12-year warranty. Chrome finish. Labor includes fasten in place only, and assumes sink is already in place. Installation times will be reduced by half if installing faucet before sink is in place. Add the cost of plumbing rough-in as required.

| | Craft@Hrs | Unit | Material | Labor | Total | Sell |
|---|---|---|---|---|---|---|
| With deck spray | P1@1.10 | Ea | 36.70 | 34.70 | 71.40 | 121.00 |
| Without spray | P1@1.00 | Ea | 31.10 | 31.50 | 62.60 | 106.00 |

|  | Craft@Hrs | Unit | Material | Labor | Total | Sell |
|---|---|---|---|---|---|---|

**Kitchen faucet, two-handle, with spray, Stanadyne.** Labor includes fasten in place only, and assumes sink is already in place. Installation times will be reduced by half if installing faucet before sink is in place. Add the cost of plumbing rough-in as required. Chrome finish.

| | | | | | | |
|---|---|---|---|---|---|---|
| Acrylic handles | P1@1.10 | Ea | 63.90 | 34.70 | 98.60 | 168.00 |
| Chrome handles | P1@1.10 | Ea | 63.90 | 34.70 | 98.60 | 168.00 |
| High arc | P1@1.10 | Ea | 90.70 | 34.70 | 125.40 | 213.00 |
| Brass accent | P1@1.10 | Ea | 96.50 | 34.70 | 131.20 | 223.00 |

**Kitchen faucet, two-handle, J A Manufacturing.** Deck mount. 8" center set. Builder's grade. Labor includes fasten in place only, and assumes sink is already in place. Installation times will be reduced by half if installing faucet before sink is in place. Add the cost of plumbing rough-in as required.

| | | | | | | |
|---|---|---|---|---|---|---|
| Brushed nickel, 8" hi-rise, with spray | P1@1.10 | Ea | 95.10 | 34.70 | 129.80 | 221.00 |
| Chrome finish | P1@1.00 | Ea | 48.30 | 31.50 | 79.80 | 136.00 |

**Kitchen faucet, two-handle, Brass Craft.** Deck mount. 8" center set. Labor includes fasten in place only, and assumes sink is already in place. Installation times will be reduced by half if installing faucet before sink is in place. Add the cost of plumbing rough-in as required.

| | | | | | | |
|---|---|---|---|---|---|---|
| Chrome finish | P1@1.00 | Ea | 10.30 | 31.50 | 41.80 | 71.10 |

**Kitchen faucet, two-handle, Hamilton American Standard.** Deck mount. 8" center set. Four-hole installation. Ceramic disc cartridge. Matching handles and deck spray. Cast brass waterways. 1/2" brass supply shanks. Labor includes fasten in place only, and assumes sink is already in place. Installation times will be reduced by half if installing faucet before sink is in place. Add the cost of plumbing rough-in as required.

| | | | | | | |
|---|---|---|---|---|---|---|
| Without spray, chrome | P1@1.00 | Ea | 84.30 | 31.50 | 115.80 | 197.00 |
| Without spray, satin nickel | P1@1.00 | Ea | 107.00 | 31.50 | 138.50 | 235.00 |
| With spray, chrome | P1@1.10 | Ea | 96.00 | 34.70 | 130.70 | 222.00 |
| With spray, satin nickel | P1@1.10 | Ea | 118.00 | 34.70 | 152.70 | 260.00 |

## Kitchen Sink Accessories

**Kitchen sink accessories, American Standard.**

| | | | | | | |
|---|---|---|---|---|---|---|
| Undercounter mounting kit | — | Ea | 10.80 | — | 10.80 | — |
| Faucet lift block | — | Ea | 6.60 | — | 6.60 | — |
| Sink mount clips, pack of 10 | P1@.050 | Ea | 2.58 | 1.58 | 4.16 | 7.07 |
| Extra long sink clip set | P1@.050 | Ea | 7.83 | 1.58 | 9.41 | 16.00 |
| J-channel installation clip set | P1@.050 | Ea | 3.69 | 1.58 | 5.27 | 8.96 |

**Sink frame, Vance Industries.** Watertight seal. No caulk bead. Prevents edge of sink from chipping, cracking or dimpling. 32" long x 21" wide.

| | | | | | | |
|---|---|---|---|---|---|---|
| Stainless steel | — | Ea | 17.80 | — | 17.80 | — |
| Stainless steel, S4 | — | Ea | 20.00 | — | 20.00 | — |
| White | — | Ea | 19.00 | — | 19.00 | — |

**Air gap, LDR Industries.**

| | | | | | | |
|---|---|---|---|---|---|---|
| Body only, stainless | P1@.400 | Ea | 3.03 | 12.60 | 15.63 | 26.60 |
| Cover only, stainless | P1@.400 | Ea | 2.81 | 12.60 | 15.41 | 26.20 |
| Chrome | P1@.400 | Ea | 4.32 | 12.60 | 16.92 | 28.80 |
| White finish | P1@.400 | Ea | 4.30 | 12.60 | 16.90 | 28.70 |

**Air gap cover, LDR Industries.**

| | | | | | | |
|---|---|---|---|---|---|---|
| Almond finish | P1@.090 | Ea | 2.42 | 2.84 | 5.26 | 8.94 |
| Chrome finish | P1@.090 | Ea | 1.68 | 2.84 | 4.52 | 7.68 |
| White finish | P1@.090 | Ea | 2.10 | 2.84 | 4.94 | 8.40 |

| | Craft@Hrs | Unit | Material | Labor | Total | Sell |
|---|---|---|---|---|---|---|

**Faucet hole cover, LDR Industries.** Blanks off deck faucet sink holes not being used.

| | Craft@Hrs | Unit | Material | Labor | Total | Sell |
|---|---|---|---|---|---|---|
| Almond finish | P1@.090 | Ea | 2.20 | 2.84 | 5.04 | 8.57 |
| Chrome finish | P1@.090 | Ea | 1.11 | 2.84 | 3.95 | 6.72 |
| Chrome finish, nipple type | P1@.090 | Ea | 2.19 | 2.84 | 5.03 | 8.55 |
| Stainless steel, bolt down | P1@.090 | Ea | 3.23 | 2.84 | 6.07 | 10.30 |
| White finish | P1@.090 | Ea | 2.18 | 2.84 | 5.02 | 8.53 |

**Deluxe sink top soap dispenser, LDR Industries.** Liquid soap or lotion dispenser. Pump action.

| | Craft@Hrs | Unit | Material | Labor | Total | Sell |
|---|---|---|---|---|---|---|
| Almond finish | P1@.250 | Ea | 17.30 | 7.88 | 25.18 | 42.80 |
| Black finish | P1@.250 | Ea | 17.20 | 7.88 | 25.08 | 42.60 |
| Chrome finish | P1@.250 | Ea | 17.30 | 7.88 | 25.18 | 42.80 |
| Stainless steel finish | P1@.250 | Ea | 17.20 | 7.88 | 25.08 | 42.60 |
| White finish | P1@.250 | Ea | 17.30 | 7.88 | 25.18 | 42.80 |

**Plastic sink top soap dispenser, LDR Industries.** Plastic. Liquid soap or lotion dispenser. Pump action. Can be refilled from above the sink.

| | Craft@Hrs | Unit | Material | Labor | Total | Sell |
|---|---|---|---|---|---|---|
| Almond finish | P1@.250 | Ea | 12.80 | 7.88 | 20.68 | 35.20 |
| Chrome finish | P1@.250 | Ea | 14.00 | 7.88 | 21.88 | 37.20 |
| White finish | P1@.250 | Ea | 14.00 | 7.88 | 21.88 | 37.20 |

## Sink Baskets and Strainer Inserts

**Kitchen sink strainer insert, Jameco.**

| | Craft@Hrs | Unit | Material | Labor | Total | Sell |
|---|---|---|---|---|---|---|
| Clip on, stainless steel | P1@.070 | Ea | 2.14 | 2.21 | 4.35 | 7.40 |
| Lock spin, stainless steel | P1@.070 | Ea | 3.23 | 2.21 | 5.44 | 9.25 |
| Long clip, stainless steel | P1@.070 | Ea | 3.22 | 2.21 | 5.43 | 9.23 |
| Plastic | P1@.070 | Ea | 1.11 | 2.21 | 3.32 | 5.64 |

**Plastic basket and strainer insert, LDR Industries.** Designer finish. Fits all standard kitchen sinks.

| | Craft@Hrs | Unit | Material | Labor | Total | Sell |
|---|---|---|---|---|---|---|
| Almond | P1@.250 | Ea | 5.45 | 7.88 | 13.33 | 22.70 |
| Almond, garbage disposal | P1@.250 | Ea | 7.72 | 7.88 | 15.60 | 26.50 |
| Stainless steel | P1@.250 | Ea | 8.63 | 7.88 | 16.51 | 28.10 |
| White | P1@.250 | Ea | 5.44 | 7.88 | 13.32 | 22.60 |
| White, garbage disposal | P1@.250 | Ea | 7.64 | 7.88 | 15.52 | 26.40 |

**Kitchen sink basket and strainer insert, Spin-N-Lock, Jameco.** Premium grade stainless steel. Positive lock-spin metal seal basket closure.

| | Craft@Hrs | Unit | Material | Labor | Total | Sell |
|---|---|---|---|---|---|---|
| Stainless steel | P1@.250 | Ea | 14.70 | 7.88 | 22.58 | 38.40 |

**Decorative kitchen sink basket and strainer insert, Jameco.** Chip and stain resistant Celcon insert and basket. Stainless steel double cup one-piece underbody. Recommended for cast iron, stainless steel and composite sinks. Watertight ball-lok mechanism basket closure.

| | Craft@Hrs | Unit | Material | Labor | Total | Sell |
|---|---|---|---|---|---|---|
| White | P1@.250 | Ea | 21.60 | 7.88 | 29.48 | 50.10 |

**Kitchen sink basket and strainer insert, Jameco.** Ball-lok basket closure for a watertight seal. Metal underbody double cup construction. Chip and stain resistant. Celcon basket and strainer insert. Threaded bushing allows color insert to be changed easily.

| | Craft@Hrs | Unit | Material | Labor | Total | Sell |
|---|---|---|---|---|---|---|
| Bone finish | P1@.250 | Ea | 21.10 | 7.88 | 28.98 | 49.30 |

**Kitchen sink basket and strainer insert, Jameco.** Polished brass finish. Heavy-duty chrome plated solid brass body. Short clip basket closure.

| | Craft@Hrs | Unit | Material | Labor | Total | Sell |
|---|---|---|---|---|---|---|
| Polished brass | P1@.250 | Ea | 31.90 | 7.88 | 39.78 | 67.60 |
| Satin nickel | P1@.250 | Ea | 32.50 | 7.88 | 40.38 | 68.60 |

| | Craft@Hrs | Unit | Material | Labor | Total | Sell |
|---|---|---|---|---|---|---|

**Kitchen sink basket and strainer insert, Plumb Shop.** Brass locknut and slip nut. Stainless steel basket. Solid cast brass body. Recommended for Corian sinks.

| | | | | | | |
|---|---|---|---|---|---|---|
| Chrome rim | P1@.250 | Ea | 17.30 | 7.88 | 25.18 | 42.80 |

**Kitchen sink basket strainer washer, Danco.** 1/8" thick rubber.

| | | | | | | |
|---|---|---|---|---|---|---|
| 1-3/4" x 1-3/8" | P1@.070 | Ea | 1.17 | 2.21 | 3.38 | 5.75 |
| 4-3/8" x 3-7/16" | P1@.070 | Ea | 1.28 | 2.21 | 3.49 | 5.93 |

**Sink strainer shank nut, Danco.** 4-3/8" outside diameter. Die cast.

| | | | | | | |
|---|---|---|---|---|---|---|
| Shank nut, 3-11/32" inside diameter | P1@.070 | Ea | 1.17 | 2.21 | 3.38 | 5.75 |
| Strainer nut, 3-3/8" inside diameter | P1@.070 | Ea | 0.63 | 2.21 | 2.84 | 4.83 |

**Flat sink drain stopper, Plumb Shop.** Universal fit. Rubber.

| | | | | | | |
|---|---|---|---|---|---|---|
| Stopper | — | Ea | 1.08 | — | 1.08 | — |

**New style drain stopper, Danco.** 5/16" stud fits Rapid Fit drains. For old style plastic and brass drain assemblies. Polished brass. 1-11/16" long

| | | | | | | |
|---|---|---|---|---|---|---|
| 3/8" diameter thread | — | Ea | 5.97 | — | 5.97 | — |

## Sink Mounted Hot Water Dispensers

**Hot water dispenser, In-Sink-Erator.** Non-swivel high-spout design. 1/3 gallon capacity. 40 cups per hour. Self-closing valve. Adjustable thermostat. Rugged, low-profile stainless steel spout features an integral copper valve body and tank for dependable performance and easy installation. One-year in-home warranty.

| | | | | | | |
|---|---|---|---|---|---|---|
| Dispenser | P1@1.25 | Ea | 113.00 | 39.40 | 152.40 | 259.00 |

**Deluxe instant hot water dispenser, In-Sink-Erator.** Near-boiling 190 degree Fahrenheit water instantly. Chrome-plated faucet completely insulated from 190-degree water. Provides 60 cups of 190-degree water per hour. Snap-action adjustable from 140 to 200 degrees, factory preset to 190 degrees. Die-molded, high-efficiency, expanded polystyrene thermal barrier surrounds tank for low heat loss. Adjustable thermostat. Convenient drain, thermally fused to prevent tank damage from dry start-up or loss of water. Tank is always open to atmospheric pressure and requires no relief valve. Instant self-closing valve. Traditional twist-handle actuation. One-year full parts and labor warranty.

| | | | | | | |
|---|---|---|---|---|---|---|
| European design, chrome | P1@1.25 | Ea | 183.00 | 39.40 | 222.40 | 378.00 |
| Standard | P1@1.25 | Ea | 160.00 | 39.40 | 199.40 | 339.00 |
| White | P1@1.25 | Ea | 183.00 | 39.40 | 222.40 | 378.00 |

## Utility Sinks, Laundry Tubs

**All-in-one standard utility tub kit, American Shower & Bath.** Made of polypropylene. Includes one standard utility tub, two 3/8" OD x 20" long faucet risers, one drain installation kit, one drain stopper, one faucet assembly, four leg levelers and Teflon® tape. Limited one-year warranty.

| | | | | | | |
|---|---|---|---|---|---|---|
| 19-gallon capacity | P1@1.75 | Ea | 64.80 | 55.10 | 119.90 | 204.00 |

**Heavy-duty utility laundry tub kit, American Shower & Bath.** Made of extra-strength structural foam. Includes ASB® heavy-duty utility tub, 2 Brass Craft® B1-20A SpeediPlumb Plus® connectors. 3/8" OD x 20" long PEX faucet risers. ASB® drain installation kit. Drain stopper. ASB® faucet assembly, 4 leg levelers and Teflon® tape. Limited one-year warranty.

| | | | | | | |
|---|---|---|---|---|---|---|
| Laundry tub | P1@1.75 | Ea | 84.00 | 55.10 | 139.10 | 236.00 |

**Utilatub® polypropylene single bowl laundry tub, E.L. Mustee.** Sturdy, one-piece molded tub. Smooth white satin finish. Easy to clean and stain-resistant. Self-draining soap and storage shelves. Includes leakproof 1-1/2" integral molded drain assembly and floor mounting hardware. Ribbed underbody for extra strength. 13" deep, 20-gallon capacity, 23" wide.

| | | | | | | |
|---|---|---|---|---|---|---|
| Laundry tub | P1@1.75 | Ea | 30.20 | 55.10 | 85.30 | 145.00 |

| | Craft@Hrs | Unit | Material | Labor | Total | Sell |
|---|---|---|---|---|---|---|

**Utilatwin® double bowl laundry tub, E.L. Mustee.** Sturdy, one-piece molded tub. Smooth white surface. Heavy gauge steel legs include levelers. 13" deep, 19-gallon capacity per tub, 40" wide. Leakproof 1-1/2" integral molded drain, hooks up to standard 1-1/2" S or P trap. Accommodates single or dual handle faucet with 4" or 8" centers. Meets or exceeds codes ANSI Z124.6, IAPMO listed.

| | Craft@Hrs | Unit | Material | Labor | Total | Sell |
|---|---|---|---|---|---|---|
| Laundry tub | P1@1.75 | Ea | 111.00 | 55.10 | 166.10 | 282.00 |

**Drop-in utility sink, American Shower & Bath.** 14.5-gallon capacity ABS utility sink. Countertop installation with self-rimming design. Built-in self-draining storage shelf and soap dish. High gloss stain-resistant finish. Easy-to-clean.

| | Craft@Hrs | Unit | Material | Labor | Total | Sell |
|---|---|---|---|---|---|---|
| Utility sink | P1@1.75 | Ea | 53.10 | 55.10 | 108.20 | 184.00 |

**Durastone Utilatub sink, E.L. Mustee.** Hi-impact molded fiberglass includes 1-1/2" drain. Extra deep 13" bowl. Hooks up to standard 1-1/2" P or S trap.

| | Craft@Hrs | Unit | Material | Labor | Total | Sell |
|---|---|---|---|---|---|---|
| Utility sink | P1@1.75 | Ea | 62.70 | 55.10 | 117.80 | 200.00 |

**Stainless steel utility bowl sink, Elkay.** 20-gauge, type-304 stainless steel. 3 faucet hole drillings. 10" extra deep bowl. 20" x 20".

| | Craft@Hrs | Unit | Material | Labor | Total | Sell |
|---|---|---|---|---|---|---|
| Utility or island top sink | P1@.600 | Ea | 137.00 | 18.90 | 155.90 | 265.00 |

**Laundry tub hose, E.L. Mustee.** Dispenses water from washing machine discharge hose into laundry tub. Eliminates draping washing machine hose over back/side wall of laundry tub and conceals washer hose. Mounts to back or side wall of laundry tub, includes mounting hardware.

| | Craft@Hrs | Unit | Material | Labor | Total | Sell |
|---|---|---|---|---|---|---|
| Hose | — | Ea | 8.50 | — | 8.50 | — |

**Laundry tub overflow tube, E.L. Mustee.** Maintains water level in laundry tub and prevents overflows. Replaces drain plug and eliminates reaching into water. 12-3/8" high; may be cut to alter height. Fits all drain openings.

| | Craft@Hrs | Unit | Material | Labor | Total | Sell |
|---|---|---|---|---|---|---|
| Overflow tube | — | Ea | 5.29 | — | 5.29 | — |

## Laundry Tub Faucets

**Laundry faucet, two-handle, Brass Craft.** Deck mount. 4" center set. Labor includes fasten in place only. Add the cost of connections and plumbing rough-in as required.

| | Craft@Hrs | Unit | Material | Labor | Total | Sell |
|---|---|---|---|---|---|---|
| Rough brass finish | P1@.900 | Ea | 14.90 | 28.40 | 43.30 | 73.60 |

**Laundry faucet, two-handle, Glacier Bay.** Deck mount. 4" center set. Drip-free washerless cartridge. Standard 3/4" hose thread on end of spout. 12-year limited warranty. Labor includes fasten in place only. Add the cost of connections and plumbing rough-in as required.

| | Craft@Hrs | Unit | Material | Labor | Total | Sell |
|---|---|---|---|---|---|---|
| Chrome finish | P1@.900 | Ea | 21.50 | 28.40 | 49.90 | 84.80 |

**Laundry faucet, two-handle, Stanadyne.** 1/2" connections. Mini lever handles. Labor includes fasten in place only. Add the cost of connections and plumbing rough-in as required.

| | Craft@Hrs | Unit | Material | Labor | Total | Sell |
|---|---|---|---|---|---|---|
| Chrome | P1@.900 | Ea | 46.10 | 28.40 | 74.50 | 127.00 |

**Laundry faucet, two-handle, wall mount, Price Pfister.** Oakland blade handles with flanges. Handicap accessible. Labor includes fasten in place only. Add the cost of connections and plumbing rough-in as required.

| | Craft@Hrs | Unit | Material | Labor | Total | Sell |
|---|---|---|---|---|---|---|
| Satin chrome finish | P1@.900 | Ea | 48.10 | 28.40 | 76.50 | 130.00 |

## Bar Sinks

**Stainless steel bar sink, Banner.** Single bowl. Standard-gauge stainless steel. Self-rimming design. Buffed finish. 2" drain. Faucet and strainer sold separately.

| | Craft@Hrs | Unit | Material | Labor | Total | Sell |
|---|---|---|---|---|---|---|
| 15" x 15" x 5-1/2" deep | P1@.600 | Ea | 47.20 | 18.90 | 66.10 | 112.00 |

|  | Craft@Hrs | Unit | Material | Labor | Total | Sell |
|---|---|---|---|---|---|---|

**Stainless steel bar sink, Neptune Series, Elkay.** Deep hospitality drop-in single bowl. Medium-thickness, 23-gauge nickel-bearing stainless steel. Buffed finish with machine-ground satin-finished bowl. Sound deadened with pads. Self-rimming installation. Square corners. Two faucet holes with 4" spacing. One-hand mounting fasteners. Includes strainer.

| 15" x 15", 5" deep | P1@.600 | Ea | 43.30 | 18.90 | 62.20 | 106.00 |

**Stainless steel bar sink, Signature Series, Elkay.** Drop-in single bowl. Self-rimming installation. Made from nickel bearing type 302 20-gauge stainless steel. Mirror-finished interior deck. Machine-ground satin-finished bowl. Highlighted outer-rim edge and bowl radius. Satin highlighted sink rim. Fully undercoated for sound deadening. Medium thickness. 3" radius vertical coved corners. Two faucet holes with 4" spacing. Complete with strainer.

| 15" x 15" x 7-1/2" deep | P1@.600 | Ea | 86.40 | 18.90 | 105.30 | 179.00 |

**Silhouette polycast island or bar sink, American Standard.** 18" x 18" x 9" deep. Porcelain enameled Americast. Lighter than cast iron. Silicone sealant supplied.

| White Heat | P1@.750 | Ea | 132.00 | 23.60 | 155.60 | 265.00 |

**Berkeley composite bar sink, CorStone.** Replaces 15" x 15" bar sinks. Large enough bowl to use as a kitchen, prep or utility sink. Full 3-1/2" drain accepts standard kitchen sized basket strainers. Faucet and drain sold separately.

| White | P1@.750 | Ea | 65.50 | 23.60 | 89.10 | 151.00 |

**Composite entertainment sink, CorStone.** Serves as a prep, bar or island sink. Accommodates all garbage disposals in its full depth of 8". Use as a stand-alone or as a "point-of-use" sink in a multi-sink kitchen.

| 18" x 18" x 8", white | P1@.750 | Ea | 85.30 | 23.60 | 108.90 | 185.00 |

**Manchester composite bar sink, International Thermocast.** 16" x 16" x 7" deep. Cutout dimensions 14.5" x 14.5" with 1" radius corners. Rich high-gloss finish. Resists stain, rust, oxidation, chipping, and scratches. Drop-in ready for quick and easy installation. No mounting clips required. Faucets, drains and accessories sold separately.

| White | P1@.750 | Ea | 63.90 | 23.60 | 87.50 | 149.00 |

**Junior basket strainer insert, Jameco.** For laundry tubs, bar sinks and mobile homes. Includes junior basket strainer, flat grid strainer and wash tray plug.

| Stainless steel | P1@.070 | Ea | 5.38 | 2.21 | 7.59 | 12.90 |

## Bar Faucets

**Bar faucet, single-handle, Waterfall, Delta, Masco.** Deck mount. 4" center set. Two-hole application. Solid brass and stainless steel construction. Designed to work on bar, preparation and laundry sinks. Labor includes fasten in place only. Add the cost of plumbing rough-in and connection as required.

| Chrome finish | P1@1.00 | Ea | 89.30 | 31.50 | 120.80 | 205.00 |
| Stainless steel | P1@1.00 | Ea | 143.00 | 31.50 | 174.50 | 297.00 |

**Bar faucet, two-handle, Watts Anderson Barrows.** 4" monoblock. Labor includes fasten in place only. Add the cost of plumbing rough-in as required.

| Chrome finish | — | Ea | 96.10 | — | 96.10 | — |

**Bar faucet, two-handle, Williamsburg, American Standard.** Deck mount. 4" center set. Porcelain handles. Labor includes fasten in place only. Add the cost of plumbing rough-in as required.

| Brass finish | P1@.906 | Ea | 134.00 | 28.50 | 162.50 | 276.00 |
| Chrome finish | P1@.906 | Ea | 86.40 | 28.50 | 114.90 | 195.00 |

**Bar faucet, Cadet Series, American Standard.** Deck mount. Labor includes fasten in place only. Add the cost of plumbing rough-in as required.

| Chrome finish | P1@.906 | Ea | 45.40 | 28.50 | 73.90 | 126.00 |

|  | Craft@Hrs | Unit | Material | Labor | Total | Sell |
|---|---|---|---|---|---|---|

**Bar faucet, two-handle, Glacier Bay.** Deck mount. 4" center set. Drip-free washerless cartridge. Solid brass waterways. 12-year limited warranty. Labor includes fasten in place only. Add the cost of plumbing rough-in as required.

| Chrome finish | P1@.906 | Ea | 35.50 | 28.50 | 64.00 | 109.00 |

**Bar faucet, two-handle, J A Manufacturing.** Deck mount. 4" center set. Builder's grade. Labor includes fasten in place only. Add the cost of connections and plumbing rough-in as required.

| Chrome finish, kit | P1@.906 | Ea | 47.10 | 28.50 | 75.60 | 129.00 |
| Chrome finish, lever handles | P1@.906 | Ea | 40.50 | 28.50 | 69.00 | 117.00 |

## Garbage Disposers

**1/3-horsepower food waste disposer, In-Sink-Erator.** Plastic 26-ounce grind chamber. Galvanized steel grinding elements with two stainless steel 360-degree swivel lugs. Stainless steel sink flange with positive seal one-piece plastic stopper. Overload protector manual reset. Permanently lubricated upper and lower bearings. Quick Lock™ mounting for fast, easy installation. Cushioned anti-splash baffle. Dishwasher drain connection. One-year parts and in-home service warranty.

| Badger 1® | P1@1.00 | Ea | 65.00 | 31.50 | 96.50 | 164.00 |

**1/2-horsepower food waste disposer, In-Sink-Erator.** Galvanized steel grinding elements with two stainless steel 360-degree swivel lugs. Plastic grind chamber. Permanently lubricated upper and lower bearings. Quick Lock™ mounting for fast, easy installation. Stainless steel sink flange with positive seal one-piece plastic stopper. Cushioned anti-splash baffle. Overload protector manual reset. Two-year parts and in-home service warranty.

| Badger 5® | P1@1.00 | Ea | 73.80 | 31.50 | 105.30 | 179.00 |

**5/8-horsepower food waste disposer, In-Sink-Erator.** Plastic 26-ounce grinding chamber. Insulated outer shell. Galvanized steel grinding elements with two stainless steel 360-degree swivel legs. Stainless steel sink flange with positive seal one-piece plastic stopper. Dishwasher drain connection. Permanently lubricated upper and lower bearings. Heavy-duty induction motor. Overload protector manual reset. Stainless steel mounting. Three-year parts and in-home service warranty. Cushioned anti-splash baffle. Quick Lock™ mounting for fast, easy installation.

| Badger 5® Plus | P1@1.00 | Ea | 96.60 | 31.50 | 128.10 | 218.00 |

**3/4-horsepower heavy-duty food waste disposer, In-Sink-Erator.** 26-ounce plastic grind chamber. Insulated outer shell. Stainless steel grinding elements with two stainless steel 360-degree swivel lugs. Stainless steel sink flange with positive seal one-piece stainless steel stopper. Dishwasher drain connection. Permanently lubricated upper and lower bearings. Overload protector manual reset. Cushioned anti-splash baffle. Quick Lock™ mounting for fast, easy installation. Four-year parts and in-home service warranty.

| Disposer | P1@1.00 | Ea | 152.00 | 31.50 | 183.50 | 312.00 |

**3/4-horsepower septic system food waste disposer, In-Sink-Erator.** Corrosion-proof grind chamber. Stainless steel grinding elements with two stainless steel 360-degree swivel lugs. Permanently lubricated upper and lower bearings. Quick Lock™ mounting for fast, easy installation. Stainless steel sink flange with positive seal one-piece plastic stopper. Overload protector manual reset. Dishwasher drain connection. Replaceable Bio-Charge™ additive cartridge. 26-ounce grind chamber capacity. Three-year parts and in-home service warranty.

| Disposer | P1@1.00 | Ea | 198.00 | 31.50 | 229.50 | 390.00 |

**3/4-horsepower batch feed disposer, In-Sink-Erator.** Extra large capacity. Locking stopper controls operation. Five-year in-home warranty. Full stainless steel construction. Labor does not include electrical connection or switch.

| Disposer | P1@1.00 | Ea | 291.00 | 31.50 | 322.50 | 548.00 |

| | Craft@Hrs | Unit | Material | Labor | Total | Sell |
|---|---|---|---|---|---|---|

**1-horsepower food waste disposer, In-Sink-Erator.** Auto-reverse grind system helps prevent jams, extends product life and provides extra sanitation. Secondary sound baffle. Stainless steel 50-ounce grind chamber. Stainless steel grinding elements with two stainless steel 360-degree swivel lugs. Corrosion protection shield. One-piece polished stainless steel stopper. Dishwasher drain connection. Permanently lubricated upper and lower bearings. Overload protector manual reset. Cushioned anti-splash baffle. Quick Lock™ mounting for fast, easy installation. Seven-year parts and in-home service warranty.

| | Craft@Hrs | Unit | Material | Labor | Total | Sell |
|---|---|---|---|---|---|---|
| Disposer | P1@1.00 | Ea | 209.00 | 31.50 | 240.50 | 409.00 |

**Garbage disposer sink stopper, In-Sink-Erator.** Economy-style sink stopper. Fits all disposers made by In-Sink-Erator. Genuine In-Sink-Erator replacement part.

| | | | | | | |
|---|---|---|---|---|---|---|
| Plastic | — | Ea | 6.45 | — | 6.45 | — |
| Polished brass | — | Ea | 26.80 | — | 26.80 | — |
| Stainless steel | — | Ea | 10.30 | — | 10.30 | — |
| Stainless steel, almond | — | Ea | 14.70 | — | 14.70 | — |
| Stainless steel, white | — | Ea | 14.70 | — | 14.70 | — |

**Garbage disposer cord kit, In-Sink-Erator.** Power cord UL-approved to resist moisture and safely connect disposer to a plug-in outlet. Includes wire nuts and retaining clamp.

| | | | | | | |
|---|---|---|---|---|---|---|
| Kit | — | Ea | 10.80 | — | 10.80 | — |

**Disposer mounting assembly, In-Sink-Erator.** Replaces existing disposer mounts. Quick Lock™ mounting assembly.

| | | | | | | |
|---|---|---|---|---|---|---|
| Kit | — | Ea | 15.10 | — | 15.10 | — |

**Dishwasher connector, In-Sink-Erator.** Connects dishwasher drain to the disposer. Standard size.

| | | | | | | |
|---|---|---|---|---|---|---|
| Hose | — | Ea | 8.32 | — | 8.32 | — |

**Sink disposer control switch, In-Sink-Erator.** Convenient and stylish new way to control disposer. Can be installed on sink or countertop. Wall-mount power module. 3' grounded power cord and 6' of air tubing.

| | | | | | | |
|---|---|---|---|---|---|---|
| Chrome or white | — | Ea | 56.40 | — | 56.40 | — |

**Deluxe disposer mounting gasket, In-Sink-Erator.**

| | | | | | | |
|---|---|---|---|---|---|---|
| Deluxe gasket A | — | Ea | 7.21 | — | 7.21 | — |
| Standard gasket B | — | Ea | 5.88 | — | 5.88 | — |

**Removable disposer sound baffle, In-Sink-Erator.** Muffles disposer noise. Prevents food from splashing up in sink.

| | | | | | | |
|---|---|---|---|---|---|---|
| Baffle | — | Ea | 6.13 | — | 6.13 | — |

**Septic disposer replacement cartridge, In-Sink-Erator.** Bio-charge 16-ounce enzyme replacement cartridge for septic disposer system. Accelerates digesting action in the septic tank. Lasts up to 4 months.

| | | | | | | |
|---|---|---|---|---|---|---|
| Cartridge | — | Ea | 14.00 | — | 14.00 | — |

**Disposer sink flange, In-Sink-Erator.** Adds a custom flair.

| | | | | | | |
|---|---|---|---|---|---|---|
| Almond | P1@.185 | Ea | 14.30 | 5.83 | 20.13 | 34.20 |
| Polished brass | P1@.185 | Ea | 26.20 | 5.83 | 32.03 | 54.50 |
| Stainless steel | P1@.185 | Ea | 14.40 | 5.83 | 20.23 | 34.40 |
| White | P1@.185 | Ea | 15.10 | 5.83 | 20.93 | 35.60 |

| | Craft@Hrs | Unit | Material | Labor | Total | Sell |
|---|---|---|---|---|---|---|

## Dishwashers

**1-cycle dishwasher, GE.** Two wash levels with normal wash cycle (dial). Heated dry (rocker switch). Hot pre-wash option. Standard upper rack. Deluxe lower rack.

| | | | | | | |
|---|---|---|---|---|---|---|
| No. HDA1200GBB, black | P1@2.40 | Ea | 172.00 | 75.60 | 247.60 | 421.00 |

**4-cycle, 9-key touch pad dishwasher, Maytag.** Space for 12 place settings. High-pressure, high water-flow wash system. Three wash arms for spray action above and below each rack. Energy Star® rated. Exceeds federal energy-efficiency standards. Microprocessor controls with angled touch pad. Sound-silencing package.

| | | | | | | |
|---|---|---|---|---|---|---|
| No. MDBH950AWQ, bisque | P1@2.40 | Ea | 321.00 | 75.60 | 396.60 | 674.00 |
| No. MDBH950AWW, white | P1@2.40 | Ea | 321.00 | 75.60 | 396.60 | 674.00 |

**2-level dishwasher, GE.** Nautilus dishwasher with the PowerScrub wash system and two wash arms. Touchpad controls. HotStart™ GE QuietPower™ motor.

| | | | | | | |
|---|---|---|---|---|---|---|
| No. GSD4000JWW, white or black | P1@2.40 | Ea | 324.00 | 75.60 | 399.60 | 679.00 |

**5-cycle, 11-key touch pad dishwasher, Maytag.** Tall tub for higher capacity. Space for 14 place settings. Adjustable upper rack allows custom configurations. High-pressure, high water-flow wash system. Three wash arms. Finer filtration. Energy Star® rated. Exceeds federal energy-efficiency standards. Microprocessor controls with angled touch pad.

| | | | | | | |
|---|---|---|---|---|---|---|
| No. MDBH970AWB, black or white | P1@2.40 | Ea | 375.00 | 75.60 | 450.60 | 766.00 |

**13-cycle stainless steel interior dishwasher, Jenn-Air.** Tall-tub design. .

| | | | | | | |
|---|---|---|---|---|---|---|
| No. JDB1080AWS | P1@2.40 | Ea | 650.00 | 75.60 | 725.60 | 1,230.00 |

## Dishwasher Accessories

**Dishwasher connector.**

| | | | | | | |
|---|---|---|---|---|---|---|
| 72" long, stainless steel braid | P1@.185 | Ea | 17.80 | 5.83 | 23.63 | 40.20 |
| 60" long, stainless steel braid | P1@.185 | Ea | 14.00 | 5.83 | 19.83 | 33.70 |
| 3/8" x 3/8" x 60", poly braid | P1@.150 | Ea | 14.80 | 4.73 | 19.53 | 33.20 |

**Dishwasher connector, compression inlet, Speedi-Plumb™, Plumb Shop.** 125 PSI maximum operating pressure. 40 to 140 degree Fahrenheit temperature range. CSA, UL and IAPMO listed.

| | | | | | | |
|---|---|---|---|---|---|---|
| 3/8" x 3/8" x 48" compression | P1@.200 | Ea | 11.60 | 6.30 | 17.90 | 30.40 |

**Garbage disposal to dishwasher drain connector, Fernco.** Connects dishwasher hose to sink drains and garbage disposals. Engineered for years of leak-free and trouble-free service.

| | | | | | | |
|---|---|---|---|---|---|---|
| Model No. PDWC-100 | P1@.250 | Ea | 2.66 | 7.88 | 10.54 | 17.90 |

| | Craft@Hrs | Unit | Material | Labor | Total | Sell |
|---|---|---|---|---|---|---|

## Cooktops

**30" electric cooktop, Maytag.** Four Insta-Heat™ radiant elements. Element-in-use and hot surface indicator lights. Frameless glass-ceramic surface. Dual-Choice™ element provides a large and small element in one location. Color-coordinated knobs with metallic accents pull off for cleaning. Overall dimensions 21" deep x 29-15/16" wide x 4-1/8" high. Frameless smooth black top. Labor cost assumes countertop is cut correctly and electrical rough-in is complete.

| No. MEC5430BDW, white | BE@.903 | Ea | 594.00 | 31.10 | 625.10 | 1,060.00 |
|---|---|---|---|---|---|---|

**30" gas cooktop, Maytag.** Four sealed surface gas burners. Two 9,200/9,100 BTU burners, one 12,500/10,500 BTU power boost burner, one 5,000/4,000 BTU simmer burner. Porcelain-on-steel surface. Pilotless electronic ignition with flame chamber. Color coordinated easy-grasp pull-off control knobs with metallic accents. Four Duraclean lift-off burner caps. Continuous Duraclean cast-iron burner grates. Labor cost assumes countertop is cut correctly and gas rough-in is complete.

| No. MGC5430BDQ, bisque | P1@1.25 | Ea | 355.00 | 39.40 | 394.40 | 670.00 |
|---|---|---|---|---|---|---|
| No. MGC5430BDB, black | P1@1.25 | Ea | 350.00 | 39.40 | 389.40 | 662.00 |
| No. MGC5430BDW, white | P1@1.25 | Ea | 370.00 | 39.40 | 409.40 | 696.00 |

**30" electric ceramic glass cooktop, GE.** Five tri-ring (12", 9" and 6") and ribbon heating elements. Frameless desing with upfront controls. Overall dimensions: 20-7/8" deep x 29-3/4" wide x 3-1/4" high. Labor cost assumes countertop is cut correctly and electrical rough-in is complete.

| No. JP970SKSS, stainless | BE@.903 | Ea | 1300.00 | 31.10 | 1,331.10 | 2,260.00 |
|---|---|---|---|---|---|---|

## Wall Ovens

**30" electric double wall oven, Maytag.** Two self-cleaning ovens with adjustable cleaning levels. Precision electronic controls. Favorite setting, cook and hold setting, adjustable keep warm setting. Delayed cook/clean options.

| Choice of finishes | BE@1.67 | Ea | 1270.00 | 57.40 | 1,327.40 | 2,260.00 |
|---|---|---|---|---|---|---|

**27" single wall oven, GE.** Preheat signal, auto oven shut-off, self-cleaning, control lock, delay bake option and TrueTemp System. Clear view window pattern.

| No. JKP20BFBB, black | BE@1.67 | Ea | 870.00 | 57.40 | 927.40 | 1,580.00 |
|---|---|---|---|---|---|---|

## Mounted Microwave Ovens

**1.4 cubic foot over-the-range microwave oven, GE.** 950 watts. cooking complete reminder, defrost, delay start, extend cook function, interior light, one-touch cooking, popcorn feature, vent/exhaust fan. 29.75" wide x 29.75" high x 15" deep. Labor cost assumes the opening has been cut and electrical rough-in is complete.

| No. JVM1443WK, white | BE@.903 | Ea | 215.00 | 31.10 | 246.10 | 418.00 |
|---|---|---|---|---|---|---|
| No. JVM1443BK, black | BE@.903 | Ea | 215.00 | 31.10 | 246.10 | 418.00 |
| No. JVM1443BK, stainless | BE@.903 | Ea | 260.00 | 31.10 | 291.10 | 495.00 |

**1.3 cubic foot Advantium™ microwave oven, GE.** Can be installed over a countertop, range, or cooktop and plugs into a standard 120-volt/15-amp outlet. Halogen cooktop lighting. Microhood two-speed, 300 CFM venting system. 15.75" deep x 16.4" high x 29.88" wide. 900 watts. Labor cost assumes the opening has been cut and electrical rough-in is complete.

| No. SCA1001KSS, stainless and black | BE@.903 | Ea | 810.00 | 31.10 | 841.10 | 1,430.00 |
|---|---|---|---|---|---|---|

|  | Craft@Hrs | Unit | Material | Labor | Total | Sell |
|---|---|---|---|---|---|---|

# Range Hoods

**Economy ducted range hood.** Under cabinet mounted 190 CFM (cubic feet per minute) exhaust. 6.0 sones. 7"
duct connector. 18" deep x 6" high. Washable charcoal filter with replaceable charcoal filter pad. Uses 75-watt
incandescent lamp. Polymeric lens. Add the cost of ducting and electrical connection.

|  | Craft@Hrs | Unit | Material | Labor | Total | Sell |
|---|---|---|---|---|---|---|
| 24" or 30" wide, enamel finish | SW@1.42 | Ea | 74.50 | 50.60 | 125.10 | 213.00 |
| 24" or 30" wide, stainless steel | SW@1.42 | Ea | 118.00 | 50.60 | 168.60 | 287.00 |
| 36" wide, enamel finish | SW@1.42 | Ea | 77.00 | 50.60 | 127.60 | 217.00 |
| 36" wide, stainless steel | SW@1.42 | Ea | 132.00 | 50.60 | 182.60 | 310.00 |
| 42" wide, enamel finish | SW@1.42 | Ea | 80.00 | 50.60 | 130.60 | 222.00 |
| 42" wide, stainless steel | SW@1.42 | Ea | 143.00 | 50.60 | 193.60 | 329.00 |
| Add for ducted hood with a duct damper | — | Ea | 10.00 | — | 10.00 | 17.00 |

**Standard quality ducted range hood.** Under cabinet mounted 190 CFM (cubic feet per minute) 3-speed blower.
3.5 sones. Either vertical or horizontal 3-1/4" x 10" duct, 7" round duct. Powder-coated 22 gauge steel. 18" deep x 6"
high. Uses one 40-watt incandescent lamp. Add the cost of ducting and electrical connection.

|  | Craft@Hrs | Unit | Material | Labor | Total | Sell |
|---|---|---|---|---|---|---|
| 24" or 30" wide, ducted, black | SW@1.42 | Ea | 126.00 | 50.60 | 176.60 | 300.00 |
| 24" or 30" wide, ducted, almond | SW@1.42 | Ea | 164.00 | 50.60 | 214.60 | 365.00 |
| 24" or 30" wide, ducted, stainless | SW@1.42 | Ea | 238.00 | 50.60 | 288.60 | 491.00 |
| 36" wide, ducted, black | SW@1.42 | Ea | 136.00 | 50.60 | 186.60 | 317.00 |
| 36" wide, ducted, almond | SW@1.42 | Ea | 173.00 | 50.60 | 223.60 | 380.00 |
| 36" wide, ducted, stainless | SW@1.42 | Ea | 248.00 | 50.60 | 298.60 | 508.00 |
| 42" wide, ducted, black | SW@1.42 | Ea | 143.00 | 50.60 | 193.60 | 329.00 |
| 42" wide, ducted, almond | SW@1.42 | Ea | 181.00 | 50.60 | 231.60 | 394.00 |
| 42" wide, ducted, stainless | SW@1.42 | Ea | 256.00 | 50.60 | 306.60 | 521.00 |
| Add for ductless range hood | — | Ea | 50.00 | — | 50.00 | 85.00 |

**Better quality range hood, Allure III NuTone.** Under cabinet mounted 400 CFM (cubic feet per minute). 0.4 sones
at 100 CFM. Teflon-coated bottom pan. Dual halogen lamps. Sensor detects excessive heat and adjusts vent
speed. 20" deep x 7-1/4" high. Four ducting options: 3 1/4" X 10" horizontal or vertical vent, 7" round vertical vent or
non-ducted. Add the cost of ducting and electrical connection.

|  | Craft@Hrs | Unit | Material | Labor | Total | Sell |
|---|---|---|---|---|---|---|
| 30" wide, polyester finish | SW@1.42 | Ea | 450.00 | 50.60 | 500.60 | 851.00 |
| 30" wide, stainless steel finish | SW@1.42 | Ea | 490.00 | 50.60 | 540.60 | 919.00 |
| 36" wide, polyester finish | SW@1.42 | Ea | 460.00 | 50.60 | 510.60 | 868.00 |
| 36" wide, stainless steel finish | SW@1.42 | Ea | 505.00 | 50.60 | 555.60 | 945.00 |
| 42" wide, polyester finish | SW@1.42 | Ea | 470.00 | 50.60 | 520.60 | 885.00 |
| 42" wide, stainless steel finish | SW@1.42 | Ea | 520.00 | 50.60 | 570.60 | 970.00 |
| Add for filter for ducted application | — | Ea | 44.00 | — | 44.00 | 75.00 |
| Add for filter for non-ducted application | — | Ea | 21.50 | — | 21.50 | 36.50 |

**Wall-mount range hood, Ballista NuTone.** Stainless steel finish with hood-mounted blower. Three-speed control.
Dual 20-watt halogen bulbs. Adjusts to fit under 8' to 10' ceiling. Stainless steel grease filter. Built-in backdraft
damper. 20" deep. 33" to 54" high. Add the cost of ducting and electrical connection.

|  | Craft@Hrs | Unit | Material | Labor | Total | Sell |
|---|---|---|---|---|---|---|
| 30" wide, 450 CFM | SW@2.50 | Ea | 1,450.00 | 89.20 | 1,539.20 | 2,620.00 |
| 36" wide, 450 CFM | SW@2.50 | Ea | 1,470.00 | 89.20 | 1,559.20 | 2,650.00 |
| 48" wide, 900 CFM | SW@2.50 | Ea | 1,640.00 | 89.20 | 1,729.20 | 2,940.00 |
| 36", Stainless steel | SW@1.42 | Ea | 167.00 | 50.60 | 217.60 | 370.00 |

**Ceiling-hung range hood, Provisa NuTone.** Stainless steel finish with hood-mounted blower. Three-speed
control. Four 20-watt halogen bulbs. Telescopic flue adjusts to fit under 8' to 9' ceiling. Stainless steel grease filter.
Built-in backdraft damper. 27" deep. 33" high. Add the cost of ducting and electrical connection.

|  | Craft@Hrs | Unit | Material | Labor | Total | Sell |
|---|---|---|---|---|---|---|
| 40" wide, 900 CFM | SW@3.50 | Ea | 2,800.00 | 125.00 | 2,925.00 | 4,970.00 |

| | Craft@Hrs | Unit | Material | Labor | Total | Sell |
|---|---|---|---|---|---|---|

**Custom range hood power package.** Exhaust system components installed in a custom-built range hood 24" to 30" above the cooling surface. Stainless steel. With variable speed control, filters and halogen or incandescent lighting. Bulbs not included. No ductwork included. Add the cost of a decorative enclosure, ducting and electrical connection.

700 CFM, dual blowers, 7" duct, 1.5 to 4.6 sones

| | Craft@Hrs | Unit | Material | Labor | Total | Sell |
|---|---|---|---|---|---|---|
| 30" wide x 22" deep | SW@1.42 | Ea | 440.00 | 50.60 | 490.60 | 834.00 |
| 32" wide x 22" deep | SW@1.42 | Ea | 480.00 | 50.60 | 530.60 | 902.00 |
| 36" wide x 28" deep | SW@1.42 | Ea | 545.00 | 50.60 | 595.60 | 1,010.00 |
| 42" wide x 28" deep | SW@1.42 | Ea | 600.00 | 50.60 | 650.60 | 1,110.00 |
| 48" wide x 28" deep | SW@1.42 | Ea | 680.00 | 50.60 | 730.60 | 1,240.00 |

1300 CFM, dual blowers, twin 7" ducts, 1.5 to 4.6 sones

| | Craft@Hrs | Unit | Material | Labor | Total | Sell |
|---|---|---|---|---|---|---|
| 30" wide x 22" deep | SW@1.42 | Ea | 715.00 | 50.60 | 765.60 | 1,300.00 |
| 32" wide x 22" deep | SW@1.42 | Ea | 787.00 | 50.60 | 837.60 | 1,420.00 |
| 36" wide x 28" deep | SW@1.42 | Ea | 885.00 | 50.60 | 935.60 | 1,590.00 |
| 42" wide x 28" deep | SW@1.42 | Ea | 974.00 | 50.60 | 1,024.60 | 1,740.00 |
| 48" wide x 28" deep | SW@1.42 | Ea | 1,110.00 | 50.60 | 1,160.60 | 1,970.00 |

**Range splash plate, Broan.** Easy to install. Mounts behind range with screws. Keeps kitchen walls clean.

| | Craft@Hrs | Unit | Material | Labor | Total | Sell |
|---|---|---|---|---|---|---|
| 30" x 24", reversible white or almond | SW@.165 | Ea | 23.00 | 5.88 | 28.88 | 49.10 |
| 30" x 24", stainless steel | SW@.165 | Ea | 51.80 | 5.88 | 57.68 | 98.10 |
| 36" x 24", reversible white or almond | SW@.165 | Ea | 28.20 | 5.88 | 34.08 | 57.90 |
| 36" x 24", stainless steel | SW@.165 | Ea | 58.70 | 5.88 | 64.58 | 110.00 |

# *Bathrooms*

# 13

There are two types of bathroom jobs. The first simply makes better use of the available space. The second is adding space, either by enlarging an existing bathroom or by adding an entirely new bathroom. Both types will be expensive when calculated on the basis of cost per square foot of floor. But real estate professionals agree that money invested in bathrooms is generally money wisely spent. A home with three or four bedrooms and only one and a half bathrooms has a major defect worth fixing. That's especially true when the home is in a neighborhood where most homes have two or more bathrooms.

## Adding a Bathroom

Older homes seldom have enough bathrooms. But the bathrooms in those homes are often much larger than the modern 5' by 7' standard for a three-fixture bath (toilet, lavatory and tub). With a large bathroom, remodeling options will be obvious: extra storage cabinets, a double-bowl vanity, more mirrors, a lighted dressing table, separate tub and shower units, or even a bathroom divided into compartments. If an entirely new bathroom is needed, the most obvious issue will be where to find the space. Keep in mind that a bathroom addition outside the existing perimeter wall is seldom a good choice of location.

Generally, bedrooms and bathrooms go together. Access to any new bathroom should be directly off either a bedroom or a central hall connecting bedrooms. Avoid any floor plan that requires walking through another room to get to the new bathroom. An exception would be if the new bathroom is a two-fixture half-bath (powder room) designed to serve a relaxation area. A good bathroom plan minimizes plumbing and electrical runs and locates plumbing fixtures on one wall. This arrangement allows the fixtures to share a common waste line and roof vent.

---

### Counting and Naming Bathrooms

Here's an explanation of what determines a bath type, if you're not familiar with bathroom naming conventions.

*Full bath* — Three or more fixtures including at least one lavatory, toilet and bathtub. Expect to see a showerhead over the tub.

*Three-quarter bath* — Three fixtures including a lavatory, toilet and stall shower (but no tub).

*Half bath* — Two fixtures, a lavatory and toilet.

*Powder room* — A two-fixture bathroom located near the front entrance.

*Mud room* — Same as a powder room but located by a back door or laundry.

*Notes:* When adding up the bathrooms in a house, round three-quarter baths up to a full bath if there are other partial bathrooms. You're sure to confuse someone if you claim a house has two and a quarter baths (full bath, three-quarter bath and half-bath). Instead, call it two and a half baths. Likewise, ignore the benefit of a second lavatory or a bidet in a bathroom. Those fixtures get no extra credit when it comes to what counts in a bathroom.

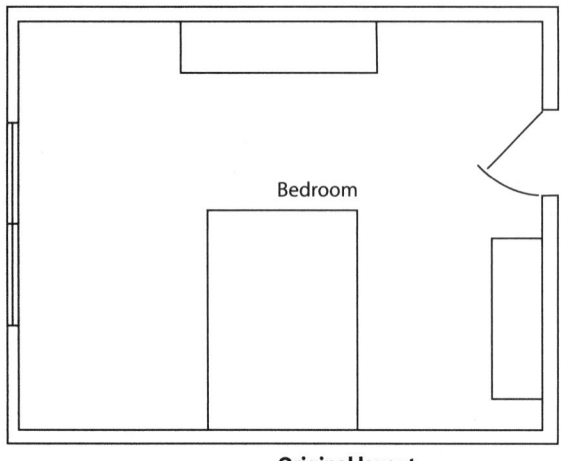 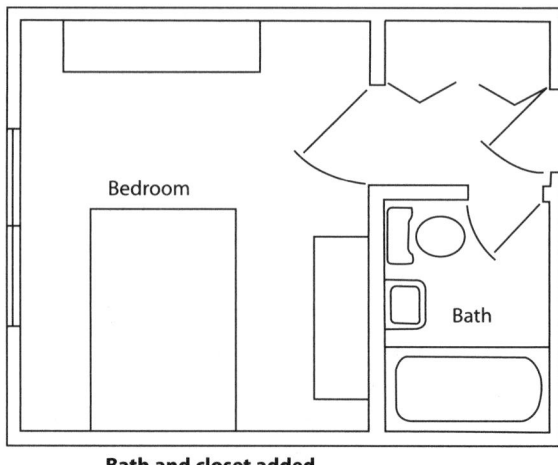

| Original layout | Bath and closet added |

**Figure 13-1**

*Portion of a large bedroom used for a bath addition*

Sometimes you'll see homes where a pantry, large closet or open area under a stairway has been converted to a bathroom. Depending on the location, converting space like this into a bathroom can be either a good choice or a poor choice. Don't convert a space into a bathroom simply because two fixtures happen to fit there. A bathroom in an entry or off the kitchen, totally removed from the bedroom area, is only good for a half bath or powder room. There's little practical advantage to a three-fixture bath that's remote from bedrooms.

## Finding Space for a Bathroom

You may find bedrooms in older homes that have a walk-in closet that's at least 5' by 7'. If a wall below that closet includes a plumbing drain, waste and vent lines, that closet is an ideal candidate for conversion to a three-fixture bathroom. Any bedroom that measures at least 16' in one direction is a good candidate to be partitioned for a new bathroom addition. See Figure 13-1. If all the bedrooms are small, consider converting the smallest bedroom into a bathroom and adding a new bedroom outside a perimeter wall.

In a one-and-one-half-story house, there's sometimes enough space under the shed dormer for another bathroom. See Figure 13-2. If you choose this area, remember that the plumbing wall in the new bathroom requires a wall directly below for pipe runs. Water supply lines to the new bathroom can be copper or poly and will snake through nearly any wall or ceiling cavity. Vent stack and waste lines aren't so flexible and may be 4" in diameter. Hiding a 4" drain pipe in an existing wall isn't easy. Any line that can't be enclosed in an existing wall will need to be framed into an enclosure within a room.

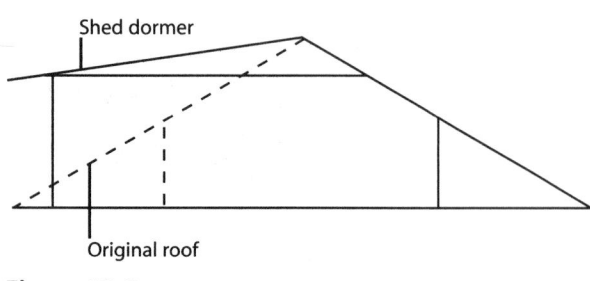

**Figure 13-2**

*Shed dormer used for additional space*

Other options may be converting the basement or garage into living space, or building on an exterior addition to the home. Each comes with its own challenges.

**Basements:** An unfinished basement offers good potential for expansion. It could include the bathroom you're adding, or be converted into a bedroom to replace the one you converted into the new bathroom. Building codes allow basements to be converted into habitable rooms if certain conditions are met. A habitable room is any space used for living, sleeping, eating, or cooking. To qualify as a habitable room, most codes require that the ground

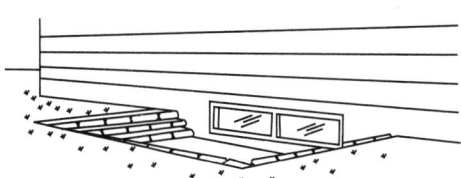

**Figure 13-3**

*Basement window areaway with sloped sides*

level outside the basement be no more than 48" above the basement floor, and the average ceiling height inside has to be at least 7'6". But bathrooms aren't considered habitable rooms under the code. So the ground level outside presents no obstacle, and the minimum ceiling height for a bathroom or other non-habitable room can usually be as low as 6'9". Your local building department will have the exact requirements for your location.

Although basements offer good space for conversions, they come with potential problems. The basement can be a difficult area in which to work. Many modifications will involve either tearing up the basement floor, or cutting through the concrete basement wall.

One obvious problem when adding a bathroom in the basement is handling sewage waste lines. If a basement floor is below the sewer main, install a sewage sump with a pump controlled by a float switch.

Another big disadvantage for living space is the lack of natural light and a view to the outside. You can add natural light with area walls or an areaway sloped so sunlight can reach basement windows. See Figure 13-3. If the slope of the lot offers an opportunity for drainage, consider creating a small sunken garden adjacent to one basement wall, as shown in Figure 13-4.

Your building code may require a fire exit in a converted basement. A basement window large enough to serve as a fire exit may meet the code requirements. Otherwise, add an areaway basement door like the one in Figure 13-5. An exterior

**Figure 13-4**

*Sunken garden forming large basement window areaway*

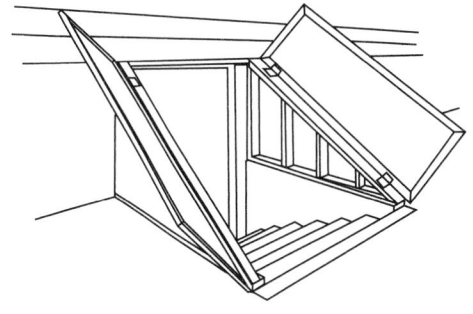

**Figure 13-5**

*Areaway type basement entrance*

entrance makes the basement a convenient place for storing lawn furniture and garden equipment. If possible, locate the basement door under the cover of a breezeway or porch roof to protect it from rain or snow.

**Garage:** If local building codes and zoning ordinances permit garage conversions, a garage attached to the house is another obvious candidate for expansion, especially if the garage is under the main roof. Considerations for the job include adding partition walls and an interior wall finish, and running ductwork for heating and air conditioning. Weigh the gain in living area against the loss of garage space and its convenience. Garages are usually close to the kitchen to save steps when unloading groceries. Even if you only use part of the garage for your addition, you may still lose that direct access convenience.

Cutting chases in a garage floor for the plumbing drain and supply lines seldom presents a serious problem. But get expert advice on recompacting the soil once access lines are in place. Fill that isn't properly compacted will have expansion characteristics different from the undisturbed soil. The result can be heaving or settling of the subsoil and an uneven floor in the converted garage.

**Exterior Additions:** Most communities have minimum setback requirements. No construction is allowed within a certain distance of lot lines and the street. The setback may be different on the two sides of the house. For example, the setback could be 5' on one side of the house and 10' on the other. If all the homes in a community have a similar setback, it's safe to assume those homes are built on the setback line. A hundred years ago, builders weren't so careful about their use of space. By modern standards, they squandered available land. You may have more possibilities for expansion when you're dealing with an older home. However, if setbacks restrict you from adding onto the front or side of a house, the only alternative may be to encroach on the back yard.

When planning exterior additions, carefully consider how to meet the needs of the owner while maintaining the overall design of the home. If the owner wants a larger bathroom, you could, as mentioned earlier, convert an existing small bedroom into a bathroom and make the exterior addition a larger new bedroom. Or partition off part of the existing living room for use as bedroom and bathroom. Then make the addition a new family room or a larger living room. In any case, the addition should be constructed to maintain the style of the house. Roofing, siding and windows should match the existing materials to preserve continuity.

The addition won't have a common foundation with the original structure. That puts stress on the point where the existing roof, wall, floor and ceiling join the addition. Cracking along the joint between the original house and the addition can be managed with control joints that allow for independent movement on each side of the joint. To minimize the length of that joint, consider building the addition as a satellite to the main house. See Figure 13-6. Connect the main house to the addition with a corridor that includes the bathroom, a new entry and a closet. Notice that satellite additions have a large perimeter in proportion to the enclosed space. Heating and cooling

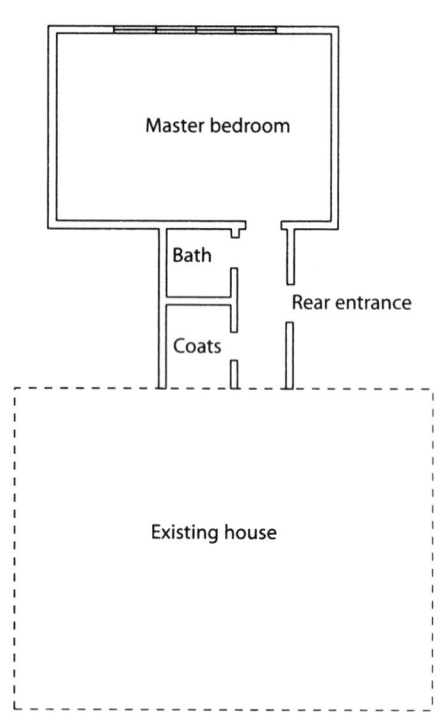

**Figure 13-6**

*Addition using the satellite concept*

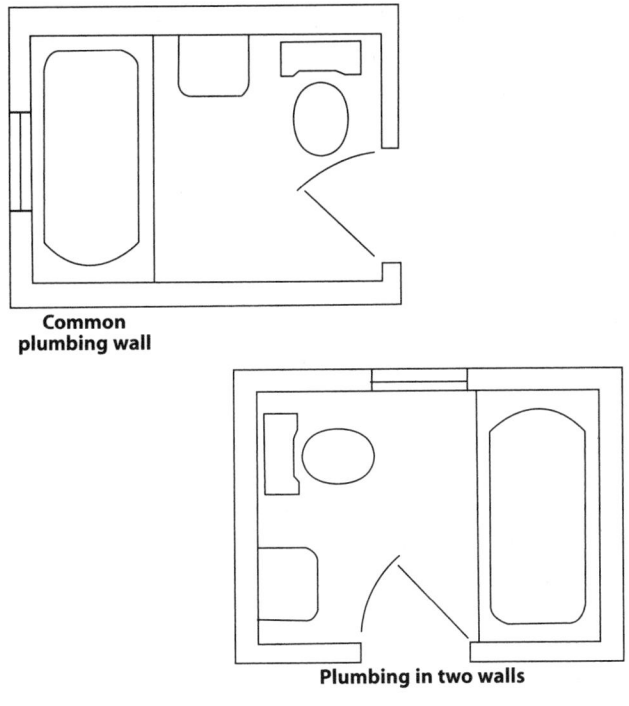

**Figure 13-7**

*Minimum size bathroom (5' by 7')*

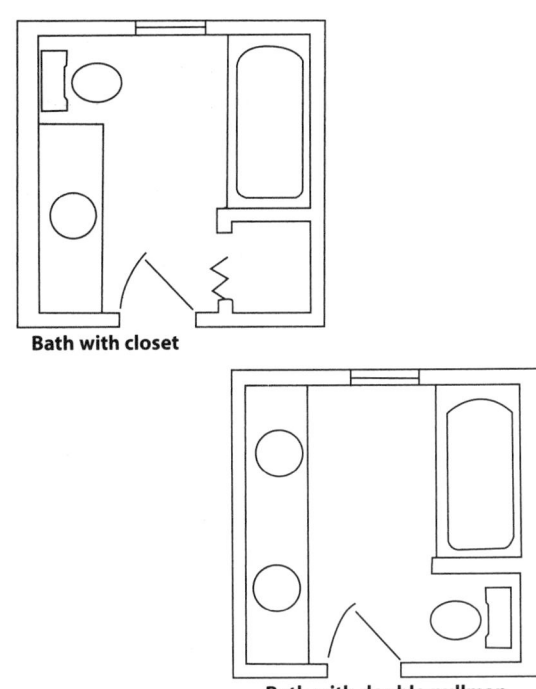

**Figure 13-8**

*Moderate size bathroom (8' by 8')*

this type of addition can be a problem. Be sure to provide good insulation and energy-efficient windows. Chapter 3 includes costs for framing the tie-in between the existing structure and the addition.

# Bathroom Floor Plans

Whether adding a new bathroom or just restyling an existing bathroom, it's important to observe some basic design rules. The minimum size for a three-fixture bathroom is 5' by 7' (Figure 13-7). Increasing that to 8' by 8' allows space for linen storage, cleaning equipment and supplies (Figure 13-8).

Figure 13-9 shows a bath divided into two compartments. That makes the bathroom suitable for use by two people, while preserving privacy. Minimum floor space for the toilet portion of a two-compartment bath is 3' by 5', assuming no tub. If a tub is included, you'll need a minimum of 5' by 5'. The lavatory compartment can be 4' x 5'. The interior dimensions of a compact two-compartment bathroom could be as little as 5' by 9' if you have a connecting door that slides or folds rather than swings. If a swinging door is planned, as in Figure 13-9, a floor area measuring 8' by 10'6" or 5' by 12' creates a more comfortable bathroom.

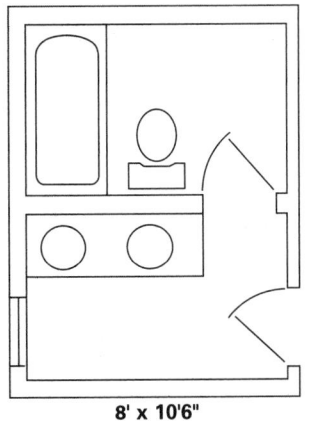

8' x 10'6"

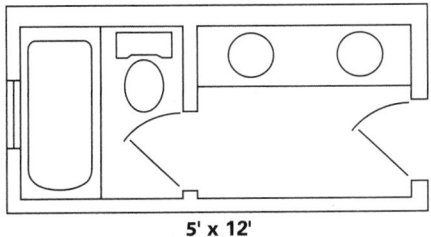

5' x 12'

**Figure 13-9**

*Compartmented bathroom*

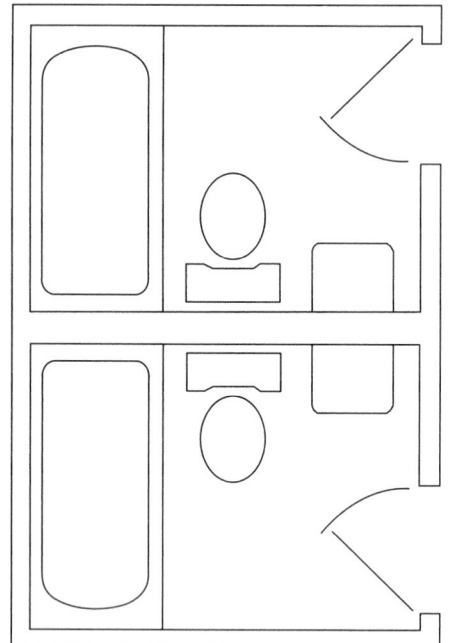

**Figure 13-10**

*Two bathrooms with economical back-to-back arrangement*

If two new bathrooms are needed on the same floor, install the plumbing fixtures back to back, as in Figure 13-10, to minimize plumbing runs. Bathrooms built on both floors of a two-story house are most economical when plumbing stacks are aligned vertically in the same wall.

Notice that all bathroom doors in Figures 13-7, 13-8, 13-9 and 13-10 open *into* the bathroom. True, that robs the bathroom of otherwise usable space, but it's better than opening a bathroom door into a hallway or bedroom. Bathroom doors are among the most used in the house and tend to open at unexpected times. A bathroom door that opens into a hallway is a constant hazard. Also, bathroom doors are generally closed when the bathroom is in use. That reclaims some of the space a swinging door would otherwise obstruct in a bathroom. Be sure to consider the swing area of the door when placing fixtures.

Every bathroom needs ventilation to prevent the accumulation of moisture and mold. If the bathroom is on an exterior wall, you should provide a window with an opening equivalent to 10 percent of the bathroom floor area. If the bathroom has only interior walls, the building code requires a fan with ducting to the exterior. Putting both the fan and light on the same switch guarantees the fan will be running when the bathroom is in use, although your customer may prefer independent switches. Don't put any switches or fixtures within arm's reach of someone using the tub or shower.

Cast iron and steel bathtubs are heavy, even when they're not filled with water. Fifty gallons of water in a tub adds another 400 pounds to the floor load. Floor joists in older homes are seldom designed to take such a concentrated load. When adding a tub, reinforce the floor joists and add cross bridging between joists. You may also need to double the studs. Even a shower stall can stress floor joists that weren't designed to support bathroom fixtures. If you plan to install ceramic tile on walls, floors or vanity tops, add extra framing to the floor and wall to support the additional weight. Ceramic tile is heavy and requires rock-solid support to resist cracking.

## Bathroom Space Minimums

❖ 5' x 4' — smallest two-fixture bathroom

❖ 5' x 7' — smallest three-fixture bathroom

❖ 12" — from the center of a toilet to the end of a tub

❖ 15" — from the center of a toilet to the side of a tub

❖ 15" — from the center of a toilet to an adjacent wall or shower stall

❖ 15" — from the center of a toilet to the center of an adjacent lavatory

❖ 15" — from the center of a lavatory to an adjacent wall or shower stall

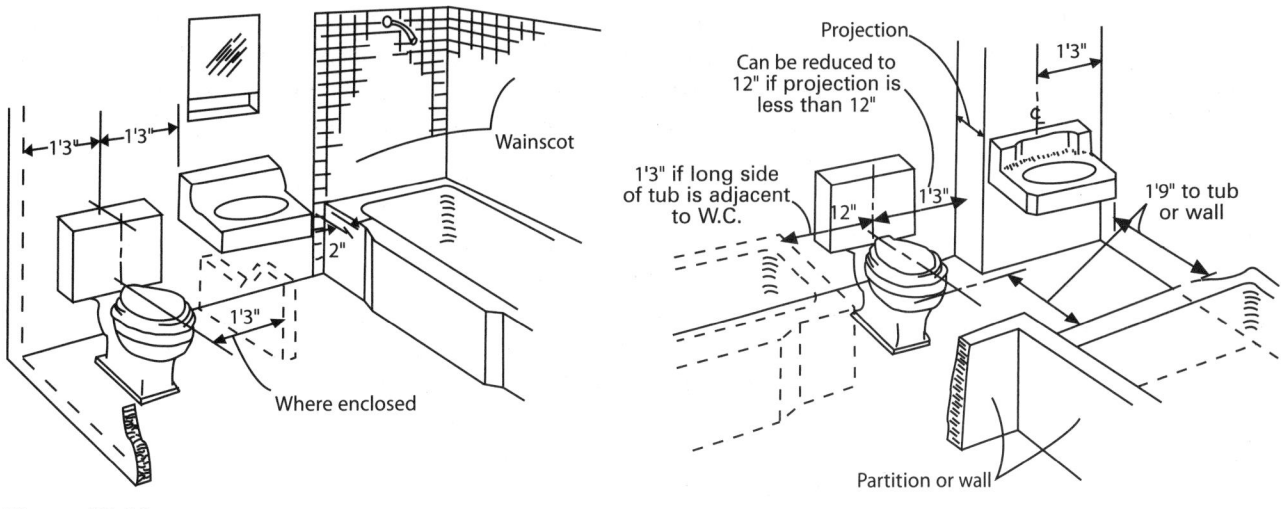

**Figure 13-11**

*Recommended dimensions for fixture spacing*

❖ 30" — distance between two lavatories mounted in the same countertop

❖ 32" to 34" — usual height for the top of vanities and lavatories

❖ 32" x 32" — width and depth of a square shower stall

❖ 30" x 60" — width and length of standard tubs

❖ 26" — height of the toilet paper roll holder above the floor

❖ 21" — depth of standing space while washing hands at a lavatory

❖ 36" — depth of standing space to open a sliding shower door

❖ 44" — depth of standing space to open a hinged shower door

Some building codes set minimum bathroom dimensions. As usual, the code is the last word. Figure 13-11 shows recommended minimum distances between and around bathroom fixtures. Remember that these are *minimums*, and more space is usually better.

| | Craft@Hrs | Unit | Material | Labor | Total | Sell |
|---|---|---|---|---|---|---|

## Plumbing Repairs

**Repair a compression faucet.** Turn valves off. Remove packing nut. Remove valve stem and cap. Remove seat washer from stem. Pry out old washer. Install new washer. Install stem. Tighten main gland valves if necessary. Turn valve on. Test faucet.

| | | | | | | |
|---|---|---|---|---|---|---|
| Per valve repaired | P1@.350 | Ea | — | 11.00 | 11.00 | 18.70 |

**Remove and replace fixture gasket.** Turn valve off. Loosen locknut. Remove spud connection. Remove gasket or washer. Clean spud seat. Install new gasket or washer. Assemble spud connection. Tighten locknut. Reassemble fixture and inspect connection.

| | | | | | | |
|---|---|---|---|---|---|---|
| Per gasket replaced | P1@.120 | Ea | — | 3.78 | 3.78 | 6.43 |

**Remove and replace toilet fill valve.** Turn water off. Empty tank. Remove valve nut and disconnect supply tube. Install new toilet fill valve. Tighten nut. Attach supply tube. Turn water on. Check operation.

| | | | | | | |
|---|---|---|---|---|---|---|
| Per fill valve | P1@.300 | Ea | — | 9.45 | 9.45 | 16.10 |

**Remove and replace lavatory trap.** Unscrew trap plug from drain. Disconnect union. Remove and inspect trap. Replace trap. Reconnect union and tighten plug. Test operation.

| | | | | | | |
|---|---|---|---|---|---|---|
| Per trap | P1@.155 | Ea | — | 4.88 | 4.88 | 8.30 |

**Remove and replace basin strainer and P-trap.** Disconnect strainer from wall nipple and packing nut. Remove strainer lock nut. Remove strainer washer and trap. Clean out trap and strainer. Install strainer nut, packing nut and connect trap to wall nipple. Test operation.

| | | | | | | |
|---|---|---|---|---|---|---|
| Per strainer and trap | P1@.446 | Ea | — | 14.00 | 14.00 | 23.80 |

**Remove and replace showerhead.** Remove showerhead from 1/2" threaded pipe. Inspect as necessary. Replace and test showerhead.

| | | | | | | |
|---|---|---|---|---|---|---|
| Per showerhead | P1@.108 | Ea | — | 3.40 | 3.40 | 5.78 |

**Reset toilet bowl.** Turn off angle valve. Remove packing nuts. Remove flange bolts. Remove bowl from flange. Replace gasket or ring. Clean bowl foundation. Reset bowl and connect flange nuts. Replace packing nuts. Turn on water and check operation.

| | | | | | | |
|---|---|---|---|---|---|---|
| Per bowl reset | P1@1.27 | Ea | — | 40.00 | 40.00 | 68.00 |

## Bathtubs

**Remove bathtub.** No salvage of materials assumed. Add for repairs to wall and floor surfaces. Debris piled on site. Labor cost does not include clearing of area or removal of wall finish.

| | | | | | | |
|---|---|---|---|---|---|---|
| Remove tub | P1@.722 | Ea | — | 22.70 | 22.70 | 38.60 |

**Ventura enameled steel bathtub.** One-piece construction for recess installations. Diagonal brace. Sound-deadening foundation pad. Straight tiling edges. Full wall flange. Slip-resistant bottom. 40-gallon capacity. 60" x 30" x 15-1/4". Labor includes setting and fastening in place only. Add for water supply and drain connections as required. White.

| | | | | | | |
|---|---|---|---|---|---|---|
| Left-hand drain | P1@2.50 | Ea | 104.00 | 78.80 | 182.80 | 311.00 |
| Right-hand drain | P1@2.50 | Ea | 104.00 | 78.80 | 182.80 | 311.00 |

**Performa™ molded bathtub, Sterling Plumbing.** Vikrell™ construction. Clear, molded color. Structural ribs for added strength. Slip-resistant bottom. Available with shower wall set. Faucet sold separately. 60" x 42" x 18". Resistant to chipping, cracking and peeling. Easy-to-clean high-gloss surface. Durable material with color molded throughout fixture. Labor includes setting and fastening in place only. Add for water supply and drain connections as required. White.

| | | | | | | |
|---|---|---|---|---|---|---|
| Left-hand drain | P1@2.50 | Ea | 142.00 | 78.80 | 220.80 | 375.00 |
| Right-hand drain | P1@2.50 | Ea | 142.00 | 78.80 | 220.80 | 375.00 |

| | Craft@Hrs | Unit | Material | Labor | Total | Sell |
|---|---|---|---|---|---|---|

**Acrylic soaking tub with skirt, Lasco Bathware.** Slip-resistant textured bottom. Integral skirt and tiling flange. Labor includes setting and fastening in place only. Add for water supply and drain connections as required. 60" x 30" x 17". White.

| | Craft@Hrs | Unit | Material | Labor | Total | Sell |
|---|---|---|---|---|---|---|
| Left-hand drain | P1@2.50 | Ea | 238.00 | 78.80 | 316.80 | 539.00 |
| Right-hand drain | P1@2.50 | Ea | 238.00 | 78.80 | 316.80 | 539.00 |

**Oval acrylic soaking tub, Lasco Bathware.** Universal drain location. Slip-resistant textured bottom. Integral armrests. Island configuration. Comfort sloped backrest. Pre-leveled base. 60" x 42" x 23". Labor includes setting and fastening in place only. Add for water supply and drain connections as required.

| | Craft@Hrs | Unit | Material | Labor | Total | Sell |
|---|---|---|---|---|---|---|
| Ariel I, white | P1@2.50 | Ea | 507.00 | 78.80 | 585.80 | 996.00 |

**Princeton recessed bathtub, American Standard.** Integral lumbar support. Slip-resistant. Faucet sold separately. Glossy porcelain finish. Americast material insulates, holding water temperature and reducing the noise of running water. Half the weight of cast iron for easy installation. 60" x 30" x 14". Labor includes setting and fastening in place only. Add for water supply and drain connections as required.

| | Craft@Hrs | Unit | Material | Labor | Total | Sell |
|---|---|---|---|---|---|---|
| Bone, left-hand drain | P1@2.50 | Ea | 332.00 | 78.80 | 410.80 | 698.00 |
| Bone, right-hand drain | P1@2.50 | Ea | 331.00 | 78.80 | 409.80 | 697.00 |
| White, left-hand drain | P1@2.50 | Ea | 221.00 | 78.80 | 299.80 | 510.00 |
| White, right-hand drain | P1@2.50 | Ea | 270.00 | 78.80 | 348.80 | 593.00 |

**Villager™ bathtub, Kohler.** Sloping back and safeguard bottom. Flat front rim provides ideal base for shower enclosure tracks. Faucet, fixture trim and accessories sold separately. Labor includes setting and fastening in place only. Add for water supply and drain connections as required.

| | Craft@Hrs | Unit | Material | Labor | Total | Sell |
|---|---|---|---|---|---|---|
| Almond, left-hand drain | P1@2.50 | Ea | 419.00 | 78.80 | 497.80 | 846.00 |
| Almond, right-hand drain | P1@2.50 | Ea | 419.00 | 78.80 | 497.80 | 846.00 |
| Bisque, left-hand drain | P1@2.50 | Ea | 414.00 | 78.80 | 492.80 | 838.00 |
| Bisque, right-hand drain | P1@2.50 | Ea | 414.00 | 78.80 | 492.80 | 838.00 |
| White, left-hand drain | P1@2.50 | Ea | 340.00 | 78.80 | 418.80 | 712.00 |
| White, right-hand drain | P1@2.50 | Ea | 341.00 | 78.80 | 419.80 | 714.00 |

## Whirlpool Tubs

**Renaissance whirlpool bath, American Standard.** 60" x 32" x 19-3/4". High-gloss acrylic. 1-HP pump. 4 body side jets, 2 lumbar back jets, 1 bubble massage control. Deck-mounted on and off switch. Faucet, fixture trim and accessories sold separately. Labor includes setting and fastening in place only. Add for water supply, drain and electrical connections as required.

| | Craft@Hrs | Unit | Material | Labor | Total | Sell |
|---|---|---|---|---|---|---|
| Bone | P1@4.00 | Ea | 491.00 | 126.00 | 617.00 | 1,050.00 |
| White | P1@4.00 | Ea | 356.00 | 126.00 | 482.00 | 819.00 |

**Cadet whirlpool bath, American Standard.** 60" x 32" x 19-3/4". High-gloss acrylic. 1-HP pump and motor. 8 multidirectional and flow-adjustable jets. Deck-mounted on and off switch. Sculpted backrest with molded-in arm and elbow rests. Faucet, fixture trim and accessories sold separately. Labor includes setting and fastening in place only. Add for water supply and drain connections as required.

| | Craft@Hrs | Unit | Material | Labor | Total | Sell |
|---|---|---|---|---|---|---|
| White | P1@4.00 | Ea | 432.00 | 126.00 | 558.00 | 949.00 |

**Cadet Elite whirlpool tub, American Standard.** Acrylic with fiberglass reinforcement. Form-fitted backrest. Molded-in arm rests with elbow supports. Includes 1.5-HP pump. Labor includes setting and fastening in place only. Add for water supply, drain and electrical connections as required.

| | Craft@Hrs | Unit | Material | Labor | Total | Sell |
|---|---|---|---|---|---|---|
| 5' x 36", white | P1@4.00 | Ea | 536.00 | 126.00 | 662.00 | 1,130.00 |
| 5' x 42", white | P1@4.00 | Ea | 723.00 | 126.00 | 849.00 | 1,440.00 |

**Williamsburg Elite whirlpool tub, American Standard.** Built-in apron model for easy, attractive installation. Jet comfort system featuring 6 interchangeable universal flex-jets. Includes 1.5-HP pump. 60" x 32" x 21-1/2". Labor includes setting and fastening in place only. Add for water supply, drain and electrical connections as required.

| | Craft@Hrs | Unit | Material | Labor | Total | Sell |
|---|---|---|---|---|---|---|
| Left-hand drain, white | P1@4.00 | Ea | 702.00 | 126.00 | 828.00 | 1,410.00 |
| Right-hand drain, white | P1@4.00 | Ea | 702.00 | 126.00 | 828.00 | 1,410.00 |

| | Craft@Hrs | Unit | Material | Labor | Total | Sell |
|---|---|---|---|---|---|---|

**Devonshire™ whirlpool tub, Sterling Plumbing.** Includes 1.25-HP pump with air switch actuator. 8 color-matched Kohler jets direct adjustable hydro-massage. Crafted of high-gloss acrylic and backed with a solid layer of fiberglass to ensure durability and resistance to chipping, cracking and flexing. Labor includes setting and fastening in place only. Add for water supply, drain and electrical connections as required.

| | Craft@Hrs | Unit | Material | Labor | Total | Sell |
|---|---|---|---|---|---|---|
| 5', drop-in | P1@4.00 | Ea | 810.00 | 126.00 | 936.00 | 1,590.00 |
| 5', left-hand drain, white | P1@4.00 | Ea | 778.00 | 126.00 | 904.00 | 1,540.00 |
| 5', right-hand drain, white | P1@4.00 | Ea | 772.00 | 126.00 | 898.00 | 1,530.00 |

**Portrait™ whirlpool tub, Kohler.** Traditional styling. 8 flexjets. Adjustable hydro-massage. Drop-in installation. High-gloss acrylic and backed with fiberglass. Labor includes setting and fastening in place only. Add for water supply, drain and electrical connections as required.

| | Craft@Hrs | Unit | Material | Labor | Total | Sell |
|---|---|---|---|---|---|---|
| 5-1/2', with heater | P1@4.00 | Ea | 2076.00 | 126.00 | 2,202.00 | 3,740.00 |

**Elite whirlpool bath EZ-Install™ heater, American Standard.** Pre-plumbed for easy installation. Maintains constant water temperature. Labor includes setting and fastening in place only. Add for electrical service and connection.

| | Craft@Hrs | Unit | Material | Labor | Total | Sell |
|---|---|---|---|---|---|---|
| EZHeat-100 | P1@1.00 | Ea | 217.00 | 31.50 | 248.50 | 422.00 |

**Whirlpool tub apron, American Standard.** Labor includes setting and fastening in place only.

| | Craft@Hrs | Unit | Material | Labor | Total | Sell |
|---|---|---|---|---|---|---|
| 5', white | P1@.500 | Ea | 119.00 | 15.80 | 134.80 | 229.00 |
| Renaissance or Cadet, bone | P1@.500 | Ea | 161.00 | 15.80 | 176.80 | 301.00 |

**Williamsburg J-spout Roman tub deck-mount tub filler, American Standard.** Lifetime finish. Will not tarnish. Ceramic disc valving. Mounts on 8" to 12" centers. Labor includes set in place and connection only. Add the cost of plumbing rough-in as required.

| | Craft@Hrs | Unit | Material | Labor | Total | Sell |
|---|---|---|---|---|---|---|
| Brushed satin nickel finish | P1@1.00 | Ea | 280.00 | 31.50 | 311.50 | 530.00 |
| Chrome finish | P1@1.00 | Ea | 215.00 | 31.50 | 246.50 | 419.00 |
| Satin chrome finish | P1@1.00 | Ea | 280.00 | 31.50 | 311.50 | 530.00 |

**Innovations® Roman tub trim kit, Delta.** Chrome and polished brass finish. Flexible deck and ledge mount. 18 GPM fill rate. Brilliant anti-tarnish finish. Installation kit included.

| | Craft@Hrs | Unit | Material | Labor | Total | Sell |
|---|---|---|---|---|---|---|
| 18 GPM fill rate | P1@1.00 | Ea | 182.00 | 31.50 | 213.50 | 363.00 |

## Tub Enclosure Walls

**One-piece Georgia tub wall with shelf.** Bath and shower wall surround. Seamless. One extra long molded soap dish. High-gloss finish.

| | Craft@Hrs | Unit | Material | Labor | Total | Sell |
|---|---|---|---|---|---|---|
| 58" to 61" x 31" x 59", white | B1@3.00 | Ea | 96.10 | 88.10 | 184.20 | 313.00 |

**One-piece Columbia tile finish tub wall.** Wall surround. High-gloss finish. With integrated soap dishes and simulated ceramic tile finish.

| | Craft@Hrs | Unit | Material | Labor | Total | Sell |
|---|---|---|---|---|---|---|
| 58" to 60" x 32" x 58", white | B1@3.00 | Ea | 128.00 | 88.10 | 216.10 | 367.00 |

**One-piece Nantucket tile finish tub wall.** Fits alcoves from 57" to 60" wide by 29" to 32" deep. Hammered tile look with 2 shelves and 1 soap dish. Flexible construction accepts wall misalignments and non-square alcove corners. Simulated tile design. Tub and faucet sold separately.

| | Craft@Hrs | Unit | Material | Labor | Total | Sell |
|---|---|---|---|---|---|---|
| White, seamless | B1@3.00 | Ea | 172.00 | 88.10 | 260.10 | 442.00 |

**Three-piece bathtub wall kit, Pro-Wall 3™.** Heavy gauge, high-impact co-polymer plastic. 2 clear towel bars, 3 molded shelves. Easy installation with few seams to caulk. Back panel can be cut for window opening. Fits alcoves 57" to 61" across the back and 30" across the ends without trimming. Permanent, waterproof, colorfast panels are easy to clean. Resistant to mold and mildew.

| | Craft@Hrs | Unit | Material | Labor | Total | Sell |
|---|---|---|---|---|---|---|
| 58" wall height, white | B1@3.00 | Ea | 183.00 | 88.10 | 271.10 | 461.00 |

|  | Craft@Hrs | Unit | Material | Labor | Total | Sell |
|---|---|---|---|---|---|---|

**Three-piece Monterey bathtub tile kit, Trayco.** Tub wall surround. Mold and mildew resistant. Limited lifetime warranty.

| 30" x 60", beige, ceramic tile look | B1@3.00 | Ea | 247.00 | 88.10 | 335.10 | 570.00 |

**Five-piece Alabama tub wall surround kit.** High-gloss finish. 5 pieces. Shower base sold separately.

| 48" to 60" x 32" x 59", white | B1@3.00 | Ea | 96.10 | 88.10 | 184.20 | 313.00 |

**Five-piece Bretton tub surround.** 2 shelves and 1 towel bar. Fits alcoves up to 60" wide by 32" to 36" deep without trimming. Transforms the traditional bathtub area into a shower alcove. Shower base sold separately.

| White | B1@3.00 | Ea | 139.00 | 88.10 | 227.10 | 386.00 |

**Five-piece overlap construction design tub surround, Tall Elite.** 2 towel bars and 6 shelves. Fits alcoves from 49" to 60-1/2" wide by 28" to 31" deep. Panels are easy to trim for window openings. White high-gloss finish.

| 80" high x 60" wide | B1@3.00 | Ea | 123.00 | 88.10 | 211.10 | 359.00 |

**Bathtub wall surround kit, Elite.** Walls only. Tub and shower faucets sold separately. 6 shelves. 2 clear towel bars. Thicker corners provide greater strength and durability.

| 58" wall height, bone | B1@3.00 | Ea | 117.00 | 88.10 | 205.10 | 349.00 |
| 58" wall height, white | B1@3.00 | Ea | 96.10 | 88.10 | 184.20 | 313.00 |

**Bathtub tile wall kit.** Rounded corners, bullnose edges, and engraved white grout for ceramic tile appearance. Groutless design. PVC finish resists stains, mold and mildew. Panels can be cut to fit window openings. 3 panels. Tub and shower faucets sold separately.

| 30" x 60" x 60" | B1@3.00 | Ea | 215.00 | 88.10 | 303.10 | 515.00 |

**Vikrell™ bathtub walls, Sterling Plumbing.** Resistant to chipping, cracking and peeling. Easy-to-clean high-gloss surface. Durable material with color molded throughout fixture. Labor includes fastening in place only.

| Bone | B1@3.00 | Ea | 215.00 | 88.10 | 303.10 | 515.00 |

**Fiberglass bath wall window trim, Swan Corporation.**

| White | B1@1.00 | Ea | 74.50 | 29.40 | 103.90 | 177.00 |

## Bypass Door Tub Enclosures

**Bypass door tub enclosure, American Shower & Bath.** 60" wide opening. With track.

| Brass, ellipse glass | BC@1.20 | Ea | 215.00 | 38.50 | 253.50 | 431.00 |
| Brass, hammertone glass | BC@1.20 | Ea | 140.00 | 38.50 | 178.50 | 303.00 |
| Brass, hammertone mirror glass | BC@1.20 | Ea | 163.00 | 38.50 | 201.50 | 343.00 |
| Silver, ellipse glass | BC@1.20 | Ea | 206.00 | 38.50 | 244.50 | 416.00 |
| Silver, frameless, clear glass | BC@1.20 | Ea | 354.00 | 38.50 | 392.50 | 667.00 |
| Silver, frameless, rainglass | BC@1.20 | Ea | 306.00 | 38.50 | 344.50 | 586.00 |
| Silver, fluted glass | BC@1.20 | Ea | 261.00 | 38.50 | 299.50 | 509.00 |
| Silver, hammertone glass | BC@1.20 | Ea | 130.00 | 38.50 | 168.50 | 286.00 |
| Silver, hammertone mirror glass | BC@1.20 | Ea | 152.00 | 38.50 | 190.50 | 324.00 |
| Silver, obscure glass | BC@1.20 | Ea | 238.00 | 38.50 | 276.50 | 470.00 |
| Silver, rainglass | BC@1.20 | Ea | 174.00 | 38.50 | 212.50 | 361.00 |
| White, ellipse glass | BC@1.20 | Ea | 217.00 | 38.50 | 255.50 | 434.00 |

| | Craft@Hrs | Unit | Material | Labor | Total | Sell |
|---|---|---|---|---|---|---|
| **Sliding tub enclosure door, Contractors Wardrobe.** 60" wide opening. With track. | | | | | | |
| Bright nickel, frameless, 1/4" glass | BC@1.20 | Ea | 332.00 | 38.50 | 370.50 | 630.00 |
| Bright silver finish, rain glass | BC@1.20 | Ea | 151.00 | 38.50 | 189.50 | 322.00 |
| Deluxe frameless, radiance glass | BC@1.20 | Ea | 289.00 | 38.50 | 327.50 | 557.00 |
| Deluxe gold, glue chip glass | BC@1.20 | Ea | 332.00 | 38.50 | 370.50 | 630.00 |
| Deluxe silver finish, rain glass | BC@1.20 | Ea | 267.00 | 38.50 | 305.50 | 519.00 |
| Deluxe silver, glue chip glass | BC@1.20 | Ea | 310.00 | 38.50 | 348.50 | 592.00 |
| Gold finish, rain glass | BC@1.20 | Ea | 162.00 | 38.50 | 200.50 | 341.00 |
| Silver, frameless, 1/4" clear glass | BC@1.20 | Ea | 215.00 | 38.50 | 253.50 | 431.00 |
| Silver, frameless, clear glass | BC@1.20 | Ea | 218.00 | 38.50 | 256.50 | 436.00 |
| Silver, frameless, clear glass | BC@1.20 | Ea | 207.00 | 38.50 | 245.50 | 417.00 |
| Silver, frameless, radiance glass | BC@1.20 | Ea | 233.00 | 38.50 | 271.50 | 462.00 |
| Silver finish, obscure glass | BC@1.20 | Ea | 105.00 | 38.50 | 143.50 | 244.00 |
| Silver finish, swan glass | BC@1.20 | Ea | 73.00 | 38.50 | 111.50 | 190.00 |
| White finish, rain glass | BC@1.20 | Ea | 151.00 | 38.50 | 189.50 | 322.00 |
| | | | | | | |
| **Framed bypass door tub enclosure, Keystone.** Fits 57" to 59" wide by 57-3/8" openings. 6'8" high. EZ Kleen® track system. Factory-installed towel bar. No exposed screws. | | | | | | |
| Gold frame, obscure glass | BC@1.20 | Ea | 161.00 | 38.50 | 199.50 | 339.00 |
| Silver frame, 3/8" rain glass | BC@1.20 | Ea | 207.00 | 38.50 | 245.50 | 417.00 |
| Silver frame, obscure glass | BC@1.20 | Ea | 193.00 | 38.50 | 231.50 | 394.00 |
| | | | | | | |
| **Frameless bypass door shower enclosure, Keystone.** 54" to 59" wide, 71" high. | | | | | | |
| Satin nickel track, 3/8" glass | BC@1.20 | Ea | 414.00 | 38.50 | 452.50 | 769.00 |
| Satin nickel track, clear glass | BC@1.20 | Ea | 344.00 | 38.50 | 382.50 | 650.00 |
| Silver track, 3/8" clear glass | BC@1.20 | Ea | 552.00 | 38.50 | 590.50 | 1,000.00 |
| Silver track, clear glass | BC@1.20 | Ea | 217.00 | 38.50 | 255.50 | 434.00 |
| Silver track, Krystal flute glass | BC@1.20 | Ea | 269.00 | 38.50 | 307.50 | 523.00 |
| White track, clear glass | BC@1.20 | Ea | 257.00 | 38.50 | 295.50 | 502.00 |

## Fiberglass Tub and Shower Combinations

**Lascoat™ tub and shower unit.** Integral toiletry shelves. Installed acrylic grab bar. Slip-resistant textured bottom. 60" x 30" x 72". White. Smooth wall. Labor includes setting and fastening in place only. Add for water supply and drain connections as required. Be sure any tub and shower unit you select for replacement purposes will fit through existing doorways.

| | Craft@Hrs | Unit | Material | Labor | Total | Sell |
|---|---|---|---|---|---|---|
| Left-hand drain | P1@4.75 | Ea | 260.00 | 150.00 | 410.00 | 697.00 |
| Right-hand drain | P1@4.75 | Ea | 260.00 | 150.00 | 410.00 | 697.00 |

**Lascoat™ two-piece fiberglass tub and shower unit.** BathLock™ front installation system. Integral toiletry shelves and acrylic grab bar. Smooth wall design. Leak-proof right angle joint flanges. Slip-resistant textured bottom. White. 60" x 30" x 72". Labor includes setting and fastening in place only. Add for water supply and drain connections as required.

| | Craft@Hrs | Unit | Material | Labor | Total | Sell |
|---|---|---|---|---|---|---|
| Left-hand drain | P1@4.75 | Ea | 336.00 | 150.00 | 486.00 | 826.00 |
| Right-hand drain | P1@4.75 | Ea | 336.00 | 150.00 | 486.00 | 826.00 |

**Tub and shower, Lasco Bathware.** 4" smooth tile design. Integral soap shelf. Slip-resistant, textured bottom. 60" x 32" x 72". White. Labor includes setting and fastening in place only. Add for water supply and drain connections as required. Be sure any tub and shower unit you select for replacement purposes will fit through existing doorways.

| | Craft@Hrs | Unit | Material | Labor | Total | Sell |
|---|---|---|---|---|---|---|
| Left-hand drain | P1@4.75 | Ea | 314.00 | 150.00 | 464.00 | 789.00 |
| Right-hand drain | P1@4.75 | Ea | 314.00 | 150.00 | 464.00 | 789.00 |

| | Craft@Hrs | Unit | Material | Labor | Total | Sell |
|---|---|---|---|---|---|---|

**Two-piece tub and shower, Lasco Bathware.** Smooth wall tile design. BathLock™ front installation system. Leak-proof right angle joint flanges. Integral soap shelves. Specially designed to hide seam. 60" x 30" x 72". White. Acrylic grab bar. Labor includes setting and fastening in place only. Add for water supply and drain connections as required.

| | Craft@Hrs | Unit | Material | Labor | Total | Sell |
|---|---|---|---|---|---|---|
| Left-hand drain | P1@4.75 | Ea | 388.00 | 150.00 | 538.00 | 915.00 |
| Right-hand drain | P1@4.75 | Ea | 389.00 | 150.00 | 539.00 | 916.00 |

**Three-piece molded fiberglass sectional tub and shower, Lasco Bathware.** 60" x 30" x 72". Smooth wall design. BathLock™ front installation system. Leak-proof right angle joint flanges. Integral toiletry shelves. Acrylic grab bar. Slip-resistant textured bottom. Labor includes setting and fastening in place only. Add for water supply and drain connections as required.

| | Craft@Hrs | Unit | Material | Labor | Total | Sell |
|---|---|---|---|---|---|---|
| Left-hand drain | P1@4.75 | Ea | 410.00 | 150.00 | 560.00 | 952.00 |
| Right-hand drain | P1@4.75 | Ea | 410.00 | 150.00 | 560.00 | 952.00 |

**Two-piece fiberglass tub and shower unit.** With towel bar, shelf, and soap dish. Smooth wall design. BathLock™ front installation system. Leak-proof right angle joint flanges. Slip-resistant textured bottom. Acrylic grab bar. Bone finish. 60" x 30" x 72". Labor includes setting and fastening in place only. Add for water supply and drain connections as required.

| | Craft@Hrs | Unit | Material | Labor | Total | Sell |
|---|---|---|---|---|---|---|
| Left-hand drain | P1@4.75 | Ea | 356.00 | 150.00 | 506.00 | 860.00 |
| Right-hand drain | P1@4.75 | Ea | 356.00 | 150.00 | 506.00 | 860.00 |

## Shower Stalls

**Demolish stall shower.** Remove 3 walls and receptor. Remove supply pipe and insulation. Remove drain with a cutting torch. Stack debris on site.

| | Craft@Hrs | Unit | Material | Labor | Total | Sell |
|---|---|---|---|---|---|---|
| Per shower stall | BL@1.55 | Ea | — | 41.30 | 41.30 | 70.20 |

**Shower stall kit, American Shower & Bath.** Free-standing shower stall. Includes heavy-gauge walls in one-piece design. Molded soap dish and shampoo holder. 2 hand rails for convenient support. Base with no-caulk drain. Faucet, showerhead and installation hardware included. Add the cost of water supply and drain connection.

| | Craft@Hrs | Unit | Material | Labor | Total | Sell |
|---|---|---|---|---|---|---|
| White | P1@2.90 | Ea | 107.00 | 91.40 | 198.40 | 337.00 |

**One-piece shower stall, Lasco Bathware.** Dimpled tile finish. Center drain location. Comfort-molded, rear corner seats. Slip-resistant textured bottom. Simulated 4" textured tile design. Integral toiletry shelves. Acrylic grab bar. Space saver design. Add for water supply, drain, valve and showerhead. White. 72" high.

| | Craft@Hrs | Unit | Material | Labor | Total | Sell |
|---|---|---|---|---|---|---|
| 32" x 32", smooth finish | P1@2.90 | Ea | 249.00 | 91.40 | 340.40 | 579.00 |
| 36" x 36", 4" tile finish | P1@2.90 | Ea | 324.00 | 91.40 | 415.40 | 706.00 |
| 48" x 35", smooth finish | P1@3.25 | Ea | 259.00 | 102.00 | 361.00 | 614.00 |

**Lascoat™ two-piece molded sectional shower stall, Lasco Bathware.** Bath Lock™ front installation system. Leak-proof right angle joint flanges. Central drain. Integral soap shelf. Smooth walls. Slip-resistant textured bottom. Specially designed to hide seam. Add the cost of water supply and drain connection, valve and showerhead. White.

| | Craft@Hrs | Unit | Material | Labor | Total | Sell |
|---|---|---|---|---|---|---|
| 32" x 32" x 72-3/4" | P1@2.90 | Ea | 324.00 | 91.40 | 415.40 | 706.00 |
| 36" x 36" x 72" | P1@2.90 | Ea | 324.00 | 91.40 | 415.40 | 706.00 |

**Smooth wall shower stall, Lasco Bathware.** Center drain. BathLock™ front installation system. Leak-proof right angle joint flanges. Integral soap shelf. Slip-resistant textured bottom. Add for water supply, drain, valve and showerhead. White. 72" high.

| | Craft@Hrs | Unit | Material | Labor | Total | Sell |
|---|---|---|---|---|---|---|
| 36" x 36", 3-piece | P1@2.90 | Ea | 356.00 | 91.40 | 447.40 | 761.00 |
| 38" x 38", 2-piece, neo-angle | P1@2.90 | Ea | 378.00 | 91.40 | 469.40 | 798.00 |
| 48" x 34", 2-piece | P1@3.25 | Ea | 334.00 | 102.00 | 436.00 | 741.00 |
| 48" x 34", 3-piece | P1@3.25 | Ea | 388.00 | 102.00 | 490.00 | 833.00 |
| 60" x 34", 3-piece | P1@3.25 | Ea | 432.00 | 102.00 | 534.00 | 908.00 |

| | Craft@Hrs | Unit | Material | Labor | Total | Sell |
|---|---|---|---|---|---|---|

**Corner entry shower stall kit, American Shower & Bath.** White aluminum frame with clear tempered safety glass. Easy-to-clean walls and corner caddy with four shelves and towel bar. Heavy-duty base. Drain, drain cover and installation hardware included. Add the cost of water supply and drain connection, valve and showerhead. White.

| | Craft@Hrs | Unit | Material | Labor | Total | Sell |
|---|---|---|---|---|---|---|
| 32" x 32" x 71" | P1@2.90 | Ea | 292.00 | 91.40 | 383.40 | 652.00 |

**Iris round shower stall kit, Maax.** Includes shower stall, door, and drain. High-luster ABS. Polycrystal™ Corinto design panels. Installs directly to studs or finished walls. Reversible sliding door. Non-slip textured base. Bottle holder and shelves. 33-3/4" x 33-3/4" base measured at walls. 78-1/4" height. Easy to clean. Add for water supply, drain connection, valve and showerhead.

| | Craft@Hrs | Unit | Material | Labor | Total | Sell |
|---|---|---|---|---|---|---|
| 34" x 34" | P1@2.90 | Ea | 408.00 | 91.40 | 499.40 | 849.00 |

**Neo-angle 38-inch shower stall kit, American Shower & Bath.** Includes anodized aluminum frame and tempered safety glass pivot door. Easy-to-clear walls. Corner caddy with 4 shelves and towel bar. Heavy-duty base, drain, drain cover and installation hardware. Add the cost of valve, water supply and drain connection.

| | Craft@Hrs | Unit | Material | Labor | Total | Sell |
|---|---|---|---|---|---|---|
| Chrome frame, obscure glass | P1@3.25 | Ea | 345.00 | 102.00 | 447.00 | 760.00 |
| Chrome frame, rainfall glass | P1@3.25 | Ea | 461.00 | 102.00 | 563.00 | 957.00 |
| Chrome with clear glass | P1@3.25 | Ea | 593.00 | 102.00 | 695.00 | 1,180.00 |
| Chrome with radiance glass | P1@3.25 | Ea | 492.00 | 102.00 | 594.00 | 1,010.00 |
| Gold frame, autumn glass | P1@3.25 | Ea | 467.00 | 102.00 | 569.00 | 967.00 |
| Gold frame, rainfall glass | P1@3.25 | Ea | 471.00 | 102.00 | 573.00 | 974.00 |
| Silver frame, obscure glass | P1@3.25 | Ea | 322.00 | 102.00 | 424.00 | 721.00 |
| White frame, rainfall glass | P1@3.25 | Ea | 259.00 | 102.00 | 361.00 | 614.00 |

**Hazel three-piece neo-angle shower stall kit, Maax.** Includes shower stall, door, and drain. Large soap dishes and clear acrylic towel bar. Reversible pivot glass panel door with left and right central opening. Magnetic watertight seal. Drip channel on door. Non-slip base. Textured 38" x 38" base. 74" high. Easy to clean. Add the cost of valve, water supply and drain connection.

| | Craft@Hrs | Unit | Material | Labor | Total | Sell |
|---|---|---|---|---|---|---|
| Chrome finish, clear glass | P1@3.25 | Ea | 307.00 | 102.00 | 409.00 | 695.00 |
| White frame, obscure glass | P1@3.25 | Ea | 280.00 | 102.00 | 382.00 | 649.00 |

**Durastall® three-piece fiberglass shower stall, E.L. Mustee.** Tongue-and-groove interlocking seams. Semi-gloss surface. Colorfast, waterproof, and mold- and mildew-resistant. Seamless corners. Swing doors. Shelf and towel bar molded-in. Molded fiberglass panels. Fits Mustee Durabase. Easy to clean. Add the cost of base, water supply and drain connection, valve and showerhead. White.

| | Craft@Hrs | Unit | Material | Labor | Total | Sell |
|---|---|---|---|---|---|---|
| 32" x 32" | P1@3.25 | Ea | 210.00 | 102.00 | 312.00 | 530.00 |
| 36" x 36" | P1@3.25 | Ea | 256.00 | 102.00 | 358.00 | 609.00 |

**Durabase® one-piece shower base, E.L. Mustee.** Underbody ribbed for added strength. Slip-resistant. Mold- and mildew-resistant. Semi-gloss surface. Labor includes setting and fastening in place only. Add for drain connections as required.

| | Craft@Hrs | Unit | Material | Labor | Total | Sell |
|---|---|---|---|---|---|---|
| 32" long, 32" wide, square, single threshold | P1@.600 | Ea | 81.60 | 18.90 | 100.50 | 171.00 |
| 36" long, 36" wide, neo-angle corner threshold | P1@.600 | Ea | 102.00 | 18.90 | 120.90 | 206.00 |
| 36" long, 36" wide, square, single threshold | P1@.600 | Ea | 93.20 | 18.90 | 112.10 | 191.00 |
| 38" long, 38" wide, neo-angle corner threshold | P1@.800 | Ea | 109.00 | 25.20 | 134.20 | 228.00 |
| 48" long, 32" wide, rectangular, single threshold | P1@.800 | Ea | 111.00 | 25.20 | 136.20 | 232.00 |
| 48" long, 34" wide, rectangular, single threshold | P1@.800 | Ea | 106.00 | 25.20 | 131.20 | 223.00 |

| | Craft@Hrs | Unit | Material | Labor | Total | Sell |
|---|---|---|---|---|---|---|

**Lascoat™ shower pan, Lasco Bathware.** Lascoat gel coat finish. Center drain location. Slip-resistant textured bottom. Integral nailing flange. 3-year warranty on Lascoat finish. White. 7-1/4" height. Labor includes setting and fastening in place only. Add for drain connections as required.

| | Craft@Hrs | Unit | Material | Labor | Total | Sell |
|---|---|---|---|---|---|---|
| 32" long, 32" wide | P1@.600 | Ea | 108.00 | 18.90 | 126.90 | 216.00 |
| 36" long, 36" wide | P1@.800 | Ea | 118.00 | 25.20 | 143.20 | 243.00 |
| 42" long, 34" wide | P1@.800 | Ea | 151.00 | 25.20 | 176.20 | 300.00 |
| 60" long, 34" wide | P1@.800 | Ea | 172.00 | 25.20 | 197.20 | 335.00 |

## Shower Enclosure Doors

**Pivot shower enclosure door, Keystone.** Obscure glass. Silver frame. Drip apron stops water drips when opening door. Reversible for left- or right-hand opening. Lifetime warranty. 3" adjustment allowance. Factory-installed handle. No exposed screws.

| | Craft@Hrs | Unit | Material | Labor | Total | Sell |
|---|---|---|---|---|---|---|
| 23-3/4" to 26-3/4" wide | BG@1.20 | Ea | 128.00 | 35.70 | 163.70 | 278.00 |
| 27-3/4" to 30-3/4" wide | BG@1.20 | Ea | 123.00 | 35.70 | 158.70 | 270.00 |
| 30-3/4" to 33-3/4" wide | BG@1.20 | Ea | 126.00 | 35.70 | 161.70 | 275.00 |
| 33-3/4" to 36-3/4" wide | BG@1.20 | Ea | 150.00 | 35.70 | 185.70 | 316.00 |

**Pivot shower enclosure door, Keystone.** Silver with rain glass.

| | Craft@Hrs | Unit | Material | Labor | Total | Sell |
|---|---|---|---|---|---|---|
| 24" wide | BG@1.20 | Ea | 150.00 | 35.70 | 185.70 | 316.00 |
| 27" wide | BG@1.20 | Ea | 151.00 | 35.70 | 186.70 | 317.00 |
| 30" wide | BG@1.20 | Ea | 160.00 | 35.70 | 195.70 | 333.00 |
| 33" wide | BG@1.20 | Ea | 157.00 | 35.70 | 192.70 | 328.00 |

**Pivot shower enclosure door, Contractors Wardrobe.**

| | Craft@Hrs | Unit | Material | Labor | Total | Sell |
|---|---|---|---|---|---|---|
| Bright clear, rain glass, 59-1/8" x 69-1/2" | BG@1.20 | Ea | 307.00 | 35.70 | 342.70 | 583.00 |
| Bright clear, rain glass, 59-1/2" x 63-1/4" | BG@1.20 | Ea | 179.00 | 35.70 | 214.70 | 365.00 |
| Frameless, bright clear, clear glass, 59-3/8" x 68-7/8" | BG@1.20 | Ea | 233.00 | 35.70 | 268.70 | 457.00 |
| Frameless, bright gold, clear glass, 59-3/8" x 68-7/8" | BG@1.20 | Ea | 259.00 | 35.70 | 294.70 | 501.00 |

**Pivot shower enclosure door, Sterling Plumbing.** Silver frame.

| | Craft@Hrs | Unit | Material | Labor | Total | Sell |
|---|---|---|---|---|---|---|
| 24" to 27-1/2" | BG@1.20 | Ea | 99.40 | 35.70 | 135.10 | 230.00 |
| 32" to 35-1/2" | BG@1.20 | Ea | 134.00 | 35.70 | 169.70 | 288.00 |
| 36" to 39-1/2" | BG@1.20 | Ea | 139.00 | 35.70 | 174.70 | 297.00 |
| 42" to 45-1/2" | BG@1.20 | Ea | 167.00 | 35.70 | 202.70 | 345.00 |
| 48" to 51-1/2" | BG@1.20 | Ea | 199.00 | 35.70 | 234.70 | 399.00 |

**Swinging shower enclosure door, Sterling Plumbing.** Standard hinged. Use in either right- or left-handed opening. Height is 64". Silver frame finish. Hammered glass texture. Self-draining, easy-to-clean bottom track. Fits most fiberglass surround units.

| | Craft@Hrs | Unit | Material | Labor | Total | Sell |
|---|---|---|---|---|---|---|
| 23-1/2" to 25" wide | BG@1.20 | Ea | 74.50 | 35.70 | 110.20 | 187.00 |
| 27" to 28-1/2" wide | BG@1.20 | Ea | 69.80 | 35.70 | 105.50 | 179.00 |
| 31" to 32-1/2" wide | BG@1.20 | Ea | 72.30 | 35.70 | 108.00 | 184.00 |

**Hinged shower enclosure door, American Shower & Bath.** Silver frame.

| | Craft@Hrs | Unit | Material | Labor | Total | Sell |
|---|---|---|---|---|---|---|
| 24", hammertone glass | BG@1.20 | Ea | 93.90 | 35.70 | 129.60 | 220.00 |
| 31", pebble glass | BG@1.20 | Ea | 165.00 | 35.70 | 200.70 | 341.00 |
| 36", pebble glass | BG@1.20 | Ea | 185.00 | 35.70 | 220.70 | 375.00 |

|  | Craft@Hrs | Unit | Material | Labor | Total | Sell |
|---|---|---|---|---|---|---|
| **Bypass door shower enclosure, American Shower & Bath.** | | | | | | |
| 48", silver, hammertone glass | BG@1.20 | Ea | 140.00 | 35.70 | 175.70 | 299.00 |
| 60", brass, hammertone glass | BG@1.20 | Ea | 172.00 | 35.70 | 207.70 | 353.00 |
| 60", silver, hammertone glass | BG@1.20 | Ea | 163.00 | 35.70 | 198.70 | 338.00 |
| 60", silver, rain glass | BG@1.20 | Ea | 239.00 | 35.70 | 274.70 | 467.00 |

## Single-Control Tub and Shower Faucets

**Single-control tub and shower faucet, Price Pfister.** Round acrylic mixing valve handle. Labor includes set in place and connection only. Add the cost of plumbing rough-in as required.

|  | Craft@Hrs | Unit | Material | Labor | Total | Sell |
|---|---|---|---|---|---|---|
| Chrome | P1@1.40 | Ea | 79.90 | 44.10 | 124.00 | 211.00 |
| Chrome, shower only | P1@1.00 | Ea | 61.30 | 31.50 | 92.80 | 158.00 |

**Monitor® single-knob tub and shower faucet set, Delta.** Pressure balance faucet prevents uncomfortable water temperature changes. Durable chrome finish. Labor includes set in place and connection only. Add the cost of plumbing rough-in as required.

|  | Craft@Hrs | Unit | Material | Labor | Total | Sell |
|---|---|---|---|---|---|---|
| Chrome handle | P1@1.40 | Ea | 87.00 | 44.10 | 131.10 | 223.00 |
| Chrome or brass handle | P1@1.40 | Ea | 148.00 | 44.10 | 192.10 | 327.00 |
| Clear handle | P1@1.40 | Ea | 77.20 | 44.10 | 121.30 | 206.00 |
| Porcelain and brass inserts | P1@1.40 | Ea | 109.00 | 44.10 | 153.10 | 260.00 |

**Single-handle tub and shower set, Stanadyne.** Pressure balanced. Classic styling. Washerless cartridge design. Labor includes set in place and connection only. Add the cost of plumbing rough-in as required.

|  | Craft@Hrs | Unit | Material | Labor | Total | Sell |
|---|---|---|---|---|---|---|
| Polished chrome finish, Villeta | P1@1.40 | Ea | 96.70 | 44.10 | 140.80 | 239.00 |
| Platinum with chrome accents | P1@1.40 | Ea | 161.00 | 44.10 | 205.10 | 349.00 |

**Reliant Plus lever-handle tub and shower faucet set, American Standard.** Pressure balanced. Metal lever handle. Ceramic disc valving resists hard and sandy water. 1/2" direct sweat or IPS FPT. High-temperature limit helps prevent scalding. Labor includes set in place and connection only. Add the cost of plumbing rough-in as required.

|  | Craft@Hrs | Unit | Material | Labor | Total | Sell |
|---|---|---|---|---|---|---|
| Chrome, shower only | P1@1.00 | Ea | 108.00 | 31.50 | 139.50 | 237.00 |
| Polished chrome | P1@1.40 | Ea | 106.00 | 44.10 | 150.10 | 255.00 |

**Single-control tub and shower faucet, Glacier Bay.** Anti-scald pressure balanced mechanism. Brass showerhead with adjustable spray pattern. Drip-free washerless cartridge design.

|  | Craft@Hrs | Unit | Material | Labor | Total | Sell |
|---|---|---|---|---|---|---|
| Chrome | P1@1.40 | Ea | 78.80 | 44.10 | 122.90 | 209.00 |

**Genesis single-control tub and shower faucet, Price Pfister.** Pforever Pfaucet™ warranty. Covers function and finish for life. Labor includes set in place and connection only. Add the cost of plumbing rough-in as required.

|  | Craft@Hrs | Unit | Material | Labor | Total | Sell |
|---|---|---|---|---|---|---|
| Chrome finish | P1@1.40 | Ea | 118.00 | 44.10 | 162.10 | 276.00 |

**Single-control tub and shower set, Moen.** Pressure balanced. Lever handle. Includes adjustable-spray showerhead, arm and flange, and diverter spout with flange. ADA compliant. Labor includes set in place and connection only. Add the cost of plumbing rough-in as required.

|  | Craft@Hrs | Unit | Material | Labor | Total | Sell |
|---|---|---|---|---|---|---|
| Chrome, brass accents | P1@1.40 | Ea | 148.00 | 44.10 | 192.10 | 327.00 |
| Chrome, chrome handle | P1@1.40 | Ea | 127.00 | 44.10 | 171.10 | 291.00 |
| Polished brass | P1@1.40 | Ea | 253.00 | 44.10 | 297.10 | 505.00 |

**Single-control tub and shower faucet, Price Pfister.** Pressure balanced. Lever handle. Washerless valves. Adjustable showerhead, arm, and flange. Diverter spout. ADA compliant. Labor includes set in place and connection only. Add the cost of plumbing rough-in as required.

|  | Craft@Hrs | Unit | Material | Labor | Total | Sell |
|---|---|---|---|---|---|---|
| Chrome, brass accent | P1@1.40 | Ea | 150.00 | 44.10 | 194.10 | 330.00 |
| Chrome, porcelain handle | P1@1.40 | Ea | 128.00 | 44.10 | 172.10 | 293.00 |
| Nickel, brass accent | P1@1.40 | Ea | 181.00 | 44.10 | 225.10 | 383.00 |

|  | Craft@Hrs | Unit | Material | Labor | Total | Sell |
|---|---|---|---|---|---|---|

**Williamsburg lever-handle tub and shower set, American Standard.** Lifetime finish, will not tarnish. Ceramic disc valving, great for hard water. Both porcelain and metal lever inserts. Pressure balanced valve prevents accidental scalding. Lifetime warranty on function and finish. Labor includes set in place and connection only. Add the cost of plumbing rough-in as required.

| | | | | | | |
|---|---|---|---|---|---|---|
| Brass and chrome finish | P1@1.40 | Ea | 160.00 | 44.10 | 204.10 | 347.00 |
| Brass with chrome and brass | P1@1.40 | Ea | 178.00 | 44.10 | 222.10 | 378.00 |
| Brass with satin and chrome | P1@1.40 | Ea | 240.00 | 44.10 | 284.10 | 483.00 |
| Brass with velvet and chrome | P1@1.40 | Ea | 201.00 | 44.10 | 245.10 | 417.00 |
| Brushed satin nickel finish | P1@1.40 | Ea | 212.00 | 44.10 | 256.10 | 435.00 |
| Chrome finish | P1@1.40 | Ea | 149.00 | 44.10 | 193.10 | 328.00 |
| Chrome and brass finish | P1@1.40 | Ea | 212.00 | 44.10 | 256.10 | 435.00 |
| Polished brass finish | P1@1.40 | Ea | 170.00 | 44.10 | 214.10 | 364.00 |
| Satin chrome finish | P1@1.40 | Ea | 314.00 | 44.10 | 358.10 | 609.00 |

**Fairfax single-control tub and shower faucet, Kohler.** Pressure balanced. Lever handle. Showerhead, arm, and flange. Rite-Temp™ pressure-balancing valves eliminate surges of hot and cold water. Nostalgic design. Polished chrome finish. Labor includes set in place and connection only. Add the cost of plumbing rough-in as required.

| | | | | | | |
|---|---|---|---|---|---|---|
| With diverter spout | P1@1.40 | Ea | 150.00 | 44.10 | 194.10 | 330.00 |

**Victorian™ lever-handle tub and shower faucet, Delta.** Individual temperature and volume control. Large, decorative showerhead delivers a good spray pattern. Single hole mount centerset. Brass. Pressure balanced with Scald-Guard. Labor includes set in place and connection only. Add the cost of plumbing rough-in as required.

| | | | | | | |
|---|---|---|---|---|---|---|
| Chrome finish | P1@1.40 | Ea | 168.00 | 44.10 | 212.10 | 361.00 |

**Pegasus 1000 series single-handle tub and shower faucet, J A Manufacturing.** Pressure balanced. 1/4 turn, anti-scald valve temperature control mechanism. Bell-shaped deluxe head with adjustable arm. Favorite temperature set. Shutoff valve. Labor includes set in place and connection only. Add the cost of plumbing rough-in as required.

| | | | | | | |
|---|---|---|---|---|---|---|
| Brushed nickel | P1@1.40 | Ea | 212.00 | 44.10 | 256.10 | 435.00 |
| Brushed nickel, shower only | P1@1.00 | Ea | 88.70 | 31.50 | 120.20 | 204.00 |
| Chrome | P1@1.40 | Ea | 182.00 | 44.10 | 226.10 | 384.00 |
| Chrome, shower only | P1@1.00 | Ea | 153.00 | 31.50 | 184.50 | 314.00 |

**Innovations single-handle tub and shower faucet, Masco.** Pressure balanced. Labor includes set in place and connection only. Add the cost of plumbing rough-in as required.

| | | | | | | |
|---|---|---|---|---|---|---|
| Pearl nickel finish | P1@1.40 | Ea | 204.00 | 44.10 | 248.10 | 422.00 |

**Two-handle tub and shower faucet, Price Pfister.** Knob handles. Includes showerhead, flange, and arm. Polished chrome. Labor includes set in place and connection only. Add the cost of plumbing rough-in as required.

| | | | | | | |
|---|---|---|---|---|---|---|
| Porcelain handle, shower only | P1@1.40 | Ea | 133.00 | 44.10 | 177.10 | 301.00 |
| Verve®, metal handle, shower only | P1@1.00 | Ea | 62.80 | 31.50 | 94.30 | 160.00 |
| Windsor, acrylic handle | P1@1.40 | Ea | 75.80 | 44.10 | 119.90 | 204.00 |

**Two-handle tub and shower faucet, Masco.** Scald-Guard® keeps water temperature within 3 degrees Fahrenheit to prevent sudden temperature changes. Touch-Clean® showerhead. Lifetime warranty. Labor includes set in place and connection only. Add the cost of plumbing rough-in as required.

| | | | | | | |
|---|---|---|---|---|---|---|
| Polished brass and chrome finish | P1@1.40 | Ea | 188.00 | 44.10 | 232.10 | 395.00 |

**Three-handle tub and shower faucet set, Price Pfister.** Chrome. Includes showerhead, arm, flanges and tub spout. Labor includes set in place and connection only. Add the cost of plumbing rough-in as required.

| | | | | | | |
|---|---|---|---|---|---|---|
| Porcelain cross handles | P1@1.40 | Ea | 156.00 | 44.10 | 200.10 | 340.00 |
| Verve® chrome handles | P1@1.40 | Ea | 75.10 | 44.10 | 119.20 | 203.00 |
| Windsor® acrylic handles | P1@1.40 | Ea | 110.00 | 44.10 | 154.10 | 262.00 |

| | Craft@Hrs | Unit | Material | Labor | Total | Sell |
|---|---|---|---|---|---|---|

**Three-handle tub and shower faucet, Moen.** Clear knobs. Includes showerhead, arm, flange, and diverter spout. Full spray, water-saving showerhead. Reliable washerless cartridge. Labor includes set in place and connection only. Add the cost of plumbing rough-in as required.

| | | | | | | |
|---|---|---|---|---|---|---|
| Chrome | P1@1.50 | Ea | 100.00 | 47.30 | 147.30 | 250.00 |

**Savannah three-handle tub and shower faucet set, Price Pfister.** Ceramic disc valving. Porcelain levers. Solid brass construction. Pressure balanced with Scald-Guard®. Labor includes set in place and connection only. Add the cost of plumbing rough-in as required.

| | | | | | | |
|---|---|---|---|---|---|---|
| White porcelain lever handles | P1@1.50 | Ea | 160.00 | 47.30 | 207.30 | 352.00 |

**Add-on shower kit for built-in tub, Brass Craft.** Includes water-saver showerhead, diverter spout, adjustable risers, and wall bracket. Labor includes set in place and connection only. Add the cost of plumbing rough-in as required.

| | | | | | | |
|---|---|---|---|---|---|---|
| Kit | P1@1.00 | Ea | 32.40 | 31.50 | 63.90 | 109.00 |

## Bath and Shower Accessories

**Power massage showerhead.** 3 massage settings: pulsating, gentle and combination. Great for low pressure. Self-cleaning, water-saving showerhead for continuous flow.

| | | | | | | |
|---|---|---|---|---|---|---|
| Chrome, brass accents | P1@.200 | Ea | 7.89 | 6.30 | 14.19 | 24.10 |
| Chrome, white accents | P1@.200 | Ea | 8.85 | 6.30 | 15.15 | 25.80 |
| White finish | P1@.200 | Ea | 7.35 | 6.30 | 13.65 | 23.20 |

**Sunflower showerhead, J A Manufacturing.**

| | | | | | | |
|---|---|---|---|---|---|---|
| 8", easy clean | P1@.200 | Ea | 129.00 | 6.30 | 135.30 | 230.00 |

## Toilets (Water Closets)

**Hot and cold bidet.** Drilled for hot and cold valves. Labor includes setting and connection only. Add the cost of valves and piping.

| | | | | | | |
|---|---|---|---|---|---|---|
| Diplomatico, small, bone | P1@1.50 | Ea | 98.70 | 47.30 | 146.00 | 248.00 |
| White | P1@1.50 | Ea | 109.00 | 47.30 | 156.30 | 266.00 |

**Bidet Valvuetwist faucet set, three-handle, J A Manufacturing.** Fits 5" to 9" center holes. Ceramic disc valves. Labor includes set in place and connection only. Add the cost of plumbing rough-in as required.

| | | | | | | |
|---|---|---|---|---|---|---|
| Chrome, brass accents | P1@1.10 | Ea | 125.00 | 34.70 | 159.70 | 271.00 |
| Chrome, porcelain handles | P1@1.10 | Ea | 102.00 | 34.70 | 136.70 | 232.00 |

**Toilet To Go kit, Jameco International.** 1.6-gallon flush. Includes tank, bowl, seat, wax ring and toilet bolts. Easy to install. Exceeds ANSI standards. Expanded 2" trapway. Large 9" x 8-1/2" water surface. Labor includes setting and connection only. Add the cost of wall stop valve and piping.

| | | | | | | |
|---|---|---|---|---|---|---|
| Round, Bel-Air, white | P1@1.10 | Ea | 70.34 | 34.70 | 105.04 | 179.00 |
| Round, Metro, stylish, white | P1@1.10 | Ea | 81.60 | 34.70 | 116.30 | 198.00 |
| Elongated, Metro | P1@1.10 | Ea | 96.30 | 34.70 | 131.00 | 223.00 |

**Express Toilet In A Box, Briggs Industries.** Includes bowl, tank, seat, wax ring and mounting bolts. Labor includes setting and connection only. Add the cost of wall stop valve and piping.

| | | | | | | |
|---|---|---|---|---|---|---|
| Round, white | P1@1.10 | Ea | 83.80 | 34.70 | 118.50 | 201.00 |

**Cadet compact toilet, American Standard.** Fits into space of a round front bowl toilet. 1.6-gallon flush. Sanitary dam. Labor includes setting and connection only. Add the cost of wall stop valve, piping and the seat.

| | | | | | | |
|---|---|---|---|---|---|---|
| Elongated bowl | P1@.550 | Ea | 67.50 | 17.30 | 84.80 | 144.00 |
| Tank | P1@.550 | Ea | 66.30 | 17.30 | 83.60 | 142.00 |

| | Craft@Hrs | Unit | Material | Labor | Total | Sell |
|---|---|---|---|---|---|---|

**Arangon IV one-piece toilet, Foremost International.** 1.6-gallon flush. Round front. Seat included. Labor includes setting and connection only. Add the cost of wall stop valve and piping.

| | Craft@Hrs | Unit | Material | Labor | Total | Sell |
|---|---|---|---|---|---|---|
| White | P1@1.10 | Ea | 162.00 | 34.70 | 196.70 | 334.00 |

**Canterbury one-piece toilet, Eljer Plumbingware.** Vitreous china. 1.6-gallon flush. Siphon action. Side-mounted flush actuator. 12" rough-in. Seat sold separately. Labor includes setting and connection only. Add the cost of wall stop valve, piping and the seat.

| | Craft@Hrs | Unit | Material | Labor | Total | Sell |
|---|---|---|---|---|---|---|
| Round, white | P1@1.10 | Ea | 70.95 | 34.70 | 105.65 | 180.00 |

**Hamilton one-piece toilet, American Standard.** Vitreous china. Low consumption. Elongated siphon action bowl. Includes color-matched Rise and Shine™ plastic seat and cover with lift-off hinge. 2-bolt cap. Labor includes setting and connection only. Add the cost of wall stop valve and piping.

| | Craft@Hrs | Unit | Material | Labor | Total | Sell |
|---|---|---|---|---|---|---|
| Bone | P1@1.10 | Ea | 281.00 | 34.70 | 315.70 | 537.00 |
| Linen | P1@1.10 | Ea | 299.00 | 34.70 | 333.70 | 567.00 |
| White | P1@1.10 | Ea | 263.00 | 34.70 | 297.70 | 506.00 |

**Cadet pressure-assist toilet, American Standard.** Labor includes setting and connection only. Add the cost of wall stop valve, piping and the seat.

| | Craft@Hrs | Unit | Material | Labor | Total | Sell |
|---|---|---|---|---|---|---|
| Round front bowl | P1@.550 | Ea | 105.00 | 17.30 | 122.30 | 208.00 |
| Tank, 1.6 gallon | P1@.550 | Ea | 181.00 | 17.30 | 198.30 | 337.00 |

**Champion toilet, American Standard.** Labor includes setting and connection only. Add the cost of wall stop valve, piping and the seat.

| | Craft@Hrs | Unit | Material | Labor | Total | Sell |
|---|---|---|---|---|---|---|
| 16-1/2" toilet bowl, white | P1@.550 | Ea | 161.00 | 17.30 | 178.30 | 303.00 |
| Toilet tank, white | P1@.550 | Ea | 112.00 | 17.30 | 129.30 | 220.00 |

**Williamsburg Antiquity one-piece toilet, American Standard.** Elongated front model fits in the space of a round toilet. 2" trapway. Seat included. Labor includes setting and connection only. Add the cost of wall stop valve and piping.

| | Craft@Hrs | Unit | Material | Labor | Total | Sell |
|---|---|---|---|---|---|---|
| Elongated, white | P1@1.10 | Ea | 302.00 | 34.70 | 336.70 | 572.00 |

**Gabrielle Comfort Height™ one-piece toilet, Kohler.** Contemporary look and superior flush. Rim height 16-1/8", consistent with standard height of chair. Vigorous flush is a result of siphon-jet flushing technology and 2" fully glazed trapway. 10" x 9" water surface, which maintains a cleaner bowl. Ingenium™ flushing system. Labor includes setting and connection only. Add the cost of wall stop valve and piping.

| | Craft@Hrs | Unit | Material | Labor | Total | Sell |
|---|---|---|---|---|---|---|
| Elongated, white | P1@1.10 | Ea | 441.00 | 34.70 | 475.70 | 809.00 |

**San Raphael™ toilet, Kohler.** One-piece design. Kohler's quietest flushing toilet. 2" fully glazed trapway. Polished chrome trip lever. Includes seat. Labor includes setting and connection only. Add the cost of wall stop valve and piping.

| | Craft@Hrs | Unit | Material | Labor | Total | Sell |
|---|---|---|---|---|---|---|
| Elongated, white | P1@1.10 | Ea | 568.00 | 34.70 | 602.70 | 1,020.00 |

**One-piece wall-mount toilet, American Standard.** Vitreous china. Full glazed trapway. Condensation channel. Flush valve. Direct-fed siphon jet action. 10" x 12" water surface area. 25" x 14-3/4" x 15". Labor includes setting and connection only. Add the cost of a flush valve, carrier and piping.

| | Craft@Hrs | Unit | Material | Labor | Total | Sell |
|---|---|---|---|---|---|---|
| Top spud, elongated, white | P1@1.10 | Ea | 121.00 | 34.70 | 155.70 | 265.00 |

**Sloan Royal flush valve.** Water saver. For floor or wall hung top spud bowls.

| | Craft@Hrs | Unit | Material | Labor | Total | Sell |
|---|---|---|---|---|---|---|
| Sloan No. 110 | P1@.600 | Ea | 125.00 | 18.90 | 143.90 | 245.00 |

**Sloan Royal urinal flush valve.** Water saver. 1.5-gallon flush. For 3/4" top spud urinals.

| | Craft@Hrs | Unit | Material | Labor | Total | Sell |
|---|---|---|---|---|---|---|
| Sloan No. 186 | P1@.600 | Ea | 124.00 | 18.90 | 142.90 | 243.00 |

| | Craft@Hrs | Unit | Material | Labor | Total | Sell |
|---|---|---|---|---|---|---|

**Dover urinal, Eljer Plumbingware.** One gallon flush. Labor includes setting and connection only. Add the cost of a flush valve, hanger and piping.

| | | | | | | |
|---|---|---|---|---|---|---|
| 3/4" top spud, white | P1@1.50 | Ea | 107.00 | 47.30 | 154.30 | 262.00 |

**Washbrook urinal, American Standard.** Vitreous china. Flushing rim. Elongated 14" rim from finished wall. Washout flush action. Includes two wall hangers and outlet connection. Labor includes setting and connection only. Add the cost of a flush valve, hanger and piping.

| | | | | | | |
|---|---|---|---|---|---|---|
| 3/4" top spud | P1@1.50 | Ea | 214.00 | 47.30 | 261.30 | 444.00 |

## Toilet Installation and Repair

**Low-Boy toilet supply hook-up kit, Brass Craft.** For iron pipe or copper tube stub outs. Kit includes 1-1/2" female pipe thread by 3/8" OD tube compression by 1/2" male pipe thread union adapter, and Teflon® tape. Includes 9" braided Speedi-Plumb water supply connector.

| | | | | | | |
|---|---|---|---|---|---|---|
| Kit | P1@.250 | Ea | 11.80 | 7.88 | 19.68 | 33.50 |

**Angle valve.**

| | | | | | | |
|---|---|---|---|---|---|---|
| 1/2" with 3/8" x 12" supply tube kit | P1@.250 | Ea | 7.47 | 7.88 | 15.35 | 26.10 |

**Decorative flexible supply stop.** 1/2" iron pipe size by 3/8" outside diameter.

| | | | | | | |
|---|---|---|---|---|---|---|
| Angle valve, polished brass | P1@.250 | Ea | 11.30 | 7.88 | 19.18 | 32.60 |

**No-Seep wax toilet bowl gasket.** Wax gasket with polyethylene flange fits 3" and 4" lines, closet bowls only. Includes 1/4" x 2-1/4" brass bolt kit with stainless washers and brass nuts.

| | | | | | | |
|---|---|---|---|---|---|---|
| With brass bolt kit | P1@.100 | Ea | 4.17 | 3.15 | 7.32 | 12.40 |

**Cast iron water closet flange.** Use on heavyweight plastic or soil pipe. Complete with body, brass ring, bolts and gasket.

| | | | | | | |
|---|---|---|---|---|---|---|
| 4" diameter, compression | P1@.250 | Ea | 15.90 | 7.88 | 23.78 | 40.40 |

**Toilet base plate cover.** For toilet replacement. Conceals ring around smaller base of new toilet. Plastic. Plates can be stacked to raise toilet. 11" wide.

| | | | | | | |
|---|---|---|---|---|---|---|
| Round nose, 22-1/2" long | P1@.080 | Ea | 9.66 | 2.52 | 12.18 | 20.70 |
| Square nose, 18-3/4" long | P1@.080 | Ea | 8.46 | 2.52 | 10.98 | 18.70 |

**Leak Sentry™ anti-siphon toilet fill valve, Fluidmaster.** Prevents automatic refill of a leaky tank. At most, only one tank of water can escape a toilet. Toilet always remains operational. Bowl refill adjustment prevents wasteful overfilling. Fits most toilets including 1.6 gallon per flush models. Made from corrosion-resistant plastic and stainless steel.

| | | | | | | |
|---|---|---|---|---|---|---|
| 1.6 gallon flush | P1@.300 | Ea | 9.71 | 9.45 | 19.16 | 32.60 |

**Complete toilet tank repair kit, Fluidmaster.** Includes flush valve with Adjust-Flush™ flapper, anti-siphon fill valve, Sure-Fit® tank lever, 3 bolts and gasket. Flapper dial regulates volume. Valve height adjusts 9" to 14". Lever trims and bends to fit. Universal gasket.

| | | | | | | |
|---|---|---|---|---|---|---|
| Kit | P1@.850 | Ea | 21.00 | 26.80 | 47.80 | 81.30 |

**One-piece toilet repair kit, Fluidmaster.** Replaces many one-piece toilet assemblies. Fewer parts. Ideal for wet or dry inlet tubes.

| | | | | | | |
|---|---|---|---|---|---|---|
| Kit | P1@.850 | Ea | 58.40 | 26.80 | 85.20 | 145.00 |

**Tank to bowl bolt kit, Fluidmaster.** Two 5/16" x 3" brass bolts with washers and wing nuts.

| | | | | | | |
|---|---|---|---|---|---|---|
| Kit | P1@.275 | Ea | 2.68 | 8.66 | 11.34 | 19.30 |

| | Craft@Hrs | Unit | Material | Labor | Total | Sell |
|---|---|---|---|---|---|---|

## Toilet Seats

**Molded wood toilet seat, Bemis Manufacturing.** Multi-coat enamel finish. Closed front and cover to fit a regular bowl. Color-matched bumpers and hinges. Top-Tite® hinges with non-corrosive, top-tightening bolts and wing nuts. 1" ring thickness including bumper. Includes lid.

| | Craft@Hrs | Unit | Material | Labor | Total | Sell |
|---|---|---|---|---|---|---|
| Round, beige | P1@.250 | Ea | 10.72 | 7.88 | 18.60 | 31.60 |
| Round, bone | P1@.250 | Ea | 11.05 | 7.88 | 18.93 | 32.20 |
| Round, gold | P1@.250 | Ea | 10.73 | 7.88 | 18.61 | 31.60 |
| Round, pink | P1@.250 | Ea | 11.80 | 7.88 | 19.68 | 33.50 |
| Round, seafoam | P1@.250 | Ea | 12.40 | 7.88 | 20.28 | 34.50 |
| Round, sky blue | P1@.250 | Ea | 11.04 | 7.88 | 18.92 | 32.20 |
| Round, silver | P1@.250 | Ea | 10.72 | 7.88 | 18.60 | 31.60 |
| Round, white | P1@.250 | Ea | 10.23 | 7.88 | 18.11 | 30.80 |
| Round, white, fits Eljer Emblem | P1@.250 | Ea | 15.05 | 7.88 | 22.93 | 39.00 |
| Round, black, chrome hinge | P1@.250 | Ea | 21.30 | 7.88 | 29.18 | 49.60 |
| Round, bone, chrome hinge | P1@.250 | Ea | 20.94 | 7.88 | 28.82 | 49.00 |
| Round, silver, chrome hinge | P1@.250 | Ea | 21.30 | 7.88 | 29.18 | 49.60 |
| Round, white, chrome hinge | P1@.250 | Ea | 16.22 | 7.88 | 24.10 | 41.00 |

**Molded wood open-front toilet seat, Bemis Manufacturing.** Multi-coat enamel finish. Color-matched bumpers and hinges. Top-Tite® hinges with non-corrosive, top-tightening bolts and wing nuts. 1" ring thickness including bumper. Includes lid. 3-year warranty.

| | Craft@Hrs | Unit | Material | Labor | Total | Sell |
|---|---|---|---|---|---|---|
| Elongated, biscuit | P1@.250 | Ea | 16.80 | 7.88 | 24.68 | 42.00 |
| Elongated, bone | P1@.250 | Ea | 16.80 | 7.88 | 24.68 | 42.00 |
| Elongated, white | P1@.250 | Ea | 15.00 | 7.88 | 22.88 | 38.90 |
| Round, biscuit | P1@.250 | Ea | 12.40 | 7.88 | 20.28 | 34.50 |
| Round, bone | P1@.250 | Ea | 11.10 | 7.88 | 18.98 | 32.30 |
| Round, white | P1@.250 | Ea | 10.85 | 7.88 | 18.73 | 31.80 |

**Plastic open-front toilet seat, Bemis Manufacturing.** Open-front with cover. Easy to install top-tightening hinges. White.

| | Craft@Hrs | Unit | Material | Labor | Total | Sell |
|---|---|---|---|---|---|---|
| Elongated, no cover | P1@.250 | Ea | 22.06 | 7.88 | 29.94 | 50.90 |
| Elongated, with cover | P1@.250 | Ea | 21.60 | 7.88 | 29.48 | 50.10 |
| Round, no cover | P1@.250 | Ea | 19.30 | 7.88 | 27.18 | 46.20 |
| Round, with cover | P1@.250 | Ea | 17.80 | 7.88 | 25.68 | 43.70 |

**Molded wood shell design toilet seat, Bemis Manufacturing.** Multi-coat enamel finish. Closed front and cover to fit a regular bowl. Color-matched bumpers and hinges. Top-Tite® hinges with non-corrosive, top-tightening bolts and wing nuts. 1" ring thickness including bumper. Includes lid.

| | Craft@Hrs | Unit | Material | Labor | Total | Sell |
|---|---|---|---|---|---|---|
| Elongated, bone | P1@.250 | Ea | 27.00 | 7.88 | 34.88 | 59.30 |
| Round, bone | P1@.250 | Ea | 21.50 | 7.88 | 29.38 | 49.90 |

**Sculpted wood toilet seat, Bemis Manufacturing.** Durable high-gloss finish. Non-corrosive hinges. Installs without tools.

| | Craft@Hrs | Unit | Material | Labor | Total | Sell |
|---|---|---|---|---|---|---|
| Elongated, bone | P1@.250 | Ea | 24.80 | 7.88 | 32.68 | 55.60 |
| Elongated, white | P1@.250 | Ea | 24.90 | 7.88 | 32.78 | 55.70 |
| Elongated, ivy, white | P1@.250 | Ea | 26.90 | 7.88 | 34.78 | 59.10 |
| Round, ivy, bone | P1@.250 | Ea | 21.60 | 7.88 | 29.48 | 50.10 |
| Round, ivy, white | P1@.250 | Ea | 21.60 | 7.88 | 29.48 | 50.10 |
| Round, rose, white | P1@.250 | Ea | 21.90 | 7.88 | 29.78 | 50.60 |
| Round, shell, black | P1@.250 | Ea | 21.60 | 7.88 | 29.48 | 50.10 |
| Round, wave, bone | P1@.250 | Ea | 21.60 | 7.88 | 29.48 | 50.10 |

| | Craft@Hrs | Unit | Material | Labor | Total | Sell |
|---|---|---|---|---|---|---|

**Soft decorator toilet seat, Bemis Manufacturing.** Cushioned for comfort. High-density padding. Durable antibacterial color-fast vinyl with molded wood core. Dial On® exclusive no-wobble hinges. Installs without tools.

| | Craft@Hrs | Unit | Material | Labor | Total | Sell |
|---|---|---|---|---|---|---|
| Round, bone | P1@.250 | Ea | 17.20 | 7.88 | 25.08 | 42.60 |
| Round, pink | P1@.250 | Ea | 17.20 | 7.88 | 25.08 | 42.60 |
| Round, sky blue | P1@.250 | Ea | 17.30 | 7.88 | 25.18 | 42.80 |
| Round, white | P1@.250 | Ea | 23.07 | 7.88 | 30.95 | 52.60 |
| Round, white, chrome hinges | P1@.250 | Ea | 21.50 | 7.88 | 29.38 | 49.90 |
| Round, white, chrome hinges | P1@.250 | Ea | 28.10 | 7.88 | 35.98 | 61.20 |
| Round, white, embroidered butterfly | P1@.250 | Ea | 21.80 | 7.88 | 29.68 | 50.50 |
| Round, white, Roman marble | P1@.250 | Ea | 21.80 | 7.88 | 29.68 | 50.50 |

**Natural maple toilet seat.** Durable, long-lasting furniture grade finish. Decorative chrome-plated brass hinge. Will not warp, crack or split.

| | Craft@Hrs | Unit | Material | Labor | Total | Sell |
|---|---|---|---|---|---|---|
| Elongated | P1@.250 | Ea | 32.70 | 7.88 | 40.58 | 69.00 |
| Elongated with diamond inlay | P1@.250 | Ea | 45.00 | 7.88 | 52.88 | 89.90 |
| Round | P1@.250 | Ea | 27.00 | 7.88 | 34.88 | 59.30 |
| Round with diamond inlay | P1@.250 | Ea | 37.80 | 7.88 | 45.68 | 77.70 |

## Pedestal Lavatories

**Bahama pedestal lavatory.** Labor includes setting and connecting only. Add the cost of faucet and accessories.

| | Craft@Hrs | Unit | Material | Labor | Total | Sell |
|---|---|---|---|---|---|---|
| Gray | P1@.800 | Ea | 60.00 | 25.20 | 85.20 | 145.00 |
| Sky rose | P1@.800 | Ea | 60.00 | 25.20 | 85.20 | 145.00 |
| White | P1@.800 | Ea | 60.00 | 25.20 | 85.20 | 145.00 |

**Williamsburg pedestal lavatory, American Standard.** Elegant turn-of-the-century styling. 24-1/2" x 19" top. Labor includes setting and connecting only. Add the cost of faucet and accessories.

| | Craft@Hrs | Unit | Material | Labor | Total | Sell |
|---|---|---|---|---|---|---|
| 4" centers, linen | P1@.800 | Ea | 124.00 | 25.20 | 149.20 | 254.00 |
| 4" centers, white | P1@.800 | Ea | 133.00 | 25.20 | 158.20 | 269.00 |
| 8" centers, linen | P1@.800 | Ea | 134.00 | 25.20 | 159.20 | 271.00 |
| 8" centers, white | P1@.800 | Ea | 130.00 | 25.20 | 155.20 | 264.00 |

**Repertoire pedestal lavatory, American Standard.** Labor includes setting and connecting only. Add the cost of faucet and accessories.

| | Craft@Hrs | Unit | Material | Labor | Total | Sell |
|---|---|---|---|---|---|---|
| 4" centers, white | P1@.800 | Ea | 143.00 | 25.20 | 168.20 | 286.00 |
| 8" centers, linen | P1@.800 | Ea | 173.00 | 25.20 | 198.20 | 337.00 |

**Seychelle pedestal lavatory, American Standard.** Labor includes setting and connecting only. Add the cost of faucet and accessories.

| | Craft@Hrs | Unit | Material | Labor | Total | Sell |
|---|---|---|---|---|---|---|
| 4" centers, white | P1@.800 | Ea | 135.00 | 25.20 | 160.20 | 272.00 |

**Memoirs™ pedestal lavatory, Kohler.** Traditional styling. Integrates with the Memoirs suite. Labor includes setting and connecting only. Add the cost of faucet and accessories.

| | Craft@Hrs | Unit | Material | Labor | Total | Sell |
|---|---|---|---|---|---|---|
| 4" centers, white | P1@.800 | Ea | 215.00 | 25.20 | 240.20 | 408.00 |
| 8" centers, white | P1@.800 | Ea | 215.00 | 25.20 | 240.20 | 408.00 |

**Memoirs™ pedestal lavatory, Kohler.** Integral backsplash. 27" x 19-3/8" x 34". Vitreous china. Drilled centers. Traditional styling integrates with the Memoirs Suite. Labor includes setting and connecting only. Add the cost of faucet and accessories.

| | Craft@Hrs | Unit | Material | Labor | Total | Sell |
|---|---|---|---|---|---|---|
| 4" centers, white, stately design | P1@.800 | Ea | 374.00 | 25.20 | 399.20 | 679.00 |

| | Craft@Hrs | Unit | Material | Labor | Total | Sell |
|---|---|---|---|---|---|---|

**Memoirs™ pedestal lavatory base, Kohler.** Traditional styling integrates with the Memoirs™ suite.

| | | | | | | |
|---|---|---|---|---|---|---|
| White | P1@.350 | Ea | 145.00 | 11.00 | 156.00 | 265.00 |

**Classic pedestal lavatory base.**

| | | | | | | |
|---|---|---|---|---|---|---|
| Bone | P1@.350 | Ea | 29.00 | 11.00 | 40.00 | 68.00 |
| Coral | P1@.350 | Ea | 29.05 | 11.00 | 40.05 | 68.10 |
| Silver | P1@.350 | Ea | 29.05 | 11.00 | 40.05 | 68.10 |
| White | P1@.350 | Ea | 29.05 | 11.00 | 40.05 | 68.10 |

**Lavatory leg set, Melard.**

| | | | | | | |
|---|---|---|---|---|---|---|
| Chrome | P1@.350 | Ea | 21.20 | 11.00 | 32.20 | 54.70 |

**Standard collection pedestal lavatory metal leg set, American Standard.**

| | | | | | | |
|---|---|---|---|---|---|---|
| Chrome | P1@.350 | Ea | 283.00 | 11.00 | 294.00 | 500.00 |
| Satin | P1@.350 | Ea | 316.00 | 11.00 | 327.00 | 556.00 |

## Lavatory Sinks

**Remove pedestal, counter or wall-mounted lavatory.** Turn off water. Disconnect hot and cold supply lines from the wall valves. Unscrew trap plug to the drain. Disconnect trap unions. Remove trap. Remove basin retaining screws or wall hangers. Remove wash basin.

| | | | | | | |
|---|---|---|---|---|---|---|
| Per lavatory basin | P1@.630 | Ea | — | 19.80 | 19.80 | 33.70 |

**Round vitreous china lavatory sink, Eljer Plumbingware.** Self-rimming drop-in counter mount. 19" diameter. Front overflow. Labor includes setting only. Add the cost of connection, faucet and accessories.

| | | | | | | |
|---|---|---|---|---|---|---|
| 4" centers, natural | P1@.850 | Ea | 43.50 | 26.80 | 70.30 | 120.00 |
| 4" centers, white | P1@.850 | Ea | 38.70 | 26.80 | 65.50 | 111.00 |

**Farmington™ drop-in lavatory sink, Kohler.** Self-rimming installation. 19-1/4" x 16-1/4" overall. Cast iron. Labor includes setting only. Add the cost of connection, faucet and accessories.

| | | | | | | |
|---|---|---|---|---|---|---|
| 4" centers, almond | P1@.850 | Ea | 108.00 | 26.80 | 134.80 | 229.00 |
| 4" centers, white | P1@.850 | Ea | 97.10 | 26.80 | 123.90 | 211.00 |
| 8" centers, white | P1@.850 | Ea | 96.20 | 26.80 | 123.00 | 209.00 |

**Renaissance drop-in lavatory sink, American Standard.** Vitreous china. Self-rimming. Round, 19" overall dimensions. Labor includes setting only. Add the cost of connection, faucet and accessories.

| | | | | | | |
|---|---|---|---|---|---|---|
| 4" centers, bone | P1@.850 | Ea | 99.00 | 26.80 | 125.80 | 214.00 |
| 4" centers, white | P1@.850 | Ea | 88.50 | 26.80 | 115.30 | 196.00 |

**Rondalyn lavatory sink, American Standard.** Tapered edges for style. Vitreous china. 19" diameter. Labor includes setting only. Add the cost of connection, faucet and accessories.

| | | | | | | |
|---|---|---|---|---|---|---|
| 4" centers, linen | P1@.850 | Ea | 88.50 | 26.80 | 115.30 | 196.00 |

**Aqualyn drop-in lavatory sink, American Standard.** Vitreous china. Tapered edges for style. Self-rimming. Front overflow. Supplied with template and color-matched sealant. 20-3/8" x 17". Labor includes setting only. Add the cost of connection, faucet and accessories.

| | | | | | | |
|---|---|---|---|---|---|---|
| 4" centers, linen | P1@.850 | Ea | 86.40 | 26.80 | 113.20 | 192.00 |

**Seychelle drop-in lavatory sink, American Standard.** Vitreous china. Self-rimming. Bowl has distinctive sculptured pattern. Front overflow. Labor includes setting only. Add the cost of connection, faucet and accessories.

| | | | | | | |
|---|---|---|---|---|---|---|
| 4" centers, white | P1@.850 | Ea | 76.70 | 26.80 | 103.50 | 176.00 |

| | Craft@Hrs | Unit | Material | Labor | Total | Sell |
|---|---|---|---|---|---|---|

**Cadet drop-in lavatory sink, American Standard.** European-style interior bowl. Self-rimming oval countertop lavatory. Front overflow. Supplied with template and color-matched sealant. Vitreous china, 20" x 17" oval. Labor includes setting only. Add the cost of connection, faucet and accessories.

| | Craft@Hrs | Unit | Material | Labor | Total | Sell |
|---|---|---|---|---|---|---|
| 4" centers, bone | P1@.850 | Ea | 97.80 | 26.80 | 124.60 | 212.00 |
| 4" centers, white | P1@.850 | Ea | 86.30 | 26.80 | 113.10 | 192.00 |
| 8" centers, white | P1@.850 | Ea | 63.80 | 26.80 | 90.60 | 154.00 |

**Standard Collection drop-in lavatory sink, American Standard.** Oval self-rimming design includes color-matched caulk for seamless, easy installation. Generous interior bowl with "invisible" front overflow. Nominal dimensions 22-7/8" x 18-1/2". Bowl size 15-3/4" wide by 11-1/4" front to back, 6-1/2" deep. Labor includes setting only. Add the cost of connection, faucet and accessories.

| | Craft@Hrs | Unit | Material | Labor | Total | Sell |
|---|---|---|---|---|---|---|
| 8" centers, white | P1@.850 | Ea | 74.70 | 26.80 | 101.50 | 173.00 |

**Williamsburg drop-in lavatory sink, American Standard.** European-style interior bowl. Self-rimming oval countertop lavatory. Front overflow. Supplied with template and color-matched sealant. 24-1/2" x 18". Vitreous china. Labor includes setting only. Add the cost of connection, faucet and accessories.

| | Craft@Hrs | Unit | Material | Labor | Total | Sell |
|---|---|---|---|---|---|---|
| 4" centers, linen | P1@.850 | Ea | 124.00 | 26.80 | 150.80 | 256.00 |
| 4" centers, white | P1@.850 | Ea | 110.00 | 26.80 | 136.80 | 233.00 |
| 8" centers, white | P1@.850 | Ea | 110.00 | 26.80 | 136.80 | 233.00 |

**Memoirs™ drop-in lavatory sink, Kohler.** Vitreous china. Glossy finish. Labor includes setting only. Add the cost of connection, faucet and accessories.

| | Craft@Hrs | Unit | Material | Labor | Total | Sell |
|---|---|---|---|---|---|---|
| 4" centers, white | P1@.850 | Ea | 146.00 | 26.80 | 172.80 | 294.00 |
| 8" centers, white | P1@.850 | Ea | 148.00 | 26.80 | 174.80 | 297.00 |

**Oval drop-in lavatory sink.** Vitreous china. 20" x 17". Self-rimming. Front overflow. Labor includes setting only. Add the cost of connection, faucet and accessories.

| | Craft@Hrs | Unit | Material | Labor | Total | Sell |
|---|---|---|---|---|---|---|
| 4" centers, natural | P1@.850 | Ea | 43.80 | 26.80 | 70.60 | 120.00 |
| 4" centers, white | P1@.850 | Ea | 34.80 | 26.80 | 61.60 | 105.00 |

**Ovalyn II undercounter-mount lavatory sink, American Standard.** Vitreous china. Unglazed rim. Rear overflow. Supplied with mounting kit. Labor includes setting only. Add the cost of connection, faucet and accessories.

| | Craft@Hrs | Unit | Material | Labor | Total | Sell |
|---|---|---|---|---|---|---|
| Bone, 17" x 14" | P1@.850 | Ea | 75.60 | 26.80 | 102.40 | 174.00 |
| Linen, 17" x 14" | P1@.850 | Ea | 76.70 | 26.80 | 103.50 | 176.00 |
| White, 17" x 14" | P1@.850 | Ea | 64.80 | 26.80 | 91.60 | 156.00 |

**Laurel round enameled steel lavatory sink.** Self-rimming counter mount. 19" diameter. Labor includes setting only. Add the cost of connection, faucet and accessories.

| | Craft@Hrs | Unit | Material | Labor | Total | Sell |
|---|---|---|---|---|---|---|
| 4" centers, natural finish | P1@.850 | Ea | 32.80 | 26.80 | 59.60 | 101.00 |

**Radiant lavatory sink, Kohler.** Vitreous china. Glossy finish. Self-rimming, counter mount. Labor includes setting only. Add the cost of connection, faucet and accessories.

| | Craft@Hrs | Unit | Material | Labor | Total | Sell |
|---|---|---|---|---|---|---|
| 4" centers, white | P1@.850 | Ea | 102.30 | 26.80 | 129.10 | 219.00 |

**Seychelle countertop lavatory sink, American Standard.** Labor includes setting only. Add the cost of connection, faucet and accessories.

| | Craft@Hrs | Unit | Material | Labor | Total | Sell |
|---|---|---|---|---|---|---|
| 4" centers, bone | P1@.800 | Ea | 97.20 | 25.20 | 122.40 | 208.00 |

**Summit acrylic lavatory sink, International Thermocast.** 20-3/4" x 18" x 6-3/4" deep. 2-gallon bowl. High-gloss finish. Resists stain, rust, oxidation, chipping, and scratches. Labor includes setting only. Add the cost of connection, faucet and accessories.

| | Craft@Hrs | Unit | Material | Labor | Total | Sell |
|---|---|---|---|---|---|---|
| 4" centers, bone | P1@.850 | Ea | 107.00 | 26.80 | 133.80 | 227.00 |
| 4" centers, white | P1@.850 | Ea | 74.50 | 26.80 | 101.30 | 172.00 |

|  | Craft@Hrs | Unit | Material | Labor | Total | Sell |
|---|---|---|---|---|---|---|

**Magnolia acrylic lavatory sink, International Thermocast.** 22" x 19" x 6-3/4" deep. 3-gallon bowl. High-gloss finish. Resists stain, rust, oxidation, chipping, and scratches. Self-rimming counter mount. Labor includes setting only. Add the cost of connection, faucet and accessories.

| 4" centers, white | P1@.850 | Ea | 86.40 | 26.80 | 113.20 | 192.00 |
|---|---|---|---|---|---|---|
| 8" centers, white | P1@.850 | Ea | 107.00 | 26.80 | 133.80 | 227.00 |

**Reminiscence lavatory sink, American Standard.** Vitreous china. Vintage-style design with raised motif backsplash. Self-rimming. Front overflow. Supplied with template and color-matched sealant. Labor includes setting only. Add the cost of connection, faucet and accessories.

| 8" centers, white | P1@.850 | Ea | 148.50 | 26.80 | 175.30 | 298.00 |
|---|---|---|---|---|---|---|

**Morning above-counter lavatory sink, American Standard.** For above counter installation. Integral faucet deck for single hole faucet. Center drain outlet. Without overflow. Supplied with template and color-matched sealant. Labor includes setting only. Add the cost of connection, faucet and accessories.

| Single hole faucet, white | P1@.850 | Ea | 231.00 | 26.80 | 257.80 | 438.00 |
|---|---|---|---|---|---|---|

**Memoirs™ lavatory sink, Kohler.** Vitreous china. Complies with Americans with Disabilities Act when installed per requirements of Accessibility Guidelines, Section 4.19 Lavatories & Mirrors. Traditional styling coordinates with the Memoirs Suite. Self-rimming, counter mount. Labor includes setting only. Add the cost of connection, faucet and accessories.

| 8" centers, white | P1@.850 | Ea | 316.00 | 26.80 | 342.80 | 583.00 |
|---|---|---|---|---|---|---|

**Miami wall-hung lavatory sink.** 19-1/2" x 17-1/4". Labor includes setting only. No area prep work included. Add the cost of connection, faucet and accessories.

| 4" centers, white | P1@.850 | Ea | 30.09 | 26.80 | 56.89 | 96.70 |
|---|---|---|---|---|---|---|

**Hydra wall-hung lavatory sink, American Standard.** Labor includes setting only. No area prep work included. Add the cost of connection, faucet and accessories.

| 4" centers, white | P1@.850 | Ea | 53.90 | 26.80 | 80.70 | 137.00 |
|---|---|---|---|---|---|---|

**Murray wall-hung lavatory sink, Eljer Plumbingware.** Labor includes setting only. No area prep work included. Add the cost of connection, faucet and accessories.

| 4" centers, white | P1@.850 | Ea | 42.20 | 26.80 | 69.00 | 117.00 |
|---|---|---|---|---|---|---|

**Lucerne wall-hung lavatory sink, American Standard.** Labor includes setting only. No area prep work included. Add the cost of connection, faucet and accessories.

| 4" centers, white | P1@.850 | Ea | 99.30 | 26.80 | 126.10 | 214.00 |
|---|---|---|---|---|---|---|

**Minette corner wall-hung lavatory sink, American Standard.** Vitreous china. Front overflow. Furnished with two wall hangers. Labor includes setting only. No area prep work included. Add the cost of connection, faucet and accessories.

| 4" centers, white | P1@.850 | Ea | 142.50 | 26.80 | 169.30 | 288.00 |
|---|---|---|---|---|---|---|

**Murro wall-hung lavatory sink, American Standard.** Vitreous china. Rear overflow. Recessed self-draining deck. For concealed arm or wall. Universal design. Labor includes setting only. No area prep work included. Add the cost of connection, faucet and accessories.

| 4" centers, white | P1@.850 | Ea | 185.00 | 26.80 | 211.80 | 360.00 |
|---|---|---|---|---|---|---|

**Lavatory hanger bracket.** Supports wall hung china lavatory basin.

| Lavatory hanger bracket | P1@.250 | Ea | 1.81 | 7.88 | 9.69 | 16.50 |
|---|---|---|---|---|---|---|

| | Craft@Hrs | Unit | Material | Labor | Total | Sell |
|---|---|---|---|---|---|---|

## Single-Handle Bath Faucets, 4" Center Set

**Single-handle lavatory faucet, Glacier Bay.** Deck mount. 4" center set. Chrome finish. Lever handle. Washerless cartridge. Includes mechanical pop-up assembly. Labor includes set in place and connection only. Add the cost of plumbing rough-in as required.

| | Craft@Hrs | Unit | Material | Labor | Total | Sell |
|---|---|---|---|---|---|---|
| Chrome with brass accents | P1@.900 | Ea | 86.10 | 28.40 | 114.50 | 195.00 |
| Polished chrome finish | P1@.900 | Ea | 32.30 | 28.40 | 60.70 | 103.00 |

**Zinc single-handle lavatory faucet.** Deck mount. 4" center set. Lever handle. Washerless cartridge. With mechanical pop-up assembly. Labor includes set in place and connection only. Add the cost of plumbing rough-in as required.

| | Craft@Hrs | Unit | Material | Labor | Total | Sell |
|---|---|---|---|---|---|---|
| Chrome finish, zinc | P1@.900 | Ea | 51.00 | 28.40 | 79.40 | 135.00 |

**Acrylic knob single-control lavatory faucet, Moen.** Includes pop-up plug and waste assembly. Washerless cartridge. 4" center set. Water- and energy-saving aerator. Labor includes set in place and connection only. Add the cost of plumbing rough-in as required.

| | Craft@Hrs | Unit | Material | Labor | Total | Sell |
|---|---|---|---|---|---|---|
| Chrome | P1@.900 | Ea | 63.00 | 28.40 | 91.40 | 155.00 |

**Reliant single-handle lavatory faucet, American Standard.** Deck mount. Lever handle. All metal construction. Ceramic disc valve cartridge with adjustable hot temperature limit safety stop. Includes mechanical pop-up assembly. 2.5 gallons per minute flow restricted aerator. 1/2" male threaded copper connector and nuts. 4" center set. Labor includes set in place and connection only. Add the cost of plumbing rough-in as required.

| | Craft@Hrs | Unit | Material | Labor | Total | Sell |
|---|---|---|---|---|---|---|
| Polished chrome | P1@.900 | Ea | 73.50 | 28.40 | 101.90 | 173.00 |

**Classic single-handle lavatory faucet, Masco.** Deck mount. 4" center set. Lever handle. Washerless cartridge. Lifetime warranty. Includes mechanical pop-up assembly. Labor includes set in place and connection only. Add the cost of plumbing rough-in as required.

| | Craft@Hrs | Unit | Material | Labor | Total | Sell |
|---|---|---|---|---|---|---|
| Chrome finish | P1@.900 | Ea | 85.00 | 28.40 | 113.40 | 193.00 |

**Genesis single-handle lavatory faucet, Price Pfister.** Lever handle. 4" center set. Metal poppet. ADA compliant. Labor includes set in place and connection only. Add the cost of plumbing rough-in as required.

| | Craft@Hrs | Unit | Material | Labor | Total | Sell |
|---|---|---|---|---|---|---|
| Chrome | P1@.900 | Ea | 74.50 | 28.40 | 102.90 | 175.00 |

**Villeta single-handle lavatory faucet.** Deck mount. 4" center set. Lever handle. Includes mechanical pop-up assembly. Lifetime limited warranties against leaks drips and finish defects. Labor includes set in place and connection only. Add the cost of plumbing rough-in as required.

| | Craft@Hrs | Unit | Material | Labor | Total | Sell |
|---|---|---|---|---|---|---|
| Chrome finish | P1@.900 | Ea | 80.00 | 28.40 | 108.40 | 184.00 |

**Chateau single-control lavatory faucet, Moen.** With waste assembly. ADA approved. Washerless cartridge. 4" center set. Includes pop-up assembly. Chrome finish. Labor includes set in place and connection only. Add the cost of plumbing rough-in as required.

| | Craft@Hrs | Unit | Material | Labor | Total | Sell |
|---|---|---|---|---|---|---|
| Acrylic knob control | P1@.900 | Ea | 79.90 | 28.40 | 108.30 | 184.00 |
| Handle control | P1@.900 | Ea | 63.80 | 28.40 | 92.20 | 157.00 |

**Innovations® single-handle lavatory faucet, Delta.** 4" center set. Deck mount. 3-hole installation. Lifetime warranty. ADA compliant. Includes mechanical pop-up assembly. Labor includes set in place and connection only. Add the cost of plumbing rough-in as required.

| | Craft@Hrs | Unit | Material | Labor | Total | Sell |
|---|---|---|---|---|---|---|
| Chrome finish | P1@.900 | Ea | 104.00 | 28.40 | 132.40 | 225.00 |
| Chrome finish, brass accents | P1@.900 | Ea | 127.00 | 28.40 | 155.40 | 264.00 |
| Polished brass, brass accents | P1@.900 | Ea | 166.00 | 28.40 | 194.40 | 330.00 |

**Parisa single-handle lavatory faucet, Price Pfister.** Solid brass. 4" center set. Single-control ceramic disc cartridge with temperature memory. High arc spout. Use with or without the deckplate for single-hole mounting. ADA approved. Add the cost of plumbing rough-in as required.

| | Craft@Hrs | Unit | Material | Labor | Total | Sell |
|---|---|---|---|---|---|---|
| Brushed nickel finish | P1@.900 | Ea | 86.40 | 28.40 | 114.80 | 195.00 |
| Chrome finish | P1@.900 | Ea | 74.90 | 28.40 | 103.30 | 176.00 |

| | Craft@Hrs | Unit | Material | Labor | Total | Sell |
|---|---|---|---|---|---|---|

**Fairfax lavatory single-handle faucet, Kohler.** Deck mount. 4" center set. Chrome finish. Latch handle. 5" spout. Ceramic disc valve cartridge. Includes mechanical pop-up assembly. Labor includes set in place and connection only. Add the cost of plumbing rough-in as required.

| Polished chrome | P1@.900 | Ea | 107.00 | 28.40 | 135.40 | 230.00 |

**Optima Plus infrared-operated hands-free lavatory faucet, Sloan Valve.** 4" center set. Deck mount. Infrared sensor automatically activates solenoid mixing valve to dispense tempered water. Labor includes set in place and connection only. Add the cost of plumbing rough-in as required.

| Battery operated | P1@.900 | Ea | 387.00 | 28.40 | 415.40 | 706.00 |

## Two-Handle Bath Faucets

**Two-handle lavatory faucet.** Deck mount. Includes mechanical pop-up assembly, except where noted. 4" center set. Labor includes set in place and connection only. Add the cost of plumbing rough-in as required.

| Cleo, satin chrome and brass | P1@.900 | Ea | 85.90 | 28.40 | 114.30 | 194.00 |
| Leonardo | P1@.900 | Ea | 42.00 | 28.40 | 70.40 | 120.00 |
| Lotus, chrome and porcelain | P1@.900 | Ea | 70.50 | 28.40 | 98.90 | 168.00 |
| Modern, chrome finish | P1@.900 | Ea | 40.75 | 28.40 | 69.15 | 118.00 |
| Porcelain lever handles | P1@.900 | Ea | 42.70 | 28.40 | 71.10 | 121.00 |
| Teapot, brushed nickel | P1@.900 | Ea | 58.90 | 28.40 | 87.30 | 148.00 |
| Without mechanical pop-up | P1@.900 | Ea | 8.21 | 28.40 | 36.61 | 62.20 |

**Acrylic two-handle lavatory faucet, Glacier Bay.** Deck mount. 4" center set. Chrome finish. Solid brass waterways. Drip-free washerless cartridge design. Labor includes set in place and connection only. Add the cost of plumbing rough-in as required.

| With pop-up drain | P1@.900 | Ea | 20.50 | 28.40 | 48.90 | 83.10 |
| Without pop-up drain | P1@.900 | Ea | 16.20 | 28.40 | 44.60 | 75.80 |

**Two-handle lavatory faucet, American Standard.** Deck mount. 4" center set. Chrome finish. Includes mechanical pop-up assembly. Ceramic disc cartridge. 1/2" brass supply shanks for easy installation. Labor includes set in place and connection only. Add the cost of plumbing rough-in as required.

| Acrylic handles | P1@.900 | Ea | 41.70 | 28.40 | 70.10 | 119.00 |
| Contemporary design | P1@.900 | Ea | 43.30 | 28.40 | 71.70 | 122.00 |
| Ravenna design | P1@.900 | Ea | 96.10 | 28.40 | 124.50 | 212.00 |

**Bedford two-handle lavatory faucet, Price Pfister.** Deck mount. 4" center set. Chrome finish. Metal verve handle. Includes brass mechanical pop-up assembly. Metal knobs. Labor includes set in place and connection only. Add the cost of plumbing rough-in as required.

| With mechanical pop-up drain | P1@.900 | Ea | 50.75 | 28.40 | 79.15 | 135.00 |
| Without pop-up drain | P1@.900 | Ea | 37.70 | 28.40 | 66.10 | 112.00 |

**Two-lever French handle lavatory faucet.** Deck mount. 4" center set. Drip-free ceramic disc cartridge. Solid brass waterways. Includes brass mechanical pop-up assembly. Labor includes set in place and connection only. Add the cost of plumbing rough-in as required.

| Chrome finish | P1@.900 | Ea | 52.70 | 28.40 | 81.10 | 138.00 |
| Polished brass | P1@.900 | Ea | 82.60 | 28.40 | 111.00 | 189.00 |

**Touch-control two-handle lavatory faucet, Moen.** Deck mount. 4" center set. Chrome finish. Acrylic handles. Limited lifetime function and finish warranty. Labor includes set in place and connection only. Add the cost of plumbing rough-in as required.

| No drain rod hole | P1@.900 | Ea | 39.70 | 28.40 | 68.10 | 116.00 |
| With mechanical pop-up drain | P1@.900 | Ea | 47.75 | 28.40 | 76.15 | 129.00 |

| | Craft@Hrs | Unit | Material | Labor | Total | Sell |
|---|---|---|---|---|---|---|

**Acrylic two-handle lavatory faucet, Delta.** Deck mount. Fits 3-hole sinks with 4" centers. Includes mechanical pop-up assembly. Machined brass cartridges. Washerless design. Lifetime warranty. Labor includes set in place and connection only. Add the cost of plumbing rough-in as required.

| | Craft@Hrs | Unit | Material | Labor | Total | Sell |
|---|---|---|---|---|---|---|
| Chrome finish | P1@.900 | Ea | 48.60 | 28.40 | 77.00 | 131.00 |

**Teapot two-handle lavatory faucet, Glacier Bay.** Deck mount. 4" center set. Teapot style. White ceramic handles. Includes mechanical pop-up assembly. Labor includes set in place and connection only. Add the cost of plumbing rough-in as required.

| | Craft@Hrs | Unit | Material | Labor | Total | Sell |
|---|---|---|---|---|---|---|
| Brass finish | P1@.900 | Ea | 42.90 | 28.40 | 71.30 | 121.00 |
| Chrome finish | P1@.900 | Ea | 32.40 | 28.40 | 60.80 | 103.00 |

**Two-handle lavatory faucet, Masco.** Deck mount. 4" center set. No pop-up included. Lifetime faucet and finish warranty. Solid brass and stainless steel construction. Labor includes set in place and connection only. Add the cost of plumbing rough-in as required.

| | Craft@Hrs | Unit | Material | Labor | Total | Sell |
|---|---|---|---|---|---|---|
| Chrome finish | P1@.900 | Ea | 87.70 | 28.40 | 116.10 | 197.00 |
| Chrome finish, brass accents | P1@.900 | Ea | 129.80 | 28.40 | 158.20 | 269.00 |

**Bathroom décor kit with two-handle lavatory faucet.** Includes chrome finish deck-mount teapot faucet, 24" towel bar, towel ring, toilet paper holder, robe hook and mounting hardware. Titanium PVD polished brass finish. Labor includes set in place and connection only. Add the cost of plumbing rough-in as required.

| | Craft@Hrs | Unit | Material | Labor | Total | Sell |
|---|---|---|---|---|---|---|
| Faucet and bath accessory kit | P1@1.90 | Ea | 75.90 | 59.90 | 135.80 | 231.00 |

**Gerber two-handle lavatory faucet.** Deck mount. 4" center set. Includes mechanical pop-up drain assembly. Traditional styling. Labor includes set in place and connection only. Add the cost of plumbing rough-in as required.

| | Craft@Hrs | Unit | Material | Labor | Total | Sell |
|---|---|---|---|---|---|---|
| Cast brass body | P1@.900 | Ea | 52.90 | 28.40 | 81.30 | 138.00 |

**Williamsburg two-handle lavatory faucet, American Standard.** Deck mount. 1/4-turn ceramic disc cartridge. 4" center set. Solid brass construction with mechanical pop-up assembly. Labor includes set in place and connection only. Add the cost of plumbing rough-in as required.

| | Craft@Hrs | Unit | Material | Labor | Total | Sell |
|---|---|---|---|---|---|---|
| Brass and chrome | P1@.900 | Ea | 116.00 | 28.40 | 144.40 | 245.00 |
| Chrome with brass accents | P1@.900 | Ea | 107.00 | 28.40 | 135.40 | 230.00 |
| Chrome, porcelain handles | P1@.900 | Ea | 96.10 | 28.40 | 124.50 | 212.00 |
| Satin brass finish | P1@.900 | Ea | 138.00 | 28.40 | 166.40 | 283.00 |
| Satin chrome finish | P1@.900 | Ea | 137.00 | 28.40 | 165.40 | 281.00 |
| Velvet and chrome finish | P1@.900 | Ea | 142.00 | 28.40 | 170.40 | 290.00 |

**Williamsburg two-handle lavatory faucet, American Standard.** Deck mount. 8" center set. Chrome finish except as noted. J-spout. 1/4-turn ceramic disc cartridge resists hard and sandy water. Complete with both metal and porcelain lever handles. Solid brass construction with metal pop-up drain. Labor includes set in place and connection only. Add the cost of plumbing rough-in as required.

| | Craft@Hrs | Unit | Material | Labor | Total | Sell |
|---|---|---|---|---|---|---|
| Brass accents | P1@.900 | Ea | 145.00 | 28.40 | 173.40 | 295.00 |
| Brass, porcelain handles | P1@.900 | Ea | 150.00 | 28.40 | 178.40 | 303.00 |
| Brass, two sets of handles | P1@.900 | Ea | 149.00 | 28.40 | 177.40 | 302.00 |
| Porcelain lever handles | P1@.900 | Ea | 117.00 | 28.40 | 145.40 | 247.00 |
| Satin chrome finish | P1@.900 | Ea | 179.00 | 28.40 | 207.40 | 353.00 |
| Two sets of lever handles | P1@.900 | Ea | 138.00 | 28.40 | 166.40 | 283.00 |

**J-spout two-handle lavatory faucet, American Standard.** Deck mount. 4" center set. 1/4-turn ceramic disc cartridge. Mechanical pop-up assembly. Labor includes set in place and connection only. Add the cost of plumbing rough-in as required.

| | Craft@Hrs | Unit | Material | Labor | Total | Sell |
|---|---|---|---|---|---|---|
| Brass, porcelain handles | P1@.900 | Ea | 116.00 | 28.40 | 144.40 | 245.00 |
| Chrome, porcelain handles | P1@.900 | Ea | 101.00 | 28.40 | 129.40 | 220.00 |

| | Craft@Hrs | Unit | Material | Labor | Total | Sell |
|---|---|---|---|---|---|---|

**Savannah two-handle lavatory faucet, Price Pfister.** Deck mount. 4" center set. High-arc swivel spout. Porcelain lever handles. Includes mechanical pop-up assembly. ADA compliant. Labor includes set in place and connection only. Add the cost of plumbing rough-in as required.

| | Craft@Hrs | Unit | Material | Labor | Total | Sell |
|---|---|---|---|---|---|---|
| Chrome finish | P1@.900 | Ea | 92.30 | 28.40 | 120.70 | 205.00 |

**Georgetown two-handle lavatory faucet, Price Pfister.** Deck mount. 4" center set. Chrome finish. Includes mechanical pop-up assembly. Labor includes set in place and connection only. Add the cost of plumbing rough-in as required.

| | Craft@Hrs | Unit | Material | Labor | Total | Sell |
|---|---|---|---|---|---|---|
| Brass accents | P1@.900 | Ea | 63.80 | 28.40 | 92.20 | 157.00 |
| Nickel, brass accent | P1@.900 | Ea | 135.00 | 28.40 | 163.40 | 278.00 |
| Porcelain cross handles | P1@.900 | Ea | 93.30 | 28.40 | 121.70 | 207.00 |
| Porcelain lever handles | P1@.900 | Ea | 85.70 | 28.40 | 114.10 | 194.00 |

**Georgetown two-handle lavatory faucet, Price Pfister.** Deck mount. 8" center set. Includes mechanical pop-up assembly. TwistPfit™ installation. Pforever Pfaucet™. Labor includes set in place and connection only. Add the cost of plumbing rough-in as required.

| | Craft@Hrs | Unit | Material | Labor | Total | Sell |
|---|---|---|---|---|---|---|
| Chrome, porcelain cross handle | P1@.900 | Ea | 151.00 | 28.40 | 179.40 | 305.00 |
| Chrome, porcelain lever handle | P1@.900 | Ea | 157.00 | 28.40 | 185.40 | 315.00 |
| Nickel, brass accent | P1@.900 | Ea | 180.00 | 28.40 | 208.40 | 354.00 |

**Innovations® two-handle lavatory faucet, Delta.** Deck mount. Includes mechanical pop-up assembly. 4" center set. Labor includes set in place and connection only. Add the cost of plumbing rough-in as required.

| | Craft@Hrs | Unit | Material | Labor | Total | Sell |
|---|---|---|---|---|---|---|
| Brass | P1@.900 | Ea | 178.00 | 28.40 | 206.40 | 351.00 |
| Chrome, brass accents | P1@.900 | Ea | 147.00 | 28.40 | 175.40 | 298.00 |
| Chrome, porcelain handles | P1@.900 | Ea | 100.00 | 28.40 | 128.40 | 218.00 |

**Traditional two-handle lavatory faucet, Delta.** Deck mount. Brilliance lifetime anti-tarnish chrome finish. Fits 3-hole sinks with 4" centers. Lifetime faucet and finish warranty. Labor includes set in place and connection only. Add the cost of plumbing rough-in as required.

| | Craft@Hrs | Unit | Material | Labor | Total | Sell |
|---|---|---|---|---|---|---|
| Brass accents | P1@.900 | Ea | 107.00 | 28.40 | 135.40 | 230.00 |

**Victorian two-handle lavatory faucet, Masco.** Deck mount. Metal lever handle. 4" center 3-hole installation. Labor includes set in place and connection only. Add the cost of plumbing rough-in as required.

| | Craft@Hrs | Unit | Material | Labor | Total | Sell |
|---|---|---|---|---|---|---|
| Chrome finish | P1@.900 | Ea | 126.00 | 28.40 | 154.40 | 262.00 |
| Satin nickel finish | P1@.900 | Ea | 247.00 | 28.40 | 275.40 | 468.00 |
| Venetian bronze finish | P1@.900 | Ea | 169.00 | 28.40 | 197.40 | 336.00 |

**Monticello two-handle lavatory faucet, Moen.** Deck mount. 4" center set. Chrome finish. Includes mechanical pop-up assembly. Reliable washerless cartridge. Labor includes set in place and connection only. Add the cost of plumbing rough-in as required.

| | Craft@Hrs | Unit | Material | Labor | Total | Sell |
|---|---|---|---|---|---|---|
| Chrome | P1@.900 | Ea | 94.70 | 28.40 | 123.10 | 209.00 |
| Chrome, brass handles | P1@.900 | Ea | 106.00 | 28.40 | 134.40 | 228.00 |

**Platinum two-handle lavatory faucet, Moen.** Deck mount. 4" center set. Includes mechanical pop-up assembly. Labor includes set in place and connection only. Add the cost of plumbing rough-in as required.

| | Craft@Hrs | Unit | Material | Labor | Total | Sell |
|---|---|---|---|---|---|---|
| Platinum finish, chrome accents | P1@.900 | Ea | 136.00 | 28.40 | 164.40 | 279.00 |

**Decorator two-handle lavatory faucet, Moen.** Metal lever handles. 4" center set. With waste assembly. ADA compliant. Labor includes set in place and connection only. Add the cost of plumbing rough-in as required.

| | Craft@Hrs | Unit | Material | Labor | Total | Sell |
|---|---|---|---|---|---|---|
| Polished brass | P1@.900 | Ea | 204.00 | 28.40 | 232.40 | 395.00 |

**Asceri Hi-Arc lavatory faucet, Moen.** Deck mount. 4" center set. Includes mechanical pop-up assembly. Lifetime limited warranty against leaks, drips and finish defects. Labor includes set in place and connection only. Add the cost of plumbing rough-in as required.

| | Craft@Hrs | Unit | Material | Labor | Total | Sell |
|---|---|---|---|---|---|---|
| Chrome with brass accents | P1@.900 | Ea | 191.00 | 28.40 | 219.40 | 373.00 |

| | Craft@Hrs | Unit | Material | Labor | Total | Sell |
|---|---|---|---|---|---|---|

**Monticello adjustable center two-handle lavatory faucet, Moen.** Deck mount. Adjustable center fits 8" to 16" centers. Includes mechanical pop-up assembly. Finish will not scratch or tarnish. Metal construction. Washerless cartridge. Comes with 1/2" high IPS connections. Labor includes set in place and connection only. Add the cost of plumbing rough-in as required.

| | Craft@Hrs | Unit | Material | Labor | Total | Sell |
|---|---|---|---|---|---|---|
| Chrome with brass accents | P1@.900 | Ea | 160.00 | 28.40 | 188.40 | 320.00 |
| Polished brass | P1@.900 | Ea | 269.00 | 28.40 | 297.40 | 506.00 |

**Adjustable center two-handle lavatory faucet, Moen.** Deck mount. Fits 8" to 16" centers. LifeShine™ non-tarnish finish. Lever handles. Includes mechanical pop-up assembly. ADA compliant. Labor includes set in place and connection only. Add the cost of plumbing rough-in as required.

| | | | | | | |
|---|---|---|---|---|---|---|
| Chrome | P1@.900 | Ea | 139.00 | 28.40 | 167.40 | 285.00 |

**Adjustable center two-handle lavatory faucet, J A Manufacturing.** Deck mount. Fits 6" to 12" centers. Energy-saving aerator. Drip-free ceramic disc cartridge. Double set screws strengthen handles. Labor includes set in place and connection only. Add the cost of plumbing rough-in as required.

| | | | | | | |
|---|---|---|---|---|---|---|
| Bronze finish | P1@.900 | Ea | 128.00 | 28.40 | 156.40 | 266.00 |
| Brushed nickel finish | P1@.900 | Ea | 168.00 | 28.40 | 196.40 | 334.00 |
| Brushed nickel and chrome | P1@.900 | Ea | 223.00 | 28.40 | 251.40 | 427.00 |
| Chrome finish | P1@.900 | Ea | 191.00 | 28.40 | 219.40 | 373.00 |

**Fairfax two-handle lavatory faucet, Kohler.** Deck mount. 8" center set. Lever handles. Includes mechanical pop-up assembly. Labor includes set in place and connection only. Add the cost of plumbing rough-in as required.

| | | | | | | |
|---|---|---|---|---|---|---|
| Polished chrome | P1@.900 | Ea | 162.00 | 28.40 | 190.40 | 324.00 |

**Fixture valve and supply lines.** Flexible hose. Cone washer for universal fit.

| | | | | | | |
|---|---|---|---|---|---|---|
| 3/8" C x 7/8" BC x 12", PVC faucet supply | P1@.250 | Ea | 3.09 | 7.88 | 10.97 | 18.60 |
| 1/2" FIP x 1/2" FIP x 16", PVC faucet supply | P1@.250 | Ea | 3.92 | 7.88 | 11.80 | 20.10 |
| 1/2" FIP x 1/2" FIP x 20", PVC faucet supply | P1@.250 | Ea | 4.30 | 7.88 | 12.18 | 20.70 |

## Vanity Cabinets Without Top

**Springfield vanity.** White finish. No assembly required. Maple finished interior with concealed hinges. Add the cost of countertop, basin and plumbing. 32" high.

| | | | | | | |
|---|---|---|---|---|---|---|
| 18" wide x 16" deep, 1 door | BC@.400 | Ea | 64.60 | 12.80 | 77.40 | 132.00 |
| 24" wide x 18" deep, 2 doors | BC@.400 | Ea | 85.80 | 12.80 | 98.60 | 168.00 |
| 30" wide x 18" deep, 3 doors | BC@.400 | Ea | 96.60 | 12.80 | 109.40 | 186.00 |

**Charleston oak vanity.** Fully assembled. Maple finished interior with concealed hinges. Includes chrome knobs. Add the cost of countertop, basin and plumbing. 32" high.

| | | | | | | |
|---|---|---|---|---|---|---|
| 24" wide x 18" deep, 2 doors | BC@.400 | Ea | 85.70 | 12.80 | 98.50 | 167.00 |
| 30" wide x 18" deep, 2 doors | BC@.400 | Ea | 96.60 | 12.80 | 109.40 | 186.00 |

**Kingston vanity.** White finish. Fully assembled. Maple finished interior with concealed hinges. Includes decorative door hardware. Add the cost of countertop, basin and plumbing. 18" deep. 34" high.

| | | | | | | |
|---|---|---|---|---|---|---|
| 24" wide, 1 door, 2 drawers | BC@.400 | Ea | 140.00 | 12.80 | 152.80 | 260.00 |
| 30" wide, 1 door, 2 drawers | BC@.400 | Ea | 162.00 | 12.80 | 174.80 | 297.00 |
| 36" wide, 2 doors, 2 drawers | BC@.400 | Ea | 183.00 | 12.80 | 195.80 | 333.00 |

| | Craft@Hrs | Unit | Material | Labor | Total | Sell |
|---|---|---|---|---|---|---|

**Arkansas white vanity cabinet.** Fully assembled, ready to install. Front surfaced in rigid thermofoil vinyl. Deluxe raised panel doors match drawer fronts. Extra deep drawers. Concealed and adjustable steel European style hinges. Chrome hardware included. Add the cost of countertop, basin and plumbing. 18" deep. 34" high.

| | Craft@Hrs | Unit | Material | Labor | Total | Sell |
|---|---|---|---|---|---|---|
| 24" wide, 1 door, 2 drawers | BC@.400 | Ea | 138.00 | 12.80 | 150.80 | 256.00 |
| 30" wide, 1 door, 2 drawers | BC@.400 | Ea | 157.00 | 12.80 | 169.80 | 289.00 |
| 36" wide, 2 doors, 2 drawers | BC@.400 | Ea | 181.00 | 12.80 | 193.80 | 329.00 |

**Monterey vanity.** White. Fully assembled. Maple finished interior with fully adjustable concealed hinges. Extra durable thick plywood drawer box. Countertop, faucet and basin sold separately. 21" deep. 32" high. Add the cost of countertop, basin and plumbing.

| | Craft@Hrs | Unit | Material | Labor | Total | Sell |
|---|---|---|---|---|---|---|
| 24" wide, 1 door, 2 drawers | BC@.400 | Ea | 201.00 | 12.80 | 213.80 | 363.00 |
| 30" wide, 1 door, 2 drawers | BC@.400 | Ea | 211.00 | 12.80 | 223.80 | 380.00 |
| 36" wide, 1 door, 2 drawers | BC@.400 | Ea | 249.00 | 12.80 | 261.80 | 445.00 |
| 48" wide, 2 doors, 4 drawers | BC@.500 | Ea | 324.00 | 16.00 | 340.00 | 578.00 |
| 60" wide, 2 doors, 4 drawers | BC@.500 | Ea | 406.00 | 16.00 | 422.00 | 717.00 |

**Monterey oak vanity.** Fully assembled. Maple finished interior with fully adjustable concealed hinges. Extra-durable thick plywood drawer box. Add the cost of countertop, basin and plumbing. 21" deep. 32" high.

| | Craft@Hrs | Unit | Material | Labor | Total | Sell |
|---|---|---|---|---|---|---|
| 24" wide, 1 door, 2 drawers | BC@.400 | Ea | 200.00 | 12.80 | 212.80 | 362.00 |
| 30" wide, 1 door, 2 drawers | BC@.400 | Ea | 211.00 | 12.80 | 223.80 | 380.00 |
| 36" wide, 1 door, 2 drawers | BC@.400 | Ea | 248.00 | 12.80 | 260.80 | 443.00 |
| 48" wide, 1 door, 4 drawers | BC@.500 | Ea | 324.00 | 16.00 | 340.00 | 578.00 |
| 60" wide, 1 door, 4 drawers | BC@.500 | Ea | 406.00 | 16.00 | 422.00 | 717.00 |

**Monterey maple vanity.** Fully assembled, ready to install. Solid maple and veneer door drawers and header. Large-capacity drawers. Solid maple face frame. Concealed adjustable hinges. Add the cost of countertop, basin and plumbing. 21" deep. 32" high.

| | Craft@Hrs | Unit | Material | Labor | Total | Sell |
|---|---|---|---|---|---|---|
| 24" wide, 1 door, 2 drawers | BC@.400 | Ea | 215.00 | 12.80 | 227.80 | 387.00 |
| 30" wide, 1 door, 2 drawers | BC@.400 | Ea | 239.00 | 12.80 | 251.80 | 428.00 |
| 36" wide, 1 door, 2 drawers | BC@.400 | Ea | 271.00 | 12.80 | 283.80 | 482.00 |
| 48" wide, 1 door, 4 drawers | BC@.500 | Ea | 378.00 | 16.00 | 394.00 | 670.00 |
| 60" wide, 1 door, 4 drawers | BC@.500 | Ea | 436.00 | 16.00 | 452.00 | 768.00 |

**Danville vanity with bottom drawer.** Fully assembled. Durable white finish. Raised arch panel door. Patented deep storage bottom drawer. Concealed adjustable hinges. 21" deep. 32" high. Add the cost of countertop, basin and plumbing.

| | Craft@Hrs | Unit | Material | Labor | Total | Sell |
|---|---|---|---|---|---|---|
| 24", 2 doors, 1 drawer | BC@.400 | Ea | 212.00 | 12.80 | 224.80 | 382.00 |
| 30", 1 door, 4 drawers | BC@.400 | Ea | 250.00 | 12.80 | 262.80 | 447.00 |
| 36", 2 doors, 4 drawers | BC@.400 | Ea | 284.00 | 12.80 | 296.80 | 505.00 |
| 48", 2 doors, 7 drawers | BC@.500 | Ea | 390.00 | 16.00 | 406.00 | 690.00 |

**Virginia bottom drawer oak vanity.** Solid oak and oak veneer. No assembly required. Maple finished interior with concealed hinges. Ball bearing glides on bottom drawer. Add the cost of countertop, basin and plumbing. 21" deep. 32" high.

| | Craft@Hrs | Unit | Material | Labor | Total | Sell |
|---|---|---|---|---|---|---|
| 24" wide, 1 door, 2 drawers | BC@.400 | Ea | 211.00 | 12.80 | 223.80 | 380.00 |
| 30" wide, 1 door, 3 drawers | BC@.400 | Ea | 251.00 | 12.80 | 263.80 | 448.00 |
| 36" wide, 2 doors, 3 drawers | BC@.400 | Ea | 283.00 | 12.80 | 295.80 | 503.00 |
| 48" wide, 2 doors, 7 drawers | BC@.500 | Ea | 390.00 | 16.00 | 406.00 | 690.00 |

| | Craft@Hrs | Unit | Material | Labor | Total | Sell |
|---|---|---|---|---|---|---|

## Cultured Marble Vanity Top

**Recessed eclipse oval bowl cultured marble vanity top.** Radius edge detail. Includes overflow. Labor includes connecting the top to the drain. Dove white.

| | Craft@Hrs | Unit | Material | Labor | Total | Sell |
|---|---|---|---|---|---|---|
| 25" wide x 22" deep | P1@.654 | Ea | 87.00 | 20.60 | 107.60 | 183.00 |
| 31" wide x 22" deep | P1@.654 | Ea | 97.50 | 20.60 | 118.10 | 201.00 |
| 37" wide x 22" deep | P1@.654 | Ea | 119.00 | 20.60 | 139.60 | 237.00 |
| 49" wide x 22" deep | P1@.654 | Ea | 152.00 | 20.60 | 172.60 | 293.00 |
| 61" wide x 22" deep, two bowls | P1@1.25 | Ea | 217.00 | 39.40 | 256.40 | 436.00 |
| 22" long side splash | P1@.200 | Ea | 27.00 | 6.30 | 33.30 | 56.60 |

**Neptune shell marble countertop.** Traditional scalloped shell bowl. 3-step edge detail. Includes overflow. Labor includes connecting the top to the drain. Glacier white.

| | Craft@Hrs | Unit | Material | Labor | Total | Sell |
|---|---|---|---|---|---|---|
| 25" wide x 22" deep | P1@.654 | Ea | 96.80 | 20.60 | 117.40 | 200.00 |
| 31" wide x 22" deep | P1@.654 | Ea | 108.00 | 20.60 | 128.60 | 219.00 |
| 37" wide x 22" deep | P1@.654 | Ea | 129.00 | 20.60 | 149.60 | 254.00 |
| 49" wide x 22" deep | P1@.654 | Ea | 162.00 | 20.60 | 182.60 | 310.00 |
| 22" long side splash | P1@.200 | Ea | 17.50 | 6.30 | 23.80 | 40.50 |

**Newport cultured marble vanity top.** White. Labor includes connecting the top to the drain.

| | Craft@Hrs | Unit | Material | Labor | Total | Sell |
|---|---|---|---|---|---|---|
| 19" wide x 17" deep | P1@.654 | Ea | 28.60 | 20.60 | 49.20 | 83.60 |
| 25" wide x 19" deep | P1@.654 | Ea | 52.00 | 20.60 | 72.60 | 123.00 |
| 31" wide x 19" deep | P1@.654 | Ea | 63.70 | 20.60 | 84.30 | 143.00 |
| 37" wide x 19" deep | P1@.654 | Ea | 74.50 | 20.60 | 95.10 | 162.00 |
| 19" long side splash | P1@.200 | Ea | 14.00 | 6.30 | 20.30 | 34.50 |

**Rectangular cultured marble vanity top.** Solid cultured marble. Heat- and stain-resistant. Labor includes connecting the top to the drain. Faucet sold separately. 19" deep.

| | Craft@Hrs | Unit | Material | Labor | Total | Sell |
|---|---|---|---|---|---|---|
| 25" wide, dove white | P1@.654 | Ea | 54.40 | 20.60 | 75.00 | 128.00 |
| 31" wide, dove white | P1@.654 | Ea | 65.20 | 20.60 | 85.80 | 146.00 |
| 37" wide, dove white | P1@.654 | Ea | 76.00 | 20.60 | 96.60 | 164.00 |
| 19" side splash, dove white | P1@.200 | Ea | 15.10 | 6.30 | 21.40 | 36.40 |
| 25" wide, pearl onyx | P1@.654 | Ea | 96.80 | 20.60 | 117.40 | 200.00 |
| 31" wide, pearl onyx | P1@.654 | Ea | 107.00 | 20.60 | 127.60 | 217.00 |
| 37" wide, pearl onyx | P1@.654 | Ea | 116.00 | 20.60 | 136.60 | 232.00 |
| 19" side splash, pearl onyx | P1@.200 | Ea | 7.75 | 6.30 | 14.05 | 23.90 |

**Rectangular premium cultured marble vanity top.** 4" backsplash. Overflow drain. 1" lip on three sides to guard against drips. Labor includes connecting the top to the drain. Solid white. Faucet sold separately.

| | Craft@Hrs | Unit | Material | Labor | Total | Sell |
|---|---|---|---|---|---|---|
| 25" wide x 22" deep | P1@.654 | Ea | 78.60 | 20.60 | 99.20 | 169.00 |
| 31" wide x 22" deep | P1@.654 | Ea | 86.80 | 20.60 | 107.40 | 183.00 |
| 49" wide x 22" deep | P1@.654 | Ea | 141.00 | 20.60 | 161.60 | 275.00 |
| 22" long side splash | P1@.200 | Ea | 15.00 | 6.30 | 21.30 | 36.20 |

**Shell bowl cultured marble vanity top.** White. Labor includes connecting the top to the drain.

| | Craft@Hrs | Unit | Material | Labor | Total | Sell |
|---|---|---|---|---|---|---|
| 25" wide x 19" deep | P1@.654 | Ea | 64.80 | 20.60 | 85.40 | 145.00 |
| 31" wide x 19" deep | P1@.654 | Ea | 75.60 | 20.60 | 96.20 | 164.00 |
| 37" wide x 19" deep | P1@.654 | Ea | 86.40 | 20.60 | 107.00 | 182.00 |
| 37" wide x 22" deep, offset left | P1@.654 | Ea | 129.00 | 20.60 | 149.60 | 254.00 |
| 37" wide x 22" deep, offset right | P1@.654 | Ea | 117.00 | 20.60 | 137.60 | 234.00 |

| | Craft@Hrs | Unit | Material | Labor | Total | Sell |
|---|---|---|---|---|---|---|
| **Shell bowl cultured marble vanity top.** White. Labor includes connecting the top to the drain. | | | | | | |
| 31" wide x 22" deep | P1@.654 | Ea | 96.10 | 20.60 | 116.70 | 198.00 |
| 37" wide x 22" deep | P1@.654 | Ea | 107.00 | 20.60 | 127.60 | 217.00 |
| 49" wide x 22" deep | P1@.654 | Ea | 139.00 | 20.60 | 159.60 | 271.00 |
| 22" long side splash | P1@.200 | Ea | 14.90 | 6.30 | 21.20 | 36.00 |
| **Aspen cultured marble vanity top and bowl.** Swirl finish. Labor includes connecting the top to the drain. | | | | | | |
| 31" wide x 22" deep | P1@.654 | Ea | 107.00 | 20.60 | 127.60 | 217.00 |
| 37" wide x 22" deep | P1@.654 | Ea | 129.00 | 20.60 | 149.60 | 254.00 |
| 49" wide x 22" deep | P1@.654 | Ea | 162.00 | 20.60 | 182.60 | 310.00 |
| 22" long side splash | P1@.200 | Ea | 17.00 | 6.30 | 23.30 | 39.60 |
| **Recessed oval bowl cultured marble vanity top.** Pearl onyx. Labor includes connecting the top to the drain. | | | | | | |
| 31" wide x 22" deep | P1@.654 | Ea | 130.00 | 20.60 | 150.60 | 256.00 |
| 37" wide x 22" deep | P1@.654 | Ea | 152.00 | 20.60 | 172.60 | 293.00 |
| 49" wide x 22" deep | P1@.654 | Ea | 206.00 | 20.60 | 226.60 | 385.00 |
| 22" long side splash | P1@.200 | Ea | 29.40 | 6.30 | 35.70 | 60.70 |

## Solid Surface Vanity Tops

| | Craft@Hrs | Unit | Material | Labor | Total | Sell |
|---|---|---|---|---|---|---|
| **Solid surface vanity sink top.** 22" deep. Oval halo bowl. 1-1/4" premium edge detail. | | | | | | |
| 25" wide, saddle | BC@.377 | Ea | 147.00 | 12.10 | 159.10 | 270.00 |
| 31" wide, saddle | BC@.468 | Ea | 151.00 | 15.00 | 166.00 | 282.00 |
| 37" wide, saddle | BC@.558 | Ea | 183.00 | 17.90 | 200.90 | 342.00 |
| 49" wide, saddle | BC@.740 | Ea | 215.00 | 23.70 | 238.70 | 406.00 |
| 61" wide, saddle, double bowl | B1@.920 | Ea | 308.00 | 27.00 | 335.00 | 570.00 |
| 22" long side splash, saddle | BC@.332 | Ea | 21.10 | 10.70 | 31.80 | 54.10 |
| 31" wide, wheat | BC@.468 | Ea | 151.00 | 15.00 | 166.00 | 282.00 |
| 37" wide, wheat | BC@.558 | Ea | 184.00 | 17.90 | 201.90 | 343.00 |
| 49" wide, wheat | BC@.740 | Ea | 216.00 | 23.70 | 239.70 | 407.00 |
| 22" long side splash, wheat | BC@.332 | Ea | 21.40 | 10.70 | 32.10 | 54.60 |
| **Granite vanity top.** 4" center. | | | | | | |
| 31" wide, beige | BC@.400 | Ea | 172.00 | 12.80 | 184.80 | 314.00 |
| 37" wide, beige | BC@.400 | Ea | 193.00 | 12.80 | 205.80 | 350.00 |

## Vanity Cabinets with Countertop and Basin

**Raised panel vanity with white marble top.** Berkeley series. Solid oak and veneer raised panel doors. 31" high. Frameless construction with no exposed hinges. Maple finish vanity interior. Assembled. Includes decorative hardware. Faucet sold separately. Labor cost includes setting in place only.

| | Craft@Hrs | Unit | Material | Labor | Total | Sell |
|---|---|---|---|---|---|---|
| 18" wide x 16" deep | BC@.400 | Ea | 88.40 | 12.80 | 101.20 | 172.00 |
| 24" wide x 18" deep | BC@.400 | Ea | 111.00 | 12.80 | 123.80 | 210.00 |
| 30" wide x 18" deep | BC@.400 | Ea | 145.00 | 12.80 | 157.80 | 268.00 |

**Raised panel vanity with white marble top.** Del Mar series. White rigid thermofoil raised panel doors. 31" high. Frameless construction with no exposed hinges. Maple finish vanity interior. Fully assembled. Includes decorative door hardware. Faucet sold separately. Labor cost includes setting in place only.

| | Craft@Hrs | Unit | Material | Labor | Total | Sell |
|---|---|---|---|---|---|---|
| 18" wide x 16" deep | BC@.400 | Ea | 89.00 | 12.80 | 101.80 | 173.00 |
| 24" wide x 18" deep | BC@.400 | Ea | 111.00 | 12.80 | 123.80 | 210.00 |
| 30" wide x 18" deep | BC@.400 | Ea | 145.00 | 12.80 | 157.80 | 268.00 |

| | Craft@Hrs | Unit | Material | Labor | Total | Sell |
|---|---|---|---|---|---|---|

**Vanity and cultured marble top.** Colorado style. Fully assembled. Faucet sold separately. 31" high. Oak finish. Labor cost includes setting in place only.

| | Craft@Hrs | Unit | Material | Labor | Total | Sell |
|---|---|---|---|---|---|---|
| 18" wide x 16" deep | BC@.400 | Ea | 85.30 | 12.80 | 98.10 | 167.00 |
| 24" wide x 18" deep | BC@.400 | Ea | 107.00 | 12.80 | 119.80 | 204.00 |
| 30" wide x 18" deep | BC@.400 | Ea | 150.00 | 12.80 | 162.80 | 277.00 |
| 30" wide, slide drawer | BC@.400 | Ea | 182.00 | 12.80 | 194.80 | 331.00 |
| 36" wide, slide drawer | BC@.400 | Ea | 204.00 | 12.80 | 216.80 | 369.00 |

**White vanity with cultured marble top.** Arkansas style. Rigid thermofoil front. Raised square panel door. Concealed 35mm European hinges. Glue and dowel construction. Full overlay frameless doors. Fully assembled. Faucet sold separately. Labor cost includes setting in place only.

| | Craft@Hrs | Unit | Material | Labor | Total | Sell |
|---|---|---|---|---|---|---|
| 18" wide x 16" deep | BC@.400 | Ea | 85.30 | 12.80 | 98.10 | 167.00 |
| 24" wide x 18" deep | BC@.400 | Ea | 107.00 | 12.80 | 119.80 | 204.00 |
| 30" wide x 18" deep | BC@.400 | Ea | 150.00 | 12.80 | 162.80 | 277.00 |
| 36" wide x 18" deep | BC@.400 | Ea | 204.00 | 12.80 | 216.80 | 369.00 |

**Vanity cabinet with top.** Cultured marble top. Solid doors and face frame. Brass hardware. Faucet sold separately. Labor cost includes setting in place only.

| | Craft@Hrs | Unit | Material | Labor | Total | Sell |
|---|---|---|---|---|---|---|
| 24" wide x 18" deep x 31" high, oak | BC@.400 | Ea | 107.00 | 12.80 | 119.80 | 204.00 |
| 24" wide x 18" deep x 31" high, white | BC@.400 | Ea | 107.00 | 12.80 | 119.80 | 204.00 |

**Vanity cabinet with cultured marble top and mirror.** 2-door, 1-drawer vanity cabinet. White melamine finish. Fully assembled bottom drawer vanity cabinet. Matching decorative Heritage-style vanity and mirror. Durable European style cultured marble top. Extended Euro top provides extra surface area for accessories. 21-1/2" wide x 23-1/2" high mirror. Faucet sold separately. Labor cost includes setting in place only.

| | Craft@Hrs | Unit | Material | Labor | Total | Sell |
|---|---|---|---|---|---|---|
| 24" wide x 13" deep x 31" high | BC@.400 | Ea | 216.00 | 12.80 | 228.80 | 389.00 |

## Bath Exhaust Fans

**ValueTest™ economy ceiling and wall exhaust fan, NuTone.** For baths, utility, and recreation rooms. Non-ducted. For baths up to 45 square feet, other rooms up to 60 square feet. Installs in ceiling or in wall. Plastic duct collar. White polymeric grille. UL listed for use in tub and shower enclosure with GFCI branch circuit wiring. Housing dimensions: 8-1/16" length, 7-3/16" width, 3-7/8" depth. Grille size, 8-11/16" x 9-1/2". 3" duct. 0.75 amp. Labor includes setting and connecting only. Add the cost of wiring and ducting. CFM (cubic feet of air per minute).

| | Craft@Hrs | Unit | Material | Labor | Total | Sell |
|---|---|---|---|---|---|---|
| 50 CFM, 2.5 sones | BE@1.00 | Ea | 14.20 | 34.40 | 48.60 | 82.60 |
| 70 CFM, 4.0 sones | BE@1.00 | Ea | 30.87 | 34.40 | 65.27 | 111.00 |

**Economy ceiling or wall bath fan, Broan Manufacturing.** Steel housing, with permanently lubricated motor and built-in damper. Torsion-spring mounted white polymeric grille. Built-in mounting ears. Housing dimensions: 7-1/4" length, 7-1/2" width, 3-5/8" depth. UL listed to use over bathtubs and showers when connected to a GFCI-protected branch circuit. For bathrooms up to 45 square feet. Labor includes setting and connecting only. Add the cost of wiring and ducting. CFM (cubic feet of air per minute).

| | Craft@Hrs | Unit | Material | Labor | Total | Sell |
|---|---|---|---|---|---|---|
| 50 CFM, 2.0 sones | BE@1.00 | Ea | 13.80 | 34.40 | 48.20 | 81.90 |

**Ceiling or wall bath fan, Broan Manufacturing.** Steel housing, with permanently lubricated motor and built-in damper. Centrifugal blower wheel. Torsion-spring mounted white polymeric grille. Built-in double steel mounting ears with keyhole slots. Fits 3" round ducts. Housing dimensions: 7-1/4" length, 7-1/2" width, 3-5/8" depth. UL listed for use over bathtubs and showers when connected to a GFCI-protected branch circuit. Labor includes setting and connecting only. Add the cost of wiring and ducting. CFM (cubic feet of air per minute).

| | Craft@Hrs | Unit | Material | Labor | Total | Sell |
|---|---|---|---|---|---|---|
| 70 CFM, 3.0 sones | BE@1.00 | Ea | 25.90 | 34.40 | 60.30 | 103.00 |

| | Craft@Hrs | Unit | Material | Labor | Total | Sell |
|---|---|---|---|---|---|---|

**Vertical discharge bath fan, Broan Manufacturing.** Galvanized steel housing, with built-in damper and spin-on white polymeric grille. Polymeric fan blade and duct connectors. Built-in double steel mounting ears with keyhole slots. Fits 8" ducts. UL listed. Labor includes setting and connecting only. Add the cost of wiring and ducting. CFM (cubic feet of air per minute).

| | Craft@Hrs | Unit | Material | Labor | Total | Sell |
|---|---|---|---|---|---|---|
| 55 CFM, 4.0 sones | BE@1.00 | Ea | 25.90 | 34.40 | 60.30 | 103.00 |
| 180 CFM, 5.0 sones | BE@1.00 | Ea | 76.70 | 34.40 | 111.10 | 189.00 |

**ValueTest™ vertical discharge bath fan, NuTone.** For 75 square foot bath. Mounts in ceiling, discharges through duct to roof or wall. White polymeric grille. Fits 7" round ducts. Housing dimensions: 6-15/16" diameter x 6-1/2" length. Grille size, 8-11/16" x 9-1/2". UL listed for use in tub/shower when used with GFI branch circuit wiring. Not recommended for kitchen use. Labor includes setting and connecting only. Add the cost of wiring and ducting. CFM (cubic feet per minute).

| | Craft@Hrs | Unit | Material | Labor | Total | Sell |
|---|---|---|---|---|---|---|
| 80 CFM, 6.0 sones | BE@1.00 | Ea | 30.50 | 34.40 | 64.90 | 110.00 |

**Vertical discharge utility fan, NuTone.** Mounts in ceiling. Discharges through duct to roof or wall. Pre-wired motor with plug-in receptacle. Adjustable hanger bars for 16" or 24" on-center joists. Silver anodized aluminum grille. Fits 7" round ducts. Housing dimensions: 11" diameter x 5-5/8" deep. UL listed. Labor includes setting and connecting only. Add the cost of wiring and ducting. CFM (cubic feet of air per minute).

| | Craft@Hrs | Unit | Material | Labor | Total | Sell |
|---|---|---|---|---|---|---|
| 210 CFM, 6.5 sones | BE@1.00 | Ea | 82.90 | 34.40 | 117.30 | 199.00 |

**Ceiling or wall bath exhaust fan, NuTone.** White polymeric grille. Torsion spring grille mounting requires no tools. Plug-in, permanently lubricated motor. Centrifugal blower wheel. Rugged, 26 gauge galvanized steel housing. Keyhole mounting brackets for quick, accurate installation. Tapered, polymeric duct fitting with built-in backdraft damper. U.L. listed for use over bathtubs and showers when connected to a GFCI-protected branch circuit. 120 volts, 0.5 amps, 2.5 sones, 80 CFM (HVI-2100 certified) and 4" round duct. Labor includes setting and connecting only. Add the cost of wiring and ducting. CFM (cubic feet of air per minute).

| | Craft@Hrs | Unit | Material | Labor | Total | Sell |
|---|---|---|---|---|---|---|
| 80 CFM, 2.5 sones | BE@1.00 | Ea | 68.00 | 34.40 | 102.40 | 174.00 |

**Bath exhaust fan, Broan Manufacturing.** Aluminum centrifugal blower wheel powered by 4-pole 1500-RPM low-sound motor. Fits 4" ducts. White polymeric designer grille, torsion-spring mounted. Housing dimensions: 8" length, 8-1/4" width, 5-3/4" depth. UL listed for use over bathtubs and showers when connected to a GFCI-protected branch circuit. Labor includes setting and connecting only. Add the cost of wiring and ducting. CFM (cubic feet of air per minute).

| | Craft@Hrs | Unit | Material | Labor | Total | Sell |
|---|---|---|---|---|---|---|
| 110 CFM, 4.0 sones | BE@1.00 | Ea | 73.50 | 34.40 | 107.90 | 183.00 |

**Solitaire™ bath exhaust fan, Broan Manufacturing.** Pre-wired outlet box for plug-in motor plate, torsion spring-mounted low-profile grille. Steel housing, damper to eliminate back drafts. Housing: 9-1/4" x 9-1/2" x 7-5/8". AMCA licensed for air and sound. UL listed to use over tub or shower enclosure with GFCI-protected branch circuit. HVI Certified. 4" duct. Labor includes setting and connecting only. Add the cost of wiring and ducting. CFM (cubic feet of air per minute).

| | Craft@Hrs | Unit | Material | Labor | Total | Sell |
|---|---|---|---|---|---|---|
| 80 CFM, 2.5 sones | BE@1.00 | Ea | 95.90 | 34.40 | 130.30 | 222.00 |
| 80 CFM, 3.5 sones | BE@1.00 | Ea | 106.00 | 34.40 | 140.40 | 239.00 |
| 90 CFM, 1.5 sones | BE@1.00 | Ea | 90.00 | 34.40 | 124.40 | 211.00 |
| 120 CFM, 2.5 sones | BE@1.00 | Ea | 156.00 | 34.40 | 190.40 | 324.00 |
| 130 CFM, 2.5 sones | BE@1.00 | Ea | 114.00 | 34.40 | 148.40 | 252.00 |

**Solitaire™ Ultra Silent® low-sound bath fan, Broan Manufacturing.** Centrifugal blower. 7-5/8" high housing. UL listed for use over bathtubs and showers when connected to a GFCI-protected branch circuit. AMCA licensed for both air and sound. 4" duct. Labor includes setting and connecting only. Add the cost of wiring and ducting. CFM (cubic feet of air per minute).

| | Craft@Hrs | Unit | Material | Labor | Total | Sell |
|---|---|---|---|---|---|---|
| 80 CFM, 0.6 sones | BE@1.00 | Ea | 124.00 | 34.40 | 158.40 | 269.00 |
| 110 CFM, 1.5 sones | BE@1.00 | Ea | 150.00 | 34.40 | 184.40 | 313.00 |

| | Craft@Hrs | Unit | Material | Labor | Total | Sell |
|---|---|---|---|---|---|---|

**Thru-the-wall utility fan, Broan Manufacturing.** Steel housing with built-in damper and white polymeric grille. Permanently lubricated motor. Rotary on and off switch. Housing dimensions: 14-1/4" long, 14-1/4" wide, 4-1/2" deep. UL listed. Labor includes setting and connecting only. Add the cost of wiring and exterior finish. CFM (cubic feet of air per minute).

| | Craft@Hrs | Unit | Material | Labor | Total | Sell |
|---|---|---|---|---|---|---|
| 180 CFM, 5.0 sones | BE@1.00 | Ea | 74.70 | 34.40 | 109.10 | 185.00 |

**Chain-operated wall exhaust fan, Broan Manufacturing.** Steel housing. Low-profile white polymeric grille. Permanently lubricated motor. Pull-chain opens door and turns on fan. Fits walls from 4-1/2" to 9-1/2" thick. Labor includes setting and connecting only. Add the cost of wiring and ducting. CFM (cubic feet of air per minute).

| | Craft@Hrs | Unit | Material | Labor | Total | Sell |
|---|---|---|---|---|---|---|
| 8", 250 CFM, 4.5 sones | BE@.850 | Ea | 111.00 | 29.20 | 140.20 | 238.00 |

**QuieTTest® low-sound bath fan, NuTone.** For 105 square foot bath. Pre-wired outlet box with plug-in receptacle. Adjustable hanger brackets. Fits 4" round ducts. Low-profile white polymeric grille, torsion-spring mounted. Housing dimensions: 9-3/8" length, 11-1/4" width, 7-7/8" depth. Grille size, 14-1/4" x 12-1/16". UL listed for use in tub and shower when used with GFI branch circuit wiring. Not recommended for kitchen use. Labor includes setting and connecting only. Add the cost of wiring and ducting. CFM (cubic feet of air per minute).

| | Craft@Hrs | Unit | Material | Labor | Total | Sell |
|---|---|---|---|---|---|---|
| 110 CFM, 2.0 sones | BE@1.00 | Ea | 130.00 | 34.40 | 164.40 | 279.00 |

**QuieTTest® low-sound ceiling blower, NuTone.** For 375 square foot room. Pre-wired outlet box with plug-in receptacle. Rounded, low-profile white polymeric grille with silver anodized trim at each end. Housing dimensions: 14-1/4" length, 10" width, 9" depth. Grille size, 16-1/2" x 12-3/92". Labor includes setting and connecting only. Add the cost of wiring and ducting. CFM (cubic feet of air per minute).

| | Craft@Hrs | Unit | Material | Labor | Total | Sell |
|---|---|---|---|---|---|---|
| 300 CFM, 4.5 sones | BE@1.00 | Ea | 154.00 | 34.40 | 188.40 | 320.00 |

## Lighted Bath Exhaust Fans

**ValueTest™ economy bath fan and light, NuTone.** Fan and light operate separately or together. Fits 4" ducts. White polymeric grille with break-resistant lens. Uses 100-watt lamp (not included). Housing dimensions: 9" length, 9" width, 5-1/2" depth. Grille size, 10-3/4" x 12-1/8". UL listed for use in tub and shower when used with GFI branch circuit wiring. Labor includes setting and connecting only. Add the cost of wiring and ducting. CFM (cubic feet of air per minute).

| | Craft@Hrs | Unit | Material | Labor | Total | Sell |
|---|---|---|---|---|---|---|
| 50 CFM, 2.5 Sones, 45 SF bath | BE@1.00 | Ea | 40.70 | 34.40 | 75.10 | 128.00 |
| 70 CFM, 4.5 Sones, 65 SF bath | BE@1.00 | Ea | 65.80 | 34.40 | 100.20 | 170.00 |

**Decorative globe bath fan and light, NuTone.** Decorative glass fixture with a powerful exhaust fan. The fan exhausts through inconspicuous openings in the base. Corrosion-resistant gloss-white finish base with frosted melon glass globe. Can be installed throughout the house in bathrooms, bedrooms or hallways. Uses two standard 60-watt bulbs (sold separately). Labor includes setting and connecting only. Add the cost of wiring and ducting. CFM (cubic feet of air per minute).

| | Craft@Hrs | Unit | Material | Labor | Total | Sell |
|---|---|---|---|---|---|---|
| 70 CFM, 3.5 sones, white | BE@.850 | Ea | 95.30 | 29.20 | 124.50 | 212.00 |

**Exhaust Air deluxe bath fan with light, NuTone.** Ventilation for baths up to 95 square feet, other rooms up to 125 square feet. 100-watt ceiling and 7-watt night-light or energy-saving fluorescent light (lamps not included). Snap-on grille assembly. Housing dimensions: 9" length, 9" width, 6" depth. Polymeric white grille, 15" diameter. UL listed for use in tub and shower when used with GFI branch circuit wiring. Labor includes setting and connecting only. Add the cost of wiring and ducting. CFM (cubic feet of air per minute).

| | Craft@Hrs | Unit | Material | Labor | Total | Sell |
|---|---|---|---|---|---|---|
| 100 CFM, 3.5 sones | BE@1.00 | Ea | 129.00 | 34.40 | 163.40 | 278.00 |

**Economy bath exhaust fan and light, Broan Manufacturing.** Type IC. Polymeric impeller. Plug-in permanently lubricated motor. 13-watt double twin tube compact fluorescent bulb provides equivalent of 60 watts of incandescent light (bulb not included). White polymeric grille. UL listed for use over bathtubs and showers when connected to GFCI-protected branch circuit. 4" duct. Labor includes setting and connecting only. Add the cost of wiring and ducting. CFM (cubic feet of air per minute).

| | Craft@Hrs | Unit | Material | Labor | Total | Sell |
|---|---|---|---|---|---|---|
| 50 CFM, 3.5 sones | BE@1.00 | Ea | 40.00 | 34.40 | 74.40 | 126.00 |
| 70 CFM, 3.5 sones | BE@1.00 | Ea | 66.90 | 34.40 | 101.30 | 172.00 |

| | Craft@Hrs | Unit | Material | Labor | Total | Sell |
|---|---|---|---|---|---|---|

**Decorative bath exhaust fan with light, Broan Manufacturing.** Corrosion-resistant finish. Frosted melon glass globe. 60-watt max. Labor includes setting and connecting only. Add the cost of wiring and ducting. CFM (cubic feet of air per minute).

| | Craft@Hrs | Unit | Material | Labor | Total | Sell |
|---|---|---|---|---|---|---|
| 70 CFM, round globe, white | BE@1.00 | Ea | 94.50 | 34.40 | 128.90 | 219.00 |

**Designer Series bath fan with light, Broan Manufacturing.** With light and 7-watt night-light. Bulbs not included. Type IC permanently lubricated motor. Centrifugal blower for high performance at low sound levels. Quit damper prevents cold back drafts. Oak frame with multi-layer moisture protection. Steel housing with adjustable brackets that span to 24". 100-watt light with glass lens and low-profile polished brass finished grille assembly. Housing dimensions: 8" x 8-1/4" x 5-3/4". Labor includes setting and connecting only. Add the cost of wiring and ducting. CFM (cubic feet of air per minute).

| | Craft@Hrs | Unit | Material | Labor | Total | Sell |
|---|---|---|---|---|---|---|
| 80 CFM, 0.6 sones | BE@1.00 | Ea | 134.00 | 34.40 | 168.40 | 286.00 |
| 100 CFM, 3.5 sones | BE@1.00 | Ea | 139.00 | 34.40 | 173.40 | 295.00 |

**Solitaire™ bath fan with light, Broan Manufacturing.** 18-watt twin-tube fluorescent bulb, built-in 7-watt night-light (bulbs not included). Type IC balanced centrifugal blower. Permanently lubricated motor. 7-5/8"-high housing. Mounts directly to ceiling joists or spans up to 24" with slide bar brackets. 4" duct. Labor includes setting and connecting only. Add the cost of wiring and ducting. CFM (cubic feet of air per minute).

| | Craft@Hrs | Unit | Material | Labor | Total | Sell |
|---|---|---|---|---|---|---|
| 110 CFM, 1.5 sones | BE@1.00 | Ea | 204.00 | 34.40 | 238.40 | 405.00 |
| 120 CFM, 2.5 sones | BE@1.00 | Ea | 209.00 | 34.40 | 243.40 | 414.00 |

## Bath Exhaust Fans with Heater

**Infrared bulb heater and fan, Broan Manufacturing.** 4-point adjustable mounting brackets span up to 24". Uses 250-watt, 120-volt R-40 infrared bulbs (not included). Heater and fan units include 70 CFM (cubic feet of air per minute), 3.5 sones ventilator fan. Damper and duct connector included. Plastic matte-white molded grille and compact housings. Labor includes setting and connecting only. Add the cost of wiring and ducting.

| | Craft@Hrs | Unit | Material | Labor | Total | Sell |
|---|---|---|---|---|---|---|
| 1-bulb heater, 4" duct | BE@1.00 | Ea | 22.50 | 34.40 | 56.90 | 96.70 |
| 2-bulb heater, 7" duct | BE@1.10 | Ea | 68.00 | 37.80 | 105.80 | 180.00 |

**Bath heater and fan, Broan Manufacturing.** Steel housing. Adjustable mounting brackets. White polymeric grille. Quiet, efficient blower removes moisture and odors. Bulb sold separately. UL listed. Labor includes setting and connecting only. Add the cost of wiring and ducting. CFM (cubic feet of air per minute).

| | Craft@Hrs | Unit | Material | Labor | Total | Sell |
|---|---|---|---|---|---|---|
| 70 CFM, 3.5 sones, 1-bulb | BE@1.00 | Ea | 54.00 | 34.40 | 88.40 | 150.00 |

**Surface-mount ceiling bath heater and fan, Broan Manufacturing.** 1250-watts. 4266 Btus (British thermal units). 120-volts. Chrome alloy wire element for instant heat. Built-in fan. Automatic overheat protection. Low-profile housing. Permanently lubricated motor. Mounts to standard 3-1/4" round or 4" octagonal ceiling electrical box. Satin-finish aluminum grille extends 2-3/4" from ceiling. 11" diameter x 2-3/4" deep. Labor includes setting and connecting only. Add the cost of wiring and ducting.

| | Craft@Hrs | Unit | Material | Labor | Total | Sell |
|---|---|---|---|---|---|---|
| 10.7 amp | BE@.750 | Ea | 56.20 | 25.80 | 82.00 | 139.00 |

**Designer Series bath heater and exhaust fan, Broan Manufacturing.** 1500-watt fan-forced heater. 120-watt light capacity, 7-watt night-light (bulbs not included). Permanently lubricated motor. Polymeric damper prevents cold back drafts. 4-point adjustable mounting brackets with keyhole slots. Torsion-spring grille mounting, no tools needed. Non-glare, light diffusing glass lenses. Fits single gang opening. Suitable for use with insulation. Includes 4-function control unit. 70 CFM (cubic feet of air per minute), 3.5 sones. Labor includes setting and connecting only. Add the cost of wiring and ducting.

| | Craft@Hrs | Unit | Material | Labor | Total | Sell |
|---|---|---|---|---|---|---|
| 4" duct, 100-watt lamp | BE@1.00 | Ea | 80.40 | 34.40 | 114.80 | 195.00 |

**Heat-A-Lamp® bulb heater and fan, NuTone.** Non-IC. Swivel-mount. Torsion spring holds plates firmly to ceiling. 250-watt radiant heat from one R-40 infrared heat lamp. Uses 4" duct. Adjustable socket for ceilings up to 1" thick. Adjustable hanger bars. Automatic reset for thermal protection. White polymeric finish. UL listed. 6-7/8" high. Labor includes setting and connecting only. Add the cost of wiring and ducting.

| | Craft@Hrs | Unit | Material | Labor | Total | Sell |
|---|---|---|---|---|---|---|
| 2.6 amp, 1 lamp, 12-1/2" x 10" | BE@1.50 | Ea | 54.30 | 51.60 | 105.90 | 180.00 |
| 5.0 amp, 2 lamp, 15-3/8" x 11" | BE@1.50 | Ea | 65.30 | 51.60 | 116.90 | 199.00 |

| | Craft@Hrs | Unit | Material | Labor | Total | Sell |
|---|---|---|---|---|---|---|

**Deluxe Heat-A-Ventlite® heater and ventilator, NuTone.** With light and night-light. Type IC. 1500-watt fan-forced heat with 55-watt exhaust motor. White enamel grille. Includes wall switch with separate on-off controls for all functions. Housing dimensions: 13-1/4" diameter x 7-1/2". Frame size, 15-3/8" diameter. 5118-Btu (British thermal unit) heater. 100-watt light capacity, 7-watt night-light (lamps not included). Labor includes setting and connecting only. Add the cost of wiring and ducting. CFM (cubic feet of air per minute).

| | Craft@Hrs | Unit | Material | Labor | Total | Sell |
|---|---|---|---|---|---|---|
| 70 CFM, 3.5 sones | BE@1.50 | Ea | 205.00 | 51.60 | 256.60 | 436.00 |
| 70 CFM, 4.0 sones | BE@1.50 | Ea | 168.00 | 51.60 | 219.60 | 373.00 |

## Bath Exhaust Fan Accessories

**Timer switch, NuTone.**

| | Craft@Hrs | Unit | Material | Labor | Total | Sell |
|---|---|---|---|---|---|---|
| 60-minute timer | BE@.250 | Ea | 22.00 | 8.60 | 30.60 | 52.00 |

**Exhaust fan switch, Broan Manufacturing.**

| | Craft@Hrs | Unit | Material | Labor | Total | Sell |
|---|---|---|---|---|---|---|
| 60-minute timer, white | BE@.300 | Ea | 22.00 | 10.30 | 32.30 | 54.90 |

**SensAire® 3-function wall control switch, Broan Manufacturing.** Fits single gang opening. Top switch for on-auto-off. Other switches for light and night-light. Use with SensAire® fans and lights.

| | Craft@Hrs | Unit | Material | Labor | Total | Sell |
|---|---|---|---|---|---|---|
| White | BE@.250 | Ea | 17.20 | 8.60 | 25.80 | 43.90 |

**Bath vent kit, Deflect-O.** UL listed. Supurr-Flex® duct, aluminum roof vent, 3" to 4" increaser and 2 clamps. Fire resistant. Labor includes working around existing drywall, feeding the duct through existing attic space, and cutting through existing roofing.

| | Craft@Hrs | Unit | Material | Labor | Total | Sell |
|---|---|---|---|---|---|---|
| 4" x 8' flexible duct | BC@2.00 | Ea | 19.50 | 64.20 | 83.70 | 142.00 |

**Roof vent kit, NuTone.** For venting kitchen or bath exhaust fan through slanted roof. Works with both 3" and 4" ducted units. Labor includes working around existing drywall, feeding the duct through existing attic space, and cutting through existing roofing.

| | Craft@Hrs | Unit | Material | Labor | Total | Sell |
|---|---|---|---|---|---|---|
| 8' length | BC@2.00 | Ea | 20.50 | 64.20 | 84.70 | 144.00 |

## Mirrors

**Vanity mirror.** Polished edges.

| | Craft@Hrs | Unit | Material | Labor | Total | Sell |
|---|---|---|---|---|---|---|
| 24" wide x 36" high | BC@.331 | Ea | 22.40 | 10.60 | 33.00 | 56.10 |
| 36" wide x 36" high | BC@.331 | Ea | 35.50 | 10.60 | 46.10 | 78.40 |
| 36" wide x 42" high | BC@.331 | Ea | 43.40 | 10.60 | 54.00 | 91.80 |
| 48" wide x 36" high | BC@.331 | Ea | 49.20 | 10.60 | 59.80 | 102.00 |
| 60" wide x 36" high | BC@.331 | Ea | 57.00 | 10.60 | 67.60 | 115.00 |

**Frosted oval mirror.** Clear etched border accents the color of wall behind it. Ready to hang vertically or horizontally.

| | Craft@Hrs | Unit | Material | Labor | Total | Sell |
|---|---|---|---|---|---|---|
| 23" wide x 29" high | BC@.331 | Ea | 45.50 | 10.60 | 56.10 | 95.40 |
| 24" wide x 36" high | BC@.331 | Ea | 63.90 | 10.60 | 74.50 | 127.00 |

## Medicine Cabinets

**Stainless steel framed bath cabinet, mirrored swing door.** Polystyrene body. Recess or surface mount. Magnetic catch. 3 fixed shelves. Fits 14" x 18" wall opening. 4-3/4" deep.

| | Craft@Hrs | Unit | Material | Labor | Total | Sell |
|---|---|---|---|---|---|---|
| 16" wide x 20" high | BC@.950 | Ea | 13.00 | 30.50 | 43.50 | 74.00 |

**Stainless steel framed door medicine cabinet.** Plastic cabinet. Stainless steel door frame. Surface or recess mount. 2 adjustable shelves. Fits 13-1/2" x 23-1/2" wall opening. 4-1/2" deep.

| | Craft@Hrs | Unit | Material | Labor | Total | Sell |
|---|---|---|---|---|---|---|
| 16-1/8" wide x 26-1/8" high | BC@.950 | Ea | 24.00 | 30.50 | 54.50 | 92.70 |

| | Craft@Hrs | Unit | Material | Labor | Total | Sell |
|---|---|---|---|---|---|---|

**Frameless swing door medicine cabinet.** Polystyrene body. Frameless beveled swing door. Recess or surface mount. 1/2" beveled mirror. Magnetic catch. 2 adjustable shelves. Reversible left- or right-hand opening. 4-3/4" deep.

| | Craft@Hrs | Unit | Material | Labor | Total | Sell |
|---|---|---|---|---|---|---|
| 16" wide x 24" high | BC@.950 | Ea | 32.50 | 30.50 | 63.00 | 107.00 |

**Frameless beveled swing door medicine cabinet.** Polystyrene body with beveled mirror edges. 2 adjustable shelves. Recess or surface mount. 4-1/2" deep.

| | Craft@Hrs | Unit | Material | Labor | Total | Sell |
|---|---|---|---|---|---|---|
| 16" wide x 26" high | BC@.950 | Ea | 45.90 | 30.50 | 76.40 | 130.00 |

**Swing door medicine cabinet, double frame.** 2 fixed shelves. Recess or surface mount. All wood construction.

| | Craft@Hrs | Unit | Material | Labor | Total | Sell |
|---|---|---|---|---|---|---|
| 16" x 26" high, oak | BC@.950 | Ea | 41.00 | 30.50 | 71.50 | 122.00 |
| 16" x 26" high, white | BC@.950 | Ea | 40.90 | 30.50 | 71.40 | 121.00 |

**Chrome framed lighted sliding door cabinet.** White stainless steel cabinet. 2 fixed shelves. Surface mount. Includes on/off light switch and electrical outlet. Bulbs sold separately. 7-3/4" deep.

| | Craft@Hrs | Unit | Material | Labor | Total | Sell |
|---|---|---|---|---|---|---|
| 24" wide x 20" high | BC@.900 | Ea | 51.00 | 28.90 | 79.90 | 136.00 |

**Frameless octagonal swing door medicine cabinet.** Polystyrene body.

| | Craft@Hrs | Unit | Material | Labor | Total | Sell |
|---|---|---|---|---|---|---|
| 16" wide x 24" high | BC@.950 | Ea | 55.70 | 30.50 | 86.20 | 147.00 |

**Etched glass medicine cabinet.** Etched mirror panel. Adjustable shelves.

| | Craft@Hrs | Unit | Material | Labor | Total | Sell |
|---|---|---|---|---|---|---|
| 24" x 25" high, oak | BC@.950 | Ea | 63.90 | 30.50 | 94.40 | 160.00 |
| 24" x 25" high, white | BC@.950 | Ea | 63.90 | 30.50 | 94.40 | 160.00 |

**Frameless corner cabinet.** 4 fixed shelves. Surface mount.

| | Craft@Hrs | Unit | Material | Labor | Total | Sell |
|---|---|---|---|---|---|---|
| 14" wide x 36" high | BC@.950 | Ea | 91.90 | 30.50 | 122.40 | 208.00 |

**Etched glass bath storage cabinet.**

| | Craft@Hrs | Unit | Material | Labor | Total | Sell |
|---|---|---|---|---|---|---|
| 24" wide x 28" high, oak | BC@.950 | Ea | 75.50 | 30.50 | 106.00 | 180.00 |
| 24" wide x 28" high, white | BC@.950 | Ea | 75.60 | 30.50 | 106.10 | 180.00 |

**Frameless beveled tri-view mirror medicine cabinet.** Wood body. Surface mount. 2 adjustable shelves. Concealed adjustable hinges.

| | Craft@Hrs | Unit | Material | Labor | Total | Sell |
|---|---|---|---|---|---|---|
| 24-1/4" wide x 26" high | BC@.950 | Ea | 86.20 | 30.50 | 116.70 | 198.00 |
| 29-1/2" wide x 25-3/8" high | BC@.950 | Ea | 97.20 | 30.50 | 127.70 | 217.00 |
| 36" wide x 29-7/8" high | BC@.950 | Ea | 126.00 | 30.50 | 156.50 | 266.00 |

**Frameless beveled edge mirrored medicine cabinet.** Recess or surface mount.

| | Craft@Hrs | Unit | Material | Labor | Total | Sell |
|---|---|---|---|---|---|---|
| 20" x 26" high | BC@.950 | Ea | 129.00 | 30.50 | 159.50 | 271.00 |
| 24" x 30" high, aluminum | BC@.950 | Ea | 146.00 | 30.50 | 176.50 | 300.00 |

**Top lighted swing door bath cabinet.** Solid hardwood frame. Concealed adjustable hinges. 3 adjustable shelves. 3 bulbs, not included. 24" wide x 36" high.

| | Craft@Hrs | Unit | Material | Labor | Total | Sell |
|---|---|---|---|---|---|---|
| Maple, nickel-plated top light | BC@1.25 | Ea | 140.00 | 40.10 | 180.10 | 306.00 |
| Oak, brass-plated top light | BC@1.25 | Ea | 125.00 | 40.10 | 165.10 | 281.00 |
| White, chrome-plated top light | BC@1.25 | Ea | 125.00 | 40.10 | 165.10 | 281.00 |

**Single door hinged medicine cabinet.** Beveled front mirror. White aluminum interior. Surface mounted or recessed.

| | Craft@Hrs | Unit | Material | Labor | Total | Sell |
|---|---|---|---|---|---|---|
| 24" x 30" high, white | BC@.950 | Ea | 128.00 | 30.50 | 158.50 | 269.00 |

**Oval beveled edge medicine cabinet.** Aluminum body. Surface or recess mount. Labor cost for recess-mounting includes installation into an existing, correctly framed opening.

| | Craft@Hrs | Unit | Material | Labor | Total | Sell |
|---|---|---|---|---|---|---|
| 24" x 36" high | BC@.950 | Ea | 204.00 | 30.50 | 234.50 | 399.00 |

| | Craft@Hrs | Unit | Material | Labor | Total | Sell |
|---|---|---|---|---|---|---|

**Frameless beveled edge bi-view medicine cabinet.** Recess or surface mount. Adjustable glass shelves. Aluminum body. Labor cost for recess-mounting includes installation into an existing, correctly framed opening.

| | Craft@Hrs | Unit | Material | Labor | Total | Sell |
|---|---|---|---|---|---|---|
| 30" x 26" high, mirrored interior | BC@.950 | Ea | 168.00 | 30.50 | 198.50 | 337.00 |
| 30" x 26" high, white interior | BC@.950 | Ea | 139.00 | 30.50 | 169.50 | 288.00 |
| 36" x 26" high, mirrored interior | BC@.950 | Ea | 191.00 | 30.50 | 221.50 | 377.00 |
| 36" x 26" high, white Interior | BC@.950 | Ea | 161.00 | 30.50 | 191.50 | 326.00 |
| 48" x 26" high, mirrored interior | B1@.950 | Ea | 275.00 | 27.90 | 302.90 | 515.00 |

**Frameless bi-view medicine cabinet.** Recess or surface mount. Labor cost for recess-mounting includes installation into an existing, correctly framed opening.

| | Craft@Hrs | Unit | Material | Labor | Total | Sell |
|---|---|---|---|---|---|---|
| 30" x 30" high, silver | BC@.950 | Ea | 258.00 | 30.50 | 288.50 | 490.00 |

**Tri-view medicine cabinet.** 36" high. Decorative crown molding and beveled mirrors. Adjustable shelves. Surface mount.

| | Craft@Hrs | Unit | Material | Labor | Total | Sell |
|---|---|---|---|---|---|---|
| 30" wide, maple | BC@.950 | Ea | 161.00 | 30.50 | 191.50 | 326.00 |
| 30" wide, oak | BC@.950 | Ea | 136.00 | 30.50 | 166.50 | 283.00 |
| 30" wide, white | BC@.950 | Ea | 136.00 | 30.50 | 166.50 | 283.00 |
| 36" wide, maple | BC@.950 | Ea | 179.00 | 30.50 | 209.50 | 356.00 |
| 36" wide, oak | BC@.950 | Ea | 159.00 | 30.50 | 189.50 | 322.00 |
| 36" wide, white | BC@.950 | Ea | 158.00 | 30.50 | 188.50 | 320.00 |
| 48" wide, maple | B1@.950 | Ea | 212.00 | 27.90 | 239.90 | 408.00 |
| 48" wide, oak | B1@.950 | Ea | 179.00 | 27.90 | 206.90 | 352.00 |
| 48" wide, white | B1@.950 | Ea | 180.00 | 27.90 | 207.90 | 353.00 |

**Tri-view medicine cabinet, Keystone.** 3 beveled front mirrors. White aluminum interior. Surface mount or recessed. Labor cost for recess-mounting includes installation into an existing, correctly framed opening.

| | Craft@Hrs | Unit | Material | Labor | Total | Sell |
|---|---|---|---|---|---|---|
| 30" x 31" high, white | BC@.950 | Ea | 182.00 | 30.50 | 212.50 | 361.00 |
| 36" x 31" high, white | BC@.950 | Ea | 215.00 | 30.50 | 245.50 | 417.00 |
| 48" x 31" high, white | B1@.950 | Ea | 258.00 | 27.90 | 285.90 | 486.00 |

**Tri-view medicine cabinet with light bar.** Recess or surface mount. Built-in top light bar. Adjustable shelves. 36" high. Labor cost for recess-mounting includes installation into an existing, correctly framed opening.

| | Craft@Hrs | Unit | Material | Labor | Total | Sell |
|---|---|---|---|---|---|---|
| 30" wide, maple | BC@.950 | Ea | 178.00 | 30.50 | 208.50 | 354.00 |
| 30" wide, oak | BC@.950 | Ea | 151.00 | 30.50 | 181.50 | 309.00 |
| 30" wide, white | BC@.950 | Ea | 151.00 | 30.50 | 181.50 | 309.00 |
| 36" wide, maple | BC@.950 | Ea | 199.00 | 30.50 | 229.50 | 390.00 |
| 36" wide, oak | BC@.950 | Ea | 178.00 | 30.50 | 208.50 | 354.00 |
| 36" wide, white | BC@.950 | Ea | 178.00 | 30.50 | 208.50 | 354.00 |
| 48" wide, maple | B1@.950 | Ea | 226.00 | 27.90 | 253.90 | 432.00 |
| 48" wide, oak | B1@.950 | Ea | 205.00 | 27.90 | 232.90 | 396.00 |
| 48" wide, white | B1@.950 | Ea | 205.00 | 27.90 | 232.90 | 396.00 |

## Medicine Cabinet Light Bars

**Light bar for beveled tri-view cabinet.** Frameless. Beveled mirror edges. Bulbs not included. Labor cost does not include electrical rough-in.

| | Craft@Hrs | Unit | Material | Labor | Total | Sell |
|---|---|---|---|---|---|---|
| 24", 4 bulb | BE@.600 | Ea | 43.40 | 20.60 | 64.00 | 109.00 |
| 30", 5 bulb | BE@.600 | Ea | 54.00 | 20.60 | 74.60 | 127.00 |
| 36", 8 bulb | BE@.600 | Ea | 59.90 | 20.60 | 80.50 | 137.00 |
| 48", 8 bulb | BE@.600 | Ea | 88.70 | 20.60 | 109.30 | 186.00 |

| | Craft@Hrs | Unit | Material | Labor | Total | Sell |
|---|---|---|---|---|---|---|

**Medicine cabinet light bar.** Surface mount. UL and CSA listed. Bulbs not included. Labor cost does not include electrical rough-in.

| | Craft@Hrs | Unit | Material | Labor | Total | Sell |
|---|---|---|---|---|---|---|
| 24", oak, 4 bulb | BE@.600 | Ea | 43.20 | 20.60 | 63.80 | 108.00 |
| 24", white, 4 bulb | BE@.600 | Ea | 43.40 | 20.60 | 64.00 | 109.00 |
| 30", oak, 5 bulb | BE@.600 | Ea | 54.00 | 20.60 | 74.60 | 127.00 |
| 30", white, 5 bulb | BE@.600 | Ea | 54.00 | 20.60 | 74.60 | 127.00 |
| 36", oak, 6 bulb | BE@.600 | Ea | 64.80 | 20.60 | 85.40 | 145.00 |
| 36", white, 5 bulb | BE@.600 | Ea | 64.80 | 20.60 | 85.40 | 145.00 |

## Bathroom Accessories

**Standard collection bath accessories, polished chrome, American Standard.** Durable solid brass with chrome finish. Easy installation on tile, wood, plaster or drywall. Concealed mounting with no exposed hardware.

| | Craft@Hrs | Unit | Material | Labor | Total | Sell |
|---|---|---|---|---|---|---|
| Robe hook | BC@.280 | Ea | 18.10 | 8.99 | 27.09 | 46.10 |
| 24" shelf with glass | BC@.280 | Ea | 79.60 | 8.99 | 88.59 | 151.00 |
| Toilet paper holder | BC@.280 | Ea | 34.09 | 8.99 | 43.08 | 73.20 |
| 18" towel bar | BC@.280 | Ea | 41.09 | 8.99 | 50.08 | 85.10 |
| 24" towel bar | BC@.280 | Ea | 43.50 | 8.99 | 52.49 | 89.20 |
| Towel ring | BC@.280 | Ea | 35.50 | 8.99 | 44.49 | 75.60 |
| Tumbler/toothbrush holder | BC@.280 | Ea | 33.50 | 8.99 | 42.49 | 72.20 |

**Standard collection bath accessories, satin chrome, American Standard.** Durable solid brass with satin chrome finish. Easy installation on tile, wood, plaster or drywall. Concealed mounting with no exposed hardware.

| | Craft@Hrs | Unit | Material | Labor | Total | Sell |
|---|---|---|---|---|---|---|
| Robe Hook | BC@.280 | Ea | 20.90 | 8.99 | 29.89 | 50.80 |
| 24" shelf with glass | BC@.280 | Ea | 99.50 | 8.99 | 108.49 | 184.00 |
| Toilet paper holder | BC@.280 | Ea | 43.90 | 8.99 | 52.89 | 89.90 |
| 18" towel bar | BC@.280 | Ea | 51.50 | 8.99 | 60.49 | 103.00 |
| 24" towel bar | BC@.280 | Ea | 53.10 | 8.99 | 62.09 | 106.00 |
| Towel ring | BC@.280 | Ea | 42.20 | 8.99 | 51.19 | 87.00 |

**Williamsburg bath accessories, satin chrome, American Standard.** Satin finish with optional chrome accent pieces included. Matches Williamsburg line of faucets and accessories. Solid brass construction. Lifetime warranty. Easy installation. Drill-free anchors. Concealed mounting hardware.

| | Craft@Hrs | Unit | Material | Labor | Total | Sell |
|---|---|---|---|---|---|---|
| Robe hook | BC@.280 | Ea | 18.30 | 8.99 | 27.29 | 46.40 |
| Soap dish | BC@.280 | Ea | 45.30 | 8.99 | 54.29 | 92.30 |
| Toilet paper holder | BC@.280 | Ea | 35.80 | 8.99 | 44.79 | 76.10 |
| 18" towel bar | BC@.280 | Ea | 41.10 | 8.99 | 50.09 | 85.20 |
| 24" towel bar | BC@.280 | Ea | 43.10 | 8.99 | 52.09 | 88.60 |
| Towel ring | BC@.280 | Ea | 35.60 | 8.99 | 44.59 | 75.80 |
| Tumbler/toothbrush holder | BC@.280 | Ea | 45.30 | 8.99 | 54.29 | 92.30 |

**Williamsburg bath accessories, chrome and brass, American Standard.** Matches Williamsburg line of faucets and accessories. Chrome finish with optional brass accent pieces included. Solid brass construction. Easy installation. Drill-free anchors. Concealed mounting hardware.

| | Craft@Hrs | Unit | Material | Labor | Total | Sell |
|---|---|---|---|---|---|---|
| Robe hook | BC@.280 | Ea | 17.30 | 8.99 | 26.29 | 44.70 |
| Soap dish | BC@.280 | Ea | 45.30 | 8.99 | 54.29 | 92.30 |
| Toilet paper holder | BC@.280 | Ea | 35.80 | 8.99 | 44.79 | 76.10 |
| 18" towel bar | BC@.280 | Ea | 41.60 | 8.99 | 50.59 | 86.00 |
| 24" towel bar | BC@.280 | Ea | 43.30 | 8.99 | 52.29 | 88.90 |
| Towel ring | BC@.280 | Ea | 35.70 | 8.99 | 44.69 | 76.00 |
| Tumbler/toothbrush holder | BC@.280 | Ea | 45.30 | 8.99 | 54.29 | 92.30 |

|  | Craft@Hrs | Unit | Material | Labor | Total | Sell |
|---|---|---|---|---|---|---|

**Williamsburg bath accessories, chrome and porcelain, American Standard.** Matches Williamsburg line of faucets and accessories. Solid brass construction. Easy installation. Drill-free anchors. Concealed mounting hardware.

| Robe hook | BC@.280 | Ea | 16.20 | 8.99 | 25.19 | 42.80 |
| Soap dish | BC@.280 | Ea | 45.30 | 8.99 | 54.29 | 92.30 |
| Toilet paper holder | BC@.280 | Ea | 35.50 | 8.99 | 44.49 | 75.60 |
| 18" towel bar | BC@.280 | Ea | 42.20 | 8.99 | 51.19 | 87.00 |
| 24" towel bar | BC@.280 | Ea | 43.20 | 8.99 | 52.19 | 88.70 |
| Towel ring | BC@.280 | Ea | 34.20 | 8.99 | 43.19 | 73.40 |
| Tumbler/toothbrush holder | BC@.280 | Ea | 45.30 | 8.99 | 54.29 | 92.30 |

**Williamsburg bath accessories, satin chrome and satin brass, American Standard.** Matches Williamsburg line of faucets and accessories. Satin finish with optional brass accent pieces included. Solid brass construction. Easy installation. Drill-free anchors. Concealed mounting hardware.

| Robe hook | BC@.280 | Ea | 17.30 | 8.99 | 26.29 | 44.70 |
| Toilet paper holder | BC@.280 | Ea | 35.80 | 8.99 | 44.79 | 76.10 |
| 18" towel bar | BC@.280 | Ea | 41.10 | 8.99 | 50.09 | 85.20 |
| 24" towel bar | BC@.280 | Ea | 43.20 | 8.99 | 52.19 | 88.70 |
| Towel ring | BC@.280 | Ea | 35.60 | 8.99 | 44.59 | 75.80 |

**Williamsburg bath accessories, polished brass, American Standard.** Matches Williamsburg line of faucets and accessories. Solid brass construction. Easy installation with drill-free anchors. Concealed mounting hardware.

| Robe hook | BC@.280 | Ea | 15.70 | 8.99 | 24.69 | 42.00 |
| Toilet paper holder | BC@.280 | Ea | 36.05 | 8.99 | 45.04 | 76.60 |
| 18" towel bar | BC@.280 | Ea | 43.80 | 8.99 | 52.79 | 89.70 |
| 24" towel bar | BC@.280 | Ea | 44.80 | 8.99 | 53.79 | 91.40 |
| Towel ring | BC@.280 | Ea | 34.90 | 8.99 | 43.89 | 74.60 |

**Edgewater bath accessories, polished chrome, Baldwin.**

| Robe hook | BC@.280 | Ea | 12.40 | 8.99 | 21.39 | 36.40 |
| Tissue roll holder | BC@.280 | Ea | 34.70 | 8.99 | 43.69 | 74.30 |
| 18" towel bar | BC@.280 | Ea | 38.80 | 8.99 | 47.79 | 81.20 |
| 24" towel bar | BC@.280 | Ea | 40.00 | 8.99 | 48.99 | 83.30 |
| Towel ring | BC@.280 | Ea | 31.80 | 8.99 | 40.79 | 69.30 |

**Laguna bath accessories, polished brass and chrome, Baldwin.** Concealed mounting. Includes mounting hardware for drywall and masonry.

| Robe hook | BC@.280 | Ea | 12.36 | 8.99 | 21.35 | 36.30 |
| Toilet paper holder | BC@.280 | Ea | 32.80 | 8.99 | 41.79 | 71.00 |
| 18" towel bar | BC@.280 | Ea | 38.00 | 8.99 | 46.99 | 79.90 |
| 24" towel bar | BC@.280 | Ea | 40.20 | 8.99 | 49.19 | 83.60 |
| Towel ring | BC@.280 | Ea | 29.60 | 8.99 | 38.59 | 65.60 |

**Best Value™ Series bath accessories, polished chrome, Franklin Brass.** Exposed screw.

| Recessed toilet paper holder | BC@.280 | Ea | 6.45 | 8.99 | 15.44 | 26.20 |
| Robe hook | BC@.280 | Ea | 1.91 | 8.99 | 10.90 | 18.50 |
| Soap dish | BC@.280 | Ea | 2.87 | 8.99 | 11.86 | 20.20 |
| Toilet paper holder | BC@.280 | Ea | 4.30 | 8.99 | 13.29 | 22.60 |
| Toothbrush holder | BC@.280 | Ea | 2.87 | 8.99 | 11.86 | 20.20 |
| Toothbrush and tumbler holder | BC@.280 | Ea | 2.67 | 8.99 | 11.66 | 19.80 |
| 18" towel bar | BC@.280 | Ea | 1.06 | 8.99 | 10.05 | 17.10 |
| 24" towel bar | BC@.280 | Ea | 2.15 | 8.99 | 11.14 | 18.90 |
| 30" towel bar | BC@.280 | Ea | 2.15 | 8.99 | 11.14 | 18.90 |
| Towel bar posts, 1 pair | BC@.280 | Ea | 2.20 | 8.99 | 11.19 | 19.00 |
| Towel ring | BC@.280 | Ea | 4.30 | 8.99 | 13.29 | 22.60 |

|  | Craft@Hrs | Unit | Material | Labor | Total | Sell |
|---|---|---|---|---|---|---|

**Futura bath accessories, polished chrome, Franklin Brass.** Zinc die-cast traditional design with concealed mounting.

| | | | | | | |
|---|---|---|---|---|---|---|
| Recessed soap dish, with grab bar | BC@.280 | Ea | 15.20 | 8.99 | 24.19 | 41.10 |
| Recessed toilet paper holder | BC@.280 | Ea | 15.20 | 8.99 | 24.19 | 41.10 |
| Robe hook | BC@.280 | Ea | 4.35 | 8.99 | 13.34 | 22.70 |
| Soap dish | BC@.280 | Ea | 6.56 | 8.99 | 15.55 | 26.40 |
| Toilet paper holder | BC@.280 | Ea | 9.78 | 8.99 | 18.77 | 31.90 |
| Toothbrush holder | BC@.280 | Ea | 6.56 | 8.99 | 15.55 | 26.40 |
| 18" towel bar | BC@.280 | Ea | 10.80 | 8.99 | 19.79 | 33.60 |
| 24" towel bar | BC@.280 | Ea | 13.00 | 8.99 | 21.99 | 37.40 |
| 30" towel bar | BC@.280 | Ea | 15.10 | 8.99 | 24.09 | 41.00 |
| Towel ring | BC@.280 | Ea | 9.75 | 8.99 | 18.74 | 31.90 |

**Classic bath accessories, Lenape.** Genuine porcelain. White. Clip-on installation.

| | | | | | | |
|---|---|---|---|---|---|---|
| Robe hook | BC@.280 | Ea | 10.73 | 8.99 | 19.72 | 33.50 |
| Soap dish | BC@.280 | Ea | 14.00 | 8.99 | 22.99 | 39.10 |
| Toilet paper holder | BC@.280 | Ea | 17.00 | 8.99 | 25.99 | 44.20 |
| Toothbrush/tumbler holder | BC@.280 | Ea | 14.00 | 8.99 | 22.99 | 39.10 |
| 18" towel bar | BC@.280 | Ea | 24.80 | 8.99 | 33.79 | 57.40 |
| 24" towel bar assembly | BC@.280 | Ea | 28.05 | 8.99 | 37.04 | 63.00 |
| 24" towel bar, double | BC@.280 | Ea | 31.20 | 8.99 | 40.19 | 68.30 |
| Towel ring | BC@.280 | Ea | 10.25 | 8.99 | 19.24 | 32.70 |

**Alexandria bath accessories, glazed ceramic and chrome, Bath Unlimited.** Wall mount.

| | | | | | | |
|---|---|---|---|---|---|---|
| Glass shelf | BC@.280 | Ea | 43.40 | 8.99 | 52.39 | 89.10 |
| Lotion dispenser | BC@.280 | Ea | 10.80 | 8.99 | 19.79 | 33.60 |
| Robe hook | BC@.280 | Ea | 16.20 | 8.99 | 25.19 | 42.80 |
| Soap dish | BC@.280 | Ea | 8.63 | 8.99 | 17.62 | 30.00 |
| Soap dish, ceramic | BC@.280 | Ea | 21.60 | 8.99 | 30.59 | 52.00 |
| Tissue cover | BC@.280 | Ea | 21.60 | 8.99 | 30.59 | 52.00 |
| Toilet paper holder | BC@.280 | Ea | 23.90 | 8.99 | 32.89 | 55.90 |
| Toothbrush/tumbler holder | BC@.280 | Ea | 21.60 | 8.99 | 30.59 | 52.00 |
| Toothbrush holder | BC@.280 | Ea | 8.62 | 8.99 | 17.61 | 29.90 |
| 18" towel bar | BC@.280 | Ea | 27.10 | 8.99 | 36.09 | 61.40 |
| 24" towel bar | BC@.280 | Ea | 29.20 | 8.99 | 38.19 | 64.90 |
| 24" towel bar, double | BC@.280 | Ea | 43.40 | 8.99 | 52.39 | 89.10 |
| Towel ring | BC@.280 | Ea | 23.90 | 8.99 | 32.89 | 55.90 |

**College Circle bath accessories, polished chrome, Bath Unlimited.**

| | | | | | | |
|---|---|---|---|---|---|---|
| Double robe hook | BC@.280 | Ea | 7.60 | 8.99 | 16.59 | 28.20 |
| Soap dish | BC@.280 | Ea | 13.00 | 8.99 | 21.99 | 37.40 |
| Toilet paper holder | BC@.280 | Ea | 15.10 | 8.99 | 24.09 | 41.00 |
| 18" towel bar | BC@.280 | Ea | 16.20 | 8.99 | 25.19 | 42.80 |
| 24" towel bar | BC@.280 | Ea | 17.40 | 8.99 | 26.39 | 44.90 |
| Towel ring | BC@.280 | Ea | 10.80 | 8.99 | 19.79 | 33.60 |
| Tumbler/toothbrush holder | BC@.280 | Ea | 13.00 | 8.99 | 21.99 | 37.40 |

| | Craft@Hrs | Unit | Material | Labor | Total | Sell |
|---|---|---|---|---|---|---|
| **Dakota bath accessories, black finish, Bath Unlimited.** Modern look. | | | | | | |
| Robe hook | BC@.280 | Ea | 8.90 | 8.99 | 17.89 | 30.40 |
| Soap dish | BC@.280 | Ea | 18.90 | 8.99 | 27.89 | 47.40 |
| Toilet paper holder | BC@.280 | Ea | 15.30 | 8.99 | 24.29 | 41.30 |
| 18" towel bar | BC@.280 | Ea | 19.90 | 8.99 | 28.89 | 49.10 |
| 24" towel bar | BC@.280 | Ea | 22.00 | 8.99 | 30.99 | 52.70 |
| Towel ring | BC@.280 | Ea | 15.40 | 8.99 | 24.39 | 41.50 |
| Tumbler/toothbrush holder | BC@.280 | Ea | 18.90 | 8.99 | 27.89 | 47.40 |
| **Greenwich bath accessories, chrome, Bath Unlimited.** | | | | | | |
| Robe hook | BC@.280 | Ea | 10.80 | 8.99 | 19.79 | 33.60 |
| Soap dish | BC@.280 | Ea | 13.00 | 8.99 | 21.99 | 37.40 |
| Tissue holder | BC@.280 | Ea | 19.50 | 8.99 | 28.49 | 48.40 |
| Toothbrush/tumbler holder | BC@.280 | Ea | 14.50 | 8.99 | 23.49 | 39.90 |
| 18" towel bar | BC@.280 | Ea | 19.90 | 8.99 | 28.89 | 49.10 |
| 24" towel bar | BC@.280 | Ea | 21.70 | 8.99 | 30.69 | 52.20 |
| Towel ring | BC@.280 | Ea | 18.40 | 8.99 | 27.39 | 46.60 |
| **Greenwich bath accessories, polished brass, Bath Unlimited.** | | | | | | |
| Robe hook | BC@.280 | Ea | 10.90 | 8.99 | 19.89 | 33.80 |
| Soap dish | BC@.280 | Ea | 14.50 | 8.99 | 23.49 | 39.90 |
| Toilet paper holder | BC@.280 | Ea | 18.60 | 8.99 | 27.59 | 46.90 |
| Toothbrush/tumbler holder | BC@.280 | Ea | 14.50 | 8.99 | 23.49 | 39.90 |
| Towel ring | BC@.280 | Ea | 19.50 | 8.99 | 28.49 | 48.40 |
| 18" towel bar | BC@.280 | Ea | 19.70 | 8.99 | 28.69 | 48.80 |
| 24" towel bar | BC@.280 | Ea | 21.80 | 8.99 | 30.79 | 52.30 |
| **Greenwich bath accessories, brass and chrome, Bath Unlimited.** | | | | | | |
| Robe hook | BC@.280 | Ea | 10.90 | 8.99 | 19.89 | 33.80 |
| Soap dish | BC@.280 | Ea | 14.50 | 8.99 | 23.49 | 39.90 |
| Tissue holder | BC@.280 | Ea | 19.50 | 8.99 | 28.49 | 48.40 |
| Toothbrush/tumbler holder | BC@.280 | Ea | 14.50 | 8.99 | 23.49 | 39.90 |
| 18" Towel bar | BC@.280 | Ea | 20.60 | 8.99 | 29.59 | 50.30 |
| 24" Towel bar | BC@.280 | Ea | 22.70 | 8.99 | 31.69 | 53.90 |
| Towel ring | BC@.280 | Ea | 18.40 | 8.99 | 27.39 | 46.60 |
| **Greenwich bath accessories, satin nickel, Bath Unlimited.** | | | | | | |
| Glass shelf | BC@.280 | Ea | 40.30 | 8.99 | 49.29 | 83.80 |
| Robe hook | BC@.280 | Ea | 13.00 | 8.99 | 21.99 | 37.40 |
| Soap dish | BC@.280 | Ea | 18.40 | 8.99 | 27.39 | 46.60 |
| Toilet paper holder | BC@.280 | Ea | 21.70 | 8.99 | 30.69 | 52.20 |
| Toothbrush/tumbler holder | BC@.280 | Ea | 17.90 | 8.99 | 26.89 | 45.70 |
| Towel bar, double | BC@.280 | Ea | 37.80 | 8.99 | 46.79 | 79.50 |
| 18" towel bar | BC@.280 | Ea | 24.90 | 8.99 | 33.89 | 57.60 |
| 24" towel bar | BC@.280 | Ea | 27.00 | 8.99 | 35.99 | 61.20 |
| Towel ring | BC@.280 | Ea | 19.50 | 8.99 | 28.49 | 48.40 |

| | Craft@Hrs | Unit | Material | Labor | Total | Sell |
|---|---|---|---|---|---|---|
| **Deluxe oak bath accessories, Bath Unlimited.** | | | | | | |
| Robe hook | BC@.280 | Ea | 6.50 | 8.99 | 15.49 | 26.30 |
| Soap dish | BC@.280 | Ea | 12.90 | 8.99 | 21.89 | 37.20 |
| Toilet paper holder | BC@.280 | Ea | 9.70 | 8.99 | 18.69 | 31.80 |
| Toothbrush/tumbler holder | BC@.280 | Ea | 12.90 | 8.99 | 21.89 | 37.20 |
| 18" towel bar | BC@.280 | Ea | 17.20 | 8.99 | 26.19 | 44.50 |
| 24" towel bar | BC@.280 | Ea | 19.40 | 8.99 | 28.39 | 48.30 |
| Towel ring | BC@.280 | Ea | 11.80 | 8.99 | 20.79 | 35.30 |
| **Monticello bath accessories, chrome and brass, Bath Unlimited.** | | | | | | |
| Robe hook | BC@.280 | Ea | 16.20 | 8.99 | 25.19 | 42.80 |
| Soap dish | BC@.280 | Ea | 20.30 | 8.99 | 29.29 | 49.80 |
| Toilet paper holder | BC@.280 | Ea | 24.90 | 8.99 | 33.89 | 57.60 |
| Toothbrush/tumbler holder | BC@.280 | Ea | 20.30 | 8.99 | 29.29 | 49.80 |
| 18" towel bar | BC@.280 | Ea | 30.30 | 8.99 | 39.29 | 66.80 |
| 24" towel bar | BC@.280 | Ea | 32.50 | 8.99 | 41.49 | 70.50 |
| Towel ring | BC@.280 | Ea | 21.70 | 8.99 | 30.69 | 52.20 |
| **Permanent shower rod.** 72" long. | | | | | | |
| Antique brass | P1@.250 | Ea | 16.90 | 7.88 | 24.78 | 42.10 |
| Bone | P1@.250 | Ea | 13.90 | 7.88 | 21.78 | 37.00 |
| Chrome | P1@.250 | Ea | 14.09 | 7.88 | 21.97 | 37.30 |
| Decor rod with brass ends | P1@.250 | Ea | 17.40 | 7.88 | 25.28 | 43.00 |
| Platinum | P1@.250 | Ea | 17.40 | 7.88 | 25.28 | 43.00 |
| White | P1@.250 | Ea | 14.10 | 7.88 | 21.98 | 37.40 |

## Bathroom Conversion Costs

**Sump pumping systems for residential wastes.** Including fiberglass sump basin, pump, 40' pipe run to septic tank or sewer line and automatic float switch. Add the cost of electrical work and excavation.

| | Craft@Hrs | Unit | Material | Labor | Total | Sell |
|---|---|---|---|---|---|---|
| To 15' head | P1@8.00 | Ea | 1,540.00 | 252.00 | 1,792.00 | 3,050.00 |
| To 25' head | P1@8.00 | Ea | 1,840.00 | 252.00 | 2,092.00 | 3,560.00 |
| To 30' head | P1@8.00 | Ea | 2,470.00 | 252.00 | 2,722.00 | 4,630.00 |
| Add for high water or pump failure alarm | P1@.500 | Ea | 380.00 | 15.80 | 395.80 | 673.00 |

**Sheet metal area wall.** Provides light well for basement windows. Galvanized corrugated steel. Labor includes the cost of setting in place only. Add the cost of excavation.

| | Craft@Hrs | Unit | Material | Labor | Total | Sell |
|---|---|---|---|---|---|---|
| 12" deep, 6" projection, 37" wide | SW@.410 | Ea | 21.00 | 14.60 | 35.60 | 60.50 |
| 18" deep, 6" projection, 37" wide | SW@.410 | Ea | 26.90 | 14.60 | 41.50 | 70.60 |
| 24" deep, 6" projection, 37" wide | SW@.410 | Ea | 35.00 | 14.60 | 49.60 | 84.30 |
| 30" deep, 6" projection, 37" wide | SW@.410 | Ea | 41.70 | 14.60 | 56.30 | 95.70 |
| Add for steel grille cover | — | Ea | 47.30 | — | 47.30 | — |

**Basement doors, Bilco.** 12 gauge primed steel, center opening basement doors. Costs include assembly and installation hardware. Add the cost of excavation, concrete, masonry, anchor placement and finish painting.
Doors (overall dimensions)

| | Craft@Hrs | Unit | Material | Labor | Total | Sell |
|---|---|---|---|---|---|---|
| 19-1/2" H, 55" W, 72" L | BC@3.41 | Ea | 365.00 | 109.00 | 474.00 | 806.00 |
| 22" H, 51" W, 64" L | BC@3.41 | Ea | 334.00 | 109.00 | 443.00 | 753.00 |
| 30" H, 47" W, 58" L | BC@3.41 | Ea | 328.00 | 109.00 | 437.00 | 743.00 |
| 52" H, 51" W, 43-1/4" L | BC@3.41 | Ea | 392.00 | 109.00 | 501.00 | 852.00 |

| | Craft@Hrs | Unit | Material | Labor | Total | Sell |
|---|---|---|---|---|---|---|
| Door extensions (available for 19-1/2" H, 55" W, 72" L door only) | | | | | | |
| 6" deep | BC@1.71 | Ea | 90.70 | 54.90 | 145.60 | 248.00 |
| 12" deep | BC@1.71 | Ea | 108.00 | 54.90 | 162.90 | 277.00 |
| 18" deep | BC@1.71 | Ea | 135.00 | 54.90 | 189.90 | 323.00 |
| 24" deep | BC@1.71 | Ea | 161.00 | 54.90 | 215.90 | 367.00 |
| Basement door stair stringers, steel, pre-cut for 2" x 10" wood treads (without treads) | | | | | | |
| 32" to 39" stair height | BC@1.71 | Ea | 76.60 | 54.90 | 131.50 | 224.00 |
| 48" to 55" stair height | BC@1.71 | Ea | 95.80 | 54.90 | 150.70 | 256.00 |
| 56" to 64" stair height | BC@1.71 | Ea | 108.00 | 54.90 | 162.90 | 277.00 |
| 65" to 72" stair height | BC@1.71 | Ea | 121.00 | 54.90 | 175.90 | 299.00 |
| 73" to 78" stair height | BC@1.71 | Ea | 166.00 | 54.90 | 220.90 | 376.00 |
| 81" to 88" stair height | BC@1.71 | Ea | 180.00 | 54.90 | 234.90 | 399.00 |
| 89" to 97" stair height | BC@1.71 | Ea | 192.00 | 54.90 | 246.90 | 420.00 |

**Extend plumbing vent through roof.** Usually required when fixtures have been installed in a dormer addition. Add the cost of refinishing the wall interior and exterior, ceiling and roof.

| | Craft@Hrs | Unit | Material | Labor | Total | Sell |
|---|---|---|---|---|---|---|
| Per roof vent | BC@5.50 | Ea | 34.70 | 176.00 | 210.70 | 358.00 |

**Relocate sink vent stack.** Usually required after a wall breakthrough. Add the cost of refinishing the wall interior and exterior, ceiling and roof.

| | Craft@Hrs | Unit | Material | Labor | Total | Sell |
|---|---|---|---|---|---|---|
| Per vent stack | BC@9.50 | Ea | 37.10 | 305.00 | 342.10 | 582.00 |

# *Plumbing and HVAC*

# 14

Not every older home needs an extra bathroom or a larger kitchen, but nearly every older home needs plumbing, heating and cooling systems brought up to modern standards. No home is truly comfortable without effective plumbing, heating and cooling. In a very old house, the entire mechanical system may have to be replaced.

## Warm Air Heating

Many old houses still have gravity heating systems. The main difference between these and a modern forced-air system is that there's no fan to move the air. It depends on the physical property of heated air to rise, drawing in unheated air to fill the vacuum. This simple circulation system moves the warm air through the house. Gravity heating has its advantages; it's quiet, and there are no fan motors to repair. But they are very large units, with so many large ducts needed to funnel air to each room in the downstairs of the house that they almost fill the basement. They also tend to be heavy gas users. That may not have been a problem 50 years ago when gas was cheap, but fuel cost is something we have to consider today.

A new high-efficiency forced air furnace will recoup the replacement cost in just a few years. And since modern furnaces are small, and use smaller ducts, eliminating the gravity system's octopus-like duct arms leaves most of the basement space available for conversion into living space. Forced air heating has the added advantage of being able to bring the house up to a comfortable temperature in a matter of minutes, rather than the hour or two a gravity system takes.

The Btu capacity of a residential heating system depends on climate, window size and orientation, insulation and square footage to be heated. For cost estimating purposes, there's an easy way to calculate the Btu capacity of the furnace needed. Multiply the square feet of heated floor area by 53; then round up to the next larger furnace size. For example, to size a furnace for a 2,000 square foot home:

2,000 times 53 equals 106,000. The next larger furnace size is 125,000 Btu.

Altitude also affects the size of the furnace needed. Reduce the stated Btu rating of a furnace by 4 percent for each 1,000 feet above sea level. For example, the capacity of a 100,000 Btu furnace installed 5,000 feet above sea level would be 20 percent less, or 80,000 Btu.

When adding a room, enclosing a porch, expanding attic space or converting a garage, the existing furnace may not have capacity to serve the addition. Even if the existing furnace has enough capacity, a long duct run may reduce the volume of heat delivered at the register to below acceptable levels. The correct way to serve a remote room addition is to install a new trunk line directly to the furnace plenum. Talk to your heating subcontractor about installing a booster fan in the existing duct run to increase the flow of warm air.

If tapping into the existing HVAC system doesn't make sense, and there are no other practical options, install a floor or wall furnace in the new room addition. Direct-venting thru-the-wall furnaces are usually an acceptable alternative.

## Hot Water Heating

One-pipe gravity steam heating systems are common in older homes. Radiators are fed by a single pipe on the entry (warm) side. The other side of the radiator has only an exhaust valve on the pipe. Once steam has cooled and condensed to water, it flows by gravity back to the boiler. This is an extremely simple system and, if properly installed and maintained, can be a reliable heat source. But again, one-pipe steam heating systems respond very slowly to requests for temperature change, and they offer far less control than modern hydronic systems. It may be possible to modernize a one-pipe system by replacing old radiators with baseboard heaters. Discuss it with your HVAC contractor.

There's a lot that can go wrong when modifying an old steam or hot water heating system. Results may not be what the owner expected. Include in your contract some language that limits both the scope of your work and your liability. For example: *Contractor makes no warranty on the condition, size or capacity of the existing boiler, radiators, or distribution piping.*

Radiant heating systems are less common, but tend to be trouble-free. Heated water flows through coils embedded in either a concrete slab or the plastered ceiling. Expert repair is required when a radiant heating system's piping develops leaks or becomes air locked. Breaks in ceiling coils can be repaired fairly easily. But repairing breaks in a floor is both difficult and expensive.

## Electric Space Heaters

Assuming the home has sufficient capacity at the service panel, it's easy to add electric space heaters in a home if you need to extend heat to a room addition. However, electric heating is expensive and may not be practical in colder climates. Also, insulation standards are much higher for a home heated with electricity, so you must take that into consideration if you plan on this type of heat for your addition.

# Water Supply

Low water pressure is usually due to an accumulation of lime or corrosion in supply lines, though it may also be that supply lines are too small. Main distribution pipes should be $^3/_4$" inside diameter. Branch lines serving a single fixture may be $^1/_2$" inside diameter.

If the house is served by a well on the premises, check the gauge on the pressure tank. A good operating pressure is 40 or 50 psi. If pressure is less than 20 psi, the pump isn't operating properly or the pressure setting is too low.

## *Water Hammer*

Water hammer is a loud banging noise in supply lines caused when water is shut off abruptly. Over a period of time, water hammer can seriously damage valves and fixtures throughout the supply system. An air chamber placed on a supply line can absorb enough shock to prevent water hammer. If there's already an air chamber on the line, the air may have escaped. A waterlogged air chamber offers no protection against water hammer. To restore air to the chamber, turn off the water at the source. Open the air chamber to drain out any excess water, then reseal the chamber and restore the water pressure.

# Plumbing Drain, Waste and Vent Lines

The drainage system consists of sewer laterals, drainage pipes and vents. It's common for pipes in older homes to break or become clogged. Most older homes have cast iron drain lines, which rust out after about 70 years. In some cases, the pipe diameter will be less than required by current codes. This is especially true for vent piping.

Sewer drain pipe in older homes may be bell-joint vitreous tile. Joints can break under pressure as the soil settles or heaves, and a broken pipe is an invitation for tree roots. Even if pipes aren't broken, tree roots can enter through the pipe joints and obstruct the flow of sewage. When this happens, remove the roots with a power-driven snake. If the system has to be snaked out every few months, you can assume there's a break or an open joint somewhere that needs repair. Tile or steel drainage pipe should last as long as the house. But these lines can accumulate sediment and need flushing out occasionally.

If you notice the smell of sewer gas, a vent stack is probably obstructed. The vent stacks you see running through the roof are designed to equalize air pressure in the drainage system. If a vent is obstructed, waste flowing through the system can create a vacuum that sucks water out of fixture traps, allowing sewer gas to enter the home. All fixtures should drain by gravity, not through suction. If you notice that water is being sucked out of the trap when a toilet is flushed, you can assume the roof vent is obstructed.

---

### *Checking Pipe Sizes*

The nominal (name) size of pipe refers to the inside diameter.

Copper pipe

$^1/_2$" inside diameter pipe has $^5/_8$" outside diameter

$^3/_4$" inside diameter pipe has $^7/_8$" outside diameter

Galvanized pipe

$^1/_2$" inside diameter pipe has $^7/_8$" outside diameter

$^3/_4$" inside diameter pipe has $1^1/_8$" outside diameter

# Water Heaters

Some older homes may still have domestic water coil systems in the furnace. That may provide enough hot water for cooking and bathing during the heating season. During summer months when the furnace isn't running, a regular water heater will be required. A gas water heater for a three-bedroom home should have at least a 30-gallon capacity, though 40 is the current standard. An electric water heater for the same home should have a 50-gallon capacity.

# Pricing Plumbing and HVAC Repairs

Home improvement contractors usually subcontract heating, air conditioning, and plumbing work to specialists. Plumbing and HVAC subcontractors usually quote the total installed prices only, without a breakdown of material and labor costs. However, to improve the data's versatility, this chapter provides a breakdown of the labor and material costs.

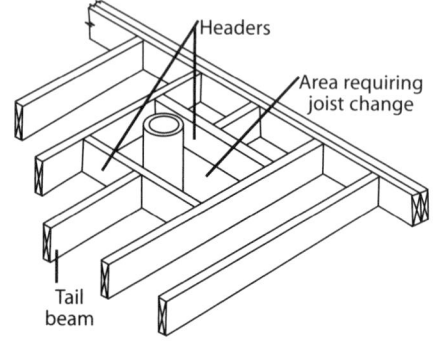

**Figure 14-1**

*Header placement to acommodate plumbing lines*

# Framing for Plumbing and HVAC Improvements

Installing new heating duct, plumbing stacks, drains or water piping will usually require alteration of the framing. Plumbers and HVAC installers will arrive on the job site with all the tools needed to cut away structural wood framing. But a carpenter may not appreciate the result. Caution your plumbing and HVAC subs to avoid cutting if possible. When critical framing members have to be cut, have a qualified carpenter add stiffeners or supports.

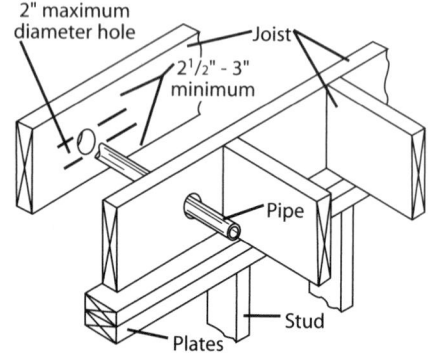

**Figure 14-2**

*Dimensions required for passing pipe through joists*

# Cutting Floor Joists

When a floor joist has to make way for the passage of a stack or vent, install headers and tail beams (as in Figure 14-1) to reinforce the area around the joist that has been cut.

Figure 14-2 shows how to pass water pipe or conduit through a joist. The hole can't be any larger than 2" in diameter. Edges of the hole must be at least $2^1/2$ to 3" from the top and bottom of the joist. You can also cut a notch at the top or bottom edge of the joist. But the notch should be in the last quarter of the span and not more than $^1/6$ the depth of the joist. If a joist has to be cut and you can't comply with these rules, add an additional joist next to the cut joist or reinforce the cut joist by nailing scabs to each side.

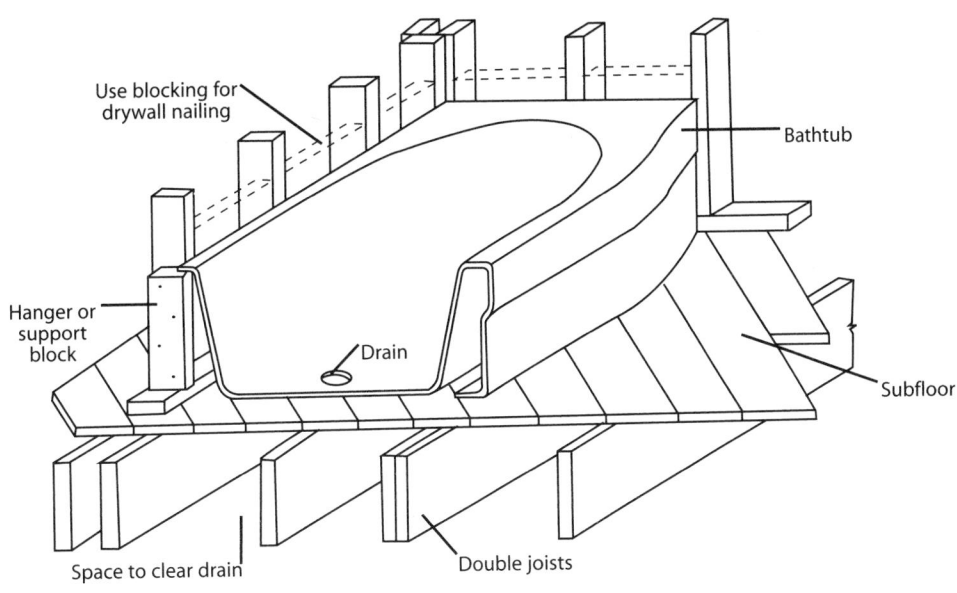

**Figure 14-3**

*Framing for bathtub*

# Bathtub Framing

When adding a bathtub, more framing may be needed to support the heavy weight of a tub filled with water. Where joists are parallel to the length of the tub, double joists under the outer edge of the tub. The inner edge is usually supported by wall framing. Use hangers or wood blocks as supports where the tub meets enclosing walls. See Figure 14-3.

# Utility Walls

Walls that enclose new plumbing stacks or venting may require special framing. A 4" soil stack won't fit in a standard 2 x 4 stud wall. If a thicker wall is needed, frame it with 2 x 6 top and bottom plates and 2 x 4 studs placed flatwise at the edge of the plates. See Figure 14-4A. This leaves the center of the wall open for running both supply and drain pipes.

A 3" vent stack will fit in a 2 x 4 stud wall. But the hole for the vent will require cutting away most of the top plate. To repair the top plate, nail scabs cut from 2 x 4s on each side of the vent. See Figure 14-4B.

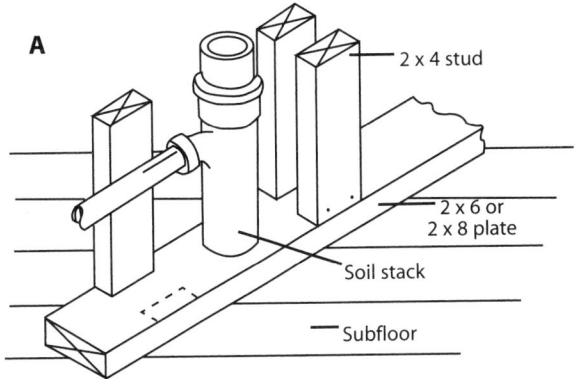

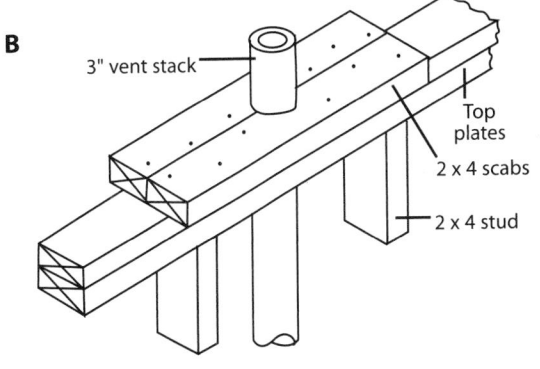

**Figure 14-4**

*Framing around a stack*

| | Craft@Hrs | Unit | Material | Labor | Total | Sell |
|---|---|---|---|---|---|---|

## Heating System Renovations

**Remove residential heating system components.** Includes hauling components from the site. No salvage value assumed. Costs will be higher if components include detectable friable asbestos or if access to the site is limited. If the furnace or ductwork is wrapped in asbestos, you may be legally required to follow asbestos abatement procedures. This may require the services of a specialized asbestos abatement contractor.

| | Craft@Hrs | Unit | Material | Labor | Total | Sell |
|---|---|---|---|---|---|---|
| Gravity feed furnace and plenum | P1@5.30 | Ea | 80.00 | 167.00 | 247.00 | 420.00 |
| Forced air furnace and plenum | P1@2.50 | Ea | 50.00 | 78.80 | 128.80 | 219.00 |
| Residential fuel oil storage tank | P1@2.50 | Ea | 50.00 | 78.80 | 128.80 | 219.00 |
| Steam or hot water boiler | P1@2.50 | Ea | 55.00 | 78.80 | 133.80 | 227.00 |

**Remove residential heating duct.** Includes hauling components from the site. No salvage value assumed. Costs will be higher if components include detectable friable asbestos or if access to the site is limited. If the furnace or ductwork is wrapped in asbestos, you may be legally required to follow asbestos abatement procedures. This may require the services of a specialized asbestos abatement contractor. Add the cost of patching wall, floor or ceiling surfaces and the cost of closing openings after duct is removed. Large register openings in a gravity system can be converted to cold air returns in a forced air system. This avoids closing and patching large wall openings.

| | Craft@Hrs | Unit | Material | Labor | Total | Sell |
|---|---|---|---|---|---|---|
| Duct runs to 30 feet, per run | P1@1.50 | Ea | 20.00 | 47.30 | 67.30 | 114.00 |
| Duct runs 30 feet to 40 feet | P1@1.75 | Ea | 25.00 | 55.10 | 80.10 | 136.00 |
| Duct runs over 40 feet | P1@2.00 | Ea | 30.00 | 63.00 | 93.00 | 158.00 |

**Gas-fired forced air furnace — duct on interior walls.** Includes a humidifier, filter, 5 to 8 galvanized sheet metal duct runs, registers and grilles, thermostat and electrical hookup, connection to gas line and gas piping, chimney vent and vent accessories, supply and return air plenums, system start-up and balancing. Installed in a one- or two-story home with conventional ducting. Add for wall, floor and ceiling patching as required.

| | Craft@Hrs | Unit | Material | Labor | Total | Sell |
|---|---|---|---|---|---|---|
| 80,000 Btu | P1@42.0 | Ea | 1,640.00 | 1,320.00 | 2,960.00 | 5,030.00 |
| 100,000 Btu | P1@43.5 | Ea | 1,820.00 | 1,370.00 | 3,190.00 | 5,420.00 |
| 125,000 Btu | P1@45.0 | Ea | 1,980.00 | 1,420.00 | 3,400.00 | 5,780.00 |
| 150,000 Btu | P1@47.0 | Ea | 2,090.00 | 1,480.00 | 3,570.00 | 6,070.00 |
| 175,000 Btu | P1@49.0 | Ea | 2,260.00 | 1,540.00 | 3,800.00 | 6,460.00 |
| Add for each second floor duct run | P1@1.50 | Ea | 35.20 | 47.30 | 82.50 | 140.00 |
| Add for trunk run to a room addition | P1@4.00 | Ea | 357.00 | 126.00 | 483.00 | 821.00 |

**Gas-fired forced air furnace — duct on exterior walls (perimeter system).** Includes a humidifier, filter, 5 to 8 galvanized sheet metal duct runs, registers and grilles, thermostat and electrical hookup, connection to gas line and gas piping, chimney vent and vent accessories, supply and return air plenums, system start-up and balancing. Installed in a one- or two-story home with perimeter ducting. Add for wall, floor and ceiling patching as required.

| | Craft@Hrs | Unit | Material | Labor | Total | Sell |
|---|---|---|---|---|---|---|
| 80,000 Btu | P1@46.0 | Ea | 1,790.00 | 1,450.00 | 3,240.00 | 5,510.00 |
| 100,000 Btu | P1@47.5 | Ea | 1,990.00 | 1,500.00 | 3,490.00 | 5,930.00 |
| 125,000 Btu | P1@49.0 | Ea | 2,130.00 | 1,540.00 | 3,670.00 | 6,240.00 |
| 150,000 Btu | P1@51.0 | Ea | 2,250.00 | 1,610.00 | 3,860.00 | 6,560.00 |
| 175,000 Btu | P1@53.0 | Ea | 2,410.00 | 1,670.00 | 4,080.00 | 6,940.00 |
| Add for each second floor duct run | P1@2.00 | Ea | 35.20 | 63.00 | 98.20 | 167.00 |
| Add for trunk run to a room addition | P1@4.00 | Ea | 357.00 | 126.00 | 483.00 | 821.00 |

**Low-boy style gas-fired forced air furnace.** When vertical space is limited, install a low-headroom gas furnace. Specs are the same as for a conventional gas furnace, but the furnace cost will be higher. Add the following to either conventional or perimeter gas furnace costs.

| | Craft@Hrs | Unit | Material | Labor | Total | Sell |
|---|---|---|---|---|---|---|
| Add for low-boy 80,000 Btu furnace | P1@1.00 | Ea | 95.20 | 31.50 | 126.70 | 215.00 |
| Add for low-boy 100,000 Btu furnace | P1@2.00 | Ea | 119.00 | 63.00 | 182.00 | 309.00 |
| Add for low-boy 125,000 Btu furnace | P1@2.00 | Ea | 131.00 | 63.00 | 194.00 | 330.00 |
| Add for low-boy 150,000 Btu furnace | P1@4.00 | Ea | 173.00 | 126.00 | 299.00 | 508.00 |
| Add for low-boy 175,000 Btu furnace | P1@4.75 | Ea | 435.00 | 150.00 | 585.00 | 995.00 |

| | Craft@Hrs | Unit | Material | Labor | Total | Sell |
|---|---|---|---|---|---|---|

**Counterflow gas-fired forced air furnace.** When ducting is run under the floor and below the level of the furnace, a counterflow circulation furnace is required. Specs are the same as for a conventional gas furnace, but the furnace cost will be higher. Add the following to either conventional or perimeter gas furnace costs.

| | Craft@Hrs | Unit | Material | Labor | Total | Sell |
|---|---|---|---|---|---|---|
| Add for counterflow furnace to 125,000 Btu | P1@3.50 | Ea | 155.00 | 110.00 | 265.00 | 451.00 |
| Add for counterflow furnace over 125,000 Btu | P1@6.00 | Ea | 459.00 | 189.00 | 648.00 | 1,100.00 |

**Horizontal flow gas-fired forced air furnace.** When space is restricted, a horizontal flow furnace may be the best option available. Specs are the same as for a conventional gas furnace, but the furnace cost will be higher. Add the following to either conventional or perimeter gas furnace costs.

| | Craft@Hrs | Unit | Material | Labor | Total | Sell |
|---|---|---|---|---|---|---|
| Add for horizontal furnace to 125,000 Btu | P1@2.50 | Ea | 119.00 | 78.80 | 197.80 | 336.00 |
| Add for horizontal furnace over 125,000 Btu | P1@4.75 | Ea | 417.00 | 150.00 | 567.00 | 964.00 |

**Oil-fired forced air furnace.** Furnace and duct costs will be the same as for a gas-fired furnace. But add the following costs, which include storage tanks, gauge filter, vent and oil supply piping to the furnace. Add the cost of excavation for buried tanks or sight-screen fencing for above-ground tanks.

| | Craft@Hrs | Unit | Material | Labor | Total | Sell |
|---|---|---|---|---|---|---|
| 275 gallon basement fuel tank | P1@3.75 | Ea | 776.00 | 118.00 | 894.00 | 1,520.00 |
| Twin 275 gallon basement tanks | P1@5.25 | LS | 1,850.00 | 165.00 | 2,015.00 | 3,430.00 |
| Concrete block foundation | P1@2.00 | Ea | 33.10 | 63.00 | 96.10 | 163.00 |
| Exterior 550 gallon fuel tank | P1@8.00 | Ea | 1,350.00 | 252.00 | 1,602.00 | 2,720.00 |
| Exterior 1,000 gallon fuel tank | P1@14.0 | Ea | 1,850.00 | 441.00 | 2,291.00 | 3,890.00 |

**Remove and replace a gas-fired furnace — heating only.** Using existing duct runs. Includes removal of the existing furnace and disposal, a new furnace with an A.G.A. certified ignition system, supply and return air plenums, filters and humidifier, connection to existing duct runs, new chimney pipe, and a new thermostat. Cost estimates for cooling assume installation with the gas furnace and include condenser, concrete pad, 20' of copper suction and liquid line refrigeration tubing, coil section and electrical connection.

| | Craft@Hrs | Unit | Material | Labor | Total | Sell |
|---|---|---|---|---|---|---|
| 80,000 Btu | P1@22.0 | Ea | 1,700.00 | 693.00 | 2,393.00 | 4,070.00 |
| 100,000 Btu | P1@22.8 | Ea | 1,800.00 | 718.00 | 2,518.00 | 4,280.00 |
| 125,000 Btu | P1@24.0 | Ea | 1,960.00 | 756.00 | 2,716.00 | 4,620.00 |
| 150,000 Btu | P1@25.0 | Ea | 2,120.00 | 788.00 | 2,908.00 | 4,940.00 |
| 175,000 Btu | P1@25.8 | Ea | 2,430.00 | 813.00 | 3,243.00 | 5,510.00 |
| Add when replacing a gravity furnace | P1@2.00 | Ea | 64.20 | 63.00 | 127.20 | 216.00 |
| Add for 24,000 Btu cooling | P1@10.0 | Ea | 1,760.00 | 315.00 | 2,075.00 | 3,530.00 |
| Add for 50,000 Btu cooling | P1@14.0 | Ea | 2,800.00 | 441.00 | 3,241.00 | 5,510.00 |
| Add for 60,000 Btu cooling | P1@16.0 | Ea | 3,940.00 | 504.00 | 4,444.00 | 7,550.00 |
| Add for a roof-mounted compressor | P1@6.00 | Ea | 1,060.00 | 189.00 | 1,249.00 | 2,120.00 |

**Install duct runs.** New work, such as for a room addition. Including diffusers and floor or wall registers. Duct suitable for either cold or warm air. Add the cost of wall, floor or ceiling refinishing, as required.

| | Craft@Hrs | Unit | Material | Labor | Total | Sell |
|---|---|---|---|---|---|---|
| Inside or outside wall, per run | P1@2.00 | Ea | 88.40 | 63.00 | 151.40 | 257.00 |
| Cold air central return duct | P1@3.00 | Ea | 82.10 | 94.50 | 176.60 | 300.00 |
| Add per run for second floor ducting | P1@1.50 | Ea | 29.50 | 47.30 | 76.80 | 131.00 |
| Add for insulation, per SF of surface | P1@.030 | SF | .58 | .95 | 1.53 | 2.60 |

**Central humidifier.** Added to an existing furnace or installed with a new furnace. Based on evaporation capacity. A 1,500 square foot house (6 rooms) needs a 10-gallon per day humidifier. A 2,000 square foot house (7 rooms) needs a 12-gallon per day humidifier. A 3,500 square foot house (8 to 10 rooms) needs a 20-gallon per day humidifier.

| | Craft@Hrs | Unit | Material | Labor | Total | Sell |
|---|---|---|---|---|---|---|
| 10 gallons per day, with new furnace | P1@3.75 | Ea | 185.00 | 118.00 | 303.00 | 515.00 |
| 10 gallons per day, existing furnace | P1@4.00 | Ea | 207.00 | 126.00 | 333.00 | 566.00 |
| 12 gallons per day, with new furnace | P1@4.00 | Ea | 196.00 | 126.00 | 322.00 | 547.00 |
| 12 gallons per day, existing furnace | P1@4.25 | Ea | 218.00 | 134.00 | 352.00 | 598.00 |
| 20 gallons per day, with new furnace | P1@4.50 | Ea | 263.00 | 142.00 | 405.00 | 689.00 |
| 20 gallons per day, existing furnace | P1@4.75 | Ea | 291.00 | 150.00 | 441.00 | 750.00 |

| | Craft@Hrs | Unit | Material | Labor | Total | Sell |
|---|---|---|---|---|---|---|

**Central air conditioning.** Added to an existing forced air furnace. Includes remote condenser and pad, coils and cabinet, refrigeration tubing, new thermostat, electric wiring and connection.

| | Craft@Hrs | Unit | Material | Labor | Total | Sell |
|---|---|---|---|---|---|---|
| 2 ton, to 1,000 square feet | P1@16.0 | Ea | 2,200.00 | 504.00 | 2,704.00 | 4,600.00 |
| 3 ton, to 1,600 square feet | P1@18.0 | Ea | 3,130.00 | 567.00 | 3,697.00 | 6,280.00 |
| 5 ton, over 2,000 square feet | P1@20.0 | Ea | 4,150.00 | 630.00 | 4,780.00 | 8,130.00 |

**Add or move steam heat or hot water radiator.** Add a new cast iron radiator or change the location of an existing cast iron radiator. Based on exposed radiator piping. Add the cost of refinishing floor and wall surfaces if piping is to be concealed. First floor work. Based on average-size rooms.

| | Craft@Hrs | Unit | Material | Labor | Total | Sell |
|---|---|---|---|---|---|---|
| Add a new bathroom radiator | P1@7.50 | Ea | 275.00 | 236.00 | 511.00 | 869.00 |
| Add a new bedroom radiator | P1@8.50 | Ea | 332.00 | 268.00 | 600.00 | 1,020.00 |
| Add a new dining room radiator | P1@7.50 | Ea | 297.00 | 236.00 | 533.00 | 906.00 |
| Add a new kitchen radiator | P1@8.50 | Ea | 384.00 | 268.00 | 652.00 | 1,110.00 |
| Add a new living room radiator | P1@9.00 | Ea | 418.00 | 284.00 | 702.00 | 1,190.00 |
| Relocate an existing radiator | P1@6.50 | Ea | 74.90 | 205.00 | 279.90 | 476.00 |
| Add for second floor installations | — | % | 20.0 | — | — | — |

**Hydronic heating for room additions.** Install up to 8 feet of baseboard convection unit with necessary supply and return lines. Based on an average-size single room.

| | Craft@Hrs | Unit | Material | Labor | Total | Sell |
|---|---|---|---|---|---|---|
| First floor work, per room | P1@9.00 | Ea | 656.00 | 284.00 | 940.00 | 1,600.00 |
| Second floor work, per room | P1@11.0 | Ea | 758.00 | 347.00 | 1,105.00 | 1,880.00 |

**Remove and replace a cast iron radiator.** Replaced on the same wall using exposed piping. Add the cost of refinishing floor and wall surfaces if required.

| | Craft@Hrs | Unit | Material | Labor | Total | Sell |
|---|---|---|---|---|---|---|
| Replace with a baseboard radiator | P1@8.00 | Ea | 605.00 | 252.00 | 857.00 | 1,460.00 |
| Replace with thin-line radiator | P1@8.00 | Ea | 599.00 | 252.00 | 851.00 | 1,450.00 |

**Remove and replace a hydronic heating system.** In an existing home. Remove and haul away the existing heating plant. Install a new gas- or propane-fired hydronic heating boiler with insulated steel jacket, new wall thermostat, electrical connections and wiring. Connect to existing gas, propane or oil supply lines and existing hydronic distribution piping. Includes electric or spark ignition, power vent controls, standard valves, gauges, and drain valves. Add the cost of radiators and refinishing floor, ceiling and wall surfaces as required.

| | Craft@Hrs | Unit | Material | Labor | Total | Sell |
|---|---|---|---|---|---|---|
| Gas fired, 100,000 Btu | P1@26.0 | Ea | 2,580.00 | 819.00 | 3,399.00 | 5,780.00 |
| Gas fired, 125,000 Btu | P1@27.5 | Ea | 2,630.00 | 866.00 | 3,496.00 | 5,940.00 |
| Oil fired, 100,000 Btu | P1@27.0 | Ea | 2,890.00 | 851.00 | 3,741.00 | 6,360.00 |
| Oil fired, 125,000 Btu | P1@28.0 | Ea | 3,080.00 | 882.00 | 3,962.00 | 6,740.00 |
| Add per bathroom radiator, copper pipe | P1@7.50 | Ea | 271.00 | 236.00 | 507.00 | 862.00 |
| Add per bedroom radiator, copper pipe | P1@5.00 | Ea | 311.00 | 158.00 | 469.00 | 797.00 |
| Add per dining room radiator, copper pipe | P1@8.50 | Ea | 328.00 | 268.00 | 596.00 | 1,010.00 |
| Add per kitchen radiator, copper pipe | P1@6.00 | Ea | 357.00 | 189.00 | 546.00 | 928.00 |
| Add per living room radiator, copper pipe | P1@7.50 | Ea | 294.00 | 236.00 | 530.00 | 901.00 |
| Add per bathroom radiator, iron pipe | P1@5.00 | Ea | 317.00 | 158.00 | 475.00 | 808.00 |
| Add per bedroom radiator, iron pipe | P1@8.50 | Ea | 379.00 | 268.00 | 647.00 | 1,100.00 |
| Add per dining room radiator, iron pipe | P1@6.50 | Ea | 470.00 | 205.00 | 675.00 | 1,150.00 |
| Add per kitchen radiator, iron pipe | P1@9.00 | Ea | 413.00 | 284.00 | 697.00 | 1,180.00 |
| Add per living room radiator, iron pipe | P1@7.00 | Ea | 436.00 | 221.00 | 657.00 | 1,120.00 |
| Add for second floor radiators | — | % | 20.0 | — | — | — |
| Add for circulating pump | P1@5.00 | Ea | 464.00 | 158.00 | 622.00 | 1,060.00 |

| | Craft@Hrs | Unit | Material | Labor | Total | Sell |
|---|---|---|---|---|---|---|

**Chimney liner (flue), aluminum.** Prefabricated, 2 ply single wall flexible aluminum gas or propane fuel chimney liner. Approved for installation in an existing masonry chimney. Maximum flue temperature of 270 degrees F. Check local codes for regulations regarding acceptable installation requirements.

| | Craft@Hrs | Unit | Material | Labor | Total | Sell |
|---|---|---|---|---|---|---|
| 3" flexible chimney liner | P1@.085 | LF | 2.16 | 2.68 | 4.84 | 8.23 |
| 4" flexible chimney liner | P1@.090 | LF | 2.63 | 2.84 | 5.47 | 9.30 |
| 5" flexible chimney liner | P1@.095 | LF | 3.09 | 2.99 | 6.08 | 10.30 |
| 6" flexible chimney liner | P1@.100 | LF | 3.76 | 3.15 | 6.91 | 11.70 |
| 7" flexible chimney liner | P1@.115 | LF | 4.29 | 3.62 | 7.91 | 13.40 |
| 8" flexible chimney liner | P1@.120 | LF | 4.85 | 3.78 | 8.63 | 14.70 |

**Chimney liner rain cap assembly, aluminum.** Approved for installation in an existing masonry chimney. Maximum flue temperature of 270 degrees F. Check local codes for regulations regarding acceptable installation requirements.

| | Craft@Hrs | Unit | Material | Labor | Total | Sell |
|---|---|---|---|---|---|---|
| 3" chimney liner cap assembly | P1@.450 | Ea | 42.90 | 14.20 | 57.10 | 97.10 |
| 4" chimney liner cap assembly | P1@.450 | Ea | 40.80 | 14.20 | 55.00 | 93.50 |
| 5" chimney liner cap assembly | P1@.450 | Ea | 47.00 | 14.20 | 61.20 | 104.00 |
| 6" chimney liner cap assembly | P1@.450 | Ea | 53.70 | 14.20 | 67.90 | 115.00 |
| 7" chimney liner cap assembly | P1@.500 | Ea | 60.60 | 15.80 | 76.40 | 130.00 |
| 8" chimney liner cap assembly | P1@.500 | Ea | 64.40 | 15.80 | 80.20 | 136.00 |

**Chimney liner (flue), stainless steel.** Prefabricated, 304/316 stainless steel single wall flexible chimney liner. Approved for installation in an existing masonry chimney. Maximum flue temperature of 2100 degrees F. Check local codes for regulations regarding acceptable installation requirements.

| | Craft@Hrs | Unit | Material | Labor | Total | Sell |
|---|---|---|---|---|---|---|
| 3" flexible chimney liner | P1@.100 | LF | 12.60 | 3.15 | 15.75 | 26.80 |
| 4" flexible chimney liner | P1@.115 | LF | 13.20 | 3.62 | 16.82 | 28.60 |
| 5" flexible chimney liner | P1@.120 | LF | 16.20 | 3.78 | 19.98 | 34.00 |
| 6" flexible chimney liner | P1@.125 | LF | 19.50 | 3.94 | 23.44 | 39.80 |
| 7" flexible chimney liner | P1@.130 | LF | 23.20 | 4.10 | 27.30 | 46.40 |
| 8" flexible chimney liner | P1@.135 | LF | 29.40 | 4.25 | 33.65 | 57.20 |

**Chimney (flue), C vent.** Prefabricated single wall 24-gauge galvanized steel snap lock gas or propane fuel chimney. Add for cutting new holes in ceiling and roof, and patching, if required. Check local codes for regulations regarding acceptable installation requirements.

| | Craft@Hrs | Unit | Material | Labor | Total | Sell |
|---|---|---|---|---|---|---|
| 5" single wall | P1@.110 | LF | 1.23 | 3.47 | 4.70 | 7.99 |
| 6" single wall | P1@.115 | LF | 1.40 | 3.62 | 5.02 | 8.53 |
| 7" single wall | P1@.130 | LF | 1.57 | 4.10 | 5.67 | 9.64 |
| 8" single wall | P1@.135 | LF | 1.69 | 4.25 | 5.94 | 10.10 |

**Chimney (flue), C vent, aluminum.** Prefabricated .020" thickness, 24-gauge single wall, aluminum snap lock gas or propane fuel chimney. Add for cutting new holes in ceiling and roof, and patching, if required. Check local codes for regulations regarding acceptable installation requirements.

| | Craft@Hrs | Unit | Material | Labor | Total | Sell |
|---|---|---|---|---|---|---|
| 3" single wall | P1@.095 | LF | .76 | 2.99 | 3.75 | 6.38 |
| 4" single wall | P1@.100 | LF | .91 | 3.15 | 4.06 | 6.90 |
| 5" single wall | P1@.110 | LF | 1.17 | 3.47 | 4.64 | 7.89 |
| 6" single wall | P1@.115 | LF | 1.38 | 3.62 | 5.00 | 8.50 |
| 7" single wall | P1@.130 | LF | 1.66 | 4.10 | 5.76 | 9.79 |
| 8" single wall | P1@.135 | LF | 1.93 | 4.25 | 6.18 | 10.50 |

| | Craft@Hrs | Unit | Material | Labor | Total | Sell |
|---|---|---|---|---|---|---|

**Chimney (flue) base tee, C vent.** Prefabricated single wall galvanized steel gas or propane fuel chimney base tee, including cap. Add for cutting new holes in ceiling and roof, and patching, if required. Check local codes for regulations regarding acceptable installation requirements.

| | Craft@Hrs | Unit | Material | Labor | Total | Sell |
|---|---|---|---|---|---|---|
| 3" single wall | P1@.150 | Ea | 18.50 | 4.73 | 23.23 | 39.50 |
| 4" single wall | P1@.155 | Ea | 18.90 | 4.88 | 23.78 | 40.40 |
| 5" single wall | P1@.165 | Ea | 23.10 | 5.20 | 28.30 | 48.10 |
| 6" single wall | P1@.180 | Ea | 26.20 | 5.67 | 31.87 | 54.20 |
| 7" single wall | P1@.195 | Ea | 30.90 | 6.14 | 37.04 | 63.00 |
| 8" single wall | P1@.220 | Ea | 42.70 | 6.93 | 49.63 | 84.40 |

**Chimney (flue) base tee, C vent, aluminum.** Prefabricated single wall aluminum gas or propane fuel chimney base tee, including cap. Add for cutting new holes in ceiling and roof, and patching, if required. Check local codes for regulations regarding acceptable installation requirements.

| | Craft@Hrs | Unit | Material | Labor | Total | Sell |
|---|---|---|---|---|---|---|
| 3" single wall | P1@.150 | Ea | 21.50 | 4.73 | 26.23 | 44.60 |
| 4" single wall | P1@.155 | Ea | 23.00 | 4.88 | 27.88 | 47.40 |
| 5" single wall | P1@.165 | Ea | 25.00 | 5.20 | 30.20 | 51.30 |
| 6" single wall | P1@.180 | Ea | 28.70 | 5.67 | 34.37 | 58.40 |
| 7" single wall | P1@.195 | Ea | 33.20 | 6.14 | 39.34 | 66.90 |
| 8" single wall | P1@.220 | Ea | 44.60 | 6.93 | 51.53 | 87.60 |

**Chimney (flue) elbow, C vent.** Prefabricated single wall galvanized steel gas or propane fuel chimney adjustable elbow. Add for cutting new holes in ceiling and roof, and patching, if required. Check local codes for regulations regarding acceptable installation requirements.

| | Craft@Hrs | Unit | Material | Labor | Total | Sell |
|---|---|---|---|---|---|---|
| 3" single wall, adjustable elbow | P1@.150 | Ea | 3.15 | 4.73 | 7.88 | 13.40 |
| 4" single wall, adjustable elbow | P1@.155 | Ea | 4.15 | 4.88 | 9.03 | 15.40 |
| 5" single wall, adjustable elbow | P1@.165 | Ea | 4.49 | 5.20 | 9.69 | 16.50 |
| 6" single wall, adjustable elbow | P1@.180 | Ea | 5.14 | 5.67 | 10.81 | 18.40 |
| 7" single wall, adjustable elbow | P1@.195 | Ea | 5.66 | 6.14 | 11.80 | 20.10 |
| 8" single wall, adjustable elbow | P1@.220 | Ea | 7.41 | 6.93 | 14.34 | 24.40 |

**Chimney (flue) elbow, C vent, aluminum.** Prefabricated single wall aluminum gas or propane fuel chimney adjustable elbow. Add for cutting new holes in ceiling and roof, and patching, if required. Check local codes for regulations regarding acceptable installation requirements.

| | Craft@Hrs | Unit | Material | Labor | Total | Sell |
|---|---|---|---|---|---|---|
| 3" single wall, adjustable elbow | P1@.150 | Ea | 3.73 | 4.73 | 8.46 | 14.40 |
| 4" single wall, adjustable elbow | P1@.155 | Ea | 4.55 | 4.88 | 9.43 | 16.00 |
| 5" single wall, adjustable elbow | P1@.165 | Ea | 5.14 | 5.20 | 10.34 | 17.60 |
| 6" single wall, adjustable elbow | P1@.180 | Ea | 6.65 | 5.67 | 12.32 | 20.90 |
| 7" single wall, adjustable elbow | P1@.195 | Ea | 9.63 | 6.14 | 15.77 | 26.80 |
| 8" single wall, adjustable elbow | P1@.220 | Ea | 13.30 | 6.93 | 20.23 | 34.40 |

**Chimney (flue) storm collar, rain cap and roof flashing assembly, C vent.** Prefabricated single wall galvanized steel gas or propane fuel chimney termination assembly. Add for cutting new holes in ceiling and roof, and patching if required. Check local codes for regulations regarding acceptable installation requirements.

| | Craft@Hrs | Unit | Material | Labor | Total | Sell |
|---|---|---|---|---|---|---|
| 3" single wall, termination assembly | P1@.600 | Ea | 30.80 | 18.90 | 49.70 | 84.50 |
| 4" single wall, termination assembly | P1@.650 | Ea | 32.90 | 20.50 | 53.40 | 90.80 |
| 5" single wall, termination assembly | P1@.700 | Ea | 35.60 | 22.10 | 57.70 | 98.10 |
| 6" single wall, termination assembly | P1@.750 | Ea | 41.60 | 23.60 | 65.20 | 111.00 |
| 7" single wall, termination assembly | P1@.900 | Ea | 49.70 | 28.40 | 78.10 | 133.00 |
| 8" single wall, termination assembly | P1@1.15 | Ea | 60.40 | 36.20 | 96.60 | 164.00 |

|  | Craft@Hrs | Unit | Material | Labor | Total | Sell |
|---|---|---|---|---|---|---|

**Chimney (flue), B vent.** Prefabricated double wall metal gas or propane fuel chimney. Aluminum inner liner with extruded aluminum outer jacket. Add for cutting new holes in ceiling and roof, and patching, if required. Check local codes for regulations regarding acceptable installation requirements.

| Item | Craft@Hrs | Unit | Material | Labor | Total | Sell |
|---|---|---|---|---|---|---|
| 3" double wall | P1@.095 | LF | 4.90 | 2.99 | 7.89 | 13.40 |
| 4" double wall | P1@.100 | LF | 5.19 | 3.15 | 8.34 | 14.20 |
| 5" double wall | P1@.115 | LF | 5.90 | 3.62 | 9.52 | 16.20 |
| 6" double wall | P1@.120 | LF | 6.19 | 3.78 | 9.97 | 16.90 |
| 7" double wall | P1@.135 | LF | 8.05 | 4.25 | 12.30 | 20.90 |
| 8" double wall | P1@.140 | LF | 9.27 | 4.41 | 13.68 | 23.30 |

**Chimney (flue) base tee, B vent.** Prefabricated double wall metal gas or propane fuel chimney base tee with cap. Aluminum inner liner with extruded aluminum outer jacket. Check local codes for regulations regarding acceptable installation requirements.

| Item | Craft@Hrs | Unit | Material | Labor | Total | Sell |
|---|---|---|---|---|---|---|
| 3" double wall | P1@.150 | Ea | 35.60 | 4.73 | 40.33 | 68.60 |
| 4" double wall | P1@.155 | Ea | 37.60 | 4.88 | 42.48 | 72.20 |
| 5" double wall | P1@.165 | Ea | 41.70 | 5.20 | 46.90 | 79.70 |
| 6" double wall | P1@.180 | Ea | 44.90 | 5.67 | 50.57 | 86.00 |
| 7" double wall | P1@.195 | Ea | 52.30 | 6.14 | 58.44 | 99.30 |
| 8" double wall | P1@.220 | Ea | 60.40 | 6.93 | 67.33 | 114.00 |

**Chimney (flue) elbow, B vent.** Prefabricated double wall metal gas or propane fuel chimney base tee with cap. Aluminum inner liner with extruded aluminum outer jacket. Check local codes for regulations regarding acceptable installation requirements.

| Item | Craft@Hrs | Unit | Material | Labor | Total | Sell |
|---|---|---|---|---|---|---|
| 3" double wall, 45 degree elbow | P1@.150 | Ea | 16.80 | 4.73 | 21.53 | 36.60 |
| 3" double wall, 90 degree elbow | P1@.150 | Ea | 22.10 | 4.73 | 26.83 | 45.60 |
| 4" double wall, 45 degree elbow | P1@.155 | Ea | 17.10 | 4.88 | 21.98 | 37.40 |
| 4" double wall, 90 degree elbow | P1@.155 | Ea | 22.30 | 4.88 | 27.18 | 46.20 |
| 5" double wall, 45 degree elbow | P1@.165 | Ea | 17.40 | 5.20 | 22.60 | 38.40 |
| 5" double wall, 90 degree elbow | P1@.165 | Ea | 22.80 | 5.20 | 28.00 | 47.60 |
| 6" double wall, 45 degree elbow | P1@.180 | Ea | 20.00 | 5.67 | 25.67 | 43.60 |
| 6" double wall, 90 degree elbow | P1@.180 | Ea | 24.50 | 5.67 | 30.17 | 51.30 |
| 7" double wall, 45 degree elbow | P1@.195 | Ea | 24.20 | 6.14 | 30.34 | 51.60 |
| 7" double wall, 90 degree elbow | P1@.195 | Ea | 28.20 | 6.14 | 34.34 | 58.40 |
| 8" double wall, 45 degree elbow | P1@.220 | Ea | 29.50 | 6.93 | 36.43 | 61.90 |
| 8" double wall, 90 degree elbow | P1@.220 | Ea | 34.90 | 6.93 | 41.83 | 71.10 |

**Chimney (flue) storm collar, rain cap and roof flashing assembly, B vent.** Add for cutting new holes in ceiling and roof, and patching, if required. Check local codes for regulations regarding acceptable installation requirements.

| Item | Craft@Hrs | Unit | Material | Labor | Total | Sell |
|---|---|---|---|---|---|---|
| 3" double wall, termination assembly | P1@.600 | Ea | 30.80 | 18.90 | 49.70 | 84.50 |
| 4" double wall, termination assembly | P1@.650 | Ea | 32.90 | 20.50 | 53.40 | 90.80 |
| 5" double wall, termination assembly | P1@.700 | Ea | 35.60 | 22.10 | 57.70 | 98.10 |
| 6" double wall, termination assembly | P1@.750 | Ea | 41.60 | 23.60 | 65.20 | 111.00 |
| 7" double wall, termination assembly | P1@.900 | Ea | 49.70 | 28.40 | 78.10 | 133.00 |
| 8" double wall, termination assembly | P1@1.15 | Ea | 60.40 | 36.20 | 96.60 | 164.00 |

**Chimney (flue), L vent.** Prefabricated double wall metal gas or propane fuel chimney. Stainless steel inner liner with extruded aluminum outer jacket. High temperature and condensing boiler or furnace applications. Maximum flue temperature of 750 degrees F. Add for cutting new holes in ceiling and roof, and patching, if required. Check local codes for regulations regarding acceptable installation requirements.

| Item | Craft@Hrs | Unit | Material | Labor | Total | Sell |
|---|---|---|---|---|---|---|
| 5" double wall | P1@.115 | LF | 18.10 | 3.62 | 21.72 | 36.90 |
| 6" double wall | P1@.120 | LF | 20.50 | 3.78 | 24.28 | 41.30 |
| 7" double wall | P1@.135 | LF | 23.50 | 4.25 | 27.75 | 47.20 |
| 8" double wall | P1@.140 | LF | 25.10 | 4.41 | 29.51 | 50.20 |

|  | Craft@Hrs | Unit | Material | Labor | Total | Sell |
|---|---|---|---|---|---|---|

**Chimney (flue) base tee, L vent.** Prefabricated double wall metal gas or propane fuel chimney base tee with cap. Stainless steel inner liner with extruded aluminum outer jacket. High temperature and condensing boiler or furnace applications. Maximum flue temperature of 750 degrees F. Check local codes for regulations regarding acceptable installation requirements.

| | | | | | | |
|---|---|---|---|---|---|---|
| 5" double wall | P1@.165 | Ea | 67.10 | 5.20 | 72.30 | 123.00 |
| 6" double wall | P1@.180 | Ea | 71.80 | 5.67 | 77.47 | 132.00 |
| 7" double wall | P1@.195 | Ea | 85.90 | 6.14 | 92.04 | 156.00 |
| 8" double wall | P1@.220 | Ea | 96.00 | 6.93 | 102.93 | 175.00 |

**Chimney (flue) elbow, L vent.** Prefabricated double wall metal gas or propane fuel chimney. Stainless steel inner liner with extruded aluminum outer jacket. High temperature and condensing boiler or furnace applications. Maximum flue temperature of 750 degrees F. Check local codes for regulations regarding acceptable installation requirements.

| | | | | | | |
|---|---|---|---|---|---|---|
| 5" double wall, 45 degree elbow | P1@.165 | Ea | 53.70 | 5.20 | 58.90 | 100.00 |
| 5" double wall, 90 degree elbow | P1@.165 | Ea | 77.80 | 5.20 | 83.00 | 141.00 |
| 6" double wall, 45 degree elbow | P1@.180 | Ea | 57.70 | 5.67 | 63.37 | 108.00 |
| 6" double wall, 90 degree elbow | P1@.180 | Ea | 79.20 | 5.67 | 84.87 | 144.00 |
| 7" double wall, 45 degree elbow | P1@.195 | Ea | 78.80 | 6.14 | 84.94 | 144.00 |
| 7" double wall, 90 degree elbow | P1@.195 | Ea | 52.50 | 6.14 | 58.64 | 99.70 |
| 8" double wall, 45 degree elbow | P1@.220 | Ea | 64.40 | 6.93 | 71.33 | 121.00 |
| 8" double wall, 90 degree elbow | P1@.220 | Ea | 92.60 | 6.93 | 99.53 | 169.00 |

**Chimney (flue) storm collar, rain cap and roof flashing assembly, L vent.** Prefabricated double wall metal gas or propane fuel chimney. Stainless steel inner liner with extruded aluminum outer jacket. High temperature and condensing boiler or furnace applications. Maximum flue temperature of 750 degrees F. Check local codes for regulations regarding acceptable installation requirements.

| | | | | | | |
|---|---|---|---|---|---|---|
| 5" double wall, termination assembly | P1@.700 | Ea | 80.50 | 22.10 | 102.60 | 174.00 |
| 6" double wall, termination assembly | P1@.750 | Ea | 93.80 | 23.60 | 117.40 | 200.00 |
| 7" double wall, termination assembly | P1@.900 | Ea | 104.00 | 28.40 | 132.40 | 225.00 |
| 8" double wall, termination assembly | P1@1.15 | Ea | 127.00 | 36.20 | 163.20 | 277.00 |

**Chimney (flue), A vent.** Prefabricated double wall insulated stainless steel all-fuels chimney. Maximum flue temperature of 1,700 degrees F. Stainless steel inner liner and outer jacket. Twist-lock connection. Add for cutting new holes in ceiling and roof, and patching, if required. Check local codes for regulations regarding acceptable installation requirements.

| | | | | | | |
|---|---|---|---|---|---|---|
| 5" double wall | P1@.120 | LF | 35.00 | 3.78 | 38.78 | 65.90 |
| 6" double wall | P1@.125 | LF | 37.10 | 3.94 | 41.04 | 69.80 |
| 7" double wall | P1@.140 | LF | 39.10 | 4.41 | 43.51 | 74.00 |
| 8" double wall | P1@.145 | LF | 41.70 | 4.57 | 46.27 | 78.70 |
| 10" double wall | P1@.160 | LF | 67.70 | 5.04 | 72.74 | 124.00 |
| Add for 2100 degree maximum temp rating | — | % | 45.0 | — | — | — |

**Chimney (flue) base tee, A vent.** Prefabricated double wall insulated all-fuels chimney base tee including cap. Maximum flue temperature of 1,700 degrees F. Stainless steel inner liner and outer jacket. Twist-lock connection. Add for cutting new holes in ceiling and roof and patching, if required. Check local codes for regulations regarding acceptable installation requirements.

| | | | | | | |
|---|---|---|---|---|---|---|
| 5" double wall | P1@.170 | Ea | 152.00 | 5.36 | 157.36 | 268.00 |
| 6" double wall | P1@.185 | Ea | 165.00 | 5.83 | 170.83 | 290.00 |
| 7" double wall | P1@.200 | Ea | 172.00 | 6.30 | 178.30 | 303.00 |
| 8" double wall | P1@.225 | Ea | 181.00 | 7.09 | 188.09 | 320.00 |
| 10" double wall | P1@.260 | Ea | 216.00 | 8.19 | 224.19 | 381.00 |
| Add for 2100 degree F. maximum temp rating | — | % | 25.0 | — | — | — |

| | Craft@Hrs | Unit | Material | Labor | Total | Sell |
|---|---|---|---|---|---|---|

**Chimney (flue) elbow, A vent.** Prefabricated double wall insulated all-fuels chimney elbow. Maximum flue temperature of 1,700 degrees F. Stainless steel inner liner and outer jacket. Twist-lock connection. Add for cutting new holes in ceiling and roof and patching, if required. Check local codes for regulations regarding acceptable installation requirements.

| | Craft@Hrs | Unit | Material | Labor | Total | Sell |
|---|---|---|---|---|---|---|
| 5" double wall, 15 degree elbow | P1@.170 | Ea | 56.80 | 5.36 | 62.16 | 106.00 |
| 5" double wall, 30 degree elbow | P1@.170 | Ea | 59.70 | 5.36 | 65.06 | 111.00 |
| 5" double wall, 45 degree elbow | P1@.170 | Ea | 85.90 | 5.36 | 91.26 | 155.00 |
| 6" double wall, 15 degree elbow | P1@.190 | Ea | 58.70 | 5.99 | 64.69 | 110.00 |
| 6" double wall, 30 degree elbow | P1@.190 | Ea | 61.50 | 5.99 | 67.49 | 115.00 |
| 6" double wall, 45 degree elbow | P1@.190 | Ea | 78.60 | 5.99 | 84.59 | 144.00 |
| 7" double wall, 15 degree elbow | P1@.225 | Ea | 67.20 | 7.09 | 74.29 | 126.00 |
| 7" double wall, 30 degree elbow | P1@.225 | Ea | 73.10 | 7.09 | 80.19 | 136.00 |
| 7" double wall, 45 degree elbow | P1@.225 | Ea | 93.70 | 7.09 | 100.79 | 171.00 |
| 8" double wall, 15 degree elbow | P1@.260 | Ea | 79.60 | 8.19 | 87.79 | 149.00 |
| 8" double wall, 30 degree elbow | P1@.260 | Ea | 87.10 | 8.19 | 95.29 | 162.00 |
| 8" double wall, 45 degree elbow | P1@.260 | Ea | 104.00 | 8.19 | 112.19 | 191.00 |
| 10" double wall, 15 degree elbow | P1@.330 | Ea | 96.50 | 10.40 | 106.90 | 182.00 |
| 10" double wall, 30 degree elbow | P1@.330 | Ea | 109.00 | 10.40 | 119.40 | 203.00 |
| 10" double wall, 45 degree elbow | P1@.330 | Ea | 133.00 | 10.40 | 143.40 | 244.00 |
| Add for 2100 degree maximum temp rating | — | % | 25.0 | — | — | — |

**Chimney (flue) storm collar, rain cap and roof flashing assembly, A vent.** Prefabricated double wall insulated all-fuels chimney. Maximum flue temperature of 1,700 degrees F. Twist-lock connection. Add for cutting new holes in ceiling and roof, and patching, if required. Check local codes for regulations regarding acceptable installation requirements.

| | Craft@Hrs | Unit | Material | Labor | Total | Sell |
|---|---|---|---|---|---|---|
| 5" double wall, termination assembly | P1@.750 | Ea | 95.70 | 23.60 | 119.30 | 203.00 |
| 6" double wall, termination assembly | P1@.800 | Ea | 107.00 | 25.20 | 132.20 | 225.00 |
| 7" double wall, termination assembly | P1@.950 | Ea | 113.00 | 29.90 | 142.90 | 243.00 |
| 8" double wall, termination assembly | P1@1.20 | Ea | 126.00 | 37.80 | 163.80 | 278.00 |
| 10" double wall, termination assembly | P1@1.30 | Ea | 147.00 | 41.00 | 188.00 | 320.00 |

**Chimney (flue) supports and brackets, A vent.** Prefabricated double wall insulated all-fuels chimney. Maximum flue temperature of 1,700 degrees F. Twist-lock connection. Add for cutting new holes in ceiling and roof, and patching, if required. Check local codes for regulations regarding acceptable installation requirements.

| | Craft@Hrs | Unit | Material | Labor | Total | Sell |
|---|---|---|---|---|---|---|
| 5" double wall, anchor plate | P1@.200 | Ea | 32.70 | 6.30 | 39.00 | 66.30 |
| 5" double wall, adjustable wall support | P1@.300 | Ea | 53.70 | 9.45 | 63.15 | 107.00 |
| 5" double wall, wall or floor support | P1@.400 | Ea | 37.30 | 12.60 | 49.90 | 84.80 |
| 6" double wall, anchor plate | P1@.225 | Ea | 36.20 | 7.09 | 43.29 | 73.60 |
| 6" double wall, adjustable wall support | P1@.325 | Ea | 67.70 | 10.20 | 77.90 | 132.00 |
| 6" double wall, wall or floor support | P1@.425 | Ea | 56.10 | 13.40 | 69.50 | 118.00 |
| 7" double wall, anchor plate | P1@.230 | Ea | 38.60 | 7.25 | 45.85 | 77.90 |
| 7" double wall, adjustable wall support | P1@.330 | Ea | 75.90 | 10.40 | 86.30 | 147.00 |
| 7" double wall, wall or floor support | P1@.430 | Ea | 59.50 | 13.50 | 73.00 | 124.00 |
| 8" double wall, anchor plate | P1@.240 | Ea | 40.80 | 7.56 | 48.36 | 82.20 |
| 8" double wall, adjustable wall support | P1@.350 | Ea | 87.50 | 11.00 | 98.50 | 167.00 |
| 8" double wall, wall or floor support | P1@.450 | Ea | 59.50 | 14.20 | 73.70 | 125.00 |
| 10" double wall, anchor plate | P1@.260 | Ea | 49.00 | 8.19 | 57.19 | 97.20 |
| 10" double wall, adjustable wall support | P1@.400 | Ea | 103.00 | 12.60 | 115.60 | 197.00 |
| 10" double wall, wall or floor support | P1@.550 | Ea | 84.00 | 17.30 | 101.30 | 172.00 |

| | Craft@Hrs | Unit | Material | Labor | Total | Sell |
|---|---|---|---|---|---|---|

**Electric baseboard heaters.** Includes a remote thermostat, a separate circuit run to the point of use and connection. Each foot of electric baseboard heater delivers approximately 925 Btu of heating capacity.

| | Craft@Hrs | Unit | Material | Labor | Total | Sell |
|---|---|---|---|---|---|---|
| 1,850 Btu, 540-watt, 2' heater | P1@4.00 | Ea | 137.00 | 126.00 | 263.00 | 447.00 |
| 2,775 Btu, 660-watt, 3' heater | P1@4.50 | Ea | 158.00 | 142.00 | 300.00 | 510.00 |
| 3,700 Btu, 1.0kW, 4' heater | P1@4.65 | Ea | 187.00 | 146.00 | 333.00 | 566.00 |
| 4,600 Btu, 1.35kW, 5' heater | P1@4.75 | Ea | 208.00 | 150.00 | 358.00 | 609.00 |
| 5,350 Btu, 1.55kW, 6' heater | P1@5.00 | Ea | 235.00 | 158.00 | 393.00 | 668.00 |
| 7,200 Btu, 2.0kW, 8' heater | P1@5.25 | Ea | 293.00 | 165.00 | 458.00 | 779.00 |
| 9,050 Btu, 2.65kW, 9' heater | P1@5.50 | Ea | 314.00 | 173.00 | 487.00 | 828.00 |
| 11,900 Btu, 3.5kW, 10' heater | P1@5.75 | Ea | 362.00 | 181.00 | 543.00 | 923.00 |

**Central coal or wood-burning furnace.** Connected to existing ductwork and chimney flue. Triple flue heat exchanger, refractory lined chamber holds logs from 22" to 32" in diameter. Includes fan lift switch, blower system, cast iron grate, automatic controlled damper and removal of the existing furnace.

| | Craft@Hrs | Unit | Material | Labor | Total | Sell |
|---|---|---|---|---|---|---|
| Furnace | P1@25.8 | Ea | 2,200.00 | 813.00 | 3,013.00 | 5,120.00 |

**Gas floor furnace.** Includes thermostat, electrical connection, gas piping, flue and valve. Add the cost of interior and exterior patching.

| | Craft@Hrs | Unit | Material | Labor | Total | Sell |
|---|---|---|---|---|---|---|
| 35,000 Btu, pilot ignition | P1@9.00 | Ea | 525.00 | 284.00 | 809.00 | 1,380.00 |
| 35,000 Btu, spark ignition | P1@9.00 | Ea | 624.00 | 284.00 | 908.00 | 1,540.00 |
| 50,000 Btu, pilot ignition | P1@9.50 | Ea | 586.00 | 299.00 | 885.00 | 1,500.00 |
| 50,000 Btu, spark ignition | P1@9.50 | Ea | 684.00 | 299.00 | 983.00 | 1,670.00 |
| 65,000 Btu, pilot ignition | P1@10.0 | Ea | 680.00 | 315.00 | 995.00 | 1,690.00 |
| 65,000 Btu, spark ignition | P1@10.0 | Ea | 790.00 | 315.00 | 1,105.00 | 1,880.00 |

**Gas wall furnace.** Includes thermostat, electrical connection, gas piping, valve, blower and debris removal. Add the cost of wall, ceiling and roof patching. Two room (rear grille) units have one thermostat and control the temperature in one room only. For a room under 300 square feet, a 25,000 Btu furnace usually has enough capacity if ceiling height is 8' or less.

| | Craft@Hrs | Unit | Material | Labor | Total | Sell |
|---|---|---|---|---|---|---|
| 25,000 Btu, pilot ignition | P1@9.00 | Ea | 443.00 | 284.00 | 727.00 | 1,240.00 |
| 25,000 Btu, spark ignition | P1@9.00 | Ea | 443.00 | 284.00 | 727.00 | 1,240.00 |
| 40,000 Btu, pilot ignition | P1@9.50 | Ea | 552.00 | 299.00 | 851.00 | 1,450.00 |
| 40,000 Btu, spark ignition | P1@9.50 | Ea | 684.00 | 299.00 | 983.00 | 1,670.00 |
| 50,000 Btu, pilot ignition | P1@9.50 | Ea | 574.00 | 299.00 | 873.00 | 1,480.00 |
| 50,000 Btu, spark ignition | P1@9.50 | Ea | 705.00 | 299.00 | 1,004.00 | 1,710.00 |
| 60,000 Btu, pilot ignition | P1@10.0 | Ea | 624.00 | 315.00 | 939.00 | 1,600.00 |
| 60,000 Btu, spark ignition | P1@10.0 | Ea | 766.00 | 315.00 | 1,081.00 | 1,840.00 |
| Add for dual wall unit (2 rooms) | P1@6.00 | Ea | 527.00 | 189.00 | 716.00 | 1,220.00 |

**Thru-the-wall gas furnace.** Includes thermostat, electrical connection, gas piping, valve, direct exterior wall vent and debris removal. Pilot ignition. Add the cost of interior and exterior wall patching if needed.

| | Craft@Hrs | Unit | Material | Labor | Total | Sell |
|---|---|---|---|---|---|---|
| 100 square foot room, 14,000 Btu | P1@8.00 | Ea | 454.00 | 252.00 | 706.00 | 1,200.00 |
| 150 square foot room, 22,000 Btu | P1@8.50 | Ea | 481.00 | 268.00 | 749.00 | 1,270.00 |
| 200 square foot room, 25,000 Btu | P1@9.00 | Ea | 515.00 | 284.00 | 799.00 | 1,360.00 |
| 250 square foot room, 30,000 Btu | P1@9.00 | Ea | 542.00 | 284.00 | 826.00 | 1,400.00 |
| 300 square foot room, 40,000 Btu | P1@9.50 | Ea | 591.00 | 299.00 | 890.00 | 1,510.00 |
| 400 square foot room, 65,000 Btu | P1@10.0 | Ea | 722.00 | 315.00 | 1,037.00 | 1,760.00 |

| | Craft@Hrs | Unit | Material | Labor | Total | Sell |
|---|---|---|---|---|---|---|

## Gas Water Heaters

**30-gallon gas water heater.** 6-year warranty. Includes temperature and pressure relief valve. Includes pipe connection labor. Add for flue, water and gas pipe, fitting materials and gas permit costs if required.

| | Craft@Hrs | Unit | Material | Labor | Total | Sell |
|---|---|---|---|---|---|---|
| 30 gallon | P1@2.50 | Ea | 314.00 | 78.80 | 392.80 | 668.00 |

**40-gallon gas-fired water heater.** 6-year warranty. Flammable Vapor Ignition Resistant (FVIR). Includes pipe connection labor. Add for flue, water and gas pipe, fitting materials and gas permit costs if required.

| | | | | | | |
|---|---|---|---|---|---|---|
| 36,000 Btu | P1@2.75 | Ea | 304.00 | 86.60 | 390.60 | 664.00 |

**60-gallon gas water heater.** 12-year warranty. Tall boy. Includes pipe connection labor. Add for flue, water and gas pipe, fitting materials and gas permit costs if required.

| | | | | | | |
|---|---|---|---|---|---|---|
| 50,000 Btu | P1@2.75 | Ea | 511.00 | 86.60 | 597.60 | 1,020.00 |

**75-gallon gas water heater.** 6-year warranty. Tall boy. Includes pipe connection labor. Add for flue, water and gas pipe, fitting materials and gas permit costs if required.

| | | | | | | |
|---|---|---|---|---|---|---|
| 70,000 Btu, Low-NOx | P1@2.75 | Ea | 644.00 | 86.60 | 730.60 | 1,240.00 |

**100-gallon commercial gas water heater.** Includes pipe connection labor. Add for flue, water and gas pipe, fitting materials and gas permit costs if required.

| | | | | | | |
|---|---|---|---|---|---|---|
| 73,000 Btu, Low-NOx | P1@3.75 | Ea | 1,190.00 | 118.00 | 1,308.00 | 2,220.00 |
| 75,000 Btu, 3-year warranty | P1@3.75 | Ea | 1,336.00 | 118.00 | 1,454.00 | 2,470.00 |
| 199,000 Btu, 3-year warranty | P1@3.75 | Ea | 2,590.00 | 118.00 | 2,708.00 | 4,600.00 |

**50-gallon high-altitude gas water heater.** 12-year warranty. Tall boy. Includes pipe connection labor. Add for flue, water and gas pipe, fitting materials and gas permit costs if required.

| | | | | | | |
|---|---|---|---|---|---|---|
| 60,000 Btu | P1@2.75 | Ea | 410.00 | 86.60 | 496.60 | 844.00 |

**75-gallon high-altitude gas water heater.** 6-year warranty. For installations above 2000 feet. Self-cleaning. Labor includes set in place only. Add for the cost of pipe and vent. Includes pipe connection labor. Add for flue, water and gas pipe, fitting materials and gas permit costs if required.

| | | | | | | |
|---|---|---|---|---|---|---|
| 75,500 Btu | P1@3.50 | Ea | 652.00 | 110.00 | 762.00 | 1,300.00 |

**Propane-fired tall boy water heater.** 6-year warranty. Flammable Vapor Ignition Resistant (FVIR) water heater. Includes pipe connection labor. Add for flue, water and gas pipe, fitting materials and gas permit costs if required.

| | | | | | | |
|---|---|---|---|---|---|---|
| 30 gallon, 27,000 Btu | P1@2.50 | Ea | 370.00 | 78.80 | 448.80 | 763.00 |
| 40 gallon, 36,000 Btu | P1@2.50 | Ea | 383.00 | 78.80 | 461.80 | 785.00 |
| 50 gallon, 36,000 Btu | P1@2.75 | Ea | 455.00 | 86.60 | 541.60 | 921.00 |

**50-gallon power-vent propane water heater.** 6-year warranty. Fan assisted flue venting. Requires positive seal chimney vent. Power-assist permits combined horizontal and vertical venting. Air intake runs can be up to 80' (including one 90-degree elbow) using 3" Schedule 40 PVC, CPVC or ABS pipe. Includes pipe connection labor. Add for flue, water and gas pipe, fitting materials and gas permit costs if required.

| | | | | | | |
|---|---|---|---|---|---|---|
| 40,000 Btu | P1@2.75 | Ea | 617.00 | 86.60 | 703.60 | 1,200.00 |

**Oil-fired water heater.** Labor includes setting and connecting only. Includes pipe connection labor. Add for flue, water and gas pipe, fitting materials and gas permit costs if required.

| | | | | | | |
|---|---|---|---|---|---|---|
| 30 gallon, TF, tank only | P1@2.25 | Ea | 539.00 | 70.90 | 609.90 | 1,040.00 |
| 30 gallon, TF, with burner assembly | P1@2.25 | Ea | 863.00 | 70.90 | 933.90 | 1,590.00 |
| 50 gallon, TF, tank only | P1@2.75 | Ea | 927.00 | 86.60 | 1,013.60 | 1,720.00 |
| 50 gallon, TF, with burner assembly | P1@2.75 | Ea | 1,156.00 | 86.60 | 1,242.60 | 2,110.00 |

| | Craft@Hrs | Unit | Material | Labor | Total | Sell |
|---|---|---|---|---|---|---|

## Electric Water Heaters

**30-gallon electric water heater.** 6-year warranty. Includes temperature and pressure relief valve. Includes pipe connection labor. Add for water pipe, fitting materials and electrical permit costs if required.

| | | | | | | |
|---|---|---|---|---|---|---|
| Two 3800-watt elements | P1@2.00 | Ea | 203.00 | 63.00 | 266.00 | 452.00 |
| Two 3800-watt elements, low boy | P1@2.00 | Ea | 180.00 | 63.00 | 243.00 | 413.00 |

**40-gallon electric water heater.** 6-year warranty. Includes temperature and pressure relief valve. 42-gallon-per-hour recovery rate. 23" diameter, 32" height. $410 estimated annual operating costs. 0.90 energy factor. 11.5 R-factor insulated jacket. Patented resistor-design heating elements prolong anode and tank life. Includes pipe connection labor. Add for water pipe, fitting materials and electrical permit costs if required.

| | | | | | | |
|---|---|---|---|---|---|---|
| Two 4500-watt elements, low boy | P1@2.25 | Ea | 228.00 | 70.90 | 298.90 | 508.00 |

**40-gallon electric water heater.** 6-year warranty. Includes temperature and pressure relief valve. 49-gallon-per-hour recovery rate. 17" diameter, 59" height. $420 estimated annual operating costs. 0.88 energy factor. 11.5 R-factor on insulated jacket. Patented resistor-design heating elements prolong anode and tank life. Includes pipe connection labor. Add for water pipe, fitting materials and electrical permit costs if required.

| | | | | | | |
|---|---|---|---|---|---|---|
| Two 4500-watt elements, tall boy | P1@2.25 | Ea | 226.00 | 70.90 | 296.90 | 505.00 |

**47-gallon electric water heater.** 6-year warranty. Includes temperature and pressure relief valve. 48-gallon-per-hour recovery rate. 27" diameter, 32" height. $425 estimated annual operating costs. 0.87 energy factor. 11.5 R-factor on insulated jacket. Patented resistor-design heating elements prolong anode and tank life. Includes pipe connection labor. Add for water pipe, fitting materials and electrical permit costs if required.

| | | | | | | |
|---|---|---|---|---|---|---|
| Two 4500-watt elements, low boy | P1@2.50 | Ea | 275.00 | 78.80 | 353.80 | 601.00 |

**50-gallon electric water heater.** 6-year warranty. Includes temperature and pressure relief valve. Includes pipe connection labor. Add for water pipe, fitting materials and electrical permit costs if required.

| | | | | | | |
|---|---|---|---|---|---|---|
| Two 4500-watt elements, low boy | P1@2.50 | Ea | 267.00 | 78.80 | 345.80 | 588.00 |
| Two 4500-watt elements, tall boy | P1@2.50 | Ea | 238.00 | 78.80 | 316.80 | 539.00 |

**80-gallon electric water heater.** 6-year warranty. Includes temperature and pressure relief valve. 23" diameter, 59" height. $425 estimated annual operating costs. 0.82 energy factor. 11.5 R-factor on insulated jacket. Includes pipe connection labor. Add for water pipe, fitting materials and electrical permit costs if required.

| | | | | | | |
|---|---|---|---|---|---|---|
| Two 4500-watt elements | P1@3.25 | Ea | 369.00 | 102.00 | 471.00 | 801.00 |

**Small electric water heater.** 6-year warranty. Labor includes set in place only. Add for the cost of piping and electrical runs. Includes pipe connection labor. Add for water pipe, fitting materials and electrical permit costs if required.

| | | | | | | |
|---|---|---|---|---|---|---|
| 6 gallons, 2000-watt element | P1@1.25 | Ea | 242.00 | 39.40 | 281.40 | 478.00 |

**Thermal Expansion Tank.** Control thermal expansion of water in domestic hot water systems. Absorbs the increased volume of water generated by the hot water heating sourcekeeping system pressure relief setting of the T&P Relief valve.

| | | | | | | |
|---|---|---|---|---|---|---|
| Thermal Expansion Tank 2.1 Gal. | P1@.300 | Ea | 31.80 | 9.45 | 41.25 | 70.10 |
| Thermal Expansion Tank 4.5 Gal. | P1@.300 | Ea | 53.70 | 9.45 | 63.15 | 107.00 |

Retrofit installation increase labor to 1 man-hour per unit

| | Craft@Hrs | Unit | Material | Labor | Total | Sell |
|---|---|---|---|---|---|---|

## Tankless Water Heaters

**Tankless gas water heater, Bosch AquaStar 125B.** Gas tankless water heater with standing pilot. Hangs on the wall to save floor space. Provides endless hot water for one major application at a time. Made of high-quality plastic, brass, and copper. 18" wide x 30" high x 9" deep. 15-year warranty on heat exchanger. Replaces 40-gallon tank-style water heater. Natural gas or propane. Requires electric connection for ignition. Includes pipe connection labor. Add for water pipe, fitting materials and permit costs if required.

| | Craft@Hrs | Unit | Material | Labor | Total | Sell |
|---|---|---|---|---|---|---|
| 117,000 Btu | P1@3.25 | Ea | 532.00 | 102.00 | 634.00 | 1,080.00 |

**Tankless gas water heater, Bosch AquaStar 125HX.** Gas tankless water heater. Hydro-generated ignition requires no electricity. Hangs on the wall to save floor space. Provides hot water for one major application at a time. Made of high-quality plastic, brass, and copper. 18" wide x 30" high x 9" deep. 15-year warranty on heat exchanger. Natural gas or propane. Replaces 40-gallon tank-style water heater. Includes pipe connection labor. Add for flue, gas, water pipe, fitting materials and permit costs if required.

| | Craft@Hrs | Unit | Material | Labor | Total | Sell |
|---|---|---|---|---|---|---|
| 117,000 Btu | P1@3.25 | Ea | 632.00 | 102.00 | 734.00 | 1,250.00 |

**Tankless gas water heater power vent kit.** For Bosch AquaStar 125B and 125HX.

| | Craft@Hrs | Unit | Material | Labor | Total | Sell |
|---|---|---|---|---|---|---|
| Power vent kit | P1@2.50 | Ea | 305.00 | 78.80 | 383.80 | 652.00 |

**Tankless electric water heater, PowerStar.** Designed to replace a 40-gallon electric storage tank water heater. Mounts on the wall. Electronic flow switch. Electronic thermostatic control. 10-year warranty. 99% efficiency rating. Glass reinforced plastic heat exchanger. 1/2" compression fitting connections. Filter screen on inlet. Thermal safety cut out. Add the cost of electrical source as required. Includes pipe connection labor. Add for pipe and fitting material costs if required.

| | Craft@Hrs | Unit | Material | Labor | Total | Sell |
|---|---|---|---|---|---|---|
| 240 volts, 17kW | P1@2.15 | Ea | 435.00 | 67.70 | 502.70 | 855.00 |
| 240 volts, 27kW | P1@2.15 | Ea | 646.00 | 67.70 | 713.70 | 1,210.00 |

**Point-of-use electric water heater, Ariston GL4 Mini-Tank.** 4-gallon capacity. 6-year warranty. Adjustable thermostatic control. Glass-lined tank. Includes temperature/pressure relief valve. Plugs into standard 110-volt outlet. Dimensions: 14" x 14" x 12". Installs independently or in-line with larger hot water source. Meets ASHRAE 90.1 Standard. UL listed. Includes pipe connection labor. Add for pipe and fitting material costs if required.

| | Craft@Hrs | Unit | Material | Labor | Total | Sell |
|---|---|---|---|---|---|---|
| 4 gallon | P1@2.15 | Ea | 140.00 | 67.70 | 207.70 | 353.00 |

**Point-of-use electric water heater, PowerStream.** For under-sink applications. 5-year warranty. Solid copper heat exchanger. Mounts in any direction. Must be hardwired. Add the cost of electrical source as required. Includes pipe connection labor. Add for pipe and fitting material costs if required.

| | Craft@Hrs | Unit | Material | Labor | Total | Sell |
|---|---|---|---|---|---|---|
| 240 volts, 9.5kW | P1@2.15 | Ea | 215.00 | 67.70 | 282.70 | 481.00 |

## Water Hammer Control

**Water hammer arrester.**

| | Craft@Hrs | Unit | Material | Labor | Total | Sell |
|---|---|---|---|---|---|---|
| 1/2" male threaded | P1@.250 | Ea | 9.35 | 7.88 | 17.23 | 29.30 |
| 3/4" male threaded | P1@.250 | Ea | 10.40 | 7.88 | 18.28 | 31.10 |

**Water surge shock absorber.** Water hammer arrestor. O-ring sealed air chamber maintains 60 PSI. Air charge absorbs high-pressure surge caused by quick closing of valves. Eliminates need for risers. Stops banging noises in pipes.

| | Craft@Hrs | Unit | Material | Labor | Total | Sell |
|---|---|---|---|---|---|---|
| 1/2" male threaded | P1@.250 | Ea | 13.03 | 7.88 | 20.91 | 35.50 |
| For washing machine line | P1@.250 | Ea | 13.10 | 7.88 | 20.98 | 35.70 |

| | Craft@Hrs | Unit | Material | Labor | Total | Sell |
|---|---|---|---|---|---|---|

## Plumbing

**Remove and replace galvanized water pipe with copper pipe.** Includes fittings and pipe clips. Based on Type M copper pipe with soft soldered joints. Includes disposal of scrap pipe and fittings. No salvage value assumed. Add for floor, wall or ceiling patching, if required.

| | Craft@Hrs | Unit | Material | Labor | Total | Sell |
|---|---|---|---|---|---|---|
| 1/2" Type M copper pipe and fittings | P1@.145 | LF | .80 | 4.57 | 5.37 | 9.13 |
| 3/4" Type M copper pipe and fittings | P1@.160 | LF | 1.22 | 5.04 | 6.26 | 10.60 |
| Prep and connect to existing 1/2" galvanized | P1@.550 | Ea | 8.50 | 17.30 | 25.80 | 43.90 |
| Prep and connect to existing 3/4" galvanized | P1@.650 | Ea | 12.00 | 20.50 | 32.50 | 55.30 |

**Remove and replace galvanized drain, waste and vent pipe (DWV) with copper pipe.** Includes fittings and hangers. Based on type DWV copper pipe with soft soldered joints. Includes disposal of scrap pipe and fittings. No salvage value assumed. Add for floor, wall or ceiling patching, if required.

| | Craft@Hrs | Unit | Material | Labor | Total | Sell |
|---|---|---|---|---|---|---|
| 1-1/2" DWV copper pipe and fittings | P1@.165 | LF | 4.55 | 5.20 | 9.75 | 16.60 |
| 2" DWV copper pipe and fittings | P1@.190 | LF | 5.70 | 5.99 | 11.69 | 19.90 |
| Prep and connect to existing 1-1/2" galvanized | P1@.650 | Ea | 24.50 | 20.50 | 45.00 | 76.50 |
| Prep and connect to existing 2" galvanized | P1@.850 | Ea | 32.00 | 26.80 | 58.80 | 100.00 |

**Remove and replace cast iron drain, waste and vent pipe (DWV) with cast iron pipe.** Includes fittings, couplings and hangers. Based on service weight hub-less cast iron pipe with mechanical joint couplings. Includes disposal of scrap pipe and fittings. No salvage value assumed. Add for floor, wall or ceiling patching, if required.

| | Craft@Hrs | Unit | Material | Labor | Total | Sell |
|---|---|---|---|---|---|---|
| 2" cast iron MJ pipe and fittings | P1@.190 | LF | 7.15 | 5.99 | 13.14 | 22.30 |
| 3" cast iron pipe and fittings | P1@.210 | LF | 7.80 | 6.62 | 14.42 | 24.50 |
| 4" cast iron pipe and fittings | P1@.245 | LF | 9.30 | 7.72 | 17.02 | 28.90 |
| Prep and connect to existing 2" cast iron | P1@.650 | Ea | 14.00 | 20.50 | 34.50 | 58.70 |
| Prep and connect to existing 3" cast iron | P1@.700 | Ea | 19.00 | 22.10 | 41.10 | 69.90 |
| Prep and connect to existing 4" cast iron | P1@.750 | Ea | 26.00 | 23.60 | 49.60 | 84.30 |

**Remove and replace cast iron drain, waste and vent pipe (DWV) with ABS pipe.** Includes fittings and hangers. Based on ABS plastic pipe with solvent weld joints. Includes disposal of scrap pipe and fittings. No salvage value assumed. Add for floor, wall or ceiling patching, if required.

| | Craft@Hrs | Unit | Material | Labor | Total | Sell |
|---|---|---|---|---|---|---|
| 1-1/2" ABS pipe and fittings | P1@.110 | LF | .92 | 3.47 | 4.39 | 7.46 |
| 2" ABS pipe and fittings | P1@.120 | LF | 1.55 | 3.78 | 5.33 | 9.06 |
| 3" ABS pipe and fittings | P1@.145 | LF | 2.45 | 4.57 | 7.02 | 11.90 |
| 4" ABS pipe and fittings | P1@.160 | LF | 3.98 | 5.04 | 9.02 | 15.30 |
| Prep and connect to existing 1-1/2" cast iron | P1@.550 | Ea | 18.00 | 17.30 | 35.30 | 60.00 |
| Prep and connect to existing 2" cast iron | P1@.250 | Ea | 22.00 | 7.88 | 29.88 | 50.80 |
| Prep and connect to existing 3" cast iron | P1@.750 | Ea | 28.00 | 23.60 | 51.60 | 87.70 |
| Prep and connect to existing 4" cast iron | P1@.850 | Ea | 36.00 | 26.80 | 62.80 | 107.00 |

**Remove and replace cast iron drain, waste and vent pipe (DWV) with PVC pipe.** Includes fittings and hangers. Based on PVC DWV plastic pipe with solvent weld joints. Includes disposal of scrap pipe and fittings. No salvage value assumed. Add for floor, wall or ceiling patching, if required.

| | Craft@Hrs | Unit | Material | Labor | Total | Sell |
|---|---|---|---|---|---|---|
| 1-1/2" PVC DWV pipe and fittings | P1@.115 | LF | 1.02 | 3.62 | 4.64 | 7.89 |
| 2" PVC DWV pipe and fittings | P1@.125 | LF | 1.78 | 3.94 | 5.72 | 9.72 |
| 3" PVC DWV pipe and fittings | P1@.150 | LF | 2.75 | 4.73 | 7.48 | 12.70 |
| 4" PVC DWV pipe and fittings | P1@.165 | LF | 4.40 | 5.20 | 9.60 | 16.30 |
| Prep and connect to existing 1-1/2" cast iron | P1@.550 | Ea | 22.00 | 17.30 | 39.30 | 66.80 |
| Prep and connect to existing 2" cast iron | P1@.250 | Ea | 26.00 | 7.88 | 33.88 | 57.60 |
| Prep and connect to existing 3" cast iron | P1@.750 | Ea | 31.00 | 23.60 | 54.60 | 92.80 |
| Prep and connect to existing 4" cast iron | P1@.850 | Ea | 38.50 | 26.80 | 65.30 | 111.00 |

# *Electrical*

**15**

**M**ost older homes have electrical systems that are in perfectly good working order — but completely inadequate by modern standards. They have too little power, too few circuits and far too few outlets. Even if the electrical service in the home has been upgraded in the last 25 years, more circuits may be needed to keep up with the demand created by today's multitude of electrical conveniences.

Few homes built before World War I were wired for electricity. Circuits were added later, usually gouged into plaster walls or run behind baseboards. In the 1920s and 1930s, most new homes were planned for 40 amps. The next jump was to four 15-amp circuits, or 60 amps total power. Many rooms had only a single duplex receptacle and a switched light fixture. By the 1950s, 100 amps was considered adequate power, unless the plan included an electric range or electric heat. Since the 1970s, 150 amps has been considered the minimum for a small home and 200 amps a better choice for most homes.

## Enough Power?

The primary electrical shortcoming in older homes is too little amperage. Your first task is to determine how much power is available at the distribution panel (or fuse box). There are three ways to find out:

❖ The first limitation is the current-carrying capacity of the service entrance wire. The wire gauge may be marked on the insulation where the wire enters the weatherhead or where the cable emerges from underground conduit. If there's no marking, an electrician can measure the wire gauge and compute the maximum amperage. If you see only two wires, rather than three, running to the service head or emerging from underground conduit, there's no hope. Three wires are required for a modern 240-volt electrical system.

❖ Second, check the manufacturer's data plate or service disconnect. It will identify the maximum amperage of the main electrical panel.

❖ And third, the main circuit breaker will be marked with an amperage rating. It's always the same or less than the service drop and electrical panel.

Unfortunately, the amperage that can be delivered to loads in the house is determined by the *smallest* number among these three identifiers.

If the ampacity available isn't enough to carry the planned loads for your home improvement, you'll need a new service panel, and possibly a larger service drop from the electric company. If there's nothing wrong with the existing electrical system, you can just leave it in place. Then upgrade by adding more circuits and outlets and the new electrical service panel. Most building departments will allow non-electricians to do minor work, such as adding an outlet or switch, but a licensed electrician will be required for service upgrades. An electrician will be able to calculate what size panel you'll need to install. While doing that math, he can also figure the most efficient way to run circuits to new light fixtures, appliances and outlets. If you're not a licensed electrician, this is a job you'll have to sub out.

### Licensing and Code Compliance

Electrical wiring can be dangerous, and a poorly-wired home is a fire hazard. Because of this, electricians have to be licensed and all their work must be done in compliance with the *National Electrical Code (NEC)*. Unfortunately, this isn't the last word on code compliance issues. Your electrical contractor will apply at the building department for a permit that covers the new work. That's easy enough. But the building department may require more. It's very unlikely that the existing electrical system in an older home complies with the current *NEC* standards. The question then becomes, how much of the old electrical system has to be brought up to code before the building inspector will sign off on the new work? If you're doing an extensive upgrade, the building department may require that you bring all the existing work up to meet current code requirements. This may involve more expense than the owner is willing, or able, to manage. And that's your best argument if you need to negotiate this issue with the building department. Too many requirements and nothing will get done — that benefits no one.

Resist the temptation to do major electrical work without a permit. You won't find a licensed electrician eager to bid on work like that, and a reputable home improvement contractor won't rely on an unlicensed electrician. A bootleg electrical job will make it hard for the owner to sell the home without first making the required code improvements. It's better for everyone to negotiate a fair compromise with the building official, use a professional electrician and get the job signed off by a building inspector.

## The Service Upgrade

The exterior portion of a service upgrade consists of replacing the meter, main service panel and the wire that connects the meter to the power grid. The lines up to the service drop belong to the utility company. The utility will probably have to do any work required on the service drop, but the homeowner will have to pay for the drop itself and the meter.

If the home has a basement, most of the service upgrade can be done there. Replace the old fuse box or boxes with a modern breaker panel. Then disconnect circuits running to the fuse box and reconnect them at the breaker box. Attach each circuit to a breaker with the correct capacity. The new breaker panel should have enough breaker spaces for existing circuits, plus a few extra for expansion. Most homes will need a 150- or 200-amp service panel.

Because the electrician's part of the work is done in the basement or outside, it can be done after all the carpentry and interior finishing is complete. However, any wiring that has to be run in the walls must be completed before drywall is installed. Otherwise, electrical wire has to be fished down between the studs.

# Wiring Residences

Nearly all new homes today are wired with non-metallic sheathed cable, but that may not be what you find when you work on an older house. Depending on its age, you may have some surprises.

## *Knob and Tube Wiring*

Homes built before about 1930 usually had knob and tube (K&T) wiring. Wires were strung between porcelain insulators driven into studs and joists. If wire had to pass through framing, a hollow porcelain tube was inserted in a hole drilled through the stud or joist. Conductors were usually single strands covered with cloth insulation. You won't find a ground wire on K&T. In those days, only lightning rods were grounded.

K&T that's given trouble-free service for nearly a century could probably do the same for another century, if no further demands were made on the system. But that's not likely. Because it isn't grounded, doesn't have enough capacity, and its insulation isn't worthy of the name, some insurance policies exclude coverage for homes with K&T wiring. For that reason alone, many of these older homes have already been upgraded.

If you find K&T wiring and the owner isn't willing to replace it, just by-pass the K&T. Work around it with new circuits. *Don't extend it.* If possible, don't even touch it unless you find an actual or impending emergency.

## *Aluminum Wiring*

Aluminum wire is another type that's no longer used for interior home electrical systems. Aluminum is a good and durable conductor and is usually less expensive than the more popular copper. However, late in the 1970s, electricians and code officials began to recognize a problem developing in homes with aluminum wiring. When aluminum wire carries current, it warms up and expands, just like copper. When it cools, the aluminum contracts, just like copper. But unlike copper, aluminum connections oxidize during the cool-down phase, creating resistance where conductors join. With time, the resistance grows into arcing — a spark that passes through the corrosion (gap) between the wire and the connector. Given the right conditions, that spark can ignite a fire.

The aluminum wire itself isn't the problem. It's the *wire connections* that are to blame. You don't need to rip out all the aluminum wire you find. But it's prudent to check connections in a home wired with aluminum. Electrical devices used with aluminum wire should be rated specifically for aluminum (usually stamped *CO/ALR* or *Al/Cu*). Look for signs of overheating, such as blackened connections or melted insulation. If you elect to extend an aluminum circuit using copper wire, your electrician will need to rent a special crimping tool made just for the purpose. When crimped, the wire connection must be covered with anti-oxidant grease.

Aluminum wire is still widely used for residential service entrance, though not for concealed wiring in walls and ceilings. Wires run between the house and the public utility grid are larger in diameter and require very few connections. That makes aluminum a good choice. For interior wiring, the price advantage of aluminum over copper usually isn't worth the risk or the extra trouble, though aluminum wire with a copper coating is used in some communities.

### Non-Metallic Sheathed Cable

Non-metallic sheathed cable, called *Romex* or "*rope*" by electricians, is the most common wire type used in homes today. The *NEC* classifies it as *Type NM cable*. (Type NM-B cable is identical but has a slightly-better temperature rating.) Romex has two or more insulated conductors and a ground wire, all covered in a plastic sheath. It's popular because it's inexpensive and easy to install. You can use NM cable in wall cavities where the wire is protected from physical damage and unlikely to get either wet or hot. When it's run through 2" x 4" stud walls, protect the cable at each stud with a metal plate to prevent damage from nails. Romex cable can be stapled to studs, rather than attached with nail-on hangers or supports. Most electrical codes permit the use of plastic (rather than metal) outlet boxes with NM cable.

You can't use Romex for exposed wiring on walls if it's within 5' of the floor. But most inspectors will approve Type AC (armored cable) for that purpose. AC is like Romex but includes a flexible aluminum cover that protects it against physical damage. If the inspector won't accept AC cable for exposed runs, you may have to install conduit. Unlike electric cable, conduit includes no wire. It's a protective tube through which wire is pulled. Conduit is used in most commercial buildings and occasionally in residences, such as in the service entrance mast where overhead wires terminate at the entrance cap. Flexible (*flex*), EMT (electric metallic tube), GRS (galvanized rigid steel) and IMC (intermediate rigid conduit) are the most common types of conduit. Flex is a hard metallic tube with enough flexibility to snake through studs. EMT is lightweight but not flexible. GRS conduit is heavier. IMC falls between EMT and GRS. Each of these types has specialized uses.

Electrical wire size is measured in American wire gauge (AWG) and usually abbreviated with the pound sign. For example, #14-3 indicates a 14-gauge wire with three conductors (and probably a separate bare ground wire). The smaller the gauge number, the bigger the wire and the greater its current-carrying capacity. Most circuits in a home are rated at 15 amps and use 14-gauge copper wire. Circuits for kitchen appliances should be rated at 20 amps and use 12-gauge copper wire. Circuits for an electric water heater, air conditioner or electric clothes dryer should be 30 amps and use 10-gauge copper wire. An electric range requires 6-gauge copper wire and a 50-amp breaker. All of these cables should include a copper ground wire.

Usually Type THHN wire is used with conduit because it has a heat-resistant thermoplastic cover. THHN is available in many colors to simplify the identification of conductors after the wire is pulled in the conduit. Small gauges are solid wire. Larger gauges are stranded wire, usually 19 strands per conductor. Stranded wire is a slightly better conductor than solid wire. It's also not as stiff, so it's easier to pull stranded wire in conduit.

Use Type UF cable for underground runs or in damp locations. It's not required by all codes, but common sense dictates that buried UF cable be trenched deep enough to make accidental damage unlikely. Either enclose the UF in PVC conduit or lay a warning tape in the trench over the wire before it's backfilled.

Protect cable with conduit wherever it's exposed, particularly where it exits the residence. For outdoor wiring, junction boxes and outlet boxes must be rated waterproof, and all receptacles must be GFCI-protected.

## Ground Wire

Electrical systems in old houses usually don't include a ground wire. If you're not sure about the house you're working on, just check any duplex receptacle. If it only has space for two contacts, you're dealing with an ungrounded electrical system. If you're not tearing into the walls, and you're working on a tight budget, it's best to simply leave this wiring as it is — assuming the building inspector doesn't intervene. An old electrical system that was properly installed is still safe and usable. But be aware that two-wire electrical cable can only support a 15-amp ungrounded circuit. That limits the number of receptacles and lighting fixtures on a circuit. If you're going to tear into the walls, plan to replace old two-wire circuits with 20-amp grounded circuits. If you disturb an old

### Fishing Electrical Cable

In new construction, wire is threaded through studs and between joists before the framing cavities are enclosed with drywall. That's the easy way to string wire. In home improvement work, wall and ceiling cavities aren't always exposed, so the wire has to be *fished* through enclosed stud spaces using an electrician's fish tape. Fishing electrical cable through existing walls is slow work.

A fish tape is a reel of springy wire that retracts into a circular case. The wire is flexible enough to bend when needed but stiff enough to be self-supporting for at least a few feet. The leading end of the tape has a hook so electrical wire is easy to attach. If possible, you'll want to fish the cable vertically — down from the attic or up from the basement. From the attic, bore down through the wall's top plate and thread fish tape down between two studs. When tape is fished from beginning point to end point, attach electric cable to the leading end of the tape. Then reel in the tape, threading new wire through the stud and joist cavities as the tape retracts.

If you have no choice but to fish cable horizontally across studs and joists, the first step is to find the shortest and easiest route between where power is available and where power is needed. With the help of an electric stud finder, count the studs and joists between those two points. When you've settled on the path of least resistance, break into the wall at each stud or joist and cut out a small notch large enough for wiring to pass through. When all the framing has been notched, start threading fish tape through the framing from notch to notch. This is slow work and requires two people, one pushing fish tape and the other finding and guiding the tape end. When the wire ends have been connected and you're sure the electrical work is done, seal the holes in the drywall. Use fiberglass mesh tape and joint compound to patch and smooth the drywall surface.

two-wire circuit, don't count on it to be entirely reliable. Insulation on old wire can become brittle and crumble when moved, leaving dangerous bare wires that you'll have to replace.

### Low Voltage Wiring

Phone lines, coax TV cable, computer network cable and speaker wire don't present the same degree of risk that comes with 120-volt alternating current circuits. Because of this, the *NEC* has only a few simple rules for low voltage wiring. Primary among these is keeping low voltage wiring out of conduit, junction and outlet boxes that include 120-volt wiring. Low voltage wiring is dangerous if it's interconnected with regular AC circuits. You should also keep video, speaker and data cable at least a foot away from AC cable. That will minimize the effect of radio frequency interference (RFI) generated by alternating current circuits.

But don't let these rules impede progress. While adding or extending AC circuits, use the opportunity to also add new connections for modern low voltage conveniences. Most of the houses you'll be upgrading were built long before the need for phone jacks in every room, multiple cable TV outlets, and computer network cabling.

### Adding Outlets and Switches

Most rooms in old houses have a single electrical outlet, and all the outlets in the house may be on the same 15-amp circuit. That's enough power for a few lamps and not much more. It certainly won't support a modern lifestyle. You may be able to turn on the lights and watch TV in the evening, but forget the microwave popcorn!

The *NEC* sets standards for residential electrical outlets. These standards may or may not be enforced in home improvement projects in your community. The degree that these regulations affect your project may be a matter for negotiation between the contractor and the building department. Select an electrical contractor with experience in negotiating with building inspectors. Remember, however, that the code exists for a reason and most of what it requires is simply good professional practice. Follow the code standards whenever possible.

**Spacing of Outlets** — In most rooms, the code requires that no point along the floor line be more than 6' from a receptacle. That means you need an outlet at least every 12' along walls. Floor outlets don't help meet this requirement unless they're near the wall. Different standards apply to kitchens, bathrooms and laundry rooms. Spacing in hallways can be 20' and closets don't need any outlets at all.

**Kitchens** — Plan at least two 20-amp small appliance circuits to serve the kitchen, pantry and dining area. These circuits are in addition to circuits used by the refrigerator, dishwasher, oven, range, garbage disposer and lighting. Every kitchen counter wider than 12" needs at least one outlet. No point on a kitchen counter can be more than 24" from an outlet. That means you need at outlet at least every 4' over counters. Outlets have to be mounted on a wall, not face-up in the counter. The outlet next to the sink must be protected with a ground-fault circuit interrupter (GFCI). Plan on dedicated circuits for the range, dishwasher, garbage disposer, refrigerator, and microwave oven. A microwave oven will trip the breaker if it's on the same circuit with another large appliance, such as a refrigerator.

**Bathrooms** — Every bathroom needs at least one GFCI-protected outlet by the sink, even if the vanity lighting fixture includes an outlet. Keep the receptacle far enough away from the bathtub and shower to prevent the use of electric shavers or hairdryers while bathing.

**Laundry** — Provide at least one GFCI-protected duplex receptacle in the laundry area. The laundry receptacle must be on a dedicated 20-amp circuit.

**Unfinished Basement or Attic** — Provide at least one outlet.

**Outdoors** — Include a GFCI-protected duplex receptacle at the front and rear of the house.

**Garage** — Provide a GFCI-protected duplex receptacle for each parking space. Detached garages may not need any outlets.

**Ground-Fault Circuit Interrupter** — GFCI protection opens the circuit when a ground-fault is detected — such as when someone gets an electric shock. Outlets over the kitchen and bathroom sinks, in the laundry room, garage and outdoors have to be GFCI-protected. Installing GFCI outlets is easy, assuming the circuit includes a ground wire. Just remove the old outlet and replace it with a GFCI outlet. You don't have to buy a GFCI outlet for every receptacle that needs ground-fault protection. Several regular outlets can be wired to a single GFCI device. Circuitry in the GFCI outlet will protect all attached receptacles. That's the good news. The bad news is that GFCI outlets require a ground to work properly. If the old circuit has no ground wire, you'll have to run grounded cable to GFCI outlets.

**Arc-Fault Interrupter** — The NEC now requires these special breakers for circuits that serve sleeping rooms. Regular breakers open when there's an overload. Arc-fault breakers open any time the circuit is creating sparks, even if there's no overload.

**Outlets per Circuit** — Plan on six duplex outlets per 20-amp circuit. Your electrician may suggest ways to put 10 or even 12 outlets on a circuit and still meet *NEC* requirements. The code doesn't prohibit mixing light fixtures and outlets on the same *home run* (connection to the breaker panel).

**Lights** — Every room needs either a switch-operated overhead light or a switch-operated outlet. The code requires the switch to be located by the door at the room's entrance. You can run the wiring to any ceiling fixture that's still in good condition. If the fixture is worn, broken, or simply unattractive, replacing it is a simple task. But don't go overboard on wattage, such as replacing a 75-watt bathroom fixture with a 500-watt heat lamp. If your choice of replacement fixtures is a fluorescent, be sure there isn't a grounding problem. Many fluorescent fixtures require a ground wire. If you install a ceiling fan or a chandelier, plan to set a ceiling fixture box specifically rated for that purpose. Allow at least 3" between a recessed (non-IC) fixture and any insulation in the ceiling. Better yet, select a recessed fixture with an insulated case (IC) and thermal cutout specifically rated for installation touching insulation.

## Smoke Detectors

Nearly all communities require some type of smoke detector. The most common are battery-powered ionization detectors that recognize products of combustion even before flame is visible. Other types of detectors recognize smoke or detect a rapid rise in temperature. If you have the opportunity, recommend an AC-powered detector rather than a battery-powered unit. Surveys show that a high percentage of battery-powered smoke detectors have a dead battery at any given time. For that reason, many building codes require AC detectors, some with a battery backup.

| | Craft@Hrs | Unit | Material | Labor | Total | Sell |
|---|---|---|---|---|---|---|

## Labor Estimates for Wiring

**Fish electrical cable through a stud wall.** Add the cost of connecting wire at both ends.

| | | | | | | |
|---|---|---|---|---|---|---|
| Per job set-up | BE@.270 | Ea | — | 9.29 | 9.29 | 15.80 |
| Per foot of cable fished | BE@.050 | Ea | — | 1.72 | 1.72 | 2.92 |
| Per electric box set | BE@.761 | Ea | — | 26.20 | 26.20 | 44.50 |
| Per stud notched and hole patched | BE@.400 | Ea | — | 13.80 | 13.80 | 23.50 |

**Cable Snake™ flat wire fish tape, Gardner Bender.** Perfect for fishing short runs of wire in existing structures. Ergonomic handle provides a sure grip for pulling. Aerodynamic tip reduces friction, requiring less force and decreased hang-ups. Reel handle allows for quick retrieval and locking for pulling.

| | | | | | | |
|---|---|---|---|---|---|---|
| 20' x 1/4" x .031" | — | Ea | 11.80 | — | 11.80 | — |

**Connect new 220-volt appliance with plug.** Does not include electrical rough-in. Includes a range, electric dryer or air conditioner using three-conductor cable with ground wire and connected plug. Add the cost of setting the appliance in place.

| | | | | | | |
|---|---|---|---|---|---|---|
| Per appliance | BE@.433 | Ea | — | 14.90 | 14.90 | 25.30 |

**Connect new 220-volt appliance to an outlet box with cable.** Includes a range, electric dryer or air conditioner to a service outlet box using three-conductor non-metallic cable with ground wire. Add the cost of setting the appliance in place.

| | | | | | | |
|---|---|---|---|---|---|---|
| Per appliance | BE@.530 | Ea | — | 18.20 | 18.20 | 30.90 |

**Disconnect and reconnect an existing 220-volt appliance to an outlet box with cable.** Includes a range, electric dryer or air conditioner to a service outlet box using three-conductor non-metallic cable with ground wire. Add the cost of setting the appliance in place.

| | | | | | | |
|---|---|---|---|---|---|---|
| Per appliance | BE@.510 | Ea | — | 17.50 | 17.50 | 29.80 |

**Connect new 220-volt appliance using flex conduit.** Includes a range, electric dryer or air conditioner connected to a service outlet box using wire pulled in flex conduit. Add the cost of setting the appliance in place.

| | | | | | | |
|---|---|---|---|---|---|---|
| Per appliance | BE@1.04 | Ea | — | 35.80 | 35.80 | 60.90 |

**Remove and replace 220-volt cable from appliance.** Includes a range, electric dryer or air conditioner. Add the cost of setting the appliance in place.

| | | | | | | |
|---|---|---|---|---|---|---|
| Per appliance | BE@.465 | Ea | — | 16.00 | 16.00 | 27.20 |

**Connect or disconnect 110-volt appliance using three 8-gauge or smaller connectors.** Includes connecting a dishwasher or garbage disposer to the supply box. Add the cost of setting or removing the appliance.

| | | | | | | |
|---|---|---|---|---|---|---|
| Per appliance | BE@.796 | Ea | — | 27.40 | 27.40 | 46.60 |

## Type NM Non-Metallic Cable

**14-gauge Type NM-B Romex® sheathed indoor cable.** White. With full-size ground wire. Labor is based on installation in a wood-frame building, including boring out and pulling cable in exposed walls. Add the cost of fishing cable through enclosed walls, when required.

| | | | | | | |
|---|---|---|---|---|---|---|
| 2 conductor, per foot | BE@.027 | LF | .49 | .93 | 1.42 | 2.41 |
| 2 conductor, 25' coil | BE@.027 | LF | .56 | .93 | 1.49 | 2.53 |
| 2 conductor, 50' coil | BE@.027 | LF | .55 | .93 | 1.48 | 2.52 |
| 2 conductor, 100' coil | BE@.027 | LF | .42 | .93 | 1.35 | 2.30 |
| 2 conductor, 250' coil | BE@.027 | LF | .25 | .93 | 1.18 | 2.01 |
| 3 conductor, per foot | BE@.030 | LF | .75 | 1.03 | 1.78 | 3.03 |
| 3 conductor, 25' coil | BE@.030 | LF | .79 | 1.03 | 1.82 | 3.09 |
| 3 conductor, 50' coil | BE@.030 | LF | .76 | 1.03 | 1.79 | 3.04 |
| 3 conductor, 100' coil | BE@.030 | LF | .63 | 1.03 | 1.66 | 2.82 |

|  | Craft@Hrs | Unit | Material | Labor | Total | Sell |
|---|---|---|---|---|---|---|
| 3 conductor, 250' coil | BE@.030 | LF | .38 | 1.03 | 1.41 | 2.40 |
| 14-2-2, per foot | BE@.032 | LF | .70 | 1.10 | 1.80 | 3.06 |
| 14-2-2, 250' coil | BE@.032 | LF | .57 | 1.10 | 1.67 | 2.84 |

**12-gauge Type NM-B Romex® sheathed indoor cable.** Yellow. With full-size ground wire. Rated at 20 amps. Labor is based on installation in a wood-frame building, including boring out and pulling cable in exposed walls. Add the cost of fishing cable through enclosed walls, when required.

|  | Craft@Hrs | Unit | Material | Labor | Total | Sell |
|---|---|---|---|---|---|---|
| 2 conductor, per foot | BE@.030 | LF | .70 | 1.03 | 1.73 | 2.94 |
| 2 conductor, 25' coil | BE@.030 | LF | .69 | 1.03 | 1.72 | 2.92 |
| 2 conductor, 50' coil | BE@.030 | LF | .69 | 1.03 | 1.72 | 2.92 |
| 2 conductor, 100' coil | BE@.030 | LF | .64 | 1.03 | 1.67 | 2.84 |
| 2 conductor, 250' coil | BE@.030 | LF | .37 | 1.03 | 1.40 | 2.38 |
| 3 conductor, per foot | BE@.031 | LF | 1.00 | 1.07 | 2.07 | 3.52 |
| 3 conductor, 25' coil | BE@.031 | LF | 1.00 | 1.07 | 2.07 | 3.52 |
| 3 conductor, 50' coil | BE@.031 | LF | .96 | 1.07 | 2.03 | 3.45 |
| 3 conductor, 100' coil | BE@.031 | LF | .85 | 1.07 | 1.92 | 3.26 |
| 3 conductor, 250' coil | BE@.031 | LF | .57 | 1.07 | 1.64 | 2.79 |
| 12-2-2, per foot | BE@.032 | LF | 1.17 | 1.10 | 2.27 | 3.86 |
| 12-2-2, 250' coil | BE@.032 | LF | .78 | 1.10 | 1.88 | 3.20 |

**10-gauge Type NM-B Romex® sheathed indoor cable.** Orange. With ground. Rated at 30 amps. Labor is based on installation in a wood-frame building, including boring out and pulling cable in exposed walls. Add the cost of fishing cable through enclosed walls, when required.

|  | Craft@Hrs | Unit | Material | Labor | Total | Sell |
|---|---|---|---|---|---|---|
| 2 conductor, per foot | BE@.031 | LF | 1.25 | 1.07 | 2.32 | 3.94 |
| 2 conductor, 25' coil | BE@.031 | LF | 1.33 | 1.07 | 2.40 | 4.08 |
| 2 conductor, 50' coil | BE@.031 | LF | 1.20 | 1.07 | 2.27 | 3.86 |
| 2 conductor, 100' coil | BE@.031 | LF | .95 | 1.07 | 2.02 | 3.43 |
| 2 conductor, 250' coil | BE@.031 | LF | .66 | 1.07 | 1.73 | 2.94 |
| 3 conductor, per foot | BE@.035 | LF | 1.45 | 1.20 | 2.65 | 4.51 |
| 3 conductor, 25' coil | BE@.035 | LF | 1.55 | 1.20 | 2.75 | 4.68 |
| 3 conductor, 50' coil | BE@.035 | LF | 1.51 | 1.20 | 2.71 | 4.61 |
| 3 conductor, 100' coil | BE@.035 | LF | 1.27 | 1.20 | 2.47 | 4.20 |
| 3 conductor, 250' coil | BE@.035 | LF | .94 | 1.20 | 2.14 | 3.64 |

**8-gauge Type NM-B Romex® sheathed indoor cable.** Black. With ground, except where noted. Rated at 40 amps. Labor is based on installation in a wood-frame building, including boring out and pulling cable in exposed walls. Add the cost of fishing cable through enclosed walls, when required.

|  | Craft@Hrs | Unit | Material | Labor | Total | Sell |
|---|---|---|---|---|---|---|
| 2 conductor, per foot | BE@.033 | LF | 1.50 | 1.13 | 2.63 | 4.47 |
| 2 conductor, 125' coil | BE@.033 | LF | 1.02 | 1.13 | 2.15 | 3.66 |
| 3 conductor, no ground, per foot | BE@.033 | LF | 1.23 | 1.13 | 2.36 | 4.01 |
| 3 conductor, per foot | BE@.037 | LF | 2.65 | 1.27 | 3.92 | 6.66 |
| 3 conductor, 125' coil | BE@.037 | LF | 1.56 | 1.27 | 2.83 | 4.81 |

**6-gauge Type NM-B Romex® sheathed indoor cable.** Black. With ground. Rated at 55 amps. Labor is based on installation in a wood-frame building, including boring out and pulling cable in exposed walls. Add the cost of fishing cable through enclosed walls when required.

|  | Craft@Hrs | Unit | Material | Labor | Total | Sell |
|---|---|---|---|---|---|---|
| 2 conductor, per foot | BE@.038 | LF | 2.42 | 1.31 | 3.73 | 6.34 |
| 2 conductor, 125' coil | BE@.038 | LF | 1.46 | 1.31 | 2.77 | 4.71 |
| 3 conductor, per foot | BE@.044 | LF | 3.23 | 1.51 | 4.74 | 8.06 |
| 3 conductor, 125' reel | BE@.044 | LF | 2.21 | 1.51 | 3.72 | 6.32 |

| | Craft@Hrs | Unit | Material | Labor | Total | Sell |
|---|---|---|---|---|---|---|

**Cable two-piece clamp connector.** Zinc-plated steel. Butterfly style, does not require a locknut. For indoor use only in a dry location, Secures 14- to 10-gauge non-metallic sheathed cable to a steel outlet box or other metal enclosure. 3/8" size fits 1/2" knockout. UL listed.

| | Craft@Hrs | Unit | Material | Labor | Total | Sell |
|---|---|---|---|---|---|---|
| 3/8" | BE@.050 | Ea | 1.77 | 1.72 | 3.49 | 5.93 |
| 3/4" to 1" | BE@.050 | Ea | 2.84 | 1.72 | 4.56 | 7.75 |
| 1-1/4" | BE@.050 | Ea | 2.30 | 1.72 | 4.02 | 6.83 |
| 1-1/2" | BE@.050 | Ea | 2.52 | 1.72 | 4.24 | 7.21 |

**Stud nail plate.** Steel. Installed at each stud to protect cable bored through framing, as defined in *NEC* Article 300-4.

| | Craft@Hrs | Unit | Material | Labor | Total | Sell |
|---|---|---|---|---|---|---|
| 1-1/2" x 2-1/2", each | BE@.050 | Ea | .24 | 1.72 | 1.96 | 3.33 |
| 1-1/2" x 2-1/2", pack of 50 | BE@2.50 | Ea | 6.97 | 86.00 | 92.97 | 158.00 |
| 1-1/2" x 5", each | BE@.100 | Ea | .35 | 3.44 | 3.79 | 6.44 |

## Type UF Non-Metallic Cable

**UF-B sheathed underground cable.** Gray. With ground. Used to supply power to landscape lighting, pumps and other loads connected from the main building. Designed for burial in a trench. Add the cost of excavation and backfill.

| | Craft@Hrs | Unit | Material | Labor | Total | Sell |
|---|---|---|---|---|---|---|
| 14 gauge, 2 conductor, per foot | BE@.010 | LF | .75 | .34 | 1.09 | 1.85 |
| 14 gauge, 2 conductor, 25' coil | BE@.010 | LF | 1.05 | .34 | 1.39 | 2.36 |
| 14 gauge, 2 conductor, 100' coil | BE@.010 | LF | .89 | .34 | 1.23 | 2.09 |
| 14 gauge, 3 conductor, per foot | BE@.010 | LF | .89 | .34 | 1.23 | 2.09 |
| 14 gauge, 3 conductor, 25' coil | BE@.010 | LF | .76 | .34 | 1.10 | 1.87 |
| 14 gauge, 3 conductor, 50' coil | BE@.010 | LF | .69 | .34 | 1.03 | 1.75 |
| 12 gauge, 2 conductor, per foot | BE@.010 | LF | .94 | .34 | 1.28 | 2.18 |
| 12 gauge, 2 conductor, 25' coil | BE@.010 | LF | .79 | .34 | 1.13 | 1.92 |
| 12 gauge, 2 conductor, 50' coil | BE@.010 | LF | .78 | .34 | 1.12 | 1.90 |
| 12 gauge, 2 conductor, 100' coil | BE@.010 | LF | .70 | .34 | 1.04 | 1.77 |
| 12 gauge, 3 conductor, per foot | BE@.010 | LF | 1.58 | .34 | 1.92 | 3.26 |
| 12 gauge, 3 conductor, 100' coil | BE@.010 | LF | .90 | .34 | 1.24 | 2.11 |
| 10 gauge, 2 conductor, per foot | BE@.010 | LF | 2.46 | .34 | 2.80 | 4.76 |
| 10 gauge, 3 conductor, per foot | BE@.010 | LF | 1.75 | .34 | 2.09 | 3.55 |

**UF non-metallic cable compression connector.** Compression type connector for underground feeder cable. Fits smaller UF cables as well as standard sizes. Box knockout size is the same as the nominal conduit size. UL listed.

| | Craft@Hrs | Unit | Material | Labor | Total | Sell |
|---|---|---|---|---|---|---|
| 1/2", 14 to 12 gauge, 2 conductor | BE@.050 | Ea | 2.18 | 1.72 | 3.90 | 6.63 |
| 3/4", 14 to 12 gauge, 3 conductor | BE@.060 | Ea | 2.72 | 2.06 | 4.78 | 8.13 |

**AC-90® steel armored flexible cable, American Flexible Conduit.** Galvanized steel. 16-gauge integral bond wire-to-armor grounding path. Sometimes referred to as "BX" cable. Paper wrap conductor insulation covering. Maximum temperature rating is 90-degrees Celsius (dry). 600 volts. UL listed. By American Wire Gauge size and number of conductors. Add the cost of fishing cable through enclosed walls when required.

| | Craft@Hrs | Unit | Material | Labor | Total | Sell |
|---|---|---|---|---|---|---|
| 14 gauge, 3 conductor, 25' coil | BE@.020 | LF | 1.04 | .69 | 1.73 | 2.94 |
| 14 gauge, 4 conductor, per foot | BE@.021 | LF | 1.87 | .72 | 2.59 | 4.40 |
| 14 gauge, 4 conductor, 250' coil | BE@.021 | LF | 1.25 | .72 | 1.97 | 3.35 |
| 12 gauge, 3 conductor, 50' coil | BE@.021 | LF | 1.13 | .72 | 1.85 | 3.15 |
| 12 gauge, 3 conductor, 100' coil | BE@.021 | LF | 1.03 | .72 | 1.75 | 2.98 |
| 12 gauge, 4 conductor, per foot | BE@.023 | LF | 1.92 | .79 | 2.71 | 4.61 |
| 12 gauge, 4 conductor, 250' reel | BE@.023 | LF | 1.28 | .79 | 2.07 | 3.52 |
| 10 gauge, 3 conductor, 125' coil | BE@.027 | LF | 1.41 | .93 | 2.34 | 3.98 |

|  | Craft@Hrs | Unit | Material | Labor | Total | Sell |
|---|---|---|---|---|---|---|

## PVC Conduit and Fittings

**PVC conduit, Schedule 40, Carlon®.** For above or below ground applications subject to physical abuse. Also for use encased in concrete or for direct burial. Sunlight resistant. For use with conductors rated to 90-degrees Celsius. Sold in 10' lengths. Nominal sizes, by outside diameter (OD), inside diameter (ID) and wall thickness:

Nominal 1/2" measures 0.840" OD, 0.622" ID and 0.109" in wall thickness.
Nominal 3/4" measures 1.050" OD, 0.820" ID and 0.113" in wall thickness.
Nominal 1" measures 1.315" OD, 1.049" ID and 0.133" in wall thickness.
Nominal 1-1/4" measures 1.660" OD, 1.380" ID and 0.140" in wall thickness.
Nominal 1-1/2" measures 1.900" OD, 1.610" ID and 0.145" in wall thickness.
Nominal 2" measures 2.375" OD, 2.067" ID and 0.154" in wall thickness.
Nominal 2-1/2" measures 2.875" OD, 2.469" ID and 0.203" in wall thickness.

| | Craft@Hrs | Unit | Material | Labor | Total | Sell |
|---|---|---|---|---|---|---|
| 1/2" nominal diameter | BE@.031 | LF | .18 | 1.07 | 1.25 | 2.13 |
| 3/4" nominal diameter | BE@.032 | LF | .23 | 1.10 | 1.33 | 2.26 |
| 1" nominal diameter | BE@.033 | LF | .33 | 1.13 | 1.46 | 2.48 |
| 1-1/4" nominal diameter | BE@.034 | LF | .56 | 1.17 | 1.73 | 2.94 |
| 1-1/2" nominal diameter | BE@.035 | LF | .67 | 1.20 | 1.87 | 3.18 |
| 2" nominal diameter | BE@.035 | LF | .79 | 1.20 | 1.99 | 3.38 |
| 2-1/2" nominal diameter | BE@.039 | LF | 1.24 | 1.34 | 2.58 | 4.39 |

**PVC conduit, Schedule 80, Carlon®.** For above or below ground applications subject to physical abuse. Also for use encased in concrete or for direct burial. Sunlight resistant. For use with conductors rated to 90-degrees Celsius. Meets specifications of Underwriters Laboratories 651 and the National Electrical Manufacturers Association TC 2.

| | Craft@Hrs | Unit | Material | Labor | Total | Sell |
|---|---|---|---|---|---|---|
| 1/2" nominal diameter | BE@.034 | LF | .38 | 1.17 | 1.55 | 2.64 |
| 3/4" nominal diameter | BE@.035 | LF | .44 | 1.20 | 1.64 | 2.79 |
| 1" nominal diameter | BE@.036 | LF | .63 | 1.24 | 1.87 | 3.18 |
| 1-1/4" nominal diameter | BE@.034 | LF | .91 | 1.17 | 2.08 | 3.54 |
| 1-1/2" nominal diameter | BE@.035 | LF | 1.08 | 1.20 | 2.28 | 3.88 |
| 2" nominal diameter | BE@.040 | LF | 1.20 | 1.38 | 2.58 | 4.39 |
| 2-1/2" nominal diameter | BE@.042 | LF | 1.86 | 1.44 | 3.30 | 5.61 |

**Snap Strap PVC conduit clamp, Carlon®.** For installation of polyvinyl chloride conduit. Can be used with rigid steel. Indoor use only. All plastic clamp allows conduit to expand and contract with temperature changes, eliminating bowing.

| | Craft@Hrs | Unit | Material | Labor | Total | Sell |
|---|---|---|---|---|---|---|
| Single mount, 1/2" | BE@.050 | Ea | .73 | 1.72 | 2.45 | 4.17 |
| Single mount, 3/4" | BE@.050 | Ea | .83 | 1.72 | 2.55 | 4.34 |
| Single mount, 1" | BE@.050 | Ea | .44 | 1.72 | 2.16 | 3.67 |
| Double mount, 1-1/4" | BE@.050 | Ea | .90 | 1.72 | 2.62 | 4.45 |
| Double mount, 2" | BE@.050 | Ea | .74 | 1.72 | 2.46 | 4.18 |

**Non-metallic reducer bushing, Carlon®.** Rigid Schedule 40 PVC conduit reducer. Socket ends. Male x female. UL listed.

| | Craft@Hrs | Unit | Material | Labor | Total | Sell |
|---|---|---|---|---|---|---|
| 3/4" x 1/2" | BE@.030 | Ea | .87 | 1.03 | 1.90 | 3.23 |
| 1" x 1/2" | BE@.030 | Ea | 1.63 | 1.03 | 2.66 | 4.52 |
| 1" x 3/4" | BE@.040 | Ea | 1.85 | 1.38 | 3.23 | 5.49 |
| 1-1/4" x 3/4" | BE@.050 | Ea | 1.90 | 1.72 | 3.62 | 6.15 |
| 1-1/4" x 1" | BE@.050 | Ea | 2.09 | 1.72 | 3.81 | 6.48 |
| 1-1/2" x 1" | BE@.050 | Ea | 1.99 | 1.72 | 3.71 | 6.31 |
| 1-1/2" x 1-1/4" | BE@.050 | Ea | 2.68 | 1.72 | 4.40 | 7.48 |
| 2" x 1-1/4" | BE@.060 | Ea | 2.85 | 2.06 | 4.91 | 8.35 |
| 2" x 1-1/2" | BE@.060 | Ea | 3.02 | 2.06 | 5.08 | 8.64 |
| 2-1/2" x 2" | BE@.060 | Ea | 8.57 | 2.06 | 10.63 | 18.10 |

|  | Craft@Hrs | Unit | Material | Labor | Total | Sell |
|---|---|---|---|---|---|---|
| **PVC conduit box adapter, Carlon®.** For non-metallic Schedule 40 or 80 conduit. | | | | | | |
| 1/2" diameter | BE@.050 | Ea | .49 | 1.72 | 2.21 | 3.76 |
| 3/4" diameter | BE@.060 | Ea | .54 | 2.06 | 2.60 | 4.42 |
| 1" diameter | BE@.080 | Ea | .58 | 2.75 | 3.33 | 5.66 |
| 1-1/4" diameter | BE@.100 | Ea | .74 | 3.44 | 4.18 | 7.11 |
| 1-1/2" diameter | BE@.100 | Ea | .83 | 3.44 | 4.27 | 7.26 |
| 2" diameter | BE@.150 | Ea | 1.06 | 5.16 | 6.22 | 10.60 |
| **Non-metallic standard PVC coupling, Carlon®.** For joining Schedule 40 PVC conduit. | | | | | | |
| 1/2" diameter | BE@.020 | Ea | .24 | .69 | .93 | 1.58 |
| 3/4" diameter | BE@.030 | Ea | .29 | 1.03 | 1.32 | 2.24 |
| 1" diameter | BE@.050 | Ea | .40 | 1.72 | 2.12 | 3.60 |
| 1-1/4" diameter | BE@.060 | Ea | .64 | 2.06 | 2.70 | 4.59 |
| 1-1/2" diameter | BE@.060 | Ea | .68 | 2.06 | 2.74 | 4.66 |
| 2" diameter | BE@.080 | Ea | .95 | 2.75 | 3.70 | 6.29 |
| 2-1/2" diameter | BE@.090 | Ea | 1.52 | 3.10 | 4.62 | 7.85 |
| **Non-metallic 45-degree elbow, plain end, Carlon®.** Standard radius. For Schedule 40 PVC conduit. | | | | | | |
| 1/2" diameter | BE@.050 | Ea | .62 | 1.72 | 2.34 | 3.98 |
| 3/4" diameter | BE@.060 | Ea | .69 | 2.06 | 2.75 | 4.68 |
| 1" diameter | BE@.080 | Ea | 1.17 | 2.75 | 3.92 | 6.66 |
| 1-1/4" diameter | BE@.080 | Ea | 1.57 | 2.75 | 4.32 | 7.34 |
| 1-1/2" diameter | BE@.100 | Ea | 2.01 | 3.44 | 5.45 | 9.27 |
| 2" diameter | BE@.150 | Ea | 2.70 | 5.16 | 7.86 | 13.40 |
| 2-1/2" diameter | BE@.150 | Ea | 4.91 | 5.16 | 10.07 | 17.10 |
| **Non-metallic 90-degree elbow with plain end, standard radius, Carlon®.** For Schedule 40 PVC conduit. | | | | | | |
| 1/2" diameter | BE@.090 | Ea | .69 | 3.10 | 3.79 | 6.44 |
| 3/4" diameter | BE@.090 | Ea | .79 | 3.10 | 3.89 | 6.61 |
| 1" diameter | BE@.090 | Ea | 1.19 | 3.10 | 4.29 | 7.29 |
| 1-1/4" diameter | BE@.090 | Ea | 1.65 | 3.10 | 4.75 | 8.08 |
| 1-1/2" diameter | BE@.090 | Ea | 2.21 | 3.10 | 5.31 | 9.03 |
| 2" diameter | BE@.090 | Ea | 3.30 | 3.10 | 6.40 | 10.90 |
| 2-1/2" diameter | BE@.150 | Ea | 5.58 | 5.16 | 10.74 | 18.30 |
| **PVC conduit body, Type T, Carlon®.** For easy access and pulling of wires through conduit runs. Will not rust or corrode. For use with Schedule 40 and 80 non-metallic conduit. Gasket included. | | | | | | |
| 1/2" | BE@.100 | Ea | 2.77 | 3.44 | 6.21 | 10.60 |
| 3/4" | BE@.150 | Ea | 3.22 | 5.16 | 8.38 | 14.20 |
| 1" | BE@.150 | Ea | 3.79 | 5.16 | 8.95 | 15.20 |
| 1-1/4" | BE@.150 | Ea | 5.85 | 5.16 | 11.01 | 18.70 |
| 1-1/2" | BE@.150 | Ea | 6.52 | 5.16 | 11.68 | 19.90 |
| 2" | BE@.150 | Ea | 9.54 | 5.16 | 14.70 | 25.00 |
| **PVC conduit body, Type LB, Carlon®.** For easy access and pulling of wires through conduit run. For use with Schedule 40 and 80 non-metallic conduit. Gasket included. UL listed. | | | | | | |
| 1/2", 4 cubic inch | BE@.100 | Ea | 2.42 | 3.44 | 5.86 | 9.96 |
| 3/4", 12 cubic inch | BE@.100 | Ea | 2.82 | 3.44 | 6.26 | 10.60 |
| 1", 12 cubic inch | BE@.150 | Ea | 3.45 | 5.16 | 8.61 | 14.60 |
| 1-1/4", 32 cubic inch | BE@.150 | Ea | 4.98 | 5.16 | 10.14 | 17.20 |
| 1-1/2", 32 cubic inch | BE@.150 | Ea | 6.30 | 5.16 | 11.46 | 19.50 |
| 2", 63 cubic inch | BE@.200 | Ea | 8.96 | 6.88 | 15.84 | 26.90 |
| 2-1/2" | BE@.215 | Ea | 30.70 | 7.39 | 38.09 | 64.80 |

| | Craft@Hrs | Unit | Material | Labor | Total | Sell |
|---|---|---|---|---|---|---|

**PVC service entrance cap.** UL listed. Plus 40 and Plus 80.

| | | | | | | |
|---|---|---|---|---|---|---|
| 3/4" | BE@.150 | Ea | 3.66 | 5.16 | 8.82 | 15.00 |
| 1" | BE@.150 | Ea | 4.65 | 5.16 | 9.81 | 16.70 |
| 1-1/4" | BE@.300 | Ea | 5.35 | 10.30 | 15.65 | 26.60 |
| 1-1/2" | BE@.300 | Ea | 6.64 | 10.30 | 16.94 | 28.80 |
| 2" | BE@.500 | Ea | 9.62 | 17.20 | 26.82 | 45.60 |

**Non-metallic two-gang conduit box, Super Blue, Carlon®.** Two-gang hard shell box with captive nails. Depth 3-1/2", width 4-1/8", length 3-7/8". 8 integral clamps, 4 each side.

| | | | | | | |
|---|---|---|---|---|---|---|
| 35 cubic inch, wood studs | BE@.200 | Ea | 2.00 | 6.88 | 8.88 | 15.10 |
| 35 cubic inch, steel studs | BE@.150 | Ea | 2.52 | 5.16 | 7.68 | 13.10 |

**Non-metallic three-gang conduit box, Super Blue, Carlon®.** Three-gang hard shell box with captive nails. Depth 3-1/2", width 3-3/4", length 5-7/8". 12 integral clamps, 6 each side.

| | | | | | | |
|---|---|---|---|---|---|---|
| 53 cubic inch, wood studs | BE@.200 | Ea | 2.73 | 6.88 | 9.61 | 16.30 |
| 53 cubic inch, steel studs | BE@.150 | Ea | 2.75 | 5.16 | 7.91 | 13.40 |

**Non-metallic ceiling fan box, Carlon®.** 4", rated for fans up to 35 pounds. Rated for lighting fixtures up to 50 pounds. Includes mounting screws.

| | | | | | | |
|---|---|---|---|---|---|---|
| 1/2" deep, 8 cubic inch | BE@.300 | Ea | 3.43 | 10.30 | 13.73 | 23.30 |
| 2-1/4" deep, 20 cubic inch | BE@.300 | Ea | 4.75 | 10.30 | 15.05 | 25.60 |

**Non-metallic old work switch box, Zip-Mount® retainers, Carlon®.** For existing construction. Non-metallic cable clamps. UL listed.

| | | | | | | |
|---|---|---|---|---|---|---|
| Single gang, 14 cubic inch | BE@.200 | Ea | 1.19 | 6.88 | 8.07 | 13.70 |
| Single gang, 20 cubic inch | BE@.200 | Ea | 1.88 | 6.88 | 8.76 | 14.90 |
| 2 gang, 25 cubic inch | BE@.200 | Ea | 2.33 | 6.88 | 9.21 | 15.70 |
| 3 gang, 55 cubic inch | BE@.200 | Ea | 3.93 | 6.88 | 10.81 | 18.40 |
| 4 gang, 68 cubic inch | BE@.300 | Ea | 6.31 | 10.30 | 16.61 | 28.20 |

**Non-metallic old work switch box, Zip-Mount® retainers, Carlon®.** For existing construction. 4 integral non-metallic cable clamps. 3 Zip-Mount® retainers. 2-3/4" deep. High temperature resistance. UL listed.

| | | | | | | |
|---|---|---|---|---|---|---|
| Round | BE@.300 | Ea | 5.89 | 10.30 | 16.19 | 27.50 |

**Non-metallic low voltage device mounting bracket, Arlington.** Low voltage device; Class 2 only. For both single- and double-gang installation on existing construction. Not to be used for AC circuits. 110 volt. 4.256" high. Non-conductive, smooth plastic construction. For communications, cable TV, computer wiring. CAT 5 listed. Adjusts to fit 1/4"- to 1"-thick wallboard, paneling, or drywall. Bracket is its own template for cutout. Specially designed wing flips up when mounting screw is tightened for a secure mount. UL listed.

| | | | | | | |
|---|---|---|---|---|---|---|
| Single gang, 2.507" wide | BE@.050 | Ea | 1.29 | 1.72 | 3.01 | 5.12 |
| Double gang, 4.185" wide | BE@.050 | Ea | 1.61 | 1.72 | 3.33 | 5.66 |

## EMT Conduit and Fittings

**Remove EMT conduit and wire.** Removed in salvage condition.

| | | | | | | |
|---|---|---|---|---|---|---|
| 1/2" to 2" conduit diameter | BE@.030 | LF | — | 1.03 | 1.03 | 1.75 |

**Electric metallic tube (EMT) conduit.** Welded and galvanized. UL listed. *NEC* approved. Meets ANSI specifications. Sold in 10' lengths. Labor assumes installation in exposed frame walls.

| | | | | | | |
|---|---|---|---|---|---|---|
| 1/2" nominal diameter | BE@.033 | LF | .21 | 1.13 | 1.34 | 2.28 |
| 3/4" nominal diameter | BE@.035 | LF | .43 | 1.20 | 1.63 | 2.77 |
| 1" nominal diameter | BE@.040 | LF | .75 | 1.38 | 2.13 | 3.62 |
| 1-1/4" nominal diameter | BE@.045 | LF | 1.42 | 1.55 | 2.97 | 5.05 |
| 1-1/2" nominal diameter | BE@.055 | LF | 1.52 | 1.89 | 3.41 | 5.80 |
| 2" nominal diameter | BE@.070 | LF | 1.72 | 2.41 | 4.13 | 7.02 |

| | Craft@Hrs | Unit | Material | Labor | Total | Sell |
|---|---|---|---|---|---|---|

**EMT 45-degree elbow.** Galvanized steel. Indoor use only. UL listed.

| | Craft@Hrs | Unit | Material | Labor | Total | Sell |
|---|---|---|---|---|---|---|
| 1/2" nominal diameter | BE@.050 | Ea | 2.55 | 1.72 | 4.27 | 7.26 |
| 3/4" nominal diameter | BE@.060 | Ea | 2.55 | 2.06 | 4.61 | 7.84 |
| 1" nominal diameter | BE@.080 | Ea | 2.90 | 2.75 | 5.65 | 9.61 |
| 1-1/4" nominal diameter | BE@.100 | Ea | 3.82 | 3.44 | 7.26 | 12.30 |
| 1-1/2" nominal diameter | BE@.150 | Ea | 4.93 | 5.16 | 10.09 | 17.20 |
| 2" nominal diameter | BE@.200 | Ea | 7.61 | 6.88 | 14.49 | 24.60 |

**EMT 90-degree elbow.** Galvanized steel. Indoor use only. UL listed.

| | Craft@Hrs | Unit | Material | Labor | Total | Sell |
|---|---|---|---|---|---|---|
| 1/2" nominal diameter | BE@.050 | Ea | 3.02 | 1.72 | 4.74 | 8.06 |
| 3/4" nominal diameter | BE@.060 | Ea | 3.30 | 2.06 | 5.36 | 9.11 |
| 1" nominal diameter | BE@.080 | Ea | 3.69 | 2.75 | 6.44 | 10.90 |
| 1-1/4" nominal diameter | BE@.100 | Ea | 4.57 | 3.44 | 8.01 | 13.60 |
| 1-1/2" nominal diameter | BE@.100 | Ea | 5.30 | 3.44 | 8.74 | 14.90 |
| 2" nominal diameter | BE@.150 | Ea | 8.70 | 5.16 | 13.86 | 23.60 |

**EMT 90-degree pulling elbow, EMT to box.** Die-cast zinc. Combination threaded/set screw. Includes cover and gasket. UL listed.

| | Craft@Hrs | Unit | Material | Labor | Total | Sell |
|---|---|---|---|---|---|---|
| 1/2" nominal diameter | BE@.050 | Ea | 2.17 | 1.72 | 3.89 | 6.61 |
| 3/4" nominal diameter | BE@.060 | Ea | 3.04 | 2.06 | 5.10 | 8.67 |
| 1" nominal diameter | BE@.080 | Ea | 6.45 | 2.75 | 9.20 | 15.60 |
| 1-1/4" nominal diameter | BE@.100 | Ea | 8.57 | 3.44 | 12.01 | 20.40 |

**EMT compression connector.** Die-cast zinc. Indoor and outdoor use. Concrete-tight. Connects EMT conduit to steel outlet box, load center, or other metal enclosure. UL and CSA listed.

| | Craft@Hrs | Unit | Material | Labor | Total | Sell |
|---|---|---|---|---|---|---|
| 1/2" | BE@.050 | Ea | .45 | 1.72 | 2.17 | 3.69 |
| 3/4" | BE@.050 | Ea | .67 | 1.72 | 2.39 | 4.06 |
| 1" | BE@.080 | Ea | 1.53 | 2.75 | 4.28 | 7.28 |
| 1-1/4" | BE@.100 | Ea | 1.68 | 3.44 | 5.12 | 8.70 |
| 1-1/2" | BE@.100 | Ea | 2.21 | 3.44 | 5.65 | 9.61 |
| 2" | BE@.150 | Ea | 3.78 | 5.16 | 8.94 | 15.20 |

**EMT insulated throat compression connector.** Die-cast zinc. Indoor and outdoor use. Concrete tight. Connects EMT conduit to a steel outlet box, load center or other metal enclosure. UL listed.

| | Craft@Hrs | Unit | Material | Labor | Total | Sell |
|---|---|---|---|---|---|---|
| 1/2" | BE@.050 | Ea | .52 | 1.72 | 2.24 | 3.81 |
| 3/4" | BE@.060 | Ea | .64 | 2.06 | 2.70 | 4.59 |
| 1" | BE@.080 | Ea | 1.64 | 2.75 | 4.39 | 7.46 |
| 1-1/4" | BE@.100 | Ea | 3.02 | 3.44 | 6.46 | 11.00 |
| 1-1/2" | BE@.100 | Ea | 3.35 | 3.44 | 6.79 | 11.50 |
| 2" | BE@.150 | Ea | 2.52 | 5.16 | 7.68 | 13.10 |

**EMT set screw connector.** Zinc plated steel. Concrete tight when taped. Connects EMT conduit to a steel outlet box, load center or other metal enclosure.

| | Craft@Hrs | Unit | Material | Labor | Total | Sell |
|---|---|---|---|---|---|---|
| 3/4" | BE@.050 | Ea | .51 | 1.72 | 2.23 | 3.79 |
| 1" | BE@.080 | Ea | .83 | 2.75 | 3.58 | 6.09 |
| 1-1/4" | BE@.100 | Ea | 1.51 | 3.44 | 4.95 | 8.42 |
| 1-1/2" | BE@.100 | Ea | 2.18 | 3.44 | 5.62 | 9.55 |
| 2" | BE@.150 | Ea | 3.28 | 5.16 | 8.44 | 14.30 |

**EMT offset screw connector.** Die-cast zinc. Concrete tight when taped. Connects EMT conduit to a metal outlet box or other enclosure where an offset is required. UL listed.

| | Craft@Hrs | Unit | Material | Labor | Total | Sell |
|---|---|---|---|---|---|---|
| 1/2" conduit | BE@.100 | Ea | 1.38 | 3.44 | 4.82 | 8.19 |
| 3/4" conduit | BE@.100 | Ea | 1.71 | 3.44 | 5.15 | 8.76 |

| | Craft@Hrs | Unit | Material | Labor | Total | Sell |
|---|---|---|---|---|---|---|

**EMT to box offset compression connector.** Die-cast zinc. Indoor and outdoor use. Concrete-tight. Connects EMT conduit to a metal outlet box or other enclosure where an offset is required. UL listed.

| | Craft@Hrs | Unit | Material | Labor | Total | Sell |
|---|---|---|---|---|---|---|
| 1/2" conduit | BE@.100 | Ea | 2.55 | 3.44 | 5.99 | 10.20 |
| 3/4" conduit | BE@.100 | Ea | 2.99 | 3.44 | 6.43 | 10.90 |

**EMT to flex combination coupling, compression to screw-in.** Die-cast zinc. Indoor or outdoor use. Joins EMT conduit to flexible metal conduit. UL listed.

| | Craft@Hrs | Unit | Material | Labor | Total | Sell |
|---|---|---|---|---|---|---|
| 1/2" | BE@.050 | Ea | 2.12 | 1.72 | 3.84 | 6.53 |
| 3/4" | BE@.060 | Ea | 2.45 | 2.06 | 4.51 | 7.67 |
| 1" | BE@.150 | Ea | 2.98 | 5.16 | 8.14 | 13.80 |

**EMT to non-metallic sheathed cable coupling.**

| | Craft@Hrs | Unit | Material | Labor | Total | Sell |
|---|---|---|---|---|---|---|
| 1/2" | BE@.100 | Ea | .52 | 3.44 | 3.96 | 6.73 |

**EMT 1-hole strap.** Zinc plated steel. Snap-on style. Indoor and outdoor use. Supports EMT conduit.

| | Craft@Hrs | Unit | Material | Labor | Total | Sell |
|---|---|---|---|---|---|---|
| 1/2" | BE@.050 | Ea | .21 | 1.72 | 1.93 | 3.28 |
| 3/4" | BE@.050 | Ea | .22 | 1.72 | 1.94 | 3.30 |
| 1" | BE@.080 | Ea | .30 | 2.75 | 3.05 | 5.19 |
| 1-1/4" | BE@.100 | Ea | .41 | 3.44 | 3.85 | 6.55 |
| 1-1/2" | BE@.100 | Ea | .67 | 3.44 | 4.11 | 6.99 |
| 2" | BE@.100 | Ea | .83 | 3.44 | 4.27 | 7.26 |

**EMT 2-hole strap.** Zinc plated steel. Indoor or outdoor use. Supports EMT conduit where greater load bearing capacity is required. UL listed.

| | Craft@Hrs | Unit | Material | Labor | Total | Sell |
|---|---|---|---|---|---|---|
| 1/2" | BE@.030 | Ea | .26 | 1.03 | 1.29 | 2.19 |
| 3/4" | BE@.040 | Ea | .30 | 1.38 | 1.68 | 2.86 |
| 1-1/4" | BE@.050 | Ea | .33 | 1.72 | 2.05 | 3.49 |
| 1-1/2" | BE@.050 | Ea | .46 | 1.72 | 2.18 | 3.71 |
| 2" | BE@.100 | Ea | .64 | 3.44 | 4.08 | 6.94 |

## Metal Conduit Boxes

**Install conduit box in confined area.** Installed in an area such as an attic, crawlspace or inserted in a wall.

| | Craft@Hrs | Unit | Material | Labor | Total | Sell |
|---|---|---|---|---|---|---|
| Junction, switch or outlet box | BE@.420 | Ea | — | 14.40 | 14.40 | 24.50 |

**Drawn handy box.** 4" x 2", 1-7/8" deep. 13 cubic inch capacity. Used to mount switches or receptacles. Bracket for wood or metal studs. Drawn construction. UL listed. KO (knockout).

| | Craft@Hrs | Unit | Material | Labor | Total | Sell |
|---|---|---|---|---|---|---|
| (3) 1/2" side KOs, (3) 1/2" bottom KOs | BE@.200 | Ea | 1.44 | 6.88 | 8.32 | 14.10 |
| (3) 1/2" side KOs, (2) 1/2" end KOs, (3) 1/2" bottom KOs | BE@.200 | Ea | 1.89 | 6.88 | 8.77 | 14.90 |

**Welded handy box, RACO.** 2-1/8" deep. 16.5 cubic inch capacity. Used to mount switches or receptacles. Bracket for wood or metal studs. UL listed. KO (knockout).

| | Craft@Hrs | Unit | Material | Labor | Total | Sell |
|---|---|---|---|---|---|---|
| 4" x 2", (2) 1/2" end KOs, (3) 1/2" bottom KOs | BE@.200 | Ea | 1.67 | 6.88 | 8.55 | 14.50 |
| 4" x 2", (3) 1/2" side KOs, (2) 1/2" end KOs, (3) 1/2" bottom KOs | BE@.200 | Ea | 2.11 | 6.88 | 8.99 | 15.30 |
| 4" x 2", (3) 1/2" side KOs, (2) 1/2" end KOs, (3) 1/2" bottom KOs | BE@.200 | Ea | 2.14 | 6.88 | 9.02 | 15.30 |

**Handy box cover, RACO.** Covers also may be used as single gang wall plates. Includes captive screws.

| | Craft@Hrs | Unit | Material | Labor | Total | Sell |
|---|---|---|---|---|---|---|
| 20-amp receptacle | BE@.030 | Ea | 1.04 | 1.03 | 2.07 | 3.52 |
| Duplex receptacle | BE@.030 | Ea | .48 | 1.03 | 1.51 | 2.57 |
| Toggle switch | BE@.030 | Ea | .48 | 1.03 | 1.51 | 2.57 |

| | Craft@Hrs | Unit | Material | Labor | Total | Sell |
|---|---|---|---|---|---|---|

**Welded square box, RACO.** Used to mount switches and receptacles. 42 cubic inch wiring capacity. Welded construction. UL listed. KO (knockout).

| | Craft@Hrs | Unit | Material | Labor | Total | Sell |
|---|---|---|---|---|---|---|
| (2) 1/2" side KOs, (7) top KOs, (1) 1/2", (2) 3/4" bottom KOs | BE@.300 | Ea | 2.35 | 10.30 | 12.65 | 21.50 |
| (2) 1/2" and (10) TKO side KOs, (2) 3/4" and (1) 1/2" TKO bottom KOs | BE@.300 | Ea | 3.40 | 10.30 | 13.70 | 23.30 |
| (4) 3/4", (4) 1" side KOs, (3) 1/2", (2) 3/4" bottom KOs | BE@.300 | Ea | 3.35 | 10.30 | 13.65 | 23.20 |

**Square surface box cover.** 30- to 50-amp receptacle. Steel. Raised 1/2".

| | Craft@Hrs | Unit | Material | Labor | Total | Sell |
|---|---|---|---|---|---|---|
| 2.156" diameter | BE@.060 | Ea | 4.15 | 2.06 | 6.21 | 10.60 |

**Multi-gang switch box with conduit KOs, RACO.** Steel. 2-1/2" deep. Combination 1/2" and 3/4" KOs. UL listed. KO (knockout).

| | Craft@Hrs | Unit | Material | Labor | Total | Sell |
|---|---|---|---|---|---|---|
| 3 gang, 5-19/32" wide, (6) top, (2) bottom, (3) back KOs | BE@.200 | Ea | 7.37 | 6.88 | 14.25 | 24.20 |
| 4 gang, 7-19/32" wide, (8) top, (2) bottom, (4) back KOs | BE@.200 | Ea | 10.13 | 6.88 | 17.01 | 28.90 |

**Drawn 2-device switch box, RACO.** 4" x 4". Used to mount switches or receptacles. 30.3 cubic inch wiring capacity. Bracket for wood or metal studs. Drawn construction. KO (knockout).

| | Craft@Hrs | Unit | Material | Labor | Total | Sell |
|---|---|---|---|---|---|---|
| 1/2" side KOs, (5) 1/2" bottom KOs | BE@.200 | Ea | 4.05 | 6.88 | 10.93 | 18.60 |
| (2) 1/2", (1) 3/4" side KOs, (3) 1/2", (2) 3/4" bottom KOs | BE@.200 | Ea | 4.20 | 6.88 | 11.08 | 18.80 |
| (2) 3/4" side KOs, (3) 1/2", (2) 3/4" bottom KOs | BE@.200 | Ea | 4.52 | 6.88 | 11.40 | 19.40 |

**Finished box cover.** Baked Cadilite finished steel. 4" square. 1/2" deep. Fully depressed corners for easier mounting. UL listed.

| | Craft@Hrs | Unit | Material | Labor | Total | Sell |
|---|---|---|---|---|---|---|
| 1 GFCI receptacle | BE@.050 | Ea | 1.26 | 1.72 | 2.98 | 5.07 |
| 1 single receptacle | BE@.050 | Ea | 1.39 | 1.72 | 3.11 | 5.29 |
| 1 duplex receptacle | BE@.050 | Ea | 1.10 | 1.72 | 2.82 | 4.79 |
| 2 duplex receptacles | BE@.050 | Ea | 1.10 | 1.72 | 2.82 | 4.79 |
| 1 toggle switch | BE@.050 | Ea | 1.39 | 1.72 | 3.11 | 5.29 |
| 2 toggle switches | BE@.050 | Ea | 1.39 | 1.72 | 3.11 | 5.29 |
| Toggle and duplex | BE@.050 | Ea | 1.39 | 1.72 | 3.11 | 5.29 |

**Ceiling fan box.** Steel. UL listed for installation of ceiling fans up to 35 pounds. Fixture supports up to 50 pounds. Affixes to joist or cross brace. 4 side cable knockouts, (2) 1/2" side conduit knockouts, 1/2" bottom conduit knockout.

| | Craft@Hrs | Unit | Material | Labor | Total | Sell |
|---|---|---|---|---|---|---|
| 4", includes fan-mounting screws with lock washers | BE@.200 | Ea | 4.29 | 6.88 | 11.17 | 19.00 |

**Ceiling fan pancake box.** 6 cubic inch capacity. Steel. Includes fan-mounting screws with lock washers. UL listed for installation of ceiling fans up to 35 pounds. Fixture supports up to 50 pounds. Affixes to joist or cross brace. Includes non-metallic connector. KO (knockout).

| | Craft@Hrs | Unit | Material | Labor | Total | Sell |
|---|---|---|---|---|---|---|
| 4", 1/2" deep, (4) 1/2" bottom KOs | BE@.200 | Ea | 3.32 | 6.88 | 10.20 | 17.30 |

**R-3 drawn octagon box, RACO.** Used to support ceiling light fixture. Sides have (4) 1/2" knockouts, bottom has 1/2" knockout. 50-pound maximum light fixture support. Not designed for ceiling fan support. UL listed.

| | Craft@Hrs | Unit | Material | Labor | Total | Sell |
|---|---|---|---|---|---|---|
| 11.8 cubic inch capacity | BE@.200 | Ea | 1.39 | 6.88 | 8.27 | 14.10 |

| | Craft@Hrs | Unit | Material | Labor | Total | Sell |
|---|---|---|---|---|---|---|

**Steel octagon box cover, RACO.** Flat. Blank. UL listed.

| | | | | | | |
|---|---|---|---|---|---|---|
| 4" x 1/2" | BE@.050 | Ea | .53 | 1.72 | 2.25 | 3.83 |

## Flex Conduit and Fittings

**Remove flexible metal conduit and wire.** Includes removing one conduit box each 10'.

| | | | | | | |
|---|---|---|---|---|---|---|
| 1/2" to 2" conduit diameter | BE@.019 | LF | — | .65 | .65 | 1.11 |

**Reduced wall flexible aluminum conduit.** By nominal trade size.

| | | | | | | |
|---|---|---|---|---|---|---|
| 1/2", 25' coil | BE@.030 | LF | .50 | 1.03 | 1.53 | 2.60 |
| 1/2", 100' coil | BE@.030 | LF | .38 | 1.03 | 1.41 | 2.40 |
| 1/2", 500' coil | BE@.030 | LF | .23 | 1.03 | 1.26 | 2.14 |
| 3/4", 100' coil | BE@.033 | LF | .55 | 1.13 | 1.68 | 2.86 |
| 1", 50' coil | BE@.033 | LF | .94 | 1.13 | 2.07 | 3.52 |

**Reduced wall flexible steel conduit.** By nominal trade size.

| | | | | | | |
|---|---|---|---|---|---|---|
| 3/8", 100' coil | BE@.025 | LF | .31 | .86 | 1.17 | 1.99 |
| 3/8", 250' coil | BE@.025 | LF | .27 | .86 | 1.13 | 1.92 |
| 1/2", per foot | BE@.028 | LF | .31 | .96 | 1.27 | 2.16 |
| 1/2", 25' coil | BE@.028 | LF | .40 | .96 | 1.36 | 2.31 |
| 1/2", 100' coil | BE@.028 | LF | .33 | .96 | 1.29 | 2.19 |
| 3/4", per foot | BE@.030 | LF | .48 | 1.03 | 1.51 | 2.57 |
| 3/4", 100' coil | BE@.030 | LF | .47 | 1.03 | 1.50 | 2.55 |
| 1", 50' coil | BE@.030 | LF | .96 | 1.03 | 1.99 | 3.38 |

**Flex screw-in connector.** Die-cast zinc. Indoor use only. Connects flexible metallic conduit to steel outlet box, load center or other metal enclosure. 3/8" size fits 1/2" knockout. UL listed.

| | | | | | | |
|---|---|---|---|---|---|---|
| 3/8" | BE@.050 | Ea | .62 | 1.72 | 2.34 | 3.98 |
| 1/2" | BE@.050 | Ea | .41 | 1.72 | 2.13 | 3.62 |
| 3/4" | BE@.050 | Ea | .94 | 1.72 | 2.66 | 4.52 |
| 1" | BE@.060 | Ea | 1.58 | 2.06 | 3.64 | 6.19 |

**Flex screw-in insulated throat connector.** For flexible metal conduit. Zinc die-cast. Knockout size is the same as the conduit trade size.

| | | | | | | |
|---|---|---|---|---|---|---|
| 3/8" conduit | BE@.050 | Ea | .43 | 1.72 | 2.15 | 3.66 |
| 1/2" conduit | BE@.050 | Ea | .44 | 1.72 | 2.16 | 3.67 |
| 3/4" conduit | BE@.060 | Ea | .75 | 2.06 | 2.81 | 4.78 |
| 1" conduit | BE@.080 | Ea | 3.45 | 2.75 | 6.20 | 10.50 |

**Flex 90-degree insulated throat connector.**

| | | | | | | |
|---|---|---|---|---|---|---|
| 1/2" conduit | BE@.050 | Ea | .75 | 1.72 | 2.47 | 4.20 |
| 3/4" conduit | BE@.050 | Ea | 3.52 | 1.72 | 5.24 | 8.91 |

**Flex 90-degree connector.** Die-cast zinc. Indoor use only. Use with flexible metal conduit, metal clad or armored cable. Use in a dry location to connect aluminum and steel flex to a steel outlet box or other metallic enclosure. 3/8" size fits 1/2" knockout. UL listed.

| | | | | | | |
|---|---|---|---|---|---|---|
| 3/8" conduit | BE@.050 | Ea | 1.00 | 1.72 | 2.72 | 4.62 |
| 1/2" conduit | BE@.050 | Ea | 1.22 | 1.72 | 2.94 | 5.00 |
| 3/4" conduit | BE@.060 | Ea | 1.89 | 2.06 | 3.95 | 6.72 |
| 1" conduit | BE@.080 | Ea | 1.09 | 2.75 | 3.84 | 6.53 |

| | Craft@Hrs | Unit | Material | Labor | Total | Sell |
|---|---|---|---|---|---|---|

**Flex squeeze connector.** Die-cast zinc. Indoor use only. In a dry location, use to connect flexible metallic conduit to a metal enclosure such as a steel outlet box or load center. 3/8" size fits 1/2" knockout. UL listed.

| | Craft@Hrs | Unit | Material | Labor | Total | Sell |
|---|---|---|---|---|---|---|
| 3/8" conduit | BE@.050 | Ea | .55 | 1.72 | 2.27 | 3.86 |
| 1/2" conduit | BE@.050 | Ea | .67 | 1.72 | 2.39 | 4.06 |
| 3/4" conduit | BE@.060 | Ea | .99 | 2.06 | 3.05 | 5.19 |
| 1" conduit | BE@.080 | Ea | 1.94 | 2.75 | 4.69 | 7.97 |

**Electrical non-metallic flex tubing (ENT) conduit, Carlon®.** Flex-Plus® Blue™ ENT. No special tools required. Highly flexible. Cuts easily and cleanly. For use with 90-degree Celsius conductors. Corrugated design for easy wire pulling and pushing. NER-290 recognized by BOCA, ICC and SBCCI. UL listed. By nominal trade size.

| | Craft@Hrs | Unit | Material | Labor | Total | Sell |
|---|---|---|---|---|---|---|
| 1/2", per foot | BE@.021 | LF | .29 | .72 | 1.01 | 1.72 |
| 1/2", 10' coil | BE@.021 | LF | .27 | .72 | .99 | 1.68 |
| 1/2", 200' coil | BE@.021 | LF | .22 | .72 | .94 | 1.60 |
| 3/4", per foot | BE@.022 | LF | .45 | .76 | 1.21 | 2.06 |
| 3/4", 10' coil | BE@.022 | LF | .38 | .76 | 1.14 | 1.94 |
| 3/4", 100' coil | BE@.025 | LF | .35 | .86 | 1.21 | 2.06 |
| 1", 10' coil | BE@.040 | LF | .33 | 1.38 | 1.71 | 2.91 |

**Carflex® non-metallic liquid-tight flexible conduit, Carlon®.** Non-conductive, non-corrosive, crush, abrasion and stain resistant. Used to connect air conditioning and heating equipment for outdoor wiring and controls.

| | Craft@Hrs | Unit | Material | Labor | Total | Sell |
|---|---|---|---|---|---|---|
| 1/2", per foot | BE@.020 | LF | .54 | .69 | 1.23 | 2.09 |
| 1/2", 100' coil | BE@.020 | LF | .49 | .69 | 1.18 | 2.01 |
| 3/4", per foot | BE@.020 | LF | .67 | .69 | 1.36 | 2.31 |
| 3/4", 100' coil | BE@.020 | LF | .67 | .69 | 1.36 | 2.31 |
| 1", per foot | BE@.025 | LF | 1.05 | .86 | 1.91 | 3.25 |
| 1", 100' coil | BE@.025 | LF | .98 | .86 | 1.84 | 3.13 |
| 2", per foot | BE@.030 | LF | 2.36 | 1.03 | 3.39 | 5.76 |

**Carflex® conduit fitting, Carlon®.** Requires no disassembly of components for installation. Locknut, O-rings and foam washer assembled onto fitting. Withstands temperatures up to 140-degrees F. UL listed.

| | Craft@Hrs | Unit | Material | Labor | Total | Sell |
|---|---|---|---|---|---|---|
| 1/2" 90-degree ell | BE@.100 | Ea | 2.99 | 3.44 | 6.43 | 10.90 |
| 3/4" 90-degree ell | BE@.100 | Ea | 3.53 | 3.44 | 6.97 | 11.80 |
| 1" 90-degree ell | BE@.150 | Ea | 6.30 | 5.16 | 11.46 | 19.50 |
| 1/2" straight coupling | BE@.100 | Ea | 2.17 | 3.44 | 5.61 | 9.54 |
| 3/4" straight coupling | BE@.100 | Ea | 2.97 | 3.44 | 6.41 | 10.90 |
| 1" straight coupling | BE@.150 | Ea | 2.84 | 5.16 | 8.00 | 13.60 |

## Weatherproof Conduit Boxes and Fittings

**1-gang rectangular steel weatherproof conduit box, Red Dot®.** Includes mounting lugs, hole plugs and screws. UL listed. Sliver.

| | Craft@Hrs | Unit | Material | Labor | Total | Sell |
|---|---|---|---|---|---|---|
| 1/2" hole, 3 holes | BE@.300 | Ea | 3.66 | 10.30 | 13.96 | 23.70 |
| 1/2" hole, 4 holes | BE@.300 | Ea | 4.86 | 10.30 | 15.16 | 25.80 |
| 1/2" hole, 5 holes | BE@.300 | Ea | 6.25 | 10.30 | 16.55 | 28.10 |
| 1/2" hole, 5 holes, 2 side holes | BE@.300 | Ea | 6.50 | 10.30 | 16.80 | 28.60 |
| 3/4" hole, 3 holes | BE@.300 | Ea | 5.90 | 10.30 | 16.20 | 27.50 |
| 3/4" hole, 4 holes | BE@.300 | Ea | 6.60 | 10.30 | 16.90 | 28.70 |
| 3/4" hole, 5 holes | BE@.300 | Ea | 7.19 | 10.30 | 17.49 | 29.70 |

**2-gang steel weatherproof conduit box, Red Dot®.** Includes mounting lugs, hole plugs and screws. UL listed. Silver.

| | Craft@Hrs | Unit | Material | Labor | Total | Sell |
|---|---|---|---|---|---|---|
| 1/2" hole, 3 holes | BE@.300 | Ea | 9.07 | 10.30 | 19.37 | 32.90 |
| 1/2" hole, 5 holes | BE@.300 | Ea | 10.48 | 10.30 | 20.78 | 35.30 |
| 1/2" hole, 7 holes | BE@.300 | Ea | 14.08 | 10.30 | 24.38 | 41.40 |
| 3/4" hole, 2 holes | BE@.300 | Ea | 9.51 | 10.30 | 19.81 | 33.70 |

| | Craft@Hrs | Unit | Material | Labor | Total | Sell |
|---|---|---|---|---|---|---|
| 3/4" hole, 5 holes | BE@.300 | Ea | 11.60 | 10.30 | 21.90 | 37.20 |
| 3/4" hole, 5 holes, round | BE@.300 | Ea | 9.86 | 10.30 | 20.16 | 34.30 |
| 3/4" hole, 7 holes | BE@.300 | Ea | 12.90 | 10.30 | 23.20 | 39.40 |
| 3/4" hole, 7 holes, 2 side holes | BE@.250 | Ea | 15.40 | 8.60 | 24.00 | 40.80 |
| 1" hole, 5 holes, deep | BE@.300 | Ea | 13.30 | 10.30 | 23.60 | 40.10 |

**Non-metallic exposed weatherproof conduit box, Carlon®.** With 1/2" threaded holes. Mounting feet included.

| | | | | | | |
|---|---|---|---|---|---|---|
| Rectangular, gray, 3 holes | BE@.300 | Ea | 6.16 | 10.30 | 16.46 | 28.00 |
| Rectangular, white, 3 holes | BE@.300 | Ea | 6.20 | 10.30 | 16.50 | 28.10 |
| Round, gray, 5 holes | BE@.300 | Ea | 7.40 | 10.30 | 17.70 | 30.10 |
| Round, white, 5 holes | BE@.300 | Ea | 8.08 | 10.30 | 18.38 | 31.20 |

**2-gang waterproof conduit box cover, Red Dot®.** Includes gasket and screws. Silver.

| | | | | | | |
|---|---|---|---|---|---|---|
| Blank cover | BE@.150 | Ea | 2.35 | 5.16 | 7.51 | 12.80 |
| Duplex receptacle cover | BE@.150 | Ea | 8.47 | 5.16 | 13.63 | 23.20 |
| Switch cover | BE@.150 | Ea | 8.00 | 5.16 | 13.16 | 22.40 |

**GFCI wet location conduit box receptacle cover, Red Dot®.** Includes cover, gasket, screws and instructions. All metal construction. Required by the *NEC*. For residential, commercial industrial or recreational use. UL listed.

| | | | | | | |
|---|---|---|---|---|---|---|
| 2 gang | BE@.150 | Ea | 20.90 | 5.16 | 26.06 | 44.30 |

**Horizontal duplex weatherproof conduit box receptacle cover, Red Dot®.** Includes gasket and screws. Die-cast construction.

| | | | | | | |
|---|---|---|---|---|---|---|
| Silver | BE@.150 | Ea | 3.73 | 5.16 | 8.89 | 15.10 |
| Silver, GFCI | BE@.150 | Ea | 3.98 | 5.16 | 9.14 | 15.50 |

**Universal non-metallic wet location conduit box cover, Red Dot®, Carlon®.** Thermoplastic cover with transparent finish. Lockable. Accommodates GFCI, single and duplex receptacles, toggle switches and single receptacle up to 1.59" diameter. Horizontal or vertical orientation. Mounts to a box or the device. All installation screws included. Device holes are keyed. Just back out the existing device screws and slip the cover over previously installed screws. Meets *NEC* requirements.

| | | | | | | |
|---|---|---|---|---|---|---|
| 1 gang, clear | BE@.150 | Ea | 11.30 | 5.16 | 16.46 | 28.00 |
| 2 gang, clear | BE@.150 | Ea | 17.60 | 5.16 | 22.76 | 38.70 |

**Non-metallic weatherproof single-gang conduit box cover, Carlon®.** Rigid PVC. Non-conductive and non-corrosive. For industrial, commercial, and residential applications. UL listed. Gray.

| | | | | | | |
|---|---|---|---|---|---|---|
| 15 amp | BE@.100 | Ea | 6.03 | 3.44 | 9.47 | 16.10 |
| 30 amp | BE@.100 | Ea | 5.40 | 3.44 | 8.84 | 15.00 |
| 50 amp | BE@.100 | Ea | 5.30 | 3.44 | 8.74 | 14.90 |
| Blank | BE@.100 | Ea | 2.35 | 3.44 | 5.79 | 9.84 |
| Single switch | BE@.100 | Ea | 5.25 | 3.44 | 8.69 | 14.80 |
| Vertical duplex receptable | BE@.100 | Ea | 6.00 | 3.44 | 9.44 | 16.00 |
| Vertical GFCI receptacle | BE@.100 | Ea | 5.50 | 3.44 | 8.94 | 15.20 |

**Weatherproof lampholder and cover, Red Dot®.** Three-hole round cover, gaskets, and screws included.

| | | | | | | |
|---|---|---|---|---|---|---|
| Single lamp, rectangular | BE@.200 | Ea | 3.40 | 6.88 | 10.28 | 17.50 |
| Dual lamps, round | BE@.800 | Ea | 9.46 | 27.50 | 36.96 | 62.80 |

**Weatherproof switch and cover, Red Dot®.** Gasket, screws, and spacers included. Switch is UL listed.

| | | | | | | |
|---|---|---|---|---|---|---|
| 1 pole, 15 amp | BE@.250 | Ea | 5.08 | 8.60 | 13.68 | 23.30 |
| 3 way, 15 amp | BE@.250 | Ea | 5.46 | 8.60 | 14.06 | 23.90 |

| | Craft@Hrs | Unit | Material | Labor | Total | Sell |
|---|---|---|---|---|---|---|

## Rigid and IMC Conduit and Fittings

**Remove rigid steel conduit and wire.** Includes conduit and wire removed in salvage condition.

| | Craft@Hrs | Unit | Material | Labor | Total | Sell |
|---|---|---|---|---|---|---|
| 1/2" to 2" conduit diameter | BE@.030 | LF | — | 1.03 | 1.03 | 1.75 |

**Install rigid steel conduit in confined area.** Working in a confined area such as an attic, crawl space or behind a wall.

| | Craft@Hrs | Unit | Material | Labor | Total | Sell |
|---|---|---|---|---|---|---|
| 1/2" to 1" conduit diameter | BE@.076 | LF | — | 2.61 | 2.61 | 4.44 |

**Galvanized rigid steel (GRS) conduit.** *NEC* approved. Meets ANSI specifications. Made of high-strength strip steel. Provides radiation protection and magnetic shielding. Sold in 10' lengths.

| | Craft@Hrs | Unit | Material | Labor | Total | Sell |
|---|---|---|---|---|---|---|
| 1/2" nominal diameter | BE@.040 | LF | 1.52 | 1.38 | 2.90 | 4.93 |
| 3/4" nominal diameter | BE@.045 | LF | 1.72 | 1.55 | 3.27 | 5.56 |
| 1" nominal diameter | BE@.050 | LF | 2.70 | 1.72 | 4.42 | 7.51 |
| 1-1/4" nominal diameter | BE@.070 | LF | 3.79 | 2.41 | 6.20 | 10.50 |
| 1-1/2" nominal diameter | BE@.080 | LF | 3.90 | 2.75 | 6.65 | 11.30 |
| 2" nominal diameter | BE@.110 | LF | 4.32 | 3.78 | 8.10 | 13.80 |

**Intermediate metal conduit (IMC).** UL listed. *NEC* approved. Meets ANSI specifications. Full-cut threads are galvanized after cutting and capped for protection. Interior coated with lubricating finish. Sold in 10' lengths.

| | Craft@Hrs | Unit | Material | Labor | Total | Sell |
|---|---|---|---|---|---|---|
| 1/2" nominal diameter | BE@.038 | LF | 1.25 | 1.31 | 2.56 | 4.35 |
| 3/4" nominal diameter | BE@.040 | LF | 1.49 | 1.38 | 2.87 | 4.88 |
| 1" nominal diameter | BE@.045 | LF | 2.04 | 1.55 | 3.59 | 6.10 |
| 1-1/4" nominal diameter | BE@.065 | LF | 2.73 | 2.24 | 4.97 | 8.45 |
| 1-1/2" nominal diameter | BE@.065 | LF | 3.15 | 2.24 | 5.39 | 9.16 |
| 2" nominal diameter | BE@.090 | LF | 3.83 | 3.10 | 6.93 | 11.80 |

**Rigid or IMC 45-degree conduit elbow.** Galvanized steel. Indoor or outdoor use. UL listed.

| | Craft@Hrs | Unit | Material | Labor | Total | Sell |
|---|---|---|---|---|---|---|
| 1/2" nominal diameter | BE@.100 | Ea | 2.73 | 3.44 | 6.17 | 10.50 |
| 3/4" nominal diameter | BE@.100 | Ea | 3.95 | 3.44 | 7.39 | 12.60 |
| 1" nominal diameter | BE@.120 | Ea | 5.09 | 4.13 | 9.22 | 15.70 |
| 1-1/4" nominal diameter | BE@.150 | Ea | 7.65 | 5.16 | 12.81 | 21.80 |
| 1-1/2" nominal diameter | BE@.150 | Ea | 8.63 | 5.16 | 13.79 | 23.40 |
| 2" nominal diameter | BE@.200 | Ea | 12.80 | 6.88 | 19.68 | 33.50 |

**Rigid or IMC 90-degree conduit elbow.** Galvanized steel. Indoor or outdoor use. UL listed.

| | Craft@Hrs | Unit | Material | Labor | Total | Sell |
|---|---|---|---|---|---|---|
| 1/2" nominal diameter | BE@.100 | Ea | 3.68 | 3.44 | 7.12 | 12.10 |
| 3/4" nominal diameter | BE@.100 | Ea | 4.33 | 3.44 | 7.77 | 13.20 |
| 1" nominal diameter | BE@.120 | Ea | 6.01 | 4.13 | 10.14 | 17.20 |
| 1-1/4" nominal diameter | BE@.150 | Ea | 8.32 | 5.16 | 13.48 | 22.90 |
| 1-1/2" nominal diameter | BE@.150 | Ea | 9.66 | 5.16 | 14.82 | 25.20 |
| 2" nominal diameter | BE@.200 | Ea | 14.70 | 6.88 | 21.58 | 36.70 |

**Rigid or IMC conduit coupling.** Galvanized steel. Indoor and outdoor use. Joins threaded rigid or IMC conduit.

| | Craft@Hrs | Unit | Material | Labor | Total | Sell |
|---|---|---|---|---|---|---|
| 1/2" conduit | BE@.050 | Ea | 1.29 | 1.72 | 3.01 | 5.12 |
| 3/4" conduit | BE@.050 | Ea | 1.77 | 1.72 | 3.49 | 5.93 |
| 1" conduit | BE@.080 | Ea | 2.47 | 2.75 | 5.22 | 8.87 |
| 1-1/4" conduit | BE@.100 | Ea | 2.74 | 3.44 | 6.18 | 10.50 |
| 1-1/2" conduit | BE@.100 | Ea | 3.13 | 3.44 | 6.57 | 11.20 |
| 2" conduit | BE@.150 | Ea | 3.74 | 5.16 | 8.90 | 15.10 |

| | Craft@Hrs | Unit | Material | Labor | Total | Sell |
|---|---|---|---|---|---|---|

**Rigid or IMC compression conduit coupling.** Steel. Concrete-tight. Indoor and outdoor use. Joins two lengths of threadless rigid or IMC conduit. UL listed.

| | | | | | | |
|---|---|---|---|---|---|---|
| 1/2" conduit | BE@.050 | Ea | 3.80 | 1.72 | 5.52 | 9.38 |
| 3/4" conduit | BE@.060 | Ea | 5.11 | 2.06 | 7.17 | 12.20 |
| 1" conduit | BE@.800 | Ea | 9.12 | 27.50 | 36.62 | 62.30 |
| 1-1/4" conduit | BE@.100 | Ea | 13.40 | 3.44 | 16.84 | 28.60 |
| 1-1/2" conduit | BE@.100 | Ea | 17.07 | 3.44 | 20.51 | 34.90 |
| 2" conduit | BE@.150 | Ea | 33.50 | 5.16 | 38.66 | 65.70 |

**Rigid or IMC insulating bushing.** High-impact thermoplastic. Flame-retardant. Indoor or outdoor. Use on threaded rigid or IMC conduit to protect wires. Required by *NEC*. UL listed.

| | | | | | | |
|---|---|---|---|---|---|---|
| 1/2" | BE@.020 | Ea | .71 | .69 | 1.40 | 2.38 |
| 3/4" | BE@.020 | Ea | .77 | .69 | 1.46 | 2.48 |
| 1" | BE@.030 | Ea | .31 | 1.03 | 1.34 | 2.28 |
| 1-1/4" | BE@.040 | Ea | .42 | 1.38 | 1.80 | 3.06 |
| 1-1/2" | BE@.040 | Ea | .42 | 1.38 | 1.80 | 3.06 |
| 2" | BE@.050 | Ea | .68 | 1.72 | 2.40 | 4.08 |

**Rigid or IMC insulated metallic grounding bushing.** Die-cast zinc. Lay-in lug. Indoor or outdoor use. Use with locknut to terminate service conduit to a cabinet. Lug provided for bonding jumper to a neutral bus bar. Thermoplastic liner rated at 150-degrees Celsius. UL listed.

| | | | | | | |
|---|---|---|---|---|---|---|
| Trade size 1/2" | BE@.100 | Ea | 3.15 | 3.44 | 6.59 | 11.20 |
| Trade size 3/4" | BE@.100 | Ea | 3.15 | 3.44 | 6.59 | 11.20 |
| Trade size 1" | BE@.100 | Ea | 3.61 | 3.44 | 7.05 | 12.00 |
| Trade size 1-1/4" | BE@.150 | Ea | 3.24 | 5.16 | 8.40 | 14.30 |
| Trade size 1-1/2" | BE@.150 | Ea | 4.40 | 5.16 | 9.56 | 16.30 |
| Trade size 2" | BE@.200 | Ea | 5.09 | 6.88 | 11.97 | 20.30 |

**Rigid or IMC reducing bushing.** Die-cast zinc. Dry locations only. Reduces the entry size of a run of threaded conduit. UL listed.

| | | | | | | |
|---|---|---|---|---|---|---|
| 3/4" x 1/2" | BE@.050 | Ea | .83 | 1.72 | 2.55 | 4.34 |
| 1" x 1/2" | BE@.075 | Ea | 1.08 | 2.58 | 3.66 | 6.22 |
| 1" x 3/4" | BE@.100 | Ea | 1.31 | 3.44 | 4.75 | 8.08 |
| 1-1/2" x 1-1/4" | BE@.100 | Ea | 2.95 | 3.44 | 6.39 | 10.90 |

**Rigid or IMC watertight conduit hub.** With insulated throat. Zinc die-cast. UL listed.

| | | | | | | |
|---|---|---|---|---|---|---|
| 1/2" nominal diameter | BE@.200 | Ea | 2.77 | 6.88 | 9.65 | 16.40 |
| 3/4" nominal diameter | BE@.250 | Ea | 3.19 | 8.60 | 11.79 | 20.00 |
| 1" nominal diameter | BE@.300 | Ea | 4.04 | 10.30 | 14.34 | 24.40 |
| 1-1/4" nominal diameter | BE@.350 | Ea | 4.89 | 12.00 | 16.89 | 28.70 |
| 1-1/2" nominal diameter | BE@.375 | Ea | 6.24 | 12.90 | 19.14 | 32.50 |
| 2" nominal diameter | BE@.400 | Ea | 7.97 | 13.80 | 21.77 | 37.00 |

**Steel sealing conduit locknut.** With PVC molded seal. UL listed.

| | | | | | | |
|---|---|---|---|---|---|---|
| 1/2" nominal diameter | BE@.020 | Ea | .94 | .69 | 1.63 | 2.77 |
| 3/4" nominal diameter | BE@.020 | Ea | 1.13 | .69 | 1.82 | 3.09 |
| 1" nominal diameter | BE@.020 | Ea | 1.63 | .69 | 2.32 | 3.94 |
| 1-1/4" nominal diameter | BE@.020 | Ea | 2.56 | .69 | 3.25 | 5.53 |
| 1-1/2" nominal diameter | BE@.020 | Ea | 3.34 | .69 | 4.03 | 6.85 |
| 2" nominal diameter | BE@.020 | Ea | 4.30 | .69 | 4.99 | 8.48 |

| | Craft@Hrs | Unit | Material | Labor | Total | Sell |
|---|---|---|---|---|---|---|

**Rigid or IMC conduit nipple.** Steel. Indoor or outdoor use. Connects two steel outlet boxes or enclosures. UL listed.

| | Craft@Hrs | Unit | Material | Labor | Total | Sell |
|---|---|---|---|---|---|---|
| 1/2" x close trade size | BE@.050 | Ea | 1.00 | 1.72 | 2.72 | 4.62 |
| 1/2" x 4" trade size | BE@.050 | Ea | 1.31 | 1.72 | 3.03 | 5.15 |
| 3/4" x close trade size | BE@.060 | Ea | 1.25 | 2.06 | 3.31 | 5.63 |
| 1" x 2" trade size | BE@.080 | Ea | 1.20 | 2.75 | 3.95 | 6.72 |
| 1" x 2-1/2" trade size | BE@.080 | Ea | 1.41 | 2.75 | 4.16 | 7.07 |
| 1" x 3" trade size | BE@.080 | Ea | 1.55 | 2.75 | 4.30 | 7.31 |
| 1" x 5" trade size | BE@.080 | Ea | 2.39 | 2.75 | 5.14 | 8.74 |
| 1-1/4" x 2" trade size | BE@.100 | Ea | 1.67 | 3.44 | 5.11 | 8.69 |
| 1-1/4" x 3" trade size | BE@.100 | Ea | 2.52 | 3.44 | 5.96 | 10.10 |
| 1-1/2" x 2" trade size | BE@.100 | Ea | 2.17 | 3.44 | 5.61 | 9.54 |
| 1-1/2" x 3" trade size | BE@.100 | Ea | 3.40 | 3.44 | 6.84 | 11.60 |
| 2" x 3" trade size | BE@.150 | Ea | 3.41 | 5.16 | 8.57 | 14.60 |
| 2" x 5" trade size | BE@.150 | Ea | 4.75 | 5.16 | 9.91 | 16.80 |
| 2" x 7" trade size | BE@.150 | Ea | 6.70 | 5.16 | 11.86 | 20.20 |
| 2" x 8" trade size | BE@.150 | Ea | 6.95 | 5.16 | 12.11 | 20.60 |

**Rigid or IMC conduit 1-hole strap.** Zinc plated malleable iron. Dry locations. Secures rigid or IMC conduit to wood, masonry, or other surfaces. UL listed.

| | Craft@Hrs | Unit | Material | Labor | Total | Sell |
|---|---|---|---|---|---|---|
| 1/2" conduit diameter | BE@.050 | Ea | .42 | 1.72 | 2.14 | 3.64 |
| 3/4" conduit diameter | BE@.060 | Ea | .53 | 2.06 | 2.59 | 4.40 |
| 1" conduit diameter | BE@.050 | Ea | 1.41 | 1.72 | 3.13 | 5.32 |
| 1-1/4" conduit diameter | BE@.075 | Ea | 2.01 | 2.58 | 4.59 | 7.80 |
| 1-1/2" conduit diameter | BE@.075 | Ea | 2.63 | 2.58 | 5.21 | 8.86 |
| 2" conduit diameter | BE@.100 | Ea | 5.56 | 3.44 | 9.00 | 15.30 |

**Rigid or IMC threaded aluminum conduit body.** Copper-free die-cast aluminum. Indoor or outdoor use. Includes cover and gasket. Used to gain access to the interior of raceway for wire pulling, splicing and maintenance (as allowed by the *NEC*). UL listed.

| | Craft@Hrs | Unit | Material | Labor | Total | Sell |
|---|---|---|---|---|---|---|
| Type C, 1/2" | BE@.100 | Ea | 6.10 | 3.44 | 9.54 | 16.20 |
| Type C, 3/4" | BE@.150 | Ea | 7.83 | 5.16 | 12.99 | 22.10 |
| Type C, 1" | BE@.200 | Ea | 8.79 | 6.88 | 15.67 | 26.60 |
| Type C, 1-1/4" | BE@.250 | Ea | 13.30 | 8.60 | 21.90 | 37.20 |
| Type C, 1-1/2" | BE@.250 | Ea | 17.80 | 8.60 | 26.40 | 44.90 |
| Type C, 2" | BE@.300 | Ea | 26.60 | 10.30 | 36.90 | 62.70 |
| Type LB, 1/2" | BE@.100 | Ea | 6.45 | 3.44 | 9.89 | 16.80 |
| Type LB, 3/4" | BE@.150 | Ea | 7.70 | 5.16 | 12.86 | 21.90 |
| Type LB, 1" | BE@.150 | Ea | 10.63 | 5.16 | 15.79 | 26.80 |
| Type LB, 1-1/4" | BE@.200 | Ea | 14.60 | 6.88 | 21.48 | 36.50 |
| Type LB, 1-1/2" | BE@.200 | Ea | 18.60 | 6.88 | 25.48 | 43.30 |
| Type LB, 2" | BE@.250 | Ea | 26.90 | 8.60 | 35.50 | 60.40 |
| Type LL or LR, 1/2" | BE@.100 | Ea | 6.65 | 3.44 | 10.09 | 17.20 |
| Type LL or LR, 3/4" | BE@.150 | Ea | 7.98 | 5.16 | 13.14 | 22.30 |
| Type LL or LR, 1" | BE@.150 | Ea | 9.21 | 5.16 | 14.37 | 24.40 |
| Type LL or LR, 1-1/4" | BE@.200 | Ea | 12.40 | 6.88 | 19.28 | 32.80 |
| Type LL or LR, 1-1/2" | BE@.250 | Ea | 17.70 | 8.60 | 26.30 | 44.70 |
| Type LL or LR, 2" | BE@.300 | Ea | 22.05 | 10.30 | 32.35 | 55.00 |
| Type T, 1/2" | BE@.150 | Ea | 8.00 | 5.16 | 13.16 | 22.40 |
| Type T, 3/4" | BE@.200 | Ea | 9.78 | 6.88 | 16.66 | 28.30 |
| Type T, 1" | BE@.250 | Ea | 11.40 | 8.60 | 20.00 | 34.00 |
| Type T, 1-1/4" | BE@.300 | Ea | 15.70 | 10.30 | 26.00 | 44.20 |
| Type T, 1-1/2" | BE@.300 | Ea | 21.60 | 10.30 | 31.90 | 54.20 |
| Type T, 2" | BE@.330 | Ea | 25.50 | 11.30 | 36.80 | 62.60 |

| | Craft@Hrs | Unit | Material | Labor | Total | Sell |
|---|---|---|---|---|---|---|

**Steel service entrance cap.** Copper-free die-cast aluminum. Clamp-on style. For overhead wiring service entrance. Mounts on top of EMT, rigid or IMC conduit. Serves as connecting point for service entrance wires. UL listed.

| | Craft@Hrs | Unit | Material | Labor | Total | Sell |
|---|---|---|---|---|---|---|
| 1/2" conduit diameter | BE@.150 | Ea | 3.71 | 5.16 | 8.87 | 15.10 |
| 3/4" conduit diameter | BE@.200 | Ea | 4.12 | 6.88 | 11.00 | 18.70 |
| 1" conduit diameter | BE@.250 | Ea | 4.52 | 8.60 | 13.12 | 22.30 |
| 1-1/4" conduit diameter | BE@.300 | Ea | 5.51 | 10.30 | 15.81 | 26.90 |
| 1-1/2" conduit diameter | BE@.300 | Ea | 8.88 | 10.30 | 19.18 | 32.60 |
| 2" conduit diameter | BE@.500 | Ea | 11.17 | 17.20 | 28.37 | 48.20 |
| 2-1/2" conduit diameter | BE@.500 | Ea | 28.00 | 17.20 | 45.20 | 76.80 |

**Rigid or IMC service entrance elbow.** Copper-free die-cast aluminum. Indoor or outdoor use. Includes cover and gasket. Use to gain access to interior of raceway for wire pulling, inspection, and maintenance. UL listed.

| | Craft@Hrs | Unit | Material | Labor | Total | Sell |
|---|---|---|---|---|---|---|
| Trade size 1/2" | BE@.100 | Ea | 4.10 | 3.44 | 7.54 | 12.80 |
| Trade size 3/4" | BE@.150 | Ea | 5.17 | 5.16 | 10.33 | 17.60 |
| Trade size 1" | BE@.150 | Ea | 7.90 | 5.16 | 13.06 | 22.20 |
| Trade size 1-1/4" | BE@.200 | Ea | 10.30 | 6.88 | 17.18 | 29.20 |
| Trade size 1-1/2" | BE@.200 | Ea | 19.09 | 6.88 | 25.97 | 44.10 |
| Trade size 2" | BE@.250 | Ea | 27.40 | 8.60 | 36.00 | 61.20 |

**Service entrance roof flashing.** Prevents leaks where conduit penetrates the roof surface. Slips over the conduit mast and mounts on the roof. Made of galvanized steel with neoprene seal.

| | Craft@Hrs | Unit | Material | Labor | Total | Sell |
|---|---|---|---|---|---|---|
| 2" conduit diameter | BE@.250 | Ea | 6.33 | 8.60 | 14.93 | 25.40 |
| 2-1/2" conduit diameter | BE@.250 | Ea | 7.65 | 8.60 | 16.25 | 27.60 |

**Service entrance conduit sill plate.** Copper-free die-cast aluminum. Helps seal the hole where service entrance cable enters the home.

| | Craft@Hrs | Unit | Material | Labor | Total | Sell |
|---|---|---|---|---|---|---|
| 1/2" to 3/4", 2 screws | BE@.100 | Ea | 1.63 | 3.44 | 5.07 | 8.62 |
| 1", 2 screws | BE@.100 | Ea | 2.18 | 3.44 | 5.62 | 9.55 |
| 1-1/4", 4 screws | BE@.200 | Ea | 2.89 | 6.88 | 9.77 | 16.60 |
| 1-1/2", 4 screws | BE@.200 | Ea | 3.53 | 6.88 | 10.41 | 17.70 |
| 2", 4 screws | BE@.200 | Ea | 5.45 | 6.88 | 12.33 | 21.00 |

**Ground rod clamp.** Bronze. Connects the grounding wire to the ground rod. Approved for direct burial. UL listed.

| | Craft@Hrs | Unit | Material | Labor | Total | Sell |
|---|---|---|---|---|---|---|
| 1/2" rod | BE@.100 | Ea | 2.51 | 3.44 | 5.95 | 10.10 |
| 5/8" rod | BE@.100 | Ea | 2.42 | 3.44 | 5.86 | 9.96 |
| 3/4" rod | BE@.100 | Ea | 2.93 | 3.44 | 6.37 | 10.80 |
| 1-1/4" to 2", 8 to 4 gauge wire | BE@.100 | Ea | 3.80 | 3.44 | 7.24 | 12.30 |

## Wire and Fittings

**Type THHN copper building wire.** Labor is based on wire pulled in conduit.

| | Craft@Hrs | Unit | Material | Labor | Total | Sell |
|---|---|---|---|---|---|---|
| 14 gauge, solid, per foot | BE@.006 | LF | .19 | .21 | .40 | .68 |
| 14 gauge, solid, 50' coil | BE@.006 | LF | .23 | .21 | .44 | .75 |
| 14 gauge, solid, 100' coil | BE@.006 | LF | .11 | .21 | .32 | .54 |
| 14 gauge, solid, 500' coil | BE@.006 | LF | .09 | .21 | .30 | .51 |
| 14 gauge, stranded, per foot | BE@.006 | LF | .21 | .21 | .42 | .71 |
| 14 gauge, stranded, 500' coil | BE@.006 | LF | .10 | .21 | .31 | .53 |
| 12 gauge, solid, per foot | BE@.007 | LF | .30 | .24 | .54 | .92 |
| 12 gauge, solid, 50' coil | BE@.007 | LF | .27 | .24 | .51 | .87 |
| 12 gauge, solid, 100' coil | BE@.007 | LF | .16 | .24 | .40 | .68 |
| 12 gauge, solid, 500' coil | BE@.007 | LF | .13 | .24 | .37 | .63 |
| 12 gauge, stranded, per foot | BE@.007 | LF | .34 | .24 | .58 | .99 |
| 12 gauge, stranded, 50' coil | BE@.007 | LF | .35 | .24 | .59 | 1.00 |
| 12 gauge, stranded, 100' coil | BE@.007 | LF | .23 | .24 | .47 | .80 |
| 12 gauge, stranded, 500' coil | BE@.007 | LF | .15 | .24 | .39 | .66 |

| | Craft@Hrs | Unit | Material | Labor | Total | Sell |
|---|---|---|---|---|---|---|
| 10 gauge, stranded, per foot | BE@.008 | LF | .45 | .28 | .73 | 1.24 |
| 10 gauge, stranded, 50' coil | BE@.008 | LF | .34 | .28 | .62 | 1.05 |
| 10 gauge, stranded, 100' coil | BE@.008 | LF | .29 | .28 | .57 | .97 |
| 10 gauge, stranded, 500' coil | BE@.008 | LF | .24 | .28 | .52 | .88 |
| 8 gauge, stranded, per foot | BE@.009 | LF | .70 | .31 | 1.01 | 1.72 |
| 8 gauge, stranded, 500' coil | BE@.009 | LF | .37 | .31 | .68 | 1.16 |
| 6 gauge, stranded, per foot | BE@.010 | LF | .93 | .34 | 1.27 | 2.16 |
| 6 gauge, stranded, 500' coil | BE@.010 | LF | .57 | .34 | .91 | 1.55 |

**20-gauge bell wire, Carol® Cable.** For alarms, thermostat controls, touch plate systems, burglar alarms, doorbells and other low voltage installations.

| | Craft@Hrs | Unit | Material | Labor | Total | Sell |
|---|---|---|---|---|---|---|
| 2 conductor, 500' coil | BE@.006 | LF | .08 | .21 | .29 | .49 |

**18-gauge thermostat cable, Carol® Cable.** For thermostat control, heating and air conditioning, touch plate systems, burglar alarms, intercom systems, doorbells, remote control units, signal systems, and other low voltage installations. Annealed solid bare copper conductors. PVC insulation and white jacket. UL listed. Type CL-2.

| | Craft@Hrs | Unit | Material | Labor | Total | Sell |
|---|---|---|---|---|---|---|
| 2 conductor, 50' roll | BE@.006 | LF | .18 | .21 | .39 | .66 |
| 2 conductor, 100' roll | BE@.006 | LF | .15 | .21 | .36 | .61 |
| 5 conductor, 50' roll | BE@.006 | LF | .32 | .21 | .53 | .90 |
| 5 conductor, 100' roll | BE@.006 | LF | .31 | .21 | .52 | .88 |

**Speaker wire, RCA.** Clear jacket. Low-noise copper wire.

| | Craft@Hrs | Unit | Material | Labor | Total | Sell |
|---|---|---|---|---|---|---|
| 16 gauge, 50' long | BE@.008 | LF | .26 | .28 | .54 | .92 |
| 16 gauge, 100' long | BE@.008 | LF | .23 | .28 | .51 | .87 |
| 18 gauge, 50' long | BE@.008 | LF | .16 | .28 | .44 | .75 |
| 18 gauge, 100' long | BE@.008 | LF | .12 | .28 | .40 | .68 |

**Coaxial cable.** For RF signal transmission, MATV and CATV, drop cable, and FM broadcast. Solid bare copperweld. AWG 18. Cellular polyethylene insulation. 100% aluminum/polyester tape shield with + 55% aluminum braid. 0.308" outside diameter. Capacitance, pF/Ft. 17.3. Vel. propagation: 78% nominal impedance: 75 ohm. UL listed.

| | Craft@Hrs | Unit | Material | Labor | Total | Sell |
|---|---|---|---|---|---|---|
| Black, per foot | BE@.007 | LF | .29 | .24 | .53 | .90 |
| Black, 500' coil | BE@.007 | LF | .14 | .24 | .38 | .65 |

**Coaxial F connector, RCA.** For connecting to RG6 coaxial cable. Lifetime warranty.

| | Craft@Hrs | Unit | Material | Labor | Total | Sell |
|---|---|---|---|---|---|---|
| Crimp-on | BE@.075 | Ea | 1.35 | 2.58 | 3.93 | 6.68 |
| Twist-on | BE@.075 | Ea | 1.60 | 2.58 | 4.18 | 7.11 |

**Coaxial cable wall plate, RCA.** Fits standard electrical outlet box or flush-mount to drywall. Ivory or white.

| | Craft@Hrs | Unit | Material | Labor | Total | Sell |
|---|---|---|---|---|---|---|
| 75 ohm, single | BE@.150 | Ea | 4.87 | 5.16 | 10.03 | 17.10 |
| 75 ohm, duplex | BE@.200 | Ea | 4.97 | 6.88 | 11.85 | 20.10 |

**Category 5E data communication cable, Carol® Cable.** UTP; Spec. 4480; Type MPR/CMR. Suitable for high-speed data/LAN transmission up to 16 Mbps; 10 BASE-T, 4 Mbps token ring, 100 VG-Any LAN, 100 Mbps TP-PMD, 100 BASE-T, and 55/155 Mbps ATM. 24 AWG solid bare annealed copper construction. Flame retardant PVC insulation. Pairing: two twists per foot. Pull-Pac® cartons. Maximum DC resistance, 9.38 (omega)/100 m at 20°C (68°F). Mutual capacitance, 4.59pF/Ft. at 1 kHz. Characteristic impedance for 772 kHz, 87-117 (omega). For 1.0-100.0 MHz, 85-115 (omega). SRL for 1.0-20.0 MHz, 23 dB. For 20.0-100.0 MHz, 23-10 log. Skew 10 ns. NVP, 70% speed of light. UL listed. cPCCFT4 listed. Meets requirements: ANSI/TIA/EIA 568A, ANSI/ICEA S-90-661 and NEMA WC63.1. Ivory. 1" outside diameter. 24 gauge.

| | Craft@Hrs | Unit | Material | Labor | Total | Sell |
|---|---|---|---|---|---|---|
| 4 conductor, per foot | BE@.007 | LF | .20 | .24 | .44 | .75 |
| 4 conductor, 500' coil | BE@.007 | LF | .15 | .24 | .39 | .66 |

**Decorator phone wall plate.** Ivory or white.

| | Craft@Hrs | Unit | Material | Labor | Total | Sell |
|---|---|---|---|---|---|---|
| Single line | BE@.270 | Ea | 5.26 | 9.29 | 14.55 | 24.70 |
| Duplex | BE@.270 | Ea | 5.26 | 9.29 | 14.55 | 24.70 |

|  | Craft@Hrs | Unit | Material | Labor | Total | Sell |
|---|---|---|---|---|---|---|

## Switches and Receptacles

**Remove switch or receptacle.** Includes removal of cover plate, disconnecting wires and taping ends.

| | | | | | | |
|---|---|---|---|---|---|---|
| Single pole switch or receptacle | BE@.110 | Ea | — | 3.78 | 3.78 | 6.43 |
| Double pole switch or receptacle | BE@.178 | Ea | — | 6.12 | 6.12 | 10.40 |

**Residential grade AC quiet toggle switch, Leviton.** 120-volt, 15-amp, with grounding screw. Side-wired and quick-wired framed toggle push-in. Quick wire accepts conductors up to 12 gauge. Large head, triple-drive combination screws. Impact-resistant. Durable thermoplastic toggle and frame. UL listed.

| | | | | | | |
|---|---|---|---|---|---|---|
| 1 pole, almond | BE@.250 | Ea | .63 | 8.60 | 9.23 | 15.70 |
| 1 pole, brown | BE@.250 | Ea | .63 | 8.60 | 9.23 | 15.70 |
| 1 pole, ivory | BE@.250 | Ea | .63 | 8.60 | 9.23 | 15.70 |
| 1 pole, white | BE@.250 | Ea | .63 | 8.60 | 9.23 | 15.70 |
| 1 pole, clear illuminated | BE@.250 | Ea | 4.62 | 8.60 | 13.22 | 22.50 |
| 3 way, clear illuminated | BE@.250 | Ea | 7.00 | 8.60 | 15.60 | 26.50 |

**Commercial grade AC quiet toggle switch, Leviton.** Side-wired 120/277-volt AC. 15-amp. Back-wiring clamps and large head. Double-drive terminal screws, backed out and staked. Accepts up to 10-gauge copper or copper-clad wire. Large silver and cadmium oxide contacts. Rust-resistant heavy-gauge steel mounting strap. Break-off plaster ears. One-piece brass alloy contact arm for reliable electrical performance. UL listed.

| | | | | | | |
|---|---|---|---|---|---|---|
| 1 pole, brown | BE@.250 | Ea | 2.21 | 8.60 | 10.81 | 18.40 |
| 1 pole, gray | BE@.250 | Ea | 2.21 | 8.60 | 10.81 | 18.40 |
| 1 pole, ivory | BE@.250 | Ea | 2.22 | 8.60 | 10.82 | 18.40 |
| 1 pole, white | BE@.250 | Ea | 2.22 | 8.60 | 10.82 | 18.40 |
| 3 way, ivory | BE@.300 | Ea | 4.51 | 10.30 | 14.81 | 25.20 |
| 3 way, white | BE@.300 | Ea | 4.51 | 10.30 | 14.81 | 25.20 |
| 4 way, framed, ivory | BE@.400 | Ea | 9.75 | 13.80 | 23.55 | 40.00 |
| 4 way, framed, white | BE@.400 | Ea | 12.20 | 13.80 | 26.00 | 44.20 |
| 4 way, grounded, ivory | BE@.400 | Ea | 12.20 | 13.80 | 26.00 | 44.20 |
| 4 way, grounded, white | BE@.400 | Ea | 12.20 | 13.80 | 26.00 | 44.20 |
| 1 pole, lighted, ivory | BE@.250 | Ea | 6.80 | 8.60 | 15.40 | 26.20 |
| 1 pole, lighted, white | BE@.250 | Ea | 6.80 | 8.60 | 15.40 | 26.20 |
| 3 way, lighted, ivory | BE@.300 | Ea | 10.08 | 10.30 | 20.38 | 34.60 |
| 3 way, lighted, white | BE@.300 | Ea | 9.08 | 10.30 | 19.38 | 32.90 |

**Decora® AC quiet rocker switch, Leviton.** 15-amp, 120/277-volt AC. Side wired with Quickwire push-in terminals accepting No.12 or No.14 copper or copper-clad wire. Switch frame shields against dust and fits in wallplate to prevent rocker binding. Sturdy construction for long service life. Full rated current capacity with tungsten, fluorescent and resistive loads. UL listed.

| | | | | | | |
|---|---|---|---|---|---|---|
| 1 pole | BE@.250 | Ea | 2.86 | 8.60 | 11.46 | 19.50 |
| 1 pole, illuminated | BE@.250 | Ea | 8.23 | 8.60 | 16.83 | 28.60 |
| 3 way | BE@.300 | Ea | 4.06 | 10.30 | 14.36 | 24.40 |
| 3 way, illuminated | BE@.300 | Ea | 11.31 | 10.30 | 21.61 | 36.70 |
| 4 way | BE@.350 | Ea | 14.50 | 12.00 | 26.50 | 45.10 |

**Residential grade duplex straight blade receptacle, Leviton.** 15 amps, 125 volts. Quickwire push-in and side wired. Large head terminal screws accept up to 12-gauge copper or copper-clad wire. Quickwire push-in terminals accept up to 14-gauge copper or copper-clad wire. Double-wide power contacts. Plated steel mounting strap. Break-off plaster ears. Captive backed-out mounting screws. Break-off fins for 2-circuit conversion. (5-15R). UL listed.

| | | | | | | |
|---|---|---|---|---|---|---|
| 15 amps, almond | BE@.200 | Ea | .50 | 6.88 | 7.38 | 12.50 |
| 15 amps, brown | BE@.200 | Ea | .50 | 6.88 | 7.38 | 12.50 |
| 15 amps, ivory | BE@.200 | Ea | .50 | 6.88 | 7.38 | 12.50 |
| 15 amps, white | BE@.200 | Ea | .50 | 6.88 | 7.38 | 12.50 |
| 15 amps, self-grounding, ivory | BE@.200 | Ea | .96 | 6.88 | 7.84 | 13.30 |
| 15 amps, self-grounding, white | BE@.200 | Ea | .96 | 6.88 | 7.84 | 13.30 |

| | Craft@Hrs | Unit | Material | Labor | Total | Sell |
|---|---|---|---|---|---|---|

**Commercial grade duplex receptacle, Leviton.** Side wired. 125 volts. Ground out. Large head triple drive terminal screws, backed out and staked. Terminals and back-wiring clamps. Accepts 10-gauge copper or copper-clad wire. Rust-resistant mounting strap. Washer type break-off plaster ears. Captive-mounting screws. 15-amp, NEMA 5-15R. 20-amp NEMA 2-20R. UL listed.

| | Craft@Hrs | Unit | Material | Labor | Total | Sell |
|---|---|---|---|---|---|---|
| 15 amps | BE@.200 | Ea | 1.73 | 6.88 | 8.61 | 14.60 |
| 20 amps | BE@.200 | Ea | 2.72 | 6.88 | 9.60 | 16.30 |

**CO/ALR duplex receptacle, Leviton.** Residential grade. 125 volts. Side-wired grounding. Fits shallow receptacle. Large head terminal screws accept up to 12-gauge copper or copper-clad wire. Quickwire push-in terminals accept 14-gauge solid copper wire only. Rust-resistant steel mounting strap. Break-off plaster ears. Captive mounting screws. Break-off tabs allow easy two-circuit conversion. 5-15R. UL listed.

| | Craft@Hrs | Unit | Material | Labor | Total | Sell |
|---|---|---|---|---|---|---|
| 15 amps, ivory | BE@.200 | Ea | 2.98 | 6.88 | 9.86 | 16.80 |
| 15 amps, white | BE@.200 | Ea | 3.25 | 6.88 | 10.13 | 17.20 |

**GFCI hospital grade receptacle.**

| | Craft@Hrs | Unit | Material | Labor | Total | Sell |
|---|---|---|---|---|---|---|
| 20 amps, ivory | BE@.300 | Ea | 16.30 | 10.30 | 26.60 | 45.20 |
| 20 amps, red | BE@.300 | Ea | 16.30 | 10.30 | 26.60 | 45.20 |

**GFCI receptacle.** 15-amp shock guard ground fault circuit interrupter. Side- and back-wired to accept 12-gauge wire. Shallow design allows for easy installation.

| | Craft@Hrs | Unit | Material | Labor | Total | Sell |
|---|---|---|---|---|---|---|
| 15 amps, ivory | BE@.300 | Ea | 8.26 | 10.30 | 18.56 | 31.60 |
| 15 amps, white | BE@.300 | Ea | 7.30 | 10.30 | 17.60 | 29.90 |

**Single receptacle, Leviton.** Commercial grade. Large head triple-drive terminal screws, backed out and staked. Terminals and back-wiring clamps accept 10-gauge copper or copper-clad wire. Heavy-gauge, rust-resistant steel mounting strap. Convenient washer-type break-off plaster ears for best flush alignment. Triple-wide power contacts. Captive-mounting screws.

| | Craft@Hrs | Unit | Material | Labor | Total | Sell |
|---|---|---|---|---|---|---|
| 15 amps, 125 volts | BE@.200 | Ea | 3.24 | 6.88 | 10.12 | 17.20 |
| 20 amps, 125 volts | BE@.200 | Ea | 4.30 | 6.88 | 11.18 | 19.00 |
| 20 amps, 250 volts | BE@.200 | Ea | 5.40 | 6.88 | 12.28 | 20.90 |

**Appliance receptacle, flush mount, Pass & Seymour/Legrand.** Straight blade receptacle. Industrial specification grade. 125/250-volt, 60 Hz, AC only. For copper or aluminum conductors. Easily identifiable color-coded terminals. Double-wipe copper alloy contacts. Terminals accept up to #4 conductors. Fits single- or 2-gang outlet boxes. Includes mounting hardware. UL listed.

| | Craft@Hrs | Unit | Material | Labor | Total | Sell |
|---|---|---|---|---|---|---|
| NEMA 14-30R, 30 amps, dryer | BE@.350 | Ea | 7.89 | 12.00 | 19.89 | 33.80 |
| NEMA 14-50R, 50 amps, range | BE@.400 | Ea | 7.83 | 13.80 | 21.63 | 36.80 |

**Single-gang plastic wall plate, Leviton.** Residential grade. Standard size. Resists fading, discoloration, grease, oils, organic solvents and scratches. Round on edges to prevent injury and wall damage. Includes matching metal mounting screws. UL listed.

| | Craft@Hrs | Unit | Material | Labor | Total | Sell |
|---|---|---|---|---|---|---|
| 1 duplex receptacle | BE@.050 | Ea | .24 | 1.72 | 1.96 | 3.33 |
| 2 duplex receptacles | BE@.100 | Ea | .89 | 3.44 | 4.33 | 7.36 |
| 1 toggle switch | BE@.050 | Ea | .24 | 1.72 | 1.96 | 3.33 |
| 1 toggle, 1 duplex receptacle | BE@.100 | Ea | .89 | 3.44 | 4.33 | 7.36 |
| 2 toggle switches | BE@.100 | Ea | .87 | 3.44 | 4.31 | 7.33 |
| Blank | BE@.050 | Ea | .52 | 1.72 | 2.24 | 3.81 |

**Single-gang Decora® wall plate, Leviton.** Smooth face and round edges that resist dust accumulation. Individual plastic wrapping to protect surfaces. Color-matched mounting screws included. UL listed.

| | Craft@Hrs | Unit | Material | Labor | Total | Sell |
|---|---|---|---|---|---|---|
| 1 rocker switch | BE@.050 | Ea | .54 | 1.72 | 2.26 | 3.84 |
| 2 rocker switches | BE@.100 | Ea | 1.08 | 3.44 | 4.52 | 7.68 |
| 1 rocker switch, 1 toggle | BE@.100 | Ea | 1.53 | 3.44 | 4.97 | 8.45 |

|  | Craft@Hrs | Unit | Material | Labor | Total | Sell |
|---|---|---|---|---|---|---|

**Incandescent slide dimmer.** Commercial specification grade. 600 watt. Compact design. Single pole. Positive on/off switching with upper-end dimming bypass for maximum brightness. Solid-state circuitry with built-in radio/TV interference filter. UL listed.

| | Craft@Hrs | Unit | Material | Labor | Total | Sell |
|---|---|---|---|---|---|---|
| Ivory | BE@.250 | Ea | 10.04 | 8.60 | 18.64 | 31.70 |
| White | BE@.250 | Ea | 10.04 | 8.60 | 18.64 | 31.70 |

**3-way illuminated incandescent slide dimmer.** Commercial specification grade. Separate heavy-duty on/off rocker for switching without disturbing preset brightness levels. 600 watt. Compact design. Solid-state circuitry with built-in radio/TV interference filter. Includes wall plate. UL listed. 120-volt, 60 hertz, AC.

| | | | | | | |
|---|---|---|---|---|---|---|
| White, preset on/off | BE@.300 | Ea | 16.10 | 10.30 | 26.40 | 44.90 |

**Spring wound timer switch, Intermatic.** Operation requires no electricity, automatically limiting length of time energy is used. Installs in most 2-1/2"-deep junction boxes. Compact design. Single pole, single throw.

| | | | | | | |
|---|---|---|---|---|---|---|
| 15 minute | BE@.200 | Ea | 18.40 | 6.88 | 25.28 | 43.00 |

## Smoke Detectors

**9-volt smoke detector, Kidde®.** With test button. Flashing red light indicates unit is receiving power. Includes batteries. 3-year warranty.

| | | | | | | |
|---|---|---|---|---|---|---|
| 3-7/8", round | BE@.300 | Ea | 7.56 | 10.30 | 17.86 | 30.40 |

**Dual-sensing smoke alarm, First Alert®.** Dual-sensing ionization and photoelectric smoke sensing technologies. Features remote-control test/silence. Unit can be silenced or tested from any location using most standard household remote controls. Microprocessor technology to reduce unwanted alarms. Low battery chirp, missing battery guard and blinking power indicator.

| | | | | | | |
|---|---|---|---|---|---|---|
| 10" diameter | BE@.300 | Ea | 21.60 | 10.30 | 31.90 | 54.20 |

**Dual-ionization smoke detector.** Battery and AC operation. Photoelectric sensor for detecting slow, smoldering fires. Dual-ionization sensor for detecting fast, flaming fires. Two test buttons to test each function, 30-day low battery signal.

| | | | | | | |
|---|---|---|---|---|---|---|
| 120-volt interconnectable | BE@.500 | Ea | 27.50 | 17.20 | 44.70 | 76.00 |

**Builder's door chime kit, Trine.** Includes front and back door black and ivory push buttons, one chime, one transformer and wire. UL listed.

| | | | | | | |
|---|---|---|---|---|---|---|
| Wired chime kit | BE@1.500 | Ea | 10.80 | 51.60 | 62.40 | 106.00 |

**Wireless door bell, front and back, Trine.** Weatherproof. Selectable front and back, push button and battery included. 128 selectable codes. 150-foot range. Westminster or ding-dong selection. Volume control and high-quality chime sound. Applications are 2 tones. Includes 12-volt remote control battery and mounting hardware. No wiring needed. Wall or ceiling mount. Coded transmission. No signal interference. Uses 3 standard D batteries. UL listed. White.

| | | | | | | |
|---|---|---|---|---|---|---|
| Wireless chime | BE@.300 | Ea | 28.70 | 10.30 | 39.00 | 66.30 |

## Electrical Distribution Panels

**Disconnect and remove distribution panel.** Add for disconnecting circuits. Single-phase breaker panel.

| | | | | | | |
|---|---|---|---|---|---|---|
| 50 to 100 amps, concrete wall | BE@.258 | Ea | — | 8.87 | 8.87 | 15.10 |
| 50 to 100 amps, frame wall | BE@.220 | Ea | — | 7.57 | 7.57 | 12.90 |
| 50 to 100 amps, steel column | BE@.416 | Ea | — | 14.30 | 14.30 | 24.30 |
| 200 amps, wood or concrete wall | BE@.300 | Ea | — | 10.30 | 10.30 | 17.50 |
| 200 amps, steel column | BE@.492 | Ea | — | 16.90 | 16.90 | 28.70 |
| Add per circuit disconnected | BE@.063 | Ea | — | 2.17 | 2.17 | 3.69 |

|  | Craft@Hrs | Unit | Material | Labor | Total | Sell |
|---|---|---|---|---|---|---|

**HomeLine® indoor main lug load center, value pack, Square D®.** Class 1170. Single-phase, 3-wire, 120/240-volt, 1-pole circuits. 10,000 RMS AIR. Factory-installed main lugs. Straight-in mains. Split branch neutral. Rotates for top or bottom feed. 10-year limited warranty. Main wire size #6 to 250 AWG aluminum/copper. Main breaker frame size QOM2. Equipment ground bar. Combination surface/flush cover. Install Square D HOM type circuit breakers only. UL listed.

| 200 amps, 20 spaces, 40 circuits, with (6) 15-amp and (6) 20-amp 1-pole and (1) 30-amp and (1) 50-amp 2-pole breakers | BE@1.50 | Ea | 105.00 | 51.60 | 156.60 | 266.00 |

**HomeLine circuit breaker, Square D®.** 1 pole, 120 volt. 10,000 AIC rated.

| 15 amp | BE@.150 | Ea | 3.64 | 5.16 | 8.80 | 15.00 |
| 20 amp | BE@.150 | Ea | 3.65 | 5.16 | 8.81 | 15.00 |
| 30 amp | BE@.150 | Ea | 3.74 | 5.16 | 8.90 | 15.10 |

**HomeLine arc-fault circuit breaker, Square D®.** 1-pole, 120-volt arc fault interrupter. 10,000 AIC rated.

| 15 amp | BE@.200 | Ea | 34.10 | 6.88 | 40.98 | 69.70 |
| 20 amp | BE@.200 | Ea | 34.00 | 6.88 | 40.88 | 69.50 |

**HomeLine ground-fault interrupter circuit breaker, Square D®.** 1 pole, 120 volt. 10,000 AIC rated. For use in Square D HomeLine load centers. UL listed.

| 15 amp | BE@.200 | Ea | 33.80 | 6.88 | 40.68 | 69.20 |
| 20 amp | BE@.200 | Ea | 34.40 | 6.88 | 41.28 | 70.20 |

**Combination service entrance all-in-one meter main load center, Square D®.** Overhead or underground. Single-phase, 3-wire, 120/240-volt, 200-amp ring utility meter socket. 200-amp main circuit breaker and 30-space, 40-circuit, HOM load center with semi-flush outdoor enclosure. Meets EUSERC standards.

| Model CSEDVP1, 200 amps, 30 spaces, 40 circuits | BE@1.30 | Ea | 138.00 | 44.70 | 182.70 | 311.00 |

**Combination ring type meter socket and main load center, Square D®.** Overhead or underground service entrance device. Dual main breaker with feed-through lugs. Single-phase, 3-wire, 120/240-volt AC. 10,000 amp short circuit current rating. Ring utility meter socket. Surface-mount convertible to semi-flush with SC200F flange kit. Flush-mount cover included with HOM load center. Enclosure is of rainproof construction and accepts "A"-type bolt-on hubs. NEMA 3R enclosure. Center mounted hub. Supplied with line side feed through lugs for #6 AWG-250 K circular mils (Al/Cu) conductors and provision for branch circuit breakers. Designed for single-family residential applications. Wire size #6 to 300 KCmil (Al/Cu). Box dimensions, 14-1/2" wide x 32-1/16" long x 6-15/16" deep. Meets EUSE standards. UL listed.

| No. SC816D200C, 200 amps, 8 spaces, 16 circuits | BE@1.30 | Ea | 143.00 | 44.70 | 187.70 | 319.00 |

**Load center flange kit, Square D®.** Converts 125- to 200-amp ring and ringless type surface mount load centers to semi-flush mount.

| Kit | BE@.050 | Ea | 4.00 | 1.72 | 5.72 | 9.72 |

| | Craft@Hrs | Unit | Material | Labor | Total | Sell |
|---|---|---|---|---|---|---|

**Combination service entrance device, Square D®.** All-in-one-main circuit breaker panel. Surface mount. Includes ring type utility meter socket, service disconnect, and integral load center. HOM load center with surface outdoor enclosure. Single-phase, 3-wire. 120/240-volt AC, 10,000 amp short circuit current rating. NEMA Type 3R enclosure. Overhead or underground service feed. Left meter location. Hub style A. Suitable only for use as service equipment. UL listed. Meets EUSERC standards. 10-year limited warranty.

| | Craft@Hrs | Unit | Material | Labor | Total | Sell |
|---|---|---|---|---|---|---|
| No. SC1624M100S, 100 amp, 16 spaces, 24 circuits | BE@1.30 | Ea | 52.00 | 44.70 | 96.70 | 164.00 |
| No. SC1624M125S2, 125 amp, 16 spaces, 24 circuits | BE@1.30 | Ea | 93.70 | 44.70 | 138.40 | 235.00 |
| No. SC2040M200C, 200 amp, 20 spaces, 40 circuits | BE@1.30 | Ea | 129.00 | 44.70 | 173.70 | 295.00 |

**Combination service entrance device, Square D®.** All-in-one-main circuit breaker panel. Surface mount. Includes ring type utility meter socket, service disconnect, and integral load center. HOM load center with surface outdoor enclosure. Single-phase, 3-wire, 120/240-volt AC, 22,000 amp short circuit current rating. NEMA Type 3R enclosure. Overhead or underground service feed. Left meter location. Hub style A. Suitable only for use as service equipment. UL listed. Meets EUSERC standards. 10-year limited warranty.

| | Craft@Hrs | Unit | Material | Labor | Total | Sell |
|---|---|---|---|---|---|---|
| No. SC2040M200S, 200 amp, 20 spaces, 40 circuit | BE@1.30 | Ea | 118.00 | 44.70 | 162.70 | 277.00 |
| No. SC3040M200SS, 200 amp, 30 spaces, 40 circuit | BE@1.30 | Ea | 125.00 | 44.70 | 169.70 | 288.00 |
| No. SC40M200S, 200 amp, 40 spaces, 40 circuit | BE@1.30 | Ea | 134.00 | 44.70 | 178.70 | 304.00 |

**Bolt-on A type universal hub, Square D®.** Bolt-on closing plate for "A" or "A-L" hub opening. Series A style. For Class 4131 meter socket. Class 4119 service entrance device. Aluminum. UL listed. Threaded for rigid conduit.

| | Craft@Hrs | Unit | Material | Labor | Total | Sell |
|---|---|---|---|---|---|---|
| 1" conduit | BE@.050 | Ea | 7.28 | 1.72 | 9.00 | 15.30 |
| 1-1/4" conduit | BE@.050 | Ea | 7.43 | 1.72 | 9.15 | 15.60 |
| 1-1/2" conduit | BE@.050 | Ea | 7.27 | 1.72 | 8.99 | 15.30 |
| 2" conduit | BE@.050 | Ea | 7.34 | 1.72 | 9.06 | 15.40 |

**Main circuit breaker, Class 1130 QO® main lugs, Rainproof, Square D®.** 2-pole, 22,000 AC rated. For convertible load centers only. Load center main rating. KCM is thousand circular mils.

| | Craft@Hrs | Unit | Material | Labor | Total | Sell |
|---|---|---|---|---|---|---|
| 100 amp, AWG #4 to 2/0 | BE@.150 | Ea | 51.00 | 5.16 | 56.16 | 95.50 |
| 125 amp, AWG #4 to 2/0 | BE@.150 | Ea | 112.00 | 5.16 | 117.16 | 199.00 |
| 200 amp, AWG #4 to 300 KCM | BE@.150 | Ea | 115.00 | 5.16 | 120.16 | 204.00 |

**QO® Indoor main lug load center, Square D®.** Surface mount. Class 736, 1130. Single-phase, 3-wire, 120/240-volt AC. Fixed mains. Factory-installed main lugs. 10,000 RMS amp short circuit current rating. Shielded copper bus. Straight-in mains. Rotates for top or bottom feed. Cover included. Ground bar kit sold separately. Uses QO, QOT, QO-EPD, QO-GFI or QO-PL branch circuit breakers separately. UL listed. 10-year limited warranty.

| | Craft@Hrs | Unit | Material | Labor | Total | Sell |
|---|---|---|---|---|---|---|
| No. QO2L30SCP, 30 amp, 2 spaces, 2 circuits, main wire #12-10 Al, #14-10 Cu | BE@.450 | Ea | 12.00 | 15.50 | 27.50 | 46.80 |
| No. QO24L70SCP, 70 amp, 2 spaces, 4 circuits, main wire #12-3 Al, #14-4 Cu | BE@.500 | Ea | 16.20 | 17.20 | 33.40 | 56.80 |
| No. QO612L100DS, 100 amp, 6 spaces, 12 circuits, main wire #8-1 Al and Cu | BE@.800 | Ea | 32.50 | 27.50 | 60.00 | 102.00 |

| | Craft@Hrs | Unit | Material | Labor | Total | Sell |
|---|---|---|---|---|---|---|
| No. QO612L100SCP, 100 amp,<br>6 spaces, 12 circuits,<br>main wire #8-1 Al and Cu | BE@.800 | Ea | 21.50 | 27.50 | 49.00 | 83.30 |
| No. QO816L100DS, 100 amp,<br>8 spaces, 16 circuits,<br>main wire #8-1 Al and Cu | BE@.900 | Ea | 49.40 | 31.00 | 80.40 | 137.00 |
| No. QO816L100SCP, 100 amp,<br>8 spaces, 16 circuits<br>main wire #8-1 Al and Cu | BE@.900 | Ea | 32.90 | 31.00 | 63.90 | 109.00 |

**QO® outdoor main lug load center, QOM1 frame size, Square D®.** Single-phase, 3-wire, 120/240-volt AC. Main wire #6-2/0 aluminum or copper. 65,000 RMS symmetrical amp short circuit current rating. Metallic enclosure. Convertible mains. Factory installed main lug. QOM1 main frame size. Convertible to main circuit breaker. Straight-in mains. Includes rainproof cover. Order ground bar kit separately. Order QO, QOT, QO-GFI, QO-EPD, QO-AFI OR QO-APL branch circuit breakers separately. Install Square D QO type circuit breakers only. 10-year limited warranty. UL Listed.

| | Craft@Hrs | Unit | Material | Labor | Total | Sell |
|---|---|---|---|---|---|---|
| No. QO11224L125GRB, 125 amp,<br>12 spaces, 24 circuits | BE@.800 | Ea | 88.10 | 27.50 | 115.60 | 197.00 |

**QO® miniature circuit breaker, Square D®.** Single pole. Square D QO miniature circuit breakers are plug-in products for use in QO load centers and NQOD panelboards. 120/240-volts AC. 10,000-amp interrupting rating (AIR) RMS symmetrical with VISI-TRIP® indicator. Class 730, 731 and 733. 1 space required. Use in Square D QO load centers only. UL listed.

| | Craft@Hrs | Unit | Material | Labor | Total | Sell |
|---|---|---|---|---|---|---|
| 15 amp, 1 pole | BE@.150 | Ea | 6.86 | 5.16 | 12.02 | 20.40 |
| 20 amp, 1 pole | BE@.150 | Ea | 6.86 | 5.16 | 12.02 | 20.40 |
| 25 amp, 1 pole | BE@.150 | Ea | 6.82 | 5.16 | 11.98 | 20.40 |
| 30 amp, 1 pole | BE@.150 | Ea | 6.86 | 5.16 | 12.02 | 20.40 |
| 40 amp, 1 pole | BE@.150 | Ea | 8.09 | 5.16 | 13.25 | 22.50 |
| 50 amp, 1 pole | BE@.150 | Ea | 8.09 | 5.16 | 13.25 | 22.50 |
| 15 amp, 2 pole | BE@.150 | Ea | 16.05 | 5.16 | 21.21 | 36.10 |
| 20 amp, 2 pole | BE@.150 | Ea | 16.30 | 5.16 | 21.46 | 36.50 |
| 25 amp, 2 pole | BE@.150 | Ea | 16.05 | 5.16 | 21.21 | 36.10 |
| 25 amp, 2 pole | BE@.150 | Ea | 16.05 | 5.16 | 21.21 | 36.10 |
| 30 amp, 2 pole | BE@.150 | Ea | 16.20 | 5.16 | 21.36 | 36.30 |
| 40 amp, 2 pole | BE@.150 | Ea | 16.30 | 5.16 | 21.46 | 36.50 |
| 50 amp, 2 pole | BE@.150 | Ea | 16.20 | 5.16 | 21.36 | 36.30 |

**Replacement tandem circuit breaker, QO®, Square D®.** 1 pole. 120-volts AC. 10,000 AIR. For use in Square D QO load centers manufactured before 1967. UL listed.

| | Craft@Hrs | Unit | Material | Labor | Total | Sell |
|---|---|---|---|---|---|---|
| 15 and 15 amp | BE@.150 | Ea | 27.30 | 5.16 | 32.46 | 55.20 |
| 20 and 20 amp | BE@.150 | Ea | 27.50 | 5.16 | 32.66 | 55.50 |

**Tandem circuit breaker, Square D®.** 1 pole. 120-volts AC. 10,000 AIR. HACR type. Visi-Trip® indicator. Requires handle tie No. QOTHT for common switching. For use in Square D QO load centers manufactured after 1967. UL listed as CTL.

| | Craft@Hrs | Unit | Material | Labor | Total | Sell |
|---|---|---|---|---|---|---|
| 15 and 15 amp | BE@.150 | Ea | 16.10 | 5.16 | 21.26 | 36.10 |
| 20 and 20 amp | BE@.150 | Ea | 16.10 | 5.16 | 21.26 | 36.10 |

| | Craft@Hrs | Unit | Material | Labor | Total | Sell |
|---|---|---|---|---|---|---|

**Ground-fault interrupter plug-on circuit breaker, QO-GFI Qwik-Gard®, Square D®.** QWIK-GARD® circuit breakers provide overload and short circuit protection, combined with Class A ground-fault protection. Class A denotes a ground-fault circuit interrupter that will trip when a ground-fault current is 6 milliamperes or greater, for people protection. Do not connect to more than 250' of load conductor for the total one-way run to prevent nuisance tripping. Class 685, 690, 730, 912 and 950. 10,000 AIR. HACR type. Visi-Trip® indicator. 1 pole are 120 volt and 2 pole are 120/240 volt. 2-pole breakers have common trip, 120-volts AC. 2 spaces required for 2-pole breakers. UL listed.

| | | | | | | |
|---|---|---|---|---|---|---|
| 20 amp, 1 pole | BE@.150 | Ea | 58.50 | 5.16 | 63.66 | 108.00 |
| 20 amp, 2 pole | BE@.150 | Ea | 96.10 | 5.16 | 101.26 | 172.00 |
| 30 amp, 2 pole | BE@.150 | Ea | 111.00 | 5.16 | 116.16 | 197.00 |
| 50 amp, 2 pole | BE@.150 | Ea | 138.00 | 5.16 | 143.16 | 243.00 |
| 60 amp, 2 pole | BE@.150 | Ea | 125.00 | 5.16 | 130.16 | 221.00 |

**QO® load center ground bar kit, Square D®.** Sold separately from the QO® load centers. Indoor, single phase, main lug, class 736 and 1130. Aluminum. UL listed.

| | | | | | | |
|---|---|---|---|---|---|---|
| No. PK3GTA, 3 terminal | BE@.200 | Ea | 3.66 | 6.88 | 10.54 | 17.90 |
| No. PK4GTA, 4 terminal | BE@.200 | Ea | 4.11 | 6.88 | 10.99 | 18.70 |
| No. PK7GTA, 7 terminal | BE@.200 | Ea | 4.40 | 6.88 | 11.28 | 19.20 |
| No. PK9GTA, 9 terminal | BE@.200 | Ea | 4.95 | 6.88 | 11.83 | 20.10 |
| No. PK12GTACP, 12 terminal | BE@.200 | Ea | 5.37 | 6.88 | 12.25 | 20.80 |
| No. PK15GTA, 15 terminal | BE@.200 | Ea | 6.02 | 6.88 | 12.90 | 21.90 |

## Recessed Lighting

**Recessed low-voltage baffle light kit, remodel, Commercial Electric.** Not for direct contact with insulation. 4" black baffle trim. Uses (1) MR16 bulb, 50 watt (included). Minimizes glare in general, accent and task lighting. Pre-wired housing. Thermally protected. UL listed for damp locations. For suspended ceilings. Installation instructions and template enclosed. Compatible with standard wall dimmer.

| | | | | | | |
|---|---|---|---|---|---|---|
| 4", with black baffle trim | BE@1.00 | Ea | 35.30 | 34.40 | 69.70 | 118.00 |

**Recessed low-voltage slot aperture light kit, remodel, Commercial Electric.** Not for direct contact with insulation. 4" white slot aperture trim. Uses (1) MR16 bulb, 50 watt (included). Pre-wired housing. Thermally protected. UL listed for damp locations. For suspended ceilings. Installation instructions and template enclosed. Compatible with standard wall dimmer.

| | | | | | | |
|---|---|---|---|---|---|---|
| 4", white slot aperture trim | BE@1.00 | Ea | 33.00 | 34.40 | 67.40 | 115.00 |

**Mini can recessed eyeball trim light kit, Commercial Electric.** 5" mini non-IC housing. 5" white eyeball trim. For new or remodel construction. Not for direct contact with insulation. For suspended ceilings or joist support (hanger bars sold separately). Pre-wired housing. Thermally protected. UL listed for damp locations. Compatible with standard wall dimmer. 75-watt maximum, PAR-30 bulb (not included).

| | | | | | | |
|---|---|---|---|---|---|---|
| 5", white | BE@1.00 | Ea | 12.50 | 34.40 | 46.90 | 79.70 |

**Mini can recessed open light kit, Commercial Electric.** 5" mini non-IC housing. 5" white open trim. For new or remodel construction. Not for direct contact with insulation. For suspended ceilings or joist support (hanger bars sold separately). Pre-wired housing. Thermally protected. UL listed for damp locations. Compatible with standard wall dimmer.

| | | | | | | |
|---|---|---|---|---|---|---|
| 5", white | BE@1.00 | Ea | 10.00 | 34.40 | 44.40 | 75.50 |

**Recessed light kit.** For use in an insulated ceiling. Kit includes housing and trim. For new construction.

| | | | | | | |
|---|---|---|---|---|---|---|
| 6", white trim | BE@.700 | Ea | 17.10 | 24.10 | 41.20 | 70.00 |

| | Craft@Hrs | Unit | Material | Labor | Total | Sell |
|---|---|---|---|---|---|---|

## Lighting Fixtures

**Remove and replace ceiling light fixture.** Disassemble and remove incandescent fixture and replace with incandescent fixture.

| | Craft@Hrs | Unit | Material | Labor | Total | Sell |
|---|---|---|---|---|---|---|
| Ceiling mounted | BE@.787 | Ea | — | 27.10 | 27.10 | 46.10 |

**Cambridge II residential fluorescent ceiling fixture, Lithonia Lighting.** Electronic ballast. 32 watt, T8 tubes.

| | Craft@Hrs | Unit | Material | Labor | Total | Sell |
|---|---|---|---|---|---|---|
| 1' x 4', 2 lamps, oak | BE@.800 | Ea | 99.60 | 27.50 | 127.10 | 216.00 |
| 1' x 4', 2 lamps, white | BE@.800 | Ea | 92.30 | 27.50 | 119.80 | 204.00 |
| 1.5' x 4', 4 lamps, oak | BE@.800 | Ea | 113.00 | 27.50 | 140.50 | 239.00 |
| 1.5' x 4', 4 lamps, dentil | BE@.800 | Ea | 118.00 | 27.50 | 145.50 | 247.00 |
| 1.5' x 4', 4 lamps, white | BE@.800 | Ea | 113.00 | 27.50 | 140.50 | 239.00 |
| 2' x 2', 2 U-lamps, oak | BE@.800 | Ea | 105.00 | 27.50 | 132.50 | 225.00 |
| 2' x 2', 2 U-lamps, white | BE@.800 | Ea | 105.00 | 27.50 | 132.50 | 225.00 |

**Low profile fluorescent diffuser fixture, Lithonia Lighting.** Suitable for use in closets, laundry areas and pantries. Acrylic white diffuser. Wall or ceiling mount. UL listed.

| | Craft@Hrs | Unit | Material | Labor | Total | Sell |
|---|---|---|---|---|---|---|
| 11", 22-watt lamp, round | BE@.900 | Ea | 31.20 | 31.00 | 62.20 | 106.00 |
| 12", 22-watt lamp, square | BE@.900 | Ea | 31.30 | 31.00 | 62.30 | 106.00 |
| 14", 2 lamps, 22/32 watts, round | BE@.900 | Ea | 41.30 | 31.00 | 72.30 | 123.00 |
| 15", 2 lamps, 22/32 watts, square | BE@.900 | Ea | 41.30 | 31.00 | 72.30 | 123.00 |
| 19", 2 lamps, 32/40 watts, round | BE@.900 | Ea | 52.00 | 31.00 | 83.00 | 141.00 |
| 20", 2 lamps, 32/40 watts, square | BE@.900 | Ea | 52.00 | 31.00 | 83.00 | 141.00 |

**Wave bath bar light fixture, Hampton Bay.** Brushed nickel finish with etched marble glass. Uses 100-watt, medium base bulb(s), sold separately. 5" and 22" wide respectively by 7-1/4" high.

| | Craft@Hrs | Unit | Material | Labor | Total | Sell |
|---|---|---|---|---|---|---|
| 1 light, brushed nickel | BE@.500 | Ea | 43.30 | 17.20 | 60.50 | 103.00 |
| 3 light, brushed nickel | BE@.500 | Ea | 86.90 | 17.20 | 104.10 | 177.00 |

**Chrome bath bar light fixture, Hampton Bay.** Chrome finish with etched opal glass. Uses 100-watt, medium base bulb(s), not included. 9-5/8" high.

| | Craft@Hrs | Unit | Material | Labor | Total | Sell |
|---|---|---|---|---|---|---|
| 1 light, chrome, 5-1/2" long | BE@.500 | Ea | 54.50 | 17.20 | 71.70 | 122.00 |
| 2 light, chrome, 13-1/4" long | BE@.500 | Ea | 76.00 | 17.20 | 93.20 | 158.00 |
| 3 light, chrome, 21-1/4" long | BE@.500 | Ea | 97.60 | 17.20 | 114.80 | 195.00 |

**Beveled mirror bath bar, Hampton Bay.** Chrome finish. Beveled mirror, flat back. Uses 100-watt (maximum) globe bulbs, not included. Mounting hardware included. 4-1/2" high. Extends 4" from wall.

| | Craft@Hrs | Unit | Material | Labor | Total | Sell |
|---|---|---|---|---|---|---|
| 24" wide, 4 light | BE@.800 | Ea | 27.80 | 27.50 | 55.30 | 94.00 |
| 36" wide, 6 light | BE@.800 | Ea | 38.10 | 27.50 | 65.60 | 112.00 |
| 48" wide, 8 light | BE@.800 | Ea | 48.20 | 27.50 | 75.70 | 129.00 |

**Opal glass wall light, Hampton Bay.** Chrome finish. Hand blown opal glass. Uses 100-watt bulbs. Chrome.

| | Craft@Hrs | Unit | Material | Labor | Total | Sell |
|---|---|---|---|---|---|---|
| 1 light, 4-3/4" wide | BE@.800 | Ea | 21.50 | 27.50 | 49.00 | 83.30 |
| 2 lights, 14" wide | BE@.800 | Ea | 32.60 | 27.50 | 60.10 | 102.00 |
| 3 lights, 15" wide | BE@.800 | Ea | 43.60 | 27.50 | 71.10 | 121.00 |
| 4 lights, 26" wide | BE@.800 | Ea | 54.30 | 27.50 | 81.80 | 139.00 |

**Two-light ceiling fixture.** Flush-mount hidden pan. Frosted glass shade. Uses two frosted A-19 or torpedo 60-watt maximum lamps. 13" deep by 6-1/4" high.

| | Craft@Hrs | Unit | Material | Labor | Total | Sell |
|---|---|---|---|---|---|---|
| Bronze patina | BE@.800 | Ea | 27.50 | 27.50 | 55.00 | 93.50 |
| Cobblestone finish | BE@.800 | Ea | 27.50 | 27.50 | 55.00 | 93.50 |

**Mushroom ceiling light fixture drum, Westinghouse Lighting.**

| | Craft@Hrs | Unit | Material | Labor | Total | Sell |
|---|---|---|---|---|---|---|
| 6", white | BE@.800 | Ea | 4.85 | 27.50 | 32.35 | 55.00 |

| | Craft@Hrs | Unit | Material | Labor | Total | Sell |
|---|---|---|---|---|---|---|

**Two-light incandescent flush-mount ceiling fixture, Sunburst.** Screw-in bubble-style glass. Medium base.

| | | | | | | |
|---|---|---|---|---|---|---|
| 9" wide, white | BE@.800 | Ea | 11.60 | 27.50 | 39.10 | 66.50 |

**Mushroom flush-mount light fixture, Hampton Bay.** 7-3/8" diameter x 4-1/2" high. Uses one medium base bulb, 60-watt maximum (not included).

| | | | | | | |
|---|---|---|---|---|---|---|
| Satin nickel finish, frosted ribbed glass | BE@.800 | Ea | 10.40 | 27.50 | 37.90 | 64.40 |

**Mushroom flush-mount light fixture, Hampton Bay.** 9-1/4" diameter x 5-5/8" high. Uses 2 medium base bulbs, 60-watt maximum (not included).

| | | | | | | |
|---|---|---|---|---|---|---|
| Polished brass finish, clear ribbed glass | BE@.800 | Ea | 12.80 | 27.50 | 40.30 | 68.50 |
| Satin nickel finish, frosted ribbed glass | BE@.800 | Ea | 14.00 | 27.50 | 41.50 | 70.60 |

**Mushroom flush-mount ceiling fixture.**

| | | | | | | |
|---|---|---|---|---|---|---|
| 8", polished brass | BE@.500 | Ea | 7.65 | 17.20 | 24.85 | 42.20 |
| 8", white on white, 75 watt | BE@.500 | Ea | 10.53 | 17.20 | 27.73 | 47.10 |
| 9-1/2", polished brass | BE@.500 | Ea | 11.60 | 17.20 | 28.80 | 49.00 |
| 10", white on white, 75 watt | BE@.500 | Ea | 11.60 | 17.20 | 28.80 | 49.00 |

**Glass globe ceiling light fixture.** 6" white dome fixture with polished brass fitter. Uses one 75-watt bulb, sold separately. UL listed. Flush mount.

| | | | | | | |
|---|---|---|---|---|---|---|
| With switch, white on white | BE@.500 | Ea | 5.60 | 17.20 | 22.80 | 38.80 |
| Without switch, polished brass | BE@.500 | Ea | 5.43 | 17.20 | 22.63 | 38.50 |
| Without pull chain, white on white | BE@.500 | Ea | 5.43 | 17.20 | 22.63 | 38.50 |

**Glass globe ceiling light fixture.** 8" white dome fixture with polished brass fitter. Uses one 75-watt bulb, sold separately. UL listed.

| | | | | | | |
|---|---|---|---|---|---|---|
| Flush mount | BE@.500 | Ea | 9.45 | 17.20 | 26.65 | 45.30 |

**Swirl flush-mount dome light ceiling fixture.** Polished brass finish. Clear swirled glass dome. Uses one 75-watt bulb, sold separately. Includes mounting hardware.

| | | | | | | |
|---|---|---|---|---|---|---|
| 11", clear | BE@.500 | Ea | 11.50 | 17.20 | 28.70 | 48.80 |
| 13", clear | BE@.500 | Ea | 15.40 | 17.20 | 32.60 | 55.40 |

**6-light crystal chandelier.**

| | | | | | | |
|---|---|---|---|---|---|---|
| Antique copper finish | BE@1.10 | Ea | 433.00 | 37.80 | 470.80 | 800.00 |
| Crystal chandelier | BE@1.10 | Ea | 375.00 | 37.80 | 412.80 | 702.00 |
| White with shades | BE@1.10 | Ea | 486.00 | 37.80 | 523.80 | 890.00 |
| With shades | BE@1.10 | Ea | 519.00 | 37.80 | 556.80 | 947.00 |

**Flair collection glass chandelier, Hampton Bay.** Brushed nickel finish chandelier with etched marble glass shades with interchangeable blue and frosted white decorative rings. Uses 100-watt medium base bulbs, not included. Fixtures are 18-1/2" and 27-1/2" wide and 20" and 23" high respectively.

| | | | | | | |
|---|---|---|---|---|---|---|
| 4 light, brushed nickel | BE@1.10 | Ea | 119.00 | 37.80 | 156.80 | 267.00 |
| 6 light, brushed nickel | BE@1.10 | Ea | 184.00 | 37.80 | 221.80 | 377.00 |

| | Craft@Hrs | Unit | Material | Labor | Total | Sell |
|---|---|---|---|---|---|---|

**Outdoor bulkhead oval light fixture, Hampton Bay.** 5" wide x 8-1/2" high x 4-1/2" extension from the wall. Cast aluminum construction. Frosted, ribbed glass lens. UL wet location listed. Uses 1 medium base bulb, 60-watt maximum (included).

| | Craft@Hrs | Unit | Material | Labor | Total | Sell |
|---|---|---|---|---|---|---|
| Black finish | BE@.800 | Ea | 21.50 | 27.50 | 49.00 | 83.30 |
| White finish | BE@.800 | Ea | 21.40 | 27.50 | 48.90 | 83.10 |

**Outdoor round bulkhead light fixture, Hampton Bay.** 8" wide x 8" high x 4-1/2" extension from the wall. Cast aluminum construction. Frosted, ribbed glass lens. UL wet location listed. Uses one medium base bulb, 60-watt maximum (included).

| | Craft@Hrs | Unit | Material | Labor | Total | Sell |
|---|---|---|---|---|---|---|
| White finish | BE@.800 | Ea | 21.50 | 27.50 | 49.00 | 83.30 |

**Outdoor wall lantern, Hampton Bay.** 7" wide x 12-1/2" high x 7-3/8" extension from the wall. Solid brass construction. Clear bent beveled glass. Install with or without tail. Uses one 100-watt maximum medium base bulb (included).

| | Craft@Hrs | Unit | Material | Labor | Total | Sell |
|---|---|---|---|---|---|---|
| Antique brass finish | BE@.800 | Ea | 32.40 | 27.50 | 59.90 | 102.00 |
| Polished brass finish | BE@.800 | Ea | 32.30 | 27.50 | 59.80 | 102.00 |

**Curved glass wall-mount lantern with photo cell, Hampton Bay.** Cast aluminum. Six sides of clear bent beveled glass panels. Photo control turns light on at dusk and off at dawn. Weather-resistant finish. Uses one medium base bulb, 100-watt maximum (not included). 7" wide x 15" high.

| | Craft@Hrs | Unit | Material | Labor | Total | Sell |
|---|---|---|---|---|---|---|
| Black | BE@.800 | Ea | 38.30 | 27.50 | 65.80 | 112.00 |
| White | BE@.800 | Ea | 38.30 | 27.50 | 65.80 | 112.00 |

**Jelly jar fixture, Catalina Lighting.** Extra thick glass. Coated aluminum screw shell. 10-1/8" high by 6" wide and 5-3/4" diameter.

| | Craft@Hrs | Unit | Material | Labor | Total | Sell |
|---|---|---|---|---|---|---|
| Black | BE@.800 | Ea | 3.30 | 27.50 | 30.80 | 52.40 |
| Polished brass | BE@.800 | Ea | 3.78 | 27.50 | 31.28 | 53.20 |
| White | BE@.800 | Ea | 3.30 | 27.50 | 30.80 | 52.40 |

**6-sided motion sensing lantern, Hampton Bay.** Cast aluminum. Six sides of clear bent beveled glass panels. Motion sensor turns on the light when motion is detected. Adjustable to 30' away with a 180-degree angle of coverage for security and convenience. Weather-resistant finish. Uses one 100-watt maximum medium base bulb (not included).

| | Craft@Hrs | Unit | Material | Labor | Total | Sell |
|---|---|---|---|---|---|---|
| Black | BE@.800 | Ea | 40.80 | 27.50 | 68.30 | 116.00 |
| White | BE@.800 | Ea | 40.90 | 27.50 | 68.40 | 116.00 |

**Motion sensor security light, Heath Zenith.** Quartz. Daylight shut-off. Selectable walk test. Automatic power outage reset. Sensitivity adjustment. Selectable light timer. Manual override. Operating temperature -22 to +120 degrees F. 150-degree angle lens. Instant-on operation. 70-foot sensor range. 2400 square foot sensor coverage area. Pulse count. Die-cast metal fixtures. Dual-Brite™ 2-level lighting. 3 and 5 hour dawn-to-dusk timer. Wall or eave mount. Two 150-watt quartz halogen bulbs included.

| | Craft@Hrs | Unit | Material | Labor | Total | Sell |
|---|---|---|---|---|---|---|
| Bronze | BE@.800 | Ea | 43.40 | 27.50 | 70.90 | 121.00 |
| White | BE@.800 | Ea | 43.40 | 27.50 | 70.90 | 121.00 |

**Heavy-duty motion sensor security light, Heath Zenith.** Metal fixtures. 150-degree angle lens. 70-foot illumination, 6400 square foot coverage. Indoor manual override wall switch. Weather-resistant die-cast aluminum construction. Bulbs not included. UL listed.

| | Craft@Hrs | Unit | Material | Labor | Total | Sell |
|---|---|---|---|---|---|---|
| Gray | BE@.800 | Ea | 21.00 | 27.50 | 48.50 | 82.50 |
| White | BE@.800 | Ea | 20.95 | 27.50 | 48.45 | 82.40 |

| | Craft@Hrs | Unit | Material | Labor | Total | Sell |
|---|---|---|---|---|---|---|

## Ceiling Fans

**San Marino™ ceiling fan with light kit, Hampton Bay.** Stylish mesh spotlight kit and finish. Powerful 153mm x 15mm motor. Triple capacitor speed control. Accu-Arm blade arms and slide-on mounting bracket make for wobble-free installation. Tri-mount installation compatible. Five reversible rosewood and black blades.

| | Craft@Hrs | Unit | Material | Labor | Total | Sell |
|---|---|---|---|---|---|---|
| 36", brushed steel | BE@1.78 | Ea | 59.30 | 61.20 | 120.50 | 205.00 |

**Gazebo™ indoor or outdoor ceiling fan, Hampton Bay.** Stainless steel hardware. UL rating for outdoor use. Lifetime warranty.

| | Craft@Hrs | Unit | Material | Labor | Total | Sell |
|---|---|---|---|---|---|---|
| 42", white | BE@1.50 | Ea | 64.80 | 51.60 | 116.40 | 198.00 |

**Gazebo Plus ™ ceiling fan with light kit, indoor or outdoor, Hampton Bay.** Special galvanized finish. Five oak wood grain weather-resistant blades with PowerMax motor. Stepped dome light kit. Stainless steel hardware. UL rating for outdoor use. Accu-Arm for accurate, easy installation. Uses two 60-watt vibration resistant bulbs, sold separately. Separate (2) pull-chain switches for fan and light. Lifetime warranty.

| | Craft@Hrs | Unit | Material | Labor | Total | Sell |
|---|---|---|---|---|---|---|
| 52", copper patina | BE@1.78 | Ea | 91.00 | 61.20 | 152.20 | 259.00 |
| 52", verde, wood grain blades | BE@1.78 | Ea | 94.20 | 61.20 | 155.40 | 264.00 |
| 52", white, white blades | BE@1.78 | Ea | 90.45 | 61.20 | 151.65 | 258.00 |

**Passport II® ceiling fan, Hunter®.** For large rooms. Five reversible blades. Light kit adaptable. 3-position Installer's Choice™ mounting system. UL listed. Lifetime warranty.

| | Craft@Hrs | Unit | Material | Labor | Total | Sell |
|---|---|---|---|---|---|---|
| 52", antique brass | BE@1.50 | Ea | 100.00 | 51.60 | 151.60 | 258.00 |
| 52", white | BE@1.50 | Ea | 100.00 | 51.60 | 151.60 | 258.00 |

# *Porches and Decks*

# 16

**A**dding a porch or a deck is a popular home improvement project. The advantage is obvious. Porches and decks add value at a lesser square foot cost than similar enclosed living area. What isn't so obvious is the disadvantage: porches and decks are more exposed to the weather and have a life expectancy considerably less than the home itself. That makes porch and deck work a common repair task. The result is a steady volume of work for porch and deck specialists. This chapter includes cost estimates for both the addition of new porches and decks, and repair work.

## Inspecting for Decay

Exposure to sun, wind and rain tends to open wood grain. The result is a higher moisture content that supports the growth of wood-destroying organisms. Nearly all wood decays if it remains moist for long periods. Check for decay and insect damage with a screwdriver. Wood has lost nearly all its load-carrying capacity when you can push a common screwdriver blade into the grain. Deck flooring needs to be replaced when it sags under foot traffic.

Check the crawl space under the porch or deck for signs of dampness. Condensation forms on the underside of an elevated porch or deck when there isn't proper ventilation. Replace decayed framing with treated lumber.

Give particular attention to posts that are in contact with concrete. Posts should be replaced when heavily decayed. Support the new post slightly above the porch floor with a post anchor, as in Figure 16-1. Embed the post anchor in concrete to help protect the porch roof from uplift wind forces.

If a decayed post is strictly ornamental, consider cutting off the decayed base. Replace the decayed portion with a wood block secured to the concrete with a pin and a washer. Add base trim to conceal the patch (see Figure 16-2). Keep trim pieces slightly above the concrete so moisture can escape.

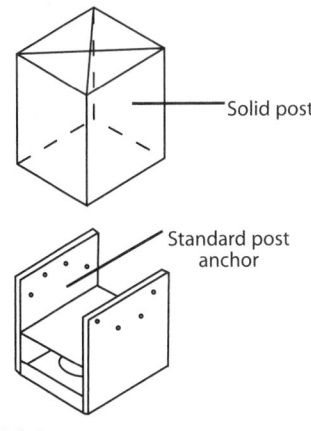

**Figure 16-1**

*Base for post; standard post anchor for resistance to uplift*

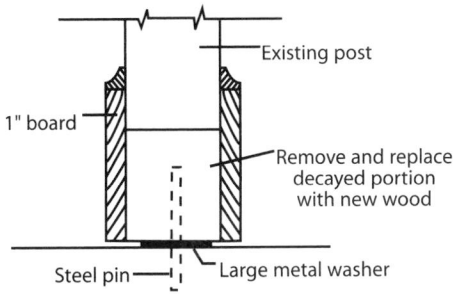

**Figure 16-2**

*Replace end of decayed porch post*

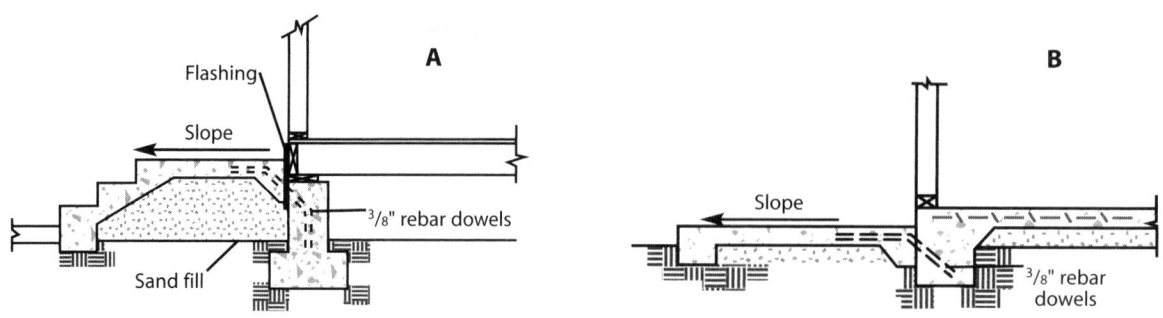

**Figure 16-3**

*Porch slab connection to house:* ***A***, *concrete porch with wood-frame floor;* ***B***, *porch with house slab*

# Adding a Porch

Figure 16-3 shows how to tie a porch slab to a house with either a conventional crawl-space foundation (Figure 16-3A) or a concrete slab (Figure 16-3B). Tie the new concrete foundation, piers or slab to the existing foundation with anchor bolts. The new footing should go at least as deep as the footing on the existing building.

Figure 16-4 shows construction details for a porch addition built on a concrete slab. In colder climates and where the porch or deck can be built at least 18" above ground level, use conventional floor framing, as shown in Figure 16-5. Provide a polyethylene soil cover to control excessive moisture. Whether wood or concrete, slope the floor away from the house at least $1/8$" per foot of width. That'll help keep water away from the existing foundation line.

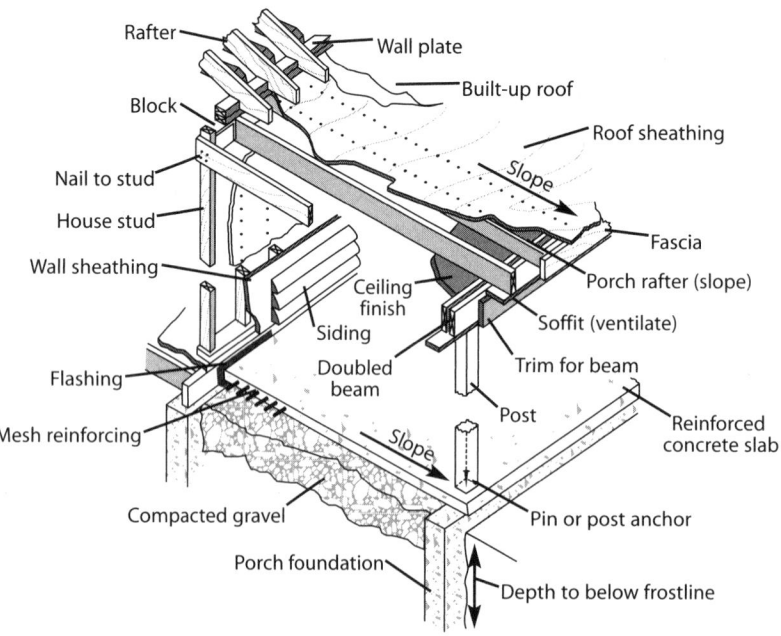

**Figure 16-4**

*Details of porch construction for concrete slab*

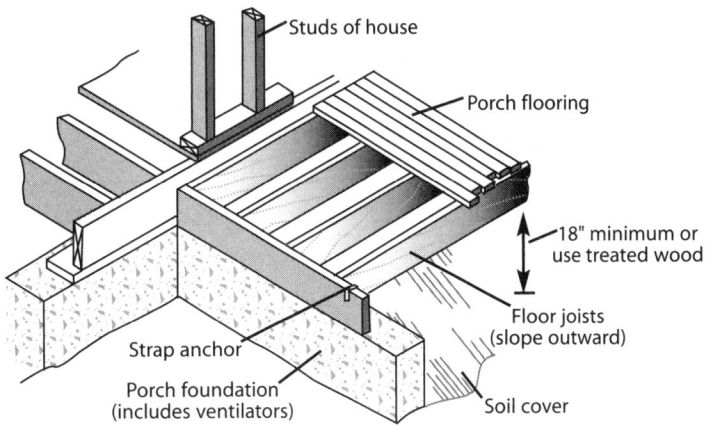

**Figure 16-5**

*Porch floor with wood framing*

## Columns and Posts

Support the porch roof with either solid or built-up posts or columns. Figure 16-6A shows a column built from doubled 2 x 4s covered with 1 x 4" casing on two opposite sides and 1 x 6" finish casing on the other sides. Solid posts should be either 4 x 4" or 6 x 6" lumber. For a traditional style home, use factory-made adjustable round columns.

Keep the post or column base out of any pocket or depression that will collect moisture and invite decay. If high winds aren't a design consideration, use a steel pin to join a one-piece post to the foundation. Set a large galvanized washer or spacer between the bottom of the post and the top of the floor or slab, as in Figure 16-6B. Apply wood preservative to cut ends at both the top and bottom of the post. For a cased post, install flashing under the base molding, as in Figure 16-6C. Build all deck posts and framing from pressure-treated lumber. Use 0.40 treated lumber for framing and 0.60 treated lumber for posts in contact with the ground.

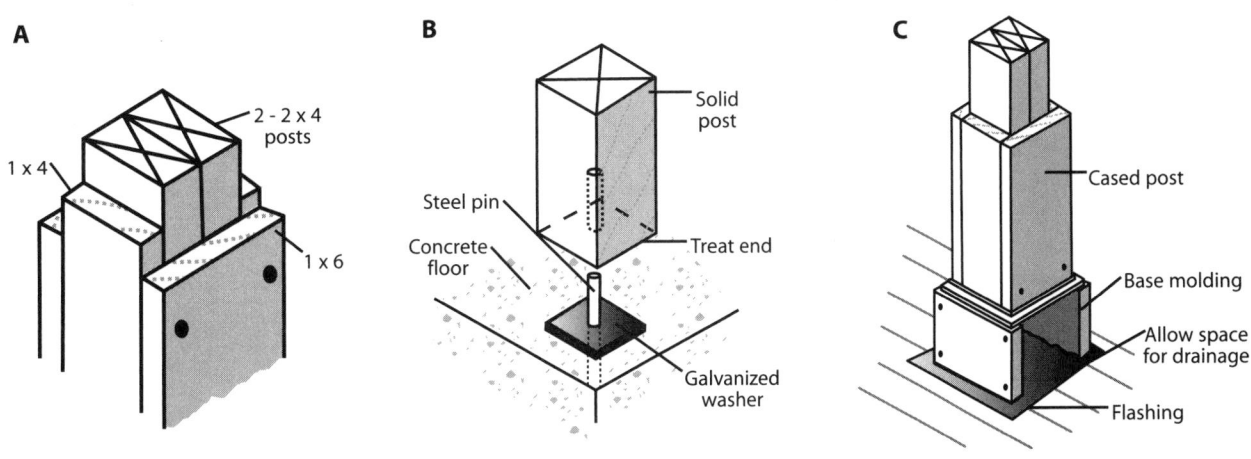

**Figure 16-6**

*Post details: **A**, cased post; **B**, pin anchor and space; **C**, flashing at base*

### Balustrades

To exclude insects and flying debris, enclose the porch with either combination windows or window screen. See Figure 16-7A. An open balustrade adds safety as well as a decorative accent to a porch that isn't enclosed. See Figure 16-7B.

All wood railing exposed to weather should be shaped to shed water. Select railing that tapers slightly at the edges. Figure 16-8A shows a balustrade with a tapered top rail, a tapered bottom rail and balusters connecting the two. Design the joint between the post and rail to avoid trapping moisture. One method is to leave a small space between the post and the end of the railing, as shown in Figure 16-8B. Treat the cut rail end with water-repellent preservative. Wood posts, balusters and railings should be made from either the heartwood of a decay-resistant wood species or from treated wood. Place a small block under the bottom railing so it doesn't contact a concrete floor. Any wood in contact with concrete should be pressure-treated.

You must provide a railing around the perimeter of the porch if it's more than 30" above the existing grade. Generally the railing on a single-family residential porch or deck must be at least 36" high. Secure all hand rails and guard rails firmly to the deck. The space between balusters must be narrow enough so that nothing larger than a 4"-diameter sphere can pass through it.

## Elevated Decks

An elevated wood deck is a good choice for a sloping lot where a concrete patio slab isn't practical. Use pressure-treated or decay-resistant lumber for support posts, beams and joists. Joist spacing can be 24", 16" or 12" on center, depending on the anticipated load and the type of deck material you use. The deck surface can be either wood that's naturally decay-resistant, pressure-treated, or a composite

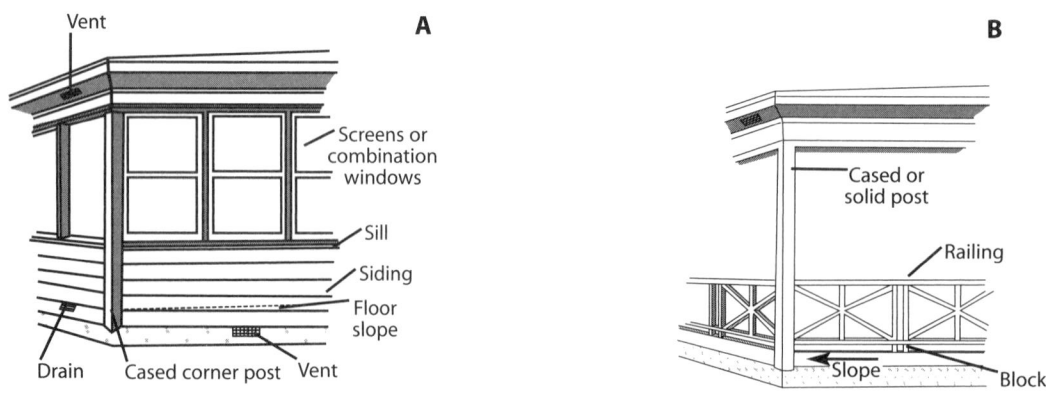

**Figure 16-7**

*Types of balustrades:* ***A***, *closed balustrades;* ***B***, *open balustrades*

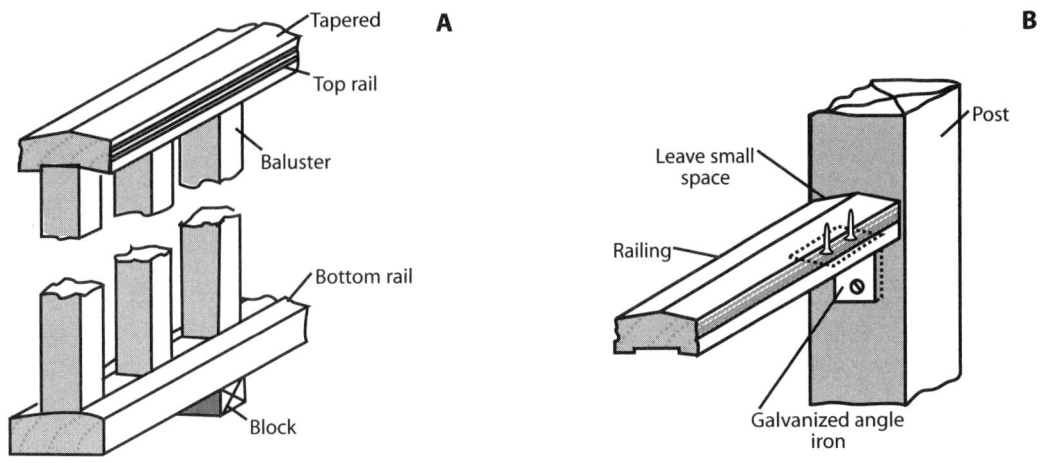

**Figure 16-8**

*Railing details: **A**, balustrade assembly; **B**, rail to post connection*

made from recycled plastics. Synthetic decking doesn't have the strength of wood decking and usually requires joists spaced 12" on center.

The bottom step of an elevated porch or deck stairway should never be in contact with the soil. Support the bottom step and stair carriage on a concrete pad or with a pressure-treated post. Figure 16-9 shows a porch stairway supported on treated posts. Posts like this should measure at least 5" in each dimension and should be embedded in the soil at least 3'. Nail and bolt a cross member between the posts. Then block the inner end of the horizontal support to the floor framing.

Where more than one step is required, use a 2 x 12 stringer at each end of the steps (see Figure 16-10). Bolt the lower end of each stringer to a treated post and attach the upper end of the stringer to the porch framing. You could also support the stairway on concrete or masonry piers.

Begin the construction of an elevated porch and deck by fastening a ledger securely to the existing sill with lag bolts (Figure 16-11). Then build a post and beam framework that'll support the joists and wood decking. Hang joists from the ledger using metal joist hangers. Joist spacing depends of the type of decking you select. As mentioned earlier, synthetic deck materials usually require joists no more than 12" on center. For treated $^5/_4$" fir and pine, you can usually space joists 16" on center. Install solid blocking between the joists at all supporting beams. Finally, fasten decking to the joists. You can face-nail, blind-nail or screw and plug wood decking in straight, diagonal, herringbone, or checkerboard patterns.

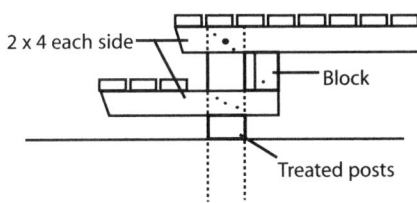

**Figure 16-9**

*Single porch step supported on a treated post*

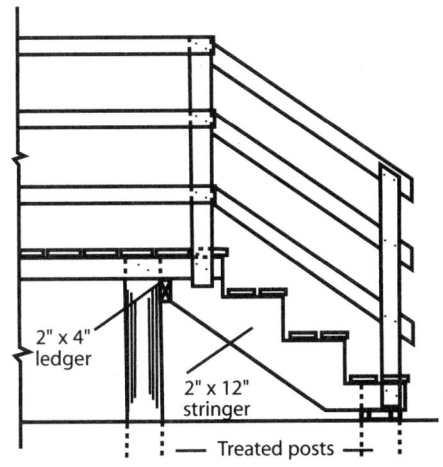

**Figure 16-10**

*Step stringer supported by porch framing and posts*

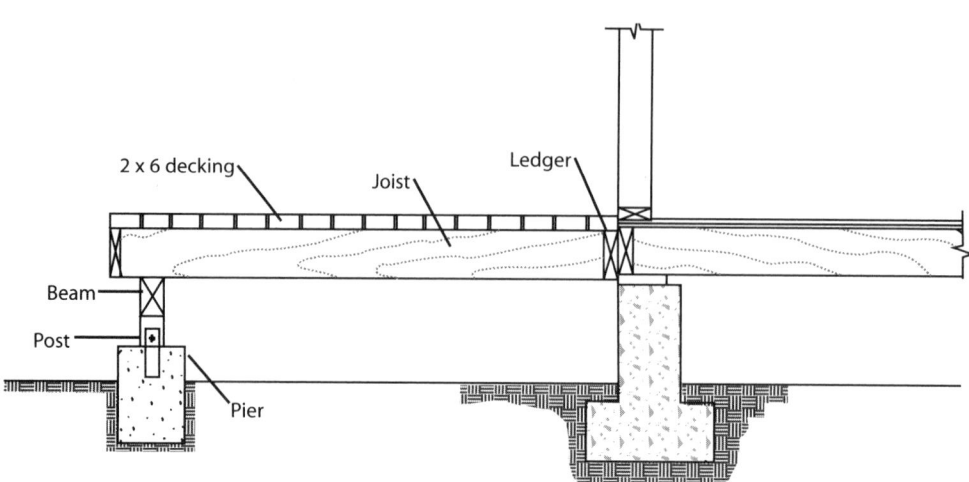

**Figure 16-11**

*Wood deck attached to house*

If you build or remodel a deck that is under a window, and the deck floor is less than 18" below the bottom edge of the window, that window may now be considered a hazardous location by the building code. If it is, the window must be special safety glass. Check for this when you estimate the job. That way if it applies, you can include the cost in your bid.

### *Choices in Porch and Deck Floor Plank*

Western red cedar and redwood are naturally decay- and insect-resistant. The cost is higher than pressure-treated lumber, especially in areas remote from the west coast of the U.S. and Canada, where western red cedar and redwood are milled. For decks, use either construction common or construction heart redwood. "Merch" redwood is a lower grade ("merchantable") and has more knots and sapwood. Heart redwood is cinnamon-red and has more natural decay-resistance. The sapwood is creamy yellow. The most common widths are 4" and 6". Lengths are usually 8', 12' and 16'. Both $5/4$" and nominal 2" thickness are available. When left unfinished, both redwood and cedar turn a rustic gray after a few years. Apply water-repellent or stain to retard the color change. Varnishes and other film-forming coatings will crack and peel on a walking surface and aren't recommended. Both western red cedar and redwood are relatively soft and can be abraded under foot traffic.

Pressure-treated lumber makes sturdy and durable deck material. Most treated lumber sold for decking measures $5/4$" ($1^1/4$") or nominal 2" thick and from 4" to 6" wide. Lengths are from 8' to 16'. Shorter lengths and narrower widths cost less per square foot of deck. Longer and wider pieces cost more but require less labor for installation. Pressure-treated wood decking costs less than other decking materials but should be fastened with galvanized nails or screws. All hangers and fasteners should be hot-dipped galvanized, such as Simpson Zmax. For 2" lumber, use 12d nails or 3" screws. For $^5/_4$" decking, use 10d nails or $2^1/_2$" screws. Most pressure-treated lumber isn't pretty. It comes from the lumberyard tinted green from the residue of the ACQ (alkaline copper quaternary) or CA (copper azol) treatment. Soft areas in the wood absorb more stain than the harder portions. The result is a mottled appearance that won't win any prizes. Exposure to the sun and rain will fade the deck from green to honey to gray after a few years. Unless deck lumber has been coated already (Thompsonized), apply a water-repellent top coat immediately after installation.

All wood deck requires occasional maintenance. Check for splits, warping, popped nails and splinters. Unless the owner likes the look of aged wood, he'll need to apply a coat of sealer every few years.

Synthetics include composites made from a combination of recycled wood fiber and recycled thermoplastics. These products (Fiberon is one name) look like real wood and have characteristics like real wood. For example, composites don't expand and contract as much as thermoplastics. But composites made from wood fiber will absorb moisture and tend to age like real wood. You can cut or drill composites with an ordinary hand drill or hand saw equipped with a carbide tip blade. Composites can be stained or painted and should never require maintenance. Deck planks made from rice hulls (Edeck®) are stronger than other composites and don't absorb moisture like deck lumber made from wood fiber. Edeck® requires installation with T-clips that permit expansion and contraction as the deck material heats and cools. Synthetics tend to be more slippery than real wood, especially when wet.

| | Craft@Hrs | Unit | Material | Labor | Total | Sell |
|---|---|---|---|---|---|---|

**Porch and deck demolition.** Includes demolition of framing, supports, flooring, posts, railing, wood stairs. Enclosed porch demolition also includes removal of roof framing, interior finish, gutters, downspouts and roofing. These figures assume use of hand tools (rather than excavation equipment). All debris piled on site. No salvage value assumed. Add the cost of debris hauling and dump fees. Add the cost of electric work, patching of exterior walls and patching of the roof, if required. Cost per square foot of floor demolished.

Demolish enclosed wood-frame porch

| | Craft@Hrs | Unit | Material | Labor | Total | Sell |
|---|---|---|---|---|---|---|
| One story | BL@.250 | SF | — | 6.66 | 6.66 | 11.30 |
| Two story | BL@.225 | SF | — | 5.99 | 5.99 | 10.20 |
| Three story | BL@.196 | SF | — | 5.22 | 5.22 | 8.87 |

Demolish a screened porch built on a concrete slab. Includes demolition of one concrete or wood frame step to grade. Per square foot of floor area.

| | Craft@Hrs | Unit | Material | Labor | Total | Sell |
|---|---|---|---|---|---|---|
| Porch and one step (no slab demolition) | BL@.100 | SF | — | 2.66 | 2.66 | 4.52 |

Demolish a wood deck with railing or kneewall. Includes demolition of the wood deck and up to 7 steps to grade.

| | Craft@Hrs | Unit | Material | Labor | Total | Sell |
|---|---|---|---|---|---|---|
| Deck to 150 SF | BL@18.5 | Ea | — | 493.00 | 493.00 | 838.00 |
| Add per SF for deck over 150 SF | BL@.100 | SF | — | 2.66 | 2.66 | 4.52 |

**Porch and deck repair.** Includes piling debris on site. No salvage value assumed.

Remove and replace structural deck posts

| | Craft@Hrs | Unit | Material | Labor | Total | Sell |
|---|---|---|---|---|---|---|
| 8' treated post, to 6" x 6" | B1@10.3 | Ea | 16.20 | 303.00 | 319.20 | 543.00 |
| 8' x 10" wood colonial column | B1@11.0 | Ea | 125.00 | 323.00 | 448.00 | 762.00 |
| Built-up 8' to 10' post | B1@10.0 | Ea | 13.00 | 294.00 | 307.00 | 522.00 |
| Cut post base and patch 6" x 6" post | B1@5.50 | Ea | 16.20 | 162.00 | 178.20 | 303.00 |

Deck and stair repairs

| | Craft@Hrs | Unit | Material | Labor | Total | Sell |
|---|---|---|---|---|---|---|
| Chip out decay in joist, reinforce with 2" x 6" to 2" x 10" sister joist | | | | | | |
| per LF of joist | B1@.333 | LF | 1.40 | 9.78 | 11.18 | 19.00 |
| Remove wood porch or deck floor and replace with 5/4" or 2" wood flooring | | | | | | |
| per SF of flooring | B1@.166 | SF | 2.20 | 4.88 | 7.08 | 12.00 |
| Remove and replace wood exterior open stairs, | | | | | | |
| per SF of tread and landing | B1@.725 | SF | 15.20 | 21.30 | 36.50 | 62.10 |

**Post and beam wood deck.** Using pressure treated framing lumber with 4" x 6" posts set in concrete. Includes concrete, pressure treated posts, beams, joists 12" on center, deck plank as described, galvanized hardware, fasteners, 4' wide stairs from deck level to ground level, 36" high railing with balusters 4" on center and stair rail. Add for deck skirt and deck finish, if needed. Per square foot of floor area.

Based on rectangular 12' x 10' deck with railing on four sides and stairway, 24" above ground level.

| | Craft@Hrs | Unit | Material | Labor | Total | Sell |
|---|---|---|---|---|---|---|
| With 5/4" pressure treated plank | B1@.260 | SF | 10.10 | 7.64 | 17.74 | 30.20 |
| With 2" Thompsonized plank | B1@.260 | SF | 10.70 | 7.64 | 18.34 | 31.20 |
| With 5/4" western red cedar plank | B1@.260 | SF | 10.70 | 7.64 | 18.34 | 31.20 |
| With 5/4" Fiberon deck plank | B1@.260 | SF | 14.10 | 7.64 | 21.74 | 37.00 |
| With 5/4" Edeck® plank | B1@.260 | SF | 17.10 | 7.64 | 24.74 | 42.10 |

Based on L-shaped 12' x 12' deck with railing on three sides and stairway, 120" above ground level.

| | Craft@Hrs | Unit | Material | Labor | Total | Sell |
|---|---|---|---|---|---|---|
| With 5/4" pressure treated plank | B1@.330 | SF | 12.60 | 9.69 | 22.29 | 37.90 |
| With 2" Thompsonized plank | B1@.330 | SF | 13.60 | 9.69 | 23.29 | 39.60 |
| With 5/4" western red cedar plank | B1@.330 | SF | 13.70 | 9.69 | 23.39 | 39.80 |
| With 5/4" Fiberon deck plank | B1@.330 | SF | 17.70 | 9.69 | 27.39 | 46.60 |
| With 5/4" Edeck® plank | B1@.330 | SF | 18.50 | 9.69 | 28.19 | 47.90 |
| Add for framing to support a spa | B1@8.50 | LS | 96.00 | 250.00 | 346.00 | 588.00 |
| Add for split level deck, 6" difference | B1@6.00 | Ea | 100.00 | 176.00 | 276.00 | 469.00 |
| Add for stain and sealer finish | B1@.006 | SF | .10 | .18 | .28 | .48 |

**Pressure treated construction select deck framing lumber.** For use above ground. .40 treatment. Nominal sizes.

| | | Unit | Material | Labor | Total | Sell |
|---|---|---|---|---|---|---|
| 2" x 4" x 8' | — | Ea | 5.49 | — | 5.49 | — |
| 2" x 4" x 10' | — | Ea | 7.83 | — | 7.83 | — |
| 2" x 4" x 12' | — | Ea | 9.90 | — | 9.90 | — |

| | Craft@Hrs | Unit | Material | Labor | Total | Sell |
|---|---|---|---|---|---|---|
| 2" x 6" x 8' | — | Ea | 9.76 | — | 9.76 | — |
| 2" x 6" x 10' | — | Ea | 12.10 | — | 12.10 | — |
| 2" x 6" x 16' | — | Ea | 19.10 | — | 19.10 | — |
| 2" x 8" x 8' | — | Ea | 12.40 | — | 12.40 | — |
| 2" x 8" x 10' | — | Ea | 14.90 | — | 14.90 | — |
| 2" x 8" x 12' | — | Ea | 19.50 | — | 19.50 | — |
| 2" x 8" x 16' | — | Ea | 24.40 | — | 24.40 | — |
| 2" x 10" x 8' | — | Ea | 15.60 | — | 15.60 | — |
| 2" x 10" x 10' | — | Ea | 19.40 | — | 19.40 | — |
| 2" x 10" x 12' | — | Ea | 24.70 | — | 24.70 | — |
| 2" x 10" x 16' | — | Ea | 32.30 | — | 32.30 | — |
| 2" x 12" x 8' | — | Ea | 19.50 | — | 19.50 | — |
| 2" x 12" x 10' | — | Ea | 23.70 | — | 23.70 | — |
| 2" x 12" x 12' | — | Ea | 28.90 | — | 28.90 | — |
| 2" x 12" x 16' | — | Ea | 38.80 | — | 38.80 | — |
| 4" x 4" x 6' | — | Ea | 23.50 | — | 23.50 | — |
| 4" x 4" x 10' | — | Ea | 15.20 | — | 15.20 | — |
| 4" x 4" x 12' | — | Ea | 18.30 | — | 18.30 | — |
| 4" x 6" x 8' | — | Ea | 18.60 | — | 18.60 | — |
| 4" x 6" x 10' | — | Ea | 23.40 | — | 23.40 | — |
| 4" x 6" x 12' | — | Ea | 27.70 | — | 27.70 | — |
| 4" x 6" x 16' | — | Ea | 37.20 | — | 37.20 | — |
| 4" x 8" x 10' | — | Ea | 32.30 | — | 32.30 | — |
| 4" x 8" x 12' | — | Ea | 38.20 | — | 38.20 | — |
| 4" x 8" x 16' | — | Ea | 50.70 | — | 50.70 | — |
| 4" x 8" x 20' | — | Ea | 60.40 | — | 60.40 | — |
| 4" x 10" x 16' | — | Ea | 63.70 | — | 63.70 | — |
| 4" x 12" x 12' | — | Ea | 61.90 | — | 61.90 | — |
| 4" x 12" x 16' | — | Ea | 84.20 | — | 84.20 | — |
| 4" x 12" x 20' | — | Ea | 97.20 | — | 97.20 | — |
| 6" x 6" x 8' | — | Ea | 34.50 | — | 34.50 | — |
| 6" x 6" x 10' | — | Ea | 45.30 | — | 45.30 | — |

**Pressure treated #2 grade deck posts**. For use in contact with the ground. .60 treatment.

| | Craft@Hrs | Unit | Material | Labor | Total | Sell |
|---|---|---|---|---|---|---|
| 4" x 6" x 8' | — | Ea | 16.20 | — | 16.20 | — |
| 4" x 6" x 10' | — | Ea | 18.30 | — | 18.30 | — |
| 4" x 6" x 12' | — | Ea | 25.90 | — | 25.90 | — |
| 4" x 6" x 14' | — | Ea | 30.20 | — | 30.20 | — |
| 4" x 6" x 16' | — | Ea | 35.60 | — | 35.60 | — |
| 6" x 6" x 8' | — | Ea | 25.90 | — | 25.90 | — |
| 6" x 6" x 10' | — | Ea | 34.50 | — | 34.50 | — |
| 6" x 6" x 12' | — | Ea | 38.80 | — | 38.80 | — |
| 6" x 6" x 16' | — | Ea | 52.90 | — | 52.90 | — |

**Western red cedar deck lumber.** Round edge, surfaced four sides.

| | Craft@Hrs | Unit | Material | Labor | Total | Sell |
|---|---|---|---|---|---|---|
| 5/4" x 4" x 8' | — | Ea | 4.15 | — | 4.15 | — |
| 5/4" x 4" x 10' | — | Ea | 5.20 | — | 5.20 | — |
| 5/4" x 4" x 12' | — | Ea | 6.21 | — | 6.21 | — |
| 5/4" x 4" x 16' | — | Ea | 8.35 | — | 8.35 | — |
| 5/4" x 6" x 8' | — | Ea | 8.01 | — | 8.01 | — |
| 5/4" x 6" x 10' | — | Ea | 10.00 | — | 10.00 | — |
| 5/4" x 6" x 12' | — | Ea | 12.20 | — | 12.20 | — |
| 5/4" x 6" x 16' | — | Ea | 16.20 | — | 16.20 | — |
| 2" x 6" x 8', select | — | Ea | 11.90 | — | 11.90 | — |

| | Craft@Hrs | Unit | Material | Labor | Total | Sell |
|---|---|---|---|---|---|---|

**Pressure treated deck lumber.** .40 treatment. Nominal sizes.

| | Craft@Hrs | Unit | Material | Labor | Total | Sell |
|---|---|---|---|---|---|---|
| 5/4" x 4" x 8', premium | — | Ea | 4.29 | — | 4.29 | — |
| 5/4" x 4" x 10', premium | — | Ea | 6.45 | — | 6.45 | — |
| 5/4" x 4" x 12', premium | — | Ea | 7.53 | — | 7.53 | — |
| 5/4" x 4" x 16', premium | — | Ea | 10.80 | — | 10.80 | — |
| 5/4" x 6" x 8', standard and better | — | Ea | 6.16 | — | 6.16 | — |
| 5/4" x 6" x 8', premium | — | Ea | 8.00 | — | 8.00 | — |
| 5/4" x 6" x 10', standard and better | — | Ea | 6.91 | — | 6.91 | — |
| 5/4" x 6" x 10', premium | — | Ea | 9.98 | — | 9.98 | — |
| 5/4" x 6" x 12', standard and better | — | Ea | 8.08 | — | 8.08 | — |
| 5/4" x 6" x 12', premium | — | Ea | 12.20 | — | 12.20 | — |
| 5/4" x 6" x 14', standard and better | — | Ea | 8.72 | — | 8.72 | — |
| 5/4" x 6" x 14', premium | — | Ea | 13.10 | — | 13.10 | — |
| 5/4" x 6" x 16', standard and better | — | Ea | 13.40 | — | 13.40 | — |
| 5/4" x 6" x 16', premium | — | Ea | 17.00 | — | 17.00 | — |

**Thompsonized outdoor decking.** Nominal sizes. Actual size is about 1/2" less when dry. Pressure treated and water repellent.

| | Craft@Hrs | Unit | Material | Labor | Total | Sell |
|---|---|---|---|---|---|---|
| 2" x 4" x 8' | — | Ea | 5.50 | — | 5.50 | — |
| 2" x 4" x 10' | — | Ea | 6.45 | — | 6.45 | — |
| 2" x 4" x 12' | — | Ea | 7.53 | — | 7.53 | — |
| 2" x 4" x 16' | — | Ea | 10.80 | — | 10.80 | — |
| 2" x 6" x 8' | — | Ea | 8.22 | — | 8.22 | — |
| 2" x 6" x 10' | — | Ea | 8.61 | — | 8.61 | — |
| 2" x 6" x 12' | — | Ea | 11.70 | — | 11.70 | — |
| 2" x 6" x 16' | — | Ea | 16.00 | — | 16.00 | — |
| Deck post surface mount kit | — | Ea | 10.80 | — | 10.80 | — |
| Deck post, 4" x 4" x 48" | — | Ea | 15.10 | — | 15.10 | — |
| End cut preservative, quart | — | Ea | 9.96 | — | 9.96 | — |

**Fiberon decking.** Composite deck made from recycled fiber and recycled polyethylene plastic. Will not split, warp, rot, cup, or crack. Slip-resistant when wet. Installs with conventional woodworking tools. Use stainless steel nails or screws for best results.

| | Craft@Hrs | Unit | Material | Labor | Total | Sell |
|---|---|---|---|---|---|---|
| 5/4" x 6" x 8' | — | Ea | 16.50 | — | 16.50 | — |
| 5/4" x 6" x 12' | — | Ea | 24.80 | — | 24.80 | — |
| 5/4" x 6" x 16' | — | Ea | 32.50 | — | 32.50 | — |
| 4" x 4" x 48" deck post | — | Ea | 20.50 | — | 20.50 | — |
| Post cap | — | Ea | 5.37 | — | 5.37 | — |
| Post bracket, joist mount | — | Ea | 15.30 | — | 15.30 | — |
| Post hardware for concrete | — | Ea | 47.00 | — | 47.00 | — |
| Post jacket top | — | Ea | 4.26 | — | 4.26 | — |
| Post sleeve, 4-1/2" x 4-1/2" x 42" | — | Ea | 15.75 | — | 15.75 | — |
| Bevel post cap | — | Ea | 1.95 | — | 1.95 | — |
| Top or bottom rail, 8' long | — | Ea | 33.80 | — | 33.80 | — |
| Square baluster, 32" long | — | Ea | 3.10 | — | 3.10 | — |
| Line rail hardware kit | — | Ea | 13.60 | — | 13.60 | — |
| Stair rail hardware kit | — | Ea | 21.20 | — | 21.20 | — |

**Edeck® composite deck.** Composite deck made from rice hulls. Doesn't absorb moisture.

| | Craft@Hrs | Unit | Material | Labor | Total | Sell |
|---|---|---|---|---|---|---|
| 5/4" x 6" x 12' Edeck® plank | — | Ea | 27.00 | — | 27.00 | — |
| 5/4" x 6" x 16' Edeck® plank | — | Ea | 35.60 | — | 35.60 | — |
| 5/4" x 6" x 20' Edeck® plank | — | Ea | 36.00 | — | 36.00 | — |
| 2" x 6" x 10' Edeck® plank | — | Ea | 26.00 | — | 26.00 | — |
| Edeck® #8 x 3/4" screws, per box | — | Ea | 6.50 | — | 6.50 | — |
| Edeck® screw hole plug | — | Ea | 9.79 | — | 9.79 | — |

| | Craft@Hrs | Unit | Material | Labor | Total | Sell |
|---|---|---|---|---|---|---|
| Edeck® rail bracket | — | Ea | 6.50 | — | 6.50 | — |
| 5" x 5" x 53" Edeck® rail post | — | Ea | 28.10 | — | 28.10 | — |
| 5" x 5" peaked Edeck® post cap | — | Ea | 2.73 | — | 2.73 | — |
| E-Clips kit, pack of 100 | — | Ea | 32.40 | — | 32.40 | — |
| 1.5" x 1.5" x 2.5' Edeck® steel bar | — | Ea | 7.57 | — | 7.57 | — |

**Turned colonial style wood porch post.**

| | Craft@Hrs | Unit | Material | Labor | Total | Sell |
|---|---|---|---|---|---|---|
| 4" x 4" x 8' | B1@1.18 | Ea | 51.50 | 34.70 | 86.20 | 147.00 |
| 5" x 5" x 8' | B1@1.18 | Ea | 73.90 | 34.70 | 108.60 | 185.00 |
| 6" x 6" x 8' | B1@1.18 | Ea | 95.60 | 34.70 | 130.30 | 222.00 |

**Plinth for turned wood porch post.**

| | Craft@Hrs | Unit | Material | Labor | Total | Sell |
|---|---|---|---|---|---|---|
| 4" plinth | B1@.250 | Ea | 5.34 | 7.34 | 12.68 | 21.60 |
| 5" plinth | B1@.250 | Ea | 5.89 | 7.34 | 13.23 | 22.50 |
| 6" plinth | B1@.250 | Ea | 6.46 | 7.34 | 13.80 | 23.50 |

**Structural aluminum porch column.** White decorative column. Includes column, cap and base. Integral interlocking device permanently holds column together. Electrostatically painted. 6" load capacity is 16,000 pounds. 8" load capacity is 20,000 pounds.

| | Craft@Hrs | Unit | Material | Labor | Total | Sell |
|---|---|---|---|---|---|---|
| 6" x 8' column, round | B1@2.00 | Ea | 87.10 | 58.70 | 145.80 | 248.00 |
| 8" x 8' column, round | B1@2.00 | Ea | 115.00 | 58.70 | 173.70 | 295.00 |
| 6" x 6" x 8' column, square | B1@2.00 | Ea | 115.00 | 58.70 | 173.70 | 295.00 |

**Round repair column.** Colonial column can wrap around existing posts or be used as load-bearing column for new construction. Fluted design. Maintenance-free baked-on enamel finish.

| | Craft@Hrs | Unit | Material | Labor | Total | Sell |
|---|---|---|---|---|---|---|
| 6" x 8' round column | B1@2.00 | Ea | 71.30 | 58.70 | 130.00 | 221.00 |
| 8" x 8' round column | B1@2.00 | Ea | 94.60 | 58.70 | 153.30 | 261.00 |

**Porch roof framing.** Conventionally framed beams and joists. Per square foot of area covered. Figures in parentheses indicate board feet per square foot of ceiling including end joists, header joists and 5% waste. Based on a 200 SF job. Add the cost of support posts and finish roof.

| | Craft@Hrs | Unit | Material | Labor | Total | Sell |
|---|---|---|---|---|---|---|
| 2" x 4", standard and better grade | | | | | | |
|   16" centers (.59 BF per SF) | B1@.020 | SF | .37 | .59 | .96 | 1.63 |
|   24" centers (.42 BF per SF) | B1@.014 | SF | .27 | .41 | .68 | 1.16 |
| 2" x 6", standard and better grade | | | | | | |
|   16" centers (.88 BF per SF) | B1@.020 | SF | .55 | .59 | 1.14 | 1.94 |
|   24" centers (.63 BF per SF) | B1@.015 | SF | .39 | .44 | .83 | 1.41 |

Add for porch support posts. Per post, including one sack of concrete per post. Add the cost of excavation and post anchors.

| | Craft@Hrs | Unit | Material | Labor | Total | Sell |
|---|---|---|---|---|---|---|
| 3" x 3" yellow pine, pressure treated, 10' long | B1@.500 | Ea | 9.85 | 14.70 | 24.55 | 41.70 |
| 4" x 4" pressure treated, 10' long | B1@.750 | Ea | 12.10 | 22.00 | 34.10 | 58.00 |
| 6" x 6" pressure treated, 10' long | B1@.750 | Ea | 16.10 | 22.00 | 38.10 | 64.80 |

**Porch roofing.** Cover over conventionally framed wood ceiling joists. Flat roof with built-up roofing. Using 1/2" CDX roof sheathing (at $29.44 per 4' x 8' sheet) with 3 ply asphalt consisting of 2 plies of 15 lb. felt with 90 lb. cap sheet and 3 hot mop coats of asphalt. Add the cost of roof framing.

| | Craft@Hrs | Unit | Material | Labor | Total | Sell |
|---|---|---|---|---|---|---|
| Total flat roof assembly cost per SF | B1@.029 | SF | 1.11 | .85 | 1.96 | 3.33 |

**Deck railing.** Typical 42" high deck rail and baluster consisting of top and bottom rail and 2" x 2" baluster 5" OC with posts lag bolted to the edge of the deck. See detailed railing material costs below. Cost per linear foot of deck railing.

| | Craft@Hrs | Unit | Material | Labor | Total | Sell |
|---|---|---|---|---|---|---|
| Pine, pressure treated | B1@.333 | LF | 3.58 | 9.78 | 13.36 | 22.70 |
| Redwood, select heart | B1@.333 | LF | 6.46 | 9.78 | 16.24 | 27.60 |
| Recycled plastic lumber | B1@.333 | LF | 17.40 | 9.78 | 27.18 | 46.20 |

| | Craft@Hrs | Unit | Material | Labor | Total | Sell |
|---|---|---|---|---|---|---|
| **Pressure treated deck posts.** 4" x 4" x 4" high. | | | | | | |
| Chamfered dado post | — | Ea | 13.20 | — | 13.20 | — |
| Treated cedar post | — | Ea | 14.40 | — | 14.40 | — |
| Laminated finial-ready post | — | Ea | 15.50 | — | 15.50 | — |
| Notched cedar post | — | Ea | 17.60 | — | 17.60 | — |
| V-groove cedar post | — | Ea | 21.60 | — | 21.60 | — |
| Turned deck post | — | Ea | 21.60 | — | 21.60 | — |
| Premium ball-top post | — | Ea | 23.20 | — | 23.20 | — |
| **Pressure treated deck post cap.** | | | | | | |
| 3-1/2" x 3-1/2" square cap | — | Ea | 1.64 | — | 1.64 | — |
| 4" x 4" plain ball cap | — | Ea | 3.36 | — | 3.36 | — |
| 4" x 4" royal coachman cap | — | Ea | 4.00 | — | 4.00 | — |
| 4" x 4" Hampton flat cap | — | Ea | 4.34 | — | 4.34 | — |
| 4" x 4" cannon ball post cap | — | Ea | 8.49 | — | 8.49 | — |
| 4" x 4" Newport pyramid cap | — | Ea | 8.98 | — | 8.98 | — |
| 4" x 4" Verona copper cap | — | Ea | 9.46 | — | 9.46 | — |
| 4" x 4" Victoria copper cap | — | Ea | 10.70 | — | 10.70 | — |
| 6" x 6" Verona copper cap | — | Ea | 15.00 | — | 15.00 | — |
| 6" x 6" Victoria copper cap | — | Ea | 15.10 | — | 15.10 | — |
| **Pressure treated deck rail.** | | | | | | |
| 2" x 3" x 8' pine rail insert | — | Ea | 6.59 | — | 6.59 | — |
| 2" x 4" x 8' molded rail | — | Ea | 7.80 | — | 7.80 | — |
| 2" x 4" x 8' combo rail | — | Ea | 11.60 | — | 11.60 | — |
| Contemporary finial | — | Ea | 3.61 | — | 3.61 | — |
| **Pressure treated deck rail baluster.** | | | | | | |
| 2" x 2" x 32", square ends | — | Ea | 1.04 | — | 1.04 | — |
| 2" x 2" x 36", square ends | — | Ea | 1.19 | — | 1.19 | — |
| 2" x 2" x 42", beveled one end | — | Ea | 1.29 | — | 1.29 | — |
| 2" x 2" x 48", beveled one end | — | Ea | 1.39 | — | 1.39 | — |
| 2" x 2" x 42", beveled two ends | — | Ea | 1.45 | — | 1.45 | — |
| **Pressure treated deck rail turned spindle.** | | | | | | |
| 2" x 2" x 32" classic spindle | — | Ea | 1.80 | — | 1.80 | — |
| 2" x 2" x 36" classic spindle | — | Ea | 2.17 | — | 2.17 | — |
| 2" x 3" x 36" classic spindle | — | Ea | 3.19 | — | 3.19 | — |
| 2" x 2" x 36" colonial spindle | — | Ea | 4.22 | — | 4.22 | — |
| 3' x 3" x 36" jumbo spindle | — | Ea | 6.25 | — | 6.25 | — |
| **Outdoor deck rail.** | | | | | | |
| 2" x 4" x 8' redwood dado rail | — | Ea | 13.10 | — | 13.10 | — |
| 2" x 5" x 8' primed handrail | — | Ea | 12.90 | — | 12.90 | — |
| 2" x 4" x 8' deluxe cedar rail | — | Ea | 16.20 | — | 16.20 | — |
| 2" x 6" x 8' western red cedar rail | — | Ea | 18.30 | — | 18.30 | — |
| 2" x 4" x 8' western red cedar rail | — | Ea | 20.40 | — | 20.40 | — |
| Rail bracket kit, 4-piece | — | Ea | 8.97 | — | 8.97 | — |
| Rail angle bracket kit, 4-piece | — | Ea | 20.50 | — | 20.50 | — |
| **Outdoor deck rail baluster.** | | | | | | |
| 2" x 2" x 24" Douglas fir baluster | — | Ea | 1.22 | — | 1.22 | — |
| 2" x 2" x 30" Douglas fir baluster | — | Ea | 1.61 | — | 1.61 | — |
| 2" x 2" x 36" Douglas fir baluster | — | Ea | 1.88 | — | 1.88 | — |
| 2" x 2" x 36", beveled redwood | — | Ea | 2.13 | — | 2.13 | — |

| | Craft@Hrs | Unit | Material | Labor | Total | Sell |
|---|---|---|---|---|---|---|
| 2" x 2" x 42", one end beveled cedar | — | Ea | 2.37 | — | 2.37 | — |
| 1-3/8" x 1-3/8" x 32", beveled redwood | — | Ea | 2.89 | — | 2.89 | — |
| 1-3/8" x 1-3/8" x 48", redwood | — | Ea | 3.24 | — | 3.24 | — |
| 3" x 3" x 32" main street spindle | — | Ea | 5.37 | — | 5.37 | — |
| 8' western red cedar fillet strip | — | Ea | 2.37 | — | 2.37 | — |

**Outdoor deck post cap.**

| | Craft@Hrs | Unit | Material | Labor | Total | Sell |
|---|---|---|---|---|---|---|
| 2" x 6" flat cedar cap | — | Ea | 1.06 | — | 1.06 | — |
| 5" cedar post cap | — | Ea | 3.21 | — | 3.21 | — |
| 4" x 4" suburban pyramid | — | Ea | 5.37 | — | 5.37 | — |
| Cedar ball top | — | Ea | 5.37 | — | 5.37 | — |
| 4" x 4" cascade peak cap | — | Ea | 7.04 | — | 7.04 | — |
| Gothic post cap | — | Ea | 7.53 | — | 7.53 | — |
| Newport classic cap | — | Ea | 8.12 | — | 8.12 | — |
| 4" x 4" Newport pyramid, cedar | — | Ea | 11.80 | — | 11.80 | — |
| 4" x 4" copper high point cap | — | Ea | 11.10 | — | 11.10 | — |
| 4" x 4" redwood ball cap | — | Ea | 14.00 | — | 14.00 | — |
| 4" x 4" cedar ball cap | — | Ea | 13.40 | — | 13.40 | — |

**Porch or deck stairs.** All stairs assume 12" tread (step) composed of two 2" x 6" lumber and riser. Add the cost of a concrete footing.

One or two steps high, box construction. Prices are per LF of step. A 4' wide stairway with 2 steps has 8 LF of step. Use 8 LF as the quantity.

| | Craft@Hrs | Unit | Material | Labor | Total | Sell |
|---|---|---|---|---|---|---|
| Pine treads, open risers | B1@.270 | LF | 4.29 | 7.93 | 12.22 | 20.80 |
| Pine treads, with pine risers | B1@.350 | LF | 7.39 | 10.30 | 17.69 | 30.10 |
| Redwood treads, open risers | B1@.270 | LF | 4.18 | 7.93 | 12.11 | 20.60 |
| Redwood treads, with redwood risers | B1@.350 | LF | 7.06 | 10.30 | 17.36 | 29.50 |
| 5/4" recycled plastic lumber treads, open risers | B1@.270 | LF | 8.32 | 7.93 | 16.25 | 27.60 |
| 5/4" recycled plastic lumber treads, with recycled plastic lumber risers | B1@.350 | LF | 17.50 | 10.30 | 27.80 | 47.30 |
| 2" x 6" recycled plastic lumber treads, open risers | B1@.270 | LF | 8.64 | 7.93 | 16.57 | 28.20 |
| 2" x 6" recycled plastic lumber treads, with recycled plastic lumber risers | B1@.350 | LF | 18.40 | 10.30 | 28.70 | 48.80 |

Three or more steps high, 36" wide. 3 stringers cut from 2" x 12" pressure treated lumber. Cost per riser.

| | Craft@Hrs | Unit | Material | Labor | Total | Sell |
|---|---|---|---|---|---|---|
| Pine treads, open risers | B1@.530 | Ea | 15.60 | 15.60 | 31.20 | 53.00 |
| Pine treads, with pine risers | B1@.620 | Ea | 24.90 | 18.20 | 43.10 | 73.30 |
| Redwood treads, open risers | B1@.530 | Ea | 15.30 | 15.60 | 30.90 | 52.50 |
| Redwood treads, with redwood risers | B1@.620 | Ea | 23.90 | 18.20 | 42.10 | 71.60 |
| 5/4" recycled plastic lumber treads, open risers | B1@.810 | Ea | 28.70 | 23.80 | 52.50 | 89.30 |
| 5/4" recycled plastic lumber treads, with recycled plastic lumber risers | B1@1.05 | Ea | 56.10 | 30.80 | 86.90 | 148.00 |
| 2" x 6" recycled plastic lumber treads, open risers | B1@.810 | Ea | 29.70 | 23.80 | 53.50 | 91.00 |
| 2" x 6" recycled plastic lumber treads, with recycled plastic lumber risers | B1@1.05 | Ea | 59.00 | 30.80 | 89.80 | 153.00 |

| | Craft@Hrs | Unit | Material | Labor | Total | Sell |
|---|---|---|---|---|---|---|
| **Precut outdoor deck stair stringers.** Cost per stringer. | | | | | | |
| 2" x 12" x 3 step, redwood | — | Ea | 17.20 | — | 17.20 | — |
| 2" x 12" x 4 step, redwood | — | Ea | 23.30 | — | 23.30 | — |
| 2" x 12" x 5 step, redwood | — | Ea | 27.90 | — | 27.90 | — |
| 2 step, treated lumber | — | Ea | 10.70 | — | 10.70 | — |
| 3 step, treated lumber | — | Ea | 8.78 | — | 8.78 | — |
| 4 step, treated lumber | — | Ea | 10.90 | — | 10.90 | — |
| 5 step, treated lumber | — | Ea | 20.40 | — | 20.40 | — |
| 2 step, steel | — | Ea | 9.69 | — | 9.69 | — |
| 3 step, steel | — | Ea | 14.00 | — | 14.00 | — |
| 4 step, steel | — | Ea | 18.30 | — | 18.30 | — |
| EZ stair building hardware kit | — | Ea | 8.37 | — | 8.37 | — |
| **Outdoor deck stair stepping.** Pressure treated lumber. | | | | | | |
| 2" x12" x 48" tread | — | Ea | 12.90 | — | 12.90 | — |

**Porch and patio misting (fog) system.** Installed 8' high on a porch exterior. Based on .008" stainless steel and brass nozzles installed each 18" along 3/8" type L copper refrigeration tubing. Includes one electric operated 1 HP, 800 PSI, 1 GPM water pump, supply valve, auto drain valve, a wireless remote control switch, 50' of underground water connection line tapped to the existing service and trenching. Add the cost of 20 amp electrical connection.

| | Craft@Hrs | Unit | Material | Labor | Total | Sell |
|---|---|---|---|---|---|---|
| Set up water pump, tap to existing water line | P1@5.00 | Ea | 1,560.00 | 158.00 | 1,718.00 | 2,920.00 |
| Add per nozzle (each 18" of tube) | P1@.160 | Ea | 6.50 | 5.04 | 11.54 | 19.60 |

**Porch enclosure screening.** Includes medium quality aluminum frame stock, screening and insertion of vinyl screen spline in the frame. Cost per square foot of screen installed, either from a kneewall to the header or from the deck to the header.

| | Craft@Hrs | Unit | Material | Labor | Total | Sell |
|---|---|---|---|---|---|---|
| With fiberglass screen wire | B1@.040 | SF | 1.08 | 1.17 | 2.25 | 3.83 |
| With aluminum screen wire | B1@.040 | SF | 1.18 | 1.17 | 2.35 | 4.00 |
| With solar bronze screen wire | B1@.040 | SF | 1.54 | 1.17 | 2.71 | 4.61 |

**Porch enclosure screen wire.** Per roll.

| | Craft@Hrs | Unit | Material | Labor | Total | Sell |
|---|---|---|---|---|---|---|
| Aluminum, 36" x 25' | — | Ea | 21.60 | — | 21.60 | — |
| Aluminum, 48" x 25' | — | Ea | 28.70 | — | 28.70 | — |
| Fiberglass, 72" x 100' | — | Ea | 122.00 | — | 122.00 | — |
| Solar screen, 36" x 25' | — | Ea | 49.50 | — | 49.50 | — |
| Solar screen, 48" x 25' | — | Ea | 62.90 | — | 62.90 | — |

**Aluminum porch screen frame stock.** Framing for screen porch enclosure. With integrated groove ready to accept screen and flat spline when framework assembly is complete. Supports the top of a kickplate coil when attached between uprights. Bronze or white finish.

| | Craft@Hrs | Unit | Material | Labor | Total | Sell |
|---|---|---|---|---|---|---|
| 1" x 2", 8' | — | Ea | 9.69 | — | 9.69 | — |
| 1" x 2", 10' | — | Ea | 11.80 | — | 11.80 | — |
| 2" x 2", 8' | — | Ea | 16.20 | — | 16.20 | — |
| 2" x 2", 10' | — | Ea | 21.60 | — | 21.60 | — |
| 2" x 3", 8' | — | Ea | 28.00 | — | 28.00 | — |

**Angle anchor clip.** Aluminum. Use to attach 2" x 3" uprights to 1" x 2" and concrete. Bronze or white finish.

| | Craft@Hrs | Unit | Material | Labor | Total | Sell |
|---|---|---|---|---|---|---|
| 2" x 2" x 1/8", bag of 25 | — | Ea | 9.15 | — | 9.15 | — |

**Capri clip angle fastener.** Aluminum. Attaches top of uprights to framework header with self-drilling screws. Bronze or white finish.

| | Craft@Hrs | Unit | Material | Labor | Total | Sell |
|---|---|---|---|---|---|---|
| Bag of 10 | — | Ea | 5.37 | — | 5.37 | — |

Porches and Decks **459**

|  | Craft@Hrs | Unit | Material | Labor | Total | Sell |
|---|---|---|---|---|---|---|

**Castle clip U-fastener.** Aluminum. Attaches kickplate rails to uprights with self-drilling screws. Bronze or white finish.

| Fastener | — | Ea | .85 | — | .85 | — |
|---|---|---|---|---|---|---|

**F-channel.** Aluminum. Attaches to soffit or fascia. Brown or white finish.

| 12' long | — | Ea | 8.37 | — | 8.37 | — |
|---|---|---|---|---|---|---|

**Kickplate channel.** Aluminum. Trims edges of kickplate coil. Slips over edges of kickplate coil and is screwed to coil and framework. Bronze or white finish.

| 10' long | — | Ea | 4.29 | — | 4.29 | — |
|---|---|---|---|---|---|---|

**Kickplate coil.** Aluminum. Covers area under kickplate rails. Helps keep dirt from blowing, and rain from splashing into the room. Bronze or white finish.

| 16" x 16' | — | Ea | 35.60 | — | 35.60 | — |
|---|---|---|---|---|---|---|

**Screen enclosure screws.** Weatherproof coating provides corrosion resistance. Use 2" screws to 1" x 2" of side wall to front wall corner upright. Use 3" screws to attach header to wood framing.

| Size 8 self-drilling hex washer head, | | | | | | |
|---|---|---|---|---|---|---|
| 9/16", bag of 100 | — | Ea | 4.29 | — | 4.29 | — |
| Size 10 hex head self-drilling hex washer head, | | | | | | |
| 2", bag of 25 | — | Ea | 2.13 | — | 2.13 | — |
| 3", bag of 25 | — | Ea | 2.45 | — | 2.45 | — |

**Screen spline.** Vinyl. Rolls into spline grooves of framework to hold screen in place.

| 100' roll, flat black | — | Ea | 5.93 | — | 5.93 | — |
|---|---|---|---|---|---|---|

# Painting and Finishing

**17**

Nothing improves the looks of an old house like a new coat of paint. But before deciding to repaint, consider washing the surface or spot painting. Excessive paint build-up increases the chance of chipping and peeling from incompatible paint layers. Paint film that's too thick is likely to crack. The only remedy for cross-grain cracking or intercoat peeling may be to strip the surface down to bare wood. Areas sheltered from the sun, such as eaves and porch ceilings, don't need to be painted as often as exposed areas. Instead of repainting the entire surface, spot paint only in areas showing the most wear. Matching colors is easy with modern optical paint-matching equipment. Most paint suppliers can match nearly any sample you bring in.

## Guidelines for Recoating

Nearly all surfaces in a home can be recoated successfully with the right surface preparation. The challenge is to identify what's on the surface so you know what treatment to use, and then to select the appropriate coating.

❖ *Kalsomine* (also called *calcimine*) is a whitewash made from zinc oxide and glue. Kalsomine turns to powder as it ages, which makes it an unsuitable surface for any type of coating. Either remove the existing surface by sandblasting or cover the surface, such as with new siding.

❖ *Grease* falls in the same category as kalsomine. No paint will adhere to walls or ceilings coated with grease. This is a common problem in kitchens. Remove all grease before recoating. If paint has been applied over grease, remove all coatings down to the bare surface.

❖ *Non-compatible paints* can result in intercoat peeling. Adding another coat on top of peeling paint will only compound the problem. You'll generally need to remove the existing paint. If the surface you're to paint is incompatible with the paint you're going to use, apply a coat of primer between the non-compatible paints. That will usually solve the problem.

❖ *Stains* caused by water damage, smoke or a foreign substance need special attention. Be sure the surface is dry and clean, then coat with a primer-sealer such as Kilz®.

❖ *Plaster cracks* should be repaired with fiberglass tape and joint compound before recoating. Don't think that paint will fill the cracks for you.

❖ *Varnish* should be cleaned with a strong tri-sodium phosphate solution and then painted soon after drying. Heavily-alligatored varnish must be removed before coating.

❖ *Metals* should be wire brushed to remove loose material and then coated with a metal primer.

❖ *Hardwood floors* of oak, birch, beech, and maple are usually finished by applying two coats of sealer. Sand lightly or buff with steel wool between coats. Then apply a coat of paste wax and buff. Maintain the finish by rewaxing. For a high-gloss finish, instead of applying two coats of sealer, make the second coat varnish.

❖ *Wood trim and paneling* are usually finished with a clear wood sealer or a stain-sealer combination and then top-coated with at least one additional coat of sealer or varnish. Sand lightly between sealing and top-coating. For a deeper finish, apply one coat of high-gloss varnish followed by a coat of semi-gloss varnish.

❖ *Painted wood trim* requires a primer or undercoat and then a coat of acrylic latex, either flat or semi-gloss. Paint with at least some gloss resists fingerprints better and can be cleaned with soap and water.

❖ *Kitchen and bathroom* walls need a coat of semi-gloss enamel. This type of finish wears well, is easy to clean, and resists moisture.

❖ *Drywall*, in rooms other than the kitchen and bathroom, needs a coat of flat latex. New drywall, since it's highly porous, should have a coat of primer before the final coat of latex paint.

# Paint Color and Gloss

Paint comes in an infinite range of colors. Your paint store probably stocks a few of the basic colors, such as white and black, but can mix any color you want. You buy the tinting base and the paint technician adds just the right amount of pigment to yield the color needed. Any color you select will be the same price. However, you'll pay more for high-gloss paints and less for low-gloss (*flat*) paints. High-gloss paint resists fingerprints and holds up better when scrubbed. Use high gloss in kitchens and bathrooms where you need a washable surface. High-gloss paint tends to reveal surface imperfections and any application defects. Flat paints cost a little less than high gloss and hide flaws better, but don't stand up as well to scrubbing. Use a flat paint in bedrooms, halls and living rooms. Eggshell paint has slightly more gloss than flat paint. Semi-gloss and satin finishes have more gloss than eggshell, but are still not as glossy as high gloss.

Paint is also classified by its reflectance value — how much light is reflected off the surface. Light colors have a light reflectance value (LRV) of 50 or more and are a good choice for interiors. Darker colors are better for exteriors.

Paint selection isn't the place to try and save a few bucks on materials. Top quality paints made by well-known and respected brands will yield better results. Lesser quality paints may require additional coats, which will end up costing more in the end.

# Exterior Trim

Expect wood trim on the exterior of an older home to be in poor condtion. If the finish is worn but the surface is smooth, another coat of paint should be enough. Trim that's damaged or badly chipped may need to be replaced. Matching old trim isn't easy. Patterns popular 50 or 100 years ago may no longer be available at your local lumberyard. You can have trim pieces custom made to order, but the cost may greater than replacing all similar trim.

# Wood Preservatives

Remove blotchy discoloration and signs of rain spatter on wood by wire brushing. Brush *with* the grain. Then treat the surface with a water-repellent preservative. Applying a coat of water-repellent preservative every few years will prevent nearly all decay.

Wood treated with water-soluble preservatives can be painted the same as untreated wood. The coating may not last as long as with untreated wood, but there's no major difference, especially when the treated wood has weathered for several months.

# Penetrating Stains

If the surface hasn't been coated previously, consider a penetrating stain. These stains make a good finish for most wood surfaces, especially rough-sawn, textured wood and knotty wood that would be difficult to paint. Penetrating stain doesn't form a film on the surface. Because the surface isn't sealed, the coating won't crack, peel or blister. Use pigmented penetrating stains to add shades of brown, green, red, or gray to the wood. To avoid lap marks, coat full lengths of siding or trim without stopping. On smooth surfaces, apply a single coat. Two or three years later, when the surface has weathered, apply another coat of penetrating stain. This time the surface will be rougher and will absorb more stain. That will protect the wood longer — up to eight years.

On rough-sawn or weathered surfaces, apply two coats of penetrating stain. Both coats should be applied within a few hours. Don't let the first coat dry before applying the second. A dry first coat will keep the second from penetrating into the wood. Rough wood that's been double-coated with penetrating stain has a finish life of ten years or more.

Refinish a stained wood surface when the color has faded and bare wood is beginning to show through. Prepare the surface by brushing with steel wool. Brush in the direction of the grain. Then hose down with water. When the surface is thoroughly dry, apply another coat of stain.

# Paint Application Methods

The three most common paint application techniques are brush, roller and spray. Brushwork takes the most time per square foot coated. Productivity is highest when a spray rig can be put to good use, but not all paints are suitable for spraying. Sprayers generally require the paint to be thinned, and some paints are designated "Do not thin." Applying paint by roller takes less time per square foot than brushing, but more time than spraying. The cost of a painting task will differ depending on the application method used. That's why labor estimates and material costs are separated in this chapter. Labor estimates come first, followed by material costs. The labor estimates include figures for each of the three application methods.

The best tool for applying paint depends on what's being coated and what's not supposed to get coated. Spray painting isn't as precise as painting with a brush or even a roller. Spray painting requires much more attention to masking and covering adjacent surfaces, such as the floor, doors, windows, hardware and trim. When painting a wall or ceiling with brush or roller, it may be enough to lay down several drop cloths. That takes only a minute or two. Coating the same surface with a spray rig will require masking off everything in the room that's not supposed to get coated — and then closing off openings to adjacent rooms. When spray painting, as much as 80 percent of the job may be masking and protecting adjacent surfaces. That's especially true in an occupied building where it's impractical

---

### *Brush, Roller or Spray?*

❖ Spray application is usually impractical (takes longer than brushing or rolling) in small rooms with multiple openings and installed hardware or fixtures.

❖ Bedrooms and bathrooms are seldom candidates for spray painting, especially if the home is occupied.

❖ Even in larger rooms, such as a kitchen or living room, spray painting isn't usually practical in an occupied home.

❖ Building exteriors are good candidates for spray painting, especially stucco, masonry and panel exteriors where less masking is required.

❖ Interiors of most occupied homes are usually painted with a brush (hand cut in) and roller. The labor tables that follow identify this work as "roll and brush."

❖ It's safe to assume that recoating railings, trim, doors, windows, metals and cabinets will be done with a brush.

to remove every ornament and piece of hardware — and where overspray poses a major risk to the dwelling contents. If spray painting a large room can be expected to take an hour, expect painters will spend two to four hours on masking and protection from overspray.

Brush, roller and spray don't necessarily produce equivalent results. For example, paint brushed on a metal surface makes better contact with the metal than paint sprayed on the same surface. Contact and adhesion affect the life of the coating, especially on metal. So most metal surfaces are coated with brush or roller, at least in home improvement work. Likewise, when applying semi-gloss to doors, most painters prefer to brush coat four sides (front, back, hinge edge and latch edge) while the door is hung in the opening. That results in a much smoother finish. Even when paint is sprayed on, some jobs are *back-rolled*. The spray coat is smoothed out with a roller immediately after application.

When a spray job is complete, good practice requires that someone inspect the surface carefully, looking for voids, imperfections and painter's holidays (areas missed). Assume that about 10 percent of the surface will require brush or roller touch-up after a spray application is complete.

When working with a roller, most jobs require hand *cut in*. Paint is applied by brush at corners, where different materials intersect and around obstructions such as hardware, vents, cases, trim and fixtures. When hand cut-in is complete, the remainder of the surface is rolled.

## Additional Costs

None of the figures in this chapter includes moving furniture, fixtures, wall hangings, objects of art, etc. All assume that loose contents are moved to the center of the room or removed entirely before painting begins. Reduce productivity by 30 to 50 percent if painters will be moving furniture, removing and rehanging pictures, and removing and replacing drapery valances.

Straight, uninterrupted walls with square corners are quick and easy paint jobs. Painting cut-up surfaces with alcoves, soffits, light wells and offsets takes longer. For a cut-up job, allow 20 to 40 percent more time than indicated in the following tables.

These estimates assume landscaping materials (shrubs and trees) are cut back or pulled away from exterior surfaces before exterior painting begins.

| | Craft@Hrs | Unit | Material | Labor | Total | Sell |
|---|---|---|---|---|---|---|

## Labor Estimates for Painting

**Cost to repaint a home interior.** Typical costs per 100 square feet of floor for roller and brush application of a single coat to walls and ceilings in all rooms of an occupied dwelling, including bathrooms and closets. Add the cost of coating cabinets, trim, doors and window trim. Includes minimum surface preparation, spackle of minor defects in wallboard or plaster, masking of adjacent surfaces and priming of stained or discolored surfaces. Use these figures for preliminary estimates and to check completed bids. These estimates equate to 3 painter-hours and 1 gallon of paint per 100 square feet of floor.

| | Craft@Hrs | Unit | Material | Labor | Total | Sell |
|---|---|---|---|---|---|---|
| Repaint a home interior, per 100 SF of floor | PT@3.00 | CSF | 16.50 | 97.10 | 113.60 | 193.00 |

**Cost to repaint a home exterior.** Typical costs per 100 square feet of floor for spray painting a single coat on exterior walls and soffits of an occupied dwelling. Add the cost of painting trim, doors and windows. Includes a water blast of the exterior, repair of minor surface defects in siding and trim, and priming of stained, cracked or peeling surfaces. Use these figures for preliminary estimates and to check completed bids. These figures equate to 2 painter-hours and 1 gallon of paint per 100 square feet of floor.

| | Craft@Hrs | Unit | Material | Labor | Total | Sell |
|---|---|---|---|---|---|---|
| Repaint a home exterior, per 100 SF of floor | PT@2.00 | CSF | 17.50 | 64.80 | 82.30 | 140.00 |

**High time difficulty factors for surface preparation and painting.** Painting takes longer when heights exceed 8' above the floor. Productivity is lower when application requires a roller pole or wand on a spray gun or when work is done from a ladder or scaffold. When painting above 8', apply the following factors.

Add 30% to the area for heights from 8' to 13' (multiply by 1.3)
Add 60% to the area for heights from 13' to 17' (multiply by 1.6)
Add 90% to the area for heights from 17' to 19' (multiply by 1.9)
Add 120% to the area for heights from 19' to 21' (multiply by 2.2)

**Labor for acid wash of gutters and downspouts.** For heights above one story, use the high time difficulty factors above.

| | Craft@Hrs | Unit | Material | Labor | Total | Sell |
|---|---|---|---|---|---|---|
| Per 100 linear feet | PT@1.05 | CLF | — | 34.00 | 34.00 | 57.80 |

**Labor to burn off paint.** Using a heat gun or torch. For heights above 8', use the high time difficulty factors above.

| | Craft@Hrs | Unit | Material | Labor | Total | Sell |
|---|---|---|---|---|---|---|
| Beveled wood siding, per 100 square feet | PT@3.33 | CSF | — | 108.00 | 108.00 | 184.00 |
| Exterior plain surfaces, per 100 square feet | PT@2.50 | CSF | — | 81.00 | 81.00 | 138.00 |
| Exterior trim, per 100 square feet | PT@5.00 | CSF | — | 162.00 | 162.00 | 275.00 |
| Interior plain surfaces, per 100 square feet | PT@4.00 | CSF | — | 130.00 | 130.00 | 221.00 |
| Interior trim, per 100 square feet | PT@6.67 | CSF | — | 216.00 | 216.00 | 367.00 |

**Labor for caulking.** Add extra time for digging out and removing old caulk and residue, protecting adjacent surfaces and masking.

| | Craft@Hrs | Unit | Material | Labor | Total | Sell |
|---|---|---|---|---|---|---|
| 1/8" gap, per 100 linear feet | PT@1.54 | CLF | — | 49.90 | 49.90 | 84.80 |
| 1/4" gap, per 100 linear feet | PT@1.82 | CLF | — | 58.90 | 58.90 | 100.00 |
| 3/8" gap, per 100 linear feet | PT@2.22 | CLF | — | 71.90 | 71.90 | 122.00 |
| 1/2" gap, per 100 linear feet | PT@2.63 | CLF | — | 85.20 | 85.20 | 145.00 |

**Labor for interior sanding.** For heights above 8', use the high time difficulty factors above.

| | Craft@Hrs | Unit | Material | Labor | Total | Sell |
|---|---|---|---|---|---|---|
| Sand flat wall areas before first coat, per 100 square feet | PT@.333 | CSF | — | 10.80 | 10.80 | 18.40 |
| Sand and putty before second flat coat, per 100 square feet | PT@.500 | CSF | — | 16.20 | 16.20 | 27.50 |
| Sand and putty before second enamel coat, per 100 square feet | PT@.800 | CSF | — | 25.90 | 25.90 | 44.00 |
| Sand and putty cabinets before second coat, per 100 square feet | PT@.667 | CSF | — | 21.60 | 21.60 | 36.70 |
| Sand and putty bookshelves before second coat, per 100 square feet | PT@.800 | CSF | — | 25.90 | 25.90 | 44.00 |

|  | Craft@Hrs | Unit | Material | Labor | Total | Sell |
|---|---|---|---|---|---|---|
| Sand bookshelves with light grit before third coat, per 100 square feet | PT@.444 | CSF | — | 14.40 | 14.40 | 24.50 |
| Sand with light grit before third enamel coat, per 100 square feet | PT@.714 | CSF | — | 23.10 | 23.10 | 39.30 |

**Labor for exterior sanding.** For heights above 8', use the high time difficulty factors above.

|  | Craft@Hrs | Unit | Material | Labor | Total | Sell |
|---|---|---|---|---|---|---|
| Sand and putty siding and trim, before second coat, per 100 square feet | PT@.500 | CSF | — | 16.20 | 16.20 | 27.50 |
| Sand and putty trim only, before second coat, per 100 square feet | PT@.909 | CSF | — | 29.40 | 29.40 | 50.00 |
| Sand trim only with light grit, before third coat, per 100 square feet | PT@.571 | CSF | — | 18.50 | 18.50 | 31.50 |

**Labor for sandblasting.** Sandblast only. Add the cost of equipment, protecting adjacent surfaces, mobilization and cleanup. For heights above 8', use the high time difficulty factors above.

|  | Craft@Hrs | Unit | Material | Labor | Total | Sell |
|---|---|---|---|---|---|---|
| Brush-off sandblast to remove cement base paint, per 100 square feet | PT@.571 | CSF | — | 18.50 | 18.50 | 31.50 |
| Brush-off sandblast to remove oil or latex paint, per 100 square feet | PT@.800 | CSF | — | 25.90 | 25.90 | 44.00 |

**Labor for stripping paint.** For heights above 8', use the high time difficulty factors above.

|  | Craft@Hrs | Unit | Material | Labor | Total | Sell |
|---|---|---|---|---|---|---|
| Strip, remove, or bleach, per 100 square feet | PT@1.43 | CSF | — | 46.30 | 46.30 | 78.70 |
| Strip varnish with light-duty liquid remover, per 100 square feet | PT@2.86 | CSF | — | 92.60 | 92.60 | 157.00 |
| Strip varnish with heavy-duty liquid remover, per 100 square feet | PT@3.33 | CSF | — | 108.00 | 108.00 | 184.00 |
| Paint removal with light-duty liquid remover, per 100 square feet | PT@3.33 | CSF | — | 108.00 | 108.00 | 184.00 |
| Paint removal with heavy-duty liquid remover, per 100 square feet | PT@4.00 | CSF | — | 130.00 | 130.00 | 221.00 |

**Labor for wash down.** For heights above 8', use the high time difficulty factors above.

|  | Craft@Hrs | Unit | Material | Labor | Total | Sell |
|---|---|---|---|---|---|---|
| Interior smooth flat wall, per 100 square feet | PT@.500 | CSF | — | 16.20 | 16.20 | 27.50 |
| Wash and touch up a smooth flat interior wall, per 100 square feet | PT@.625 | CSF | — | 20.20 | 20.20 | 34.30 |
| Wash down a rough surface flat interior wall, per 100 square feet | PT@.667 | CSF | — | 21.60 | 21.60 | 36.70 |
| Wash and touch up a rough surface flat interior wall, per 100 square feet | PT@1.05 | CSF | — | 34.00 | 34.00 | 57.80 |
| Wash down an interior enamel wall, per 100 square feet | PT@.465 | CSF | — | 15.10 | 15.10 | 25.70 |
| Wash down and touch up an interior enamel wall, per 100 square feet | PT@.885 | CSF | — | 28.70 | 28.70 | 48.80 |
| Wash down interior enamel trim, per 100 square feet | PT@.667 | CSF | — | 21.60 | 21.60 | 36.70 |
| Wash and touch up interior enamel trim, per 100 square feet | PT@.833 | CSF | — | 27.00 | 27.00 | 45.90 |
| Wash interior varnish trim, per 100 square feet | PT@.513 | CSF | — | 16.60 | 16.60 | 28.20 |

| | Craft@Hrs | Unit | Material | Labor | Total | Sell |
|---|---|---|---|---|---|---|
| Wash and touch up interior varnish trim, per 100 square feet | PT@.714 | CSF | — | 23.10 | 23.10 | 39.30 |
| Wash interior varnished floors, per 100 square feet | PT@.476 | CSF | — | 15.40 | 15.40 | 26.20 |
| Wash and touch up interior varnish floors, per 100 square feet | PT@.654 | CSF | — | 21.20 | 21.20 | 36.00 |
| Wash interior smooth plaster, per 100 square feet | PT@.571 | CSF | — | 18.50 | 18.50 | 31.50 |
| Wash and touch up interior smooth plaster, per 100 square feet | PT@.714 | CSF | — | 23.10 | 23.10 | 39.30 |
| Wash interior sand finish plaster, per 100 square feet | PT@.741 | CSF | — | 24.00 | 24.00 | 40.80 |
| Wash and touch up interior sand finish plaster, per 100 square feet | PT@.909 | CSF | — | 29.40 | 29.40 | 50.00 |

**Labor for waterblasting (pressure washing).** Waterblast to remove deteriorated, cracked or flaking paint from wood, concrete, brick, block, plaster or stucco surfaces. Add the cost of waterblast equipment, protecting adjacent surfaces and mobilization. For heights above 8', use the high time difficulty factors. These production rates will apply on larger jobs, such as an entire home. For smaller jobs, use a minimum charge of 3 or 4 hours to move on, set up, waterblast, clean up and move off. Production rates are based on medium-pressure equipment (2500 PSI). High-pressure equipment (3000 PSI) will increase productivity 10% to 50%. Low-pressure equipment (1700 PSI) will reduce productivity by 10% to 50%. Rates assume a 1/4"-diameter nozzle. Jobs will be more difficult where the surface is badly deteriorated or contaminated with grime or grease. Note that the surface must dry thoroughly before coating begins. Per 100 square feet of surface cleaned.

| | Craft@Hrs | Unit | Material | Labor | Total | Sell |
|---|---|---|---|---|---|---|
| Light blast, 1500 SF per hour, per 100 square feet | PT@.067 | CSF | — | 2.17 | 2.17 | 3.69 |
| Most jobs, 1000 SF per hour, per 100 square feet | PT@.100 | CSF | — | 3.24 | 3.24 | 5.51 |
| Difficult jobs, 750 SF per hour, per 100 square feet | PT@.133 | CSF | — | 4.31 | 4.31 | 7.33 |

**Labor for painting baseboard.** Paint application only. Add the cost of surface preparation, protecting adjacent surfaces, mobilization, cleanup and callbacks.

| | Craft@Hrs | Unit | Material | Labor | Total | Sell |
|---|---|---|---|---|---|---|
| Brush 1 coat, 700 linear feet per gallon, per 100 linear feet | PT@.833 | CLF | — | 27.00 | 27.00 | 45.90 |
| Roll 1 coat, 750 linear feet per gallon, per 100 linear feet | PT@.083 | CLF | — | 2.69 | 2.69 | 4.57 |

**Labor for painting bookcases and shelves.** Paint application only. Add the cost of surface preparation, protecting adjacent surfaces, mobilization, cleanup and callbacks.

| | Craft@Hrs | Unit | Material | Labor | Total | Sell |
|---|---|---|---|---|---|---|
| Roll and brush undercoat, 280 square feet per gallon, per 100 square feet | PT@3.33 | CSF | — | 108.00 | 108.00 | 184.00 |
| Roll and brush finish coat, 340 square feet per gallon, per 100 square feet | PT@2.50 | CSF | — | 81.00 | 81.00 | 138.00 |
| Spray undercoat, 133 square feet per gallon, per 100 square feet | PT@.606 | CSF | — | 19.60 | 19.60 | 33.30 |
| Spray finish coat, 158 square feet per gallon, per 100 square feet | PT@.400 | CSF | — | 13.00 | 13.00 | 22.10 |

| | Craft@Hrs | Unit | Material | Labor | Total | Sell |
|---|---|---|---|---|---|---|

**Labor for painting cabinets.** Paint application only. Add the cost of surface preparation, protecting adjacent surfaces, mobilization, cleanup and callbacks.

| | Craft@Hrs | Unit | Material | Labor | Total | Sell |
|---|---|---|---|---|---|---|
| Brush cabinet back, 275 square feet per gallon, per 100 square feet | PT@.769 | CSF | — | 24.90 | 24.90 | 42.30 |
| Roll and brush cabinet face, 250 square feet per gallon, per 100 square feet of face | PT@1.08 | CSF | — | 35.00 | 35.00 | 59.50 |
| Spray cabinet face, 113 square feet per gallon, per 100 square feet of face | PT@.714 | CSF | — | 23.10 | 23.10 | 39.30 |

**Labor for applying acoustic ceiling texture.** Application only. Add the cost of surface preparation, protecting adjacent surfaces, mobilization, cleanup and callbacks. For work more than 8' above floor level, use the high time difficulty factors.

| | Craft@Hrs | Unit | Material | Labor | Total | Sell |
|---|---|---|---|---|---|---|
| Spray prime coat for acoustic textured ceiling, 90 square feet per gallon, per 100 square feet | PT@.333 | CSF | — | 10.80 | 10.80 | 18.40 |
| Spray finish coat for acoustic textured ceiling, 170 square feet per gallon, per 100 square feet | PT@.222 | CSF | — | 7.19 | 7.19 | 12.20 |
| Spray stipple finish texture paint, light coverage, dry powder mix, 7.5 square feet per gallon, per 100 square feet | PT@.400 | CSF | — | 13.00 | 13.00 | 22.10 |

**Labor for painting tongue-and-groove ceiling — brush.** Paint application only. Add the cost of surface preparation, protecting adjacent surfaces, mobilization, cleanup and callbacks. For heights above 8', use the high time difficulty factors.

| | Craft@Hrs | Unit | Material | Labor | Total | Sell |
|---|---|---|---|---|---|---|
| Brush 1st coat of water-base flat latex, 288 square feet per gallon, per 100 square feet | PT@1.54 | CSF | — | 49.90 | 49.90 | 84.80 |
| Brush additional coats of water-base flat latex, 338 square feet per gallon, per 100 square feet | PT@1.33 | CSF | — | 43.10 | 43.10 | 73.30 |
| Brush prime coat of water-base sealer, 288 square feet per gallon, per 100 square feet | PT@1.43 | CSF | — | 46.30 | 46.30 | 78.70 |
| Brush 1st finish coat of water-base enamel, 288 square feet per gallon, per 100 square feet | PT@1.33 | CSF | — | 43.10 | 43.10 | 73.30 |
| Brush additional coats of water-base enamel, 363 square feet per gallon, per 100 square feet | PT@1.11 | CSF | — | 35.90 | 35.90 | 61.00 |

**Labor for painting tongue-and-groove ceiling — roll.** Paint application only. Add the cost of surface preparation, protecting adjacent surfaces, mobilization, cleanup and callbacks. For heights above 8', use the high time difficulty factors.

| | Craft@Hrs | Unit | Material | Labor | Total | Sell |
|---|---|---|---|---|---|---|
| Roll 1st coat of water-base flat latex, 263 square feet per gallon, per 100 square feet | PT@.769 | CSF | — | 24.90 | 24.90 | 42.30 |
| Roll additional coats of water-base flat latex, 313 square feet per gallon, per 100 square feet | PT@.645 | CSF | — | 20.90 | 20.90 | 35.50 |

| | Craft@Hrs | Unit | Material | Labor | Total | Sell |
|---|---|---|---|---|---|---|
| Roll prime coat of water-base sealer, 263 square feet per gallon, per 100 square feet | PT@.833 | CSF | — | 27.00 | 27.00 | 45.90 |
| Roll 1st finish coat of water-base enamel, 313 square feet per gallon, per 100 square feet | PT@.690 | CSF | — | 22.30 | 22.30 | 37.90 |
| Roll additional coats of water-base enamel, 338 square feet per gallon, per 100 square feet | PT@.526 | CSF | — | 17.00 | 17.00 | 28.90 |

**Labor for painting tongue-and-groove ceiling — spray.** Paint application only. Add the cost of surface preparation, protecting adjacent surfaces, mobilization, cleanup and callbacks. For heights above 8', use the high time difficulty factors.

| | Craft@Hrs | Unit | Material | Labor | Total | Sell |
|---|---|---|---|---|---|---|
| Spray 1st coat of water-base flat latex, 155 square feet per gallon, per 100 square feet | PT@.278 | CSF | — | 9.00 | 9.00 | 15.30 |
| Spray additional coats of water-base flat latex, 225 square feet per gallon, per 100 square feet | PT@.213 | CSF | — | 6.90 | 6.90 | 11.70 |
| Spray prime coat of water-base sealer, 155 square feet per gallon, per 100 square feet | PT@.263 | CSF | — | 8.52 | 8.52 | 14.50 |
| Spray 1st finish coat of water-base enamel, 225 square feet per gallon, per 100 square feet | PT@.222 | CSF | — | 7.19 | 7.19 | 12.20 |
| Spray additional coats of water-base enamel, 300 square feet per gallon, per 100 square feet | PT@.182 | CSF | — | 5.89 | 5.89 | 10.00 |

**Labor for staining tongue-and-groove ceiling.** Stain application only. Add the cost of surface preparation, protecting adjacent surfaces, mobilization, cleanup and callbacks. For heights above 8', use the high time difficulty factors.

| | Craft@Hrs | Unit | Material | Labor | Total | Sell |
|---|---|---|---|---|---|---|
| Roll and brush water-base semi-transparent stain, 275 square feet per gallon, per 100 square feet | PT@.417 | CSF | — | 13.50 | 13.50 | 23.00 |

**Labor for painting closet shelf and pole.** Paint application only. Add the cost of surface preparation, protecting adjacent surfaces, mobilization, cleanup and callbacks.

| | Craft@Hrs | Unit | Material | Labor | Total | Sell |
|---|---|---|---|---|---|---|
| Brush 1 coat of water-base undercoat, 70 linear feet per gallon, per 100 linear feet | PT@4.55 | CLF | — | 147.00 | 147.00 | 250.00 |
| Brush split coat of water-base undercoat and enamel, 70 linear feet per gallon, per 100 linear feet | PT@4.76 | CLF | — | 154.00 | 154.00 | 262.00 |
| Brush finish coat of water-base enamel, 70 linear feet per gallon, per 100 linear feet | PT@5.00 | CLF | — | 162.00 | 162.00 | 275.00 |

| | Craft@Hrs | Unit | Material | Labor | Total | Sell |
|---|---|---|---|---|---|---|

**Labor for painting door frame and trim only.** Paint application only. Opening refers to both sides of a doorway. Add the cost of surface preparation, protecting adjacent surfaces, mobilization, cleanup and callbacks.

| | Craft@Hrs | Unit | Material | Labor | Total | Sell |
|---|---|---|---|---|---|---|
| Brush 1 coat of water-base undercoat,<br>28 openings per gallon, per opening | PT@.063 | Ea | — | 2.04 | 2.04 | 3.47 |
| Brush split coat of water-base undercoat and enamel,<br>31 openings per gallon, per opening | PT@.071 | Ea | — | 2.30 | 2.30 | 3.91 |
| Brush water-base enamel,<br>31 openings per gallon, per opening | PT@.077 | Ea | — | 2.49 | 2.49 | 4.23 |

**Labor for painting flush exterior door, frame and trim.** Paint application only. Add the cost of surface preparation, protecting adjacent surfaces, mobilization, cleanup and callbacks.

| | Craft@Hrs | Unit | Material | Labor | Total | Sell |
|---|---|---|---|---|---|---|
| Roll and brush 2 coats of water-base exterior enamel,<br>6 doors per gallon, per door | PT@.400 | Ea | — | 13.00 | 13.00 | 22.10 |
| Roll and brush 2 coats of oil-base exterior enamel,<br>6 doors per gallon, per door | PT@.400 | Ea | — | 13.00 | 13.00 | 22.10 |
| Brush 2 coats of polyurethane,<br>4.5 doors per gallon, per door | PT@.600 | Ea | — | 19.40 | 19.40 | 33.00 |

**Labor for painting exterior French door, frame and trim.** Paint application only. Add preparation time for sanding, puttying, protecting adjacent surfaces and masking or waxing glass panes.

| | Craft@Hrs | Unit | Material | Labor | Total | Sell |
|---|---|---|---|---|---|---|
| Roll and brush 2 coats of water-base exterior enamel,<br>10 doors per gallon, per door | PT@.800 | Ea | — | 25.90 | 25.90 | 44.00 |
| Roll and brush 2 coats of oil-base exterior enamel,<br>12 doors per gallon, per door | PT@.800 | Ea | — | 25.90 | 25.90 | 44.00 |
| Brush 2 coats of polyurethane,<br>7.5 doors per gallon, per door | PT@1.30 | Ea | — | 42.10 | 42.10 | 71.60 |

**Labor for painting exterior louver door, frame and trim.** Paint application only. Add the cost of surface preparation, protecting adjacent surfaces, mobilization, cleanup and callbacks.

| | Craft@Hrs | Unit | Material | Labor | Total | Sell |
|---|---|---|---|---|---|---|
| Roll and brush 2 coats of water-base exterior enamel,<br>6 doors per gallon, per door | PT@1.10 | Ea | — | 35.60 | 35.60 | 60.50 |
| Brush 2 coats of polyurethane,<br>4.5 doors per gallon, per door | PT@1.50 | Ea | — | 48.60 | 48.60 | 82.60 |

**Labor for painting exterior panel (entry) door, frame and trim.** Paint application only. Add the cost of surface preparation, protecting adjacent surfaces, mobilization, cleanup and callbacks.

| | Craft@Hrs | Unit | Material | Labor | Total | Sell |
|---|---|---|---|---|---|---|
| Roll and brush 2 coats of water-base exterior enamel,<br>3 doors per gallon, per door | PT@1.10 | Ea | — | 35.60 | 35.60 | 60.50 |
| Brush 2 coats of polyurethane,<br>3 doors per gallon, per door | PT@1.50 | Ea | — | 48.60 | 48.60 | 82.60 |

**Labor for painting interior flush door, frame and trim — roll and brush.** Paint application only. Add the cost of surface preparation, protecting adjacent surfaces, mobilization, cleanup and callbacks.

| | Craft@Hrs | Unit | Material | Labor | Total | Sell |
|---|---|---|---|---|---|---|
| Roll and brush 1 coat of water-base undercoat,<br>13 doors per gallon, per door | PT@.300 | Ea | — | 9.71 | 9.71 | 16.50 |
| Roll and brush 1st finish coat of water-base enamel,<br>12.5 doors per gallon, per door | PT@.250 | Ea | — | 8.10 | 8.10 | 13.80 |
| Roll and brush additional finish coats<br>of water-base enamel,<br>13.5 doors per gallon, per door | PT@.200 | Ea | — | 6.48 | 6.48 | 11.00 |

| | Craft@Hrs | Unit | Material | Labor | Total | Sell |
|---|---|---|---|---|---|---|

**Labor for painting interior flush door, frame and trim — spray.** New installations only. Includes paint application only. Add the cost of surface preparation, protecting adjacent surfaces, mobilization, cleanup and callbacks.

| | Craft@Hrs | Unit | Material | Labor | Total | Sell |
|---|---|---|---|---|---|---|
| Spray 1 coat of water-base undercoat, 16 doors per gallon, per door | PT@.090 | Ea | — | 2.91 | 2.91 | 4.95 |
| Spray 1 coat of oil-base undercoat, 16 doors per gallon, per door | PT@.090 | Ea | — | 2.91 | 2.91 | 4.95 |
| Spray 1st finish coat of water-base enamel, 17 doors per gallon, per door | PT@.080 | Ea | — | 2.59 | 2.59 | 4.40 |
| Spray additional coats of water-base enamel, 18 doors per gallon, per door | PT@.070 | Ea | — | 2.27 | 2.27 | 3.86 |

**Labor for painting interior French door, frame and trim.** Paint application only. Add preparation time for sanding, puttying, protecting adjacent surfaces and masking or waxing glass panes.

| | Craft@Hrs | Unit | Material | Labor | Total | Sell |
|---|---|---|---|---|---|---|
| Roll and brush 1 coat of water-base undercoat, 13 doors per gallon, per door | PT@.380 | Ea | — | 12.30 | 12.30 | 20.90 |
| Roll and brush 1st finish coat of water-base enamel, 14 doors per gallon, per door | PT@.350 | Ea | — | 11.30 | 11.30 | 19.20 |
| Roll and brush additional finish coats of water-base enamel, 15 doors per gallon, per door | PT@.330 | Ea | — | 10.70 | 10.70 | 18.20 |

**Labor for painting interior louver door, frame and trim — roll and brush.** Paint application only. Add the cost of surface preparation, protecting adjacent surfaces, mobilization, cleanup and callbacks.

| | Craft@Hrs | Unit | Material | Labor | Total | Sell |
|---|---|---|---|---|---|---|
| Roll and brush 1 coat of water-base undercoat, 7 doors per gallon, per door | PT@.540 | Ea | — | 17.50 | 17.50 | 29.80 |
| Roll and brush 1st finish coat of water-base enamel, 8 doors per gallon, per door | PT@.420 | Ea | — | 13.60 | 13.60 | 23.10 |
| Roll and brush additional finish coats of water-base enamel, 9 doors per gallon, per door | PT@.300 | Ea | — | 9.71 | 9.71 | 16.50 |

**Labor for painting interior louver door, frame and trim — spray.** New installations only. Includes paint application only. Add the cost of surface preparation, protecting adjacent surfaces, mobilization, cleanup and callbacks.

| | Craft@Hrs | Unit | Material | Labor | Total | Sell |
|---|---|---|---|---|---|---|
| Spray 1 coat of water-base undercoat, 11 doors per gallon, per door | PT@.140 | Ea | — | 4.53 | 4.53 | 7.70 |
| Spray 1st finish coat of water-base enamel, 12 doors per gallon, per door | PT@.110 | Ea | — | 3.56 | 3.56 | 6.05 |
| Spray additional finish coats of water-base enamel, 13 doors per gallon, per door | PT@.090 | Ea | — | 2.91 | 2.91 | 4.95 |

**Labor for painting interior panel door, frame and trim — roll and brush.** Paint application only. Add the cost of surface preparation, protecting adjacent surfaces, mobilization, cleanup and callbacks.

| | Craft@Hrs | Unit | Material | Labor | Total | Sell |
|---|---|---|---|---|---|---|
| Roll and brush 1 coat of water-base undercoat, 7 doors per gallon, per door | PT@.330 | Ea | — | 10.70 | 10.70 | 18.20 |
| Roll and brush 1st finish coat of water-base enamel, 10 doors per gallon, per door | PT@.300 | Ea | — | 9.71 | 9.71 | 16.50 |
| Roll and brush additional finish coats of water-base enamel, 11 doors per gallon, per door | PT@.250 | Ea | — | 8.10 | 8.10 | 13.80 |

| | Craft@Hrs | Unit | Material | Labor | Total | Sell |
|---|---|---|---|---|---|---|

**Labor for painting interior panel door, frame and trim — spray.** New installations only. Includes paint application only. Add the cost of surface preparation, protecting adjacent surfaces, mobilization, cleanup and callbacks.

| | Craft@Hrs | Unit | Material | Labor | Total | Sell |
|---|---|---|---|---|---|---|
| Spray 1 coat of water-base undercoat, 14 doors per gallon, per door | PT@.110 | Ea | — | 3.56 | 3.56 | 6.05 |
| Spray 1st finish coat of water-base enamel, 15 doors per gallon, per door | PT@.080 | Ea | — | 2.59 | 2.59 | 4.40 |
| Spray additional finish coats of water-base enamel, 16 doors per gallon, per door | PT@.080 | Ea | — | 2.59 | 2.59 | 4.40 |

**Labor for staining 2" x 4" fascia.** Front side and bottom edge only. Stain application only. Add the cost of surface preparation, protecting adjacent surfaces, mobilization, cleanup and callbacks. For heights above 8', use the high time difficulty factors.

| | Craft@Hrs | Unit | Material | Labor | Total | Sell |
|---|---|---|---|---|---|---|
| Roll and brush each coat of solid-body stain, 130 linear feet per gallon, per 100 linear feet | PT@.488 | CLF | — | 15.80 | 15.80 | 26.90 |
| Roll and brush each coat of semi-transparent stain, 150 linear feet per gallon, per 100 linear feet | PT@.444 | CLF | — | 14.40 | 14.40 | 24.50 |
| Spray each coat of solid-body stain, 100 linear feet per gallon, per 100 linear feet | PT@.308 | CLF | — | 9.97 | 9.97 | 16.90 |
| Spray each coat of semi-transparent stain, 115 linear feet per gallon, per 100 linear feet | PT@.286 | CLF | — | 9.26 | 9.26 | 15.70 |

**Labor for staining 2" x 6" or 2" x 10" fascia.** Front side and bottom edge only. Stain application only. Add the cost of surface preparation, protecting adjacent surfaces, mobilization, cleanup and callbacks. For heights above 8', use the high time difficulty factors.

| | Craft@Hrs | Unit | Material | Labor | Total | Sell |
|---|---|---|---|---|---|---|
| Roll and brush each coat of solid-body stain, 110 linear feet per gallon, per 100 linear feet | PT@.571 | CLF | — | 18.50 | 18.50 | 31.50 |
| Roll and brush each coat of semi-transparent stain, 130 linear feet per gallon, per 100 linear feet | PT@.513 | CLF | — | 16.60 | 16.60 | 28.20 |
| Spray each coat of solid-body stain, 80 linear feet per gallon, per 100 linear feet | PT@.333 | CLF | — | 10.80 | 10.80 | 18.40 |
| Spray each coat of semi-transparent stain, 95 linear feet per gallon, per 100 linear feet | PT@.319 | CLF | — | 10.30 | 10.30 | 17.50 |

**Labor for staining 2" x 12" fascia.** Front side and bottom edge only. Stain application only. Add the cost of surface preparation, protecting adjacent surfaces, mobilization, cleanup and callbacks. For heights above 8', use the high time difficulty factors.

| | Craft@Hrs | Unit | Material | Labor | Total | Sell |
|---|---|---|---|---|---|---|
| Roll and brush each coat of solid-body stain, 75 linear feet per gallon, per 100 linear feet | PT@.769 | CLF | — | 24.90 | 24.90 | 42.30 |
| Roll and brush each coat of semi-transparent stain, 95 linear feet per gallon, per 100 linear feet | PT@.667 | CLF | — | 21.60 | 21.60 | 36.70 |

| | Craft@Hrs | Unit | Material | Labor | Total | Sell |
|---|---|---|---|---|---|---|
| Spray each coat of solid-body stain, 50 linear feet per gallon, per 100 linear feet | PT@.380 | CLF | — | 12.30 | 12.30 | 20.90 |
| Spray each coat of semi-transparent stain, 65 linear feet per gallon, per 100 linear feet | PT@.347 | CLF | — | 11.20 | 11.20 | 19.00 |

**Labor for staining picket fence.** Area is the fence length times the fence height times two. Stain application only. Add the cost of surface preparation, protecting adjacent surfaces, mobilization, cleanup and callbacks.

| | Craft@Hrs | Unit | Material | Labor | Total | Sell |
|---|---|---|---|---|---|---|
| Roll and brush 1st coat of solid-body or semi-transparent stain, 343 square feet per gallon, per 100 square feet | PT@.690 | CSF | — | 22.30 | 22.30 | 37.90 |
| Roll and brush additional coats of solid-body or semi-transparent stain, 388 square feet per gallon, per 100 square feet | PT@.444 | CSF | — | 14.40 | 14.40 | 24.50 |
| Spray 1st coat of solid-body or semi-transparent stain, 275 square feet per gallon, per 100 square feet | PT@.200 | CSF | — | 6.48 | 6.48 | 11.00 |
| Spray additional coats of solid-body or semi-transparent stain, 325 square feet per gallon, per 100 square feet | PT@.167 | CSF | — | 5.41 | 5.41 | 9.20 |

**Labor for painting concrete floor — roll.** Paint application only. Add the cost of surface preparation, protecting adjacent surfaces, mobilization, cleanup and callbacks.

| | Craft@Hrs | Unit | Material | Labor | Total | Sell |
|---|---|---|---|---|---|---|
| Roll and brush 1st coat of water-base masonry paint, 263 square feet per gallon, per 100 square feet | PT@.459 | CSF | — | 14.90 | 14.90 | 25.30 |
| Roll and brush 2nd coat of water-base masonry paint, 325 square feet per gallon, per 100 square feet | PT@.373 | CSF | — | 12.10 | 12.10 | 20.60 |
| Roll and brush 3rd or additional coats of water-base masonry paint, 350 square feet per gallon, per 100 square feet | PT@.333 | CSF | — | 10.80 | 10.80 | 18.40 |
| Roll and brush each coat of 1-part water-base epoxy, 488 square feet per gallon, per 100 square feet | PT@.444 | CSF | — | 14.40 | 14.40 | 24.50 |
| Roll and brush each coat of 2-part water-base epoxy, 488 square feet per gallon, per 100 square feet | PT@.481 | CSF | — | 15.60 | 15.60 | 26.50 |
| Roll and brush 1st coat of penetrating oil stain, 425 square feet per gallon, per 100 square feet | PT@.400 | CSF | — | 13.00 | 13.00 | 22.10 |
| Roll and brush 2nd coat of penetrating oil stain, 475 square feet per gallon, per 100 square feet | PT@.290 | CSF | — | 9.39 | 9.39 | 16.00 |

| | Craft@Hrs | Unit | Material | Labor | Total | Sell |
|---|---|---|---|---|---|---|
| Roll and brush 3rd or additional coats of penetrating oil stain, 500 square feet per gallon, per 100 square feet | PT@.261 | CSF | — | 8.45 | 8.45 | 14.40 |

**Labor for painting concrete floor — spray.** Paint application only. Add the cost of surface preparation, protecting adjacent surfaces, mobilization, cleanup and callbacks.

| | Craft@Hrs | Unit | Material | Labor | Total | Sell |
|---|---|---|---|---|---|---|
| Spray 1st coat of water-base masonry paint, 163 square feet per gallon, per 100 square feet | PT@.111 | CSF | — | 3.59 | 3.59 | 6.10 |
| Spray 2nd coat of water-base masonry paint, 263 square feet per gallon, per 100 square feet | PT@.100 | CSF | — | 3.24 | 3.24 | 5.51 |
| Spray 3rd or additional coats of water-base masonry paint, 313 square feet per gallon, per 100 square feet | PT@.091 | CSF | — | 2.95 | 2.95 | 5.02 |

**Labor for painting wood deck or floor — roll and brush.** Paint application only. Add the cost of surface preparation, protecting adjacent surfaces, mobilization, cleanup and callbacks.

| | Craft@Hrs | Unit | Material | Labor | Total | Sell |
|---|---|---|---|---|---|---|
| Roll and brush water-base prime undercoat, 425 square feet per gallon, per 100 square feet | PT@.228 | CSF | — | 7.38 | 7.38 | 12.50 |
| Roll and brush porch and deck enamel, 450 square feet per gallon, per 100 square feet | PT@.216 | CSF | — | 6.99 | 6.99 | 11.90 |
| Brush 1-part water-base epoxy, 400 square feet per gallon, per 100 square feet | PT@.400 | CSF | — | 13.00 | 13.00 | 22.10 |
| Roll 2-part epoxy, 225 square feet per gallon, per 100 square feet | PT@.444 | CSF | — | 14.40 | 14.40 | 24.50 |

**Labor for staining wood deck or floor — brush.** Stain application only. Add the cost of surface preparation, protecting adjacent surfaces, mobilization, cleanup and callbacks.

| | Craft@Hrs | Unit | Material | Labor | Total | Sell |
|---|---|---|---|---|---|---|
| Brush 1st coat oil-base wiping stain, wipe and fill wood floor, 475 square feet per gallon, per 100 square feet | PT@.400 | CSF | — | 13.00 | 13.00 | 22.10 |
| Brush 2nd coat oil-base stain, wipe and fill wood floor, 500 square feet per gallon, per 100 square feet | PT@.235 | CSF | — | 7.61 | 7.61 | 12.90 |
| Brush 3rd or additional coats of oil-base stain, wipe and fill wood floor, 525 square feet per gallon, per 100 square feet | PT@.222 | CSF | — | 7.19 | 7.19 | 12.20 |
| Brush 1st coat of sanding sealer on maple or pine floor, 450 square feet per gallon, per 100 square feet | PT@.250 | CSF | — | 8.10 | 8.10 | 13.80 |
| Brush additional coats of sanding sealer on maple or pine floor, 525 square feet per gallon, per 100 square feet | PT@.222 | CSF | — | 7.19 | 7.19 | 12.20 |

| | Craft@Hrs | Unit | Material | Labor | Total | Sell |
|---|---|---|---|---|---|---|
| Brush 1st coat of sanding sealer on oak floor, 500 square feet per gallon, per 100 square feet | PT@.235 | CSF | — | 7.61 | 7.61 | 12.90 |
| Brush additional coats of sanding sealer on oak floor, 600 square feet per gallon, per 100 square feet | PT@.190 | CSF | — | 6.15 | 6.15 | 10.50 |
| Brush 1st coat of penetrating stain, wax and wipe wood floor, 525 square feet per gallon, per 100 square feet | PT@.400 | CSF | — | 13.00 | 13.00 | 22.10 |
| Brush additional coats of penetrating stain, wax and wipe wood floor, 525 square feet per gallon, per 100 square feet | PT@.333 | CSF | — | 10.80 | 10.80 | 18.40 |
| Wax and polish wood floors by hand, per 100 square feet | PT@.500 | CSF | — | 16.20 | 16.20 | 27.50 |
| Wax and polish wood floors by machine, per 100 square feet | PT@.222 | CSF | — | 7.19 | 7.19 | 12.20 |

**Labor for painting gutters and downspouts.** Paint application only. Add the cost of surface preparation, protecting adjacent surfaces, mobilization, cleanup and callbacks. For heights above 8', use the high time difficulty factors.

| | Craft@Hrs | Unit | Material | Labor | Total | Sell |
|---|---|---|---|---|---|---|
| Brush prime coat rust inhibitor on metal gutters, 375 linear feet per gallon, per 100 linear feet | PT@1.11 | CLF | — | 35.90 | 35.90 | 61.00 |
| Brush 1st finish coat on metal gutters, 400 linear feet per gallon, per 100 linear feet | PT@.909 | CLF | — | 29.40 | 29.40 | 50.00 |
| Brush additional finish coats on metal gutters, 425 linear feet per gallon, per 100 linear feet | PT@.769 | CLF | — | 24.90 | 24.90 | 42.30 |
| Brush prime coat rust inhibitor on metal downspouts, 225 linear feet per gallon, per 100 linear feet | PT@2.86 | CLF | — | 92.60 | 92.60 | 157.00 |

**Labor for applying masonry block filler on brick masonry.** Filler application only. Add the cost of surface preparation, protecting adjacent surfaces, mobilization, cleanup and callbacks. For heights above 8', use the high time difficulty factors.

| | Craft@Hrs | Unit | Material | Labor | Total | Sell |
|---|---|---|---|---|---|---|
| Brush one coat of masonry block filler, 65 square feet per gallon, per 100 square feet | PT@.800 | CSF | — | 25.90 | 25.90 | 44.00 |
| Roll one coat of masonry block filler, 60 square feet per gallon, per 100 square feet | PT@.465 | CSF | — | 15.10 | 15.10 | 25.70 |
| Spray one coat of masonry block filler, 55 square feet per gallon, per 100 square feet | PT@.190 | CSF | — | 6.15 | 6.15 | 10.50 |

| | Craft@Hrs | Unit | Material | Labor | Total | Sell |
|---|---|---|---|---|---|---|

**Labor for painting smooth brick masonry — brush.** Paint application only. Add the cost of surface preparation, protecting adjacent surfaces, mobilization, cleanup and callbacks. For heights above 8', use the high time difficulty factors.

| | Craft@Hrs | Unit | Material | Labor | Total | Sell |
|---|---|---|---|---|---|---|
| Brush 1st coat of water-base masonry paint, 275 square feet per gallon, per 100 square feet | PT@.444 | CSF | — | 14.40 | 14.40 | 24.50 |
| Brush additional coats of water-base masonry paint, 300 square feet per gallon, per 100 square feet | PT@.364 | CSF | — | 11.80 | 11.80 | 20.10 |

**Labor for painting smooth brick masonry — roll.** Paint application only. Add the cost of surface preparation, protecting adjacent surfaces, mobilization, cleanup and callbacks. For heights above 8', use the high time difficulty factors.

| | Craft@Hrs | Unit | Material | Labor | Total | Sell |
|---|---|---|---|---|---|---|
| Roll 1st coat of water-base masonry paint, 213 square feet per gallon, per 100 square feet | PT@.286 | CSF | — | 9.26 | 9.26 | 15.70 |
| Roll additional coats of water-base masonry paint, 250 square feet per gallon, per 100 square feet | PT@.250 | CSF | — | 8.10 | 8.10 | 13.80 |
| Roll 1st coat of clear hydro sealer, 150 square feet per gallon, per 100 square feet | PT@.444 | CSF | — | 14.40 | 14.40 | 24.50 |
| Roll additional coats of clear hydro sealer, 190 square feet per gallon, per 100 square feet | PT@.400 | CSF | — | 13.00 | 13.00 | 22.10 |

**Labor for painting smooth brick masonry — spray.** Paint application only. Add the cost of surface preparation, protecting adjacent surfaces, mobilization, cleanup and callbacks. For heights above 8', use the high time difficulty factors.

| | Craft@Hrs | Unit | Material | Labor | Total | Sell |
|---|---|---|---|---|---|---|
| Spray 1st coat of water-base masonry paint, 225 square feet per gallon, per 100 square feet | PT@.133 | CSF | — | 4.31 | 4.31 | 7.33 |
| Spray additional coats of water-base masonry paint, 238 square feet per gallon, per 100 square feet | PT@.121 | CSF | — | 3.92 | 3.92 | 6.66 |
| Spray 1st coat of clear hydro sealer, 100 square feet per gallon, per 100 square feet | PT@.125 | CSF | — | 4.05 | 4.05 | 6.89 |
| Spray additional coats of clear hydro sealer, 138 square feet per gallon, per 100 square feet | PT@.111 | CSF | — | 3.59 | 3.59 | 6.10 |

**Labor for painting concrete masonry units (CMU) — brush.** Paint application only. Add the cost of surface preparation, protecting adjacent surfaces, mobilization, cleanup and callbacks. For heights above 8', use the high time difficulty factors.

| | Craft@Hrs | Unit | Material | Labor | Total | Sell |
|---|---|---|---|---|---|---|
| Brush 1st coat of water-base masonry paint, 88 square feet per gallon, per 100 square feet | PT@.769 | CSF | — | 24.90 | 24.90 | 42.30 |
| Brush additional coats of water-base masonry paint, 168 square feet per gallon, per 100 square feet | PT@.476 | CSF | — | 15.40 | 15.40 | 26.20 |

| | Craft@Hrs | Unit | Material | Labor | Total | Sell |
|---|---|---|---|---|---|---|
| Brush 1st coat of 2-part clear epoxy, 98 square feet per gallon, per 100 square feet | PT@.870 | CSF | — | 28.20 | 28.20 | 47.90 |
| Brush additional coats of 2-part clear epoxy, 188 square feet per gallon, per 100 square feet | PT@.526 | CSF | — | 17.00 | 17.00 | 28.90 |
| Brush 1st coat of clear hydro sealer, 80 square feet per gallon, per 100 square feet | PT@.667 | CSF | — | 21.60 | 21.60 | 36.70 |
| Brush additional coats of clear hydro sealer, 110 square feet per gallon, per 100 square feet | PT@.364 | CSF | — | 11.80 | 11.80 | 20.10 |

**Labor for painting concrete masonry units (CMU) — roll.** Paint application only. Add the cost of surface preparation, protecting adjacent surfaces, mobilization, cleanup and callbacks. For heights above 8', use the high time difficulty factors.

| | Craft@Hrs | Unit | Material | Labor | Total | Sell |
|---|---|---|---|---|---|---|
| Roll 1st coat of water-base masonry paint, 78 square feet per gallon, per 100 square feet | PT@.333 | CSF | — | 10.80 | 10.80 | 18.40 |
| Roll additional coats of water-base masonry paint, 143 square feet per gallon, per 100 square feet | PT@.308 | CSF | — | 9.97 | 9.97 | 16.90 |
| Roll 1st coat 2-part epoxy coating, 88 square feet per gallon, per 100 square feet | PT@.364 | CSF | — | 11.80 | 11.80 | 20.10 |
| Roll additional costs of 2-part epoxy coating, 160 square feet per gallon, per 100 square feet | PT@.333 | CSF | — | 10.80 | 10.80 | 18.40 |

**Labor for painting concrete masonry units (CMU) — spray.** Paint application only. Add the cost of surface preparation, protecting adjacent surfaces, mobilization, cleanup and callbacks. For heights above 8', use the high time difficulty factors.

| | Craft@Hrs | Unit | Material | Labor | Total | Sell |
|---|---|---|---|---|---|---|
| Spray 1st coat of water-base masonry paint, 78 square feet per gallon, per 100 square feet | PT@.143 | CSF | — | 4.63 | 4.63 | 7.87 |
| Spray additional coats of water-base masonry paint, 133 square feet per gallon, per 100 square feet | PT@.125 | CSF | — | 4.05 | 4.05 | 6.89 |
| Spray 1st coat of 2-part clear epoxy, 68 square feet per gallon, per 100 square feet | PT@.167 | CSF | — | 5.41 | 5.41 | 9.20 |
| Spray additional coats of 2-part clear epoxy, 130 square feet per gallon, per 100 square feet | PT@.143 | CSF | — | 4.63 | 4.63 | 7.87 |
| Spray 1st coat of clear hydro sealer, 50 square feet per gallon, per 100 square feet | PT@.143 | CSF | — | 4.63 | 4.63 | 7.87 |
| Spray additional coats of clear hydro sealer, 75 square feet per gallon, per 100 square feet | PT@.125 | CSF | — | 4.05 | 4.05 | 6.89 |

| | Craft@Hrs | Unit | Material | Labor | Total | Sell |
|---|---|---|---|---|---|---|

**Labor for painting molding or trim.** Paint application only. Add the cost of surface preparation, protecting adjacent surfaces, mobilization, cleanup and callbacks. For heights above 8', use the high time difficulty factors.

| | Craft@Hrs | Unit | Material | Labor | Total | Sell |
|---|---|---|---|---|---|---|
| Brush prime coat, 600 linear feet per gallon, per 100 linear feet | PT@.488 | CLF | — | 15.80 | 15.80 | 26.90 |
| Brush split coat of 1/2 undercoat and 1/2 enamel, 675 linear feet per gallon, per 100 linear feet | PT@.741 | CLF | — | 24.00 | 24.00 | 40.80 |
| Brush 1st finish coat of enamel, 675 linear feet per gallon, per 100 linear feet | PT@.625 | CLF | — | 20.20 | 20.20 | 34.30 |
| Brush 2nd or additional coats of enamel, 675 linear feet per gallon, per 100 linear feet | PT@.667 | CLF | — | 21.60 | 21.60 | 36.70 |
| Brush stipple finish enamel on smooth exterior molding, per 100 linear feet | PT@1.11 | CLF | — | 35.90 | 35.90 | 61.00 |
| Brush glazing or mottling over enamel on smooth molding, 900 linear feet per gallon, per 100 linear feet | PT@1.54 | CLF | — | 49.90 | 49.90 | 84.80 |

**Labor for painting roof overhang to 30" wide — roll and brush.** Paint application only. Add the cost of surface preparation, protecting adjacent surfaces, mobilization, cleanup and callbacks. See the overhang difficulty factors that follow.

| | Craft@Hrs | Unit | Material | Labor | Total | Sell |
|---|---|---|---|---|---|---|
| Roll and brush 1st coat of solid-body stain, 185 square feet per gallon, per 100 square feet | PT@.800 | CSF | — | 25.90 | 25.90 | 44.00 |
| Roll and brush 2nd coat of solid-body stain, 240 square feet per gallon, per 100 square feet | PT@.541 | CSF | — | 17.50 | 17.50 | 29.80 |
| Roll and brush additional coats of solid body stain, 270 square feet per gallon, per 100 square feet | PT@.444 | CSF | — | 14.40 | 14.40 | 24.50 |

**Labor for painting roof overhang to 30" wide — spray.** Paint application only. Add the cost of surface preparation, protecting adjacent surfaces, mobilization, cleanup and callbacks. See the overhang difficulty factors that follow.

| | Craft@Hrs | Unit | Material | Labor | Total | Sell |
|---|---|---|---|---|---|---|
| Spray 1st coat of semi-transparent stain, 125 square feet per gallon, per 100 square feet | PT@.286 | CSF | — | 9.26 | 9.26 | 15.70 |
| Spray 2nd coat of semi-transparent stain, 163 square feet per gallon, per 100 square feet | PT@.222 | CSF | — | 7.19 | 7.19 | 12.20 |
| Spray 3rd or additional coats of semi-transparent stain, 213 square feet per gallon, per 100 square feet | PT@.190 | CSF | — | 6.15 | 6.15 | 10.50 |

| | Craft@Hrs | Unit | Material | Labor | Total | Sell |
|---|---|---|---|---|---|---|

**Labor for painting large roof overhang over 30" wide — roll and brush.** Paint application only. Add the cost of surface preparation, protecting adjacent surfaces, mobilization, cleanup and callbacks. See the overhang difficulty factors below.

| | Craft@Hrs | Unit | Material | Labor | Total | Sell |
|---|---|---|---|---|---|---|
| Roll and brush 1st coat of solid-body stain, 210 square feet per gallon, per 100 square feet | PT@.556 | CSF | — | 18.00 | 18.00 | 30.60 |
| Roll and brush 2nd coat of solid-body stain, 275 square feet per gallon, per 100 square feet | PT@.444 | CSF | — | 14.40 | 14.40 | 24.50 |
| Roll and brush 3rd coat of solid-body stain, 280 square feet per gallon, per 100 square feet | PT@.357 | CSF | — | 11.60 | 11.60 | 19.70 |

**Labor for painting large roof overhang over 30" wide — spray.** Paint application only. Add the cost of surface preparation, protecting adjacent surfaces, mobilization, cleanup and callbacks. See the overhang difficulty factors below.

| | Craft@Hrs | Unit | Material | Labor | Total | Sell |
|---|---|---|---|---|---|---|
| Spray 1st coat of solid-body or semi-transparent stain, 90 square feet per gallon, per 100 square feet | PT@.167 | CSF | — | 5.41 | 5.41 | 9.20 |
| Spray 2nd coat of solid-body or semi-transparent stain, 163 square feet per gallon, per 100 square feet | PT@.154 | CSF | — | 4.99 | 4.99 | 8.48 |
| Spray 3rd or additional coats of solid-body or semi-transparent stain, 213 square feet per gallon, per 100 square feet | PT@.143 | CSF | — | 4.63 | 4.63 | 7.87 |

**Overhang difficulty factors for eaves and cornices.** The labor estimates above are for roof overhang with boxed eaves (no exposed rafter tails) on new single-story buildings.

Add 50% for repainting overhang or painting overhang with exposed rafter tails or second color (multiply the area by 1.5).

Add 50% for painting second-story overhang from scaffolding (multiply the area by 1.5).

Add 100% for second story with no scaffolding or exposed rafter tails (multiply the area by 2.0)

**Labor for painting exterior plaster or stucco — brush.** Paint application only. Add the cost of surface preparation, protecting adjacent surfaces, mobilization, cleanup and callbacks. For heights above 8', use the high time difficulty factors.

| | Craft@Hrs | Unit | Material | Labor | Total | Sell |
|---|---|---|---|---|---|---|
| Brush water-base masonry paint, 213 square feet per gallon, per 100 square feet | PT@.613 | CSF | — | 19.80 | 19.80 | 33.70 |
| Brush 2nd coat water-base masonry paint, 230 square feet per gallon, per 100 square feet | PT@.613 | CSF | — | 19.80 | 19.80 | 33.70 |
| Brush 3rd or additional coat water-base masonry paint, 245 square feet per gallon, per 100 square feet | PT@.578 | CSF | — | 18.70 | 18.70 | 31.80 |
| Brush 1st coat of clear hydro sealer, 163 square feet per gallon, per 100 square feet | PT@.667 | CSF | — | 21.60 | 21.60 | 36.70 |
| Brush 2nd coat of clear hydro sealer, 188 square feet per gallon, per 100 square feet | PT@.500 | CSF | — | 16.20 | 16.20 | 27.50 |

| | Craft@Hrs | Unit | Material | Labor | Total | Sell |
|---|---|---|---|---|---|---|

**Labor for painting exterior plaster or stucco — roll.** Paint application only. Add the cost of surface preparation, protecting adjacent surfaces, mobilization, cleanup and callbacks. For heights above 8', use the high time difficulty factors.

| | Craft@Hrs | Unit | Material | Labor | Total | Sell |
|---|---|---|---|---|---|---|
| Roll 1st water-base masonry prime coat, 175 square feet per gallon, per 100 square feet | PT@.366 | CSF | — | 11.90 | 11.90 | 20.20 |
| Roll 2nd water-base masonry coat, 200 square feet per gallon, per 100 square feet | PT@.313 | CSF | — | 10.10 | 10.10 | 17.20 |
| Roll 3rd or additional water-base masonry coats, 225 square feet per gallon, per 100 square feet | PT@.294 | CSF | — | 9.52 | 9.52 | 16.20 |
| Roll 1st coat of clear hydro sealer, 138 square feet per gallon, per 100 square feet | PT@.275 | CSF | — | 8.90 | 8.90 | 15.10 |
| Roll 2nd or additional coats of clear hydro sealer, 163 square feet per gallon, per 100 square feet | PT@.235 | CSF | — | 7.61 | 7.61 | 12.90 |

**Labor for painting exterior plaster or stucco — spray.** Paint application only. Add the cost of surface preparation, protecting adjacent surfaces, mobilization, cleanup and callbacks. For heights above 8', use the high time difficulty factors.

| | Craft@Hrs | Unit | Material | Labor | Total | Sell |
|---|---|---|---|---|---|---|
| Spray 1st water-base masonry prime coat, 120 square feet per gallon, per 100 square feet | PT@.148 | CSF | — | 4.79 | 4.79 | 8.14 |
| Spray 2nd water-base masonry coat, 150 square feet per gallon, per 100 square feet | PT@.125 | CSF | — | 4.05 | 4.05 | 6.89 |
| Spray 3rd or additional water-base masonry coat, 165 square feet per gallon, per 100 square feet | PT@.118 | CSF | — | 3.82 | 3.82 | 6.49 |
| Spray 1st coat of clear hydro sealer, 113 square feet per gallon, per 100 square feet | PT@.143 | CSF | — | 4.63 | 4.63 | 7.87 |
| Spray 2nd or additional coats of clear hydro sealer, 138 square feet per gallon, per 100 square feet | PT@.121 | CSF | — | 3.92 | 3.92 | 6.66 |

**Labor for painting decorative wood handrail — brush.** Paint application only. Add the cost of surface preparation, protecting adjacent surfaces, mobilization, cleanup and callbacks.

| | Craft@Hrs | Unit | Material | Labor | Total | Sell |
|---|---|---|---|---|---|---|
| Brush undercoat or enamel, 110 linear feet per gallon, per 100 linear feet | PT@2.86 | CLF | — | 92.60 | 92.60 | 157.00 |

**Labor for painting 2' x 4' shutters or blinds — brush.** Paint application only. Add the cost of surface preparation, protecting adjacent surfaces, mobilization, cleanup and callbacks. For heights above 8', use the high time difficulty factors.

| | Craft@Hrs | Unit | Material | Labor | Total | Sell |
|---|---|---|---|---|---|---|
| Brush undercoat, 11 shutters per gallon, per shutter | PT@.333 | Ea | — | 10.80 | 10.80 | 18.40 |
| Brush split coat (1/2 undercoat + 1/2 enamel), 14 shutters per gallon, per shutter | PT@.286 | Ea | — | 9.26 | 9.26 | 15.70 |
| Brush exterior enamel, 14 shutters per gallon, per shutter | PT@.400 | Ea | — | 13.00 | 13.00 | 22.10 |

| | Craft@Hrs | Unit | Material | Labor | Total | Sell |
|---|---|---|---|---|---|---|

**Labor for painting 2' x 4' shutters or blinds — spray.** Paint application only. Add the cost of surface preparation, protecting adjacent surfaces, mobilization, cleanup and callbacks. For heights above 8', use the high time difficulty factors.

| | Craft@Hrs | Unit | Material | Labor | Total | Sell |
|---|---|---|---|---|---|---|
| Spray undercoat, 9 shutters per gallon, per shutter | PT@.111 | Ea | — | 3.59 | 3.59 | 6.10 |
| Spray split coat (1/2 undercoat + 1/2 enamel), 11 shutters per gallon, per shutter | PT@.100 | Ea | — | 3.24 | 3.24 | 5.51 |
| Spray exterior enamel, 11 shutters per gallon, per shutter | PT@.125 | Ea | — | 4.05 | 4.05 | 6.89 |

**Labor for painting aluminum siding — brush.** Paint application only. Add the cost of surface preparation, protecting adjacent surfaces, mobilization, cleanup and callbacks. For heights above 8', use the high time difficulty factors.

| | Craft@Hrs | Unit | Material | Labor | Total | Sell |
|---|---|---|---|---|---|---|
| Brush metal primer, 420 square feet per gallon, per 100 square feet | PT@.426 | CSF | — | 13.80 | 13.80 | 23.50 |
| Brush 1st or additional finish coats on aluminum siding, 465 square feet per gallon, per 100 square feet | PT@.351 | CSF | — | 11.40 | 11.40 | 19.40 |

**Labor for staining rough wood siding — roll and brush.** Stain application only. Add the cost of surface preparation, protecting adjacent surfaces, mobilization, cleanup and callbacks. For heights above 8', use the high time difficulty factors.

| | Craft@Hrs | Unit | Material | Labor | Total | Sell |
|---|---|---|---|---|---|---|
| Roll and brush 1st coat of water-base stain, 213 square feet per gallon, per 100 square feet | PT@.500 | CSF | — | 16.20 | 16.20 | 27.50 |
| Roll and brush 2nd coat of water-base stain, 275 square feet per gallon, per 100 square feet | PT@.400 | CSF | — | 13.00 | 13.00 | 22.10 |
| Roll and brush 3rd or additional coats of water-base stain, 325 square feet per gallon, per 100 square feet | PT@.347 | CSF | — | 11.20 | 11.20 | 19.00 |
| Roll and brush 1st coat of penetrating oil stain, 125 square feet per gallon, per 100 square feet | PT@.444 | CSF | — | 14.40 | 14.40 | 24.50 |
| Roll and brush 2nd coat of penetrating oil stain, 240 square feet per gallon, per 100 square feet | PT@.364 | CSF | — | 11.80 | 11.80 | 20.10 |
| Roll and brush 3rd or additional coats of penetrating oil stain, 330 square feet per gallon, per 100 square feet | PT@.299 | CSF | — | 9.68 | 9.68 | 16.50 |

**Labor for staining rough wood siding — spray.** Stain application only. Add the cost of surface preparation, protecting adjacent surfaces, mobilization, cleanup and callbacks. For heights above 8', use the high time difficulty factors.

| | Craft@Hrs | Unit | Material | Labor | Total | Sell |
|---|---|---|---|---|---|---|
| Spray 1st coat of water-base stain, 115 square feet per gallon, per 100 square feet | PT@.200 | CSF | — | 6.48 | 6.48 | 11.00 |
| Spray 2nd coat of water-base stain, 125 square feet per gallon, per 100 square feet | PT@.167 | CSF | — | 5.41 | 5.41 | 9.20 |

| | Craft@Hrs | Unit | Material | Labor | Total | Sell |
|---|---|---|---|---|---|---|
| Spray 3rd or additional coats of water-base stain, 170 square feet per gallon, per 100 square feet | PT@.154 | CSF | — | 4.99 | 4.99 | 8.48 |
| Spray 1st coat penetrating oil stain, 180 square feet per gallon, per 100 square feet | PT@.200 | CSF | — | 6.48 | 6.48 | 11.00 |
| Spray 2nd coat of penetrating oil stain, 245 square feet per gallon, per 100 square feet | PT@.182 | CSF | — | 5.89 | 5.89 | 10.00 |
| Spray 3rd or additional coats of penetrating oil stain, 360 square feet per gallon, per 100 square feet | PT@.154 | CSF | — | 4.99 | 4.99 | 8.48 |
| Spray 1st coat of clear hydro sealer, 113 square feet per gallon, per 100 square feet | PT@.174 | CSF | — | 5.63 | 5.63 | 9.57 |
| Spray 2nd coat of clear hydro sealer, 150 square feet per gallon, per 100 square feet | PT@.148 | CSF | — | 4.79 | 4.79 | 8.14 |
| Spray 3rd or additional coats of clear hydro sealer, 175 square feet per gallon, per 100 square feet | PT@.133 | CSF | — | 4.31 | 4.31 | 7.33 |

**Labor for painting gypsum drywall walls — roll and brush.** Paint application only. Add the cost of surface preparation, protecting adjacent surfaces, mobilization, cleanup and callbacks. For heights above 8', use the high time difficulty factors.

| | Craft@Hrs | Unit | Material | Labor | Total | Sell |
|---|---|---|---|---|---|---|
| Roll and brush 1st coat of water-base flat latex, 275 square feet per gallon, per 100 square feet | PT@.186 | CSF | — | 6.02 | 6.02 | 10.20 |
| Roll and brush 2nd coat of water-base flat latex, 313 square feet per gallon, per 100 square feet | PT@.167 | CSF | — | 5.41 | 5.41 | 9.20 |
| Roll and brush additional coats of water-base flat latex, 338 square feet per gallon, per 100 square feet | PT@.154 | CSF | — | 4.99 | 4.99 | 8.48 |
| Roll and brush prime coat of water-base sealer, 263 square feet per gallon, per 100 square feet | PT@.200 | CSF | — | 6.48 | 6.48 | 11.00 |
| Roll and brush 1st finish coat of water-base enamel, 263 square feet per gallon, per 100 square feet | PT@.222 | CSF | — | 7.19 | 7.19 | 12.20 |
| Roll and brush 2nd finish coat of water-base enamel, 288 square feet per gallon, per 100 square feet | PT@.211 | CSF | — | 6.83 | 6.83 | 11.60 |

**Labor for painting gypsum drywall walls — spray.** Paint application only. Add the cost of surface preparation, protecting adjacent surfaces, mobilization, cleanup and callbacks. For heights above 8', use the high time difficulty factors.

| | Craft@Hrs | Unit | Material | Labor | Total | Sell |
|---|---|---|---|---|---|---|
| Spray 1st coat of water-base flat latex, 225 square feet per gallon, per 100 square feet | PT@.125 | CSF | — | 4.05 | 4.05 | 6.89 |
| Spray 2nd coat of water-base flat latex, 275 square feet per gallon, per 100 square feet | PT@.111 | CSF | — | 3.59 | 3.59 | 6.10 |

| | Craft@Hrs | Unit | Material | Labor | Total | Sell |
|---|---|---|---|---|---|---|
| Spray additional coats of water-base flat latex, 300 square feet per gallon, per 100 square feet | PT@.105 | CSF | — | 3.40 | 3.40 | 5.78 |
| Spray prime coat of water-base sealer, 225 square feet per gallon, per 100 square feet | PT@.136 | CSF | — | 4.40 | 4.40 | 7.48 |
| Spray 1st finish coat of water-base enamel, 238 square feet per gallon, per 100 square feet | PT@.148 | CSF | — | 4.79 | 4.79 | 8.14 |
| Spray 2nd finish coat of water-base enamel, 263 square feet per gallon, per 100 square feet | PT@.143 | CSF | — | 4.63 | 4.63 | 7.87 |
| Spray additional finish coats of water-base enamel, 275 square feet per gallon, per 100 square feet | PT@.129 | CSF | — | 4.18 | 4.18 | 7.11 |

**Labor for painting gypsum drywall ceiling — roll and brush.** Paint application only. Add the cost of surface preparation, protecting adjacent surfaces, mobilization, cleanup and callbacks. For heights above 8', use the high time difficulty factors.

| | Craft@Hrs | Unit | Material | Labor | Total | Sell |
|---|---|---|---|---|---|---|
| Roll and brush 1st coat of flat latex, 275 square feet per gallon, per 100 square feet | PT@.286 | CSF | — | 9.26 | 9.26 | 15.70 |
| Roll and brush 2nd coat of flat latex, 313 square feet per gallon, per 100 square feet | PT@.267 | CSF | — | 8.65 | 8.65 | 14.70 |
| Roll and brush 3rd coat of flat latex, 338 square feet per gallon, per 100 square feet | PT@.235 | CSF | — | 7.61 | 7.61 | 12.90 |
| Roll and brush prime coat of water-base sealer, 275 square feet per gallon, per 100 square feet | PT@.267 | CSF | — | 8.65 | 8.65 | 14.70 |
| Roll and brush 1st finish coat of water-base enamel, 313 square feet per gallon, per 100 square feet | PT@.286 | CSF | — | 9.26 | 9.26 | 15.70 |
| Roll and brush additional coats of water-base enamel, 338 square feet per gallon, per 100 square feet | PT@.250 | CSF | — | 8:10 | 8.10 | 13.80 |
| Roll and brush 1st coat of white epoxy coating, 288 square feet per gallon, per 100 square feet | PT@.308 | CSF | — | 9.97 | 9.97 | 16.90 |
| Roll and brush additional coats of white epoxy coating, 288 square feet per gallon, per 100 square feet | PT@.267 | CSF | — | 8.65 | 8.65 | 14.70 |

**Labor for painting gypsum drywall ceiling — spray.** Paint application only. Add the cost of surface preparation, protecting adjacent surfaces, mobilization, cleanup and callbacks. For heights above 8', use the high time difficulty factors.

| | Craft@Hrs | Unit | Material | Labor | Total | Sell |
|---|---|---|---|---|---|---|
| Spray 1st coat of flat latex, 200 square feet per gallon, per 100 square feet | PT@.133 | CSF | — | 4.31 | 4.31 | 7.33 |
| Spray 2nd coat of flat latex, 225 square feet per gallon, per 100 square feet | PT@.114 | CSF | — | 3.69 | 3.69 | 6.27 |

| | Craft@Hrs | Unit | Material | Labor | Total | Sell |
|---|---|---|---|---|---|---|
| Spray additional coats of flat latex, 250 square feet per gallon, per 100 square feet | PT@.108 | CSF | — | 3.50 | 3.50 | 5.95 |
| Spray prime coat of water-base sealer, 200 square feet per gallon, per 100 square feet | PT@.125 | CSF | — | 4.05 | 4.05 | 6.89 |
| Spray 1st finish coat of water-base enamel, 225 square feet per gallon, per 100 square feet | PT@.121 | CSF | — | 3.92 | 3.92 | 6.66 |
| Spray additional coats of water-base enamel, 250 square feet per gallon, per 100 square feet | PT@.114 | CSF | — | 3.69 | 3.69 | 6.27 |

**Labor for painting wood paneled walls — roll and brush.** Paint application only. Add the cost of surface preparation, protecting adjacent surfaces, mobilization, cleanup and callbacks. For heights above 8', use the high time difficulty factors.

| | Craft@Hrs | Unit | Material | Labor | Total | Sell |
|---|---|---|---|---|---|---|
| Roll and brush water-base undercoat, 263 square feet per gallon, per 100 square feet | PT@.333 | CSF | — | 10.80 | 10.80 | 18.40 |
| Roll and brush water-base split coat (1/2 undercoat + 1/2 enamel), 263 square feet per gallon, per 100 square feet | PT@.364 | CSF | — | 11.80 | 11.80 | 20.10 |
| Roll and brush 1st finish coat of water-base enamel, 313 square feet per gallon, per 100 square feet | PT@.235 | CSF | — | 7.61 | 7.61 | 12.90 |

**Labor for painting wood paneled walls — spray.** Paint application only. Add the cost of surface preparation, protecting adjacent surfaces, mobilization, cleanup and callbacks. For heights above 8', use the high time difficulty factors.

| | Craft@Hrs | Unit | Material | Labor | Total | Sell |
|---|---|---|---|---|---|---|
| Spray water-base undercoat, 150 square feet per gallon, per 100 square feet | PT@.235 | CSF | — | 7.61 | 7.61 | 12.90 |
| Spray water-base split coat (1/2 undercoat + 1/2 enamel), 150 square feet per gallon, per 100 square feet | PT@.250 | CSF | — | 8.10 | 8.10 | 13.80 |
| Spray 1st finish coat of water-base enamel, 225 square feet per gallon, per 100 square feet | PT@.182 | CSF | — | 5.89 | 5.89 | 10.00 |
| Spray additional finish coats of water-base enamel, 325 square feet per gallon, per 100 square feet | PT@.154 | CSF | — | 4.99 | 4.99 | 8.48 |

**Labor for staining wood paneled walls.** Stain application only. Add the cost of surface preparation, protecting adjacent surfaces, mobilization, cleanup and callbacks. For heights above 8', use the high time difficulty factors.

| | Craft@Hrs | Unit | Material | Labor | Total | Sell |
|---|---|---|---|---|---|---|
| Roll and brush each coat of solid or semi-transparent stain, 450 square feet per gallon, per 100 square feet | PT@.380 | CSF | — | 12.30 | 12.30 | 20.90 |
| Spray solid or semi-transparent stain, 250 square feet per gallon, per 100 square feet | PT@.250 | CSF | — | 8.10 | 8.10 | 13.80 |

| | Craft@Hrs | Unit | Material | Labor | Total | Sell |
|---|---|---|---|---|---|---|

**Labor for painting windows to 15 square feet, 1, 2 or 3 panes, per side painted.** Paint application only. Add preparation time for sanding, puttying, protecting adjacent surfaces and masking or waxing window panes. For heights above 8', use the high time difficulty factors.

Brush undercoat,

| | | | | | | |
|---|---|---|---|---|---|---|
| 14.5 windows per gallon, per window | PT@.200 | Ea | — | 6.48 | 6.48 | 11.00 |

Brush split coat (1/2 undercoat + 1/2 enamel),

| | | | | | | |
|---|---|---|---|---|---|---|
| 16.5 windows per gallon, per window | PT@.250 | Ea | — | 8.10 | 8.10 | 13.80 |

Brush finish coat of enamel,

| | | | | | | |
|---|---|---|---|---|---|---|
| 15.5 windows per gallon, per window | PT@.300 | Ea | — | 9.71 | 9.71 | 16.50 |

**Labor for painting windows to 15 square feet, 4, 5 or 6 panes, per side painted.** Paint application only. Add preparation time for sanding, puttying, protecting adjacent surfaces and masking or waxing window panes. For heights above 8', use the high time difficulty factors.

Brush undercoat,

| | | | | | | |
|---|---|---|---|---|---|---|
| 13.5 windows per gallon, per window | PT@.300 | Ea | — | 9.71 | 9.71 | 16.50 |

Brush split coat (1/2 undercoat + 1/2 enamel),

| | | | | | | |
|---|---|---|---|---|---|---|
| 15.5 windows per gallon, per window | PT@.350 | Ea | — | 11.30 | 11.30 | 19.20 |

Brush finish coat of enamel,

| | | | | | | |
|---|---|---|---|---|---|---|
| 14.5 windows per gallon, per window | PT@.450 | Ea | — | 14.60 | 14.60 | 24.80 |

**Labor for painting windows to 15 square feet, 7 or 8 panes, per side painted.** Paint application only. Add preparation time for sanding, puttying, protecting adjacent surfaces and masking or waxing window panes. For heights above 8', use the high time difficulty factors.

Brush undercoat,

| | | | | | | |
|---|---|---|---|---|---|---|
| 13 windows per gallon, per window | PT@.400 | Ea | — | 13.00 | 13.00 | 22.10 |

Brush split coat (1/2 undercoat + 1/2 enamel),

| | | | | | | |
|---|---|---|---|---|---|---|
| 15 windows per gallon, per window | PT@.500 | Ea | — | 16.20 | 16.20 | 27.50 |

Brush finish coat of enamel,

| | | | | | | |
|---|---|---|---|---|---|---|
| 14 windows per gallon, per window | PT@.620 | Ea | — | 20.10 | 20.10 | 34.20 |

**Labor for painting windows to 15 square feet, 9, 10 or 11 panes, per side painted.** Paint application only. Add preparation time for sanding, puttying, protecting adjacent surfaces and masking or waxing window panes. For heights above 8', use the high time difficulty factors.

Brush undercoat,

| | | | | | | |
|---|---|---|---|---|---|---|
| 12 windows per gallon, per window | PT@.500 | Ea | — | 16.20 | 16.20 | 27.50 |

Brush split coat (1/2 undercoat + 1/2 enamel),

| | | | | | | |
|---|---|---|---|---|---|---|
| 14 windows per gallon, per window | PT@.620 | Ea | — | 20.10 | 20.10 | 34.20 |

Brush finish coat of enamel,

| | | | | | | |
|---|---|---|---|---|---|---|
| 13 windows per gallon, per window | PT@.730 | Ea | — | 23.60 | 23.60 | 40.10 |

**Labor for painting windows to 15 square feet, 12 panes, per side painted.** Paint application only. Add preparation time for sanding, puttying, protecting adjacent surfaces and masking or waxing window panes. For heights above 8', use the high time difficulty factors.

Brush undercoat,

| | | | | | | |
|---|---|---|---|---|---|---|
| 11 windows per gallon, per window | PT@.620 | Ea | — | 20.10 | 20.10 | 34.20 |

Brush split coat (1/2 undercoat + 1/2 enamel),

| | | | | | | |
|---|---|---|---|---|---|---|
| 13 windows per gallon, per window | PT@.670 | Ea | — | 21.70 | 21.70 | 36.90 |

Brush finish coat of enamel,

| | | | | | | |
|---|---|---|---|---|---|---|
| 12 windows per gallon, per window | PT@.800 | Ea | — | 25.90 | 25.90 | 44.00 |

| | Craft@Hrs | Unit | Material | Labor | Total | Sell |
|---|---|---|---|---|---|---|

**Labor for painting windows, per square foot of window, per side painted.** Use figures in the following table to estimate labor for windows larger than 15 square feet (length times width). To calculate the window area, add 1 foot to each dimension (top, bottom, left and right side). Then multiply the width times length. For example, a window measuring 4' x 4' with 1 foot added to the top, bottom, right side and left side would be 6' x 6' or 36 square feet. Then, add 2 square feet for each window pane to allow time for painting the mullions, muntins and sash. The calculation for a 4' x 4' double-hung window with two panes would be 6' x 6' + 4', or 40 square feet. Add preparation time for sanding, puttying, protecting adjacent surfaces and masking or waxing window panes. For heights above 8', use the high time difficulty factors.

| | Craft@Hrs | Unit | Material | Labor | Total | Sell |
|---|---|---|---|---|---|---|
| Brush undercoat, 450 square feet per gallon, per 100 square feet | PT@.606 | CSF | — | 19.60 | 19.60 | 33.30 |
| Brush split coat (1/2 undercoat + 1/2 enamel), 500 square feet per gallon, per 100 square feet | PT@.741 | CSF | — | 24.00 | 24.00 | 40.80 |
| Brush finish coat of enamel, 480 square feet per gallon, per 100 square feet | PT@.885 | CSF | — | 28.70 | 28.70 | 48.80 |

## Exterior Solid Stain

**Deck Plus® solid color deck and siding stain, Behr.** 100% acrylic latex. Opaque. For decks (including pressure-treated), siding, fences and patio furniture. Water-repellent and mildew-resistant. Water cleanup. Resists scuffing, cracking and peeling. Covers 400 square feet per gallon. Dries to touch in 2 to 3 hours; to recoat in 24 hours. Full cure in 30 days.

| | Craft@Hrs | Unit | Material | Labor | Total | Sell |
|---|---|---|---|---|---|---|
| Gallon | — | Ea | 23.70 | — | 23.70 | — |
| 5 gallons | — | Ea | 107.00 | — | 107.00 | — |

**Plus 10® solid color exterior stain, Behr.** Oil and latex formula. Water cleanup. Protects against fading, weathering and chalking for up to 10 years on vertical surfaces. Prevents cracking, peeling and blistering. Opaque finish covers wood grain and allows wood texture to remain. For exterior shakes, shingles, wood siding, fencing, trim, stucco and masonry surfaces.

| | Craft@Hrs | Unit | Material | Labor | Total | Sell |
|---|---|---|---|---|---|---|
| Gallon | — | Ea | 20.40 | — | 20.40 | — |
| 5 gallons | — | Ea | 90.20 | — | 90.20 | — |

**Acrylic exterior and interior stain, Woodpride.**

| | Craft@Hrs | Unit | Material | Labor | Total | Sell |
|---|---|---|---|---|---|---|
| Gallon | — | Ea | 18.30 | — | 18.30 | — |
| 5 gallons | — | Ea | 79.90 | — | 79.90 | — |

**Preserva-Wood penetrating finish, Preservaproducts.** Will not crack or peel. Tinted natural finishes to enhance the natural beauty and color of various types of wood. Ultraviolet protection, environmentally responsible. Clear or tinted.

| | Craft@Hrs | Unit | Material | Labor | Total | Sell |
|---|---|---|---|---|---|---|
| Quart | — | Ea | 10.80 | — | 10.80 | — |
| Gallon | — | Ea | 24.80 | — | 24.80 | — |
| 5 gallons | — | Ea | 105.00 | — | 105.00 | — |

**Wood bleach kit, Klean-Strip.** 2-part solution. Lightens stripped or unfinished wood. Prepares bare wood for staining. Pint covers 15 to 20 square feet.

| | Craft@Hrs | Unit | Material | Labor | Total | Sell |
|---|---|---|---|---|---|---|
| Pint | — | Ea | 9.35 | — | 9.35 | — |

## Interior Oil Wiping Stain

**Wood finish wiping stain, Minwax®.** Use on interior bare or stripped wood surface. Penetrates into wood fibers to highlight grain. Available in 20 wood-tone colors.

| | Craft@Hrs | Unit | Material | Labor | Total | Sell |
|---|---|---|---|---|---|---|
| Quart | — | Ea | 6.33 | — | 6.33 | — |
| Gallon | — | Ea | 21.60 | — | 21.60 | — |

| | Craft@Hrs | Unit | Material | Labor | Total | Sell |
|---|---|---|---|---|---|---|

**Watco™ Danish Oil™, Rust-Oleum®.** Natural, hand-rubbed appearance. Penetrating stain soaks into pores of wood. Won't crack, chip or peel. Walnut, fruitwood or natural tint..

| | Craft@Hrs | Unit | Material | Labor | Total | Sell |
|---|---|---|---|---|---|---|
| Quart | — | Ea | 10.70 | — | 10.70 | — |
| Gallon | — | Ea | 28.50 | — | 28.50 | — |

## Waterborne Stain

**Water-based pickling stain, Minwax®.** Leaves a unique, subtle shade of white while allowing the natural wood grain to show through.

| | Craft@Hrs | Unit | Material | Labor | Total | Sell |
|---|---|---|---|---|---|---|
| Quart | — | Ea | 9.60 | — | 9.60 | — |

**Water-based wood stain, Minwax®.** Provides rich, even stain penetration. Fast drying and easily cleans up with soap and water.

| | Craft@Hrs | Unit | Material | Labor | Total | Sell |
|---|---|---|---|---|---|---|
| Half pint | — | Ea | 4.84 | — | 4.84 | — |
| Quart | — | Ea | 9.58 | — | 9.58 | — |

## Gel Stain

**Gel stain, Minwax®.** Oil-based, gelled stain for finishing vertical surfaces and non-wood surfaces like fiberglass, metal, veneer and fiberboard. Non-drip formula. Suitable for fine woods. Controlled penetration for uniform color.

| | Craft@Hrs | Unit | Material | Labor | Total | Sell |
|---|---|---|---|---|---|---|
| Quart | — | Ea | 12.80 | — | 12.80 | — |

**Polyurethane stain, Minwax®.** Stain color and tough polyurethane protection in one easy step. Available in nine colors.

| | Craft@Hrs | Unit | Material | Labor | Total | Sell |
|---|---|---|---|---|---|---|
| Half pint | — | Ea | 5.37 | — | 5.37 | — |
| Quart | — | Ea | 10.60 | — | 10.60 | — |

## Exterior Clear Wood Coating

**Marine varnish, Valspar.** For outdoor furniture, doors, entryways, finish, and trim. Resists water, salt, and sun. Low volatile organic compounds (VOCs).

| | Craft@Hrs | Unit | Material | Labor | Total | Sell |
|---|---|---|---|---|---|---|
| Gloss, quart | — | Ea | 15.10 | — | 15.10 | — |
| Gloss, gallon | — | Ea | 43.20 | — | 43.20 | — |
| Satin, quart | — | Ea | 15.10 | — | 15.10 | — |
| Satin, gallon | — | Ea | 43.20 | — | 43.20 | — |

**Man-O-War marine spar varnish, Valspar.** Protection against weathering and salt water. Resists cracking and peeling. Better than polyurethane. Expands and contracts with weather conditions. For fences and outdoor furniture. Gloss or satin finish.

| | Craft@Hrs | Unit | Material | Labor | Total | Sell |
|---|---|---|---|---|---|---|
| Gloss, quart | — | Ea | 13.90 | — | 13.90 | — |
| Gloss, gallon | — | Ea | 38.70 | — | 38.70 | — |

**Helmsman® spar urethane, Minwax.** Clear finish protects wood exposed to sunlight, water or temperature changes. Ultraviolet absorbers help protect wood from graying and fading in the sun. Contains oils that help the finish expand and contract with wood as the seasons and temperature change. 24-hour drying time. 400 square feet per gallon. 2-3 coat application. Clean up with mineral spirits. High gloss, semi-gloss or satin finish.

| | Craft@Hrs | Unit | Material | Labor | Total | Sell |
|---|---|---|---|---|---|---|
| Quart | — | Ea | 12.90 | — | 12.90 | — |
| Gallon | — | Ea | 32.40 | — | 32.40 | — |

**Clearshield™ weather-resistant wood coating, Minwax®.** Clear protective topcoat with ultraviolet absorbers to help protect wood from weather damage. Satin or semi-gloss.

| | Craft@Hrs | Unit | Material | Labor | Total | Sell |
|---|---|---|---|---|---|---|
| Quart | — | Ea | 14.00 | — | 14.00 | — |

| | Craft@Hrs | Unit | Material | Labor | Total | Sell |
|---|---|---|---|---|---|---|

**Varathane® exterior classic clear finish, water-based, Rust-Oleum®.** For outdoor patio furniture, garage doors and front doors. Excellent scratch and impact resistance. Transparent, non-yellowing, and low odor. Fast drying and easy water clean up. Gloss, semi-gloss or satin finish.

| | Craft@Hrs | Unit | Material | Labor | Total | Sell |
|---|---|---|---|---|---|---|
| 12-ounce aerosol | — | Ea | 7.35 | — | 7.35 | — |
| Quart | — | Ea | 17.20 | — | 17.20 | — |
| Gallon | — | Ea | 48.70 | — | 48.70 | — |

**Varathane® exterior classic clear finish, oil-based, Rust-Oleum®.** For outdoor furniture, garage doors and front doors. Excellent scratch and impact resistance. Expands and contracts with weather conditions. Fast drying and self-leveling. Gloss or satin finish.

| | Craft@Hrs | Unit | Material | Labor | Total | Sell |
|---|---|---|---|---|---|---|
| Quart | — | Ea | 13.90 | — | 13.90 | — |
| Gallon | — | Ea | 40.70 | — | 40.70 | — |

**Log home gloss finish.** For log and timber frame homes, siding, fencing and railings. Contains mildewcide and offers UV protection. Water clean-up. Dries to the touch in 8 to 12 hours. Recoat after 24 hours. 3 to 4 coats provide maximum protection. Clear or cedar finish.

| | Craft@Hrs | Unit | Material | Labor | Total | Sell |
|---|---|---|---|---|---|---|
| Gallon | — | Ea | 32.40 | — | 32.40 | — |
| 5 gallons | — | Ea | 145.00 | — | 145.00 | — |

## Oil-based Polyurethane Interior Clear Finish

**Fast-drying clear polyurethane, Minwax®.** A clear, oil-based, durable protective finish for wood. Provides long-lasting protection and beauty to interior wood surfaces such as furniture, doors, cabinets and floors. Gloss, semi-gloss or satin finish.

| | Craft@Hrs | Unit | Material | Labor | Total | Sell |
|---|---|---|---|---|---|---|
| Half pint | — | Ea | 4.85 | — | 4.85 | — |
| Quart | — | Ea | 9.02 | — | 9.02 | — |
| Gallon | — | Ea | 26.80 | — | 26.80 | — |
| 2-1/2 gallons | — | Ea | 64.80 | — | 64.80 | — |

**Varathane® oil-based floor finish, Rust-Oleum®.** High impact resistance. Self-leveling to eliminate brush marks. Scratch and mar resistance. High solids content. Gloss, semi-gloss or satin.

| | Craft@Hrs | Unit | Material | Labor | Total | Sell |
|---|---|---|---|---|---|---|
| Gallon | — | Ea | 40.60 | — | 40.60 | — |

**Varathane® interior classic clear finish, oil-based, Rust-Oleum®.** Extra hard finish resists scuffing, chipping, and abrasion. Self-leveling. Outlasts varnish. Gloss, semi-gloss or satin.

| | Craft@Hrs | Unit | Material | Labor | Total | Sell |
|---|---|---|---|---|---|---|
| 12-ounce aerosol | — | Ea | 6.87 | — | 6.87 | — |
| Quart | — | Ea | 12.10 | — | 12.10 | — |
| Gallon | — | Ea | 38.00 | — | 38.00 | — |

## Waterborne Clear Finish

**Polycrylic® protective finish, Minwax®.** A hard, clear, ultra fast-drying protective finish. Non-flammable and very little odor. Cleans up with water. For use over light woods, pastel stained woods, painted surfaces, both latex- and oil-based and well-bonded wall coverings. Available in clear gloss, clear semi-gloss and clear satin. Covers 125 square feet per quart. Dries clear in 2 hours.

| | Craft@Hrs | Unit | Material | Labor | Total | Sell |
|---|---|---|---|---|---|---|
| 12-ounce aerosol | — | Ea | 9.68 | — | 9.68 | — |
| Half pint | — | Ea | 6.44 | — | 6.44 | — |
| Quart | — | Ea | 14.10 | — | 14.10 | — |
| Gallon | — | Ea | 43.20 | — | 43.20 | — |

**Parks PRO Finisher waterborne sanding sealer, Parks Corporation.** A fast-drying water-based sanding sealer to be used on unfinished wood before applying a waterborne top coat. Thoroughly sand sealer to retard grain raise.

| | Craft@Hrs | Unit | Material | Labor | Total | Sell |
|---|---|---|---|---|---|---|
| Gallon | — | Ea | 28.00 | — | 28.00 | — |

| | Craft@Hrs | Unit | Material | Labor | Total | Sell |
|---|---|---|---|---|---|---|

## Shellac

**Bulls Eye shellac, Zinsser.** Protects cabinetry, trim work, and furniture. Seals in stain to protect from water leaks, grease marks, smoke, and soot. Adheres to glossy surfaces without sanding. Seals knots and tannins. May be tinted to produce fast-drying deep color primer. Interior use only. Clear or amber color.

| | | | | | | |
|---|---|---|---|---|---|---|
| Amber, quart | — | Ea | 10.80 | — | 10.80 | — |
| Amber, gallon | — | Ea | 32.40 | — | 32.40 | — |
| Clear, quart | — | Ea | 10.80 | — | 10.80 | — |
| Clear, gallon | — | Ea | 32.30 | — | 32.30 | — |

## Water Seal for Wood

**WaterSeal® clear wood protector, Thompson's.** Enhance and maintain wood's natural color. Protects against damage from water and sun and provides a mildew-resistant coating. Oil-enriched formula provides long-lasting protection for all types of wood.

| | | | | | | |
|---|---|---|---|---|---|---|
| Gallon | — | Ea | 14.80 | — | 14.80 | — |
| 5 gallons | — | Ea | 64.60 | — | 64.60 | — |

**WaterSeal® clear multi-surface waterproofer, Thompson's.** For wood, brick, concrete, stucco and masonry. Can be applied to ACQ pressure-treated wood. Protects against warping, cracking, swelling, and water damage. Paintable. One-coat coverage. Withstands abrasion. Allows wood to weather naturally. Prevents moisture from penetrating by setting up barrier beneath. Goes on milky white, dries clear.

| | | | | | | |
|---|---|---|---|---|---|---|
| 12-ounce aerosol | — | Ea | 5.36 | — | 5.36 | — |
| Gallon | — | Ea | 9.56 | — | 9.56 | — |
| 5 gallons | — | Ea | 43.00 | — | 43.00 | — |

**Waterproofing wood protector, Behr.** For decks, fences, siding and furniture. Lasts up to one year on decks and up to two years on fences and siding.

| | | | | | | |
|---|---|---|---|---|---|---|
| Gallon | — | Ea | 14.00 | — | 14.00 | — |
| 5 gallons | — | Ea | 62.60 | — | 62.60 | — |

**Advanced wood protector, Sherwin-Williams.** Clear or tinted.

| | | | | | | |
|---|---|---|---|---|---|---|
| Gallon | — | Ea | 21.50 | — | 21.50 | — |
| 5 gallons | — | Ea | 94.40 | — | 94.40 | — |

## Wood Preservative

**Dock and fence post preservative, No. 90, Behr.** Exterior, green, water-repellent. For marine or below-ground use. Deep-penetrating formula. For wood in contact with soil and water. Protects against wood rot, warping, swelling, termite damage, mildew and moisture. Water clean up.

| | | | | | | |
|---|---|---|---|---|---|---|
| Gallon | — | Ea | 19.50 | — | 19.50 | — |

**Seasonite® new wood treatment, The Flood Company.** Protects pressure-treated and untreated wood against moisture. Provides a gradual seasoning process to reduce warping, cupping, splitting, and checking. Readies surface for painting, staining, or finishing. Ideal for new deck and patio lumber. Mildew resistant. Soap and water clean up.

| | | | | | | |
|---|---|---|---|---|---|---|
| Gallon | — | Ea | 13.90 | — | 13.90 | — |

**Termin-8 wood preservative, green, Jasco.** Oil-based. Preserves and protects wood from carpenter ants, termites, and powder post beetles. Protects against rot, warping, mildew, swelling, fungus, and moss. Slow drying for deep absorption. Exterior or marine use. Gallon covers 100 to 300 square feet. Paintable after 2 to 7 days.

| | | | | | | |
|---|---|---|---|---|---|---|
| Quart | — | Ea | 8.16 | — | 8.16 | — |
| Gallon | — | Ea | 16.90 | — | 16.90 | — |
| 5 gallons | — | Ea | 69.10 | — | 69.10 | — |

|  | Craft@Hrs | Unit | Material | Labor | Total | Sell |
|---|---|---|---|---|---|---|
| **Copper wood preservative, Jasco.** |  |  |  |  |  |  |
| Brown, quart | — | Ea | 7.61 | — | 7.61 | — |
| Brown, gallon | — | Ea | 16.30 | — | 16.30 | — |
| Clear, quart | — | Ea | 7.60 | — | 7.60 | — |
| Clear, gallon | — | Ea | 17.50 | — | 17.50 | — |

**ZPW clear wood preservative, Jasco.** Kills termites, carpenter ants, protects decks and fences. Use inside or outside house above or below ground. Paintable and tintable. Controls warping, swelling and cracking.

|  | Craft@Hrs | Unit | Material | Labor | Total | Sell |
|---|---|---|---|---|---|---|
| Quart | — | Ea | 7.67 | — | 7.67 | — |
| Gallon | — | Ea | 16.30 | — | 16.30 | — |

## Wood Conditioners

|  | Craft@Hrs | Unit | Material | Labor | Total | Sell |
|---|---|---|---|---|---|---|
| **Tung oil.** |  |  |  |  |  |  |
| Quart | — | Ea | 8.09 | — | 8.09 | — |
| Gallon | — | Ea | 19.40 | — | 19.40 | — |

**Boston polish, Butcher's.** Like White Diamond bowling alley wax except for color. Amber tint hides kinks and scratches on wood.

|  | Craft@Hrs | Unit | Material | Labor | Total | Sell |
|---|---|---|---|---|---|---|
| Amber, pound | — | Ea | 7.14 | — | 7.14 | — |

**White Diamond bowling alley wax, Butcher's.** Clear wax in turpentine and other select solvents. Cleans as it waxes. Polishes easily to a deep luster. Also recommended for shoes, boots, leather, copper, brass, fiberglass, marble, sealed brick and flagstone, linoleum and cork floors.

|  | Craft@Hrs | Unit | Material | Labor | Total | Sell |
|---|---|---|---|---|---|---|
| Clear, pound | — | Ea | 7.14 | — | 7.14 | — |

**Teak oil, Valspar.**

|  | Craft@Hrs | Unit | Material | Labor | Total | Sell |
|---|---|---|---|---|---|---|
| Quart | — | Ea | 10.10 | — | 10.10 | — |

**Finishing paste wax, Minwax®.** Protects and adds luster to any stained or finished wood surface.

|  | Craft@Hrs | Unit | Material | Labor | Total | Sell |
|---|---|---|---|---|---|---|
| 1 pound | — | Ea | 8.08 | — | 8.08 | — |

**Wood conditioner, Minwax®.** Ensures even stain penetration on soft or porous woods such as pine, fir, alder and maple. When working with soft or porous woods, wood conditioner helps prevent stains from blotching and streaking.

|  | Craft@Hrs | Unit | Material | Labor | Total | Sell |
|---|---|---|---|---|---|---|
| Pint | — | Ea | 5.38 | — | 5.38 | — |
| Quart | — | Ea | 8.62 | — | 8.62 | — |
| Gallon | — | Ea | 23.10 | — | 23.10 | — |

## Interior Primers

**Interior enamel undercoater.** Improves sheen, hide, adhesion, flow and leveling of enamel topcoats. Blocks stains. Conceals surface blemishes. For use with wood, drywall, plaster, stucco, masonry, brick and metal. Use prior to top-coating with satin, semi-gloss and high-gloss topcoats.

|  | Craft@Hrs | Unit | Material | Labor | Total | Sell |
|---|---|---|---|---|---|---|
| Quart | — | Ea | 9.16 | — | 9.16 | — |
| Gallon | — | Ea | 19.40 | — | 19.40 | — |
| 5 gallons | — | Ea | 85.30 | — | 85.30 | — |

**Interior polyvinyl acrylic latex primer, Speed-Wall®.** General purpose. Easy application, fast-drying and high-hiding. Covers 400 square feet per gallon.

|  | Craft@Hrs | Unit | Material | Labor | Total | Sell |
|---|---|---|---|---|---|---|
| Gallon | — | Ea | 9.70 | — | 9.70 | — |
| 5 gallons | — | Ea | 37.80 | — | 37.80 | — |

| | Craft@Hrs | Unit | Material | Labor | Total | Sell |
|---|---|---|---|---|---|---|

**KILZ® Premium sealer-primer-stainblocker, Masterchem.** Interior/exterior, water-base formula. Superior stainblocker. Great for mildew-prone areas. Excellent enamel undercoater. High hide and durability.

| | | | | | | |
|---|---|---|---|---|---|---|
| Quart | — | Ea | 8.62 | — | 8.62 | — |
| Gallon | — | Ea | 19.40 | — | 19.40 | — |
| 5 gallons | — | Ea | 85.30 | — | 85.30 | — |

## Interior Paint

**Interior latex ceiling paint.** Minimal splatter and excellent coverage. Hides minor surface imperfections and dries quickly to a smooth, even finish. Clean up with water. Covers 300 square feet per gallon.

| | | | | | | |
|---|---|---|---|---|---|---|
| Gallon | — | Ea | 14.00 | — | 14.00 | — |
| 5 gallons | — | Ea | 57.20 | — | 57.20 | — |

**Interior acrylic latex flat wall and trim paint.** Dries to a smooth, even finish. Hides minor surface imperfections. Easy to clean up and dries quickly. Covers 400 square feet per gallon.

| | | | | | | |
|---|---|---|---|---|---|---|
| Gallon | — | Ea | 10.80 | — | 10.80 | — |
| 5 gallons | — | Ea | 50.20 | — | 50.20 | — |

**Ultra-Hide® interior latex flat finish, Glidden.** For use on interior walls, ceilings, and low-traffic areas. High-hiding with excellent touch-up ability when applied by spray, brush, or roller. Quick drying and recoat. Covers 400 square feet per gallon.

| | | | | | | |
|---|---|---|---|---|---|---|
| Gallon | — | Ea | 15.10 | — | 15.10 | — |
| 5 gallons | — | Ea | 62.60 | — | 62.60 | — |

**Premium Plus® interior flat paint, Behr.** 100% acrylic latex. Hides minor surface imperfections. For walls, woodwork and ceilings. Lifetime guarantee. Matte, flat finish. Easy water clean up.

| | | | | | | |
|---|---|---|---|---|---|---|
| Pastel base, gallon | — | Ea | 21.60 | — | 21.60 | — |
| Pastel base, 5 gallons | — | Ea | 96.50 | — | 96.50 | — |
| Ultra white, quart | — | Ea | 10.20 | — | 10.20 | — |
| Ultra white, gallon | — | Ea | 21.60 | — | 21.60 | — |
| Ultra white, 5 gallons | — | Ea | 96.20 | — | 96.20 | — |

**Latex eggshell wall and trim paint.** Acrylic latex. 15-year durability.

| | | | | | | |
|---|---|---|---|---|---|---|
| Quart | — | Ea | 11.90 | — | 11.90 | — |
| Gallon | — | Ea | 24.80 | — | 24.80 | — |
| 5 gallons | — | Ea | 111.00 | — | 110.00 | — |

**Ultra-Hide® interior latex eggshell, Glidden.**

| | | | | | | |
|---|---|---|---|---|---|---|
| Gallon | — | Ea | 17.20 | — | 17.20 | — |
| 5 gallons | — | Ea | 59.00 | — | 59.00 | — |

**Latex satin wall and trim enamel, Behr.** 100% acrylic latex. For kitchens and bathrooms. Exceptional washability, scrubbability, stain and mildew-resistance. Pearl-like finish. Hides minor surface imperfections. Superior adhesion. Lifetime guarantee.

| | | | | | | |
|---|---|---|---|---|---|---|
| Quart | — | Ea | 12.40 | — | 12.40 | — |
| Gallon | — | Ea | 25.90 | — | 25.90 | — |
| 5 gallons | — | Ea | 117.00 | — | 117.00 | — |

**Ultra-Hide® interior latex semi-gloss wall and trim enamel, Glidden.** High-hiding. Good adhesion and moisture resistance. Easy application. Fast-drying. Covers 400 square feet per gallon.

| | | | | | | |
|---|---|---|---|---|---|---|
| Gallon | — | Ea | 18.30 | — | 18.30 | — |
| 5 gallons | — | Ea | 82.00 | — | 82.00 | — |

| | Craft@Hrs | Unit | Material | Labor | Total | Sell |
|---|---|---|---|---|---|---|

**Speed-Wall® interior latex semi-gloss wall and trim enamel, Glidden.** Good dry hide. Hard and durable. Easy application. Fast-drying. Covers 400 to 450 square feet per gallon.

| | | | | | | |
|---|---|---|---|---|---|---|
| Gallon | — | Ea | 11.90 | — | 11.90 | — |
| 5 gallons | — | Ea | 56.10 | — | 56.10 | — |

**Spred Enamel semi-gloss interior latex paint, Glidden.** For walls, ceilings, trim, previously painted or primed masonry, previously painted or primed metal. Scrubbable finish. Resists scuffs and stains. Quick-drying.

| | | | | | | |
|---|---|---|---|---|---|---|
| Quart | — | Ea | 9.66 | — | 9.66 | — |
| Gallon | — | Ea | 21.20 | — | 21.20 | — |
| 5 gallons | — | Ea | 103.00 | — | 103.00 | — |

**Premium Plus® interior and exterior high-gloss enamel, Behr.** 100% acrylic latex. Mildew-proof. Glass-like finish. For kitchen, bath, railings, trim, doors, cabinets, and woodwork. Exceptional washability and stain removal.

| | | | | | | |
|---|---|---|---|---|---|---|
| Quart | — | Ea | 14.00 | — | 14.00 | — |
| Gallon | — | Ea | 29.00 | — | 29.00 | — |
| 5 gallons | — | Ea | 130.00 | — | 130.00 | — |

**Ultra-Hide® interior and exterior alkyd gloss enamel, Glidden.** Durable high-gloss finish. High solids and low VOC. Good abrasion resistance. Excellent resistance to grease, oil and water. Covers 350 to 400 square feet per gallon.

| | | | | | | |
|---|---|---|---|---|---|---|
| Quart | — | Ea | 11.30 | — | 11.30 | — |
| Gallon | — | Ea | 23.80 | — | 23.80 | — |
| 5 gallons | — | Ea | 111.00 | — | 110.00 | — |

**Interior texture paint, Behr.** Decorative finish for walls and ceilings. Alternative to wallpaper, borders, and stenciling. Reinforced with interwoven fibers to easily repair old textured surfaces. Usable with topcoat. Brightest white, ready to use. 2-gallon container.

| | | | | | | |
|---|---|---|---|---|---|---|
| Trowel-on Mediterranean ceiling | — | Ea | 21.30 | — | 21.30 | — |
| Sand stucco knockdown | — | Ea | 21.20 | — | 21.20 | — |
| Roll-on smooth acoustic | — | Ea | 21.20 | — | 21.20 | — |

**Roll-on stucco texture, Litex.** Interior and exterior.

| | | | | | | |
|---|---|---|---|---|---|---|
| 3.5 gallons | — | Ea | 34.30 | — | 34.30 | — |

**Sand-Texture™, Homax®.** A classic wall texture for a sandstone-like finish. Mixes with any paint, won't change the color of the paint. Use one package per gallon

| | | | | | | |
|---|---|---|---|---|---|---|
| 6 ounces | — | Ea | 2.13 | — | 2.13 | — |

**Perlite texturing paint additive, Litex.** Add to any paint to create texture and sand finishes.

| | | | | | | |
|---|---|---|---|---|---|---|
| Fine, quart | — | Ea | 3.32 | — | 3.32 | — |
| Medium, quart | — | Ea | 3.48 | — | 3.48 | — |
| Coarse, quart | — | Ea | 3.35 | — | 3.35 | — |

## Exterior Primer

**Premium Plus® oil-based exterior primer sealer, Behr.** Universal undercoat for use on wood, hardboard, aluminum and wrought iron. Prevents extractive bleeding.

| | | | | | | |
|---|---|---|---|---|---|---|
| Quart | — | Ea | 11.90 | — | 11.90 | — |
| Gallon | — | Ea | 24.80 | — | 24.80 | — |

**Water-based concrete bonding primer, No. 880, Behr.** Clear epoxy treatment for porous concrete, garage floors, driveways, walkways, patios and stucco. Penetrates and bonds. Resistant to water, alkali, and efflorescence. Creates a sound surface for painting or staining. Promotes uniform finish of topcoats. Dries to touch in 2 to 4 hours. Approximately 350 to 400 square feet per gallon coverage.

| | | | | | | |
|---|---|---|---|---|---|---|
| Clear, gallon | — | Ea | 17.30 | — | 17.30 | — |

| | Craft@Hrs | Unit | Material | Labor | Total | Sell |
|---|---|---|---|---|---|---|

**Premium Plus® exterior water-base primer and sealer, Behr.** 100% acrylic latex. Multi-purpose stain blocking. Use on exterior wood, hardboard, masonry, stucco, brick, galvanized metal, and aluminum.

| | Craft@Hrs | Unit | Material | Labor | Total | Sell |
|---|---|---|---|---|---|---|
| Quart | — | Ea | 9.70 | — | 9.70 | — |
| Gallon | — | Ea | 20.50 | — | 20.50 | — |
| 5 gallons | — | Ea | 90.80 | — | 90.80 | — |

**KILZ® exterior sealer primer stainblocker, Masterchem.** Exterior oil-base formula. Resists cracking and peeling. Mildew-resistant coating. Blocks tannin bleed on redwood and cedar.

| | Craft@Hrs | Unit | Material | Labor | Total | Sell |
|---|---|---|---|---|---|---|
| Gallon | — | Ea | 17.20 | — | 17.20 | — |
| 5 gallons | — | Ea | 77.40 | — | 77.40 | — |

## Exterior Paint

**Evermore® exterior latex house and trim paint, 20 year, Glidden.** 100% acrylic to protect against UV rays, color fading and severe weather conditions. Hides minor surface imperfections, gives a mildew-resistant coating. Non-chalking. Good weather resistance against dirt, blistering and peeling. For siding, wood, stucco, brick and masonry surfaces. Use the satin finish on windows and doors. Covers 300 to 400 square feet per gallon.

| | Craft@Hrs | Unit | Material | Labor | Total | Sell |
|---|---|---|---|---|---|---|
| Flat, quart | — | Ea | 9.70 | — | 9.70 | — |
| Flat, gallon | — | Ea | 20.50 | — | 20.50 | — |
| Flat, 5 gallon | — | Ea | 92.90 | — | 92.90 | — |
| Satin, quart | — | Ea | 10.80 | — | 10.80 | — |
| Satin, gallon | — | Ea | 22.70 | — | 22.70 | — |
| Satin, 5 gallons | — | Ea | 103.00 | — | 103.00 | — |
| Semi-gloss, quart | — | Ea | 11.30 | — | 11.30 | — |
| Semi-gloss, gallon | — | Ea | 23.80 | — | 23.80 | — |
| Semi-gloss, 5 gallons | — | Ea | 108.00 | — | 108.00 | — |
| Gloss, quart | — | Ea | 11.80 | — | 11.80 | — |
| Gloss, gallon | — | Ea | 25.90 | — | 25.90 | — |

**Premium Plus® exterior paint, Behr.** 100% acrylic latex. Mildew and fade resistant. For use on wood, masonry, brick, stucco, vinyl and aluminum siding, and metal.

| | Craft@Hrs | Unit | Material | Labor | Total | Sell |
|---|---|---|---|---|---|---|
| Flat, quart | — | Ea | 11.30 | — | 11.30 | — |
| Flat, gallon | — | Ea | 23.70 | — | 23.70 | — |
| Flat, 5 gallons | — | Ea | 106.00 | — | 106.00 | — |
| Satin, gallon | — | Ea | 25.90 | — | 25.90 | — |
| Satin, 5 gallons | — | Ea | 117.00 | — | 117.00 | — |
| Semi-gloss, quart | — | Ea | 12.90 | — | 12.90 | — |
| Semi-gloss, gallon | — | Ea | 27.00 | — | 27.00 | — |
| Semi-gloss, 5 gallons | — | Ea | 120.00 | — | 120.00 | — |

## Masonry and Stucco Paint

**Plus 10® latex flat masonry and stucco paint, Behr.** For interior or exterior use. Acrylic latex paint for smooth, rough and textured masonry surfaces. Weather- and temperature-resistant.

| | Craft@Hrs | Unit | Material | Labor | Total | Sell |
|---|---|---|---|---|---|---|
| Gallon | — | Ea | 18.30 | — | 18.30 | — |
| 5 gallons | — | Ea | 74.50 | — | 74.50 | — |

**Plus 10® elastomeric waterproofing paint, Behr.** Acrylic latex for masonry, stucco, and concrete. Exterior, flexible, high-build coating. Expands and contracts, bridging hairline cracks. Stretches up to 600%. Resists mildew and dirt. Retains color. Passes Federal Specification TT-C-555B sec. 3.3.3.

| | Craft@Hrs | Unit | Material | Labor | Total | Sell |
|---|---|---|---|---|---|---|
| Gallon | — | Ea | 25.90 | — | 25.90 | — |
| 5 gallons | — | Ea | 117.00 | — | 117.00 | — |

| | Craft@Hrs | Unit | Material | Labor | Total | Sell |
|---|---|---|---|---|---|---|

## Concrete Coatings

**Waterproofing sealer, Seal-Krete.** Waterborne, high-binding acrylic base. Penetrating sealer formulated for use on all vertical concrete, cement, masonry and stucco surfaces. Non-toxic.

| | | | | | | |
|---|---|---|---|---|---|---|
| Gallon | — | Ea | 13.30 | — | 13.30 | — |
| 5 gallons | — | Ea | 56.50 | — | 56.50 | — |

**Exposed aggregate sealer.**

| | | | | | | |
|---|---|---|---|---|---|---|
| Brown or clear, gallon | — | Ea | 17.50 | — | 17.50 | — |
| Brown or clear, 5 gallons | — | Ea | 75.50 | — | 75.50 | — |

**Concrete and masonry waterproofer, Behr.** Penetrating water-based silicone formula reduces water damage and deterioration for up to 10 years. Forms a breathable barrier that reduces corrosion, freeze-thaw damage, mildew and algae staining.

| | | | | | | |
|---|---|---|---|---|---|---|
| Gallon | — | Ea | 20.50 | — | 20.50 | — |
| 5 gallons | — | Ea | 90.70 | — | 90.70 | — |

**Thoroseal® protective coating, Thoro.** For walls, pool floors, cisterns, tanks, and masonry reservoirs. Finishes masonry of all types. Seals water and dampness out of walls, above or below grade, inside or out. Leakproofs water containers. Under ordinary water conditions, 2 pounds cover 1 square yard as a base coat; 1 pound covers 1 square yard as a second coat. Under extreme water pressures, 1 brush coat at 2 pounds per square yard and 1 trowel coat. White or gray.

| | | | | | | |
|---|---|---|---|---|---|---|
| 50 pound bag | — | Ea | 23.70 | — | 23.70 | — |

**Epoxy floor coating, 2 part, Litex.** Prevents hot tire pickup. Resists gasoline, oil, grease, acid, transmission fluid, saltwater, bleach, and alkali.

| | | | | | | |
|---|---|---|---|---|---|---|
| Clear, gallon | — | Ea | 27.80 | — | 27.80 | — |
| Gray, gallon | — | Ea | 22.20 | — | 22.20 | — |
| Most colors, gallon | — | Ea | 33.10 | — | 33.10 | — |
| Tint base, gallon | — | Ea | 30.60 | — | 30.60 | — |

**Cure-Seal wet-look concrete sealer, Jasco.** Protects exposed aggregate, brick, flagstone, and concrete from oil, rust, and water staining. Dries clear. Covers 150 to 400 square feet per gallon. Ready for traffic in eight to 24 hours.

| | | | | | | |
|---|---|---|---|---|---|---|
| Gallon | — | Ea | 16.20 | — | 16.20 | — |
| 5 gallons | — | Ea | 66.90 | — | 66.90 | — |

**Plus 10® concrete stain.** Durable, water-repellent. Interior and exterior. Forms tough, longwearing film on all properly-prepared surfaces such as patios, basements, pool decks, pillars, bricks and sidewalks. Natural slate color. 100% acrylic resins.

| | | | | | | |
|---|---|---|---|---|---|---|
| Gallon | — | Ea | 22.10 | — | 22.10 | — |
| 5 gallons | — | Ea | 98.70 | — | 98.70 | — |

**Premium Plus® gloss porch and floor enamel, Behr.** 100% acrylic. Mildew-resistant. Dries to touch in 2 hours. Recoat in 24 to 48 hours. Full cure in 30 days.

| | | | | | | |
|---|---|---|---|---|---|---|
| Gloss, quart | — | Ea | 10.50 | — | 10.50 | — |
| Gloss, gallon | — | Ea | 24.20 | — | 24.20 | — |
| Slate gray, low luster, gallon | — | Ea | 22.00 | — | 22.00 | — |
| Slate gray, low luster, 5 gallons | — | Ea | 98.70 | — | 98.70 | — |

**Porch and floor polyurethane oil gloss, Glidden.** For interior or exterior wood and concrete floors, stairs, porches, railings and steps. Covers 400 to 500 square feet per gallon.

| | | | | | | |
|---|---|---|---|---|---|---|
| Gallon | — | Ea | 20.50 | — | 20.50 | — |
| 5 gallons | — | Ea | 92.90 | — | 92.90 | — |

| | Craft@Hrs | Unit | Material | Labor | Total | Sell |
|---|---|---|---|---|---|---|

## Rust Preventative

**Clean metal primer.** White. For use on bare, lightly rusted or previously-painted surfaces. Creates a strong surface for better topcoat adhesion and weather resistance.

| | Craft@Hrs | Unit | Material | Labor | Total | Sell |
|---|---|---|---|---|---|---|
| Quart | — | Ea | 8.47 | — | 8.47 | — |
| Quart, low VOC | — | Ea | 8.45 | — | 8.45 | — |

**Rust converting metal primer, Rust Destroyer®.** Non-toxic, USDA and FDA approved. Heat resistant up to 800 degrees F. Five-year guarantee. Works on clean or rusty steel, zinc galvanized metal, core tan steel, aluminum, tin, previously painted surfaces and barbeques. When applied over new metal, will prevent rusting.

| | | | | | | |
|---|---|---|---|---|---|---|
| Quart | — | Ea | 11.80 | — | 11.80 | — |
| Gallon | — | Ea | 39.90 | — | 39.90 | — |

**Professional oil-based enamel, Rust-Oleum®.** Tough industrial enamel. Prevents rust, resists cracking, peeling, chipping, and fading. Scuff-, abrasion- and moisture-resistant. Lead-free. Not low-volatile-organic-compound compliant.

| | | | | | | |
|---|---|---|---|---|---|---|
| Gallon | — | Ea | 27.50 | — | 27.50 | — |
| 5 gallons | — | Ea | 102.00 | — | 102.00 | — |

**Stops Rust, Rust-Oleum®.** Resists moisture and corrosion. For metal, wood, concrete and masonry.

| | | | | | | |
|---|---|---|---|---|---|---|
| Half pint | — | Ea | 3.95 | — | 3.95 | — |
| Quart | — | Ea | 7.93 | — | 7.93 | — |

**Rusty metal primer.** Fish oil based formula penetrates rust to bare metal. Drives out corrosive air and moisture. Bonds tightly to rust to form a surface that topcoats can adhere to. Volatile organic compound (VOC) compliant.

| | | | | | | |
|---|---|---|---|---|---|---|
| Half pint | — | Ea | 4.10 | — | 4.10 | — |
| Quart | — | Ea | 7.42 | — | 7.42 | — |

## Specialty Coatings

**Chlorinated rubber swimming pool paint, Insl-X.** For new or existing pools, indoor or outdoor. Semi-gloss, self-priming for gunite and marcite pools. Durable in salt or fresh water. Resists fading, fungus, algae, abrasion and alkalis.

| | | | | | | |
|---|---|---|---|---|---|---|
| Gallon | — | Ea | 25.90 | — | 25.90 | — |

**Barn and fence paint.**

| | | | | | | |
|---|---|---|---|---|---|---|
| Gallon | — | Ea | 11.90 | — | 11.90 | — |
| 5 gallons | — | Ea | 52.90 | — | 52.90 | — |

**Tru-Glaze WB™ epoxy gloss coating, Devoe.** Waterborne. Hard, durable high-performance architectural coating for use on interior concrete, concrete block, drywall, metal and wood.

| | | | | | | |
|---|---|---|---|---|---|---|
| Gallon | — | Ea | 35.60 | — | 35.60 | — |

**Tru-Glaze WB™ epoxy primer, Devoe.** High-performance waterborne epoxy rust-inhibitive primer for steel, aluminum and certain galvanized metal surfaces. Can also be used over concrete, masonry, glazed brick or ceramic tile. Excellent adhesion, chemical resistance, and good resistance to splash, spillage or fumes for a wide range of chemicals. May be top-coated with epoxy, urethane, alkyd or latex coatings.

| | | | | | | |
|---|---|---|---|---|---|---|
| Quart | — | Ea | 12.90 | — | 12.90 | — |
| Gallon | — | Ea | 35.60 | — | 35.60 | — |

# Index

**A**

ABS pipe .........................408
Absorber
  water surge ...................407
ABTCO ..........................251
AC cable ........................419
AC quiet rocker switch
  Decora .......................434
AC quiet toggle switch ...........434
Access door .....................248
Access doors
  foundation ....................89
Accessories
  ridge vents ....................91
Acid cleaning
  masonry .....................116
  wash ......................50, 116
Acid wash
  manhours .....................466
Acoustic ceiling texture
  labor ........................469
Acoustic roll-on .................493
Acoustic tile
  demolition .....................34
Acoustical
  tile, demolition ...............315
Acoustical ceiling
  demolition ...................255
  panel .....................256-257
  patch ........................245
  suspended ...................256
  texture touch-up kit ..........244
  tile .........................254
  tile adhesive .................254
  touch-up kit ..................244
Acousti-Tex ceiling texture .......244
Acrylic
  mirrored sheet ...............172
  mortar admix .................307
  paint .........................40
  sheet ........................172
  skylights .....................174
  tape .........................291
  wood filler ...................281
Acrylic latex paint ..............492
Acrylic roof coating .............137
Acrylic stain ...................487
Actuating devices, garage doors ..225
Adapter ........................421
  box ..........................421
Addition, second story ...........57
Additive
  feather edge ..................286
  resilient flooring ..............285
Additives
  coloring .......................47
  concrete ......................47
  perlite paint texturing ........493
Adhesive
  acoustical ceiling tile ..........254
  bond .........................133
  cold process ..................133
  cove base .....................290
  field .........................134
  flooring ......................282
  marble & granite ..............306
  modified bitumen .............133
  nozzle .......................291
  primer .......................282
  resilient tile ..................290
  sheet flooring .................290
  spray ........................249
  tile ......................293-294
  vinyl tile .....................290
  white silicon .................172

wood flooring ...................282
Adhesive caulk
  ceramic tile ...................307
Adhesive remover ...............282
  flooring ...................284-285
Admix
  acrylic mortar ................307
Admixtures
  coloring .......................47
  concrete ......................47
Aggregate
  base .......................40, 46
  roofing .......................132
Aggregate sealer ...............495
Air compressor
  rental .........................28
Air conditioner .................398
Air gap ........................334
  cover ........................334
Air hose
  rental .........................28
Alkyd gloss enamel .............493
All-season storm doors .......204-205
Aluminum
  door bottom ..................196
  drip edge ....................141
  fascia .......................115
  foil ..........................89
  patio doors ..................215
  roll flashing .................141
  roof coating .................137
  screen rolls ..................169
  siding .......................114
  siding removal ...............107
  siding repair .................106
  soffit .......................115
  storm windows ...............166
  thermal door sill ............195
  windows, fixed ..............162
Aluminum conduit
  threaded .....................431
Aluminum siding
  painting .....................482
Amber shellac .................490
American Flexible Conduit .......419
American Marazzi ......296-298, 303
American Molding ..............262
American Shower & Bath ....337, 355,
  ........................357-360
American Standard .....330, 332-333,
  ...335, 338-339, 353-354, 360-364,
  ....................366-372, 385-386
Anchor clip ....................458
Andersen ..................214, 216
Angle
  flashing .....................142
Angle bay windows .............165
Angle valve ...................364
Ann Arbor birch paneling .......252
Anti-fracture
  membrane kit ................307
  membrane waterproofing .....305
  tile membrane ...............305
Anti-siphon toilet fill valve .....364
Appliance receptacle ..........435
Appliances
  connect .....................417
  connecting ..................417
  connection ..................417
  disconnect ..................417
  kitchen, removal .............315
  new connection ..............417
  remove & replace cable ......417
Apron
  roof .........................143

whirlpool tub ..................354
Aqua Mix ...................307-308
Aqua-Touch faucet .............331
Aqua-Tough tile underlayment ...295
Arch
  bullnose drywall .............247
Architectural laminate shingles ...130
Architectural shingles ..........131
Area walls, sheet metal .........389
Arlington Industries ...........422
Armstrong ..256-257, 279, 285-291, 293
Asceri Hi-Arc lavatory faucet .....373
Ascot brass doors ..............199
Asphalt
  base .........................51
  paving ........................51
  paving, demolition .........32, 51
  recycling ......................29
Asphalt fiberglass shingles ......130
Asphalt roofing ................132
  built-up .....................132
  emulsion ....................133
  fabric .......................131
  saturated cotton roof patch ...136
  SEBS ........................135
Asphalt shingles
  fiberglass ....................131
  repair .......................130
Astragal, T ....................194
Atomix ........................334
Attachments
  skid steer equipment ..........39
Attic conversion .................57
Attic fans
  belt drive .....................92
Attic vents .....................92
Awning
  remove and replace ...........107
Awning windows
  aluminum ...................162
  operators ...................167
  vinyl-clad wood .............160

**B**

Bacharach kitchen faucet ........332
Backer
  tile .........................245
Backer tape
  tile .........................249
Backerboard
  ceramic tile ..................295
  RhinoBoard ..................295
  tile .........................295
  Wonderboard ...............295
Backfill
  concrete excavation ..........40
Backhoe
  rental .........................28
Backing
  carpentry .....................74
Backplate
  knob ........................326
Backsplash ....................322
Baffle
  sound .......................341
Baffle light kit ................440
Balance, sash ..................167
Baldwin .......................386
Baluster .......................455
  deck rail ....................456
Balustrade
  deck .........................456
Bamboo
  flooring trim .................280
  hardwood ...................280
  plank flooring ...............280

Banner Sterling ................338
Bar
  light ........................385
  towel ..........385-386, 388-389
Bar doors .....................209
Bar faucet ....................338
  single handle ................338
  two-handle ..................338
Bar kit
  QO load center ground .......440
Bar light fixture ...............441
Bar locks .....................224
Bar sink
  Berkeley .....................338
  composite ...................338
  Manchester composite .......338
  polycast .....................338
  single bowl ..................338
  stainless steel ...............338
Bar, towel ....................305
Barn & fence paint .............496
Barrier, foil ....................89
Base
  asphalt .......................51
  ceramic tile ..................295
  lavatory .....................367
  rubber wall ..................292
  self-stick wall ...............293
  shower ...................358-359
  slab ..........................49
  vinyl wall ................292-293
  wall, thermoplastic ..........292
Base adhesive .................290
Base cabinets ..................318
  end panel ...................317
  oak .........................320
  prefinished ...............320-321
  unfinished birch .............317
  unfinished oak ..............318
Base cap molding
  primed finger joint ...........260
Base corner molding ...........259
  MDF primed .................261
Base molding ..................263
  colonial .....................263
  combination .................264
  MDF colonial ................261
  paperwrap colonial ..........263
  paperwrap ranch ............264
  polystyrene colonial .........264
  primed finger joint ...........260
  shoe ........................264
  Victorian hardwood ..........259
Base plate cover, toilet .........364
Base sheet
  fiberglass ...................133
Base shoe molding
  primed finger joint ...........260
Baseboard
  painting labor ...............468
Baseboard heater, electric ......404
Baseboard molding
  beaded ......................259
  MDF ........................261
Basement
  doors .......................389
  leaks .........................37
Basement post removal ..........64
Basement storm windows .......166
Basin strainer
  remove & replace .............352
Basket strainer
  garbage disposal .............336
  junior .......................339
  kitchen sink ..............335-336

Basket strainer washer
kitchen sink .336
Basswood molding .259
Bath accessories .385-388, 389
brass .386, 388
brass & chrome .386, 388
ceramic & chrome .387
ceramic tile .294
chrome .386-388
chrome & brass .386, 389
chrome & porcelain .386
economy .386
nickel .388
oak .389
porcelain .387
Bath accessory set
porcelain .305
Bath cabinet
framed .382
stainless steel .382
storage .383
swing door .383
top lighted .383
Bath ceiling blower .378
Bath fan .378-381
ceiling .381
ceiling & wall .378
ceiling or wall .379
chain-operated .380
Designer Series .381
exhaust .380
exhaust with heater .381
exhaust with light .380
Heat-A-Lamp .381
QuieTTest .380
solitaire .379, 381
Solitaire Ultra Silent .379
timer switch .382
vertical discharge .379
with heater, infrared .381
with light .380-381
Bath heater .381
exhaust fan .381
Heat-A-Lamp .381
infrared bulb .381
Bath light bar .441
Bath Unlimited .387-389
Bath vent kit .382
Bath wraps .355
Bathrobe hook .385-386, 388-389
Bathroom accessories
ceramic tile soap dish .304
towel bar, ceramic .305
Bathroom sink .367-369
Aqualyn .367
Bayside .368
cadet .368
drop-in .367-368
Farmington .367
Lucerne .369
Memoirs .369
Miami .369
Minette .369
Murray .369
Murro .369
oval .368
Ovalyn .368
Radiant .368
Renaissance .367
Rondalyn .367
round .368
Seychelle .367-368
steel .368
undermount .368
wall hung .369
wall hung, corner .369
Bathtub .352-353
acrylic soaker .353
ceramic tile look .355
enameled steel .352
Performa .352
Princeton .353
recessed .353
removal .352
soaker .353
soaker with skirt .353
surround .355
three-piece .355
tile wall kit .355
Ventura .352
Villager .353
wall kit .355
wall surround .355
wall, Vikrell .355
whirlpool .353-354
Bathtub & shower unit .357
Bathtub & shower units .356-357
Bathtub door .356
bypass .355
sliding .356
Bathtub wall .354
Georgia .354
one-piece with shelf .354
three-piece .354
Batt insulation .85
Batten tileboard .251
Bay vent channel
rafter .89
Bay window roof .165
Bay windows .158, 164
Bead
bullnose corner .247
corner .248
drywall .248
drywall corner .247
glazing .171
inside corner .247
L trim .248
metal J trim .248
outside corner .247
plastic .248
spray adhesive .249
tape on .247
Bead molding .263
primed finger joint .260
Beaded baseboard molding .259
Beaded Cape Cod plank .253
Beaded paneling .251
Beadex flex metal tape .248
Beadex paper faced metal .247
Beams
ceiling .69
collar .76
grade .48
needle .63
repair .53
Bed molding .263
Bedding mortar
CustomFloat .306
Behr .487, 490, 492-495
Bell wire .433
Bemis Manufacturing .365-366
Bestview Ltd .298
Bevel siding
cedar .109
pine .109
western red cedar .109
Beveled mirror bath bar .441
Bidding
wood framing jobs .53
Bidet
hot & cold .362
Bidet faucet set
Valvuetwist .362
Bi-fold doors .219-221
French .221
hardware .221
mirrored .221-222
Bin, trash .29
Birch doors .192, 206-207, 210
Birch paneling .252
Birdstop .140
Bitumen adhesive, modified .133
Bitumen cold process adhesive,
modified .133
Bituminous
roofing .132
Blank cover
2 gang .428
handy box .424
weatherproof .428
Bleach kit
wood .487
Blinds
painting .481-482
Block masonry
painting .476
Block molding
corner .262
rosette .259
Blocking
fire .67
fireblocks .69
Blow-in cellulose insulation .86
Blue Board insulation .86
Blueboard .245
Board
blue .245
cement .245
forming .44
sheathing .67
Board insulation .86
Board siding
cedar .109
pine .109
repair .106
Bobcats .39
Boiler
hydronic heating system .398
Boiler breeching .401-403
Bolt kit .324
toilet tank .364
Bolt-on A type universal hub .438
Bolts
double-hung window .167
Bond adhesive
permanent .133
Bonding mortar .305-306
flexible .306
Bonding primer
concrete .493
Bookcase
painting labor .468
Boone International .293
Boston polish .491
Bottle jacks .63
Bow window roof .164-165
Bowl
lavatory .366
Bowl sink
utility .337
Bowling alley wax .491
Box .428
2 gang weatherproof electrical .427
ceiling fan .425
ceiling fan pancake .425
drawn 2-device switch .425
drawn handy .424
drawn octagon .425
multi gang switch .425
non-metallic old work switch .422
non-metallic round conduit .428
rectangular weatherproof .427
Super Blue non-metallic
three gang .422
Super Blue non-metallic
two gang .422
welded square .425
Zip Box non-metallic ceiling fan .422
Box adapter .421
Box cover .426
blank handy .424
finished .425
square surface .425
Bracing
carpentry .68
temporary .68
Bracket
lavatory hanger .369
non-metallic low voltage
mounting .422
Brass
storm .205
Brass Craft .334, 338, 362, 364
Breaker
arc fault, HomeLine .437
GFCI .437
ground fault interrupter plug-on
circuit .440
HomeLine .437
main circuit .438
QO miniature circuit .439
tandem circuit .439
Breaker box
ground bar kit .440
Breaker, paving
rental .28
Bretton .355
Brick
cleaning .116
demolition .30
painting .477-478
pointing .116
repointing .116
Brick masonry
painting labor .476
Brick molding .195
Brick molding set
primed finger joint .260
Brick paver tile .300
Bridging
carpentry .66
Briggs Industries .362
Broan Manufacturing .344, 378-382
Brookfield cast iron kitchen sink .330
Brookstone birch paneling .252
Broom finish, concrete .49
Brown coat plaster .115
Bruce .278, 282
Brush
painting labor .468
wire .50
Brush chipper
rental .28
Brushes
crow's foot .250
drywall texture .250
Buggies
rental .39
Building paper .88, 134
Building wire
type THHN .432
Built-up girders
carpentry .63
Built-up roofing .132, 134
Cold-Ap cement .133
demolition .33
insulation .131-132
patch .131
tear-off .129
torch and mop system .134

Bulb heater & fan
  Heat-A-Lamp .................381
Bulkhead light fixture
  outdoor ....................443
Bull float
  rental .......................39
Bullnose
  ceramic tile ................301
  corner bead ................247
  corner tile .................302
  drywall arch ...............247
  plastic corner .............247
Bullnose outside corner .......247
Bulls Eye shellac .............490
Buried cable ..................419
Burn off paint labor ..........466
Bush removal ...................29
Bushing
  insulated metallic grounding ....430
  non-metallic reducer ........420
  rigid or IMC reducing ........430
  thermoplastic insulating .....430
Butcher's .....................491
BX cable
  flexible armored ...........419
Bypass door
  bathtub ....................355
  shower .....................356
Bypass door hardware ..........222
Bypass doors ..................218
  accessories ................219
  beveled mirror .............219
  mirrored ...............218, 219

**C**

Cabinet crown mold ............322
Cabinet door ..................316
Cabinet drawer front ..........316
Cabinet hinge
  overlay ....................327
  self-closing ...............328
  semi-concealed .............328
Cabinet knob .............324-326
Cabinet trim ..................320
Cabinets
  base ...........317-318, 320-321
  bath ...................382-383
  bathroom ...............374-375
  corner .....................383
  end panel ..............316-318
  filler strip ............318-319
  frameless ..................383
  medicine ...............382-384
  oak ...................320-321
  painting labor .............469
  pantry .................319-321
  prefinished ................320
  removal ....................315
  trim .......................320
  utility ................317-321
  utility ....................321
  vanity .................375, 378
  wall ..................317-321
Cable
  AC .........................419
  appliance connection .......417
  appliance re-connect .......417
  appliance, remove & replace ...417
  BX .........................419
  coaxial ....................433
  data communication .........433
  fish through stud wall .....417
  flexible armored BX ........419
  thermostat .................433
  type NM-B Romex sheathed .417-418
  type NM-B Romex sheathed ...418
  type UF-B sheathed .........419
  underground ................419

Cable coupling
  EMT to non-metallic sheathed ..424
Cable snake flat wire fish tape ....417
Cadet kitchen faucet ..........333
Cadet Series bar faucet .......339
Cafe doors ....................209
Calcium chloride ...............47
Cal-Flor accessory systems .....291
Cambray vinyl sheet flooring ...287
Cambria countertop ............323
Can
  ceiling light ..............440
Canopy
  remove and replace .........107
Cant strips ...................137
  roofing .....................87
Cap ...........................146
  ceramic surface ............302
  chimney ....................146
  PVC service entrance .......422
  service entrance ...........432
Cap molding
  colonial ...................264
Cap sheet
  roofing ....................133
Cap, post ................456-457
Cape Cod plank ................253
Carborundum rub ................50
Carflex liquid-tight flexible
  conduit ....................427
  fitting ....................427
Carlon ............420-422, 427-428
Carol Cable ...................433
Carpentry
  backing and nailers .........74
  beams and girders ...........63
  bridging or blocking ........66
  ceiling beams ...............69
  ceiling joists ..............74
  door openings ...............70
  dormer studs ................77
  fascia board ................77
  fireblocks ..................69
  floor joists .............65-66
  floor repairs ...............63
  furring .....................73
  plates ......................68
  posts .......................64
  rafters .....................76
  roof sheathing ..............77
  roof trusses ................77
  room additions ..............65
  sheathing ...................71
  sill plates .................65
  studs .......................67
  subflooring .................67
  wall assemblies .............71
  wall framing ................67
  window openings .............70
Carpet ........................308
  demolition ..................35
  removal ....................278
  roller, extendable .........291
  transition strip ...........292
Cartridge
  disposal ...................341
Casement windows ..............165
  lock handle ................167
  operator handle ............167
  vinyl-clad wood ............160
  wood .......................164
Casing ........................263
  aluminum siding ............115
  colonial ...................260
  colonial door ..............195
  door ...................191, 206
  economy ....................265
  MDF ........................261

paperwrap colonial ............264
paperwrap ranch ...............265
polystyrene ...................266
polystyrene colonial ..........264
polystyrene fluted ............265
polystyrene ranch .............265
prefinished ...................265
prefinished colonial ......264-265
prefinished fluted ............265
ranch .....................260, 265
Caspian sheet vinyl flooring ...286
Cast iron
  pipe, remove & replace .....408
  radiator, remove & replace ...398
  water closet flange ........364
Cast iron kitchen sink .....329-330
Cast molding ..................262
  chair rail .................262
Cast-in-place concrete .........47
Cat doors .....................217
Catalina Lighting .............443
Caulk
  ceramic tile ...............307
  ceramic tile adhesive ......307
  removal ....................157
  roof .......................136
  window exterior ............157
Caulking
  manhours ...................466
Cedar
  panels, sidewall shingle ...114
  roofing ....................137
  shingles ...................138
  siding .................109-110
Cedar louvered vents ...........91
Cedarmill .....................113
Ceiling
  beams .......................69
  demolition .................242
  joists ......................74
  painting ...................470
  painting manhours ......469-470
  repair .....................244
  staining manhours ..........470
  texture manhours ...........469
  tile adhesive ..............253
  tile, acoustical ...........254
  tile, wood fiber ...........254
Ceiling and wall panel
  FRP ........................253
Ceiling fan ...................444
  bath .......................378
  bath heater & fan ..........381
  gazebo .....................444
  QuieTTest ..................380
  San Marino .................444
Ceiling fan box ...............425
  pancake ....................425
  Zip Box non-metallic .......422
Ceiling fixture ...............440
  drum .......................441
  fluorescent ................441
  flush mount ................441
  glass globe ................442
  incandescent flush mount ...442
  mushroom flush mount .......442
  remove & replace ...........441
  Sunburst flush mount .......442
  swirl dome light ...........442
Ceiling latex paint ...........492
Ceiling lighting panel
  acrylic ....................258
  styrene ....................258
Ceiling panel .................257
  acoustical .................256
  classic fine-textured ......256
  fiberglass .............256-257
Firecode Fifth Avenue .........257

fissured ......................257
fissured fire guard ...........257
Grenoble ......................257
gypsum lay-in .................257
mineral fiber .................257
plain white ...................256
random textured ...............256
sahara ........................256
suspended .................256-258
textured ......................258
Ceiling repair ................245
Ceiling texture
  Acousti-Tex ................244
  cottage cheese .............244
  patch ......................245
  popcorn patch ..............244
  spray ......................244
  touch-up kit ...............244
Cellar doors, steel ...........194
Cement
  coloring ....................47
  drywall ....................245
  joint compound .............245
  tile .......................139
  underlayment ...............285
  white ......................115
Cement board ..................245
Cement mixer
  rental ......................39
Cement siding .................114
Cement soffit
  fiber ......................113
Cement, roofing
  cold process ...............133
  flashing ...................135
  Kool Patch .................135
  lap roof ...................133
  rubber wet patch ...........135
  Wet Stick ..................135
Center
  HomeLine indoor main lug load ..437
Central air conditioning .......398
Central humidifier ............397
Ceramic
  floor & wall ...............297
Ceramic tile
  accessories ................304
  accessory set ..............305
  adhesive ...................293
  adhesive caulk .............307
  backer .....................295
  backerboard ................295
  bullnose ...................301
  bullnose corner ............302
  caulk ......................307
  demolition .............35, 315
  floor ..................295-298
  floor warming system .......305
  glazed .....................303
  installation ...............294
  mosaic accent ..............304
  mosaic octagon .............304
  Natura .....................296
  removal ....................278
  Saltillo ...................299
  sink rail ..................302
  soap dish ..................304
  surface cap ................302
  surface cap corner .........302
  threshold ..................301
  toilet tissue holder .......304
  tool .......................291
  toothbrush & tumbler holder ....305
  wall ...................301, 303
  window sill ................301
Ceramic wall tile .............303
CertainTeed ...................131

Chain link fence
demolition .29
Chain saw
rental .28
Chain-operated wall exhaust fan .380
Chair rail
cast polyurethane .262
Clearwood PS .266
MDF .261
paperwrap .266
polymer .263
polystyrene .266
polystyreneg .266
prefinished .266
primed finger joint .260
Chalk line, self chalking .291
Chandelier .442
crystal .442
Channel .89
vent, rafter bay .89
Channel siding
cedar .109
Chesapeake floor tile .289
Chime kit, door .436
Chimney .399-403
A vent .402-403
all fuels .402-403
aluminum .399-400
B vent .401
base tee .400-402
C vent .399-400
cap .146
double wall .401-402
double wall insulated .402-403
elbow .400-403
flexible .399
flue .399- 403
flue rain cap .399
gas or propane .399-402
L vent .401-402
liner .399
liner rain cap .399
rain cap .400-403
roof flashing .400-403
single wall .399-400
stack .401-402
storm collar .400-403
supports & brackets .403
termination assembly .400-403
Chlorinated rubber paint .496
Chop saw
rental .28
Chrome light fixture .441
Circuit breaker
arc fault, HomeLine .437
GFCI .437
ground fault interrupter plug-on .440
HomeLine .437
main .438
main, all-in-one .438
QO miniature .439
tandem .439
Circular windows
half circle, vinyl .159
Civic Square .290
Clamp
ground rod .432
Snap Strap PVC conduit .420
wood floor .281
Clamp connector
two-piece .419
Classic faucet .370
Clay
roofing tile .139
Cleaning
brick .116
masonry .116
pressure wash .468
stone .116
terra cotta .116
Clear
cedar siding .109
Clear wood
preservative .491
protector .490
Clearshield .488
Clip, anchor .458
Clips
drywall repair .249
Clips & screws .329
Closers
door .217
Closet
painting labor .470
Closet doors .210
bi-fold .219-220
bypass .218
mirrored bypass .218-219
Closet pole molding .263
Closure strips, horizontal .141
CMU
painting .477
CO/ALR duplex receptacle .435
Coal furnace .404
Coated glass fabric .136
yellow resin .136
Coating
fibered roof .137
floor, epoxy .495
foundation and roof .129
roof .137
roof and foundation .129
weather-resistant .488
wood .488
Coating kit
seam .291
Coatings
concrete .495
epoxy gloss .496
epoxy primer .496
specialty .496
Coaxial cable .433
wall plate .433
Coaxial F-connector
for RG6 .433
Cobra Rigid ridge vent .145
Cofair Products .135
Coil, kickplate .459
Cold process adhesive
modified bitumen .133
Cold process cement .133
Collar beams
carpentry .76
Collar ties .76
Colonial base molding .263
MDF .261
paperwrap .263
polystyrene .264
Colonial cap paperwrap molding .264
Colonial casing
finger jointed .260
paperwrap .264
polystyrene .264
prefinished .264-265
Colonial crown paperwrap
molding .268
Colonial door casing .195
Colonial outside corner molding .267
Colonial stop molding .269
Colonial storm doors .205
Colonnade kitchen faucet .332
Colorant
grout .307
Coloring agents
concrete .47, 49
Columbia tile finish tub wall
one-piece .354
Column, porch .455
Combination base molding .264
Combination coupling .424
Combination service entrance
device .438
Combination service entrance load
center .437
Commercial Electric .440
Communication cable
category 5E .433
Composite kitchen sink .338
60/40 bowl .331
bar .338
Beaumont .331
Berkeley .338
double bowl .330
Forsyth .330
Greenwich .330
single bowl .330
Composition tile
vinyl .289-290
Compound
glazing .171
latex glazing .171
Compression conduit coupling,
rigid or IMC .430
Compression connector
EMT .423
EMT-to-box offset .424
non-metallic UF .419
Compression faucet
repair .352
Compression inlet
dishwasher connector .342
Compressor
rental .28
Concrete
additives .47
architectural .47
bonding primer .493
coatings .495
coloring .47
curb .44
delivery charges .46
driveways .46
embossed finish .50
enamel .495
epoxy floor coating .495
expansion joints .46
exposed aggregate .49
finishing .50
flatwork .40
floor painting labor .474-475
footings, demolition .31
form stripping .43
forms .42, 48
foundations .48
grade beams .48
granite aggregate .46
grout mix .46
grouting .50
high early strength .47
joints .46
keyway .48
lightweight .46
mix .46
patching .50
plasticized .47
protective coating .495
pump mix .46
pumping .39
ready-mix .46
recycling .29
reinforcing steel .44
retaining walls .47
sawing .50
sealer .495
shingles .140
sidewalk .49
slab demolition .32
slabs .49
specialty finishes .50
stain .495
stamped finish .50
tile .139
tile roofing .140
topping .50
walks .49
wall demolition .32
wall form .43
walls .47
waterproofer .495
waterproofing sealer .495
Concrete & masonry waterproofer . .495
Concrete block
demolition .31
painting .476-478
Concrete conveyor
rental .39
Concrete deck
patch built-up roofing .131
Concrete pier replacement .64
Concrete saws
rental .39
Conditioner
wood .491
Conduit
Carflex liquid-tight .427
EMT .422, 426
ENT .427
flex appliance connection .417
flexible aluminum .426
flexible steel .426
galvanized IMC .429
GRS .429
IMC .429
installation, rigid .429
removal, rigid .429
Schedule 40 PVC .420
Schedule 80 PVC .420
Conduit body .421
PVC .421
threaded aluminum .431
Conduit box
adapter .421
cover, steel .424, 426
installation .424
non-metallic round .428
steel .425
Conduit bushing
rigid .430
Conduit clamp
Snap Strap PVC .420
Conduit connector
90-degree .426
flex .427
Conduit coupling
rigid .429-430
Conduit elbow
45-degree rigid steel .429
rigid or IMC steel .429
Conduit fitting
Carflex .427
non-metallic .421
Conduit hub
watertight .430
Conduit locknut .430
Conduit nipple .431
Conduit sill plate
service entrance .432
Conduit strap .431
Connector
90-degree flexible .426
dishwasher .341-342
EMT compression .423

EMT offset screw ..............423
EMT set screw ...............423
EMT-to-box offset compression .424
flex screw-in ...............426
flex squeeze ...............427
non-metallic UF compression ...419
two-piece clamp ...........419
Continuous vents ................90
Contractors Wardrobe .......356, 359
Control switch
fan .......................382
Convection heating, baseboard ...398
Convertible range hood .........343
Cooktop
electric ................342-343
Cooling system, patio ...........458
Copper
formed flashing ...............142
roll flashing ................141
window roofs ...............165
Copper pipe
DWV ......................408
water ....................408
Copper step flashing ...........142
Copper wire
type THHN ...............432
Copper wood preservative ......491
Cord
disposal accessory kit .........340
Corian....................322
countertops .................322
Cork
roll ........................293
Cork tile
natural .....................293
Corner
MDF primed base .............261
plastic bullnose ..............247
Corner and cove molding
inside ...............266-267
paperwrap ................266
prefinished ................267
Corner bead
bullnose ...................247
drywall ...................247
flex .......................248
inside ....................247
outside ...................248
spray adhesive ..............249
Corner block
polyurethane ................262
Corner cabinet
frameless ..................383
Corner entry shower kit .........358
Corner molding
inside .....................267
outside ...................261
paperwrap outside ...........267
polystyrene inside .............267
polystyrene outside ...........267
prefinished inside ............267
prefinished outside ..........267
Corner shelf, ceramic ...........304
Corner tape
flexible metal ...............249
Corner trowel ................250
Cornices
painting ...................480
Corrugated
fiberglass ..................140
metal roofing ...............138
metal siding ...............114
CorStone ............330-331, 338
Cottage cheese ceiling finish ....244
Cotton insulation ...............88
Cotton roof patch
asphalt saturated .............136
Counter edge

oak .......................268
Counterflashing .................144
Countertops ...................322
Corian ...................322
granite ..................323
laminated plastic .........323-324
marble vanity .............376
miter bolt kit ..............324
plastic laminate .............323
setting with adhesive .........294
setting with mortar ..........294
solid surface ...............322
stone ...................323
tile ......................295
Coupling
combination ................424
EMT to non-metallic sheathed
cable ....................424
non-metallic standard ........421
rigid conduit ...............429
rigid or IMC compression
conduit .................430
Cove and corner molding
inside .................266-267
paperwrap ................266
Cove base
adhesive .................290
adhesive nozzle ..............291
Cove molding .................268
spring ...................268
Cover
air gap ...................335
faucet hole ...............335
Covers
2 gang blank .............428
blank handy box ............424
dual weatherproof lampholder ..428
GFCI wet location receptacle ....428
horizontal duplex weatherproof
receptacle ...............428
receptacle ...............428
steel conduit box .........424, 426
switch ....................428
weatherproof blank ..........428
Crack filler
siliconizer ................136
Crack repair
mesh ...................248
Cracks, foundation
repair .....................37
Cribbing ....................63
Cross bridging ...............66
Cross head cast molding ........262
Crossbuck doors ..............193
Crown molding
paperwrapcolonial .............268
Crown molding .........259, 263
cabinet .................320, 322
MDF ...................262
MDF ultralite ..............262
polystyrene .................268
prefinished .................268
primed finger joint ...........261
Crow's foot brush .............250
Crystalite countertop ..........323
Curb mount fixed skylights ......173
Curbs
concrete ....................44
Cure-Seal concrete sealer .......495
Curved glass wall lantern .......443
Custom Building Products ....305-307
CustomBlend Thin-Set Mortar ...306
CustomFloat bedding mortar ....306
Cutter
vinyl tile ................291
Cutters
glass .....................171
Cutting

slabs .......................50

**D**

Dado partition
demolition ................316
Dal-Tile .............297, 303-304
Danco ...................336
Danish Oil ................488
DAP ....................171
Data communication cable
category 5E ...............433
Deadbolts ...............202
Debris removal .............29
Deck
demolition ................452
lumber ...................453
lumber ...................452
lumber ...................454
lumber, Fiberon .............454
lumber, Thompsonized ........454
misting system .............458
painting ..................475
post & beam ...............452
post cap ..................457
posts ................453, 456
railing ...............455-456
railing, turned spindle ........456
repair ...................452
roof ....................455
staining ..................475
stairs ...................458
Deck & siding stain ...........487
Deck Plus stain .............487
Deck, roof
removal ..................129
repair ...................129
walking ..................140
Deck, roofing
patching built-up .........131-132
Decor Bone kitchen sink basket
strainer .................336
Decora AC quiet rocker switch ....434
Decora wall plate
single gang ...............435
Deflect-O .................382
Delta .....332-334, 339, 354, 360-361,
...................370, 372-373
Demolition
acoustical ceiling .............315
asphalt paving ...............51
brick sidewalk ...............31
brick wall ................30
building components ...........30
carpet ...................35
ceiling ................34, 242
concrete block ..............31
concrete sidewalk ...........31
concrete slab ..............32
concrete wall ..............32
decks, wood .............32, 452
doors ....................35
equipment rental .............28
estimating .................27
fence ....................29
floor cover .................315
flooring ..................278
framing ...................33
garage ...................30
guardrail ..................29
joists ................32-33
masonry ..................30
paneling ..................253
plaster ceiling .............315
porch ................32, 452
shower stall ...............357
sidewalks .................31
skylight, poly dome ..........173
slab .....................30

stairs .....................35
suspended ceiling ............315
wall .....................270
walls ..................30-31
windows ..............35, 156
wood framing ..............30
Demolition checklist ...........27
DensShield tile backer .........245
Dentil cast molding ..........262
Dentil chair rail .............262
Designer panel wallcovering .....252
Designer Series bath fan ........381
Detaching membrane ..........308
Detector
dual-ionization smoke .........436
smoke ...................436
Devoe Coatings .............496
DeWalt ..................249
DeWitts ...................135
Difficulty factors, painting
manhours .................466
overhang .................480
Diffuser fixture
fluorescent ...............441
Diffusers
duct, installation ............397
Dimmer
illuminated slide ............436
incandescent slide ..........436
Dirt, recycling ..............29
Disconnect
appliances .................417
distribution panel ............436
Dish, soap .................304
Dishwasher ..............341-342
installation ................342
removal ..................315
Dishwasher connector .......341-342
compression inlet ...........342
FIP inlet ................342
Dishwasher supply line .........342
Dispenser
hot water ..............336-337
soap ....................335
water cooler ...............337
Disposal
basket strainer ..............336
batch feed ................340
cartridge .................341
food waste .............339-340
mounting assembly ..........340
septic system .............340
sink flange ................341
sink stopper ..............340
waste .....................29
Disposer
garbage ...............339-340
Disposer cord accessory kit ......340
Disposer sink stopper ..........340
Distribution panel
disconnect & remove ..........436
Ditching ...................41
Ditra ....................308
Dock & fence post preservative ...490
Dome light ceiling fixture ........442
Door
cabinet ...................316
framing ...................70
hinges, brass ...............196
kick plates ................213
knockers .................213
openers ..................225
painting labor ...........471-473
peep sights ...............214
pulls ....................213
push plates ...............213
removal ...................35

Door accessories
  bypass ...........................219
Door bar .........................226
Door bell
  wireless ........................436
Door bottom
  aluminum & vinyl ...............196
  drip cap .......................196
Door casing ................191, 206
  colonial ...................195, 260
  ranch ..........................195
  ranch, pine ....................195
Door chime builder's kit .........436
Door closer installation .........217
Door closers ....................217
  hinge pin ......................217
  hydraulic ......................217
  pneumatic ......................217
Door frame
  painting labor .................471
Door frames .....................194
  fire-rated .....................194
  pocket .........................222
  removal ........................197
Door grille .....................216
Door handles, patio .............216
Door hangers ...................222
Door hardware ..................216
  track ..........................222
Door hinge, remove .............196
Door installation ........191, 206, 209
  interior .......................206
Door jambs
  exterior .......................194
  garage .........................225
  interior .......................209
  patio door .....................217
  pocket .........................222
  remove & replace .........191, 206
Door kit ........................222
Door lever
  privacy ........................213
Door locks ......................218
  patio ..........................216
Door locksets ..............201-202
Door push bar ..................226
Door roller .....................216
Door stop .......................213
Door sweep
  Flex-O-Matic ..................196
Door threshold ..................217
Door track hardware ............222
Door trim .......................194
  painting labor .................471
Doors
  4-panel ........................193
  6-panel ........................192
  bar ............................209
  basement ......................389
  bathtub ........................356
  bi-fold ....................219-222
  bi-fold, louvered ..............221
  birch .....................192, 210
  birch, prehung ................197
  bypass ........................218
  bypass, mirrored .........218-219
  cafe ...........................209
  camber top ....................199
  casing .........................195
  cellar .........................194
  closers ........................217
  closet ...............210, 218-219
  crossbuck ......................193
  demolition ......................35
  entrance .......................194
  entry ...............193, 198-200
  entry, decorative fan ..........200
  entry, decorative half lite ....200

exterior ..........192-193, 197-198
exterior fan lite ................197
exterior flush ..................197
exterior prehung .........197-199
exterior, installation ............191
fiberglass .................200-201
fir .............................193
fire ............................192
fire-rated ......................197
flush ......................219-220
flush closet ....................219
folding ....................217-218
foundation access ...............89
French .........................193
French, interior ................208
garage .....................223-224
garage, extension spring .......225
garage, jamb hinge .............225
garage, wood ..................224
handle sets ....................202
hardboard ................192, 209
hardware .......................221
hardware, exit .................226
hemlock .......................197
installation ........191, 197, 206, 209
interior ..........207-208, 211-212
interior cafe ...................209
interior slab ...................206
interior, colonial ..............211
interior, colonist ..............211
interior, fir ...................208
interior, French ...........212-213
interior, full louver ...........212
interior, hardboard ............207
interior, hollow core ..........207
interior, oak ..............208, 212
interior, pine ..................208
interior, prehung .......209-210, 212
interior, solid core ............207
lauan ..........................210
lever latchset .................213
lockset repair .................201
metal, reversible ..............194
mirrored ..................221-222
mirrored, frameless ...........222
mobile home ..............222-223
molded face ..............207-208
oak .......................192, 210
pantry .........................212
patio ..........................214
patio grille ...................216
patio handle ..................216
patio lock ....................216
patio roller ..................216
patio threshold ...............217
patio, aluminum ..............215
patio, double .................215
patio, fiberglass .............216
patio, gliding ................214
patio, hinged .................214
patio, sliding ............214-215
patio, steel ..............215-216
patio, swinging ..........214, 216
patio, vinyl ..................215
pet ...........................216
pocket ........................222
prehung .......................197
repair ........................191
screen ........................203
screen, installation ..........202
screen, replacement .......203, 216
screen, vinyl .................203
screen, wood .................203
screens, patio ...............215
security ......................205
security with screen .........205
shower ..............356, 359-360
sidelites .....................198

slab .......................192-193
solid core ................192, 197
steel ..........................198
steel exterior .................199
steel, 6-panel ................198
steel, entry ..................198
steel, exterior ...............199
steel, prehung ...............198
stile & rail ...................208
stile and rail ................207
storm .....................204-205
storm, aluminum .............204
storm, removal ..............204
storm, self-storing ..........204
weatherstripping .............196
Dormer
  studs ......................76-77
  vent ........................144
  vents ........................91
Double bowl kitchen sink .......329
  cast iron .....................329
  stainless steel ...............328
Double cylinder door handle ....202
Double doors ..............211-212
Double oven
  wall ..........................343
Double-hung windows
  insulated .....................159
  replacement ..................157
  screens .......................160
  vinyl-clad ....................159
  wood .........................164
Double-track storm windows .....166
Downlight
  ceiling fixture ................440
Downspouts
  acid wash .....................466
  galvanized steel ..............146
  painting labor ................476
  remove and replace ...........107
  residential ...................146
Drain connector
  garbage disposal to dishwasher ..342
Drain stopper ...................336
  flat sink .....................336
Drainboards, countertops ........295
Drawer front
  cabinet .......................316
Drawn box
  2-device switch ...............425
  handy .........................424
  octagon .......................425
Drip cap ........................141
  door bottom seal .............196
  molding, wood ................115
Drip edge .......................142
  aluminum .....................141
Driveway
  forms .......................43, 49
Driveways
  concrete ......................49
Drop ceiling
  panel .....................256-258
  soffit ........................316
Drop-in lavatory sink ...........367
Drum mushroom ceiling fixture ...441
Dry tile grout
  non-sanded ...................307
Dryback vinyl wall base .........292
Drywall
  arch, bullnose ...............247
  ceiling, demolition ...........315
  corner bead ..................247
  corner tape ..................249
  demolition ...........34, 242, 316
  dustless sanding system ......251
  finish ........................244
  fire code type X .............243

fire-resistant ..................243
flexible .......................244
greenboard ....................243
hammer ........................250
hand sander ...................250
installation ...............242, 244
J bead ........................248
J trim ........................248
joint compound ...........245-246
joint tape .....................246
L trim ........................248
moisture resistant ............243
moisture-resistant ...........243
nails .........................249
outside corner trowel .........250
painting ..................483-484
panel lift .....................250
patch .........................242
primer ........................246
repair ........................242
repair clips ...................249
repair kit .....................248
repair patch ..................248
repair sheet ..................243
sander ........................251
sander kit ....................250
sanding respirator ............250
sanding sponge ...............250
screwdriver ...................249
screwdriver bit ...............250
screws ........................249
screws, collated ..............249
sound deadening .............243
spray texture .................244
stilts ........................250
tape ..........................246
texture .......................244
texture brush .................250
texture spray .................244
tool kit ......................251
topping compound ...........246
trowel ........................250
T-square ......................250
vinyl outside corner bead ......248
Dual sensing smoke alarm .......436
Dual weatherproof
  lampholder & cover ...........428
Dual-ionization smoke detector ...436
Duct
  heating, removal .............396
  runs, installation .............397
Ductless range hood ...........344
Dumont kitchen sink ...........329
Dump fees ......................29
Dump trucks
  rental .........................28
Dumpster
  rental .........................29
  trash ..........................29
Duplex receptacle
  CO/ALR .......................435
  commercial grade ............435
  handy box cover ..............424
  straight blade ................434
  waterproof cover .............428
Durabase ......................358
Durastall ......................358
Durock backerboard ...........295
Durock cement board ..........245
Dustless drywall sanding system ..251
DWV pipe
  remove & replace .............407

**E**

E.L. Mustee ............337-338, 358
Easysills ......................166
Eave
  vents .........................91

Eave closure .....140
Eaves
painting .....480
Edeck .....452
composite deck .....454
Edge
counter .....268
drip .....141-142
flashing, roof .....143
Edge forms, concrete .....40, 43
Edge protection tile .....303
Edge pulls .....222
Edge V plank .....253
Efficiency cast iron kitchen sink .....329
Egg & dart cast molding .....262
Egg knob locksets .....202
Eggshell wall paint, latex .....492
Elastocaulk .....136
Elastomastic .....136
Elastomeric
roof coating .....137
Elastomeric waterproofing paint .....494
Elastotape .....136
Elbow .....429
45-degree plain end .....421
90-degree .....423
90-degree standard radius .....421
EMT 45-degree .....423
EMT 90-degree .....423
pulling .....423
rigid or IMC steel conduit .....429
rigid steel conduit .....429
service entrance .....432
Electric baseboard heaters .....404
Electric cooktop .....342-343
Electric metallic tube conduit .....422, 426
Electric service panel .....439
Electric wall oven .....343
Electric water heater .....406-407
PowerStar .....407
tankless .....407
Electrical non-metallic tubing
conduit .....427
Eliane ceramic tile .....297
Eljer .....329
Eljer Plumbingware .363-364, 367, 369
Elkay .....337
Elkay Manufacturing ..328-329, 337-338
Embossed basswood molding .....259
Embossed concrete .....50
Embossed decorative molding .....259
Embossed polystyrene casing .....265
Embossed polystyrene molding
base .....263
chair rail .....266
crown .....268
Embossed prefinished molding
casing .....265
chair rail .....266
Embossed tileboard .....252
Embossing leveler .....285
Emser International .....298
EMT conduit .....422, 426
45-degree elbow .....423
90-degree elbow .....423
combination coupling .....424
compression connector .....423
coupling .....424
offset compression connector .....424
offset screw connector .....423
set screw connector .....423
strap .....424
EMT strap .....424
Emulsion
asphalt .....133
Enamel
floor & porch .....495
interior & exterior .....493
latex .....492
oil-based .....496
painting .....483-484
wall & trim .....492
Enamel undercoater
interior .....491
Encapsulated fiberglass roll
insulation .....86
Enclosure, shower .....356
pivot door .....359
End panel
base cabinets .....317
cabinet .....316
cabinets .....318
wall cabinets .....319
ENT conduit .....427
Entertainment sink .....338
Entrance cap
PVC .....422
Entry doors .....193, 198-200
deco fan lite .....193
fiberglass .....200-201
oak .....194
Epoxy floor coating .....495
Epoxy gloss coating .....496
Epoxy mortar .....306
Epoxy primer .....496
Equipment rental
demolition .....28
foundations and slabs .....39
Escape window .....173
Estate Nordic Pine paneling .....252
Estimating
demolition .....27
Estwing .....250
Eternit .....139
Excavation .....41
backfilling .....40
fine grading .....40
footings .....40
rock .....40
spreading .....41
tamping .....41
topsoil .....41
Excelon vinyl composition tile .289-290
Exhaust fan .....380
bath .....378-381
bath, vertical discharge .....379
ceiling .....379
QuieTTest .....380
switch .....382
timer switch .....382
utility, vertical discharge .....379
ValueTest with light .....380
vertical discharge .....379
wall .....379
with heater .....381
with infrared heater .....381
Exhauster, roof .....145
Exit bar .....226
Expansion joints
concrete .....46
Exposed aggregate finish .....49
Exposed aggregate sealer .....495
Exposed round conduit box
non-metallic .....428
Express Toilet In A Box .....362
Extendable roller .....291
Extension kit
jamb .....217
Exterior & interior stain
acrylic .....487
Exterior clear finish .....489
Exterior door jambs .....194
Exterior doors .....193
fir .....192
flush .....192
prehung .....197
Exterior jambs
kerfed .....194
Exterior paint
satin flat .....494
Exterior painting .....466
Exterior sanding
manhours .....467
Exterior stain .....487
Plus 10 .....487
Exterior wall assemblies .....72
Exterior window shutters .....168
Extruded polymer molding .....263
Eyeball trim light kit .....440


## F


F channel .....459
F connector
for RG6 coaxial .....433
Fabback acrylic mirrored sheet .....172
Fabric
asphalt roofing .....131
polyester stress-block .....136
Factors
high time painting .....466
painting difficulty .....480
Fairfax faucet
single-handle .....371
two-handle .....374
Fairfax kitchen faucet .....332, 334
Fan
ceiling .....444
gazebo .....444
outdoor ceiling .....444
Fan lite doors .....197, 201
Fans
attic exhaust .....92
bath .....378-381
bath exhaust .....378, 380-381
bath exhaust & heater .....381
bath exhaust, QuieTTest .....380
bath exhaust, vertical discharge .379
bath with heater .....381
bath, premium .....379
SensAire control switch .....382
utility exhaust, vertical discharge .379
utility, thru-the-wall .....380
wall exhaust, chain-operated .....380
Fascia
aluminum .....115
board .....77
staining .....473
vinyl .....109
Fashion Series .....252
Fasteners .....458-459
Fast-setting patch .....285
Faucet hole cover .....335
Faucets
bar .....338
bar, single handle .....338
bar, two-handle .....338
bidet, Valvuetwist .....362
compression, repair .....352
kitchen .....331-333
kitchen, two-handle .....333-334
laundry, two-handle .....337
lavatory .....370-372
lavatory, adjustable center .....374
lavatory, center set .....371-374
lavatory, infrared .....371
lavatory, two-handle .....372
pressure balanced .....361
Roman tub .....354
shower .....360
shower & tub .....360
supply .....374
tub & shower .....360-362
Feather edge additive .....286
Fees
recycling .....29
Felt
roofing .....133
roofing, glass fiber .....134
scribing .....291
Fence
demolition .....29
staining labor .....474
Fence & barn paint .....496
Fence post preservative .....490
Fiber cement
lap siding .....113
siding .....114
siding panel .....113
soffit .....113
trim for siding .....113
Fiber-cement
roofing .....139
Fibered aluminum roof coating .....137
Fibered roof coating .....137
Wet-Stick .....137
Fiberglass
base sheet .....133
bathtubs .....357
ceiling panel .....256-257
drywall tape, white .....246
panels .....140
reinforcing membrane .....136
screen rolls .....169
shower stall .....358
window trim, bath .....355
Fiberglass asphalt shingles .....131
Fiberglass ceiling panel .....257
Fiberglass doors
entry .....200-201
Fiberglass insulation
batt .....85
roll .....85-86
unfaced .....85
Fiberglass mat roof patch .....135
Fiberglass reinforced plastic
panels .....253
Fiberon .....452, 454
Field adhesive
SBS .....134
Field tile
setting with adhesive .....294
setting with mortar .....294
Fill
excavation .....40
Filler
crack .....136
Filler strip .....317-319
oak .....321
Filler strip, wood .....138
Film
polyethylene .....88
Finish carpentry
countertops .....322
siding .....108
Finish sealer
high-gloss .....307
Finished box cover .....425
Finishes
exterior clear .....489
flat latex .....492
floor, oil-based .....489
interior classic clear .....489
log home gloss .....489
rust-resistant .....496
stain gel, non-wood .....488
wood .....487
Finishing
concrete .....49-50
Finishing compound
wallboard .....246

Finishing putty
  Pergo .........................283
Fink truss .......................77
FIP inlet
  dishwasher connector .........342
Fir doors ...................192-193
Fire doors ......................192
Fireblocks .......................69
Firebrick .......................116
Fireplaces
  demolition ...................116
  prefabricated .................117
  repair .......................116
Fire-rated door frames ...........194
Fire-rated doors ................197
Fire-resistant drywall ...........243
First Alert ......................436
Fish tape
  cable snake ..................417
Fissured ceiling panel ...........257
Fixed skylights .............173-174
Fixture gasket removal ...........352
Fixture supply lines .............374
Fixture valve lines .............374
Fixture, light
  ceiling ......................442
  chandelier ...................442
  chrome bath bar ..............441
  fluorescent ceiling ..........441
  flush mount ceiling ..........441
  jelly jar ....................443
  motion sensing ...............443
  mushroom flush mount .........442
  outdoor bulkhead .............443
  outdoor wall lantern .........443
  remove & replace .............441
  wave bath ....................441
Fixtures
  troffer fluorescent ..........258
  U-lamp troffer fluorescent .......258
Flange kit
  load center ..................437
Flashing
  aluminum roll ................141
  aluminum siding ..............114
  angle ........................142
  copper formed ................142
  counterflash .................144
  galvanized ...................142
  gravel stop ..................142
  lead .....................143, 145
  patch ........................135
  pipe .....................143, 145
  removal ......................129
  repair .......................129
  roll valley ..............141, 143
  roof .........................134
  roof edge ................143-144
  roof, removal ................134
  roof-to-wall .................144
  service entrance roof ........432
  shingle ......................144
  step .........................144
  w-valley .....................144
Flashing cement
  pro-grade ....................135
Flashing paper
  roof ..........................89
Flat masonry & stucco paint .....494
Flat paint .......................492
  exterior, satin ..............494
  interior .....................492
  latex ........................494
Flat sink drain stopper ..........336
Flex conduit
  appliance connection .........417
Flex screw-in connector ..........426
FlexBond bonding mortar .......306

Flexible aluminum conduit .......426
Flexible armored BX cable .......419
Flexible bonding mortar .........306
Flexible connector
  90-degree ....................426
  90-degree insulated ..........426
Flexible metal corner tape ......249
Flexible steel conduit ...........426
Flip locks
  window .......................172
Float finish
  concrete ......................49
Floating floor
  installation kit .............280
Flood Company, The ..............490
Floor
  beam .........................63
  beam, shimming ...............63
  clamp ........................281
  decking, subflooring ..........67
  demolition ................33, 35
  furnace, gas .................404
  furring ......................73
  joists ....................65-66
  repairs, framing .............63
  sagging ......................54
  slab, concrete ...............49
  staining .....................475
Floor & porch enamel ............495
Floor & porch oil gloss .........495
Floor & wall tile
  ceramic ......................297
Floor coating
  epoxy ........................495
Floor enamel ....................495
Floor finish
  oil-based ....................489
Floor molding
  hardwood .....................278
Floor primer ....................285
Floor squeaks .....................56
Floor tile
  ceramic ..................295-298
  granite ......................299
  marble .......................298
  natural stone ................299
  parquet ......................282
  Place n Press ................289
  Saltillo ceramic .............299
  slate ........................299
  travertine ...................299
  vinyl ....................288-289
  wood .........................282
Floor trim
  wood .........................279
Floor warming system ...........305
  thermostat ...................305
Flooring
  adhesive for rubber tile .......290
  adhesive for vinyl sheet .......290
  adhesive for vinyl tile .......290
  adhesive for wood ............282
  adhesive remover .............285
  bamboo plank .................280
  carpet .......................308
  ceramic tile .................294
  countersink ...................56
  demolition ...........35, 278, 315
  hardwood plank ...............280
  hardwood, sanding ............281
  heart pine plank .............280
  Hevea parquet ................282
  installation spacer ..........283
  laminate floor underlayment ...284
  laminate strip ...............284
  maple hardwood ...............279
  molding, laminate ............284
  molding, wood laminate .......283

Northwoods hardwood .........278
oak .............................279
oak parquet .....................282
oak self-stick parquet ..........282
oak, solid hardwood .............278
Pergo ...........................283
Pergo underlayment ............283
red oak strip ...................281
removal .........................278
repairs ..........................63
rustic oak plank ................280
sealant, laminate ...............283
setting with adhesive ...........294
setting with mortar ............294
sheet vinyl ................286-287
strips, natural wood ............279
surface preparation .............284
teak ............................279
Traffic Master laminate .........284
transition reducer strip ........292
transition strip ................292
trim, hardwood ..................281
underlayment ....................285
underlayment & patch ............286
underlayment skimcoat .........286
vinyl roll .................286-287
vinyl sheet .....................287
vinyl wall base .................293
white oak strip .................281
wood ............................281
wood laminate ...................283
wood plank ......................280
wood, installation tools ........281
Flooring installation system
  laminate floor ...............284
Flooring trim ...................284
  bamboo .......................280
  heart pine ...................281
  rosewood flooring ............279
Flue
  chimney ......................399
  furnace ..................399-403
Fluidmaster .................342, 364
Fluorescent ceiling fixture .......441
Fluorescent diffuser fixture
  low profile ..................441
Fluorescent fixtures
  parabolic troffer ............258
  troffer ......................258
  U-lamp troffer ...............258
Flush door
  painting labor ..............471472
Flush doors .....................219
  bi-fold closet ...........219-220
  birch exterior ...........192, 197
  birch interior ......206-207, 210
  fire-rated ...................192
  hardboard ....................192
  hardboard exterior ...........197
  hardboard interior ....206-207, 209
  hardwood exterior ........192, 197
  lauan exterior ...............192
  lauan interior .......206-207, 210
  oak entry ....................194
  oak interior .............206, 210
  red oak interior .............210
  steel exterior ...........198-199
  steel utility ................198
Flush mount ceiling fixture ...441-442
Flush mount light fixture ........442
Flush valve
  Sloan Royal ..................363
  urinal, Sloan Royal ..........363
Flute & reed molding ............261
Fluted casing embossed polystyrene
  molding ......................265
Fluted casing prefinished molding .265
Foam insulation ..................86

Foam underlayment .........283-284
Foam weatherstripping ..........196
Foamular XAE ....................86
Foil
  aluminum ......................89
  barrier .......................89
Folding doors ...............217-218
  locks ........................218
Footing
  demolition ....................31
Footings .........................48
  concrete ..................46, 48
  continuous ....................48
  excavation ....................40
  keyway ........................48
Forced air furnace ..............397
  removal ......................397
Foremost International ..........363
Formica countertops .............322
Forms
  board ........................43
  concrete ......................48
  curb ..........................44
  footings ......................44
  foundation ................43-44
  keyway ........................44
  plywood .......................43
  stripping .....................44
  wall ..........................43
Formwork
  stripping .....................44
Forsyth composite sink ..........330
Foundation
  access doors ..................89
  checklist, pre-construction ....37
  cracked .......................37
  demolition ....................31
  deterioration .................37
  repair ........................37
  vent ..........................89
  vents .........................89
Foundation and roof coating
  non-fibered ..................129
Foundation insulation ...........87
Foundations
  concrete ..................46-47
  equipment rental ..............39
  footings ......................48
  forms .........................43
Frame
  painting labor ...............471
  sink .........................335
Framed bypass door
  bathtub ......................356
Frameless corner cabinet .........383
Frames
  door .........................194
Framing
  cut window opening ...........156
  demolition ...........32–33, 270
  interior doors ...........206, 209
  partitions, wood ..........67–68
  repairs .......................63
  room additions ................65
  wall studs ....................67
Framing lumber ..................452
Franklin Brass ..............386-387
French door
  painting labor ...........471-472
French doors ................193, 208
  bi-fold ......................221
  interior, double .........212-213
Frenchwood .................214, 216
Frieze cast molding .............262
FRP panel molding ...............254
FRP wall and ceiling panel ......253
Full-view storm doors ..........205

Furnace
coal ....404
counterflow, gas ....397
gas floor ....404
gas wall ....404
gas, duct ....396
gas, thru-wall ....404
horizontal flow, gas ....397
low boy, gas ....396
oil fired ....397
removal ....397
wood burning ....404
Furnace flue ....399-403
Furring
floor ....73
wood ....73

## G

Gable
truss ....77
Gable louver, metal ....91
Gable vents ....90
metal ....91
power attic ....93
redwood ....91
Gabrielle Comfort Height toilet ....363
GAF ....131, 134, 145
GAFGLAS
ply 4 glass felt ....134
Torch and Mop System ....134
Galvanized
bullnose corner bead ....247
Galvanized flashing ....142
Galvanized metal
soffit vents ....90
Galvanized rigid steel conduit ....429
Galvanized roofing ....138
louver ....145
Galvanized steel
rain gutters ....146
Gap, air ....335
Garage
demolition ....30
Garage doors ....223
bonded steel ....224
bottom seal ....223
extension springs ....225
hardware ....224-225
installation ....223
jamb hinge ....225
locks ....224
low headroom kit ....225
operators ....225
raised panel ....224
replacement ....223
steel ....223
steel, non-insulated ....224
torsion spring ....223
trollies, installation ....223
weather seal ....224
wood ....224
Garage slabs ....42
Garbage disposal
basket strainer ....336
connector, dishwasher drain ....342
Garbage disposer ....339-340
cord kit ....340
mounting assembly ....340
mounting gasket ....341
removal ....315
sink flange ....341
sink stopper ....340
switch, sink top ....341
Garden windows ....158
Gardner ....135-136
Gardner Bender ....417
Gas breaker
rental ....28
Gas furnace
counterflow ....397
exterior wall duct ....396
horizontal flow ....397
interior wall duct ....396
low boy ....396
oil fired ....397
removal ....397
Gas water heater ....405
AquaStar ....406
power vent ....407
tankless ....406
Gasket
fixture, remove & replace ....352
mounting ....341
wax toilet bowl ....364
Gates
security ....205
Gazebo ceiling fan ....444
GE Appliances ....341, 343
Gel stain ....488
General requirements ....29
Genesis kitchen faucet ....332
Georgetown lavatory faucet ....373
Georgia-Pacific ....245, 251-253
GFCI circuit breaker ....437
plug-on ....440
GFCI receptacle ....425, 435
duplex waterproof ....428
hospital grade ....435
vertical ....428
wet location cover ....428
GFI plug-on circuit breaker ....440
Girders
carpentry ....63
Glacier Bay ....332, 334, 338-339, ....360, 370-372
Glacier Extensa kitchen faucet ....332
Glaslock interlocking shingles ....131
Glass
jalousie ....170
measure & cut ....170
single strength ....170
Glass block
mortar mix ....306
Glass cutters ....171
Glass felt
ply 4 ....134
Glass globe ceiling fixture ....442
Glazing bead ....170
Glazing compound ....170
latex ....170
Glazing points ....170
Glazing putty
remove and replace ....170
Glazing tool ....171
Glidden ....492-495
Gloss
polyurethane oil ....495
Gloss & seal, tile ....308
Gloss enamel
alkyd ....493
Gloss finish, log home ....489
Glue
Pergo laminate floor ....283
Grade beams
concrete ....48
formwork ....44
Grand Sequoia Shingles ....131
Granite
aggregate concrete ....46
floor tile ....299
mortar mix ....306
setting adhesive ....306
tile, natural ....299
Granite countertop ....323
Granite vanity top ....377
Granules
roofing ....136
Gravel stop ....142
Gray Board insulation ....86
Grecian cast molding ....262
Greek key chair rail ....262
Greenboard ....72, 243
Greenhouse
panels ....140
Greenwaste
recycling ....29
Greenwich kitchen sink ....330
Grilles
door, patio ....216
Ground bar kit
breaker box ....440
QO load center ....440
Ground fault interrupter
plug-on circuit breaker ....440
Ground rod clamp ....432
Grounding bushing
insulated metallic ....430
Grout
colorant ....307
Saltillo ....307
sanded ....307
sealer ....308
tile ....294
tile, non-sanded dry ....307
Grout mix concrete ....46
Grouting
concrete ....50
GRS conduit ....429
installation ....429
removal ....429
Guard
window ....168
Guardrail
demolition ....29
Gutters
acid wash ....466
aluminum ....146
flashing cement ....135
galvanized steel ....146
painting labor ....476
patching material ....135
remove and replace ....107
roof caulk ....136
vinyl ....146
Gutting, demolition ....30
Gypboard
painting ....483-484
Gypsum
demolition ....34
greenboard ....72
type X ....72
Gypsum base
Imperial ....245
Gypsum lay-in ceiling panel ....257
Gypsum panel lift ....250
Gypsum wallboard
demolition ....242
fire code ....243
flexible ....243
installation ....242, 244
joint cement ....245-246
joint tape ....246
primer ....246
repair ....242
spray texture ....244
water resistant ....243

## H

Hammer
drywall ....250
Hampton Bay ....441-444
Hand sander ....251
drywall ....250
Handiflo ....337
Handle sets ....216
double cylinder ....202
entry door ....202
Handrail
painting ....481
patio deck ....455
Handy box
cover, blank ....424
drawn ....424
welded ....424
Hanger bracket
lavatory ....369
Hangers
pocket door ....222
Harbour Collection ....289
Hardboard
lap siding ....113
panel siding ....112
Hardboard doors ....192, 207, 209
fire-rated ....197
Hardibacker ....295
Hardie ....113
Hardipanel ....113
Hardiplank ....113
Hardisoffit ....113
Hardware
bi-fold door ....221
door, patio ....216
garage door ....224-225
pocket door ....222
screen door ....203
Hardwood
flooring demolition ....35, 315
molding, embossed ....259
molding, Victorian base ....259
paneling ....251
Hardwood doors ....192
fire-rated ....192
prehung ....197
Hardwood flooring
maple ....279
molding ....278, 280
oak ....278-279
removal ....278
rosewood ....279
sanding ....281
strip ....278
TapTight ....280
trim ....281
Hardwood molding ....278, 280
bruce parquet ....282
Harris-Tarkett ....278-281
Hartland cast iron kitchen sink ....330
Headers
carpentry ....69
Headroom kit ....225
Heart pine
flooring trim ....281
plank flooring ....280
Heat-A-Lamp bulb heater & fan ....381
Heat-A-Ventlite heater ....382
Heaters
baseboard, electric ....404
Heat-A-Ventlite ....382
water ....405
water, electric ....406-407
water, gas-fired ....405
water, oil-fired ....405
water, propane-fired ....405
whirlpool bath EZ-Install ....354
Heath Zenith ....443
Heating duct
removal ....396
Heating system
component removal ....396
Heating, hydronic ....398

Helmsman
    spar urethane . . . . . . . . . . . . . . . . .488
Hemlock doors . . . . . . . . . . . . . . . . . .212
Henry . . . . .133, 135-137, 285-286, 290
Hevea parquet flooring . . . . . . . . . .282
High boy threshold, oak . . . . . . . . . .195
High early strength
    concrete . . . . . . . . . . . . . . . . . . . . .47
High time factors . . . . . . . . . . . . . . .466
High-gloss enamel . . . . . . . . . . . . . .493
High-gloss finish sealer . . . . . . . . . .307
Hilti . . . . . . . . . . . . . . . . . . . . . . . . .250
Hinge
    cabinet . . . . . . . . . . . . . . . . . . . . .328
    cabinet, Euro . . . . . . . . . . . . . . . .327
Hinges
    door . . . . . . . . . . . . . . . . . . . . . . .196
    garage door . . . . . . . . . . . . . . . . .225
    solid brass . . . . . . . . . . . . . . . . . .196
Hip and ridge shingles . . . . . . . . . . .130
Hip shingles . . . . . . . . . . . . . . . . . . .131
Hole cover
    faucet . . . . . . . . . . . . . . . . . . . . .335
Hollow core doors . . . . . . . . . . . . . .207
Hollow metal doors . . . . . . . . . . . . .194
Homax . . . . . . . . . . . . . . . . . . .244, 493
HomeLine
    circuit breaker . . . . . . . . . . . . . . .437
    indoor main lug load center . . . .437
HomeStyle suspended ceiling . . . .256
Hood
    range . . . . . . . . . . . . . . . . . .343-344
    range, ductless . . . . . . . . . . . . . .344
    range, quiet . . . . . . . . . . . . . . . . .344
    range, vented . . . . . . . . . . . . . . .344
Hooks
    bathrobe . . . . . . . . .385-386, 388-389
    robe . . . . . . . . . . . .385-386, 388-389
Hopper gun
    ceiling finish . . . . . . . . . . . . . . . .244
Hopper vinyl window . . . . . . . . . . . .159
Hospital grade receptacle
    GFCI . . . . . . . . . . . . . . . . . . . . . . .435
Hot mix, spreading . . . . . . . . . . . . . .51
Hot water dispenser . . . . . . . . .336-337
Hotmopping
    roofing . . . . . . . . . . . . . . . . . . . .132
House numbers . . . . . . . . . . . . . . . .214
House wrap . . . . . . . . . . . . . . . . . . . .89
Hub
    bolt-on A type universal . . . . . . .438
    watertight conduit . . . . . . . . . . .430
Humidifier, central . . . . . . . . . . . . . .397
Hunter . . . . . . . . . . . . . . . . . . . . . . .444
HVAC
    central air conditioning . . . . . . . .398
    component removal . . . . . . . . . . .396
    duct . . . . . . . . . . . . . . . . . . . . . . .397
    hydronic heating . . . . . . . . . . . . .398
    hydronic heating system . . . . . . .398
    radiator, cast iron . . . . . . . . . . . .398
    radiator, relocation . . . . . . . . . . .398
    removal of components . . . . . . . .396
Hydraulic door closers . . . . . . . . . . .217
Hydraulic jacks . . . . . . . . . . . . . . . . .63
Hydronic heating . . . . . . . . . . . . . . .398

**I**
Ice dam shield . . . . . . . . . . . . . . . . . .89
Illuminated incandescent slide dim-
mer . . . . . . . . . . . . . . . . . . . . . . . . .436
IMC conduit . . . . . . . . . . . . . . . . . . .429
    body . . . . . . . . . . . . . . . . . . . . . . .431
    bushing . . . . . . . . . . . . . . . . . . . .430
    coupling . . . . . . . . . . . . . . . .429-430
    elbow . . . . . . . . . . . . . . . . . . . . . .429
    hub . . . . . . . . . . . . . . . . . . . . . . .430
    nipple . . . . . . . . . . . . . . . . . . . . .431

service entrance elbow . . . . . . . .432
    strap . . . . . . . . . . . . . . . . . . . . . .431
Imperial
    gypsum base . . . . . . . . . . . . . . . .245
Incandescent ceiling fixture . . . . . .442
Incandescent slide dimmer . . . . . . .436
Indoor & outdoor cable
    type NM-B Romex sheathed . . . .418
Indoor cable
    type NM-B Romex sheathed .417-418
Indoor main lug load center . . . . . .438
    Homeline . . . . . . . . . . . . . . . . . .437
Infiltration barrier . . . . . . . . . . . . . . .89
Infrared bulb heater & fan . . . . . . . .381
Ingenium Flushing System . . . . . . .363
Innovations lavatory faucet . . . . . . .373
Innovations Series Faucet . . . . . . . .370
Inside corner & cove molding . . . . .266
Inside corner bead
    metal . . . . . . . . . . . . . . . . . . . . . .247
Inside corner molding
    polystyrene . . . . . . . . . . . . . . . . .267
    prefinished . . . . . . . . . . . . . . . . .267
In-Sink-Erator . . . . . . .336-337, 339-341
Insl-X . . . . . . . . . . . . . . . . . . . . . . . .496
Installation kit
    floating floor . . . . . . . . . . . . . . . .280
    Pergo . . . . . . . . . . . . . . . . . . . . . .283
Installation system
    laminate flooring . . . . . . . . . . . . .284
Insulated grounding bushing,
metallic . . . . . . . . . . . . . . . . . . . . . .430
Insulating bushing
    high-impact thermoplastic . . . . . .430
Insulation
    blow-in cellulose . . . . . . . . . . . . . .86
    board . . . . . . . . . . . . . . . . . . . . . . .86
    board, demolition . . . . . . . . .34, 316
    built-up roofing, patch . . . . . .131-132
    FBX . . . . . . . . . . . . . . . . . . . . . . . .87
    fiberglass . . . . . . . . . . . . . . . . .85-86
    fiberglass batt . . . . . . . . . . . . . . . .85
    fiberglass, unfaced . . . . . . . . . . . .85
    foam . . . . . . . . . . . . . . . . . . . . . . .86
    foundation . . . . . . . . . . . . . . . . . .87
    masonry . . . . . . . . . . . . . . . . . . . .88
    natural fiber . . . . . . . . . . . . . . . . .88
    polystyrene . . . . . . . . . . . . . . . . . .86
    pouring . . . . . . . . . . . . . . . . . . . . .88
    R-Gard . . . . . . . . . . . . . . . . . . . . . .87
    rockwool . . . . . . . . . . . . . . . . .87-88
    roof, perlite . . . . . . . . . . . . . . . . . .87
    SAFB . . . . . . . . . . . . . . . . . . . . . . .88
    sill gasket . . . . . . . . . . . . . . . . . . .87
    sound control . . . . . . . . . . . . . . . .87
    urethane sheathing . . . . . . . . . . . .87
    vermiculite . . . . . . . . . . . . . . . . . .88
Intellisensor . . . . . . . . . . . . . . . . . . .249
Interior & exterior stain
    acrylic . . . . . . . . . . . . . . . . . . . . .487
Interior acrylic latex paint . . . . . . . .492
Interior doors . . . . . . . . . . . . . .212-213
    birch, flush . . . . . . . . . . . . . .206-207
    cafe . . . . . . . . . . . . . . . . . . . . . . .209
    fir . . . . . . . . . . . . . . . . . . . . . . . .208
    fire-rated . . . . . . . . . . . . . . . . . . .211
    flush . . . . . . . . . . . . . . . . . . .209-210
    flush . . . . . . . . . . . . . . . . . . . . . .207
    half louver . . . . . . . . . . . . . . . . . .208
    hardboard . . . . . . . . . . . . . . . . . .207
    hardboard, flush . . . . . . . . . .206-207
    heater closet . . . . . . . . . . . . . . . .210
    hemlock, prehung . . . . . . . . . . . .212
    knotty alder . . . . . . . . . . . . . . . . .212
    lauan . . . . . . . . . . . . . . . . . . . . . .207
    lauan, flush . . . . . . . . . . . . . .206-207
    oak . . . . . . . . . . . . . . . . . . . . . . .208
    oak . . . . . . . . . . . . . . . . . . . . . . .212

oak, flush . . . . . . . . . . . . . . . . . . .206
    pine . . . . . . . . . . . . . . . . . . . .208, 212
    prehung . . . . . . . . . . . . . . . . . . . .211
    prehung . . . . . . . . . . . . . . . . . . . .212
Interior flat paint . . . . . . . . . . . . . . .492
Interior latex
    eggshell . . . . . . . . . . . . . . . . . . . .492
    flat finish . . . . . . . . . . . . . . . . . . .492
    paint, semi-gloss . . . . . . . . . . . . .493
Interior paint . . . . . . . . . . . . . . . . . .492
Interior painting . . . . . . . . . . . . . . . .466
Interior sanding
    labor . . . . . . . . . . . . . . . . . . . . . .466
Interior wall assemblies . . . . . . . . . .71
Interlocking shingles . . . . . . . . . . . .131
Intermatic . . . . . . . . . . . . . . . . . . . .436
Intermediate metal conduit . . . . . . .429
International Thermocast . . . .331, 339,
    . . . . . . . . . . . . . . . . . . . . . . .368-369
In-wall spring wound timer . . . . . . .436
Island sink
    polycast . . . . . . . . . . . . . . . . . . . .338

**J**
J A Manufacturing . . . . . . . . . .334, 339,
    . . . . . . . . . . . . . . . . .361-362, 371, 374
J bead
    metal . . . . . . . . . . . . . . . . . . . . . .248
    plastic . . . . . . . . . . . . . . . . . . . . .248
J trim
    metal . . . . . . . . . . . . . . . . . . . . . .248
Jack
    roof . . . . . . . . . . . . . . . . . . . . . . .145
Jack posts . . . . . . . . . . . . . . . . . . . . .64
Jackhammer
    rental . . . . . . . . . . . . . . . . . . . . . . .28
Jalousie windows
    aluminum . . . . . . . . . . . . . . . . . .163
    operators . . . . . . . . . . . . . . . . . .167
    replacement glass . . . . . . . . . . . .169
    vinyl . . . . . . . . . . . . . . . . . . . . . .159
Jamb extension kit . . . . . . . . . . . . . .217
Jamb sets
    finger joint . . . . . . . . . . . . . . . . . .209
    interior . . . . . . . . . . . . . . . . . . . . .209
    solid clear pine . . . . . . . . . . . . . .209
    with hinges . . . . . . . . . . . . . . . . .209
Jambs
    exterior . . . . . . . . . . . . . . . . . . . .194
    exterior door . . . . . . . . . . . . . . . .194
    exterior, remove & replace . . . . .191
    interior, remove & replace . . . . . .206
    pocket door . . . . . . . . . . . . . . . . .222
Jameco . . . . . . . . . . . . . . .335-336, 339
Jameco International . . . . . . . . . . . .362
James Hardie . . . . . . . . . . . . . .113, 295
Jasco . . . . . . . . . .282, 308, 490-491, 495
J-channel
    thermoplastic . . . . . . . . . . . . . . . .109
    vinyl . . . . . . . . . . . . . . . . . . . . . .108
Jeffrey Court . . . . . . . . . . . . . . . . . .300
Jelly jar fixture . . . . . . . . . . . . . . . . .443
Jenn-Air . . . . . . . . . . . . . . . . . . . . . .343
Joint cement
    cartridge . . . . . . . . . . . . . . . . . . .245
    lightweight . . . . . . . . . . . . . . . . .245
Joint compound . . . . . . . . . . . .245-246
Joint tape . . . . . . . . . . . . . . . . . . . . .246
Joints
    expansion . . . . . . . . . . . . . . . . . . .46
Joists
    bridging . . . . . . . . . . . . . . . . . . . .66
    ceiling . . . . . . . . . . . . . . . . . . . . . .74
    demolition . . . . . . . . . . . . . . . . . .33
    floor . . . . . . . . . . . . . . . . . . . . .65- 66
    TJI . . . . . . . . . . . . . . . . . . . . . . . . .66
Jubilee White Ice beaded paneling .251
Junction box . . . . . . . . . . . . . . . . . .424

Junior basket strainer . . . . . . . . . . .339

**K**
Karnak . . . . . . . . . . . . . . . . . . . . . . .136
Key locks
    sliding window . . . . . . . . . . . . . . .172
Keystone . . . . . . . . . . . . . .356, 359, 384
Keyway form . . . . . . . . . . . . . . . . . . .44
Kick plates, door hardware . . . . . . .213
Kickplate channel . . . . . . . . . . . . . .459
Kickplate coil . . . . . . . . . . . . . . . . . .459
Kidde . . . . . . . . . . . . . . . . . . . . . . . .436
KILZ
    sealer, primer & stainblocker . . . .492
Kitchen
    appliance removal . . . . . . . . . . . .315
    cabinets, removal . . . . . . . . . . . .315
    countertop . . . . . . . . . . . . . .323-324
Kitchen cabinet doors . . . . . . . . . . .316
Kitchen cabinets
    base . . . . . . . . . . . . . . . . . . . .317-318
    wall . . . . . . . . . . . . . . . . . . . .317-319
Kitchen countertop
    miter bolt kit . . . . . . . . . . . . . . . .324
Kitchen faucet
    deck spray . . . . . . . . . . . . . . . . . .333
    pull-out spout . . . . . . . . . . . .331-332
    single-handle . . . . . . . . . . . . .331-333
    two-handle . . . . . . . . . . . . . .333-334
Kitchen sink . . . . . . . . . . . . . . .329-330
    Beaumont . . . . . . . . . . . . . . . . . .331
    cast iron . . . . . . . . . . . . . . . .329-330
    Charleston . . . . . . . . . . . . . . . . . .329
    composite . . . . . . . . . . . . . . .330-331
    double bowl . . . . . . . . . . . . .328-329
    Emory . . . . . . . . . . . . . . . . . . . . .329
    Iveness . . . . . . . . . . . . . . . . . . . .331
    Newport . . . . . . . . . . . . . . . . . . .331
    polycast . . . . . . . . . . . . . . . . . . . .330
    Risotto . . . . . . . . . . . . . . . . . . . .329
    stainless steel . . . . . . . . . . . . . . .328
Kitchen sink basket strainer . .335-336
    Decor Bone . . . . . . . . . . . . . . . . .336
    Spin-N-Lock . . . . . . . . . . . . . . . . .336
    washer . . . . . . . . . . . . . . . . . . . .336
Klean-Strip . . . . . . . . . . . . . . . . . . . .487
Knob
    cabinet . . . . . . . . . . . . . . . . .324-326
Knock down door frames . . . . . . . . .194
Knockdown
    stucco . . . . . . . . . . . . . . . . . . . . .493
Knockdown ceiling . . . . . . . . . . . . . .245
Knockdown spray texture . . . . . . . .244
Knockers, door . . . . . . . . . . . . . . . . .213
Kohler . . . . .328-330, 332, 334, 353-354,
    . . . . . . . . .361, 363, 366-369, 371, 374
Kool Patch
    acrylic roof patching cement . . . .135
Kool Seal . . . . . . . . . . . . . . . . . . . . .135
Kraft-faced
    fiberglass roll insulation . . . . . . . .85

**L**
L trim
    drywall . . . . . . . . . . . . . . . . . . . . .248
Labor
    high time painting . . . . . . . . . . . .466
Lag screws
    drill tip suspension . . . . . . . . . . . .256
Laminate flooring
    finishing putty . . . . . . . . . . . . . . .283
    glue . . . . . . . . . . . . . . . . . . . . . . .283
    installation spacer . . . . . . . . . . . .283
    installation system . . . . . . . . . . . .284
    molding, Perpetual . . . . . . . . . . .284
    Pergo Presto . . . . . . . . . . . . . . . .283
    Pergo Presto molding . . . . . . . . .283
    sealant . . . . . . . . . . . . . . . . . . . .283

strip . . . . . . . . . . . . . . . . . . . . . . .284
   Traffic Master . . . . . . . . . . . . . . . . . .284
   trim . . . . . . . . . . . . . . . . . . . . . . . . .284
   underlayment . . . . . . . . . . . . . .283-284
Laminate shingles . . . . . . . . . . . . . . .131
   architectural . . . . . . . . . . . . . . . . . .130
Laminated plastic
   countertops . . . . . . . . . . . . . . .322-323
Laminated plastic countertop . . . . .324
Lampholder & cover
   dual weatherproof . . . . . . . . . . . . .428
Lantern
   curved glass wall . . . . . . . . . . . . . .443
   motion sensing . . . . . . . . . . . . . . . .443
   outdoor wall . . . . . . . . . . . . . . . . . .443
Lap
   cement . . . . . . . . . . . . . . . . . . . . . .133
Lap siding
   fiber cement . . . . . . . . . . . . . . . . . .113
   hardboard . . . . . . . . . . . . . . . . . . . .113
   OSB . . . . . . . . . . . . . . . . . . . . . . . .112
   spruce . . . . . . . . . . . . . . . . . . . . . . .111
   textured . . . . . . . . . . . . . . . . . . . . .112
Lasco Bathware . . . . .353, 356-357, 359
Lascoat . . . . . . . . . . . . . . . . . . .356, 359
Latch
   passage . . . . . . . . . . . . . . . . . . . . .213
Latchsets
   privacy . . . . . . . . . . . . . . . . . . . . . .213
   privacy lever . . . . . . . . . . . . . . . . . .213
Latex
   painting . . . . . . . . . . . . . . . . . .483-484
Latex eggshell paint . . . . . . . . . . . . .492
Latex glazing compound . . . . . . . . .171
Latex paint
   eggshell . . . . . . . . . . . . . . . . . . . . .492
   enamel . . . . . . . . . . . . . . . . . .492-493
   exterior . . . . . . . . . . . . . . . . . . . . . .494
   flat finish, interior . . . . . . . . . . . . .492
   masonry & stucco . . . . . . . . . . . . .494
   semi-gloss . . . . . . . . . . . . . . . . . . .493
   waterproofing . . . . . . . . . . . . . . . .494
Latex primer
   PVA . . . . . . . . . . . . . . . . . . . . . . . .491
   resilient flooring . . . . . . . . . . . . . .285
Lattice molding . . . . . . . . . . . . . . . . .269
Lauan doors . . . . . . . . . . . .206-207, 210
   flush . . . . . . . . . . . . . . . . . . . . . . . .192
Laundry faucet
   two-handle . . . . . . . . . . . . . . . . . .337
   wall mount . . . . . . . . . . . . . . . . . . .337
Laundry tub
   double bowl . . . . . . . . . . . . . . . . . .337
   Handiflo . . . . . . . . . . . . . . . . . . . . .337
   overflow tube . . . . . . . . . . . . . . . . .337
   single bowl . . . . . . . . . . . . . . . . . . .337
   Utilatub . . . . . . . . . . . . . . . . . . . . .337
   Utilatwin . . . . . . . . . . . . . . . . . . . .337
   utility kit . . . . . . . . . . . . . . . . . . . . .337
Lavatory base . . . . . . . . . . . . . . . . . .367
Lavatory bowl . . . . . . . . . . . . . . . . . .366
Lavatory faucet
   adjustable center . . . . . . . . . . . . . .374
   decorator . . . . . . . . . . . . . . . . . . . .373
   hands free . . . . . . . . . . . . . . . . . . .371
   hi-arc . . . . . . . . . . . . . . . . . . . . . . .373
   single-handle . . . . . . . . . . .370-371
   two-handle . . . . . . . . . . . . . .371-374
Lavatory hanger bracket . . . . . . . . .369
Lavatory leg set . . . . . . . . . . . . . . . .367
Lavatory sink . . . . . . . . . . . . . . .367-369
   above counter . . . . . . . . . . . . . . . .369
   Aqualyn . . . . . . . . . . . . . . . . . . . . .367
   Bayside . . . . . . . . . . . . . . . . . . . . .368
   cadet . . . . . . . . . . . . . . . . . . . . . . .368
   drop-in . . . . . . . . . . . . . . . . . .367-368
   Farmington . . . . . . . . . . . . . . . . . .367
   Miami . . . . . . . . . . . . . . . . . . . . . . .369
   Murro . . . . . . . . . . . . . . . . . . . . . . .369

oval . . . . . . . . . . . . . . . . . . . . . . . . .368
Ovalyn . . . . . . . . . . . . . . . . . . . . . . . .368
pedestal . . . . . . . . . . . . . . . . . . . . . .366
Radiant . . . . . . . . . . . . . . . . . . . . . . .368
Reminiscence . . . . . . . . . . . . . . . . . .369
removal . . . . . . . . . . . . . . . . . . . . . . .367
Renaissance . . . . . . . . . . . . . . . . . . .367
Rondalyn . . . . . . . . . . . . . . . . . . . . . .367
round . . . . . . . . . . . . . . . . . . . . . . . .368
Seychelle . . . . . . . . . . . . . . . . .367-368
undermount . . . . . . . . . . . . . . . . . . .368
wall hung . . . . . . . . . . . . . . . . . . . . .369
wall hung, corner . . . . . . . . . . . . . . .369
Lavatory trap
   remove & replace . . . . . . . . . . . . . .352
Lay-in ceiling panel
   gypsum . . . . . . . . . . . . . . . . . . . . . .257
LB conduit body
   PVC . . . . . . . . . . . . . . . . . . . . . . . . .421
LDR Industries . . . . . . . . . . . . . .335-336
Lead flashing . . . . . . . . . . . . . .143, 145
Leak Sentry toilet fill valve . . . . . . .364
Leak Stopper
   rubberized roof patch . . . . . . . . . .135
Leaks
   roof . . . . . . . . . . . . . . . . . . . . . . . . . .58
Leaks, basement . . . . . . . . . . . . . . . . .37
Ledgers . . . . . . . . . . . . . . . . . . . . . . . .74
Leg set
   pedestal lavatory . . . . . . . . . . . . . .367
Lenape . . . . . . . . . . . . . . . . .304-305, 387
Lens
   suspended lighting panel . . . . . . .258
Let-in bracing . . . . . . . . . . . . . . . . . . .68
Let-in ribbons . . . . . . . . . . . . . . . . . . .74
Leveler
   embossing . . . . . . . . . . . . . . . . . . .285
Lever locksets . . . . . . . . . . . . . . . . . .202
Leviton . . . . . . . . . . . . . . . . . . .434-435
Lexan XL-10 . . . . . . . . . . . . . . . . . . .172
L-flashing . . . . . . . . . . . . . . . . . . . . .143
Lift sash . . . . . . . . . . . . . . . . . . . . . .167
Light bar
   beveled . . . . . . . . . . . . . . . . . . . . .384
   medicine cabinet . . . . . . . . . .384-385
Light fixture
   ceiling . . . . . . . . . . . . . . . . . . . . . . .440
   chandelier . . . . . . . . . . . . . . . . . . .442
   chrome bath bar . . . . . . . . . . . . . .441
   drum, mushroom . . . . . . . . . . . . . .441
   jelly jar . . . . . . . . . . . . . . . . . . . . . .443
   mushroom flush mount . . . . . . . . .442
   outdoor bulkhead . . . . . . . . . . . . .443
   remove & replace . . . . . . . . . . . . . .441
   wave bath . . . . . . . . . . . . . . . . . . . .441
Light kit
   mini can . . . . . . . . . . . . . . . . . . . . .440
   mini can recessed eyeball . . . . . . .440
   recessed . . . . . . . . . . . . . . . . . . . . .440
   recessed non-IC baffle . . . . . . . . .440
   slot aperture . . . . . . . . . . . . . . . . . .440
Lighting
   motion sensor . . . . . . . . . . . . . . . .443
   recessed . . . . . . . . . . . . . . . . . . . . .440
Lighting fixtures
   soffit, kitchen . . . . . . . . . . . . . . . . .316
Lighting panel
   acrylic . . . . . . . . . . . . . . . . . . . . . . .258
   styrene . . . . . . . . . . . . . . . . . . . . . .258
Lights
   suspended troffer fixture . . . . . . .258
Lightweight concrete . . . . . . . . . . . . .46
Liquid underlayment . . . . . . . . . . . .285
Liquid-tight flexible conduit
   Carflex . . . . . . . . . . . . . . . . . . . . . .427
Listello
   tumbled marble tile . . . . . . . . . . .300
Litex . . . . . . . . . . . . . . . . . . . . .493, 495
Lithonia Lighting . . . . . . . . . . . . . . .441

Load center
   combination ring type . . . . . . . . .437
   flange kit . . . . . . . . . . . . . . . . . . . .437
   ground bar kit . . . . . . . . . . . . . . . .440
   HomeLine indoor main lug . . . . .437
   indoor main lug . . . . . . . . . . . . . .438
   outdoor . . . . . . . . . . . . . . . . . . . . .439
Loading trucks . . . . . . . . . . . . . . . . . .40
Locknut
   steel sealing conduit . . . . . . . . . . .430
Locks
   bar . . . . . . . . . . . . . . . . . . . . . . . . .224
   deadbolt . . . . . . . . . . . . . . . . . . . .202
   door, folding . . . . . . . . . . . . . . . . .218
   garage door . . . . . . . . . . . . . . . . . .224
   installation . . . . . . . . . . . . . . . . . . .201
   patio door . . . . . . . . . . . . . . . . . . .216
   sash . . . . . . . . . . . . . . . . . . . . . . . .167
   sliding window . . . . . . . . . . . . . . . .172
   window flip . . . . . . . . . . . . . . . . . .172
Locksets
   and deadbolt . . . . . . . . . . . . . . . . .202
   colonial knob . . . . . . . . . . . . . . . . .202
   egg knob . . . . . . . . . . . . . . . . . . . .202
   installation . . . . . . . . . . . . . . . . . . .201
   keyed entry . . . . . . . . . . . . . .201-202
   lever . . . . . . . . . . . . . . . . . . . . . . . .202
Log cabin siding . . . . . . . . . . . . . . . .111
Log home gloss finish . . . . . . . . . . .489
Log lap siding . . . . . . . . . . . . . . . . . .111
Longstrip oak flooring . . . . . . . . . . .279
Louver
   gable . . . . . . . . . . . . . . . . . . . . . . . .91
   roof . . . . . . . . . . . . . . . . . . . . . . . .145
   roofing . . . . . . . . . . . . . . . . . . . . . .145
Louver doors
   painting . . . . . . . . . . . . . . . . .471-472
Louver windows . . . . . . . . . . . . . . . .163
Louvered doors . . . . . . . . . . . . . . . .212
Louvered exterior window
   shutters . . . . . . . . . . . . . . . . . . . . .168
Louvered vents . . . . . . . . . . . . . . . . . .91
Louvers
   midget . . . . . . . . . . . . . . . . . . . . . . .92
   roof . . . . . . . . . . . . . . . . . . . . . . . . .92
   wall . . . . . . . . . . . . . . . . . . . . . . . . .92
Low-Boy toilet kit . . . . . . . . . . . . . . .364
Low-boy, trash disposal . . . . . . . . . . .29
Low-sound bath fan
   QuieTTest . . . . . . . . . . . . . . . . . . .380
Low-voltage light kit . . . . . . . . . . . .440
Lug load center
   HomeLine indoor . . . . . . . . . . . . .437
Lumber
   alternative building materials . . .457
   deck . . . . . . . . . . . . . . . . . . .453-454
   pressure treated . . . . . . . . . . . . . .452
   recycled plastic . . . . . . . . . . . . . . .457
   Thompsonized . . . . . . . . . . . . . . .454

**M**

Maax . . . . . . . . . . . . . . . . . . . . . . . . .358
Magna . . . . . . . . . . . . . . . . . . . . . . . .251
Magnetic weatherstrip . . . . . . . . . . .196
Magnolia Brick and Tile . . . . . . . . . .300
Mahogany doors . . . . . . . . . . . . . . . .193
Main circuit breaker . . . . . . . . . . . . .438
Main load center
   combination . . . . . . . . . . . . . . . . . .437
Main lug load center . . . . . . . . . . . .439
   HomeLine indoor . . . . . . . . . . . . .437
   indoor . . . . . . . . . . . . . . . . . . . . . .438
Mallet, rubber . . . . . . . . . . . . . . . . . .291
Man-O-War marine spar varnish . .488
Maple
   hardwood flooring, solid . . . . . . .279
   plank flooring . . . . . . . . . . . . . . . .280
   strip flooring . . . . . . . . . . . . . . . . .281

Marble
   floor tile . . . . . . . . . . . . . . . . . . . . .298
   listello tumbled tile . . . . . . . . . . .300
   mortar mix . . . . . . . . . . . . . . . . . . .306
   setting adhesive . . . . . . . . . . . . . .306
   threshold . . . . . . . . . . . . . . . . . . . .300
   tile . . . . . . . . . . . . . . . . . . . . . . . . .298
Marble countertop . . . . . . . . . . . . . .376
Margin notch trowel . . . . . . . . . . . .291
Marine spar varnish
   Man-O-War . . . . . . . . . . . . . . . . . .488
Marine varnish . . . . . . . . . . . . . . . . .488
Masco . . . . .333, 339, 361, 370, 372-373
Mask
   safety . . . . . . . . . . . . . . . . . . . . . . .250
Masonite Corp . . . . . . . . . . . . . . . . .253
Masonry
   cleaning . . . . . . . . . . . . . . . . . . . . .116
   demolition . . . . . . . . . . . . . . .30, 116
   fireplaces . . . . . . . . . . . . . . . . . . . .116
   furring . . . . . . . . . . . . . . . . . . . . . . .73
   insulation . . . . . . . . . . . . . . . . . . . .88
   joints . . . . . . . . . . . . . . . . . . . . . . .116
   painting labor . . . . . . . . . . . .476-478
   pointing . . . . . . . . . . . . . . . . . . . . .116
   protective coating . . . . . . . . . . . . .495
   recycling . . . . . . . . . . . . . . . . . . . . .29
   repointing . . . . . . . . . . . . . . . . . . .116
Masonry & concrete waterproofer . .495
Masonry & stucco paint . . . . . . . . . .494
Masonry walls
   waterproofing . . . . . . . . . . . . . . . .130
Masterchem . . . . . . . . . . . . . . . .492, 494
Mastercure . . . . . . . . . . . . . . . . . . . . .49
Mastic
   tile . . . . . . . . . . . . . . . . . . . . . . . . .294
Mayfield sink . . . . . . . . . . . . . . . . . .329
Maytag . . . . . . . . . . . . . . . . . . .341-343
MDF
   casing . . . . . . . . . . . . . . . . . . . . . . .261
   molding, base corner . . . . . . . . . .261
   molding, chair rail . . . . . . . . . . . . .261
   molding, colonial base . . . . . . . . .261
   molding, crown . . . . . . . . . . . . . . .262
   molding, plinth block . . . . . . . . . .262
   molding, ultralite crown . . . . . . . .262
   paneling . . . . . . . . . . . . . . . . . . . . .251
Medicine cabinet
   beveled edge . . . . . . . . . . . . .383-384
   bi-view . . . . . . . . . . . . . . . . . . . . . .384
   etched glass . . . . . . . . . . . . . . . . . .383
   framed . . . . . . . . . . . . . . . . . . . . . .383
   framed door . . . . . . . . . . . . . . . . . .382
   frameless . . . . . . . . . . . . . . . .383-384
   light bar . . . . . . . . . . . . . . . . .384-385
   lighted . . . . . . . . . . . . . . . . . . . . . .383
   mirrored . . . . . . . . . . . . . . . . . . . . .383
   octagonal . . . . . . . . . . . . . . . . . . . .383
   oval . . . . . . . . . . . . . . . . . . . . . . . . .383
   single door . . . . . . . . . . . . . . . . . . .383
   sliding door . . . . . . . . . . . . . . . . . .383
   swing door . . . . . . . . . . . . . . . . . . .383
   tri-view . . . . . . . . . . . . . . . . . . . . . .384
Medium density fiberboard
   molding . . . . . . . . . . . . . . . . . . . . .261
Medley sheet vinyl flooring . . . . . . .286
Megatrade . . . . . . . . . . . . . . . . . . . . .297
Melard . . . . . . . . . . . . . . . . . . . . . . . .367
Membrane
   anti-fracture . . . . . . . . . . . . . . . . . .305
   fiberglass reinforcing . . . . . . . . . .136
   polyethylene . . . . . . . . . . . . . . . . .308
   tile, anti-fracture . . . . . . . . . . . . . .305
Memoirs pedestal lavatory
   bowl . . . . . . . . . . . . . . . . . . . . . . . .366
Mesh
   crack repair . . . . . . . . . . . . . . . . . .248
Metal conduit
   flexible . . . . . . . . . . . . . . . . . . . . . .426
   galvanized . . . . . . . . . . . . . . . . . . .429

Metal corner bead . . . . . . . . . .247, 249
Metal doors . . . . . . . . . . . . . . . . . . .194
Metal primer . . . . . . . . . . . . . . . . . .496
  rust converting . . . . . . . . . . . . . . .496
Metal screen doors . . . . . . . . . . . . .203
Metal soffit vents
  galvanized . . . . . . . . . . . . . . . . . . .90
Metallic grounding bushing
  insulated . . . . . . . . . . . . . . . . . . .430
Meter main load center . . . . . . . . . .437
Meter socket . . . . . . . . . . . . . . . . . .437
Metro sheet vinyl flooring . . . . . . .287
Microwave oven
  Advantium . . . . . . . . . . . . . . . . . .343
  over-the-range . . . . . . . . . . . . . . .343
Midget louvers . . . . . . . . . . . . . . . . .92
Mineral fiber ceiling panel . . . . . . .257
Mineral surface roll roofing . . . . . .133
Mini can recessed light kit . . . . . . .440
Miniature circuit breaker
  QO . . . . . . . . . . . . . . . . . . . . . . . .439
Minwax . . . . . . . . . . . . . .487-489, 491
  pickling stain . . . . . . . . . . . . . . . .488
  Polycrylic . . . . . . . . . . . . . . . . . . .489
  Polyshades . . . . . . . . . . . . . . . . . .488
  polyurethane . . . . . . . . . . . . . . . .489
  spar urethane . . . . . . . . . . . . . . . .488
Miracle Sealants Co . . . . . . . . . . . .307
Mirror bath light bar . . . . . . . . . . .441
Mirrored doors
  bi-fold . . . . . . . . . . . . . . . . .221-222
  bypass . . . . . . . . . . . . . . . . . . . . .219
Mirrored sheet . . . . . . . . . . . . . . . .172
Mirrors
  frosted . . . . . . . . . . . . . . . . . . . . .382
  oval . . . . . . . . . . . . . . . . . . . . . . .382
  vanity . . . . . . . . . . . . . . . . . . . . . .382
Mission tile roofing . . . . . . . . . . . . .139
Miter bolt kit . . . . . . . . . . . . . . . . . .324
Mobile home doors . . . . . . . . .222-223
Modified bitumen adhesive . . . . . . .133
Moen . . .332, 360, 362, 370-371, 373-374
Moisturbloc polyethylene film . . . .284
Moisture barrier
  Moisturbloc . . . . . . . . . . . . . . . . .284
  polyethylene . . . . . . . . . . . . . . . . .284
Moisture-resistant drywall . . . . . . .243
Molded face closet doors . . . . . . . .220
Molded sectional shower stall . . . .357
Molding
  base . . . . . . . . . . . . . . . . . . .263-264
  base cap . . . . . . . . . . . . . . . . . . . .260
  base corner . . . . . . . . . . . . . . . . . .259
  base shoe . . . . . . . . . . . . . . .260, 264
  bead . . . . . . . . . . . . . . . . . . . . . . .260
  beaded baseboard . . . . . . . . . . . .259
  brick . . . . . . . . . . . . . . . . . . . . . . .195
  casing . . . . . . . . . . . . . . . . . .264-265
  casing paperwrap . . . . . . . . . . . . .265
  casing polystyrene . . . . . . . . . . . .265
  cast . . . . . . . . . . . . . . . . . . . . . . . .262
  chair rail . . . . . . . . . . . . .260, 262, 266
  chair rail paperwrap . . . . . . . . . . .266
  chair rail polystyrene . . . . . . . . . .266
  chair rail prefinished . . . . . . . . . . .266
  colonial crown paperwrap . . . . . .268
  cove . . . . . . . . . . . . . . . . . . . . . . .268
  crown . . . . . . . . . . . . . . . . . .259, 261
  crown embossed polystyrene . . .268
  crown polystyrene . . . . . . . . . . . .268
  crown prefinished . . . . . . . . . . . . .268
  embossed . . . . . . . . . . . . . . . . . . .265
  embossed basswood . . . . . . . . . .259
  embossed decorative hardwood .259
  extruded polymer . . . . . . . . . . . . .263
  hardwood . . . . . . . . . . . . . . . . . . .280
  hardwood floor . . . . . . . . . . . . . .278
  hardwood parquet . . . . . . . . . . . .282

inside corner . . . . . . . . . . . . . . . . . .267
inside corner & cove . . . . . . .266-267
inside corner & cove paperwrap . .266
inside corner polystyrene . . . . . . .267
inside corner prefinished . . . . . . .267
laminate flooring . . . . . . . . . . . . .284
lattice . . . . . . . . . . . . . . . . . . . . . .269
MDF colonial base . . . . . . . . . . . .261
MDF crown . . . . . . . . . . . . . . . . . .262
MDF ultralite crown . . . . . . . . . . .262
medium density fiberboard . . . . . .261
outside corner . . . . . . . . . . .261, 267
outside corner paperwrap . . . . . . .267
outside corner polystyrene . . . . . .267
outside corner prefinished . . . . . .267
painting labor . . . . . . . . . . . . . . . .479
panel . . . . . . . . . . . . . . . . . . . . . . .254
paperwrap . . . . . . . . . . . . . .263-264
polystyrene . . . . . . . . . . . . . .263-265
prefinished casing . . . . . . . . .264-265
primed finger joint base . . . . . . . .260
quarter round . . . . . . . . . . . . . . . .261
reversible flute & reed . . . . . . . . .261
rosette block . . . . . . . . . . . . . . . . .259
round edge stop . . . . . . . . . . . . . .261
spring cove . . . . . . . . . . . . . . . . . .268
spring cove polystyrene . . . . . . . .268
stool . . . . . . . . . . . . . . . . . . . . . . .261
stop . . . . . . . . . . . . . . . . . .195, 269
stop paperwrap . . . . . . . . . . . . . .269
stop polystyrene . . . . . . . . . . . . . .269
stucco . . . . . . . . . . . . . . . . . . . . . .196
T-astragal . . . . . . . . . . . . . . . . . . .194
Victorian hardwood base . . . . . . .259
vinyl tileboard . . . . . . . . . . . . . . .270
wall base . . . . . . . . . . . . . . . . . . . .284
wall, suspended ceiling . . . . . . . .255
Molding set
brick . . . . . . . . . . . . . . . . . . . . . . .260
Monterey bathtub tile
three-piece kit . . . . . . . . . . . . . . .355
Monticello lavatory faucet . . . . . . .373
Mop, roof . . . . . . . . . . . . . . . . . . . .136
Mortar
acrylic admix . . . . . . . . . . . . . . . .307
bedding . . . . . . . . . . . . . . . . . . . .306
bonding . . . . . . . . . . . . . . . . .305-306
CustomBlend thin-set . . . . . . . . .306
epoxy . . . . . . . . . . . . . . . . . . . . . .306
flexible bonding . . . . . . . . . . . . . .306
porcelain . . . . . . . . . . . . . . . . . . .306
thin-set . . . . . . . . . . . . . . . . . . . . .306
thin-set tile . . . . . . . . . . . . . . . . . .293
tile . . . . . . . . . . . . . . . . . . . . . . . .294
tile repair . . . . . . . . . . . . . . . . . . .306
underwater tile repair . . . . . . . . .307
Mortar mix
glass block . . . . . . . . . . . . . . . . . .306
marble & granite . . . . . . . . . . . . .306
Mosaic ceramic tile . . . . . . . . . . . .304
Motion sensor security light . . . . .443
heavy-duty . . . . . . . . . . . . . . . . . .443
lantern . . . . . . . . . . . . . . . . . . . . .443
Mount Vernon paneling . . . . . . . . .251
Mounting assembly
disposal . . . . . . . . . . . . . . . . . . . .340
Mounting bracket
non-metallic low voltage . . . . . . .422
Mounting kit
sink . . . . . . . . . . . . . . . . . . . . . . . .335
Mud
joint cement . . . . . . . . . . . . . . . . .245
joint cement cartridge . . . . . . . . .245
wallboard . . . . . . . . . . . . . . .245-246
Mullions
window . . . . . . . . . . . . . . . . . . . .167
Multi gang switch box . . . . . . . . . .425
Mushroom light fixture . . . . . . . . .442

**N**

Nail plate . . . . . . . . . . . . . . . . . . . .419
Nailer strip, carpentry . . . . . . . . . . .74
Nailers . . . . . . . . . . . . . . . . . . . . . . .74
Nailite siding . . . . . . . . . . . . . . . . .108
Nails
  drywall . . . . . . . . . . . . . . . . . . . .249
  vinyl siding . . . . . . . . . . . . . . . . . .108
Narroline . . . . . . . . . . . . . . . . . . . .214
National Gypsum . . . . . . . . . . . . . .246
Natural Images floor tile . . . . . . . .289
Natural stone floor tile . . . . . . . . .299
Needle beams . . . . . . . . . . . . . . . . .63
Neo-angle shower kit . . . . . . . . . . .358
Neptune . . . . . . . . . . . . . . . . . . . . .328
Neptune marble countertop . . . . .376
New wood treatment . . . . . . . . . . .490
Nipple
  rigid conduit . . . . . . . . . . . . . . . .431
NM-B Romex sheathed cable
  indoor . . . . . . . . . . . . . . . . . . . . .418
  indoor . . . . . . . . . . . . . . . . . . . . .418
  indoor . . . . . . . . . . . . . . . . . . . . .418
  indoor & outdoor . . . . . . . . . . . .418
NM-B Romex sheathed indoor
cable . . . . . . . . . . . . . . . . . . . . . . .417
No-burst
  dishwasher connector . . . . . . . . .342
Non-fibered roof and foundation
coating . . . . . . . . . . . . . . . . . . . . .129
Non-metallic
  cable . . . . . . . . . . . . . . . . . . . . . .417
  ceiling fan box . . . . . . . . . . . . . . .422
  conduit fitting . . . . . . . . . . . . . . .421
  low voltage device mounting
   bracket . . . . . . . . . . . . . . . . . . .422
  old work switch box . . . . . . . . . .422
  reducer bushing . . . . . . . . . . . . . .420
  sheathed cable . . . . . . . . . . . . . . .418
  standard coupling . . . . . . . . . . . .421
  UF compression connector . . . . .419
  Zip Box . . . . . . . . . . . . . . . . . . . .422
Northwoods hardwood flooring . .278
Notched trowels . . . . . . . . . . . . . . .291
Numbers, house . . . . . . . . . . . . . . .214
NuTone . . . . . . . . .343-344, 378-382
Nylon carpet . . . . . . . . . . . . . . . . .308
Nylon paneling rivets . . . . . . . . . . .254

**O**

Oak
  counter edge molding . . . . . . . . .268
  hardwood flooring . . . . . . . . . . . .278
  molding, unfinished . . . . . . . . . . .281
  paneling . . . . . . . . . . . . . . . . . . . .252
  parquet flooring . . . . . . . . . . . . .282
  plank flooring . . . . . . . . . . . . . . .280
  strip flooring . . . . . . . . . . . .278, 281
  wood trim . . . . . . . . . . . . . . . . . .279
Oak doors . . . . . . . . . . . .192, 206, 210
  entry . . . . . . . . . . . . . . . . . . . . . .194
Oak fill strip . . . . . . . . . . . . . . . . . .321
Oak threshold . . . . . . . . . . . . . . . . .195
OakCrest Products . . . . . . . . . . . . .282
Octagon box
  drawn . . . . . . . . . . . . . . . . . . . . .425
Octagonal windows . . . . . . . . . . . .164
Offset compression connector
  EMT-to-box . . . . . . . . . . . . . . . . .424
Offset screw connector
  EMT . . . . . . . . . . . . . . . . . . . . . . .423
Oil
  Danish . . . . . . . . . . . . . . . . . . . . .488
  teak . . . . . . . . . . . . . . . . . . . . . . .491
  tung . . . . . . . . . . . . . . . . . . . . . . .491
Oil gloss
  polyurethane . . . . . . . . . . . . . . . .495
Oil-based enamel . . . . . . . . . . . . . .496

Oil-based floor finish
  Varathane . . . . . . . . . . . . . . . . . . .489
Oil-based primer sealer . . . . . . . . .493
Oil-fired water heater . . . . . . . . . . .405
Old work switch box
  non-metallic . . . . . . . . . . . . . . . . .422
Olefin carpet . . . . . . . . . . . . . . . . .308
Opal glass wall light . . . . . . . . . . . .441
Open light kit . . . . . . . . . . . . . . . . .440
Openers, garage door . . . . . . . . . . .225
Opening, breakthrough wall . . . . . .73
Operators, window . . . . . . . . . . . . .167
Orange peel ceiling . . . . . . . . . . . .245
Orange peel spray texture . . . . . . .244
Oriented strand board
  siding . . . . . . . . . . . . . . . . . . . . . .112
OSB
  roof sheathing . . . . . . . . . . . . . . .77
  sheathing . . . . . . . . . . . . . . . .67, 71
  siding . . . . . . . . . . . . . . . . . . . . . .112
  subfloor . . . . . . . . . . . . . . . . . . . .67
Outdoor ceiling fan . . . . . . . . . . . .444
Outdoor wall lantern . . . . . . . . . . .443
Outlet box
  connecting appliance . . . . . . . . . .417
  re-connect appliance . . . . . . . . . .417
Outside corner bead, drywall . . . .248
Outside corner molding . . . . . .263, 267
  colonial . . . . . . . . . . . . . . . . . . . .267
  paperwrap . . . . . . . . . . . . . . . . . .267
  polystyrene . . . . . . . . . . . . . . . . .267
  polystyrene colonial . . . . . . . . . . .267
  prefinished . . . . . . . . . . . . . . . . . .267
  primed finger joint . . . . . . . . . . . .261
Outside corner trowel . . . . . . . . . . .250
Oval light fixture
  outdoor bulkhead . . . . . . . . . . . .443
Oven
  double wall . . . . . . . . . . . . . . . . .342
  microwave . . . . . . . . . . . . . . . . . .342
  microwave, above range . . . . . . .342
  wall . . . . . . . . . . . . . . . . . . . . . . .342
Overhang
  painting . . . . . . . . . . . . . . . .479-480
Overhang painting, difficulty . . . . .480
Overhead doors
  garage . . . . . . . . . . . . . . . . . . . . .223
Owens Corning . . . . . . . . .130-131, 144

**P**

Paint
  acrylic latex . . . . . . . . . . . . . . . . .492
  alkyd . . . . . . . . . . . . . . . . . . . . . .493
  barn & fence . . . . . . . . . . . . . . . .496
  ceiling . . . . . . . . . . . . . . . . . . . . .492
  elastomeric . . . . . . . . . . . . . . . . .494
  flat . . . . . . . . . . . . . . . . . . . .492, 494
  floor & porch . . . . . . . . . . . . . . . .495
  gloss . . . . . . . . . . . . . . . . . . . . . . .493
  house & trim . . . . . . . . . . . . . . . .494
  interior . . . . . . . . . . . . . . . . . . . . .492
  latex . . . . . . . . . . . . . . . . . . . . . . .494
  latex ceiling . . . . . . . . . . . . . . . . .492
  latex interior . . . . . . . . . . . . . . . . .492
  masonry & stucco . . . . . . . . . . . .494
  polyurethane oil gloss . . . . . . . . .495
  satin flat . . . . . . . . . . . . . . . . . . . .494
  semi-gloss . . . . . . . . . . . . . . . . . . .493
  semi-gloss latex . . . . . . . . . . . . . .493
  swimming pool . . . . . . . . . . . . . .496
  texture . . . . . . . . . . . . . . . . . . . . .493
  wall & trim . . . . . . . . . . . . . .492-493
  waterproofing . . . . . . . . . . . . . . .494
Paint additive
  Perlite texturing . . . . . . . . . . . . . .493
Paint seal removal . . . . . . . . . . . . .156
Paint stripping
  labor . . . . . . . . . . . . . . . . . . . . . .467

Painting
  burn off, manhours ...........466
  high-time difficulty factors ......466
Painting labor, manhours
  acid wash ....................466
  baseboard ...................468
  blinds .................481-482
  bookcases ..................468
  brick masonry ...............477
  burn off paint ...............466
  cabinets ...................469
  caulking ...................466
  ceiling texture ..............469
  ceilings ................469-470
  closet shelf and pole ..........470
  concrete floor ...........474-475
  concrete masonry units ....477-478
  door frame and trim ..........471
  drywall ceiling ..............484
  drywall walls ...............483
  eaves and cornices ...........480
  entry doors ................471
  exterior doors ..............471
  French doors ............471-472
  gutters and downspouts .......476
  handrail ...................481
  house exterior ..............466
  house interior ..............466
  interior doors ...........471-473
  louver doors ...............471
  molding ...................479
  paneled walls ...............485
  plaster or stucco ..........480-481
  pressure washing ............468
  roof overhang ...........479-480
  shutters .................481-482
  siding ....................482
  stripping paint ..............467
  trim ......................479
  windows ...............486-487
  wood deck .................475
Paling, rail ...................455
Pan
  shower ....................359
Pancake box
  ceiling fan .................425
Panel molding
  FRP .......................254
Panel siding
  cedar shingle ...............114
  fiber cement siding ...........113
  hardboard .................112
  OSB .......................112
  plywood ...................112
Panel wallcovering
  designer ...................252
Panel, ceiling .................253
  acoustical .................256
  lay-in .....................257
  lighting ...................258
  suspended .............256-258
Panel, distribution
  disconnect & remove .........436
Panel, electric service ..........439
Paneling
  beaded ....................251
  birch .....................252
  demolition .................253
  fiberboard .................251
  FRP .......................253
  hardboard .................251
  hardwood .................252
  hardwood plank .........251, 253
  MDF .......................251
  oak .......................252
  oak print ..................251
  pine ......................252
  pine plank .................253

plywood ..............251-253
prefinished ..................252
rivets ......................254
tileboard ................251-252
trim pack ...................253
wainscot ...................253
Panelized shingles ..............114
Panels
  fiberglass ..................140
  greenhouse .................140
  sun .......................140
Panic bar
  door ......................226
Panic door hardware ...........226
Pantry
  cabinet ....................319
  cabinets ...............320-321
Pantry doors ..................212
Paper
  Aquabar ....................88
  building ...................134
  Moistop .....................89
  Pyro-Kure ...................89
  roof flashing .................89
  roof underlay ................89
  Seekure .....................89
  vapor barrier ................88
Paper holder
  toilet tissue ................304
Paper-faced metal trim .........247
Paperwrap molding
  base ......................263
  casing ..................264-265
  chair rail ..................266
  colonial cap ................264
  colonial crown ..............268
  inside corner & cove .........266
  outside corner ..............267
  stop ......................269
Parabolic troffer fluorescent
  fixture .....................258
Parisa faucet .................370
Parks Corporation .............489
Parquet
  molding ...................282
Parquet floor tile ..............282
Parquet flooring
  Hevea ....................282
  oak .......................282
  oak, self-stick ..............282
Parting stop molding ...........269
Partition walls
  load bearing .................57
  removal .....................71
Partition walls, adding .........270
Pass & Seymour/Legrand .......435
Passage latch .................213
Paste wax ....................491
Patch
  built-up roofing .........131-132
  drywall ...................248
  Elastocaulk roof patch .......136
  Elastomastic roof patch .......136
  flashing ...................135
  Kool Patch cement ...........135
  PondPatch .................136
  popcorn ...................244
  roof caulk ..................136
  roof cement ...............135
  roofing .................134-135
  roofing, cotton .............136
  roofing, wood shingle .........137
  underlayment ...........285-286
  vinyl .....................285
  Wet Patch roof cement .......135
  Wet Stick roof cement .......135
Patch kit .....................246
Patching
  concrete ....................50

Patching cement, roof ..........135
Patio doors ................214-216
  aluminum .................214
  aluminum, double ...........214
  fiberglass, swinging ..........216
  gliding, vinyl ...............214
  grille .....................216
  handles ...................216
  hardware ..................216
  hinged, vinyl-clad ...........214
  installation ................214
  jamb extension kit ..........217
  locks .....................216
  roller .....................216
  screen ....................216
  screen & track ..............216
  sliding wood ...............214
  steel, double ...............215
  swinging, double ...........214
  threshold ..................217
  vinyl .....................215
Patio roof ....................455
Patio screens
  replacement ...............216
Pavement
  concrete ...............46, 49
  demolition .................32
Paver tile ....................299
  brick .....................300
Paving
  asphalt .....................51
  concrete ...............46, 49
  demolition .................32
Paving breaker
  rental .....................28
Pea gravel
  concrete ....................46
Pedestal lavatory
  Bahama ...................366
  Classic base ................367
  leg set ....................367
  Memoirs ..................366
  Memoirs base ..............367
  metal leg set ...............367
  removal ...................367
  Repertoire .................366
  Seychelle ..................366
  Williamsburg ...............366
Peel & Patch waterproofing ....135
Peep sights ..................214
Penetrating wood finish
  Preserva-Wood .............487
Pergo .......................283
  finishing putty ..............283
  flooring ...................283
  flooring sealant .............283
  flooring underlayment .......283
  installation ................283
  installation kit ..............283
  laminate glue ...............283
  Presto floor molding .........283
  Presto laminate flooring ......283
Perimeter wall insulation ........87
Perlite texturing paint additive ...493
Permanent Bond Adhesive ......133
Perma-Shield ................216
Perpetual laminate flooring
  molding ...................284
Pet doors ...................217
Phone wall plate .............433
Picket fence
  staining labor ..............474
Pickling stain
  white wash ................488
Picture windows .............163
Pier, concrete .................64
Pine
  door casing ................195
  door casing, fluted ..........195

flooring trim ..................281
plank flooring ...............280
siding ..................109-111
Pine doors ...................208
Pine jamb sets ................209
Pine paneling .................252
Pine planking ................253
Pine square rosette molding ......259
Pipe
  ABS ......................408
  copper replacement .........407
  PVC DWV .................408
Pipe flashing .............143, 145
Pitch pan ....................129
Pitch pocket .................129
Pits, excavation ...............40
Place n Press floor tile ..........289
Plain end 45-degree elbow .......421
Plank floor clamp .............281
Plank flooring
  bamboo ...................280
  heart pine .................280
  rustic oak .................280
  wood .....................280
Plank paneling ...............253
  beaded ...................253
  Edge V ...................253
Plaster
  demolition ..........34, 242, 316
  painting ...............480-481
  veneer ...................246
Plaster of Paris ...............246
Plastic
  gutters ...................146
  siding ....................108
Plastic conduit ...............420
Plastic lumber ...............457
Plasticized concrete ............47
Plate
  nail ......................419
  service entrance conduit sill ...432
  single gang Decora wall ......435
  single gang plastic wall ......435
  splash ...................344
Plate cover, toilet base ..........364
Plates
  carpentry ..................65
  carpentry piecework rates .....68
  door hardware .............213
Plexiglas ....................172
Plinth ......................455
Plinth block .................259
  MDF primed ...............262
Plug
  connecting appliance .........417
Plug-on circuit breaker
  ground fault interrupter .......440
Plumb bob ...................291
Plumb Shop .............336, 342
Plumbing vent ...............390
  relocate ..................390
Plus 10 Exterior Stain ..........487
Ply 4 glass felt ...............134
Plywood
  roof sheathing ..............77
  sheathing ................67, 71
  siding ....................112
  siding, T 1-11 ..............112
  subfloor ...................67
Plywood cap molding .........263
Pneumatic breaker
  rental ....................28
Pocket doors ................222
  edge pull .................222
  hangers ..................222
  hardware .................222
  jamb .....................222
  kit, converging ............222

Pointing trowel ...291
Pointing, masonry ...116
Points
glazing ...171
Pole sander ...251
Polish
Boston ...491
Polyblend
ceramic tile caulk ...307
Polycarbonate sheet ...172
Polycast kitchen sink ...330
Polycrylic
protective finish ...489
Polyester
stress-block fabric ...136
Polyester carpet ...308
Polyethylene
film ...40
Polyethylene film ...88
moisture barrier ...284
Polyethylene membrane
Schluter Ditra ...308
Polymer molding ...263
Polyshades polyurethane stain ...488
Polystyrene casing ...264
reversible ...266
Polystyrene insulation ...86-87
Polystyrene molding
base ...264
base embossed ...263
chair rail ...266
crown ...268
crown embossed ...268
embossed ...265
inside corner ...267
lattice ...269
outside corner ...267
spring cove ...268
stop ...269
Polystyrenecasing ...265
Polyurethane
fast-drying ...489
oil gloss ...495
stain ...488
Polyurethane cast molding ...262
chair rail ...262
corner block ...262
cross head ...262
frieze ...262
rope ...262
PondPatch ...136
Pool paint ...496
Popcorn ceiling
patch ...244-245
spray ...244
spray texture ...244
Porcelain
bath accessory set ...305
mortar ...306
towel bar ...305
Porch
columns ...455
columns, aluminum ...455
demolition ...452
enclosure screen ...458
misting system ...458
repair ...452
roof framing ...455
screen frame ...458
support posts ...455
Porch & floor enamel ...495
Portland cement
stucco ...115
Portrait whirlpool bath ...354
Post
base, plinth ...455
Posts
basement ...64
cap ...456-457
carpentry ...64
deck ...452-453, 456
jack posts ...64
porch ...452, 455
wood ...455
Pouring
insulation ...88
Power attic gable vents ...93
Power massage shower head ...362
Power vent
gas water heater ...407
water heater ...405
Power ventilators
attic gable ...93
roof-mount ...93
Power washing ...468
Prefinished molding
casing ...264, 265
chair rail ...266
crown ...268
embossed ...265
inside corner ...267
inside corner & cove ...267
outside corner ...267
Prefinished paneling ...252
Prehung doors ...209, 210, 211, 212
double ...212
exterior ...198
fire-rated ...197, 211
French ...212, 213
hardboard ...197
heater closet ...210
installation ...197
oak ...212
pantry ...212
patio, steel ...215
steel ...199, 200
steel, exterior ...198
steel, half lite ...198
steel, installation ...197
Premium Plus ...495
interior flat paint ...492
primer sealer ...493
satin flat exterior paint ...494
Preservaproducts ...487
Preservative
dock & fence post ...490
wood ...490, 491
Preserva-Wood penetrating finish .487
Pressure washing
manhours ...468
Price Pfister ...332, 333, 338, 360, 36
362, 370, 371, 373
Primed finger joint casing
colonial ...260
ranch ...260
Primed finger joint molding
base ...260
base cap ...260
base shoe ...260
bead ...260
brick ...260
chair rail ...260
corner ...261
crown ...261
flute and reed ...261
quarter round ...261
stool ...261
Primed tilt double-hung windows ..164
Primer
adhesive ...282
bonding ...493
clean metal ...496
concrete ...493
epoxy ...496
flooring ...285
gypsum wallboard ...246
latex ...491
metal ...496
resilient flooring ...285
rusty metal ...496
water-based ...492
Primer sealer ...494
oil-based ...493
stainblocker ...494
Privacy latchset ...213
Pro Finisher ...489
ProForm patch kit ...246
Propane-fired water heater ...405
Protection
tile edge ...303
Protector
wood ...490
ProtectoWrap ...307
Pro-Wall 3 ...354
P-trap
remove & replace ...352
Pull
brass-plated spoon-foot ...327
cabinet ...324, 326-327
S swirl ...327
satin nickel ...327
wire ...327
wire, solid brass ...327
Pull bar ...281
Pull plates, door ...213
Pulling elbow, 90-degree ...423
Pulls
edge, pocket door ...222
Pump mix concrete ...46
Pumps
sewer ...389
Pure-Tex spray texture ...246
Purity siding ...114
Purlins
carpentry ...77
Push bar
door ...226
Push plates, door ...213
Putty
finishing ...283
PVA latex primer ...491
PVC
faucet supply ...374
PVC conduit
coupling ...421
elbow ...421
Schedule 40 ...420
Schedule 80 ...420
PVC conduit body ...421
PVC conduit clamp
Snap Strap ...420
PVC pipe ...408
PVC service entrance cap ...422
Pyro-Kure paper ...89

**Q**

QO circuit breaker
GFI plug-on ...440
main ...438
miniature ...439
tandem ...439
QO load center ground bar kit ...440
QO main lug load center ...438-439
Quarry tile ...299
Quarrybasics ...299
Quarter round
bamboo ...280
floor molding ...278-280, 284
Quarter round molding ...263
primed finger joint ...261
Quick Roof ...135
Quiet range hood ...344
QuieTTest bath fan ...380
Quietzone ...87
Quik-Tape ...248
Qwik-Gard ...440

**R**

RACO ...424-426
Radar Illusion ceiling panel ...257
Radiator
cast iron ...398
hot water ...398
relocating ...398
Rafter
bay vent channel ...89
spacing ...89
Rafters
demolition ...33
rough carpentry ...76
Rail
deck ...456
painting ...481
sink ...302
Rail molding
chair ...266
embossed polystyrene ...266
embossed prefinished ...266
paperwrap ...266
prefinished chair ...266
Railing
patio deck ...455
Rain cap, chimney ...400
Rain gutters ...146
aluminum ...146
galvanized steel ...146
Ranch base paperwrap molding ..264
Ranch casing ...195, 265
Clearwood PS ...265
paperwrap ...265
polystyrene ...265
primed finger joint ...195, 260
Ranch stop molding ...269
Random textured ceiling panel ...256
Range hood
convertible ...343
custom ...344
ductless ...344
quiet ...344
vented ...344
RCA ...433
Ready-mix concrete ...46
Rebar ...44
Rebond ...308
Receptacle
appliance flush mount ...435
CO/ALR duplex ...435
duplex ...435
duplex straight blade ...434
GFCI ...435
GFCI, hospital grade ...435
remove ...434
side-wired single ...435
Receptacle cover
GFCI, wet location ...428
horizontal duplex weatherproof ..428
Recessed bathtub
Princeton ...353
Recessed light kit ...440
Recessed lighting ...440
Rectangular weatherproof box ...427
Recycler fees ...29
Recycling
fees ...29
Red cedar
western, siding ...109
Red Dot ...427-428
Red oak
plank flooring ...280
strip flooring ...278, 281
Red rosin sheathing ...88

RedGard waterproofing
  membrane ...................305
Reducer
  bamboo ......................280
  carpet ........................284
  floor molding ...........278-280
  laminate flooring ...........284
Reducer strip
  tile ..........................304
  vinyl flooring ................292
Reducing
  window opening size ...........156
Reducing bushing
  non-metallic .................420
  rigid or IMC .................430
Redwood
  siding .......................110
  vents .........................91
Reed & flute molding ...........261
Reinforcing
  bars ..........................44
  membrane, fiberglass ..........136
  slab ..........................47
  steel .........................44
  steel, galvanized .............45
Reliant kitchen faucet ..........332
Removal
  glazing putty .................170
  skylight ......................173
  window ........................156
  window glass .............170-171
  windows .......................156
Remover
  adhesive .....................282
  flooring adhesive .........284-285
Rental
  gypsum panel lift .............250
Repainting a home ..............466
Repair
  shingles, tin .................144
Repair clips
  drywall .......................249
Repair kit
  drywall .......................248
  toilet ........................364
Repair mesh
  crack .........................248
Repair patch ...................248
Repairs
  door ..........................191
Replace-A-Vent ..................90
Replacement screen doors .......203
Replacement windows
  vinyl .........................157
  vinyl double-hung .............157
  vinyl single-hung .............158
Resilient tile
  adhesive .....................290
  demolition ....................35
Resin siding ...................108
Respirator
  sanding drywall ...............250
Respray texture ................246
Retaining wall forms ............47
Reversible doors ...............194
Reversible flute & reed molding
  primed finger joint ...........261
Reversible polystyrene casing ....266
R-Gard insulation ...............87
RhinoBoard .....................295
Ribbands .......................74
Ribbons, carpentry ..............74
Ridge and hip shingles .........130
Ridge shingles .................131
Ridge vent .....................144
  Cobra Rigid ..................145
  rigid .........................145
  shingle-over .................145

VentSure ......................144
Ridge vent coil .................91
Ridge vents, aluminum ...........91
Rigid conduit
  coupling .....................429
  strap ........................431
Rigid insulation ................86
Rigid steel conduit
  elbow ........................429
  galvanized ...................429
  installation .................429
  removal ......................429
Ring type meter socket .........437
Rings
  towel ...............385-386, 388
Risotto kitchen sink ...........329
Rivets .........................256
  paneling .....................254
Robe hook .........385-386, 388-389
Roberts ........................282
Rock
  recycling .....................29
Rocker switch
  Decora AC quiet ..............434
Rockwool insulation .............88
Rod clamp
  ground .......................432
Roll
  screen .......................169
Roll flashing
  aluminum .....................141
  copper .......................141
Roll roofing ...................134
  mineral surface ..............133
Roll valley
  flashing .....................143
Roller
  extendable ...................291
  painting labor ...............468
Roller spline tool .............169
Rollers
  door, patio ..................216
Roll-on
  texture, acoustic ............493
  texture, stucco ..............493
Rollup doors
  garage .......................223
Roman tub
  filler .......................354
  trim & rough-in ..............354
Romex sheathed cable
  indoor ..................417-418
  indoor & outdoor .............418
Roof
  bay window ...................165
  bow window ...................165
  built-up .....................132
  cant strips ...................87
  deck .........................455
  defects .......................58
  demolition ....................33
  flashing ......................89
  flashing, removal ............134
  flashing, self-adhesive ......134
  insulation ....................87
  leaks .........................58
  louver ...................92, 145
  membrane .....................132
  overhang, adding ..............61
  patch ........................135
  patch, shingle ...............137
  patio ........................455
  sheathing, OSB .............71, 77
  sheathing, plywood ............77
  tile, clay ...................139
  tile, concrete ...............140
  trusses .......................77
  trusses, repair ...............77

underlay .......................89
urethane spray foam ...........140
Roof and foundation coating
  non-fibered ..................129
Roof apron .....................143
Roof caulk .....................136
Roof cement ....................135
  wet patch ....................135
  Wet-Stick ....................135
Roof coating
  acrylic ......................137
  aluminum .....................137
  elastomeric ..................137
  fibered ......................137
  Snow Roof ....................137
Roof covers ....................165
Roof deck
  removal ......................129
  repair .......................129
  walking ......................140
Roof edge flashing .........143-144
Roof exhaust vent
  T-top ........................145
Roof flashing
  chimney ..................400-401
  service entrance .............432
Roof framing ....................76
  jacking .......................76
Roof jack ......................145
  rubber .......................145
Roof loading ...................140
Roof openings
  cut for skylights ............173
Roof overhang
  painting .................479-480
Roof patch
  cotton .......................136
  rubberized ...................135
Roof patching cement
  white ........................135
Roof sheathing
  repair ........................60
Roof turbine vent ..............146
Roof turbine vents ..............92
Roof vent
  repair, waterproofing ........130
  waterproofing ................130
Roof vent kit
  bath .........................382
Roof waterproofing .............135
  repair .......................129
  replace ......................130
Roofing
  asphalt ......................132
  base sheet ...................133
  broom ........................136
  built-up ................132, 134
  built-up, patch ..........131-132
  built-up, tear-off ...........129
  cement trowel ................136
  clay tile ....................139
  concrete tile ................140
  demolition ....................33
  felt .........................133
  fiber-cement slate ...........139
  fiberglass panels ............140
  flashing, removal ............134
  flashing, self-adhesive ......134
  galvanized ...................138
  granules .....................136
  gravel .......................132
  membrane .....................132
  mop ..........................136
  patch ........................135
  roll .....................133-134
  sheet metal ..................138
  sheets .......................138
  shingle remover ..............136

shingle starter strip ..........131
shingles, algae-resistant ......130
shingles, architectural ........131
shingles, asphalt ..............131
shingles, asphalt fiberglass ...130
shingles, fiberglass asphalt ...131
shingles, hip and ridge ....130-131
shingles, laminate .............131
slate ..........................139
tar ............................132
tear-off .......................129
tile ...........................140
tile, clay .....................139
tools ..........................136
Roofing cement .................133
  built-up .....................133
Roofing fabric
  asphalt ......................131
Roofing louver
  galvanized ...................145
Roof-mount
  power ventilator ..............93
Roof-to-wall flashing ..........144
Rope cast molding ..............262
Rope crown cast molding ........262
Roppe ..........................292
Rosette molding ................259
  square .......................259
Rosette set ....................195
Rosewood
  flooring trim ................279
  hardwood flooring ............279
Roto hammer
  rental ........................28
Rough carpentry
  backing and nailers ...........74
  bracing .......................68
  ceiling joists ................74
  ceilings and joists ...........74
  collar beams ..................76
  collar ties ...................76
  curbs .........................76
  dormer studs ..................76
  floor joists ..................65
  floor trusses .................66
  furring .......................73
  girders .......................63
  joists ........................69
  ledger strips .................74
  plates ........................68
  posts .........................64
  purlins .......................77
  rafters .......................76
  ribbons .......................74
  roof trusses ...............76-77
  sill plates ...................65
  studding ...................67-68
  subflooring ...................67
  trimmers ......................76
  wall openings .................73
  wall studs ....................67
  window openings ...............73
Round conduit box
  non-metallic .................428
Round edge stop molding .......261
Royal Mouldings ............263-270
Royelle vinyl sheet flooring ...287
Rub, carborundum ................50
Rubber
  mallet .......................291
  roof jack ....................145
  studded tile .................290
  tile adhesive ................290
  wall base ....................292
  wet patch roof cement ........135
Rubberized
  roof patch ...................135
Rust Destroyer .................496

Rust resistant finish ..............496
Rustic oak plank flooring ........280
Rustic siding
cedar ..........................109
pine ...........................111
Rust-Oleum ............488-489, 496

## S

Sack finish ........................50
Saddle & sill .....................300
Safety
respirator mask ................250
Sahara ceiling panels ............256
Saltillo
ceramic floor tile ...............299
grout mix ......................307
Salvage, demolition ................35
San Marino ceiling fan ...........444
Sand
fill ...........................40, 49
Sand & Kleen .....................251
Sandblasting
manhours .....................467
masonry .......................116
Sanded grout ....................307
Sander
drywall ........................250
gypsum wallboard .............251
hand ..........................251
Sanding
hardwood floor ................281
Sanding sealer
waterborne ....................489
Sanding sponge .................250
Sanding system
drywall ........................251
dustless .......................251
Sanding, paint prep
manhours .................466-467
Sand-Texture wall texture ........493
Sash
balance ........................167
cord repair .....................156
lift .............................167
locks ..........................167
springs,wood window .........167
window, repair .................156
Saturated cotton roof patch
asphalt ........................136
Savannah lavatory faucet .........373
Sawing
concrete slab ....................50
SBS field adhesive ................134
ScapeWel window ................173
Schedule 40 PVC conduit .........420
Schedule 80 PVC conduit .........420
Schluter Ditra polyethylene membrane ...........................308
Schluter Systems .......303-304, 308
Scratch coat, plaster ..............115
Screed
rental ...........................39
Screen
aluminum rolls .................169
fiberglass rolls .................169
frame kit .......................168
patch ..........................169
porch enclosure ................458
screws .........................459
soffit ............................90
spline .....................169, 459
Screen doors .....................203
fan lite .........................203
hardware ......................203
installation ................202-203
replacement ...................203
security ........................203
steel, replacement .............203
vinyl ...........................203
Screen frame ....................458
Screen vents .......................90
Screen wire ......................458
Screens
patio door .....................216
patio door .....................215
window ........................160
Screw gun
drywall ........................249
Screwdriver
drywall ........................249
Screwdriver bit
drywall ........................250
Screw-in connector, flex .........426
Screws
drywall ........................249
drywall to steel ................249
drywall to wood ...............249
lag, drill tip ....................256
screen enclosure ...............459
Screws & clips, sink mount .......329
Screws, self-tapping .............141
Scribing felt .....................291
Seal
garage door ...................223
Seal & tile gloss .................308
Sealant
laminate flooring ..............283
Sealdon 25 shingles .............131
Sealer
concrete .......................495
exposed aggregate ............495
grout ..........................308
high-gloss finish ...............307
oil-based ......................493
painting ...................483-484
primer .........................493
sanding .......................489
sill ..............................87
tile ............................308
tile & stone ....................307
water-based ...................492
waterproofing .................495
Sealer primer
stainblocker ...................494
water-base ....................494
Sealer's Choice 15 Gold ..........307
Sealing conduit locknut
steel ...........................430
Seal-Krete .......................495
Seam coating kit ................291
Seats
toilet ......................365-366
toilet, shell design .............365
SEBS styrene-ethylene-butylene-styrene
asphalt ........................135
Second story addition ............57
Security doors ...................205
heavy-duty ....................205
Security gates, folding ...........205
Security light
motion sensor .................443
Security storm doors .............205
Seekure paper .....................89
Self-stick
parquet flooring, oak ...........282
vinyl wall base .................293
wall base ......................293
Self-storing storm doors .........204
Self-tapping screws ..............141
Semi-gloss
latex paint .....................493
wall enamel ...................492
Senco ............................249
SensAire wall control switch ......382
Sentinel Products ...........284-285
Septic disposal cartridge ........341
Sequentia .......................253
Service entrance
cap ............................432
cap, PVC .......................422
device, combination ...........438
elbow .........................432
meter main load center ........437
roof flashing ...................432
Service entrance conduit sill plate ..432
Service panel
electric ........................439
Set screw connector
EMT ...........................423
Setting adhesive
marble & granite stone .........306
Seychelle pedestal lavatory bowl .366
Shake felt .......................137
Shakertown siding ...............114
Shakes
siding, replacement ............106
Shakes, cedar ....................137
Shank nut
sink strainer ...................336
Shaping slopes ....................41
Shaw HardSurfaces ..............284
Sheathed cable
indoor .........................418
indoor & outdoor Romex .......418
type NM-B Romex .........417-418
Sheathing
board .......................67, 71
carpentry .......................67
demolition ......................34
insulated .......................86
plywood ...................71, 78
roof ........................71, 77
T 1-11 .........................112
Sheathing paper ...................88
Sheathing, roof
repair ...........................60
Sheet
acrylic .........................172
mirrored .......................172
Plexiglas .......................172
Sheet flooring
adhesive .......................290
removal .......................278
vinyl ......................286-287
Sheet metal
gutters, downspouts ...........146
roofing ........................138
Sheetrock
ceiling panel ...................257
demolition .....................242
fire code .......................243
flexible ........................243
installation ...............242, 244
painting ...................483-484
repair ..........................242
water resistant .................243
Sheets, roofing ..................138
Shelf
bath, with glass ................385
bathroom ......................388
corner .........................304
glass ...........................388
painting .......................468
Shellac
amber .........................490
Shelves
painting labor ..................468
Shim shingles ...................138
Shims, flooring ....................63
Shims, stud .......................68
Shingle
flashing .......................144
Shingle panels ..................114
Shingle siding ...................114
cedar ..........................114
Shingle starter strip .............131
Shingles
algae-resistant .................130
architectural ...................131
asphalt ....................130-131
asphalt, repair .................130
cedar ..........................138
demolition ......................33
fiberglass asphalt ..............131
hip and ridge ..............130-131
interlocking ....................131
laminate ..................130-131
panelized ......................114
roofing ........................130
siding .........................114
siding, replacement ............106
slate ...........................139
tin repair ......................144
wood ..........................137
wood, repair ...................137
Shoe molding
base ...........................264
Shower
corner entry kit ................358
fiberglass ......................357
neo-angle kit ..................358
stall ...........................357
stall kit ........................357
stall kit, add-on ................362
Shower & bathtub units ..........356
Shower & tub faucet
single control ..................360
Shower base .....................358
Shower curtain rod ..............389
Shower door .....................356
bypass .........................360
hinged .........................359
pivot ...........................359
swinging .......................359
Shower faucet ..............360-362
lever handle ...................361
single-handle ..................360
Shower head
power massage ................362
remove & replace ..............352
sunflower ......................362
Shower pan ......................359
Shower stall .....................357
demolition .....................357
round ..........................358
smooth wall ...................357
three-piece ....................358
Showers ...................356-357
Shutters
painting ...................481-482
remove and replace ............107
window ........................168
Sidelites
steel door ......................198
Sidewalk
demolition ......................31
Sidewalks
concrete ....................46, 49
Side-wired receptacle, single ......435
Siding ......................108, 113
accessories, mounting .........108
aluminum ......................115
aluminum corrugated ..........114
cedar ..........................109
cedar rustic ....................109
cedar T&G .....................110
corrugated ....................114
demolition ......................34
fiber cement ...................113
fiber cement shingles ..........114
hardboard lap .................113

hardboard panel ...............112
lap ........................111
log cabin ...................111
mounting block, fixture ........108
Nailite .....................108
OSB .......................112
painting ...................482
panelized shingles ...........114
pine .......................109
pine rustic .................111
pine shiplap ................111
plywood ...................112
redwood ...................110
removal ...................107
repair .....................106
resin ......................108
shingle ....................114
shingles, replacement ........106
southern yellow pine ......110-111
staining ...................482
T 1-11 plywood .............112
thermoplastic ..............108
touch up paint .............109
trim .......................115
vinyl ......................108
western red cedar ..........109
Siding & deck stain ...........487
Siding and roofing
  metal ....................247
Signature kitchen faucet .......332
Silent Step .................283
Silestone countertop .........323
Silhouette polycast kitchen sink ..330
Silicone sealant
  window caulk ..............157
Siliconizer
  crack filler ...............136
Sill
  door, thermal .............195
  window ..................166
  window, ceramic ...........301
Sill & saddle ...............300
Sill plate
  service entrance conduit .....432
Sill plate gasket. .............87
Sill plates ..................65
  removal ..................64
Sill sealer ..................87
Single gang wall plate ........435
Single receptacle
  side-wired ...............435
Single strength glass .........170
Single-hung windows ........158, 161
Single-texture drywall brush .....250
Sink
  bathroom ..............367-369
  frame ...................335
  hardware ................329
  lavatory ...............367-369
  mounting kit .............335
  removal ................315
  strainer shank nut .........336
  under counter ............368
Sink rail, ceramic ............302
Sink trap
  remove & replace ..........352
Sinks
  bar ...................338-339
  bowl ...................337
  Breckenridge .............331
  Cambridge ...............331
  cast iron .............329-330
  composite .........330-331, 338
  double bowl ...........328, 330
  double bowl composite .....331
  entertainment ............338
  kitchen ...............329-331
  kitchen, basket strainer .....335

mounting clips ..............329
polycast ...................330
Silhouette polycast ...........330
single bowl .............328, 330
stainless steel ...........328-329
stainless steel kitchen .......328
stainless steel single bowl .....328
Utilatub ...................336
utility ....................336
Skid steer loaders
  rental ...................39
Skimcoat patch & underlayment ..286
Skylight removal .............172
Skylights
  acrylic ...................172
  cut roof opening ...........172
  fixed ....................172
  trimmers .................76
  tubular ..................172
  venting ..................173
Slab doors .........192-194, 206-207
  installation ...............191
Slabs
  base ....................49
  calculating area ...........42
  concrete .................46
  finishes ..................49
  forms ...................43
  garage ..................42
  on grade ................49
Slate
  floor tile, natural ..........299
  roofing ..................139
  tile, tumbled .............300
Sleepers, carpentry ...........74
Slide dimmer
  illuminated ..............436
  incandescent .............436
Sliding windows ..........161-162
Sloan Royal flush valve .........363
Sloan Valve ................371
Slopes, shaping .............41
Slot aperture light kit .........440
Smoke alarm, dual sensing ......436
Smoke detector
  basic ...................436
  dual sensing .............436
  dual-ionization ...........436
Snap Strap PVC conduit clamp ...420
Soaker bathtub ..............353
  acrylic ..................353
  with skirt ...............353
Soap dish ...............386-389
  ceramic .................304
  porcelain ...............305
Soap dispenser .............335
Soffit
  painting ...............479-480
  wood ...................74
Soffit screen wire .............90
Soffit vents
  galvanized metal ...........90
  louvered .................90
  stucco ..................90
Soffits
  aluminum ...............115
  fiber cement ..............113
  kitchen ..................316
  pendent lighting ..........316
  refrigerator, under counter .....316
  removal ................107
  stucco ..................115
  vinyl ...................109
Soffits, demolition ............34
SoftSeal underlayment .........283
Soil compaction .............41
Solar lens .................173

Solid core doors
  birch exterior ..........192, 197
  entryr ...................193
  fir entry .................193
  hardboard exterior ......192, 197
  hardboard interior .........207
  hardwood exterior .........192
  lauan exterior .............192
  oak .....................192
  prehung exterior ..........197
  red oak exterior ...........192
Solid surface countertop ........322
Solitaire bath fan .........379, 381
  Ultra Silent ..............379
Sound baffle, removable ........341
Sound control ..............87
Soundbloc foam underlayment ...283
Southern yellow pine
  siding ................110-111
Spacer
  laminate flooring ..........283
Spacing
  rafter ...................89
Spanish tile roofing ...........139
Spar urethane
  Helmsman ...............488
Spar varnish
  marine ..................488
Spatter ceiling ..............245
Spatter texture .............244
Speaker wire ...............433
Speed-Wall
  interior latex enamel .......493
  PVA latex primer ..........491
Spin-N-Lock sink strainer ......336
Splash plate ...............344
Spline
  screen, porch .............459
  tool ....................169
  window screen ...........169
Sponge
  sanding .................250
Spray adhesive .............249
Spray painting
  labor ...................468
Spray texture ...........244, 246
  ceiling ..................244
  wallboard ...............244
Spreading soil ..............41
Spring cove molding ..........268
Spring wound timer
  in-wall ..................436
Springs
  garage door .............223
  garage, extension .........225
  window sash .............167
Spruce
  lap siding ...............111
Square
  drywall .................250
Square box
  surface cover ............425
  welded .................425
Square D ..............437-440
Square rosette molding .........259
Squeaks, floor ..............56
Squeeze connector, flex ........427
Stain
  acrylic interior & exterior .....487
  concrete ................495
  deck ...................487
  deck & siding ............487
  Deck Plus ...............487
  exterior .................487
  gel ....................488
  Plus 10 .................487
  polyurethane .............488
  siding ..................487

white wash ................488
wood ....................488
Stainblocker
  primer sealer .............494
  water-based ..............492
Staining labor, manhours
  ceiling ..................470
  fascia ..................473
  paneled walls ............485
  picket fence ..............474
  siding ..................482
  wood deck ..............475
Stainless steel sink ........328-329
  double bowl .............328
  single bowl ..............328
Stair nose .................284
  bamboo .................280
  molding .............278-280
Stair stringers, basement ........390
Stairs
  basement ...............390
  deck .................457-458
  demolition ..............35
  porch ................457-458
  stringers ................458
Stall, shower .............357-358
Stamped concrete ............50
Stanadyne ...........334, 338, 360
Standard coupling
  non-metallic .............421
Standard radius elbow ...........421
Stanley ...................249
Starter strip ................131
Starter strip, aluminum siding .....115
Steam cleaning, masonry ........116
Steam heat
  radiator, hot water ..........398
Steel conduit
  elbow ..................429
  elbow, rigid or IMC ........429
  galvanized rigid ...........429
Steel doors
  entry .................199-200
  exterior ..............198-199
  frames .................194
  garage .................223
  installation ..............197
  screen .................203
Steel octagon box cover ........426
Steel sealing conduit locknut ....430
Stem wall .................48
Step flashing .............142, 144
  copper .................142
Sterling Plumbing ..352, 354-355, 359
Stile & rail doors .............208
Stilts
  drywall .................250
Stone
  base ...................49
  cleaning ................116
  floor tile ................299
  roofing .................139
  sealer ..................307
  setting adhesive ..........306
  slate ...................139
  tile, travertine ...........298
  tile, tumbled ............300
Stone countertop ............323
Stool molding
  primed finger joint .........261
Stop
  door ...................213
  fixture supply ............364
  gravel ..................142
Stop molding ..............195
  colonial .................269
  paperwrap colonial .........269
  parting .................269

polystyrene colonial ...........269
polystyrene ranch ............269
ranch .....................269
round edge .................261
Stopper, drain ...............336
Stops Rust ..................496
Storage cabinet
  bath ....................383
  etched glass .............383
Store-in-Door ...............204
Storm collar
  chimney ..............400-401
Storm doors ............204-205
  full-view .................205
  removal .................204
  security .................205
  self-storing ..............204
Storm panels ................164
Storm windows ...............165
  remove and replace .........107
StormGuard ..................89
Straight blade receptacle
  duplex ..................434
Strainer
  junior basket .............339
  sink shank nut ............336
Strainer, basin
  remove & replace ..........352
Strap
  EMT ...................424
  rigid conduit .............431
Stress-block fabric
  polyester .................136
Stringers
  deck stair ...............458
Strip
  transition ................292
Strip flooring
  laminate ................284
  maple, unfinished ..........281
  natural wood .............279
  red oak .................281
  white oak ...............281
Stripping
  forms ....................44
  soil ....................41
Stripping paint
  labor ...................467
Strips
  asphalt roofing starter .......131
  cant ...................137
  shingle starter ............131
Stucco
  knockdown ..............493
  paint ...................494
  painting ...........480-481
  plastering ...............115
  portland cement ..........115
  removal ................107
  repair ..................106
  texture .................493
  wall assembly, wood framed ...72
Stucco molding, redwood ......196
Stucco soffit vents .............90
Stud sensor
  electronic ...............249
Stud straightening .............68
Stud wall
  demolition ...............33
Studding ...................67
Studs
  carpentry piecework rates ....67
  dormer .................77
  wall assemblies ...........67
Stump grinder
  rental ...................28
Stump grinding ...............29
Stylistik II floor tile ............289

Styrene-ethylene-butylene-styrene
  asphalt ..................135
Subfloor
  board sheathing ..........67
  OSB ...................67
  plywood ................67
Subflooring
  installation ..............67
  piecework rates ..........67
Sump pump ................389
Sundial vinyl sheet flooring ......287
Sunflower shower head ........362
Supply line
  dishwasher ..............342
Supply lines .................374
Supply stop .................364
Surface box cover, square ......425
Surface preparation
  flooring .................284
  labor ..................466
  manhours ..............466
Surround
  bathtub .................355
  bathtub wall .............355
  five-piece bathtub ........355
  tub wall ................355
Suspended ceiling
  demolition .........34, 254, 315
  installation ..............254
  kit ....................256
  light kit .................440
  lighting panel ............258
  panel .............256-258
  panel, acoustical .........256
  systems ................255
  zero clearance system .....255
Swan Corporation ...........355
Sweep, door ................196
Swimming pool paint .........496
Switch
  AC quiet toggle ..........434
  Decora AC quiet rocker ....434
  remove .................434
  sink top ................341
Switch & cover ..............428
Switch box
  drawn 2-device ..........425
  multi gang ..............425
  non-metallic old work .....422
Switches
  exhaust fan .............382
  fan control ..............382
  timer, exhaust fan ........382

**T**

T 1-11 plywood siding ..........112
T molding ..............280, 284
  bamboo ................280
  wood floor ..........278-279
Tall boy water heater ..........405
Tall Elite ...................355
Tamping, excavation ..........41
Tandem circuit breaker ........439
Tape
  acrylic ..................291
  corner bead .............249
  drywall .................246
  drywall joint .............246
  fish ....................417
  tile backer ..............249
Tape-on bead ...............247
Tapping block ...............281
TapTight
  hardwood flooring ........280
  hardwood molding ........280
Tar
  roofing .................132
T-astragal ..................194

Teak
  flooring .................279
Teak oil ....................491
Tear-off, roofing ..............129
Temporary
  wall bracing .............68
Termin-8 wood preservative .....490
Termite damage .............53
Terrazzo
  demolition ...............35
Textolite counters ............322
Texture
  paint ...................493
  paint additive ............493
  roll-on .................493
  Sand-Texture wall ........493
  stucco .................493
Texture 1-11 plywood siding ....112
Texture brushes
  drywall .................250
Texture spray
  gypsum board ...........244
Texture touch-up kit
  acoustic ceiling ..........244
Texture, drywall .............244
  ceiling .................244
  spray ..............244, 246
Textured ceiling panels ........258
Textured lap siding ...........112
Texturing paint additive
  Perlite ..................493
Themes sheet vinyl flooring ....287
Thermal sill .................195
Thermoplastic siding ..........108
Thermostat
  floor warming system .......305
Thermostat cable ............433
THHN copper wire
  gauge .................432
Thin-set mortar ..............306
  CustomBlend ............306
Thompson's ................490
Thompsonized decking lumber ..454
Thoro .....................495
Thoroseal protective coating ....495
Threaded conduit body ........431
Three gang box
  Super Blue, non-metallic ....422
3M Company ...............249
Threshold
  ceramic .................301
  high boy ................195
  marble .................300
  patio door ..............217
Thru-the-wall utility fan .........380
Thru-wall
  furnace, gas .............404
Ties
  collar ...................76
Tile
  accessories, ceramic .......304
  accessories, toilet tissue holder .304
  accessories, toothbrush holder .305
  accessories, towel bar .......305
  acoustic, demolition ........34
  anti-fracture membrane .....305
  anti-fracture membrane kit ....307
  backerboard .............295
  bullnose, ceramic ......301-302
  carpet ..................308
  ceiling, acoustical .........253
  ceramic caulk ............307
  ceramic floor .........295-298
  ceramic floor & wall ........297
  ceramic wall ........301, 303
  ceramic, demolition ........35
  clay, roofing .............139
  concrete roof ............140

  demolition ..............315
  edge protection ..........303
  finish ..................307
  floor ...................299
  glazed ceramic ..........303
  gloss & seal .............308
  granite .................299
  granite floor .............299
  grout ..................294
  grout, sanded ...........307
  installation ..............294
  marble .................298
  marble floor .............298
  mortar .................306
  mosaic ceramic ..........304
  natural cork .............293
  non-sanded dry grout .....307
  parquet floor ............282
  paver ..................299
  Place n Press floor ........289
  polyethylene membrane ....308
  quarry .................299
  reducer strip ............304
  repair, mortar ...........306
  resilient, demolition ........35
  roofing .................140
  rubber studded ..........290
  Saltillo ceramic floor .......299
  sealer ..................307
  sealer, high-gloss .........307
  setting with adhesive ......294
  setting with mortar ........294
  slate floor ...............299
  thin-set mortar ...........293
  travertine floor ...........299
  travertine stone ..........298
  tumbled marble listello ....300
  tumbled slate ............300
  tumbled stone ...........300
  vinyl composition .....289-290
  vinyl floor ...........288-289
  wall ...................303
  wall, ceramic ............303
  wood flooring ...........282
Tile accessory set
  porcelain ...............305
Tile adhesive
  ceramic .................293
  resilient ................290
  rubber .................290
  stone ..................306
  thin-set ................306
  vinyl ..................290
Tile adhesive caulk
  ceramic .................307
Tile backer .................245
Tile backer tape .............249
Tile backerboard ............295
Tile cutter
  vinyl ..................291
Tile forms ..................355
Tile Perfect .................307
Tile repair
  mortar .................306
  mortar, underwater .......307
Tile tool
  ceramic .................291
Tile tub wall ............354-355
Tile underlayment
  Aqua-Tough ............295
Tile wall kit ................355
Tileboard
  batten ..................251
  embossed ..............252
Tileboard molding
  vinyl ..................270
Tilt double-hung windows ......164

Timberline
Ultra ... 131
Timbertex ... 131
Timer
in-wall spring wound ... 436
Timer switch ... 382
Tin
repair shingles ... 144
Tissue holder ... 387-388
TJI
joists ... 66
Toggle switch
AC quiet ... 434
Toilet base plate cover ... 364
Toilet bowl
gasket ... 364
reset ... 352
Toilet fill valve
anti-siphon ... 364
remove & replace ... 352
Toilet flange
cast iron ... 364
Toilet paper holder .. 385-386, 388-389
recessed ... 386
Toilet repair kit ... 364
Toilet seat ... 365
cushioned vinyl ... 366
molded wood ... 365
natural maple ... 366
open-front ... 365
wood ... 365
Toilet tank bolt kit ... 364
Toilet tissue paper holder ... 304-305
Toilet To Go Kit ... 362
Toilets
cadet ... 362-363
Canterbury ... 363
Champion ... 363
Comfort Height ... 363
compact ... 362
Gabrielle ... 363
Hamilton ... 363
hook-up kit ... 364
one-piece ... 363
pressure assist ... 363
San Raphael ... 363
urinals ... 364
wall-mount ... 363
Williamsburg ... 363
Tongue and groove siding
cedar ... 110
Tool kit
drywall ... 251
Tools
ceramic tile ... 291
glazing ... 171
roof repair ... 136
spline ... 169
Toothbrush & tumbler holder ... 305
Toothbrush holder .. 385-386, 388-389
porcelain ... 305
Top
laminated plastic ... 323-324
Topping compound
wallboard ... 246
Topping, concrete ... 50
Topsoil, excavation ... 41
Torsion springs ... 223
Touch control lavatory faucet ... 371
Touch-up kit
spray texture ... 244
wood floor ... 278
Towel bar ... 385-389
ceramic ... 305
porcelain ... 305
Towel ring ... 385-389
Track, patio door ... 216
Traffic Master laminate flooring ... 284
Transit mix concrete ... 46
Transition reducer strip
vinyl flooring ... 292
Transition strip
carpet to tile ... 292
oak to tile ... 292
Trap, lavatory
remove & replace ... 352
Trash
bins ... 29
Trash compactor
removal ... 315
Travertine
floor tile ... 299
stone tile ... 298
Trayco ... 355
Treads
concrete, finishing ... 49
Treatment
wood ... 490
Tree removal ... 29
Trenchers
rental ... 39
Trenching ... 41
hand ... 40
Trim
aluminum siding ... 115
bamboo flooring ... 280
door ... 194
fiber cement siding ... 113
hardwood flooring ... 281
heart pine flooring ... 281
oak ... 279
painting labor ... 471, 479
window ... 355
wood flooring ... 279
Trim & rough-in
Roman tub ... 354
Trim enamel
latex satin ... 492
Trim pack
wood paneling ... 253
Trim paint ... 492
eggshell ... 492
flat ... 494
Trimless door frame ... 194
Trimmers ... 76
Trimming, trench bottom ... 40
Trine ... 436
Triple-track storm doors ... 204-205
Troffer fluorescent fixtures ... 258
parabolic ... 258
U-lamp ... 258
Trowel
drywall ... 250
outside corner ... 250
Trowel finishing, concrete ... 49
Troweling machine
rental ... 39
Trowels
margin notch ... 291
notched ... 291
pointing ... 291
Tru-Glaze WB epoxy gloss coating .. 496
Trusses
fink ... 77
gable ... 77
roof ... 77
T-square
drywall ... 250
T-top roof exhaust vent ... 145
Tub
laundry ... 337
overflow tube ... 338
utility laundry, kit ... 337
Tub & shower faucet
single control ... 360
single-handle ... 360
Tub door
bypass ... 355
sliding ... 356
Tub faucet ... 360-362
lever handle ... 361
Tub, bath
removal ... 352
Tubular skylights ... 173
Tumbled tile
marble listello ... 300
slate ... 300
stone ... 300
Tumbler & toothbrush holder ... 305
Tung oil ... 491
Turbine vent
roof ... 146
Turbine vents
roof ... 92
Turned spindle deck rail ... 456
Two gang box
Super Blue, non-metallic ... 422
Two-handle lavatory faucet ... 372
Type X gypsum board ... 243
Tyvek house wrap ... 89

**U**

U.S. Ceramic Tile ... 299, 301-303
UF compression connector
non-metallic ... 419
UF-B sheathed cable ... 419
U-lamp troffer fluorescent fixture . 258
Ultra Silent bath fan ... 379
Ultra Touch ... 88
Ultra white elastomeric roof
coating ... 137
Ultra-Hide
alkyd gloss enamel ... 493
interior latex eggshell ... 492
interior latex paint ... 492
interior paint ... 492
Undercoater
interior enamel ... 491
Undercounter refrigerator ... 316
Underground cable
UF-B sheathed ... 419
Underground feeder compression
connector
non-metallic ... 419
Underlay, roof ... 89
Underlayment ... 140
cement-based ... 285
fast-setting patch ... 285
foam ... 283-284
laminate flooring ... 283-284
liquid ... 285
roof ... 134
SoftSeal ... 283
tile ... 295
tile backerboard ... 295
Underlayment & floor patch ... 286
Underwater tile repair mortar ... 307
Unfaced fiberglass batt insulation .. 85
Universal hub
bolt-on A type ... 438
Urethane
roof, spray foam ... 140
Urethane sheathing ... 87
Urinal flush valve
Sloan Royal ... 363
Urinals
Dover ... 364
Washbrook ... 364
USG ... 245
USG Interiors ... 257-258, 295
Utilatub laundry tub ... 337
Utilatub sink ... 337
Utility access door ... 248
Utility bowl sink ... 337
Utility cabinets
birch ... 317
finished ... 320
maple ... 321
oak ... 319, 321
red oak ... 318
Utility fan
thru-the-wall ... 380
vertical discharge ... 379
Utility laundry tub kit ... 336
Utility sink ... 336
Utility tub kit ... 336

**V**

Valance
cabinet ... 316
cabinets ... 319
Valley flashing
aluminum ... 141
roll ... 143
Valspar ... 488, 491
ValueTest exhaust fan ... 378
ValueTest range hood ... 344
ValueTest vertical discharge
bath fan ... 379
ValuSeries
pointing trowel ... 291
Valve lines ... 374
Valves
angle ... 364
flush, Sloan Royal ... 363
flush, urinal ... 363
toilet, remove & replace ... 352
wall stop ... 364
Valvuetwist faucet set ... 362
Vance Industries ... 335
Vanguard hardwood flooring ... 280
Vanity
Charleston ... 374
Danville ... 375
Kingston ... 374
maple ... 375
Monterey ... 375
oak ... 374-375
solid surface sink top ... 377
Springfield ... 374
Virginia ... 375
with drawer ... 375
with marble top ... 377-378
Vanity cabinet ... 378
mirrored with marble top ... 378
Vanity mirror ... 382
Vanity top
cultured marble ... 377
granite ... 377
Neptune shell marble ... 376
Newport cultured marble ... 376
oval bowl cultured marble .. 376-377
rectangular cultured marble ... 376
shell bowl cultured marble . 376-377
Vapor barrier ... 40
paper ... 88
SoftSeal ... 283
Varathane
exterior classic clear ... 489
exterior clear finish ... 489
interior classic clear ... 489
oil-based floor finish ... 489
Varnish
marine ... 488
marine spar ... 488
Velux ... 173-174
Veneer base board ... 245
Veneer plaster ... 246
Vent
plumbing ... 390
Vent channel
rafter bay ... 89

Vent kit
bath ....382
Vent, roof
dormer ....144
ridge ....144-145
T-top exhaust ....145
turbine ....146
waterproofing ....130
Vented range hood ....344
Ventilator
bath ....382
Ventilators
power ....93
Venting skylights ....174
Ventlite doors
fiberglass exterior ....201
steel exterior ....199
Vents
attic ....92
cedar ....91
continuous ....90
dormer ....91
eave ....91
foundation ....89
gable ....90-91
galvanized metal ....90
louvered ....91
redwood ....91
ridge ....91
roof turbine ....92
screen ....90
soffit ....90
stucco soffit ....90
VentSure ridge vent ....144
Vermiculite
insulation ....88
Vertical discharge bath fan ....379
ValueTest ....379
Vertical discharge utility fan ....379
Vibrators
rental ....39
Victorian base molding
hardwood ....259
Victorian lavatory faucet ....373
Villeta faucet ....370
Vinyl
bay windows ....158
circular windows ....159
door bottom ....196
double-hung replacement
windows ....157
fascia ....109
garden windows ....158
hopper windows ....159
jalousie windows ....159
patch ....285
screen doors ....203
siding ....108
siding removal ....107
siding repair ....106
single-hung windows ....158
soffit ....109
Vinyl composition tile ....289-290
Vinyl drywall
outside corner bead ....248
Vinyl floor tile ....288-289
Vinyl flooring
removal ....278
roll ....286-287
roller, extendable ....291
sheet ....286-287
transition reducer strip ....292
Vinyl sheet flooring ....287
Vinyl tile
adhesive ....290
cutter ....291
Vinyl tile cutter ....291
Vinyl tileboard molding ....270
Vinyl wall base
dryback ....292
self-stick ....293
Vinyl-clad
wood awning windows ....160
wood casement windows ....160
wood windows ....159

**W**

Waferboard ....112
Wainscot
tile ....294
Wainscot paneling ....253
Walkway
demolition ....31
forms ....49
Walkway forms ....43
Walkways
concrete ....49
Wall
bathtub ....354-355
braces ....68
fireblocks ....69
furnace, gas ....404
opening, breakthrough ....73
plates ....68
tub ....354-355
Wall & floor tile
ceramic ....297
Wall and ceiling panel
FRP ....253
Wall base
molding ....284
rubber ....292
self-stick ....293
strip, vinyl ....293
thermoplastic ....292
vinyl ....292
Wall cabinets ....317-319
end panel ....319
oak ....321
prefinished ....321
square ....320
washer dryer ....319
Wall control switch ....382
Wall demolition ....31
Wall enamel
latex satin ....492
Wall exhaust fan
chain-operated ....380
Wall forms, concrete ....43, 47
Wall framing
replacement ....57
Wall kit
bathtub ....355
tile ....355
Wall lantern ....443
Wall light
opal glass ....441
Wall louvers ....92
Wall oven ....342
double ....342
Wall paint ....492
eggshell ....492
Latex ....492
Wall plate
coaxial cable ....433
decorator phone ....433
single gang Decora ....435
single gang plastic ....435
Wall registers, installation ....397
Wall stop ....364
Wall texture
Sand-Texture ....493
Wall tile
ceramic ....301, 303
Wallaceburg bar faucet ....339
Wallboard
cement ....245
demolition ....34, 316
L trim ....248
painting ....483-484
primer ....246
sander ....250
stilts ....250
texture spray ....244
Wallcovering
designer panel ....252
Wall-mount lantern
curved glass ....443
Wall-mounted lavatory
removal ....367
Walls
concrete ....47
demolition ....30
painting ....480-481, 483-485
partition, adding ....270
setting with adhesive ....294
setting with mortar ....294
staining ....485
tile ....294
Warming system
ceramic tile floor ....305
thermostat ....305
Wash down
manhours ....467
Washer
kitchen sink basket strainer ....336
Washer dryer wall cabinets ....319
Washing
masonry ....116
pressure ....468
Washing, paint prep
manhours ....467
Waste disposal ....29
Watco ....488
Water
blasting ....116
Water closet
flange, cast iron ....364
toilet ....362-363
Water cooler ....337
Water dispenser ....336-337
Water hammer
arrester ....407
Water heater ....405
Ariston ....407
electric ....406-407
electric, tankless ....407
gas fired ....405
gas, tankless ....406-407
high altitude ....405
oil fired ....405
point-of-use ....407
PowerStream ....407
propane fired ....405
tall boy ....405
Water pipe
remove & replace ....408
Water resistant gypsum board ....243
Water Seal waterproofer ....490
Water shield ....89
Water surge shock absorber ....407
Water table ....37
Water-based primer ....493
Water-based wood stain ....488
Waterblasting
manhours ....468
masonry ....116
Waterborne sanding sealer ....489
Waterfall kitchen faucet ....334
Waterproofer
concrete & masonry ....495
Water Seal ....490
Waterproofing
masonry walls ....130
material ....135
metal roof ....135
paint, elastomeric ....494
paper ....89
roof vent ....130
roof, remove & replace ....130
roof, repair ....129
sealer ....495
wood protector ....490
Waterproofing membrane ....305
Watertight conduit hub ....430
Watts Anderson Barrows 339, 342, 374
Watts Radiant ....305
Wave bath light fixture ....441
Wax
bowling alley ....491
paste ....491
Wax toilet bowl gasket ....364
Weather head ....432
PVC ....422
Weatherproof blank cover
non-metallic ....428
Weatherproof conduit box ....427
Weatherproof electrical box
2 gang ....427
Weatherproof lampholder & cover .428
Weatherproofing ....89
Weather-resistant wood coating ..488
Weatherstripping
aluminum, adjustable ....196
foam ....196
garage door ....223
magnetic ....196
vinyl, remove & replace ....196
window seal sets ....172
Weco ....307
Welded handy box ....424
Welded square box ....425
Welded wire mesh ....45
Western red cedar
siding ....109
Westinghouse Lighting ....441
Westminster Ceramics ....300
Wet location conduit box cover ..428
Wet location receptacle cover
GFCI ....428
Wet patch roof cement ....135
rubber ....135
Wet-Stick fibered roof coating ....137
Wet-Stick roof cement ....135
Wheel glass cutters ....171
Wheel loader
rental ....28
Wheel loaders
rental ....39
Whirlpool bath
Cadet ....353
Cadet Elite ....353
Devonshire ....354
Elite EZ-Install heater ....354
Portrait ....354
Renaissance ....353
Williamsburg Elite ....353
Whirlpool tub apron ....354
WhispAire range hood ....343
White cement ....115
White Diamond bowling
alley wax ....491
White oak
strip flooring ....281
White wash pickling stain ....488
Whole house fans ....92
Williamsburg kitchen faucet ....333
Williamsburg lavatory faucet ....372
Williamsburg Roman tub filler ....354
Wind clips ....140

Window
  drip cap . . . . . . . . . . . . . . . . . . .141
  framing . . . . . . . . . . . . . . . . . . . . .70
Window bolt
  double-hung . . . . . . . . . . . . . . .167
Window glass
  remove & replace . . . . . . . . . .169-170
Window grille systems . . . . . . . . . . .169
Window guards . . . . . . . . . . . . . . .167
Window locks
  flip lock . . . . . . . . . . . . . . . . . . .172
  key, sliding . . . . . . . . . . . . . . . . .172
  thumbscrew bar . . . . . . . . . . . . .172
Window mullions . . . . . . . . . . . . . . .166
Window openings
  closure, masonry wall . . . . . . . . .156
  closure, wood frame . . . . . . . . . .156
  cut new . . . . . . . . . . . . . . . . . . .156
Window operators
  awning . . . . . . . . . . . . . . . . . . . .166
  jalousie . . . . . . . . . . . . . . . . . . .166
Window painting
  labor . . . . . . . . . . . . . . . . .486-487
Window roof covers, bay . . . . . . . . .165
Window roofs . . . . . . . . . . . . . .165-166
Window screens . . . . . . . . . . . . . . .160
  patch . . . . . . . . . . . . . . . . . . . . .168
  spline . . . . . . . . . . . . . . . . . . . .168
Window shutters . . . . . . . . . . . . . . .167
Window sill . . . . . . . . . . . . . . . . . . .166
  ceramic . . . . . . . . . . . . . . . . . . .301
Window trim
  bath wall . . . . . . . . . . . . . . . . . .355
Windows
  aluminum single hung . . . . . . . . .161
  aluminum sliding . . . . . . . . . .161-162
  aluminum, horizontal . . . . . . . . . .161
  awning . . . . . . . . . . . . . . . . . . . .160
  awning, aluminum . . . . . . . . . . . .162

bay . . . . . . . . . . . . . . . . . . . . . . . .158
bay, angle . . . . . . . . . . . . . . . . . . .164
casement . . . . . . . . . . . . . . . . . . .160
casement bow . . . . . . . . . . . . . . . .164
casement, bay . . . . . . . . . . . . . . . .158
casement, wood . . . . . . . . . . . . . . .164
caulk . . . . . . . . . . . . . . . . . . . . . . .157
caulk removal . . . . . . . . . . . . . . . . .157
circular, vinyl . . . . . . . . . . . . . . . . .159
cut new opening . . . . . . . . . . . . . .156
double-hung . . . . . . . . . . . . .157, 164
egress window . . . . . . . . . . . . . . . .173
escape wells . . . . . . . . . . . . . . . . .173
fixed aluminum . . . . . . . . . . . . . . .163
fixed octagonal . . . . . . . . . . . . . . .163
garden . . . . . . . . . . . . . . . . . . . . .158
glass setting labor . . . . . . . . . . . . .171
grille sets, tilt sash . . . . . . . . . . . . .169
jalousie . . . . . . . . . . . . . . . . .159, 163
louvered, aluminum . . . . . . . . . . . .163
measure & cut glass . . . . . . . . . . . .170
octagonal, wood . . . . . . . . . . . . . .164
picture . . . . . . . . . . . . . . . . . . . . .163
removal . . . . . . . . . . . . . . . . . .35, 156
replacement . . . . . . . . . . . . . .157-158
replacement glass . . . . . . . . . . . . .170
sash repair . . . . . . . . . . . . . . . . . .156
sash springs . . . . . . . . . . . . . . . . .166
sill, ceramic . . . . . . . . . . . . . . . . . .301
size, reducing . . . . . . . . . . . . . . . .156
storm . . . . . . . . . . . . . . . . . . . . . .165
trim removal . . . . . . . . . . . . . . . . .156
vinyl bay . . . . . . . . . . . . . . . . . . . .158
vinyl double hung . . . . . . . . . . . . .157
vinyl garden . . . . . . . . . . . . . . . . .158
vinyl hopper . . . . . . . . . . . . . . . . .159
vinyl jalousie . . . . . . . . . . . . . . . . .159
vinyl single hung . . . . . . . . . . . . . .158
vinyl, custom . . . . . . . . . . . . . . . . .157

vinyl, replacement . . . . . . . . . . . . .157
vinyl-clad awning . . . . . . . . . . . . . .160
vinyl-clad casement . . . . . . . . . . . .160
vinyl-clad double-hung . . . . .159-160
vinyl-clad wood . . . . . . . . . . . . . . .159
weatherstripping . . . . . . . . . . . . . .172
wood, replacement . . . . . . . . . . . .156
Wire
  bell . . . . . . . . . . . . . . . . . . . . . . .433
  screen . . . . . . . . . . . . . . . . . . . . .458
  speaker . . . . . . . . . . . . . . . . . . . .433
  thermostat cable . . . . . . . . . . . . .433
  type THHN copper . . . . . . . . . . . .432
  UF cable . . . . . . . . . . . . . . . . . . .419
Wire brush . . . . . . . . . . . . . . . . . . . .50
Wire mesh, welded . . . . . . . . . . . . . .45
Wireless door bell . . . . . . . . . . . . . .436
Wonderboard tile backer . . . . . . . . .295
Wood
  shakes . . . . . . . . . . . . . . . . . . . . .137
  shingles . . . . . . . . . . . . . . . . . . .137
  siding removal . . . . . . . . . . . . . . .107
Wood beams
  repair . . . . . . . . . . . . . . . . . . . . . .53
Wood bleach kit . . . . . . . . . . . . . . .487
Wood coating
  weather-resistant . . . . . . . . . . . . .488
Wood conditioner . . . . . . . . . . . . . .491
Wood deck
  patch built-up roofing . . . . . . . . .132
  staining . . . . . . . . . . . . . . . . . . . .475
Wood doors
  entry . . . . . . . . . . . . . . . . . . . . . .193
  patio . . . . . . . . . . . . . . . . . . . . . .214
  screen . . . . . . . . . . . . . . . . . . . . .203
  screen, installation . . . . . . . . . . . .202
  screen, remove & replace . . . . . . .202
Wood filler
  acrylic . . . . . . . . . . . . . . . . . . . . .281

Wood finish . . . . . . . . . . . . . . . . . . .487
  penetrating . . . . . . . . . . . . . . . . .487
Wood flooring . . . . . . . . . . . . . . . . .281
  adhesive . . . . . . . . . . . . . . . . . . .282
  installation tools . . . . . . . . . . . . . .281
  plank . . . . . . . . . . . . . . . . . . . . . .280
  strips . . . . . . . . . . . . . . . . . . . . . .279
  tile . . . . . . . . . . . . . . . . . . . . . . . .282
  touch-up . . . . . . . . . . . . . . . . . . .278
  trim . . . . . . . . . . . . . . . . . . . . . . .279
Wood paneling trim pack . . . . . . . .253
Wood preservative . . . . . . . . .490-491
  copper . . . . . . . . . . . . . . . . . . . . .491
  Termin-8 . . . . . . . . . . . . . . . . . . .490
Wood protector
  clear . . . . . . . . . . . . . . . . . . . . . .490
  waterproofing . . . . . . . . . . . . . . .490
Wood shingle
  patch . . . . . . . . . . . . . . . . . . . . . .137
Wood stain
  water-based . . . . . . . . . . . . . . . . .488
Wood toilet seat . . . . . . . . . . . . . . .365
Wood treatment . . . . . . . . . . . . . . .490
Wood trim
  oak . . . . . . . . . . . . . . . . . . . . . . .279
Wood-burning furnace . . . . . . . . . .404
Woodcore storm doors . . . . . . . . . .205
Woodpride . . . . . . . . . . . . . . . . . . .487
Wool carpet . . . . . . . . . . . . . . . . . .308

**X**
X type gypsum board . . . . . . . . . . .243

**Z**
Zinsser . . . . . . . . . . . . . . . . . . . . . .490
Zip Box
  non-metallic ceiling fan box . . . . .422
ZPW clear wood preservative . . . . .491

# Practical References for Builders

## Construction Forms & Contracts

125 forms you can copy and use — or load into your computer (from the FREE disk enclosed). Then you can customize the forms to fit your company, fill them out, and print. Loads into Word for Windows, Lotus 1-2-3, WordPerfect, Works, or Excel programs. You'll find forms covering accounting, estimating, fieldwork, contracts, and general office. Each form comes with complete instructions on when to use it and how to fill it out. These forms were designed, tested and used by contractors, and will help keep your business organized, profitable and out of legal, accounting and collection troubles. Includes a CD-ROM for WindowsTM and Mac. **432 pages, 8½ x 11, $41.75**

## CD Estimator

If your computer has *Windows*TM and a CD-ROM drive, CD Estimator puts at your fingertips over 135,000 construction costs for new construction, remodeling, renovation & insurance repair, home improvement, framing & finish carpentry, electrical, concrete & masonry, painting, and plumbing & HVAC. Monthly cost updates are available at no charge on the Internet. You'll also have the *National Estimator* program – a stand-alone estimating program for *Windows*TM that *Remodeling* magazine called a "computer wiz," and *Job Cost Wizard*, a program that lets you export your estimates to QuickBooks Pro for actual job costing. A 60-minute interactive video teaches you how to use this CD-ROM to estimate construction costs. And to top it off, to help you create professional-looking estimates, the disk includes over 40 construction estimating and bidding forms in a format that's perfect for nearly any *Windows*TM word processing or spreadsheet program. **CD Estimator is $78.50**

## Contractor's Guide to QuickBooks Pro 2006

This user-friendly manual walks you through QuickBooks Pro's detailed setup procedure and explains step-by-step how to create a first-rate accounting system. You'll learn in days, rather than weeks, how to use *QuickBooks Pro* to get your contracting business organized, with simple, fast accounting procedures. On the CD included with the book you'll find a *QuickBooks Pro* file for a construction company (You open it, enter your own company's data, and add info on your suppliers and subs.) You also get a complete estimating program, including a database, and a job costing program that lets you export your estimates to *QuickBooks Pro*. It even includes many useful construction forms to use in your business. **352 pages, 8½ x 11, $51.50**

Also available: **Contractor's Guide to QuickBooks Pro 2001, $45.25**
**Contractor's Guide to QuickBooks Pro 2003, $47.75**
**Contractor's Guide to QuickBooks Pro 2004, $48.50**
**Contractor's Guide to QuickBooks Pro 2005, $49.75**

## Basic Engineering for Builders

If you've ever been stumped by an engineering problem on the job, yet wanted to avoid the expense of hiring a qualified engineer, you should have this book. Here you'll find engineering principles explained in non-technical language and practical methods for applying them on the job. With the help of this book you'll be able to understand engineering  functions in the plans and how to meet the requirements, how to get permits issued without the help of an engineer, and anticipate requirements for concrete, steel, wood and masonry. See why you sometimes have to hire an engineer and what you can undertake yourself: surveying, concrete, lumber loads and stresses, steel, masonry, plumbing, and HVAC systems. This book is designed to help the builder save money by understanding engineering principles that you can incorporate into the jobs you bid. **400 pages, 8½ x 11, $36.50**

## Roofing Construction & Estimating

Installation, repair and estimating for nearly every type of roof covering available today in residential and commercial structures: asphalt shingles, roll roofing, wood shingles and shakes, clay tile, slate, metal, built-up, and elastomeric. Covers sheathing and underlayment techniques, as well as secrets for installing leakproof valleys. Many estimating tips help you minimize waste, as well as insure a profit on every job. Troubleshooting techniques help you identify the true source of most leaks. Over 300 large, clear illustrations help you find the answer to just about all your roofing questions. **432 pages, 8½ x 11, $38.00**

## National Construction Estimator

Current building costs for residential, commercial, and industrial construction. Estimated prices for every common building material. Provides man-hours, recommended crew, and gives the labor cost for installation. Includes a CD-ROM with an electronic version of the book with *National Estimator*, a stand-alone *Windows*TM estimating program, plus an interactive multimedia video that shows how to use the disk to compile construction cost estimates. **656 pages, 8½ x 11, $52.50. Revised annually**

## Electrician's Exam Preparation Guide

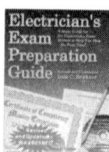

Need help in passing the apprentice, journeyman, or master electrician's exam? This is a book of questions and answers based on actual electrician's exams over the last few years. Almost a thousand multiple-choice questions — exactly the type you'll find on the exam — cover every area of electrical installation: electrical drawings, services and systems, transformers, capacitors, distribution equipment, branch circuits, feeders, calculations, measuring and testing, and more. It gives you the correct answer, an explanation, and where to find it in the latest *NEC*. Also tells how to apply for the test, how best to study, and what to expect on examination day. Includes a FREE CD-ROM with all the questions in the book in interactive test-yourself software that makes studying for the exam almost fun! Updated to the 2005 *NEC*. **352 pages, 8½ x 11, $39.50**

## Concrete Countertops

At last, a complete how-to instruction manual on creating countertops from one of the most versatile building materials available. Here you'll find step-by-step instructions for making concrete countertops, from planning and building the mold, mixing and pouring the concrete, to curing, grinding, finishing, polishing, and finally installing the finished countertop and backsplash. You'll even learn a potential side business of maintaining concrete countertops. Over 300 full-color photos show you combinations of granite, wood, and brass as well as colored, stamped, and stained concrete. **202 pages, 9 x 11, $29.95**

## 2005 National Electrical Code

This new electrical code incorporates sweeping improvements to make the code more functional and user-friendly. Here you'll find the essential foundation for electrical code requirements for the 21st century. With hundreds of significant and widespread changes, this 2005 *NEC* contains all the latest electrical technologies, recently developed techniques, and enhanced safety standards for electrical work.  This is the standard all electricians are required to know, even if it hasn't yet been adopted by their local or state jurisdictions. **784 pages, 8½ x 11, $65.00**

## National Painting Cost Estimator

A complete guide to estimating painting costs for just about any type of residential, commercial, or industrial painting, whether by brush, spray, or roller. Shows typical costs and bid prices for fast, medium, and slow work, including material costs per gallon; square feet covered per gallon; square feet covered per manhour; labor, material, overhead, and taxes per 100 square feet; and how much to add for profit. Includes a CD-ROM with an electronic version of the book with *National Estimator*, a stand-alone *Windows*TM estimating program, plus an interactive multimedia video that shows how to use the disk to compile construction cost estimates. **440 pages, 8½ x 11, $53.00. Revised annually**

## 2006 International Residential Code

Replacing the *CABO One-* and *Two-Family Dwelling Code*, this book has the latest technological advances in building design and construction. Among the changes are provisions for steel framing and energy savings. Also contains mechanical, fuel gas and plumbing provisions that coordinate with the *International Mechanical Code* and *International Plumbing Code*. **578 pages, 8½ x 11, $68.00**

Also available: **2003 International Residential Code, $62.00**
**2000 International Residential Code, $59.00**
**2000 International Residential Code on interactive CD-ROM, $48.00**

## Guide to Stamped Concrete

Become a concrete stamping expert with the practical tips in this book. Here you'll find hundreds of ideas and sources for stamping designs. You'll learn of site conditions affecting stamped concrete and how to avoid problems. Shows everything from preparing the subgrade to erecting forms, installing reinforcement, and placing the concrete. Gives different methods of stamping concrete, coloring stamped concrete, applying color hardener, release agents, installing expansion joints, and special techniques for finishing and sealing the concrete, Shows how to improve the durability of stamped concrete, and how to fix minor flaws. Explains how to write a fair contract for the work you perform and how to determine what to charge your customers for the finished job. **142 pages, 8½ x 11, $45.00**

## How to Succeed With Your Own Construction Business

Everything you need to start your own construction business: setting up the paperwork, finding the work, advertising, using contracts, dealing with lenders, estimating, scheduling, finding and keeping good employees, keeping the books, and coping with success. If you're considering starting your own construction business, all the knowledge, tips, and blank forms you need are here. **336 pages, 8½ x 11, $28.50**

## Wood-Frame House Construction

Step-by-step construction details, from the layout of the outer walls, excavation and formwork, to finish carpentry and painting. Packed with clear illustrations and explanations updated for modern construction methods. Everything you need to know about framing, roofing, siding, interior finishings, floor covering and stairs — your complete book of wood-frame homebuilding. **320 pages, 8½ x 11, $25.50. Revised edition**

## National Electrical Estimator

 This year's prices for installation of all common electrical work: conduit, wire, boxes, fixtures, switches, outlets, loadcenters, panelboards, raceway, duct, signal systems, and more. Provides material costs, manhours per unit, and total installed cost. Explains what you should know to estimate each part of an electrical system. Includes a CD-ROM with an electronic version of the book with *National Estimator*, a stand-alone *Windows*™ estimating program, plus an interactive multimedia video that shows how to use the disk to compile construction cost estimates. **552 pages, 8½ x 11, $52.75. Revised annually**

## Finish Carpentry: Efficient Techniques for Custom Interiors

Professional finish carpentry demands expert skills, precise tools, and a solid understanding of how to do the work. This new book explains how to install moldings, paneled walls and ceilings, and just about every aspect of interior trim — including doors and windows. Covers built-in bookshelves, coffered ceilings, and skylight wells and soffits, including paneled ceilings with decorative beams. **288 pages, 8½ x 11, $34.95**

## National Repair & Remodeling Estimator

The complete pricing guide for dwelling reconstruction costs. Reliable, specific data you can apply on every repair and remodeling job. Up-to-date material costs and labor figures based on thousands of jobs across the country. Provides recommended crew sizes; average production rates; exact material, equipment, and labor costs; a total unit cost and a total price including overhead and profit. Separate listings for high- and low-volume builders, so prices shown are specific for any size business. Estimating tips specific to repair and remodeling work to make your bids complete, realistic, and profitable. Includes a CD-ROM with an electronic version of the book with *National Estimator*, a stand-alone *Windows*™ estimating program, plus an interactive multimedia video that shows how to use the disk to compile construction cost estimates. **328 pages, 8½ x 11, $53.50. Revised annually**

## Building Contractor's Exam Preparation Guide

Passing today's contractor's exams can be a major task. This book shows you how to study, how questions are likely to be worded, and the kinds of choices usually given for answers. Includes sample questions from actual state, county, and city examinations, plus a sample exam to practice on. This book isn't a substitute for the study material that your testing board recommends, but it will help prepare you for the types of questions — and their correct answers — that are likely to appear on the actual exam. Knowing how to answer these questions, as well as what to expect from the exam, can greatly increase your chances of passing. **320 pages, 8½ x 11, $35.00**

## Estimating With Microsoft Excel

 Most builders estimate with *Excel* because it's easy to learn, quick to use, and can be customized to your style of estimating. Here you'll find step-by-step instructions on how to create your own customized automated spreadsheet estimating program for use with *Excel*. You'll learn how to use the magic of *Excel* to create detail sheets, cost breakdown summaries, and links. You'll put this all to use in estimating concrete, rebar, permit fees, and roofing. You can even create your own macros. Includes a CD-ROM that illustrates examples in the book and provides you with templates you can use to set up your own estimating system. **148 pages, 7 x 9, $39.95**

## Illustrated Guide to the International Plumbing & Fuel Gas Codes

A comprehensive guide to the *International Plumbing* and *Fuel Gas Codes* that explains the intricacies of the code in easy-to-understand language. Packed with plumbing isometrics and helpful illustrations, it makes clear the code requirements for the installation methods and materials for plumbing and fuel gas systems. Includes code tables for pipe sizing and fixture units, and code requirements for just about all areas of plumbing, from water supply and vents to sanitary drainage systems. Covers the principles and terminology of the codes, how the various systems work and are regulated, and code-compliance issues you'll likely encounter on the job. Each chapter has a set of self-test questions for anyone studying for the plumber's exam, and tells you where to look in the code for the details. Written by a former plumbing inspector, this guide has the help you need to install systems in compliance with the *IPC* and the *IFGC*. **312 pages, 8½ x 11, $37.00**

## Steel-Frame House Construction

Framing with steel has obvious advantages over wood, yet building with steel requires new skills that can present challenges to the wood builder. This new book explains the secrets of steel framing techniques for building homes, whether pre-engineered or built stick by stick. It shows you the techniques, the tools, the materials, and how you can make it happen. Includes hundreds of photos and illustrations, plus a CD-ROM with steel framing details, a database of steel materials and manhours, with an estimating program. **320 pages, 8½ x 11, $39.75**

## Contractor's Plain-English Legal Guide

For today's contractors, legal problems are like snakes in the swamp – you might not see them, but you know they're there.  This book tells you where the snakes are hiding and directs you to the safe path. With the directions in this easy-to-read handbook you're less likely to need a $200-an-hour lawyer. Includes simple directions for starting your business, writing contracts that cover just about any eventuality, collecting what's owed you, filing liens, protecting yourself from unethical subcontractors, and more. For about the price of 15 minutes in a lawyer's office, you'll have a guide that will make many of those visits unnecessary. Includes a CD-ROM with blank copies of all the forms and contracts in the book. **272 pages, 8½ x 11, $49.50**

## Troubleshooting Guide to Residential Construction

How to solve practically every construction problem — before it happens to you! With this book you'll learn from the mistakes other builders made as they faced 63 typical residential construction problems. Filled with clear photos and drawings that explain how to enhance your reputation as well as your bottom line by avoiding problems that plague most builders. Shows how to avoid, or fix, problems ranging from defective slabs, walls and ceilings, through roofing, plumbing & HVAC, to paint. **304 pages, 8½ x 11, $32.50**

## Markup & Profit: A Contractor's Guide

In order to succeed in a construction business, you have to be able to price your jobs to cover all labor, material and overhead expenses, *and* make a decent profit. The problem is knowing what markup to use. You don't want to lose jobs because you charge too much, and you don't want to work for free because you've charged too little. If you know how to calculate markup, you can apply it to your job costs to find the right sales price for your work. This book gives you tried and tested formulas, with step-by-step instructions and easy-to-follow examples, so you can easily figure the markup that's right for *your* business. Includes a CD-ROM with forms and checklists for your use. **320 pages, 8½ x 11, $32.50**

## Residential Structure & Framing

With this easy-to-understand guide you'll learn how to calculate loads, size joists and beams, and tackle many common structural problems facing residential contractors. It covers cantilevered floors, complex roof structures, tall window walls, and seismic and wind bracing. Plus, you'll learn field-proven production techniques for advanced wall, floor, and roof framing with both dimensional and engineered lumber. You'll find information on sizing joists and beams, framing with wood I-joists, supporting oversized dormers, unequal-pitched roofs, coffered ceilings, and more. Fully illustrated with lots of photos.
**272 pages, 8½ x 11, $34.95. Published by JLC**

## Profits in Buying & Renovating Homes

Step-by-step instructions for selecting, repairing, improving, and selling highly profitable "fixer-uppers." Shows which price ranges offer the highest profit-to-investment ratios, which neighborhoods offer the best return, practical directions for repairs, and tips on dealing with buyers, sellers, and real estate agents. Shows you how to determine your profit before you buy, what "bargains" to avoid, and how to make simple, profitable, inexpensive upgrades. **304 pages, 8½ x 11, $24.75**

## Professional Kitchen Design

Remodeling kitchens requires a "special" touch — one that blends artistic flair with function to create a kitchen with charm and personality as well as one that is easy to work in. Here you'll find how to make the best use of the space available in any kitchen design job, as well as tips and lessons on how to design one-wall, two-wall, L-shaped, U-shaped, peninsula and island kitchens. Also includes what you need to know to run a profitable kitchen design business. **176 pages, 8½ x 11, $24.50**

## Roof Framing

Shows how to frame any type of roof in common use today, even if you've never framed a roof before. Includes using a pocket calculator to figure any common, hip, valley, or jack rafter length in seconds. Over 400 illustrations cover every measurement and every cut on each type of roof: gable, hip, Dutch, Tudor, gambrel, shed, gazebo, and more. **480 pages, 5½ x 8½, $24.50**

## Builder's Guide to Accounting Revised

Step-by-step, easy-to-follow guidelines for setting up and maintaining records for your building business. This practical guide to all accounting methods shows how to meet state and federal accounting requirements, explains the new depreciation rules, and describes how the Tax Reform Act can affect the way you keep records. Full of charts, diagrams, simple directions and examples, to help you keep track of where your money is going. Recommended reading for many state contractor's exams. Each chapter ends with a set of test questions, and a CD-ROM included FREE has all the questions in interactive self-test software. Use the Study Mode to make studying for the exam much easier, and Exam Mode to practice your skills. **360 pages, 8½ x 11, $35.50**

## Renovating & Restyling Older Homes

Any builder can turn a run-down old house into a showcase of perfection — if the customer has unlimited funds to spend. Unfortunately, most customers are on a tight budget. They usually want more improvements than they can afford — and they expect you to deliver. This book shows how to add economical improvements that can increase the property value by two, five or even ten times the cost of the remodel. Sound impossible? Here you'll find the secrets of a builder who has been putting these techniques to work on Victorian and Craftsman-style houses for twenty years. You'll see what to repair, what to replace and what to leave, so you can remodel or restyle older homes for the least amount of money and the greatest increase in value. **416 pages, 8½ x 11, $33.50**

## Craftsman's Construction Installation Encyclopedia

Step-by-step installation instructions for just about any residential construction, remodeling or repair task, arranged alphabetically, from *Acoustic tile* to *Wood flooring*. Includes hundreds of illustrations that show how to build, install, or remodel each part of the job, as well as manhour tables for each work item so you can estimate and bid with confidence. Also includes a CD-ROM with all the material in the book, handy look-up features, and the ability to capture and print out for your crew the instructions and diagrams for any job. **792 pages, 8½ x 11, $65.00**

---

**Craftsman** Craftsman Book Company
6058 Corte del Cedro
P.O. Box 6500
Carlsbad, CA 92018

☎ 24 hour order line
**1-800-829-8123**
Fax (760) 438-0398

### In A Hurry?
We accept phone orders charged to your
○ Visa, ○ MasterCard, ○ Discover or ○ American Express